Microscopy of Semiconducting Materials 1995

Other titles in the Series

The Institute of Physics Conference Series regularly features papers presented at important conferences and symposia highlighting new developments in physics and related fields. Recent titles include:

Microscopy of Semiconducting Materials 1995

Proceedings of the Institute of Physics Conference held at Oxford University, 20–23 March 1995

Edited by A G Cullis and A E Staton-Bevan

S
M M
IX

Institute of Physics Conference Series Number 146
Institute of Physics Publishing, Bristol and Philadelphia

CODEN IPHSAC 146 1–795 (1995)

British Library Cataloguing in Publication Data

A catalogue record for this book is available from the British Library.

ISBN 0 7503 0347 6

Library of Congress Cataloging-in-Publication Data are available

Conference Chairmen
    A G Cullis and A E Staton-Bevan

Honorary Editors
    A G Cullis and A E Staton-Bevan

Scientific Sponsors
    The Institute of Physics
    The Royal Microscopical Society
    The Materials Research Society

This work relates to Department of Navy Grant N00014-95-1-9012 issued by the Office of Naval Research European Office. The United States has a royalty-free licence throughout the world in all copyrightable material contained herein.

Published by Institute of Physics Publishing, wholly owned by The Institute of Physics, London
Institute of Physics Publishing, Techno House, Redcliffe Way, Bristol BS1 6NX, UK
US Editorial Office: Institute of Physics Publishing, The Public Ledger Building, Suite 1035, 150 South Independence Mall West, Philadelphia, PA 19106, USA

Printed in the UK by Galliard (Printers) Ltd, Great Yarmouth, Norfolk

Contents

Dislocations and grain boundaries

Epitaxial layers

Bulk gallium arsenide and other compounds

Processed silicon and diamond

Silicides and metal–semiconductor contacts

Device structures

STM and AFM applications

Advanced SEM and SOM applications

Preface

This volume contains the invited and contributed papers presented at the conference on 'Microscopy of Semiconducting Materials' held at the University of Oxford on 20–23 March 1995. The event was organized with scientific sponsorship by the Electron Microscopy and Analysis Group of the Institute of Physics, the Royal Microscopical Society and the Materials Research Society. This conference was the ninth in a series focusing on the most recent international developments in semiconductor studies carried out by all forms of microscopy. Delegates from 20 countries gave a comprehensive account of work progressing at the present state-of-the-art across the entire field.

Following the tradition of this conference series, these Proceedings report fundamental studies of the microstructure and defect content of semiconductor crystals using sophisticated microscopical techniques which provide information down to the atomic scale. Another important feature of the volume is the device-related work which is reported. Indeed, the semiconductor structures required for advanced electronic devices often incorporate novel materials and geometrical configurations. This leads almost inevitably to extended device development periods needed to overcome problems which can occur during fabrication. Under these circumstances, electron and other forms of microscopy may well be the principal analytical methods by which detailed information on the very small-scale device elements can be obtained. There is much evidence in the present Proceedings of the amount of device-oriented work in progess.

Each camera-ready manuscript submitted for publication in this volume has been reviewed by two referees and modified accordingly. The editors are very grateful to the following scientific referees for their rapid and careful work:

R M Anderson, P D Augustus, L J Balk, R Beanland, M R Brozel, G R Booker, J C Bravman, P D Brown, H Cerva, A Cornet, B Cunningahm, K Durose, D J Eaglesham, D M Follstaedt, C Frigeri, D Gerthsen, P J Goodhew, A Gustafsson, A J Harvey, V Higgs, D B Holt, R Hull, C J Humphreys, J L Hutchison, A Jakubowicz, K G F Janssens, S Mahajan, R E Mallard, C D Marsh, A J McGibbon, S Mil'shtein, I J Murgatroyd, A G Norman, C E Norman, D D Perovic, P Pirouz, F A Ponce, P Pongratz, K A Prior, A M Rocher, F M Ross, G A Rozgonyi, J L Sudijono, G C Weatherly, M E Welland and G M Williams.

The conference organizers are particularly pleased to acknowledge the financial support provided by the Royal Society, the Office of US Naval Research, European Office and the International Science Foundation, Washington DC. S Lippmann (Institute of Physics) is gratefully acknowledged for the smooth running of the administration which underpinned the whole conference. Special thanks are due to P Z A Desouza (Imperial College), P J Hull and R Nayak (Oxford University) and C Reeves (DRA, Malvern) for assistance in correcting proof copies of many manuscripts.

A G Cullis
A E Staton-Bevan
September 1995

Inst. Phys. Conf. Ser. No 146
Paper presented at Microsc. Semicond. Mater. Conf., Oxford, 20–23 March 1995

High-resolution transmission electron microscopy of semiconductor interfaces

Koichi Ishida

Analysis and Evaluation Technology Center, NEC Corporation
1753 Shimonumabe, Nakaharaku, Kawasaki, 211, Japan

ABSTRACT: Imaging conditions of <110> cross-sectional high-resolution transmission electron microscopy were investigated to observe interfacial structures on an atomic scale. The {111} beam amplitudes of GaAs(Ge) exhibit a minimum at the extinction distance, while those of AlAs(Si) show appreciable values. This leads to a marked contrast between HREM images of GaAs(Ge) and AlAs(Si), allowing edge-on observation of the interfacial structures. An artifact-free TEM specimen preparation technique is also presented. This HREM method enabled us to observe the step structures and the ordered interfacial structure at AlAs/GaAs and Si/Ge interfaces, respectively.

1. INTRODUCTION

Heterostructures of III-V semiconductor epitaxial layers, such as quantum wells and quantum wires, have attracted much attention because of their novel optical and electrical properties (Sakaki 1983). Si/Ge superlattices have also been extensively studied, since it has been proposed that direct band gap semiconductors might be produced from originally indirect band gap semiconductors by forming superlattices (Gnutzmann and Clausecker 1974). Since the properties of epitaxial layers depend directly on the qualities of interfaces, the characterization of epitaxial layers on an atomic scale is required.

Cross-sectional high-resolution transmission electron microscopy (HREM) has been widely used to investigate the quality of epitaxial layers. Especially in a lattice-matched system such as AlAs/GaAs, much effort has been devoted to investigate the compositional abruptness at the interfaces and the step structures. In order to observe the interfacial structures on an atomic scale, the interfacial structure must be observed edge-on under HREM imaging conditions where the two layers are distinguished clearly. It has been noticed in <100> cross-sectional HREM of AlAs/GaAs interfaces that the two layers are recognized by the difference in their HREM image patterns owing to the contribution of the compositionally sensitive four {200} beams to the image formation (Hetherington et al. 1985, Ourmazd et al. 1989). However, this HREM method cannot be applied to Si/Ge interfaces, since the 200-reflection is forbidden in Si and Ge. Furthermore, it is difficult to observe the interface structure, since the interfaces may have strongly anisotropic structures along the <110> directions.

Several attempts have been made to observe the interfacial structures edge-on in the <110> HREM (de-Jong et al. 1987, Poudoulec et al. 1990). However, the step structures have not yet been clearly observed since the thick specimen for the conventional <110> HREM imaging conditions and the artifacts of ion-milling have prevented detailed observation of the interfacial structures. In the present paper, we propose new <110> HREM imaging conditions whose principles can be applied to both AlAs/GaAs and Si/Ge interfaces, together with an artifact-free preparation technique for cross-sectional specimens. Using the HREM method, the step structures at AlAs/GaAs interfaces and the ordered interfacial structure at Si/Ge interfaces were observed.

2. IMAGING CONDITIONS AND SPECIMEN PREPARATION

In the <110> cross section, the four {111} beams make a dominant contribution to the HREM image formation. It has been considered that the two layers are difficult to distinguish since the {111} beams are strong in both layers and the HREM images are dominated by the {111}-fringes which have less intensity difference at the interfaces. However, the amplitudes of diffracted beams vary differently in the two layers with sample thickness because of dynamical interaction and may exhibit large differences at certain thicknesses. Therefore, we examined the specimen thickness dependence of the {111} beam amplitudes using the multislice method (Cowley and Moodie 1957).

First, we examine imaging conditions for AlAs/GaAs interfaces (Ikarashi et al. 1993b). The thickness dependence of the {111} beam and (000) beam amplitudes for GaAs and AlAs are shown in Fig. 1(a) and 1(b), respectively. It is seen that the (111) and ($\bar{1}\bar{1}\bar{1}$) beam amplitudes have different thickness dependencies. This is because of the failure of Friedel's law under dynamical scattering conditions in non-centrosymmetric structures such as zincblende crystals. Figures 1(a) and 1(b) show that, at a specimen thickness of around 14.4nm, the amplitudes of the {111} beams for GaAs are at a minimum, while those of AlAs have appreciable values. This remarkable difference in the amplitudes suggests the possibility of distinguishing GaAs from AlAs by the difference in their lattice fringe intensities in the <110> HREM images. Figure 1(a) shows that the (000) beam amplitude in GaAs exhibits minima at sample thicknesses of 7.2nm and 22.0nm, indicating that the first and the second thickness fringes appear at those values, and shows a maximum at 14.4nm; the extinction distance, ζ_{0GaAs}, of the transmitted beam for GaAs is thus 14.4nm. Hence, marked differences in {111} fringe intensities are expected to be observed around a specimen thickness of ζ_{0GaAs} between the thickness fringe.

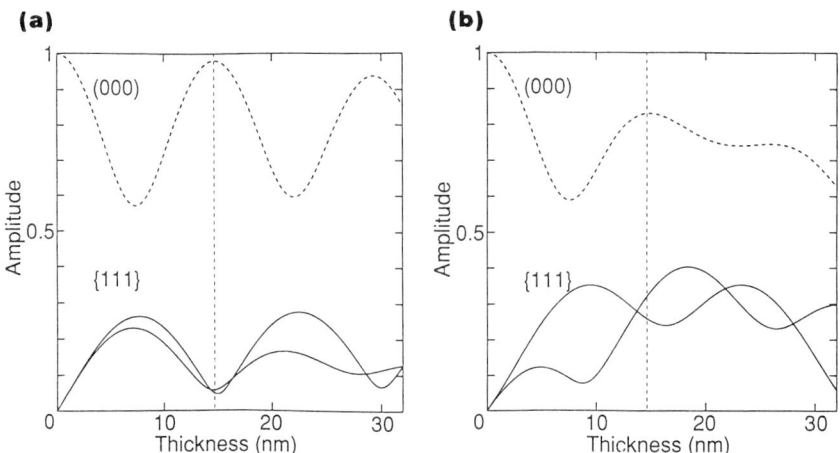

Fig. 1 Variation of {111} and (000) beam amplitudes with specimen thickness for GaAs (a) and AlAs (b) at an acceleration voltage of 200kV. The {111} beams of GaAs show minima at a thickness of 14.4nm.

In order to confirm the above arguments, we made through-thickness and through-focal image simulations near a specimen thickness of 14.4nm. The parameters used for the calculation were chosen to fit the imaging conditions of a TOPCON EM002B electron microscope used for the present HREM observations at an acceleration voltage of 200kV. The results are shown in Fig. 2. An atomic model of [(AlAs)1ML/(GaAs)19ML] was used for the calculation. Arrowheads in the figure indicate the positions of 1ML AlAs layers. As expected from Fig. 1(a), the GaAs lattice fringes are very weak near 14 nm. In order to find an appropriate focus-setting, defocus values were varied from 10nm (overfocus) to -40nm (underfocus) for each specimen thickness. Note that AlAs is most clearly observed around the defocus value of -10nm for a specimen thickness of 14.4nm, while the GaAs lattice gives very weak lattice fringes, which allows <110> HREM of AlAs/GaAs interface structures at these imaging conditions. The comparison between the atomic model and the simulated images found the Al atomic column to appear black in the simulated images; in addition, the black dot intensities of $Al_xGa_{1-x}As$ alloy layers increased linearly with x for $0 < x < 0.7$ and saturated for $x > 0.7$.

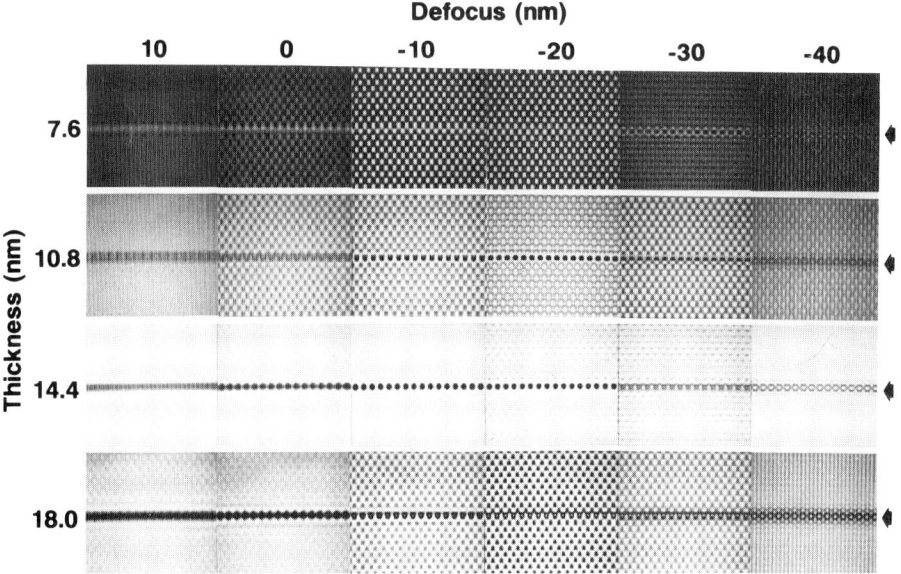

Fig. 2 Through-focal and through-thickness simulated images of [(AlAs)1ML/(GaAs)19ML]. Images of GaAs and AlAs show highest contrast at 14.4nm and defocus of -10nm.

Next, the <110> imaging conditions for Si/Ge interfaces are considered (Ikarashi et al. 1994a). Figures 3(a) and 3(b) show the variation of {111} and (000) beam amplitudes with specimen thickness for Ge and Si, respectively. The amplitudes of the {111} beams and those of the (000) beams show clear periodic oscillation with increasing specimen thickness in both Ge and Si. Note that around the Ge extinction distance (14.8nm), the amplitudes of the {111} beams for Ge are at a minimum, while those for Si have substantial values. As was shown for AlAs/GaAs interfaces, this large difference in the {111} beam amplitudes indicates the possibility of distinguishing Ge from Si by the difference in the lattice image intensities around a specimen thickness of ζ_{0Ge}. The regions of this specimen thickness are bright background regions between the thickness fringes, since the Ge(000) beam shows a minimum at thickness of 7.2nm and 22.0nm and a maximum at 14.8nm.

4

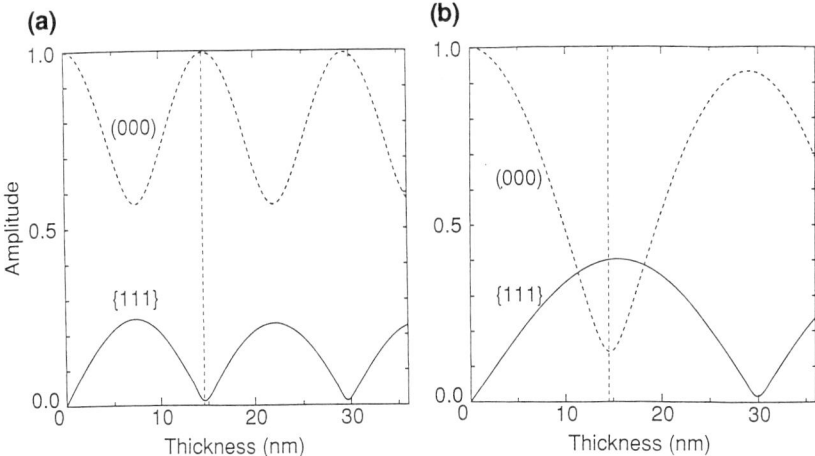

Fig. 3 Variation of {111} and (000) beam amplitudes with specimen thickness for Ge (a) and Si (b) at an acceleration voltage of 200kV. The Ge {111} beam amplitudes show a minimum at 14.8nm, while the Si {111} beams are quite intense at the same thickness.

As expected from the thickness dependence of the {111} beam amplitudes of Ge and Si, the through-thickness and through-focal simulated images of the Si/Ge interface around a specimen thickness of 14.8nm have shown that the {111}-fringes of Ge are almost invisible at this thickness and thus the Si/Ge interface is easily recognized by the difference in {111} fringe intensities. At this thickness, the maximum contrast was expected when the defocus value was -20nm. However, the simulated images of Si showed half-period spacing, which prevented the observation of the details of the interface structure. Thus, the defocus value of -10nm, where the Si lattice image consists of crossed {111} fringes with black contrast located above Si atomic column pairs, is preferable.

Ion milling is usually used as a thinning technique in cross-sectional TEM specimen preparation. However, ion-milling introduces various artifacts on the specimen surface to a depth of several nm, causing noise in the HREM images of specimens whose thicknesses are less than 20nm. On the other hand, chemical etching does not produce such artifacts, but because of the differences in etching rates for materials, chemical etching is not commonly used for the preparation of cross-sectional specimens of interfaces. Thus, in order to prepare artifact-free TEM specimens, the samples are chemically etched after ion-milling. Selective etching effects are reduced by chemically etching only the thin damaged layer on the specimen surface. Furthermore, an etching solution having equal etching rate for AlAs(Si) and GaAs(Ge) is used. The specimen with the AlAs/GaAs interface is etched by dipping the specimen into $30H_3PO_4$-$3H_2O_2$-$100CH_3OH$ for several seconds at room temperature. The specimen with the Si/Ge interface is similarly etched using $1H_2O_2$-$1HF$-$500CH_3COOH$ solution.

The validity of the present imaging conditions and of the specimen preparation technique is demonstrated by the observation of an AlAs/GaAs superlattice. The superlattice was grown by molecular beam epitaxy (MBE) at 600°C. Figure 4 shows an experimental image of a typical TEM specimen having a wedge shape. The imaging orientation is [1$\bar{1}$0] and the defocus value is about -10 nm. Note that in the bright background region (14 nm thickness) between the dark regions of the first and second thickness fringes, the AlAs layers of about 1ML thickness exhibit remarkable contrast with respect to the very weak GaAs image in agreement with the simulated images shown in Fig. 2. Thus, in the present HREM, the specimen thickness and the defocus value for each micrograph are precisely known, since through-focal images are taken at a specimen thickness where the {111} GaAs fringes exhibit minimum intensities. It is also noteworthy that noise from ion milling artifacts is not observed in the image. This indicates that the damaged layer of the specimen was successfully removed by the chemical etching process.

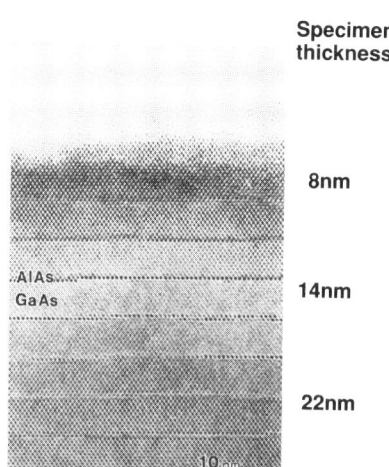

Specimen thickness

8nm

14nm

22nm

Fig. 4 An experimental $[1\bar{1}0]$ HREM image of an AlAs/GaAs superlattice containing 1 ML AlAs. Specimen thicknesses estimated from the GaAs thickness fringes are indicated in the figure.

3. STEP STRUCTURE AT THE AlAs/GaAs INTERFACE

We first examine the visibility of 1 ML-high steps and the straightness of step edges (Ikarashi et al. 1993). The superlattices examined were grown on GaAs (001) substrates tilted 2° towards the (111)A (sample A) and 2° towards the (111)B (sample B). Here, A (Ga-terminated) steps (parallel to the $[1\bar{1}0]$ direction) were introduced on the surface of sample A, while B (As-terminated) steps (parallel to the [110] direction) were present for the sample B. The average interval between neighboring 1 ML-high steps on these surfaces is 8 nm. We grew a multilayer structure on these substrates, by alternately depositing 1 ML of AlAs and 14 ML of GaAs a total of 20 times. The substrate temperature was 520°C. Before growing this superlattice structure, a 100 nm-thick GaAs buffer layer was deposited. At the growth temperature of the present and related specimens (500~600°C), the estimated interdiffusion constant of the AlAs/GaAs interfaces is very small (~10^{-21} cm^2 sec^{-1}) (Ourmazd et al. 1990) so that the majority of Al atoms diffuse only 0.28nm (1ML) into the GaAs layer and constitute only 0.6% of its composition, indicating that the interdiffusion of the Al atoms is negligible. Thus, the contrast variation at the AlAs/GaAs interface represents the interfacial structures.

Figures 5(a) and 5(b) show typical HREM images of sample A and sample B, respectively. The layer thicknesses of AlAs and GaAs are about 1 ML and about 13-14 MLs, as intended. As indicated by arrowheads, the ends of black dot rows of an AlAs layer clearly show the positions of 1 ML-high steps at AlAs-on-GaAs interfaces. Staircases of 1 ML-high steps are observed; the average slope of the staircases agrees with the tilt angles of the substrates in both samples.

In Fig. 5, the difference in the sharpness of step edge images is also noticeable. High-magnification images of typical steps in samples A and B are shown in Figs. 6(a) and 6(b), respectively. In Fig. 6(a), the step edges exhibit sharp intensity variations, while in Fig. 6(b), the contrast changes gradually at step edges over a few nm wide. Simulated images of atomic models containing steps showed that the straight step parallel to the incident beam display abrupt contrast changes at step edges, while steps running obliquely to the incident beam direction or ragged step edges lead to gradual intensity changes. Therefore, the above results show that the A steps are virtually straight through the thickness of the TEM specimen (about 14 nm), while the B steps are ragged with kinks of a few nm in depth.

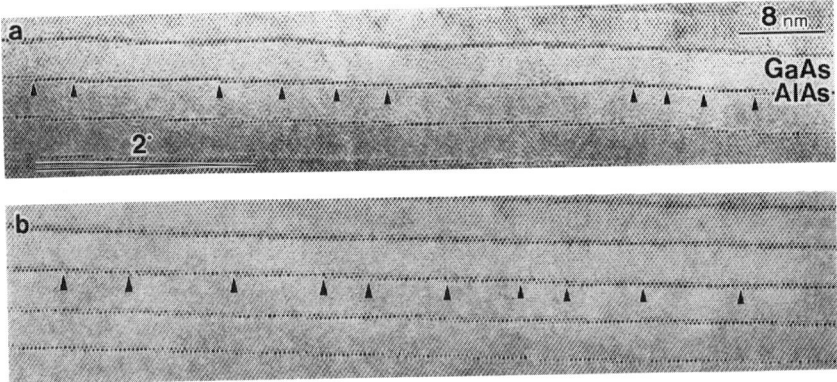

Fig. 5 HREM image of sample A (a) and sample B (b). Arrowheads indicate position of AlAs-on-GaAs interfacial steps.

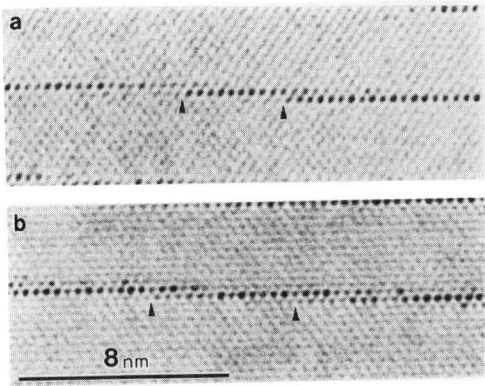

Fig. 6 High magnification images of step edges:(a) sample A, (b) sample B. Across the step edge, the intensity of black dots changes abruptly in (a) but gradually in (b).

Next, in order to study the difference between the step intervals along [110] and [1̄10], (i.e., anisotropy of 1 ML-high islands at AlAs-on-GaAs interfaces), we carried out edge-on observations of both in the [110] and the [1̄10] cross-section (Ikarashi 1993b). An AlAs/GaAs superlattice having an alternating [(AlAs)1-2ML/(GaAs)10ML] structure was grown on GaAs (001)0.1°substrates by MBE at 600°C. Representative [1̄10] and [110] images of the AlAs-on-GaAs interface are shown in Fig. 7. It is noticeable that the step intervals are quite different between the two directions; that is, the interval of B steps is several times wider than that of A steps. On the other hand, the straightness of the step edges showed the same tendency as that observed for the vicinal interfaces; that is, the A step is straighter than the B-step. The above observed difference in the straightness of step edges and the anisotropy of island shapes at AlAs-on-GaAs interfaces characterizes the structure of the GaAs surface during MBE growth and agrees with the results obtained by reflection high-energy electron diffraction (RHEED) (Pukite et al. 1989) and scanning tunneling microscopy (STM) (Pashley et al. 1991, Ide et al. 1992) studies of the GaAs (001) surface.

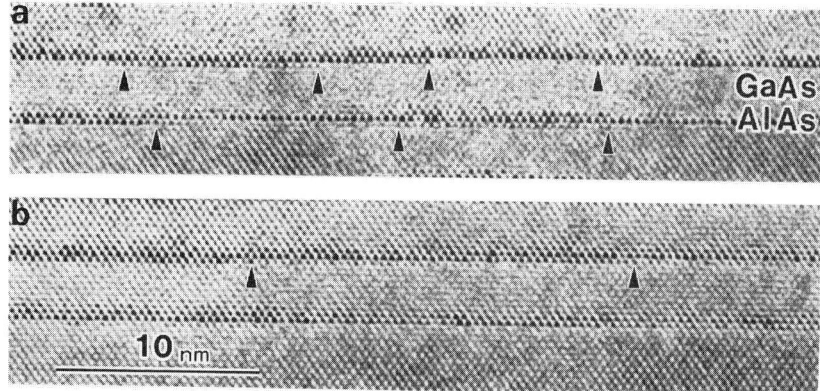

Fig. 7 HREM images of the superlattice in the [1$\bar{1}$0] (a) and [110] (b) projections. Arrowheads indicate the positions of AlAs-on-GaAs interfacial steps.

On the other hand, across the GaAs-on-AlAs interfaces, the image intensities of AlAs layer change gradually, and no interfacial steps were observed at the interfaces, contrary to those present at the AlAs-on-GaAs interfaces. This suggests the presence of interfacial roughness, such as atomic islands and valleys with sizes much smaller than the TEM specimen thickness, since these small structures are not clearly observed in the cross-section.

It was noticed by photoluminescence (PL) and RHEED observations that the steps are smoothed out by growth interruption before the formation of interfaces during MBE growth (Sakaki et al. 1985, Fukunaga et al. 1985) resulting in wide step intervals. This effect was examined directly by the present HREM method. The growth condition was the same as that above, but the growth was interrupted for 120s at both the AlAs-on-GaAs and GaAs-on-AlAs interfaces. Figure 8(a) is a typical [1$\bar{1}$0] image of an AlAs/GaAs interface grown by MBE with growth interruption. Figure 8(b) shows an AlAs/GaAs interface grown by conventional MBE, for comparison. At the AlAs-on-GaAs interface, no steps are observed as shown in Fig. 8(a), while there are several steps in Fig. 8(b). At an interface formed with growth interruption,

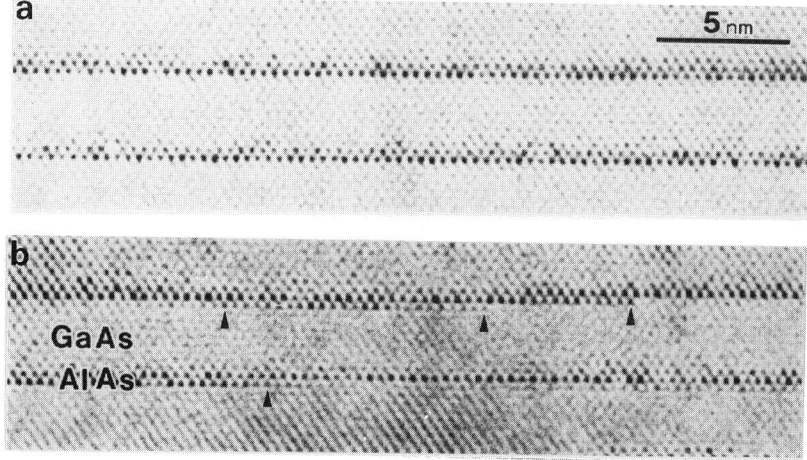

Fig. 8 [1$\bar{1}$0] HREM images of a superlattice grown with growth interruption (a) and without growth interruption (b). Arrowheads indicate the positions of AlAs-on-GaAs interfacial steps.

the step intervals are usually larger than several tens of nm. Thus, it is evident that the steps are dramatically smoothed out by growth interruption. On the other hand, no clear differences were observed between GaAs-on-AlAs interfaces grown by conventional MBE and by MBE with growth interruption. Thus, growth interruption of the GaAs growing surface smooths out surface steps, but it does not have a noticeable effect on the AlAs surface. These results provide direct evidence for the effect of growth interruption as observed by PL and RHEED.

4. ORDERED INTERFACIAL STRUCTURE AT THE Si/Ge INTERFACE

The sample examined has a $(Si_5/Ge_{44})_5$ superlattice structure grown by solid-source MBE. The superlattice was grown on a Ge (001) substrate at a substrate temperature of 400°C. The growth rates were 0.10 nm/s for Si and 0.07 nm/s for Ge. The <110> image of the Si/Ge superlattice in the present HREM method revealed an ordered interfacial structure at the Si/Ge interface (Ikarashi et al. 1994a, 1994b). Figure 9(a) shows an experimental image of the Si/Ge interface. The specimen thickness, as estimated from the thickness fringes in the Ge lattice image, is about 15nm. Note that noise caused by ion milling artifacts is not observed in the image, indicating that the chemical etching during the TEM specimen preparation removed them. The Si layer shows a clear lattice image, while the Ge lattice image is very weak, as expected from the variation of the {111} beam amplitudes with sample thickness shown in Fig. 3. This marked contrast between the lattice images of Ge and Si allows clear observations of the interfacial structure. It should be noted in Fig. 9(a) that the image of the Si-on-Ge interface shows a periodicity of twice the (110) spacing, as indicated by the arrowheads in Fig. 9(a). The doubling of the periodicity indicates the presence of an ordered structure at the interface in the <110> projection, caused by Si and Ge chemical ordering. The ordered regions often extend over 10 nm in width. These clear intensity changes are confined to a bilayer thickness at the interface. On the other hand, no interfacial ordering was observed at the Ge-on-Si interface. The transmission electron diffraction pattern shown in Fig. 9(b) indicates that the periodicity of the ordered interfaces is 2x1. Fundamental diffraction spots from the bulk crystal appear at the [20](equivalent to 220) and the [02]($2\bar{2}0$) reciprocal point. Diffraction spots of fractional indices (indicated by arrows) originate from the superposition of 2x1 and 1x2 structures. The periodicity of the ordered interface is not 2x2 because no <h/2,k/2,> (h,k odd) spots are observed.

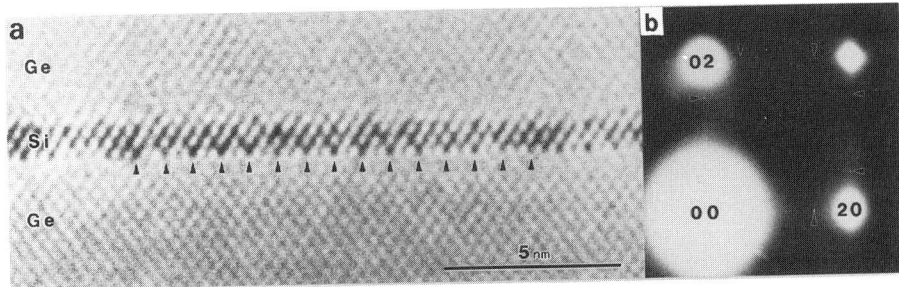

Fig. 9 HREM image of the Si/Ge interface (a) and transmission electron diffraction pattern (b). Arrowheads indicate an ordered interfacial structure in (a) and diffraction spots of fractional indices showing a 2x1 periodicity in (b).

A magnified experimental image, an atomic model of the ordered structure and the resulting simulated image are shown in Fig. 10(a). The simulated image is in good agreement with the experimental image. The refinement of the ordered interfacial structure shown in Fig, 10(a) was carried out by using grazing incidence X-ray diffraction; the alloying in the interfacial dumbbell and small displacements of atoms were determined. On the other hand, Jesson et al. (1991) presented a different atomic model for the ordered interfacial structure and proposed a Ge atom pump model for the ordering. Although the present data show that the ordered interfacial structure is a single phase, their model has several variants, all of which have two strongly ordered atomic layers at the interface. Fig. 10(b) shows the atomic structure of Jesson's model and the simulated image which shows smeared out regions for columns of Si-Ge pairs in disagreement with our experimental image. Furthermore, the periodicity of their model is 2x2 in contradiction with our diffraction data.

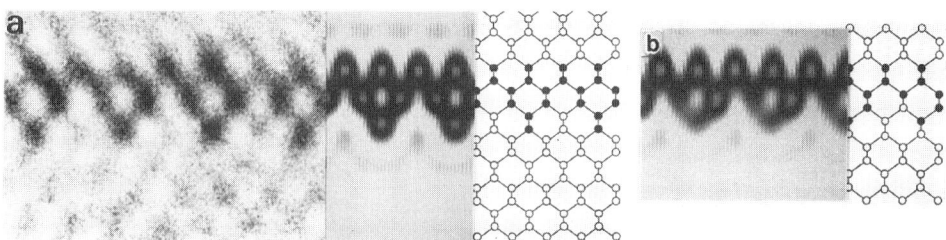

Fig. 10 A magnified image, an atomic model of the ordered structure and the simulated image of our model (a), and the atomic model and simulated image of Jesson's model (b). Open (solid) circles show Ge (Si) atomic columns.

The ordered interfacial structure determined above has several features in common with the ordered structures proposed for the SiGe alloy surface (Kelires 1989) and for bulk SiGe alloys (LeGoues 1990). In these structures, the formation of ordering is attributed to relaxation of the surface stress inherent to a 2x1 dimerized (001) surface; that is, atoms having different covalent radii tend to segregate so as to release the surface stress leading to the 2x1 ordered structure. Thus, the same mechanism may be operating for the formation of the present ordered interfacial structure; in other words, the ordered interface may have been originally formed at the growing surface because of surface stress and subsequently buried at the interface. However, our finding that no ordering was observed at the Ge-on-Si interface points out an important role of Ge segregation in the formation of the ordering. Our results suggest that Ge segregation and surface stress play different roles in the ordered interface formation; the main cause of the atom replacement is Ge segregation due to its lower surface energy, while the ordering occurs in order to relax the stress caused by 2x1 surface reconstruction at the growing surface.

5. CONCLUSION

HREM imaging conditions for studying interfacial structures in AlAs/GaAs and Si/Ge superlattices are presented together with an artifact-free TEM specimen preparation technique. We have shown that for optimized imaging conditions, AlAs(Si) and GaAs(Ge) interfaces are clearly distinguishable in the <110> cross-sectional imaging direction. The present <110> HREM method allows us to directly investigate the AlAs/GaAs and Si/Ge interfacial structures on an atomic scale. Moreover, since this HREM technique utilizes the difference between the extinction distances of the materials making up an interface, it can be applied to observation of other hetero-epitaxial interfaces.

10

6. ACKNOWLEDGMENTS

The present HREM work is the result of collaboration with N. Ikarashi. The author would like to thank him for his cooperation. Thanks are also due to Professor H. Sakaki, M. Tanaka, T. Baba, and T. Tatsumi for providing us with specimens and for useful discussions on MBE growth.

REFERENCES

Cowley J M and Moodie A F, 1957 Acta Crystallogr.10 607
de Jong A F, Bender H, and Coene W, 1987 Ultramicroscopy 21 373
Fukunaga T, Kobayashi K L I, and Nakashima H 1985 Jpn.J.Appl.Phys. 24 L510
Gnutzmann U and Clausecker K, 1974 Appl. Phys.3 9
Hetherington C J, Bi J, Barry J C, Humphreys C J, Grange J and Wood C, 1989 Mater. Res.Symp.Proc. 37 201
Ide T, Yamashita A and Mizutani T 1992 Phys. Rev.46 1905
Ikarashi N, Tanaka M, Sakaki H, and Ishida K, 1992 Appl.Phys.Lett. 60 1360
Ikarashi N, Baba T, and Ishida K, 1993a Appl. Phys. Lett. 62 1632
Ikarashi N, Tanaka M, Baba T, Sakaki H, Ishida K, 1993b Jpn. J. Appl. Phys. 32 2824
Ikarashi N, Tatsumi T and Ishida K 1994 Jpn. J Appl. Phys. 33 1228
Ikarashi N, Akimoto K, Tatsumi T and Ishida K 1994 Phys.Rev.Lett. 72 3198
Jesson D E, Pennycook S J, and Baribeau J-M, 1991 Phys.Rev.Lett. 66 750
Kelires P C and Tersoff J, 1989 Phys. Rev. Lett. 63 1164
LeGoues F K, Kesan V P, Iyer S S, Tersoff J, and Tromp R, 1990 Phys. Rev. Lett. 64 2038
Ourmazd A, Taylor D W, Cunningham J C and Tu C W, 1989 Phys.Rev.Lett. 62 933
Ourmazd A, Kim Y and Bode M, 1990 Mater. Res. Soc. Symp. Proc. 163 639
Pashley M D, Haberern K W, and Gaines J M, 1991 Appl.Phys.Lett. 58 40
Poudoulec A, Guenais B, and D'anterroches C, and Regreny A, 1990J. Cryst. Growth 100 529
Pukite P R, Petrich G S, Batra S and Cohen P I 1989 J.Cryst.Growth 95 269
Sakaki H Proc. Int. Symp. Foundation of Quantum Mechanics Tokyo 1983 (Physical Society of Japan, Tokyo, 1984) pp 94
Sakaki H,Tanaka M, and Yoshino J 1985 Jpn.J.Appl.Phys. 24 L417

Inst. Phys. Conf. Ser. No 146
Paper presented at Microsc. Semicond. Mater. Conf., Oxford, 20–23 March 1995
© *1995 IOP Publishing Ltd*

Strain relaxation induced local crystal tilts at Si/SiGe interfaces in cross-sectional transmission electron microscope specimens

T Walther, C B Boothroyd and C J Humphreys

Department of Materials Science and Metallurgy, University of Cambridge, Pembroke Street, Cambridge CB2 3QZ, UK

ABSTRACT: High-resolution electron micrographs of strained Si/SiGe interfaces can exhibit local variations in the image pattern that appear to be due to local crystal tilts resulting from strain relaxation during the preparation of the cross-sectional specimen. In this study we measured the crystal tilt on a nanometre scale around the interface region using convergent beam micro-diffraction in a scanning transmission electron microscope. In our specimen, regions about 8nm apart can differ in orientation by up to 7mrad. These crystal tilts significantly affect HREM image patterns and must be taken into account in any image pattern analysis if reliable quantitative compositional information is to be obtained.

1. INTRODUCTION

Pattern recognition algorithms have recently been used to study the interface quality in strained layer systems of semiconductor quantum wells by high-resolution transmission electron microscopy (HREM) with near-atomic spatial resolution (Stenkamp and Jäger 1993, Schwander et al 1993). Similar approaches have proved partially successful applied to lattice matched heterostructures (Hetherington et al 1985, de Jong and van Dyck 1990, Ourmazd et al 1990, Thoma and Cerva 1991, Walther and Gerthsen 1993), although their validity inherently assumes that the pattern in an experimental image matches that of the image simulations, which detailed comparisons show to be a very crude approximation (Boothroyd and Stobbs 1989, Anstis et al 1993, Hytch and Stobbs 1994). These algorithms will only yield correct results if the experimental image pattern depends on the imaging parameters in exactly the same way as predicted from the calculations for the whole range of relevant image frequencies (i.e. from the background intensity to the higher spatial frequencies constituting the pattern). Effects like bending of the lattice planes near the surface of the thin foil specimens (Treacy et al 1985, Treacy and Gibson 1986) and amorphous surface layers (Boothroyd et al 1994) will introduce errors in the compositional maps obtained which can be taken into account if the effects are known to contribute to the pattern. We think that local crystal tilt may be a parameter which has been widely ignored in the these approaches, probably because it has not been recognised as an important parameter in HREM, although strain contrast due to tilted lattice planes is well known from conventional bright and dark field imaging.

In this study we measured the crystal tilt on a nanometre scale around the interface region using convergent beam electron micro-diffraction (CBED) in a scanning transmission electron microscope (STEM) by comparing the observed diffraction patterns with simulations. The estimated values of the local crystal tilt were then used as input for HREM image simulations.

2. MICRO-DIFFRACTION EXPERIMENTS

The specimen investigated was an Ar⁺ ion milled <110> cross section of a Si/SiGe multi quantum well structure grown directly (without buffer) on a (001) Si substrate by low pressure chemical vapour deposition and had been studied before by HREM (Walther et al 1994). The nominal width of the SiGe layers is 3nm, the Ge content 20at% and the spacing

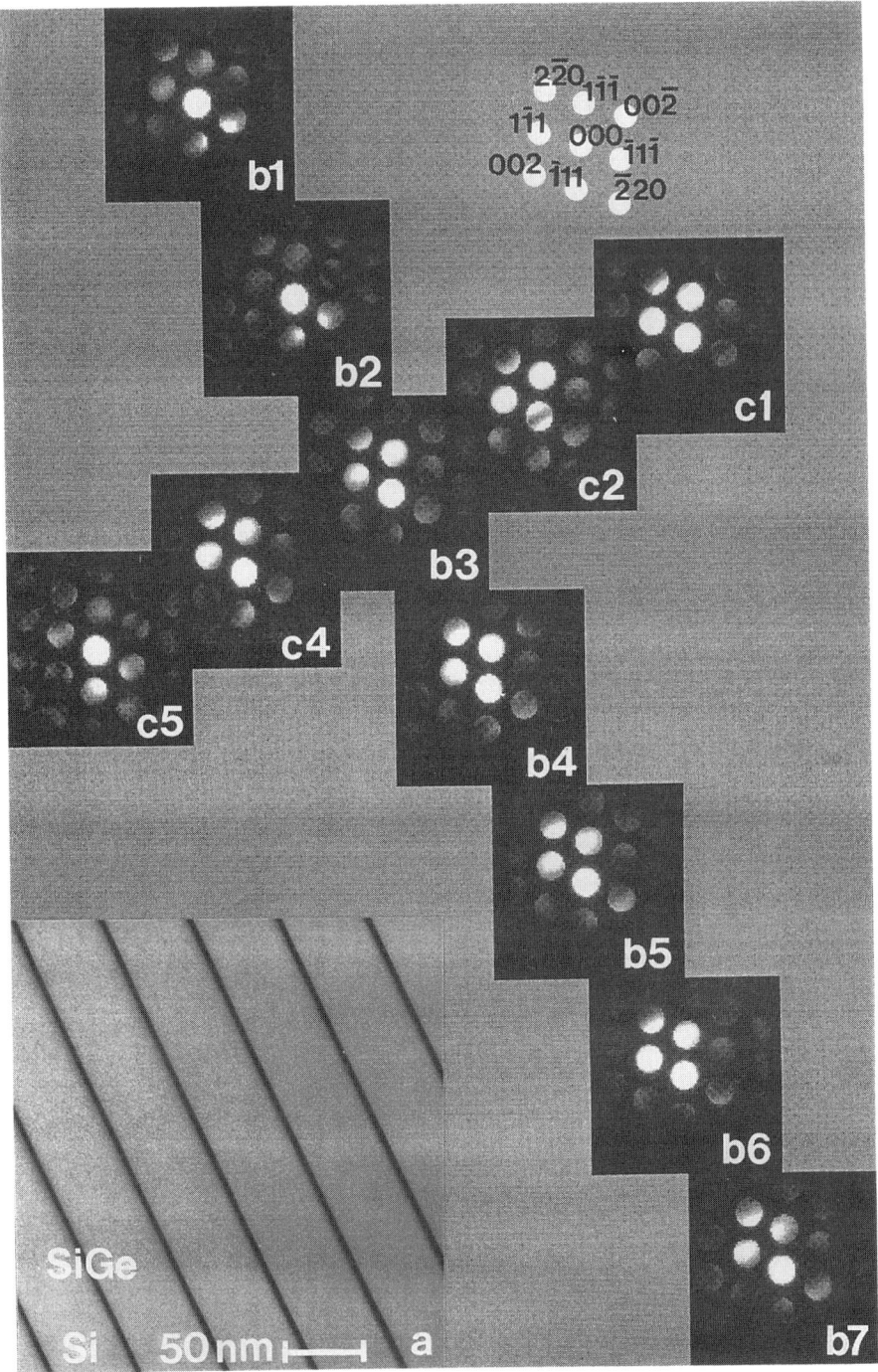

Fig. 1: Bright field image of the multi quantum well structure (a) and montage of CBED
patterns obtained parallel (series b) and perpendicular (series c) to one of the quantum
wells slightly more than 50nm from the specimen edge

between the quantum wells 47nm. Fig. 1a is a conventional bright field electron microscope image confirming the overall uniformity of the structure. The CBED patterns were obtained in a VG HB501 STEM operated at 100keV which is equipped with a GATAN energy filter to exclude inelastically scattered electrons and were recorded directly on a CCD. The probe size of the instrument was about 2nm in this mode, and the CBED patterns shown in Fig. 1b,c were taken close to the <110> zone axis from areas about 8±3nm apart. Seven images were acquired along and five perpendicular to one of the quantum wells slightly more than 50nm away from the specimen edge. Figs. 1a and 1b,c are aligned such that the growth direction is the same in both. The diffraction pattern c3 (from the series taken along the growth direction) was acquired very close to the quantum well position and is not shown since it was almost identical with the pattern b3 from the series along the layer as can be seen from the intensity values in table 1. This demonstrates the good reproducibility of the data obtained. Visual comparison of the CBED patterns already reveals a varying degree of asymmetry, indicating that the crystal tilt with respect to the electron beam direction is not constant over the area analysed. In this context it has to be admitted, however, that eight of the twelve diffraction patterns (indexed b1 to c1 in Fig. 1) suffer from an experimental artefact in form of a partial cut-off of the $(\bar{2}20)$ disk caused by a drift of the energy slit during the data acquisition. This is not very obvious in the figure since the intensity diffracted into that beam is rather low compared to those of the five central beams. We did not try to correct the values for this reflection, but have simply chosen not to include this disk in the least-square fit procedure described in the following paragraph. Care was taken to acquire the diffraction patterns with a moderate electron beam flux and in reasonably short time in order to avoid bending of the crystal planes because of heating up or even amorphisation. An HREM analysis of the specimen area afterwards exhibited no amorphous regions though a lot of speckle contrast indicating typical beam damage probably due to the HREM work at 400keV rather than to the micro-diffraction study at 100keV.

image	I(000)	I($1\bar{1}1$)	I($1\bar{1}\bar{1}$)	I($\bar{1}1\bar{1}$)	I($\bar{1}11$)	I(002)	I($00\bar{2}$)	I($2\bar{2}0$)	I($\bar{2}20$)
b1	0.518	0.090	0.084	0.119	0.071	0.026	0.033	0.033	(0.004)
b2	0.464	0.036	0.058	0.108	0.081	0.019	0.021	0.035	(0.008)
b3	0.272	0.143	0.186	0.053	0.025	0.032	0.029	0.075	(0.002)
b4	0.216	0.182	0.217	0.067	0.033	0.018	0.021	0.158	(0.003)
b5	0.360	0.162	0.179	0.081	0.022	0.019	0.023	0.153	(0.001)
b6	0.249	0.247	0.204	0.040	0.012	0.022	0.028	0.117	(0.001)
b7	0.323	0.163	0.136	0.073	0.024	0.014	0.025	0.128	(0.001)
c1	0.297	0.233	0.212	0.030	0.005	0.034	0.027	0.094	(0.001)
c2	0.167	0.225	0.237	0.050	0.044	0.042	0.044	0.072	0.013
c3	0.303	0.140	0.179	0.054	0.033	0.021	0.019	0.122	0.008
c4	0.300	0.176	0.168	0.056	0.067	0.018	0.026	0.111	0.010
c5	0.330	0.079	0.066	0.085	0.124	0.028	0.029	0.037	0.038

Table 1: Experimental average intensities of the CBED disks, scaled to I(000)=0.33 for c5. The values in brackets are invalid because of a partial cut-off of the $(\bar{2}20)$ CBED disk.

3. QUANTIFICATION OF THE TILTS USING IMAGE SIMULATIONS

The agreement between experimental and simulated intensities of diffracted beams depends critically on a knowledge of the specimen thickness. We intended to analyse a region having a thickness below 40nm suitable for HREM work. From the specimen geometry we could exclude regions much more than 50nm away from the edge. A rather precise value of the thickness could be obtained from the absolute value of the intensity of the transmitted beam in the diffraction patterns, however, Fig. 2a indicates that this value also is a sensitive function of the crystal tilt which we want to determine. We have therefore chosen the experimental diffraction pattern with the lowest apparent asymmetry (c5 in Fig. 1) as the best approximation to an on-axis pattern. Simulations show that for thicknesses between 28nm and 31nm, the intensity of the transmitted beam changes by less than 0.05 on a scale where the incident intensity is unity, if the crystal is tilted away from the zone axis up to 3mrad in amplitude in any direction, and by less than 0.02 for thicknesses between 21nm and 33nm if the tilt is smaller than 1mrad. Thus it should be possible for small tilts to measure the specimen thickness rather reliably from the transmitted beam intensity, assuming that the calculated extinction distance for the undiffracted beam is in agreement with the experimental one. Integrating the intensity in the transmitted beam and comparing it to the total intensity integrated over the whole image (assumed to be unity) yields an intensity of the transmitted beam of 0.33±0.02. Comparing this value with image simulations for Si, using the EMS program package by Stadelmann (1987) with absorption constants V'_g/V_g up to 0.02 for all $\mathbf{g}$ vectors, shows that in the thickness interval of 20nm up to 40nm this value is only in agreement with a specimen thickness of 29±2nm. Further calculations therefore focused on the thickness interval between 27nm and 31nm. This approach must of course be an approximation because it neglects all higher order elastically scattered beams not recorded in the images. The intensity values given in Table 1 will therefore be upper estimates, and this systematic error can be taken into account by multiplying all experimental intensity values with a scaling constant $c \le 1$. For values $0.8 \le c \le 1$ we checked that the fit procedure yielded the same results for the crystal tilts and indeed did not change the fit quality by more than a few percent.

In the following we shall use the symbols I for the intensity, $\mathbf{g}$ for the vectors characterising the diffracted beams and $(\mathbf{k_x}, \mathbf{k_y})$ to denote the crystal tilt components as given in reciprocal space by the shift of the Ewald sphere along the [001] and [$\bar{1}$10] direction, respectively (i.e. perpendicular to the tilt axes in real space), though we usually quote their magnitudes in mrad rather than in nm^{-1}.

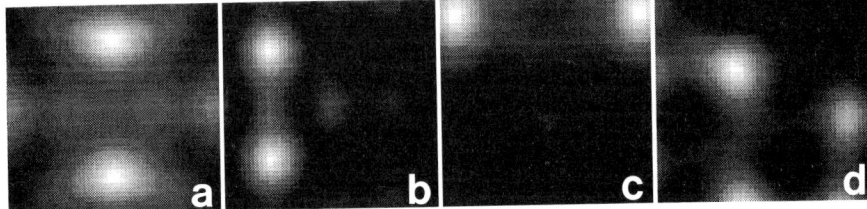

Fig. 2: Calculated beam intensities I($\mathbf{g}$) as a function of crystal tilt for tilts between (-12mrad, -12mrad) and (12mrad,12mrad) and a specimen thickness of 30nm for $\mathbf{g}$=(000) (a; max=0.78), $\mathbf{g}$=(002) (b; max=0.27), $\mathbf{g}$=(2$\bar{2}$0) (c; max=0.43) and $\mathbf{g}$=($\bar{1}$11) (d; max=0.37); directions (perpendicular to tilt axes): x-axis ‖ [001] and y-axis ‖ [$\bar{1}$10]; grey levels: black=0, white =max

Fig. 2 depicts the intensities I($\mathbf{g}$) calculated for $\mathbf{g}$=(000), (002), (2$\bar{2}$0) and ($\bar{1}$11) as a function of the tilt $(\mathbf{k_x}, \mathbf{k_y})$ in the range from (-12mrad,-12mrad) to (12mrad,12mrad) in steps of 0.5mrad. The corresponding plots for the other $\mathbf{g}$ vectors not shown here are related to the ones given in this figure by the 2m symmetry of the projected structure with respect to the x and y axes. In order to speed up the computation using the multi-slice algorithm for slice thicknesses of 0.19nm and a shifted phase grating to model the tilt, the values were calculated for the sector (0,0) to (12mrad,12mrad) only, using increments of 1mrad. The other values were obtained using the symmetry and linear interpolation of the 1mrad grid onto a grid with 0.5mrad mesh size. Thus a higher sampling density for the tilt array was achieved which is important for a good fit to the experimental data. No absorption was taken into account in these calculations. A first approximation for the absorption values V'_g/V_g of the lower order reflections in Si at 100keV would be a constant of the order of 0.01 (Humphreys and Hirsch 1968) and - for an orientation on the zone axis - lead to a decrease of I($\mathbf{g}$) of $\le$0.03 for all $\mathbf{g}$ vectors considered here. We have further neglected any intensity variations across the disks which should be a good approximation for thin crystals. A matching of the faint patterns in the disks with simulations would require significantly more computing time and for the small thicknesses here be inaccurate anyway. The scope of this work is not to simulate CBED patterns but only to get an idea of the amount of crystal tilt necessary to explain the observed asymmetries in the diffraction patterns. However, we made sure that taking the intensities at the centres of the disks (corresponding to a semi-angle of incidence of ≈0.5mrad) instead of the average values did not change the results obtained for the tilts by more than 0.5mrad.

An analysis of the asymmetry with regard to the **g** vectors of the diffraction pattern in form of a signal defined by $\Delta(\mathbf{g})=[I(\mathbf{g})-I(-\mathbf{g})]/[I(\mathbf{g})+I(-\mathbf{g})]$ which effectively describes the difference between the intensities diffracted into the **g** and the -**g** reflection, normalised by their average value, looks promising at first glance because of its independence from I(000) which is strongly thickness dependent. However, this value seems to be useful only for extremely small tilts since it tends to unity as the asymmetry between I(**g**) and I(-**g**) increases and then does not measure the crystal tilt reliably anymore. Moreover, as the sample thickness in our case is quite small and the diffracted beam intensities thus are comparably low, we can only reliably extract I(**g**) values for nine beams (or eight if we refrain from the $(\bar{2}\,2\,0)$disk), yielding only four (or three) values for a fit. We have therefore chosen to perform a least-square fit of the calculated diffracted beam intensities to the experimental values, taking into account the different orders of magnitude of the intensity values of the diffracted beams by dividing the difference between experimental and calculated intensity values by the value calculated for the corresponding tilt. This normalisation seems to be necessary if the transmitted beam is included, the intensity of which is much larger than the intensities of the diffracted beams and would otherwise dominate the least-square fit result. Hence, we used

$$f(\boldsymbol{k}_{\mathrm{x}},\boldsymbol{k}_{\mathrm{y}}) = \left[\sum_{g} \left(\frac{I_{g,\exp} - I_{g,\mathrm{cal}}(\boldsymbol{k}_{\mathrm{x}},\boldsymbol{k}_{\mathrm{y}})}{I_{g,\mathrm{cal}}(\boldsymbol{k}_{\mathrm{x}},\boldsymbol{k}_{\mathrm{y}})} \right)^{2} \right]^{1/2}$$

as a suitable fit function to be minimised for **g**=(000), $(1\bar{1}1)$, $(1\bar{1}\,\bar{1})$, $(\bar{1}11)$, $(\bar{1}1\bar{1})$, (002), $(00\bar{2})$ and $(2\,\bar{2}\,0)$. The results are shown in Fig. 3 for the four images indexed b2, b3, c4 and c5 in Fig. 1 in form of the inverse f^{-1} for a crystal thickness of 30nm which gave the best fit for all diffraction patterns. All maps are scaled to the same minimum and maximum values for black and white. The lightest positions correspond to the best fits which are obtained for crystal tilts of (0,4mrad) for b2, (0,-3mrad) for b3 and c4 and (-0.5mrad,3mrad) for c5. The crystal tilt of the regions corresponding to the CBED patterns b3-b7 and c1-c3 for which the fitted maps are not shown stays at a nearly constant value around (0,-2.5mrad). All tilt values were reproduced within 0.5mrad precision for all thicknesses between 27nm and 31nm though sometimes spurious peaks at tilt values above 10mrad amplitude were obtained as well. Hence, we have demonstrated the existence of crystal tilts along a cross-sectional TEM specimen of a strained quantum well of 3-4mrad from the average zone axis orientation and in particular that regions about 8nm apart can differ in orientation relative to each other up to 7±1mrad, with the tilts being in the $[\bar{1}10]$ direction, i.e. about the growth direction as tilt axis. Assuming the fraction in the equation above to be a constant independent from **g**, the summation can be replaced by a factor 8 for the eight different reflections considered and the fit value becomes approximately three times the relative error in % for fitting the average intensity of each CBED disk. This value then is between 15% (for c5) and 50% (for c4), indicating that our fit procedure is sufficient for extracting the tilt values, but far from optimum.

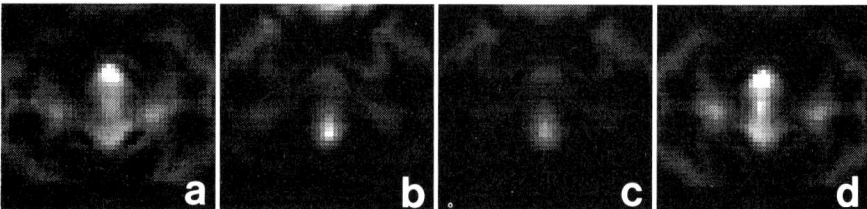

Fig. 3: Calculated fits f^{-1} between experimental and simulated diffracted intensities as a function of crystal tilt for tilts between (-12mrad,-12mrad) and (12mrad,12mrad) for the CBED patterns b2 (a), b3 (b), c4 (c) and c5 (d); grey levels: black=0, white =1; best fit values: f=0.6 (a), 1.0 (b), 1.4 (c) and 0.4 (d)

4. INFLUENCE OF THE CRYSTAL TILT ON HIGH-RESOLUTION IMAGES

Fig. 4 is a patchwork of HREM images simulated for 21nm thick <110> Si in a typical 400keV high-resolution electron microscope (JEOL 4000EX-II with a spherical aberration of C_s=0.9mm, a defocus spread of s=10nm and a semi-angle of beam convergence of α=0.9mrad; both rms values) for crystal tilts about the [001] tilt axis up to 8mrad in amplitude with increments of 2mrad. It is obvious that tilts of the order determined experimentally by CBED above change the image pattern drastically. A tilt of 4mrad yields a half-spacing contrast similar to that observed for SiGe at smaller thicknesses. For tilts of 6mrad to 8mrad complete contrast reversals (column contrast to tunnel contrast) are possible though a closer inspection in this case shows an additional slight shift of the intensity extrema with regard to the on-axis pattern. In particular, we would expect the thin foil buckling to increase as the specimen thickness decreases. Hence, the values for the local crystal tilt determined here are lower estimates since HREM work is often performed on crystals significantly thinner than 30nm.

Fig. 4: HREM image simulations for
    <110> Si of 21nm thickness
    with crystal tilts about the [001]
    axis (horizontally) with ampli-
    tudes of 0mrad (a), 2mrad (b),
    4mrad (c), 6mrad (d), 8mrad (d);
    grey levels: black=0, white=2

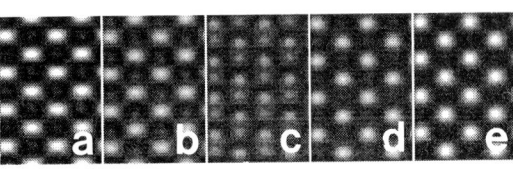

5. SUMMARY AND CONCLUSION

In cross-sectional specimens of (001) SiGe/Si structures prepared by standard ion milling procedures, sample areas of ≈30nm thickness and 5nm to 11nm apart have been demonstrated to exhibit differences in orientation with respect to the electron beam along the [1 10] direction, with [001] as tilt axis. These crystal tilts significantly influence the pattern in an HREM image. Given the magnitude of the local crystal tilt found here and its effect on HREM images, it appears questionable whether any reliable quantitative compositional information can be obtained from a pattern analysis of single HREM images of strained semiconductor heterostructures unless an upper estimate of the local crystal tilt is obtained by CBED from the same specimen region and then taken into account in the HREM image analysis.

ACKNOWLEDGEMENT

Thanks are due to DJ Robbins of the DRA Malvern for growing the wafers, AR Preston for comments and the EPSRC UK and DRA Malvern for a CASE award studentship to TW.

REFERENCES

Anstis GR, Auchterlonie GJ and Barry JC 1993 Ultramicroscopy 52 167-178
Boothroyd CB and Stobbs WM 1989 Ultramicroscopy 31 259-274
Boothroyd CB, Dunin-Borkowski RE, Stobbs WM and Humphreys CJ 1994, paper
    presented at the MRS Fall Meeting, Boston, Dec 1994
de Jong AF and van Dyck D 1990 Ultramicroscopy 33 269-279
Hetherington CJD , Barry JC, Bi JM, Humphreys CJ, Grange J and Wood C 1985 Mat Res
    Soc Symp Proc 37 41-46
Humphreys CJ and Hirsch PB 1968 Phil Mag 18 115-122
Hytch MJ and Stobbs WM 1994 Ultramicroscopy 53 191-203
Ourmazd A, Baumann FH, Bode M and Kim Y 1990 Ultramicroscopy 34 237-255
Schwander P, Kisielowski C, Seibt M, Baumann FH, Kim Y and Ourmazd A 1993 Phys Rev
    Lett 71 4150-4153
Stadelmann PA 1987 Ultramicroscopy 21 131-146
Stenkamp D and Jäger W 1993 Ultramicroscopy 50 321-354
Thoma S and Cerva H 1991 Ultramicroscopy 38 265-289
Treacy MMJ, Gibson JM and Howie A 1985 Phil Mag A 51 389-417
Treacy MMJ and Gibson JM 1986 J Vac Sci Technol B 4 1458-1466
Walther T, Boothroyd CB, Humphreys CJ and Cullis AG 1994 Proc ICEM-13 1 365-366
Walther T and Gerthsen D 1993 Appl Phys A 57 393-400

Inst. Phys. Conf. Ser. No 146
Paper presented at Microsc. Semicond. Mater. Conf., Oxford, 20–23 March 1995
© 1995 IOP Publishing Ltd

Direct determination of defocus, thickness and composition from high-resolution electron microscopy lattice images of group-IV semiconductors

D Stenkamp and H P Strunk

Institut für Werkstoffwissenschaften VII, Universität Erlangen-Nürnberg, Cauerstr. 6
D-91058 Erlangen, Germany

ABSTRACT: A quantitative method for determination of defocus $\triangle f$ and local thickness t from high-resolution electron microscopy lattice images of wedge-shaped Si, Ge and $Si_x Ge_{1-x}$ samples in [100], [110] and [111] projections is proposed, which also allows a restriction-free determination of local compositions x at coherent $Si/Si_x Ge_{1-x}$ interfaces. The method relies on the analytically derived dependence of the first-order linear and non-linear image Fourier coefficients J_1 and J_2 on $\triangle f$, t and x. By plotting J_1 versus J_2 for varying t, ellipses with defocus- and composition-specific geometry are obtained. By reconstructing the appropriate ellipse for image regions of constant composition, $\triangle f$ and t can be determined independently. At interfaces, local compositions x can be determined by utilizing a systematic "rotation" of the ellipses with varying composition.

1. INTRODUCTION

High-resolution electron microscopy (HREM) reveals the atomic structure of crystal defects by matching experimental images with simulated images of structure models, for which a precise knowledge of sample thickness t and defocus $\triangle f$ is of critical importance. Compared to indirect determination methods for $\triangle f$ and t such as thickness fringe and amorphous phase contrast analysis, direct methods usually yield precise values for $\triangle f$ and t by contrast matching of defect-free image regions close to the image area of interest (Thust and Urban 1992, King and Campbell 1993). However, these methods principally fail for highly-symmetric materials with small lattice parameters for which identical image contrast patterns occur for many different combinations of $\triangle f$ and t. As an important example, this case occurs for lattice images of the group-IV semiconductors Si, Ge and $Si_x Ge_{1-x}$ alloys taken along [100] and [111] zone axes.

Similar difficulties occur for the composition determination at $Si/Si_x Ge_{1-x}$ interfaces by analysis of local image contrast patterns. Here, contrast variations due to variations in $\triangle f$ and t are a priori not descernible from variations in x, which is why sophisticated methods have been proposed in order to allow a quantitative determination of x on an atomic-scale (Stenkamp and Jäger 1993, Schwander et al. 1993).

In this paper we describe a novel quantitative approach for the independent determination of defocus $\triangle f$ and local sample thickness t from single HREM lattice images of wedge-shaped Si, Ge and $Si_x Ge_{1-x}$ alloy samples. Moreover, the method allows a restriction-free determination of the local composition x at coherent $Si/Si_x Ge_{1-x}$ interfaces for which only one image region of known composition is required as a reference.

2. THEORY

The proposed method basically relies on the functional dependence of the image intensity distribution on the parameters defocus Δf, sample thickness t and composition x. In this chapter, this dependence is systematically derived by an analytical description of the electron scattering and imaging process.

From theoretical investigations of the scattering behaviour of Si, Ge and $Si_x Ge_{1-x}$ alloys for low-indexed zone axes it is known that the electron wave field is dominated by only two Bloch waves (Glaisher et al. 1989, Stenkamp and Jäger 1993). In order to obtain analytical expressions for the amplitude U_0 and phase θ_0 of the undiffracted beam g_0 and for the amplitudes U_1 and phases θ_1 of the equivalent first-order diffracted beams g_1 for zone axes $\mathbf{B} = [100]$, $[110]$ and $[111]$, we formulate the dynamical electron scattering process as a Bloch wave eigenvalue problem including 5 Bloch waves. Comparisons with numerical 71 Bloch wave eigenvalue calculations reveal that the inclusion of only 5 Bloch waves is sufficient to describe the thickness dependence of g_0 and g_1 correctly. In a second step, we further reduce the effective number of Bloch waves contributing to the wave field down to 2 by exploiting the centro-symmetry of the crystal potential and applying matrix reduction methods. By using the [100] zone axis as example, this leads to eigenvalue equations

$$
\begin{pmatrix} \gamma^{(j)} & 4\,W_{220}(x) \\ W_{220}(x) & \gamma^{(j)} - g_{220}^2 + 2W_{400}(x) + W_{440}(x) \end{pmatrix} \times \begin{pmatrix} C_{000}^{(j)} \\ C_{220}^{(j)} \end{pmatrix} = 0,
$$

in which $\gamma^{(j)}$ denotes the eigenvalue and the $C_g^{(j)}$ denote the eigenvector components of the j-th Bloch wave ($j = 1, 2$), and the $W_g(x)$ denote scaled Fourier coefficients of the crystal potential for a given composition x. Superposition of these two Bloch waves leads to a thickness dependence of the beam amplitudes U_0, U_1 and of the relative beam phase $\Delta\theta_1 = \theta_1 - \theta_0$ given by

$$
U_0(t) = \sqrt{1 - 4\,C_0^{(1)2}\,C_0^{(2)2}\,\sin^2[\frac{\Delta k_B}{2}\,t]}, \qquad U_1(t) = C_0^{(1)}\,C_0^{(2)}\,\left|\sin[\frac{\Delta k_B}{2}\,t]\right|
$$

$$
\Delta\theta_1(t) = \frac{\Delta k_B}{2}\,t - \arctan\frac{C_0^{(1)2}\,\sin[\Delta k_B\,t]}{C_0^{(1)2}\,\cos[\Delta k_B\,t] + C_0^{(2)2}} + \frac{\pi}{2}\,\text{sign}\{\sin[\frac{\Delta k_B}{2}\,t]\}
$$

Here, Δk_B denotes the wave vector difference $(k_B^{(1)} - k_B^{(2)})$ between Bloch wave #1 and #2 which is directly correlated to the extinction distance ξ by $\Delta k_B = 2\pi/\xi$.

Using the non-linear imaging theory, the spatially varying part of the image intensity distribution under coherent 5- ($\mathbf{B} = [100]$, $[110]$) and 7-beam ($\mathbf{B} = [111]$) illumination conditions can be expressed by $I(\mathbf{r}) = \sum_{i=1,2} J_i \cos[\,\mathbf{g}_i \cdot \mathbf{r}\,]$. Here, J_1 denotes a linear image Fourier coefficient describing a single-periodic image contrast modulation with period g_1 and J_2 denotes a non-linear image Fourier coefficient describing a double-periodic image contrast modulation with period $\mathbf{g}_2 = 2\mathbf{g}_1$. J_1 and J_2 are given by

$$
J_1(t, \Delta f) = 4\,U_0(t)\,U_1(t)\,\cos[\,\Delta\theta_1(t) + \chi(g_1, \Delta f)\,] \quad \text{and} \quad J_2(t) = 2\,U_1^2(t),
$$

where $\chi(\Delta f) = \pi\lambda\,\Delta f\,g^2 + 0.5\,\pi\,C_s\,\lambda^3 g^4$ denotes the objective lens wave aberration function (λ electron wavelength, C_s spherical aberration). By substituting U_0, U_1 and $\Delta\theta_1(t)$ by their previously derived equivalents, final equations for $J_1(t, \Delta f)$ and $J_2(t)$ are obtained

$$
J_1(t, \Delta f) = (C_0^{(1)2} - C_0^{(2)2})\cos\chi(\Delta f) - \sqrt{1 - 4\,C_0^{(1)2}\,C_0^{(2)2}\cos^2\chi(\Delta f)}\,\cos[\Delta k_B\,t + \phi(\Delta f)]
$$

$$
J_2(t) = 0.5\,C_0^{(1)}C_0^{(2)}\,(1 - \cos[\Delta k_B\,t]),
$$

with $\sin\phi(\triangle f) = -\sin\chi(\triangle f)/(1 - 4\,C_0^{(1)2}\,C_0^{(2)2}\cos^2\chi(\triangle f))^{1/2}$. Both $J_1(t,\triangle f)$ and $J_2(t)$ are periodic functions in t with a period length given by the extinction distance ξ. Moreover, $J_1(t,\triangle f)$ is also periodic in $\triangle f$ with a period length given by $\delta f = 2/(\lambda\,g^2)$.

3. ELLIPTIC REPRESENTATION

Inspection of the final equations for J_1 and J_2 directly reveals that these can be rewriten in a more convenient form as

$$J_1(t,\triangle f) = M_1(\triangle f) - R_1(\triangle f)\,\cos[\,2\,\pi\,t/\xi + \phi(\triangle f)\,]$$
$$J_2(t) = M_2 - R_2\,\cos[\,2\,\pi\,t/\xi\,],$$

(note that $M_2 = R_2$). In this representation, the equations for J_1 and J_2 directly remind of those for a well-known geometrical curve: that of a 2-dimensional ellipse. By taking (J_1, J_2) as the (x, y) components of the ellipse, its "variable" phase angle is given by $2\pi t/\xi$, its geometrical center by $(M_1(\triangle f), M_2)$ and its eccentricity is determined by the ratio $R_1(\triangle f)/R_2$ and by the phase angle offset $\phi(\triangle f)$. Moreover, all possible ellipses cross the origin $(0, 0)$ for thicknesses $t = 0$ and $t = \xi$, independent of $\triangle f$ and x. As an example, Fig. 1 depicts such an ellipse for Si at $400\,\text{keV}$ which is composed of all possible combinations of "contrast patterns" (J_1, J_2) that can occur for the choosen defocus of $\triangle f = -57\,\text{nm}$. As is indicated by the arrow, for this defocus the path of the ellipse is traversed in a clockwise sense for increasing t.

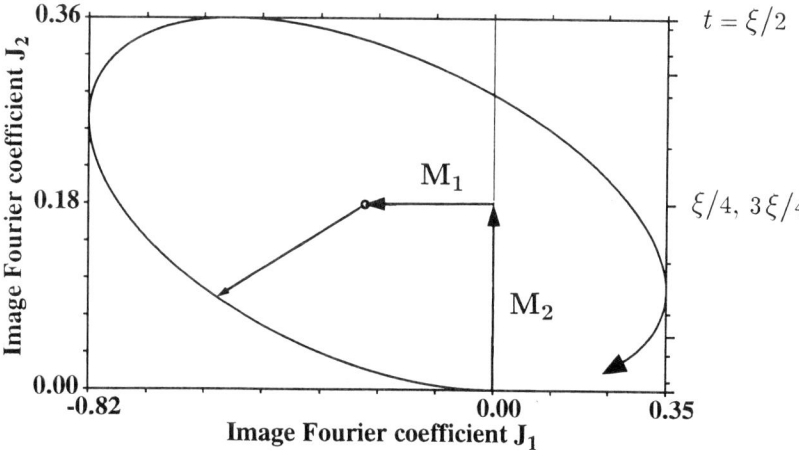

Fig. 1: Plot of image Fourier coefficients J_2 vs. J_1 as a function of thickness t (Si, $400\,\text{keV}$, $\mathbf{B} = [100]$, $\triangle f = -57\,\text{nm}$).

An advantage of this elliptic representation is the possibility to directly read out the appropriate absolute thickness to each individual "contrast pattern" (J_1, J_2). This becomes possible due to the access to the maximum and minimum of J_2, which correspond to the "highest", respectively "deepest" points of the ellipse and which correspond to $t = 0$ and $t = \xi/2$. From this knowledge and by additionally taking into account the ellipse's rotation sense, absolute values for t can be determined. Thereby it is possible to define a correspondingly scaled thickness axis parallel to the J_2 axis from which t can be read out (Fig. 1). As an important aspect, this thickness determination is not influenced by any linear scaling of J_1 or J_2 since for determination of t only the scale-invariant ratio J_2/M_2 is used. This is of importance concerning the effect of partially coherent illumination which leads to a linear damping of J_1 and J_2 (Ishizuka 1980).

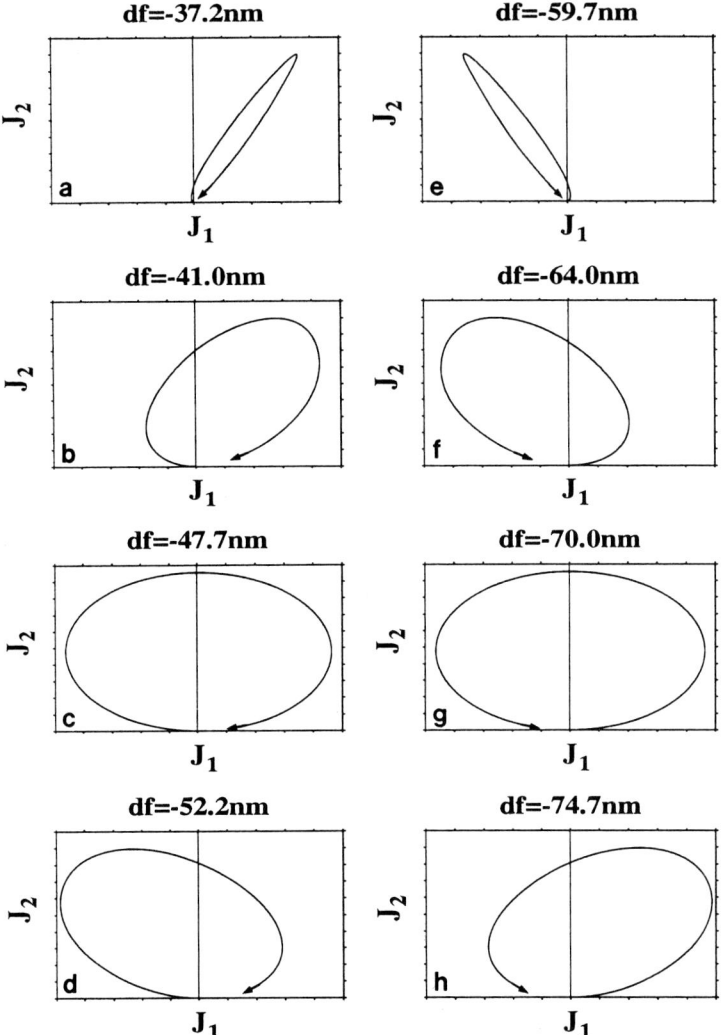

Fig. 2: Plots of image Fourier coefficients J_2 vs. J_1 as a function of thickness t for different defoci $\triangle f$ (Si, 400 keV, $\mathbf{B} = [100]$).

The defocus-dependence of the ellipse is shown in Fig. 2, again using the example of Si at 400 keV. Figures (a) to (h) show the variation of the eccentricity and rotation sense over a full defocus period δf. Important aspects are the systematic in- and decrease of the eccentricity and the change in the rotation sense beetween $\triangle f = -52.2$ (Fig. 2d) and $\triangle f = -59.7$ (Fig. 2e). Based on simple geometrical arguments, it is possible to directly determine the appropriate value of the wave aberration function from the ellipse's geometry

$$\chi \;=\; \arctan\{\, \frac{J_1^*(\xi/4)}{M_1}\, [C_0^{(2)^2} - C_0^{(1)^2}]\,\},$$

where $J_1^*(\xi/4) = J_1(\xi/4) - M_1$. Since χ itself depends linearly on $\triangle f$, this allows an unambiguous determination of $\triangle f$. Similar than for the thickness determination, also the determination of $\triangle f$ is invariant against any linear scaling of J_1 or J_2 since the resulting effects are automatically canceled out by using the ratio $J_1^*(\xi/4)/M_1$.

The composition-dependence of the ellipse is shown in Fig. 3, which depicts the corresponding ellipses for Si, a $Si_{0.5}Ge_{0.5}$ alloy and Ge at 400 keV, $\mathbf{B} = [100]$ and $\triangle f = -52.2$ nm. As becomes directly obvious, an increase in the Ge content leads to a systematic "rotation" of the ellipse for Si, similar to the case of varying defocus (Fig. 2). The reason for this "rotation" are systematic changes in the ratio of the eigenvector components $C_0^{(1)}/C_0^{(2)}$ and in the length of g_1 with increasing Ge content, which lead to substantial variations in M_1, M_2, R_1 and ϕ. It is important to note that parallel to the directly obvious "rotation" effect there also occurs a "shrinking" of the ellipse's size with increasing Ge content. A detailed study of the composition dependence of ellipses for Si_xGe_{1-x} alloys within the range $0 \leq x \leq 1$ has revealed that the demonstrated example is representative also for the zone axes [110] and [111] and for different acceleration voltages.

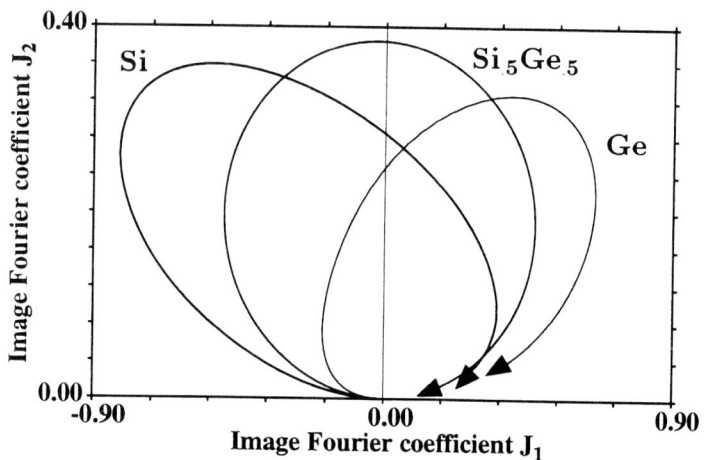

Fig. 3 : Plot of image Fourier coefficients J_2 vs. J_1 as a function of thickness t for Si, a $Si_{0.5}Ge_{0.5}$ alloy and Ge (400 keV, $\mathbf{B} = [100]$, $\triangle f = -52.2$ nm).

4. DEFOCUS, THICKNESS AND COMPOSITION DETERMINATION

Basing on the derived theoretical fundament (Sect. 2) and utilizing the described ellipse represention for J_1 and J_2 (Sect. 3), we now describe an approach for the direct determination of defocus and thickness from single experimental HREM lattice images of wedge-shaped Si, Ge and Si_xGe_{1-x} alloy samples. The approach basically relies on the justified assumption that within limited regions of a lattice image $\triangle f$ is nearly constant. Moreover, typical HREM samples must be assumed as wedge shaped what leads to variations of the thickness and therefore of the image contrast pattern within the field of view. With respect to the introduced elliptic representation, each lattice image therefore naturally contains a number of different "contrast patterns" (J_1, J_2) which fall within a segment of the ellipse for the choosen defocus (Fig. 1). The principal idea of our approach is to utilize this segment in order to reconstruct the complete ellipse from it and then to determine defocus and local thickness values separately in the described way (Sect. 3).

Practically this is achieved by first deviding the digitized image into small image cells and measuring J_1 and J_2 within each cell by Fourier transform (see also Stenkamp and Jäger 1993). Afterwards, an appropriate ellipse is reconstructed from this data, taking additionally into account that the ellipse has to cross the origin. Finally, $\triangle f$ and local t values for each image cell are determined.

For the additional determination of local composition values at coherent Si/Si_xGe_{1-x} interfaces, we first measure data points for (J_1, J_2) within the total image, including at least one image region of known composition as reference. From this image region we then reconstruct the appropriate ellipse and determine the image defocus. With the knowledge of the defocus and the composition of this region, we then perform a systematic ellipse scan as function of x over all data points (J_1, J_2) of unknown composition. By application of this procedure, a restriction-free composition determination within the full composition range $0 \leq x \leq 1$ becomes feasible which takes possible thickness fluctuations automatically into account.

5. SUMMARY AND CONCLUSIONS

In this paper we present a novel approach for the direct determination of defocus and local sample thickness from lattice images of group-IV semiconductors. The method fundamentally relies on the dependence of the first-order image Fourier coefficients J_1 and J_2 on $\triangle f$ and t, as is analytically derived by application of the Bloch wave scattering formalism and application of the non-linear imaging theory. By introduction of an elliptic representation for J_1 and J_2, a direct determination of local t values as a function of the appropriate "contrast pattern" (J_1, J_2) becomes possible and $\triangle f$ can be obtained from simple geometrical parameters of the ellipse. By application of the approach to lattice images of crystal defects, precise values for $\triangle f$ and t can be obtained for defect-free image regions, thus providing an optimum start position for the atomic structure determination of the defect by image matching techniques.

For the analysis of coherent interfaces in Si/Si_xGe_{1-x} heterostructures, the method uses a systematic "rotation" of the ellipse with varying x. By using only one image region of known x as a reference, a quantitative composition determination over the full composition range $0 \leq x \leq 1$ becomes feasible which takes variations of the sample thickness automatically into account. It thereby becomes possible to quantitatively analyse also interfaces for which the composition is known only for one side of the interface or for which the compositional transition region is very broad (e.g. of the order of 10 nm).

REFERENCES

Glaisher R W, Spargo A E C and Smith D J 1989 Ultramicroscopy **27** 19, 35
Ishizuka K 1980 Ultramicroscopy **5** 55
King W E and Campbell G H 1993 Ultramicroscopy **51** 128
Schwander P, Kisielowski C, Seibt M, Baumann F H, Kim Y and Ourmazd A 1993 Phys. Rev. Lett. **71** 4150
Stenkamp D and Jäger W 1993 Ultramicroscopy **50** 321
Thust A and Urban K 1992 Ultramicroscopy **45** 23

Inst. Phys. Conf. Ser. No 146
Paper presented at Microsc. Semicond. Mater. Conf., Oxford, 20–23 March 1995
© *1995 IOP Publishing Ltd*

Effect of the static atomic displacements in quantitative HREM of III-V alloys

F Glas

France Telecom, Centre National d'Etudes des Télécommunications, Paris B, Laboratoire de Bagneux, 196 avenue Henri Ravéra, BP 107, 92225 Bagneux Cedex, France

ABSTRACT: We investigate the effect upon HREM images of the static atomic displacements (SD) present in most III-V alloys, by comparing pairs of images of the same distributions of atoms calculated with and without taking the SD into account. Large differences usually exist between these images. We assess them quantitatively and demonstrate that proper consideration of the SD is essential to achieve reliable composition analysis by HREM.

1. INTRODUCTION

Although any III-V alloy possesses an average cubic sphalerite structure, each particular specimen deviates from perfect periodicity (except in the highly hypothetical case of perfect ordering), for two reasons: (i) any site of the mixed sublattice(s) is occupied by a real atom, not an average atom; (ii) with the single exception of Ga and Al, the atoms of the mixed sublattice(s) have very different covalent radii. Consequently, each atom is displaced from the site to which it belongs. We demonstrated previously that, even if the alloy is ideally homogeneous, these static atomic displacements (SD) induce a typical diffuse intensity distribution in the diffraction patterns (Glas et al 1990) and a previously unrecognized and characteristic 'atomic size effect contrast' (ASEC) in the conventional Transmission Electron Microscopy images (Glas 1993, 1995a). Here, we investigate the effect of the SD in High Resolution Electron Microscopy (HREM).

Extracting compositional data from the images (ideally, the composition of each atomic column) has recently become a major task for quantitative HREM of alloys. Several schemes have been proposed and applied, albeit nearly exclusively to the SD-free $Ga_xAl_{1-x}As$ alloy (Ourmazd et al 1989, Ikarashi et al 1989, de Jong and Van Dyck 1990, Thoma and Cerva 1991). Our primary aim is not to judge the applicability of these various schemes to alloys with SD, but to assess quantitatively the effect of the SD in HREM. To this end, we compare pairs of simulated images of the same distribution of atoms, one of them calculated with the SD, the other without.

2. SPECIMENS AND IMAGES: SIMULATIONS AND COMPARISON CRITERIA

2.1 Specimen simulation

We simulate ternary $A_xB_{1-x}C$ crystalline alloy specimens as detailed previously (Glas et al 1990, Glas 1995a). First, the lattice parameter a of the average crystal of composition x is obtained from Vegard's law. We place A and B atoms at the sites of the average mixed sublattice and C atoms on the other sublattice. To study the intrinsic effect of the SD (aside from that of any genuine composition variation), we simulate ideally homogeneous non-ordered alloys, where the occupancies of the sites of the mixed sublattice are totally uncorrelated. We let all the atoms relax by minimizing the Valence Force Field excess energy (see references and justifications in Glas 1995a). The atoms are allowed to move slightly away from the average sites, but not to exchange sites. We thus get the SD field associated with the initial atomic distribution. Since, even in ideally disordered alloys, the SD are (partially) correlated over distances of several nm (Glas 1995a), we simulate crystals with sides L equal to 5 and 10 nm. Surface relaxation effects have very little effect on the following results.

24

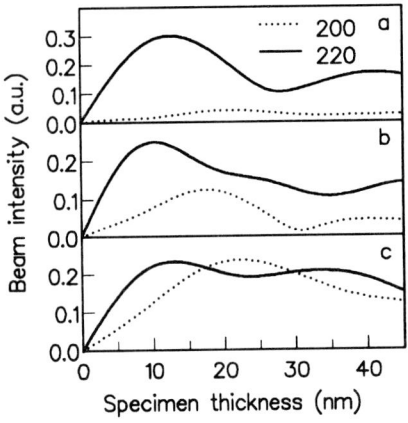

Fig. 1: Variation with specimen thickness of the intensities of the 200 and 220 beams in GaAs (a), InAs (b) and AlAs (c); $E_0 = 400$ keV.

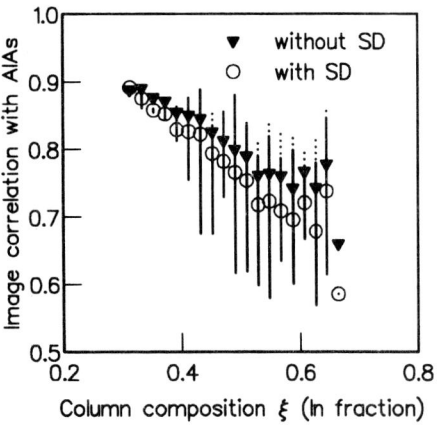

Fig. 3: As Fig. 1 for an $In_{0.5}Al_{0.5}As$ alloy specimen 30 nm thick and cross-correlation with a reference AlAs. $E_0 = 400$ keV, $\Delta f = -50$ nm, $\Phi = 12$ nm^{-1}.

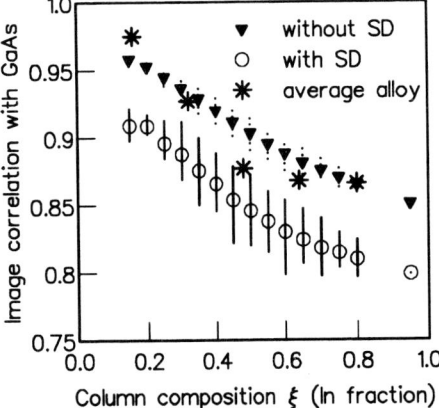

Fig. 2: Variation with group III column composition, in a disordered $In_{0.5}Ga_{0.5}As$ alloy specimen 12 nm thick, of the cross-correlation C between the image unit cell centered on the column and the image of a reference GaAs specimen. Symbols: averages; bars: ranges. $E_0 = 400$ keV, $\Delta f = -50$ nm, $\Phi = 8$ nm^{-1}.

Fig. 4: Variation with specimen thickness (10.6, 15.8, 21.1, 26.6 nm from a to d) of the cross-correlations between the image unit cells of an $In_{0.5}Al_{0.5}As$ alloy specimen and of a reference AlAs specimen of the same thickness. $E_0 = 400$ keV, $\Delta f = -50$ nm, $\Phi = 12$ nm^{-1}. Same symbols as in Figs. 2 and 3.

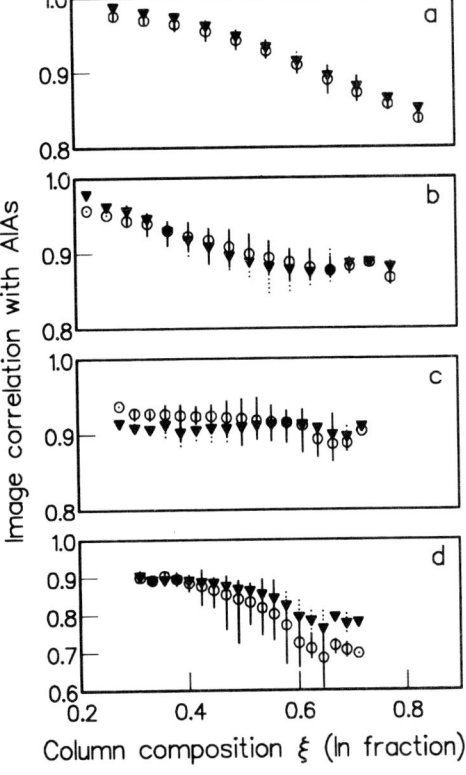

2.2 Image calculation

The whole simulated crystal is treated as a giant supercell, whose images are calculated by using the standard multislice routine of the EMS package (Stadelmann 1987). The slices, all different, have a thickness $a/2$. The images are calculated in the [001] orientation usually chosen for quantitative

HREM in these alloys: the atomic columns then contain either only group III or only group V atoms. In this orientation, the diffracted beams used to form the image of the III-V binaries are the 200 beams, and, depending on the resolution of the microscope and the choice of the aperture, the 220 beams. In the alloys, these beams, now correponding to the average lattice, are still used. However, since the specimen is not perfectly periodic (see section 1), there exists, in addition to the average beams, a continuous distribution of diffusely scattered beams, which we sample at intervals $1/L$ along reciprocal directions [100] and [010]. Such a fine sampling is essential, since the information about any difference in intensity between columns and any local change of pattern is entirely contained in these diffuse beams. Moreover, the differences between the calculations without and with SD are mainly contained in the differences of diffusely scattered intensity. 512x512 beams are used in the calculation. The imaginary part of the potential is taken equal to 10% of its real part. The images are calculated for primary beam energies E_0 of 200 or 400 keV, and for a microscope with a spherical aberration constant $C_s = 1$ mm. The aperture diameter Φ is either 8 nm^{-1} or 12 nm^{-1}; this choice is discussed below.

2.3 Quantitative image comparison

Although our specimens are ideally homogeneous, each mixed column contains so few atoms that, due to statistical fluctuations, its actual composition ξ in element A may vary widely. To compare the images calculated with and without SD, we first divide the image into unit cells of sides $a/2$ along [100] and [010] and centered on a mixed column (Glas 1995b). Each cell is then subdivided into n pixels, and, following Ourmazd et al (1989), we calculate the cross-correlation C between this cell and a corresponding cell in a reference binary (AC or BC):

$$C = \left(\sum_{i=1}^{n} x_i \, y_i \right) / \left(\sum_{i=1}^{n} x_i^2 \, \sum_{i=1}^{n} y_i^2 \right)^{1/2}$$

where the intensities of corresponding reference and alloy pixels are x_i and y_i. The results will be presented as a comparison of the variation of C with ξ between specimens with and without SD.

2.4 Choice of the alloys studied

Two types of methods have been proposed to analyse quantitatively the HREM micrographs of the III-V alloys. The first type relies on the 'chemical sensitivity' of the 200 beams: the relative variation with composition of the structure factor is usually much larger for the latter than for the 220 beams. In the simplest of these methods, one forms the image with an aperture retaining only the 200 beams (Ikarashi et al 1989, de Jong and Van Dyck 1990). A more refined method uses a large aperture, but filters the image to select an information associated with the 200-type beams (Thoma and Cerva 1991). On the other hand, the methods of the second type (Ourmazd et al 1989) use the sensitivity to composition of the relative weights of the 200 and 220 beams to evaluate the changes of composition through the changes of patterns in HREM images formed with these two types of beams.

These methods have been mainly applied to Ga$_x$Al$_{1-x}$As. It has not been noticed that usually, for each III-V alloy, one of these methods is better suited than the other. Fig. 1 illustrates this point in the cases of In$_x$Ga$_{1-x}$As and In$_x$Al$_{1-x}$As. The latter alloy is adapted to a treatment of the second type, because thickness ranges exist where the 200 beams are more intense than the 220 beams in AlAs, but not in GaAs. On the contrary, at any reasonable thickness, the 220 beams are stronger than the 200 beams in GaAs as well as in InAs; the image of any In$_x$Ga$_{1-x}$As alloy will thus be dominated by the 220 fringes pattern. To illustrate this difference of behaviour, we give results for both alloys.

3. RESULTS

3.1 Using composition sensitive beams

Fig. 2 shows the variation with alloy column composition ξ of the cross-correlation C between the image unit cells of an In$_{0.5}$Ga$_{0.5}$As specimen and of a reference GaAs specimen of the same thickness $t = 12$ nm, calculated at Scherzer defocus $\Delta f = -50$ nm for $E_0 = 400$ keV and with an aperture excluding the 220 beams ($\Phi = 8$ nm^{-1}). Comparing calculations with and without SD, we notice that:
- For any column composition ξ, there is a range of C values, which is larger with SD than without SD.
- In both cases, as expected, C decreases (on average) as the deviation of ξ from pure GaAs increases.
- However, the average of C over columns of a given composition is systematically lower with SD

than without SD. This has important consequences for quantitative HREM analysis because, in order to find the composition of an unknown alloy column, even if experimental images of alloys of known composition can be obtained, calibration curves such as these must be used to interpolate the behaviour of C between these references. Then, the omission of the SD has disastrous consequences: for instance, a measured $C = 0.9$ corresponds to a column composition $\xi = 0.52$ $(-0.07/+0.06)$ if the SD are omitted, but to $\xi = 0.24$ $(-0.04/+0.09)$ only if they are correctly taken into account.
- Not only the values of C are different with and without SD, but the local slopes $dC/d\xi$ as well.

Calculating the images by using 'average' atoms (a common practice) produces results differing from the calculations made with as well as without SD, although they are closer to the latter (Fig. 2).

3.2 Using composition sensitive image patterns

Fig. 3 is the analogue of Fig. 2 for a disordered $In_{0.5}Al_{0.5}As$ specimen with $t = 30$ nm (chosen from Fig. 1). To include the $\bar{2}20$ beams, we select $\Phi = 12$ nm^{-1}. The reference binary is now AlAs. Even in such conditions, very favourable for this method (C varies indeed faster with ξ than in Fig. 2, at least for $\xi \le 0.55$), omitting the SD leads to errors in quantitation if one uses the average values of the correlations to estimate a particular column composition, although these errors are smaller here than in the previous case. These average values are however of little practical use, because the correlations calculated for each column composition are widely scattered: for instance, any correlation between $C = 0.6$ and $C = 0.8$ may correspond to any composition between $\xi = 0.48$ and $\xi = 0.66$.

4. DISCUSSION

These results prove that the SD usually affect strongly the HREM images of III-V alloys. Many calculations, where we varied E_0, Δf, Φ and t, show that a significant difference between images with and without SD is a very common feature (see also Glas 1995b). We nevertheless find rare cases, in particular near Scherzer focus for alloys of the $In_{0.5}Al_{0.5}As$ type (Fig. 4), where the SD have less effect on the HREM images (or at least on the averages of C for each ξ). However, even then, C can hardly be used for quantitative analysis: either it varies little with ξ (Fig. 4 b,c) or its range for each ξ is broad (Figs. 3 and 4 d), except in narrow thickness intervals (Fig. 4 a).

This effect of the SD on the images can be understood in several ways. The projected potential is changed because of the SD. Since the latter are correlated over distances of several nm (Glas 1995a), the local diffraction conditions are altered. In reciprocal space terms, the disorder-induced distributions of diffusely scattered beams are different with and without SD.

The difference between calculations with and without SD manifests itself by several features. Firstly, the cross-correlations C for a given column are different. C is usually lower with SD than without SD: indeed, the SD are an extra source of difference between the alloy and the binary. Secondly, for a given ξ, the values of C are systematically more scattered with SD than without. The scatter without SD is due to two effects: the potential at a given point does not depend only on the nearest atom but also on its neighbours; moreover, two columns having the same composition ξ contain the same atoms, but very often stacked differently. With the SD, a third effect appears: the SD of each atom is the net result of the strains induced by all its neighbours, and thus depends on its environment. This third effect usually dominates (Figs. 2, 3).

Our results show that it is essential to take properly into account the SD in any quantitative HREM study of the III-V alloys, and probably of any alloy, with atomic size effects.

REFERENCES

de Jong A F and Van Dyck D 1990 Ultramicroscopy $\underline{33}$ 269
Glas F 1993 Proc. Micr. Semicond. Mater. 1993, eds A G Cullis, A E Staton-Bevan and J L Hutchison, Inst. Phys. Conf. Ser. No 134 (Bristol: Institute of Physics) pp 269-278
Glas F 1995a Phys. Rev. B $\underline{51}$ 825
Glas F 1995b accepted for publication in Ultramicroscopy
Glas F, Gors C and and Hénoc P 1990 Phil. Mag. B $\underline{62}$ 373
Ikarashi N, Sakai A, Baba T and Ishida K 1989 Appl. Phys. Lett. $\underline{55}$ 2509
Ourmazd A, Taylor D W, Cunningham J and Tu C W 1989 Phys. Rev. Lett. $\underline{62}$ 933
Stadelmann P A 1987 Ultramicroscopy $\underline{21}$ 131
Thoma S and Cerva H 1991 Ultramicroscopy $\underline{38}$ 265

Inst. Phys. Conf. Ser. No 146
Paper presented at Microsc. Semicond. Mater. Conf., Oxford, 20–23 March 1995
© *1995 IOP Publishing Ltd*

Self-assembled monolayers of semiconductor quantum dots characterised by high resolution electron microscopy

P J Hull, O V Salata*, J L Hutchison and P J Dobson*

Department of Materials, University of Oxford, Parks Road, Oxford OX1 3PH
*Department of Engineering Science, University of Oxford, Parks Road, Oxford OX1 3PJ

ABSTRACT: Quantum dots are nanometre scale semiconductor crystallites predicted to have properties desirable for optoelectronic applications. Colloidal chemistry techniques, though able to form quantum dots in the required size range, are inadequate if uniform, dense layers of dots are required. In this investigation, bifunctional molecules have been used to selectively bind colloidal quantum dots onto a substrate. HREM has proved vital in the analysis of the resulting systems.

1. INTRODUCTION

There has recently been a great deal of interest in semiconductor quantum dots (QD) for potential optoelectronic applications (Brus 1991, Weller 1993). They represent an extension of the principles behind the more familiar quantum wells. By confining the charge carriers to a volume less than the size of the bulk exciton, new physical properties arise. In particular, the parabolic bulk density of states is replaced by a series of sharp peaks, whose size depends strongly on the crystallite diameter. This has potential applications as monochromatic light emitters or modulators.

Conventional lithography and etching are well suited to producing ordered arrays of quantum dots, but are limited to lateral dimensions of the order of 100nm. To achieve measurable quantum confinement effects at room temperature it is necessary to fabricate structures less than 10nm in size, and various novel techniques have been proposed to achieve this (see for example, Oshinowo 1995). This work concentrates on colloidal techniques that have many advantages over other, more elaborate schemes.

2. COLLOIDAL TECHNIQUES

2.1 Formation of quantum dots

The essence of the colloidal technique is the direct reaction of two dilute solutions to form a product semiconductor that is insoluble. Under appropriate conditions of temperature, pH and the presence of suitable surface stabilisers, the resultant semiconductor will be in the form of nanometre-scale quantum dots. Colloid chemistry is robust, scalable and a mature technology.

For this investigation, cadmium sulphide, a direct band gap semiconductor, was chosen. It has a bulk band gap in the orange region of the spectrum, which can be blue-shifted into the ultra-violet by quantum confinement effects. The particles were grown at room temperature, using solutions of cadmium nitrate and sodium sulphide, at a pH above 9 (Weller 1993).

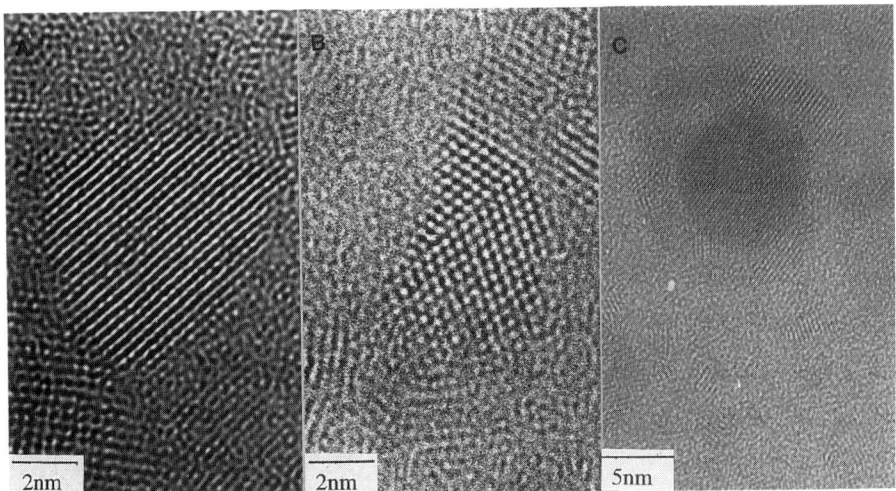

Figure 1 Colloidal CdS quantum dots: a) Perfect crystal showing 2.9Å {200} lattice planes. b) A {111} cubic-to-hexagonal stacking fault. c) Composite quantum dot with a core of CdO ({200} planes, 2.4Å) and a shell of CdS ({111} planes, 3.4Å).

Using high resolution electron microscopy it is possible to characterise a sample of quantum dots, at the atomic level. Figure 1 shows micrographs of various colloids taken with a JEOL 4000EX operating at 400KV. The point resolution under these conditions is 1.6Å. Note that HREM reveals information that would not be apparent in conventional TEM. The presence of the stacking fault (Figure 1b) is interesting; bulk CdS is hexagonal, whereas most quantum dots possess the cubic zinc blende structure. This type of defect represents exactly the insertion of one plane of hexagonal stacking into a cubic crystal. It is supposed that as the crystallite grows, the incidence of these faults increases until the hexagonal form dominates

2.2 Problems with colloids

While suspended in a solvent, the QD are stable over a long period of time. Once dried however, adjacent dots will fuse together due to the large areas of reactive surface present. In addition, the moving solvent front of an evaporating liquid film will tend to pull the dots together by surface tension forces, exacerbating the problem.

3. SELF-ASSEMBLED MONOLAYERS

One possible solution is to use self-assembled monolayers (SAM) as a "molecular glue". These layers are composed of bifunctional molecules, the head of which binds to a substrate, with the tail binding to the quantum dots (Colvin 1991, Fendler 1994). When a solution of such molecules is applied to a substrate, a monolayer *spontaneously* forms. (Figure 2) Subsequent application of the colloid to the monolayer results in the quantum dots binding chemically to the surface, and are thus unable to aggregate or fuse together. By

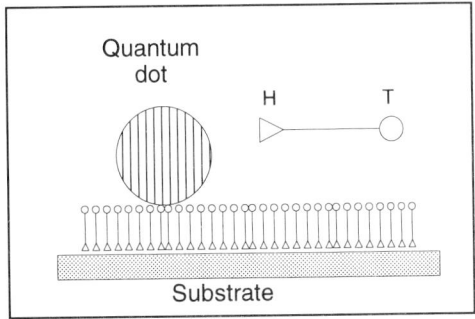

Figure 2 Schematic formation of a quantum dot/monolayer film

Substrate	Head	Body	Tail	Semiconductor
C, Au	HS—			
Al_2O_3	COOH—	$(CH_2)_n$	—SH	CdS
SiO_2	$Si(OCH_3)_3$—			

Table I Some functional groups used in self-assembled monolayers

choosing the head and tail groups it is possible to bind a variety of semiconductors and substrates — see Table I for examples. Note that it is also possible to use silicon and aluminium substrates, given the native oxides which invariably form on these materials. Two systems have been investigated here — mercaptopropyltrimethoxysilane (MPMS, $Si(OCH_3)_3(CH_2)_3SH$) on silica, and hexanedithiol (HDT, $HS(CH_2)_6SH$) on amorphous carbon.

4. EXPERIMENTAL AND RESULTS

4.1 MPMS on Silica

A film of silica, nominally 10nm thick was sputtered onto a carbon-coated copper grid. Due to the deposition conditions used, the film was not continuous but consisted of several silica islands on the carbon. This grid was placed in an ethanol solution of MPMS for approximately ten hours. This was followed by a rinse in ethanol, and then a droplet of aqueous CdS colloid was placed on the grid. After a few hours exposure the excess colloid was rinsed off in ethanol and dried.

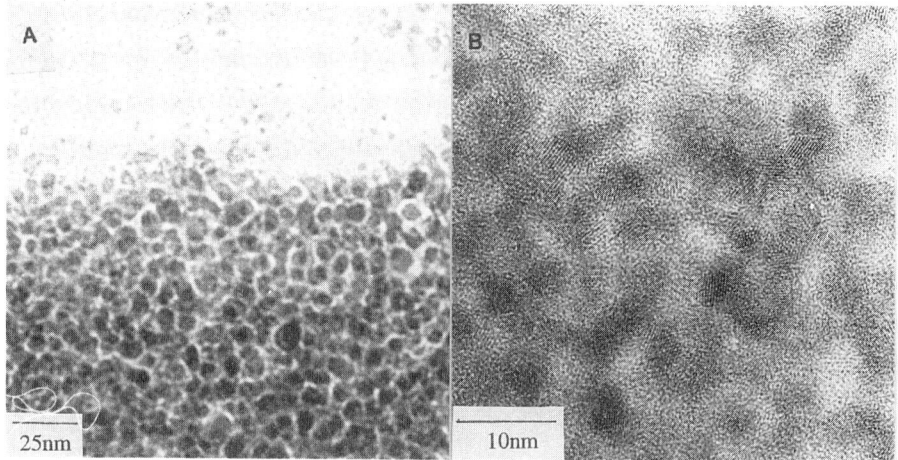

Figure 3a) CdS quantum dots bound to a silica island with MPMS b) At high magnification, showing lattice spacings

In Figure 3 the quantum dots can clearly be seen, localised to a silica island. The —$Si(OCH_3)_3$ group binds strongly to the silica and poorly, if at all, to the carbon. The dots appear densely packed, but *not* overlapping. Some dots can be seen on the carbon; this may be due to very small stray pieces of silica or a less favourable reaction.

4.2 HDT on Carbon

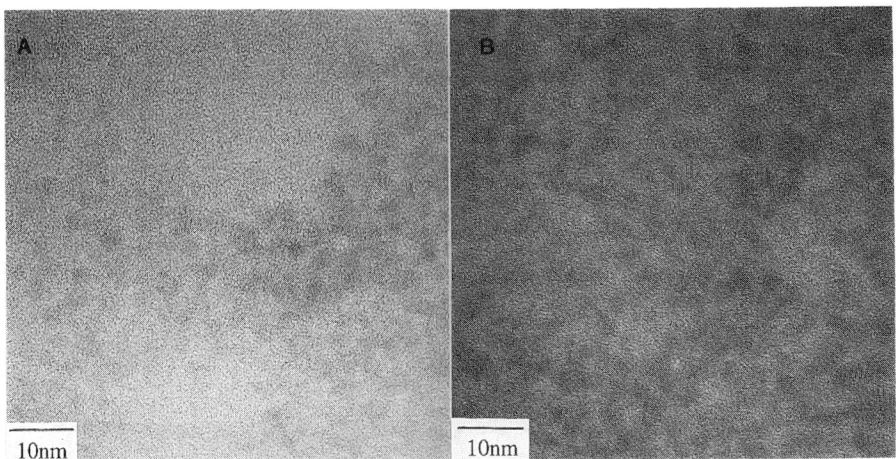

Figure 4 CdS bound to hexanedithiol (HDT) on carbon a) 1 hour exposure to HDT b) 19 hours exposure to HDT

The method used was similar to that used for MPMS, except that no silica film was deposited. In Figure 4, micrographs representing HDT treatment times of 1 hour and 19 hours are presented. It is clear that the longer time produces a higher coverage area of quantum dots. However, within the covered area the density of dots appears fairly constant, which may imply that an incoming HDT molecule is more likely to attach to the substrate if another HDT molecule is already bound nearby.

5. CONCLUSIONS

It been demonstrated that self-assembled monolayers can be effective as a 'molecular glue' to bind semiconductor quantum dots to a variety of surfaces, thus improving the localisation and stability of the sample. The binding has also been shown to be selective, which opens up the possibility of patterning a substrate and producing ordered arrays of quantum dot patches.

The role of electron microscopy is critical, since it allows characterisation of individual crystallites and their spatial location. By contrast, spectroscopic and diffraction techniques yield results averaged over the whole sample. Furthermore, HREM is uniquely able to determine the lattice structure and reveal stacking faults or composite particles.

REFERENCES

Brus L E 1991 Appl. Phys. A **53** 465
Colvin V L, Goldstein A N, Alivisatos A P 1991 J. Am. Chem. Soc **114** 5221
Fendler J H 1994 Advances in polymer science **113**, Springer-Verlag
Oshinowo J, Nishioka M, Ishida S, Arakawa Y 1994 J. Cryst. Growth **145** 986
Weller H 1993 Adv. Mater. **5** 88

Inst. Phys. Conf. Ser. No 146
Paper presented at Microsc. Semicond. Mater. Conf., Oxford, 20–23 March 1995
© 1995 IOP Publishing Ltd

TEM/HREM characterization of self-organized (In,Ga)As quantum dots

S Ruvimov, P Werner, K Scheerschmidt, U Richter, U Gösele, J Heydenreich, N N Ledentsov*, M Grundmann*, D Bimberg*, V M Ustinov[†], A Yu Egorov[†], P S Kop'ev[†] and Zh I Alferov[†]

Max-Planck-Institut für Mikrostrukturphysik, Weinberg 2, 06120 Halle/Salle, Germany
* Technische Universität Berlin, Hardenbergstrasse 36, 10623 Berlin, Germany
[†]A F Ioffe Physical-Technical Institute, Politekhniczeskaya 26, 194021 St Petersburg, Russia

ABSTRACT: The morphology evolution of molecular beam epitaxy grown InAs and $In_{0.5}Ga_{0.5}As$ layers as a function of deposition thickness, ranging from 1 to 10 monolayers (ML), is studied by transmission electron microscopy (TEM) to characterize the formation and the self-organization of pseudomorphic quantum dots. The luminescence of all samples with coherent dots exhibits a high quantum efficiency.

1. INTRODUCTION

Currently, self-organization phenomena in crystalline semiconductors have increased in research interest (see e.g. Leonard et al. 1993, Nötzel et al. 1994, Ledentsov et al 1994) as a way possible to form nanoscale islands, namely quantum dots (QDs), without any patterning process. QDs are expected to exhibit unique properties like δ-function density of states leading to novel and/or strongly improved properties for photonic and electronic devices, e.g. lasers. Recently, this δ-function density of states (Ledentsov et al. 1994, Grundmann et al. 1995) and QD lasers (Kirstaedter et al. 1994) with unique properties were demonstrated.

Islanding growth in the (In,Ga)As-GaAs system is considered to result from a morphology evolution of the two-dimensional (2D) layer after the growth of 1 - 2 monolayers, explained by a Stranski-Krastanov model of coherent island growth. Either the elastic relaxation of the strain in heterostructures, i.e. a local minimum in the total energy (Eaglesham & Cerullo 1990), or the kinetics of strain-induced surface roughness (Snyder et al. 1991, Madhukar et al. 1994) has been assumed to be responsible for the self-organized formation of nm-size 3D islands. However, a conclusive picture for the 2D-3D morphology transformation has not yet been established. The results of the present work demonstrate the importance of both strain and temperature, which implies that there are energetic and kinetic effects.

Here we report on the nucleation of (In,Ga)As QDs on GaAs during the growth of thin pseudomorphic layers. Our experimental results as to dot size and shape differ from those reported in other papers on InAs-GaAs (Moison et al. 1994)) and InGaAs-GaAs dots (Leonard et al. 1993, Mo et al. 1990) grown under similar conditions. Here pyramid-shaped dots, as known for Ge on Si (Mo et al. 1990), are reported for InAs and $In_{0.5}Ga_{0.5}As$ for the first time. In addition, a self-organized short-range ordering of dots into rows along <100> is revealed, forming a primitive two-dimensional cubic lattice.

32

2. RESULTS AND DISCUSSION

InAs and InGaAs dots were grown by molecular beam epitaxy (MBE) on GaAs (001) using an EP1201 system (see, e.g., the details in Ledentsov et al. 1994). Their evolution was studied *in situ* by reflection high energy electron diffraction (RHEED), and *ex situ* (after the deposition of a cap layer) by transmission electron microscopy (TEM) using JEOL JEM1000 (1MV) and JEM4000EX (400kV) microscopes. Low-temperature (T = 8 K) photoluminescence (PL) and cathodoluminescence (CL) were used to characterize the optical properties of the dots.

In situ RHEED experiments show that the critical thickness for the formation of 3D islands (quantum dots) depends on the layer composition, i.e. elastic energy. Indeed, the critical thickness of $In_xGa_{1-x}As$ increases from 1.7 to 3 ML with decreasing In content x from 1.0 to 0.5. TEM studies generally confirm this observation. Figs. 1 (a)-(d) show typical plan-view TEM images of InGaAs and InAs layers grown at T_d=480°C and differing in thickness (which is above the critical value in all four cases). At the initial stages of (In, Ga)As dot formation with slashes in the RHEED pattern just appearing, the corresponding TEM images (see, e.g., Fig. 1 (a) for 3.3 ML of $In_{0.5}Ga_{0.5}As$) demonstrate a fine-scale, black-and-white granular contrast composed of round-shaped dots having a diameter of approxiamtely 6 nm. Locally connected, the dots appear in agglomerates. Depositing further material causes to the formation of well-developed coherent islands of increasingly uniform size, most probably owing to the higher growth rate of relatively small islands (see Figs. 1 (b); (d)).

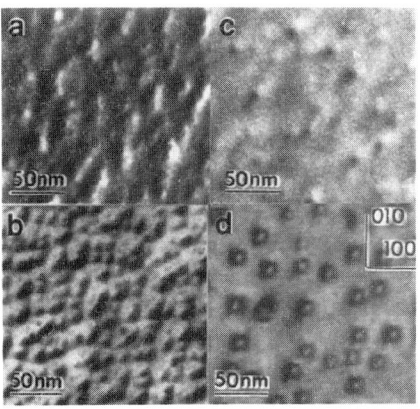

At that stage diffraction spots appear in the RHEED pattern. At a coverage of 5.3 ML of $In_{0.5}Ga_{0.5}As$, coherent islands of about 5-15 nm in size are observed. Further increasing the nominal thickness of $In_{0.5}Ga_{0.5}As$ up to 7.3 ML results in the formation of well-developed dots with size and spacing less varing (Fig. 1 (b)).

Fig.1 a-d. Plan-view bright-field electron micrographs of (In,Ga)As quantum dots corresponding to different nominal thicknesses of In-containing layers: 3.3 ML (a) and 7.3 ML (b) of $In_{0.5}Ga_{0.5}As$, and 2 ML (c) and 4 ML (d) of InAs. Images (a)-(c) are taken at **g**=220, Fig. (d) is a symmetrical [001] zone-axis image.

The formation of well-developed InAs dots (at T_d = 480°C) already begins at an average thickness of less than 2 ML. Figs. 1 (c) and (d) show dots formed during InAs deposition of nominal thicknesses of 2 and 4 ML, respectively.

Comparing the results of the InAs and $In_{0.5}Ga_{0.5}As$ deposition indicates that the onset of dot formation is governed by the strain energy per interface area. Figs. 1 (b) and (d) indicate that the dots are of pyramidal shape with the square base of those principal axes being close to the two orthogonal <100> directions and the average length being about 12±1 nm.

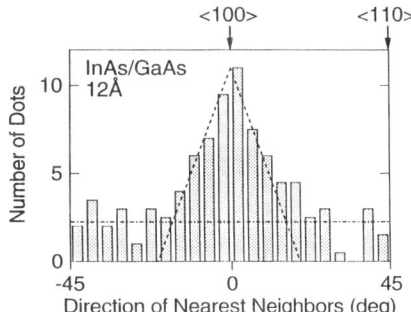

Fig.2. Histogram of next neighbour dot direction (modulus 90°) for 4 ML InAs, obtained from a TEM image as, e.g., Fig. 1 (d). The dashed line denotes the partially disordered square lattice, the dash-dot line illustrates the superimposed random dot distribution.

The size of our InAs dots is similar to that reported by Leonard et al (1994) but about half of that reported by Moison et al. (1994), obtained by atomic force microscopy (AFM), for dots of an average coverage of 2.3 ML grown at 500°C, under rather similar growth conditions. Thus either AFM overvalues the size of the dots, or the dot formation is extremely sensitive to the growth conditions. The size distribution of the dots in InAs samples of 2 and 4 ML is rather narrow (< 20%) and similar to results reported by Leonard et al (1994). The variations of dot size and interdot distance at the initial stages of dot formation for $In_{0.5}Ga_{0.5}As$ deposition are larger than that for InAs deposition.

The dots align in rows along <100>, demonstrating the symmetry of a two-dimensional primitive cubic lattice as Fig. 1 (d) reveals. A histogram of the direction of adjacent dots (Fig. 2) was obtained from a statistical analysis of a TEM image. It clearly has a maximum in <100> direction confirming the impression from the image of Fig. 1 (d). The dashed line in Fig. 2 marks the dependence expected of dots in a partially disordered square lattice. The self-ordering is superimposed by some randomness, resulting in an offset (dash-dot line) independent of the azimuth.

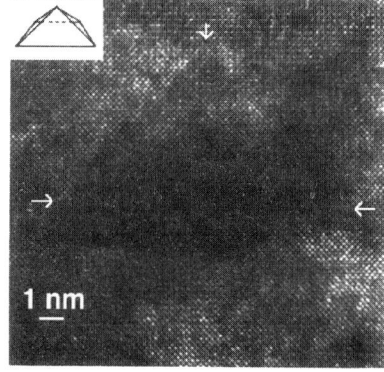

Fig. 3. Experimental 400 kV HREM micrograph of 3 ML InAs QD in GaAs. Arrows indicate the boundary facets.

The dot geometry is additionally confirmed by cross section HREM images as in Fig. 3. The height of the dots is about 6 nm. The sidewalls are close to {110}. Weakly pronounced is the top of the pyramid which together with the strong strain-induced contrast around the dot is shown in the simulated HREM micrograph of Fig.4 calculated by use of CERIUS Programme package (Molecular Simulation Inc.), with the pyramid always appearing to be truncated.

Being rather typical dot size and shape seem to be energetically favourable (Tersoff & Tromp 1993), corresponding to the energy decrease arising from the strain relaxation similar to surface faceting (Marchenko 1981). Growth kinetics are likely to strongly affect the initial stage of layer transformation either at extremely low substrate temperatures or under the changing the optimal As pressure. The regular distribution of the dots may be due to two alternative processes, viz. the ordering of nucleation centres on the surface followed by equalization of the dot sizes during further material deposition and/or by strain field induced ordering with increasing coverage.

Indeed, periodic lateral modulations of the layer composition have been observed in highly strained heterostructures (Glas et al 1990) resulting from the spinodal decomposition of ternary or quaternary solid solutions. Similar effects may also result in the formation of quasi-periodic dot arrays.

Fig. 4. Simulated HREM image of pyramidal InAs QD in GaAs (400 kV, C_s=1 mm, a_0=0.6 A^{-1}, def=-70 nm, foil thickness t=22.6 nm).

The optical properties of the dots are correlated with the morphology revealed by TEM. In general, at low temperatures the dots exhibit a high internal quantum efficiency close to 100% . Fig. 4 depicts PL spectra of the structures with 2 ML and 4 ML of InAs. They differ in peak position and intensity in accordance with the TEM data on size and distribution of the dots. The smaller dots with larger dispersion give rise to a weaker and broader PL line which is also closer to the GaAs bandgap energy as compared to the PL peak of the larger dots showing a more homogeneous size distribution.

34

The low density of quantum-dot induced states in 2 ML InAs dots is not distinguishable in the absorption spectrum, which is dominated by the large density of states in the residual non-transformed 2D layer. For a dense array of uniform InAs dots (Fig. 1 (d)), however, the absorption peak of the dot ground state is clearly identified, and coincides with the PL maximum (the absorption of the wetting layer additionally occuring here).

The corrugation of the $In_{0.5}Ga_{0.5}As$ layer (Fig. 1 (a)) revealed by TEM at the initial stage of layer transformation has also a strong effect on luminescence properties resulting in an intense PL peak which is shifted from the position expected for a uniform InGaAs layer (~1.4 eV for a 1 nm thick $In_{0.5}Ga_{0.5}As$ layer) to lower energies. The structures with 3.3 to 4 ML of $In_{0.5}Ga_{0.5}As$ deposited at 400 - 450°C exhibit the PL peak at 1.23-1.26 eV with 40 - 60 meV as the full width at half maximum (FWHM).

Here we observe a large Stokes shift between the PL line (1.24 eV) and the absorption (1.31 eV) indicating a non-uniform size distribution of InGaAs dots.

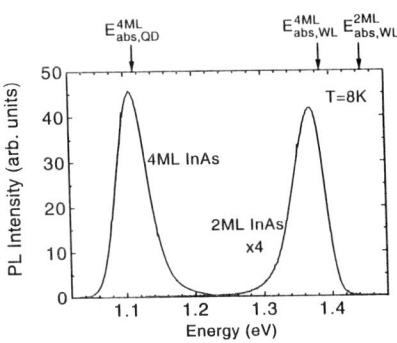

Fig. 4. Low-temperature (T=8 K) PL spectra of quantum dots in samples with 2 ML and 4 ML of InAs of an excitation density D=1 Wcm^{-2}. The peak energies in the absorption spectra of the wetting layer (WL) in both samples and the quantum dots (QD) in the 4 ML sample are marked by arrows.

In conclusion, the formation of ordered arrays (forming a primitive two-dimensional cubic lattice) of pseudomorphic dots of pyramid-like shape, typically 6 nm in height and about 12 nm in base diameter, has been observed on a residual two-dimensional layer above a nominal coverage of 2 ML for InAs, and of 6 ML for $In_{0.5}Ga_{0.5}As$ on GaAs. The dispersion of dot size and interdot distance was remarkably low (< 20 %). Efficient luminescence from dot states has been observed for all samples. For dense dot arrays (~10^{11} dots/cm^2) grown under optimum conditions, the quantum dot ground state photoluminescence and absorption were found to coincide energetically.

ANKNOWLEDGEMENT
This work was partly supported by the Volkswagen Stiftung and INTAS.

REFERENCES

Eaglesham D J & Cerullo M 1990 Phys. Rev. Lett. <u>64</u> 1943
Glas F, Gors C & Henoc P 1990 Phil. Mag. <u>B62</u> 373
Grundmann M, Ledentsov N, Christen J et al 1995 phys. stat. sol. *in press*
Kirstaedter N, Ledentsov N N, Grundmann M et al 1994 Electr. Lett. <u>30</u> 1416
Ledentsov N N, Grundmann M, Kirstaedter M et al 1994 Proc. of the 22nd Int. Conf. Phys. Semicon., (Vancouver, Canada, 1994) *in press*.
Leonard D, Krishnamurthy M, Reaves C M, Denbaars S P & Petroff P M 1993 Appl. Phys. Lett. <u>63</u> 3203
Leonard D, Krishnamurty M, Fafard S, Merz J L & Petroff P M 1994 J. Vac. Sci. Technol. <u>B12</u> 1063
Madhukar A, Xie Q, Chen P & Konkar A Appl. Phys. Lett. 1994 <u>64</u> 2727
Marchenko V I 1981 Sov. Phys. JETP <u>54</u> 605
Mo Y-W, SavageD E, Swartzentruber B S and. Lagally M G 1990 Phys. Rev. Lett. <u>65</u> 1020
Moison J M, Houzay F, Barthe F, Leprice L, Andre E & Vatel O 1994 Appl. Phys. Lett. <u>64</u> 196
Nötzel R, Temmyo J & Tamamura T 1994 Nature <u>369</u> 131
Snyder C W, Orr B G, Kessler D & Sander L M 1991 Phys. Rev. Lett. <u>66</u> 3032
Tersoff J & Tromp R M 1993 Phys. Rev. Lett. <u>70</u> 2782

Inst. Phys. Conf. Ser. No 146
Paper presented at Microsc. Semicond. Mater. Conf., Oxford, 20–23 March 1995
© 1995 IOP Publishing Ltd

High-resolution electron microscope analysis of $(AlAs)_n(GaAs)_m$ short-period superlattices in <110> and <100> projections

H Cerva, H Riechert and D Bernklau

Siemens AG, Research Laboratories, Otto Hahn Ring 6, D-81730 München, Germany

ABSTRACT: $(AlAs)_3(GaAs)_7$ short period superlattices and $Al_{0.6}Ga_{0.4}As/GaAs$ layer structures have abrupt and flat $Al_xGa_{1-x}As$ on GaAs interfaces whereas the GaAs on $Al_xGa_{1-x}As$ interfaces show a transition 2-3 cation layers thick. These results are obtained by visual inspection from <110> high resolution images for the imaging conditions: $-10nm \le f \le 5nm$ and $10nm \le t \le 14nm$ at 200kV in a JEOL 4000EX microscope. Quantitative high resolution images in <100> projection at 400kV from the same specimen corroborate these results.

1. INTRODUCTION

The growth of thick layers of $(AlAs)_n(GaAs)_m$ short-period superlattices (SPS) by molecular beam epitaxy (MBE) is an alternative to the growth of $Al_xGa_{1-x}As$ layers. Growing SPS at reduced temperatures (<700°C) results in very good optical quality and improved laser performance (Riechert et al 1994). Various SPS samples grown at temperatures between 580°C and 680°C with nominal values n=3 and m=7 reveal the following structural and optical properties as obtained by high-resolution X-ray diffraction and photoluminescence (Riechert et al 1994): While the thickness of the AlAs layers remains constant the thickness of the GaAs layers decreases with increasing growth temperature. This reflects the effect of Ga desorption which is known to lead to thickness fluctuations and compositional inhomogeneities. The full width at half maximum of the photoluminescence peak measured in the various samples increases with growth temperature indicating structurally less perfect layers. Quantitative high-resolution electron microscopy (HREM) carried out on <100> oriented 90°-wedge specimens (Thoma and Cerva 1991) of SPS layers grown at 610°C and 680°C yields layer thicknesses which are in agreement with the X-ray diffraction results. At both temperatures a spreading of all the interfaces over 2 - 3 cation layers is observed (Cerva 1993).

The present work reports on a more detailed comparative HREM characterisation of the SPS interface quality in <100> and <110> projections. Earlier quantitative HREM in <100> projection of much thicker $Al_xGa_{1-x}As/GaAs$ layer structures with x=0 and x=0.6 reveal that $Al_xGa_{1-x}As$ on GaAs interfaces are more abrupt than the reverse ones (Thoma et al 1992). It will be shown that this is also true for SPS interfaces even when they are grown at a high temperature of 680°C. Moreover, it is demonstrated by comparison with quantitatively evaluated <100> images that imaging in <110> projection at similar conditions as proposed by Ikarashi et al (1993) yields reliable information on the interface structure by visual inspection.

2. EXPERIMENTAL

The layer structures are grown by conventional MBE on untilted (100) GaAs substrates. The growth temperatures and rates of the SPS and the thicker $Al_{0.6}Ga_{0.4}As/GaAs$ structure are 680°C, 0.28nm/s and 600°C, 0.15nm/s ($Al_{0.6}Ga_{0.4}As$), 0.06nm/s (GaAs), respectively.

Conventional <100> and <110> oriented cross sections are prepared by grinding and ionmilling (3 kV, angle of incidence 12°, liquid N_2 cooling). Quantitative HREM images in <100> projection

36

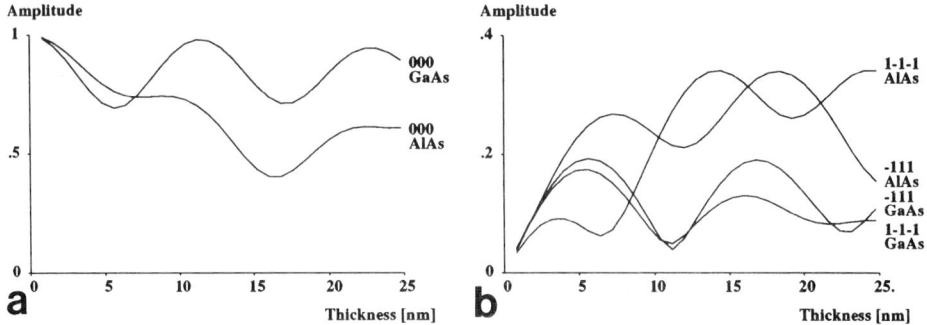

a

b

Fig. 1. Amplitude/thickness curves of (a) the transmitted and (b) the {111} beams for the <110> projection at 200 kV in GaAs and AlAs.

are obtained at 400 kV and are analysed as described by Thoma and Cerva (1991). While the <100> HREM image of the $Al_{0.6}Ga_{0.4}As$/GaAs structure is obtained from the as-ion milled specimen, the specimen containing the SPS is given a short dip in a 0.5% bromine-ethanol solution prior to imaging which leads to a better signal to noise ratio. The <110> oriented specimens are all briefly etched for a few seconds in $H_3PO_4/H_2O_2/C_2H_5OH$ (30:3:100) before investigation which removes the ionmilling-induced amorphous layers on the specimen surfaces (Ikarashi et al 1993). A JEOL 4000EX microscope is used operated at either 200 or 400 kV. Though similar results are obtained at 400 kV for particular imaging conditions, operation at 200 kV avoids rapid irradiation damage.

3. <110> IMAGE SIMULATIONS

Multislice calculations for the <110> projection at 200 kV are carried out with the EMS program of Stadelmann (1987). The amplitude/thickness curves in GaAs (Fig. 1b) show a distinct minimum for

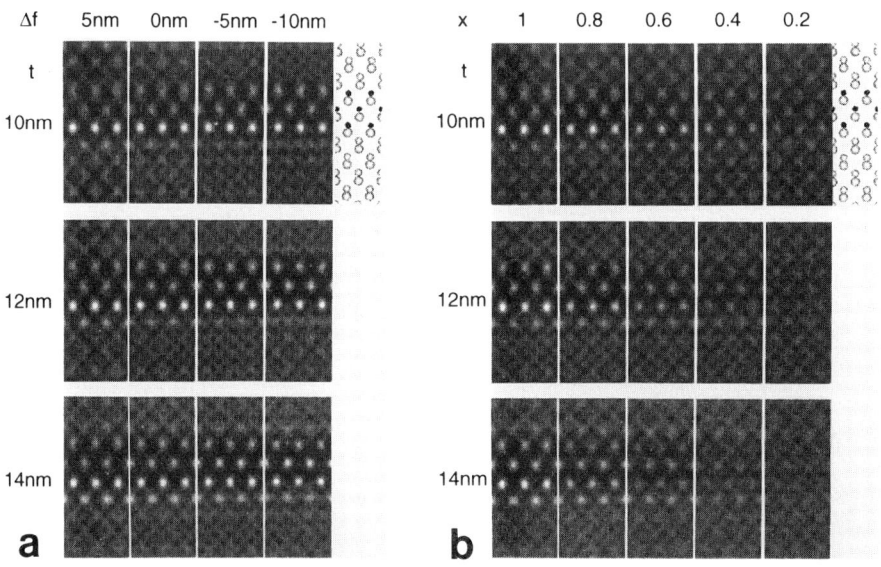

a

b

Fig. 2. Calculated <110> images of (a) an $(AlAs)_3(GaAs)_7$ SPS and (b) a series of $(Al_xGa_{1-x}As)_3/$ $(GaAs)_7$ SPSs for $\Delta f = 0nm$. 200 kV, $C_S = 1.2$ nm, $\Delta=13nm$, and $\alpha= 0.9mrad$, Debye-Waller factors $M=0.006$, $f_{abs}/f_{el}=0.07$ (Ga, As), $f_{abs}/f_{el}=0.03$ (Al) Schematic: Al black, Ga large, As medium atoms.

both types of {111} beams at t=11nm whereas in AlAs the {111} beams have a comparatively high value at this thickness. Therefore, at suitable imaging conditions, the AlAs layers should give a strong contrast while the contrast of the GaAs layers is suppressed. The amplitude/thickness curve of the transmitted beam in GaAs (Fig. 1a) shows that the second maximum coincides with the minimum of the {111} beams. Thus, in practice, experimental images have to be recorded in specimen areas where the second bright thickness fringe of GaAs appears.

With the TEM parameters U=200kV, C_S=1.2mm, Δ=13nm, α=0.9mrad the AlAs layers are strongly visible for -10 nm $\leq \Delta f \leq$ 5 nm and 10 nm $\leq t \leq$ 14 nm (Fig. 2a). Then, the Al-atom positions correspond to bright dots inbetween very dark tunnels. At the AlAs on GaAs interface the first row of Al-atoms appears as a row of extremely intense bright dots. This is experimentally confirmed in the images of Fig. 5. Figure 2b shows that at these imaging conditions the contrast is also sensitive to the Al-content x. The tunnels become less dark with decreasing x. The image in Fig. 3c is an example that these imaging conditions are also sensitive to $Al_xGa_{1-x}As$ layers with a lower Al-content.

4. EXPERIMENTAL IMAGES

Figure 3b shows the quantitative evaluation of the <100> image in Fig. 3a. The GaAs on $Al_{0.6}Ga_{0.4}As$ interface reveals a transition 3 cation layers wide while the $Al_{0.6}Ga_{0.4}As$ on GaAs interface is sharp within 1 cation layer. The same result may be extracted visually from the <110> image in Fig 3c. Note that both images give the same mean layer thickness.

The <100> projection of the SPS and its quantitative evaluation are displayed in Fig. 4. Due to Ga desorption the layer periodicity is reduced from 10 to 8 cation layers with the GaAs layers being only 5 cation layers thick. Obviously the transition layers at the AlAs on GaAs and the reverse interfaces are 1 and 2 cation layers wide, respectively. According to Ikarashi et al (1993) steps along [110] and [1-10] have a different shape which may be distinguished. In such images of our SPS, however, the AlAs on GaAs interfaces appear to be flat over a distance of at least 35 nm despite the normal growth

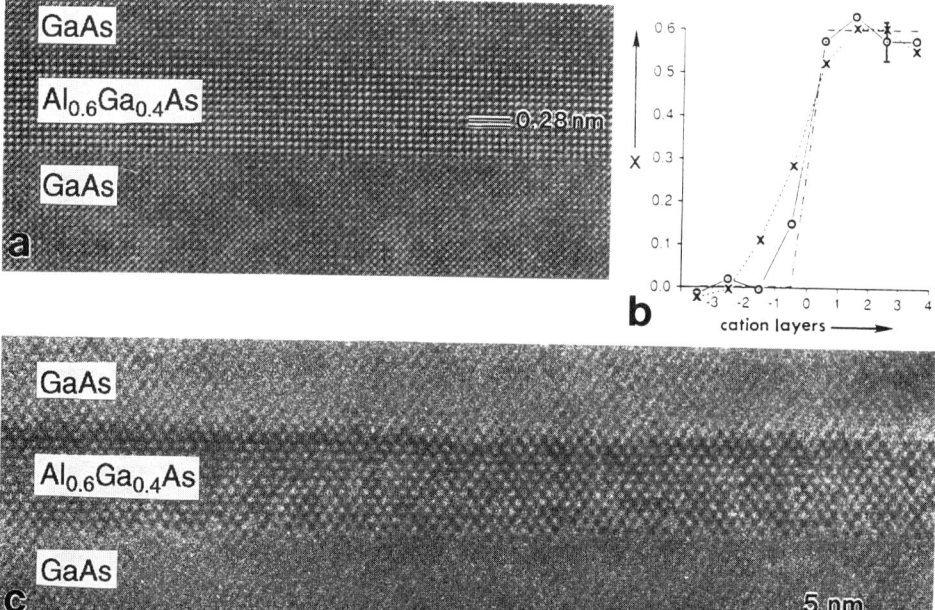

Fig. 3. The same $Al_{0.6}Ga_{0.4}As$/GaAs structure in (a) <100> and (c) <110> projection. (a) 400 kV, Δf=-20 nm, t$\approx$10 nm, (b) 200 kV, Δf=0 nm, t$\approx$12 nm). (b) Quantitative evaluation of the interfaces in the <100> image: (—) AlAs on GaAs, (----) GaAs on AlAs, (-.-.-.-) perfectly abrupt interface.

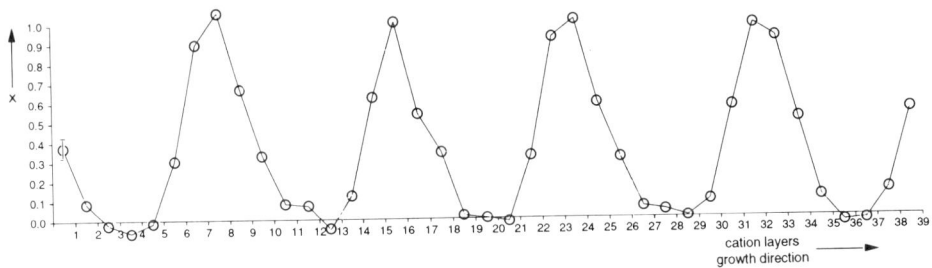

Fig. 4. Al-composition profile of the $(AlAs)_3(GaAs)_7$ SPS obtained from a <100> HREM image (not shown due to restricted space). Average over 64 cells parallel to the interface.

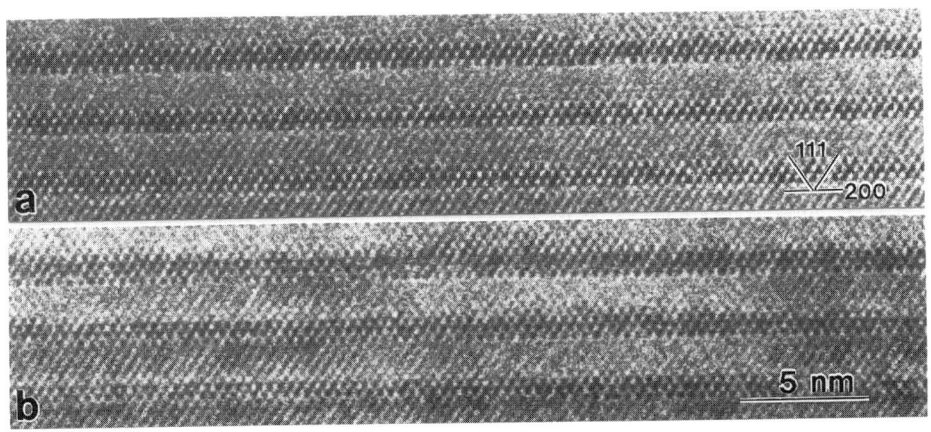

Fig. 5. HREM images of the $(AlAs)_3(GaAs)_7$ SPS in two <110> projections perpendicular to each other. Imaging conditions: 200 kV, $\Delta f \approx 0$ nm and $t \approx 12$ nm.

rate (Fig. 5). Both images confirm the results obtained in the <100> image: the lower interface is abrupt while there is a 1-2 cation layers wide transition at the upper interface of the AlAs layers.

3. CONCLUSION

HREM imaging of $Al_xGa_{1-x}As/GaAs$ layer structures and $(AlAs)_n(GaAs)_m$ SPS in <110> projection at particular imaging conditions gives reliable results as verified by quantitatively evaluated <100> images obtained from the same layer structures. Calculated images show an unambiguous relationship between contrast and Al-content. Thus, a quantification algorithm may be set up and applied. $(AlAs)_3(GaAs)_7$ SPS layers grown at a high temperature of 680°C and normal growth rate have abrupt and flat AlAs on GaAs interfaces and diffuse GaAs on AlAs interfaces.

REFERENCES

Cerva H 1993 Solid-State Electronics 37 1045
Ikarashi N, Tanaka M, Baba T, Sakaki S and Ishida K 1993 Jpn. J. Appl. Phys. 32 2824
Riechert H, Bernklau D, Milde A, Schuster M, Cerva H, Hoyler C and Wolf H-D
    1994 21st Int. Symp. on Compound Semiconductors, San Diego, USA, Sept., IOP Publishing
Thoma S and Cerva H 1991 Ultramicroscopy 38 265
Thoma S, Riechert H, Mitwalsky A, Oppolzer H 1992 J. Crystal Growth 123 287
Stadelmann P 1987 Ultramicroscopy 21 131

Inst. Phys. Conf. Ser. No 146
Paper presented at Microsc. Semicond. Mater. Conf., Oxford, 20–23 March 1995
© *1995 IOP Publishing Ltd*

HREM - analysis of interfaces on the basis of molecular dynamic structure relaxations

Kurt Scheerschmidt, Sergej Ruvimov and Dirk Timpel

Max Planck Institute of Microstructure Physics, 06120 Halle, Germany

ABSTRACT: Interfaces are investigated by high resolution electron microscopy (HREM) providing local atomic information on the defects. To analyse the micrographs images are calculated on the basis of relaxed atomic structures of different interfaces modelled by molecular dynamics (MD) and static energy minimization. Semiconductor multi- layers and silver particles in glasses are discussed to study limitations in revealing the compositional variations and the elastic deformations at the interfaces by HREM.

1. INTRODUCTION

Solid interfaces have a strong influence on material properties because of inhomogeneities of their mechanical, chemical and electrical behaviour. Hence, the structural characterization of interfaces and their relaxations are of great importance. Two types of interfaces were investigated by high resolution electron microscopy (HREM) and used to discuss the relevance of the molecular dynamics (MD) model generation for the structure analysis applying HREM image simulations: First, silver particles in sodium-silicate glasses generated after sodium-silver ion exchange by thermally activated migration [Scheerschmidt, Timpel 1994]. Second, HREM imaging of multi-layers in the binary and ternary systems based on GaAs and GaSb, as, e.g., quantum-well structures and $In_xGa_{1-x}As/Al_yGa_{1-y}As$ with small misfits, but a high stoichiometric variability [Ruvimov et al. 1994, Scheerschmidt et al. 1995]. The systems are characterized by different misfits and varyious scattering factors resulting in a different elastic behaviour at the boundaries and in respective imaging effects.The interpretation of the micrographs requires methods of image analysis and the computer simulation of the HREM contrast by calculating the electron beam/specimen interaction of a theoretical structure model and by subsequently consi- dering the electron-optical process.

2. EXPERIMENTS AND COMPUTATIONS

The semiconductor samples are grown by molecular beam epitaxy. Cleaved 90° wedge- shaped, planar and cross-sectional samples are used for HREM investigations in a JEOL-JEM 4000 EX electron microscope. The sodium-silicate glass consists of $70\%SiO_2$-$30\%Na_2O$ with stabilizing and reducing components; the ion exchange in a $NaNO_3/AgNO_3$ melt and subsequent annealing create Ag particles of different diameters. The relaxed atomic structures of the multi- layers and the particles as well as their interfaces are modelled by MD and static energy minimiza- tion, with the bonding forces and the three-body potentials being varied. For the computer-aided generation of structure models the CERIUS program package is applied [Scheerschmidt 1994]. The structure of glasses are modelled by relaxing disturbed and topologically rearranged sodium silicate crystals and including differently shaped silver particles. The atomic arrangements of the epilayer interfaces are modelled assuming coherent adjacent homogeneously strained crystals. The total energy thereby has to include the valence bonding, non-bonded forces as well as topological constraints. The force field of the many-particle system is calculated by the energy gradient owing to two-, three- or four-body interactions, i.e. distance, angle, torsion and inversion of the bonds. Special many-body forces are known for covalent structures and glasses; the description of details of surfaces and interfaces, however, requires refined potentials reproducing more experimentally observed features. Some of the structures are refined using Born-Mayer- Huggins pair potentials in combination with Stillinger-Weber angular terms [Garofalini 1990]. Of great importance are the correct setting of covalent bonds and the choice of the atomic distances of the crystal structure instead of the molecular ones.The misfit is determined by the different equilibrium interatomic distances, which are chosen to generate stable crystals.

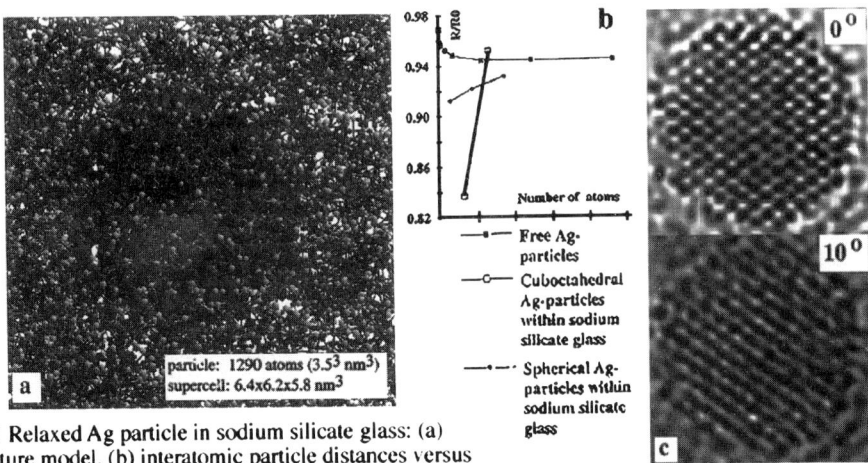

Fig.1 Relaxed Ag particle in sodium silicate glass: (a)
structure model, (b) interatomic particle distances versus
radius, (c) HREM simulations U=400kV, Cs=1mm,
δ=8nm, α =.5mrad , Δ = -37nm, beam=011

Fig.2: Relaxed atomic model of an
InAs/AlSb heterostructure (a) and
contrast simulations (b) using different
defoci Δ for 4 different atomic sequences
across the InAs/AlSb interface (Al/As or
In/As interface with an AsAl or InSb
interlayer); U=400kV, Cs=1mm,δ=8nm,
α_D=.5mrad,α=16nm^{-1}, t=11.3nm

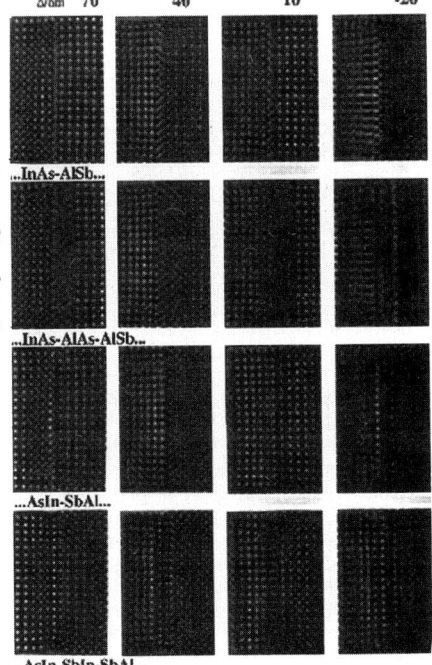

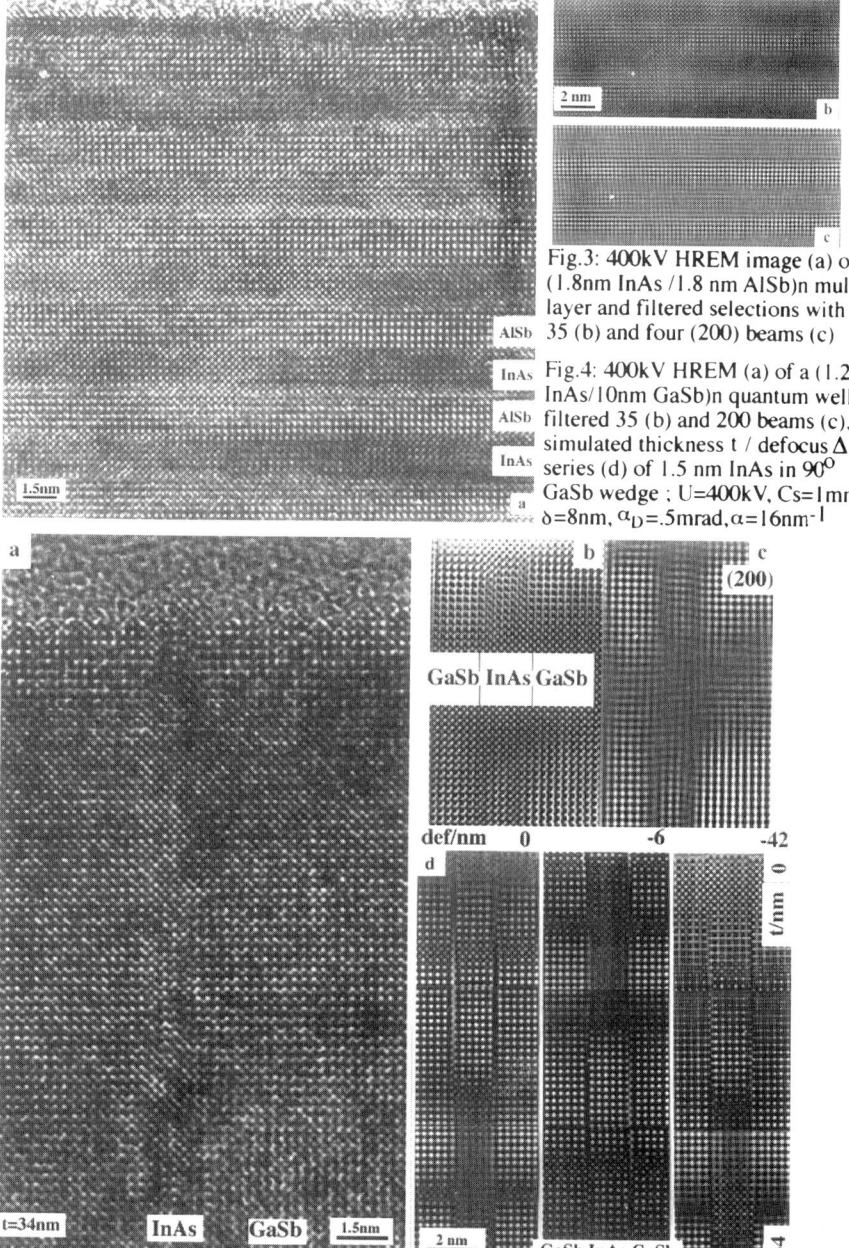

Fig.3: 400kV HREM image (a) of a (1.8nm InAs /1.8 nm AlSb)n multi-layer and filtered selections with 35 (b) and four (200) beams (c)

Fig.4: 400kV HREM (a) of a (1.2nm InAs/10nm GaSb)n quantum well, filtered 35 (b) and 200 beams (c), simulated thickness t / defocus Δ series (d) of 1.5 nm InAs in 90° GaSb wedge ; U=400kV, Cs=1mm, δ=8nm, α_D=.5mrad, α=16nm^{-1}

3. RESULTS AND DISCUSSION

Fig. 1 shows the relaxed structure of a spherical Ag-particle in a sodium silicate glass (a), the relative interatomic distances of the particle as a function of its radius (b) and the calculated image contrast for two different particle orientations (c). Besides the Fresnel fringes at the particle borders and the speckled contrast of the amorphous matrix, both the lattice disturbances at the surface and the compression of the particle can be revealed.

Fig. 2 demonstrates the model generation using MD calculations (a) and the resulting HREM image simulations (b) for a relaxed coherent interface of an InAs/AlSb heterostructure assuming no further restrictions of the MD relaxation. The different layers are matched, with the bonds being recalculated before the energy is minimized. Different configurations were used as geometrical start models resulting in a different relaxation behaviour at the interface and in a different corresponding interface HREM contrast (see Fig. 2b): InAs is stacked onto AlSb, and vice versa, thus generating Al/As and In/Sb interfaces, respectively. Furthermore, an additional layer is inserted replacing one of the components by that of the adjacent layers thus creating twice the Al/As and In/Sb interface, respectively. Different contrasts at the boundary arise from the four possibilities of InAs on AlSb, or vice versa, and In/Sb or Al/As at the interface. The different species are clearly revealed. In addition to this structure factor contrast and the delocalization effects, inclined striations occur with the homogeneous strains increasing at the interfaces. The InAs/AlSb system considered is characterized by a large misfit and a structure factor asymmetry as well as by two most different interfaces: InSb having a small asymmetry, AlAs having a large one. Fig. 3 shows an experimental HREM micrograph (a) of the InAs/AlSb multi-layer system and its contrast enhancement by Fourier processing using 35 beams (b) or solely the 200 reflexions (c). The analysis enables one to discriminate the layer sequence (InSb and AlAs interface) and to identify the roughness of the interfaces (steps). The interfaces are no longer abrupt, and local variations of the orientation create virtual boundary structures, which cannot be analysed in a phenomenological manner. In some places finer fringes occur, which can be interpreted as "half spacings". The different assumptions of the local interface structure in the InAs/AlSb system imply different contrasts at the boundaries where the interface region is blurred and extended if additional interlayers are considered.

Fig.4 shows an experimental 400kV HREM image (a) of a (1.2nm InAs/10nm GaSb)$_n$ quantum well structure grown on (001)-oriented GaSb prepared as a 90° wedge-shaped sample. The filtered image (b) includes 35 beams, whereas a micrograph image-processed by using the 200 beams is shown in (c). The HREM contrast simulations in Fig. 4d present a defocus series of a 1.5 nm wide InAs layer in a 90° wedge-shaped GaSb matrix, thickness t of the wedge ranges from 0 to 34nm. The imaging parameters chosen (see Figure caption) are that of the JEM4000EX with defocus values def =0,-6, and-42nm.The influence of strains in addition to effects of the scattering factor contrast are illustrated in both the image-processed micrographs (Figs. 4 b,c) and the corresponding simulated HREM images (Fig. 4d), which is a result of the small structure factor asymmetry of InAs/GaSb (approximately between 0.0 and 0.03) for the perfect regions and independent of the actual interface stacking. For systems having a comparable medium misfit but a higher degree of structure factor asymmetry, however, the structure factor contrast prevails over the strain contrast. The strains at the interfaces in Fig. 5 create the delocalized defect contrast and cause the spreading of the interface region.

The HREM image contrast is determined by the geometrical interface structure as well as by the imaging conditions.The main contrast features result from the differences of the structure factors of the projected atom columns, their asymmetry in the bulk and at the interfaces controls the defocus/thickness dependencies and image delocalization at the defects. Different atomic species at the interfaces can be revealed by the contrast features as well as by the relaxation of the lattices owing to the lattice misfit and the stoichiometry of mixed compounds. The interface structure has to be refined by using more realistic pair potentials, enabling the rearrangement of bonds and including semi-coherent interfaces in the molecular dynamics simulation.

Garofalini S H 1990 Journ. Non-Cryst. Solids **120** 1
Ruvimov S, et al. 1994 Electron Microscopy: Proc. 13. ICEM Paris, Vol.1, p. 403-405
Scheerschmidt K, Timpel D 1994 Electron Microscopy: Proc. 13. ICEM Paris, Vol.2A, p.395-396
Scheerschmidt K 1994 Proc. MRS Spring Meeting San Francisco, Eds.: Borgesen P et al., **338** 121
Scheerschmidt K et al. 1995, Journal of Microscopy, accepted

Inst. Phys. Conf. Ser. No 146
Paper presented at Microsc. Semicond. Mater. Conf., Oxford, 20–23 March 1995
© 1995 IOP Publishing Ltd

The structure of interfaces in semiconductor low-dimensional systems grown by MBE

A K Gutakovski, A L Aseev and J L Hutchison*

Institute of Semiconductor Physics, Russian Academy of Sciences, Siberian Branch, 630090 Novosibirsk, Russia
* Department of Materials, University of Oxford, Parks Road, Oxford OX1 3PH, UK

ABSTRACT: Results are presented of high resolution electron microscopy (HREM) investigations of semiconductor low-dimensional systems of III-V superlattices, quantum wires formed on facetted surfaces at GaAs/AlAs epitaxial interfaces, quantum dots formed by epitaxial growth of Ge on Si, Sb delta-doped Si as well as InSb epitaxial layers on GaAs grown by molecular beam epitaxy.

1. INTRODUCTION

Studies of the atomic structure of interfaces and defects in semiconductor low-dimensional structures remain an important task in spite of numerous results obtained for this field of research reviewed, for example, by Smith and McCartney (1994), Heydenreich (1993) and Hutchison (1990). This is due to the wide variety of low-dimensional semiconductor structures with unusual quantum properties fabricated mainly by molecular beam epitaxy (MBE) and electron beam lithography methods. The HREM is a technologically important tool for investigation of the atomic structure of interfaces and defect nucleation in artificially grown semiconductor epitaxial systems. The aim of the present work is to experimentally study the structures of semiconductor epitaxial systems grown by MBE, using the HREM. The JEM-4000EX electron microscopes installed both in the Department of Materials at Oxford University and in the Siberian Branch of Russian Academy of Sciences were used.

2. RESULTS AND DISCUSSION

2.1. Modulated GaAs/AlAs epitaxial system

Fig.1 shows typical transmission electron microscopy (TEM) and HREM images of a GaAs/AlAs superlattice grown on the MBE machine "Katun" developed by Pchelyakov et al.(1994). Increased atomic scale roughness of GaAs/AlAs interface compared to AlAs/GaAs interface (upper layer is indicated first) is clearly seen in the HREM image, Fig.1a. This difference indicates a lower migration coefficient of Al adsorbed atoms compared to that of Ga adsorbed atoms for the given growth conditions. The lower migration coefficient of Al results in multi-level nucleation of two-dimensional islands and increased roughness of the AlAs growing surface. Contrary to this, high mobility of Ga adatoms provides conditions for layer-by-layer growth with formation of a flat growing surface of GaAs. With decreasing growth temperature both migration coefficients decrease and result in a drastic increase of roughness of both growing surfaces. Experimental TEM images of cross-sectional specimens illustrate the appearance of

44

morphological instability of growing surface for this case (Fig.1c) compared to optimal growth conditions (Fig.1b). The typical height of corrugated interfaces exceeds the thickness of a separated layer at further stages of growth. So, a laterally modulated structure is formed in the upper part of the epitaxial system as seen in Fig.1c.

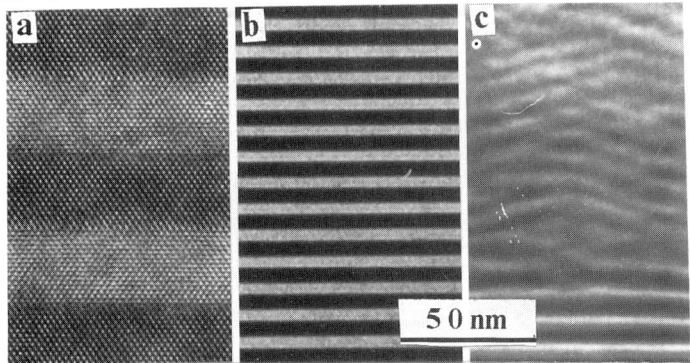

Fig.1. HREM (a) and TEM (b,c) images of cross-sectional specimens of GaAs/AlAs super-lattices grown at optimal (a,b) and non-optimal (c) substrate temperature.

As proposed by Ploog et al.(1991) the alternate growth of GaAs and AlAs on (311)A oriented GaAs substrate results in the formation of ordered arrays of quantum wires oriented along [233]. Fig.2 shows an HREM image of a cross-sectional specimen with epitaxially grown arrays of quantum wires. Interpretation of the image contrast is complicated because the specimen is oriented on a (233) plane perpendicular to the quantum wires' direction. Nevertheless, a weak modulation of the contrast with periodicity of about 10 nm is observed in Fig.2. This value of perodicity differs significantly from that predicted by Ploog et al. (1991).

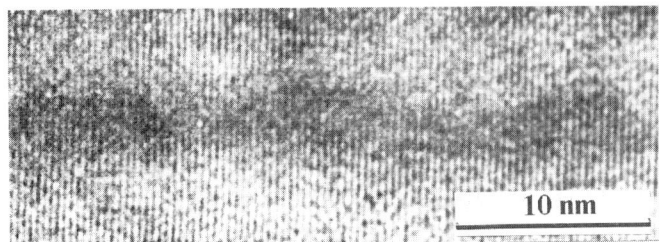

Fig.2. HREM image of cross-sectioned AlAs/GaAs specimen projected along [233] direction. The substrate plane is oriented on (311).

2.2. Quantum dots in Ge/Si heterostructure

According to previous investigations by Yakimov et al.(1992) the initial stage of Ge epitaxy on Si substrates includes the formation of Ge islands in the size range 10 -100 nm following the Stranski-Krastanov growth mechanism. Such islands must represent quantum dots containing zero-dimensional electron gas. Ge/Si heterostructures were grown by MBE after preliminary thermal cleaning of the substrate. After growth of about 1 nm thick Ge layer, morphological transformation of the continuous strained Ge layer into separate islands was observed by Yakimov et al.(1992).

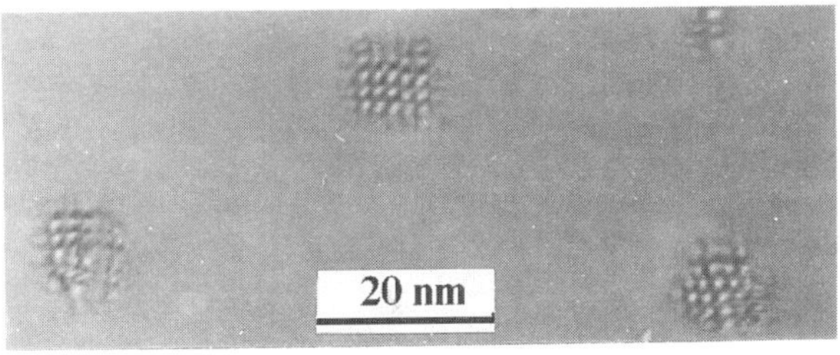

Fig.3. Plan-view TEM image of Ge islands on Si (100) substrate grown by MBE at 400 C.

This is confirmed by the plan view TEM images of Ge layers (Fig.3). The Ge islands are revealed easily due to the presence of a network of misfit dislocations at the interface.
The average distance between islands is approximately constant and their sizes vary in the range 10 - 20 nm (Fig.3). The conductance measurement of Ge/Si heterostructures containing such Ge islands indicates hole tunneling into quantum dots associated with the islands (Yakimov et al. 1992, 1994).

2.3. InSb/GaAs epitaxial structure

This system is characterized by a large lattice mismatch of 13.6%. Fig.4 shows TEM and HREM images of a cross-sectional specimen. Threading dislocations with density increasing at the interface are observed on the TEM image (Fig.4a). The region of InSb layer at a distance beyond 1000 nm from the interface is characterized by the density of threading dislocations of about 10^8 cm^{-2}. Measurement of charge carrier life-times gives a value of about 10 ns in a thin layer of InSb with a high density of threading dislocations. The life-time value increases up to 500 ns for epitaxial layers of InSb 1 μm thick. So, a strong correlation between the dislocation density and charge carrier life-time was detected. The HREM images (Fig.4b) show a network of Lomer-type misfit dislocations located in the interface plane. The lateral spacing of the misfit dislocations, having Burgers vector a/2<110>, corresponds to the average periodicity of approximately seven interatomic distances in the interface plane. This correlates with the misfit dislocation spacing predicted for the complete accommodation of the lattice mismatch between GaAs and InSb. Fig.4b shows an HREM image of a misfit dislocation core after digital Fourier-filtering. The superimposed interatomic bonds show complete linking of bonds in the core. This observation corresponds to that previously obtained for the core structure of the Lomer dislocation in crystals with diamond-cubic lattice, Bourret et al.(1982).

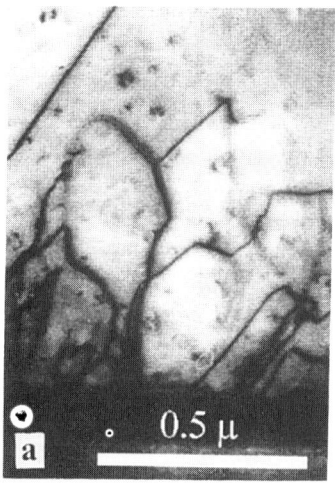
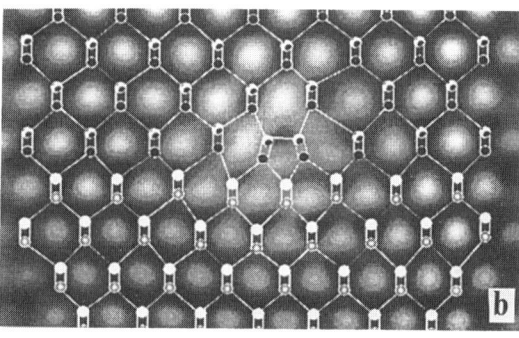

Fig.4. (a) TEM image of cross-sectional specimen InSb/GaAs. (b) filtered HREM image with schematic representation of atomic rows.

2.4. Sb δ -doped silicon

Fig.5a shows a TEM image of a Si cross-sectional specimen with Sb δ-doped layers, grown by MBE as described by Nikiforov et al.(1992). The δ-doped layers differ in Sb concentration as indicated in the caption. Each layer is revealed by the diffraction contrast increasing with increasing Sb concentration. As seen in Fig.5a, threading defects nucleate directly on the δ-doped layers. An HREM image corresponding to one δ-doped layer is shown in Fig.5b. This image shows the formation of multiple microtwins propagating from the δ-doped layer along the growth direction. The increase in microtwin density and microtwin areas with increasing Sb content suggests the promotion of microtwin nucleation by the presence of Sb atoms.

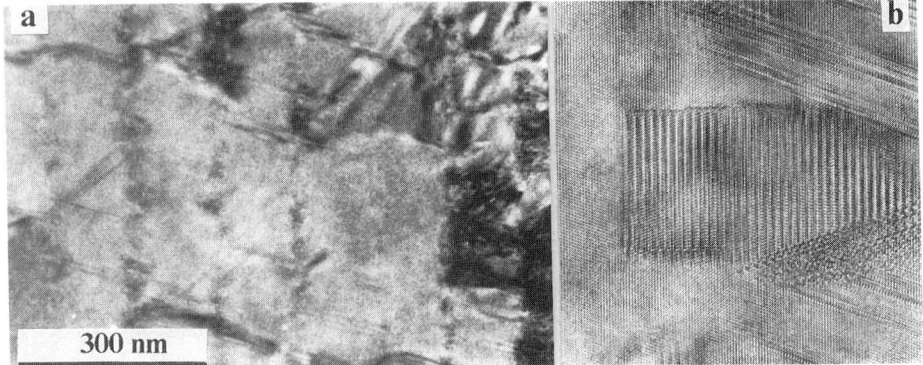

Fig.5. TEM (a) and HREM (b) images of cross-sectional specimen of Sb δ-doped Si. The concentration of Sb is 1×10^{11}, 2×10^{12}, 3×10^{13}, 4×10^{13}, 5×10^{13}, 6×10^{14}, 7×10^{15} cm^{-2} from lower (left side of the TEM image) to upper layers (right side of the TEM image). HREM image corresponds to a layer with Sb content 5×10^{13} cm^{-2}.

3. CONCLUSION

The morphological instability of the growing surface resulted in lateral patterning of the epitaxial system GaAs/AlAs. The transformation of the continuous strained Ge layer into separate islands provides the possibility to form the system of quantum dots. The formation of Lomer dislocations in the InSb/GaAs interface is demonstrated. Finally, the nucleation of multiple microtwins induced by the Sb atoms is observed for MBE grown δ-doped layers.

Acknowledgments

The authors are indebted to colleagues from the Institute of Semiconductor Physics in Novosibirsk engaged in the MBE research work: O P Pchelyakov, L V Sokolov, A I Toropov, V V Preobrazhenski, A A Velichko, B Z Kanter, A I Nikiforov and S V Rubanov. This work was supported partly by the Program of the Russian Ministery of Sciences and Technological Policy "Physics of Solid State Nanostructures", grant 3-007/2. ALA is grateful to the Royal Society for supporting his visit to the Department of Materials at Oxford University.

REFERENCES

Bourret A, Desseaux J and Renault A 1982 Phil.Mag. A 45, 1

Heydenreich J 1993 Solid State Phenomena 32-33 511

Hutchison J L 1990 JEOL News 28E 4

Nikiforov A I, Kanter B Z, Stenin S I and Rubanov S V 1992 Poverchnost.Phys.Khim.Mekh 10-11 95 (in Russian)

Pchelyakov O P, Markov V A and Sokolov L V 1994 Brazilian Journ. Phys. 24 77

Ploog K, Brandt O and Notzel R 1991 10th Symp.Rec.of Alloy Semicond.Phys. and Electronics, eds A Sasaki and T Nishinaga (Nagoya) pp 47-56

Smith D J and McCartney M R 1994 Mat.Res.Soc.Symp.Proc. 332 43

Yakimov A I, Markov V A, Dvurechenski A V and Pchelyakov O P 1992 Phil.Mag.B65 701

Yakimov A I, Markov V A, Dvurechenski A V and Pchelyakov O P 1994 J.Phys.Condens. Matter 6 2573

Inst. Phys. Conf. Ser. No 146
Paper presented at Microsc. Semicond. Mater. Conf., Oxford, 20–23 March 1995
© *1995 IOP Publishing Ltd*

Strain determination in bilayer systems using HREM

M D Robertson,* J M Corbett,* J E Currie* and J B Webb**

*Guelph-Waterloo Program for Graduate Work in Physics, University of Waterloo, Waterloo, Ontario, Canada N2L 3G1
**Institute for Microstructural Sciences, National Research Council of Canada, Montreal Road, Bldg. M-50, Ottawa, Ontario, Canada K1A 0R6

ABSTRACT: A new technique is presented to directly measure strains in epitaxial systems from HREM images. This method involves the calculation of the cumulative sum of deviations in lattice fringe spacings from the average value. By utilizing the cumulative sums of deviations as opposed to individual lattice fringe spacings, a very sensitive method for measuring strains in the presence of noise has been developed. This technique has been applied to a simulated as well as an experimental bilayer HREM image and the strains determined agree with the expected values to within experimental error.

1. INTRODUCTION

The quantification of local strains in semiconductor epilayer systems is of technological importance since strain can affect the physical, electrical and optical properties of the material. For the direct measurement of strains in epitaxial structures on the nanometer scale, high resolution electron microscopy (HREM) images recorded using slow scan charge-coupled device (CCD) cameras and analyzed using Fourier filtering techniques can provide reliable, quantitative results that are independent of image contrast. Bierwolf et al. (1993) have reported the direct measurement of local lattice distortions within semiconductor crystals by superimposing two two-dimensional lattices and investigating the resulting moiré structure. Another approach, reported by Hytch and Bayle (1994) involved studying the behavior of the amplitude and phase from processed CCD images of individual <111> and <002> lattice fringes from a thin buried layer of Ni in Au. Jouneau et al. (1994 a,b) presented a third technique where local distortions between (002) atomic planes were measured and the morphology of the interfaces determined.

In this paper we present an alternative technique for the quantification of elastic strains in epitaxial layers by HREM. Our method is based on the analysis of the cumulative sum of deviations (CUSUM) of lattice fringe spacings from a target value. Page (1961) and Goldsmith and Whitfield (1961) have shown the CUSUM technique to be a sensitive method for detecting small changes from target conditions when the data is subject to noise. A description of the CUSUM method, and its application to a simulated strained bilayer and an experimental $In_{.89}Ga_{.11}Sb$ on (001) GaAs bilayer will be discussed.

2. DESCRIPTION OF METHOD

The HREM image of the interfacial region of interest is Fourier transformed to obtain its digital diffractogram. Each reflection carries information concerning the strain in the direction of the corresponding g-vector. Ideally, one is interested in studying the lattice fringe information from two reflections, one parallel and one perpendicular to the interface normal. From the lattice fringes running parallel to the interface normal, information on the degree of misfit strain relaxation can be obtained. Also, if the lattice fringes perpendicular to the interfacial normal are studied, then the lattice misfit between the two layers can be obtained. Although this information may be obtained from a single reflection with its g-vector containing both parallel and perpendicular components, analysis is simpler when two orthogonal reflections are employed. For a semiconductor sample cross sectioned in $[1\bar{1}0]$ zone axis projection with the growth direction along [001], the [002] and [220] reflections work well. All unnecessary reflections are filtered out using elliptical masks of approximately the first Brillouin zone in size. Performing the inverse-Fourier-transform then yields lattice fringes on which the strain analysis is to be performed. One must be careful when interpreting results near the edges of the Fourier filtered image due to the effects of Fourier periodic continuation in these regions.

Next, the pixel spacing between two consecutive lattice fringes is measured. It has been found that using polynomial fits to the intensity data to determine the location of the intensity maxima and minima is a more accurate method than sinusoidal fitting. This is believed to be a result of the presence of the many Fourier frequencies inside the elliptical mask when the inverse Fourier-transform is performed. In order to evaluate the state of strain across the interface, the individual lattice fringe spacings are required in a direction parallel to the interface normal. For the (002) lattice fringes this is straightforword when the interface normal is along [001]. Let the spacing of each of the n individual fringes of the image be Z_i, where $i = 1, \ldots , n$. The average fringe spacing is denoted by $\overline{Z}$. For the (220) lattices fringes a slightly different procedure is required. The (220) lattice fringes run parallel to [001], hence the individual fringe spacings along [001] cannot be determined directly. Instead, let the average fringe spacing across a row pixels be denoted by Z_i where $i = 1, \ldots, N$ and N is the pixel dimension of the image in the direction of the interface normal. Then, $\overline{Z}$ is again defined as the average of the fringe spacings across the interface.

The CUSUM technique requires a sum of deviations of the individual lattice fringe spacings from a target value. Let this target value be the average fringe spacing, $\overline{Z}$. Thus, the sum $\sum \left(Z_i - \overline{Z} \right)$ is plotted vs pixel position across the interface. $\overline{Z}$ is a useful choice for the target value since regions of the CUSUM plot where the lattice fringe spacing was less than the average value will have negative slope and regions where the lattice fringe spacing was greater than the average will have a positive slope. The vertices of the CUSUM plot indicate the location where the lattice fringe spacings have changed and in strained epilayer systems, this marks the position of the interface between the two materials.

The slopes, m, of the CUSUM plots can be related to the strains present in the image (Robertson et al. 1995). The amount of lattice misfit parallel to the interface normal, e, and the relaxed portion of the natural lattice misfit, δ, are given by

$$e = \frac{m_{film}^{002} - m_{sub}^{002}}{1 - m_{film}^{002}}, \quad \text{and} \quad \delta = \frac{m_{film}^{220} - m_{sub}^{220}}{m_{sub}^{220} + \overline{Z}} \qquad (1),\ (2)$$

respectively. In an epitaxial system, e is related to δ by

$$e = \delta + p(f - \delta), \qquad (3)$$

where f is the natural lattice misfit of the system and p is the factor of extension / contraction in a direction parallel to the interface normal when the dimensions perpendicular to the interface normal are contracted / extended as predicted by continuum elasticity theory. Ignoring surface relaxation effects, p approaches the bulk value of $(1 + v)/(1 - v)$, but in thin materials p approaches the two dimensional limit of $(1 + v)$ where v is Poisson's ratio. Equation (3) can be solved to give the natural lattice misfit of the system.

Fig. 1 illustrates a hypothetical interface where the lattice fringe spacings have been subjected to considerable noise introduced by a random number generator. The first 350 points are normally distributed with a mean of zero. From points 351 to 449 the mean is increased linearly from 0.0 to 0.2 and for points 450 to 800 a constant mean of 0.2 is maintained. A variance of 1.0, 5 times greater than the difference in means, was maintained throughout the data. This series of data was meant to simulate a strained bicrystal system with some surface roughness and with the interface located at point 400. It is apparent that the CUSUM plot provides a significant improvement to the visual interpretation of the data. The position of the hypothetical interface is readily observed to be near point 400 with the CUSUM plot but cannot be easily determined using the raw data plot. The slopes for the points ranging from 0 to 350 and 450 to 800 were -0.102(0.001) and 0.111(0.001) respectively. This gives a difference in means of 0.213(0.001), a value within 5% of the expected difference of means.

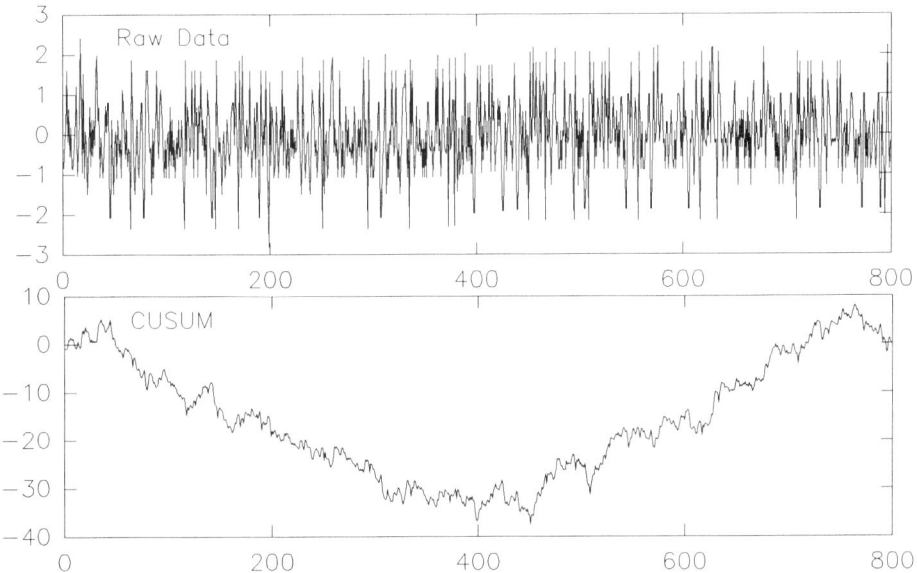

Fig. 1. Raw data and CUSUM plots of a hypothetical interface where the variance in lattice fringe spacings is 5 times the difference in mean lattice spacings between the two layers. The interface is located at x coordinate 400.

3. SURFACE RELAXATION EFFECTS

It is critical to include the effects of surface relaxation if a quantitative value of the interfacial strains is to be determined by high resolution methods. Surface relaxation effects can be included by modifying equation (3)

$$
c = \frac{\delta}{1 + \varepsilon_{sub}^{\parallel}} + \left[\frac{\left(\dfrac{1+v}{1-v}\right) + \dfrac{\varepsilon_{film}^{\parallel} - \varepsilon_{sub}^{\parallel}}{f - \delta}}{1 + \varepsilon_{sub}^{\parallel}} \right](f - \delta) \tag{4}
$$

where $\varepsilon_{film}^{\parallel}$ and $\varepsilon_{sub}^{\parallel}$ are the strains in the structure parallel to the interface normal due to the surface relaxation of the film and substrate respectively. Faux (1994) gives expressions for these surface

50

relaxation effects for bilayer and superlattice structures with rectangular geometry. Note that for the case where the misfit strains have been completely relieved through the introduction of misfit dislocations, $\delta = f$, and the effects of surface relaxation due to coherency stresses becomes zero.

4. IMAGE SIMULATION

An important question that may be asked is "Does this CUSUM Fourier filtering technique operate effectively on HREM images?". To answer this question, image simulations were performed and the results compared to the known initial conditions. Fig. 2a is a HREM image simulation for a coherently strained ($\delta=0$) InSb bilayer where surface relaxation effects have been neglected. The lattice misfit between the upper and lower halves of the image was $f = -0.02$ and was incorporated using the deformable ion approximation (Humphreys 1979). The image simulation was performed for a 20 nm thick crystal using the eigenvalue approach with 200 beams. Transfer characteristics were for a Philips CM20ST electron microscope operated at 200 kV (C_s=1.2 mm, objective aperture radius of 4.4 nm^{-1} and a defocus of -65.8 nm). Fig. 2b is the Fourier filtered image for the [002] reflection and Fig. 2c is the CUSUM curve plotted against the pixel position. The vertex of the CUSUM plot correctly determined the position of the interface at the centre of the plot. The slopes of the plot were found to be -0.018(0.0001) and 0.018(0.0001) for the strained layer and substrate respectively. Using equation (1) and a value of 0.28 for Poisson's ratio, the lattice misfit was determined to be f=-0.0203(0.0002), a value within two standard deviations of the actual misfit of $f = -0.02$.

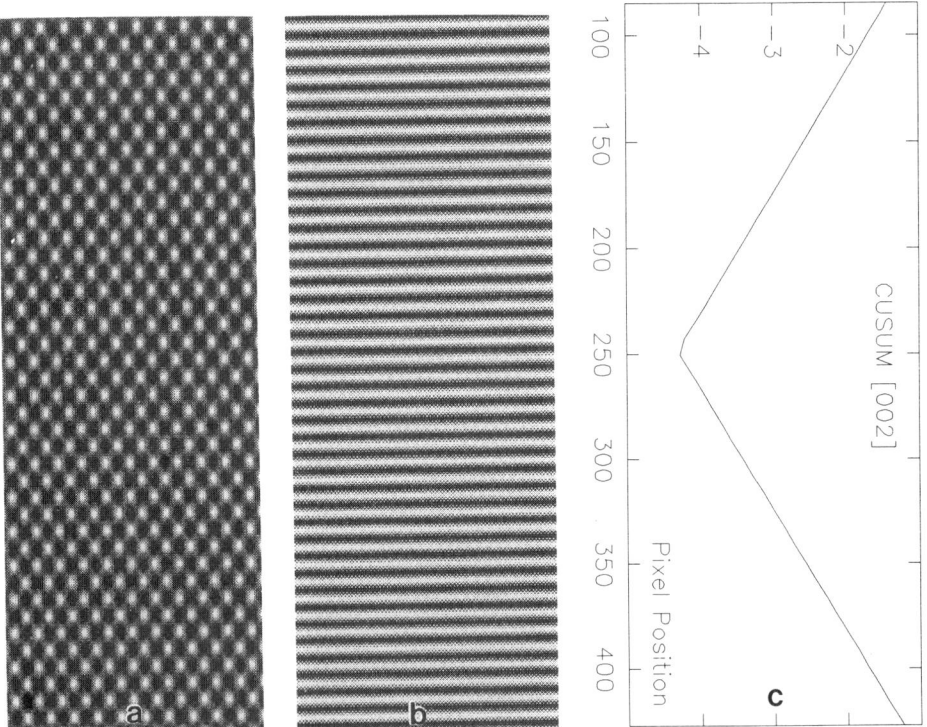

Fig. 2. (a) A HREM image simulation of an InSb bilayer where the upper half of the imaged is coherently strained by $f = -0.02$. (b) The Fourier filtered (002) lattice fringe image. (c) CUSUM plot of the (002) lattice fringe spacings. The interface is located at the vertex of this plot.

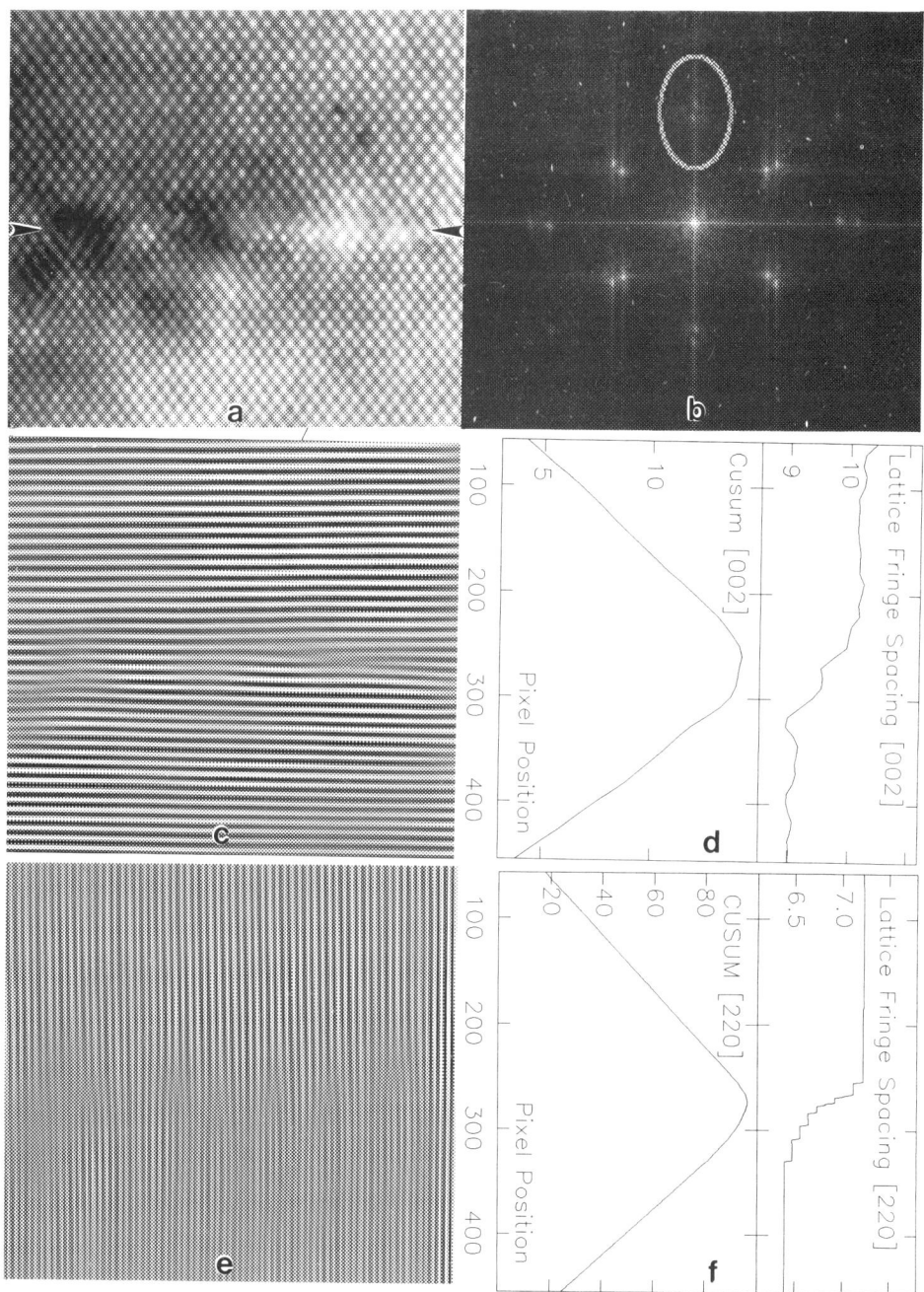

Fig. 3. (a) A HREM image of an $In_{.89}Ga_{.11}Sb$ on (001) GaAs bilayer. The interface is highlighted by the arrow-heads and there is an array of misfit dislocations of edge type at the interface. (b) The digital diffractogram of (a) displaying an elliptical Fourier filter mask. (c),(d) (220) lattice fringe image and CUSUM and lattice fringe spacing plots. (e),(f) (002) lattice fringe image and CUSUM and lattice fringe spacing plots.

5. EXPERIMENTAL BILAYER

The CUSUM technique was applied to an $In_{.89}Ga_{.11}Sb$ on (001) GaAs bilayer grown by metalorganic magnetron sputtering (Rousina et al. 1990). The HREM images were obtained using a Philips CM20ST transmission electron microscope operated at 200 kV and fitted with a Gatan slow scan CCD camera. Fig. 3a displays a 445 x 398 pixel subset of a 512x512 pixel HREM image of the bilayer in the $[1\bar{1}0]$ zone axis projection and Fig. 3b is the associated digital diffractogram. A sample elliptical mask is also displayed. From electron diffraction studies, the lattice misfit of this system has been determined to be f = 0.13(0.001) and the system was found to be completely relaxed (Robertson et al. 1992). The lattice misfit was accommodated by an array of edge dislocations at the interface. The location of the interface is highlighted by the two arrow-heads in Fig. 3a. Fig. 3c is the (002) Fourier filtered lattice fringe image and Fig. 3d is the associated CUSUM and raw lattice fringe spacing plots. Figs. 3e and 3f are the same images and plots repeated for the [220] reflection. Note that both the [002] and [220] plots are averages over the entire 512 pixel width. The vertices of the CUSUM plots for both the [002] and [220] reflections accurately represented the location of the interface. The slopes of the [002] CUSUM plots, avoiding the regions affected by Fourier continuation near the edges, were $m_{film}^{002} = 0.0556(0.0008)$ and $m_{sub}^{002} = -0.0672(0.001)$ giving e = 0.130(0.001). From the [220] CUSUM plot, the slopes were $m_{film}^{220} = 0.37948(0.00005)$ and $m_{sub}^{220} = -0.45367(0.00004)$ and the average lattice fringe spacing was $\overline{Z}^{220} = 6.846(0.017)$, giving δ = 0.1303(0.0003). Using equation (3), we see that f=δ=0.0130(0.001) as expected. Recall that surface relaxation effects are negligible for completely relaxed systems. Near the interface, there will be additional strains due to the presence of the edge dislocation array. However, these become insignificant at distances greater than about one dislocation spacing away from the interface due to cancellation by the misfit strains.

6. ACKNOWLEDGEMENTS

Two of the authors (M.D.R. and J.M.C) would like to acknowledge the support of this research by the Natural Sciences and Engineering Research Council of Canada. Also, one of the authors (M.D.R.) thanks the National Research Council of Canada for the generous support provided by the visiting research graduate program.

REFERENCES

Bierwolf R, Hohenstein M, Phillipp F, Brandt O, Crook G E and Ploog K 1993 Ultramicroscopy 49 273
Faux 1994 J. Appl. Phys. 75 186
Goldsmith P L and Whitfield H 1961 Technometrics 3 11
Humphreys C J 1979 Rep. Prog. Phys. 42 1825
Hytch M J and Bayle 1994 Proc. 13th Int. Conf. on Electron Microscopy 2A, eds B Jouffrey and C Colliex (Les Ulis: Les Editions de Physique) pp 129-130
Jouneau P H, Tardot A, Feuillet G, Mariette H and Cibert J 1994a J. Appl. Phys. 75 7310
Jouneau P H, Cibert J, Feuillet G, Mariette H and Tardot A 1994b Proc. 13th Int. Conf. on Electron Microscopy 1, eds B Jouffrey and C Colliex (Les Ulis: Les Editions de Physique) pp 399-400
Page E S 1961 Technometrics 3 1
Robertson M D, Corbett J M, Webb J B and Rousina R 1992 Can. J. Phys. 70 866
Robertson M D, Currie J E, Corbett J M and Webb J B 1995 Ultramicroscopy, in press
Rousina R, Halpin C and Webb J B 1990 J. Appl. Phys. 68 2181

Inst. Phys. Conf. Ser. No 146
Paper presented at Microsc. Semicond. Mater. Conf., Oxford, 20–23 March 1995
© 1995 IOP Publishing Ltd

A contribution to the quantitative comparison of experimental high–resolution electron micrographs and image simulations

T Walther, C J D Hetherington and C J Humphreys

Department of Materials Science and Metallurgy, University of Cambridge, Pembroke Street, Cambridge CB2 3QZ, UK

ABSTRACT: For the assessment of the reliability of any measurements from high–resolution transmission electron microscope (HREM) images it will be necessary to match quantitatively experimental and calculated images. Using a focal series from a cleaved GaAs/AlGaAs specimen we compared the Fourier spectra of the experimental and calculated image intensities for various thicknesses and spatial frequencies at Scherzer focus. The result is a significant discrepancy between the experimental and calculated data, depending on the spatial frequency and the specimen thickness, although the overall form of the thickness dependence of the Fourier components could be rather well reproduced.

1. INTRODUCTION

With the progressing sophistication of image simulation and analysis programs on the one hand and the availability of CCDs for image recording with high dynamic range and linearity on the other, there obviously is the need for quantifying the information content of a high–resolution transmission electron micrograph. The quantified data will allow atom positions and local stoichiometry to be investigated, but it will be important to assess the accuracy and reliability of any measurements. Several attempts to match quantitatively experimental and calculated images failed partially because the imaging parameters were not precisely known (Anstis et al 1993, Hytch and Stobbs 1994). In both cases the image patterns of the experimental HREM images could only be matched assuming specimen thicknesses that differed from the values determined otherwise by more than a factor of 2.

A defocus series from cleaved GaAs/AlGaAs wedge specimens of known geometry allows a reliable measurement of specimen thickness and defocus, assuming that the crystalline edge is perfectly sharp without any rounding. The Fourier spectra of the image intensities of the experimental and the simulated images are preliminarily compared for one image at Scherzer defocus. Any discrepancy between experimental and simulated images can be attributed to a combination of effects not yet taken into account in the image simulations, such as inappropriate modelling of the microscope transfer function, electron beam convergence on the top specimen surface and the role of inclined entrance and exit surfaces of the electron beam. A separation of the influence of these effects on the image would have to involve a comparison of the whole focal series with image simulations covering all reasonable ranges for the relevant parameters and must be subject to further work.

2. EXPERIMENTAL

We acquired a focal series consisting of 11 negatives with equidistant focal steps from a cleaved (001) GaAs/AlGaAs wedge specimen taken along the <100> zone axis in a JEOL 4000EX-II electron microscope operated at 400keV, along with an exposure time series developed under identical conditions. The negatives were digitised with a sampling of 0.014nm/pixel with an EKTRON EIKONIX scanner having a linear 4096 diode array. The effects of inhomogeneous illumination and different gain of the diodes resulting in streaking were compensated by convoluting the images along both directions with one dimensional

54

correction functions obtained from the negatives exposed to the electrons without specimen. The exposure series was used as a calibration to convert the grey levels into absolute electron intensities on a scale where the incident electron intensity is one, with an absolute error of only a few percent. Thus, the output can directly be compared with image simulations. The defoci of the images were determined at the crystal edges by matching the power spectra of the amorphous material with the microscope transfer function. Fig. 1 shows an HREM image of a GaAs/Al$_{0.3}$Ga$_{0.7}$As interface taken close to Scherzer focus and the corresponding power spectrum of the amorphous edge (underfocus $\Delta f=-40\pm3$nm, with an astigmatism such that its axis lies close to the vertical direction and the value is about 4nm lower in this than in the horizontal direction). We have chosen this image from the focal series because the broad pass band of the contrast transfer function of the microscope allows many spatial frequencies to contribute to the image, and at the same time the sensitivity to small focus variations is low, rendering the focus change experienced by the exit wavefunction along the inclined lower surface not critical at least for small thicknesses.

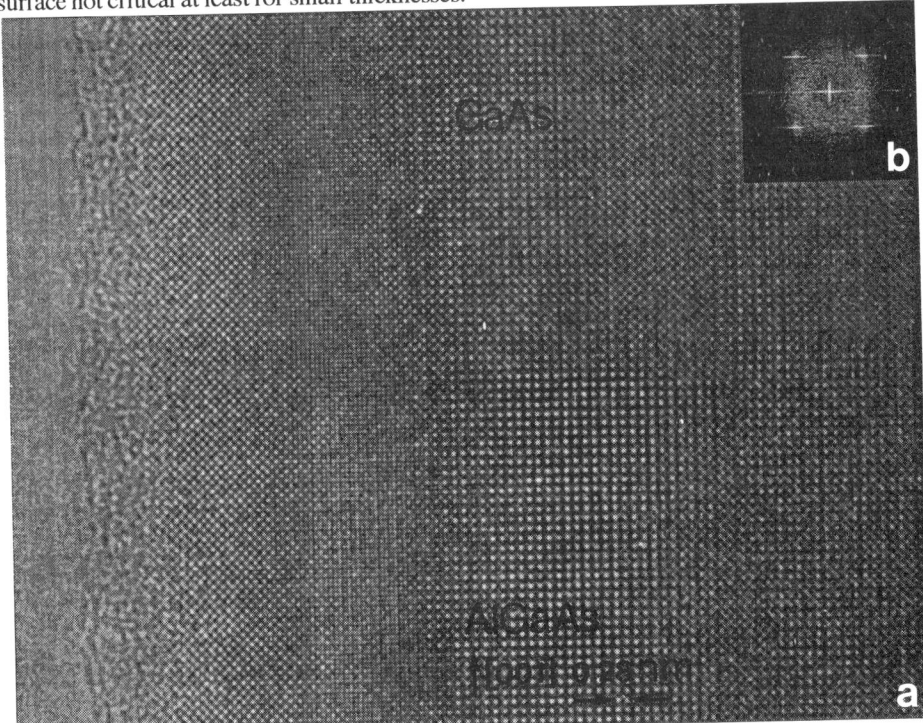

Fig. 1: Experimental <100> HREM image of cleaved Al$_{0.3}$Ga$_{0.7}$As/GaAs with the specimen thickness given as twice the distance from the crystal edge (a). Black=0.5, white=1.5. Inset: power spectrum of the amorphous region, indicating Scherzer focus (b).

3. IMAGE ANALYSIS

The HREM image was analysed in a way similar to that described by Walther and Gerthsen (1993). A least–square fit to the intensity extrema yielded the positions of the atomic columns. The absence of reference points meant that the group III and V sites could not be distinguished reliably but since we were not analysing the phases of the Fourier components of the image intensity this could be ignored. The intensity in the image unit cells of $(0.283\text{nm})^2$ thus obtained was resampled onto a 16^2 pixel square grid and Fourier trans–formed. The amplitudes U($\mathbf{g}$) of the image intensity (eqivalent to the square root of the com–ponents of the local power spectrum) were calculated for $\mathbf{g}$=(000), (002), (022) and (004) for every image unit cell and averaged along the [001] growth direction, separately for GaAs and AlGaAs. Plots of U($\mathbf{g}$) as a function of specimen thickness are shown in Fig. 2.

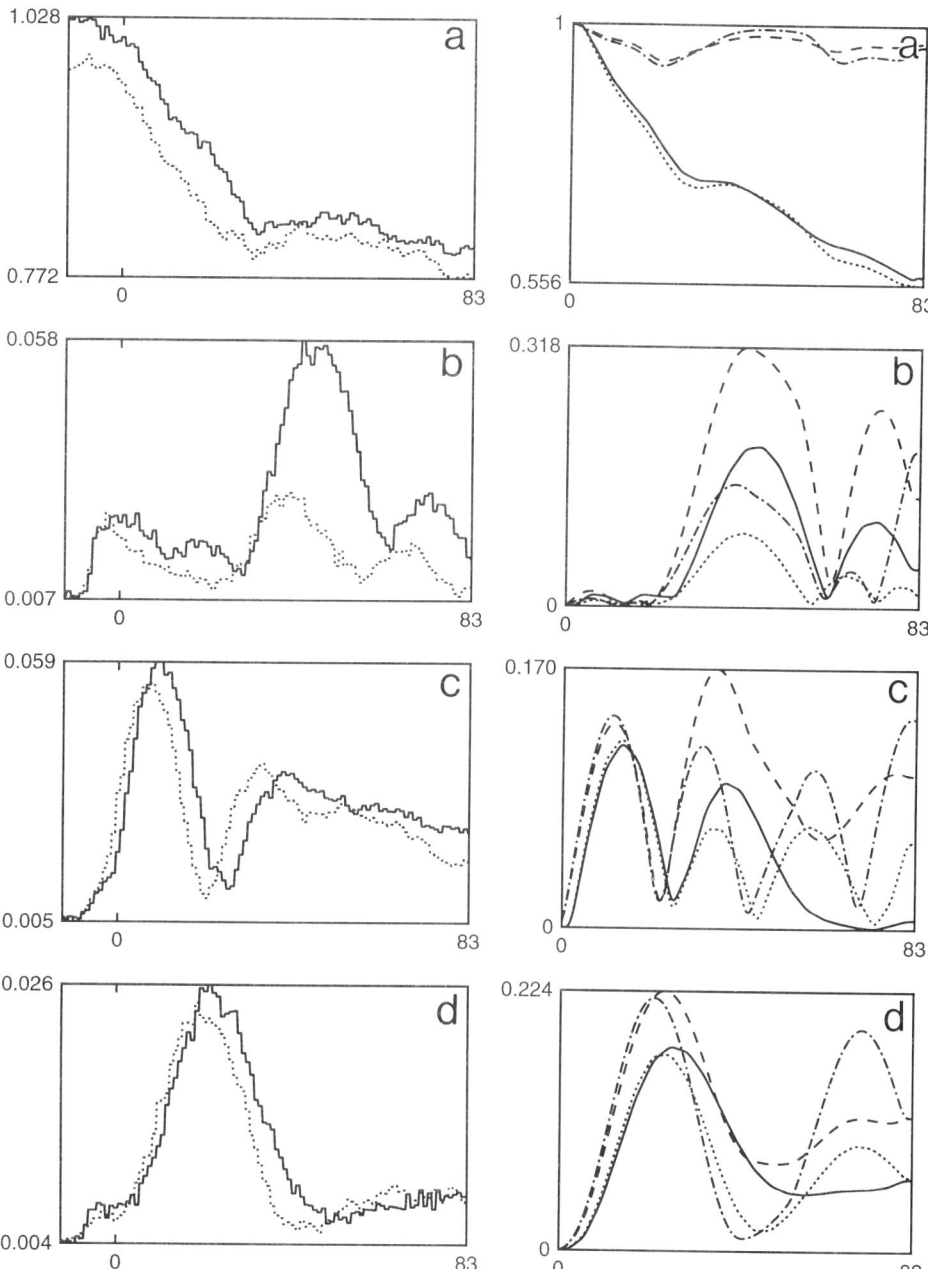

Fig. 2: Experimental values of U($\mathbf{g}$) as ave–
raged along [001] as a function of the
distance from the crystalline edge [mono-
layers; 0=edge] for AlGaAs (solid line)
and GaAs (dotted); $\mathbf{g}$=(000) (a), (002)
(b), (022) (c) and (004) (d)

Fig. 3: Values of U($\mathbf{g}$) as function of specimen
thickness [unit cells] calculated without
absorption for AlGaAs (dashed) and
GaAs (dot-dashed) and with absorption
for AlGaAs (solid) and GaAs (dotted);
$\mathbf{g}$=(000) (a), (002) (b), (022) (c), (004) (d)

The projected width of the amorphous surface layer is <2nm throughout the defocus series. The drop of U(000) in this layer is <0.02, yielding a loss in background intensity in the amorphous region under 4% of the total intensity. This might be considered a drastic improvement compared to specimens prepared by ion milling for which drops of the background intensity in the amorphous surface layers of the order of 20% are common and even values up to 44% have been reported (Dobson et al 1991). It also indicates that the inelastic scattering by the amorphous surface layer of our specimen is probably not responsible for the large discrepancy between experimental and simulated data discussed in the following.

4. IMAGE SIMULATIONS

We calculated the Fourier spectra of HREM images without and with absorption. The calculation ignoring absorption was performed with a multi–slice algorithm based on the work by Cowley and Moodie (1957), using the scattering factors provided by Doyle and Turner (1968). We checked that multi–slice and Bloch wave calculations yielded the same result if no absorption parameters were included. The imaging part was carried out in SEMPER, using as parameters a spherical aberration constant of 0.9mm, a focus spread of 10nm, a semi-angle of beam divergence of 0.9mrad, a defocus of –40nm, a lattice constant of 0.5653nm and a Debye–Waller factor of $0.006nm^{-2}$. The calculation including absorption used the Bloch wave program 'cb0' of the EMS software package (Stadelmann 1987) in a new version with the absorption parameters according to Weickenmeir and Kohl (1991). The exit wavefunction was then transferred into SEMPER for the imaging part. Both calculations were carried out for GaAs and AlGaAs and are shown in Fig. 3 on the same scale.

It is obvious that the thickness oscillations of U(000) as obtained by the calculations without absorption (Fig. 3a) are very different from the observed nearly monotonic decrease of U(000) with increasing specimen thickness (Fig. 2a). Taking into account inelastic scattering by a complex scattering potential improves the fits, and the position of the extrema as a function of the specimen thickness can be fitted quite well for all spatial frequencies considered, demonstrating that the specimen thickness increases indeed linearly with the distance from the crystal edge. However, the absolute values of the experimental and the calculated data strongly disagree. The maxima of U($\mathbf{g}$) in the experiment are smaller than the corresponding values simulated with absorption by a factor of 3 for $\mathbf{g}$=(002), 2 for $\mathbf{g}$=(022) and 6 for $\mathbf{g}$=(004). It is thus interesting to note that the disagreement between experiment and simulation does not increase monotonically with the spatial frequency and can therefore not properly be modelled by simply including Gaussian vibration parameters in the simulations that would preferably damp finer image details.

5. CONCLUSION

Even for a specimen with a minimum amount of amorphous surface contamination and known specimen geometry the dependence of the Fourier components of the image intensity on the specimen thickness could only be modelled qualitatively. The thickness values for which certain patterns occurred were in good agreement, but the experimental contrast compared to the background intensity was a factor of 2–6 smaller than calculations suggested.

ACKNOWLEDGEMENT

Thanks are due to WO Saxton and RE Schäublin for help with the image analysis and the EPSRC UK and the DRA Malvern for a CASE award studentship to TW.

REFERENCES

Anstis GR, Auchterlonie GJ and Barry JC 1993 Ultramicroscopy 52 167-178
Cowley JM and Moodie AF 1957 Acta Cryst 10 609-619
Dobson AS, Preston AR and Stobbs WM 1991 Inst Phys Conf Ser 119 449-452
Doyle PA and Turner PS 1968 Acta Cryst A 24 390-397
Hytch MJ and Stobbs WM 1994 Ultramicroscopy 53 191-203
Stadelmann PA 1987 Ultramicroscopy 21 131-146
Walther T and Gerthsen D 1993 Appl Phys A 57 393-400
Weickenmeir A and Kohl H 1991 Acta Cryst A 47 590-597

Inst. Phys. Conf. Ser. No 146
Paper presented at Microsc. Semicond. Mater. Conf., Oxford, 20–23 March 1995
© *1995 IOP Publishing Ltd*

A fuzzy logic approach to quantitative HREM

R Hillebrand, P Werner, H Hofmeister and U Gösele

Max-Planck-Institut für Mikrostrukturphysik, Weinberg 2, D-06120 Halle, Germany

ABSTRACT: Analysing HREM images of III-V compounds, a fuzzy logic approach is developed to study interdiffusion phenomena in layered structures by using the monotonous relation of "image similarity" and "chemical composition" under appropriate imaging conditions. Simulated and experimental HREM images of GaAs/P (As/P variation) and Al/GaAs (Al/Ga variation) are analysed by fuzzy logic image processing to extract chemical reliefs.

1. INTRODUCTION

Contrast features in HREM micrographs of III-V semiconductors can be correlated with the variation of the chemical composition of the crystals. In contrast to the methods of chemical mapping, developed by Ourmazd et al. (1990) and extended by Schwander et al. (1993), the present paper introduces a fuzzy logic approach, initiated by a paper of Tyan and Wang (1993). It makes use of the one-to-one relation of similarity and chemical composition under appropriate experimental conditions by constructing triangular types of fuzzy sets. Contrary to the group III sublattice of GaAs, see, e.g., Tan et al. (1991), very little is known of diffusion processes and point defects governing in the As sublattice. In addition to SIMS and cathodoluminescence, HREM will be utilized on the basis of the simulation of contrast tableaus and their "chemical analysis" to gain the desired information from actual micrographs.

2. METHOD

In the following, simulated HREM images of GaAs crystals of a variable P content in the As sublattice are used to illustrate both the physical and the mathematical concept of the fuzzy logic approach. The electron diffraction of a statistical mixture of As/P in the As sublattice can be described by an interpolation. "Multi-slice" calculations using the Stadelmann (1987) program EMS showed that the intensity of the (200) reflections strongly varies with the increase of the P content. Besides the chemical composition of the crystals studied, specimen thickness t and defocus Δ strongly influence the image patterns. Appropriate experimental conditions (t / Δ window) are checked by computer simulations.

Fig. 1a shows HREM patterns (400kV) calculated for (100)-oriented $GaAs_{1-x}P_x$. The lateral size of one crystallographic image cell is 32x32 pixels. Here, in the vertical direction the specimen thickness increases linearly from 9 to 16 nm. According to pre-studies an underfocus of 30 nm is applied. In the horizontal direction of the figure the P/As ratio is varied in steps of 10%. The influence of additive amorphous noise, e.g. from a carbon film, is included. The systematic changes in the crystalline patterns have to be quantified by image processing.

As pointed out by Diday and Simon (1976), a measure of similarity (resemblance) has to be a real, symmetric function; values close to 0 imply a high similarity. To compare two images, we have chosen the standard deviation (SD) of the difference cell pattern for the sake of linearity. The

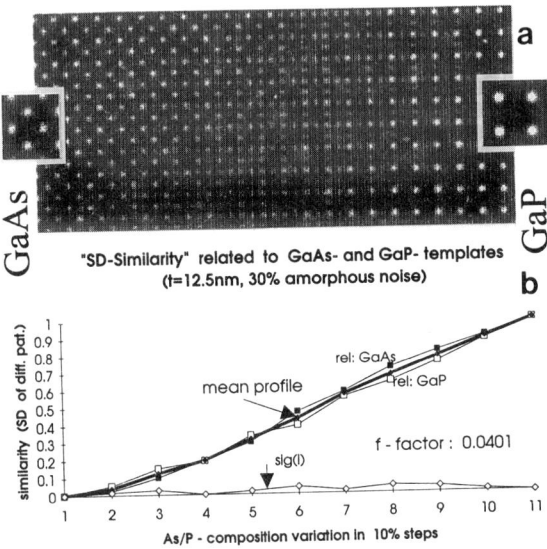

GaAs   GaP

"SD-Similarity" related to GaAs- and GaP- templates
(t=12.5nm, 30% amorphous noise)

a

b

Fig. 1. a) HREM contrast tableau of $GaAs_{1-x}P_x (100)$
b) Profile of similarity with f-factor (σ).

closest similarity of compared images is influenced by additive amorphous noise and a lateral cell shift in the step of image subtraction. If the lateral grid shift can be compared to the size of the asymmetric unit the resulting SD-values can no longer be associated with only one composition.

The insets of Fig. 1a show HREM images typical of pure GaAs(100) and GaP(100) of a crystal thickness around 12 nm. At the next stage the simulated GaAs/P contrast tableau is compared cell by cell with the two templates. The two profiles of similarity, normalized and inverted respectively, can be combined resulting in a mean function, the standard deviations σ_i of which define the factor $f = \max(\sigma_i)$, with i = 1...11 being the number of As/P-steps here. This factor controls the classifications in the subsequent analysis.

Fig. 2 introduces the mathematical approach, starting with image comparisons and resulting in the formation of fuzzy sets. The nearly straight increase of the SD-distributions (cf. Fig. 1b) allows tri-angular graphs of similarity to be constructed by inversion (comple-ment). Each of such functions I_n expresses the degree of membership of any image in the set of images, being similar to the prototype image selected. Extending the considerations to a sequence of scale images I_1, I_2, etc., yields Fig. 2a, which is part of a series of triangular fuzzy sets (see Seraphin 1994). The size of the triangular base is determined by the number of images that are members,

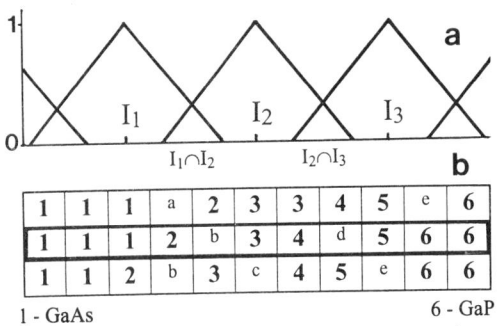

a

b

1	1	1	a	2	3	3	4	5	e	6
1	1	1	2	b	3	4	d	5	6	6
1	1	2	b	3	c	4	5	e	6	6

1 - GaAs   6 - GaP

Fig. 2. a) Triangular intersecting fuzzy sets of similarity
b) Chemical fuzzy map (NClass = 6).

called the support *supp* of the fuzzy set. The value of f (here 0.04), limits the number of useful classes (NClass) defining their degree of intersection. The values of the linguistic variable "similarity of images" are stored in SD-matrices. They are sorted and classified into: [very similar .... very different], with the number of classes and the steps of aggregation varying. According to Fig. 2a:

-    The interference (overlap) region between two different fuzzy sets is the intersection denoted by *inter* $I_1 \cap I_2$ (see Pedrycz 1993), which is determined by f.
-    If I_1, I_2 and I_3 are neighboring fuzzy sets, then the expression

$$supp(I_2) \setminus (supp(I_1 \cap I_2) \cup supp(I_2 \cap I_3)) \tag{1}$$

describes the set of images, unambiguously related to prototype 2, called the *unique* of 2, where $\setminus$ is the set theoretic difference.
-    The ratio between the interference and unique classifications reads:

$$R_{fuzzy} = inter \,/\, unique = f \times (NClass - 1) \,/\, (1 - f \times (NClass - 1)). \qquad (2)$$

A small value of R_{fuzzy} means a high level of confidence in the triangular constructions.

Based on these classifications, the physical property P, which here is the chemical composition of III-V crystals on the scale of images X, can be concluded from fuzzy-inference (Seraphin 1994). The higher the membership of any image X in a set of images A, the more their physical properties coincide, symbolically:

$$\text{THE MORE} \quad I_A(X) \quad \text{THE LESS} \quad \| P(A) - P(X) \| \qquad \text{etc.,} \qquad (3)$$

with norm $\| \ \|$ describing the deviation of the physical properties.

3. RESULTS

In the chemical map of Fig. 2b the bold numbers (1-6) characterize the As/P classes whereas the small letters (a-e) mark the fuzzy-$\cap$ (intersection) of neighbouring sets. A useful number of classes detectable in the GaAs/P example is NClass=6; the corresponding value of R_{fuzzy}=0.251 is a reasonable limit. As a result 79% of all cells analysed in Fig. 2b (11nm < t < 14nm) provides unique results of similarity and As/P-ratios. A much higher number of classes, e.g. NClass=15, would lead to a very low percentage of unique images in the defuzzification.

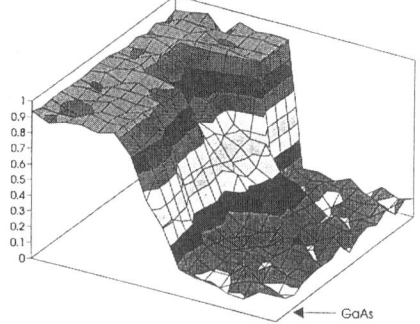

The contribution presented lays special emphasis on P-poor GaAs/P (P < 50%). In Fig. 3 four different "chemical barrier models" (t=11nm) between an abrupt As/P step and a linear variation of the As/P ratio are analysed in the above way. The similarity profiles obtained allow the clear distinction between the different slopes of chemical transition even for an amorphous noise of 30%. The fuzzy chemical classification of the barriers follows the mapping of Fig. 2.

Fig. 3. Relief of similarity of four simulated $GaAs_{0.5}P_{05}$/GaAs barriers (HREM, U=400kV).

Aspects of analysing experimental Al/GaAs(110) micrographs are indicated in Fig. 4, in particular the non-linearity of image formation. The imaging conditions in Fig. 5a were determined by image matching (best fit of simulated patterns). Then, in simulations the Al/Ga ratio was varied linearly for the crystal thickness and defocus value identified. The resulting set of chemical-tuned images showed similarity profiles (cf. Fig. 4), clearly non-linear but monotonous. The resulting fuzzy sets may be non-equidistant for stronger deviations from linearity.

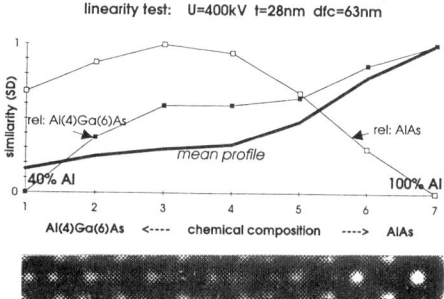

Fig. 5a shows part of an experimental HREM micrograph of an AlAs / $Al_{0.4}Ga_{0.6}As$ interface in cross section. The image of the MBE-grown material was taken in a JEM 4000EX. For studying the quality of the interface as well as for extracting the Al/Ga-profiles the reference templates were obtained from image areas far from the

Fig. 4. Non-linearity between Al/As ratio and similarity for $Al_{0.4}Ga_{0.6}As$/AlAs (110).

interface. These patterns are averaged by Fourier filtering and/or accumulation techniques. Their chemical composition is known by growth and physical reference methods (e.g., SIMS).

The relief of similarity presented in Fig. 5b is calculated cell by cell (16x16 pixels). In determining the f-factor of experimental images outliers have to be excluded (σ_i <0.07 for 89.3% of

60

the cells), which result from image distortions, noise and grid misfit. In this experimental example the f-factor of 0.07 allows 4 useful intervals (R_{fuzzy}=0.266), i.e. the step width of the Al/Ga - scale is 15% with fully considering the experimental noise. In the chemical map of Fig. 5c, the bold numbers (1-4) characterize the unambiguous Al/As classes whereas (a-c) denote the intersections.

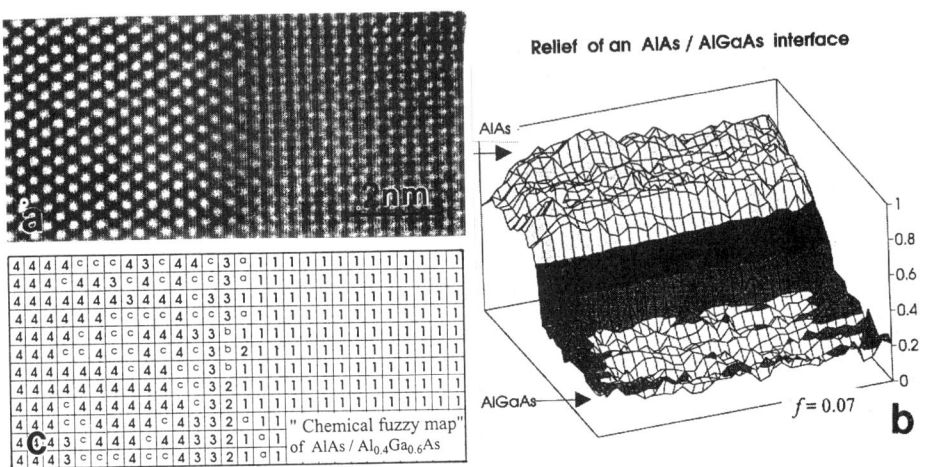

Fig. 5. a) Micrograph of an AlAs/Al$_{0.4}$Ga$_{0.6}$As (110) interface, U = 400kV b) Relief of similarity
c) Part of the analysed Al/Ga-classes.

The parameters being of physical interest for diffusion studies, i.e. the slope, can clearly be derived from the chemical profile of the interface. Improving the sensitivity of the approach to experimental applications requires the individual fit of the templates cell by cell, i.e. the tweaking of the grid, to compensate geometrical imperfections of the crystalline images.

4. CONCLUSIONS

The fuzzy logic approach proposed makes use of the correlation between image similarity in HREM and crystal composition. The triangular fuzzy sets of similarity derived, interfering according to the experimental conditions, allow HREM micrographs of layered structures of III-V compounds to be interpreted in terms of their chemical composition.

REFERENCES

Diday E and Simon J C 1976 Digital Pattern Recognition, ed KS Fu (Berlin: Springer-Verlag) pp 51
Ourmazd A, Baumann F H, Bode M and Kim Y 1990 Ultramicroscopy 34 237
Pedrycz W 1993 Fuzzy Control and Fuzzy Systems (New York: J. Wiley) p 14
Schwander P, Kisielowski C, Seibt M , Baumann F H, Kim Y and Ourmazd A 1993 Phys. Rev. Lett. 71 4150
Seraphin M 1994 Neuronale Netze und Fuzzy-Logik (München: Franzis') pp 115, 134
Stadelmann P 1987 Ultramicroscopy 2 131
Tan T Y, Gösele U and Yu S 1991 Crit. Rev. Solid State Mater. Sci. 17 47
Tyan C-Y and Wang P P 1993 Proc. 2nd IEEE Conf. on Fuzzy Systems (San Francisco) p 660

Acknowledgements:

The authors are grateful to Ch. Bauer and U. Gromann for helpful mathematical discussions. Thanks of one of us (R. H.) are due to A. Ourmazd (AT&T Bell Labs., US) for the opportunity to visit his group and to study chemical mapping.

Inst. Phys. Conf. Ser. No 146
Paper presented at Microsc. Semicond. Mater. Conf., Oxford, 20–23 March 1995

Ultra high resolution quantitative analysis of nano-structures

U Bangert[1], A J Harvey[1], S Gardelis[1], R J Keyse[2], C Dieker[3] and A Hartmann[3]

1) Department of Pure and Applied Physics, UMIST, Manchester M60 1QD, UK
2) Department of Materials Science and Engineering, University of Liverpool, Liverpool L69 3BX, UK
3) Institut fuer Schicht- und Ionentechnik, Forschungszentrum Juelich, D-52425 Juelich, Germany

ABSTRACT: Results showing the potential of energy dispersive X-ray analysis (EDX) and electron energy loss spectrometry (EELS) with ultra high spatial resolution, in conjunction with high angle dark field (HDF) imaging, performed with a Fisons VG601UX HB FEGSTEM on quantum wells (QW) and quantum wires (QWR) of the III-V compound and the SiGe system and on nano-crystals in porous silicon are presented and discussed.

1. INTRODUCTION

A new generation of scanning transmission electron microscopes (STEM) enables highly spatially resolved compositional analysis. By modelling the X-ray production from an electron beam passing through a sample of known geometry it is possible to calculate the detection limit and to match calculated composition profiles of X-ray line scans taken across nano features. Results using this approach will be presented. Results of ultra high resolution point analysis using the STEM will also be presented.

2. EXPERIMENTAL

Investigations have been conducted on semiconductor nano-structures. The geometry and composition of the samples used in this study are described in the individual figures. Cross-sectional samples of the heterostructures were made by dimpling and Ar-milling at LN temperature, in the case of porous silicon cleaved edges were used. EDX and EELS analysis were performed with a VG HB601UX FEG STEM operated at 100 keV.

3. RESULTS AND DISCUSSION

3.1 Compositional inhomogeneities in III-V semiconductor alloys

3-D island formation and wavy growth in strained multiple quantum well stacks of the ternary and quaternary III-V compound system has been extensively reported but so far only been indirectly related to compositional fluctuations of the alloy [e.g. Treacy et al 1985, Ponchet et al 1993, Zhou et al 1993].

In order to estimate the possibility of direct detection we modelled the detection limit of clusters of a certain size of element A in an otherwise uniform matrix AB, by means of a simple model, which neglects absorption effects, the mathematical derivation of which can be found elsewhere [Bangert et al 1995]. In our case this corresponds to the detection of Ga-rich regions (=A) in the group III alloy constituent GaIn (=AB) of the quaternary alloy GaInAsP. We assume the electron excitation volume to be a truncated cone (see fig.1). Fig.1 models the case in which cylindrical clusters of three different radii rc of element A in a matrix AB, where $conc._A/conc._B = 0.333$, are placed at different depths xt from the probe entrance surface. The sample thickness t=50 nm. The curves a-c show the fractional increase in counts of element A, as a function of cluster length tc. The detection limit in our experiments, due to statistics was generally found to be 8%. In the case of a cluster placed at the probe entrance surface (curve c) only 40 atoms (corresponding to a cluster length of 1.3 nm) of element A are sufficient to raise the fractional count rate above the statistical limit. If the clusters lie dense, they can hardly be spatially resolved with a scan point separation of 3 nm, which is a value adjusted to the beam broadening, and would appear as large background scatter. If the clusters are less frequent but distributed throughout the sample, the ones near the surface will be detected, if they contain more than 40 atoms.

X-ray count fluctuations near and above the detection limit of 8% are frequently observed in III-V compound alloys. In order to ascertain the existence of clusters in quaternary alloys, X-ray line scans have been undertaken in a strained GaInAsP multiple quantum well structure with changing As/P ratio in wells and barriers

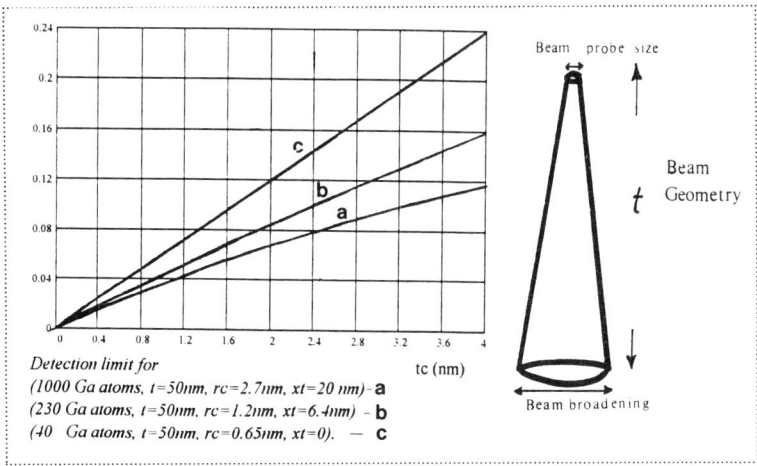

Detection limit for
(1000 Ga atoms, $t=50$nm, $rc=2.7$nm, $xt=20$ nm) - **a**
(230 Ga atoms, $t=50$nm, $rc=1.2$nm, $xt=6.4$nm) - **b**
(40 Ga atoms, $t=50$nm, $rc=0.65$nm, $xt=0$). — **c**

Fig.1 Fractional X-ray count of cluster A as function of cluster length. For explanations see text.

(+/-1% strain), but nominally constant Ga/In ratio of 0.333. Fig.2 shows line scans of the Ga/In ratio along subsequent wells and barriers of a strained GaInAsP multiple quantum well structure, together with the corresponding HDF image.

10 nm

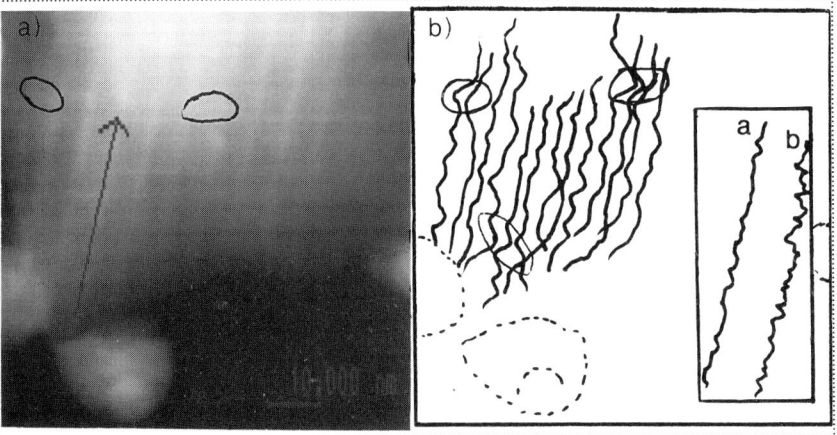

Fig.2 a) High angle DF image of a $Ga_{0.26}In_{0.74}As_{0.87}P_{0.13}/Ga_{0.26}In_{0.74}As_{0.5}P_{0.5}$ MQW structure. b) Line scans of the Ga/In count ratio along successive wells and barriers. The circles mark larger fluctuations in the scans and the corresponding features in the HADF image.

There are fluctuations (of ~ 3 nm size), which are marginally above the 8% limit in each linescan, their repetition in a similar place in a neighbouring linescan, however, lends more weight to their existence (see encircled regions). Some fluctuations coincide with 'wavy' layer growth (see circles in the HDF image). An In-line scan (a) taken in the InP substrate as well as a scan of the Ga/In ratio (b) taken in a lattice matched thick layer of GaInAs show statistical fluctuations, which are less than those in the strained quaternary MQW. We conclude therefore that fluctuations in the Ga/In ratio do occur and that the corresponding cluster sizes are of the order of 3 nm.

3.2 Composition profiles of quantum structures

It can be expected that the composition achieved during epitaxial growth of two dimensional structures and in particular growth on patterned surfaces deviates from that of thick, bulk-like layers. Fig.3a shows Ge_{Ka} X-ray counts obtained in scans along the arrows across a quantum well of 1 nm width, and fig.3b across a quantum wire of 120 nm width in a Si/Ge hetero-structure (see paper by Dieker et al in these proceedings). Distinct asymmetries in the scans of both features can be observed. Due to beam broadening substantial sampling outside the feature, in particular the well, can be expected, so that the above values, do not reflect the true composition.

A line scan can be simulated which requires (the known) values of the sample thickness t, the beam probe diameter and the beam broadening. The bottom sections show how the sample thickness t influences the simulated scan profiles. In curves (1) t=80 nm, in curve (2) t=120 nm and in curves (3) t=180 nm. The numbers represent the percentage fraction of the 'true' concentration detected by the beam. The 'true' concentration thus obtained was 31% for the well and 16% for the wire (nominal: 40%). The Ge profiles are both asymmetrical: the Ge incorporation in the well (i.e. on the <100> surface) decreases as the layer thickness increases, and it increases in the wires, indicating preferred capture in the wires, at first of Si and, as the radius of curvature increases, of Ge. The Ge

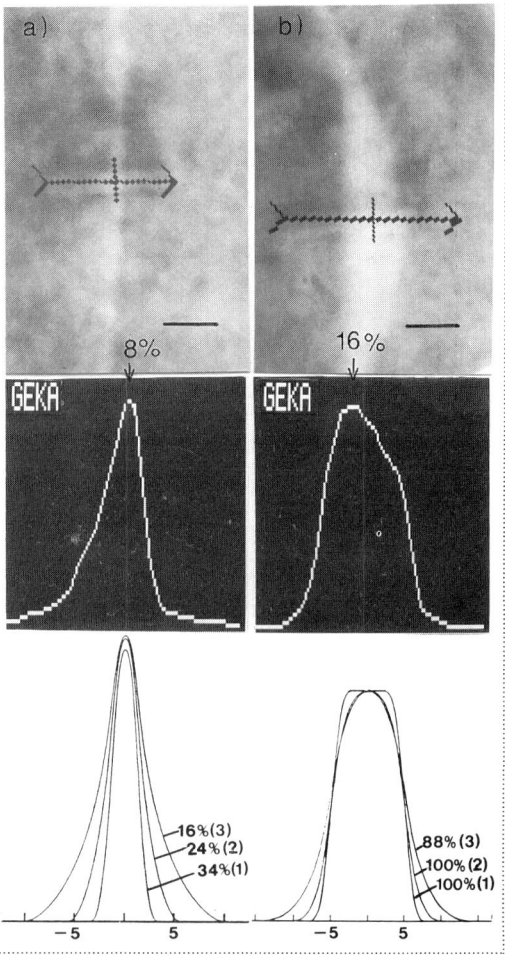

Fig.3 a) HADF image of a <100> SiGe QW with a nominal width of 1 nm and nominal Ge-content of 40%. The arrow indicates the X-ray scan line and corresponds to the GeK_a counts below. At the bottom are simulated profiles for sample thicknesses t= 80, 120 and 180 nm (curves 1, 2 & 3), resulting from a composition step of width w=1 nm and of 32% Ge; b) HADF image of a wire cross-section in a groove defined by the <100> and <111> planes. Below is the GeK_a linescan and at the bottom simulated profiles for w=10 nm, t= 80(1), 120(2) &180(3) nm, resulting from a 18% composition step. The % numbers show the fractional height of the composition step, represented in the linescan. Markers are 10 nm, units on the bottom axis are in nm.

64

concentration in the wire does not, however reach the initial value of the <100> face. There is indication that surface diffusion towards the wires during growth plays an important role (see paper by Dieker et al.). The scatter in the results from repeated scans was found to be in the range of 15-20%.

3.3 Si-oxide content of nano-crystals in porous silicon

Fig.4 shows beam point EELS spectra, taken at different positions within Si of 80% porosity on ultra thin regions sticking out at the edge of cleaved wedges. Bright field lattice images as well as HDF field images were taken simultaneously, in order to guide us, as to which regions are crystalline and which are amorphous. On the left is a bright field lattice image. The letters mark the points at which the EELS spectra, shown on the right, were taken: in a nano-crystal (B) and in an amorphous part (A). The size of the lattice give an indication of the volume sampled by the beam. The crystalline spectrum shows the Si L-edge fine structure (modified due to the nano-size), whereas the 'sponge' spectrum shows a shift of 2 eV similar to that of SiO_2, and is presumed to be a sub-oxide (see paper by Gardelis et al in these proceedings). The detailed analysis of the fine structure of the spectra, is currently being explored.

During X-ray line sacns, which typically lasted for 5 min (i.e. 0.1 s per point) in total, the specimen drift was usually less than 4 Angstroems. Drifts were even less for X-ray and EELS point analyses, which took typically 1 min. Contamination problems became unnoticable after 'fixing' the contaminants, i.e. after subjecting the sample to a large area electron beam illumination for about 30 min, prior to analysis.

In conclusion it can be said that the above applications show that it is possible to undertake quantitative analysis in quantum structures within features on the nano-meter scale.

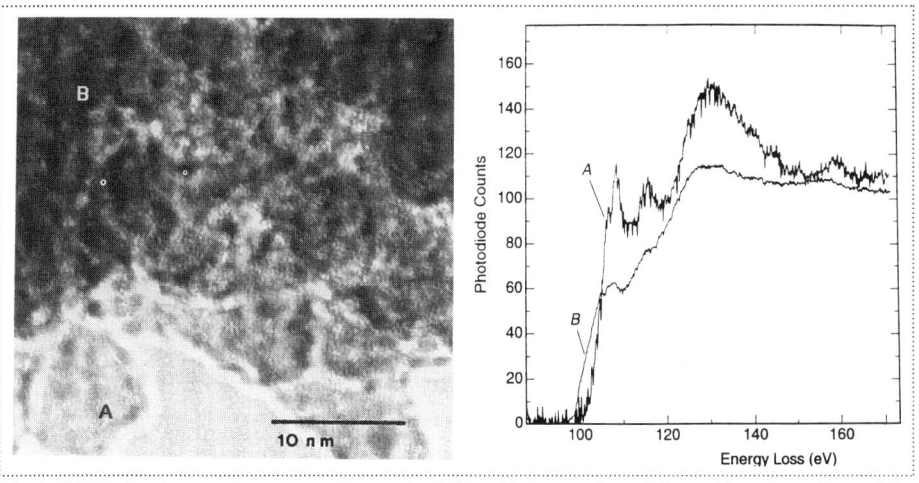

Fig.4 BF lattice image of a porous Si sample (left) with letters A and B marking the points of the EELS spectra show on the right.

REFERENCES

Bangert U, Harvey A J, Dieker C and Hardtdegen H 1995, accepted by J. Cryst. Growth
Ponchet A, Rocher A, Emery J Y, Starck C and Goldstein L 1993 Inst. Phys. Conf Ser. 134 485
Treacy M, Gibson J M and Howie A 1985 Phil. Mag. A51 389
Zhou X, Charsley P and Bangert U 1993 Inst. Phys. Conf. Ser. 134 489

Inst. Phys. Conf. Ser. No 146
Paper presented at Microsc. Semicond. Mater. Conf., Oxford, 20–23 March 1995

Structure and conductivity of films from various kinds of oriented carbon nanotubes

N A Kiselev[1], O I Lebedev[1], A N Kiselev[1], V I Bondarenko[1], L A Chernozatonskii[23], Z J Kosakovskaja[3], O E Omelianovskii[4],V I Tzebro[4] and E A Fedorov[3]

[1] Institute of Crystallography RAS, Moscow 117333, Russia
[2] Institute of Chemical Physics RAS, Moscow 117334, Russia
[3] Institute of Radio Engineering and Electronics RAS, Moscow 103907, Russia
[4] Lebedev Physical Institute RAS, Moscow 117924, Russia

ABSTRACT: Carbon nanotubes films obtained on glass and (111)Si by electron-beam evaporation of graphite were investigated by REM, STM and HREM. Nanotubes were oriented perpendicular to the substrate. At the "medium" density carbon particle flow, HREM revealed multilayered nanotubes. At the "high" flow density the bundles of single layered nanotubes with 0.72 nm diameter were revealed. Transverse to the tubes axis resistivity was $\rho_\perp(T)=\rho_o\exp(T_o/T)^{1/n}$. This is typical of amorphous semiconductors with hopping conductivity. We have found that for multilayered nanotubes films n=4 and $T_o=10^6$ K (T=25-150 K), for single layered ones n=2 and T_o=15-25K (T=4-40 K).

1.INTRODUCTION

Recently much attention has been paid to carbon nanotubes, which are a striking new solid state object (Dresselhaus et al., 1992; Chernozatonskii, 1992). Textured films on different substrates were obtained by electron beam evaporation. STM revealed that surface of the films on Si and quartz was formed by 1 nm carbon "rods" at the "high" density flow (Kosakovskaja et al.,1992; Chernozatonskii et al, 1993). and dome-shaped and conical tips at the "medium" density flow (Chernozatonskii et al., 1995). HREM investigation of the films obtained on KBr, glass, quartz and (111)Si at the medium density flow revealed that the films consist of oriented multilayered carbon nanotubes (Chernozatonskii et al., 1994).In this paper HREM investigation of the films on (111)Si and glass obtained correspondingly at "medium" and high density flow were investigated. It was shown for the first time that films obtained at " high" density flow consist of oriented single layered nanotubes. Correlation between the type of the nanotubes and film resistivity was established.

2.EXPERIMENTAL

Carbon films were obtained on glass and (111)Si by electron beam evaporation of graphite in 10^{-5}-10^{-6} Torr vacuum at the "medium" and "high" carbon density flow. Films with the texture oriented normal to the substrate were obtained by orienting the carbon particle flow perpendicular to the substrate. The films thickness was about 0.1 μm.

66

The specimens were prepared by removing the film from the substrate. Fragments of the films were dispersed in an ultrasonic bath using acetone. The specimens were investigated in a Philips EM-430ST operated at 200 kV. Experimental images were interpreted using computer simulation based on the multi-slice method. The measurement of resistivity was performed using standart four-terminal technique in the temperature range 4.2-250 K.

3.RESULTS AND DISCUSSION

HREM revealed that the main components of the nanotube films obtained at medium carbon particle flow were multilayered nanotubes (MLT) (Fig.1), similar to Iijima (1991) nanotubes. Usually the observed number of layers was 11-32. There were two types of tips: dome-shaped and conical. These tube tips form a surface relief of the films as was confirmed by STM (Chernozatonskii et al, 1994). In many cases outer layers of such tubes were destroyed. The tubes sometimes were covered by an amorphous layer.

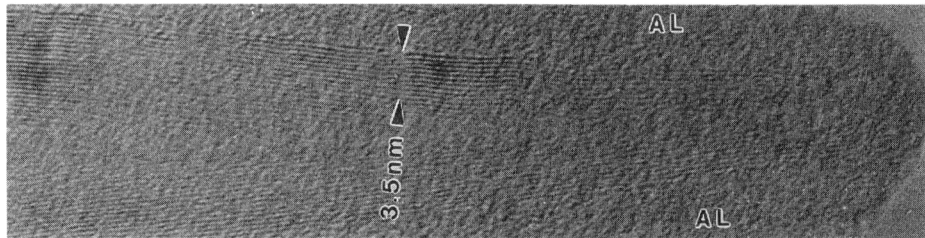

Fig.1 HREM image of MLT from the film on (111)Si covered by an amorphous layer (AL).

Fig.2 HREM image of a bundle of SLTs from the film on the glass.

In carbon films obtained on the glass for a "high" flow rate of carbon particles different kinds of structures were observed. A typical view of the main type of structure is shown in Fig.2. This structure could be interpreted as a bundle of single layered nanotubes (SLT). Judging by the appearance of thin parts of such bundles the diameter of the tubes was 0.72 nm and distance between the axes was 1.07 nm, maximal observed length was up to 135 nm. If the cross-section of such bundle is isometric, the large bundles must include up to 1000 nanotubes. These results were in agreement with STM data (Kosakovskaja et al.,1992; Chernozatonskii et al, 1993). The analysis of many bundle images shows that individual SLTs did not have any intermediate layers and, apparently, formed close hexagonal packing (Fig.3). For image simulation, tubes of 0.72 nm in diameter were used. Different kinds of tube parameters were tried. The best results were obtained for zero-start tubes with mutual orientation shown in Fig 3. The gallery of different kinds of bundle images and corresponding simulated images are shown in Fig.4. Bundles are usually slightly twisted. This could be concluded, for example, from the contrast changing along the bundle (Fig.2). Along

Fig. 3. Schematic representation of the close hexagonal packing SLTs viewed along the axis. The full dots represent the atoms in the paper plane and the open dots are projected positions of the atoms off the paper plane.

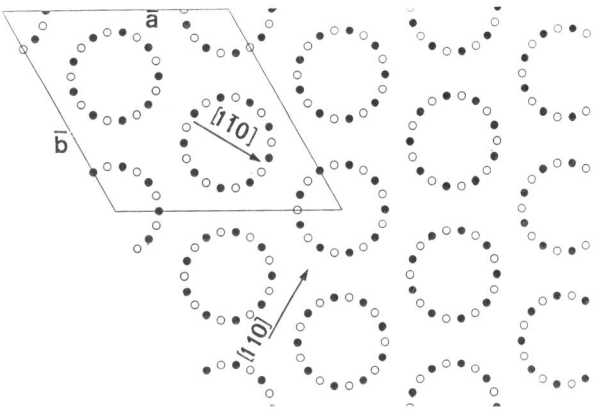

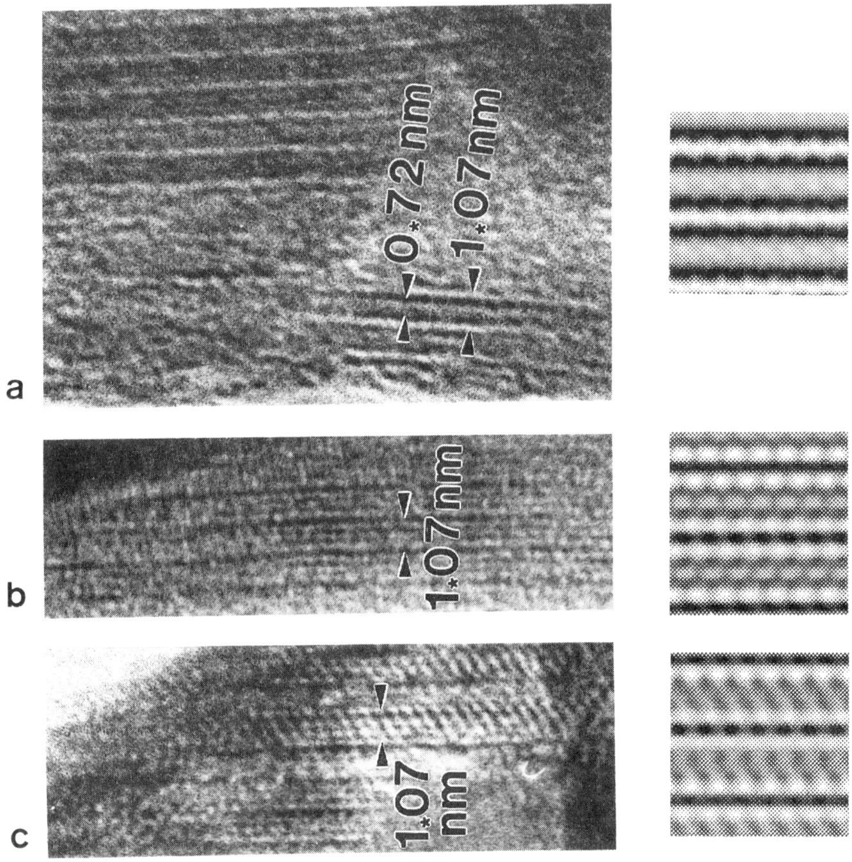

Fig. 4. Different kinds of SLT bundle images. On the right side simulated images along [110] are shown. (a) t = 1,3nm, Δf = -20nm; (b) t = 0,5nm, Δf = -5nm; (c) t = 0,8nm, Δf = -5nm

68

with SLTs a small amount of C_{60} 3D crystals were revealed as well as the "onions" (Ugarte, 1992). The latter were probably formed at the first stage of film formation, that is on the lower boundary of the film.

As shown in a large number of theoretical studies, the carbon nanotubes are expected to exhibit a noticeable variation in their electronic properties ranging from metallic through semimetallic to semiconducting depending on their diameter and helical arrangement. It is known from two experimental studies (Song et al., 1994, Langer et al, 1994) that both SLTs and MLTs bundles show to some extent a semimetallic behaviour of conductivity with evidence of weak localization. It was of particular interest to investigate the resistivity of carbon nanotube films in the direction perpendicular to the axes of tubes. It should be noted that these investigations have been carried out on the same films as HREM studies. It was found that for all films investigated, the temperature dependence of resistivity at low temperatures is strictly described by the equation: $\rho_\perp(T)=\rho_0 \exp(T_0/T)^{1/n}$. However, the character of this equation depends on the structure the film. Thus, the films formed from SLTs show that n=2 and T_0=15-25 K (for 4-40 K), whereas for the MLT films n=4 and $T_0=10^6$ K (for 25-150 K). For this reason the films investigated can be considered as semimetallic granular materials. On this basis SLT structure probably could be described by the Coulomb gap variable-range hopping model for granular and porous materials (Fung et al,1994). In the films formed from MLT`s the contacts between tubes are realized by outer layers which are often damaged and covered by amorphous layers (this is probably the most important aspect). For this reason, the equation with n=4 for MLT structure could be explained by 3D conductivity with a variable range hopping described by the Mott law.

4.CONCLUSION

MLTs and SLTs films were structurally characterized by HREM. A film resistivity dependence $exp(T_0/T)^{1/n}$ was observed. The direct correlation between the film structures and power of n was shown. The tentative explanation of this behaviour is based on the variable-range hopping conductivity model.

ACKNOWLEDGEMENTS

The investigation was performed under grant No 4 of the Russian Program "Fullerenes and Atomic Clusters".

REFERENCES

Chernozatonskii L A, Fedorov E A, Kosakovskaja Z J, Panov V I and Savinov S V 1993 JETP Letters 57 35

Chernozatonskii L A, Kosakovskaja Z J, Kiselev A N and Kiselev N A 1994 Chem. Phys. Letters 228 94.

Chernozatonskii L A, Kosakovskaja Z J, Fedorov E A and Panov V I 1995 Phys.Lett.A, 194 40

Dresselhaus M S, Dresselhaus G D and Saito R 1992 Phys.Rev. B45 6234

Fung A W P, Wang Z H, Dresselhaus M S et.al 1994 Phys.Rev.B49 17325.

Iijima S 1992 Nature 359 56.

Langer L, Stockman L, Heremans J P et al 1994 J.Mater.Res. 9 927

Song S N, Wang X K, Chang R P H et al 1994 Phys.Rev.Lett. 72 697

Ugarte D 1992 Nature 359 707

Inst. Phys. Conf. Ser. No 146
Paper presented at Microsc. Semicond. Mater. Conf., Oxford, 20–23 March 1995

Partial dislocations in semiconductors: structure, properties and their role in strain relaxation

P Pirouz and X J Ning

Department of Materials Science and Engineering, Case Western Reserve University, Cleveland, Ohio 44106, USA

ABSTRACT: The consideration of the structure of semiconductors as an assembly of tetrahedra leads to the identification of the 'glide plane' rather than the 'shuffle plane' as the slip plane in these materials. The core nature of partial dislocations and reconstruction of the broken bonds in the core gives rise to a difference in mobility of leading and trailing partials. This asymmetry in mobility can, in turn, give rise to twinning under certain conditions of temperature and stress. It is shown that twin formation in thin heteroepitaxial films, or during indentation, can take place by surface nucleation of partial dislocations on adjacent (1̄11) planes.

1. INTRODUCTION

In this paper, the importance of partial dislocations in semiconductors and their role in certain phenomena, e.g. twinning, will be reviewed. The discussion focuses mainly on some of the work carried out in the authors' laboratory. In general, a twinned region in a cubic semiconductor is related to the untwinned matrix by some symmetry operation (e.g. reflection in a {111} plane, or rotation about a <111> axis). Geometrically, the twin may be considered in terms of overlapping stacking faults on adjacent (111) slip planes of the untwinned matrix. This applies to both growth twins (obtained, e.g., during crystal growth from the vapor phase), and deformation twins produced by mechanical deformation of the crystal. There are various dislocation mechanisms that have been proposed for deformation twinning (for a recent review, see, e.g. Christian and Mahajan (1995)). The discussion in the present paper pertains to a mechanism proposed in 1987 for Si and later extended to compound semiconductors. The application of this mechanism to bulk semiconductors has been recently reviewed (Pirouz 1994) and will not be considered here. Instead a different mechanism for twin formation in thin films of semiconductors grown heteroepitaxially on a dissimilar substrate, and the application of this mechanism to describe twin formation during indentation, is discussed and some experimental results in its support are presented.

2. CRYSTAL STRUCTURE OF A SEMICONDUCTOR AS AN ASSEMBLY OF TETRAHEDRA

Since all semiconductors are tetrahedrally bonded, a convenient and instructive way to describe the structure of a semiconductor XY, consisting of elements X and Y, is by considering it as an assembly of tetrahedra (X=Y for an elemental semiconductor). Each tetrahedron contains four atoms of X (or Y) at its corners and a Y (or X) atom at its centroid (Fig. 1a). Every corner X atom is directly bonded to the centroid Y atom and the two atoms are nearest neighbors in the structure. If we consider a cubic semiconductor then all four faces of the tetrahedron are parallel to {111} planes and its edges are parallel to <110> directions. The tetrahedron in Fig. 1(a) has 3-fold rotational symmetry about any of the <111> axes joining the centroid Y atom to a corner X atom; one of these is labeled as the "c-axis" in Fig. 1(a). We take the face of the tetrahedron normal to the c-axis as the (111) (basal) plane.

Rotating the tetrahedron by 180° around the c-axis breaks the 3-fold symmetry and produces a different variant which is shown in Fig. 1(b). This may be called a *twinned* variant as opposed to the *normal* variant shown in Fig. 1(a). For ease of representation, it is more convenient to use the projections of these two variants as shown in Figs. 1(c) and 1(d), respectively. In these figures, the 3-D tetrahedra are projected along one of the edges of the basal plane, specifically the edge parallel to the $[\bar{1}10]$ direction.

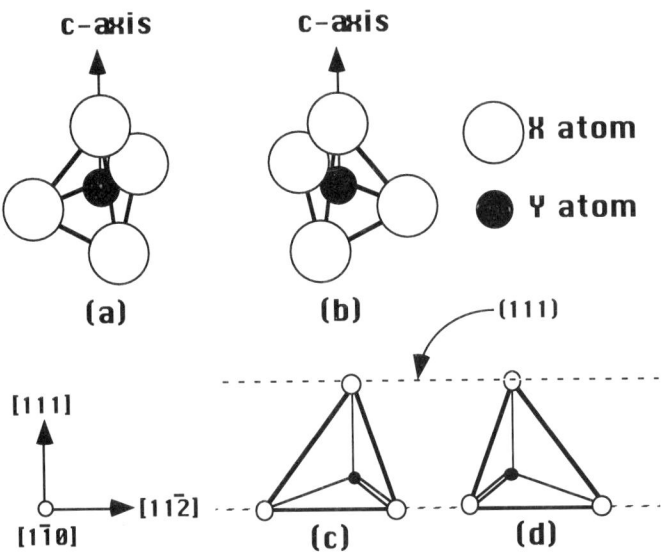

Fig. 1. The two variants of a tetrahedron as the building blocks of a XY semiconductor. The X atoms are located at the corners and the Y atom is at the centroid of the tetrahedron. The faces of the tetrahedron are parallel to {111} planes while its edges are along <110> directions, (a) the normal variant, (b) the twinned variant, (c) and (d) projections of the normal and twinned variants along the $[\bar{1}10]$ edge of the tetrahedron, respectively. The thick solid lines in (c) and (d) are projections of the {111} faces in (a) and (b), while the thin solid lines are X-Y bonds.

The complete structure of an XY semiconductor is built up by corner sharing of the XY tetrahedra. Thus, the projection of the cubic XY (zincblende) structure is shown in Fig. 2; note that only one variant of the tetrahedron, normal or twinned, is used in the cubic structure. On the other hand, to build up the crystal structure of *non-cubic* semiconductors, e.g. wurtzite, both variants are necessary (Pirouz and Yang 1993). The (111) interplanar spacing, d_{111}, in the zincblende structure is equal to the height of the tetrahedron, h. Since the Y atom is at the centroid of an equilateral tetrahedron, it divides h in a ratio of 3:1, i.e. $XY=\frac{3}{4}h$ and Y is located at a distance of $\frac{1}{4}h$ from the (111) base of the tetrahedron. Note that there are three (111) planes in every tetrahedral row (Fig. 2): a (111) plane passing through X atoms at the base of the tetrahedra, a (111) plane passing through Y atoms at a distance $\frac{1}{4}h$ above the base, and finally another (111) plane passing through X atoms at a height $h=d_{111}$.

In terms of the ...ABC... stacking sequence along a <111> direction, commonly used to represent crystals based on an fcc lattice, it should be noted that the centroid Y atom and the top corner X atom are vertically above one another and occupy the same spatial site in the $[\bar{1}11]$ projection. For this reason, the plane passing through the centroid Y atoms is denoted by, say, γ (or α or β), and the plane passing through the top corner X atoms is denoted by C (or A or B). The stacking sequence of the zincblende structure is then ...αAβBγC... as shown on the right hand side of Fig. 2.

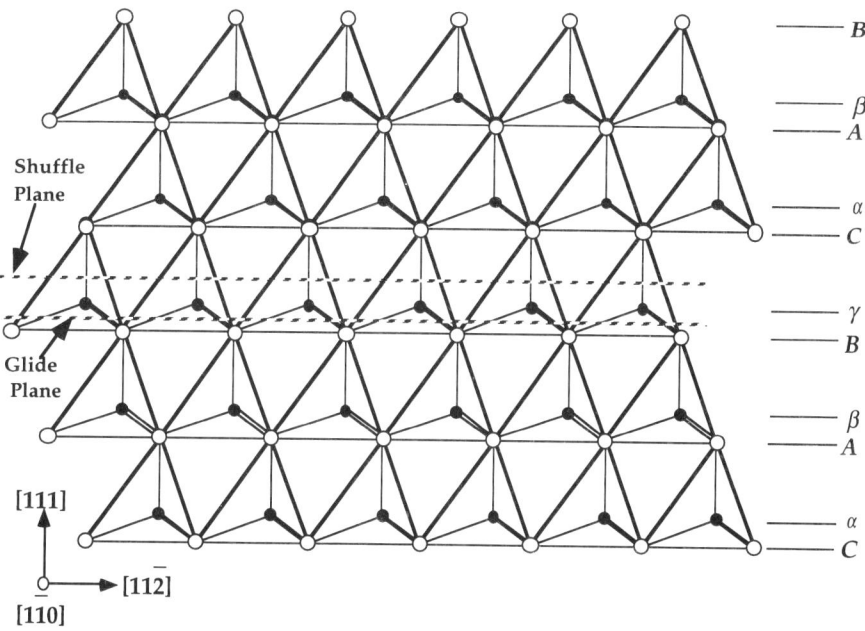

Fig. 2. The zincblende structure common to most cubic semiconductors.

3. SLIP IN SEMICONDUCTORS

3.1 Shuffle or Glide Plane

Since the diamond cubic and zincblende structures are based on the fcc lattice, they have the same slip system, {111}<1$\bar{1}$0>. However, because the centroid Y atoms divide the tetrahedra in a 3:1 ratio, two different {111} slip planes can be conceived in these structures; one in between the narrowly spaced B and γ (or C and α, or A and β) planes, and one in between the widely spaced γ and C (or α and A, or β and B) planes. These two planes are respectively called the *shuffle* and *glide* planes (Hirth and Lothe 1968), and an example of each is shown in Fig. 2. As in the fcc structure, the perfect dislocations in diamond cubic or zincblende structures have a $\frac{1}{2}$<110> Burgers vector. In addition, they have been found to be dissociated into $\frac{1}{6}$<11$\bar{2}$> partial dislocations in all the semiconducting materials that have been investigated (for a review see, e.g. George and Rabier (1987)).

The question as to whether shear in a diamond cubic or zincblende structure takes place by the motion of dislocations on the shuffle or glide plane has been a subject of discussion for a long time. In our opinion, slip takes place on the glide plane because the motion of a partial dislocation on the shuffle plane produces a much higher energy stacking fault than that produced by the corresponding motion on a glide plane. This can be seen by a comparison of the bond configuration between the centroid Y atoms and the top corner X atoms in the tetrahedra situated on the plane of the stacking fault (cf. Figs. 3(a) and 3(b)). Each distorted bond across the plane of the stacking fault (Fig. 3(a)) contributes to its higher energy. Note the conversion of normal to twinned tetrahedra after the motion of the partial dislocation (Fig. 3(b)).

72

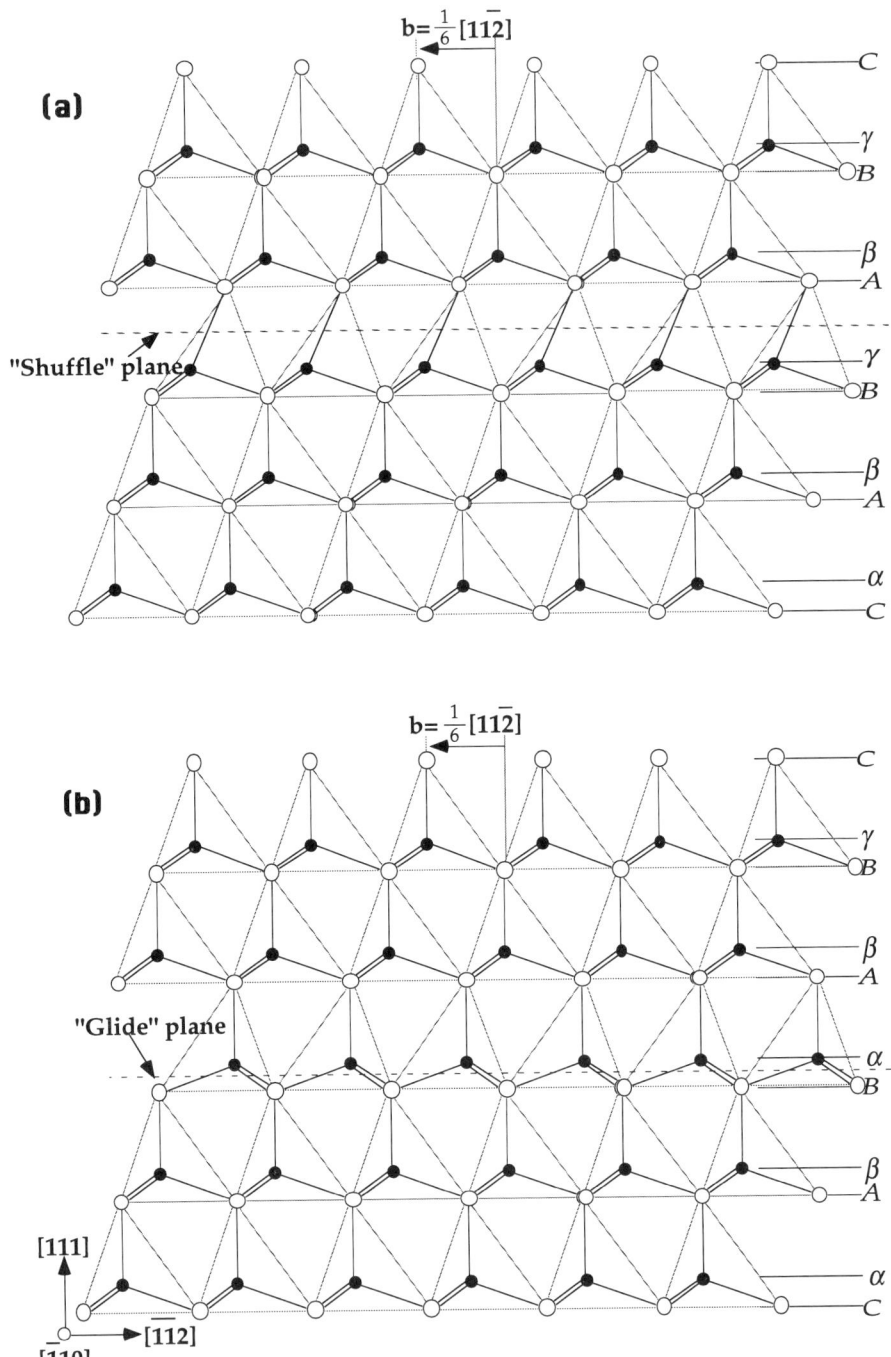

Fig. 3. The upper part of the crystal (above the slip plane) has been sheared by $\frac{1}{6}[11\bar{2}]$. Note the (a) distorted versus (b) undistorted bonds in the twinned tetrahedra.

3.2 Core Structure of Partial Dislocations

A stacking fault, e.g. as in Fig. 3, terminates on both sides at a partial dislocation (not shown in the figure). Simple topological considerations show that the core of all partial dislocations consist of only one atomic species, i.e. the atoms at the core are either all X or all Y. In fact, a 60° dislocation dissociates into a 90° and a 30° partial the cores of both of which consist of the same atom (both X or both Y) whereas a screw dislocation dissociates into two 30° partials the cores of which consist of different species (one X and one Y). The core configuration of the two 30° partials on the (111) glide plane is shown in Fig. 4. Here it has been assumed that the partials are lying along a <110> Peierls valley and both are reconstructed with no dangling bonds present at either core. Since the dislocations are assumed to lie on the *glide* plane, the notation X(g) (or Y(g)) is used to indicate a dislocation whose core consists of X (or Y) atoms (Alexander et al. 1979). The nature of partial dislocations can be experimentally determined using the contrast asymmetry exhibited by a non-screw dislocation under **g** and -**g** reflections in electron microscopy provided that the polarity of the TEM specimen surface facing the gun is known (Feuillet, 1981). However, this technique does not work for low stacking fault energy materials (e.g. SiC). Instead, recently, the LACBED technique has successfully been applied to determine the nature of partial dislocation cores in SiC (Ning and Pirouz 1995).

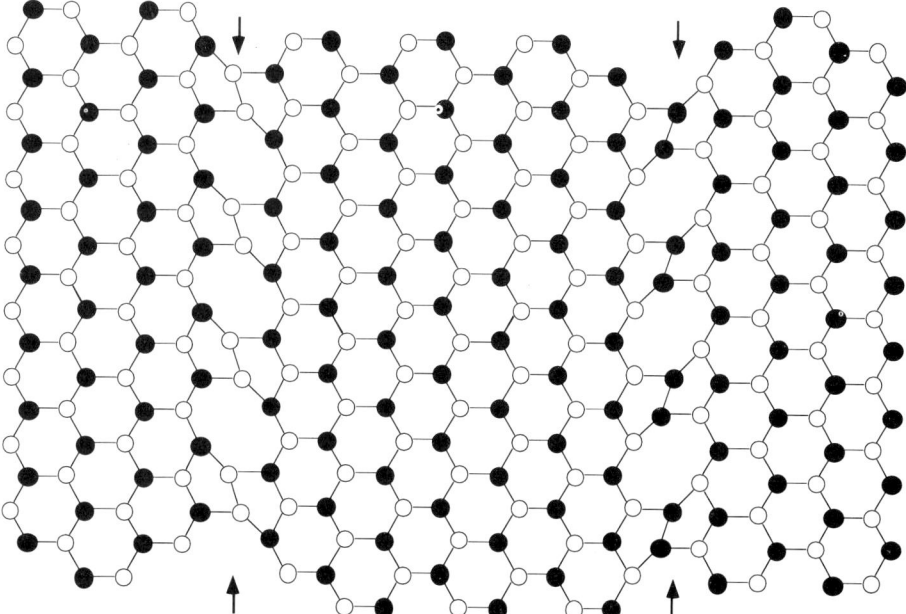

Fig. 4. The reconstructed core of 30° partial dislocations (arrowed) lying on the {111} glide plane of a zincblende structure. Note that the core of each partial consists of only one type of atom (X on the left and Y on the right). The region between the two partials defines a stacking fault ribbon which is shifted by $\frac{1}{6}$<11$\bar{2}$> with respect to the unfaulted region.

The accepted mechanism for the motion of a dislocation segment in semiconductors is by its sequential transference to neighboring Peierls valleys. The transfer from one Peierls valley to a neighboring one takes place by the generation of kink pairs on the dislocation segment and the sideways motion (or migration) of kinks to the end of the segment (e.g., Hirth and Lothe (1968)). Both kink formation and kink migration are thermally activated processes and require the overcoming of activation barriers with heights E_f and E_m, respectively. In addition, both kink-pair formation and kink migration involve breaking bonds (see Fig. 4) which are different for the X(g) and Y(g) partials. Consequently, the mobility of partial dislocations in compound semiconductors are expected to be different. This has been experimentally verified in most compound semiconductors investigated so far (where the two different types of dislocations are often designated α and β; see, e.g., George and Rabier (1987)). Interestingly, the mobility of partial dislocations is also different in elemental semiconductors where the core structures are expected to be the same (Wessel and Alexander 1977).

4. DEFORMATION TWINNING IN THIN FILM SEMICONDUCTORS

In thin films grown heteroepitaxially on a substrate, twin bands are often observed that extend from the film/substrate interface into the film, or all the way to the film surface. This phenomenon is common to both small and large lattice-mismatched systems. In the former case, the misfit parameter, f (defined as $\Delta a/a_s$ where a_s and a_f are the lattice parameters of substrate and film, respectively, and $\Delta a = a_s - a_f$), is less than about 5% whereas in the latter case $f > 5\%$. One significant difference between these two systems is that while there is considerable elastic strain energy in the initial stages of film growth in small lattice-mismatched systems, the films are effectively fully relaxed from the start in large lattice-mismatched systems.

Provided that the film thickness is below a critical value, h_c, the coherency strains in small lattice-mismatched systems can be accommodated elastically by the film. At larger thicknesses, $h > h_c$, the strain energy is released by the formation of misfit dislocations which accommodate the lattice mismatch and relax the film (see, e.g., Matthews (1975)). On the other hand, in large lattice-mismatched systems, the films are effectively strain-free from the very initial stages of film growth; the mismatch here is accommodated by a set of "geometrical misfit" dislocations which exist from the very start of film deposition (Ikuhara et al. 1994). We believe that twin bands in large lattice-mismatched systems predominantly form by stacking errors during deposition, i.e. they are basically growth twins (Pirouz et al. 1988, Ernst and Pirouz 1989, Pirouz 1989).

In small lattice-mismatched systems, where coherency stresses can be very large in the strained film, twins may form by nucleation of partial dislocations from the surface. In general, any heterogeneity on the film surface, e.g. a scratch, an impurity particle, or a surface step, are likely sites for the nucleation of dislocation half-loops under coherency stresses in the film. A mechanism proposed in this case is shown in Fig. 5 (Pirouz 1994). This is a schematic view of the reconstructed (001) surface of a silicon film with the two terraces separated by a single-layer step. The step is of the so-called SA type where the reconstructed dimers on the upper terrace are perpendicular to the step edge. STM observations have shown that such steps are rough and contain a high density of kink-pairs (see, e.g., Lagally (1993)). Fig. 5(a) shows the nucleation of a partial dislocation half-loop at a kink-pair site on a SA type step. Of course, as mentioned above, other surface defects could also be the nucleation site for dislocation half-loops.

In Fig. 5, it has been assumed that only the leading partial has been nucleated and formed a faulted half-loop. Depending on the mobility difference of the two partials, temperature, and orientation of the film (which determines the relative magnitude of the resolved shear stresses on the leading and trailing partials), it may be possible for the leading partial to expand without the trailing partial ever nucleating. An example of such a case is shown in Fig. 6 where a single partial dislocation half-loop has nucleated from the (110) surface of a GaSb crystal which has been indented at 200°C (Ning et al. 1995).

If the half-loop is a perfect dislocation, a surface site, such as the one shown in Fig. 5(a), can act as a recurrent source of dislocation half-loops all generated to relax the coherency stresses. However, if the half-loop is faulted, i.e. is produced by the leading partial, then only one half-loop can form on the same (111) plane at any time. Moreover, the nucleation of a dislocation half-loop from the surface results in a step with a height equal to the component of the Burgers vector along the normal to the surface. Thus, the faulted half-loop produced at

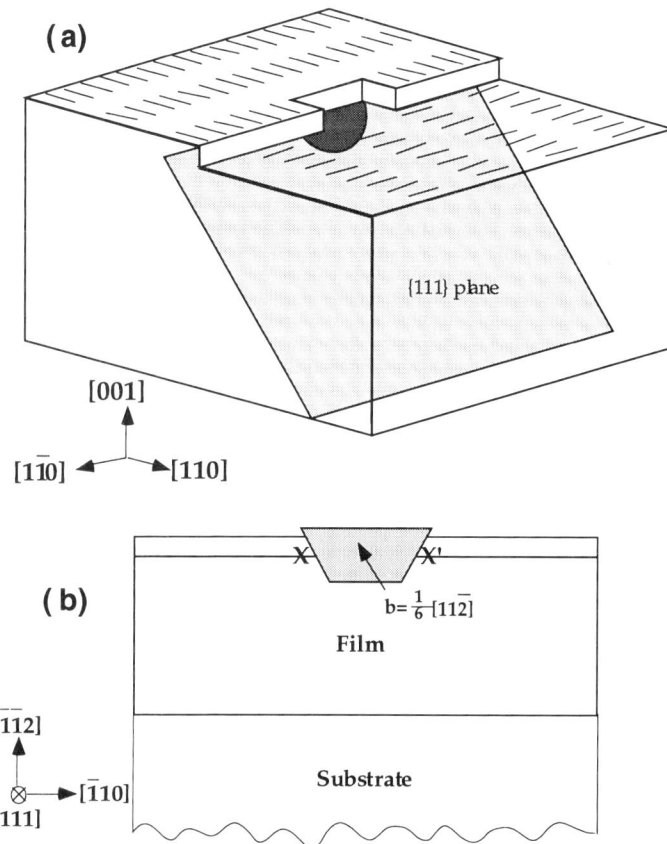

Fig. 5. (a) Nucleation of a dislocation half-loop at a kink-pair along a SA-type step, (b) [111] projection of the (111) glide plane on which the half-loop (hexagonal-shaped here) has nucleated. The faulted loop has increased the height of the surface step by $a/3$.

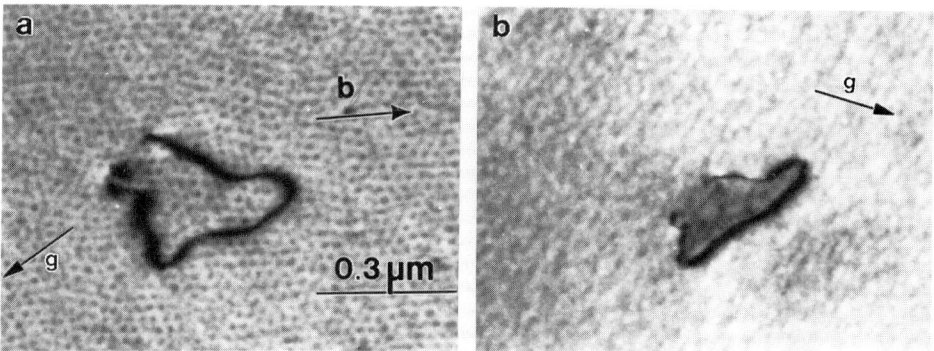

Fig. 6. A single partial dislocation nucleated by $200°C$ indentation of GaSb, (a) $\mathbf{g}=\bar{2}\bar{2}0$ near the $[1\bar{1}\bar{1}]$ zone axis, showing dislocation contrast, (b) $\mathbf{g}=20\bar{2}$ near the $[1\bar{1}1]$ zone axis showing stacking fault contrast.

the SA-type surface step (Fig. 5(a)) increases the step height by $a/3$ (Fig. 5(b)). The interesting point about this situation is that the presence of the step produced by the faulted half-loop makes the adjacent (111) plane a low-energy site for the nucleation of the next partial dislocation half-loop. This is more clearly shown in Figs. 7(a-d) for the nucleation of partial dislocation half-loops from the (001) surface of a cubic semiconductor; successive nucleation has produced an intrinsic stacking fault (ISF, Fig. 7(b)), an extrinsic stacking fault (ESF, Fig. 7(c)), and a 3-layer twin (Fig. 7(d)). While there is sufficient (coherency) stress in the film, and no better sources are available, successive faulted loops will be produced on adjacent (111) planes and the twin thickness will increase.

Notice that, assuming that the extra half-plane of a non-screw dislocation is toward the surface, the dislocations on the $(\bar{1}11)$ and $(1\bar{1}1)$ planes have different core natures, i.e. one is X(g) and the other Y(g), respectively (Fig. 7(a)). Since the activation barrier for the nucleation and motion of these two dislocations can be significantly different in compound semiconductors, an asymmetry in twin formation may result. Thus, assuming that E_f and/or E_m for an X(g) partial is less than the corresponding values for a Y(g) partial, the majority of twins bands will have their habit plane parallel to the $(\bar{1}11)$ planes.

Actually similar considerations also apply for growth twins except that, in this case, the asymmetry does not result from the different formation energies of the partial dislocations but rather from the different surface energies of $(1\bar{1}1)$ and $(\bar{1}11)$ planes that are terminated in X and Y atoms. This asymmetry results in different probabilities for the occurrence of stacking errors on the two planes (Ernst and Pirouz 1989).

A similar mechanism as the one described above may operate in the formation of indentation twins (Ning et al. 1995). Fig. 8 show a twin band formed by indentation of (110) GaSb at 200°C. Successive partial dislocation half-loops are shown nucleating from the surface of the indented crystal on adjacent (111) planes. The larger half-loops, which have nucleated first, have already expanded such that their segments which intersect the surface are nearly parallel; the segments of the half-loops parallel to the surface have been cut off by the bottom surface of the TEM foil. The micrograph thus shows an array of parallel partial dislocations which define the twin band.

A similar situation is shown in Fig. 9(a) for two twin bands, labeled A and B, produced by a 200°C indentation of a GaAs crystal. The surface of the indented crystal has been labeled. A schematic of the partial dislocation half-loops for twin band B is shown in Fig. 9(b). In this figure, because of the difficulty in drawing the configuration, all the loops are shown to lie on the same (111) slip plane, whereas in the micrograph (Fig. 9(a)), they actually lie on *adjacent* (111) planes.

5. CONCLUSION

A common source for perfect dislocations in a bulk crystal is sequential multiplication of dislocation loops from a pinned dislocation segment by the Frank-Read mechanism. The same mechanism may also operate for partial dislocations to produce twins in bulk semiconductors. The reason is that while perfect dislocations can successively multiply on the same glide plane by the Frank-Read mechanism, formation of a partial dislocation loop is restricted to only one time. However, if the undissociated pinned dislocation segment has a screw character, then cross-slip to neighboring glide planes can take place and multiplication of faulted loops on adjacent planes can produce a twinned region in the semiconductor.

In thin films grown heteroepitaxially on a substrate, twin formation can take place in two ways. In large lattice-mismatched systems (misfit parameter>4-5%) where growth takes place in a three-dimensional mode, stacking errors on the faceted islands in the initial stages of growth can give rise to stacking faults and microtwin bands. As a result, such twins may be considered to be *growth twins* whose formation is independent of shear stresses in the film. On the other hand, in small lattice-mismatched systems (misfit parameter<4%), or in indentation of a bulk crystal, deformation twins form by the nucleation of partial dislocations from the film surface under shear stresses in the film. If a partial dislocation half-loop is nucleated from a heterogeneity on the crystal surface, then the surface step generated by the dislocation can cause the successive formation of additional loops on adjacent glide planes. Thus, a twin band may form which originates from the crystal surface.

ACKNOWLEDGMENT

The support of DoE (grant number DE-FG02-93ER45496) and NASA (grant number NAG3-1702) are gratefully acknowledged.

REFERENCES

Alexander H., Haasen P., Labusch R. and Schröter W. 1979 Foreword to J. Phys. (Paris) **40** Colloque C6

Christian J. W. and Mahajan S. 1995 Progress in Materials Science **39** 1-157

Ernst F. and Pirouz P. 1989 J. Mater. Res. **4** 834-842

Feuillet G. 1982 M.Sc. Thesis, University of Oxford.

George A. and Rabier J. 1987 Rev. Phys. Appl. **22** 941-966

Hirth J. P. and Lothe J. 1968 Theory of Dislocations (New York: McGraw-Hill)

Ikuhara Y., Pirouz P., Heuer A., Yadavalli S. and Flynn C. P. 1994 Phil. Mag. A **70** 75-97

Lagally M. G. 1993 Jpn. J. Appl. Phys. **32** 1493-1501

Matthews J. W. 1975, in *Epitaxial Growth*, edited by Matthews J. W. (New York: Academic Press) *pp* 559-609

Ning X. J., Perez T. and Pirouz P. 1995 Phil. Mag. In press.

Ning X. J. and Pirouz P. 1995 Submitted to Phil. Mag.

Pirouz P., Ernst F. and Cheng T. T. 1988, in *Heteroepitaxy on Silicon: Fundamentals, Structures, and Devices*, eds Choi H. K., Ishiwara H., Hull R. and Nemanich R. J. (Pittsburgh: Materials Research Society) *pp* 57-70

Pirouz P. 1989, in *Polycrystalline Semiconductors*, edited by Werner J. H., Möller H.-J. and Strunk H. P. (Berlin: Springer-Verlag) *pp* 200-212

Pirouz P. and Yang J. W. 1993 Ultramicroscopy **51** 189-214

Pirouz P. 1994, in *Twinning in Advanced Materials*, edited by Yoo M. H. and Wuttig M. (Pittsburgh: The Minerals, Metals, and Materials Society) *pp* 275-295

Wessel K. and Alexander H. 1977 Phil. Mag. **35** 1523-1536

Inst. Phys. Conf. Ser. No 146
Paper presented at Microsc. Semicond. Mater. Conf., Oxford, 20–23 March 1995
© *1995 IOP Publishing Ltd*

Direct atomic resolution imaging of dislocation core structures in a 300kV STEM

A J McGibbon* and S J Pennycook

Solid State Division, Oak Ridge National Laboratories, Oak Ridge, TN 37831, USA
*Present address: Department of Physics and Astronomy, University of Glasgow, Glasgow G12 8QQ, Scotland, UK

ABSTRACT: By employing the technique of Z-contrast imaging in a 300kV STEM, we show that it is possible to provide directly interpretable, atomic resolution images of the sublattice in compound semiconductors. Using this approach, analysis of dislocations at an interface in the CdTe(001)/GaAs(001) system reveals unexpected core structures at Lomer dislocations.

1. INTRODUCTION

The electronic properties of a wide range of semiconductor materials rely strongly on the precise atomic structure at interfaces, grain boundaries and dislocations. Consequently, if growth processes and device characteristics are to be understood, controlled and, ultimately improved, a direct knowledge of atomic arrangements on the column-by-column level is essential. Such an aim can be achieved by employing the technique of atomic resolution Z-contrast imaging on a newly-developed 300kV scanning transmission electron microscope (STEM). In this paper, we apply the technique to the study of dislocation core structures and show that it is possible to retrieve important information on atomic arrangements without the need for pre-conceived structural models.

2. SUBLATTICE IMAGING

Pennycook and Jesson (1992) demonstrated that Z-contrast imaging in a STEM is an incoherent imaging technique in which, when observing a crystalline specimen oriented along a principle zone axis, the recorded image can be interpreted as a convolution between an object function (the Z-sensitive columnar scattering cross-section into the high-angle annular detector) and a point spread function (the effective electron probe). Consequently, unlike data acquired using high resolution electron microscopy (HREM) in which contrast reversals can occur as a function of beam defocus, Z-contrast images can be interpreted directly. Previously carried out at 100kV, with a probe limited spatial resolution of 2.2Å, the technique can now be implemented on a VG Microscopes HB603 300kV STEM at a spatial resolution of 1.3Å. At this resolution, it is possible to resolve the nearest

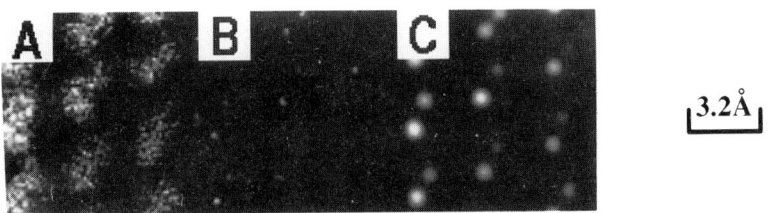

Fig. 1. A: As-acquired Z-contrast image of CdTe viewed along the [110] projection, directly revealing sublattice polarity. B: 'Most likely' Z-contrast object function of A obtained by maximum entropy analysis. C: Maximum entropy image of A.

neighbour spacing in all compound and elemental semiconductors viewed in the [110] projection (McGibbon et al. 1994). The benefit of enhanced spatial resolution is demonstrated by the as-acquired Z-contrast image of a region of CdTe shown in fig. 1A which, in addition to resolving individual atomic column positions, the sublattice polarity (Te brighter than Cd) can be seen. A further advantage of incoherent imaging in STEM is that it is ideally suited to the application of maximum entropy analysis techniques (e.g. Gull and Skilling, 1984) in which, given a knowledge of the electron probe current density distribution, a 'most likely' object function is calculated (McGibbon et al. 1995). In this way, it is possible to retrieve numerical information on both column position and composition from each Z-contrast image acquired. Using the combined Z-contrast/maximum entropy approach, the most likely object function (an array of narrow spikes located at column sites, each with a strength related to the columnar scattering power) of the region of specimen analysed in Fig. 1A is shown in Fig. 1B. This information is more readily interpreted through the convolution with a small Gaussian probe to give a 'maximum entropy image' (Fig. 1C). In this image, quantitative object function is preserved, whilst it is still possible to observe structures intuitively as in Fig. 1A, but in the absence of shot noise.

3. ANALYSIS OF DISLOCATIONS IN CdTe(001)/GaAs(001)

The full power of Z-contrast imaging can be fully utilised by the analysis of materials in which local atomic arrangements have a strong effect on bulk characteristics. An example of this is the CdTe(001)/GaAs(001) system grown by molecular beam epitaxy (MBE), which is an ideal model system in the investigation of the inter-relation between growth conditions, materials structure and electronic properties in highly mismatched III-V/II-VI heteroepitaxial materials. In this material, the lattice mismatch between substrate and epilayer is -14.6%. Previous analysis by Angelo et al. (1993) carried out using high resolution electron microscopy (HREM) has provided important and detailed information on, for example, the relative density of different defect structures and the angle of substrate tilt. However, it has not until now proved possible to provide direct structural information with Z-sensitivity on the precise sublattice arrangements at dislocations.

The dislocations shown in Figs. 2A and 2B, with schematic representations in 2C and 2D respectively are core structures in CdTe viewed along $[1\bar{1}0]$ (a) and [110] (b). From the information presented in these images, the atomic arrangements correspond to a perfect 60° dislocation, the core of which is located in the CdTe epilayer. Furthermore, from the image data alone, it is clear that both dislocations shown here are of the glide set, indicating the

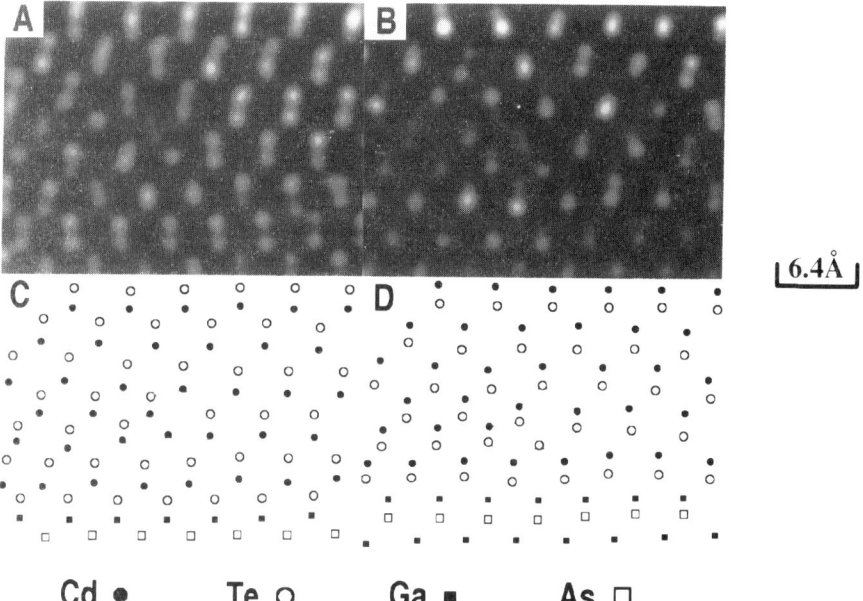

Fig. 2. 60° dislocations at the interface in CdTe(001)/GaAs(001) viewed in the [1$\bar{1}$0] (A) and [110] (B) orientations, with schematic representations given in C and D respectively.

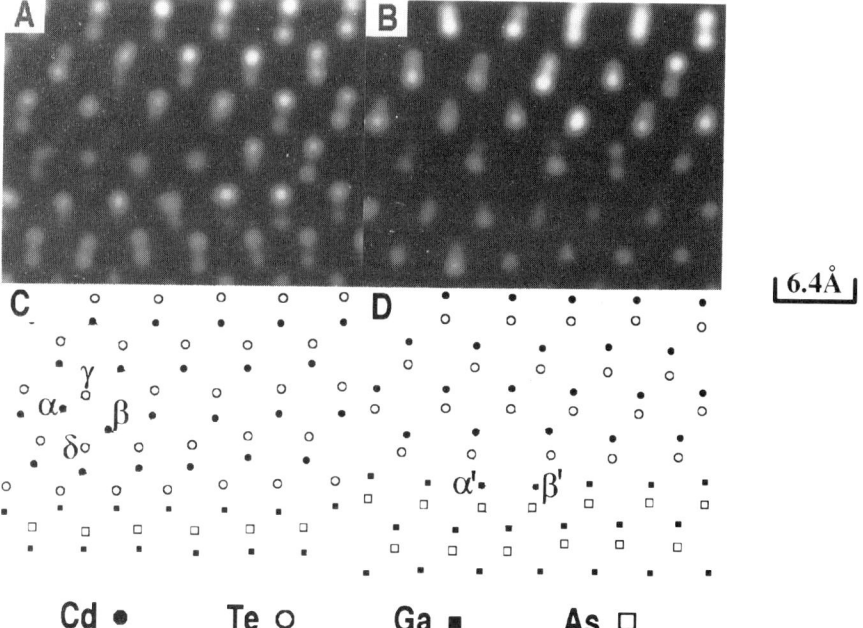

Fig. 3: Lomer dislocations at the interface in CdTe(001)/GaAs(001) viewed in the [1$\bar{1}$0] (A) and [110] (B) orientations, with schematic representations given in C and D respectively.

presence of Cd-terminated dislocations along $[1\bar{1}0]$ and Te-terminated dislocations along $[110]$.

In a manner similar to that of Fig. 2, Lomer dislocations viewed along $[1\bar{1}0]$ and $[110]$ are shown in Figs. 3A and 3B respectively, with their corresponding core structures in Figs. 3C and3D. In this case, it can be seen that the core structure viewed along $[1\bar{1}0]$ is situated above the interface and is asymmetric in nature, best described as consisting of 5 irregular six-fold rings surrounding a 4-fold ring. Clearly such a structure is unlike that predicted by Hornstra (1958) which can be described as a 7-fold ring coupled to a 5-fold ring. However, atomic arrangements similar to the Hornstra model were observed along $[110]$, but with the dislocation located exactly at the interface, implying that the two columns common to both the 5- and 7-membered rings (marked α' and β') are occupied by Ga atoms. The most likely rationalisation of the observed atomic arrangements in the asymmetric structure is that Cd columns α and β each possess 1 dangling bond per atom and that there is also a small shear (positive for α and negative for β or vice versa) of each column along $[1\bar{1}0]$ (parallel to the dislocation line direction) to accommodate a 'skewed' tetrahedral bonding configuration for atoms in Te columns γ and δ. These data suggest that, possibly as a result of the highly polar nature of the CdTe, the Hornstra structure is not energetically favoured when a dislocation occurs entirely within the material.

4. CONCLUSIONS

Through the application of Z-contrast imaging in a 300kV STEM in conjuction with maximum entropy image analysis, it is possible to retrieve information on core structures of dislocations on the atomic level without the need for pre-conceived structural models. In the analysis of the CdTe(001)/GaAs(001) system, it has been possible to determine the exact nature (glide or shuffle) of 60° dislocations and to observe unexpected core structures at Lomer dislocations.

5. ACKNOWLEDGEMENTS

We would like to thank J. E. Angelo (Purdue University) for the provision of specimens. This research was sponsored by the Division of Material Sciences, US Department of energy, under contract No. DE-AC05-840R21400 with Martin Marietta Energy Systems, Inc., and supported in part by the Oak Ridge Institute for Science and Education.

REFERENCES

Angelo J E, Gerberich W W, Bratina G, Sorba L, and Franciosi A 1993 J. Cryst. Growth 130 459
Gull S F and Skilling J 1984 IEE Proc. 131F 646
Hornstra J 1958 J. Phys. Chem. Solids 5 129
McGibbon A J, Pennycook S J and Wasilewski Z 1994 Mat. Res. Soc. Symp. Proc. 326 299
McGibbon A J, Pennycook S J and Jesson D E 1995 Submitted to J. Miscrosc.
Pennycook S J and Jesson D E 1992 Acta Metall. Mater. 40 S149

Inst. Phys. Conf. Ser. No 146
Paper presented at Microsc. Semicond. Mater. Conf., Oxford, 20–23 March 1995

Models for the Lomer dislocation core at the GaAs/Si interface

A Vilà[1], A Cornet[1], J R Morante[1], P Ruterana[2], R Bonnet[3] and M Loubradou[3]

[1]LCMM-FAE, Facultat de Física, Universitat de Barcelona, Diagonal 645-647, 08028-BARCELONA (España)
[2]LERMAT, Université de Caen, 6 Bld Maréchal Juin, 14050-CAEN (France)
[3]Institut National Polytechnique, LPTCAM-ENSEEG, Domaine Universitaire, BP 75, 38042-SAINT MARTIN D'HERES (France)

ABSTRACT: In this work we focus on the core structure of the Lomer dislocation and we propose a new asymmetrical description in which the atomic positions are determined on the basis of anisotropic elasticity calculations. The validity of this description has been tested both by superimposing the generated atomic array onto the experimental image and by extensive image simulation.

1. INTRODUCTION

The knowledge of the structure of crystallographic defects is a subject of great interest in microelectronics. Elasticity has shown its capability in modelling the stress field around dislocations in bulk material and at interfaces (Bourret et al. 1983, Mills and Stadelmann 1989). However, the singularity taking place at the dislocation core makes it difficult for a detailed description. Although a large amount of work has used HREM for analysis of grain boundaries and defects, only a few papers have been published on the atomic structure of interface defects in heteroepitaxial materials (Gerthsen 1990).

In this work, we analyze the inner structure of the Lomer dislocation in GaAs/Si layers obtained by Atomic Layer Molecular Beam Epitaxy (ALMBE). From anisotropic elasticity models, High Resolution Electron Microscopy (HREM) and extensive image simulation, some asymmetrical models are found to describe best the dislocation core.

2. EXPERIMENTAL DETAILS

The HREM samples were prepared by the standard techniques of mechanical polishing and ion milling until perforation. Observations were carried out on a Philips EM-430ST microscope operating at 300 kV. The images were simulated using the EMS program of Stadelmann (1987), taking the atomic positions described by the anisotropic elasticity calculations (Bonnet 1981) as input data.

84

3. RESULTS AND DISCUSSION

As it can be seen in Fig. 1, observations show compact and well defined contrast in most dislocation cores, which are asymmetrical referring to the {220} medial plane (Vilà et al. 1994). In each case, the asymmetry can be related to the different distances to neighbouring defects, with the core displaced towards the most distant neighbour.

The first important model for the core of the Lomer dislocation is Hornstra's, proposed in 1958 for the cubic diamond bulk structure. Two variants were identified: the core of the first one corresponds to the glide set, and consists of an eight-atom ring with an inner atom and two dangling bonds (Fig. 2a). The other structure, describing the shuffle type, can be deduced from the first by omitting the inner atom and connecting the resulting free bonds, as shown in Fig. 2b. Both are symmetrical about the medial {220} plane.

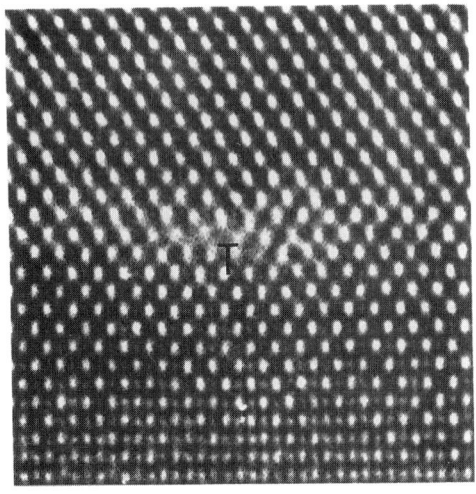

Fig.1. Experimental image of a typical Lomer dislocation at the ALMBE GaAs/Si interface. $\Delta f \sim 34$ nm.

On the other hand, in the work of Bourret et al. (1982) it was shown that in bulk Ge and Si Lomer dislocation core images are not symmetrical. They proposed a new core model which has an atomic pair in a stacking fault position (Fig.2c).

a   b

c

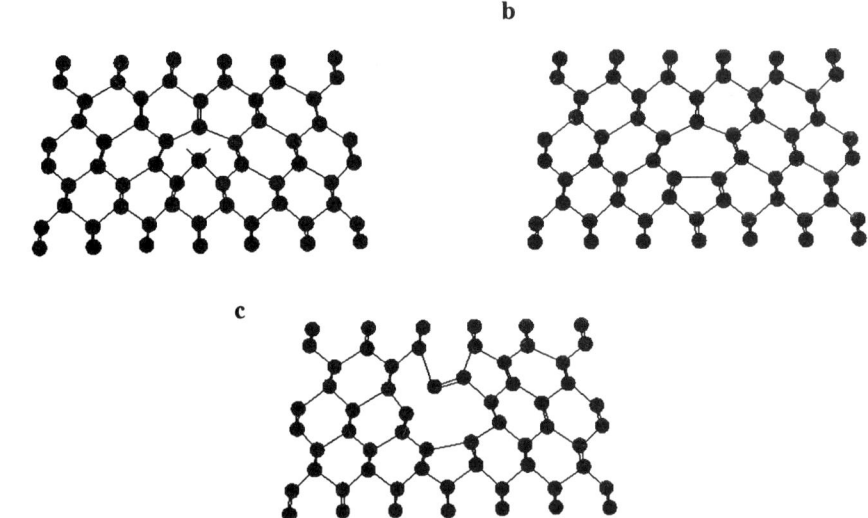

Fig. 2. *Hornstra's glide (a) and shuffle (b) and Bourret (c) models for the Lomer dislocation in bulk material.*

Unfortunately, none of these models could describe the majority of the Lomer dislocations at the GaAs/Si interface. In order to determine the displacement field associated with these dislocations, we have carried out anisotropic elasticity calculations by assuming a periodic array of dislocations located at the interface following the Somigliana (1914) formalism using the elastical constants given by Hearmon (1979). After a trial-and-error method among several possibilities, two 6 and 8 atomic rings were found in these cores, as in the Hornstra model. However, they are asymmetrical to the {220} medial plane. So, we propose a new description for the core of the Lomer dislocation, corresponding to the four diagrams shown in Fig. 3:

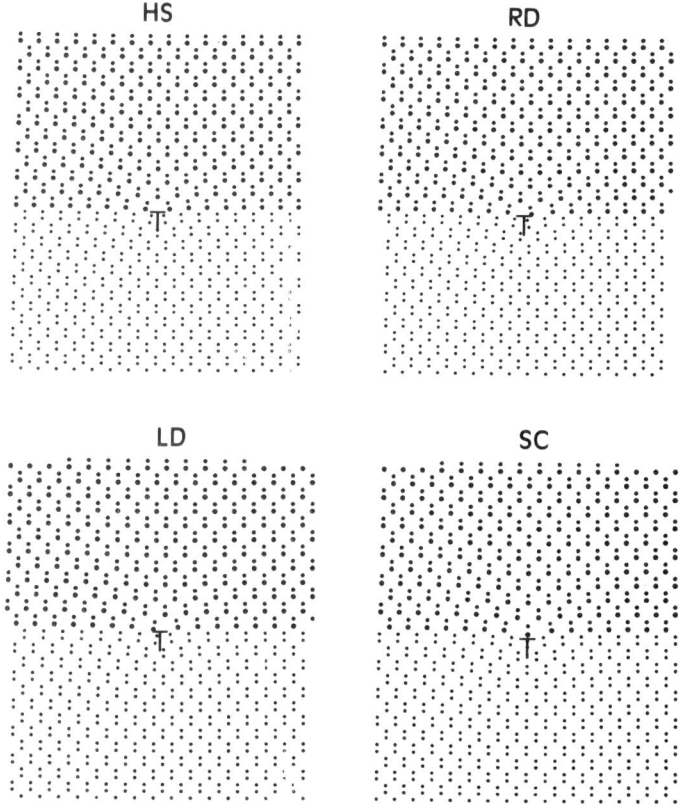

Fig.3. Core models for the Lomer dislocation: the symmetrical model (HS), with the core displaced to the right RD, to the left (LD) and an atypical symmetrical-centered model (SC).

Outside the dislocation core, the difference in the generated atomic positions between these models is negligible, and all of them can be used for comparison with experiment. The above four models have been used to fit the experimental images, and they could describe the majority (72%) of the Lomer dislocations analyzed (Vilà et al. 1994). As shown in Fig. 4, both the superimposition of the atomic array to the experimental image and the image simulation coincide well with the images. Fig. 4 shows calculated images using these models as well as the experimental image of the core of the dislocation in Fig. 1 in which calculated atomic positions were superimposed. It can be seen that this model gives a fair description of this Lomer dislocation.

a   b

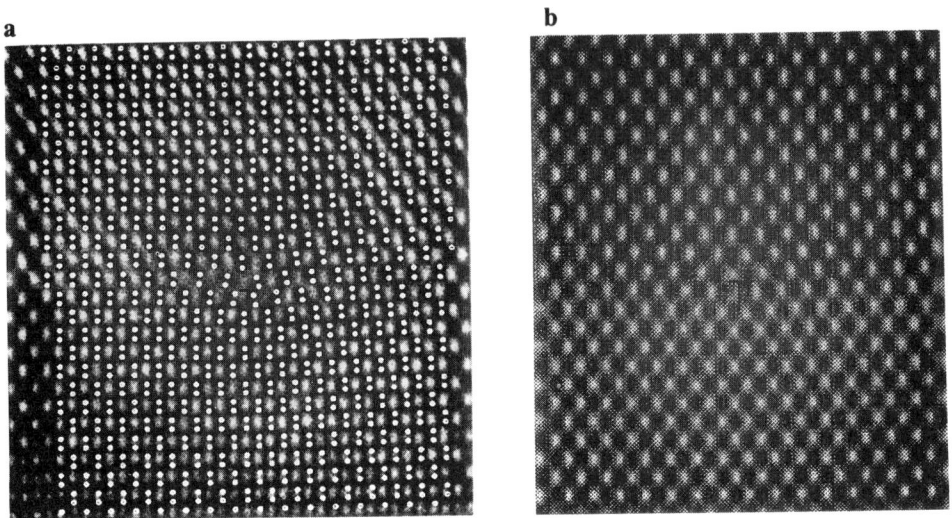

Fig.4. *Comparison between the experimental image of the dislocation in Fig.1 and the RD model shown in Fig. 3. In (a) the calculated atomic positions are superimposed to the image. In (b) multislice simulation from these positions.*

4. CONCLUSIONS

Use of the HREM technique combined with image simulation using elastic models has been found to be useful to analyze the core structure of Lomer dislocations at ALMBE GaAs/Si interface. Interfacial Lomer dislocations are made of two 8 and 6 atomic rings which are in most times asymmetrical about the medial {220} plane. The displacement of the core takes place towards the most distant neighbouring defect.

REFERENCES

Bonnet R 1981 Phil. Mag. A 44 625
Bourret A, Desseaux J and Renault A 1982 Phil. Mag. A 45 1
Bourret A, Desseaux-Thibault J and Lancon F 1983 J. Physique (Paris) sup.n.9 vol.44 C4-15
Gerthsen D 1990 Proc. 12th International Congress for Electron Microscopy p. 342
Hearmon R F S 1979 *Landolt-Börnstein numerical data and functional relationships in science and technology*, New Series Vol. 11, edited by K H Hellwege and A M Hellwege (Springer: Berlin) p.27
Hornstra J 1958 J. Phys. Chem. Solids 5 129
Mills M J and Stadelmann P 1989 Phil. Mag. A 60 355
Somigliana C 1914 Atti. Acad. Nazl. Lincei. Rond. Classe Sci. Fis. Mat. Nat. 23 463
Stadelmann P 1987 Ultramicroscopy 21 131
Vilà A, Cornet A, Morante J R, Ruterana P, Loubradou M, Bonnet R, González Y and González L 1995 Phil. Mag. 71 85

Inst. Phys. Conf. Ser. No 146
Paper presented at Microsc. Semicond. Mater. Conf., Oxford, 20–23 March 1995

87

Faulted dipoles in Indium-doped GaAs

I Yonenaga*, P D Brown, W G Burgess and C J Humphreys

Department of Materials Science and Metallurgy, University of Cambridge, Pembroke Street, Cambridge, CB2 3QZ, UK.
*On leave from the Institute for Materials Research, Tohoku University, Sendai 980, Japan.

ABSTRACT   Faulted dipoles generated in In-doped GaAs were observed using HREM. Shockley partial and stair-rod dislocations show no differences with dislocation (α or β) type.   Contrary to expectation, no precipitates arising from the In impurity, neither on the partial dislocations nor on the stacking faults, were observed.   The stacking fault energy was estimated to be 57 mJ/m^2.

1. INTRODUCTION

α and β dislocations in III-V and II-VI sphalerite compounds, which are distinguished by a reversal of the sign of the Burgers vector (**b**), exhibit quite different dynamic properties and additionally show differences in impurity segregation effects upon them (Yonenaga and Sumino 1989 and 1993).   Faulted dipoles are known to be generated following the interaction of dissociated dislocations having opposite **b** moving on parallel slip planes.   The high resolution electron microscopy (HREM) observation of such defects provides valuable information on the nature of dislocation core structures and the effect of impurities upon them. Previous studies on mobilities of dislocations within In-doped GaAs demonstrate that α dislocations are easily locked by In-dopant species as compared with β dislocations (Yonenaga and Sumino 1989).   It is of interest to know whether such macroscopical effects can be related to the details of the dislocation microstructure.   We report on a HREM investigation of faulted dipoles in In-doped GaAs.

2. EXPERIMENTAL DETAILS

GaAs single crystals grown by the liquid encapsulated Czochralski technique doped with In to a concentration of 2×10^{20} cm^{-3} were deformed in compression at 450°C along [011] as previously described (Yonenaga and Sumino, 1991).   Thin plates parallel to (01$\bar{1}$) were cut from the deformed crystals.   3-mm-diameter disks were then cut and thinned to electron transparency by mechanical dimpling followed by chemical polishing using 3 : 1 : 1 H$_2$SO$_4$: H$_2$O$_2$: H$_2$O at ≈ 80°C.   HREM observations were performed using a Jeol 4000EX-II electron microscope operating at 400keV, while conventional TEM images and convergent-beam electron diffraction (CBED) patterns were obtained using a Jeol 2000FX microscope operating at 200keV.

3. RESULTS

Figure 1 shows a bright-field image of In-doped GaAs.   In addition to some dipoles, a small Z-type defect can be seen in the second bright pendellösung fringe (arrowed).   The thickness of the foil in this vicinity was estimated to be 20 nm.   Figure 2 is a high-resolution image of the Z-type defect, and shows the stacking faults and individual partial dislocations.

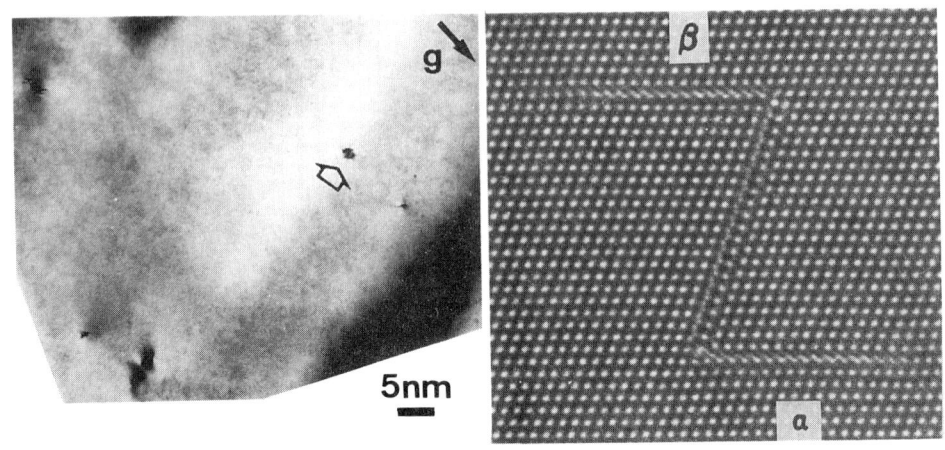

Fig. 1 Bright field image (g=022)
of In-doped GaAs.

Fig. 2 HREM image of the same
Z-type defect.

The contrast from the stacking fault demonstrates it to be intrinsic, while a Burgers circuit around the defect as a whole shows no closure error, which confirms the defect to be a faulted dipole. Extra half-planes located above the top of the stacking fault and below the bottom of the stacking fault also imply this to be a vacancy-type defect.

The assessment of absolute crystal polarity of such samples using CBED (Burgess et al. 1993) is essential for the determination of dislocation type. The CBED pattern obtained for the [01$\bar{1}$] projection of this sample is shown in Fig. 3 and is correctly oriented with respect to the TEM images presented in Figs. 1 and 2. It is apparent that fine structure is present within the four {111} diffraction discs around the central (000) disc. The asymmetry of this fine structure coupled with simulations allows {111}Ga and {$\bar{1}\bar{1}\bar{1}$}As planes to be distinguished, and this result is shown schematically in Fig. 3 (b). Thus, in Fig. 2, the partial dislocations are α and β at the top and bottom of this defect respectively.

Two faulted dipoles of different size imaged within In-doped GaAs by HREM are shown in Figures 4 and 5(a). Again, both are of the vacancy type with intrinsic stacking faults. In particular, there is no difference in the core structure of the α and β dislocations of Shockley and stair-rod type, nor in the associated stacking faults widths.

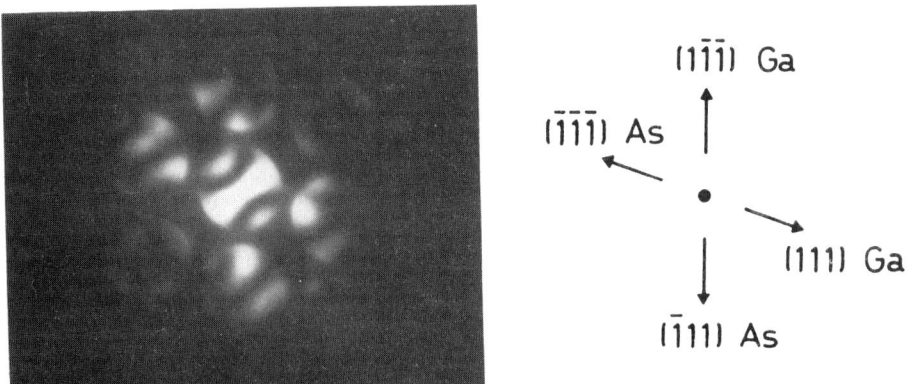

Fig. 3 [01$\bar{1}$] CBED pattern used to determine absolute crystal polarity.

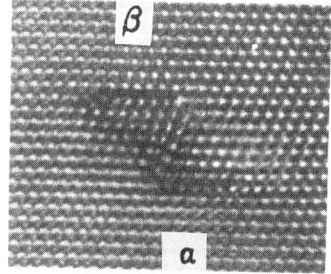

Fig. 4 HREM image of a smaller faulted
dipole in In-doped GaAs.

Figure 5(b) shows an HREM image of the faulted dipole shown in Fig. 5(a) after extended
irradiation under the electron beam.   It is evident that both α and β-types of 90° Shockley
partial have climbed from the stacking fault plane through absorption of interstitials.   From
Fig. 5(b) it also appears that the β-type partial has climbed further than the α-type partial.

4. DISCUSSION

All the faulted dipoles in In-doped GaAs were found to be of vacancy and Z-type.   Partial
Shockley and of stair-rod dislocations show no differences with α or β-type.   In addition, no
precipitates related to the In dopant were observed.

It is noted that De Cooman and Carter (1987 and 1988) found faulted dipoles only in p-type
deformed GaAs and not in n-type and undoped material, and concluded that faulted dipoles are
only formed within material within which α-, β- and screw type dislocations all move with
comparable velocities. However, in the present work, dipoles were imaged within highly In-
doped GaAs, within which α-dislocations exhibit velocities one or two orders of magnitude
greater than β- or screw dislocations, and this characteristic of dislocation velocity is similar
within n-type and undoped material (Yonenaga and Sumino 1989).   Thus, it seems that the
formation of faulted dipoles does not require dislocations with opposite b to move with
comparable velocities, and that these dislocations decrease their mobilities and cease their
motion through mutual attraction.

The configuration of the faulted dipoles allows an estimation of the stacking fault energy
(SFE).   We recognise that estimation of the true value of the SFE using dissociated widths of
extended dislocations within weak-beam images or by HREM is not easy and leads to much
confusion as demonstrated by Gerthsen and Carter (1993).   Thus, the comparison of stacking

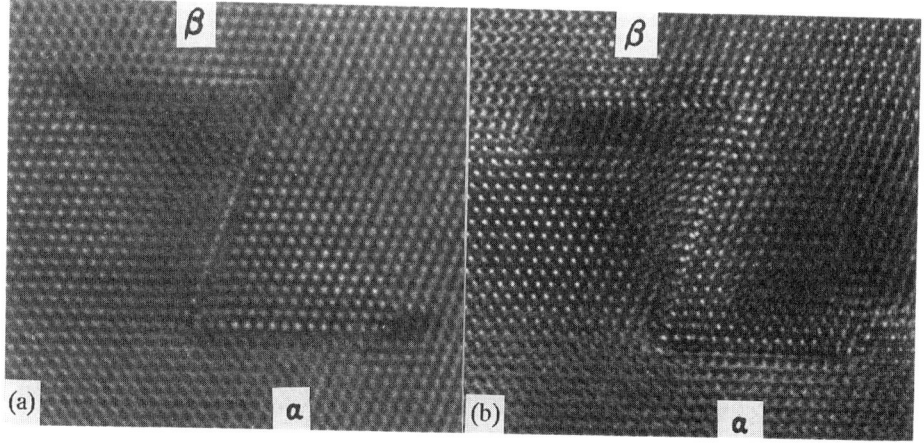

Fig. 5 (a) HREM image of a faulted dipole and (b) same region showing the effect of
prolonged electron irradiation.

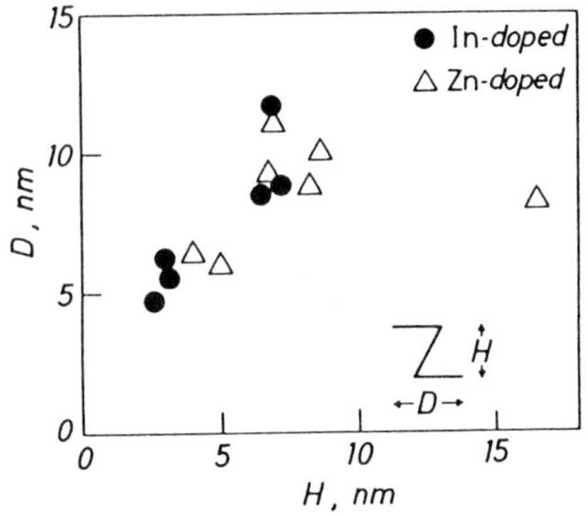

Fig. 6 Variation in width vs. height of faulted dipoles in In-doped and Zn-doped GaAs.

fault widths of dislocations within Z-type faulted dipoles from HREM is likely to be more meaningful.

Figure 6 shows the height vs. width of the faulted dipoles observed within In-doped GaAs. The data points from Zn-doped GaAs obtained by De Cooman and Carter (1987, 1988) are also shown for comparison. For a dislocation dipole in an equilibrium configuration, the net force (interaction energy and SFE) per unit length on each dislocation constituting the faulted dipole is zero. Accordingly, the SFE is estimated to be 57 ± 17 mJ/m^2 for the larger three dipoles, which is comparable to that determined for p-type material by De Cooman and Carter (1987,1988). Jimenez-Melendo et al. (1988) reported a decrease of SFE to 27 mJ/m^2 from weak beam measurements in high temperature deformed In-doped GaAs, and related this to the segregation of In onto dislocation stacking fault due to the Suzuki effect. Such a change of SFE by In-doping was not detected in the present investigation.

We find no difference in the core structure of α and β-type dislocations with the exception of the extent of climb following electron irradiation. Moreover, the macroscopic behaviour of dislocation locking due to the In dopant with GaAs (Yonenaga and Sumino 1989) cannot be correlated with differences in the microscopic core structure, nor with the dissociation widths.

5. CONCLUSIONS

Faulted dipoles within deformed In-doped GaAs were observed using HREM. Shockley partial and stair-rod dislocations showed no differences with dislocation type (α or β). In addition, and contrary to expectation, no precipitates were found to be associated with these defects. The SFE was estimated to be 57 mJ/m^2.

6. Acknowledgements

IY wishes to thank the Ministry of Education, Science and Culture of Japan for funding to visit Cambridge. PDB wishes to acknowledges SERC for funding under contact NO. GR/J37966.

REFERENCES

Burgess W G, Saunders M, Bird D and Humphreys C J 1993 *Microbeam Analysis '93*, 5222
De Cooman B C and Carter C B 1987 Appl. Phys. Lett. 50 40
De Cooman B C and Carter C B 1988 Phys. Stat. Sol. (a) 112 41
Yonenaga I and Sumino K 1989 J Appl. Phys. 65 85
Yonenaga I and Sumino K 1991 Inst. Phys. Conf. Ser. 117 129
Yonenaga I and Sumino K 1993 J Appl. Phys. 74 917
Jimenez-Melendo M, Djemel A, Riviére A P, Castaing J, Thomas C and Duseaux M Revue Phys. Appl. 23 251(1988).
Gerthsen D and Carter C B 1993 Phys. Stat. Sol. (a) 136 29

Inst. Phys. Conf. Ser. No 146
Paper presented at Microsc. Semicond. Mater. Conf., Oxford, 20–23 March 1995
© 1995 IOP Publishing Ltd

Can transmission electron holography reveal the electric potential of linear charged dislocations?

D Cavalcoli, G Matteucci and M Muccini*

Dept. of Physics, University of Bologna, via Irnerio 46, I-40126 Bologna, Italy.
* Institute of Molecular Spectroscopy- CNR, via Gobetti 101, I-40129 Bologna, Italy.

ABSTRACT: The possibility of revealing the electric potential distribution arising from straight charged dislocations in silicon by means of electron holography techniques in the transmission mode is considered. The experimental arrangement with the line charge parallel to the electron beam is investigated. Beam injection conditions are also discussed.

1. INTRODUCTION

Over many years several investigations have been carried out to evaluate the electrical properties of deformed crystals to understand dislocation behaviour. As their formation is accompanied by the introduction of a high density of point defects and impurity atoms, the investigated electrical properties must be related not only to the dislocations but to all the modifications the plastic deformation has induced in the crystal. The explanation of the experimental results is very complex (Alexander and Teichler, 1991).

The aim of this work is to point out a theoretical study of the possibility of applying transmission electron holography to reveal the electric potential distribution generated by a straight charged dislocation (Cavalcoli et al, 1995). This technique has already been successfully applied in the study of the electrostatic field of p-n junctions (Frabboni et al, 1987), of charged dielectric spheres (Chen et al, 1989) and of charged microtips (Matteucci et al, 1992). The potential of electron holography is particularly evident when phase changes due to the specimen are not large enough to be displayed in a standard contour map (Tonomura, 1992). In such cases, optical or digital phase amplification techniques can be employed to reveal small phase differences (Hasegawa et al, 1992). The possibility of detecting the dislocation charge is based on these methods. The evaluation of the electric potential of a straight dislocation from the Read model will be recalled in the next section (Read, 1954), with the electron holographic technique. The simulations of the electronic wave phase shift due to the dislocation potential distribution and the beam injection conditions will be reported in the following ones.

2. ELECTRON HOLOGRAPHIC METHOD

Let us recall a few essential facts about electron holography which can be considered as a two-step procedure. In the first step, the object under investigation is illuminated by a high brightness electron beam emitted by a field emission gun, as shown in Fig.1. The wave-front scattered by the specimen and a reference wave-front R, travelling outside its border, is overlapped by the electrostatic field produced by the thin charged wire W placed between two earthed plates P (Moellenstedt-Dueker electron biprism). This interferometry device is inserted at the selected aperture level. An interference pattern (hologram) is formed in the observation plane OP. The projector lenses PL of the microscope provide a further magnification of the hologram H on the final viewing screen. The phase shift $f(x,y)$ of the electron wave function, caused by a localized electrostatic field described by the potential Φ, is given by:

$f(x,y) = \dfrac{\pi}{\lambda V} \int_l \Phi(x,y,z)dz$, where λ is the electron wavelength and V the accelerating potential. The integral is taken along the electron optical path l parallel to the optic axis z inside and outside the object.

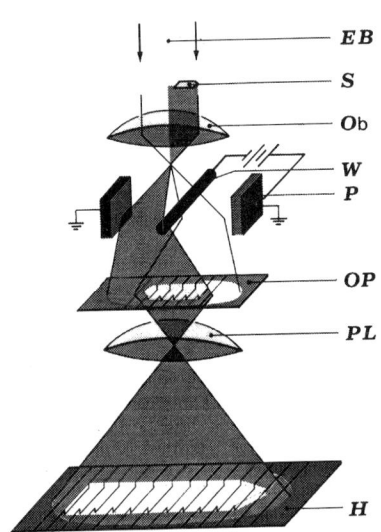

Fig. 1. Schematic arrangement for hologram formation with an electron biprism. EB: plane wave illuminating the specimen; S: specimen placed off-axis; Ob: objective lens; W: wire and P: earthed plates respectively of the electron biprism; OP: observation plane; PL projector lenses; H: hologram.

Since the field under investigation is strictly confined within the specimen, the reference wave is not affected, therefore the absolute phase shift caused by the electric potential can be determined. In order to extract the phase information, holograms can be processed, by using slow-scan CCD cameras, or an optical bench. The phase variations can be displayed as a set of contour fringes superimposed on the in-focus reconstructed specimen image. The fringe trend can be strictly related to the projected equipotential distribution of the electric field under investigation.

An approximate method for the evaluation of the electronic charge at the dislocation can be deduced by the Read model (Read, 1954). A neutral dislocation is assumed to have n_d [cm^{-1}] rechargeable acceptor centers per unit length. When it becomes negatively charged, the conduction electrons will be scattered away from the dislocation line. Consequently, a cylinder of ionized, positive atoms will be formed around the dislocation. The relation between the line charge $q = n_d e$ localized at the dislocation and the potential Φ at a distance r from the screened charged line is given by:

$$\Phi(r) = \frac{q}{\varepsilon\varepsilon_0 \pi R^2}\left[\frac{R^2 - r^2}{4} + \frac{R^2}{2}\ln\left(\frac{R}{r}\right)\right]$$

where $\varepsilon\varepsilon_0$ is the Si dielectric constant times the vacuum dielectric constant and R is the radius of the Read cylinder $R = [q/(e\,N_D)]^{1/2}$ with N_D the doping density. Since q and R are interrelated, the line charge determines Φ. A local measure of Φ can thus give an evaluation of the line charge q at the dislocation, i.e. of the linear density of dislocation-related centers.

3. SIMULATIONS OF CHARGED DISLOCATIONS

Within the framework of the Read model and by using electron holographic principles, computer simulations can be performed of the phase shift suffered by the electron wave which crosses the potential distribution generated by a charged dislocation. The simulations have been made by means of a personal computer equipped with a video board able to display 512x512 pixels at 256 grey levels.

In fig. 2 a linear charged dislocation is sketched, represented by the dark central line D confined within the cylinder C. The defect is perpendicular to the sample surface and parallel to the electron beam direction EB.

The phase shift induced by the dislocation is given by $f(r)=\pi\Phi(r)L/(\lambda V)$, where $r=(x^2+y^2)^{1/2}$, L is the line charge length that equals the specimen thickness. In this case, taking experimental values for $\Phi=0.35V$ (Fell et al , 1993), and using 100kV electrons, the maximum phase shift is about $\pi/5$. Since this phase shift is rather weak, phase difference amplification techniques are needed.

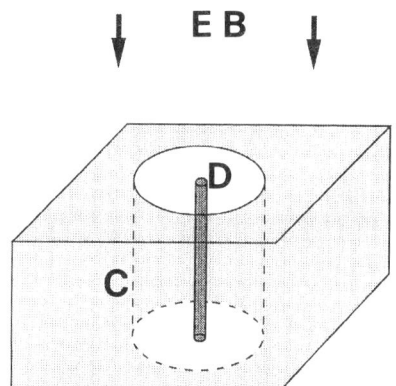

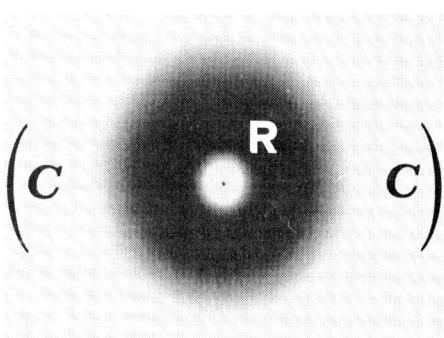

Fig.2 Schematic drawing of a linear charged dislocation parallel to the electron beam direction

.Fig.3 Calculated contour map (fifteen times phase amplification) of the potential distribution for the dislocation oriented as in Fig.2.

A potential map of a linear charged dislocation with L=200 nm and R 100 nm, amplified fifteen times, is shown in Fig.3. The dark central spot corresponds to the in - plane projection of the vertical charged dislocation core D of Fig.2. The border of the charged cylinder is represented by the two vertical solid lines *C - C*. The electric field has a radial distribution with respect to the charged line. Therefore the dark ring R can be interpreted as the set of projected equipotential cylindrical surfaces around the dislocation core.

4. BEAM INJECTION CONDITIONS

An important point to be discussed here is the analysis of the beam injection conditions during the electron hologram recording step. As the Read model is based on the calculation of the equilibrium line charge, the concentration of electron - hole pairs induced by the electron beam should be low enough to avoid perturbation effects on the dislocation potential distribution. The Bethe - Bloch formula for the energy loss in the Si sample has been used to evaluate the electron - hole generation rate. The obtained values are reported in Table 1, together with the data relative to an EBIC (Electron Beam Induced Current) experiment, a typical low injection technique. Both the generation rates refer to the same value of the beam current of about 10^{-10}A, which is a reasonably low value for hologram recording. For the same values of the beam current, the TEM generation rates result to be two orders of magnitude lower than the EBIC ones. Moreover, as is expected, the generation rate in the TEM experiment is a decreasing function of the beam energy, this being due to the small energy loss of the impinging electrons in the silicon sample.

TABLE I.   Evaluation of the generation rate G (electron - hole pairs generated per second) in the TEM   and   SEM-EBIC   mode. The evaluation refers to a beam current $I_b \approx 10^{-10}$ A, and electron - hole pair formation energy of 3.6 eV. E_b is the beam energy.

TEM		EBIC	
E_b (keV)	G (s⁻¹)	E_b (keV)	G (s⁻¹)
100	4.0×10^{10}	10	3.0×10^{12}
200	2.5×10^{10}	20	5.5×10^{12}
300	2.0×10^{10}	30	8.0×10^{12}

5. CONCLUSIONS

Until today, a technique able to map and measure the electronic properties of a single extended defect has been lacking, so the possibility of imaging the potential distribution of a single dislocation represents an important tool in the understanding of its electrical behaviour. The problem of imaging projected potentials associated with charged dislocations by means of electron holographic contour mapping techniques has been considered. By using the Read model for charged dislocations, an analytical expression for the phase shift has been calculated and the expected phase contour map simulated. The experimental conditions suitable for revealing the potential distribution generated by a charged dislocation have been discussed.

Finally, it should be pointed out that the values of the radius of the Read cylinder, the linear charge density and the height of the potential barrier refer to literature data on "clean" dislocations, the possibility of imaging decorated dislocations (i.e. with higher values of linear charge density and with a higher potential barrier at the dislocation) requires, therefore, less stringent experimental conditions and lower phase amplification factors.

REFERENCES

Alexander H and Teichler H 1991 Material Science and Technology, 4 Electronic Structure and Properties of Semiconductors, eds Cahan RW, Haasen P, Kramer EJ, Vol. ed. Schroeter W, (Weinheim, Germany: VCH) pp 249-71.
Cavalcoli D, Matteucci G and Muccini M 1995 Ultramicroscopy (in press).
Chen J W, Matteucci G, Migliori A, Missiroli G F, Nichelatti E, Pozzi G and Vanzi M 1989 Phys. Rev. A 40 3136.
Fell T S, Wilshaw P R and de Coteau M 1993 Phys. Stat. Sol. (a) 138 695.
Frabboni S, Matteucci G, Pozzi G 1987 Ultramicroscopy 23 2196.
Hasegawa S Kawasaki T, Endo J, Tonomura A, Honda Y, Futamoto M, Yoshida K, Kugiya F Koizumi M, 1989 J.Appl. Phys. 65 2000.
Matteucci G, Missiroli G F, Muccini M and Pozzi G 1992 Ultramicroscopy 45 77.
Read W T Jr. 1954 Phil. Mag. 45 775.
Tonomura A 1992 Advances in Physics 41 59.

Inst. Phys. Conf. Ser. No 146
Paper presented at Microsc. Semicond. Mater. Conf., Oxford, 20–23 March 1995
© *1995 IOP Publishing Ltd*

Segregation and precipitation of iron in a Σ=25 silicon bicrystal

X Portier, R Rizk, G Nouet and G Allais

LERMAT, URA CNRS 1317, ISMRA, 6 Bd Marechal Juin, 14050 Caen Cedex, France

ABSTRACT: Segregation and precipitation of iron in a Σ=25 silicon bicrystal have been carefully investigated by means of high resolution electron microscopy (HREM) and energy dispersive X-ray (EDX) analysis. After the intentional incorporation of iron in the bicrystal by a conventional heating procedure, it was shown that the non-equilibrium segregation of iron has been obtained after rapid cooling whereas its precipitation has been produced by a low cooling rate. The iron silicide crystals have been formed at the grain boundary (GB) plane and they have been found to belong either to the ε-FeSi cubic phase or to the α-FeSi$_2$ tetragonal phase.

1. INTRODUCTION

Because of their technological importance as interconnects in microelectronics, metal silicides have been the object of numerous studies. Beside the successful achievement of high quality films of these materials such as CoSi$_2$ (Bulle Lieuwma et al 1991) and NiSi$_2$ (Chen and Chen 1993) which exhibit nearly perfect interfaces with the crystalline silicon substrate, particular interest is attached these days to the study of buried layers. In this connection, we have carried out recently (Portier et al 1994) a careful examination of the interfaces between silicon and NiSi$_2$ precipitates at the tilt GB of a Σ=25 silicon bicrystal. The ability of the silicides, such as NiSi$_2$ and Cu$_3$Si, formed at the GB, to act as buried Schottky diodes has been demonstrated by our recently published work (Rizk et al 1994). Considering the recent interest accorded to the growth of iron silicides (Berbezier et al 1994), we report in this paper the results of an HREM study of the segregation and precipitation at the GB plane of a Σ=25 silicon bicrystal, of the moderately diffusing iron, in combination with energy dispersive X-ray (EDX) analyses.

2. EXPERIMENTAL PROCEDURE

The samples used in this investigation were low-doped with phosphorus (N_d=3.1x10^{14} at.cm^{-3}) Σ=25 silicon bicrystals ({710} GB plane, [001] tilt axis and 16.26° tilt angle). The specimens were chemically etched in an acid solution (16HNO$_3$:3CH$_3$CO$_2$:1HF) before being covered electrochemically on all their faces by iron. They were then annealed in a double wall quartz tube furnace under a continuous flow of pure argon at 1200°C for 3 hours before being cooled down either by rapid withdrawal from the furnace to the cold end of the tube or by only switching off the electrical power of the furnace. The cooling rates can be estimated to be 50°C.s^{-1} or 0.5°C.s^{-1}, respectively. HREM images were obtained in a [011] silicon grain direction which is close to the GB plane in order to minimize the projected area of the GB plane on the micrographs. The electron microscope used was a Topcon 002B (200 kV) equipped with Kevex EDX microanalysers. The EDX analyses were carried out with a probe of 5.9nm in diameter.

3. RESULTS

The formation of iron precipitates in the GB zone has been found after slow cooling whereas iron seems to segregate upon fast cooling. In this latter case, no precipitate is observed but the reproducible EDX profile (Fig. 1) shows a relatively high local iron content at the GB level that is at least twice higher than that in the neighbourhood, in spite of the great uncertainty inherent to the detected low rate.

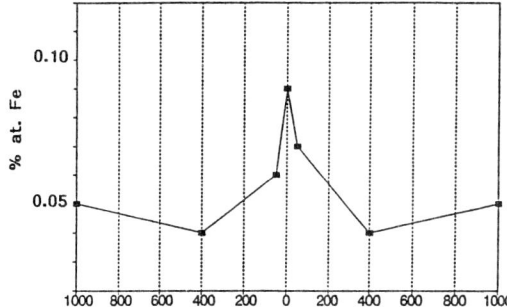

Fig.1 EDX profile of the iron concentration across the GB plane. The distance from the GB is in nanometers. Owing to the detection limit of the technique (~0.1%), the values obtained must be comparatively considered, regardless their absolute value.

On the other hand, a low cooling rate leads to the formation of isolated (average separating distance: ~1000nm) and small (size < 50nm) precipitates located in the GB region. Figs. 2 and 3 show two examples of both kinds of the formed precipitates.

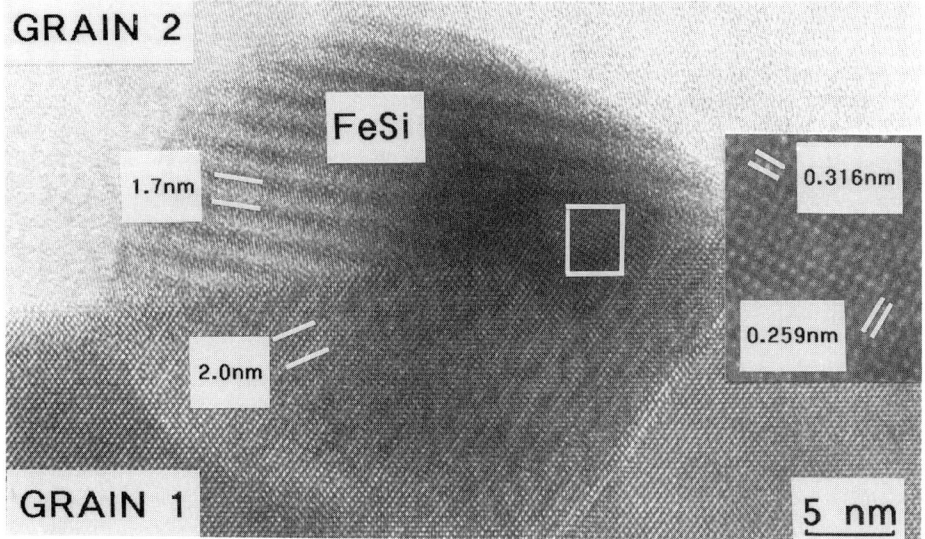

Fig.2 HREM image of a ε-FeSi precipitate located at the GB plane.

Considering the domain where the precipitate is not superposed on the silicon matrix, as delimited by the white square in Fig. 2 and magnified in the right hand side, three families of planes are visible and associated with the following interplanar distances, namely 0.259nm, 0.316nm and 0.200nm. This configuration attests that we are dealing with a cubic ε-FeSi phase with the lattice parameter a=0.446nm. The above-mentioned planes are (111), ($\bar{1}$01) and (012) respectively. The precipitate is oriented with respect to the grain 1 and its orientation relationship is Si[011]/[1$\bar{2}$1]ε-FeSi and Si(1$\bar{1}$1)/(111)ε-FeSi with a slight misorientation of about 1° for this latter. Such an orientation has been already observed by Le Thanh Vinh et al (1992) for an ε-FeSi layer deposited on a (111) oriented silicon surface. Furthermore, one can recognize two sets of Moiré fringes. The first, lying in the grain 1, is characterized by an interfringe spacing of about 2.00nm. It results from the superposition of the (012)ε-FeSi and the (02$\bar{2}$) grain 1 planes. The second set of Moiré fringes, spaced by 1.74nm and located in the grain 2, is due to the overlapping of the (012)ε-FeSi and (02$\bar{2}$) grain 2 planes.

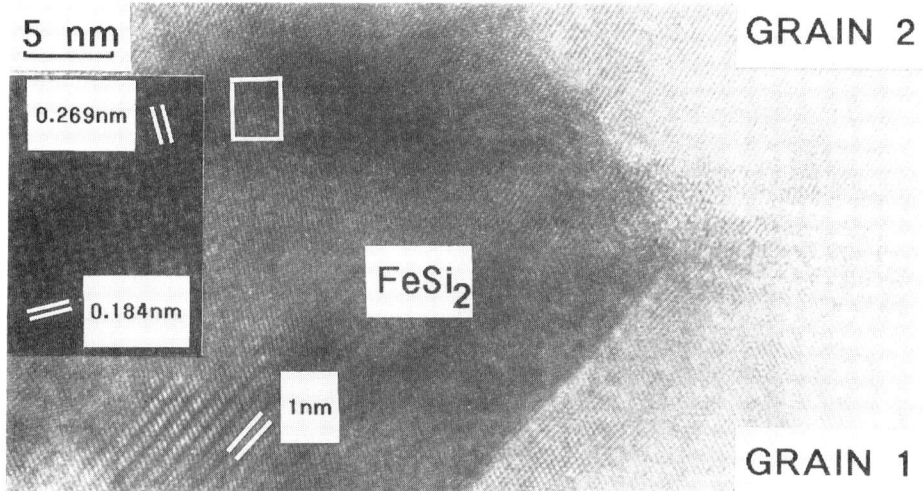

Fig.3 HREM image of a α-FeSi$_2$ crystal at the GB plane.

Similar analyses of the iron precipitate shown in Fig. 3 leads to the recognition of the α-FeSi$_2$ tetragonal phase. It results from the observation of the two families of planes depicted in the inset (an enlargement of the framed area). There, we are dealing with the perpendicular (010) and (102) families of planes, with spacings of 0.269nm and 0.184nm, respectively. The orientation relationship with respect to the grain 1, is Si[011]/[$\bar{2}$01]α-FeSi$_2$ and Si(1$\bar{1}$1)/(1$\bar{1}$2)α-FeSi$_2$. This orientation has been observed also in the case of epitaxially grown α-FeSi$_2$ on (111) silicon oriented wafers by Berbezier et al (1993). The Moiré fringes in the grain 1 which are spaced by about 1 nm are due to the overlapping of the (11$\bar{1}$) grain 1 and the (010) α-FeSi$_2$ planes.

Additionally, EDX analyses of the precipitates have been performed. The results in atomic percent are: 65% Si, 35% Fe for ε-FeSi and 79% Si, 21% Fe for α-FeSi$_2$. The exact composition of the particles cannot be determined in spite of the small size of the probe (5.9nm), because the analyzed area is never limited to the precipitate only but it extends to the silicon matrix. Nevertheless, the relative concentrations of iron, as measured on both kinds of precipitates, are consistent with the iron content in each of FeSi and FeSi$_2$ compounds.

4. DISCUSSION

The formation of silicide by precipitation requires an elevated annealing temperature in order to ensure a sufficient solubility level of the impurity in the silicon matrix, as suggested by Cullis and Katz (1974) who have used an annealing temperature as high as 1200°C for the growth of α-FeSi$_2$ precipitates in the bulk. The other determining factor for iron precipitation lies in the cooling rate. Considering that iron diffuses via interstitial sites (Weber and Riotte 1980), it seems that rapid cooling "freezes" the iron atoms in interstitial sites (Fe_i) throughout the sample, and also leads to non-equilibrium segregation of iron atoms at the GB level as demonstrated by the EDX profile of Fig. 1. The first feature is, however, confirmed by the electron beam induced current (EBIC) measurements performed on the same and similarly treated samples and reported by Ihlal et al (1995) where Fe_i acts as a recombination center leading to a general decrease of EBIC as well as the appearance of narrow denuded zones on both sides of the GB. In contrast, the slow cooling favours a sufficient local segregation around the nucleation sites resulting in the so-called precipitation. The present observations contrast strongly with those reported for the case of nickel or that of copper, where the precipitation has been obtained only after rapid cooling and where no segregation has been detected after slow cooling (Rizk et al 1994). These results are explained by the higher diffusivity of both nickel and copper in silicon in comparison with that of iron. Consequently, a cooling rate of $50°C.s^{-1}$ leads to the precipitation of nickel and/or copper whereas a cooling rate as low as $0.5°C.s^{-1}$ rejects these metallic elements to the sample surfaces.

The two observed phases (ε-FeSi and α-FeSi$_2$), as well as the orthorombic β-FeSi$_2$ phase, belong to the binary phase diagram Fe/Si (Kubaschewsky 1982), in contrast to the two cubic γ- and s-FeSi$_2$ phases which are considered by some authors (Le Thanh Vinh et al 1992, Gerthsen et al 1993, Onda et al 1992) as transitional structures whose role is to accommodate elastically the lattice parameter mismatch between the β-FeSi$_2$ and the silicon matrix. The β-FeSi$_2$ phase appears unstable for temperatures higher than 940°C (Berbezier et al 1994) in such a way that the absence of the transition structures in our work is quite conceivable. Furthermore, on the basis of the binary phase diagram exhibiting an eutectic point at 1212°C (Kubaschewsky 1982), the experimental conditions employed here appear rather favourable to an equilibrium between each of the ε-FeSi and α-FeSi$_2$ phases and the silicon matrix.

Unlike the case of nickel precipitates in a Σ=25 silicon bicrystal (Rizk et al 1994), one can emphasize a low density of iron particles and the absence of large precipitates in the present study. This is probably due to the lower solubility and diffusivity of iron compared to nickel in silicon.

REFERENCES

Berbezier I, Chevrier J and Derrien J 1994 Surf. Sci. **315**, 27

Berbezier I, Regolini J L and D'Anterroches C 1993 Microsc. Microanal. Microstruct. **4**, 5

Bulle Lieuwma C W T, De Jong A P and Vandenhoudt D E W 1991 Phil. Mag. A **64**, 255

Chen W J and Chen F R 1993 Phil. Mag. A **68**, 605

Cullis A G and Katz L E 1974 Phil. Mag. **30**, 1419

Derrien J, Chevrier J, Le Thanh Vinh, Berbezier I, Giannini C, Lagomardino S and Grimaldi M G 1993 Appl. Surf. Sc. **70**, 90

Gerthsen D, Schäfer H Ch, Rösen B, Rizzi A, Moritz H and Lüth H 1993 Inst. Phys.Conf. Ser. **134**, 185

Ihlal A Rizk R Voivenel P and Nouet G, this conference

Kubaschewsky O 1982 Iron-Binary phase diagrams (Springer-Verlag, Berlin)

Le Thanh Vinh, Chevrier J and Derrien J 1992 Phys. Rev. B **46**, 15946

Müller E, Grindatto D P, Nissen H U, Onda N and Von Känel H 1994 Appl. Phys. Lett. **64**, 1938

Onda N, Henz J, Müller E, Mader K and Von Känel H 1992 Appl. Surf. Sc. **56-58**, 421

Portier X, Rizk R, Nouet G and Allais G to be published in Phil. Mag. A.

Rizk R, Portier X, Allais G and Nouet G 1994 J. Appl. Phys **76**, 952

Weber E and Riotte H G 1980 J. Appl. Phys. **51**, 1484

Inst. Phys. Conf. Ser. No 146
Paper presented at Microsc. Semicond. Mater. Conf., Oxford, 20–23 March 1995
© *1995 IOP Publishing Ltd*

The contrast of grain boundary dislocations

M Joksch, P Wurzinger and P Pongratz

Institut für Angewandte und Technische Physik, TU Wien, Wiender Hauptstrasse 8-10, A-1040 Vienna, Austria

ABSTRACT: The contrast formed in TEM strong beam images by grain boundary dislocations in CVD-diamond was studied in detail. For certain diffraction conditions new contrast criteria were found, which can be used to determine the sign of $\mathbf{g}\cdot\mathbf{b}$ and to estimate its absolute value.

1. INTRODUCTION

Grain boundaries and their interactions with dislocations have an influence on the mechanical and electrical properities of polycrystalline materials. Therefore, dislocations at twin boundaries in polycrystalline CVD (chemical vapour deposition) diamond films were investigated with the TEM. The most accurate method to analyze Burgers vectors of grain boundary dislocations is to match computer simulations to experimental images (Humble and Forwood 1975). This requires a large number of calculations, because of numerous possible Burgers vectors. Contrast criteria (Marukawa and Matsubara 1979) based on certain geometrical and diffraction conditions are useful to reduce the number of Burgers vectors which have to be considered for the computer simulations. The present work was performed to extend the small number of such criteria for grain boundary dislocations.

2. EXPERIMENTAL

TEM (JEOL 200 CX, 200 keV) bright field images of dislocations at twin boundaries were compared to images computed with a computer programm for the contrast simulation of grain boundary dislocations which is based on the ideas of Humble and Forwood (1974). Two diffraction conditions were used: (a) Two beam diffraction is operating in one grain no diffracted beams are strongly excited in the other grain (single-$\mathbf{g}$ case). (b) Different two-beam conditions are excited in the two crystals (two-$\mathbf{g}$ case).

3. RESULTS AND DISCUSSION

3.1 Contrast in the single-g case

Marukawa and Matsubara (1979) developed contrast criteria for dislocations at grain boundaries in isotropic crystals for a single-$\mathbf{g}$ case, for which the beam direction $\mathbf{k}$, the dislocation line $\mathbf{u}$ and the grain boundary normal $\mathbf{n}$ are coplanar. For $\mathbf{g}\cdot\mathbf{b}=0$ ($\mathbf{b}$ Burgers vector) the image of the dislocation appears as a symmetric dark ($\mathbf{g}\cdot(\mathbf{b}\mathbf{x}\mathbf{u})>0$) or bright ($\mathbf{g}\cdot(\mathbf{b}\mathbf{x}\mathbf{u})<0$) line if the upper grain is in contrast (fig. 1a). For $\mathbf{g}\cdot\mathbf{b}>0$ the dislocation contrast is asymmetric. When looking along the dislocation line (from bottom to top) the image is dark on the left side and bright on the right side of the dislocation (fig. 1c). For the opposite sign of **g·b** the bright and dark sides are exchanged.

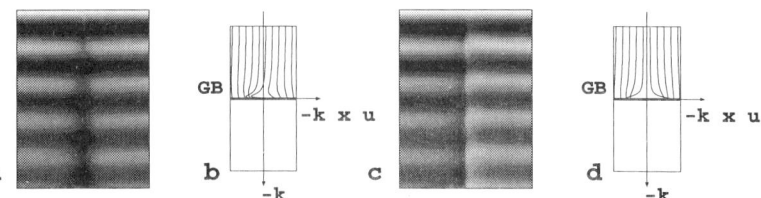

Figure 1: Computed images of a dislocation at a twin boundary in CVD diamond and the values of d(**g·R**)/dz in the generalized cross-section (in arbitrary units) for the case of coplanar **u**, **k** and **n**. The contrast is symmetric for **g·b**=0 (a) and asymmetric for **g·b**=1 (c). The parameters used are **g**=(022), **b**=(1/2)[$0\bar{1}1$](a) (1/2)[$\bar{1}10$](c), **u**=[$1\bar{1}0$], **k**=[$11\bar{1}$] and **n**=($11\bar{1}$); no diffraction was assumed for the lower grain.

Bright and dark features in images of dislocations are caused by positive and negative values of d(**g·R**)/dz (Marukawa and Matsubara 1979) which describes the effective local deviation from the Bragg condition in the Howie-Whelan equations (Howie and Whelan 1961). **R** is the lattice displacement vector, **g** the diffraction vector and z the coordinate along the beam direction. Figure 1b,d illustrate the values of d(**g·R**)/dz in a cross section of the crystal with the axes **k** and -(**kxu**) and the dislocation in the origin (generalized cross-section in Head et al 1973).

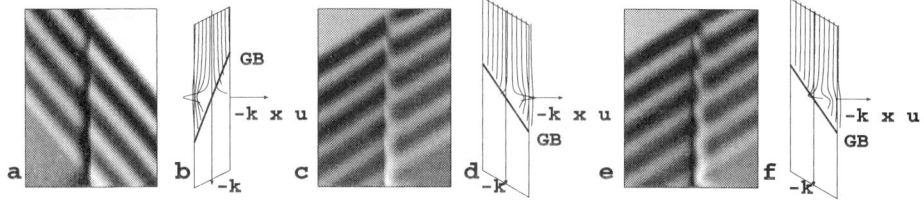

Figure 2: Computed images of a dislocation at a twin boundary in CVD diamond and the values of d(**g·R**)/dz for non-coplanar **u**, **k**, and **n**. Dependent on the inclination of **n** the dislocation appears as dark (a) or bright (c) line for **g·b**=1. For **g·b**=0 the contrast is asymmetric (e). The parameters used are **g**=($0\bar{2}2$), **b**=(1/2)[101](a,c) (1/2)[$01\bar{1}$](e), **u**=[101], **k**=[111](a) [$4\bar{1}\bar{1}$](c,e) and **n**=($11\bar{1}$); no diffraction was assumed for the lower grain.

When **n** deviates substantially from the plane of **u** and **k** the contrast changes fundamentally. This may be demonstrated for a case with **g·b**=1 (fig. 2a). Two-beam diffraction is assumed only in the upper grain. Therefore, the grain boundary which is now inclined to the beam, precludes the strain field at the right side of the dislocation from being effective for the dislocation contrast. The main contrast features thus appear at the left side of the dislocation which is the side of larger grain thickness (fig 2b). Because of the negative d(**g·R**)/dz on this side the dislocation contrast is a dark line. Under the same diffraction condition the crystal may be tilted so that the grain boundary is inclined to the other side (fig. 2c), the dislocation gives now bright contrast lying at the left side, because positive values of d(**g·R**)/dz are effective for the image. Changing the sign of **g·b** the contrast is reversed (fig. 3). As the symmetry of d(**g·R**)/dz for **g·b**=0 is broken by the grain boundary (cf. fig. 1a & fig. 2f) the image contrast becomes asymmetric for **g·b**=0. The mean features of the weak contrast are a dark and a bright line lying side by side.

The contrast features presented above can be generalized and compiled to contrast criteria for non-coplanar **u**, **k** and **n** in the single-**g** case (Table 1). They are independent of the grain in which two-beam diffraction is excited and describe the additional contrast features caused by the dislocation. In particular the dark features of the contrast depend on the absolute value of **g·b** as shown in fig. 3. This is useful for an estimate of the magnitude of **g·b** and can be used to confine the number of possible Burgers vectors without computer simulation.

	g·b>0	**g·b**<0	**g·b**=0
LGT left	dark	bright	bright and dark parallel
LGT right	bright	dark	weak contrast

Tabel 1: Additional contrast caused by a dislocation at a grain boundary in the single-**g** case with non-coplanar **u**, **k** and **n**. LGT: Larger thickness of the grain in which the two-beam condition is excited.

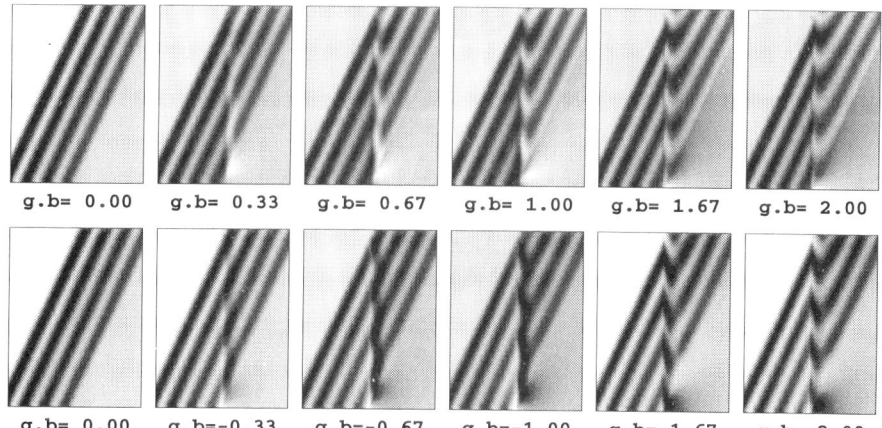

Figure 3: Computed images of a dislocation showing characteristic features for different **g·b**. The Burgers vectors are: (1/2)[011], (1/6)[1$\overline{2}$1], (1/6)[$\overline{2}$1$\overline{1}$], (1/2)[101], (1/6)[$\overline{1}$14] and (1/2)[0$\overline{1}$1]. Other parameters: **g**=(0$\overline{2}$2), **k**=[111], **u**=[011], **n**=[11$\overline{1}$]

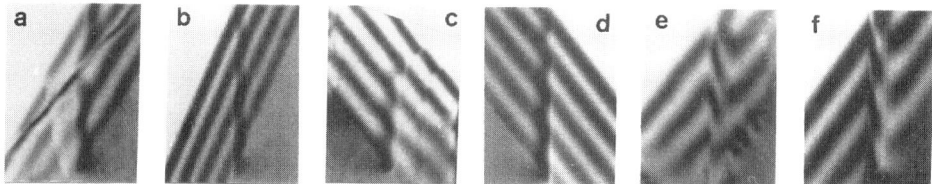

Figure 4: Experimental and computed images of dislocations at a twin boundary in CVD diamond in the single-**g** case: a,b) **g**=(02$\overline{2}$), **b**=(1/6)[112], **u**=[$\overline{1}$43], **k**=[111], **n**=[11$\overline{1}$]; c,d) **g**=($\overline{2}$02), **b**=(1/6)[$\overline{1}$21], **u**=[101], **k**=[232], **n**=[11$\overline{1}$]; e,f) **g**=(0$\overline{2}$2), **b**=(1/2)[10$\overline{1}$], **u**=[011], **k**=[111], **n**=[11$\overline{1}$]

Figure 4 gives three examples for the contrast in experimental images compared to computer simulations. All three micrographs are taken with non-coplanar **u**, **k** and **n**. The contrast features discussed above can clearly be recognized.

3.2. Contrast in the two-g case

In the two-**g** case the contrast of bright field images is formed by the multiplication of the amplitudes of the transmitted beam of the two crystal parts (Humble and Forwood, 1975). The complex formation of the contrast hinders the compilation of contrast criteria. However, the contrast of some cases can be unterstood from the criteria in the previous section, if it is considered that the image in the two-**g** case is a combination of two single-**g** images. An example is given in fig. 5(a,b) for almost coplanar **k**, **u** and **n**: **g·b** has the same sign in both grains. Therefore, similar contrast features are added and the contrast is strong. On the contrary, when **g·b** has the oposete sign in the two grains, the contrast becomes weak, because opposed contrast features are overlaid (fig. 5c).

Figure 5: Experimental (a) and computed (b,c) images of a dislocation in the two-**g** case. Used Parameters are: $\mathbf{g}_1=(1\bar{1}\bar{1})$, $\mathbf{g}_2=(\bar{2}20)$, $\mathbf{u}_1=[101]$, $\mathbf{k}_1=[232]$, $\mathbf{n}_1=[1\bar{1}\bar{1}]$, $\mathbf{b}_1=(1/6)[\bar{1}21]$(a,b), $\mathbf{b}_1=(1/6)[\bar{2}1\bar{1}]$(c), $_1$...upper grain, $_2$...lower grain.

5. CONCLUSION

In conclusion, we have found new contrast criteria for strong beam images to estimate the value and sign of **g·b** of grain boundary dislocations in the single-**g** case. The criteria were found from computer contrast simulations and checked on images of dislocations at grain boundaries in CVD diamond. The applicability of the criteria is not restricted by geometrical conditions. They can also be used to interpret features of the contrast in the two-**g** case.

ACKNOWLEDGEMENTS

The authors gratefully acknowledge financial support of the present work by the Austrian Science Foundation FWF (project S5903) carried out under the auspicies of the trinational "D-A-CH" cooperation of Germany, Austria and Switzerland on the "Synthesis of Superhard Materials".

REFERENCES

Head A K Humble P Clarebrough L M Morton A J and Forwood C T 1974 Computed Electron Micrographs and Defect Identification (North-Holland, Amsterdam)
Howie A and Whelan M 1961 Proc. R. Soc. A. <u>263</u> 217-237
Humble P and Forwood C T 1975 Phil. Mag. <u>31</u> 1011-1023
Marukawa K 1979 Phil. Mag. A <u>40</u> 303-312
Marukawa K and Matsubara Y 1979 Trans. JIM <u>20</u> 560-568

Inst. Phys. Conf. Ser. No 146
Paper presented at Microsc. Semicond. Mater. Conf., Oxford, 20–23 March 1995

Silicon wafer bonding technology: TEM study of low-angle twist boundaries

M Benamara[1], A Rocher[1], A Laporte[2,3], G Sarrabayrouse[2], L Lescouzères[3], A PeyreLavigne[3] and A Claverie[1]

1 CEMES-LOE/CNRS, 29 rue Jeanne Marvig, BP 4347, 31055 Toulouse Cedex, France
2 LAAS/CNRS, 7av du Colonel Roche, 31077 Toulouse Cedex, France
3 Motorola Semiconducteurs S A, av Général Eisenhower, 31023 Toulouse Cedex, France

ABSTRACT: Low-angle ($0<\theta<13.5°$) [001] twist boundaries prepared using the silicon wafer bonding technique are studies by transmission Electron Microscopy techniques. They consist of square networks of pure screw dislocations, the spacing of which depends on the twist misorientation angle. In addition, they always exhibit an uncontrolled small tilt component ($<0.5°$) compensated by a periodic array of $60°$ dislocation lines.

1. INTRODUCTION

The direct wafer bonding (DWB) technology where two materials of different compositions or doping levels are bonded together at the atomic level, opens new possibilities for the realization of electronic devices (Bengtsson, 1992). It is also an attractive technique for the study of the electrical properties of polycrystalline semiconductors as it permits one to produce grain boundaries with controlled geometries over very large areas. In this paper, several [001] twist boundaries were prepared using this technique with the misorientations covering the angular range $0<\theta<15°$. Then, the resulting grain boundary dislocation (GBD) networks were extensively studied by plan-view and cross-section transmission electron microscopy (TEM) techniques. Perfect arrays of screw dislocations with Burgers vectors $\mathbf{b}=a/2<110>$ were found in the vicinity of all the resulting bonding interfaces. This study consists of a considerable outgrowth of previous investigations of (001) twist boundaries in silicon (Föll and Ast 1979; Carter et al 1981; Gafiteanu et al 1993; Benamara et al 1994).

2. EXPERIMENTAL PROCEDURE

Polished 4-inch, (001) oriented, 525 μm-thick, lightly boron-doped (10^{15} cm^{-3}) CZ silicon wafers were used. Pairs of hydrophobic wafers were put in contact by manual assembly in a class 10 clean room with the wafer flats intentionally misoriented from aligned to about $15°$. Afterwards, the bonded wafers pairs were annealed at $1200°C$ for 50mn in an N_2 ambient. TEM studies were performed with JEOL 200 CX and Philips CM20 microscopes operating at 200 kV, using both images and diffraction patterns, on both cross-section and plan-view samples.

3. RESULTS

3.1. General observations

Figure 1.a is a bright field image showing the general view of a bonding interface when the electron beam is close to the [001] direction. Large domains are periodically separated by dark lines. Amorphous precipitates of silicon oxide are often seen decorating these interfaces. Their facets are also bounded by the {111} and {100} planes of silicon. The overlapping of the two misoriented crystals gives rise to double-diffraction effects (Fig 1.c). Moiré fringes result from this double-diffraction effect and arise owing the interference essentially between the 220 reflections in the two crystals. They appear in the imaging mode as shown in Fig 1.b. These fringes give no direct information about the boundary structure but they show that the two crystals are misoriented around their common <001> axis. In this case, the Moiré spacing is measured to be d=23 Å $\pm$ 1Å. From this value, one can deduce the corresponding twist angle θ=4.8° $\pm0.2°$.

104

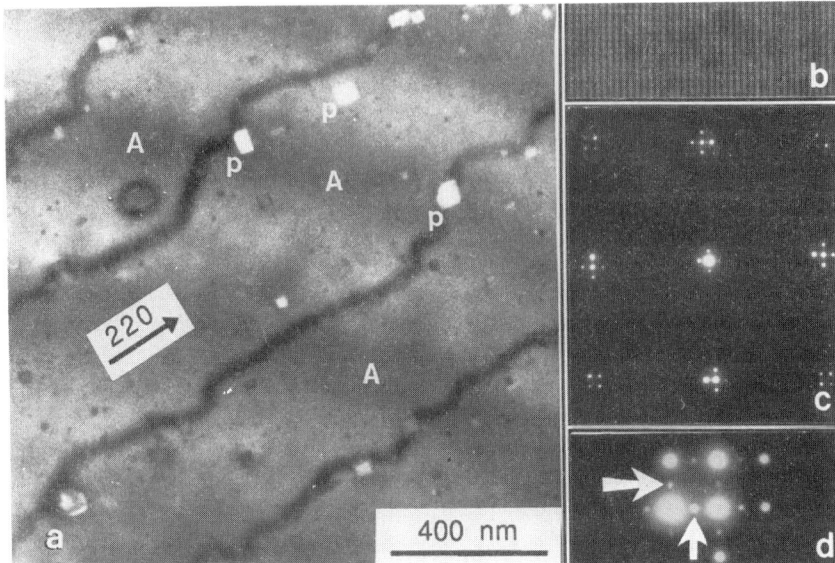

figure 1: a) BF plan-view micrograph taken under [001] zone axis. Note the formation of terraces (A) resulting from a small tilt misorientation and oxide precipitates (P) mostly connected to the 60° dislocation lines. **b)** the moiré fringes, seen within terraces and parallel to <110> directions, permit one to deduce the twist misorientation angle **c)** diffraction pattern along the [001] zone axis. The sharp extra-satellite spots that are pointed by the arrows in fig 1.d attest of the periodicity of the dislocation network on a large scale.

3.2. GBD accommodating low-angle twist components

As expected from cristallography of low-angle twist boundaries, the structure of the interface consists of two orthogonal sets of screw dislocations. Their Burgers vectors are $b_1 = a/2[110]$ and $b_2 = a/2[\bar{1}10]$, respectively. The dislocation spacing is given by the relation, $Dd = b/(2.\sin(\theta/2))$ where θ is the twist misorientation angle.

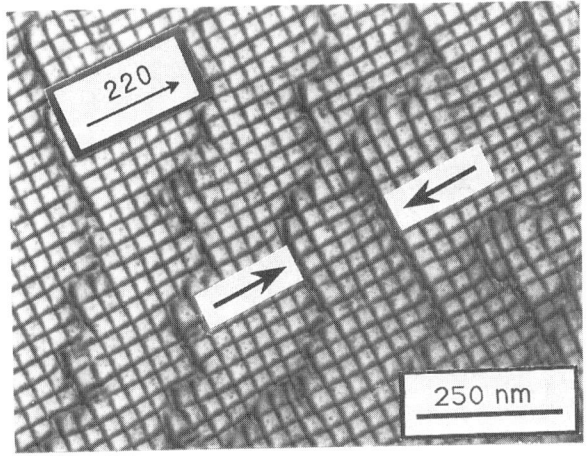

figure 2: Typical BF multibeam plan-view image from the interface of two lightly misoriented wafers. The multiple diffraction condition is satisfied by using simultaneously the 220, $\bar{2}20$ and 040 reflections. The square array of screw dislocations is disrupted by the 60° dislocation lines (pointed by the arrows). Note the perfect grating of such spaced dislocations.

a) $\theta < 1°$

For such angles, the dislocation spacing is still large enough ($Dd > 220$Å) so that the entire network can be easily imaged in multibeam conditions. Figure 2 is a typical multibeam plan-view image from one of these low-angle boundaries. The corresponding twist angle was found from the moiré spacing to be $0.58° \pm 0.03°$. The separation distance between dislocations can be directly measured from this image. It is measured to be Dd=380Å. No correction needs to be carried out since dislocations of the same set possess the same Burgers vector.

b) $1°<\theta < 7°$

As the misorientation is increased, the grain boundary dislocation spacing becomes smaller and the network becomes more regular because of the attractive forces tending to hold them together in an uniform array. In this case, the structure cannot be observed in multibeam nor in two beam conditions because of the strong contrast arising from the moiré patterns. This moiré contrast may also induce confusion as its periodicity is half the separation distance between dislocations. Thus, the best way to determine the structure is to image it using the weak-beam technique because moiré contrasts rapidly fade as the deviation from the exact Bragg condition increases. Figures 3.a & b are "weak-beam" images of the sample shown in fig. 1: each of them shows one set of screw dislocations, the other one being out-of-contrast. The dislocations appear in the images as continous and straight white lines. Their spacing is measured to be equal to 46 Å. This is in good agreement with a twist misorientation angle of 4.8° (see 3.1). The interfacial stress is minimized by this two sets of screw dislocations.

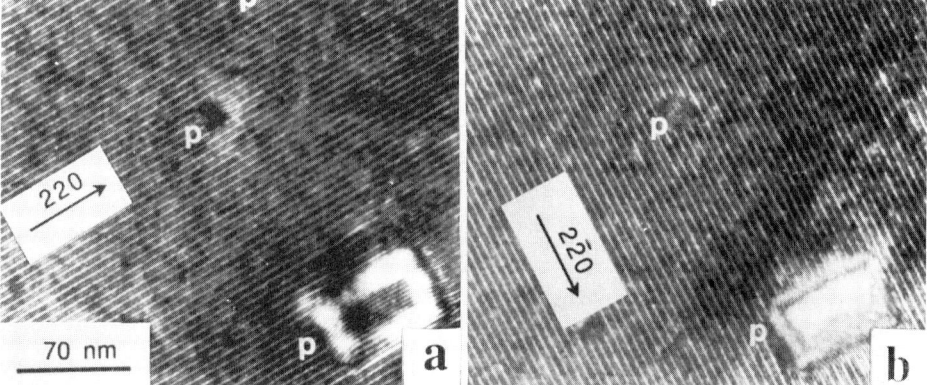

figure 3: Weak-Beam plan-view TEM images showing each of them one set of screw dislocations.

It is known that planar periodic grain boundary dislocation networks examined by electron diffraction give rise to specific reflections. These extra-spots are arranged in a array whose characteristics are directly connected to the GBD network. For a twist boundary consisting of a square grid of screw dislocation with spacing Dd examined under normal incidence, the diffraction pattern consists of a square array of spots spaced 1/Dd centered around the two cristal reflections and in the same orientation as the square GBD's grid (Guan and Sass, 1973). This square grid of extra reflections is shown in fig 1.d.

c) $\theta > 7°$

The analysis of the diffraction pattern becomes a precious tool when the GBD structure cannot be determined from the image. It usually happens for dislocations rendered invisible because too closely spaced: when the corresponding misorientation angle is above 7° (Carter et al, 1981). In this case, it is easier to detect the periodic network by the presence of the array of the screw dislocation reflections though their intensity was shown to decrease when decreasing Dd. Thus, a misorientation angle of 13.5° leads to the formation of a screw dislocation array detectable from the diffraction pattern. This diffraction pattern is shown in fig 4. The corresponding weak spots midway from the 220 reflections are the only proof of the GBD existence. Note that this sample corresponds to the maximum misorientation angle used in this work and we still do not know wether angles larger can also produce the same type of GBD network.

106

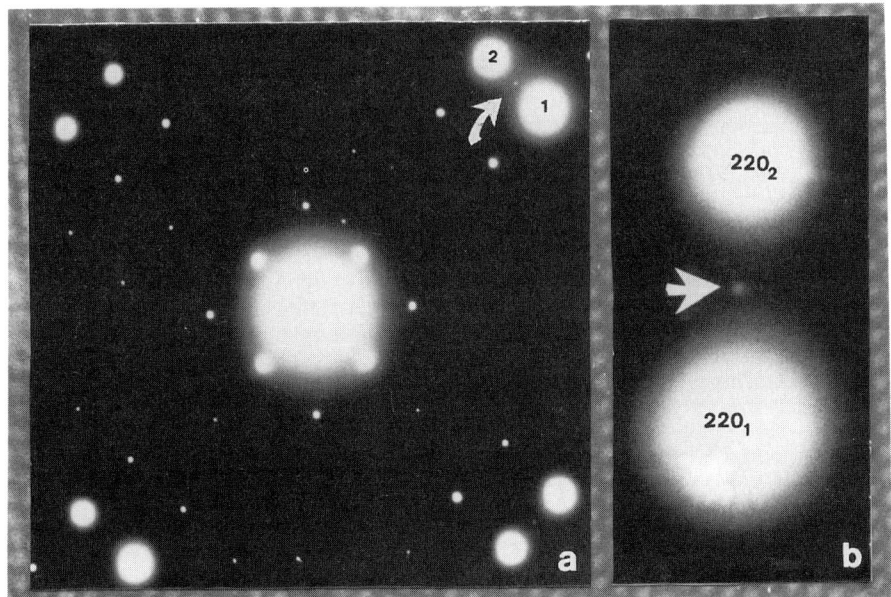

figure 4: a) diffraction pattern of the [001] zone axis. **b)** enlargement of the image of the arrowed region in a). The weak spots midway from the 220 reflections are the only proofs of the GBD existence.

3.3) GBD accommodating low-angle tilt components

The interfacial sructures always exibited an uncontrolled small tilt component. This tilt component (<0.5°) corresponds to small rotations around ramdom axes of the boundary plane. As a result, the interface consists of flat (001) terraces separated by steps lying along the direction perpendicular to the tilt direction. These steps are formed by the addition of families of 60° dislocation lines periodically located at the interface and   all have the same component perpendicular to the interface, a/2=2,7 Å (M.Benamara et al. 1994). They are clearly observed in all samples even when the screw dislocation network cannot be detected when too closely spaced. Their spacings do not depend on the screw dislocation network. In the sample corresponding to fig 1, they are the dark lines 400 nm   spaced. This allows us to precisely determine the tilt angle from which they originate to be 0.04°.

4. SUMMARY AND CONCLUSION

The general structure of these interfaces is that of a perfect grain boundary and evidently depends on the misorientation between the two bonded wafers. A twist component $0<\theta<13.5°$ creates a square network of pure screw dislocation whereas an uncontrolled tilt component (<0.5°) is compensated by a periodic array of 60° dislocation lines perpendicular to the tilt direction. All these dislocations are located at the interface and allow the structure to be fully relaxed with no extended volume defects.

Benamara M, Rocher A, Laânab L, Claverie A, Laporte A, Sarrabayrouse G, Lescouzères L
    and PeyreLavigne A 1994 CRAS Paris 318 (2) 1459
Bengtsson S 1992 J. of Electronic Materials 21 (8) 841
Carter C B, Föll H, Ast D G and Sass S L 1981 Phil. Mag. A 43 (5) 441
Föll H and Ast D 1979 Phil. Mag. A 40 (5) 589.
Gafiteanu R, Chevacharoenkul S, Gösele U M, Tan T Y 1993 Microscopy of Semiconducting
    Materials 134 87
Guan D Y and Sass S L 1973 Phil. Mag A 27 1211

Inst. Phys. Conf. Ser. No 146
Paper presented at Microsc. Semicond. Mater. Conf., Oxford, 20–23 March 1995
© 1995 IOP Publishing Ltd

EBIC studies of grain boundaries

B Raza and D B Holt

Department of Materials, Imperial College, London SW7 2BP

Abstract: Electrical property analysis by EBIC (electron beam induced current) and structure determination by EBSP (electron backscattered pattern) methods were applied to grain boundaries (GBs) in polycrystalline Si solar cells and II-VI compounds. It was found that in Si solar cells large-angle, large-Σ GBs have high and temperature independent EBIC contrast while low-angle and especially coherent $\Sigma = 3$ twin boundaries have low contrast with a negative temperature dependence. The GBs in ZnSe exhibit strong trapped-charge REBIC contrast.

1. INTRODUCTION

Grain boundaries (GBs) are of importance in large-area, low-cost devices. Polycrystalline silicon solar cells are competitors for terrestrial solar power generation. Relatively low cost per unit area is balanced against reduced power conversion efficiency for polycrystalline as compared with monocrystalline cells. The lower efficiency is largely due to the presence of electrically active GBs. This paper reports the results of the use of EBIC to measure the interfacial recombination velocity of the GBs and its variation with temperature and the crystallographic character of the boundaries. There are a number of other EBIC techniques which give information on other electrical properties of grain boundaries (Holt 1994) and one of these, REBIC (remote EBIC) revealing the charge on grain boundaries in ZnSe, will be reported here.

2. EXPERIMENTAL

A JEOL JSM-840A SEM fitted with a Matelect ISM-5 and a Link EBSP (electron backscattered pattern) analysis system was used. The solar cells were made from Wacker SILSO polycrystalline Si and kindly supplied by Dr. T. Bruton of BP Solar International. The large-grained polycrystalline samples of II-VI materials were kindly suplied by Dr. K. Durose and the late Dr. G.J. Russell of the Physics Department of Durham University.

For quantitative EBIC the solar cells were mechanically polished to minimise the backscattering effects of surface roughness.

3. RESULTS

3.1 Grain Boundaries in Polycrystalline Si Solar Cells

The cells contained boundaries nearly normal to the p-n junction and the top surface of the cells. This was established by pairs of images: EBIC micrographs of the top and

108

secondary electron images of the lightly etched grain structure of the bottom surface in which the patterns of the grains corresponded accurately. The grain size was sufficiently large (Figure 1) to allow properties to be measured both at boundaries and inside grains. Quantitative EBIC line scan profiles were recorded across GBs and analysed using the phenomenological theory of dark contrast (Donolato 1983, 1985). This gives values for the interfacial recombination velocity of the GBs and the minority carrier diffusion lengths in the neighbouring grains.

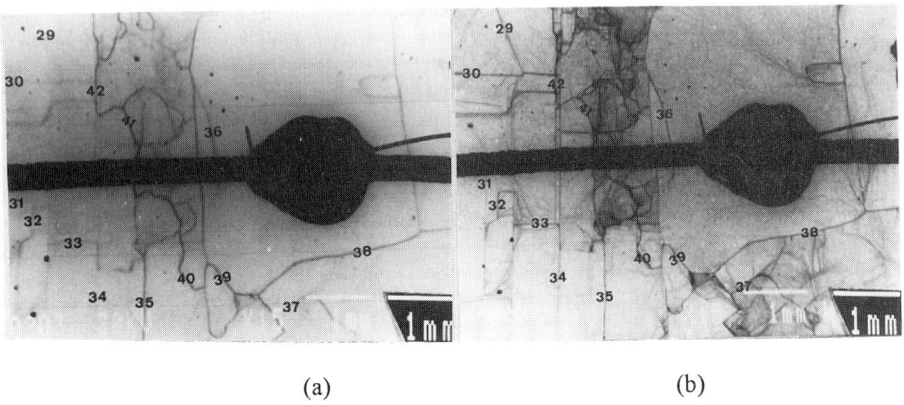

(a)   (b)

Figure 1. An area of an unpassivated polycrystalline Si solar cell shown in SEM EBIC micrographs at (a) room temperature and (b) liquid nitrogen temperature. In (b) in addition to the grain boundaries seen in (a), a large number of additional subgrain boundaries are in contrast.

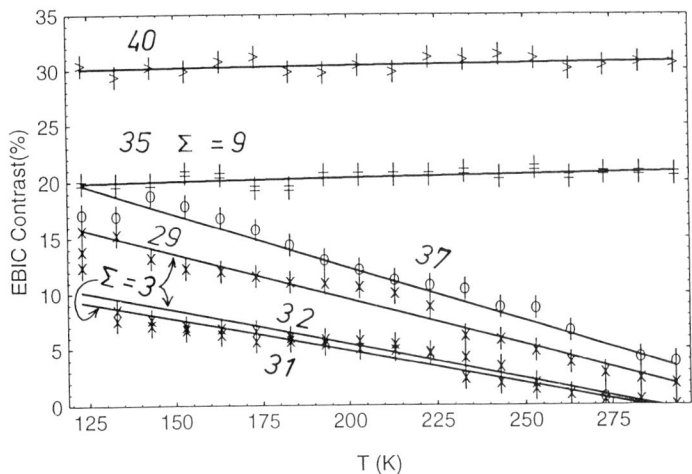

Figure 2. Variation of the EBIC contrast of six of the numbered grain boundaries in Figure 1. All the many boundaries studied had either high constant C vs T plots or low ones with negative slopes.

It was found that reducing the specimens from room temperature to liquid nitrogen temperature made many additional GBs visible in EBIC (Figure 1) as also found by Cheng (1985). Measurements were made of the variation of the EBIC contrast of large numbers of grain boundaries. The results of Table 1 and Figure 2 are typical. All the boundaries fell into two classes of EBIC contrast behaviour: (i) those having a strong contrast that was constant over the range (visible in Figure 1a) and (ii) those with a low contrast falling toward invisibility at room temperature (visible only in Figure 1b). EBSPs were obtained for the grains on either side of a large number of boundaries for which the temperature dependence of the EBIC contrast had been measured. From the EBSPs the orientations of both grains are obtained as well as the rotation axis and misorientation angle characterizing the grain boundary. By consulting tables, in many cases the Friedel index, Σ, of the boundary could also be found.

Table 1. Crystallography and Temperature Dependence of the EBIC Contrast of Numbered Grain Boundaries in Figure 1.

GB no.	Geometry	of	the	Grain	Boundary	T Dependence
	[hkl]	θ°		Σ	Description	EBIC Contrast change from RT to liq. N_2 T
29	[111]	60		3	first order twin	increase
30	[111]	60		3	"	"
31	[1$\bar{1}$1]	59.4		3	"	"
32	[111]	60		3	"	"
33	[1$\bar{1}$1]	59.7		3	"	"
34	[$\bar{1}$ $\bar{1}$ 1]	59		3	"	"
36	[$\overline{44}$1]	2		-	small angle	"
37	[$\bar{1}$ $\bar{1}$ 0]	1		-	"	"
35	[101	42.5		9	2nd order twin	no change
38	[$\bar{2}$ $\bar{3}$ 0]	40		-	large angle	"
39	[221]	43		-	"	"
40	[211]	45		21	"	"
41	[231]	40		-	"	"
42	[100]	28.7		17	"	"

It can be seen in Figure 2 that (i) the coherent first order twin ($\Sigma = 3$) boundaries (nos. 29, 31, 32)have the lowest contrast at all temperatures and a negative temperature coefficient, (ii) the small angle boundary (no. 37) has higher values of contrast with a negative coefficient while the second order twin ($\Sigma = 9$) boundary (no 35) and a large angle boundary (no. 40) have EBIC contrast that is larger and temperature independent.

3.2 Grain Boundaries in ZnSe

Grain boundaries that are electrostatically charged due to trapped carriers will produce band bending and charge collecting electrical fields in the depletion region on either side. When an electron beam sweeps across such a barrier the charge collection in opposite directions on either side will result in EBIC signals of opposite sign which will appear in micrographs as bright/dark contrast (see e.g. Holt 1994). Such bright/dark GB contrast was found in early OBIC (optical beam induced current) linescans across GBs in Ge (Figielski 1960) and seen in SEM REBIC (remote EBIC) micrographs in Ge (Matare and Laakso 1969), GaP (Ziegler et al 1982) and ZnSe (Russell et al 1980).

Direct evidence that the contrast observed at grain boundaries in ZnSe is due to GB charge and band bending is shown by the bright/dark form of EBIC line scan profiles recorded across boundaries at high magnification (Figure 3).

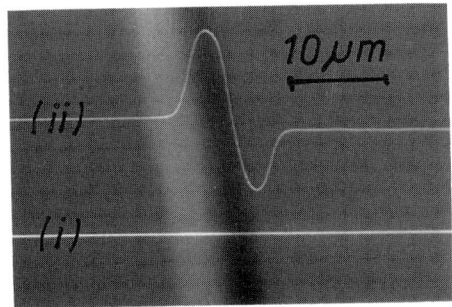

Figure 3. REBIC micrograph of a grain boundary in ZnSe plus (i) the superimposed line scanned during the recording of (ii) the EBIC contrast profile also superimposed.

Low magnification, room temperature REBIC pictures of polycrystalline ZnSe showed strong grain boundary contrast and weak contrast at some of the numerous straight twin interfaces. Observations on boundaries that are joined to twins (Figure 4) show that where the joins occur and, therefore, the crystallographic character of the boundary changes the contrast also changes.

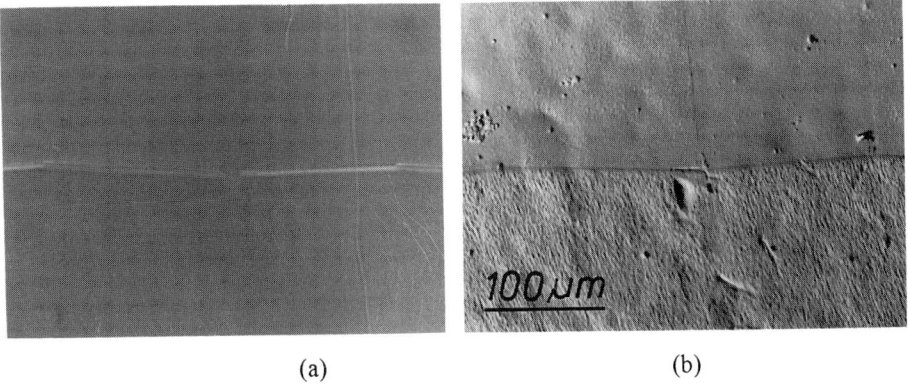

(a)   (b)

Figure 4. (a) REBIC and (b) secondary electron images of a nearly horizontal grain boundary intersected by a number of parallel twin boundaries coming in from above.

4. DISCUSSION

The results on the variation of contrast and its temperature dependence (Table 1) with GB type show that temperature independent EBIC contast was exhibited by large angle, large Σ boundaries and weak, temperature dependent contrast by small angle, small Σ (especially $\Sigma = 3$ i.e. first order coherent twin) boundaries. There are a number of reports that twin boundaries are electrically inactive (Queisser 1963) or have electrical effects depending on dislocation content (Cunningham et al 1982). Our results on both Si and ZnSe show that coherent first order twin interfaces have small but non-zero electrical effects.

The temperature dependence of contrast can best be explained using concepts borrowed from the Wilshaw physical theory of the EBIC contrast of dislocations. Wilshaw's model of charge controlled recombination shows that two dislocations (or grain boundaries) with the same recombination centre energy level can exhibit zero ($dC/dT = 0$) or negative ($dC/dT < 0$) temperature coefficients of EBIC contrast over some range of temperatures. However, on this model the $dC/dT = 0$ boundaries must have the smaller values of contrast and recombination centre density while the $dC/dT < 0$ boundaries have the larger values. But the zero coefficient boundaries are the large angle ones with the larger contrast and, it is believed, larger recombination centre densities, whether the latter centres are intrisic e.g. dangling bond centres or extrinsic, impurity decorated ones. Thus on charge-controlled recombination theory the difference between the 0 and -ve temperature dependence boundaries is that their recombination centres have energy levels E_{d1} and E_{d2}, at different depths in the band gap. The different behaviour of the two types of boundary is then due to the movement of E_F from room temperature (0.195 eV above E_V) to liquid nitrogen temperature (0.070 eV above E_V) taking it through the E_{d2} (shallower, small-angle, small Σ) levels but not through the E_{d1} (deeper, large-angle, large Σ) ones. (These values for the position of E_F are those for the $N_A = 4.5 \times 10^{15}$ cm^{-3} p-type material of most of the thickness of the solar cells.) Thus the charge on the $dC/dT < 0$ boundaries changes with T, changing the charge-controlled recombination, whereas the charge on the $dC/dT = 0$ boundaries is unchanged over this range.

The dark contrast of GBs in Si solar cells is due to enhanced minority carrier recombination by centres associated with the GB. This effect is exhibited (e.g. in Figure 2) in experiments in which the carriers are collected in a direction parallel to the GB by a p-n junction. The brignt/dark contrast effect of Figures 4 and 5 occurs in experiments in which charge collection is due to the GB and not to an independent junction and the collection is in a direction perpendicular to the GB. The two effects reveal different properties of GBs. There are still other experimental situations in which still other properties such as enhanced conduction along GBs can be seen in EBIC experiments (Palm et al 1994, Holt 1994).

Bright/dark contrast profiles like that in Figure 3 can be simulated by a computer program which models the grain boundary as two Schottky barriers back to back (Holt 1994). The fitting parameters are the widths of the depletion (field) regions and the minority carrier diffusion lengths. Thus curve fitting to experimental data can be used to evaluate these quantitites. Evidence that GBs in Si can also trap charge to produce potential barriers was presented by Stutzler and Queisser (1986) and Palm et al (1994).

ACKNOWLEDGEMENTS

We would like to thank Dr. E. Napchan and Dr. M. Mazzer for help with programs for simulation and the analysis of EBIC linescans using the Donolato (1985) theory.

112

REFERENCES

Cheng L J 1985 in Proc. 13th Internat. Conf. Defects in Semicond., eds L C Kimerling and J M Parsey (Warrendale, Penn.: Met. Soc. AIME) pp. 403 - 407

Cunningham B Strunk H P and Ast D G 1982 in Mat. Res. Soc. Symp. Proc. Vol. **5,** eds H J Leamy, G E Pike and C H Seager (New York: Elsevier) pp. 51 - 56

Donolato C 1983 J. Appl. Phys. 54, 1314-1322

Donolato C 1985 in Polycrystalline Semiconductors, ed G Harbecke (Berlin: Springer-Verlag) pp. 138 - 154

Figielski T 1960 Acta Phys. Polonica 19, 607 - 630

Holt D B 1994 in Polycrystalline Semiconductors III. Sol. State Phenomena Vols. 37 - 38 (Zurich: Scitech Publishers) pp. 171 - 182

Matare H F and Laakso C W 1969 J. Appl. Phys. 40, 476 - 482

Palm J Steinbach D and Alexander A 1994 in Polycrystalline Semiconductors III. Sol. State Phenomena Vols. 37 - 38 (Zurich: Scitech Publishers) pp. 183 - 188

Raza B and Holt D B 1991 in Polycrystalline Semiconductors II. Proc. in Physics 54 eds J H Werner and H P Strunk (Berlin: Springer-Verlag) pp. 72 -76

Russell G J Waite P and Woods J 1981 in Microscopy of Semiconducting Materials 1981. Conf. Series No. 60 eds. A G Cullis and D C Joy (Bristol: Inst. Phys) pp. 371-376

Russell G J Robertson M J Vincent B and Woods J 1980 J. Mater. Sci. 15, 939-944

Stutzler F J and Queisser H J 1986 J. Apl. Phys. 60, 3910 - 3915

Ziegler E Siegel W Blumtritt H and Breitenstein O 1982 Phys. Stat. Sol. a 72, 593-604

Inst. Phys. Conf. Ser. No 146
Paper presented at Microsc. Semicond. Mater. Conf., Oxford, 20–23 March 1995
© *1995 IOP Publishing Ltd*

Electron beam induced current studies of a Ni- and Fe-contaminated Σ=25 silicon bicrystal

A Ihlal*, R Rizk, P Voivenel and G Nouet

LERMAT, URA CNRS 1317, ISMRA, 6 Bd du Maréchal Juin, F-14050 Caen cedex, France
*Permanent address: Département de Physique, Faculté des Sciences, B.P. 28/S, Agadir, Morocco

ABSTRACT: A scanning electron microscope operating in the induced current mode has been used to study the recombination activity of Ni and Fe in a Σ=25 silicon bicrystal. With the fast diffuser Ni, the silicides ($NiSi_2$) have been formed upon rapid cooling which have been found to be at the origin of both the electrical activity in the bulk and the enhanced EBIC contrast at the boundary. By contrast, the moderately diffusing Fe precipitated only after slow cooling whereas the quenching resulted in the "freezing" of iron in interstitial sites (Fe_i) acting as recombining centers that led to the decrease of the EBIC signal.

1. INTRODUCTION

The problems connected with the electrical activity induced by transition metals (TMs) in silicon have been the subject of great interest (Weber 1983). The effects of these contaminants in silicon, whether confined as point defects in the matrix or precipitated at the grain boundaries (GBs) of a polycrystal, for instance, depend strongly on the diffusivity and solubility of the elements. Beside the numerous studies devoted by many authors (Maurice and Colliex 1989, Broniatowski and Haut 1990) and recently by us (Rizk et al 1994, 1995) to the impact of the easy contaminants (Cu and Ni) on the electronic behaviour of a typical GB, such as a Σ=25 silicon bicrystal, there are very few or no data concerning Fe, which is considered an efficient contaminant. In this paper we report a comparative study using a scanning electron microscope (SEM) operating in the induced current mode, of the electrical activity of both Ni and Fe, i.e. fast and moderate diffuser TMs, respectively, deliberately incorporated in a Σ=25 silicon bicrystal.

2. EXPERIMENTAL

The samples used in this study were cut from low-doped (N_d=3.1×10^{14} P/cm^3) Σ=25 silicon bicrystals characterized by a {710} twin plane, a [001] tilt axis and a 16.26° tilt angle. They were electrochemically covered by nickel or iron before being annealed in a double wall quartz tube furnace at 1100-1200°C for 3 hours under a continous flow of pure argon. The samples were subsequently subjected either to fast cooling (50°C.s^{-1}) by their rapid withdrawal from the furnace or to slow cooling (0.5°C.s^{-1}) by simply switching off the furnace.

For electron beam induced current (EBIC) measurements, three diodes were fabricated on each sample, one at the GB and one on each of the adjacent grains, by deposition of gold on the mirror-like polished and chemically etched surface. The EBIC experiments have been performed in a SEM operating with an electron beam intensity of 1nA and an energy as high as 35 keV which enabled us to explore the neutral zone of the semiconductor.

3. RESULTS AND DISCUSSION

For the as-grown and heat-treated samples, no defect has been detected in the bulk grains, although the GB showed either a weak (1%) or a noticeable (40%) contrast (not shown) for the slowly cooled or quenched specimens, respectively. The electrical activity recorded at the GB in the rapidly cooled sample has been assigned recently by us (Ihlal et al 1995) to native impurities and/or oxide precipitates.

Concerning the Ni-contaminated sample, while the slow cooling did not lead to a significant increase in the EBIC signal in any part of the material, the EBIC image obtained after quenching [Fig. 1(a)] reveals an intense activity in both the bulk and the GB. The EBIC profile of Fig. 1(b) indicates a contrast attaining a value of ~ 40% on some precipitates in the bulk, in good agreement with that reported by Kittler et al (1991), and a value exceeding 50% at the GB.

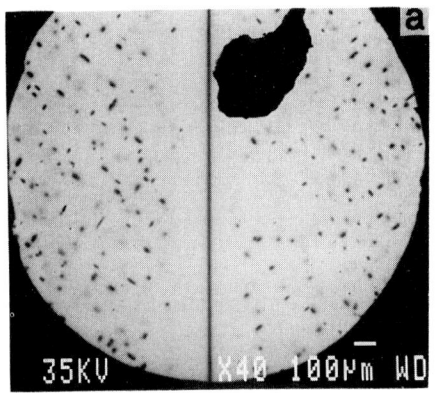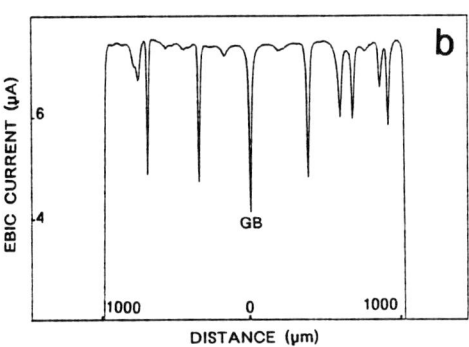

Fig. 1 EBIC image (a) and EBIC scan profile across the grain boudary (b) of a Ni-contaminated silicon bicrystal (Σ=25) that is quenched after annealing at 1100°C.

Owing to the transmission electron microscopy (TEM) observations and energy dispersive x-ray (EDX) analyses, which were reported in our lately published works (Rizk et al 1994,1995), the Ni contaminants in the quenched bicrystals have been found to precipitate in $NiSi_2$ plates growing on Si {111} planes. At the same time, no defect or precipitate has been observed either in the bulk or at the GB in the corresponding slow cooled sample. Thus, whether it be the electrical activity detected in the matrix or the substantial increase in the EBIC contrast noticed at the boundary (Fig. 1), they seem both to be due to $NiSi_2$ silicides. Fig. 1 reveals also the existence on both sides of the GB of a denuded zone (DZ), free from any defect or detectable recombination, extending to ~150µm on each side of the GB. The occurrence of such a DZ is originating from the well-known gettering effect of the boundaries (Schröter et al 1991), but its extent can be correlated with the efficiency of this gettering as well as with the high diffusivity of Ni.

Figs. 2 and 3 show the EBIC images and the superimposed scan profiles of the slowly cooled and quenched bicrystals precontaminated with Fe, respectively. Unlike the case of other similarly treated samples (as-received or Ni-contaminated) the contrast at the GB in Fig. 2 approaches 8% instead of 1%. One can notice also the appearance of a narrow (~50µm) and a slightly perceptible DZ in the GB neighbourhood. The contrast increase can be assigned straightforwardly to the formation of iron silicides (ε-FeSi and α-FeSi$_2$) in the GB near-region upon slow cooling, as evidenced in the work of Portier et al (this conference). These observations are in contrast with those reported above for nickel or elsewhere for copper (Rizk et al 1994) where the precipitation is obtained only after rapid cooling. This discrimination may be easily explained by the differences in the diffusion coefficients between both elements, approximately one order of magnitude (Weber 1983). Thus, the highly mobile Ni would require a relatively high cooling rate (say, 50°C. s^{-1}) to precipitate, whereas the lower cooling rate (0.5°C.s^{-1}) could let it   finally segregate at or near the surface. In this connection, the behaviour of iron appears quite consistent with its moderate diffusivity which leads it to either precipitate after slow cooling, as reported above, or to "freeze" in interstitial sites upon rapid cooling, as stated below.

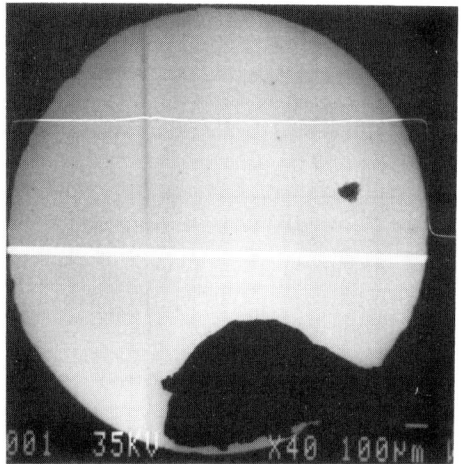 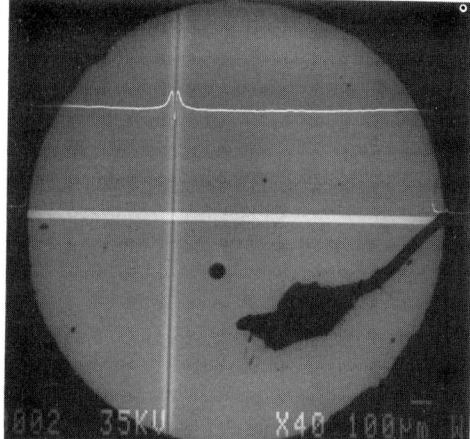

Fig. 2 EBIC image and scan profile of the slowly cooled sample precontaminated with Fe.

Fig. 3 EBIC image and scan profile of the rapidly cooled sample containing Fe.

For the case of rapid cooling, Fig. 3 reveals a relatively marked DZ on both sides of the GB extending to ~50μm, as the preceding one, as well as a GB-related contrast which appears less important than its counterpart in the similarly treated but nonintentionally contaminated sample. From the comparison between these profiles in Fig. 4(a), it can be inferred that this "apparent" reduction in the GB contrast is due to an overall decrease of the EBIC signal in the bulk originating from the "freezing" of the moderate   diffuser iron in interstitial sites (Fe$_i$) in all the sample during the rapid cooling process. Considering that Fe$_i$ atoms induce a deep donor level in the bandgap at E$_v$ + 0.4 eV (Feichtinger et al 1978), it is quite conceivable that these defects act as recombining centers leading to the above-mentioned reduction in EBIC. This feature is totally absent for the slow cooled sample where the level of the EBIC signal in the bulk [curve 1 in Fig. 4(b)] remains unchanged, in comparison with the corresponding profile recorded after quenching [curve 2 in Fig. 4(b)]. Furthermore, the EBIC signals recorded at the GB level in both curves 1 and 2 [Fig.4(a)] seem to indicate that the native impurities remain the main origin of the activity of these GBs. According to Portier et al (this conference) the quenching would lead simply to a segregation of Fe at the GB resulting from the "clearing" of the adjacent regions from Fe$_i$ defects which gives rise to a relatively pronounced DZ. The ratio of the DZ widths for Ni and Fe cases (150μm/50μm ~3) is quite comparable to that of the square roots of their diffusivities [(D$_{Ni}$/D$_{Fe}$)$^{1/2}$ ~(10)$^{1/2}$ ~3].

In order to support the aforementioned suggestion relating to the recombining activity of Fe$_i$ atoms, we have proceeded to the reannealing and subsequent slow cooling of the previously quenched sample precontaminated with Fe. Such a reanneal is expected to dissolve

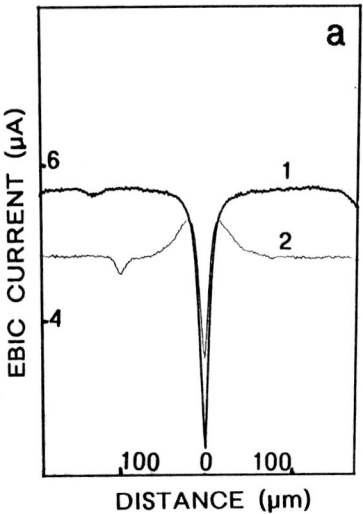
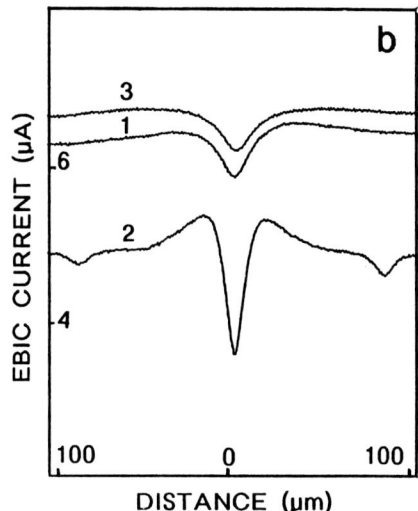

Fig. 4 (a) EBIC scan profiles across the GB of an as-grown (curve 1) and Fe-contaminated (curve 2) bicrystals recorded after quenching from 1200°C. (b) EBIC scan profiles for the slowly (curve 1) and rapidly (curve 2) cooled samples precontaminated with Fe; curve 3 has been recorded after reannealing at 1200°C and subsequent slow cooling of the sample corresponding to curve 2.

the interstitial iron atoms (Fe_i) which might precipitate at the GB during the subsequent slow cooling, as already observed upon such a heat cycle. The concomitant segregation of Fe near the surface during the same thermal process can also be considered. Indeed, the EBIC profile taken after such a heat treatment and displayed in Fig. 4(b) (curve 3) clearly shows that, apart from an understandable improvement of the overall signal due to the just considered gettering effect of the surface, it recovers almost identically the shape of that initially recorded for the primarily slow cooled sample containing Fe. This finding is indicative of the recombining ability of Fe_i defects in the quenched sample which disappears with the precipitation of iron at the GB after the above-mentioned additional heat treatment. This latter seems to dissolve also the native impurities segregating at the GB that are thought to be responsible for the noticeable contrast at the boundary level after quenching, since the contrasts at the GB in both curves 1 and 3 are quite comparable.

REFERENCES

Broniatowski A and Haut C 1990 Phil. Mag. Lett. 62 407
Feichtinger H, Waltl J and Gschwandtner A 1978 Solid State Commun. 27 867
Ihlal A, Rizk R, Voivenel P and Nouet G 1995 J. Phys. III, in press
Kittler M, Lärz J, Seifert W, Seibt M and Schröter W 1991 Appl. Phys. Lett. 58 911
Maurice J L and Colliex C 1989 Appl. Phys. Lett. 55 241
Portier X, Rizk R, Nouet G and Allais G This conference
Rizk R, Portier X, Allais G and Nouet G 1994 J. Appl. Phys. 76 952
Rizk R, Ihlal A and Portier X 1995 J. Appl. Phys. 77 1875
Schröter W, Seibt M and Gilles D 1991 Mater. Sci. Technol. 4 539
Weber E 1983 Appl. Phys. A 30 1

Inst. Phys. Conf. Ser. No 146
Paper presented at Microsc. Semicond. Mater. Conf., Oxford, 20–23 March 1995
© *1995 IOP Publishing Ltd*

The introduction of dislocations in low misfit epitaxial systems

D D Perovic and D C Houghton†

Department of Metallurgy and Materials Science, Univ. of Toronto, Toronto M5S 3E4 Canada
† Institute for Microstructural Sciences, National Research Council, Ottawa K1A 0R6 Canada

ABSTRACT: Macroscopic and microscopic analyses have been combined to give a comprehensive description of the misfit dislocation generation process in the early stages of strain relief. A generalized model has been developed to explain the relaxation behaviour in all stages of strained layer epitaxy and provides a consistent description of various results reported in the literature.

1. INTRODUCTION

The structural stability and the kinetics of elastic strain relaxation in lattice mismatched semiconductor heterostructures continues to receive significant attention by many research groups worldwide. It has been frequently demonstrated that coherently strained epitaxial layers thicker than the equilibrium critical thickness can be obtained by low temperature growth processes such as molecular beam epitaxy (MBE) and chemical vapour deposition (CVD). However, the unique optical and transport properties associated with band structure differences arising from elastic strain accommodation across heteroepitaxial interfaces, may only be realized if structural integrity and low defect densities are maintained during epitaxial growth and subsequent thermal processing treatments for device application. In view of the above there are a number of questions that continue to motivate intense research activity in this area. Is the equilibrium critical thickness theory correct? What is the limiting factor in strain relaxation (nucleation or growth)? Where do misfit dislocations originate? Can strain relaxation be controlled?

In order to answer the above questions we have carried out extensive experimental studies on a wide range of low misfit Ge_xSi_{1-x}/Si ($0.03 < x < 0.25$) heterostructures, grown by MBE and CVD techniques. By combining both macroscopic and microscopic analyses we have developed a comprehensive semi-empirical predictive model to describe the early stages of strain relaxation. Moreover, this has led to a novel description of the dislocation nucleation and propagation process which has fundamental implications for strained layer epitaxial growth in general.

2. EQUILIBRIUM CRITICAL THICKNESS

Following the seminal paper by Frank and van der Merwe (1949) there have been several theoretical treatments developed, based on either energy or force balance considerations, to predict the critical thickness at which it becomes favourable to accommodate elastic misfit strain by the introduction of interfacial misfit dislocations. If properly formulated, the force and energy balance treatments are equivalent; (see Freund 1994, Jain 1994 for recent reviews). Nevertheless, the force-balance model first introduced by Matthews *et al.* (1970) has proved to be the most popular in view of its simplicity. Moreover, unlike the energy minimization approaches, the treatment by Matthews and co-workers considers a mechanism for the generation of misfit dislocations, provided a *sufficient* density of pre-existing sources (e.g.

118

substrate threading dislocations) are available. In this way one can relate the equilibrium critical thickness to the *first* misfit dislocation that is introduced at a misfitting interface. Recent work has extended the Matthews critical thickness theory to include dislocation interaction effects explicitly for periodic and non-periodic misfit arrays which considers the *last* (i.e. total) misfit dislocation required to stabilize a strained epitaxial layer (Freund 1994, Jain 1994). Although the inclusion of dislocation interaction effects lowers the critical thickness relative to the simple Matthews result, the difference is almost negligible (e.g. a few Å for 60° dislocations). Therefore, using simple isotropic continuum elasticity, a well-defined theoretical critical thickness exists to describe both the initial and final state of strain relaxation under *equilibrium* conditions.

Alternatively, many experimental studies of strained layer critical thickness have been reported which do not lead to a consistent result; the numbers of such papers are far too numerous to quote here (for a complete review see Perovic 1995). Fig. 1 is a typical example taken from the GeSi/Si(100) system. The lowest curve corresponds to the equilibrium critical thickness as defined above, however, most of the experimental data is displaced to greater thicknesses. Accordingly there have been many reports to date claiming that experimental data correspond to a *different* critical thickness. People and Bean (1985) first attempted to describe the larger "observed" critical thicknesses in GeSi/Si using an energy balance model that considered spontaneous homogeneous interfacial nucleation of misfit dislocations in the absence of pre-existing dislocations. The model defines an *areal energy density* for an isolated dislocation which is physically invalid. Nevertheless, this model continues to be used in the modelling of experimental critical thicknesses that apparently exceed equilibrium predictions.

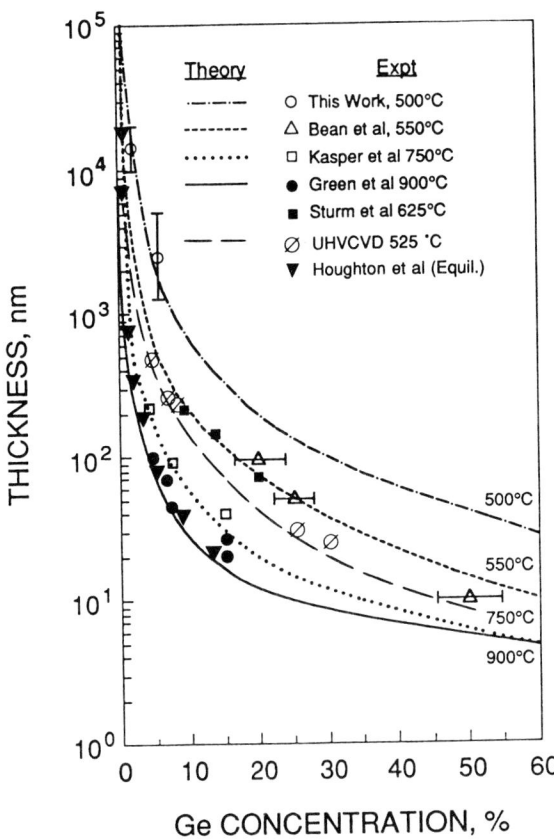

Fig. 1: Kinetically limited "critical" thickness for Ge_xSi_{1-x}/Si (100) growth at various temperatures. The extent of metastability beyond the equilibrium critical thickness (solid line) is simulated by integration of Eq. (1) for: MBE growth at 500, 550 and 750 °C (growth rate ~0.5 nm/s), RTCVD growth at 625 °C (0.1 nm/s) and 900 °C (1.5 nm/sec) and UHVCVD growth at 525 °C (0.01 nm/s). Experimental data for MBE and RTCVD growth is included from: J.C. Bean *et al.*, 1986 J. Vac. Sci. Technol. A, 2, 436; E. Kasper and H-J Herzog, 1975, Thin Solid Films, 44, 357; M.L. Green *et al.*, 1991, J. Appl. Phys., 69, 745; J.C. Sturm and D.C. Houghton, (unpublished); Houghton *et al.* (1990).

Two simple factors can explain the above discrepancy. Firstly, the precise point at which plastic deformation begins is affected by kinetic (i.e. Peierls) barriers that limit the extent of strain relaxation in metastable structures for a given temperature and time. Fig. 1 shows that the *apparent* critical thickness increases at lower growth temperatures. Exceeding the equilibrium critical thickness is only a *necessary* condition but not *sufficient* to predict observable strain relaxation. Therefore, it is not possible to compare data from metastable structures with equilibrium theory. In the case of metallic systems, the Peierls barriers are relatively small such that kinetic effects do not restrict the relaxation process and experimentally measured critical thicknesses can be predicted from equilibrium theory (Kuk *et al.* 1983, Fox and Jesser 1990)

The second factor that has led to much confusion in the literature concerns the experimental sensitivity for detecting the onset of strain relaxation. Analytical techniques frequently used to study dislocation-induced strain relief (e.g. x-ray diffraction, TEM, ion channelling, Raman spectroscopy, photoluminescence spectroscopy, laser light scattering, reflection high-energy electron diffraction and wafer curvature/scanning laser reflection) are not able to detect $< 10^4$-10^5 dislocations/cm^2. Accordingly the *apparent* critical thicknesses measured using these techniques are resolution limited by experiment. Sensitive techniques that can reveal individual dislocation segments over a large area (e.g. chemical etching/Nomarski microscopy, x-ray topography, photoluminescence microscopy and charge collection/cathodoluminescence SEM) will allow for the *first* misfit dislocation segment to be detected. In a previous study on low-misfit GeSi/Si heterostructures (Houghton *et al.* 1990) we showed that upon eliminating kinetic barriers via post-growth annealing, and detecting the first stages of misfit dislocation formation using sensitive techniques, the experimental critical thickness was in excellent agreement with equilibrium predictions.

3. NON-EQUILIBRIUM EFFECTS

In view of the discussion above it is clear that equilibrium stability theory is incapable of generally describing strain relaxation effects in metastable strained layer heterostructures grown at relatively low temperatures. Accordingly several non-equilibrium models have been developed in attempt to provide descriptions of misfit dislocation nucleation, propagation, multiplication and interaction in strained layer heterostructures. The fundamental basis for all these models comes from the classic work on plastic flow in bulk covalent materials by the groups of Alexander and Haasen (1968, 1986).

The first attempt to include dislocation dynamics in the modelling of heterostructure strain relaxation was given by Matthews *et al.* (1970). The kinetics of strain relief were included via a thermally activated friction force (i.e. Peierls stress) which opposed dislocation glide of pre-existing threading segments. Fox and Jesser (1990) extended the Matthews model to include a 'static' Peierls stress and point defect pinning. However, the exponential time dependence predicted by these models cannot properly describe the evolution of strain relaxation as observed experimentally.

Dodson and Tsao (1987) used Alexander and Haasen plastic flow theory to develop a phenomenological kinetic model which considered the thermally activated glide and multiplication of misfit dislocations by the *effective* (or excess) stress associated with lattice misfit strains exceeding the equilibrium limit (i.e. metastability). The Dodson and Tsao theory is appropriate for the later stages of strain relaxation when misfit dislocation interactions become important; an *ad hoc* "background" dislocation source density term was included to account for misfit dislocation nucleation in the early stages. Analogous to Dodson and Tsao, Nix *et al.* (1990) proposed a similar phenomenological model which considered the multiplication of *threading* dislocations using a "breeding factor" but did not include an effective stress dependence. Recently Jain (1994) has extended the Dodson and Tsao and Nix *et al.* models to correctly include dislocation interaction effects which enhances the predicted strain relaxation by up to a factor of two. In any case both phenomenological models employ adjustable fitting parameters which must be varied to explain various experimental results (Jain 1994) and thus cannot provide useful quantitative predictions. Moreover these models do not consider dislocation nucleation.

Hull *et al.* (1989) developed a phenomenological kinetic model for strain relaxation including the effects of misfit dislocation nucleation, propagation and interaction; multiplication effects were intentionally excluded. Misfit dislocation velocities and nucleation rates were measured from metastable GeSi/Si(100) heterostructures annealed *in situ* in the TEM, giving activation energy barriers of $Q_v = 1.1$ eV and $Q_n = 0.3$ eV for misfit dislocation glide and nucleation respectively. Dislocation interactions were included using a rudimentary calculation, which incorrectly defines the interaction rate. Recently Gosling *et al.* (1994) refined the approach by Hull *et al.* by correctly incorporating dislocation interaction effects for both single segments and dislocation clusters based on the ideas of dislocation blocking first developed by Freund (1994). Secondly they used the velocity expression obtained by Tuppen and Gibbings (1990) with $Q_v \approx 2.2$ eV. Thirdly the nucleation rate was described by a stress dependent Arrhenius-type relationship which was fitted to the experimental data of Hull *et al.* giving $Q_n = 0.3$ eV. Interestingly the refined theory of Gosling *et al.*, including a much larger Q_v, yields essentially the same strain relaxation predictions as the model of Hull *et al.* Moreover, the total activation energy for relaxation (Q_r), determined independently from misfit dislocation spacing measurements, was found to be $Q_r \approx 0.3$ eV! Accordingly *in situ* TEM measurements appear to be "nucleation-controlled" and hence not a true representation of strain relaxation in bulk material where glide-limited processes are known to be important. In view of the limited sensitivity for detection of misfit dislocations (> 10^5 cm^{-2}), it will be shown below that *in situ* TEM data do not represent the *early* stage of strain relaxation (hereafter referred to as Stage I) as suggested by Hull *et al.* but in fact represent dislocation interaction/multiplication effects in the later stages (Stage II). The aforementioned models illustrate the unreliability of predictive modelling in Stage II where the complex mechanistic processes of misfit dislocation nucleation, interaction and multiplication must be understood if sophisticated theory is to be developed.

4. MISFIT DISLOCATION GENERATION

It is evident from our review of existing kinetic strain relaxation models, all of which are appropriate only in the later stages (Stage II), that considering our current level of theoretical and experimental sophistication, it is difficult to accurately model Stage II strain relaxation in a predictive manner. Alternatively we concentrated on developing a semi-empirical predictive model, based on sensitive experimental techniques, to describe the *first* stages of strain relaxation (Stage I) where the misfit dislocation density varies from 0 to 10^5 cm^{-2} since it is in this crucial regime where the optical and transport properties are compromised in metastable strained layer device structures.

4.1 Quantitative Measurements

A comprehensive strain relaxation model has been developed (Houghton 1991, Perovic and Houghton 1993) based on *bulk* measurements of misfit dislocation nucleation and glide for a range of metastable (100) oriented Ge$_x$Si$_{1-x}$/Si (0.03< x <0.25) heterostructures grown by MBE, RTCVD and UHVCVD. The composition (x) and strained layer thicknesses (t) were determined by double-crystal x-ray diffraction and cross-sectional TEM. The various heterostructures provided a range of effective stresses (τ_{eff} = 0-750 MPa) for study. Following post-growth annealing (5-200 s; 450-900 °C) of metastable, coherently strained heterostructures (i.e. initially misfit dislocation-free) in an inert gas atmosphere, quantitative nucleation rate measurements from regions free from scratches or edge effects were obtained by scanning representative areas (several cm^2) of an etched wafer surface, using Nomarski interference microscopy. The number of new misfit dislocation segments generated *intrinsically* per unit area, per unit time, and *independently*, the misfit dislocation length per unit time (i.e. velocity), were measured for a given driving force (i.e. effective stress) and relaxation temperature. It is important to note that these measurements were made in the low misfit dislocation density regime (0 - 10^5 cm^{-2}) during the initial stage of elastic strain relief (Stage I) where dislocation interactions and possible multiplication effects can be ignored.

Ge$_x$Si$_{1-x}$/Si (100) misfit dislocation velocities (60° type) were found to obey the classic Alexander and Haasen Arrhenius-type relationship, with a glide activation energy of $Q_v = 2.25 \pm 0.05$ eV, a unique exponential prefactor (V_o) and a power-law dependence on effective stress ($v \propto \tau_{eff}^m$, $m=2$), for all heterostructure geometries (e.g. capped/uncapped single layers and superlattices) and misfit strains ($\varepsilon_o < 1\%$). A number of other studies have used various experimental methods to measure misfit dislocation velocities in the GeSi/Si system. Data from Tuppen and Gibbings (1990), Nix *et al.* (1990) and Yamashita *et al.* (1993) can all be accurately described by the same Q_v and τ_{eff} dependence as found in our work. The more recent *in situ* TEM measurements by Hull *et al.* (1991) resulted in a range of Q_v (1.55-2.08 eV) values depending on composition (x) and heterostructure geometry (capped vs. uncapped). Although the raw velocity data of Hull *et al.* can be described by a $(\tau_{eff})^2$ dependence, it is also possible to normalize all dislocation velocity data upon introducing an effective stress-dependence of Q_v, such that $1 < m < 2$, as found in bulk Si and Ge (Alexander 1986).

The nucleation of misfit dislocations has received much less attention and remains a poorly understood mechanism in strain relaxation. Figure 2 compares the misfit dislocation nucleation rates measured from a representative set of (100)-oriented MBE-, RTCVD- and UHVCVD-grown Ge_xSi_{1-x}/Si heterostructures possessing a range of effective stress (τ_{eff}). A number of important features can be noted from our data. Firstly, the exponential evolution of misfit dislocation density ($N(t)$) as a function of temperature (T) reveals a remarkably consistent activation energy for nucleation ($Q_n = 2.5 \pm 0.5$ eV), *independent of glide* for all heterostructure types independent of growth technique; it should be noted that our first report of quantitative nucleation data (Perovic *et al.* 1990) suffered from incorrect temperature calibration resulting in an anomalously low value for Q_n. Secondly, the nucleation rate ($dN(t)/dt$) increases markedly with τ_{eff} through a power-law dependence ($n = 2.5$). Thirdly, the misfit dislocation nucleation rate varies approximately linearly with the *initial* misfit dislocation source density (N_o) present in the as-grown structure at $t = 0$ (i.e. substrate threading dislocations, residual substrate-buffer layer precipitates; see Perovic *et al.* 1989).

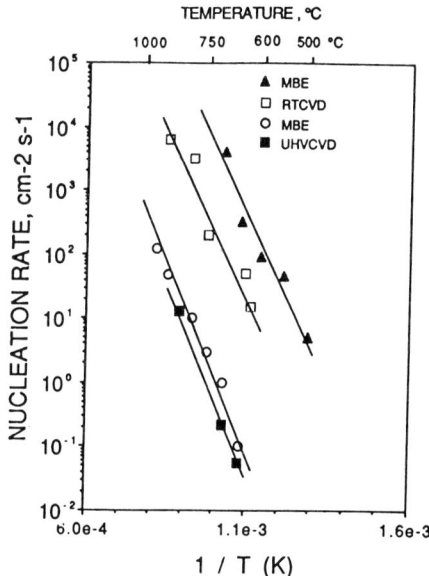

Fig. 2: Misfit dislocation nucleation rate data from post-growth annealing of Ge_xSi_{1-x}/Si (100) heterostructures grown by MBE, RTCVD and UHVCVD: (▲) 190 nm $Ge_{0.17}Si_{0.83}$ layer (τ_{eff} = 575 MPa), (□) 75 nm $Ge_{0.2}Si_{0.8}$ layer (τ_{eff} = 587 MPa), (○) 50 nm $Ge_{0.035}Si_{0.965}$ layer (τ_{eff} = 99 MPa), (■) 10-period (15 nm $Ge_{0.1}Si_{0.9}$/21 nm Si) multilayer/ (τ_{eff} = 113 MPa).

This effect can also be observed upon comparing heterostructures possessing identical effective stresses but grown by either MBE ($N_o \geq 10^3\,cm^{-2}$), RTCVD ($N_o \leq 10^2\,cm^{-2}$) or UHVCVD ($N_o \leq 10^2\,cm^{-2}$) wherein the magnitude of N_o varies due to different substrate cleaning procedures prior to growth. Most importantly, the linear increase in misfit dislocation density with time proceeds to densities well beyond N_o indicating the thermal activation of *new* sources in the early stages.

Relatively few other experiments have been carried out to quantify misfit dislocation nucleation energies. Firstly, the one earlier measurement of $Q_n = 0.3$ eV, obtained from *in situ* TEM studies by Hull *et al.* (1989), was about an order of magnitude lower than our bulk measurement. Secondly, recent work by LeGoues *et al.* (1993) compared the relaxation of Ge$_x$Si$_{1-x}$/Si strained layers grown by UHVCVD on vicinal substrates near <100> with either uniform or graded composition (*x*); (N.B. thus far all layers have been considered to be of uniform composition). The variation in τ_{eff} with substrate misorientation produces an imbalance in the number of dislocations nucleated on different slip systems giving rise to a strain induced tilt in relaxed layers as measured by x-ray diffraction. A phenomenological model was developed which allowed for the determination of the activation energy for 60° misfit dislocation nucleation. For uniform composition layers with 0.2%< ε_o < 0.9% and compositionally graded layers with 0.2%< ε_o < 1.9%, strain relaxation was said to be "nucleation-limited" with an activation energy of Q_n= 4.0 ± 0.5 eV. However, uniform composition layers with ε_o ~> 1% were found to be "glide limited" with Q_n<~ 0.3 eV. Thirdly, using the same technique as LeGoues *et al.*, Chen *et al.* (1995) have determined nucleation barriers of 1.5 eV and 1.4 eV for α and β-character 90° partial dislocations respectively, for the In$_{0.2}$Ga$_{0.8}$As/GaAs system. The significance of the various nucleation energies measured by other workers will be discussed in §6 as part of a generalized treatment of strain relaxation.

Returning to our measurements, the independent descriptions of misfit dislocation nucleation and glide during Stage I relaxation can be combined to determine the overall strain relaxation rate ($d\varepsilon/dt$). Using the classic Orowan plasticity equation ($d\varepsilon/dt = N(t)V(t)b_{eff}$), where b_{eff} is the effective Burgers vector magnitude of a 60° dislocation projected in the strained interface, we obtain (Houghton 1991):

$$\frac{d\varepsilon}{dt} = V_o B N_o b_{eff} t \left(\frac{\tau_{eff}}{\mu}\right)^{m+n} exp-\left(\frac{Q_n+Q_v}{kT}\right) \tag{1}$$

where all parameters are either material constants (μ is the shear modulus) or experimentally determined for the given materials system. It is important to note that independent measurements of *overall* strain relaxation (e.g. misfit dislocation spacing) in Stage I (Houghton 1991, Zhao 1993) revealed an activation energy for relaxation (Q_r = 4.2 ±0.5 eV) which is consistent with combining the nucleation and glide energies (i.e. $Q_r = Q_n + Q_v$) in series.

Upon integrating Eq. (1) the total strain relaxed for any thermal cycle (*T*, *t*), heterostructure geometry and growth technique is obtained from which the kinetically-limited "critical" thickness during growth is determined (see §2 and 3); here "critical" is a misnomer. We have tested the model against experimental results from other groups by simulating growth conditions upon converting growth rate to time and using suitable values for N_o. Where necessary, the integration of Eq. (1) was carried out using a finite integration limit, ε_{min}= 10^{-5}, to account for the minimum plastic strain detectable by various structural probes used to determine the onset of relaxation (see §2). Fig. 1 illustrates the excellent agreement obtained with data from various laboratories. It is evident that CVD-grown material is generally more stable compared to MBE material, primarily because of the higher N_o in the latter. For sufficiently large values of *t* and *T*, the kinetically-limited apparent "critical" thickness approaches the Matthews equilibrium stability limit, as expected.

4.2 Microscopic Mechanisms

The detailed mechanisms of misfit dislocation introduction in low misfit (ε_o< 1%) strain relaxation have been studied to some extent, primarily in the GeSi/Si system. Nevertheless the *initial* source of misfit dislocations has yet to be described comprehensively. In our previous work (Perovic *et al.* 1989), it was demonstrated that nanometer-sized, coherent β-SiC precipitate plates, localized at the Ge$_x$Si$_{1-x}$/Si epitaxial layer/substrate interface following incomplete substrate cleaning, can act as efficient sources of 60° misfit dislocations. Similarly, Tuppen *et al.* (1989) studied precipitate-induced dislocation nucleation effects which were suggested to originate from large (micron-sized) oxide particles, each of which could produce multiple misfit dislocation segments. The group of Humphreys *et al.* (1991) observed the so-called 'diamond defect', a faulted dislocation loop (*b*= a/6<411>) which acts as a multiply regenerative, heterogeneous source of 60° misfit dislocations; this source was suggested to originate from interstitial precipitation based on bond-breaking arguments. Recently, a new form of 'diamond defect' (*b*= a/3<111>; vacancy character) has been observed in ZnTe/GaSb by Cherns *et al* (1995) and has been shown to act as a source of interfacial misfit dislocations. Interestingly, from studies involving metallic contamination of GeSi/Si and Si, the nucleation of both surface half-loops (Higgs *et al*. 1991) and *b*= *a*/6<411> dislocation loops (de Coteau *et al*. 1991) have been observed. Unfortunately all of the above examples are unable to *generally* explain existing strain relaxation data for a range of misfit strains because: (i) some of the heterogeneous sources can be controlled to a certain degree resulting in an active nucleation site density which is well below that required to sufficiently relax misfit strains in Stage I and (ii) some of the sources appear to be unique to specific materials systems and/or growth conditions.

Several workers have considered the "nucleation" of misfit dislocations following the interaction of pre-existing misfit segments in both the GeSi/Si and InGaAs/GaAs systems. Hagen and Strunk (1978) first proposed a multiplication mechanism demonstrating the importance of certain orthogonal misfit dislocation intersection events. Subsequently, most workers concluded that the Hagen-Strunk mechanism was not valid and a variety of different multiplication models based on orthogonal misfit segment annihilation reactions have been developed (Lefebvre *et al*. 1991, LeGoues *et al*. 1992, Shiryaev 1993, Beanland 1995) Alternatively, a number of other models consider the combination of misfit dislocation pinning and cross-slip giving rise to multiplication via spiral or Frank-Read sources (Tuppen *et al*. 1990, Washburn and Kvam 1990, Capano 1992, Beanland 1992, Shiryaev 1993). In the present context these multiplication mechanisms are not considered as "nucleation" processes that affect Stage I relaxation but, as will be discussed, are the dominant mechanism governing strain relaxation in Stage II.

We first reported on the *interfacial* nucleation of 60° misfit dislocations which was associated with the formation of sub-nanometre sized Ge-rich platelets due to strain-induced surface segregation during growth (Perovic and Houghton 1992, 1993). More recent work on strain-induced surface instabilities has concentrated on the nucleation of misfit dislocations from stress concentrations associated with surface cusps as predicted theoretically (Jesson *et al*. 1993, Yang and Srolovitz 1993) and observed by TEM in InGaAs (Cullis *et al*. 1995, Androussi *et al*. 1995) and GeSi/Si (Perovic 1995, unpublished). In the remainder of this paper we will only consider the nucleation of dislocations from low misfit heterostructures maintaining planar interfaces during growth.

Several GeSi/Si specimens used for quantitative nucleation studies (§4.1) were carefully analyzed using cross-sectional and plan-view TEM and photoluminescence (PL) spectroscopy in order to elucidate the nature of active dislocation sources during the initial stages of strain relaxation. Since the heterostructures studied here possess misfit dislocation densities <~ 10^5 cm^{-2}, several thin foils of each sample were required to improve imaging statistics. Moreover, the use of plan-view imaging from wedge-profile specimen regions, where the heterostructure interfaces are exposed to a free surface and clearly delineated due to thin foil surface relaxation effects (Perovic and Weatherly 1991), allows for representative imaging of defect configurations within specific areas of a given heterostructure without the need for cross-sectional analysis. In fact, cross-sectional imaging of 3-D dislocation arrangements can be problematic since dislocation segments may be thinned away during specimen preparation. This fact explains a previous result where incomplete nodal balances of Burgers vectors from particulate-induced dislocation sources were observed (Perovic *et al*. 1989).

Fig. 3 shows two dislocation nucleation configurations which we have commonly observed in metastable MBE-grown GeSi/Si heterostructures during post-growth annealing in the Stage I regime. Fig. 3a is a 60° half-loop segment that has internally nucleated at a Ge$_{0.1}$Si$_{0.9}$/Si superlattice interface. Fig. 3b gives examples of: (i) complete 60° shear dislocation loops (arrowed) buried at the strained layer interface, (ii) the characteristic paired configuration known as the *'double half-loop'* dislocation source (Perovic and Houghton 1992, 1993) with threading segments intersecting the wafer surface and (iii) straight <011>-oriented interfacial 60° misfit segments. The *'double half-loop'* possesses misfit segments on different {111} variants whose lines of intersection with the <100> interface are orthogonally opposed; the Burgers vectors of each half-loop can be of the same or different *a*/2<110> 60° types. Given sufficient thermal activation, the single and double half-loops can glide (Perovic and Houghton 1992) leaving single or orthogonal misfit segments respectively, *directly* at the interface. Under these conditions the original dislocation source is masked following misfit

124

dislocation extension thus leaving no evidence on the microscopic scale following operation of the source. Accordingly the detection of such sources by TEM is difficult during the initial stages of relaxation.

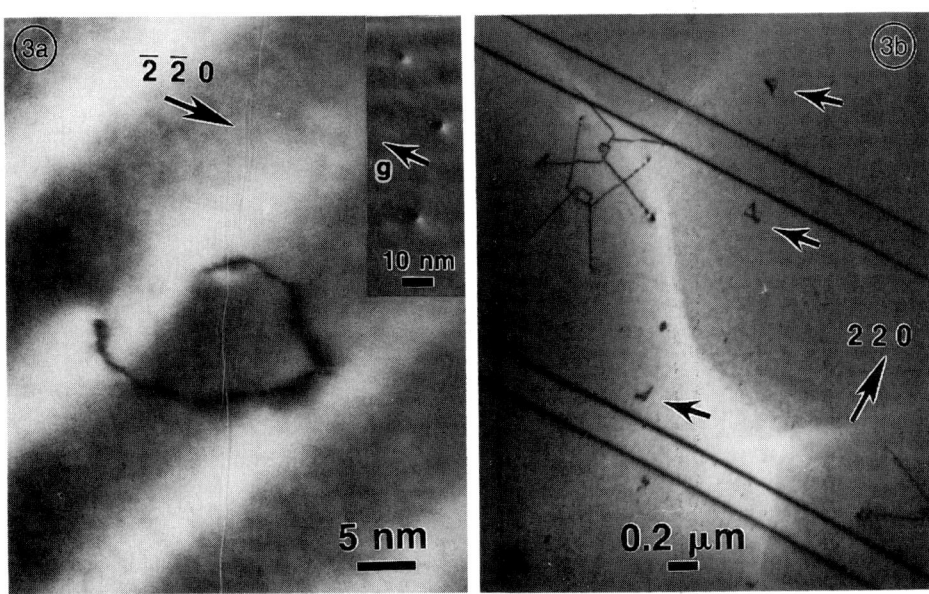

Fig. 3: (a) Oblique section plan-view TEM image of a MBE-grown 10-period Ge$_{0.25}$Si$_{0.75}$/Si superlattice showing interfacial nucleation of a 60° shear loop following post-growth annealing; inset shows examples of interstitial platelet strain field contrast. (b) Plan-view image of a single Ge$_{0.15}$Si$_{0.85}$ strained layer showing various interfacial misfit dislocation sources following post-growth annealing (see text for details).

In addition to post-growth annealed metastable heterostructures, we have studied heterostructures possessing strained layers well above the equilibrium critical thickness which undergo extensive strain relaxation during growth, typical of the Stage II regime. Fig. 4 (a-c) is an example of a single Ge$_{0.25}$Si$_{0.75}$ layer that has generated a very high density of 60° misfit dislocations, which have nucleated at the interface and are forced deep into the Si buffer layer with relatively few threading dislocations extending through the strained layer. It should be noted that although there is minimal stress in the buffer layer/substrate, dislocation injection below the strained interface can relieve strain energy, (Houghton et al. 1990). Interestingly, the dislocation configurations in Fig. 4a appear very similar to the observations from MBE (Fitzgerald et al. 1992) and CVD-grown (LeGoues et al. 1992) graded layer GeSi/Si heterostructures which LeGoues et al. have been attributed to Frank-Read multiplication. It is obvious that multiple slip has occurred on single glide planes resulting in dislocation pileups below the interface. In plan-view imaging (Fig. 4b,c) many of the buffer layer loops are removed during specimen thinning. Nevertheless, Fig. 4b clearly shows extensive interaction (annihilation and pinning) between orthogonal misfit segments at the strained layer interface indicating that various multiplication processes are possible (Shiryaev 1993). It is important to note that in the absence of cross-sectional views, conventional plan-view imaging of complicated dislocation structures as shown in Fig. 4 can lead to erroneous results. In our first study of such layers (Perovic et al. 1990) we used only plan-view stereomicroscopy and concluded that the half-loop segments were nucleated at the surface and extended to the strained interface. Fig. 4a shows that this is clearly not the case. Secondly, the heterostructure shown in Fig. 4 is the same as that studied by Rajan and Denhoff (1987) who claimed that dislocation multiplication via the Hagen-Strunk mechanism was operative. However, it is clear from Fig. 4 that multiplication proceeds from orthogonal dislocation annihilation-induced inclined tips extending into the substrate (Lefebvre *et al.* 1991) and not towards the wafer surface as required by the Hagen-Strunk mechanism and as claimed by Rajan and Denhoff (1987).

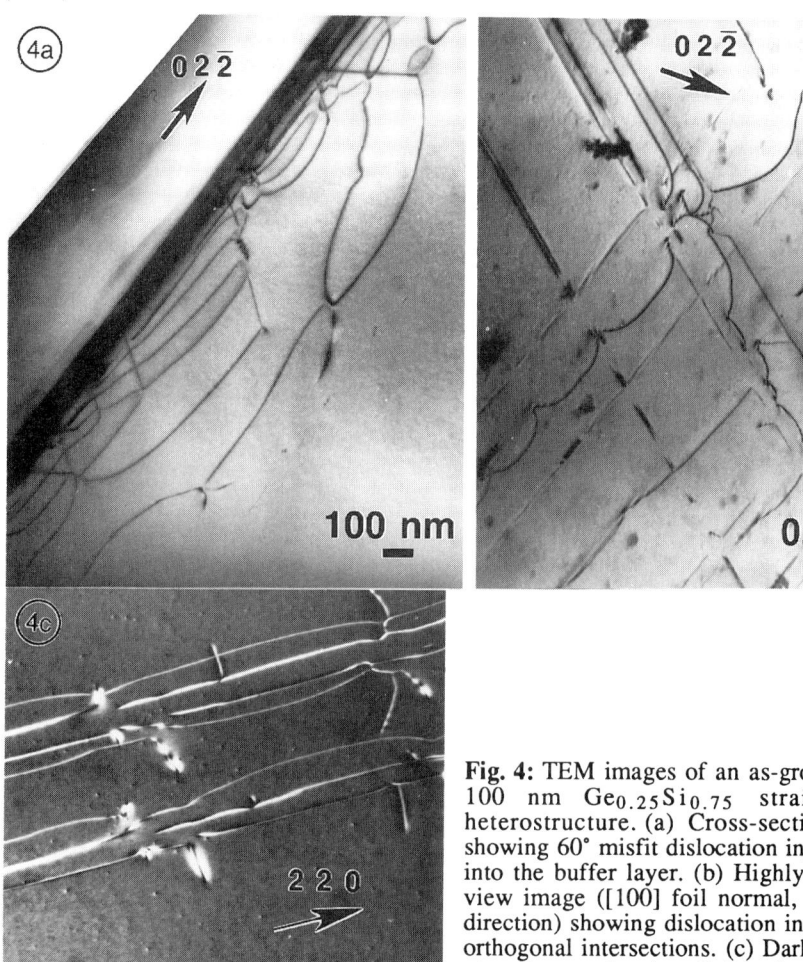

Fig. 4: TEM images of an as-grown (MBE), 100 nm $Ge_{0.25}Si_{0.75}$ strained layer heterostructure. (a) Cross-sectional image showing 60° misfit dislocation injection deep into the buffer layer. (b) Highly tilted plan-view image ([100] foil normal, [011] beam direction) showing dislocation interactions at orthogonal intersections. (c) Dark-field plan-view image showing dislocation pileups on single {111} planes and a high density of interstitial platelets

In our attempt to identify the *initial* source of the misfit dislocations, we used various kinematic (weak-beam) and dynamic diffraction contrast imaging conditions. However, no detectable heterogeneity could be observed at the apex of the single or *'double-half loop'* sources. Interestingly, several defects (<1.5 nm) possessing interstitial strain fields with (100)-oriented plate-shaped symmetry are commonly observed at Ge_xSi_{1-x}/Si interfaces (see Figs. 3a (inset), 4c and Dynna and Weatherly 1994). The interstitial platelets were never observed in the thinnest foil regions away from the heterostructure (i.e. in Si capping layers) which indicates that they are intrinsic to the strained layer structure. The density of these defects can be as high as $\sim 10^9$ cm^{-2} and have been identified as the origin of intense, broad-band photoluminescence (PL) in MBE-grown GeSi/Si layers (Noël *et al.* 1992). The change in behaviour from phonon-resolved, near band-gap PL to broad-band PL has been attributed to the observed superlinear increase in platelet density with increasing strained layer thickness. The occurrence of the broad PL band, which is shifted lower in energy than the phonon-resolved PL, was attributed to exciton localization in lower band-gap Ge-rich platelets. Since the platelets can vary in size and Ge fraction, a broad PL band is thus expected reflecting the distribution of Ge-profiles in the platelet regions. Accordingly, by combining TEM and PL characteristics we concluded that the interstitial platelet defects are Ge-rich, only a few monolayers thick in the (100) growth direction and nominally less than ~1.5 nm in diameter. As mentioned above and discussed in more detail elsewhere (Rowell *et al.* 1993), the platelet formation has been attributed to the onset of 2D islanding under step flow growth conditions wherein Ge preferentially accumulates at the periphery of atomic terrace kink sites and is subsequently covered (i.e. buried) by successive atomic layers during growth. A recent atomic force microscopy study by Chen *et al.* (1995) provides direct evidence for Ge clustering at (100) Si surface step instabilities under step flow growth conditions. Therefore, we propose that the true source of the dislocation half-loops must be associated with atomic-scale interfacial heterogeneities intrinsic to the strained layer growth process.

5. DISLOCATION NUCLEATION THEORY

Following the classic works of Frank and Hirth, several workers have considered the homogeneous nucleation of misfit dislocations at a free surface. Initially, theoretical models were developed to consider nucleation of perfect and partial dislocation half loops within a single strained epitaxial layer. More recently, these models have been revised considering different dislocation configurations, core energies and/or heterostructure geometries for both Si-Ge and III-V-based systems; further references can be found elsewhere (Perovic and Houghton 1992). In any case, existing theoretical calculations based on homogeneous theory still predict relatively large activation energy barriers (>100 eV) for misfit dislocation nucleation in low misfit (< 1%) systems. Based on the experimental results discussed thus far it is clearly evident that homogeneous nucleation is not applicable, even in relatively "perfect" Si-based material.

In view of the low nucleation activation energies observed experimentally and the electron microscopic evidence for interfacial nucleation at low misfit strains, a refined nucleation theory has been developed in order to substantiate our conjecture that misfit dislocation nucleation occurs at localized strain perturbations inherent in strained layer growth.

In any discussion of dislocation nucleation theory there are two conditions (the Orowan criteria) that must be satisfied. The *necessary* condition requires that a dislocated interface is a lower energy state relative to the fully coherent interface. The *sufficient* condition requires that the interfacial dislocation can in fact be nucleated. Accordingly, the nucleation of misfit dislocations in strained layer epitaxy will be considered as a two-step process (Fig. 5):

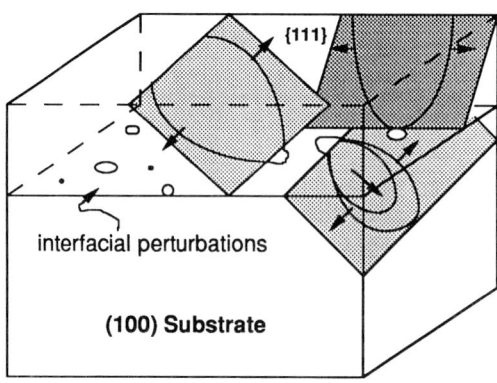

interfacial perturbations

(100) Substrate

Fig. 5: Schematic diagram illustrating the nucleation of prismatic dislocation loops (vacancy-type) at interstitial perturbations localized at the strained layer interface. Prismatic loops generate 60° shear loops forming half-loops in the epilayer and/or multiple loops in the buffer layer (see text for details).

(i) the nucleation of the initial (i.e. incipient) dislocation loop and (ii) the expansion of the incipient loop to form a stable misfit dislocation. The theory is based on linear isotropic elasticity in the continuum approximation. Although the main results will be presented here, a more detailed description will be given elsewhere (Perovic and Houghton, to be published).

5.1 Incipient Dislocations

The starting point in our treatment of the nucleation problem considers the generation of *interfacial* prismatic dislocation loops via the loss of coherency at interstitial perturbations (e.g. Ge-rich platelets). The theory is essentially identical to previous calculations developed to explain coherency breakdown at precipitates in metallic systems (Weatherly 1968, Ashby and Johnson 1969 and Brown and Woolhouse 1970). Using an energy balance approach wherein the elastic energy (E_p) of a coherent platelet is balanced against the self-energy of a dislocation loop (E_d), we obtain the total energy $(E_t = E_\varepsilon - E_d)$:

$$E_t = 4\pi\mu b\varepsilon_o' R^2 - \frac{\mu b^2 R(2-v)}{4(1-v)} \ln\left(\frac{8\alpha R}{e^2 b}\right) \cdot \Theta \qquad (2)$$

where μ is the shear modulus, v is Poisson's ratio, e is the Naperian base, b is the magnitude of the Burgers vector and ε_o' is the unconstrained lattice misfit strain between the precipitate and matrix.

The elastic strain energy of the interstitial platelet is obtained from Eshelby (1957) which allows for a range of possible morphologies to be treated; for simplicity we have approximated the interstitial plate as a constrained spherical inclusion. The inclusion radius is set equal to the dislocation loop radius (R) where the inclusion strain energy is a minimum. The dislocation self energy is obtained from the recently developed *exact* expression for a dislocation loop in an isotropic half-space (Beltz and Freund 1993, Gosling and Willis 1994) which contains a significant correction factor (Θ) for the approximate infinite body self-energy expression used in all previous dislocation calculations.

The use of continuum elasticity in such calculations suffers primarily from the difficulty of accurately accounting for the dislocation core energy through the core parameter, α. Although previous calculations normally use a value of $\alpha = 4$, which is believed to be appropriate for covalent crystals (Hirth and Lothe 1982), Perovic and Houghton 1992 argued that, based on a number of more recent non-continuum treatments, the core parameter $\alpha \leq 1$ for Si and Ge. This was confirmed by Rajan (1992) who estimated the core parameter based on TEM measurements of extended nodes from misfit dislocations in GeSi/Si. Accordingly we have used a core parameter value $\alpha = 1$ in all calculations.

From the total energy expression of Eq. (2), a critical radius condition (R^*) can be determined wherein the strained perturbation can lower its energy upon creating a prismatic loop at its interface. Fig. 6 shows the variation in R^* as a function of lattice misfit. It can be seen that a critical radius is non-existent for $\varepsilon_o > 4\%$. In other words, pure Ge clusters embedded in Si should spontaneously (i.e. homogeneously) generate an interfacial dislocation loop without an energy barrier. At lower misfit strains the critical radius becomes significant with an associated activation energy that must be overcome by some external driving force. It should be noted that the above calculation describes equilibrium conditions. For perturbation sizes just above the critical strain condition, dislocation nucleation is possible, however, kinetic limitations may prevent coherency loss at low growth temperatures. Therefore, at larger radii the perturbations become metastable and will only generate interfacial dislocations in the presence of some external driving force (e.g. post-growth annealing).

Our result compares favourably with the nucleation calculations of Brown and Woolhouse (1970) who obtained an expression for the critical misfit strain (ε_{crit}) required for spontaneous dislocation nucleation by shear at a spherical inclusion. They developed a simple relationship where $\varepsilon_{crit} = (6\pi\alpha)^{-1}$ such that $\varepsilon_{crit} \approx 5\%$ for a core parameter of $\alpha = 1$.

The same result can be obtained upon considering if the maximum shear stress at the interface of a misfitting perturbation can exceed the theoretical shear strength of the matrix crystal (Weatherly 1968). Such a calculation is independent of the inclusion size since the critical nucleation step occurs at the inclusion surface where the stress is independent of inclusion size. Previously (Perovic and Houghton 1992) we calculated the ε_{crit} required for spontaneous nucleation of a complete dislocation loop for various inclusion morphologies and stress states in an elastic matrix of Si. For the case of a spherical perturbation, ε_{crit} = 5.3% in fair agreement with the previous energy minimization calculations. Moreover, for the case of a plate-shaped perturbation (flat-oblate spheroid), which is more appropriate for the interstitial platelets of interest here, we calculated ε_{crit} = 2.4%. As expected the energy calculations based on a spherical perturbation morphology (Fig. 6) will underestimate R^* for a (100)-oriented platelet which possess stress concentrations at the periphery of the plate. The size-independent critical stress condition based on our energy calculation above has been included in Fig. 6.

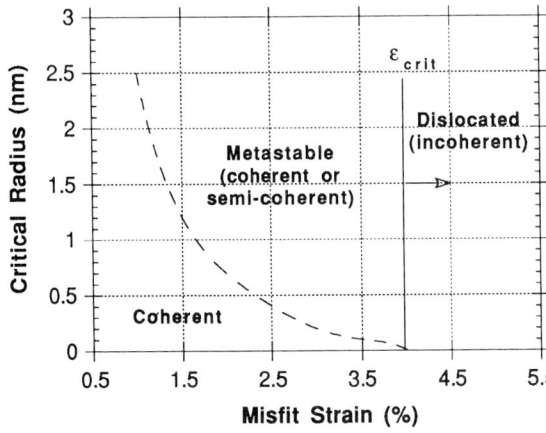

Fig. 6: Equilibrium critical radius for nucleation of a prismatic dislocation loop at an interstitial perturbation as a function of perturbation misfit strain (dashed line). Critical strain for spontaneous dislocation nucleation (solid line).

Therefore, it is evident that relatively low misfit strains are required to nucleate an *incipient* interfacial dislocation loop at an interstitial platelet such that the *sufficient* stress condition for misfit dislocation nucleation can be satisfied. For the GeSi/Si system such a condition can easily be achieved for Ge platelets ~> 0.5 nm. However, TEM diffraction contrast imaging cannot resolve the incipient dislocation loop since at these dimensions the strain field image contrast (Figs. 3a, 4c) from prismatic dislocation loops and plate-shaped precipitates is identical (Ashby and Brown 1963).

5.2 Misfit Dislocations

Having shown that sub-nanometre sized Ge-rich clusters constrained in a matrix of Si can be an efficient source of prismatic vacancy-character dislocation loops, the second step of the nucleation calculation considers whether such prismatic loops can lead to misfit dislocation generation at strained layer interfaces. The schematic diagram of Fig. 5 shows a number of (100)-oriented interfacial platelets. A few platelets are located at the apex of inclined misfit dislocation half loops. Following the suggestion of Nabarro (Hirth and Lothe 1982), it is possible that prismatic vacancy loops on {100} planes can operate as sources of dislocation shear loops on {111} planes in the presence of a resolved stress on the glide plane (i.e. τ_{eff}). Depending on the local stress state, segments of the {100} prismatic loops can bow out onto the {111} planes, analogous to a Frank-Read source, towards the free surface and eventually form a half-loop. In the event of multiple nucleation events on the same glide plane, interdislocation stresses can force misfit segments into the substrate (see Fig. 5). Such a mechanism is capable of producing the observed single and double half-loop segments emanating from a strained layer interface in both the epilayer (Fig. 3) and substrate (Fig. 4).

The calculation of the critical condition for the expansion of a *pre-existing* shear loop into a misfit segment is analogous to the calculation described in the previous section. In this case the total Helmholtz energy balance becomes: $E_t = E_\varepsilon - E_d \pm E_s - TS$ where E_d is dislocation half-loop self-energy including the appropriate correction factor (Θ) for a semi-infinite body (Gosling and Willis 1994) and $\alpha = 1$. E_s represents the energy gained or lost during the creation or removal of a surface step and TS is the entropy. The governing equation for (100)-oriented interfaces in the diamond-cubic slip system ($\{111\}<110>$) can be written as:

$$E_t = \frac{\mu b^2 R(2-\nu)}{8(1-\nu)} \ln\left(\frac{8\alpha R}{e^2 b}\right) - \Theta' - \frac{\pi R^2 \mu b \varepsilon_{eff}(1+\nu)}{\sqrt{6}(1-\nu)} \pm \frac{\mu b^2 R}{4\sqrt{2}} - \frac{3.6\pi R \, k_B T}{b} \qquad (3).$$

Here the E_ε term has been modified by replacing the average lattice misfit strain (ε_o) by an effective strain (ε_{eff}) which is a superposition of the strain field of the perturbation with that of the strained layer. Using Eshelby theory it is possible to incorporate various inclusion morphologies. Once again we assume a spherical perturbation of radius R_o, but unlike the previous calculation the distance between the perturbation and the loop will be sufficiently larger than the perturbation radius. Under these conditions E_ε is effectively shape-independent. From the total energy expression one can calculate the critical radius for nucleation (R^*) upon maximizing Eq. (3) with respect to R. Subsequently, the activation energy for nucleation (E^*)

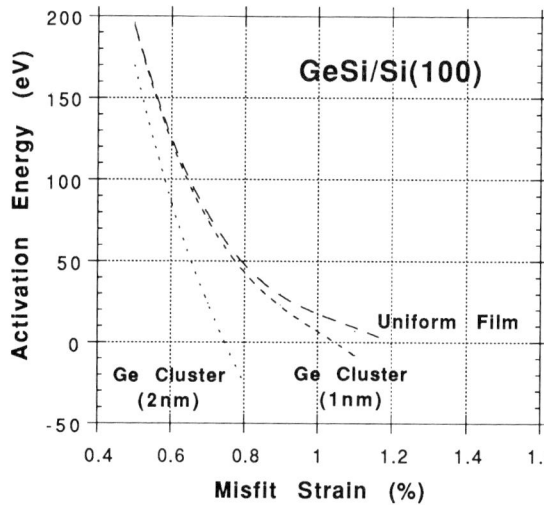

Fig. 7: Activation energy for interfacial nucleation of a 60° misfit dislocation half-loop as a function of misfit strain for the GeSi/Si (100) system for a uniform interface ($R_0 = 0$) and interfaces containing Ge clusters with $R_0 = 1$ nm and 2 nm.

is obtained by substituting R^* into Eq. (3). Fig. 7 illustrates the effect of including a perturbation strain field for the case of spherical Ge clusters embedded in Si. The activation energy reduction for heterogeneous half-loop nucleation is a sensitive function of the perturbation size and misfit strain resulting in a continuum of energy states bounded by the homogeneous nucleation condition (i.e. uniform film).

Interestingly, two recent nucleation calculations have also considered heterogeneous internal nucleation of misfit dislocations. Zhang and Yang (1994) developed a 3D boundary element method to numerically calculate the misfit strain required to nucleate a misfit dislocation loop from a pre-existing shear loop surrounding a spherical cavity. The absence of a stress field from the cavity results in significantly greater critical radii for nucleation relative to the strained perturbation case considered here. Jain et al. (1995) used an energy calculation analogous to that developed here by introducing an interaction energy term between the nucleating dislocation loop and a perturbation of either spherical or plate-shaped morphology; the activation energies calculated are in accord with Fig. 7.

It is important to note that the misfit strain of a critical-sized perturbation can be fully relaxed following the formation of an incipient dislocation loop. Under these conditions ε_{eff} tends to ε_o and the activation energy is given by the curve for the uniform film. However, as discussed in §5.1, there will be a distribution of metastable perturbations, of various size and misfit, where coherency strains are not relieved following nucleation of the incipient loop. Analogous to the Matthews theory, there will exist a range of effective stresses, increasing with perturbation size that facilitate a driving force for reduction of the activation barrier for misfit dislocation injection. It is these sources that become misfit dislocations during growth or following post-growth annealing. Nevertheless, the results of Fig. 7 indicate that there still exists a substantial nucleation barrier at very low misfits (< 0.5%) where the stress concentrating effect of Ge-rich clusters is insufficient even for relatively large cluster sizes (> 3 nm). Accordingly in this very low misfit strain regime the nucleation sources must be generated from more "intense" stress concentrations (e.g. β-SiC plates, large oxide particles), where multiple nucleation from a single source is possible (see Zhang and Yang 1994), or from other pre-existing defects (e.g. substrate dislocations). In the absence of a sufficient density (N_o) of pre-existing sources required for strain relief at very low misfits, the material becomes more stable against strain relaxation (c.f. CVD vs. MBE, Fig. 1).

6. GENERALIZED STRAIN RELAXATION MODEL

Based on the experimental and theoretical considerations presented thus far, a generalized description of the strain relaxation process has been developed. The time evolution of misfit strain relaxation in metastable heterostructures, during post-growth annealing in the early stages, is analogous to static deformation (i.e. creep) behaviour in bulk materials where the change in plastic strain is measured under conditions of constant applied stress and temperature (Alexander and Haasen 1968). However, in the later stages of strain relaxation (Stage II and beyond) the effective stress decreases significantly since it is not maintained as a constant. In addition, the effective stress is further reduced as the dislocation density (N) increases. Haasen employed a stress correction factor for bulk deformation, where $\tau_{eff}' = \tau_{eff} - AN^{1/2}$; here A is an elastic term to account for dislocation interactions.

Fig. 8 is a schematic representation describing the time evolution of strain relaxation as applied to strained layer epitaxy. Thermal expansion effects are negligible and thus are not included. Fig. 8a plots the increase in plastic strain (ε) as a function of time at constant T and τ_{eff}. The 'S'-shaped" curve is characteristic of creep behaviour in materials with a relatively low N_0 (Alexander and Haasen 1968). Curves obtained at increasing T and τ_{eff} are of similar shape but shifted towards shorter times. Fig. 8 has been divided into three characteristic regimes. The three stages will be generally described in terms of the Q and τ_{eff} dependence observed in various strain relaxation experiments. It will be seen that, analogous to the experimental determination of equilibrium critical thickness (§2), the strain relaxation behaviour observed depends on the resolution of the diagnostic technique.

6.1 Stage I

Following our extensive modelling of Stage I relaxation (c.f. Eq. (1)), misfit dislocation nucleation was observed to increase linearly with time, beyond N_o, whereas misfit dislocation velocity was effectively constant resulting in a linear increase in the overall strain relaxation rate with time (Fig. 8b); this regime is said to be *nucleation-limited*.

The relaxation curves of Fig. 8 are not identical to creep deformation in bulk Si and Ge in Stage I. In the initial stages of creep in pure, nearly dislocation-free bulk material ($N_0 < 10^2$ cm^{-2}), the dislocation density must evolve through multiplication of pre-existing sources giving rise to an incubation period exponential in time. Practically, the study of this regime is difficult in creep deformation experiments since the deformation is usually inhomogeneous due to Lüders band formation at the specimen grips (Alexander and Haasen 1968). Alternatively, strained layer relaxation is ideally suited for study of Stage I behaviour. It has been shown to be dominated by the presence of interfacial perturbations and other heterogeneities which enhance the nucleation rate stress dependence (i.e. $N(t) \propto (\tau_{eff})^{2.5}$) resulting in a linear increase in the thermally activated misfit dislocation density with time ($t \geq 0$). Accordingly the presence of heterogeneous nucleation sources in strained epitaxial layers precludes the necessity for multiplication processes in the early stages of strain relaxation thus eliminating the incubation period. This is consistent with creep studies in bulk Si and Ge where the initial deformation behaviour scales linearly with increasing N_0.

Although the activation energy for misfit dislocation glide ($Q_v \approx 2$ eV) is well understood in terms of double-kink nucleation and glide processes (see §4.1), the activation energy for Stage I nucleation has yet to be explained. Based on the theoretical calculations in §5.2 it was shown that the presence of interstitially strained perturbations can give rise to a range of low activation energies depending on perturbation size. However, the quantitative measurements indicate a universal value of $Q_n \approx 2.5$ eV. For the first time we attribute the dislocation nucleation process to be controlled by vacancy formation (E_f) and/or migration (E_m) during the generation of the incipient loop at an interstitial perturbation as discussed in §5.1. Analogous to coherency loss in classical precipitation hardening systems, the elastic stress-field of the interstitial platelets can be relaxed by condensation of vacancies at the platelet leaving a vacancy-type prismatic dislocation loop. In the presence of a sufficiently large vacancy supersaturation, dislocation loop nucleation will be controlled solely by vacancy migration, otherwise, the formation of vacancies is required.

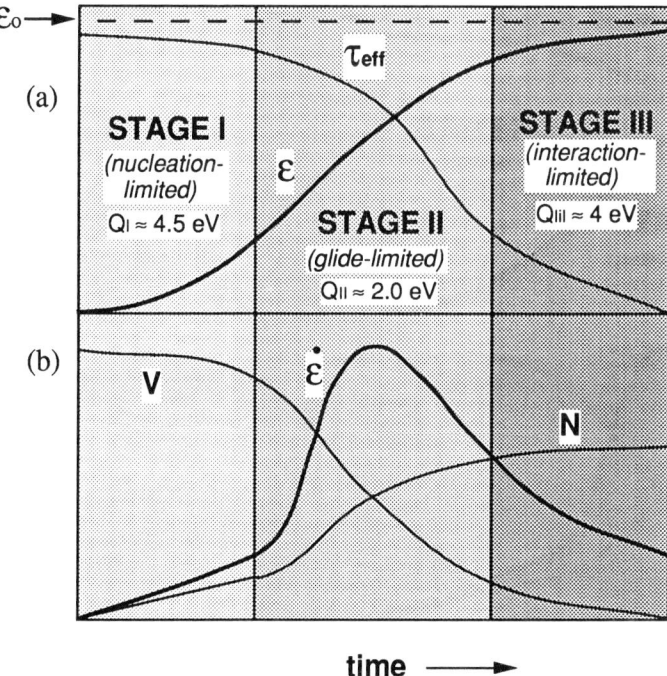

Fig. 8: Schematic representation of three strain relaxation regimes showing the time evolution of strain relaxation rate ($\dot{\varepsilon}$), relaxed strain (ε), misfit dislocation velocity (v), effective stress (τ_{eff}) and misfit dislocation density (N). Activation energies quoted are for Ge_xSi_{1-x}/Si(100).

Following the vast number of theoretical and experimental atomic diffusion studies in Si and Ge there exist well-defined values for $E_f^{Ge} = 2.0$ eV, $E_m^{Ge} = 1.0$ eV and $E_m^{Si} = 0.33$ eV (Van Vecheten 1980, Fahey *et al.* 1989). However there is still much controversy regarding the magnitude of E_f^{Si}. Nevertheless most experiments and thermodynamic theory give a value in the range $E_f^{Si} = 2.3$-2.5 eV (Van Vecheten 1980). Accordingly the $Q_n = 2.5$ eV observed during Stage I relaxation (Fig. 2) is consistent with either: (i) vacancy formation in Si or (ii) vacancy formation in Ge followed by migration in Si, consistent with the incipient dislocation loop mechanism described in §5.1; further details will be given elsewhere (Perovic and Houghton, to be published).

6.2 Stage II

Beyond Stage I the plastic strain increases dramatically resulting in steady-state relaxation behaviour. The inflection point in Fig. 8a corresponds to the highest strain relaxation rate as shown in Fig. 8b. This is a consequence of the marked increase in misfit dislocation density concomitant with the onset of multiplication processes. The velocity is seen to decrease in Stage II because of the reduction in τ_{eff} (Fig. 8a) and increase in dislocation blocking as relaxation proceeds. This regime is thus considered to be *glide-limited*.

Stage II strain relaxation is again not identical to creep deformation in bulk materials. In this regime multiplication processes are dominant with a parabolic stress dependence (i.e. $d\varepsilon/dt \propto (\tau_{eff})^2$) as compared to the $(\tau_{eff})^3$ dependence observed in bulk creep studies (Alexander and Haasen 1968). However, in either case the steady-state strain relaxation rate is controlled by misfit dislocation glide with the same activation energy. The difference in stress dependence is most likely due to the decreasing applied τ_{eff} during strain relaxation which is not the case in creep deformation. Dodson and Tsao (1987) accounted for this difference by assuming a linear relationship between the multiplication rate and τ_{eff} whereas Alexander and Haasen (1968) used a parabolic dependence.

Unlike the case of Stage I relaxation behaviour several workers have quantitatively studied Stage II behaviour in GeSi/Si using diagnostic techniques appropriate for areal dislocation densities > 10^4 - 10^5 cm^{-2} (i.e. x-ray diffraction, ion channeling, wafer bending and Raman spectroscopy). Hauenstein *et al.* (1989), Volkert *et al.* (1991), Gillard *et al.* (1992) Bai *et al.* (1994), Sardela and Hansson (1995) have confirmed that $Q_r \approx 2.0$ eV and $d\varepsilon/dt \propto (\tau_{eff})^2$ with a time dependency as depicted in Fig. 8. However, it is important to note that the incubation period observed in the above studies is not a true representation of a nucleation barrier but simply an experimentally-limited resolution limit for the detection of strain relaxation.

The *in situ* TEM measurements of Hull *et al* (1989) should also represent Stage II relaxation, however, the values reported for Q_r and Q_n are anomalously low. Based on our vacancy-controlled nucleation description it is possible to reconcile the $Q_n = 0.3$ eV measured by Hull *et al.* (1989) using *in situ* TEM. Considering the relatively small dimensions of a plan-view thin foil, and the exposed strained layer(s) at the perforation, oxidation-induced vacancy injection at the surface (LeGoues *et al.* 1989) can occur in GeSi alloys. Under these conditions, vacancy migration in the Si substrate would be the rate-limiting process and can account for the $Q_n = 0.3$ eV measured by Hull *et al.*

6.3 Stage III

Stage III corresponds to the onset of strain hardening where the relaxation eventually ceases because the diminishing effective stress is balanced by the interaction stress associated with dislocation blocking (Gillard *et al.* 1994) which impedes further dislocation extension and multiplication. The limiting degree of strain relieved is less than the initial misfit strain (ε_o) corresponding to the residual strain normally observed in strained layers of finite thickness (Jain 1994). Stage III is considered to be *interaction-limited*.

The recent experimental results (see §4.1) of LeGoues *et al.* (1993) are believed to represent dislocation interaction effects (i.e. Stage III behaviour), not dislocation nucleation. Unlike the previous studies described above, where the relaxation behaviour is observed following post-growth annealing of metastable structures, LeGoues *et al.* studied strain relaxation *during* growth where τ_{eff} is not decreasing as shown in Fig. 8, analogous to creep deformation. This has the effect of shifting the curves of Fig. 8 to the left such that Stage III behaviour occurs at earlier times. Under these conditions the deformation reaches another quasi-steady-state condition where the relaxation rate is related to the activation energy for cross-slip of screw dislocations (Siethoff 1983) which is in contrast to metals where diffusional processes dominate. Based on the TEM evidence of LeGoues *et al.* (c.f. Fig. 4) it is clear that following extensive dislocation multiplication the structure adopts a low energy dislocation sub-structure (i.e. misoriented sub-grains separated by high dislocation density walls), which accounts for the mosaic structure observed in x-ray diffraction reciprocal space mapping (Mooney *et al.* 1993, Sardela and Hansson 1995). Accordingly the strain relaxation behaviour will depend on the ability of mobile misfit dislocations to penetrate the dislocation walls. Siethoff (1983) used such an interpretation to yield an expression which can describe Stage III relaxation by a stress-dependent activation energy: $(d\epsilon/dt) \propto exp(-Q_{III}(1-\tau_{eff}/\tau_o)/kT)$, where $\tau_o \approx$ 300-400 MPa for Ge and Si and $Q_{III}{}^{Ge} \approx$ 4 eV and $Q_{III}{}^{Si} \approx$ 5 eV. These values are in agreement with the activation energy measurements of LeGoues *et al.* At sufficiently large τ_{eff}, the activation energies for cross-slip tend to zero since the effective stress is large enough to drive the misfit dislocation through the walls precluding the necessity for cross-slip. In other words, the strain relaxation curve shifts to the left in Fig. 8 such that the overall relaxation rate becomes glide-limited, with a negligible nucleation barrier typical of Stage II behaviour. This explains the two activation energies observed by LeGoues *et al.*

7. SUMMARY

Extensive experimental studies have been carried out on a variety of low misfit GeSi/Si ($0.03 < x < 0.25$) heterostructures. By combining both macroscopic and microscopic analyses we have developed a comprehensive semi-empirical predictive model for strain relaxation to quantitatively describe the early stages of misfit strain relief. The introduction of misfit dislocations has been shown to be dominated by interfacial perturbations which can give rise to a vanishingly small activation energy for nucleation. A generalized strain relaxation description has been formulated which can consistently explain all stages of relaxation and thus reconcile a wide range of strain relaxation data from the literature.

ACKNOWLEDGMENTS

We are grateful to Profs. L.M. Brown, P.B. Hirsch, J.P. Hirth and G.C. Weatherly who have participated in valuable discussions over the years.

REFERENCES

Alexander H 1986 Dislocations in Solids Vol. 7 (Amsterdam, North Holland) pp. 113-234
Alexander H and Haasen P 1968 Solid State Phys. 22 27
Androussi Y, Lefebvre A, Delamarre C, Wang L P, Dubon A, Courboules B, Deparis C and
    Massies J 1995 Appl. Phys. Lett. 66 3450
Ashby M F and Brown L M 1963 Phil. Mag. 8 1649
Ashby M F and Johnson L 1969 Phil. Mag. 20 1009
Bai G, Nicolet M-A, Chern C H and Wang K L 1994 J. Appl. Phys. 75 4475
Beanland R 1992 J. Appl. Phys. 72 4031; 1995 J. Appl. Phys. 77 6217
Beltz G E and Freund L B 1993 Phys. Stat. Sol. 180 303
Brown L M and Woolhouse G R 1970 Phil. Mag. 21 329
Capano M A 1992 Phys. Rev. B 45 11768
Chen K M, Jesson D E, Pennycook S J, Mostoller M, Kaplan T, Thundat T and Warmack R J
    1995 (to be published)
Chen Y, Liliental-Weber Z, Washburn J, Klem J F and Tsao J 1995 Appl. Phys. Lett. 66 499
Cherns D, Mylonas S, Chou C T, Wu J, Ashenford D A and Lunn B 1995 Scanning
    Microscopy 8 841
Cullis A G, Pidduck A J and Emeny M T 1995 Phys. Rev. Lett. (in press)
deCoteau M D, Wilshaw P R and Falster R 1991 Solid State Phenomena 19-20 27
Dodson B W and Tsao J Y 1987 Appl. Phys. Lett. 51, 1325; 52, 852 (E)
Dynna M and Weatherly G C 1994 J. Appl. Phys. 76 4625
Eshelby J D 1957 Proc. Roy. Soc. A 241 376
Fahey P M, Griffin P B and Plummer J D 1989 Rev. Mod. Phys. 61 289
Fitzgerald E A, Xie Y-H, Monroe D, Silverman P J, Kuo J M, Kortan A R, Thiel F A and
    Weir B E 1992 J. Vac. Sci. Technol. B 10 1807
Fox B A and Jesser W A 1990 J. Appl. Phys. 68 2801
Frank F C and Van der Merwe J 1949 Proc. Roy. Soc. Lond. A 198 216
Freund L B 1994 Adv. Appl. Mech. 30 1
Gillard V T, Noble D B and Nix W D 1992 Mat. Res. Soc. Symp. Proc. 239 395
Gillard V T, Nix W D and Freund L B 1994 J. Appl. Phys. 76 7280

Gosling T J and Willis J R 1994 J. Mech. Phys. Solids 42 1199

Gosling T J, Jain S C and Harker A H 1994 Phys. Stat. Sol. (a) 146 713

Hagen W and Strunk H 1978 Appl. Phys. 17 85

Hauenstein R J, Clemens B M, Miles R H, Marsh O J, Croke E T and McGill T C 1989 J. Vac. Sci. Technol. B 7 757

Higgs V, Kightley P, Goodhew P and Augustus P 1991 Appl. Phys. Lett. 59 829

Hirth J P and Lothe J 1982 Theory of Dislocations, 2nd ed., (New York: Wiley) p.231; p. 757

Houghton D C 1991 J. Appl. Phys. 70 2136

Houghton D C, Perovic D D, Baribeau J-M and Weatherly G C 1990 J. Appl. Phys. 67 1850

Hull R, Bean J C and Buescher C 1989 J. Appl. Phys. 66 5837

Hull R, Bean J C, Bahnck D, Peticolas L J, Short K T and Unterwald F C 1991 J. Appl. Phys. 70 2052

Humphreys C J, Maher D M, Eaglesham D J, Kvam E P and Salisbury I G 1991 J. de Phys. III 1 1119

Jain S C 1994 Germanium-Silicon Strained Layers and Heterostructures, Adv. Electron. Electron Phys. Suppl., Vol. 24, (London:Academic)

Jain U, Jain S C, Harker A H and Bullough R 1995 J. Appl. Phys. 77 103

Jesson D E, Pennycook S J, Baribeau J-M and Houghton D C 1993 Phys. Rev. Lett. 71 1744

Kuk Y, Feldman L C and Silverman P J 1983 Phys. Rev. Lett. 50 511

Lefebvre A, Herbeaux C, Bouillet C and Di Persio J 1991 Phil. Mag. Lett. 63 23

LeGoues F K, Rosenberg R, Nguyen T, Himpsel F and Meyerson B 1989 J. Appl. Phys. 65 1724

LeGoues F K, Meyerson B S, Morar J F and Kirchner P D 1992 J. Appl. Phys. 71 4230

LeGoues F K, Mooney P M and Tersoff J 1993 Phys. Rev. Lett. 71 396; 3234 (E)

Matthews J W, Mader S and Light T B 1970 J. Appl. Phys. 41 3800

Mooney P M, LeGoues F K, Chu J O and Nelson S F 1993 Appl. Phys. Lett. 62 3464

Nix W D, Noble D B and Turlo J F 1990 Mat. Res. Soc. Symp. Proc. 188, 315

Noël J-P, Rowell N , Houghton D , Wang A and Perovic D D 1992 Appl. Phys. Lett. 61 690

People R and Bean J C, 1985 Appl. Phys. Lett. 47 322; 1986 49 229

Perovic D D 1995 Control of Dislocations in Epitaxial Structures, Rep. Prog. Phys (in prep.).

Perovic D D and Houghton D C 1992 Mat. Res. Soc. Symp. Proc. 263 391

Perovic D D and Houghton D C 1993 Phys. Stat. Sol. (a) 138 425

Perovic D D and Weatherly G C 1991 Ultramicroscopy 35 271

Perovic D D, Weatherly G C and Houghton D C 1990 Mat. Res. Soc. Symp. Proc. 160, 65

Perovic D D, Weatherly G C, Baribeau J- and Houghton D C 1989 Thin Solid Films 183 141

Rajan K 1992 J. Appl. Phys. 71 5853

Rajan K and Denhoff M 1987 J. Appl. Phys. 62 1710

Rowell N L, Noël J-P, Houghton D C, Wang A, Lenchyshyn L C, Thewalt M L W and Perovic D D 1993 J. Appl. Phys. 74 2790

Sardela Jr. M R and Hansson V 1995 J. Vac. Sci. Technol. A 13 314

Shiryaev S Yu 1993 Phil. Mag. Lett. 68 195

Siethoff H 1983 Phil Mag. A 47 657

Tuppen C G , Gibbings C J and Hockly M 1989 J. Cryst. Growth 94 392

Tuppen C G and Gibbings C J 1990 J. Appl. Phys. 68 1526

Tuppen C G, Gibbings C J, Hockly M and Roberts S G 1990 Appl. Phys. Lett. 56 54

Van Vecheten J A 1980 Handbook on Semiconductors, Vol 3 ed T S Moss (Amsterdam, North Holland) p. 1

Volkert C A, Fitzgerald E A, Hull R, Xie Y H and Mii Y J 1991 J. Elect. Mat. 20 833

Washburn J and Kvam E P 1990 Appl. Phys. Lett. 57 1637

Weatherly G C 1968 Phil Mag. 17 791

Yamashita Y, Maeda K, Fujita K, Usami N, Suzuki K, Fukatsu S, Mera Y and Shiraki Y 1993 Phil. Mag. Lett. 67 165

Yang W H and Srolovitz D J 1993 Phys. Rev. Lett. 71 1593

Zhang H and Yang W 1994 J. Mech. Phys. Solids 42 913

Zhao A 1993 B.A.Sc. Thesis, Dept. of Metallurgy and Materials Science, Univ. of Toronto

Inst. Phys. Conf. Ser. No 146
Paper presented at Microsc. Semicond. Mater. Conf., Oxford, 20–23 March 1995
© *1995 IOP Publishing Ltd*

Relaxation of large lattice-mismatched (001) heterostructures by Lomer dislocations

André M Rocher and Joon M Kang*

Centre d'Elaboration des Matériaux et d'Etudes Structurales, CEMES/ CNRS
29, rue Jeanne Marvig, BP 4347, F- 31055 Toulouse, France.

ABSTRACT: The morphology of the interface controls the crystalline quality of hetero-structures grown by epitaxy. For GaAs/Si, with poor interfacial organization, the density of threading dislocations in the top of 3μm thick layer of GaAs is about $10^8/cm^2$. When the surface is ideal, as for GaSb/(001)GaAs, the misfit dislocation network is a perfect grid of Lomer dislocations. The density of threading dislocations, generated at the imperfections of the interface, is then limited.

1. INTRODUCTION

Extensive efforts are being devoted to both theoretical and experimental studies on the growth and relaxation of semiconductor heterostructures. The epitaxy of large lattice mismatch systems was developed at the end of the 80's, mainly on the GaAs/Si, a very promising combination for many optoelectronic and electronic applications, associating specific properties of both GaAs and Si. This system can only be successful if the density of extended defects propagating into the epilayer is limited. In order to make progress in this direction, it is necessary to determine the origin of the defects that propagate into the epitaxial layer.

In order to produce an epitaxial layer without defects, a growth process that enables a 2-D network to be formed at the interface must be found (Bourret and Fuoss 1992). The basic interface structure between two perfect crystals consists of a periodic lattice. A lattice-mismatch between two crystals is ideally accommodated by an array of dislocations. Many parameters such as the growth temperature, the growth rate, the type of surface, the chemical composition, etc... are responsible for the crystalline quality of both the interface and the epilayer. The aim of this paper is to discuss the role of the surface quality on the interface organization and the creation of defects propagating into the epitaxial layers in systems such as GaAs/Si and GaSb/GaAs.

2. POSITION OF THE PROBLEM

Much work, reviewed by Fang et al. (1990), has been performed on GaAs/Si with the view to obtaining defect-free layers of GaAs, acting as substrates on which active layers of III-V or II-VI compounds are grown. Such GaAs layers, being a few μm thick, must be relaxed by mechanisms involving point defects and/or dislocations when the material can be deformed plastically.

The best GaAs/Si growth has involved a two step process. The first step is performed at low temperature and hence the initial interface is not at all well organized; many defects are created during the relaxation induced by subsequent thermal annealing. A XTEM image at 2 MeV (Fig. 1) shows a typical configuration of a 3μm GaAs layer on (001)Si: i) the silicon is defect-free; ii) in the region of the interface a very high density of threading dislocations is observed; iii) the density of defects decreases when the thickness increases. Beyond 3 μm of GaAs, this density is about $10^8/cm^2$ (Rocher and Charasse 1989), which corresponds to the

136

value given by Knall et al. (1994) on a patterned silicon substrate after post thermal annealing. The lower density of defects propagating into a GaAs layer grown on silicon seems to be limited to this value: $10^8/cm^2$.

The density of defects appears to be largest at the interface. The best way to investigate defects related to the relaxation of the misfit stress is therefore to image by plan view sample with a thin epilayer. The advantage of this procedure has been discussed earlier by Rocher et al. (1991).

GaAs

1μm   Si

Fig. 1: 2MeV XTEM of 3μm GaAs/(001)Si. Note the high density of defects near the interface.

3. ROLE OF THE SURFACE ON THE QUALITY OF THE INTERFACE

GaAs/Si is not ideal for studying the origin of defects related to the growth and the relaxation of epilayer since a well organized interface is obtained only after   thermal annealing. On the other, GaSb/(001)GaAs with its 8% lattice mismatch shows a well organized misfit dislocation network when the growth conditions are optimized (Aindow et al. 1993, Rocher et al. 1990). Many problems related to the growth and the accommodation of heterostructures with a large lattice mismatch can then be explained. Two aspects of the growth of lattice mismatched heterostructures are to be discussed: i) the periodic mechanism of misfit dislocation creation appearing at the interface; ii) the role of the growth temperature on the defect density in the epilayer.

3.1 Perfect structure of the interface: the GaSb/(001)GaAs

The GaSb/GaAs interface exhibits a quasi-perfect square array of Lomer dislocations characterized by $\{u\,[110]\,, b\,[1\bar{1}0]\}$ as shown on Fig. 2. The measured spacing, Δ, between misfit dislocations is 54±3Å, which is in good agreement with the 8% lattice mismatch between GaSb and GaAs.

The regularity of the misfit dislocation network observed in GaSb/(001)GaAs is due to a very well defined mechanism of generation of misfit dislocations. This mechanism is related to the 3-D GaSb growth observed on Fig. 3 through the misfit dislocation network appearing as a grid under the islands. The shape of the GaSb islands is defined as a trapezoidal pyramid limited by {111} planes. The GaSb grows homoepitaxially on the {111} facets of the islands. The growth process is well adapted to the creation of Lomer dislocations at the edge of the islands where the {111} planes of GaSb intersect with the (001) GaAs surface. The creation mechanism of this misfit dislocations has been described by Rocher et al. (1991). It involves the movement of a small number of atoms (Kiely et al. 1989). It is at the origin of the perfect organization of the dislocation network.

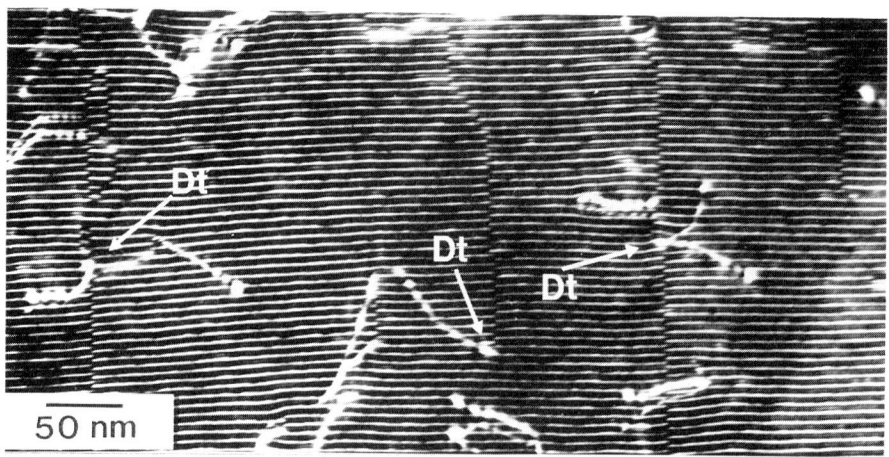

Fig. 2: Weak-beam image of a 770Å thick GaSb film illustrating the relation between the threading dislocations and the half-period-shift of 90° dislocations. Some threading dislocations are marked as D_t.

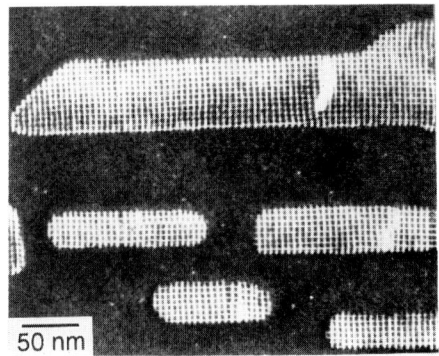

Fig. 3: Double weak beam image of a 30 Å thick GaSb deposit showing the two families of 90° misfit dislocation arrays. Note the formation of GaSb islands elongated in the [$\bar{1}$10] direction.

3.2 Misfit Dislocations relative to the (001) interface

In large lattice-mismatched systems such as GaSb/GaAs, InSb/GaAs (Zhang et al. 1990) and annealed GaAs/Si (Rocher and Charasse, 1989), the misfit dislocations are mainly of Lomer type. This is a general result for large lattice mismatch systems. Several physical arguments can explain why the misfit dislocations are of Lomer type.

The first concerns the geometry of the coincidence site lattice between the two lattice-mismatched crystals. The superimposition of the two crystals on a <110> cross-section (Fig. 4) shows that the coincidence is obtained with a periodicity of D_{110}. The coincidence between the {220} planes observed by the moiré fringes is only partial. The {220} planes of both GaAs and GaSb are not really superimposed at the anticoincidence site A: an a/2 [001] translation needs to be introduced for a perfect superimposition of these planes.

138

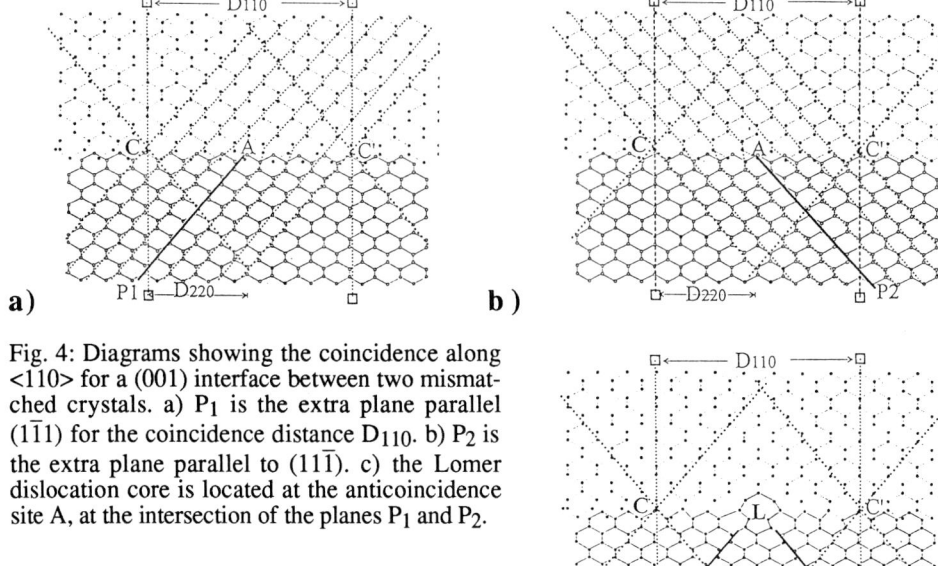

Fig. 4: Diagrams showing the coincidence along <110> for a (001) interface between two mismatched crystals. a) P_1 is the extra plane parallel $(1\overline{1}1)$ for the coincidence distance D_{110}. b) P_2 is the extra plane parallel to $(11\overline{1})$. c) the Lomer dislocation core is located at the anticoincidence site A, at the intersection of the planes P_1 and P_2.

If we follow the continuity of the {111} planes at the level of the interface we observe that two planes, labelled as P1 and P2, of the smaller crystal appear to be in excess at the anticoincidence site, A, as shown Fig.4a and Fig.4b. They are parallel respectively to $(1\overline{1}1)$ and $(11\overline{1})$. The misfit dislocations consist of two 60° dislocations with opposite [001] components of their Burgers vectors. Their association constitutes the Lomer dislocation.

The geometrical configuration shown in Fig.4a or Fig.4b does not correspond to a 60° dislocation array: Such an array cannot accommodate a lattice mismatch because the [001] component of the Burgers vector would introduce a rotation of the film with respect to the substrate. Only the alternance of the two configuration can accommodate the lattice mismatch.

In addition, the elastic energy of the Lomer dislocation array is smaller than that of the 60° dislocations, as discussed by Kang et al. (1994a). The Burgers vector of Lomer dislocation is twice as efficient as a 60° dislocation. The energy of alternative 60° dislocation configuration would be larger than that of the Lomer dislocations, owing to the interactions between the vertical components of their Burgers vectors (Kang 1993).

3.3 Origin of the threading dislocations

An interfacial network with Lomer dislocations allows one to obtain the highest interface quality. Indeed, the two perpendicular sets of Lomer dislocations do not interact elastically: dislocation intersections are clean. The large value of the misfit dislocation length, larger than 300nm as shown in Fig. 2, indicates that the coalescence of GaSb islands involves specific interaction mechanisms. The dislocations perpendicular to the coalescence line are able to line up by elastic interactions.

Nevertheless, imperfections are seen at the interface. The most commonly observed configuration of interfacial dislocation reaction, seen on Fig. 2, is a $\Delta/2 = D_{110}/2$ shift between Lomer dislocations subarrays. Two mechanisms can explain this shift:

(1) It could result from the interaction between the set of Lomer dislocations and a 60° dislocation as discussed by Zhu and Carter (1990). One of the 60° dislocation origins is due to the coalescence mechanism itself. Dislocations parallel to the line of coalescence do not interact

strongly and the distances between misfit dislocations from adjacent islands correspond to the the nucleation sites which are randomly distributed on the surface, at least for nominal orientations. When the minimum distance x between misfit dislocations in two adjacent islands is of the order of $2*\Delta$, the accommodation process introduces an extra dislocation. Its Burgers vector is a function of the distance x. If $x = \Delta$, no extra dislocation is needed to accommodate the misfit. If $x = 2\Delta$, an extra Lomer dislocation has to be introduced. In this case, The dislocations perpandicular to the coalescence line are able to line up by elastic interaction (Fig. 5a). If x is nearly equal to $3\Delta/2$, the accommodation requires only the introduction of a 60° dislocation to relax the residual lattice mismatch. As shown by Zhu and Carter (1990), this 60° dislocation introduces a $\Delta/2$ shift between the two subarrays of dislocations. The dislocation lines are then curved in the plane to give the $\Delta/2$ shift as shown Fig. 5b.

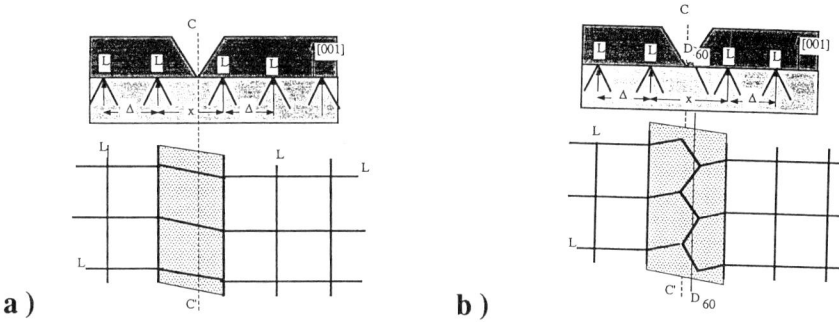

a)   b)

Fig. 5: Diagrams showing the interaction mechanisms involved during coalescence: a) $x<1.5\Delta$ self alignement of Lomer dislocations. b) $x > 1.5\Delta$: introduction of a 60° dislocation, which interacts with the Lomer dislocations in order to introduce a $\Delta/2$ shift.

(2) When the (001) interface contain steps, the scheme of Fig. 3 shows that the coincidence between the {110} planes is translated by half a period Δ. The dislocation site is translated by $\Delta/2$. In this case, as shown by Komninou et al. (1994), the shift experimentally seen is not explained by an interaction with a 60° dislocation but only by the presence of an atomic step on the surface. The continuity between the two shifted dislocations is obtained by a screw part of the dislocation.

In Fig. 2, each threading dislocation is seen at the end of $\Delta/2$ line shift. It is created by bending its line into the growth direction, whereupon it becomes a super-numerary dislocation at the level of the interface. Dislocations can be of either 60° or 90° types.

The effect of surfaces that have not been optimized before GaSb growth by a GaAs buffer layer is shown in Fig. 6: the islands are irregulary shaped, and the misfit dislocation network is not well organized as that for the best surface obtained by the deposition of a buffer layer seen Fig. 2.

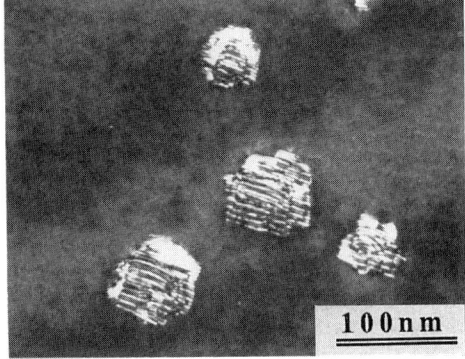

Fig. 6: Weak beam image of the 6Å GaSb layer without GaAs buffer layer: islands and misfit dislocation network are irregulary shaped.

140

The origin of the defects propagating into the epilayer is clearly established: for good growth conditions, the threading dislocations are created at the imperfections of the interface due mainly to the coalescence between islands. The imperfections are also related to the surface quality, which plays a role in both the nucleation site and the likelihood of developing a periodic growth process favorable to the organization of a well defined misfit dislocation networks.

The growth temperature is also a very important parameter (Kang et al. 1994a). At low temperature the number of nucleation sites generally increases (Biegelsen et al. 1990) and so does the number of islands for a given quantity of deposition, which in turn causes coalescence at low thickness. At 420°C the layer presents a uniform thickness after only 150Å. The dislocation arrays are well organized but with smaller perfect subarrays than those formed at 470°C. The density of defects measured in the region of the interface is $3x10^{10}/cm^2$ (Fig.7). At lower temperatures, the 2-D growth is attained more rapidly but the defect density is increased. At 470°C, uniform thickness was reached after coalescence for the deposition of at least 400Å. The interfacial defect density was $10^{10}/cm^2$ (Fig. 2). For a temperature of 520°C, the 150Å film consists of discrete islands with loss of geometric shape and appeerence of 60° dislocations network as shown in Fig. 8 (Kang et al. 1994a).

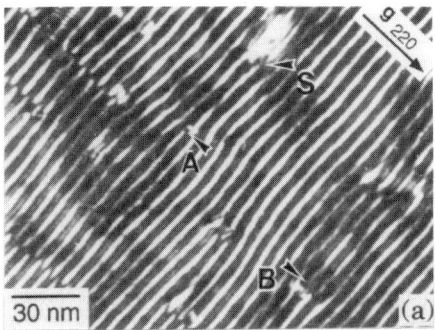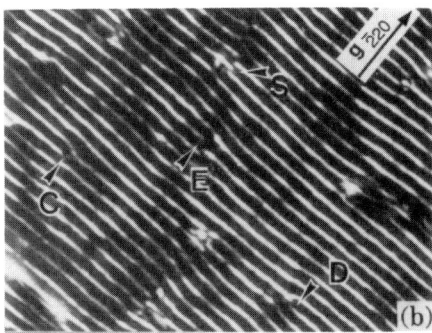

Fig. 7: Weak-beam image of 150Å thick layer of GaSb grown on (001)GaAs at 420°C: a) (220) and b) ($\overline{2}$20) DF images showing dislocation network. Note the threading dislocations A, B, C, D, E comming from 90° dislocations and S from 60° dislocation.

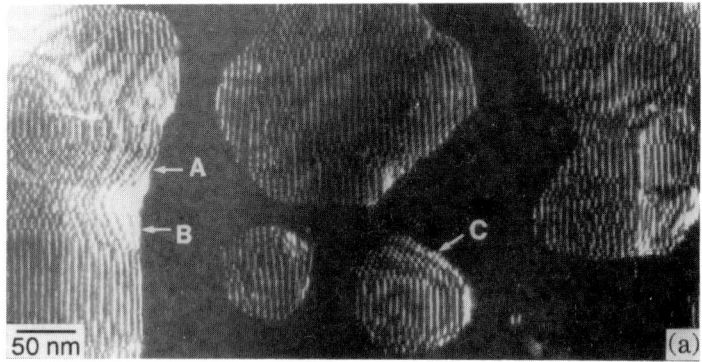

Fig. 8: Weak-beam image of 150Å thick layer of GaSb grown on (001)GaAs at 520°C: (220) DF image showing island shape and dislocation network poorly organized. Note that the A, B and C areas where the interfaces consist of 60° dislocation network.

At higher temperatures, the quality of the growth process decreases and the layer becomes more perturbated (Kang et al. 1994a). These results indicate that the density of defects in the interface could vary with the growth temperature as shown in Fig. 9. This function passes through a minimum at a temperature close to the conventional homo-epitaxy temperature of GaSb. The minimum interfacial defect density appears to be $10^{10}/cm^2$, about two orders of magnitude lower than that of GaAs/Si measured at the interface.

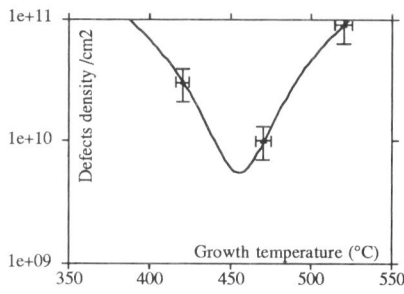

Fig. 9: Variation of the threading dislocation density at the GaSb/(001)GaAs interface with the growth temperature.

4. DISCUSSION

The common point between the GaSb/GaAs and GaAs/Si heterostructures is that in both cases, we hope to obtain a perfect epitaxial layer. As we have seen, when the growth conditions are optimized, the maximum density of defects appears to be created at the imperfections of the interface.

In the case of GaAs/Si, the density of defects in the epilayer is very high. Owing to the thermal annealing process, Lomer dislocation arrays are created with sizes limited to 60 nm. At the level of the interface, the density of threading dislocations is higher than $10^{12}/cm^2$. The areal density of dislocations decreases rapidly just above the interface and reaches a value of $10^8/cm^2$ for layers thicker than a few µm. The reduction is related to the elastic interaction between dislocations, which can be annihiled together. The optimum density of defects is limited by the efficiency of this interaction which is determined by the distance between dislocations. For $10^8/cm^2$, the mean distance between dislocations is about one µm and their interaction is limited. More investigation and modeling are needed to clarify this reduction effect.

The defect density of the relaxed heterostructure needs to be lower than the expected value to be compatible with practical applications in optoelectronics, i.e. between 10^4 to $10^6/cm^2$. This maximum value is about two order of magnitude lower than the density observed in GaAs/Si after annealing. The conventional annealing process is not capable of reducing the density of defects from 10^8 to $10^6/cm^2$: the annealing interaction between threading dislocations is governed by the dislocation distance, which in this case is larger than 1µm. The reduction of defect density by the thermal annealing process is certainly more efficient for GaAs on silicon than for the well organized system GaSb/GaAs, where the density of defects at the level of the interface is at least two orders of magnitude lower. It is not obvious that the final quality of a thick layer of GaSb on GaAs would be better than the GaAs/Si.

If we assume that the defect density needs to be lower than $10^6/cm^2$ for electronic applications, the hetero-interface must be perfect over a surface of about 100 µm². In the case of GaSb/GaAs, this interface must contain 4×10^6 dislocation intersections free of defects. The best experimental value is about 10^3 intersections free of defects in GaSb/GaAs (Rocher and Raisin 1991).

5. CONCLUSION

We have shown in this paper the importance of the quality of the initial surface of the substrate on the crystalline quality of the heterostructure. In order to obtain a perfect layer, the initial surface and the interface need to be perfect and uniform. The two examples discussed here show the role played by the surface on the crystalline quality of both the interface and the deposited layer. The perfect surface allows a well organized interface to be obtained in the GaSb/GaAs system by a periodic creation mechanism of misfit dislocations, for which the random effect is minimized.

The fundamental problem of heteroepitaxy is to obtain directly an epitaxial layer system with a defect density lower than $10^6/cm^2$. This goal requires a direct creation mechanism of misfit dislocations. In order to limit the coalescence effect which may introduce threading dislocations, the crystal growth conditions should minimize the density of nucleation sites, in the limit only one as for monocrystals which require only one seed. The conditions for the formation of a defect-free, (i.e. less than $10^6/cm^2$), hetero-epitaxial film on top of a substrate with a large lattice mismatch are difficult to determine.

Acknowledgements

The authors are indebted to P. Hawkes, A Claverie and M Benamara for helpful discussions.

REFERENCES

Aindow M, Cheng T T, Mason N J, Seong T-Y & Walker P J 1993, J Cryst Growth **133**, 168

Biegelsen D K, Bringans R D, Northrup J E & Swartz LE 1990, Appl. Phys. Lett. 57, 2419.

Bourret A & Fuoss 1992 Appl Phys Lett. **61** (9) , 1034.

Fang S F, Adomi K, Iyer S, Morkoç H & Zabel H 1990 J Appl Phys **68** (7), R31.

Hull R, Rosner S J, Koch S M &Harris 1986 Appl Phys Lett, **49**, 1714.

Kiely C J, Chyi J I, Rockett & Morkoç H, 1989 Phil Mag A, **60** (3), 321.

Kang J M, Thèse de l'Université Paul Sabatier, Toulouse, 1993.

Kang J M, Nouaoura M, Lassabatère L & Rocher A 1994a J. Cryst Growth **143**, 115

Kang J M, Min S K & Rocher A 1994b Appl. Phys. Let. **65**, 2954.

Knall J, Romano L T, Krusor BS, Biegelsen D K & Bringans R D 1994 J. Vac Sci Technol B **12**(6), 3069.

Komninou P, Stoemenos J, Dimitrikopoulos G P & Karakostas T, J Appl. Phys. 1994 **75** (1), 143.

Rocher A & Charasse M N , Sol. Stat. Phenomena 1989 **Vol 6&7**, 547.

Rocher A, Da Silva F W & Raisin C 1990, Rev Phys Appl **25**, 957

Rocher A & Raisin C, Springer Proceedings in Physics **Vol 54**, 1991 Polycrystalline Semiconductor II, Editors Werner J H and Strunk H P, 483

Zhang X, Staton-Bevan A E, Pashley D W, Parker SD, Droopad & Williams RL 1990 J Appl Phys **67** (2), 800.

Zhu J G &Carter C B 1990 Phil. Mag. A, **62**, 3, 319.

* Present address: Semiconductor Materials Laboratory, Korea Institue of Science and Technology, PO Box 131, Cheongryang, Seoul, South Korea.

Inst. Phys. Conf. Ser. No 146
Paper presented at Microsc. Semicond. Mater. Conf., Oxford, 20–23 March 1995
© *1995 IOP Publishing Ltd*

Growth mechanisms of semiconductor heterostructures with large misfits

D Gerthsen and K Tillmann*

Laboratorium für Elektronenmikroskopie, Universität Karlsruhe, Kaiserstr.12, 76128 Karlsruhe, Federal Republic of Germany
*Institut für Festkörperforschung, Forschungszentrum Jülich GmbH, Postfach 1913, 52425 Jülich, Federal Republic of Germany

ABSTRACT: The MBE growth of epitaxial layers with a mismatch of more than approximately 2.5 % often occurs in the Stranski-Krastanow mode which is investigated by transmission electron microscopy. The study focuses on the growth of $In_xGa_{1-x}As$ on GaAs(001) substrates. For indium contents $\geq$ 60 % a large fraction of the misfit is relaxed within layer thicknesses of less than 20 nm. A detailed analysis of the residual strain of the epitaxial islands was carried out to understand the mechanism of the misfit relaxation. The influence of the substrate temperature on the layer morphology was also examined.

1. INTRODUCTION

A large number of studies have been carried out about the epitaxial growth of lattice-mismatched semiconductor heterostructures. The emphasis was often put on the investigation of the generation of defects to optimize the growth process regarding minimal defect densities which are required for the fabrication of electronic and optoelectronic devices.

More recently, the interest was drawn towards heteroepitaxial systems with larger misfits where the three-dimensional growth modes prevail. The Stranski-Krastanow (SK) and the Volmer-Weber (VW) growth modes offer the opportunity to investigate quantum effects in self-assembled structures (quantum dots) if the growth is terminated during the early stage of the growth which was shown by Leonard et al (1993) and Nabetani et al (1994). For the SK growth mode epitaxial islands are nucleated on a few complete monolayers in contrast to the VW mode where the islands immediately grow on the substrate.

The present study concentrates on the growth of $In_xGa_{1-x}As$ on GaAs(001) substrates. The misfit m can be varied between 0 % and 7 % depending on the indium content. Despite the large number of reports about the InGaAs/GaAs system the results are often difficult to compare because the results of only one set of growth parameters is described. Therefore a large number of wafers was grown with a systematic variation of the parameters indium concentration, layer thickness and growth temperature with a particular emphasis on $In_{0.6}Ga_{0.4}As$/GaAs(001) with m = 4.1 % whose misfit is comparable to GaAs/Si(001) and to Ge/Si(001). A particular effort was made to characterize and to understand the strain state of the epitaxial layers in the early stage of the growth where a large amount of the misfit is already relaxed although a film of homogeneous thickness does not yet exist.

2. EXPERIMENTAL TECHNIQUES

The $In_xGa_{1-x}As$ layers were grown on GaAs(001) substrates using a Varian MOD Gen II system. To provide a clean surface, a 0.1 μm GaAs buffer layer was always deposited at a substrate temperature of 610 °C prior to the $In_xGa_{1-x}As$ growth.

Three different series of wafers are considered in this article. For the first wafer series the indium concentration was varied between 20 % and 100 % at a constant substrate temperature T_s = 500 ^oC and at constant deposition rates between 0.5 μm/hour and 1 μm/hour which are considered as standard growth conditions. The nominal layer thickness t_n was kept constant at 5 nm. For the second series, $In_{0.6}Ga_{0.4}As$ was grown with different nominal layer thicknesses of 2 nm, 3.5 nm, 5 nm, 10 nm, 20 nm, 200 nm and 1 μm under standard growth conditions. In a third series of wafers $In_{0.6}Ga_{0.4}As$ was grown with a constant nominal thickness of 3.5 nm corresponding to 12 monolayers at different growth temperatures ranging from 480 ^oC to 560 ^oC in intervals of 20 degrees. To compensate for the indium desorption at increasing temperatures the indium beam equivalent pressure was adjusted from 1.5 x 10^{-7} torr for 480 ^oC ≤ T_s ≤ 520 ^oC to 1.8 x 10^{-7} torr at 540 ^oC to 2.1 x 10^{-7} torr at 560 ^oC. The indium content was determined by X-ray diffractometry on 1 μm thick layers grown under identical conditions which yielded lattice parameters for $In_xGa_{1-x}As$ with x = 0.6 ± 0.02.

The layers were analysed by transmission electron microscopy (TEM). Plan-view TEM was carried out under Laue and two-beam diffraction conditions to study the layer morphology and the density of misfit dislocations. High-resolution transmission electron microscopy (HRTEM) was performed on cross-section specimens along the <110>-projection using a JEOL 4000EX electron microscope to examine the morphology and the residual strain of the layers on an atomic scale.

3. EXPERIMENTAL RESULTS

Cross-sectional images of the first series of wafers with a constant nominal layer thickness t_n = 5 nm grown at T_s = 500 ^oC reveal a gradual transition from the two-dimensional to the three-dimensional growth modes for indium concentrations $0.2 \le x \le 0.6$. For $In_{0.4}Ga_{0.6}As$ a closed layer with a significant thickness variation is observed which changes into a pronounced SK growth mode for $In_{0.6}Ga_{0.4}As$. The thickness of the continuous layer under the epitaxial islands for $In_{0.6}Ga_{0.4}As$ is in the order of 1 nm.

In the following the results of the $In_{0.6}Ga_{0.4}As$ layers from the second wafer series with different nominal thicknesses are presented. Plan-view images of the thinnest film with t_n = 2 nm did not reveal any contrast related to misfit dislocations. The HRTEM on cross-section specimens shows that small coherently strained epitaxial islands exist (Fig.1).

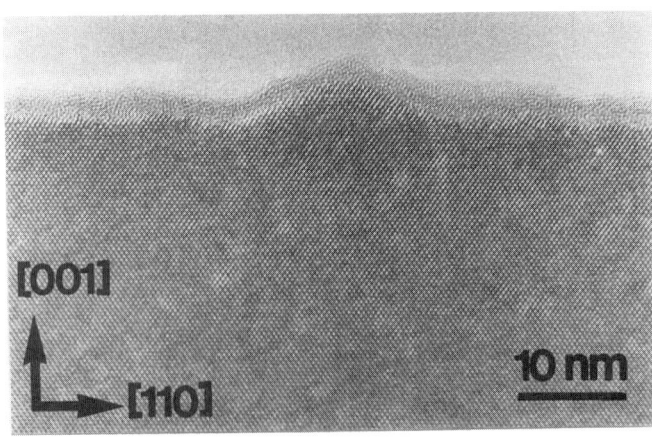

Fig.1: HRTEM cross-section image along the <110>-projection of a coherently strained epitaxial island of the layer with t_n = 2 nm

Fig. 2 depicts a plan-view image of the layer with t_n = 3.5 nm along the [001]-zone axis. The islands which are relaxed by misfit dislocations are characterized by moiré fringes along the [110]- and the [$\bar{1}$10]-directions. Weak-beam images show that the misfit dislocations are oriented along the <110>-directions with Burgers vectors perpendicular to the dislocation lines. The arrows mark islands without moiré fringes which are coherently strained.

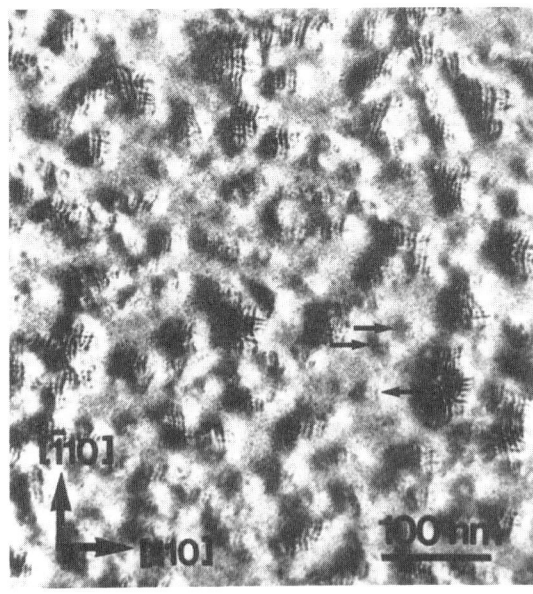

Fig.2: Plan-view image of the layer with $t_n = 3.5$ nm. Moiré fringes are visible for islands with misfit dislocations. The arrows mark coherently strained islands.

The residual strain in the islands, which is not relaxed by misfit dislocations, was determined along the two <110>-directions. The extension of a single island l was measured from the extension of the area covered by the moiré fringes. The number of dislocations in single islands was obtained by weak-beam TEM. The residual strain ε is given by Eq. (1) where b_e is the edge component of the misfit dislocation Burgers vector projected into the (001)-plane which amounts to $a/\sqrt{2}$ (a: lattice parameter of the InGaAs).

$$\varepsilon_n = m - \frac{nb_e}{l} \tag{1}$$

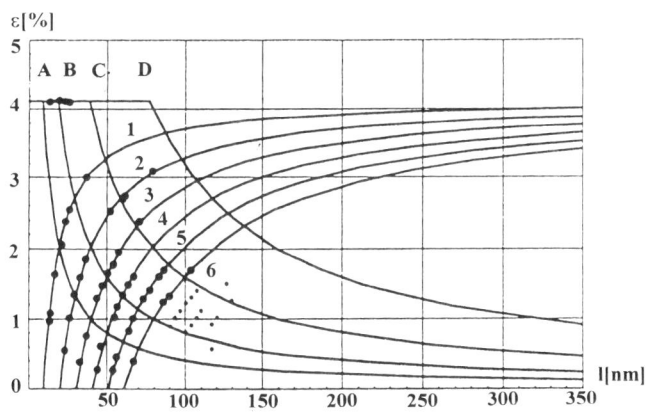

Fig.3: Measured residual strain of single islands ε as a function of the island extension l along the <110>- directions for $t_n = 3.5$ nm. The numbered curves are calculated according to Eq.(1) for $1 \le n \le 6$ misfit dislocations. The curves labeled A to D describe the expected relaxation of the misfit for different island height-to-extention ratios p = 1/8 (A), 1/15 (B), 1/30 (C), 1/60 (D).

Each point in Fig.3 represents the residual strain ε of a single island. The strain states of the islands cover a wide range from unrelaxed to almost completely relaxed. The curves labeled 1 to 6 are calculated using Eq.(1) for a constant number of dislocations $1 \leq n \leq 6$. The further analysis of the strain states and the curves A to D which describe the expected relaxation of the lattice-parameter mismatch will be presented in section 4.

For a more complete description of the island morphologies the ratios of the island height/island extention p were measured along the <110>-projection from HRTEM cross-section images. The results are plotted in Fig.4 for $t_n = 3.5$ nm. The height-to-extension ratio decreases with inceasing island size 1 because large island are often the result of the coalescence of small islands.

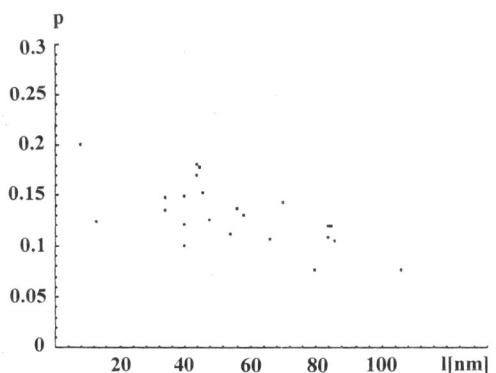

Fig.4: Measured height-to-extention ratios p for single islands as a function of the island extension along the <110>-directions for $t_n = 3.5$ nm.

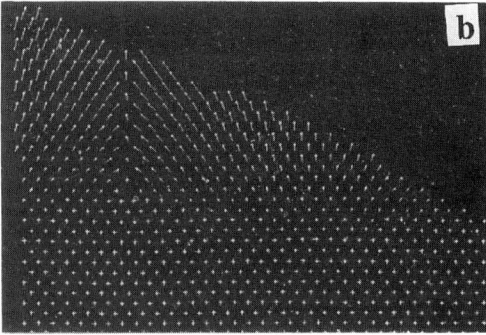

Fig.5:(a) <110> HRTEM cross-section image of an island of the layer with $t_n = 3.5$ nm. The arrows mark the positions of Lomer dislocations viewed end-on. (b) Processed image of the area within the frame of (a). The crosses mark the positions of the intensity minima with negligible deviation from the GaAs reference lattice. The directions and the lengths of the arrows indicate the deviations of the intensity minima in the InGaAs.

Fig.5a displays a HRTEM cross-section image of a typical island observed for the layer with t_n = 3.5 nm or t_n = 5 nm. The cores of three Lomer misfit dislocations viewed end-on are marked by the arrows. The Burgers vectors of the length $a/\sqrt{2}$ are oriented perpendicular to the dislocation lines and parallel to the interface. Two-beam $\vec{g} \bullet \vec{b}$ analyses verified that the Lomer dislocation is the dominating type of misfit dislocation.

The section of the image within the frame of Fig.5a was evaluated by the image processing routine DALI developed by Rosenauer (1995a, 1995b). The main steps of the routine involve the detection of the intensity minima and the generation of a reference lattice in the substrate which is superimposed on the epitaxial island. The crosses mark the positions of the intensity minima with a negligible deviation from the reference lattice which are located in the substrate. The deviations of the positions of the intensity minima in the InGaAs from the reference lattice are visualized by the lengths and the directions of the arrows. Assuming that the sample thickness does not significantly change in the investigated area the intensity minima are in a constant spatial relationship with the positions of the atomic columns. Although the original image does not provide any chemical contrast the location of the interface can be quite accurately determined in the region where the distortions first appear. The change of the directions of the arrows in the island shows that an ideal tetragonal distortion is only present in small regions of the island. An ideal tetragonal distortion would require a parallel alignment of the arrows. The deviation can be attributed to the elastic relaxation which is particularly obvious in the right part of the island where the arrows point outwards in contrast to the strain field to the right side of the dislocation core. With respect to the analysis of the lattice distortions on an atomic scale the residual strains plotted in Fig.3 must be considered as approximate values which are reduced by the elastic relaxation.

The distribution of the residual strains in the islands does not significantly change for t_n = 5 nm. At a nominal thickness of 10 nm the coalescence of the islands begins to dominate. Cross-section images show that threading dislocations are often generated in the regions where islands coalesce. At t_n= 20 nm (Fig.6) an almost closed layer has developed. The measurement of the residual strain in island-like areas yields that the strain state of the layer has homogenized at $\varepsilon \approx 1.4$ %.

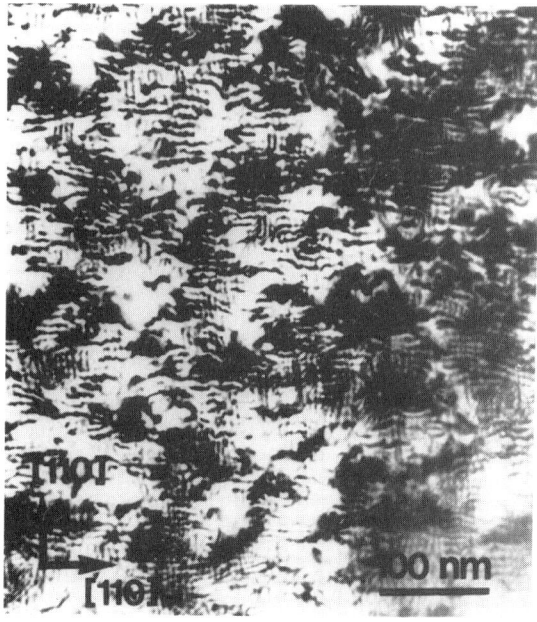

Fig.6: Plan-view image of the $In_{0.6}Ga_{0.4}As$ layer with t_n = 20 nm.

148

Fig. 7 shows plan-view images of the third series of wafers with constant $t_n = 3.5$ nm which were grown at different temperatures between 480 °C and 560 °C in intervals of 20 degrees. Apart from an increase of the average island size the situation does not change between 480 °C and 520 °C where islands in different states of relaxation are observed. At 540 °C small elongated islands appear in addition to the "conventional" islands. At 560 °C a complete transformation of the island shapes into the elongated islands has occured. The islands are exclusively oriented along one of the <110>-directions which will be denoted as the [$\bar{1}$10]-direction. The average extension of the cigar-shaped islands is 230 nm ± 25 nm along the [$\bar{1}$10]-direction. The lack of moiré fringes shows that the islands are not relaxed by misfit dislocations despite the large island sizes. The HRTEM on cross-section specimens yields island widths of 33 nm ± 3nm which can be associated with the narrow structures observed in Fig. 7e while the wider islands are likely to be generated by the coalescence of the narrow islands.

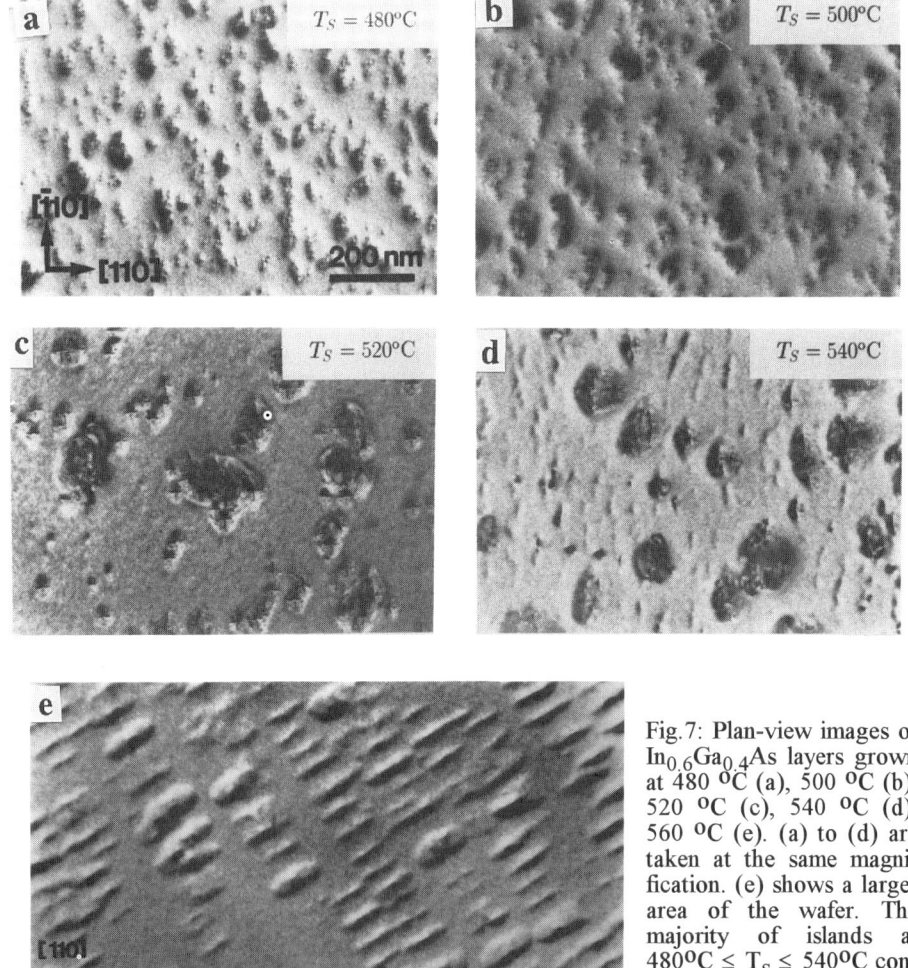

Fig. 7: Plan-view images of $In_{0.6}Ga_{0.4}As$ layers grown at 480 °C (a), 500 °C (b), 520 °C (c), 540 °C (d), 560 °C (e). (a) to (d) are taken at the same magnification. (e) shows a larger area of the wafer. The majority of islands at $480°C \leq T_S \leq 540°C$ contain moiré fringes. Only strain contrast is observed for the islands at $T_S = 560$ °C.

4. DISCUSSION

At the beginning of the MBE growth of $In_xGa_{1-x}As$ on GaAs(001) with $x \geq 0.3$ a coherently strained Stranski-Krastanow layer develops. There are considerable difficulties to study the deposition of the first monolayers (MLs) by TEM. However, in-situ scanning tunneling microscopy (STM) and reflection high-energy electron diffraction (RHEED) studies by Snyder et al (1991) have clearly demonstrated the transition from the two-dimensional growth to the nucleation of islands for indium concentrations $\geq$ 30 %. The thickness of the two-dimensional layer under the islands strongly depends on the indium content and on the growth temperature. The thickness varies between 1 ML for InAs (Nabetani et al 1994) and more than 10 MLs for $In_{0.35}Ga_{0.65}As$ (Grandjean and Massies 1993). For $In_{0.35}Ga_{0.65}As$ it was also shown that the transition from the two- to the three-dimensional growth occurs at 7 MLs for T_s = 530 °C and at 30 MLs for T_s = 480 °C. For $In_{0.6}Ga_{0.4}As$ small coherently strained islands can be observed by HRTEM on cross-section samples (Fig.1) at t_n = 2 nm which corresponds to 7 MLs.

The first misfit dislocations are generated for $In_{0.6}Ga_{0.4}As$ at nominal thicknesses between 2 nm and 3.5 nm. However, the strain state of the layer which consists of small epitaxial islands with a considerable variation of sizes is inhomogeneous ranging from completely strained to almost fully relaxed (Fig.3). Is is therefore more appropriate to define a critical island size instead of a critical layer thickness. To understand the measured residual strain values in the islands as a function of the island size two factors must be considered. In analogy to the growth of two-dimensional layers a model for the critical island size and the relaxation of the lattice-parameter mismatch in the islands can be established on the basis of the energy balance principle (review for the two-dimensional growth e.g. by van der Merve 1991). The second factor is the dramatic change of the residual strain in a small island if one dislocation is generated.

To calculate the relaxation of the residual strain in an island the strain energy E_ε and the energy of the dislocation network E_d are expressed as a function of the island size which is in the following characterized by the island volume V, the interface area A and the island height H. The strain energy of an epitaxial island is given by Eq.(2) where G_l is the shear modulus of the layer and v the Poisson number. The energy of the network of the two orthogonal sets of misfit dislocations is approximately given by Eq.(3) where e_d is the dislocation energy per unit length. The expression for e_d is a reasonable approximation during the early stage of the growth if the island height H is smaller than the average distance between the misfit dislocations d. During the later stages, H must be replaced by d/2. The length of the Burgers vector is denoted by b and the angle between the dislocation line and the Burgers vector by φ. The reduced shear modulus is given by $G_r = G_sG_l/(G_s+G_l)$ (G_s: shear modulus of the substrate). The dislocation density is expressed in terms of the misfit m, the residual strain ε and the edge component of the Burgers vector projected into the interface plane b_e. The minimization of the sum of E_ε and E_d with respect to ε yields the equilibrium residual strain Eq.(4) as a function of the island morphology.

$$E_\varepsilon = B\varepsilon^2 V \qquad \text{with} \qquad B = \frac{2G_l(1+v)}{(1-v)} \qquad (2)$$

$$E_d = \frac{2(m-\varepsilon)}{b_e} e_d A \qquad \text{with} \qquad e_d = \frac{G_r b^2}{4\pi}\left(\cos^2\varphi + \frac{\sin^2\varphi}{1-v}\right)\left(\ln\frac{H}{b}+1\right) \qquad (3)$$

$$\varepsilon = \frac{e_d}{Bb_e}\frac{A}{V} \qquad (4)$$

The island morphology is taken into account in Eq.(4) by the geometry factor A/V. Matthews (1972) already considered square tile-shaped islands with $V = H \, l^2$ which are assumed to grow with a constant height-to-extension ratio $p = l/H$. More realistic island shapes can be used for the calculation of $\varepsilon(l)$, e.g. a semiellipsoid with $V = \pi/6 \, l_l \, l_s \, H$ and $A = \pi/4 \, l_l \, l_s$ where l_l and l_s denote the large and the small axes of the semiellipsoid. The curves labeled A to D in Fig.3 were computed using Eq.(4) for semiellipsoidal islands containing Lomer dislocations with $l_l = l_s$ and constant height-to-extension ratios of 1/8 (A), 1/15 (B), 1/30 (C) and 1/60 (D). The critical island sizes l_c for the generation of the first misfit dislocation strongly depend on p. The majority of the experimentally determined height-to-extension ratios (Fig.4) lie between 1/5 and 1/15. The calculated curves reasonably describe the range of the observed residual strains. However, larger residual strains are also observed.

The deviations can be attributed to a number of simplifications in the approach presented above. The surface energy is completely neglected. The effect of the image force which strongly acts on dislocations close to a surface and the interaction between different dislocations are also not taken into account. The expression for E_ε is based on the assumption of an ideal tetragonal distortion of the unit cells in the epitaxial islands while Fig.5b show considerable deviations due to the elastic relaxation. The tendency towards larger residual strains could be a result of the elastic relaxation of the tetragonal distortion which leads to a reduction of E_ε given by Eq.(2) and therefore to larger residual strains.

The postponement of the generation of misfit dislocations due to the dislocation kinetics, which was shown to occur for SiGe on Si(001) with low Ge concentration at relatively low growth temperatures by Dodson and Tsao (1987), is not expected for $In_{0.6}Ga_{0.4}As$ where the large misfit induces high stresses in the order of several 100 MPa or even GPa in the coherently strained islands. The investigation of the dislocation velocities in InAs and GaAs by Yonenaga and Sumino (1992) showed that the dislocation mobilities are already high around 500 ^oC.

A second effect must be taken into account to understand the large variation of the residual strain in small epitaxial islands. It was already pointed out by Vincent (1969) that a discontinuous behaviour of $\varepsilon(l)$ occurs in small epitaxial islands which becomes very pronounced for islands with sizes of less than 100 nm. To demonstrate the effect $\varepsilon(l)$ was calculated for an $In_{0.6}Ga_{0.4}As$ island on a GaAs substrate (Fig.8). A tile-shaped geometry with $p = 1/20$ was adopted where the side edges of the island with the extention l and the dislocations are oriented along the two <110>-directions. The minimization of the sum of E_ε and E_d yields the discrete island extensions l where the misfit dislocations are generated. The residual strain changes from ε_{n-1} to ε_n according to Eq.(1) if the nth misfit dislocation is generated.

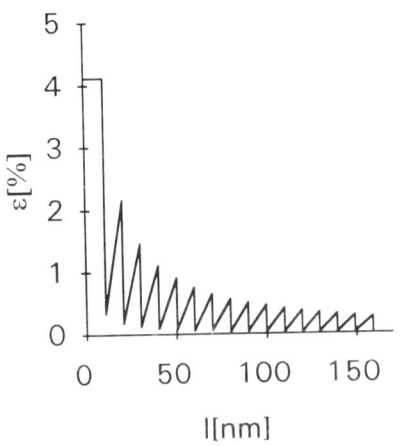

Fig.8: Calculated residual strain ε for one tile-shaped $In_{0.6}Ga_{0.4}As$ island with increasing extension l on GaAs(001) with a height-to-extension ratio of 1/20. The abrupt reductions of ε are induced by the generation of a new misfit dislocation.

According to Fig.8 the critical island size l_c is exceeded at 10 nm. The generation of the first misfit dislocation induces an abrupt drop of ε from 4.1 % to 0.4 %. If the islands size increases the strain also rises up to 2.2 % until the second misfit dislocation is generated at $l = 20$ nm. As a consequence a sawtooth-like behaviour of the strain is obtained in contrast to the two-dimensional layers. A reversal from a compressive to a tensile stress is theoretically possible for island sizes below 9 nm if the first dislocation is introduced.

With the concepts discussed above the measured broad distribution of strain states (Fig.3) in epitaxial islands can be well understood. The residual strain of the layer homogenizes when the islands coalesce which occurs around $t_n = 20$ nm under standard growth conditions for $In_{0.6}Ga_{0.4}As$. The details of the growth of thicker layers are given by Tillmann et al (1995a).

The third wafer series shows the influence of the growth temperature at a constant nominal thickness of 3.5 nm. The increase of the average island size results in a reduction of the island density. The results of the evaluation of the island density and the influence of vicinal substrates, which was also studied, are described by Tillmann et al (1995b). The preferential orientation of the islands at 560 °C along the $[\bar{1}10]$-direction on wafers which were accurately cut with a very small tolerance $0.02^0 \leq \Delta\alpha \leq 0.04^0$ parallel to the (001)-plane could be ascribed to the surface of the homoepitaxially grown GaAs buffer layer. STM investigations by Heller and Lagally (1992) show that homoepitaxially grown GaAs layers have a tendency to form terraces separated by straight steps along the $[\bar{1}10]$-direction and by highly kinked steps along the [110]-direction. This anisotropy is explained by the step edge energy anisotropy. It is reasonable to assume that the kink density will also affect the nucleation of the InGaAs islands which is in general expected to preferably occur at steps. The effect is only observed at high growth temperatures because a high adatom mobility is required to allow a sufficient number of the atoms to diffuse to the nearest step.

The second unexpected observation at 560 °C is the lack of misfit dislocations despite the large island size which can be explained by two effects. The elastic relaxation could occur in a more efficient manner at higher temperatures closer to the thermodynamical equilibrium. As a consequence the strain energy would be reduced resulting in an increase of the critical island size. A strong elastic relaxation was observed in $Ge_{0.85}Si_{0.25}$ islands grown by liquid phase epitaxy (LPE) on Si(001) by Christiansen et al (1994) which lead to very large coherently strained islands. The LPE operates considerably closer to the thermodynamical equilibrium than the MBE. A similar effect was observed by Eaglesham and Cerullo (1990) for Ge on Si(001) for a strongly reduced MBE growth rate of 0.02 μm/hour. The lowering of the growth rates also leads to an approach of the thermodynamical equilibrium resulting in a very efficient elastic relaxation.

The second effect is the loss of indium by the desorption at high temperatures. The indium loss was detected using the image processing routine DALI by the comparison of the distances of the (002)-planes in $In_{0.6}Ga_{0.4}As$ islands grown at 500 °C and at 560 °C where the tetragonal distortion was approximately reduced to one third. However, care must be taken to quantitatively determine the indium concentration because it cannot be excluded that the HRTEM specimens relax along the direction of the electron beam depending on the local thickness. A complete elastic relaxation along the electron beam direction would result in a reduction of the tetragonal distortion by a factor of approximately 1.3 compared to the bulk case. The effect of the elastic relaxation of the thin HRTEM samples still needs further investigation. The observation of an indium loss is in contradiction to the results of the X-ray diffractometry. However, the desorption could affect only the region close to the surface of the InGaAs layer which would only be a small fraction of the 1 μm thick layer. An additional driving force for the desorption in thin strained layers could be the reduction of the misfit which is achieved by the decrease of the indium concentration.

5. CONCLUSIONS

The MBE growth of InGaAs was investigated by transmission electron microscopy which an emphasis on higher indium concentrations where the Stranski-Krastanow growth mode prevails. A large number of wafers was grown with systematical variations of the parameters indium concentration, layer thickness and growth temperature.

152

It was shown that the growth temperature has a significant impact on the morphology of the layer during the early stages of the growth. Apart from the expected increase of the average island size a morphological transformation from isotropically-shaped into elongated epitaxial islands is observed which are exclusively oriented along one of the <110>-directions on nominal singular substrates. The shape transformation could be the result of a more efficient elastic relaxation. The preferential orientation along the [$\bar{1}$10]-direction could be related to the anisotropic   step structure of the homoepitaxially grown GaAs buffer layer. The desorption of indium at 560 °C was detected by measuring the (002)-plane distances.

For $In_{0.6}Ga_{0.4}As/GaAs(001)$ whose misfit of 4.1 % is comparable to Ge/Si or GaAs/Si a significant fraction of the lattice-parameter mismatch is already relaxed within nominal layer thicknesses of less than 20 nm where a layer of homogeneous thickness does not yet exist. A wide distribution of island sizes is observed under standard growth conditions during the early stage of the growth. The analysis of the residual strains in the islands yields values from completely strained to almost fully relaxed. Due to the inhomogeneous strain state of the Stranski-Krastanow layer it is more appropriate to define a critical island size instead of a critical layer thickness. The range of the relaxation states of the islands as a function of the island size is reasonably described by an energy balance model. Deviations towards larger residual strains could result from the elastic relaxation of the tetragonal distortion which could be visualized on an atomic scale.

The second effect that contributes to the inhomogeneous strain states of the epitaxial islands is the dramatic change of the residual strain if dislocations are generated in small islands. This effect will also be present under growth conditions where a narrow island size distribution is obtained.

ACKNOWLEDGMENTS

We are indebted to A Förster (Institute for Thin Film and Ion Technology, Research Center Jülich GmbH, FRG) for his expertise in growing a large number of MBE samples for our study. The help of A Rosenauer who developed the image processing program DALI (Digital Analysis of Lattice Images), and who installed the program at the Laboratory for Electron Microscopy in Karlsruhe is also greatly acknowleged.

REFERENCES

Christiansen S, Albrecht M, Strunk H P and Maier H J 1994 Appl. Phys. Lett. 64 3617
Dodson B W and Tsao J Y 1987 Appl. Phys. Lett. 51 1325
Eaglesham D J and Cerullo M 1990 Phys. Rev. Lett. 64 1943
Grandjean N and Massies J 1993 J. Cryst. Growth 134 51
Heller E J and Lagally M G 1992 Appl. Phys. Lett. 60 2675
Matthews J W 1972 Surface Science 31 241
Leonard D, Krishnamurthy M, Reaves C M, Denbaars S P and Petroff P M 1993 Appl. Phys. Lett. 63 3203
van der Merve 1991 Crit. Reviews in Solid State and Materials Science 17 187
Nabetani Y, Ishikawa T, Noda S and Sasaki A 1994 Appl. Phys. Lett. 76 347
Rosenauer A, Reisinger T, Steinkirchner E and Gebhardt W 1995a J. Cryst: Growth in press
Rosenauer A 1995b private communication, Institute of Solid State Physics, University of Regensburg
Snyder C W, Orr B G, Kessler D and Sander L M 1991 Phys. Rev. Lett. 66 3032
Tillmann K, Gerthsen D, Förster A and Urban K 1995a Thin Solid Films in press
Tillmann K, Gerthsen D, Förster A and Urban K 1995b presented at the Conference of Semiconducting Materials Oxford 20.3.-23.3.95
Vincent R 1969 Phil. Mag. 19 1127
Yonenaga I and Sumino K 1992 phys. stat. sol.(a) 131 663

Inst. Phys. Conf. Ser. No 146
Paper presented at Microsc. Semicond. Mater. Conf., Oxford, 20–23 March 1995
© 1995 IOP Publishing Ltd

The origin of misfit dislocations and the surface cross-hatch pattern in low-misfit strained epitaxial layers

R Beanland and A R Boyd*

GEC-Marconi Materials Technology, Caswell, Towcester, Northants, NN12 8EQ
*Department of Materials Science, The University of Liverpool, P.O. Box 147, Liverpool, L69 3BX

ABSTRACT: The relaxation of $In_xGa_{1-x}As$ layers on GaAs occurs through two 'waves' of dislocation introduction. The first occurs at the equilibrium critical thickness for misfit dislocation introduction, h_c. The second occurs when the layer thickness reaches $\sim 4h_c$. We present a study of this relaxation process using transmission electron microscopy and in-situ laser-light scattering. It is concluded that the second wave of misfit dislocation introduction is due to dislocation multiplication. This process also produces a 'cross-hatch' pattern on the surface of the layer.

1. INTRODUCTION

Strained epitaxial semiconductor layers are routinely grown on substrates which are so perfect that significant strain relaxation should never occur (e.g. Fitzgerald 1991, Beanland 1992). The density of dislocations threading from the substrate is typically below 10^4 cm^{-2} in GaAs, and below 10^2 cm^{-2} in Si. If each misfit dislocation arising from these sources extends to the edge of the wafer, the strain that can be relieved is only 0.05% in layers grown on 5 cm GaAs substrates, and 100 times less in layers grown on Si. Nevertheless, in layers which are sufficiently thick, strains of several percent are relieved by misfit dislocations. The origin of misfit dislocations in such systems is thus of great interest. In this paper we are concerned with $In_xGa_{1-x}As$ on GaAs, although the conclusions should apply to all systems where misfit relief is known to occur by the movement of glissile threading dislocations.

Several studies of $In_xGa_{1-x}As$ on GaAs (x < 0.25) have found that the strain in the layer at a given layer thickness is reproducible (Dunstan et al. 1994, Drigo et al. 1989). The strain of several $In_xGa_{1-x}As$/GaAs layers measured by double-crystal X-ray diffraction are shown in Fig. 1. Significant relaxation begins at a thickness $h_r = (0.8\pm0.1)/\varepsilon_0$ nm, where ε_0 is the misfit strain, and has been found to be independent of deposition technique and growth temperature (Dunstan et al. 1994). The strain in layers immediately above this thickness is also described by this law, i.e. $\varepsilon = (0.8\pm0.1)/h$, but at much greater thicknesses the strain tends towards a finite value. The critical thickness is roughly given by $h_c = (0.2\pm0.1)/\varepsilon_0$ nm, which is also shown in Fig. 1. This behaviour seems to be most consistent with dislocation multiplication rather than the effect of finite dislocation velocities or dislocation nucleation (Beanland et al. 1995a). Here, we describe a study of the relaxation processes which occur at h_r and the development of a distinctive surface topography - the well-known 'cross-hatch' pattern (e.g. Olsen 1975, Beanland 1995b) - which appears to be closely linked to significant relaxation.

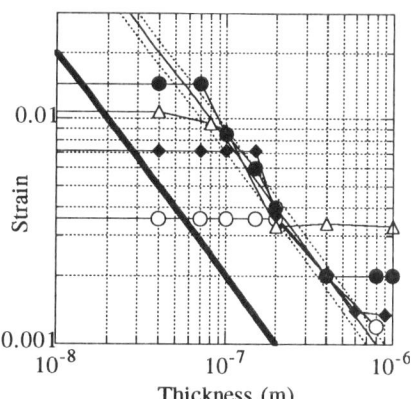

Figure 1. The relaxation of $In_xGa_{1-x}As$ layers on GaAs. The equilibrium strain according to the Matthews and Blakeslee model is shown by the thick solid line; the thin line with dashed bounds represents an empirical fit to the onset of relaxation h_r and the strain in layers with thicknesses just above h_r. Experimental points are x=0.05 ○, 0.10 ◆, 0.15 △, 0.20 ● and 0.25 ◇

2. IN-SITU LASER LIGHT SCATTERING OBSERVATIONS

Laser light scattering (LLS) is an emerging technique for the study of the evolution of surface roughness during epitaxial growth (e.g. Pidduck et al. 1992, Celii et al. 1993) and is straightforward to apply. In-situ LLS was performed during growth of $In_xGa_{1-x}As$ on (001) GaAs by chemical beam epitaxy (CBE) at a substrate temperature of 500°C. Ar^+ laser light was used ($\lambda = 514$ nm) with the incident beam normal to the surface and scattered light collected by a photomultiplier at 23° from normal. A typical LLS trace is shown in Fig. 2a. The growing GaAs surface is anisotropic, consisting of elliptical islands a few monolayers high with major axes along the $[\bar{1}10]$ direction (Fatt 1993, Beanland et al. 1995c), which produces the finite width of the LLS trace as the substrate rotates during growth (Fig. 2b). As soon as InGaAs growth commences, the islands become more elongated, as can be seen from the increase in the width of the LLS trace (Fig. 2b). At a thickness of about 110 nm, the cross-hatch pattern begins to appear, giving sharp peaks of intensity when the substrate is aligned along the <110> directions (Fig. 2c). The onset of the cross-hatch pattern measured from such traces is shown in Fig. 2d. The similarity between this graph and Fig. 1 suggests a strong link between significant relaxation and the surface cross-hatch pattern (Beanland et al. 1995c).

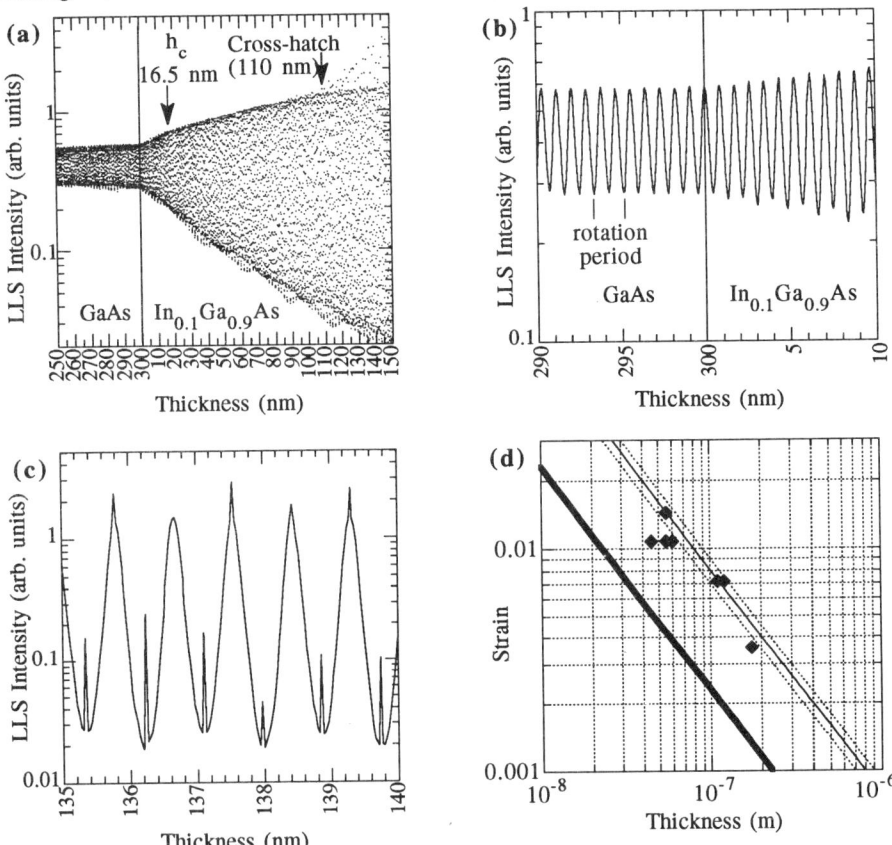

Figure 2. a) A LLS trace obtained during the growth of a GaAs buffer layer and a $In_{0.1}Ga_{0.9}As$ layer by chemical beam epitaxy at 500°C. b) and c) are extracts from the data set showing the oscillation of the intensity of scattered light due to the rotation of the substrate. (d) shows the onset of the cross-hatch pattern as measured by LLS in several such layers.

3. TRANSMISSION ELECTRON MICROSCOPE OBSERVATIONS

Plan-view transmission electron microscopy (PVTEM) was used to investigate the nature of misfit dislocation sources and any correlation between the misfit dislocation structure and the cross-hatch pattern in samples grown by CBE. Following the deposition of a 300nm thick GaAs buffer layer,

$In_xGa_{1-x}As$ layers were grown whilst being monitored in-situ using LLS. In samples which were required to be just above h_r, growth was stopped once the cross-hatch pattern appeared, resulting in a LLS trace similar to Fig. 2c. TEM specimens were prepared in two ways. Conventional specimens were prepared by mechanical back thinning to a thickness of about 100 μm followed by jet etching to electron transparency. Lift-off specimens were prepared by growing a 100 nm AlAs layer 100 nm below the $In_xGa_{1-x}As$ layer. Small pieces of the structure were immersed in HF for 24 hours, which removed the AlAs layer and allowed large electron transparent regions of the epitaxial layer to be picked up on a grid. Specimens were examined in JEOL 2000FX and 2000EX microscopes, the latter being equipped with a video recorder.

Misfit dislocations were found to be present in all layers with thicknesses greater than the equilibrium critical thickness h_c. Layers with thicknesses between h_c and h_r contained a low density of misfit dislocations, almost all of which were found to be of 60° type and to lie in the interface between the $In_xGa_{1-x}As$ layer and the GaAs buffer layer. Other studies have also shown (e.g. Whitehouse et al. 1993) that dislocations present in layers with thicknesses just above h_c arise from the threading dislocations in the substrate, and that once these dislocations are introduced there is essentially no misfit relief until h_r is reached.

An interesting phenomenon was observed in PVTEM specimens of layers with thicknesses just above h_r; the movement of a considerable proportion of dislocations in the specimen under the electron beam. Three types of moving dislocation were identified, which we call here type I and type IIa and IIb. All involve dislocations lying out of, but parallel to, the interface between $In_xGa_{1-x}As$ and GaAs. We shall call the dislocations residing in the epitaxial layer 'internal' dislocations in order to distinguish them from interfacial misfit dislocations.

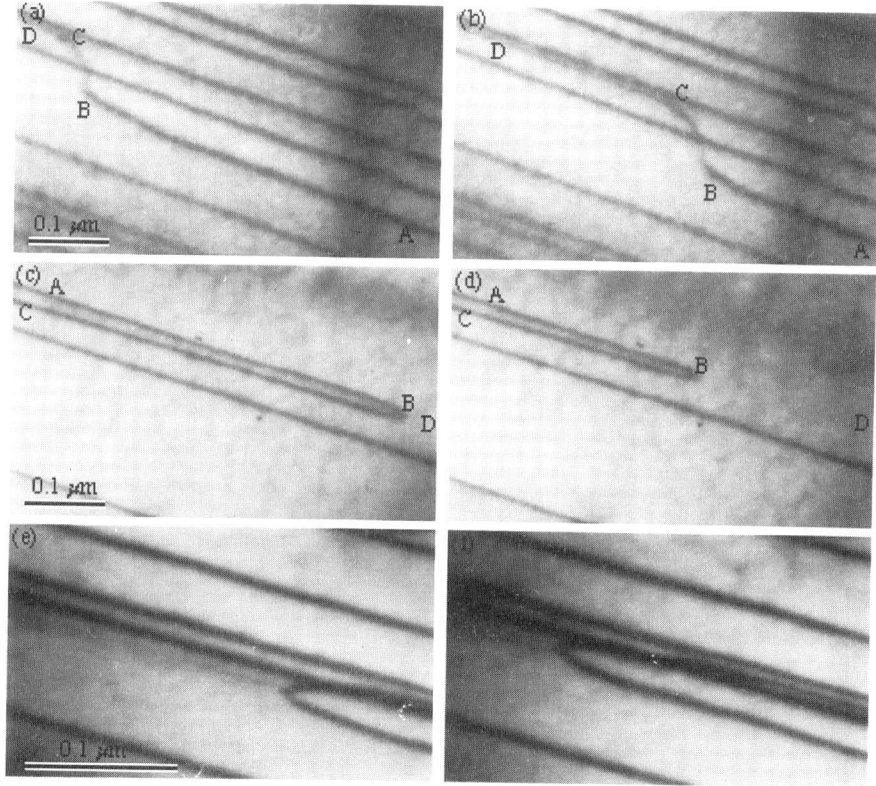

Figure 3. Dislocation movements in a lift-off PVTEM specimen of a layer of $In_{0.1}Ga_{0.9}As$ with thickness just above h_r. a) and b) Type I movement; A-B lies out of, but parallel to, the interface, whereas the segment C-D lies in the interface. c) and d) Type IIa movement, where two 60° dislocations in the interface (A-B and C-B) combine to form an edge dislocation lying out of the interface (B-D). The edge dislocation is invisible in the imaging conditions used. e) and f) Type IIb movement, which is the opposite reaction to type IIa.

156

A typical type I movement is shown in Figs. 3a and b. A glissile 60° dislocation, lying out of, but parallel to, the interface, moves into the interface when the electron beam is focused close to, or on, the dislocation. In all cases observed, the dislocation segment lying in the interface is very close to a parallel 60° dislocation. (We note that in three of the four possible configurations where two 60° dislocations are parallel the dislocations experience a mutual repulsion, since no energetically favourable reaction product can be formed.) Although most type I movements were of dislocations which initially lay below the interface, some involved dislocations lying inside the $In_xGa_{1-x}As$ layer as determined from stereo pairs. Type IIa movements involved the reaction $1/2[101] + 1/2[01\bar{1}] \rightarrow 1/2[110]$, and consists of two 60° dislocations in the interface which combine to form an edge dislocation lying out of the interface (Figs. 3c and d). Type IIb movements involve the opposite reaction (Fig. Figs. 3e and f). In all cases observed, Type IIa movements only occurred when the final pair of dislocations were relatively closely spaced, whereas Type IIb movements occurred when the initial pair of dislocations were further apart.

In order to investigate any correlation of dislocation microstructure with surface topography, atomic force microscopy (AFM) was performed on a lift-off sample followed by TEM of the same area. Correlation of the two techniques can be difficult due to the lack of accuracy in placing the AFM probe on a given area. Unfortunately, it has not been possible to image the long linear features responsible for the cross-hatch pattern (Beanland et al. 1995b) in both AFM and TEM. However, the feasibility of the technique is shown by Fig. 4a and 4b, which show the elliptical islands responsible for the log(sin) oscillation of LLS intensity with substrate rotation (Fig. 2b) and the underlying dislocation structure. Fig. 4c shows a pair of islands on the as-grown surface in greater detail.

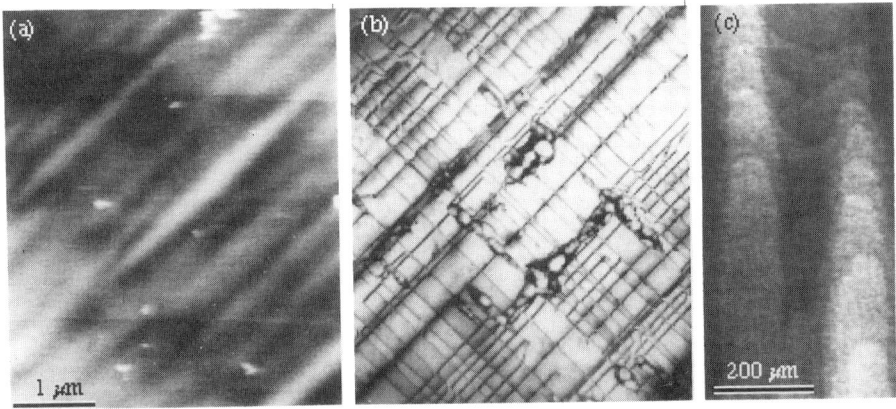

Figure 4. (a) and (b) are AFM and TEM images of the same region. Little correlation can be found between the two images, indicating that the dislocation structure has little influence on the elliptical islands. (c) An AFM image of two islands on the as-grown surface. Monolayer steps are easily seen.

In layers much thicker than h_r, internal 60° and edge dislocations are very common, as can be seen in the $g = 220$ PVTEM micrograph of Fig. 5a. Here, the specimen is wedge-shaped; the substrate and GaAs buffer layer have been removed completely above the line x-x, allowing the internal dislocation structure of the 400 nm thick $In_{0.17}Ga_{0.83}As$ layer to be seen in the upper part of the image. Many internal dislocations can be seen, both of edge type, which give double images and are invisible when $g = \bar{2}20$, and 60° type, which give single images. Some of these internal dislocations can be induced to move under the electron beam, although the details of their movement are difficult to distinguish due to the contrast of the interfacial dislocation array. In the very thickest layers grown, it was possible to examine the internal dislocation structure of the layer more easily, since the GaAs was removed from all electron transparent material. Long internal 60° and edge dislocations could be seen, and occasionally dislocation loops, or pile-ups, such as those shown in Fig. 5b, were observed.

4. DISCUSSION

These observations seem to indicate that internal dislocations are the cause - or the result - of the dislocation multiplication mechanisms responsible for significant plastic relaxation. Also, because of the link established between the cross-hatch pattern and the onset of relaxation, they are probably also responsible for the presence of the cross-hatch pattern. These proposals are made tentatively, since, despite the large number of PVTEM specimens examined, very little direct evidence of dislocation multiplication - apart from the presence of pile-ups such as shown in Fig. 5b - was seen.

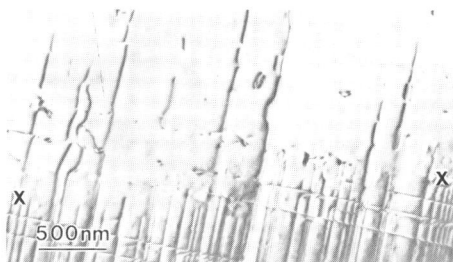 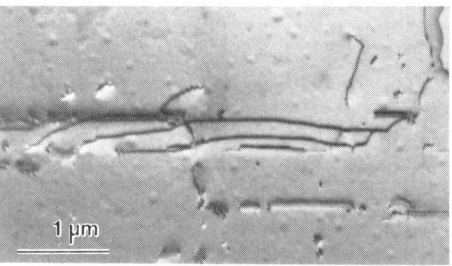

Figure 5a. A PVTEM micrograph of a 400nm thick $In_{0.17}Ga_{0.83}As$ layer. The substrate and buffer layer have been removed above x-x.

Figure 5b. A PVTEM micrograph of a dislocation source in a 1000nm thick $In_{0.17}Ga_{0.83}As$ layer.

It may be argued that this lack of direct observation is a simple consequence of the density of sources that are present and the contrast due to the interfacial dislocation array, which obscures the view of dislocations lying out of the interface. Even in fully relaxed $In_xGa_{1-x}As$ layers ($0.05 < x < 0.25$), which have misfit strains between 0.35 to 1.8%, the dislocation source density required is only 7×10^4 cm^{-2} to 3.6×10^5 cm^{-2}, assuming that dislocations travel to the edges of the wafer, and that each source produces only one new misfit dislocation. This is close to the lower detection limit of PVTEM; it is estimated that only one source is likely to be present in an entire PVTEM sample. Furthermore, in layers which experience plastic strains of 1% or more the interfacial and internal defect density is so high that it is difficult to observe dislocation configurations in detail.

The presence of internal dislocations indicates that there are many potential pinning points which lie out of the interface, which is a necessary criterion which must be satisfied if multiplication by spiral or Frank-Read sources is to take place (Beanland 1992). In particular, it seems worthwhile to examine the effect that internal dislocations will have on the mobile threading dislocations which produce the dislocation array (Fig. 6a). An internal 60° dislocation will repel or attract a threading dislocation which attempts to pass over it. The possible configurations have been described by Freund (1990), who considered the passage of a threading dislocation over an interfacial dislocation. In the case of a 60° internal dislocation, there are four possible configurations which are not symmetrically equivalent (Beanland 1995). In three of these four configurations, a repulsion occurs between the dislocations. In the fourth, the dislocations attract and an internal edge dislocation may be formed by cross-slip (Fig. 6b). Further movement of the threading dislocation segments gives a pair of spiral sources (Beanland 1995) (Figs. 6c, d). Configurations similar to those proposed were observed in layers with thicknesses just above h_r (Fig. 7). It is also interesting to note that the minimum layer thickness that is required for such a source to operate is calculated to be roughly $4h_c$ (Beanland 1992), which is in good agreement with the observed onset of significant relaxation (Fig. 1).

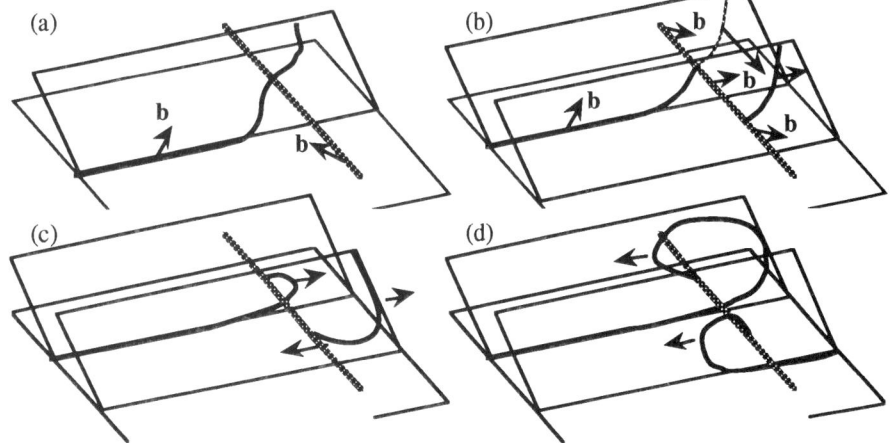

Figure 6. (a) A repulsion between a mobile threading dislocation and an internal dislocation may lead to the 'blocking' of the threading dislocation. (b) When an attraction exists between the two dislocations, the threading dislocation may cross-slip and run along the internal dislocation, converting it to edge type. (c) and (d) A second cross-slip event may result in a pair of spiral sources.

158

Figure 7. A PVTEM micrograph of a configuration consistent with a spiral source in a 140 nm thick In$_{0.1}$Ga$_{0.9}$As layer.

As noted above, the correlation between the onset of the cross-hatch pattern and the onset of significant relaxation implies that internal dislocations are in some way responsible for the change in surface topography. The most widely favoured explanation of the origin of the cross-hatch pattern is that a local change in growth rate results from the varying strain at the surface when misfit dislocations are present (e.g. Fitzgerald et al. 1993). If this is indeed the case, dislocations which lie out of the interface may be expected to have a stronger effect on the surface structure than interfacial dislocations. Indeed, there is some evidence that there is a direct link between internal dislocations and surface ridges (Beanland et al. 1995c, 1995d).

However, it should be noted that there are alternative explanations which explain the correlation equally well. For example, since the density of misfit dislocations is much larger than that of internal dislocations, variations in the spacing of misfit dislocations may produce an effect just as large as that of internal dislocations. Furthermore, when significant plastic relaxation occurs, a high density of slip steps is produced on the surface parallel to the <110> directions which may have a significant effect on the surface topography.

5. SUMMARY AND CONCLUSIONS

We have described TEM and LLS studies aimed at determining the nature of dislocation introduction mechanisms responsible for the relaxation of low-misfit strained layers. No direct evidence of multiplication was found in layers which had experienced small amounts of plastic relaxation. However, it was found that dislocations which lie outside, but parallel to, the interface were present, many of which move under the electron beam when examined in PVTEM. Layers which have experienced significant amounts of relaxation contain a high density of dislocations which lie parallel to the interface but inside the strained layer, which we call 'internal' dislocations. It is proposed that internal dislocations can act as pinning points for mobile threading dislocations, and a simple multiplication mechanism, which consists of a pair of spiral sources, has been described. Configurations which are consistent with the operation of such sources have been observed.

In-situ laser-light scattering observations of the surface topography show that the 'cross-hatch' pattern begins to form at the same thickness as the onset of significant relaxation. However, the origin of this surface roughness is unclear at present. The most likely explanation appears to be that the varying strain field at the surface produced by internal dislocations and/or misfit dislocations produces a local change in growth rate, although other mechanisms seem equally likely at present.

REFERENCES

Beanland R 1992 J. Appl. Phys. 72 4031
Beanland R 1995 J. Appl. Phys. 77 6217
Beanland R Dunstan D J and Goodhew P J 1995a Scanning Microscopy 8 859
Beanland R Boyd A R and Joyce T J 1995b J. Cryst. Growth, to be submitted
Beanland R Aindow M Kidd P J Lourenço M A Joyce T J and Goodhew P J 1995c J. Cryst. Growth 149 1
Beanland R Sacedon A Calleja E and Aindow M 1995d J. Appl. Phys., in press
Celii F G Kao Y C Liu H Y Files-Sesler L A and Beam E A 1993 J. Vac. Sci. Tech. B11 1014
Dunstan D J Kidd P Fewster P F Andrew N L Grey R David J P R González L González Y Sacedon A and González-Sanz F 1994 Appl. Phys. Lett. 65 839
Drigo A V Aydinli A Carnera A Genova F Rigo C Ferrari C Franziosi P and Salviati G 1989 J. Appl. Phys. 66 1975
Fatt Y S 1993 Semicond. Sci. Tech. 8 509
Fitzgerald E A 1991 Mat. Sci. Reports 7 87
Freund L B 1990 J. Mech. Phys. Solids 35 657
Olsen G H 1975 J. Cryst. Growth 31 223
Pidduck A J Robbins D J Cullis A G Leong W Y and Pitt A M 1992 Thin Solid Films 222 78
Whitehouse C R Barnett S J Usher B F Cullis A G Keir A M Johnson A D Clark G F Tanner B K Spirkl W Lunn B Hagston W E and Hogg J C H 1993 Microscopy of Semiconducting Materials VIII Inst. Phys. Conf. Ser. No. 134 563

Inst. Phys. Conf. Ser. No 146
Paper presented at Microsc. Semicond. Mater. Conf., Oxford, 20–23 March 1995

Diamond defects and the nucleation of misfit dislocations in ZnTe/GaSb

S Mylonas[*], D Cherns[*], D E Ashenford[**] and B Lunn[**]

[*] H. H. Wills Physics Laboratory, University of Bristol, Tyndall Avenue, Bristol BS8 1TL, UK
[**] Department of Engineering Design and Manufacturing, University of Hull, Cottingham Road, Hull, North Humberside HU6 7RX, UK

ABSTRACT: Transmission Electron Microscopy (TEM) studies of ZnTe/(001)GaSb films, grown by Molecular Beam Epitaxy (MBE), have shown the nucleation of 1/2 <110> misfit dislocations of 60° character at a new type of diamond defect lying on {111} planes. Analysis showed that the diamond defects were of vacancy type with 1/3 <111> Burgers vectors. Misfit dislocations were generated by an unfaulting mechanism, occasionally observed under the action of the electron beam. A new nucleation mechanism is proposed which explains the experimental observations.

1. INTRODUCTION

The introduction of 60° 1/2 <110> misfit dislocations into (001) semiconductor heterostructures to relieve misfit strain has been widely studied. However, although the role of pre-existing 1/2 <110> dislocations as sources, and the subsequent growth of 60° dislocations segments by glide, are quite well understood, few new sources of these dislocations have been identified. An exception is the diamond defect found by Eaglesham *et al.* (1989) in some Si/SiGe foils. These defects were identified as 1/6 <114> dislocations of interstitial character and are believed to act as regenerative sources for the nucleation of 1/2 <110> dislocation loops (Humphreys *et al.* (1989)).

In this paper we show that a new type of diamond defect is present in ZnTe/(001)GaSb foils grown by molecular beam epitaxy and acts as a non-regenerative source of 1/2 <110> misfit dislocations. It is shown that this diamond defect, although identical in geometry to those in Si/SiGe, has, in contrast, opposite character (vacancy type) and a different Burgers vector (1/3 <111>). Our observations show that many of the new diamond defects are associated with 1/2 <110> misfit dislocations. Based on the observed configurations, a nucleation mechanism is proposed whereby the 1/2 <110> dislocations arise by unfaulting of the diamond defect. It is shown that this reaction proceeds occasionally under the action of the electron beam.

2. THE MATERIAL

The samples were grown by molecular beam epitaxy in a VG V80H system at the University of Hull, UK. Growth was carried out using GaSb(001) substrates, with a substrate temperature of about 300°C and at growth rates of 1-2Åsec⁻¹. Studies were carried out on plan view TEM specimens and for ZnTe layer thicknesses in the range 0.15-0.8µm. The back-thinning of the samples was performed using mechanical thinning followed by Argon ion thinning at 4kV at an angle of incidence of 18°. For the final stages of the ion thinning the voltage was lowered to 3kV, at an angle of incidence of 15°, in order to minimise the damage of the foils. The experiments were carried out on a Philips EM430 electron microscope at 250kV.

3. EXPERIMENTAL OBSERVATIONS AND ANALYSIS

The observations carried out on ZnTe/GaSb films, which have a low natural mismatch (0.07%), showed the presence of a low density of 1/2<110> interfacial dislocations, as expected. In addition, stacking fault pyramids and small faulted loops, which lay on {111} planes and had edges along <110> directions, were observed. Fig. 1 shows an area from a 0.22µm ZnTe/GaSb sample containing several such loops. These loops were named "diamond defects" (Cherns *et al.*, 1995), by analogy with similarly shaped loops found in Si/SiGe foils (Eaglesham *et al.*, 1989).

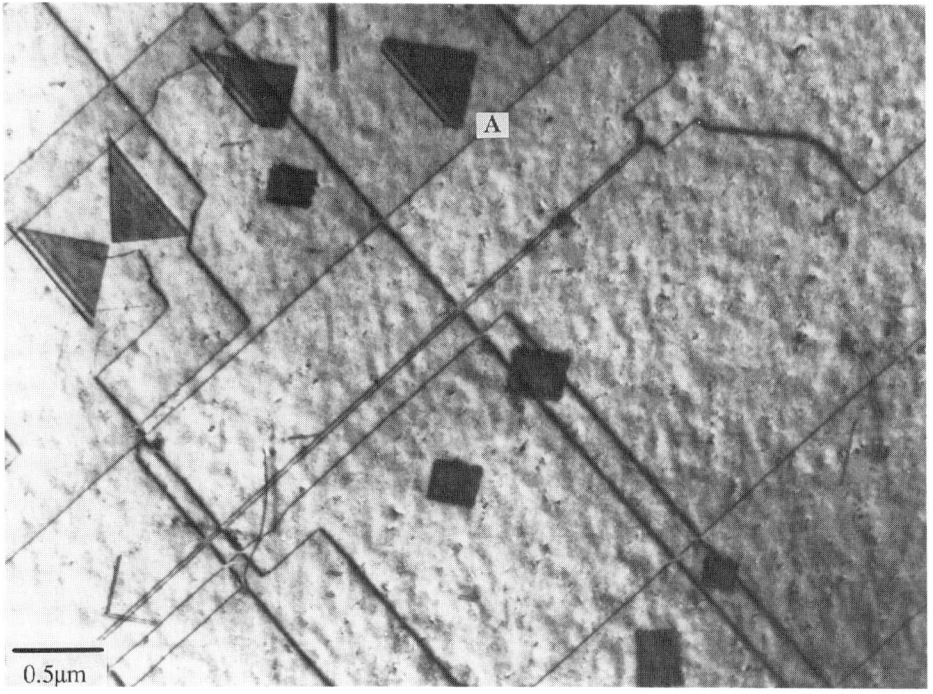

Figure 1. Diamond defects in a 0.22µm ZnTe/GaSb bicrystal. The diamond defects lie on either of the inclined {111} planes. The defect marked A intersects the growth surface.

The diamond defects were analysed using standard methods and some details of the analysis have been described elsewhere (Cherns *et al.*, 1995). Briefly, examination of the contrast under standard two-beam imaging conditions suggested a Burgers vector perpendicular to the loop plane. The sign of the Burgers vector was examined using inside-outside contrast and found to be of vacancy type. Moreover, the fact that some diamond defects intersected the growth surface of the foil (for example, see feature marked A in Fig. 1) allowed us to use the standard dark field imaging method (Gevers *et al.* (1963)) to confirm that the stacking fault was intrinsic. These results imply that the diamond defects were standard vacancy Frank loops.

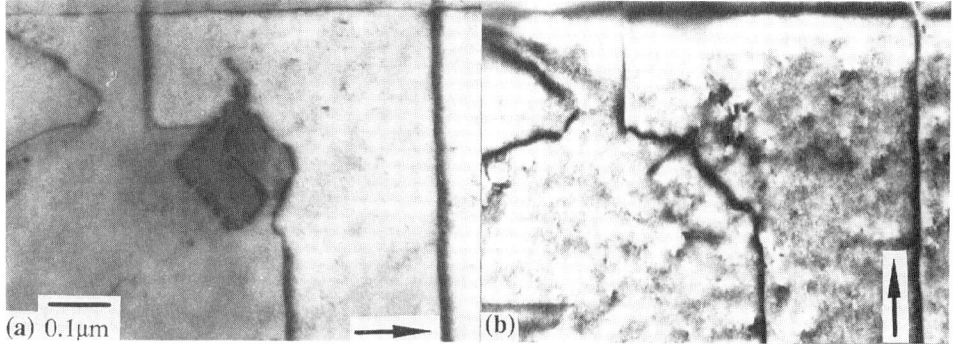

Figure 2. A diamond defect with an associated dislocation imaged under (a) **g**=220 and (b) **g**=2$\bar{2}$0. The directions of g are shown.

Examination of a large number of diamond defects showed that many were associated with 1/2 <110> misfit dislocations. A typical example is shown in Fig. 2. The diamond defect itself is in contrast in **g**=220, and out of contrast in **g**=2$\bar{2}$0, consistent with the Burgers vector of 1/3 <111>. A full analysis gave the Burgers vectors of the diamond defect and the attached misfit dislocation, as summarised in Fig. 3(a) using Thompson notation.

Stereo pairs of micrographs showed that the segments 1-2 and 6-7 were in the ZnTe/GaSb interface and that the remainder of DA looped up over the diamond defect. A close examination showed that the segment 3-4 was of type Aδ, produced by a reaction

$$AD + D\delta \rightarrow A\delta \tag{1}$$

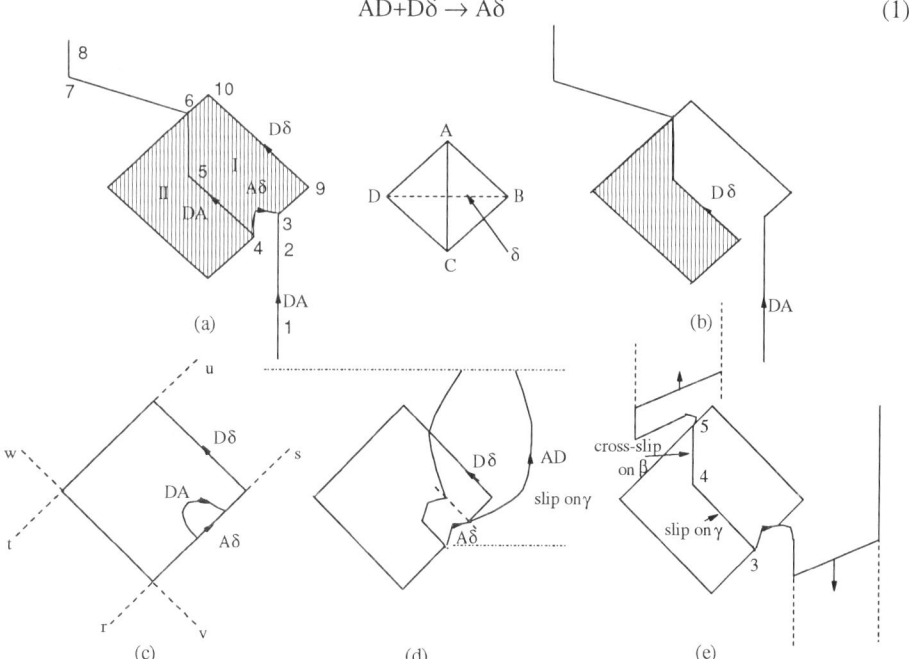

Figure 3. Schematic representation of the unfaulting mechanism.

162

During examination of the diamond defect in Fig. 2, a sudden unfaulting reaction was observed. The final configuration is shown in Fig. 4; this was reanalysed and the results summarised in Fig 3(c).

4. NUCLEATION MODEL

The dislocation density in the examined samples varied with the thickness of the epilayer. The interfacial dislocations were present, and with a dislocation spacing of about $1\mu m$, when the ZnTe layer had a thickness of $0.22\mu m$, but were seldom observed when the ZnTe thickness was less than $0.18\mu m$. At the same time, diamond defects were present in both cases but decreased in number with increasing ZnTe thickness. The inverse proportionality of diamond defects and misfit dislocations, together with the observed connection between faulted and unfaulted diamond defects and misfit dislocations, can be used as the basis for a nucleation model. Figure 3 summarises the proposed nucleation mechanism, based on the configurations in Figures 2 and 4. It is assumed that the fault on δ is an intrinsic Frank dislocation, $D\delta$ in Thompson notation. Dissociation of the segment $D\delta$, which was assumed to lie originally along the inclined <110> direction at the intersection of slip planes δ and α (i.e. the direction BC), can thus take place by the reaction

$$D\delta \rightarrow DA+A\delta \qquad (2)$$

This produces a perfect dislocation DA which can glide on slip plane γ and a Shockley dislocation $A\delta$ which is glissile on δ. Glide of DA on γ and then cross-slip on β is then needed to produce the observed interfacial dislocation. Finally glide of $A\delta$ on δ can take place to partially remove the stacking fault as was observed.

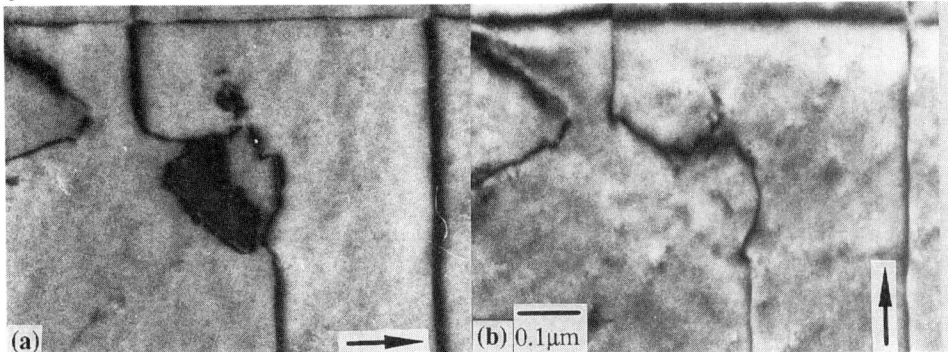

Figure 4. The diamond defect with the associated dislocation of Fig. 2 **after** the unfaulting reaction imaged under (a) $\mathbf{g}=220$ and (b) $\mathbf{g}=2\bar{2}0$.

5. CONCLUSIONS

The proposed nucleation mechanism was consistent with most diamond defect/misfit dislocation interactions. A closely similar nucleation mechanism was found to operate at stacking fault pyramids, and details will be given in a forthcoming publication (Cherns *et al.*, in preparation).

REFERENCES

Humphreys C J, Maher D M, Eaglesham D J and Salisbury I G 1989 Inst. Phys. Conf. Ser. No. 100 241-252.
Cherns D, Mylonas S, Chou C T, Wu J, Ashenford D E and Lunn B 1995 Scanning Microscopy, in press.
Eaglesham D J, Mayer D M, Kvam E P, Bean J C and Humphreys C J 1989 Phys. Rev. Letters 62 187-190.
Gevers R, Art A and Amelinckx S 1963 Physica Status Solidi 3, 1563-1593.

Inst. Phys. Conf. Ser. No 146

Paper presented at Microsc. Semicond. Mater. Conf., Oxford, 20–23 March 1995

Morphology and strain relief in the InGaAs/GaAs epitaxial system

A G Cullis, A J Pidduck and M T Emeny

DRA Malvern, St Andrews Road, Malvern, Worcs WR14 3PS

ABSTRACT: In the present work, morphological undulations which form upon heteroepitaxial $In_xGa_{1-x}As$ alloy grown on GaAs are examined by use of complementary transmission electron microscopy and atomic force microscopy. For an In x-value of 0.25, the manner in which final ripple arrays evolve from isolated islands is described and the stress interaction between islands is highlighted. The magnitude of the periodic elastic stress field which accompanies the formation of the ripple structures is microscopically measured and dilatation within the ripple crests is shown to yield essentially complete local misfit relief. Increased compressive stress present at ripple troughs is shown to lead to misfit defect source behaviour, which is expected to be of wide-ranging importance for defect generation in strained, undulating epitaxial films in general

1. INTRODUCTION

For heteroepitaxial materials exhibiting large mismatch stress, the formation of surface undulations of specific types can lead to elastic stress relief and a lowering of the system free energy: under these conditions surface growth fluctuations are favoured. This situation occurs in certain growth regimes for SiGe deposition on Si, such that ordered arrays of surface ripples can be produced on continuous layers (Srolovitz 1989, Cullis et al 1991, Robbins et al 1991, Cullis et al 1992, Pidduck et al 1992, Cullis et al 1994), while islands of deposit can be formed upon initial growth (Cullis and Booker 1971, Eaglesham and Cerullo 1990, LeGoues et al 1990). When layers of InGaAs alloy are grown on GaAs, the levels of compressive mismatch stress which can be encountered are similar to those occurring in the SiGe/Si system. A number of studies have shown (Berger et al 1988, Whaley and Cohen 1990, Guha et al 1990, Snyder et al 1991, Spencer et al 1991) that island formation in the early stages of growth can lead to elastic relaxation of the film lattice parameter. Defect evolution at later stages of growth in relatively thick single layers and multilayers (Salviati et al 1993) and compositionally-graded layers (Kavanagh et al 1995) has also been studied in detail. Furthermore, growth undulations in GaInP/InAsP multilayers have been analysed by Ponchet et al (1994).

In the present work, the growth of InGaAs alloy layers on (100) GaAs has been carried out by molecular beam epitaxy (MBE). *Ex situ* layer investigations have relied upon the complementary use of atomic force microscopy (AFM) and transmission electron microscopy (TEM) measurements. The combined results have allowed an understanding to be obtained of the regimes of morphological stability during the early stages of growth in the InGaAs/GaAs system.

2. EXPERIMENTAL DETAILS

The growth of $In_xGa_{1-x}As$/GaAs was performed in a VG V80H MBE system. All layers were deposited on (100) oriented GaAs substrates. Growth rates were maintained constant at $1.0ml.s^{-1}$ for the $In_xGa_{1-x}As$ alloys, with correction for any In desorption at the growth temperature (580°C) employed. The V/III beam equivalent pressure ratio was 5:1 for the alloy growth. Each wafer had 200nm of GaAs grown as a buffer layer, followed by a pause, before continuing with alloy growth. After the latter terminated, the substrate was cooled to room temperature as quickly as possible.

Layers were studied *ex situ* using AFM and TEM observations. The former employed a

Nanoscope III instrument using either tapping mode or contact mode with either a silicon or silicon nitride stylus. TEM studies were carried out using a JEOL JEM 4000EX microscope operated at an electron accelerating voltage of 400kV: specimens were thinned to electron transparency in cross-sectional configuration using sequential mechanical polishing and Ar^+ ion milling.

3. RESULTS AND DISCUSSION

3.1 Morphology and strain relief

Although for low $In_xGa_{1-x}As$ alloy x-values the epitaxial surface is planar (see eg Cullis et al 1995a), when the x-value approaches 0.25, layer growth rapidly becomes nonplanar. In fact, as has been found previously (Snyder et al 1992), layers of this type exhibit roughening only after a very thin initial flat film has been deposited in a classic Stranski-Krastanow mode. This is clear from Fig. 1a, which shows the formation of discrete islands, up to 3-4nm in height, for an $In_{0.25}Ga_{0.75}As$ alloy at a layer thickness of 5nm (after an initial ~3nm of planar growth). It is likely that this delay is associated, at least in part, with the build-up by segregation of the In concentration in the mobile atom component upon the layer surface. Over a critical surface In concentration range, it appears (Cullis et al 1995a) that changes in surface step configurations then can lead to the formation of randomly-positioned small islands. These are initially very elongated (Fig. 1a), yielding a shape that has a relatively high boundary length per unit volume. Since, as will be seen below, the unconstrained edges of such islands experience lattice relaxation, such 3-dimensional islands will receive additional stabilization.

The islands, once formed, grow in size and begin to interact through surrounding stress fields. This interaction leads to alignment of the islands with, for the growth temperatures examined, the ultimate formation of ordered arrays, which take on the form of surface ripples. This sequence is illustrated in Figs 1b and c which show layers of 10nm and 20nm, respectively, of the same alloy. For 10nm of growth, it is evident that the surface morphological structure has transformed into an array of linear ridges. Furthermore, for 20nm of growth the layer exhibits a dense array of island-like mounds and ridges of 10-20nm height and up to 100nm spacing. The undulations upon the latter thicker layers have the appearance of interacting ripples and run in a chevron pattern about the [011] axis, often with near <001> alignment. These types of surface morphology are typical of layers with high In x-values, but they can occur at different layer thicknesses.

The cross-sectional TEM image of Fig. 2a demonstrates, once again, that the surface undulations present upon the $In_{0.25}Ga_{0.75}As$ layer described above have the characteristics of an array of facetted ripples. The ripple troughs do not penetrate through to the underlying GaAs and, indeed, there are many similarities to rippled growth in the SiGe/Si system (Cullis et al 1992, Cullis et al 1994). The directionality of the ripple rows is particularly evident in the plan-view TEM image of Fig. 2b, where the local angular appearance of the ripple contrast is due to ripple facetting. Although only short, isolated misfit dislocations have been introduced for this layer thickness, ultimately, for an increase in thickness up to 40nm, many dislocations are introduced (Fig. 2c).

The manner in which the TEM image contrast exhibited by the surface ripple array varies with the operating Bragg reflection is shown in Fig. 3. All sections of the ripple structure give black/white contrast in either $0\overline{2}\,\overline{2}$ or $0\overline{2}2$ reflections (Figs. 3a,b) to yield images dominated by chevron patterns. However, with either $00\overline{4}$ or 040 reflections (Figs. 3c,d) the appearance of the ripple changes significantly and is dominated by linear black/white features lying normal to the g-vector of the reflection. The ripple row contrast in each of these images is due to lattice plane bending resulting from elastic relaxation in the ripple crests. The observed contrast variations can be accounted for by consideration of this lattice strain, as described elsewhere (Cullis et al 1992, Cullis et al 1995a). The image contrast observed particularly for 040-type reflections can be employed to microscopically measure the lattice distortion in the ripples by examination of the manner in which the contrast varies with angular deviation from the exact Bragg orientation. A study of the image contrast variation (Cullis et al 1995a) shows that contrast just disappears for a sample (positive) deviation of ~1.7° from the exact Bragg condition. This measurement can be related to the maximum lattice relaxation in the crests by use of an Airy stress function analysis (Cullis et al 1994) and indicates that the maximum strength of the periodic stress in the layer approaches the basic mismatch stress (compare the earlier large area

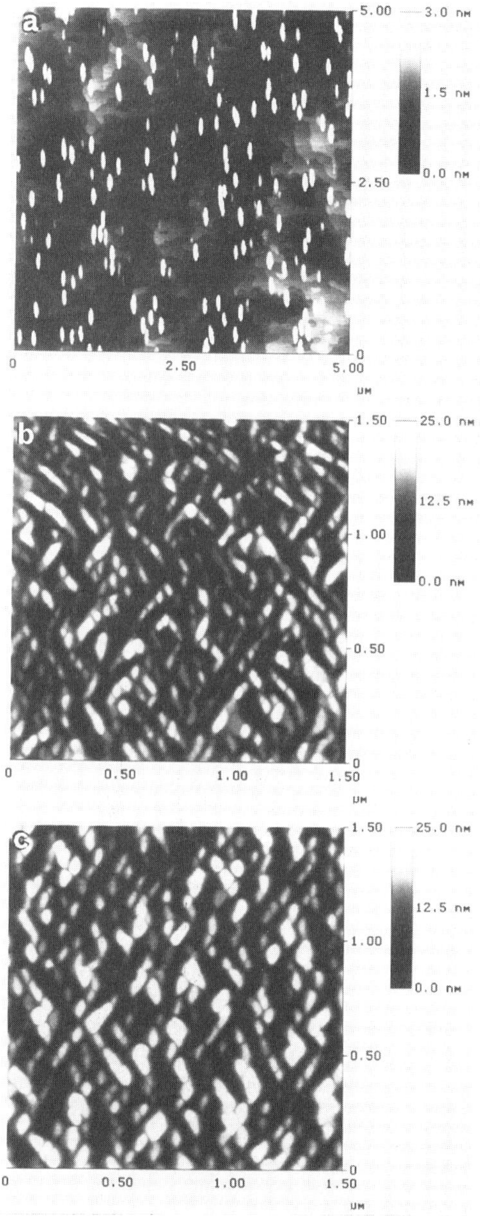

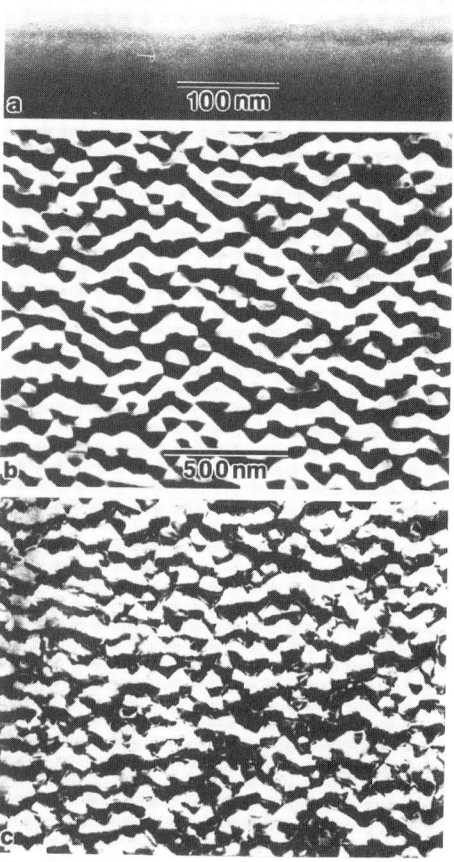

Fig. 2 TEM images of $In_{0.25}Ga_{0.75}$ As layers grown at 580°C a) cross-sectional 400 image of 20nm mean thickness layer, b) and c) plan-view $\mathbf{g}$=022 images of layers with, respectively, 20nm and 40nm mean thicknesses.

Fig. 1 AFM images of $In_{0.25}Ga_{0.75}As$ layers grown at 580°C a) 5nm thickness showing onset of roughening by formation of separated discrete growth islands b) 10nm thickness showing aligned linear undulations and c) 20nm mean thickness showing dense interacting array of island-like ripples.

RHEED studies by Berger et al (1988)). Alternatively expressed, the present work demonstrates that at the ripple peaks, where lattice expansion is a maximum, the mismatch stress is largely relaxed in the direction of the periodic field. This relaxation provides the substantial driving force for the formation of the ripple structures in the present epitaxial system.

It is important to note that there is a lattice compression of substantial magnitude in the immediate vicinity of ripple troughs. Indeed, due

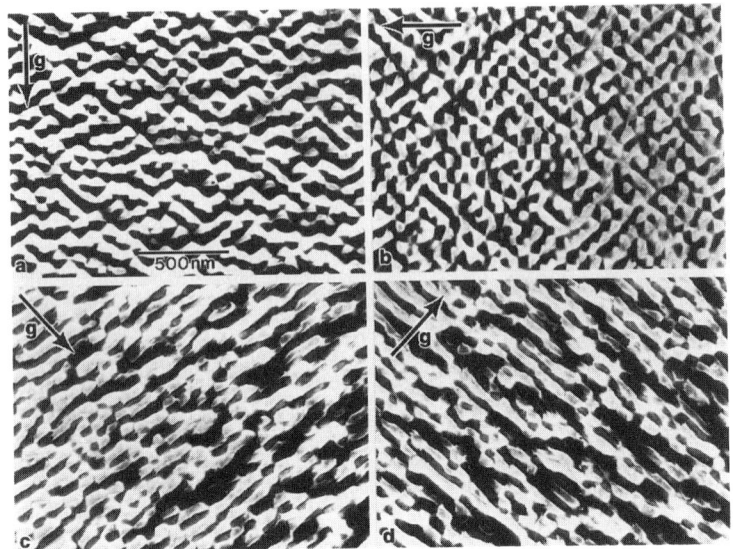

Fig. 3 TEM images of the same area of an $In_{0.25}Ga_{0.75}As$ layer grown at 580°C to a thickness of 20nm a) $g=0\overline{2}\,\overline{2}$, b) $g=0\overline{2}2$, c) $g=00\overline{4}$ and d) $g=040$.

to the observed facetting of the ripple slopes, the radius of curvature of the troughs is significantly smaller than that of the crests. Therefore, since the magnitude of the stress depends directly upon the local effective periodicity (Cullis et al 1995a), the compressive stress at a typical trough will be larger in magnitude than, although opposite in sign to, the expansion stress at a crest. In this way, although quite locally constrained, the total compressive stress at a trough may achieve values several times the basic mismatch stress. This behaviour is potentially important for defect source operation during misfit defect formation, as described in the next section.

3.2 Misfit dislocation nucleation

The initial misfit defects observed to form beneath smooth $In_xGa_{1-x}As$ layers undoubtedly result from the bending over of threading dislocations originally present in the GaAs substrate. This is the standard basic mechanism proposed by Matthews et al (1970) and, subsequently, dislocation multiplication processes may become important.

For morphologically undulating layers, the evolution of the misfit dislocation network is strongly determined by the configuration of the island or ripple arrays. The stress introduced by the individual elements of the ripple ensures that the final dislocation network structure conforms to the symmetry of the undulating array (see Fig. 2c). Of particular importance is the existence of new dislocation sources in the morphologically-distorted layers.

The 20nm thick, $x=0.25$ alloy layer shown in Figs. 2b and 3 contains only a distribution of individual, short misfit defect segments, since defect introduction had only just begun. The defects are of two distinct types: dislocations which vary their line direction to lie along the valleys formed by the ripple troughs and short straight defect segments lying along only one of the in-plane $<011>$-type directions. It is possible to examine the origin of the latter structures, but particularly short defect lengths are required. A suitable area of the layer was found near the edge of the substrate wafer where growth had taken place at a slightly reduced temperature. Figure 4 shows such an area of this layer which exhibits discontinuous ripple arrays. Within the rippled areas, short straight misfit defect segments are present and these are, respectively, in contrast in Fig. 4a and out of contrast in Fig. 4b (except for their terminations, visible due to surface strain-field relaxation). The latter image removes the defect/ripple contrast interactions and allows the two types of structural feature to be separately

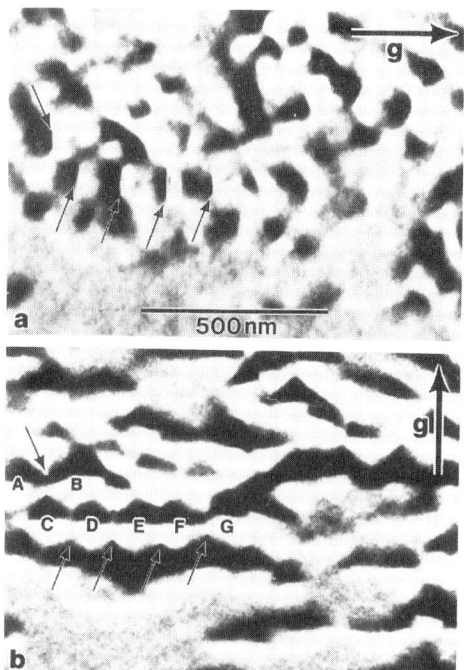

Fig. 4 TEM bright-field, plan-view images of 20nm $In_{0.25}Ga_{0.75}As$ layer showing misfit defects (arrowed) formed in troughs between surface ridges (A-G): a) $\mathbf{g}=0\overline{2}2$ and b) $\mathbf{g}=022$.

distinguished (the defects being located by their termination points). It is evident that the defect segments (arrowed) lie exclusively in the troughs between ripple ridge sections A-G, the ridges exhibiting dark/bright contrast inversions (with horizontal boundaries) as in the image of Fig. 4b. These locations are, therefore, identified as the sources of the defects (Cullis et al 1995a, 1995b). As was shown above, such troughs are the sites of greatly increased local compressional stress, which exceeds the basic misfit stress by a substantial factor. Therefore, as previously predicted (Freund et al 1989, Jesson et al 1993, Cullis et al 1994), this elevated level of stress is likely to be responsible for the way in which the troughs behave as misfit defect sources. Other misfit dislocations which exhibit various line directions are also seen (Cullis et al 1995a) to be pinned along interconnected ripple troughs. (Compare related observations for SiGe/Si by Albrecht et al (1995).)

Diffraction contrast analysis and high resolution imaging (Cullis et al 1995b) indicates that the straight defects have edge character and are likely to be faulted Frank dislocation half-loops on inclined {111} planes. These loops may form by vacancy aggregation in the compressive stress field of a trough, accompanied by the loss of interstitials at the free surface and the associated partial relaxation of the local stress. Such faulted half-loops have also been seen in epitaxial metal systems (Cherns and Stowell 1973). Furthermore, it is interesting that misfit relief in high-stress heteroepitaxial systems is commonly seen ultimately to involve the formation of sessile Lomer dislocations. In this regard, a reaction scheme (Cullis et al 1995b) by which an initial Frank half-loop can transform into such a Lomer dislocation involves combination of the Frank partial dislocation with a Shockley partial dislocation which moves in from the surface and annihilates the fault:

$$a/3[\overline{1}1\overline{1}] + a/6[21\overline{1}] \rightarrow a/2[01\overline{1}].$$

This formation mechanism would account for the common observation that sessile Lomer dislocations generally occur well beneath the surface of an epitaxial layer, despite the fact that they are unable to move by a glide process.

4. CONCLUSIONS

For $In_xGa_{1-x}As$ alloy layers on GaAs, as the In x-value approaches 0.25 there is a layer growth transition to a roughened structure due to 3-dimensional island formation after a small amount of smooth growth. Once formed, the 3-dimensional islands grow in size and, due to in-built strain, they interact with one another to form an ordered array which can have many of the characteristics of facetted surface ripples. The ripple rows lie preferentially along <010> directions and are strained with a periodic lattice expansion at the ripple crests. Direct microscopic measurement demonstrates that he local lattice dilatation in the outer regions of the ripple crests achieves a magnitude corresponding to essentially complete relaxation at these locations. This expansion, overall, provides partial elastic relaxation of the misfit and is the driving force for undulation formation through the establishment of surface adatom diffusion current.

168

When rippled alloy surfaces form, in addition to the lattice expansion at the ripple crests, there is a complementary additional lattice compression at the troughs. Although this is very localised, it can achieve a magnitude considerably greater than the misfit stress. As a direct result, the ripple troughs become sources of misfit defects, such that both edge and mixed dislocations can form in the trough regions.

REFERENCES

Albrecht M, Christiansen S, Hansson P O, Michler J, Strunk H P and Bauser E 1995 This Proceedings Volume, pp. 177-182

Berger P R, Chang K, Bhattacharya P, Singh J and Bajaj K K 1988 Appl. Phys. Lett. **53**, 684

Cherns D and Stowell M J 1973 Scripta Metall. **7**, 489

Cullis A G and Booker G R 1971 J. Crystal Growth **9**, 132

Cullis A G, Robbins D J, Pidduck A J and Smith P W 1991 in: Microscopy of Semiconducting Materials 1991, Eds A.G. Cullis and N.J. Long (IOP Publishing, Bristol) pp. 439-444

Cullis A G, Robbins D J, Pidduck A J and Smith P W 1992 J. Crystal Growth **123**, 333

Cullis A G, Robbins D J, Barnett S J and Pidduck A J 1994 J. Vac. Sci. Technol. A **12**, 1924

Cullis A G, Pidduck A J and Emeny M T 1995a J. Crystal Growth *in the press*

Cullis A G, Pidduck A J and Emeny M T 1995b Phys. Rev. Lett. **75**, 2368

Eaglesham D J and Cerullo M 1990 Phys. Rev. Lett. **64**, 1943

Freund L B, Bower A and Ramirez J C 1989 Mater. Res. Soc. Symp. Proc. **130**, 139

Guha S, Madhukar A and Rajkumar K C 1990 Appl. Phys. Lett. **57**, 2110

Jesson D E, Pennycook S J, Baribeau J-M and Houghton D C 1993 Phys. Rev. Lett. **71**, 1744

Kavanagh K L, Goldman R S and Chang J C P 1995 Scanning Microscopy **8**, 905

LeGoues F K, Copel M and Tromp R M 1990 Phys. Rev. B **42**, 690

Matthews J W, Mader S and Light T B 1970 J. Appl. Phys. **41**, 3800

Pidduck A J, Robbins D J, Cullis A G, Leong W Y and Pitt A M 1992 Thin Solid Films **222**, 78

Ponchet A, Rocher A, Ougazzaden A and Mircea A 1994 J. Appl. Phys. **75**, 7881

Robbins D J, Cullis A G and Pidduck A J 1991 J. Vac. Sci. Technol. B **9**, 2048

Salviati G, Ferrari C, Lazzarini L, Nasi L, Norman C E, Bruni M R, Simeone M G and Martelli F 1993 J. Electrochem. Soc., **140**, 2422

Snyder C W, Mansfield J F and Orr B G 1992 Phys. Rev. B **46**, 9551

Snyder C W, Orr B G, Kessler D and Sander L M 1991 Phys. Rev. Lett. **66**, 3032

Spencer B J, Voorhees P W and Davis S H 1991 Phys. Rev. Lett. **67**, 3696

Srolovitz D J 1989 Acta Metall. **37**, 621

Whaley G J and Cohen P I 1990 Appl. Phys. Lett. **57**, 144

Inst. Phys. Conf. Ser. No 146
Paper presented at Microsc. Semicond. Mater. Conf., Oxford, 20–23 March 1995
© 1995 IOP Publishing Ltd

TEM Observations of relaxation in $In_xGa_{1-x}As$ on (111)B GaAs

R Beanland, A Sacedon* and E Calleja*

GEC-Marconi Materials Technology, Caswell, Towcester, Northants, NN12 8EQ
*ETSI Telecomunicación, Departamento Ingenieria Electrónica, Universidad Politecnica Madrid, Ciudad Universitaria, 28040 Madrid, Spain

ABSTRACT: The relaxation of $In_xGa_{1-x}As$ layers on (111)B GaAs, offcut by 2° towards [2$\bar{1}\bar{1}$], is investigated. Most misfit dislocations are 60° type, gliding on {111} planes. The relaxation is highly inhomogeneous, with regions of both low and high dislocation density coexisting in the same sample. Large pile-ups of dislocations are present in the thicker layers. Contrast due to bunches of interfacial steps are also observed, which change in character close to dislocations. This indicates that diffusion across the interface can occur at a growth temperature of 500°C.

1. INTRODUCTION

Strained epitaxial III-V semiconductor layers grown on (111) substrates have recently become of interest because of the built-in field that can be produced due to the piezoelectric effect (e.g. Pabla et al. 1993, Smith and Maillhiot 1990 and references therein). The strain relief process is also of interest, since the difference in geometry from the more usual (001) growth provides a different set of misfit dislocations (e.g. Mitchell and Ubal 1991). In this paper, we report a transmission electron microscopy (TEM) study of misfit dislocations in $In_{0.1}Ga_{0.9}As$ and $In_{0.2}Ga_{0.8}As$ grown on (111)B GaAs offcut by 2° towards [2$\bar{1}\bar{1}$], and compare the relaxation processes with growth on (001) substrates. We also describe anomalous contrast at the interface between the $In_xGa_{1-x}As$ layer and the underlying GaAs, which we ascribe to the presence of bunches of interfacial steps.

2. EXPERIMENTAL

Layers were grown by MBE at a substrate temperature of 500°C. A 200nm thick GaAs buffer layer was deposited prior to growth of the $In_xGa_{1-x}As$ layer; further details are given elsewhere (Sacedon et al. 1994). Plan-view transmission electron microscope (PVTEM) specimens were prepared by mechanical back thinning to a thickness of about 100μm followed by etching to perforation using Cl in methanol. It was found that specimens could not be prepared by jet etching, since this resulted in very small amounts of thin area. However, it was found that etching using a solution in a rotating beaker, with the sample mounted on a PTFE disk using Lacomit varnish, provided good samples. All samples were examined on a JEOL 2000FX operating at 200kV.

3. DISLOCATION MICROSTRUCTURE

Although the most efficient misfit dislocations in (111) layers have Burgers vectors $a/2<\bar{1}10>$ and lie along $<11\bar{2}>$ directions, there is no easy mechanism of introduction to the interface. By analogy with (001) layers, misfit relief may be expected to take place by the movement of glissile $a/2<110>$ dislocations on {111} planes inclined to the interface. This produces a triangular array of 60° misfit dislocations with $<110>$ line directions, with two possible Burgers vectors for each line direction. Each dislocation can lie on two glide planes. Despite the similarity of these dislocations with those in (001) layers, the critical thickness in (111) layers is larger than in (001) because of the anisotropic nature of the elastic constants (Hornstra and Bartels 1978, Sacedon et al. 1994).

All the layers examined in this study were above the critical thickness, and all were found to contain misfit dislocations. A straightforward **g.b** analysis showed that most dislocations in layers just above the critical thickness were of $a/2<110>$ type, as expected. Typical observations of misfit dislocations in a PVTEM specimen of a 350nm thick layer of $In_{0.1}Ga_{0.9}As$ are shown in Fig. 1. In Fig. 1a, the dislocation spacing varies between 100 nm and 2 μm. In a different region of the same specimen, the dislocation spacing is closer to 20 nm (Fig. 1b). This kind of variation was found to occur in all specimens, irrespective of the layer thickness or composition; relaxation in these (111) layers appears to proceed by the localised production of high densities of dislocations, rather than a distributed array as in (001) layers. The coexistence of relaxed and unrelaxed regions has also been

170

detected by Raman spectroscopy (Sacedon et al. 1994). In both the images of Fig. 1, the specimen is wedge-shaped, with thicker material to the bottom of the picture. In the region of low dislocation density, interfacial roughness can be seen as a background texture at the bottom of the image. The line where the interface between the $In_{0.1}Ga_{0.9}As$ layer and the underlying GaAs intersects the wedge-shaped PVTEM specimen surface is easily visible, and it is clear that all dislocations lie in the interfacial plane. However, in the region of high dislocation density, several dislocations can be seen to lie inside the layer, since they extend past the line where the interfacial dislocations end. We call these 'internal' dislocations in order to distinguish them from interfacial misfit dislocations. The presence of internal dislocations in regions of high dislocation density was found to occur for all specimens examined.

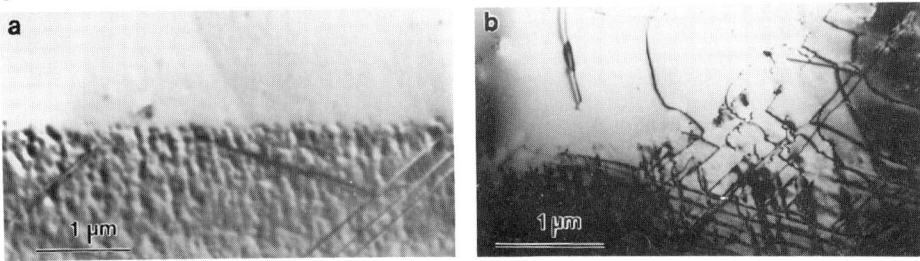

Figure 1. Bright field, $\mathbf{g}$ = 220 PVTEM micrographs of dislocations in a 350 nm thick layer of $In_{0.1}Ga_{0.9}As$. a) a region of low dislocation density, b) a region of high dislocation density. Both images are from the same PVTEM specimen, and are from regions only a few tens of μm apart.

Internal dislocations were often observed to have the form of complex pile-ups lying on the same inclined {111} plane. A typical example of this microstructure is shown in Fig. 2a, which shows several pile-ups in a 290 nm thick $In_{0.2}Ga_{0.8}As$ layer. Figs. 2b-d show two pile-ups in three different imaging conditions, illustrating that stacking faults and dislocations of more than one Burgers vector are present. In the thickest layers examined, it was generally found that dislocation pile-ups occurred in bunches, separated by regions of relatively low dislocation density where most dislocations lay in the interface. However, it should be noted that the dislocation density between the bunches of pile-ups was significantly higher in thick layers than in thin layers.

Dislocations in a pile-up were often observed to cross-slip, resulting in a distinctive geometry where a curved dislocation in a pile-up, bowing towards the layer surface, cross-slips onto an alternate glide plane and threads back down towards the interface. This can be seen in the stereo pair of Fig. 3, and is also present in Fig. 2.

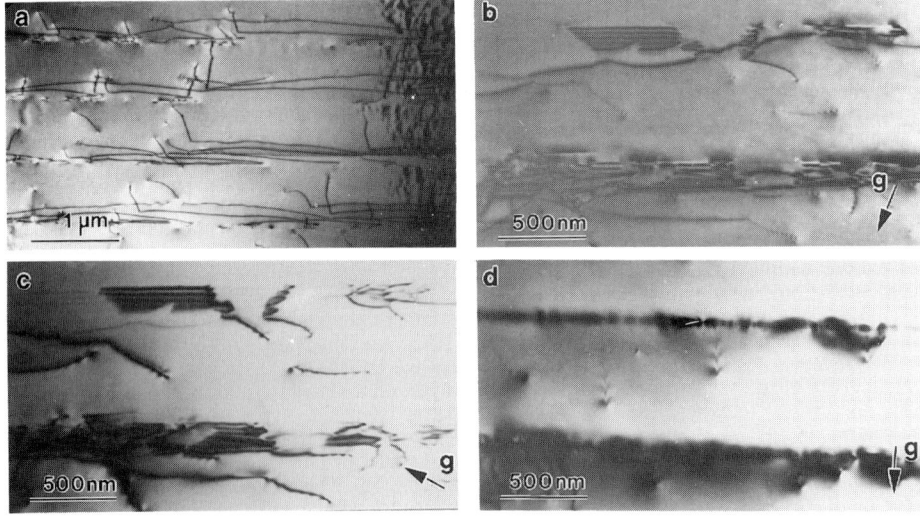

Figure 2. Dislocation pile-ups in a 290 nm thick layer of $In_{0.2}Ga_{0.8}As$. a) Several pile ups on parallel {111} planes. b), c) and d) Two pile-ups imaged in $\overline{2}20$, $22\overline{2}$ and $\overline{2}22$ two-beam bright-field conditions respectively. The arrows indicate the direction of $\mathbf{g}$.

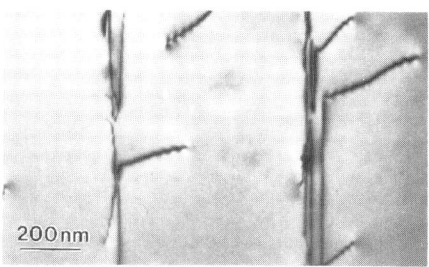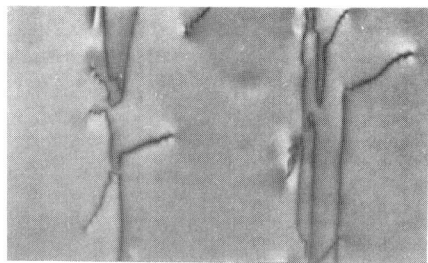

Figure 3. Stereo pair of a distinctive cross-slip configuration, found to occur in all complex pile-ups. Curved dislocations, bowing towards the layer surface, thread back towards the interface on an alternative glide plane. The layer is $In_{0.2}Ga_{0.8}As$, 290 nm thick.

The above observations show that the dislocation microstructure in (111)B $In_xGa_{1-x}As$ layers is rather different to that of (001) layers. In (001) layers, the dislocation density tends to increase in all regions of the material. Internal dislocations are observed in (001) layers which have relaxed significantly (Beanland and Boyd 1995), but they tend to be present alone rather than in pile-ups, and when pile-ups do occur they contain fewer dislocations and contain dislocations of only one Burgers vector. In (111) layers, dislocation pile-up formation appears to be considerably easier than in (001) layers. In both cases, however, the mechanism of pile-up formation is probably the same; spiral and Frank-Read dislocation sources. In both cases, the onset of significant relaxation is close to four times the critical thickness (Sacedon et al. 1995), which agrees very well with the minimum thickness required for this type of source to operate (Beanland 1992).

The primary cause of the difference in the relaxation behaviour of (111) and (001) layers is difficult to ascertain, since the exact nature of dislocation sources is not clear at present. However, Frank-Read and spiral sources seem most likely to form at intersections between dislocations which glide on different {111} planes, as this may provide a pinning point outside the interface about which cross-slip can occur (Beanland and Boyd 1995). When this occurs in (001) layers, both glide planes are occupied by dislocations, which will restrict the movement of any new dislocations produced. However, in (111) layers cross-slip may occur onto a third glide plane, which contains no pre-existing dislocations. This may allow large pile-ups to develop in (111) layers which cannot form in (001) layers.

3. INTERFACIAL STEP STRUCTURE

As noted above, contrast was observed in all samples which appeared to be associated with the interfacial plane. In some samples, the contrast had a textured appearance, whilst very strong linear contrast was observed in others. The strength of the contrast tended to increase with layer thickness and misfit strain. The most likely interpretation of this contrast appears to be bunches of steps in the interfacial plane. A step in an interface has similar topological properties to a dislocation in a bulk crystal; it must either form a closed loop or end at a surface. When the crystals on either side of the interface are different, a dislocation will generally be associated with the step (Pond 1989). In the case where the crystals have the same structure and are in parallel orientation, but have different lattice parameter, an interfacial step must have Burgers vector

$$\mathbf{b} = \varepsilon \, \mathbf{t} \tag{1}$$

where $\mathbf{t}$ is the translation vector which characterises the interfacial step and ε is the misfit strain. In the case of an offcut substrate, many interfacial steps will be present which have line directions parallel to the offcut axis. In the present case, the contrast due to an interfacial step one monolayer high would be expected to be too small to be observed by TEM, since it corresponds to a Burgers vector of only 0.0029 or 0.0057 nm for compositions of x=0.1 and 0.2 respectively. However, if several steps combine to form a single large step, they may become visible. This hypothesis is also consistent with the observation that the lines are not continuous. The contrast also varies with diffraction conditions, which implies that the lines have characteristics somewhat similar to edge dislocations (Fig. 4).

In this image, the lines have a mean spacing of ~170 nm, whereas interfacial steps due to the offcut of 2° are expected to have a spacing of about 9nm. If the interfacial step density is determined mainly by the offcut from (111), it appears that steps of height less than 5 nm, with associated Burgers vectors of less than 0.09nm, are clearly visible. Furthermore, since fine structure can be seen between the strong lines, it seems that monolayer height steps can also be seen. If this is the case, it is not clear why the contrast is so strong.

172

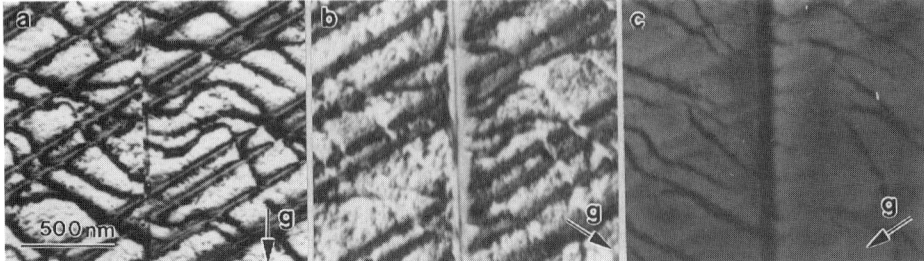

Figure 4. Interfacial contrast in a 47 nm layer of $In_{0.2}Ga_{0.8}As$, imaged under a) $g = 02\bar{2}$, b) $g = \bar{2}20$, and c) $g = 20\bar{2}$ bright-field conditions. The arrows indicate the direction of g. Several misfit dislocations can also be seen, most running from bottom left to top right and one running vertically

Interaction between the step bunches and misfit dislocations appears to be present in Fig. 4, with some step bunches terminating where they meet dislocations. In some layers, it was found that the step contrast became more disjointed close to the dislocation lines, especially where several

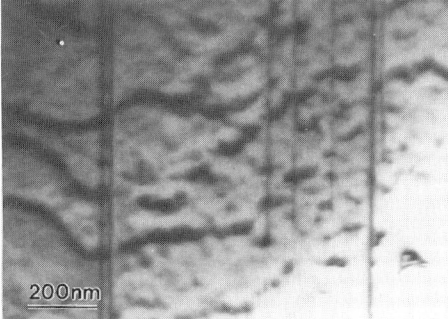

dislocations were present (Fig. 5). This seems to indicate that the interfacial step structure is not determined solely by the surface of the GaAs buffer layer, and that diffusion across the interface can occur during growth. If this is the case, local changes in composition must occur close to the $GaAs/In_xGa_{1-x}As$ interface.

Figure 5. Changes in the contrast from interfacial step bunches close to dislocations in a 88 nm layer of $In_{0.2}Ga_{0.8}As$. At the left of the image, the step bunches are relatively long, but at the right of the image where several dislocations are present, the lines become more disjointed.

4. SUMMARY AND CONCLUSIONS

We have described the microstructure of $In_xGa_{1-x}As$ layers grown on (111)B GaAs grown by MBE. We find that the relaxation is highly inhomogeneous. In layers just above the critical thickness, regions of low and high dislocation density are present. As the layer thickens, the dislocation density increases only slowly in regions of low dislocation density, whereas complex dislocation pile-ups, consisting of dislocations with different Burgers vectors and stacking faults, develop inside the $In_xGa_{1-x}As$ layer in regions of high dislocation density. It is proposed that basis of this relaxation process - the formation of spiral and Frank-Read sources - is generic to the relaxation of strained semiconductor layers. Anomalous contrast, emanating from the interface between the $In_xGa_{1-x}As$ layer and the underlying GaAs buffer layer, is ascribed to the formation of step bunches at the interface, although the contrast is far stronger than expected. From the observation that interactions may occur between these step bunches and the misfit dislocations, it is proposed that diffusion across the interface may occur during growth at 500°C.

REFERENCES

Beanland R 1992 J. Appl. Phys. 72 4031
Beanland R and Boyd A R 1995 these proceedings
Hornstra J and Bartels W J 1978 J. Cryst. Growth 44 513
Mitchell T E and Ubal O 1991 J. Electron. Mats. 20 723
Pabla A S, Sanchez-Rojas J L, Woodhead J, Grey R, David J P R, Rees J T, Hill G, Pate M A, Robson P N, Hogg R A, Fisher T A, Wilcox A R K, Whittaker D M, Skolnic M S and Mowbray D 1993 J Appl. Phys. Lett. 63 752
Pond R C 1989 in Dislocations in solids, ed F R N Nabarro (North Holland: Amsterdam), Chapter 38
Sacedon A, Calle F, Alvarez A L, Calleja E, Munoz E, Beanland R and Goodhew P J 1994 Appl. Phys. Lett. 65 3323
Sacedon A, Beanland R, Colson H and Calleja E 1995 J. Appl. Phys. to be submitted
Smith D L and Maillhiot C (1990) Rev. Mod. Phys. 62 173

Inst. Phys. Conf. Ser. No 146
Paper presented at Microsc. Semicond. Mater. Conf., Oxford, 20–23 March 1995
© *1995 IOP Publishing Ltd*

Study of relaxation in strained InGaAs/GaAs and (AlGa)InAs/GaAs heterostructures by TEM and HRXRD

N A Bert, N N Faleev and Y G Musikhin

A F Ioffe Physical-Technical Institute, St Petersburg, 194021, Russia

ABSTRACT: Transmission Electron Microscopy and High Resolution X-ray Diffraction have been used to study the strain relaxation in lattice-mismatched GaAs/InGaAs/GaAs and (AlGa)InAs/GaAs heterostructures. It has been shown that the density of defects at the top InGaAs/GaAs interface, as well as in the InGaAs layer, increases with increasing thickness of the top GaAs capping layer. The investigation of (AlGa)InAs/GaAs heterostructures has shown that the elastic strain relaxation depends on the initial surface conditions which are affected by the presence of a buffer layer.

1. INTRODUCTION

InGaAs/GaAs heterostructures still continue to attract much attention due to their importance for use in electronic and optoelectronic devices. Additionally, InGaAs layers incorporating Al, i.e. (AlGa)InAs, can be used as a source of polarised electrons. The strain relaxation behaviour of lattice-mismatched InGaAs/GaAs has been reported in many papers, and the increasing thickness of a capping GaAs layer has been shown to suppress the release of strain by Wang et al. (1994). Far less work has been carried out on the structural features of (AlGa)InAs/GaAs. In this paper we report the results of a study of elastic strain relaxation in both systems.

2. EXPERIMENTAL

Both GaAs/In$_x$Ga$_{1-x}$As/GaAs double heterostructures and (AlGa)$_{1-x}$In$_x$As layers were grown by MOCVD at 750°C on GaAs (001) substrates. The In content in the GaAs/In$_x$Ga$_{1-x}$As/GaAs heterostructures was approximately x = 0.20. The InGaAs layer thicknesses varied from 8nm to 31nm and the GaAs capping layers were 0.15μm or 1.3μm thick. The In content in the (AlGa)$_{1-x}$In$_x$As/GaAs was approximately x = 0.12. (AlGa)InAs layers, 130nm thick, were grown either directly on a GaAs substrate or on a complex buffer stack, consisting of a 10 x (AlAs)(GaAs) superlattice of period 105nm, plus a 1μm AlGaAs layer. A double-crystal, high-resolution, X-ray diffractometer TRS-1 and a Philips EM420 TEM were used in this study. X-ray diffraction patterns were taken around the (004) GaAs reflection. An asymmetrically cut Ge(001) crystal served as the X-ray incident beam conditioner. Plan-view and cross-sectional TEM specimens were prepared using conventional procedures of wet etching, mechanical treatment and Ar$^+$ ion-beam milling. TEM observations were performed, in Bright Field (BF) and Dark Field (DF) modes, together with selected area diffraction.

3. RESULTS

X-ray diffraction showed that the relaxation of misfit strain started at an InGaAs layer thickness of approximately 16nm, when the oscillation fringes on the diffraction curve began to blur, (see Fig.1a).

This may have resulted from surface roughness and appeared to be associated with the first stage of relaxation. X-ray diffraction analysis of the InGaAs/GaAs QW heterostructure with 31.5nm-thick InGaAs layers, (Fig.1b), indicated the possibility of the presence of clusters of point defects and dislocations, similar to previous observations of Faleev et al. (1994)).

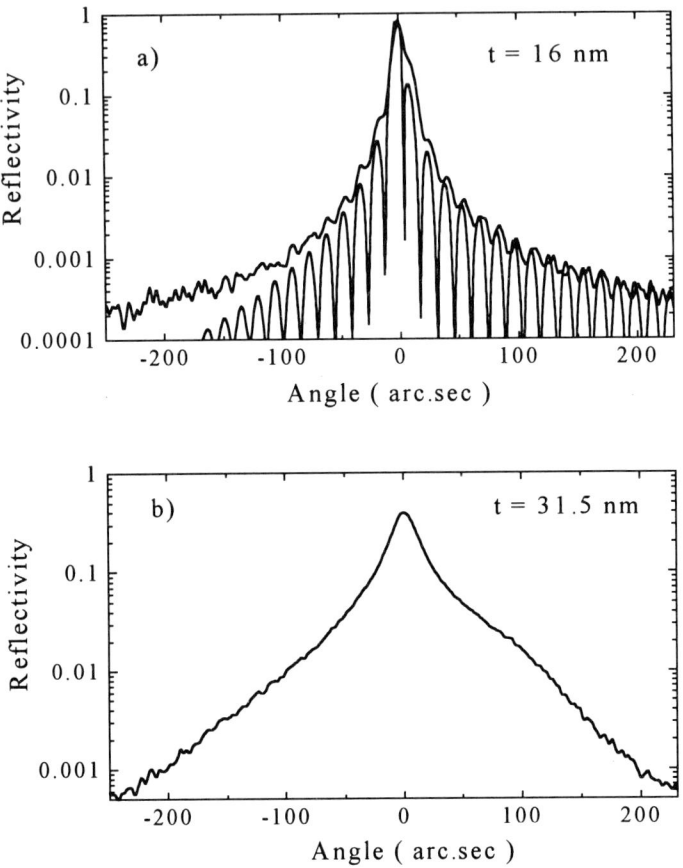

Fig. 1   X-Ray diffraction patterns of GaAs(1.3μm)/InGaAs/GaAs QW heterostructures in the vicinity of the (004) GaAs reflection. a) Thickness of InGaAs layer approximately 16nm, b) Thickness of InGaAs layer approximately 31.5nm.

TEM investigations of GaAs (0.15μm)/InGaAs(31.5nm)/GaAs structures showed the existence of dislocations at the bottom interface, as well as within the layer of InGaAs. The density of dislocations at the top interface was negligible (see Fig.2a).

An increase in thickness of the GaAs capping layer from 0.15μm to 1.3μm significantly increased the dislocation density, both at the top interface and within the InGaAs layer. This may be seen by comparing Figs. 2a and 2b. These results contradict the results reported by Wang et al. (1994). This may be explained by the fact that the dislocation mobility would be higher in the present work because of the higher growth temperature. Consequently the strain in the layer would not significantly affect dislocation glide to the interface.

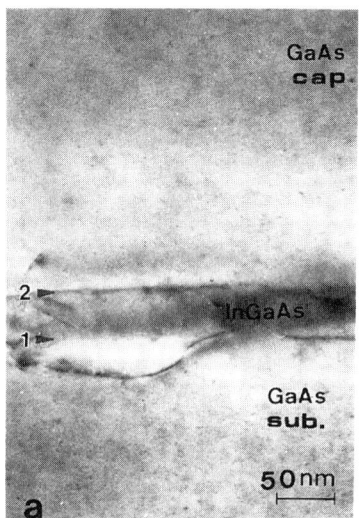
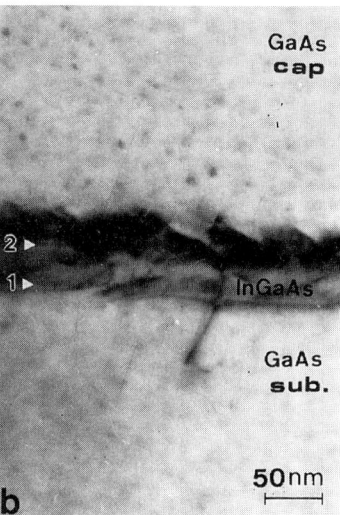

Fig. 2 BF (220) micrographs of GaAs/InGaAs/GaAs QW heterostructures, with GaAs cappping layer thicknesses of: a) 0.15μm and b) 1.3μm. Arrows indicate: 1-bottom interface of the InGaAs layer and 2- top interface of the InGaAs layer.

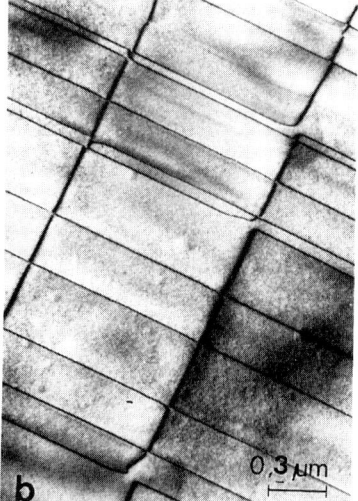

Fig.3 BF plan view image of (AlGa)InAs/GaAs heterostructures: a) grown immediately on the GaAs substrate, b) Grown on the complex buffer stack.

HRXRD investigations showed that the (AlGa)InAs layers grown directly on GaAs substrates relaxed to a considerable extent. The layers appeared to consist of a separate block for which the average angle of misorientation, determined from the FWHM of the curve, was approximately 10 arc min. The coherence length for diffracted the X-ray beams was approximately 40nm, which was much shorter than the layer thickness. The shape of the XRD curves indicated that dislocations were the main type of defect in the layer. This was supported by plan view TEM, (not shown), which revealed a dense misfit dislocation network localised at the interface. The misfit dislocations extended along the two orthogonal <110> directions in the interface.

The similar (AlGa)InAs layers grown on the buffer stacks appeared to be less relaxed. HRXRD curves showed the existence of perfect (unrelaxed) and imperfect (relaxed) parts. The coherence length for diffracted X-ray beams in the layer was approximately 110nm. Clusters of point defects and dislocations were the main types of defect in the unrelaxed and relaxed regions of the layers respectively. In the relaxed regions, TEM showed the misfit dislocation density to be approximately one order of magnitude lower than in the samples with an (AlGa)InAs layer grown directly on the substrate (see Figs. 3a and 3b). The analysis of contrast elimination showed the Burgers vectors of the dislocations to be $1/2[101]$ indicating dislocations of the 60° type.

4. SUMMARY

A study of lattice-mismatch relaxation in GaAs/InGaAs/GaAs and (AlGa)InAs/GaAs heterostructures using TEM and HRXRD has shown that the density of dislocations at the top InGaAs/GaAs interface and within the InGaAs layer increases with increasing GaAs cap layer thickness. An investigation of (AlGa)InAs/GaAs heterostructures has shown that the elastic strain relaxation depends on the initial growth conditions, which are determined by the nature of the buffer layer.

This work was supported, in part, by the ISF Grant NU 6000.

REFERENCES

Faleev N, Stabenow R and Sinitsyn M 1994 Mater. Sci. Forum 293 166

Wang S M, Anderson T G, Lai Z H and Thordson J V 1994 Semicond. Sci. Technol. 9 1230

Inst. Phys. Conf. Ser. No 146
Paper presented at Microsc. Semicond. Mater. Conf., Oxford, 20–23 March 1995
© *1995 IOP Publishing Ltd*

Surface ripples, Crosshatch pattern, dislocation formation: interplay of elastic and plastic strain relaxation mechanisms

M Albrecht[1], S Christiansen[1], P O Hansson[2], J.Michler[1], H P Strunk[1] and E Bauser[2]

[1]Universität Erlangen-Nürnberg, Institut für Werkstoffwissenschaften, Lehrstuhl Mikrocharakterisierung, Cauerstr.6, D-91058 Erlangen, FRG
[2]Max-Planck-Institut für Festkörperforschung, Heisenbergstr.1, 70569 Stuttgart, FRG

ABSTRACT: We investigate by transmission electron microscopy and atomic force microscopy the interaction of elastic relaxation (due to roughening) and dislocation formation in $Si_{0.97}Ge_{0.03}$ grown from In solution onto Si(001). We show that formation of an undulated layer and dislocation formation are cooperating mechanisms in strain relaxation that mutually influence each other: First ripples aligned along <001> directions form, then dislocations nucleate at the rim of the substrate and cause enhanced growth at strain relaxed parts on top of these dislocations, such that ridges form. These ridges build a crosshatch pattern. In the next stage of relaxation dislocations preferably nucleate near the troughs due to locally enhanced shear stresses. Based on a three dimensional finite element calculations of shear stresses in the undulated layers, we give a criterion for preferred dislocation nucleation at these troughs.

1. INTRODUCTION

Strain in heteroepitaxial layers can be released by essentially two mechanisms: elastic strain relaxation by surface roughening in form of undulated surfaces (Grinfeld 1986, Srolovitz 1989) and by misfit dislocation formation (Matthews and Blakeslee 1974). The interaction of these two basic mechanisms in the process of strain relaxation is currently under investigation by several groups (Cullis et al. 1992, Tersoff and LeGoues 1994, Freund et al. 1993, Jesson et al. 1994). While Tersoff and LeGoues (1994) show that dislocation formation and strain induced roughening are competing mechanisms for strain relaxation, other authors emphasize that either dislocation formation results in undulated surfaces (Jonsdottir and Freund 1993) or that undulations result in stress concentrations at troughs or surface cusps (Gao 1994, Jesson et al. 1994) which then cause misfit dislocations to nucleate.

We in this paper show that at low misfits, dislocation formation and surface roughening are cooperating mechanisms, that mutually affect each other. Therefore we investigate $Si_{0.97}Ge_{0.03}$ layers grown from In solution onto Si(001) substrates, by transmission electron microscopy (TEM) and atomic force microscopy (AFM). The as-grown surface topology is revealed in the TEM by strain contrast of surface steps that mediate the undulations. We analyse the dislocation formation process based on the stress distribution in the layer as calculated by the finite element method, assuming homogeneous dislocation nucleation.

2. EXPERIMENTAL

$Si_{0.97}Ge_{0.03}$ layers are grown at $900°C$ from In solution on Si(001) substrates using a cooling rate of $10°C/h$ and an initial supercooling of $\Delta T=1°C$. The Si substrate is cleaned by an RCA treatment, followed by a (2.5%) HF-dip and in-situ oxide desorption. The solvent In is saturated with Ge and Si at the saturation temperature $T_{sat}=900°C$. The solution is transported on and off the substrate by gravitational force using a combined tilting and sliding boat (for details see Hansson et al.1990). Different growth stages under otherwise identical growth conditions are obtained on the same sample by slowly withdrawing the solution. Thereby we obtain epitaxial layers that show a thickness gradient between 200 nm (where the solution is removed first) and 4 μm (where the solution is removed last).

The surface topology of the layers is characterised by optical microscopy in Nomarski differential interference contrast (NDIC). We measure the amplitude A and the wavelength λ of the surface undulations by AFM (for details see Dorsch et al. 1995). The arrangement of misfit dislocations in these layers is analysed by transmission electron microscopy of plane-view and cross-sectional samples in a Philips CM 20 electron microscope operated at 200 kV. Completely relaxed layers exhibit a thickness of about 4μm. Therefore, plane view samples of these layers are investigated in a JEOL high voltage electron microscope (HVTEM) operated at 1 MV. To preserve the as-grown surface, plane-view samples are thinned exclusively from the substrate side. Cross-sectional samples are prepared by standard techniques starting with mechanical grinding and polishing from both sides down to a thickness of 10 μm, followed by Ar^+ ion beam etching to electron transparency. The depth distribution of dislocations of plane-view samples is analysed by stereomicroscopy.

3. RESULTS

In the following we present our results concerning the different stages of heteroepitaxial growth in the $Si_{0.97}Ge/Si(001)$ system, focussing on the interaction between dislocation formation and formation of undulated surfaces.

Up to a layer thickness of about 300 nm the surface of the epitaxial layer is formed by a square arrangement of troughs and ridges aligned along <001> directions (Fig.1) (ripples according to Cullis et al. (1992)). The ripples have a wavelength λ of about 11.5±0.7 μm and an amplitude A increasing from A=5.4 ±0.4 nm at a layer thickness d of 150 nm to A=52±4 nm at a layer thickness d of 300 nm. TEM investigations show these layers to be completely free of dislocations. In the next stage ridges, caused by 60°dislocations (for details see Albrecht et al. 1995a,b), extend from the rim of the substrate along <110> directions (Fig.2). No indications for dislocation nucleation at trough sites have been found in TEM plane views of samples that represent this growth stage. An AFM micrograph of a ridge is shown in Fig.3a. The corresponding height profile in Fig.3b shows that the ridge has a width of about 10μm and a height of 80 nm. Adjacent to the ridge, troughs have formed. Note that the ripples (visible at the right hand side in the height profile in Fig. 3b) and the ridge have the same "wavelength" but an amplitude A, which is smaller by

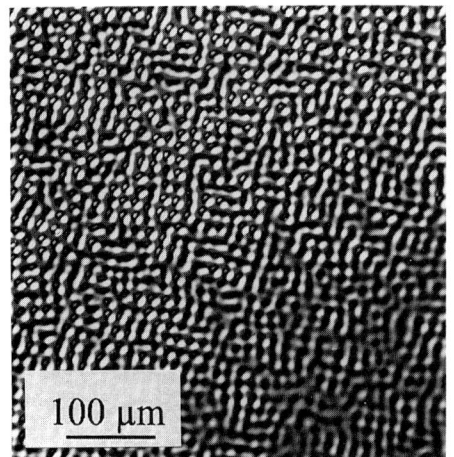

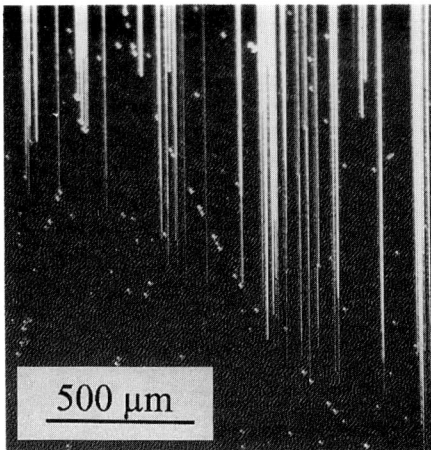

Fig. 1: Surface ripples aligned along [100] and [010] with a wavelength λ of about 10μm. Optical micrograph in Nomarski Differential interference contrast.

Fig. 2: Formation of ridges correlated with dislocations. Nomarski differential interference contrast micrograph showing ridges to extend from the rim into the centre of the substrate. The ridges are the longer the thicker the layer is.

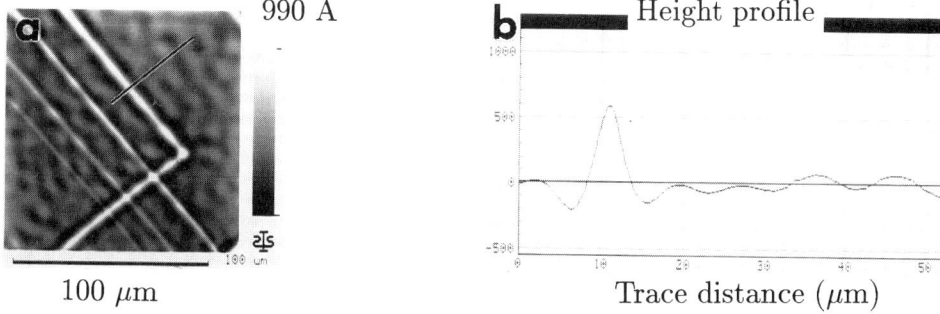

Fig. 3: (a) Formation of a cross hatch pattern by intersection of two perpendicular sets of ridges (AFM micrograph). (b) Height profile along the line in Fig. 2.

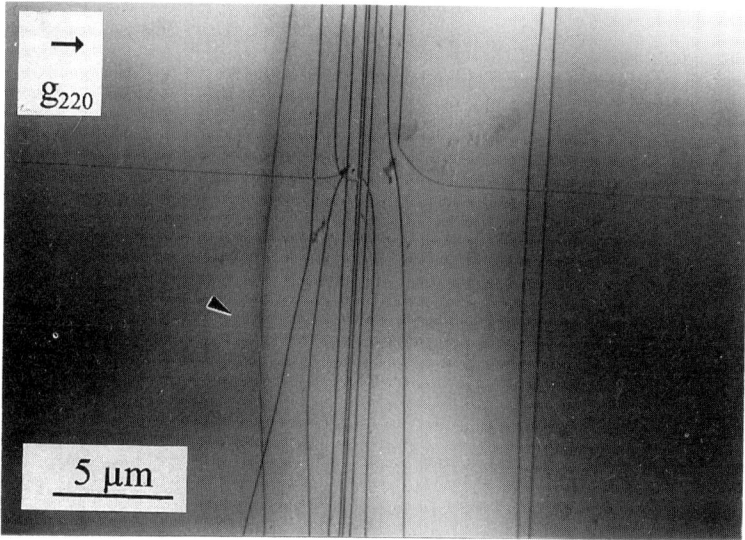

Fig. 4: Preferred dislocation nucleation in the troughs of the cross hatched pattern. Transmission electron micrograph taken at 1MV. The ridge can be revealed by dark absorption contrast, the contours of the ridge by the strain contrast of surface steps (one of them marked by arrows). A high density of 60°-dislocations is visible in the trough.

a factor of 10. As can be seen in Fig. 3a ridges have an equilibrium interdistance of 33±1μm and finally form a crosshatch pattern with even this wavelength.

Fig. 4 shows a HVTEM micrograph of a part of the sample, which exhibits a thickness of 4 μm. Dark regions (high absorption contrast) represent ridges, bright regions troughs. Details of the surface topology are easily revealed by strain contrast at surface steps (one of them is indicated by an arrow). In the trough a high density of 60°-misfit dislocations has formed. Stereomicroscopy reveals that some of these dislocations extend into the substrate.

4. DISCUSSION

In previous work we discussed, how strain relaxation by surface undulations and dislocation formation influences the growth morphology by a local strain dependent variation of the chemical potential at the growth surface. Here we focus on how dislocation formation is influenced by the inhomogeneous strain distribution in the undulated layers. We start from the model of homogeneous halfloop nucleation (Matthews and Blakeslee 1974) and therefore consider the shear stress distribution in the epitaxial layers. Fig. 5 shows the shear stress distribution in the {111} glide plane for the example of a layer with an aspect ratio $\lambda/A=10$. Maximum shear stresses occur in the trough. Ridges are approximately shear stress free (the top of the ridge shear stresses are reduced by 80%). Shear stresses in part of the layer adjacent to the interface are almost constant and comparable to that of a two dimensional layer. Fig. 6 shows the shear stress concentration S at the trough relative to that of a twodimensional layer, as dependent on the aspect ratio λ/A of the undulation, calculated by the finite element method (for details see Christiansen et al. 1995). For aspect ratios λ/A of 5 which have been observed in single cases for the crosshatch pattern, the shear stresses on {111} glide planes are enhanced by a factor 2.5 as compared to a two dimensional layer. For aspect ratios of about 100 which have been observed for both, ripples and crosshatch pattern, stress concentrations assume a value of 1.5. This means: for a given wavelength λ stress concentrations at troughs increase with increasing amplitude A.

Thus we should expect that dislocations preferably nucleate at troughs where S is maximum. However, in parallel to an increasing amplitude A that leads to an increasing S, the layer thickness d at trough sites decreases by the amplitude A to (d-A). If the critical radius for halfloop nucleation is considered as a criterion for dislocation formation in these layers, the stress concentration has to assume a threshold value to obtain prefered dislocation nucleation at troughs. According to Matthews and Blakeslee (1974) the critical radius (thickness) for dislocation nucleation in a twodimensional layer ($h_c^{2D}=150$ nm for the present misfit of $f=0.0012$) depends on the shear stress τ in a certain glide system (here the $a/2<110>\{111\}$ glide system) by

$$h_c^{2D} \propto \frac{1}{\tau} \tag{1}$$

Since shear stresses τ at troughs increase by a factor S compared to shear stresses of a two dimensional layer, the critical thickness decreases according to $h_c^{3D}(S)=h_c^{2D}/S$. Dislocations nucleate preferentially at troughs if

$$h_c^{3D}(S) < (d - A) \tag{2}$$

i.e.

$$S \geq \frac{h_c^{2D}}{(d - A)} \tag{3}$$

This rough assessment easily explains the experimental observation that dislocations nucleate preferentially at troughs of the crosshatch pattern, but not at troughs of the rippled surface undulation: surface ripples occur at layer thicknesses closely to the the theoretical critical thickness h_c^{2D} of these layers ($h_c^{2D}=150$ nm) and exhibit an aspect ratio of $\lambda/A>100$. Furthermore the undulation amplitude is comparable to the layer thickness. In consequence the stress concentration is small (S<1.5) and does not compensate the reduced layer thickness (d-A). In contrast crosshatch pattern according to our observations are related to layer thicknesses d far beyond the critical thickness (d>2μm) and exhibit aspect ratios λ/A up to $\lambda/A=5$. Thus the critical radius and correspondingly the activation energy is reduced at the troughs due to the shear stress concentration S and results in preferred dislocation nucleation.

6. CONCLUSIONS

In conclusion, we have shown that surface ripples that occur in the first pseudomorphic growth stage exclusively contribute to elastic strain relaxation, which is in good accordance with recent work of different authors (Cullis et al. 1992, Duratre et al. 1994). First misfit dislocations in our samples nucleate at the rim of the substrate and not in the troughs of the rippled surface since enhanced shear stress concentrations at trough sites do not compensate the reduced layer thickness (d-A) at the troughs. However, once dislocations

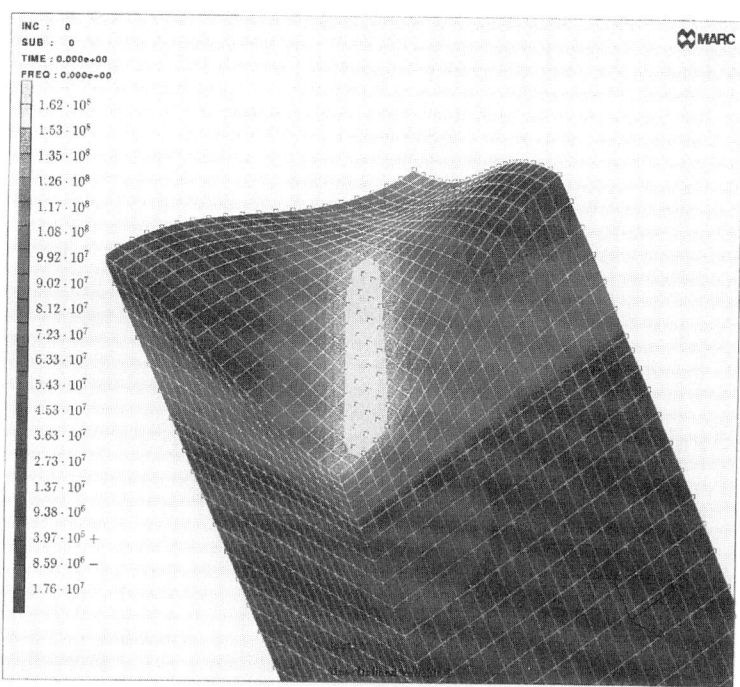

Fig. 5: Shear stress distribution in a undulated layer as calculated by the finite element dislocation. The undulation has a aspect ratio $\lambda/A = 10$. High shear stresses occur at troughs. Ridges are almost shear stress free.

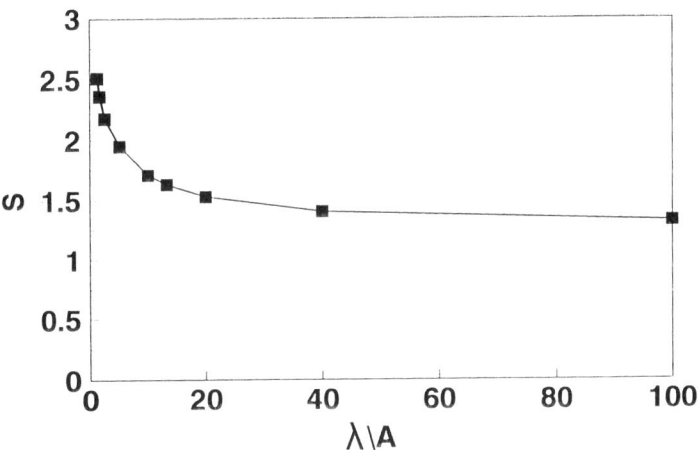

Fig. 6: Shear stress concentration at the trough of a undulated layers standardised on that of a two dimensional layer as dependent on the aspect ratio λ/A.

have nucleated at the rim of the substrate, they relax part of the strain at the surface and cause ridges to form. These ridges result in a crosshatch pattern, that exhibits an aspect ratio $\lambda/A \leq 5$, which is by a factor 20 lower than that of the ripples. The corresponding shear stress concentrations in the troughs of the crosshatch pattern then are sufficiently high to yield preferrential dislocation nucleation.

ACKNOWLEDGEMENTS

We thank J.Heydenreich and Ch. Dietzsch for giving us the oppurtunity to use the high voltage transmission electron microscope at the Max-Planck-Institut für Mikrostrukturphysik, Halle/Saale. We acknowledge financial support by the Bundesministerium für Forschung und Technologie under contract Numbers 01M1920 A and C.

REFERENCES

Albrecht M, Christiansen S, Michler J, Dorsch W, Strunk H P, Hansson P O, and Bauser E 1995a
    accepted for publication in Appl.Phys.Lett.
Albrecht M, Christiansen S, Michler J, Strunk H P, Hansson P O, and Bauser E 1995b submitted to
    J.Cryst.Growth
Christiansen S, Albrecht M, Michler J, and Strunk H P 1995 submitted to J.Appl. Phys.
Cullis A G, Robbins D J, Pidduck A J and Smith P W 1992, J.Crystal Growth 123 333
Dorsch W, Christiansen S, Albrecht M, Strunk H P, Hansson P O and Bauser E 1995 accepted for
    publication in Surf. Science 1995
Dutartre D, Warren P, Chollet F, Gisbert F, Bérenguer M, and Berbézier I 1994 J.Cryst.Growth 142 78
Freund L B, Beltz G E, Jonsdottir F 1993 Mat. Res. Soc. Symp. Proc. 308 308.
Gao H 1994 J.Mech.Phys.Solids 42 741
Grinfeld M A 1986, Dokl.Akad.Nauk SSR 290 1358
Hansson P O, Werner J H, Tapfer L, Tilly L P, and Bauser E 1990 J.Appl.Phys.68 2158
Jesson D E, Pennycook S J, Baribeau J M, Houghton D C 1994 Mat.Res.Soc.Symp.Proc.317 297
Jonsdottir F and Freund L B 1993 Mat.Res.Soc.Symp.Proc. 317 309
Matthews J W and Blakeslee A E 1974 J Cryst. Growth 27 118
Srolovitz D J 1989 Acta metall. 37 621
Tersoff J and LeGoues F K 1994 Phys. Rev Lett. 72 3570

Inst. Phys. Conf. Ser. No 146
Paper presented at Microsc. Semicond. Mater. Conf., Oxford, 20–23 March 1995
© 1995 IOP Publishing Ltd

Relaxation of (001) Si/Si$_{1-x}$Ge$_x$/Si heterostructures

Y Xin*, P D Brown, R E Schäublin and C J Humphreys

Department of Materials Science and Metallurgy, University of Cambridge, Pembroke Street, Cambridge, CB2 3QZ, UK;
* on leave from Beijing Laboratory of Electron Microscopy, CAS, Beijing, P.R. China.

ABSTRACT: As-grown and annealed (001)Si/Si$_{1-x}$Ge$_x$/Si heterostructures have been studied with x=0.12 corresponding to 0.49% strain. It is found that there are misfit dislocations at the two interfaces and also dislocations deep into the Si substrate and in the Si cap. It is tentatively suggested that the dislocation sources are small dislocation loops, the formation of which is possibly related to boron doping and subsequent annealing. The dissociation of misfit dislocations is observed.

1. INTRODUCTION

The strain relaxation mechanisms within SiGe/Si heterostructures have attracted much interest because thermal processing as used in device fabrication induces relaxation of metastable structures through the introduction of dislocations which can deleteriously affect device performance. In particular, the source of dislocations within low x content Si$_{1-x}$Ge$_x$/Si has attracted much debate. Several authors (e.g. Matthews et al 1976) note that homogeneous nucleation of dislocation loops or half loops is only possible under conditions of high stain, i.e. for more than 2% lattice mismatch. Possible heterogeneous dislocation sources in low strain, low x content Si$_{1-x}$Ge$_x$/Si include (i) pre-existing threading dislocations in the substrate, although the density of these is normally too low to provide sufficient misfit dislocations for substantial relaxation of lattice mismatch; (ii) growth induced defects, such as the "diamond defect", within the epilayer (Humphreys et al 1991); (iii) undesorbed particles (e.g. oxide, carbide) from the substrate (Perovic et al 1989). In this paper, the dislocations within as-grown and annealed Si/Si$_{0.88}$Ge$_{0.12}$/Si heterostructures are examined and the processes of relaxation are discussed.

2. EXPERIMENTAL

Samples comprising 0.8μm Si / 0.35μm Si$_{0.88}$Ge$_{0.12}$ / 0.3μm Si buffer were grown on (001) oriented Si substrates by MBE at 550°C. This system is of practical interest for 2-d hole gas device structures (Smith et al 1992). The Si substrate was doped with boron to a concentration of 5x10^{18}cm^{-3}. The epilayer was also uniformly doped with boron to about 10^{16}cm^{-3} during growth. Some samples were treated at 830°C for 1 hour. Plan view and cross-sectional samples of both as-grown and annealed samples were prepared using conventional specimen preparation procedures before ion-milling and examined using Philips CM30, Jeol 2000FX and Jeol 4000-EXII microscopes.

3. RESULTS

Fig. 1 shows [001] plan view images of the as-grown sample which was characterised by bands of dislocations separated by a few microns. Calculations show that these widely spaced misfit dislocations provide a very small amount of relaxation of the epilayer. All the misfit dislocations were of 60° character as determined using **g.b** analysis, and two features distinguished this initial stage of relaxation. The first, as shown in Fig. 1a, is the dipole

184

configuration typically formed within capped heterostructures which has fixed spacing as determined by the location of the two <110> segments on {111} at the two interfaces and the thickness of the SiGe layer (Hockly et al 1991). Few such hair-pin features were seen as the dipoles were extended over considerable distances and have low density. Dislocation half loops were the other distinguishing feature, as illustrated by Fig.1b,

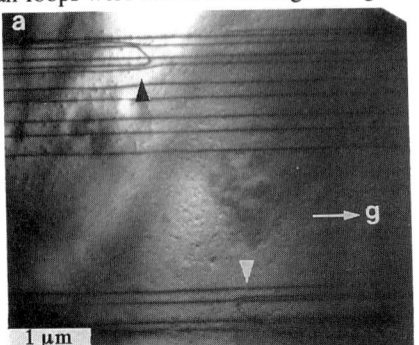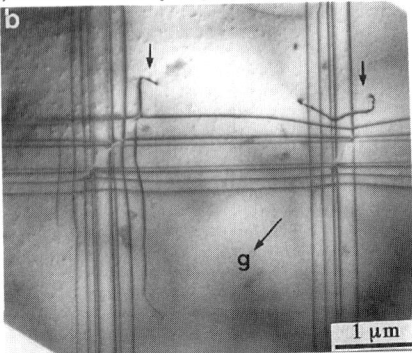

Fig. 1 [001] plan view image of the as-grown Si/SiGe/Si heterostructure showing (a) dipole hair-pins ($\mathbf{g}=\bar{2}20$) and (b) dislocation half-loops ($\mathbf{g}=400$), as indicated.

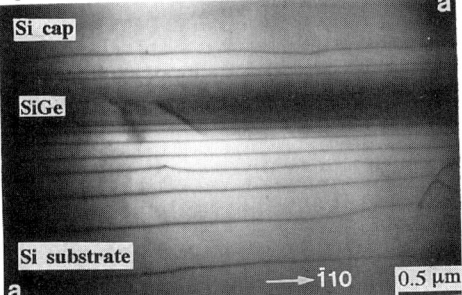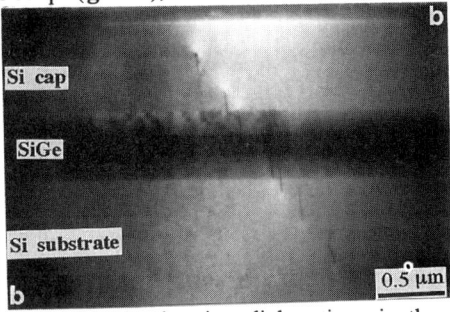

Fig. 2 [110] cross-sectional view of the as-grown sample showing dislocations in the substrate (a) along [$\bar{1}$10] and (b) along [110] ($\mathbf{g}=\bar{2}20$).

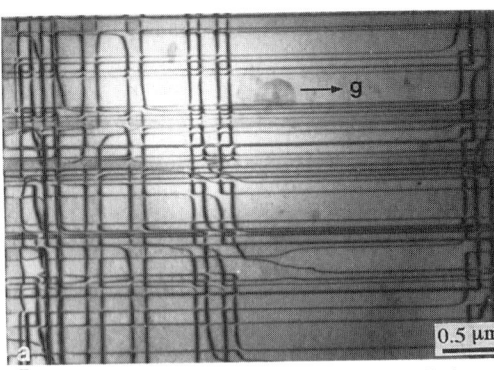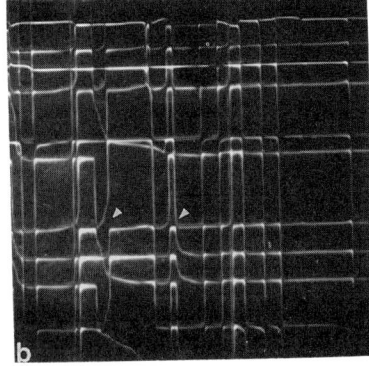

Fig. 3 [001] plan view image of the annealed sample showing the dislocation network (a) bright field ($\mathbf{g}=\bar{2}20$) and (b) weak beam ($\mathbf{g},4\mathbf{g}$) dark field.

observed within thick regions of material. Stereo-microscopy demonstrates that such half loops reside below the lower interface of the sample in this case.

Fig. 2a shows a [110] cross-sectional view of the as-grown sample and dislocations are located within the Si cap and also deep (>1μm) into the Si substrate. Such dislocations were observed to run for several tens of microns along [$\bar{1}$10]. Fig. 2b shows segments of dislocations running along [110].

Fig. 3a is an [001] plan view image of an annealed sample, showing the formation of an extensive network of misfit dislocations, comprising long, straight orthogonal arrays of 60° dislocations running along [110] and [1̄10]. These more densely spaced misfit dislocations provide a greater epilayer relaxation relative to the as-grown samples.

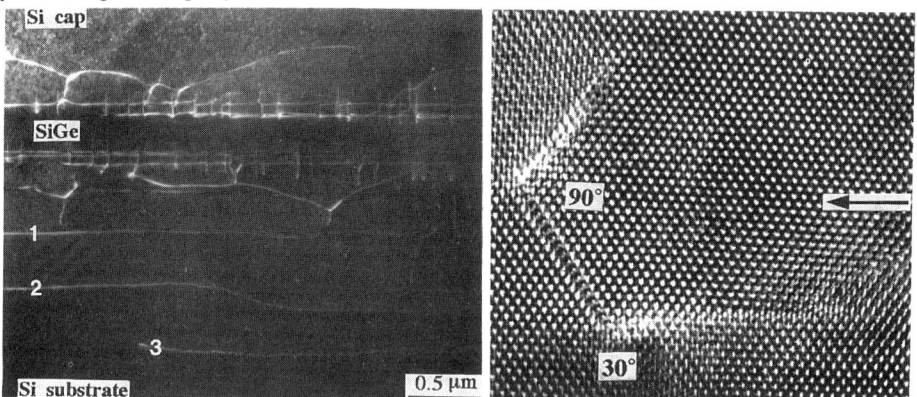

Fig. 4 Weak beam cross-sectional image of the annealed sample.

Fig. 6 HREM image of a dissociated misfit dislocation with two tails.

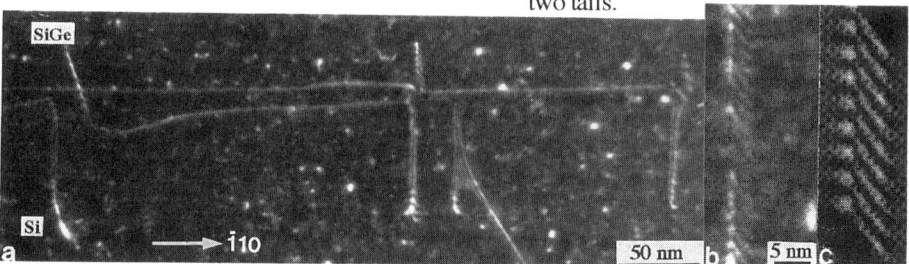

Fig. 5 Weak beam image of (a) dissociated misfit dislocations at the SiGe/Si interface; (b) (**g**,4**g**) image showing stacking fault fringes bounded by partials showing strong and weak contrast with (c) simulation.

Fig. 3b is a weak beam image of such a network and shows the commonly observed configurations arising from dislocation interactions.

Fig. 4 is a weak beam image of an annealed sample viewed in cross-section, tilted around [1̄10] to show the misfit dislocations at the Si/SiGe and SiGe/Si interfaces. Again dislocations were present at the Si cap and deep in the substrate, but few or zero dislocations were observed within the SiGe layer. CBED patterns allowed the determination of the Burgers vectors of the dislocations shown in the substrate (1 to 3). These were all found to have **b** in the opposite sense to that required to relieve misfit strain at the interface, indicating that these dislocation segments have been pushed away into the substrate. In addition, they are found to glide on different {111} planes. The deepest dislocations in the substrate (2 and 3) exhibit a kinked structure, indicative of a reduction of the driving force pushing them into the substrate. It is noted that dislocations in the as-grown sample showed no such features at the same depth. Small native dislocation loops were also present in the substrate and cap of the annealed sample.

Fig. 5a is a (**g**,4**g**) weak beam image (**g**=2̄20) of the misfit dislocations located at the lower interface taken from a cross-sectional sample tilted around [1̄10] towards [111]. The misfit dislocations along [110] are dissociated. One example is enlarged in Fig. 5b and shows strong and weak contrast partials bounding a stacking fault. The stacking fault fringes indicate that this dissociated dislocation lies on (1̄11), with possible total **b** of 1/2[011̄] or 1/2[1̄01̄]. Simulations (Schäublin and Stadelmann 1993) were made to identify the dislocation shown in Fig. 5b. A qualitative match of this experimental image was obtained using $1/2[01\bar{1}] = 1/6[12\bar{1}] + 1/6[\bar{1}1\bar{2}]$, showing that the 90° partial exhibits stronger contrast, while the 30° partial is very weak.

Fig. 6 is an HREM image of such a dissociated misfit dislocation. The dissociation width of the stacking fault on $(\bar{1}11)$ is 6nm, consistent with the projected width of the weak beam image. The leading 90° partial is at the SiGe/Si interface (arrowed) with the 30° partial located in the Si. Sometimes both partials are located within the Si but with the 90° partial directed towards this interface. Similarly, dissociated dislocations located at the Si cap/SiGe interface had the 90° partial located at or close to the interface with the 30° partial in the Si cap. The other distinguishing feature of this HREM image is the two tails always found at the ends of the stacking fault. The tails associated with the 90° partials generally lie on {111} while those associated with the 30° partials lie on (001). The reason for this defect configuration remains unclear.

4. DISCUSSION

The misfit strain at both interfaces of this $Si/Si_{0.88}Ge_{0.12}/Si$ heterostructure is 0.49% using $(a_{SiGe}-a_{Si})/a_{Si}$, (a_{SiGe} is lattice parameter of SiGe alloy and a_{Si} parameter of Si) with the SiGe layer in compressive strain. The experimental evidence of the samples studied here demonstrates the as-grown layer to be slightly relaxed (estimated 0.07%), while the annealed material was substantially relaxed (estimated 0.3%), assuming 90° partials relieving the misfit strain. The suggestion is that the initial stage of relaxation proceeds from pre-existing threading dislocations in the substrate. However, there are insufficient pre-existing dislocations in the Si substrate to provide sufficient misfit dislocations to substantially relieve the lattice mismatch, and hence other sources must become operative during annealing to introduce the dislocations needed to relax the misfit strain. While such sources are difficult to identify, and no other defects except small dislocation loops were observed in Si cap and substrate, one may speculate that these small dislocation loops are a possible source and that their formation is possibly related to boron doping and subsequent annealing. The dislocation half loops observed in the as-grown sample may originate from such sources. In view of the absence of dislocations in the SiGe layer, these secondary dislocation sources are considered to predominate in the Si substrate and cap which is possibly due to the different diffusivity of boron in Si from that in SiGe alloy, and components of the extended dislocation loops favourable for relieving strain would glide towards the interface to form misfit dislocations and relieve misfit strain, while opposite segments would be pushed away into substrate or glide out of the free surface. In situ annealing experiments are in progress to try and clarify this issue. Weak beam and HREM observations presented here for our material show that misfit dislocations are dissociated and of comparable width to dissociated dislocations in elemental Si.

ACKNOWLEDGEMENTS

YX thanks the Cambridge Overseas Trust and ORS Award scheme for funding and PDB acknowledges SERC for funding under contact number No. GR/J37966. We are also grateful to Prof. EHC Parker and Dr. RA Kubiak of Warwick University for provision of material used in this study.

REFERENCES

Hockly M, Tuppen CG, Gibbings CJ, Martin ASR, Shafi ZA and Ashburn P 1991 Inst. Phys. Conf. Ser. 117 445
Humphreys CJ, Maher DM, Eaglesham DJ, Kvam EP and Salisbury IG 1991 J. Phys. 1 1119
Matthews JW, Blakslee AE and Mader S 1976 Thin Solid Film 33 253
Perovic DD, Weatherly GC, Baribeau JM and Houghton DC 1989 Thin Solid Films 183 141
Schäublin R, Stadelmann P 1993 Mater. Sci. and Eng. A164 373
Smith DW, Emeleus CJ, Kubiak RA, Parker EHC and Whall TE 1992 Appl. Phys. Lett. 61 1453

Inst. Phys. Conf. Ser. No 146
Paper presented at Microsc. Semicond. Mater. Conf., Oxford, 20–23 March 1995
© 1995 IOP Publishing Ltd

A formation of three-dimensional misfit dislocation networks in SiGe/Si and SiGe/Ge heterostructures

V I Vdovin

Institute of Rare Metals, B. Tolmachevsky per. 5, Moscow 109017, Russia

ABSTRACT: Formation of three-dimensional misfit dislocation (MD) networks in single layer SiGe/Si,Ge heterostructures occurs due to the MD multiplication at the interface through the Frank-Read mechanism. As the layer thickness increases the MD network changes progressively from flat, wide-meshed and regular to dense three-dimensional. This process is accompanied by substrate plastic deformation in the near-interface region. The regularities of the dislocation distribution along the thickness of relevant heterostructures are discussed.

1. INTRODUCTION

Three-dimensional misfit dislocation (3D-MD) networks were observed in different hetero-structures (HSs) such as InGaP/GaP (Abrahams et al 1975), GaAsP/GaP (Ahearn and Laird 1977), InGaAs/GaAs (Fitzgerald et al 1988, Herbeaux et al 1989), SiGe/Si (LeGoues et al 1991, Vdovin et al 1994) and SiGe/Ge (Vdovin et al 1993) which were single layer HSs as well as strained superlattices. For all of these HSs, the inherent characteristics are relatively high lattice mismatch ($f = n \cdot 10^{-3}$) and large layer thickness ($0.2 < h < 2$ μm). A propagation of dislocation half loops deeply downwards into the buffer layer and/or substrate (hereafter referred to as the substrate) is the characteristic property of 3D-MD networks. Along with this characteristic, a multilevel distribution of MDs is observed over a relatively thin layer covering near-interface regions both of the layer and the substrate.

Authors of the papers mentioned above explain the formation of 3D-MD networks as a result of MD multiplication through the Frank-Read mechanism. In some papers (Ahearn and Laird 1977, Matthews et al 1976), dislocation multiplication from Frank-Read sources arising at the growth surface was considered. In contrast, Abrahams et al (1975) and Herbeaux et al (1989) have proposed that classical Frank-Read sources may operate at the interface. In this case, the source of dislocation multiplication is a MD segment clamped in the neighboring dislocation nodes arising at the crossing points of perpendicular MDs. This idea obtained more development at a later time. Recently, LeGoues et al (1991) have observed such effects in SiGe/Si HSs with graded layers as well as superlattice and have described in detail a possible mechanism.

In the present work, a study was carried out on the regularities of 3D-MD network formation in single layer SiGe/Si and SiGe/Ge HSs with constant alloy composition layers.

2. EXPERIMENT

The 3D-MD networks were studied by transmission electron microscopy on the plan-view and cross-sectional specimens of HSs grown by different techniques. $Si_{1-x}Ge_x$ layers were grown on Si (001) substrates by molecular beam epitaxy (MBE) at 850 °C and by hybrid epitaxy (HbE) at 700 °C. In the hybrid technique, epitaxial growth was carried out from a Si molecular beam and GeH_4 gas flow. $Ge_{1-x}Si_x$ layers were grown on Ge (111) substrates by hydride epitaxy (HdE) at 600 °C. Hereafter we will mark the appropriate HSs as MBE-HSs, HbE-HSs and HdE-HSs in accordance with their growth technique. For all three types of HSs, approximately similar alloy composition range (0.02<x<0.25) was examined, within which the lattice mismatch varies from 0.05 to ≈1%. The layer thickness varied from 0.1 to 3.5 μm. The dislocation density in these substrates was $(1-5) \cdot 10^3$ cm^{-2} or less.

3. RESULTS

Shown in Fig.1 are cross-sectional transmission electron microscope images typical of HSs with developed 3D-MD networks. As follows from Fig.1, the 3D-MD network seems to be a relatively thin region with multilevel distribution of MDs, from which dislocation half loops spread into the substrate to high depth. The dislocation density (N_S) and depth (h_S) of their propagation into the substrates depends on alloy composition and have been plotted by curves showing a maximum (Vdovin et al 1993). Appropriate characteristics, including threading dislocation (TD) density (N_{TD}) at the layer surface, of the HSs with maximum substrate plastic deformation are shown in part A of the table.

The layer thickness and alloy composition strongly influence the formation of 3D-MD networks, which is illustrated in the case of the HbE-HSs (part B of the table). As follows from a comparison of the characteristics of the appropriate samples from parts A and B, the increase of misfit strain due to the layer growth leads to an increase in the dislocation density and their propagation depth into the substrate (x=0.09). However, the influence of the increased misfit strain due to greater lattice mismatch (or increase of Ge content in alloy) is not so definite. For example in the sample with x=0.11 and h=1.1 μm the depth of dislocation half loop propagation proved to be considerably less (≈0.4 μm) than that in the sample with x=0.09 and h=1.2 μm, whereas the dislocation densities in the substrate and the layer of these two samples are comparable in magnitude. The thickness of the multilevel part of the 3D-MD network also depends upon layer thickness and alloy composition and varies from 0.03 to 0.2 μm.. However, it is difficult to distinguish between the contribution from each of these factors.

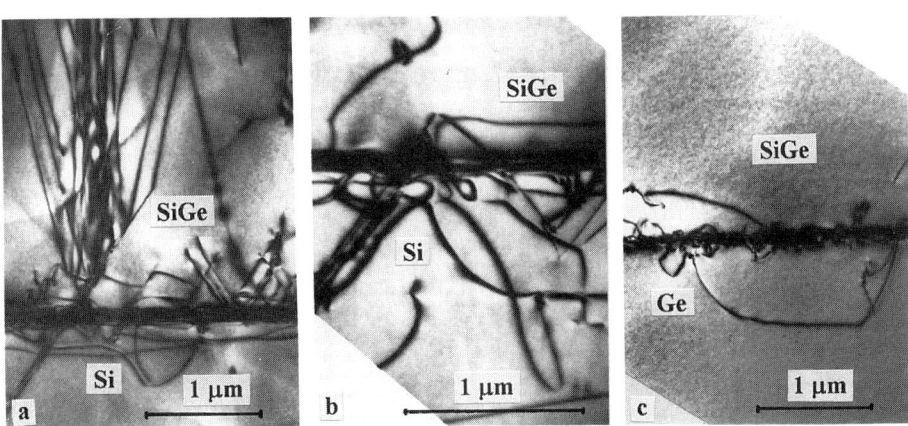

Fig.1. Three-dimensional MD networks in HSs with constant alloy composition layers, in which maximum substrate plastic deformation is attained: a - MBE-HS (x=0.11; h=1.8 μm); b - HbE-HS (x=0.09 h=1.2 μm); c - HdE-HS (x=0.08; h=3.5 μm).

A more detailed study performed on an example HdE-HSs of the dependence of the 3D-MD network characteristics upon layer thickness and alloy composition has shown that the formation of such networks occurs progressively through distinct stages as the layer thickness is increased. The initial stage is a formation of a flat regular MD network lying strictly at the interface (Fig.2, a). In such networks, local bending of dislocation lines into the substrate is often observed at the crossing points of perpendicular MDs where dislocation reactions occur to split the dislocation node or to form a dislocation junction. Such MD networks were observed in the following HdE-HSs, for example, where the ratio of alloy composition to layer thickness are as follows: x/h = 0.016/3.5; 0.025/2.1; 0.03/1.35;

0.042/0.4; 0.064/0.25 (at. sh./μm). In these HSs, bending of substrate dislocations into the interface was often observed. However along with the MDs lying strictly at the interface, dislocations parallel to the interface were also observed in the substrate at an arbitrary distance from the interface over the depth of about 0.2 μm. The ratio of these dislocations to the MDs at the interface is as similar as 1:10.

Table. Structural features of single layer heterostructures.

	Type of heterostructure	Growth method	T, °C	x, at.sh.	h, μm	N_{TD}, cm^{-2}	N_S, cm^{-2}	h_S, μm
A	$Si_{1-x}Ge_x$/Si	MBE	850	0.11	1.8	$8 \cdot 10^8$	$8 \cdot 10^5$	2
	$Si_{1-x}Ge_x$/Si	HbE	700	0.09	1.2	$8 \cdot 10^5$	$1 \cdot 10^7$	2.5
	$Ge_{1-x}Si_x$/Ge	HdE	600	0.08	3.5	$2 \cdot 10^5$	$2 \cdot 10^6$	2
B	$Si_{1-x}Ge_x$/Si	HbE	700	0.09	0.2	$2 \cdot 10^5$	$1 \cdot 10^6$	0.2
	$Si_{1-x}Ge_x$/Si	HbE	700	0.11	1.1	$4 \cdot 10^5$	$5 \cdot 10^7$	0.4

At the second stage (Fig.2, b), a MD multiplication occurs through the Frank-Read mechanism. This process leads to the formation of a multilevel network of parallel MDs, covering near-interface regions both of the layer and substrate, and propagation of large dislocation half loops down into the substrate. The TD density in the epitaxial layer remains at a similar level of 10^5 cm^{-2}. For the alloy compositions relevant to that mensioned above, such types of 3D-MD networks form in HSs with thick layers (for example, the HdE-HS with ratio x/h=0.035/2.7 μm). For 0.08<x<0.20, the appropriate 3D-MD networks were already observed in HbE-HSs and HdE-HSs with thickness of 0.1-0.2 μm. The third stage occurs as an intensive development of the 3D-MD network, so that a dense network of the dislocation half loops forms in the substrate. These dislocation half loops lose their ability to spread along the interface by glide because of numerous intersections between each other. Simultaneously, dislocation half loops bow out from the MD network into the layer where they arrange close to the interface, causing insignificant increasing TD density at the layer surface (Fig.1, b and c).

In all types of HSs with x ≥ 0.20 and thick layers (h>1 μm), a dense network of TDs (h ≥ 10^8 cm^{-2}) is usually observed along with highly developed 3D-MD network at the interface.

4. SUMMARY AND CONCLUSIONS

Experimental data obtained indicate that 3D-MD networks form in the single layer HSs with constant alloy composition in much the way as in HSs with graded layers or superlattices (LeGoues et al 1991). Moreover a comparative analysis of structural features of the HSs grown by different techniques indicates that this process occurs, on the whole, similarly in them. It concerns both the kinetics of 3D-MD network formation during the epitaxial growth and the dependence of this effect upon the alloy composition. The most surprising is that the plastic deformation of Si and Ge substrates is characterized by practically the same depth of the dislocation propagation down into the substrate and the same magnitude of the dislocation density in this region. As one can see, there is no distinguishing correlation between HS features and substrate orientation, epitaxial growth technique, and temperature.

A distinguishing difference between HSs studied is observed in TD density in the layers, but more precisely in its dependence upon alloy composition. In MBE-HSs, a relationship N_{TD}>N_S is true over the whole composition range. In contrast, in HbE-HSs and HdE-HSs, a relationship N_{TD}<N_S similar for both heterosystems is observed in the composition range x<(0.18-0.20). If x is outside these values, TD density in these layers increases up to 10^8 cm^{-2}. This fact appears to be explained by difference in the MD generation mechanisms at the initial stage of MD network formation. In MBE-HSs, MDs predominantly generate through the dislocation half loops nucleated at the layer surface (Vdovin et al 1994). In two other types of HSs, the bending of substrate dislocations into the interface was observed.

190

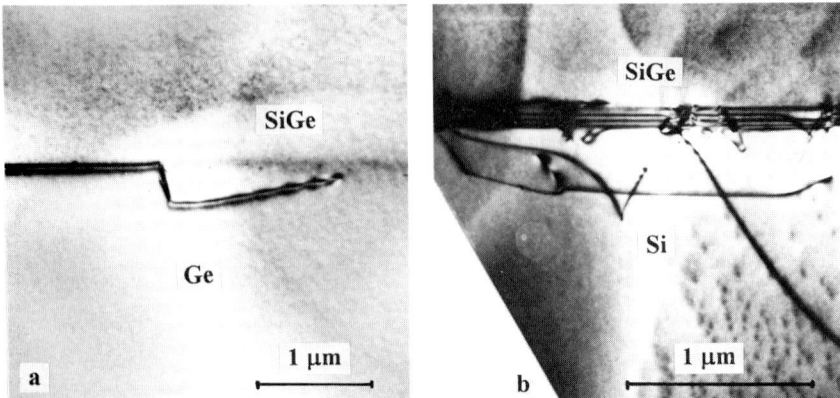

Fig.2. Intermediate stage of three-dimensional MD network formation in HSs: a - HdE-HS
(x=0.016; h=3.5 μm); b - HbE-HS (x=0.044; h=1.5 μm).

However, the initial substrate dislocation density is too small for considerable misfit strain relief. Thus, we may suppose that   additional MDs introduce into the interface and substrate close to it through some different mechanisms. Similar generation of MD network in strained $Si_{0.7}Ge_{0.3}$ HSs has recently been observed by Godbey et al (1990) during sample annealing. In addition, at high alloy composition (large lattice mismatch) a high TD density may be caused by peculiarities of the initial stages of epitaxial growth. In either case, it implies that generation of high density of TDs propagating through the whole epitaxial layer from the interface to the surface close to the growth direction is not connected with 3D-MD network formation.

As for the kinetics of 3D-MD network formation, the model proposed by LeGoues et al (1991) seems to be fairly good. However, some experimental results allow us to suggest that the real process is more complicated. In particular, the distribution of MDs over the multilevel part of 3D-MD network seems to be more symmetrical in relation to the interface rather than being shifted into the layer. The dislocation half loops which are highly extended along the interface are often observed in the near-interface substrate region at the second stage of MD network formation, whereas similar dislocation loops in the epitaxial layer are still absent. In addition, at relatively small alloy composition, active MD multiplication occurs in layers of more than 1 μm thick. In this case, one should expect more complex dislocation behaviour than the the the model suggests. Different mechanisms describing propagation of the dislocation half loops in the substrate along the interface should, perhaps, be examined.

The author is grateful to Prof M.G.Mil'vidskii and Dr T.G.Yugova for useful assistance and discussions. Thanks to Prof E.H.C.Parker and Dr O.Mironov for their interest to this work and stimulating discussion. This work was supported in part by ISF-MGE000 and INTAS-93-1403 grants.

REFERENCES

Abrahams M S, Buiocchi C J and Olsen G H 1975 J. Appl. Phys. 46 4259
Ahearn J C and Laird C 1977 J. Mat. Sci. 12 699
Fitzgerald E A, Ast D G, Kirchner P D, Pettit G H and Woodall J M 1988 J. Appl. Phys. 63 693
Godbey D, Hughes H, Kub F, Twigg M, Palkuti L and Leonov P 1990 App. Phys. Lett. 56 373
Herbeaux C, Persio J D and Lefebvre A 1989 Appl. Phys. Lett. 54 1004
LeGoues F K, Meyerson B S and Morar J F 1991 Phys. Rev. Lett. 66 2903
Matthews J W, Blakeslee A E and Mader S 1976 Thin Solid Films 33 253
Vdovin V I, Mil'vidskii M G, Yugova T G and Lyutovich K L 1994 J. Crystal Growth 141 109
Vdovin V I, Mil'vidskii M G and Yugova T G 1993 Solid State Phenomena 32-33 345

Inst. Phys. Conf. Ser. No 146
Paper presented at Microsc. Semicond. Mater. Conf., Oxford, 20–23 March 1995
191

Effects of strain and temperature on the roughness transition in epitaxial growth

A J Harvey[1], U Bangert[1], C Dieker[2] and H Hardtdegen[2]

[1]Department of Pure and Applied Physics, UMIST, Manchester M60 1QD, U.K.
[2]Institut fuer Schicht- und Ionentechnik, Forschungszentrum Juelich, D- 52425 Juelich, Germany

ABSTRACT: The experimental evidence for the strain dependence of the roughness transition temperature and of the surface modulation wavelength is reviewed. It has been found that the   region of planar growth can be extended in III-V compound semiconductors by the favourable   choice of the partial pressure ratio of group V and group III gas phase reactants.

1. INTRODUCTION

It has been found that quaternary and ternary layers grown epitaxially on InP substrates can grow with a distorted morphology , far from the planar morphology expected and desired. Similar phenomena had been noticed in other systems, notably the SiGe/Si system. The relationship between growth temperature and the square of the 'misfit strain squared thickness product' ($\sqrt{sum}$), which is proportional to the square of the areal misfit strain energy, established for the GaInAs/InP system, is shown in the curve below. The curve separates 2D and 3D growth regimes.

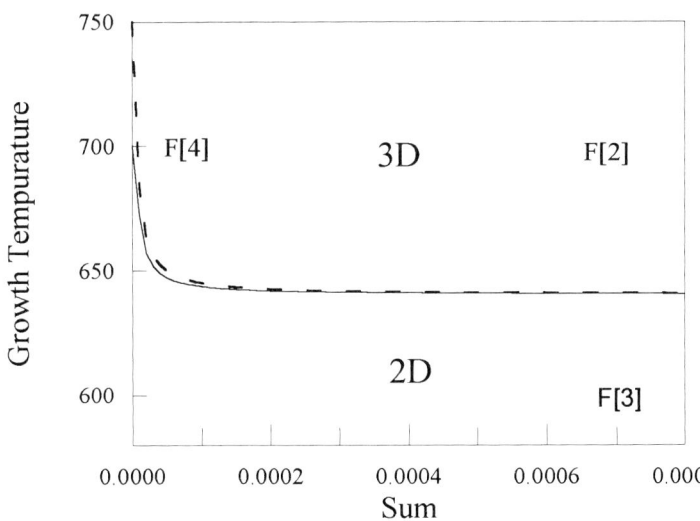

Figure 1
Roughness Transition Temperature against Sum ~ ϵ_f^2. The solid curve is for a 'low' and the dashed curve is for a 'high' V-III gas presure ratio. The dashed curve is shifted slightly upwards extending the 2D region.

This relationship can be viewed as the strain energy dependence of the roughness transition temperature, as such it should be a phenomenon which is common to all materials growth. In this paper we present evidence to support this assumption. The interesting point about this approach is that it should be possible, by appropriate choice of the growth parameters, to shift the curve so as to extended the 2D growth region to higher temperature for a given strain.

2. THEORY

The idea behind the model for the roughness transition temperature T_R is to equate the energy for planar growth to the energy to form spherical islands. This gives an expression for the transition temperature [Bangert et al 1995a].

$$T_R = \frac{A\gamma^3}{\epsilon_f^2} + T_C \qquad \frac{\epsilon_f}{p} = \left(\sum f_i^2 h_i \right)^2 = \sqrt{Sum} \tag{1}$$

where A is a constant depending on atomic volume, the angle of contact of the spherical cap with the growth surface and γ is the surface energy. T_C represents a constant temperature. ϵ_f is the misfit strain squared times layer thickness or $\sqrt{Sum}$. This expression is a first order approximation to the empirical expression which fits the experimental data.

$$T_R = \left(1 - \exp\left(-\frac{s}{\epsilon_f^2} \right) \right) T_o + T_C \tag{2}$$

where s is a constant, which amongst other parameters, depends on the surface energy.
The modulation wavelength of the surface can be related to the stress σ by a relationship developed by Asaro and Tiller [1972]

$$\frac{1}{\delta} \frac{d\delta}{dt} = r = \frac{\Omega^2 n_s D_s}{kT} \left(\frac{\omega^2 \sigma^2}{3E^*} - 4\gamma\omega^3 \right) \tag{3}$$

This expresses the growth rate of different instabilities with amplitude δ as a function of n_s the adatom density, D_s the surface diffusion coefficient, E^* Young's modulus and ω the modulation frequency. The mode with the fastest growth rate is the unstable mode and will become established. This is found from

$$\frac{dr}{d\omega} = 0 \tag{4}$$

which gives the relationship $\sigma^2 \sim \omega$ or $\sigma^2 \sim \lambda^{-1}$ If it is assumed that the actual strain in the layers is proportional to the misfit strain f, it is expected that $f^2 \sim \lambda^{-1}$.

3. EXPERIMENTAL

The samples were all MOCVD grown GaInAs/InP superlattices on (001) substrates with strained GaInAs wells and InP barriers. Wells of misfit strains ranging up to -1.63% were grown by varying the In concentration from 53 to 77%. Samples of each strain value were grown at several temperatures. The table below gives data about samples used in this study, concerning their growth temperature, their *sum* values and their group V - group III partial pressure ratio during growth. Plan-view and cross-sectional TEM samples were prepared in the conventional way from the samples listed in the table and TEM carried out in a EM430 Philips machine at 300 keV.

Sample	F[2]	F[3]	F[4]	F[6]	F[7]
Figure	2 and 5	3	4	6	7
Growth Temp $^\circ C$	700	600	700	700	700
Sum	7×10^{-4}	7×10^{-4}	4.5×10^{-5}	0.0	0.0
V-III Ratio	150	150	150	75	300

4. RESULTS AND DISCUSSION

Three samples, which are marked on the graph in figure [1] by their micrograph figure numbers were chosen to illustrate the results. Sample marked as F[2] is seen in plan view in figure (2). It was grown at 700 °C and has the same misfit strain as the sample F[3] which was grown at 600 °C. There is a different strain relaxation mechanism at work in these two samples. F[2] grown above the curve shows wavy growth and F[3] grown below the curve shows the usual misfit dislocation array. Sample F[4], which was also grown at 700 °C but with a lower misfit strain than sample F[2], can be seen in figure (4) to also show wavy layer growth. However the wave length of the undulations is larger in sample [F4] than in sample [F2] as expected from the $\sigma^2 \sim \lambda^{-1}$ relationship discussed in the theory section.

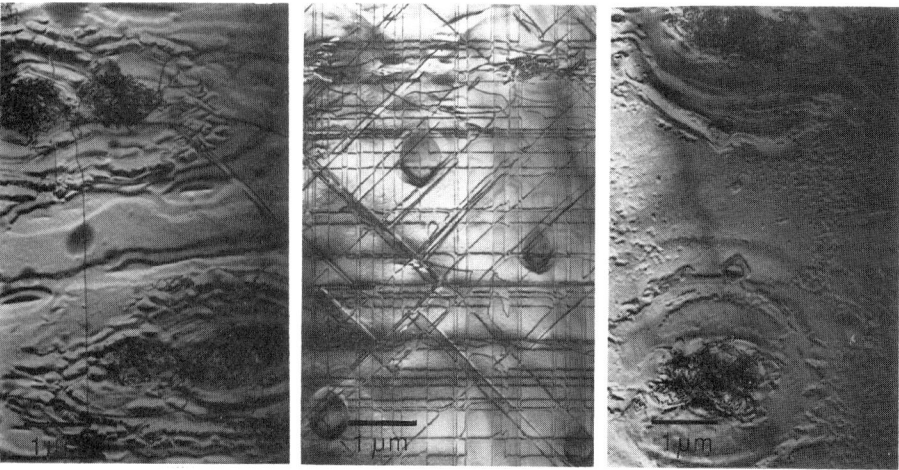

Figure 2 (001) plan view of F[2] showing wavy growth. Figure 3 (001) plan view of F[3] showing dislocation network. Figure 4 (001) plan view of F[4] showing wavy growth with a longer wave length than F[2].

194

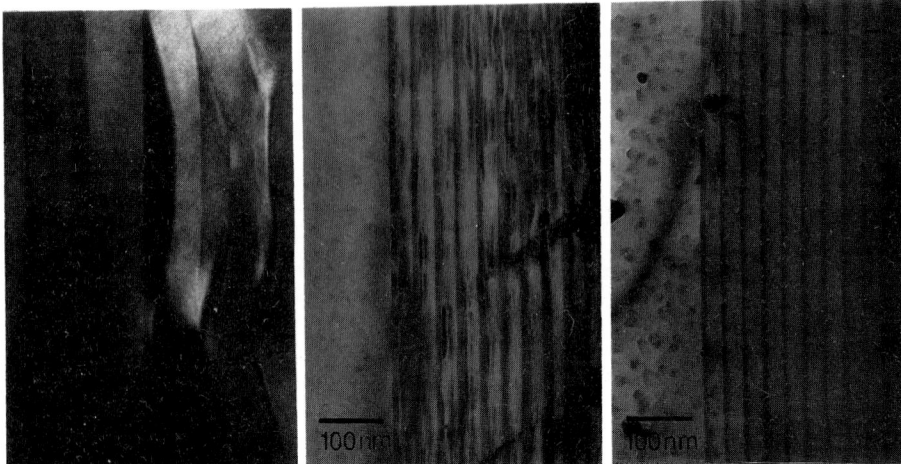

Figure 5 (110) cross-section of F[2] showing the wavy layers.

Figure 6 (110) cross-section of F[6] showing wavy growth at the low V-III ratio.

Figure 7 (110) cross-section of F[7] not showing wavy growth at the high V-III ratio.

In order to demonstrate that the transition curve can be shifted by varying the growth parameters, samples F[6] and F[7] were grown at 'low' and 'high' V-III ratios respectively with the same growth temperature of 700 °C and the same strain of 0%. A lower V-III ratio results in group III elements covering the surface, leading to considerable surface diffusion. A higher V-III ratio results in saturation of the surface with group V elements and little surface diffusion. This affects the surface energy γ. From equations 1 and 2 it is then expected that the curve is shifted downwards at low V-III ratios and upwards at high V-III ratios. The position of samples F[6] and F[7] in figure 1 is the same. The solid curve relevant to sample F[6] is seen to meet the temperature axis at about 700 °C. A cross-section of this sample is shown in figure 6 where the layers are seen to be wavy. The micrograph in figure 7 shows the sample grown with a higher V-III ratio and no wavy layers are visible, which indicates that the curve has indeed been shifted upwards, extending the region of 2D growth, as shown by the dashed curve in figure 1.

5. CONCLUSION

Using selected samples, we have demonstrated the strain dependence of the roughness transition temperature in III-V compound semiconductor growth. We also illustrated the strain dependency of the modulation wavelength. This has been presented in greater theoretical and experimental detail elsewhere [Bangert et al 1995a, 1995b]. Furthermore in this paper we give initial evidence for the possibility of shifting the curve. Further research may make it possible to controllably extend the 2D growth regime.

REFERENCES

Asaro R and Tiller W A, 1972 Met. Trans. 3 1789
Bangert U, Harvey A J, Dieker C, Hardtdegen H, Vescan L and Smith A 1995a J. Appl. Phys. (accepted for publ.)
Bangert U, Harvey A J, Dieker C, and Hardtdegen H 1995b J. Crystal Growth (accepted for publ.)

Inst. Phys. Conf. Ser. No 146
Paper presented at Microsc. Semicond. Mater. Conf., Oxford, 20–23 March 1995
© 1995 IOP Publishing Ltd

Morphological transformations of MBE-grown $In_{0.6}Ga_{0.4}As$ islands on GaAs(001) substrates

K Tillmann, D Gerthsen*, P Pfundstein*, A Förster[†] and K Urban

Institut für Festkörperforschung and [†]Institut für Schicht- und Ionentechnik, Forschungszentrum Jülich GmbH, D-52425 Jülich, Federal Republic of Germany

*Laboratorium für Elektronenmikroskopie, Universität Karlsruhe, Kaiserstraße 12, D-76128 Karlsruhe, Federal Republic of Germany

ABSTRACT: Electron microscopic techniques were employed to investigate the influence of the growth temperature on the morphology and on the lattice-parameter mismatch relaxation mechanisms of thin $In_{0.6}Ga_{0.4}As$ layers, deposited by molecular beam epitaxy (MBE) on nominal singular and on vicinal GaAs(001) wafers. For nominal layer thicknesses in the order of 3.5 nm it is shown that an increase of the growth temperature yields a shape transformation from small circular islands to larger, cigar-like structures and an indium desorption preventing the formation of misfit dislocations. The use of vicinal wafers is conducive to the nucleation of symmetrical, coherent dot-like structures.

1. INTRODUCTION

As well known, the epitaxial growth of highly lattice mismatched $In_xGa_{1-x}As$ layers on GaAs(001) takes place in the Stranski-Krastanow (1939) growth mode, starting with layer-by-layer growth followed by the nucleation of three-dimensional islands. Lately, island formation is regarded as a possibility of great promise for the production of self-assembled quantum dots, whose electronic properties are determined by their microscopic structure.

In the present study the influences of the growth temperature T_S and the substrate tilt angle α on the island structure are studied in the case of the $In_{0.6}Ga_{0.4}As$ layers on GaAs(001) substrates, which are characterized by a lattice-parameter mismatch of $f = 4.1\%$.

2. EXPERIMENTAL DETAILS

$In_{0.6}Ga_{0.4}As$ layers with a constant nominal thickness of 3.5 nm were grown at a depostion rate of 0.2 μm/h by MBE on vicinal wafers tilted by $\alpha = 2.0^o \pm 0.25^o$ around the [010]-axis and on nominal singular ($\alpha = 0^o \pm 0.04^o$) substrates. The growth temperature was varied between 480°C and 560°C. From cross-section HRTEM images it is known that the growth of a GaAs buffer layer of 100 nm in thickness does not influence the nominal tilt angles, i.e. $\alpha = 0^o$ and $\alpha = 2^o$ are valid parameters for the $In_{0.6}Ga_{0.4}As$ epitaxy.

The layer morphology and defect structure were investigated by the application of the moiré- and the weak-beam technique on plan-view TEM samples. The amount of the elastic relaxation and the indium desorption were estimated by the measurement of the lattice-plane distances on HRTEM cross-section samples. The specimens were prepared by standard mechanical preparation procedures.

3. EXPERIMENTAL RESULTS AND DISCUSSION

At lower growth temperatures ($T_S = 480^\circ$C, 500°C) the layer morphology on the vicinal wafers is dominated by a high density of small islands which appear circular in the [001]-projection, e.g. island (A) in Fig. 1a. The dark lines observed in these islands are caused by strong lattice distortions leading to bend contours due to the local fulfilment of the Bragg condition. A quantitative explanation for the formation of these image contrasts, also observed in a cross-shaped variant in Fig. 2a, has been given recently by Androussi et al (1994). Additionally performed HRTEM investigations on cross-section samples (Fig. 1b) confirm the absence of misfit dislocations and give clear evidence that an average of at least $\Delta\varepsilon \approx 0.9$ % of the lattice-parameter mismatch induced strain of $f = 4.1$ % is accomodated by elastic relaxation, i.e. by a lateral bending of the islands lattice planes.

A lower fraction of larger, anisotropic islands (labeled (B) in Fig. 1a) shows preferential alignment along the $[\bar{1}10]$-direction. Two sets of moiré-fringes running along the [110]- and the $[\bar{1}10]$-direction indicate that these islands are relaxed in the conventional manner by the formation of Lomer misfit dislocation with a Burgers vectors of type $\mathbf{b} = a/2 <110>$ perpendicular to the dislocation line (plastic relaxation mechanism). However, in these larger islands, presumabbly formed by island coalescence, a distinctly smaller part $\Delta\varepsilon \approx 0.5$ % of the misfit strain f is accomodated by elastic relaxation. This decrease may be explained by a reduced possibility of lateral expansion in a larger island.

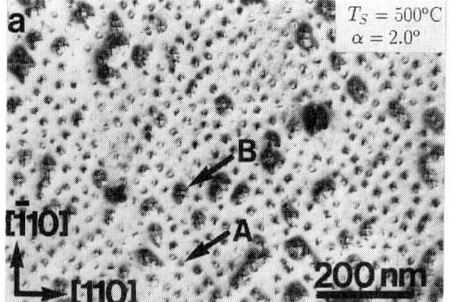
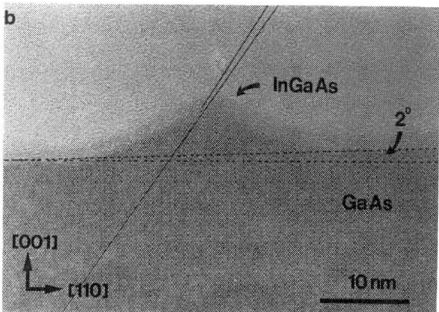

Fig. 1 Plan-view TEM image (a) of a specimen grown at $T_S = 500^\circ$C on a vicinal wafer. Small coherent (A) and larger (B) islands, which show image contrast dominated by two orthogonal sets of moiré-fringes, are visible. HRTEM images of cross-section samples (b) confirm the coherence of the small islands nucleated on vicinal wafers. While the upper dashed line indicates the interface between the substrate and the layer, the lower one indicates the (001)-plane tilted by $\alpha = 2^\circ$.

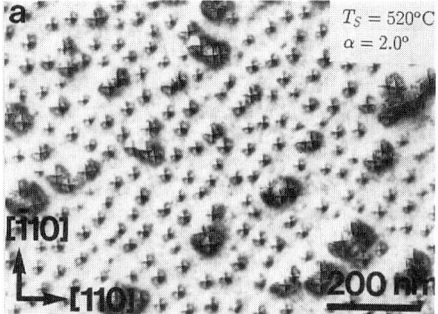
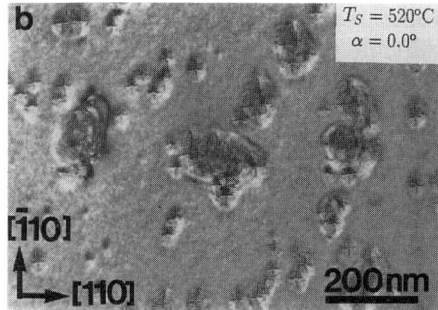

Fig. 2 Plan-view TEM images of layers grown on a vicinal (a) and a nominal singular (b) wafer at $T_S = 520^\circ$C. Most islands nucleated on the vicinal wafer are circular in shape and coherently strained (cross-shaped image contrast), while the morphology of the islands grown on the well-oriented wafer is in general irregular.

Apart from the substrate tilt angle, the images shown in Fig. 2 were taken from two samples grown under identical conditions at $T_S = 520°C$ on a vicinal substrate (a) and a well-oriented substrate (b). Generally, islands grown on the vicinal wafers are smaller and exhibit a circular circumference in contrast to a more irregular shape observed on the nominal singular substrates. This irregularity may be a consequence of the island coalescence, occuring easier on well-oriented substrates due to the absence of substrate steps. Therefore the use of vicinal wafers strongly increases the island density. This behaviour was found to be independent of the growth temperature, cf. also Fig. 5.

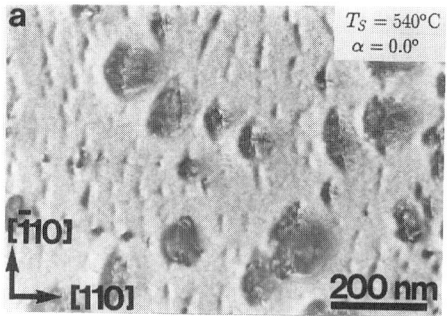 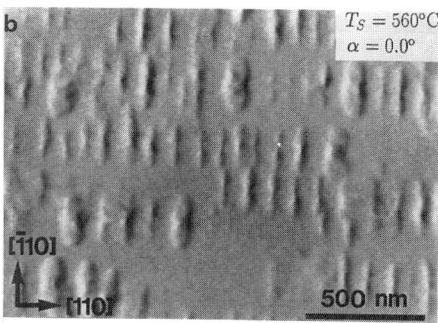

Fig. 3 Plan-view TEM images of layers grown at $T_S = 540°C$ (a) and $T_S = 560°C$ (b) on nominal singular wafers. The cigar-shaped structures only show strain contrast.

At $T_S = 540°C$ a distinction must be made between incoherent, almost circularly shaped islands and cigar-shaped structures between these islands, which are exclusively aligned along the $[\bar{1}10]$-direction, cf. Fig. 3a. In these islands moiré fringes are not observed. They only show strain contrast. Weak-beam images confirm the absence of dislocations.

Finally, at $T_S = 560°C$, the morphology is completely dominated by this type of island (Fig. 3b). From an analysis of the (002)-plane distances of cross-section HRTEM-images it is known that these cigar-shaped $In_xGa_{1-x}As$ islands are characterized by a reduced indium concentration $x < 0.6$. Furthermore, contrast changes sometimes observed in cross-section HRTEM images at some distance from the interface (Fig. 4) indicate that indium segregation near the growth front, formerly observed in MBE grown $In_xGa_{1-x}As$ quantum wells by Kawai et al (1993), is likely to occur at higher growth temperatures. On the other hand, additionally performed X-ray diffraction measurements carried out on thicker layers grown under identical conditions yielded $x = 0.60 \pm 0.02$.

It is therefore proposed that an indium desorption occurs during the very last stages of epitaxial growth, possibly not before the termination of MBE growth. The formation of elongated islands may be understood by a more effective elastic strain relaxation in anisotropic structures as shown by Tersoff and Tromp (1993). The coherence itself may be a consequence of compositional modifications along the growth direction avoiding the generation of large stress fields near the interface and therefore postponing the formation of misfit dislocations. The observed strictly parallel alignment of the cigar shaped islands along the $[\bar{1}10]$-direction without the formation of domains rotated by 90° is possibly caused

Fig. 4 HRTEM cross-section image in the $[\bar{1}10]$-projection of a cigar shaped island grown at $T_S = 560°C$ on a nominally singular wafer. A change in the image contrast is observed six to seven monolayers above the interface (dashed line).

by the inequivalence of the steps along the [110]- and the $[\bar{1}10]$-direction of the homoepitaxially grown GaAs(001) buffer layer (Heller and Lagally (1992)).

Despite the observed morphological and compositional transformations, the island density N as a function of the growth temperature shows an Arrhenius-like behaviour with an activation energy E_D, which is higher for the tilted substrates (Fig. 5). This may be understood by the presence of the substrate steps, which act as barriers to the surface diffusion of adatoms on the vicinal wafers. The average diffusion length of adatoms on the surface $\lambda = 1 / \sqrt{\pi N}$ of some 10 nm is larger than the average plateau extension of 8 nm between steps of monolayer height for $\alpha = 2.0°$. In contrast the step distance on the nominal singular substrates amounts to at least 430 nm, so adatom climbing processes across steps are unlikely to occur.

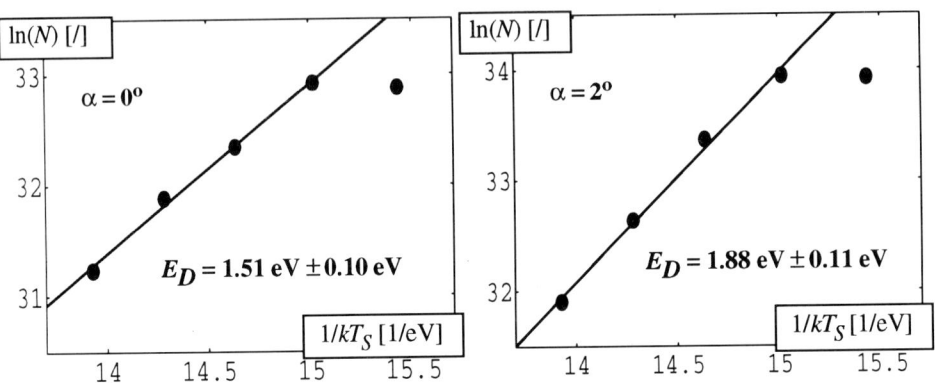

Fig. 5 Arrhenius plots of the islands densities N vs. $1/kT_S$ for layers grown on nominal singular ($\alpha = 0°$) and vicinal ($\alpha = 2°$) wafers. Both N and E_D are strongly dependent on the wafers tilt angle.

Deviations from the Arrhenius behaviour at $T_S < 500°C$ may be explained by energetic considerations. The island formation in highly mismatched systems is driven by the ability of a misfit strain energy reduction E_{elast} by the elastic relaxation at the expense of an increased surface energy E_{surf}. Below a critical volume, which is equivalent to a maximum island density, the islands will become unstable, independent of the growth temperature, when E_{elast} does not outweigh the additional contribution E_{surf}.

4. CONCLUSIONS

It has been demonstrated that the growth temperature of the MBE process has a significant influence on the layer morphology, the composition and the defect structure of $In_{0.6}Ga_{0.4}As$ islands on GaAs(001) substrates. An increasing growth temperature yields a transformation from round shaped smaller islands to larger, coherent cigar-shaped islands. This transformation goes along with an indium desorption, reducing the misfit and therefore postponing the formation of misfit dislocations.

At lower growth temperatures, the use of vicinal substrates strongly supports the formation of regularly shaped small, coherent $In_{0.6}Ga_{0.4}As$ islands, while islands grown on nominal singular wafers are incoherent and irregular in shape.

REFERENCES

Androussi Y, Lefebvre A, Courboules B, Grandjean N, Massies J, Bouhacina T and Aime J P 1994 Appl. Phys. Lett. 65 1162
Heller E J and Lagally M G 1992 Appl. Phys. Lett. 60 2675
Kawai T, Yonezy H, Ogasawara Y, Saito D and Pak K 1993 J. Appl. Phys. 74 55
Stranski I N and Krastanow L 1939 Akad. Wiss. Wien, Math.-Naturwiss. Kl IIb 146 797
Tersoff J and Tromp R M 1993 Phys. Rev. Lett. 70 2782

Inst. Phys. Conf. Ser. No 146
Paper presented at Microsc. Semicond. Mater. Conf., Oxford, 20–23 March 1995
© 1995 IOP Publishing Ltd

Relationship between the growth temperature and the 3D morphology of GaInP/InAsP strained heterostructures grown by Metal-Organic Vapour Phase Epitaxy

A Ponchet, A Rocher, A Ougazzaden[1] and A Mircea[1]

CEMES-CNRS, 29 rue Jeanne Marvig, BP 4347, 31055 Toulouse, FRANCE
[1] France Télécom, CNET-PAB, BP 107, 92225 Bagneux, FRANCE

ABSTRACT: Zero-net strained multilayers of alternating GaInP and InAsP with lattice mismatches of -1% and 1% respectively have been grown on (001)InP by Metal-Organic Vapour Phase Epitaxy. They form a laterally modulated structure: rich-GaInP and rich-InAsP ribbons alternate along the lateral [110] direction. The effect of growth temperature on the morphology is reported and discussed. For the highest temperature (650°C), the structure is perfectly modulated and free of crystalline defects. The lateral period is of 0.28μm while the so formed ribbons are several μm long in the [1$\bar{1}$0] direction. At 610°C, the lateral period of the ribbons is reduced to 0.12μm but their length is lowered by at least one order of magnitude. Considering the possible effects of surface reconstruction, it is proposed that the length along [1$\bar{1}$0] is directly limited by the surface diffusion and tends to be infinite to stabilize the structure.

1. INTRODUCTION

III-V strained and zero-net strained multilayers are often submitted to elastic 3D growth modes. In particular, using tensile layers of any composition of GaInAsP alloy with lattice mismatches from -0.5% to -2% with InP(001) substrate induces lateral thickness modulations which are systematically parallel to [1$\bar{1}$0]. This equally occurs for Molecular Beam Epitaxy (MBE) (Ponchet 1993, Ponchet 1995) and for Metal-Organic Vapour Phase Epitaxy (MOVPE) (Ponchet 1994, Zhou 1994). It is clear that the driving force is elastic relaxation, and that the morphology results from the competition between elastic and surface energies (with kinetic limitations). This could also be supported by instability models (Grilhé 1993 and ref. therein).

However, some differences have been observed in structures grown by MBE and MOVPE. Independent of the lattice mismatches, the modulation period always ranges around 20-50nm for MBE and above 0.1μm for MOVPE. Also the anisotropy is often more developed for MOVPE than for MBE. This raises the question of the different mechanisms involved in the modulation development. We have suggested that the period could be controlled by kinetics, the surface diffusion length being usually larger for MOVPE than for MBE (Ponchet 1994). However, these techniques involve different growth mechanisms and surface phenomena and it is not sure that the energetical balance is similar. This limits the validity of the comparison which is based only on different surface diffusion lengths.

The work reported here concerns tensile/compressive multilayers grown by MOVPE at different temperatures. We will focus the discussion on the role of growth temperature on the modulation development.

2. EXPERIMENTS

The multilayer consists of tensile $Ga_{0.15}In_{0.85}P$ alternated with compressive $InAs_{0.3}P_{0.7}$ with lattice mismatches of -1% and 1% respectively with the InP substrate. The layer thicknesses are 5nm. Epitaxy was carried out in a T-shaped reactor at atmospheric pressure using Trimethylindium and Trimethylgallium as sources for group III elements and pure phosphine and pure arsine as sources for group V elements. Several samples have been grown at two different temperatures of 650°C and 610°C.

The Transmission Electron Microscopy study was performed at 200kV on a Philips CM20 microscope using both cross-sectional and plan view observations. Thin cross-sectional and plan view specimens were obtained by mechanical polishing followed by Ar^+ beam milling. Dark field images were taken with g=002, which is highly sensitive to chemical composition variations, and with g=220 which is highly sensitive to strain fields.

3. RESULTS

The behaviour of samples grown at 650°C has been reported in detail and discussed (Ponchet 1994). They form remarkably regular and self-induced modulated structures alternating GaInP-rich and InAsP-rich ribbons parallel to [1$\bar{1}$0]. GaInP and InAsP form "valleys" and "mesas" built with (117)A or (118)A and (114)A facets respectively. The modulation period along [110] is 0.28μm (+/- 10%) for all samples (Fig. 1a). Modulations remain coherent over very large distances along the [1$\bar{1}$0] direction: elongated ribbons of about 2-3μm and sometimes greater than 5μm have been achieved (Fig, 2a).

The samples grown at 610°C differ in some points:
- the modulation is much less regular
- the modulation period along [110] is systematically decreased to 0.12-0.14μm (Fig. 1b)
- the length of modulation along [1$\bar{1}$0] is lowered to typically 0.2-0.3μm (Fig. 2b)
- the facets are steeper (Fig. 3)

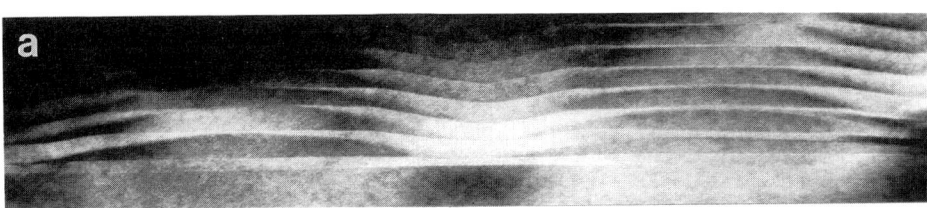

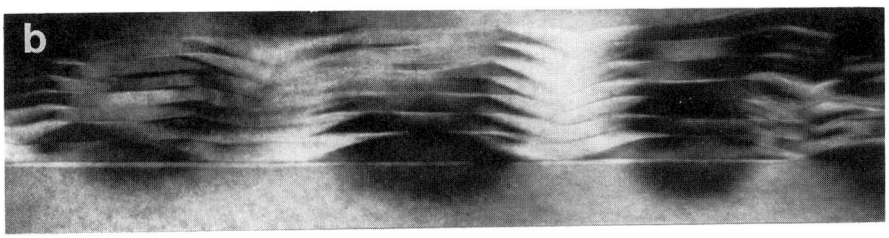

0.1 μm

Fig. 1. ($\bar{1}$ 10) cross-sectional micrographs (g=002 dark field): (a) 650°C; (b) 610°C GaInP and InAsP are bright and dark layers respectively.

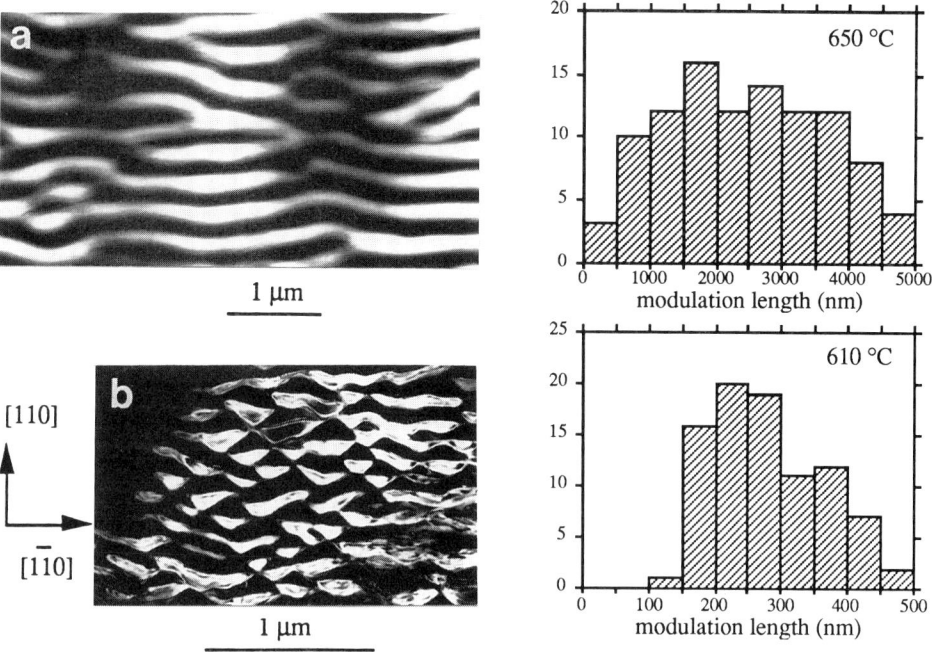

Fig. 2. Impact of growth temperature on the modulation length along [1 1̄ 0]: plan views (dark field, g=220) and length histograms; (a) 650°C; (b) 610°C.

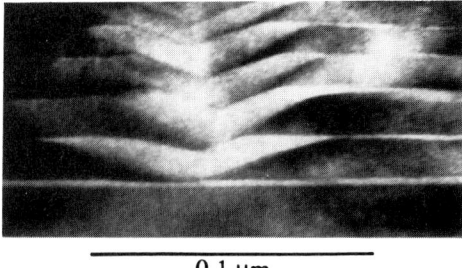

Fig. 3. Detail of one sample grown at 610°C. Facets are around (114)A for GaInP valleys and (113)A type for InAsP mesas.

0.1 μm

4. DISCUSSION

When the surface energy is negligible with regard to the elastic energy, the surface characteristics play an important role. Since in semi-conducting compounds, the surface energy is mainly determined by electronic dangling bonds, the modulation morphology should be examined through the surface reconstruction. In III-V compounds for (001) surfaces under arsenic or phosphorus flux, the group V elements form dimers parallel to [1 1̄ 0] and are the basis of any surface reconstruction which is anisotropic and periodic (see for instance Larsen 1988). It was proposed a few years ago that stable 3D islands should be infinite in the dimers direction, otherwise the surface reconstruction is not satisfied (Pashley 1988 and 1989). It has been suggested that this was the reason why these kinds of modulations are generally oriented along [1 1̄ 0] (Ponchet 1995). These hypotheses have been developed for MBE, for which the surface reconstructions can be studied in-situ and are therefore well known. Although they are not well known during MOVPE process, it can be assumed that the group V elements form dimers at surface and that this explanation may also apply to MOVPE strained multilayers.

Two factors however can reduce the modulation length: (1) kinetic limitation (surface diffusion length) (2) if the gain in elastic energy provided by a bi-dimensional modulation with regard to a one-dimensional one is larger than the cost in surface energy due to unsatisfied reconstruction, the modulation will be isotropic (this is observed in highly strained layers which form islands). Our work shows that the modulation length could reach several μm at the highest temperature (650°C). First this supposes that the surface diffusion length is as large as these values. Secondly, the modulation length is infinite with regard to the lateral period (ratio of 10). Thirdly, it is decreased with the temperature. It is concluded that the modulation length is only limited by the surface diffusion and should be infinite in a more stable state.

The modulation period is also lowered by decreasing the surface diffusion length. However, the variations of the period and the length with the growth temperature are different. At 610°C, the ratio length/period is only about 3: period and length tend to be relatively close values limited by the surface diffusion (this is also true for MBE grown structures). But unlike the length which tends to become infinite when this is kinematically possible, the period becomes much smaller than the surface diffusion length at high temperature. It cannot be stated if the period will continue to increase or is close to a maximal value. It is rather expected that the elastic relaxation is efficient only for a finite period.

Finally it not impossible that other mechanisms which have been neglected should play a role in the energy balance, in particular the distortion of the substrate which participates in the strain relaxation and induces long range elastic interactions.

5. CONCLUSIONS

The impact of growth temperature on MOVPE grown GaInP/InAsP layers has been reported. It has been established that the modulation period and the modulation length decrease with the temperature. The period is less affected that the length. The later appears to be directly related to the surface diffusion length, while the former results from complex competition involving energy minimization and kinetic factors. The modulation anisotropy can be accounted for by surface reconstruction and it is found that the most stable modulation should be infinite along $[1\bar{1}0]$. For the range of lattice mismatch, the surface energy (through surface reconstruction) plays an important role in determining the equilibrium shape of III-V compounds.

REFERENCES

Grilhé J 1993 Acta Metall. 41 909
Larsen P K and Chadi D J 1988, Phys. Rev. B37 8282
Pashley M D, Haberern K W and Woodall J M 1988 J. Vac. Sci. Technol. B6 1468
Pashley M D 1989 Phys. Rev. B40 10481
Ponchet A, Rocher A, Emery J-Y, Starck C and Goldstein L 1993 J. Appl. Phys. 74 3778
Ponchet A, Rocher A, Ougazzaden A and Mircea A 1994 J. Appl. Phys. 75 7881
Ponchet A, Le Corre A, Godefroy A, Salaün S and Poudoulec A 1995 J. Cryst. Growth (submitted)
Zhou X, Charsley P, Smith A D and Briggs A T R 1994 Mat. Res. Soc. Symp. Proc. 340 337

Inst. Phys. Conf. Ser. No 146
Paper presented at Microsc. Semicond. Mater. Conf., Oxford, 20–23 March 1995

Interfacial defect control of highly mismatched (100)InAs/GaAs heterostructures grown by MBE

A Trampert, E Tournié[1] and K H Ploog[2]

Max-Planck-Institut für Festkörperforschung, Heisenbergstr. 1, D-70569 Stuttgart, Germany
[1] Centre Recherche sur l'Hétéro-Epitaxie et ses Applications, CNRS, 06560 Valbonne, France
[2] Paul-Drude-Institut für Festkörperelektronik, D-10117 Berlin, Germany

ABSTRACT: Using transmission electron microscopy we demonstrate that the initial plastic relaxation of highly mismatched InAs layers grown on (001) GaAs is governed by the growth mode. During molecular beam epitaxy in the Stranski-Krastanov mode, 60°-type dislocations are generated at the island edges, which glide to the interface to relieve the strain. When InAs is forced to grow layer-by-layer, pure edge-type dislocations are introduced lying exactly at the interface. Our results can be explained by the different dislocation nucleation mechanism imposed by the surface morphology of the strained layer.

1. INTRODUCTION

The most important prerequisite for a device application of lattice mismatched heterostructures based on III/V semiconducting materials is the understanding of defect generation during the strained layer growth. Thereby it is well established that in low mismatched systems the strain relaxation is caused by the generation of misfit dislocations which are observed to be of 60°-type. This kind of misfit dislocation can be simply formed by the elongation of pre-existing threading dislocations into the interfacial plane, requiring the least possible energy, or by the nucleation of dislocation half-loops at instrinsic growth defects and subsequent gliding to the interface.

At high lattice mismatches, where in principle homogenous dislocation half-loop nucleation at the planar surface would become energetically allowable (Hull and Bean 1992), a strain induced morphological transition normally takes place from two-dimensional (2D) layer-by-layer growth to three-dimensional (3D) island nucleation (Stranski-Krastanov growth mode). Here, the strain is first relieved by coherent island formation and then by a 60°-dislocation generation mechanism at the island edges. Furthermore during coalescence a variety of defect types can be created when islands merge. Also pure edge-type (Lomer) misfit dislocations can be formed by a reaction between two suitable oriented glissile 60°-dislocations and subsequent climb to the interface. The Lomer dislocations are most efficient in strain relief but cannot move by simple glide processes.

In this work we demonstrate the occurrence of a new relaxation mode in the highly mismatched InAs/GaAs-system (f_0 = 7.2 %) forced to grow in 2D along the [001] direction. By modifying the surface stabilisation from anion- to cation-stable conditions during molecular beam epitaxy (MBE) we are able to inhibit the 2D-3D transition which normally occurs in this system. The relaxation induced by the two growth modes results in a totally different strain relieving dislocation generation mechanism and interfacial defect morphology.

2. EXPERIMENTAL

The InAs films were grown by solid source MBE on GaAs buffer layers deposited at 580°C on (001) oriented GaAs substrates (EPD < 3000 cm^{-2}). The growth temperature during the InAs layer deposition was kept at 430°C and the growth rate at 0.4 monolayers (ML) per second. The various surface stabilization conditions, As- or In-stablility, during growth were adjusted by realizing different As$_4$/In flux ratios and the obtained growth mode was controlled by *in-situ* reflection high-energy electron diffraction (RHEED). Details are given by Tournié et al (1994). As-grown epilayers with thicknesses between 4 and 300 ML were taken to study the different relaxation states by transmission electron microscopy (TEM). Cross-sectional and plan-view samples were prepared by conventional argon ion milling using the cold stage specimen holder. The observations were carried out in a JEOL 4000 FX microscope.

3. RESULTS AND DISCUSSION

3.1 Interfacial Structure

Island formation during the Stranski-Krastanov growth of the InAs on (001) GaAs leads to a degradation of the heterostructure. Using conventional cross-section TEM, Zhang et al (1992) recently reported on perturbations at the interface attributed to the high stresses at the edges of the coherent InAs islands deforming the substrate surface. On the atomic level the interfacial morphology of the plastically relaxed layer also appears very rough as observed by cross-sectional high resolution TEM (HREM) by Trampert et al (1994). The misfit dislocations, which consist of 60°-type, are distributed in a wide band of about 1 nm around the interface. A high linear dislocation density can be estimated from the micrographs indicating an inefficient misfit strain relief.

In contrast, the HREM analysis of the hetero-interface formed by 2D-growth of 300 ML InAs reveals a very planar interfacial morphology. The misfit dislocations are situated along one (001) plane distributed in a very periodic manner (Fig. 1). All dislocations are of pure edge-type with $\mathbf{b}=a/2[110]$ parallel to the interface. The average spacing of the misfit dislocation cores in the experimental images amounts to 30 ± 5 {220} lattice planes, which corresponds to a distance D = (6.0 ± 1.0) nm. Weak beam dark field images of tilted cross-sections confirm that such a regular array with the distance D is present over a large interfacial region. The HREM images taken from the perpendicular direction results in the same interfacial microstructure. Therefore the 2D misfit dislocation network consists of a nearly square arrangement of pure edge-type dislocations as schematically illustrated in figure 1. In addition, a few stacking faults are detected along the [110] direction, whereas no stacking faults are found in the perpendicular direction pointing out an asymmetric behaviour of the strain relaxation mechanism (Trampert et al 1995).

3.2 Defect Generation

To study the defect generation mechanisms we investigate the very early stage of relaxation by characterising epilayers just above the critical thickness. During the 3D growth the plastic relaxation starts by misfit dislocation nucleation at the island edges (Trampert et al 1994) which are zones of highest shear stress favouring the 60°-dislocation type. But in the

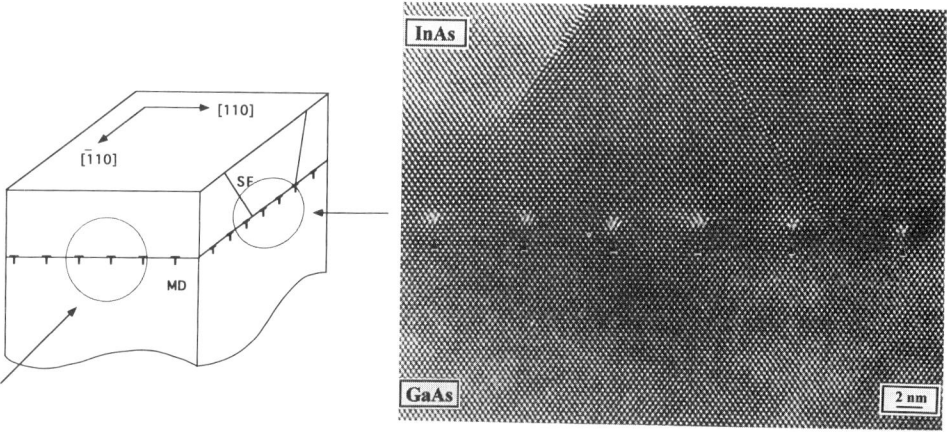

Fig. 1: The interface structure of InAs/(001)GaAs system in the case of the layer-by-layer growth mode: a schematic diagram showing the regular 2D network of Lomer misfit dislocations (MD) and the corresponding cross-sectional HREM image of a 300 ML thick film.

2D case, where the strain is homogeneously contributed over the epilayer, no specific nucleation site exists. The interfacial structure of the 2D grown 8 ML-thick InAs appears completely different compared to the case of Stranski-Krastanov growth. Figure 2a represents the [110] cross-sectional HREM images of the heterostructure demonstrating that the lattice mismatch is only relaxed by interfacial dislocations of pure edge-type. A generation mechanism based on the reaction of pre-existing 60°-dislocations as described above is not very likely, because all misfit dislocations are placed exactly in the interfacial plane. In addition, no 60°-dislocations could be found in cross-sectional as well as in the plan-view TEM using the **g·b**-criterion (Fig. 2b). Therefore we expect a homogenous surface half-loop nucleation mechanism (Fig. 2c). If we apply the formalism given by Hull et al (1992) for calculating the activation barrier for the critical loop-radius, the Lomer dislocation half-loop will be favoured against 60°-dislocation loops. In addition, because of the high lattice mismatch of the InAs/GaAs system the energy barrier can be overcome by thermal activation. However, the energetic conditions for half-loop nucleation is only a necessary condition for loop formation. The growth of a supercritical loop must also be kinetically allowed. In general, glide motion is more likely than climb motion, because dislocation climb needs the diffusion of point defects. However, diffusion-mediated climb processes become important at temperatures higher than half the melting temperatures, which is in fact nearly the case during the MBE growth of InAs at 430 °C. Furthermore, during the condensation process, a transient supersaturation of point defects might exist supporting the climb loop formation. Therefore, the nucleation and propagation of misfit dislocations should only occur during the deposition and no post-growth relaxation should happen. In fact, we find an excellent agreement between *in-situ* RHEED analysis of the in-plane lattice constant and the *ex-situ* TEM observations of the relaxation state of the epilayer (Trampert et al 1994).

In conclusion, we have demonstrated that the early stage of plastic relaxation in the highly mismatched InAs/GaAs system is governed by the growth mode. In contrast to the normally found Stranski-Krastanov growth mode, where glissile 60°-dislocations are

206

introduced at the island edges, the generation of misfit dislocations during 2D growth is determined by the surface half-loop nucleation of pure edge-type dislocations.

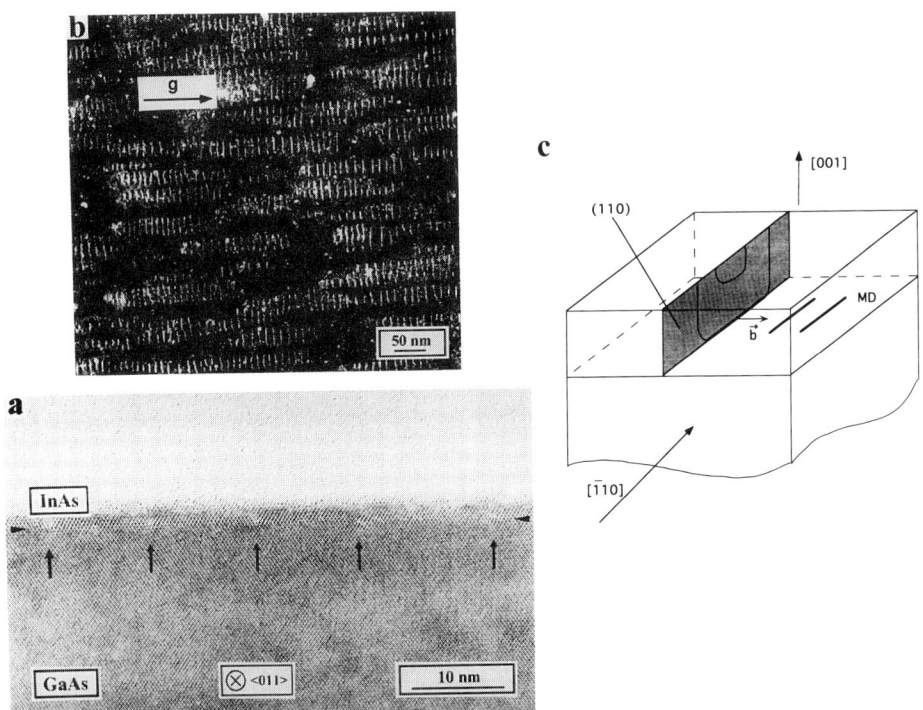

Fig. 2: (a) Cross-sectional HREM image of 2D-grown 8 ML InAs containing Lomer dislocations at the interface (arrows), (b) the corresponding plan-view dark field micrograph (g = 220) of the sample indicating the short segments of misfit dislocations and (c) a schematic diagram of the dislocation generation mechanism.

ACKNOWLEDGEMENTS

Part of this work was sponsored by the Bundesministerium für Forschung und Technologie of the Federal Republic of Germany. We thank the Max-Planck-Institut für Metallfoschung for the possibility to use the HREM facilities.

REFERENCES

Hull R and Bean J C 1992, Crit. Rev. Sol. Stat. Mat. Sci. **17**, 507
Tournié E, Trampert A and Ploog K H 1994, Europhys. Lett. **25**, 663, **26**, 315
Zhang X, Pashley D W, Neave J H, Zhang J and Joyce B A 1992, J. Cryst. Growth **121**, 381
Trampert A, Tournié E and Ploog K H 1994, phys. stat. sol. (a) **145**, 481
Trampert A, Tournié E and Ploog K H 1995 to be published

Inst. Phys. Conf. Ser. No 146
Paper presented at Microsc. Semicond. Mater. Conf., Oxford, 20–23 March 1995

Crack formation in tensile strained III-V epilayers grown on InP substrates

R T Murray, C J Kiely, M Hopkinson* and P J Goodhew

Department of Materials Science and Engineering, University of Liverpool, UK
*Department of Electronic and Electrical Engineering, University of Sheffield, UK

ABSTRACT: Layers of $In_x(GaAl)_{1-x}As$ with x<0.53 grown on InP(001) are under tension. Beyond a critical thickness, governed by the lattice misfit, such layers prefer to relieve their stored energy by crack formation. In this paper, experimental observations of this critical thickness over a range of misfits are compared with the predictions of a simple theory. Measurements of crack profiles and separations have allowed us to estimate the amount of stored elastic energy in such layers and explain why cracks penetrate beyond the epilayer into the substrate.

1. INTRODUCTION

If a III-V semiconductor alloy layer is deposited pseudomorphically onto a substrate which has a slightly different lattice parameter, the misfit strain generated results in elastic energy being stored in the epilayer. Matthews (1975) has shown that when the stored energy density exceeds some critical level, interfacial misfit dislocations will be introduced to relax some of the strain. If the misfit f (which equals $[a_e-a_s]/a_s$) is negative, corresponding to a tensile epilayer strain, then a second process for energy release can come into play. Matthews (1972) demonstrated that this involves the formation of cracks which penetrate from the free surface down to the substrate. Murray et al (1993) have shown that for $In_x(GaAl)_{1-x}As$ epilayers grown on InP(001) with |f| > -0.01, strain relaxation is brought about by a combination of 60° interfacial misfit dislocations, segments of edge-type interfacial misfit dislocations *and* an array of vertical cracks on (110) and (1$\bar{1}$0) planes. Franzosi et al (1988) have deduced that similar cracks in ternary III-V alloys grown on InP actually penetrate through the interface plane into the underlying substrate. In this study we seek to model the conditions under which cracks initiate and to explain their penetration into the substrate.

2. EXPERIMENTAL

A systematic series of $In_x(GaAl)_{1-x}As$ layers have been grown epitaxially on nominally (001) InP substrates held at 500±20°C by Molecular Beam Epitaxy. The value of x has been varied from 0.11 up to 0.53 which corresponds to adjusting the lattice mismatch between the epilayer and substrate from -3% to 0%. Crack density measurements as a function of mismatch and layer thickness were performed on as-grown wafers using Nomarski interference microscopy. This method was favoured because it allows one to obtain statistically significant data with minimum specimen disruption. Transmission electron microscopy observations on plan view and cross-sectional specimens allowed us to characterise details of crack geometry and penetration depth.

3. CRITICAL THICKNESS FOR THE ONSET OF CRACKING

3.1 Theoretical model

According to the simple model proposed by Matthews (1975), the critical thickness, h_c, for misfit dislocation introduction to relieve epilayer strain is given by the hyperbolic expansion

$$h_c = \frac{b}{8\pi(1+\upsilon)f} \ln\left(\frac{h_c}{b} + 1\right) \qquad (1)$$

where v is Poissons ratio and b is the Burgers vector of the dislocation.

For a crack to form, the stored elastic energy must exceed the energy needed to create the two new surfaces which constitute the crack. Following the treatment of Matthews (1975) for dislocation introduction, let us assume that only energy stored within a range t_c of the crack can be accessed, (where t_c is the layer thickness at which cracks are first observed). Thus, the stored energy, E_e, per unit length of incipient crack is

$$E_e = 2G \frac{(1+\upsilon)}{(1-\upsilon)} t_c^2 \, f^2 (1 - \beta) \qquad (2)$$

where G is the shear modulus and $2t_c^2$ is the volume from which energy is dissipated. The parameter β accounts for the fraction of energy relieved by dislocations introduced prior to the onset of cracking. The required surface energy, E_s, is given by

$$E_s = 2 \Gamma t \qquad (3)$$

where Γ is the (110) surface energy of the semiconductor epilayer. Since the minimum crack depth able to relieve interfacial strain during growth is given by $t = t_c$, we obtain the critical thickness condition (for the simple case when $\beta = 0$) as

$$t_c = \frac{\Gamma(1-\upsilon)}{G(1+\upsilon)} \frac{1}{f^2} \qquad (4)$$

The best numerical values available are ; $\Gamma = 0.86$ Jm^{-2} for GaAs, G = 5×10^{10} Nm^{-2} and $v = 0.33$ (taken from Adachi (1992)). Substitution of these values into (4) yields

$$t_c = 8.6 \times 10^{-12} f^{-2} \qquad (5)$$

In practice, for any value of f we always find that t_c is considerably larger than h_c

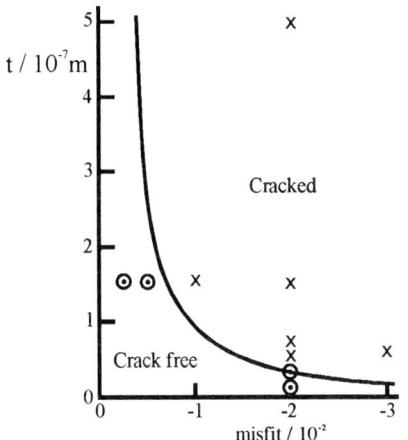

Figure 1   Observed crack zone.
The boundary shown has been calculated from equation (5).

X indicates cracked wafers
⊙ indicates crack free wafers

3.2 Experimental observations

Presented in Figure 1 are our observations using Nomarski interference microscopy on the existence of cracks as a function of lattice misfit and epilayer thickness. Also plotted on this figure is the theoretical critical thickness curve for cracking as given by equation (5). Our experimental observations on the samples grown to date are in good agreement with the theory in that all wafers to the left of the curve were uncracked whereas those to the right exhibited crack contrast. This correspondance is remarkably good considering the fact that β has been set equal to zero.

When the wafer is well to the right of the boundary, cracks will be found running parallel to both the [110] and [1$\bar{1}$0] directions (Figure 2(a)). However a wafer close to the critical thickness will normally show an asymmetry in crack population (Figure 2(b)) and in extreme cases may exhibit cracks along the [110] direction only. Athough many orthogonal cracks are observed to cross each other unimpeded, crack terminations have been observed at T-junctions (as indicated Figure 2(a)).

For layers of InGaP grown on GaAs(001), Olsen et al(1974) observed very high asymmetry in crack density between [110] and [1$\bar{1}$0]. They invoked the Cottrell (1958) model for crack initiation and ascribed this to the known difference in velocity between dislocations on $\{111\}_A$ and $\{111\}_B$ planes.

Taking their model a stage further one may postulate that in situations of low asymmetry the stress levels have exceeded the higher critical level required for motion of the slower dislocations and hence permitted both crack directions to initiate.

Figure 2(a) Nomarski micrograph of orthogonal cracks terminating at T junctions

Figure 2(b) Nomarski micrograph of cracks running along [110]only

Figure 3 'V' shaped cracks penetrating into the substrate. Note the concave free surfaces.

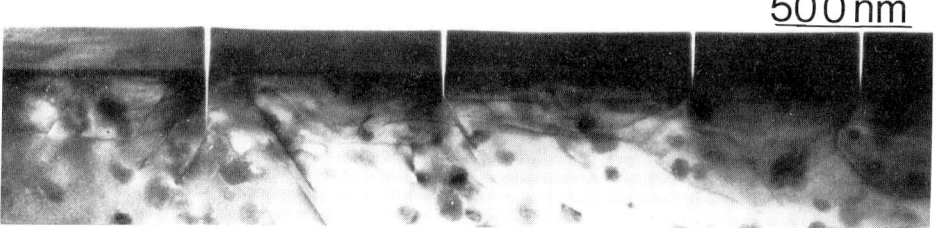

4. RESIDUAL STRAIN IN A CRACKED EPILAYER

4.1 Observations of crack profile

Figure 3 shows a cross-sectional image, taken with $\mathbf{B}= [1\bar{1}0]$ of a 150nm thick epilayer of $In_{.25}(GaAl)_{.75}As$ on InP(001). The following generic features may be observed;

(i). The cracks which have V-shaped profiles, lie on {110} planes and propagate a distance of 50-70nm into the substrate. The tips of the cracks are extremely sharp and in several cases dislocations initiate at the crack tip. In extreme cases, the cracks have been observed to deviate onto inclined {111}-type planes thus initiating a process of delamination.

(ii). Intercrack mesas adopt a concave profile. By comparing equivalent triangles, the radius of curvature, R,is found to be given by

$$R \sim \frac{ld}{w} \qquad (6)$$

where l is the intercrack spacing,, d is the crack depth and w is the crack width at the free surface. For the mesa indicated in Figure 3, $R \sim 8\mu m$.

4.2 Calculation of residual elastic energy.

In this section we shall attempt to compare the stored elastic energy in unrelieved layers with layers containing various crack profiles. The stored energy, E_0, in the psuedomorphic case is

$$E_0 = G \frac{(1+\upsilon)}{(1-\upsilon)} f^2 t \qquad (7)$$

For layers with a non-uniform strain distribution (along z) due to the presence of cracks, a more general expression for the stored energy, E_c, is

$$E_c = G \frac{(1+\upsilon)}{(1-\upsilon)} \int_0^t \varepsilon^2(z)\, dz \qquad (8)$$

where t is the film thickness and $\varepsilon(z)$ is the strain at level z. Consider the situation shown in Fig. 4(a)

210

Figure 4(a) Crack penetrates to the interface
$E_c/E_o = 1/3$

Figure 4(b) Crack penetrates into the
substrate; d=225nm, t=150nm; N=65nm:
w=15nm: l=590nm.

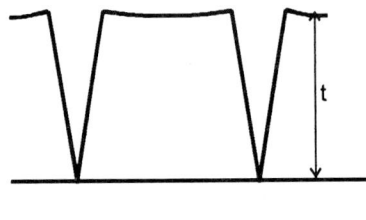

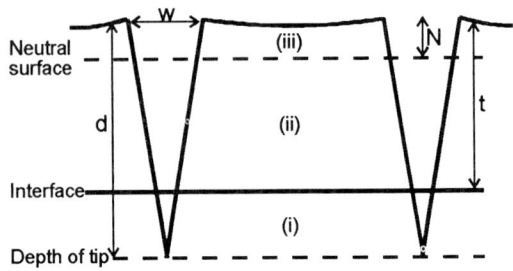

where the cracks terminate at the interface. The strain varies as

$$\varepsilon(z) = f\left(1 - \frac{z}{t}\right) \qquad (9)$$

and if one evaluates this integral the result obtained is

$$E_c = G \frac{(1+\upsilon)}{(1-\upsilon)} \frac{f^2 t}{3} \qquad (10)$$

Hence the final stored energy, E_c, is one third of E_o.

Let us now consider the situation depicted in Figure 4(b), where the cracks are allowed to penetrate beyond the interface plane, as is the experimentally observed situation. The mesa can now be divided into three distinct regions whose energies must be summed. Zone 1 is compressed substrate, zone 2 is tensile strained epilayer and zone 3 is a cap of compressed epilayer. The strain dependency on z in each of these zones is;

$$\varepsilon_I(z) = \frac{wz}{ld} \quad ; \quad \varepsilon_{II}(z) = f + \frac{wz}{ld} \quad ; \quad \varepsilon_{III}(z) = \frac{w}{ld}(z-d+N)$$

The integration can now be performed between the appropriate limits and the three contributions summed. If the experimental values for t, d, N and w which have been measured from Figure 3 are incorporated we find that the residual energy, E_c, in this case is equal to $0.12E_o$. Thus surprisingly the situation where the crack penetrates into the compressed substrate involves less stored energy than that for the crack terminating at the interface. It should be noted that our simple models do not take into account either the energetics of the convex free surface or of partial strain relief through the presence of some interfacial misfit dislocations.

5. CONCLUSIONS

A critical thickness condition exists for the introduction of cracks into tensile strained epilayers. The application of a highly idealised energy conservation model has allowed us to predict this critical thickness. Slight disparities between experimental results and theoretical predictions are well within experimental uncertainties. Cracks once initiated tend to propagate into the substrate because in doing so more strain can be relieved.

6. ACKNOWLEDGEMENTS

Funding for this work have been shared by EPSRC and Research Development fund of the University of Liverpool. Useful discussions with Dr J.David are gratefully acknowledged.

REFERENCES

Adachi S, (1992), Physical Properties of III-V Semiconductor Compounds, New York, Wiley
    Interscience
Cottrell A H, (1958),Trans. AIME , 212 192.
Franzosi P, Salviati G and Scaffardi M, (1988), J Cryst.Growth , 88 ,135.
Matthews J W, and Klokholm E (1972) Mat.Res..Bul, 7,213.
Matthews J W (1975) J Vac. Sci. Tech, 12 ,126
Murray R T, Kiely C J, Goodhew P J and Hopkinson M, (1993), Inst.Phys.Conf.Ser. 138, 309-312

Inst. Phys. Conf. Ser. No 146
Paper presented at Microsc. Semicond. Mater. Conf., Oxford, 20–23 March 1995
© *1995 IOP Publishing Ltd*

Strain relaxation in step and linearly-graded InGaAs buffer layers on (001) GaAs

F J Pacheco, D González, S I Molina, M P Villar, A Sacedón[#], E Calleja[#] and R García

Dpto. de Ciencia de los Materiales e I.M. y Q.I., Universidad de Cádiz. 11510-Puerto Real (Cádiz). Spain.
[#]Dpto. de Ingeniería Electrónica. E.T.S.I.T., U.P.M., Ciudad Universitaria sn., 28040-Madrid. Spain.

ABSTRACT: The strain distribution through three different graded InGaAs buffer structures (step, inverse-step and linearly-graded) deposited by Molecular Beam Epitaxy on (001) GaAs substrates and capped by InGaAs single layers is studied using Transmission Electron Microscopy and Photoluminescence. The cap layer relaxation state and the efficiency of these buffers when filtering threading dislocations are discussed.

1. INTRODUCTION

The growth of device-quality relaxed III-V epitaxial layers on highly mismatched substrates represents a current challenge for the III-V semiconductor community. Particularly interesting is the strained system of InGaAs grown on GaAs, which has many technological applications. However, due to the lattice mismatch between InGaAs and GaAs, misfit dislocations are generated when the epilayer thickness passes a certain critical value. As a consequence of misfit dislocation interactions, they can thread up to the epilayer free surface. A usual approach to avoid these threading dislocations and to grow the heterostructures successfully is to interpose a buffer layer between the substrate and the active layer. The buffers based on step (Krisnamoorthy et al. 1992, González et al. 1994a) and continuously graded (Chang et al. 1992, Molina et al. 1994) layers are particularly promising. In this paper, we study the strain distribution through step, inverse-step and linearly-graded buffer layers grown by Molecular Beam Epitaxy (MBE) using Transmission Electron Microscopy (TEM) and Photoluminescence (PL). Moreover, we compare the cap layer relaxation state and the efficiency of the different buffer structures to filter threading dislocations.

2. EXPERIMENTAL

A series of samples consisting of InGaAs step and linearly-graded buffer layers, respectively, followed by an InGaAs cap layer, were grown on (001) GaAs substrates by MBE at 500°C (Fig. 1). The defect densities and morphologies were studied by Cross Section TEM (XTEM) and Plan View TEM (PVTEM) on a JEOL 1200-EX and a JEOL 2000-EX transmission electron microscopes operating at 120 and 200 kV respectively. PVTEM specimens were prepared by chemical etching methods (Br_2 + CH_3OH and H_2SO_4 + H_2O_2 + H_2O) and XTEM specimens by Ar^+ ion milling. The dislocation Burgers vector analysis was performed using the conventional invisibility criteria. PL measurements were performed at low temperature, 15K. The samples were excited with the 514.5 nm line of an He-Ne laser. The PL signal was recorded by a locking system using a Ge-detector.

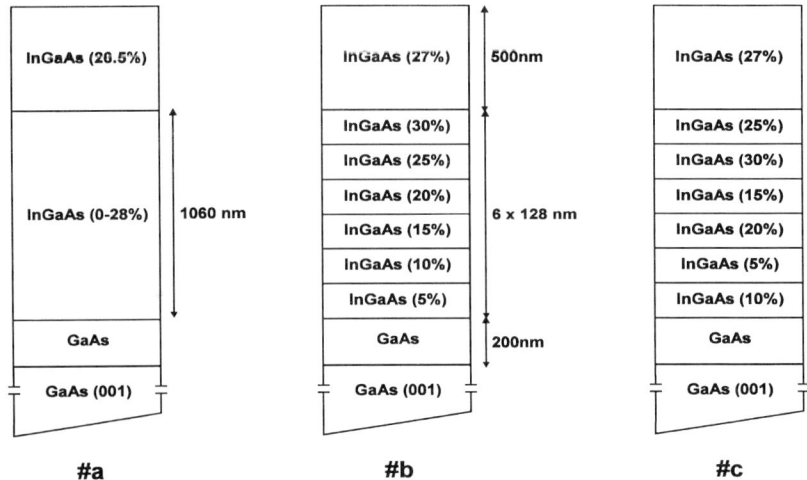

Fig. 1. Sample design scheme

3. RESULTS AND DISCUSSION

In sample #a, with a linearly-graded buffer layer, TEM observations reveal the existence of a three-dimensional network of dislocations in the buffer layer. Misfit dislocations (MDs) running mostly along the [110] and [1$\bar{1}$0] directions are observed in the region between the InGaAs/GaAs interface and a critical thickness z_2. The presence of two areas containing different cross-sectional dislocation densities is evidenced in the XTEM micrographs (Fig. 2.a). In the first area, from the graded layer/substrate interface up to a thickness z_1, the measured MD density is $<5 \cdot 10^8 \text{cm}^{-2}$ while in the second one, that reaches up to z_2, is $8.7 \pm 1 \cdot 10^9 \text{cm}^{-2}$.

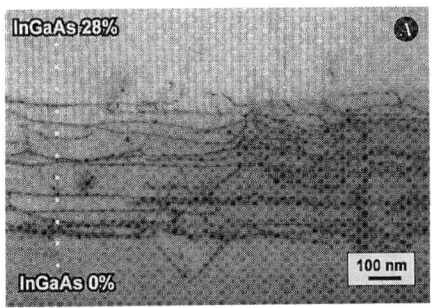

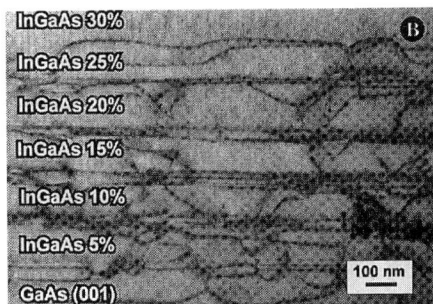

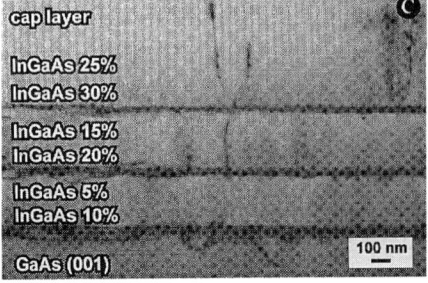

Fig. 2. BF (220) XTEM images of the buffer structures showing three different dislocation distributions in the linearly (#a), direct-step (#b) and inverse-step (#c) graded InGaAs layers.

Some threading segments between the misfit segments are also visible in the XTEM and PVTEM images of the graded layer. However, we have found no evidence of threading dislocations (TDs) at the top surface of the buffer layer and throughout the cap layer.

TEM micrographs of the step-graded sample with monotonically increasing In content (#b) show MDs lying along the [110] and [1$\bar{1}$0] directions at the interfaces. TDs are also present between the interfaces through the layers of the buffer. Table I displays the linear MD densities measured by XTEM (Fig. 2.b and 2.c). The densities are roughly similar for all the interfaces excepting the last one where no dislocations are observed. Moreover, there is no evidence of dislocations in this cap layer by XTEM.

In sample #c, with reversal steps, we only found MDs at the interfaces. However, the densities at the interfaces are quite dependent on the position in the buffer, as expected. In the interfaces with increasing In content the densities are considerably higher than those in sample #b while at the interfaces with decreasing In content the densities are much lower (Table I). The main difference with the other studied structures is the presence of threading dislocations in the cap layer.

	Misfit dislocation density ($\rho \pm 1 \cdot 10^4$ cm^{-1})					
Sample	1st	2nd	3rd	4th	5th	6th
#b	9.0	10.7	9.5	11.5	8.0	3.5
#c	12.5	1.0	21.0	2.5	21.0	2.0

Table I. Misfit dislocation densities at all the interfaces in the step-graded buffers.

From our knowledge, the comparison of the strain distribution through the layers for these three kinds of buffer structures has never been reported. Here, we estimate the residual strain along these kinds of structures from TEM measured dislocation densities.

To evaluate the capability of each kind of buffer to relax the lattice mismatch up to a determined value it is essential to locally probe the strain as a function of height. In order to calculate the strain it is necessary to know the ratio between edge and 60° dislocations. For step graded buffers, we assume our estimated value obtained previously comparing TEM and X-Ray measurements in samples consisting of InGaAs multilayers (2 and 3 layers) (González et al. 1994a). On the other hand, in the linearly-graded layers, the percentage of edge dislocations has been assessed using the conventional invisibility criteria in TEM observations.

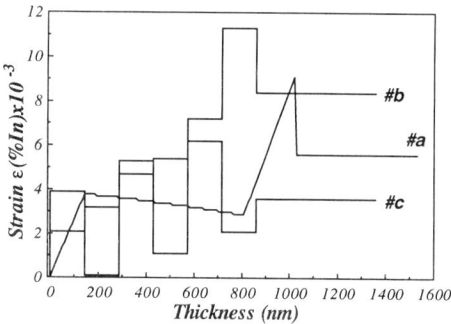

Fig. 3. Strain distribution through the layers of the buffers and the cap layers.

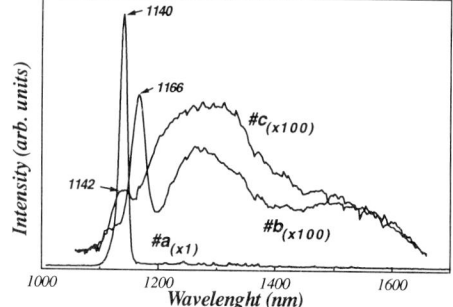

Fig. 4. Low temperature (15K) PL spectra.

As it is very difficult to determine the strain of each layer in heterostructures with more than three layers by using any technique, we carry out a previously published simple calculation procedure (González et al. 1994b) and we obtain the results presented in Fig. 3.

214

PL measurements at low temperature were performed in order to verify our TEM estimated values for the cap layer residual strain. The PL spectra are shown in Fig. 4. From the PL peaks positions it is possible to calculate the strain in the layer. Table II displays the strain values estimated by PL and TEM for the cap layers, which are in good agreement.

Samples	In(%) X-ray	Strain(TEM)	Strain(PL)
#a	26.5±1	0.0057±7	0.0051±5
#b	27±1	0.0084±7	0.0087±5
#c	27±1	0.0038±7	0.0038±5

Table II. Strain of the cap layers calculated from PL and TEM measurements.

The strain increases monotonically with the buffer layer thickness in sample #b but fluctuates in sample #c (see Fig. 3). However, in the graded layer, we can distinguish three different regions containing unequal dislocation distributions. At the beginning the strain increases quickly due to the low dislocation density in this area. In the dislocation-rich region the strain undergoes a slow reduction. Finally, in the dislocation free area, a continuous increment of strain again occurs.

Sample #c presents the lowest residual strain in the cap layer. However, this structure suffers from a poor efficiency as a threading dislocation filter. On the other hand, the structure #b shows the most strained cap layer. Therefore, the best graded buffer structure is the linearly-graded layer (sample #a) since this system relaxes approximately as much as sample #c and possesses a good capability to filter TDs. These different TD filtering efficiencies are related to the strain behaviour through the buffer structures. In samples #a and #b pronounced increments of strain at the top area of the buffer take place, while in sample #c the strain at the end of the buffer   is   too low to avoid the dislocations threading upward.

Nevertheless, it would be of high interest to investigate the behaviour of a buffer structure consisting of a combination of an inverse and monotonously step-graded layers.

ACKNOWLEDGEMENTS

The authors would like to thank Dr. Araújo and Dr. Aragón for fruitful discussions. The present work has received financial support from the Spanish *Comisión Interministerial de Ciencia y Tecnología*, CICYT, BLES Proyect (ESPRIT Program, ref. 6854) and the *Junta de Andalucía*, under the group 6020. This study was carried out at the *División de Microscopía Electrónica de la Universidad de Cádiz*.

REFERENCES

Chang J C P, Chen J, Fernández J M, Wieder H H. and Kavanagh K L 1992 Appl. Phys. Lett. 60 1129.
Molina S I, Pacheco F J, Araújo D, García R, Sacedón A, Calleja E, Yang Z and Kidd P 1994 Appl. Phys. Lett. 65 2460.
González D, Araújo D, Molina S I, Pacheco F J, Aragón G, González L, González Y, Kidd P and García R 1994a Proceedings of the 13th International Congress on Electron Microscopy (ICEM) vol. 2a, 607.
González D, Araújo D, Molina S I, Sacedón A, Calleja E and García R 1994b Mat. Sci. Eng. B28 515.
Krisnamoorthy V, Lin Y W and Park R M 1992 J. Appl. Phys. 72 1752.

Inst. Phys. Conf. Ser. No 146
Paper presented at Microsc. Semicond. Mater. Conf., Oxford, 20–23 March 1995
© 1995 IOP Publishing Ltd

The possibility of low edge dislocation densities in InGaAs/GaAs step-graded layers

G MacPherson, R Beanland and P J Goodhew

Department of Materials Science and Engineering, University of Liverpool, L69 3BX, United Kingdom

ABSTRACT: Step-graded layers up to a nominal indium composition of 30% have been examined using cross-sectional transmission electron microscopy (TEM). The layers were grown according to a new design method, with the aim of ensuring that 60° dislocations within the structure are separated to such an extent that glide along {111} plane to form an edge dislocation necessitates the traversing of another InGaAs/GaAs interface. This results in the reaction of gliding 60° dislocations with the dislocation array present at the interface, reducing the density of edge dislocations.

1. INTRODUCTION

For materials such as InGaAs/GaAs which are lattice-mismatched the epitaxial growth is pseudomorphic for layers that are less than the critical thickness, and misfit relieving dislocations are generated once this critical thickness has been exceeded (Matthews and Blakeslee 1974). For the case of single layers with indium concentrations less than ~ 18% (Krishnamoorthy, Lin and Park 1992b) the strain is relieved by an orthogonal array of 60° dislocations and the layer then forms by 2D growth. Above ~ 18% the growth becomes strain induced 3D growth leading to a high density of threading dislocations within the structure. The growth of step-graded layers up to indium compositions of 30% by 2D layer-by-layer growth is possible due to the successive relief of strain by dislocations formed at the interfaces within the structure. 2D growth prevails provided the strain at the surface never exceeds that associated with ~ 18% indium.

The main problem in step-graded layers grown by 2D growth is the threading dislocations that remain within the structure. Threading dislocations may remain because their glide path to the edge of the wafer has been blocked. The glide path can be blocked due to the inability of the dislocation to overcome orthogonal dislocation strain fields (Freund 1990). The effectiveness of blocking can be increased if the orthogonal dislocations are sessile edge dislocations found <u>within</u> each of the layers contributing to the step-graded structure, rather than at the interfaces. The presence of edge dislocations within the structure is therefore undesirable and we describe here an attempt to reduce the density of edge dislocations.

2. THEORY

Only a brief discussion of the theory will be presented here with a more detailed discussion found elsewhere (MacPherson, Beanland and Goodhew 1995b). Initially the theory for single layers will be outlined and the ideas extended to that of a step-graded layer.

2.1 SINGLE LAYERS

For two correctly oriented 60° dislocations within a single layer, they must glide along {111} planes. For a layer of thickness h, simple geometric considerations limit the separation of the two 60° dislocations at the interface to $\sqrt{2}h$ if they were to form an edge dislocation within the epilayer. Krishnamoorthy et al (1992a) showed that there are two distinct regimes associated with strain relief within a layer and provides two accurate equations describing the percentage strain relief in the two regimes, corresponding to $h_c < h < 30 h_c$ and $30 h_c < h < \approx 450 h_c$ where h_c is the critical thickness derived by Matthews and Blakeslee. It is possible to produce equations which provide an estimate of what the magnitudes of h_c and h should be to ensure a mean separation equal to $\sqrt{2}$ h at the interface for each of the two regimes:

$$h = \frac{h_c}{2}\left[1+\sqrt{\left(1+\frac{\sqrt{2}b}{h_cAf}\right)}\right] \quad h_c<h\leq30h_c \quad (1)$$

$$h = \frac{h_c(C-B)}{2B}\left[\sqrt{\left(1+\frac{\sqrt{2}Bb}{h_cf(C-B)^2}\right)}-1\right] \quad 30h_c<h<\approx450h_c \quad (2)$$

Where $A=0.02\pm 10^{-3}$, $B=2\times10^{-4} \pm 5\times10^{-5}$ and $C= 0.66 \pm 0.01$. Equations (1) and (2) will only be correct i.e., suppress any edge dislocation formation in the idealised situation of all the dislocations being separated by a distance $\sqrt{2}$ h. In the real situation there is a distribution of separations with a mean that can be more than 50% greater than the modal value (MacPherson, Beanland and Goodhew, (1995)), so it is still likely that there will be some edge dislocation formation from dislocations that lie below $\sqrt{2}$ h in the distribution of separations. This should result in a reduction in the proportion of observed edge dislocations rather than a complete absence.

2.2. STEP-GRADED LAYERS

In the case of single layers there is no such barrier for the formation of edge dislocation in the substrate. For the interfaces in a step-graded layer this is not a problem since, apart from the substrate/epitaxial layer interface, all other interfaces have interfaces above and below them at the correct positions for suppression of edge dislocations.

2.2.1. BARRIERS FOR EDGE DISLOCATION FORMATION

INTERFACIAL DISLOCATION ARRAYS

Since we have concentrated on step-graded layers there will be arrays of misfit dislocations at the interfaces in the designed structures. Because these interfaces are separated so that two 60° dislocations originating from one interface that wish to form an edge dislocation have to pass through a dislocation array then the dislocation interactions will be important. Exactly how the formation of the edge dislocations will be affected is not obvious since the interactions depend on the particular Burgers vectors of the dislocations involved. Some possible interactions are shown in fig. 1.

CHANGE IN ELASTIC MODULUS

Another possibility is the change in elastic modulus that occurs when traversing an interface where the indium composition changes within the structure. For a continuous interface, where the lattice constants are almost identical but have different elastic moduli the crossing of such an interface by a dislocation will be affected by this change in elastic modulus. If the two materials are simplified to be 'hard' and 'soft' then for a dislocation located in the hard material there is an attraction to the interface (Benlahsen, Lépinoux and Grilhé (1993)). For a dislocation located in the soft material the opposite is the case. Therefore, if this is extended to two 60° dislocations located in the soft material gliding to form an edge dislocation, then it would be expected there would be some barrier for the dislocations to overcome due to the interfacial repulsion.

For the lower layers of a step-grade the relaxation is significantly greater than that of a single layer and can be almost complete. Therefore, for lower layers, another approach can be taken giving rise to the equation:

$$\Delta f = \frac{b}{2\sqrt{2}h} \quad (3)$$

Where Δf and h correspond to the misfit and thickness of the individual layers in the step-grade. Equation (3) gives a direct relationship between Δf and h and will govern the relationship of Δf and h for lower layers. Equations (1) and (2) may still be used for the top layer of the step-grade which can be approximated as a single layer. An example of the thicknesses involved for the lower layers, an indium composition step of 3% gives rise to layer thicknesses of 64nm.

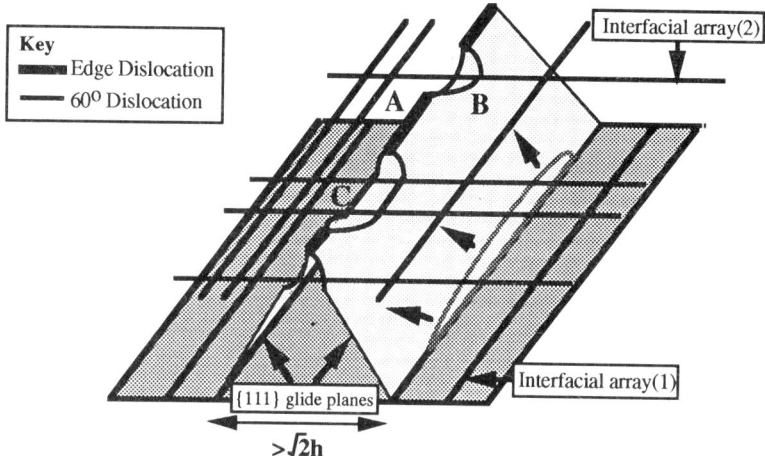

Fig. 1 Two 60° dislocations (which are correctly oriented) gliding along their respective {111} planes interacting with another interfacial array within the step-grade. A number of possibilities can be envisaged; A) completion of the reaction to form a small edge dislocation segment, B) repulsion between like-signed dislocations preventing formation, and C) closely spaced dislocations preventing continued glide of the 60° dislocations.

3. EXPERIMENT

Three InGaAs/GaAs step graded layers were grown by metal organic molecular beam epitaxy(MOMBE) at a temperature of $550^{\circ}C$ on semi-insulating GaAs substrates. A GaAs buffer layer of approximately 250nm was grown prior to InGaAs growth. Thicknesses were measured by cross-sectional TEM.

The top layer thickness was calculated using equation (1) but with the Matthews and Blakeslee value for the residual strain rather than the value of strain relief obtained from Krishnamoorthys' equations. The only effect of this was to slightly reduce the thickness of the top layer, leading to a small increase in the calculated separation of dislocations in the top interface. The sample details are shown in table 1.

Table 1

Sample	$\Delta x(\%)$	Lower Layers (nm)	Top Layer (nm)
a	5	78	120
b	3	64	140
c	3	32	105

4. RESULTS

Measurements of the mean dislocation density at the lower interfaces show that in all the samples these lower layers were almost completely relaxed indicating that subsequent growth had further relaxed the layers. This shows that equation (3) can be used for predicting the mean dislocation density of the lower layers.

The TEM micrograph of fig. 2 shows the distribution of dislocations within the structure showing 60 ° dislocations lying generally in or near to the interfaces. As expected there was a high density of edge dislocations in the substrate, but this was not the case for the epitaxial layers in which few edge dislocations could be observed. Of these edge dislocations the vast majority tended to lie in or near to the interfaces where the indium composition is increased. Fig. 3 shows a cross-sectional TEM micrograph of sample b viewed in the g004 reflection. When viewed in the g004 reflection, edge dislocations show a residual contrast ($\mathbf{g}.(\mathbf{bxu}) \neq 0$) and this residual contrast can be observed in or near two of the interfaces within the structure. One possible explanation for the position of these edge dislocations is that 60° dislocations from one interface have glided to form an edge dislocation at an adjacent interface.

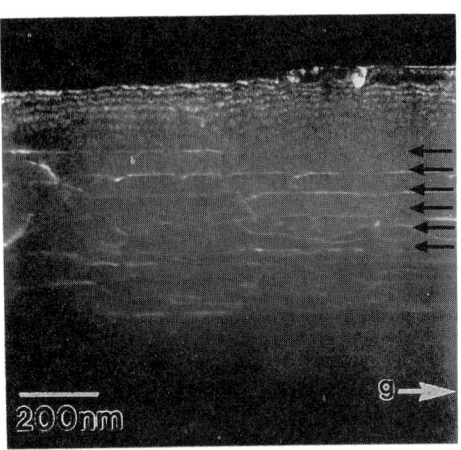

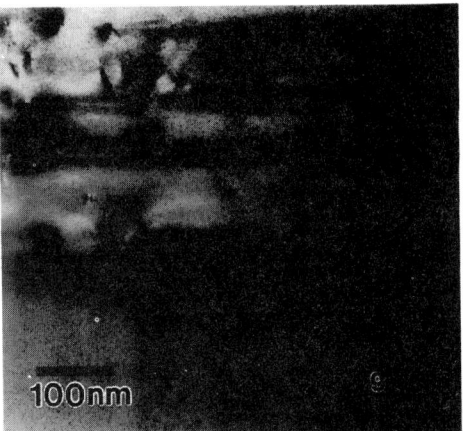

Fig. 2 Cross-sectional TEM micrograph of sample b showing the dislocation array. Note 60° dislocations generally lying in or near to the interfaces(denoted by arrows) were the indium composition has been increased. Imaging condition: dark field, diffraction condition $\mathbf{g}$, $3\mathbf{g}(\bar{2}20)$

Fig. 3 Cross-sectional TEM micrograph of sample b showing edge dislocations identified by residual contrast in or near to the indium step increase interfaces, with the interfaces denoted by arrows. Imaging condition: bright field, diffraction condition $\mathbf{g}(004)$

5. CONCLUSION

Cross-sectional TEM evidence has shown that a reduction in the density of edge dislocations has been achieved by using the described model. The densities are less than recently predicted estimates of 14% (González et al (1994)). The limited number of edge dislocations that were observed in the structure appeared to be confined to positions in or near to the interfaces with only a very small fraction observed at any significant distance from the interfaces.

REFERENCES

Benlahsen M, Lépinoux J and Grilhé J 1993 Mat.Sci.Eng. A164 428
Freund L B 1990 J.Appl.Phys. 68(5) 2073
González D, Araujo D, Molina S I, Pacheco F J, Aragón G, González L, González Y, Kidd P and García R 1994 ICEM 13 Paris 607
Krishnamoorthy V, Lin Y W, Calhoun L, Liu L H, and Park R M, 1992a Appl.Phys.Lett. 61(22) 2680
Krishnamoorthy V, Lin Y W and Park R M 1992b Mat.Res.Soc.Symp.Proc. 263 439
MacPherson G, Beanland R and Goodhew P J 1995a Scripta.Metall.Mater (in press)
MacPherson G, Beanland R and Goodhew P J 1995b Phil.Mag (submitted)
Matthews J W and Blakeslee A E 1974 J.Cryst.Growth 27 118

Inst. Phys. Conf. Ser. No 146
Paper presented at Microsc. Semicond. Mater. Conf., Oxford, 20–23 March 1995
© *1995 IOP Publishing Ltd*

The growth of InAs layers on vicinal GaAs (110) substrates by MBE

X Zhang, D W Pashley[†], J H Neave, I Kamiya[‡] and B A Joyce

Interdisciplinary Research Centre for Semiconductor Materials, The Blackett Laboratory, Imperial College, Prince Consort Road, London SW7 2BZ, U.K.
[†] Department of Materials, Imperial College, Prince Consort Road, London SW7 2BP, U.K.
[‡] Also the Research Fellow of Research Development Corporation of Japan (JRDC).

ABSTRACT: InAs layers of different layer thicknesses grown on GaAs (110) with an off cut of 1.5° towards (111)A were studied by TEM. It was found that at the early stage of growth, InAs preferentially deposited along the step edges forming elongated strips along [1$\bar{1}$0] direction on the surface. The cross-sections showed that as the layer thickness increased, the width of InAs strips expanded on both the upper and lower terraces, although predominantly the latter. The complete coverage of the GaAs surface occurred when the InAs strips coalesced laterally.

1. INTRODUCTION

A large amount of effort has been put towards the study of growth mechanisms of epitaxial hetero-structures because of its essential importance in understanding and improving the behaviour of existing as well as potential electronic devices. The majority of such studies were carried out on {100} oriented surfaces and the modes of growth (Frank and van der Merwe, Volmer-Weber or Stranski-Krastanov) of hetero-epitaxial layers have been often related to the interfacial energy and the degree of strain caused by the mismatch of lattice spacing between the two materials. In our previous study of InAs on GaAs, with a misfit strain $\varepsilon = 7.1\%$, it was found that on (001) substrate the growth mode could change from Volmer-Weber (2D) to Stranski-Krastanov (2D to 3D) depending on the growth conditions, but on singular (110) substrates 2D island nucleation always occurred (Zhang et al 1993). In this paper, we demonstrate that on vicinal (110) substrates the growth behaviour of InAs on GaAs changes to preferential step decoration.

2. EXPERIMENTAL PROCEDURES

The GaAs substrate used in this study was (110) type with a misorientation of 1.5° towards (111)A. This configuration would be expected to give rise to surface steps of monolayer height with an average step separation of ≈85Å, with step edges running parallel to the [1$\bar{1}$0] direction as shown in fig.1.
All the samples studied were grown by MBE under nominally identical growth conditions. Before the InAs deposition, a GaAs buffer layer (≈2000Å thick) was grown with an As$_2$/Ga flux ratio of 2.6 at 580°C. The InAs growth was carried out at 420°C with As$_2$/In flux ratio of 2.94. The InAs layer thicknesses were equivalent to 5Å, 15Å, 60Å and ≈200Å respectively. At the end of growth, each sample was capped with an amorphous As layer (≈50Å thick) so that during TEM foil preparation any direct contact to the sample surface could be avoided.

Fig.1. The surface steps of a (110)
substrate with an off cut at 1.5°
towards (111)A.

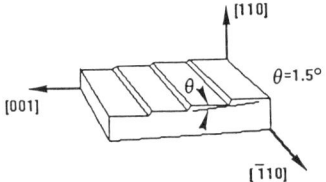

TEM studies were carried out in both plan view and [1$\bar{1}$0] cross-sections. Chemical jet thinning and standard Ar ion milling were used for the plan view and cross-sectional thin foils respectively. The instruments used were JEOL 2000FX and JEM-2010 microscopes, both operated at 200KeV.

3. RESULTS AND DISCUSSION

It has been shown that monolayer surface steps due to the misorientation redistribute during the growth of GaAs buffer layer on GaAs substrates (Krishnamurthy et al 1993, Zhang et al 1995). In this case, at the end of ≈2000Å buffer layer growth, the average step height was of the order of 65±25Å and step separation (or terrace width) was 3000±800Å. The resultant bunched surface steps were aligned approximately along the [1$\bar{1}$0] direction on the buffer layer surface.

Figures 2(a) to (d) show the plan view (upper half) and the cross-sectional (lower half) images of 5Å, 15Å, 60Å and ≈200Å InAs 'layers' respectively on the misoriented GaAs (110) substrate. At 5Å, Fig.2(a), InAs formed strips outlining the contour of surface steps. The cross-section showed that the InAs deposit accumulated at the GaAs step edges and mostly on the lower terrace side. Several InAs deposits showed evidence of forming facets which had acute angles to the (110) surface, varying from 30° to 35° (shown by the arrow in Fig.1.(a)), indicating that the facets were likely to be {11h} type (where h=1, 2, or 3).

As the thickness increased to 15Å, Fig.2(b), the width of the InAs strips expanded in the lateral direction on both the upper and lower terraces, with the latter process dominating. The height of the strips remained more or less unchanged. The facet feature of the InAs deposit on the lower terrace had become quite pronounced by this stage.

At 60Å, Fig.2(c), the InAs strips continued their lateral expansion, and by this stage the growth on the upper terraces became very noticeable as indicated by the arrow in Fig.2(c). At ≈200Å, the surface coverage of InAs on GaAs was 100%, Fig.2(d). The surface steps on the InAs layer resembled those on the GaAs buffer layer before the InAs deposition.

The behaviour of strain relaxation of the InAs strips on the misoriented GaAs (110) substrate is very similar to that observed on the singular GaAs (110) (Zhang et al 1993). The strain along the [1$\bar{1}$0] direction was relieved via the formation of Lomer type misfit dislocations (observed by TEM). The relaxation of InAs in [1$\bar{1}$0] direction on singular GaAs (110) had been observed by reflection high energy electron diffraction (not shown) to occur mainly during the first monolayer of InAs deposition and was essentially complete by 2 monolayers.

Figure 3(a) shows a plan view (2$\bar{2}$0) dark field image of a 15Å thick InAs on misoriented GaAs (110). The light line contrast along [001] direction corresponds to Lomer type misfit dislocations. Figure 3(b) is the (004) dark field image from the same sample. In this case, the

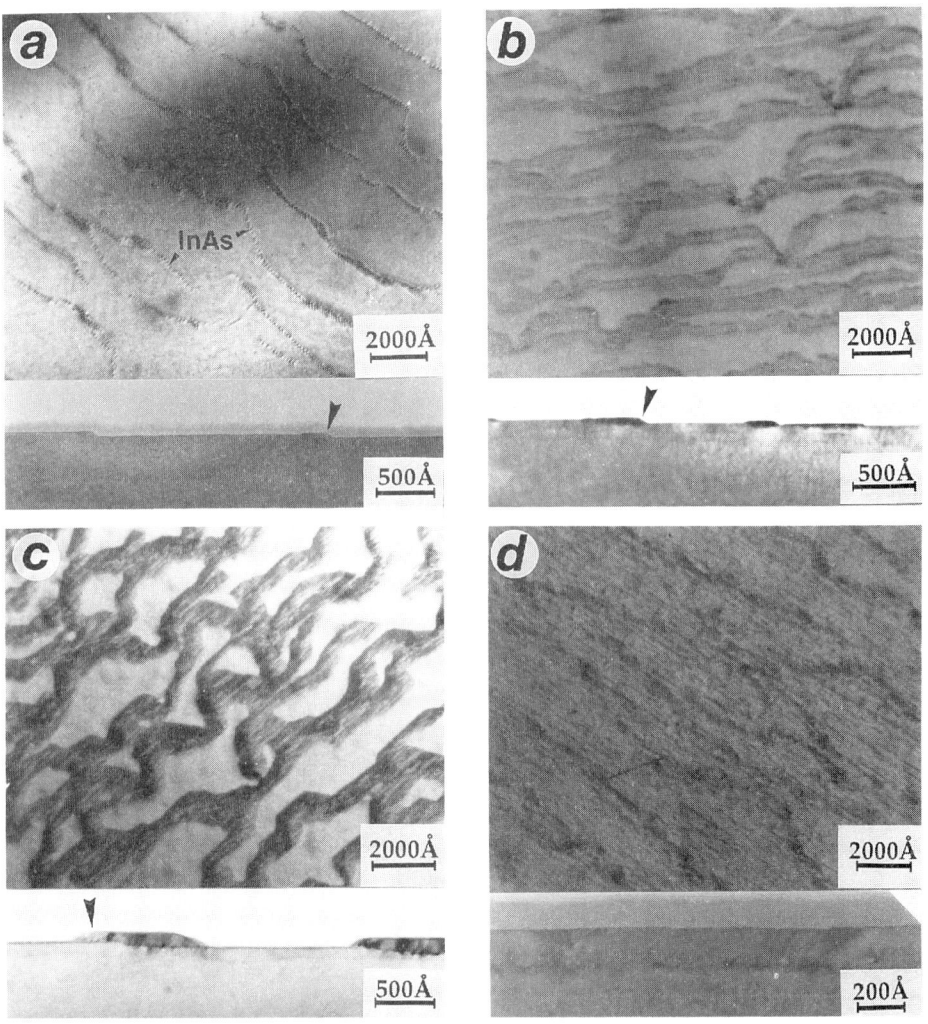

Fig.2. Plan view and cross-sectional images of InAs layers with nominal layer thicknesses equivalent to (a) 5Å, (b) 15Å, (c) 60Å and (d) ≈200Å on misoriented GaAs (1$\bar{1}$0).

line contrast along the [1$\bar{1}$0] direction is due mainly to the partial dislocations bounding stacking faults on (111) and (11$\bar{1}$) planes intersecting at the interface. A few 60° type misfit dislocations are also present. The transmission electron diffraction patterns of the areas shown indicated that, on average, the relaxation along the [1$\bar{1}$0] is ≈89% and that along [001] ≈51%. This asymmetrical strain relaxations in the two orthogonal directions decreases as the layer thickness increases so that by ≈200Å, the relaxation in the [1$\bar{1}$0] and [001] directions is approximately 97% and 87% respectively.

It is not clear at this stage whether the bunched GaAs surface steps (due to rearrangement during the buffer layer growth) consisted monolayer, double layer or other

222

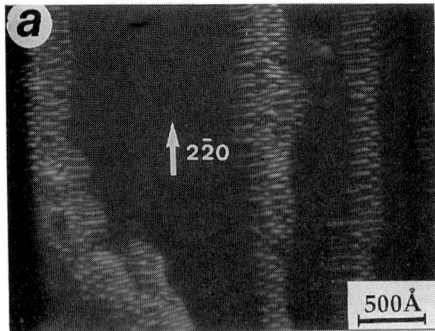 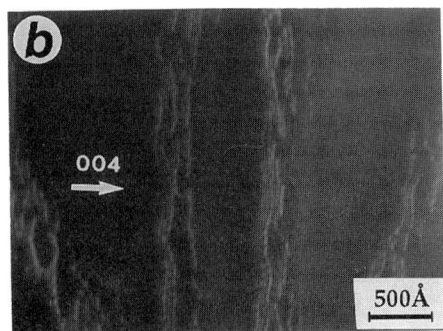

Fig.3. Dark field images of 15Å InAs on misoriented GaAs (110), (a) g=2$\bar{2}$0 and (b) g=004.

combination of steps. They certainly provided lower energy sinks on the surface compared to the terrace regions for InAs to deposit preferentially at the step edges. In some recent studies of MBE growth on patterned substrates, it was demonstrated that under identical ambient conditions the growth rate of the epilayer on the different exposed surfaces (mesa surface, groves or the side walls) was different and was closed related to the sticking coefficient as well as the surface migration length of the group III atoms on these surfaces (e.g. Konkar et al 1994). We might speculate that the surface structure at the bunched steps resembles a minute half-size 'mesa' where two different surfaces are coexisting (terrace and step edge) and growth of InAs is controlled by the competition between the two surfaces.

4. CONCLUSION

InAs layers were deposited by MBE on misoriented GaAs (110) substrate after $\approx$2000Å of GaAs buffer layer growth. At the initial stage of growth, InAs was found to deposit preferentially on the GaAs surface step edges, forming elongated strips along the [1$\bar{1}$0] direction. Further InAs deposition involved expansion of InAs strips laterally on both upper and lower terraces. Complete surface coverage occurred at the point where the strips joined laterally. The strain relaxation behaviour of the InAs strips was similar to that on the singular GaAs (110) substrate where rapid strain relief was found initially along the [1$\bar{1}$0] direction via the formation of Lomer dislocations with slower relief along the [001] direction via 60° type dislocations and partial dislocations.

REFERENCES

Konkar A, Rajkumar K C, Xie Q, Chen P, Madhukar A, Lin H T and Rich D H 1994 8th Int. Conf. on MBE, Japan.
Krishnamurthy M, Lorke A, Massermeier, Williams M D R and Petroff P M 1993 J. Vac. Sci. Technol. B 11 1384.
Zhang X, Pashley D W, Fawcett P N, Neave J N, and Joyce B A 1993 Inst. Phys. Conf. Ser. 134 300
Zhang X, Pashley D W, Neave J N, Kamiya I and Joyce B A 1995 J. Crystal Growth 147 234

Inst. Phys. Conf. Ser. No 146
Paper presented at Microsc. Semicond. Mater. Conf., Oxford, 20–23 March 1995

Microstructure of Ge/Si(001) heteroepitaxy with surfactants

H Matsuhata, K Sakamoto, K Kyoya*, K Miki and T Sakamoto

Electrotechnical Laboratory, 1-1-4, Umezono, Tsukuba, 305, Japan,
*On leave from Meiji University

Abstract: The surfactant effect of Sb and Bi on the growth of Ge/Si heterostructure by MBE was investigated using TEM and *in situ* RHEED. Suppression of island formation and an improvement in the critical thickness for pseudomorphic growth were observed with both surfactants, particularly by Sb. However, roughening of the (001) growth surface of Si was observed with both surfactants. We conclude that the effect of surfactants on Ge/Si(001) growth is not only to reduce surface free energy, but also to reduce surface diffusion.

1. INTRODUCTION

In the growth of Ge on Si (001) by molecular beam epitaxy(MBE), Ge grows in the Stranski-Krastanov (S-K) mode, which is to change from two-dimensional layer-by-layer growth mode to three-dimensional island formation after an initial 3 to 5 layer growth. Zalm *et al.* (1989) and Copel *et al.* (1989) reported that one monolayer (ML) of Ga and As always segregates towards the growing surface of Si or Ge, reducing the surface segregation of Ge at a $Si/Si_{1-x}Ge_x$ heterointerface, as well as suppressing the tendency of three-dimensional island formation. Copel *et al.* (1989) named such elements surfactants. To date, various elements have been reported to act as surfactants (e.g. Sb, Bi, Te, Sn and H, see Copel *et al.*, 1990, Sakamoto *et al.*, 1993, Higuchi and Nakanishi, 1991, Iwanari and Takayanagi, 1991, Copel and Tromp, 1991, respectively). Some differences in behavior of those surfactants were reported, but the detailed differences still have not been investigated satisfactorily.

In a previous paper (Matsuhata *et al.*, 1993), we reported that 1 ML of either Bi or Sb behaves as a surfactant and improves the abruptness of the Si/Ge/Si heterointerface. However, the improved abrupt-heterointerfaces showed interface-steps more clearly than the case without surfactants. For this report we investigated further details of the effect of surfactant, particularly on the tendency to form a flat growing surface of Si(001). We also investigate the difference in behavior of Bi and Sb. Lastly we discuss the mechanism of the surfactant effect on the Si(001) surface.

2. EXPERIMENTAL PROCEDURE

A Si substrate with a clean (001) surface was loaded into the growth chamber of an MBE system, which has an Si electron-beam evaporator and a Ge effusion cell. After removal of the protective oxide film on the surface at 1073K, a Si buffer layer was deposited at 963K. Various specimens with heterostructures were fabricated either with or without a 1 ML of Sb or Bi. Growth rates of Si and Ge were 0.3 MLs^{-1} and 0.1 MLs^{-1}, respectively at 673 K. The growing surface was monitored using *in situ* RHEED at 30 KV. The structures of those individual specimens are described in detail in the following section. The cross-sectional specimens for electron microscopy were prepared using a standard ion-milling technique. The electron microscopy was carried out using a JEM-4000FX and 4000EX, operated at 200 KV to reduce radiation damage.

3. EXPERIMENTAL RESULTS

The surfactant effect of suppressing the formation of three-dimensional island of Ge on Si was observed clearly in the RHEED intensity oscillation as shown in Fig.1(a). The RHEED intensity oscillations of the specular spot during Ge growth without surfactant decayed after five oscillation periods, suggesting that the onset of three-dimensional growth mode occurs after a 5 ML layer-by-layer growth. The other RHEED intensity oscillation graph in the figure shows the case of a 1 ML deposition of Sb before the growth of Ge, where the initial two oscillation periods in the graph correspond to 1 ML of Sb. The intensity oscillations lasted 12 periods during Ge growth, suggesting that 1 ML of Sb improves the two-dimensional growth mode up to a 12 ML growth of Ge. Figure 1(b) shows the evolution of surface lattice constants during Ge growth, that we estimated from the *in situ* RHEED spot spacing. The critical thicknesses of the Ge pseudomorphic growth on Si were 10 ML and 5 ML for Sb and Bi, respectively. We also found that changes in the surface lattice constant and the three-dimensional island formation take place at different Ge thicknesses.

Figure 2 (a),(b),(c) show cross-sectional TEM images of Si/10MLGe/Si without surfactant, with 1 ML Sb before Ge growth, and with 1 ML Bi before 10 ML of Ge growth, respectively. Fig. 2(a) shows the three-dimensional islands of Ge. This image was taken at a small tilt from the [110] orientation to avoid the strain contrast caused by dislocations. Figure 2(b) and Fig. 2(c) both taken at the [110] orientation show a flatly grown 10 ML Ge layer. They show stacking faults and Lomer-Cottrell dislocations, which originated in the 10 ML Ge and continued into the Si overlayer. The V-shaped Lomer-Cottrell dislocations seen in Fig. 2(b) are due to elongation of lattice parameter perpendicular to the growth direction, and Λ–shaped Lomer-Cottrell dislocations are due to shrinkage of the lattice parameter. The defect density seen in Fig. 2(c) is higher than that seen in Fig. 2(b). The increase in the lattice parameters shown in the RHEED patterns as seen in Fig. 1(b) corresponds to that of the defect densities observed in the TEM images. These experimental results show that Sb is more effective than Bi in suppressing the introduction of lattice defects and in increasing the lattice parameter, although both suppress island formation.

In order to observe the effect of surfactants on the flatness of a Si(001) growth surface, we fabricated the specimens shown schematically in Fig. 3(a). At first we deposited 10 ML of Ge on a clean Si(001) surface, and then intentionally formed a rough surface by S-K mode growth of Ge. After formation of this rough-surface, we then deposited 1 ML of Bi as a surfactant for one specimen, and no Bi for another specimen. After an additional deposition of 50 nm Si on both specimens, we deposited 2 ML Ge as a marker to indicate the growing surface morphology, and then we deposited a Si cap-layer. TEM images at locations without threading dislocations and defects are shown in Fig. 3(b) and 3(c) for the specimens without Bi and with 1ML Bi, respectively. Although the marker in the specimen with surfactant in Fig. 3(c) shows sharper contrast, its Ge marker layer is not as flat as that in the specimen without the surfactant in Fig.3(b). This indicates that the surfactant reduces the tendency to form a flat growth Si (001) surface, and suggests that the surface diffusion on Si(001) is suppressed by the surfactant.

4. DISCUSSION

It is known that each Si or Ge atom has one dangling bond on a clean Si(001) 1x2 surface as shown in Fig. 4(a). In contrast, 1 ML Sb or Bi forms a new 2x1 dimer structure at the top surface and can terminate the surface without a dangling bond, see Fig. 4(b). This is the reason why Bi and Sb are stabilized at the surface and reduce the surface free energy.

We saw in Fig. 1(b) and Fig. 2 that Sb has a larger effect than Bi in suppressing the introduction of lattice defects and surface lattice parameter changes. This difference in surfactant effect is considered to be due to the difference in the covalent radii of these atoms, namely, Bi has a larger radius than Sb. Thus, at the surface of the growing Ge layer, Bi enhances the stress effect caused by Ge/Si heteroepitaxy. In our previous paper (Matsuhata *et al.*, 1993) we reported that a small amount of Sb is incorporated in Si, whereas most of Bi is segregated at the top surface. This phenomenon can also be interpreted in terms of the

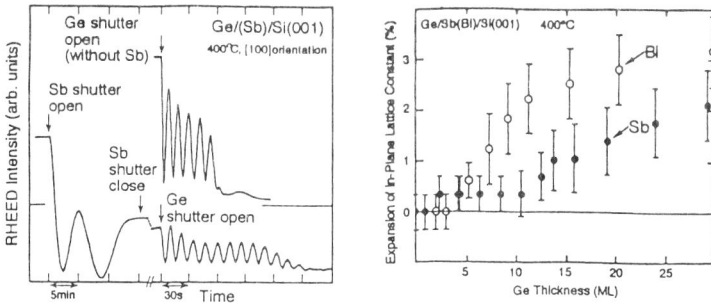

Fig. 1(a) RHEED intensity oscillations during Ge growth with 1 ML of Sb as a surfactant and without surfactant. (b) Evolution of surface lattice parameter during Ge growth measured from the spacing of spots in the *in situ* RHEED.

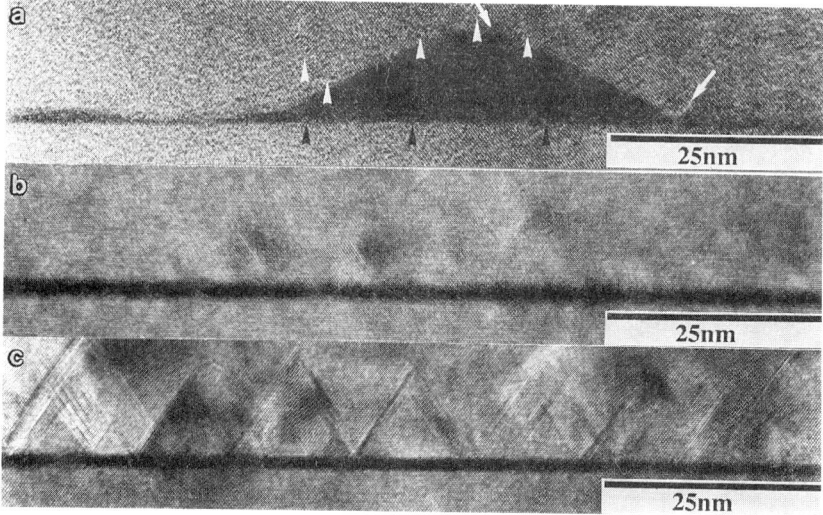

Fig. 2 Cross-sectional TEM images of Si/10MLGe/Si(001) heterostructure, (a) without surfactant: defects are indicated by arrows, (b) with 1 ML of Sb before 10 ML of Ge, (c) with 1 ML Bi before 10 ML of Ge.

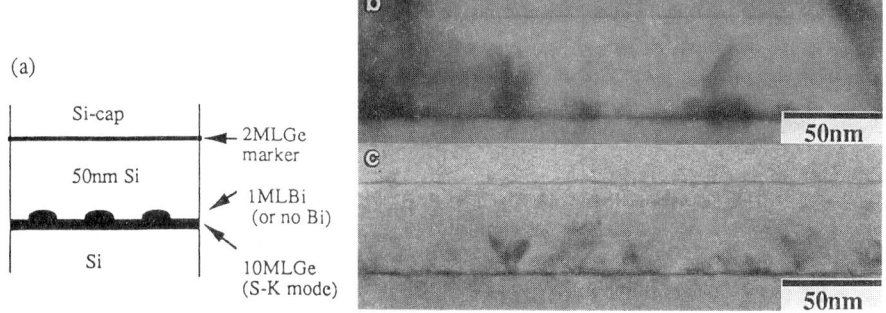

Fig. 3 (a)Schematic of the fabricated structure, (b)Cross-sectional TEM image of the 2 ML Ge marker without surfactant, and (c) with 1 ML of Bi.

226

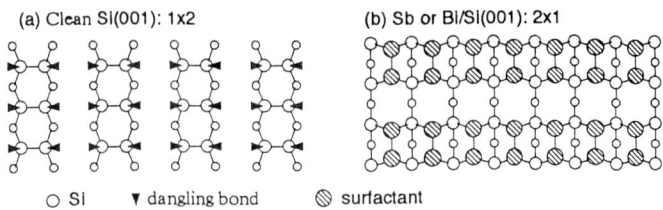

(a) Clean Si(001): 1x2   (b) Sb or Bi/Si(001): 2x1

○ Si   ▼ dangling bond   ◎ surfactant

Fig. 4 (a) Plan-view of a clean Si(001) 1x2 surface, (b) With 1 ML of either Sb or Bi on the surface. Black triangles denote the dangling bonds, and dark spheres are the surfactant atoms.

difference in size of these elements; the smaller Sb atom is incorporated in Si more easily than the larger Bi atom.

Since Bi and Sb are stable at the surface, the impinging molecules of Si and Ge on the surface are considered to move quickly below the surfactant, and diffusion on the top surface is suppressed. Because the tendency to form a flat growing surface of Si(001) is due to the step-flow growth mode caused by surface diffusion of Si, the suppression of the tendency to form flat (001) growth surfaces seen in Fig. 3 suggests that these surfactants reduce surface diffusion of Si. The formation of three-dimensional islands of Ge after the growth of several ML on Si(001) is known to be due to the surface diffusion of Ge. Thus, we conclude that the suppression in three-dimensional island formation by surfactant as seen in Fig. 2 is due to the suppression of surface diffusion. Copel *et al.* (1989) interpreted the mechanism of S-K mode suppression in terms of the reduction of surface free energy by the surfactant, but the effect is a very complicated phenomenon, and it can also be interpreted in terms of the suppression of surface diffusion. Though it has been reported that a surfactant enhances surface diffusion on the Si(111) 7x7 surface (Nakahara and Ichikawa 1992), this is not the case for the Si(001) 1x2 surface.

5. CONCLUSION

The effect of 1 ML of either Bi or Sb as a surfactant on Ge/Si(001) heteroepitaxy was investigated in detail. The RHEED and the TEM images showed that both surfactants reduce S-K mode growth of Ge on Si. This tendency was discussed in terms of its effect in reducing the surface diffusion. In our experiments, Sb had a stronger effect than Bi in suppressing the introduction of defects and the change in the lattice parameter at the surface. We attributed this property of Sb to the size difference of these two elements.

ACKNOWLEDGMENT

Portions of the high-resolution work were carried out at Kyushu University. Thus we gratefully acknowledge Prof. Tomokiyo and Mr. Manabe for the opportunity to use their microscope.

REFERENCES

Copel M, Router C, Kaxiras E and Tromp R M 1989 Phys Rev. Lett. **63** 632
Copel M, Router C,Von Hoegen H and Tromp R M 1990 Phys. Rev. **B42** 11682
Copel M and Tromp R M 1991 Appl. Phy. Lett. **58** 2648
Higuchi S and Nakanishi Y 1991 Surf. Sci. lett. **254** L465
Matsuhata H, Sakamoto K, Kyoya K, Miki K and Sakamoto T 1993 IOP series No. **134** Proc. Conf. Micro. Semicon. Mat. pp395-399
Nakahara H and Ichikawa M 1992 Appl. Phys. Lett **61** 1531
Iwanari S and Takayanagi K 1991 Jpn. J. Appl. Phys. **30** L1978
Sakamoto K, Miki K, Sakamoto T, Matsuhata H and Kyoya K 1993 J. Cryst. Growth **127** 392
Zalm P C, Van De Walle F A, Gravesteijin D J and Van Gorkum A A 1989 Appl. Phys. Lett. **55** 2520

Inst. Phys. Conf. Ser. No 146
Paper presented at Microsc. Semicond. Mater. Conf., Oxford, 20–23 March 1995
© *1995 IOP Publishing Ltd*

Study of 2D nucleation at silicon homoepitaxy by means of UHV REM

A V Latyshev, A B Krasilnikov and A L Aseev

Institute of Semiconductor Physics, Russian Academy of Sciences, Siberian Branch, 630090 Novosibirsk, Russia

ABSTRACT: Two-dimensional nucleation during silicon homoepitaxy has been investigated. A direct determination of the critical interstep distance for the appearance of the first growth islands has been made. Quantitative data about diffusion parameters and their dependence on various conditions have been obtained and adatom migration on reconstructed and unreconstructed surfaces is discussed.

1. INTRODUCTION

The present work deals with direct observations of the nucleation of two-dimensional growth islands in the initial stage of Si epitaxy on Si(111) surface using the in-situ method of ultra high vacuum reflection electron microscopy (UHV REM) which was described by Latyshev et al (1992). UHV REM allows one to obtain directly data about the change of critical distance h_c for two-dimensional nucleation at various temperatures and deposition fluxes. In this work we used mainly the silicon surface with step bands and anti-bands for the determination of h_c. The anti-band formation during long-time annealing was studied by Latyshev et al (1994).

2. RESULTS AND DISCUSSION

Fig. 1a shows an REM image of the initial silicon surface with a step band and an anti-band. The inset shows schematically the morphology of this surface. Step band b-b' represents a region with a high initial density of monoatomic steps and anti-band a-a' represents a region with a high initial density of monoatomic steps having opposite sign. There is almost a singular (111) surface region b'-a' between the step band and the anti-band. Deposition of silicon leads to the slow movement of the step band and anti-band in opposite directions due to step-layer growth. It results in an increase of the distance between step band and anti-band, i.e. in an increase of the width of terrace b'-a'. When the distance b'-a' reaches a critical value (h_c), the formation of two-dimensional islands takes place (Fig. 1b). The advantage of this method lies in the possibility of obtaining data about critical distances up to very large values.

According to the classical Einstein relation, the diffusion length of adsorbed surface atoms λ_s at the temperature T is defined as

$$\lambda_s^2 = 2D\tau_s, \tag{1}$$

where τ_s is the surface lifetime of an adsorbed atom, $D = D_0 \exp(-E_s/k_BT)$ is the diffusion constant for adatoms, E_s is the activation energy for adatom diffusion, and $D_0 = \nu a^2$ is the factor proportional to the adatom vibration frequency ν and the nearest neighbour hopping distance a.

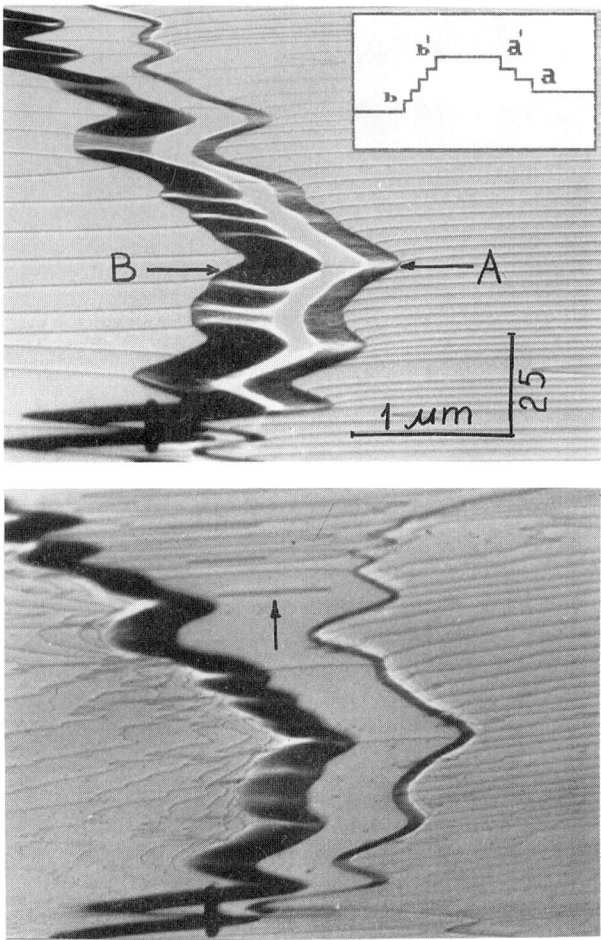

Fig.1. REM images of the Si(111) surface containing a step band and an anti-band before (a) and after (b) silicon deposition. The arrow marks a two-dimensional island.

In the case where the adatom diffusion length is sufficiently large for adatom migration from the central part of the terrace to the monoatomic step, all adatoms incorporate at the steps giving rise to the steps movement. When the terrace width reaches the value of $2\lambda_s$, there takes place formation of islands of growth or, in other words, there is a change in the mechanism of growth. It means that the critical distance between steps for the two-dimensional nucleation to take place can be evaluated as $h_c = 2\lambda_s$, and equation (1) can be written in the form

$$h_c^2 = 8D\tau_s \tag{2}$$

UHV REM data showing the dependence of $h_c^2(1/T)$ for various fluxes are shown in Fig. 2. The values of activation energies obtained from the slopes of straight lines drawn through the experimental data are shown in Fig. 3.

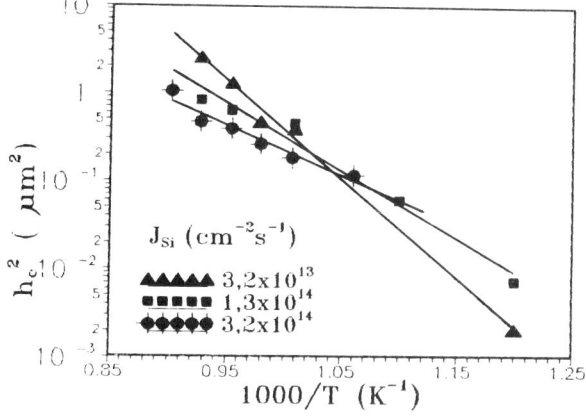

Fig.2. Dependence of h_c^2 on the absolute temperature for various fluxes of deposition.

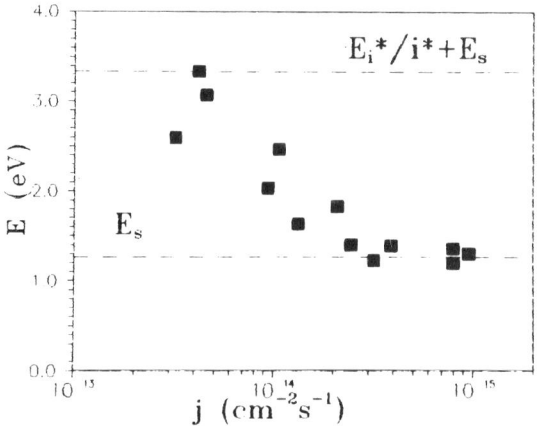

Fig.3. The activation energies obtained from the experimental data.

To interpret the dependence of E on J obtained from our experimental observations consider limiting case of large fluxes. In this case $\tau_s > \tau_i$, where $\tau_i = N_s/J$ is the average interval between arrival times of atoms per site and N_s is the atom density on the Si(111) surface. Given the inequality $\tau_s > \tau_i$, the adatom lifetime is determined by τ_i. Assuming that the size of the critical nucleus is equal to that of a single atom equation (2) can be transformed to

$$h_c^2 = 8 N_s \nu a^2 \exp(-E_s/k_B T)/J. \qquad (3)$$

According to this equation, the activation energy of surface diffusion is independent of the flux. This is confirmed experimentally (Fig. 3). One can see that when $j > J^* = 3 \times 10^{14} cm^{-2} s^{-1}$, the activation energy $E = E_s = 1.4 \pm 0.1$ eV which is consistent with previous REM observations by Latyshev et al (1989). Fluxes less than J^* lead to an increase of E. It means that $\tau_s < \tau_i^* = N_s/J^*$ and that the Einstein relation is not applicable to this case.

230

At standard epitaxial temperatures the adatom desorption may be disregarded. It seems likely that the decrease in adatom lifetime is due to the increase of critical nucleus size (the number of atoms), so adatom interaction is the main sink for the adatoms. The relation between critical interstep distance for two-dimensional nucleation and the critical nucleus size was derived by Nakahara et al (1993)

$$\ln(h_c) = \frac{i^*}{(i^*+1)} \left[\frac{\ln(1/J) - E_i^*/i^* + E_s}{k_B T} \right] + \alpha, \tag{4}$$

where E_i^* is the binding energy of a critical nucleus of size i^* and α is the pre-exponential term. One can see that at $i^* = 1$ equation (4) transforms to equation (3), and that the increase of i^* leads to the increase of E because $E_i^*/i^* = E_b(i^* - 1)/i^*$ (where E_b is the Si-Si binding energy) which is consistent with experimental data (Fig. 3).

A number of other researchers suggest a possible influence of surface reconstruction on adatom diffusion properties. In the present paper we have attempted to investigate the influence of superstructural domains on the activation energies of surface diffusion. To do this the data on h_c^2 were obtained over a wide temperature range, including the temperature of the $(7 \times 7) \Leftrightarrow (1 \times 1)$ transition (Fig. 4).

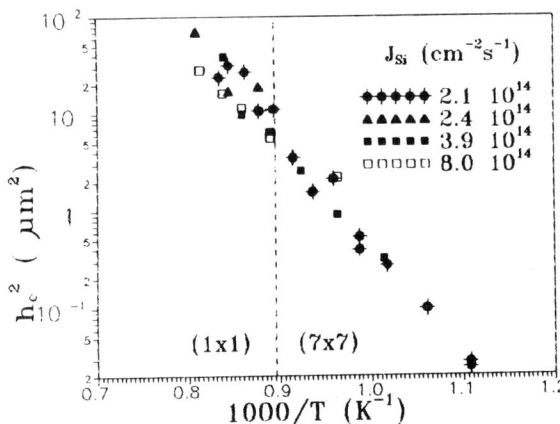

Fig.4. The data characterizing the behaviour of h_c^2 for various fluxes of deposition over a wide range of temperatures including the temperature of the $(7 \times 7) \Leftrightarrow (1 \times 1)$ transition

One can see that the data obtained did not indicate any significant changes in the displacement or slope of the straight lines h_c^2 vs. $1/T$ in the Arrhenius plots, and further investigation is required to determine the effect of surface reconstruction.

ACKNOWLEDGMENTS: this work was supported by International Science Foundation (grant U86000) and Russian Fund of Fundamental Researches (grant 94-02-06138).

REFERENCES

Latyshev A V, Aseev A L and Krasilnikov A B (1989) Phys. Stat. Sol. (a) 113 421
Latyshev A V, Krasilnikov A B and Aseev A L (1992) Mic. Res. and Technique 20 341
Latyshev A V, Krasilnikov A B and Aseev A L (1994) Surf. Sci. 311 341
Nakahara H, Ichikawa M and Stoyanov S (1993) Ultramicroscopy 48 417

Inst. Phys. Conf. Ser. No 146
Paper presented at Microsc. Semicond. Mater. Conf., Oxford, 20–23 March 1995

An investigation into gas source molecular beam epitaxy deposition onto non-planar InP patterned substrates

C A Mullan, B J Robinson, D A Thompson and G C Weatherly

Centre for Electrophotonic Devices and Materials, McMaster University, Hamilton, Ontario, Canada, L8S 4L7

ABSTRACT: We have compared deposition of InP, and lattice matched InGaAs and InGaAsP onto an InP substrate patterned as a diffraction grating. The planarisation necessary for device fabrication occurred after 300Å of InP was deposited, but the surfaces retained the shape of the substrate after much thicker deposits of the ternary and quaternary compounds. In addition, these exhibited phase separation which has been investigated by a combination of transmission electron microscopy and X-ray analysis.

1. INTRODUCTION

Epitaxial growth on patterned substrates has generated a great deal of interest over the past few years, primarily motivated by the fabrication of laser structures and other tuned optoelectronic devices. The substrates are patterned prior to growth through a combination of photolithography and etching. This paper will consider deposition by gas source molecular beam epitaxy(GS-MBE) of lattice matched III-V ternary and quaternary compounds onto InP wafers patterned as diffraction gratings. This study addresses two questions; how much deposited material is required to planarise the substrate (which is necessary for subsequent device fabrication) and whether decomposition of the ternary and quaternary compounds occurs during growth, resulting in compositional fluctuations. Transmission electron microscopy has been used to examine the samples.

2. EXPERIMENTAL DETAILS

Epitaxial layers of lattice matched $In_{0.53}Ga_{0.47}As$ and InGaAsP (E_g=0.954eV) were grown on a non-planar InP substrate. The substrate had previously been patterned as a second order diffraction grating running in the [110] direction with a pitch of 2000Å and a grating height of 412Å. The grating consisted of (001) mesas (approximately 600Å long) and {311} side walls. Epitaxial layers were deposited onto the sample in a gas source MBE system at 450-460°C. The sample was rotated during deposition. After growth the layers were characterised by a JEOL 2010F, operated at 200kV for TEM examination and chemical analysis. Cross-sectional samples were prepared by mechanical polishing and subsequent ion beam thinning at a low angle of incidence (~13°).

3. RESULTS

3.1 Deposition of the ternary compound InGaAs

Figure 1a is a [1$\bar{1}$0] axial image of a ternary layer structure deposited onto the grating. It consists of 3 periods of <220ÅInGaAs/30Å InP> and a terminal 250Å InGaAs layer (InP is employed to act as a marker layer only). Although this image was taken ~2° off axis, it can be seen that depositing almost 1000 Å of InGaAs has not planarised the grating and the top surface has an undulating profile. The waves are periodic and their maxima are positioned above the centre of the (001) mesa with their minima occurring directly above the bottom of the grating. The maximum to minimum in the undulation is 190 Å and less than half of the grating has been filled. It is interesting to note that, after TEM examination, we calculated that deposition of just 250Å InP will planarise this substrate.

Figure 1a Axial [1$\bar{1}$0] image of InGaAs deposited onto the substrate

Figure 1b is a dark field image of this structure imaged using **g**=022. Bands of alternating black and white contrast are seen in the ternary layer and their wavelength is approximately 1000 Å. This is half the period of the underlying substrate. If we compare this micrograph to the axial image in figure 1a, it can be seen that over the 2000 Å period of the InP grating there are four contrast reversals; (i) half way along the (001) mesa, (ii) half way along a {311} side wall, (iii) at the position corresponding to the bottom of the grating and, (iv), half way along the opposite {311} side wall. No diffraction evidence was found for CuPt ordering in this layer.

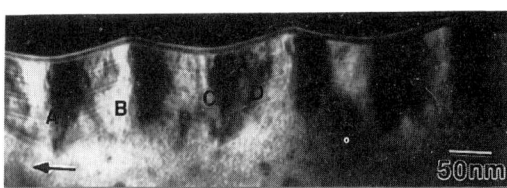

Figure 1b DF (**g**=022) image of InGaAs deposited onto the substrate

This specimen was then analysed by EDX. We took over 20 spectra from the sample in four sets of equivalent positions. These are labelled A (an interface between dark and bright contrast), B (the middle of a bright area), C (an interface between a bright and dark contrast) and D (the middle of a dark area) in figure 1b. Our preliminary results indicate that the In/Ga ratio altered with position. At both A and C (the interfaces) the relative amounts of In/Ga were the same. However, regions which are bright in figure 1b have a lower In/Ga ratio than those which are dark.

3.2 Deposition of the quaternary compound InGaAsP

Figure 2a is an axial [1$\bar{1}$0] image of a structure consisting of 1000Å InGaAsP, a 2 period <75ÅInGaAs/100ÅInGaAsP> structure, a 75Å InGaAs layer and a final 800ÅInGaAsP layer. It can be seen that in the quaternary layer, directly above the substrate mesa, there is a variation in contrast which indicates that facetting occurs during growth. The contrast lines intersect to form the apex of a triangle whose angle is ~111°. This is the angle between {311} planes when viewed along the [110] direction. Growth of such facets are often seen at an intermediate growth stage when III-V layers are deposited onto non planar substrates (Kapon 1994). This quaternary is in the range of compositions which are unstable at this growth temperature (Stringfellow 1982). Theoretical and experimental evidence indicates that decomposition results in the formation of InAs and GaP-rich regions and we propose that the InGaAsP has decomposed to form alternating {311} layers of slightly different compositions.

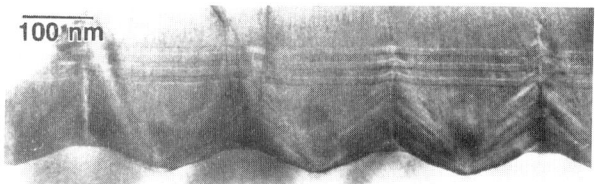

Figure 2a   An axial [1$\bar{1}$0] image of the InGaAsP deposition

After the 800 Å quaternary deposition the surface is not perfectly flat, although it does not exhibit the undulations seen for the ternary overgrowth (figure 1a). There are facets above the middle of the (001) mesa but it appears that most of the substrate has been filled in. The height difference of the facet maximum and the minimum is only 100Å.

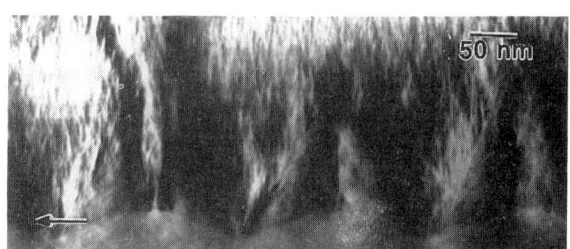

Figure 2b   DF (g=022) image of InGaAsP deposited onto the substrate

Figure 2b is a DF(g=022) image of this structure. There are alternating bands of dark-white contrast through the quaternary layer (there are also faint traces of the banding seen in Figure 2a). Near the substrate this has a wavelength of 1000 Å and it scales with the grating in a manner similar to that seen in the ternary (figure 1b). After depositing 800 Å of InGaAsP only fine modulations are visible and their wavelength decreases to between 100-150 at the surface. Both the coarse and fine contrast modulations behaved in the same way when examined under different ±g vectors.

234

4. DISCUSSION

We have found that this InP patterned substrate could only be planarised easily with InP. After depositing 300Å of InP, the grating was filled and a flat surface was achieved, whereas, after 1000 Å of InGaAs or 800 Å of InGaAsP, the pattern of the substrate was retained on the surface.

During growth we are effectively depositing onto two different planes, each of which will have different sticking and diffusion coefficients for the different atomic species (Cotta 1994). This will translate into localized variations of the growth rates and could result in the formation of line or planar defects, as seen by Sugiura et al.(1993). However, no such defects were observed in our samples which leads us to believe that the lattice constant variations are due to compositional modulations which are accommodated completely by elastic strain in the ternary alloy. Our TEM micrographs are consistent with phase separation and the EDX data shows that coarse diffraction contrast modulations are associated with small changes in the In/Ga ratio. Similar contrast modulations have also been noted by Okada et al.(1995) where tensile strained InGaAs and InGaAsP was deposited onto planar InP substrates and these have also been ascribed to compositional variations. We thus conclude that when both ternary and quaternary III-V compounds are deposited onto non-planar substrates, phase separation can occur. We believe this must be governed, in a yet unknown manner, by surface effects.

REFERENCES

Cotta MA, Hamm RA, Chu SNG and Harriott LR 1994, J Applied Physics, 75(1), 630 Kapon E, Semiconductors and Semimetals 1994, Volume 40, Ed Arthur C Gossard, (Academic Press Inc.) Chapter 4
Okada T, LaPierre R, Mullan CA, Weatherly GA, Robinson BJ and Thompson DA
        1995, in this proceedings
Stringfellow GB 1982, J. Electronic Materials ,11, 903
Sugiura H, Rudra A, Carlin JF, Buhlmann HJ, Araugo D and M Ilegems 1993,
        Semiconductor Science and Technology, 8(6), 1063

ACKNOWLEDGMENTS

We would like to thank the NERC and Bell Northern Research for supporting this work

Inst. Phys. Conf. Ser. No 146
Paper presented at Microsc. Semicond. Mater. Conf., Oxford, 20–23 March 1995
© *1995 IOP Publishing Ltd*

Origin of coarse contrast modulations in $In_{0.53}Ga_{0.47}As$ layers

K Lee, W C Johnson* and S Mahajan

Department of Materials Science, Carnegie Mellon University, Pittsburgh, Pennsylvania 15213, USA
*Currently at: Department of Materials Science, University of Virginia, Charlottesville, Virginia 22903, USA

ABSTRACT: The influence of growth temperature on the formation of coarse contrast modulations in (001) InGaAs layers, grown by liquid phase epitaxy, has been examined using transmission electron microscopy, Hall and photoluminescence (PL) measurements. These modulations are only observed in layers grown at low temperatures. These layers also exhibit lower carrier mobility and broader PL peaks. This behaviour is attributed to the increased compositional difference between the phase separated regions constituting the fine scale modulations. This enhanced difference may increase the magnitude of the two-dimensional strains associated with the fine scale modulations. It is suggested that in the presence of these strains, undulations develop on the growing surface. The coarse modulations are identified with these undulations.

1. INTRODUCTION

Since the initial observation by Henoc et al. (1982) of coarse contrast modulations in InGaAsP layers grown by liquid phase epitaxy (LPE) on (001) InP substrates, several investigators have observed these features (Launois et al. 1982 Mahajan et al. 1984, Chu et al. 1985, Norman and Booker 1985, Treacy et al. 1985, Mahajan et al. 1989, and McDevitt et al. 1992). Two explanations have emerged for the formation of the coarse modulations. Henoc et al. (1982), Launois et al. (1982) and Norman and Booker (1985) envisage that they evolve by surface spinodal decomposition during the layer growth. On the other hand, Mahajan et al. (1984), Mahajan et al (1989) and McDevitt et al. (1992) attribute them to the co-existing fine scale modulations. They argue that the coarse modulations evolve to accommodate two-dimensional strains associated with the fine scale speckle structure, implying that the two features are coupled.

McDevitt et al. (1991) have carried out reversion experiments to demonstrate the coupling between the two types of features. Results indicate that on annealing the coarse modulations and the speckle structure tend to simultaneously disappear, suggesting a strong coupling between the two. If the coarse modulations were true composition modulations, then they would take considerably longer to revert because the wavelength is ten times larger than that of the fine scale modulations.

We have carried out additional experiments to demonstrate this coupling. This study was based on the premise that two-dimensional strains associated with the speckle structure can be enhanced by depositing the same lattice matched composition at different temperatures. This is inferred because the composition difference between the phase separated regions is accentuated with the decreasing growth temperature. In addition, the stability of both features to the in-diffusion of zinc has been investigated. Results of these studies constitute the present paper.

236

2. EXPERIMENTAL DETAILS

InGaAs layers, emitting at 1.67 μm, were grown in the temperature range of 480 to 705°C by LPE on (001) InP substrates. The compositional differences between the phase separated regions were qualitatively monitored using Hall and photoluminescence (PL) measurements. Zinc was in-diffused in some of the layers using ampoule diffusion; the diffusion temperature was varied between 390 and 543°C. Different samples were examined in plan-view and cross-section using transmission electron microscopy.

3. RESULTS

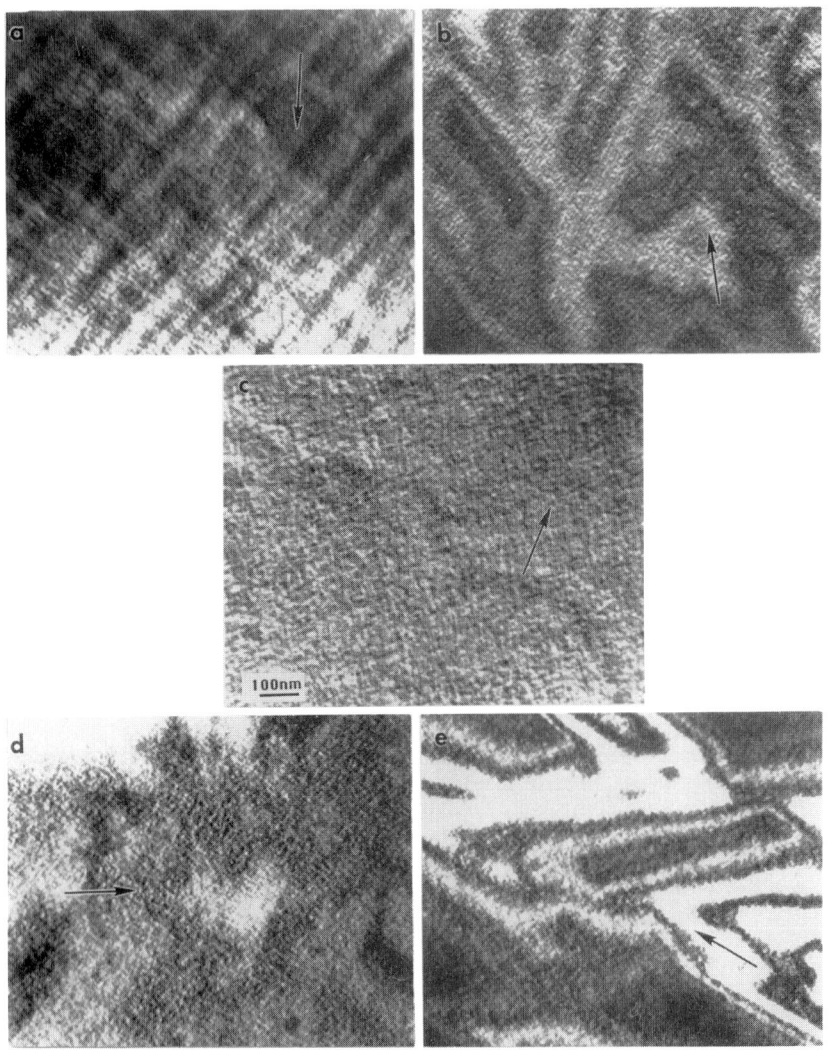

Fig. 1 Typical dark-field electron micrographs obtained from (001) InGaAs layers deposited on InP substrates by liquid phase epitaxy at different temperatures: (a) 480, (b) 540, (c) 600, (d) 660 and (e) 705°C. The reflection used to form the image in each case was 220.

Figure 1 shows typical dark-field electron micrographs obtained from layers grown at different temperatures; the growth temperatures in (a), (b), (c), (d) and (e) are, respectively, 480, 540, 600, 660 and 705°C. It is clear that the fine scale speckle structure is observed in all the micrographs, whereas the coarse modulations are only seen in the layer grown at 480°C. The presence of fine scale modulations was borne out by the [001] electron diffraction patterns. The <200> spots were elongated and <400> spots showed satellites. By measuring the spacing of the satellite spots from the fundamental reflection, the wavelength of the modulated structure (λ) has been computed using the following expression:

$$\lambda = \frac{|\vec{g}|}{|\Delta\vec{g}|} \frac{a}{\sqrt{h^2 + k^2 + l^2}} \, ,$$

where $\vec{g}$ is the fundamental reflection, $|\Delta\vec{g}|$ is the spacing between the fundamental reflection and the satellite, a is the lattice parameter and h, k, l are indices of the planes giving rise to the fundamental reflection. The measured values of λ as a function of growth temperatures are shown in Fig. 2 and are compared with that of McDevitt (1990) on InGaAs

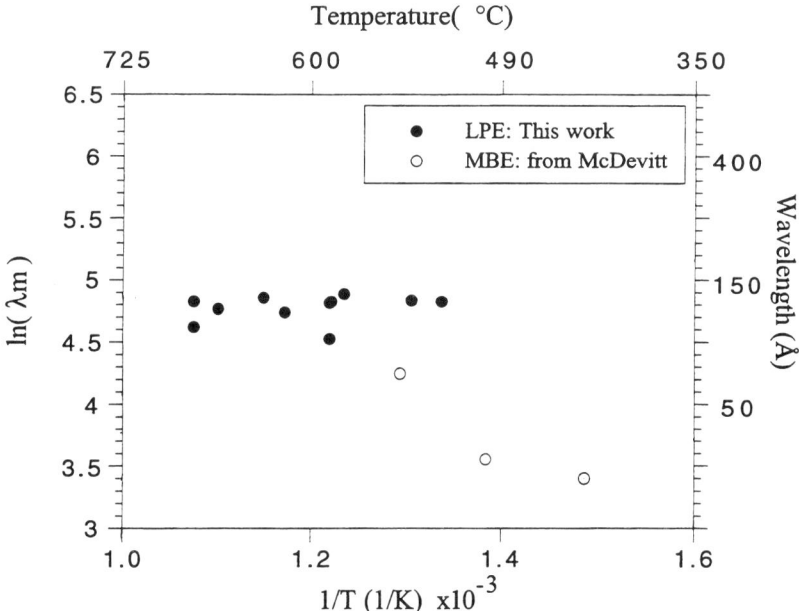

Fig. 2 Plot shows the variation in wavelength of the fine scale modulations versus the growth temperature; results of McDevitt et al (1992) on MBE grown InGaAs layers are included for the comparison. Note that the wavelength of the fine modulations in LPE grown InGaAs layers is essentially independent of the growth temperature.

layers grown by molecular beam epitaxy (MBE). It is clear that in the temperature range of 480 to 705°C, the wavelength of the fine scale speckle structure in InGaAs layers is independent of the growth temperature, whereas the wavelength of the speckle structure in MBE grown InGaAs layers exhibits strong dependence on the growth temperature. Lee (1994) has argued that when diffusion is very fast, even at low temperatures the maximum wavelength predicted by the Cahn-Hilliard theory can develop in a phase separated microstructure, i.e., the wavelength does not depend on the temperature.

238

Figure 3 shows the carrier mobility at 77K of InGaAs layers grown at different temperatures. The mobility ranges from 5,000 to 20,000 cm^2/v-sec., and is low at the growth temperature where the coarse modulations are observed, see Fig. 1(a). Furthermore, results show that the full width of half maximum of PL peaks decreases with the growth temperature as depicted in Fig. 4.

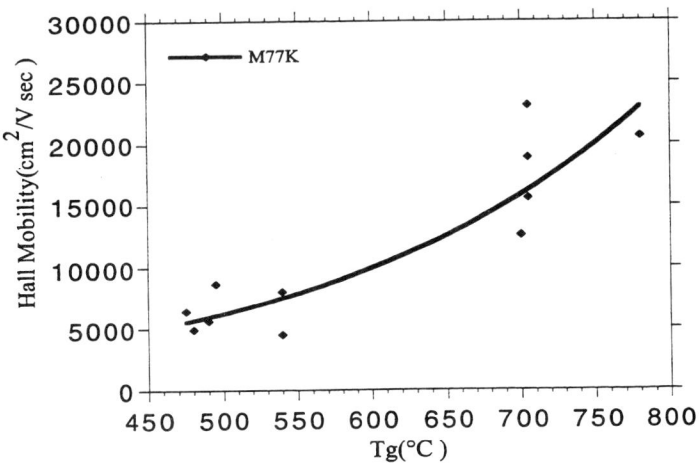

Fig.3   Plot showing the variation in 77 K carrier mobility of InGaAs layers as a function of the growth temperature.

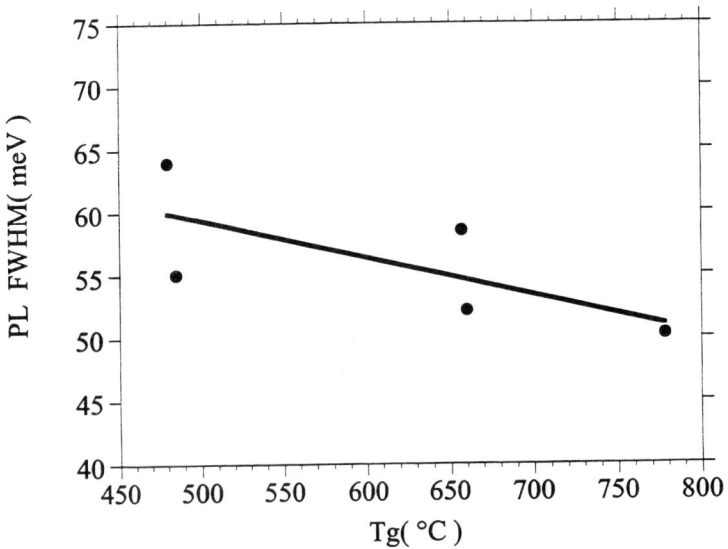

Fig. 4   Plot showing the variation in full width at half maximum of PL peaks observed from InGaAs layers grown at different temperatures.

A cross-sectional electron micrograph, obtained from a layer grown at 485°C and in which Zn was subsequently diffused at 390°C for 50 hrs., is shown in Fig. 5. Three different regions are observed: (i) near the surface, the microstructure is essentially homogeneous, (ii) both types of modulations are present near the layer-substrate interface, and (iii) in the region defined by the arrows, only coarse modulations are present. These results imply that the correlation between the two types of modulations may be a bit more complicated than that envisaged by McDevitt et al. (1991).

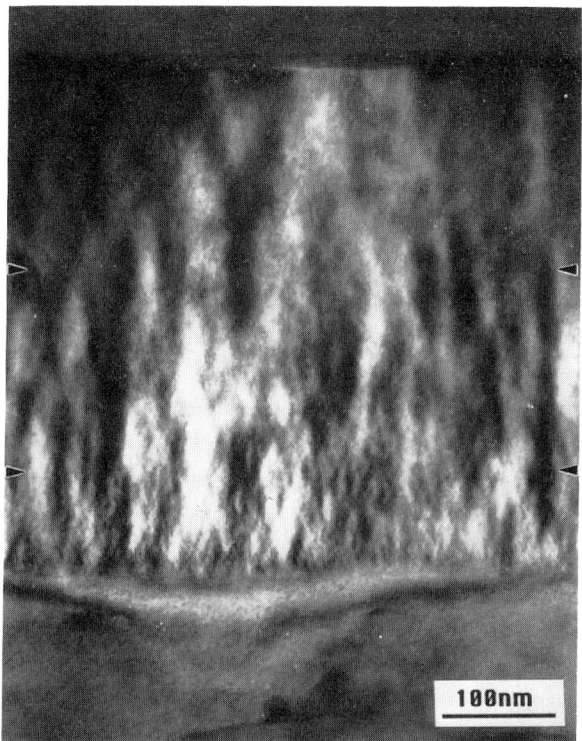

Fig. 5 Cross-section electron micrograph obtained from an InGaAs layer grown at 485°C in which Zn was diffused at 390°C for 50 hours. Note the presence of three regions: (i) near the surface, the microstructure is more or less homogeneous, (ii) near the layer-substrate interface both the coarse and fine modulations are present, and (iii) in the region defined by the arrows the fine modulations have been eliminated by the in-diffusion of zinc, whereas the less distinct coarse modulations are still present.

4. DISCUSSION

Several interesting observations have emerged from this study. First, the coarse modulations are observed only in layers grown at low temperatures and they co-exist with the fine scale speckle structure. Second, the layers exhibiting coarse modulations have a lower carrier mobility at 77 K and the full width at half maximum of the PL peaks is also larger. Third, when the zinc in-diffusion-induced homogenization of the speckle structure occurs, the coarse modulations are still observed, see Fig. 5.

When the lattice matched layers of InGaAs are grown at different temperature by LPE, the wavelength of the speckle structure is essentially invariant. The observed decrease in the carrier mobility can be rationalized if it is invoked that the compositional difference between the In- and Ga-rich regions is accentuated at the lower growth temperatures. Since the occurrence of phase separation creates mini-hetero barriers in the layer (Blood and Grassie 1984), the enhanced compositional differences would lead to larger band offsets. As a result, the carriers would tend to stay in the lower band gap regions. Under the action of an applied field, the carriers must overcome the mini barriers and propagate in the layer. If the barriers are larger in the case of the low temperature growth, then the mobility will be lower. Likewise, the enhanced compositional difference between the In- and Ga-rich regions would broaden the PL peaks because the range of wavelengths covered by the composition modulations will be larger.

Chu et al. (1985) and McDevitt et al. (1992) have shown that phase separation in (001) InGaAsP layers is two-dimensional in nature, occurs on the surface while the layer is growing and produces two-dimensional strains along the [100] and [010] directions. Lee's studies (1994) have demonstrated identical results for the growth of InGaAs layers on (001) InP substrates. It is envisaged that the magnitudes of the two-dimensional strains in the InGaAs layers grown at lower temperatures will be higher because of the enhanced compositional differences between the phase separated regions.

A number of investigators have observed that in the presence of misfit-induced strains the planar growth is unstable (Srolovitz 1989, Cullis et al. 1992 and Jesson et al. 1993). The surface becomes undulated, and the wavelength and amplitude of the undulations depend on several material parameters. We visualize that the presence of higher two-dimensional strains in the InGaAs layers grown at lower temperatures destablizes planar growth and produces undulations. The coarse contrast modulations are identified with these undulations.

If a layer is grown for an extended period after the formation of the undulations, the atomic species constituting the layer and having different tetrahedral radii may segregate to the tension and compression regions; the larger atoms would tend to segregate to the tension regions, whereas the smaller ones would go to the compression regions. The possibility of stress-induced segregation to the undulations may explain the original result of Henoc et al. (1982) on composition variations across the coarse modulations. In addition, within the framework of the above suggestion, the validity of the results of Treacy et al. (1985) is maintained. The only difference being that the two-dimensional strains and not the composition modulations as assumed by Treacy et al. (1985) are responsible for the formation of tension and compression regions at the surface.

There are two plausible explanations for our in-diffusion results shown in Fig. 5. First, the stress-induced segregation occurs at the coarse modulations during growth. Therefore, they are more difficult to eliminate during the diffusion anneal because the wavelength is large. Second, the zinc atoms may segregate to the tension regions during diffusion and stabilize the coarse modulations. With the available results, it is not possible to distinguish between the two possibilities.

In summary, the coarse modulations in (001) InGaAs layers are only observed at low growth temperatures. Concomitantly, these layers exhibit lower carrier mobility and broader PL peaks. This is attributed to the enhanced compositional difference between the In- and Ga-rich regions. It is proposed that this increased compositional difference leads to higher two-dimensional strains which destabilize the growing surface, resulting in undulations. The coarse contrast modulations are identified with these undulations.

REFERENCES

Blood P and Grassie A D C 1984 J. Appl. Phys. 56 1866
Chu S N G, Nakahara S, Strege K E and Johnston Jr W D 1985 J. Appl. Phys. 57 4610
Cullis A G, Robbins D J, Pidduck A J and Smith P W 1992 J. Cryst. Growth 123 333
Henoc P Izrael A, Quillec M and Launois H 1982 Appl. Phys. Lett. 40 963
Hillard J E 1970 Phase Transformations (Metals Park: ASM)
Jesson D E, Pennycook S J, Baribean J-M and Houghton D C 1993 Phys. Rev. Lett. 71 1744
Launois H, Quillec M, Glas F and Treacy M M J 1982 Inst. Phys. Conf. Ser. No 65 537
Lee K 1994 Ph.D. Dissertation Carnegie Mellon University
Mahajan S, Dutt B V, Temkin H, Cava R J and Bonner W A 1984 J Cryst. Growth 68 589
Mahajan S, Shahid M A and Laughlin D E 1989 Inst. Phys. Conf. Ser. No 100
McDevitt T L, Mahajan S, Laughlin D E, Bonner W A and Keramidas V G 1992 Phys. Rev. B 45 6614
McDevitt T L, Mahajan S, Laughlin D E, Bonner W A and Keramidas V G Inst. Phys. Conf. Ser. No 117 477
Norman A G and Booker G R 1985 J. Appl. Phys. 57 4715
Srolovitz D J 1989 Acta Metall. 37 621

Inst. Phys. Conf. Ser. No 146
Paper presented at Microsc. Semicond. Mater. Conf., Oxford, 20–23 March 1995
© *1995 IOP Publishing Ltd*

Correlation between surface topography and atomically ordered domains in MOCVD InGaAs layers

T-Y Seong*, G R Booker, A G Norman, P J F Harris[1] and A G Cullis[2]

Department of Materials, University of Oxford, Oxford, OX1 3PH, UK.
[1]Chemical Crystallography Laboratory, University of Oxford, Oxford, OX1 3PD, UK.
[2]DRA Malvern, St. Andrews Road, Malvern, Worcs., WR14 3PS, UK.

ABSTRACT: MOVPE InGaAs layers grown on (001) InP substrates at 550°C are examined by TEM/TED to investigate the CuPt-type ordered domains present and by atomic force microscopy (AFM) to observe the surface topography. There is a direct correlation between domains ordered on $(\bar{1}11)$ and $(1\bar{1}1)$ planes and [110] surface steps facing in opposite directions. These results explain the shapes and sizes of the domains present in the layers and give strong support to the correctness of the atomic models recently proposed for CuPt - type atomic ordering.

1. INTRODUCTION

In recent studies (Seong 1991, Seong et al 1991, Seong et al 1994) we used transmission electron microscopy (TEM) to examine $In_{0.53} Ga_{0.47}$ As layers grown by metal-organic chemical vapour deposition (MOCVD) on (001) InP substrates so as to investigate the CuPt-type atomic ordering present. Particular attention was given to two layers grown at 550°C, one at 0.2 nm/s and the other at 2.0 nm/s, termed here layers X and Y respectively. Both layers were ~ 1 μm thick.

For both layers, transmission electron diffraction (TED) patterns from [110] cross-section (CS) specimens, and from [001] plan-view (PV) specimens after tilting to other poles, showed $1/2$ $[\bar{1}11]$ and $1/2$ $[1\bar{1}1]$ type superlattice spots as well as the main lattice spots, indicating the presence of CuPt-type atomic ordering. Dark-field (DF) images obtained using individual superlattice spots revealed well defined domains corresponding to ordering on either the $(\bar{1}11)$ or $(1\bar{1}1)$ lattice planes, termed ordered variants I and II respectively.

For both layers, CS specimens showed that the domains were columnar and extended through the full layer thickness, and were typically ~ 0.5 μm wide. For layer X, PV specimens showed that the domains were crescent-shaped (Fig. 1). In general, a variant I crescent domain oriented in one sense was joined to a variant II crescent domain oriented in the other sense, with unordered material between the two crescents. For layer Y, PV specimens showed that the domains were approximately triangular-shaped (Fig. 2). In general, a variant I triangular domain oriented in one sense was joined to a variant II triangular domain oriented in the other sense, with a small area of unordered material between the two triangles. In the PV images, the line along which pairs of crescents or pairs of triangles were joined together corresponded to the [110] direction. The irregular dark lines within the crescent-shaped domains of Figs 1a and 1b, and the more regular dark lines within the triangular domains of Figs 2a and 2b, are antiphase boundaries (APBs).

---

*Now at Department of Materials Science and Engineering, Kwangju Institute of Science and Technology, 572 Sangam-dong, Kwangsan-ku, Kwangju, 506-303 Korea.

242

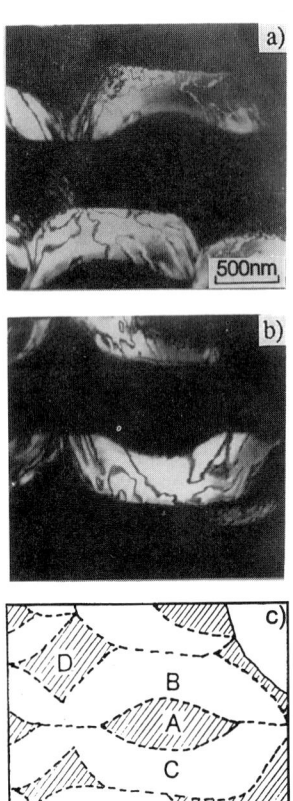

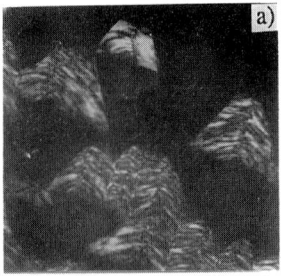

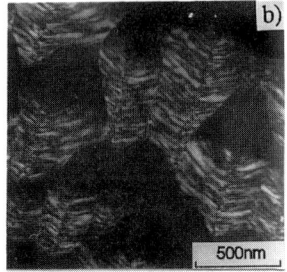

Fig. 2. Similar layer to that of Fig. 1.
grown at 550°C and 2.0 nm/s.
a) ($\bar{1}11$) ordered domains, b) ($1\bar{1}1$)
ordered domains.

Fig. 1. MOCVD InGaAs layer grown on
(001) InP substrate at 550°C and 0.2 nm/s.
a) Plan-view TEM DF g 1/2 <331>,
showing ($\bar{1}11$) ordered domains, b) similar
showing ($1\bar{1}1$) ordered domains,
c) composite diagram, B - ($\bar{1}11$) domain,
C - ($1\bar{1}1$) domain, A & D - unordered.

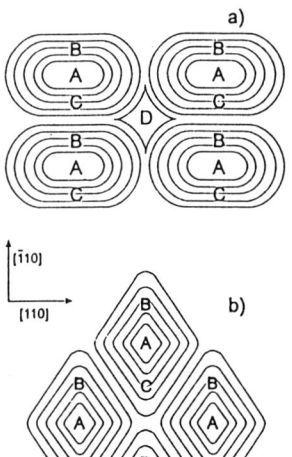

Fig. 3. Diagrams showing our previously suggested
atomic step distrubutions for surface undulations.
a) Corresponds to Fig. 1., b) corresponds to Fig. 2.
Areas B and C slope downwards, areas A and D
are parallel to surface.

Many investigators now consider that CuPt-type ordering grows in at the layer surface. Surface bond reconstruction orders the atoms in the surface monolayers Atomic steps moving across the surface order the atoms in successive monolayers. When atomic steps along the [110] direction are present, a series of such steps facing in one sense produces a variant I ordered domain, and a series facing in the other sense produces a variant II ordered domain (Suzuki et al 1988, Bellon et al 1989, Murgatroyd et al 1990, Seong 1991, Seong et al 1991, Chen et al 1991, Suzuki and Gomyo 1991 and 1993, Norman et al 1993a, Norman et al 1993b, Philips et al 1994). On this basis we proposed that the crescent and triangular shaped domains could have arisen as follows (Seong et al 1994).

For specimen X, we suggested that there were surface undulations as idealy shown in the PV diagram of Fig. 3a. Areas A correspond to the tops of hills which are relatively flat, i.e. there are few surface steps. Areas B slope downwards from areas A, i.e. there are many surface steps facing in one sense. Areas C are similar with many surface steps facing in the other sense. Areas D are relatively flat. Growth on areas B gives variant I domains and on areas C gives variant II domains. Growth on areas A and D gives either no ordered material (no steps) or very small randomly distributed variant I and variant II domains (Seong et al 1994). Hence, the crescent shaped domains observed would arise.

For specimen Y, we suggested that there were surface undulations as idealy shown in Fig. 3b. Areas A again correspond to the flat tops of hills and areas B and C slope downwards. The main differences between these undulations and those of Fig. 3a are that the areas A are individually smaller and have a different shape, and are collectively aligned in a diamond array, rather than a rectangular array. Hence, the triangular shaped domains observed would arise.

2. RESULTS AND DISCUSSIONS

Since our previous TEM work described above, we have used the atomic force microscope (AFM) to obtain quantitative data concerning the surface undulations for these layers. Two-dimensional PV images for layers X and Y are shown in Fig. 4, with the brighter regions corresponding to the higher regions of the layer surface.

For layer X (Fig. 4a), the elliptical bright areas are hills ~ 2 to ~ 3 μm long and ~ 0.5 μm wide, with the length direction corresponding to [110]. Analogous three-dimensional images showed that the hills are ~ 50 nm above the valleys which are present in between. For individual hills, there is a central area approximately parallel to the [001] substrate plane. The sides of the hill slope downwards, the surface making angles of ~ 2° for the length direction and ~ 5° for the width direction with the [001] plane. These hills correspond closely to those proposed above (Fig. 3a) for layer X, each ellipse comprising A, B and C areas.

For layer Y (Fig. 4b), the diamond-shaped bright areas are hills, often with facetted surfaces. The hills range from 0.7 μm long by 0.5 μm wide to 2 μm long by 1.5 μm wide, with the width direction corresponding to [110]. Analogous three-dimensional images showed that the large hills are ~ 150 nm, and the small hills ~ 50 nm, above the valleys. For individual hills, the sides slope downwards, the surface making angles of ~ 10° for both the length and width directions. The central area approximately parallel to the [001] substrate plane is now smaller than for the elliptical hills for Fig. 3a. Groups of the diamond-shaped hills often form a close-packed array. These hills correspond closely to those proposed above (Fig. 3b) for layer Y, each diamond comprising A, B and C areas.

The heights and angles of the hills determined above by the AFM are for the surfaces of the 1 μm thick layers. It is possible that the heights and angles were smaller than these when the layers began to grow and that the heights and angles progressively increased during growth.

Work previously performed on CuPt-type ordering in III-V ternary semiconductors has shown that such ordering occurs on vicinal [001] substrates for angles with off-cuts towards [11̄1] of up to ~ 15° (Suzuki and Gomo 1993). Consequently, in the present work the angles determined for the sides of the hills are in the correct range for ordered domains to occur.

244

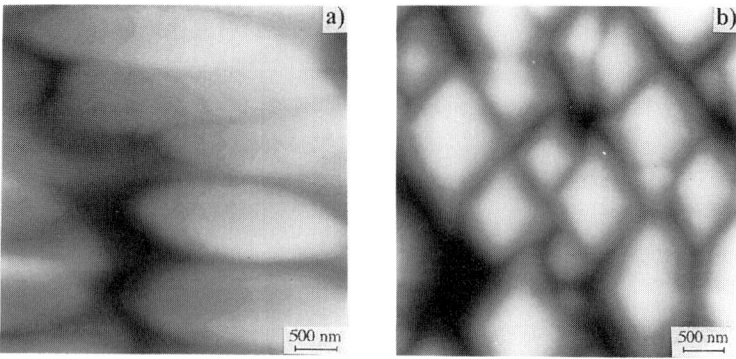

Fig. 4. Plan-view AFM images. a) elliptical hills correspond to Figs 1 and 3a, b) diamond hills correspond to Figs 2 and 3b.

The present AFM surface results strikingly confirm the proposals that we put forward previously for layer surface topography to explain our earlier TEM structural results for CuPt-type ordered domains in two markedy different MOCVD InGaAs epitaxial layers (Seong 1991, Seong et al 1994). At the same time they give strong support to the general correctness of the ordering mechanism on which the proposals were based. Our experimental results are analogous to observations (Friedman et al 1993) performed on abnormally large CuPt - type ordered domains (up to 20 μm long and 5 μm wide) in 10 μm thick MOVPE GaInP layers. Combined Nomarski optical microscopy and TEM studies showed that surface facets facing in opposite directions correlated with $(1\bar{1}1)$ and $(11\bar{1})$ ordered domains.

3. ACKNOWLEDGEMENTS

We wish to thank S J Bass and L L Taylor (DRA) for growing the layers and the British Council and SERC (UK) for financial support.

REFERENCES

Bellon P, Chevalier J P, Augard E, André J P, Martin G P, 1989 J. Appl. Phys. **66** 2388
Chen G S, Jaw D H and Stringfellow G B, 1991 J. Appl. Phys. **69** 4263
Friedman D J, Zhu J G, Kibbler A E, Olsen J M and Moreland J, 1993 Appl. Phys. Lett. **63** 1774
Murgatroyd I J, Norman A G and Booker G R, 1990 J. Appl. Phys. **67** 2310
Norman A G, Seong T-Y, Ferguson I T, Booker G R and Joyce B A, 1993a Semicon. Sci. & Tech. **8** S9
Norman A G, Seong T-Y, Philips B A, Booker G R and Mahajan S, 1993b Inst. Phys. Conf. Ser. **134** 279
Philips B A, Norman A G, Seong T-Y, Mahajan S, Booker G R, Skowronski M, Harbison J P, and Keramidis V G, 1994 J. Crystal Growth, **140** 249
Seong T-Y, 1991 D.Phil. thesis Oxford University
Seong T-Y, Norman A G, Hutchison J L, Booker G R, Cullis A G, Bass S J and Taylor L L, 1991 Inst. Phys. Conf. Ser. **117** 463
Seong T-Y, Norman A G, Booker G R and Cullis A G, 1994 J. Appl. Phys. **75** 7852
Suzuki T, Gomyo A and Iijima S, 1988 J. Cryst. Growth **93** 396
Suzuki T and Gomyo A, 1991 J. Cryst. Growth **111** 353
Suzuki T and Gomyo A, 1993 Semiconductor Interfaces at the Sub-Nanometre Scale, edited by H W M Salemink and M D Pashley (Netherlands: Kluwer Academic Publishers) p11

Inst. Phys. Conf. Ser. No 146
Paper presented at Microsc. Semicond. Mater. Conf., Oxford, 20–23 March 1995
© *1995 IOP Publishing Ltd*

Disordering of CuPt-type ordered structure in GaInP under electron irradiation

N Noda and S Takeda

Department of Physics, Faculty of Science, Osaka University, 1-16 Machikane-yama, Toyonaka, Osaka 560, Japan

ABSTRACT: We have found disordering of the CuPt-type ordered structure in GaInP under electron irradiation. In order to study the disordering process in detail, electron diffraction intensities have been measured under various irradiation conditions. It is found that the intensities of the ordered reflections exponentially decrease with irradiation time. Based on the experimental data, we have estimated the threshold electron energy for disordering to be 150 keV and the correlation factor of the displaced Ga and In atoms to be 0.23.

1. INTRODUCTION

It is well known that ordered structures of III-V compound semiconductors are invariably grown by various growth techniques such as molecular beam epitaxy (MBE) and metalorganic vapour phase epitaxy (MOVPE) (Gomyo, Suzuki and Iijima 1988). During characterization of the ordered structure in GaInP, we have found that the ordered structure gradually disappears in an electron-irradiated area. This finding itself is not surprising, since electron bombardment induces mixing of constituent atoms and extensive studies have indicated that the ordered structures are not the thermodynamic equilibrium phases; in other works, there is no driving force to recover the ordered structure under electron irradiation. However, by applying a theory on disordering of metallic ordered structures under electron irradiation, it is possible to obtain precise information on point defects by transmission electron microscopy (TEM). Such information has been difficult to obtain in semiconducting materials by TEM. We present in this report the measurement of transmission electron diffraction (TED) intensities of the ordered structure as functions of flux and energy of incident electrons, irradiation time and temperature. The experimental results are analyzed, based on the disordering theory.

2. EXPERIMENTAL

The sample was un-doped $Ga_{0.5}In_{0.5}P$, which was gown at 660°C on a GaAs substrate by MOVPE. The surface of the substrate was 2° off (001) towards [110]. The grown specimen was 0.5 μm thick. the [110] cross-sections of the specimens were observed by TEM. The disordering process was observed both in high resolution TEM images and TED patterns. High resolution TEM images were recorded at room temperature by a Hitachi H9000 microscope, which was operated at 300 kV and equipped with a video cassette recorder. TED patterns were taken with a JEOL JEM 2000EX microscope. Electron diffraction was observed under various irradiation conditions: electron flux (1.6, 4.7 and 7.8 x 10^{18} e cm^{-2}s^{-1}), incident electron energy (160-200keV) and temperature (150-300K). Diffraction intensities were measured using a photo-densitometer.

3. HIGH RESOLUTION ELECTRON MICROSCOPY

Fig. 1 shows a series of high resolution TEM images taken from video. The initial structure in our observation is CuPt-type ordered structure (Gomyo et al 1988), in which

Ga and In atoms are located, respectively, on two different sublattices called α and β while P atoms are located on the regular sites of the zincblend-type structure. In other words, Ga and In atoms segregate on alternate {111} planes of the zincblend-type structure. Disordering means that Ga and In atoms exchange their locations. Just after the irradiation commenced (Fig 1(a)), the ordered lattice fringes (indicated by the arrows) were clearly seen. The fringes in different directions are due to the coexistence of several equivalent ordered grains. As the irradiation proceeded, the ordered lattice fringes faded away gradually and uniformly (Fig. 1(b) to (c)) .

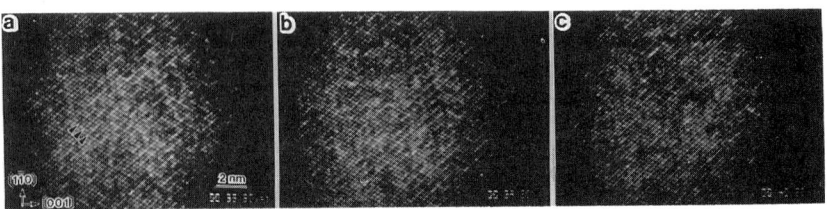

Fig. 1. Disordering observed in high resolution TEM images. Irradiation time is (a) 0 s, (b) 30 s and (c) 90 s. The arrows indicate the ordered lattice fringes.

4 TRANSMISSION ELECTRON DIFFRACTION

4.1 Dependence on Electron Flux and Dose

We measured intensities of electron diffraction under various irradiation conditions such as electron flux, ϕ, incident electron energy, E and specimen temperature, T. Fig. 2 (a) demonstrates the change of TED patterns under irradiation, in which the electron flux was estimated to be 7.8×10^{18} electrons $cm^{-2}s^{-1}$ and the specimen was observed at room temperature. It is clearly seen that the intensities of the ordered reflections (e.g., 111) decrease with irradiation time while those of the fundamental reflections (e.g., 004, 222) remain the same.

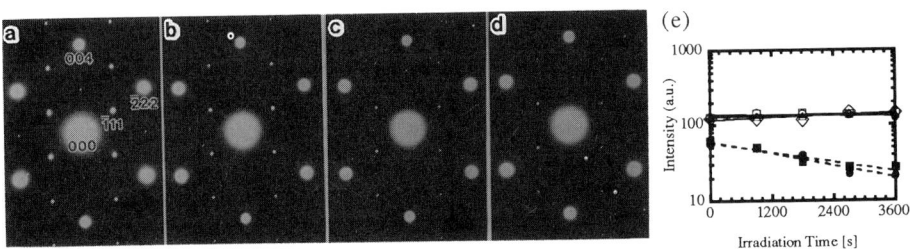

Fig. 2. Change of TED patterns. After irradiation for (a) 0 s, (b) 900 s, (c) 1800 s and (d) 2700 s. (e) Change of the observed intensities with irradiation time. The solid and dotted lines represent, respectively, the intensities of the fundamental and the ordered reflections.

Fig. 2 (e) shows the quantitative measurement of the TED intensities as a function of irradiation time. The measurements were performed at the condition (T=room temperature, E=200 keV and ϕ=4.6×10^{18} electrons $cm^{-2}s^{-1}$). The intensities of the ordered reflections, I^{SL} (dotted lines) exponentially decrease with irradiation time, t. Regardless of the irradiation conditions, the decay of I^{SL} obeys the exponential law. Therefore, we introduce a parameter, μ in the following formula, in order to express the experiment results.

$$I^{SL}=I_0\exp(-\mu t) \tag{1}$$

in which I_0 represents the intensity of the ordered reflections just after irradiation commences. Of course, μ itself depends on the irradiation condition. Fig. 3 (a) shows that μ is proportional to electron fluxes when the electron energy and temperature were kept constant; $E=200$ keV, $T=300$ K. Hence, the rate of disordering is simply proportional to the total dose.

4.2 Dependence on Incident Electron Energy

The disordering depends on incident electron energy, E. It is found that the disordering was never observed when E was smaller than about 150 keV. Fig. 3 (b) depicts the parameter μ as a function of E, in which ϕ and T were set constant, i.e., 7.8×10^{18} electrons cm^{-2}s^{-1} and room temperature, respectively. The disordering needs a certain threshold energy, and we estimate it to be 150 keV.

4.3 Dependence on Temperature

Disordering was observed in the temperature range between 150~300 K by utilizing available specimen containers. It is found in Fig. 3 (c) that the parameter, μ is constant in the temperature range ($\phi=7.8 \times 10^{18}$ e cm^{-2}s^{-1}, $E=200$ kV).

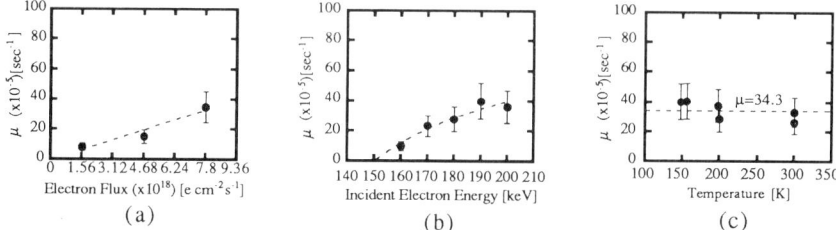

(a) (b) (c)

Fig. 3. Dependence of parameter μ on the irradiation condition; (a) electron flux, (b) incident electron energy and (c) temperature.

5. DISCUSSION

We have found that the disordering of the CuPt-type GaInP occurs under electron irradiation. The phenomenon is certainly due to atomic displacement under electron irradiation. The minimum electron energy for disordering was estimated to be 150 keV. The value is commensurate to those determined by electronic measurement in InP and GaP, which are, respectively, 120 keV (Massarani and Bourgion 1986) and 150 keV (Esposit and Loferki 1968).

We consider the behavior of displaced atoms in the crystal, based on the experimental data. It is natural to assume that an atom displaced by incident electrons either scatters in the crystal lattice and then recombines with vacancies at another site or returns to the initial site. We describe the probability of the former event as f and that for the latter event as $1-f$, in which f is called the correlation factor on the analogy of diffusion of atoms in a crystal. Next we assume that the scattered Ga or In atom eventually recombines with a vacancy at the sites for the group III element atoms. Since the ordered structure possesses the two sublattices for the group III element atoms, i.e., α, β and the disordered structure is most likely the thermal equilibrium phase, one can postulate the equal probability for Ga (or In) atoms settled in the α and β sublattices. Based on these assumptions, we set up the standard rate equations such as

$$\frac{dc_{Ga}^{\alpha}}{dt} = -G_{Ga}^{\alpha} \frac{c_{v}^{\beta}}{c_{v}^{\alpha}+c_{v}^{\beta}} \frac{f}{2} + G_{Ga}^{\beta} \frac{c_{v}^{\alpha}}{c_{v}^{\alpha}+c_{v}^{\beta}} \frac{f}{2} \qquad (2)$$

where c_δ^γ is the concentration of δ (Ga atoms or vacancies) at γ sites (α or β) and G_{Ga}^γ is the atomic displacement rate per lattice site for Ga atoms on γ sites. We have reached the expression below for the kinematical electron diffraction intensities of the ordered reflections,

$$I^{SL} = S_0 \exp\left\{-\left(\frac{2\sigma_d^{Ga}\sigma_d^{In}}{\sigma_d^{Ga}+\sigma_d^{In}}\phi f\right)t\right\}$$
(3)

where S_0 is the order parameters before irradiation and σ_d is the cross-section for atomic displacement of Ga and In atoms. The formula in the parenthesis above corresponds to the parameter μ. The cross-sections are calculated in the standard procedure (Mckinley and Feshbach 1948), in which it is assumed that the threshold energy of an In atom to be 120 keV (Massarani and Bourgion 1986) and that of a Ga atom 150 keV (Esposit and Loferki 1968), respectively. Based on the dynamical electron diffraction theory, we confirmed that I^{SL} is approximately proportional to S^2 when a wedge-shaped specimen is observed. By comparing with experimental results, we have therefore obtained the correlation factor, f in GaInP. Fig. 4 shows the change of the correlation factor depending on incident electron energy and temperature. It is concluded that the correlation factor is constant independent of the incident electron energy and temperature, i.e., 0.23. The result indicates that over 70 percent of the displaced atoms return to the same position.

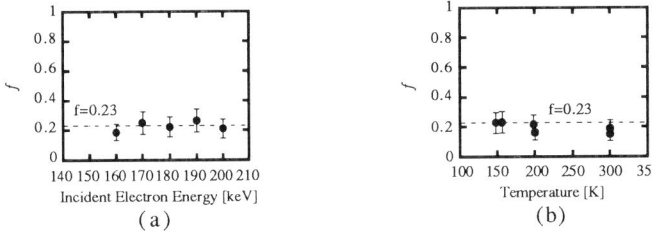

Fig. 4. Dependence of the correlation factor f on irradiation condition; (a) incident electron energy and (b) temperature. The correlation factor is independent of irradiation condition

6. CONCLUSION

We have found that the disordering of the CuPt-type order structure in GaInP occurs in a TEM and studied the behavior of point defects introduced by electron irradiation. The intensities of the ordered reflections exponentially decrease with irradiation time. We have estimated the threshold energy for disordering to be 150 keV and the correlation factor to be about 0.23 during electron irradiation.

ACKNOWLEDGMENTS

The authors are very indebted to Mitsubishi Kagaku Co. for offering the specimens. High resolution TEM was performed at the Ultra-High Voltage Electron Microscope Center of Osaka University. This work was supported in part by the Grant-in-Aid from the Ministry of Science, Culture and Education.

REFERENCES

Esposit R M, and Loferki J J 1968 Proc. 9th Int. Conf. on the Physics of Semiconductors, ed S M Ryvkin (Moscow), p.p 1105-1109
Gomyo A, Suzuki T and Iijima S 1988 Phys. Rev. Lett. 60 2645
Massarani B, and Bourgion J C 1986 Phys. Rev. B 34 2470
Mckinley W A, and Feshbach H 1948 Phys. Rev. 74 1759

Inst. Phys. Conf. Ser. No 146
Paper presented at Microsc. Semicond. Mater. Conf., Oxford, 20–23 March 1995
© *1995 IOP Publishing Ltd*

TEM and photoluminescence characterization of ordered domains in $Ga_{0.51}In_{0.49}P$ alloys

L Nasi, L Lazzarini, G Salviati, F Fermi* and G Lenzi*

C.N.R.-MASPEC Institute, Via Chiavari 18/A, 43100 Parma, Italy
*INFM-Phyisics Department, University of Parma, Viale delle Scienze, 43100 Parma, Italy

ABSTRACT: Transmission electron microscopy (TEM), photoluminescence (PL) and polarized photoluminescence excitation (PLE) spectroscopy, have been performed on MOCVD grown $GaInP_2$ layers differently misoriented on (001) GaAs. The layers spontaneously order in $CuPt_B$ variants. The more misosoriented samples show a higher structural order and a sharper distribution of the ordering degree in the CuPt domains. The more relevant parameters suggesting a direct relation between CuPt structure and electronic energy gap are the crystal field split-off energy of the valence band and the band-to-band transition onset in the PLE spectra.

1. INTRODUCTION

CuPt-type ordered $GaInP_2$ alloys are of technological interest in the fabrication of optoelectronic devices, due to the possibility of tuning the band gap at constant composition. Experimental evidences of the occurrence of spontaneous ordering in $GaInP_2$ are reported by many authors (Zunger and Mahajan 1994), however the ordering effects on the absorption and emission spectra of these structures are so far an open problem. In particular the correlation between ordering parameter and the optical gap of the materials is currently not satisfactory. In this paper the relation between structural order, ordering degree and optical band gap is discussed.

2. EXPERIMENTAL DETAILS

The lattice matched $GaInP_2$ layers have been grown by LP-MOCVD on (001) GaAs substrate at exact orientation and $2°$ and $6°$ toward (1-10) at temperature of 660 °C. The V/III ratio and the growth rate are 120 and 2.5 μm/h respectively.
TEM observations have been performed using a Jeol 2000FX working at 200 kV. The PL measurements have been made by means of a typical photoluminescence apparatus equipped with two monochromators, one for the photoexcitation and one for the analysis of the emitted light. The excitation source was an XBO 150W high pressure Xe lamp having an excitation power less than 1μW.

3. TEM RESULTS

The layers spontaneously order in CuPt-like structure. Of the four possible {111} ordering variants only the (-111) and (1-11) are observed. The doubling of the real-space periodicity is revealed by Transmission electron diffraction (TED) observing the extra reflections at {1/2,1/2,1/2} positions in the [110] cross section.

Due to the important role of the surface steps in the development of the ordered domains (Chen 1991), the misorientation of the substrate is fundamental in determining the distribution of the two ordered variants in the layers. In the 0° off sample the two variants are equally distributed in plate-like macro-domains, inclined at a shallow angle and in an opposite sense relative to the interfacial plane. The tilting of the substrate around the [110] axis toward (1-10) breaks down the symmetry of the distribution of the two ordered variants: the (1-11) ordered regions are enhanced, while the ordered domains relative to the (-111)-type variant gradually disappear. The enlargment of the (1-11) strongest ordered variant is shown in figure 1 by comparing the (1/2,-1/2,1/2) dark field images of the two limit 0° and 6° off samples.

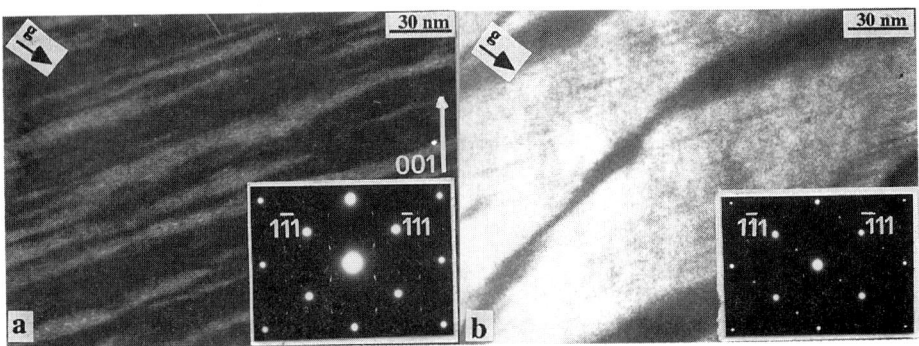

Figure 1 - (a) and (b) - (1/2,-1/2,1/2) dark-field images of 0° and 6° off sample respectively. TED patterns of the [110] cross section of the samples are shown in the insets.

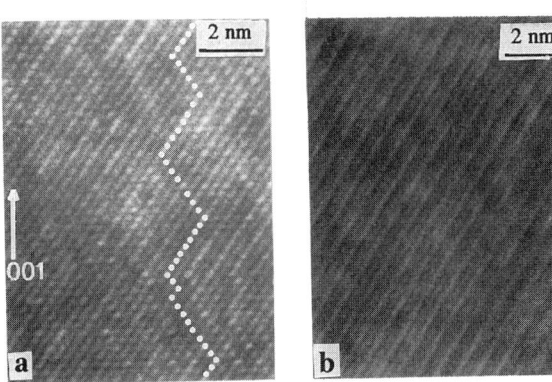

Figure 2 - (a) and (b) - HREM images of the 0° and 6° off samples respectively, taken (a) in the zone axis (b) by tilting the sample about (1-11) planes.

In the 6° off sample the (1-11) CuPt ordered structure is distributed over almost the entire epilayer, the dark bands being domain boundaries between adjacent regions in antiphase relationship. As shown in the inset of each micrograph, the shape and intensity of the extra spots in the TED patterns are clearly indicative of the morphology of the ordered domains in real space.

In the 0° and 2° misoriented samples the frequent interlocking of the two variants gives rise to a complex structure of (-111) and (1-11) ordered micro-domains. This fine-scale structure is evidenced in the lattice image of the 0° sample (fig. 2a) showing the doubling of the lattice fringes periodicity in both directions. On the contrary, due to the absence of one of the two variants, the 6° off sample exhibits a more homogeneous structure consisting of large single ordered domains. This is clearly evidenced in the HREM image of figure 2b showing no interruption of ordering on a large scale.

4. PHOTOLUMINESCENCE

The main features of the PL and PLE spectra are collected in figures 3, 4, 5 and 6.

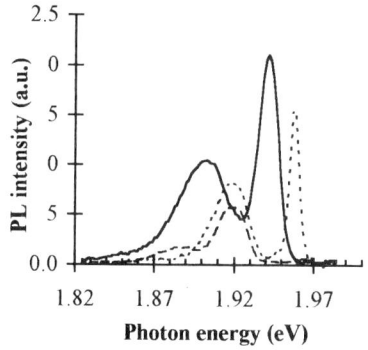

Figure 3 - PL observed at T=18 K of GaInP₂ MOCVD grown on 0°, 2° and 6° miscut GaAs substrates: —— 0°, − − − 2°, ----- 6°.
Excitation: λ= 620 nm (2 eV), Δλ = 3 nm.

Figure 4 - Linear polarized PLE observed at T=18K of GaInP₂ MOCVD grown on 0°, 2° and 6° miscut GaAs substrates. Exciting light field parallel to [110] : —— 0°, − − − 2°, ----- 6°.

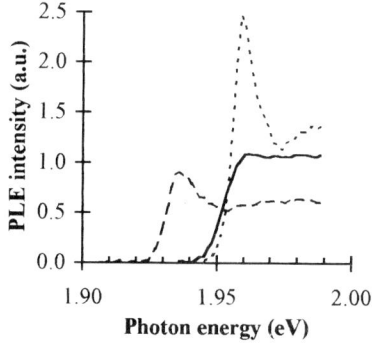

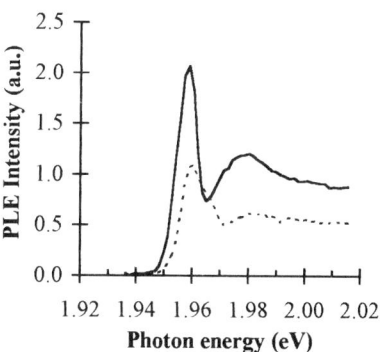

Figure 5 - Linear polarized PLE observed at T=18 K of GaInP₂ MOCVD grown on 0°, 2° and 6° miscut GaAs substrates. Exciting light field parallel to [1-10] : —— 0°, − − − 2°, ---- 6°.

Figure 6 - Linear polarized PLE observed at T=18K of GaInP₂ MOCVD grown on 6° miscut GaAs substate.
Light electric field: —— [110], ----- [1-10].
Emission: λ=648nm (1.9 eV), Δλ=2nm.

Every sample shows two emission bands (fig.3). The high energy PL bands are strictly related to the excitonic transitions detected in the PLE spectra while those at lower energy can be attributed to localized states. The excitonic PL bands show substantial red shift only for the $0°$ (15 meV) and $2°$ off sample (10 meV). These bands have large full width (FW) and change their peak energy and FW increasing the temperature. On the contrary, the excitonic PL band of the $6°$ off sample is very sharp (FW=6 meV), overlaps the excitonic PLE band and is insensitive to temperature increase at least up to T=45 K.

Following Horner et al (1994), the linear polarized PLE spectra (fig.4 and 5) show a sharp increase of the excitonic band in the sequence of $0°$, $2°$ and $6°$ off samples. Moreover these bands are slightly blue shifted for light polarization along the [1-10] direction showing also an increase of the excitonic band width for the $2°$ off sample, and the rise of a second band on the high energy side of the excitonic band for the $6°$ off sample (fig.5). Again, the PLE spectra of the $6°$ off sample show (fig.6) the clear onset of the band to band transitions.

5. CONCLUSIONS

The PL and PLE spectra in ordered crystals have been explained by many authors (Horner 1994) as follows: the crystals present a statistical distribution of order parameters whose mean value is higher the lower the optical band gap; the red shift of the PL excitonic peaks is due to thermalization of the photoexcited excitons in the CuPt domains with lower band gap. Our results confirm this interpretation, however comparison between TEM and PL measurements allow more detailed considerations. TEM observations show a distribution of small CuPt domains in both the (-111) and (1-11) variants in the $0°$ off sample. The $2°$ off sample is similar but has larger CuPt domains, while the $6°$ off sample presents very large domains of (1-11)-type only. PL and PLE spectra seem to reflect well these structural situations. Samples with higher structural order show sharper excitonic PLE peaks indicating a sharper distribution of order parameters in the CuPt domains. Consequently it seems to exist a relation between structural order and homogeneity of the ordering degree. However contributions of localized states, quantum dimensional effects and elastic local energy in $0°$ and $2°$ off samples, can significantly contribute to the position of the PLE onset. As a consequence, conclusions about the order parameter value, on the basis of the energy onset of the PLE, must be carefully evaluated. In practice the more relevant parameters suggesting a direct relation between CuPt structure and electronic energy gap are the crystal field split-off energy of the valence band and the band-to-band transition onset in the PLE spectra. The $6°$ misoriented sample seems to show all of these characteristics as evidenced in fig.6. In this figure the onset of the band-to-band transition is quite evident in the [110] polarized PLE, while a second excitonic peak emerging in the [1-10] polarized PLE could be attributed to the exciton associated to the crystal field split-off band ($\Delta E \cong 9$ meV) in agreement with the optical selection rules for the CuPt structures (Horner 1993). The $6°$ off sample is, to our knowledge, one of the best examples of partially ordered GaInP$_2$. Its order parameter is low owing to the low band gap reduction ($\cong 20$ meV) but its distribution must be very sharp.

ACKNOWLEDGEMENTS: Thanks are due to Dr. F. Vidimari and Dr. S. Pellegrino of TELETTRA-ALCATEL (Vimercate-MI, Italy) for growing the samples.

REFERENCES

Chen G S, Stringfellow G B 1991 Appl. Phys. Lett. 59 3258
Horner G S, Mascarenhas A, Alonso R G, Froyen S, Bartness K A and Olson J. M 1994 Phys. Rev. B 49 1727
Horner G S, Mascarenhas A, Froyen S, Alonso R G, Bartness K A and Olson J. M 1993 Phys. Rev. B 47 4041
Zunger A and Mahajan S 1994 Handbook on Semiconductors vol 3 (19) 1399

Inst. Phys. Conf. Ser. No 146
Paper presented at Microsc. Semicond. Mater. Conf., Oxford, 20–23 March 1995
© *1995 IOP Publishing Ltd*

Transmission electron microscopy and photoluminescence studies of $In_{1-x}Ga_xAs_yP_{1-y}$ films grown on $<100>$ InP substrates

T Okada, R LaPierre, C Mullan, G C Weatherly, B J Robinson and D A Thompson

Centre for Electrophotonic Materials and Devices, McMaster University, Hamilton, Canada L8S 4L7

ABSTRACT: The growth of lattice matched and strained $In_{1-x}Ga_xAs_yP_{1-y}$ layers on $<001>$ InP substrates has been studied by TEM and PL. Both studies point to phase separation. The PL signal shows temperature dependent intensity variations and peak shifts consistent with phase separation for layers whose composition lies within the miscibility gap. TEM studies show that the separation occurs along [110], with a modulation of 10nm in lattice matched layers. In layers under tension, phase separation is linked to a change in growth mode from $<001>$ to faceted growth.

1. INTRODUCTION

Ternary or quaternary compounds of the family $In_{1-x}Ga_xAs_yP_{1-y}$ display a wide variety of ordering or phase separation phenomena when grown as thin epitaxial films on InP or GaAs substrates. The thermodynamic properties of the bulk alloys predict that phase separation, rather than ordering, should be favoured in these systems (Stringfellow 1982), although the role of elastic strain energy in stabilizing a random solid solution has to be considered (Glas 1987; Ipatova et al 1993). Ordering, when observed, is believed to be a surface mediated process, controlled by the alternating, short-range, tensile and compressive stresses associated with the (2x4) reconstruction on (001) surfaces (Philips et al 1994). Phase separation can also be coupled to a change in growth mode, i.e. from planar to island or cusp-like growth morphologies (Spencer et al 1993; Guyer and Voorhees 1995). Such processes are driven by a reduction in the free energy of a system which remains elastically coupled to the substrate without the introduction of dislocations. In addition, phase separation or ordering may degrade the photoluminescence (PL) from the layer, causing peak shifts and/or line broadening in the spectra.

In this contribution we describe the correlation between the microstructural, growth and PL characteristics of a series of lattice matched and strained layers of $In_{1-x}Ga_xAs_yP_{1-y}$ grown by gas source molecular beam epitaxy (MBE). We find no diffraction evidence for ordering. However both cross sectional and plan view transmission electron microscopy (TEM), as well as the PL results, provide evidence for phase separation, although this does not occur by "classical" spinodal decomposition.

2. EXPERIMENTAL PROCEDURE

Gas-source MBE was used to grow lattice-matched, tensile or compressively strained ($\pm0.5\%$) $In_{1-x}Ga_xAs_yP_{1-y}$ layers on n-type (001) InP substrates. A 50nm thick InP buffer layer was first deposited, prior to a 150-300nm thick alloy layer, using a growth rate of 1μm/hr at temperatures in the range of 450-490°C. The structure was then capped with a 25nm InP layer. [001] plan view and $<110>$ cross sectional TEM (XTEM) samples were prepared by chemical thinning (2 vol% Br_2 in methanol) and Ar ion milling in a liquid N_2 cooled stage. TEM observations were done with a Philips CM12 microscope operating at 120 kV. PL spectra were collected from 11 to 300K using the 488nm line of a 20mW cw Ar ion laser. All the PL data presented here were obtained from samples having a 150nm thick alloy layer grown at 450°C. The compositions of the layers were established by combining X-ray diffraction measurements of lattice mismatch and PL bandgap determinations (Macrander and Lau 1991).

3. RESULTS

3.1 TEM Observations

[1$\overline{1}$0] XTEM observations using g = 220 showed a modulated contrast in dark field, the period of the modulations being about 10nm (Fig.1). No modulations were observed in the [001] growth direction. The modulations were strongest for the sample having a composition x = 0.285, y = 0.630, but were very weak for x = 0.117, y = 0.309, and less pronounced for the lattice matched InGaAs layer (x = 0.467 and y = 1). These trends correlate with the position of the bulk (strain-free) miscibility gap for In$_{1-x}$Ga$_x$As$_y$P$_{1-y}$ alloys calculated using the procedure described by Stringfellow (1982). At 450°C, a lattice matched composition having x = 0.285, y = 0.630, lies well within, while a composition of x = 0.117, y = 0.309 lies outside the miscibility gap. No diffraction evidence was found for CuPt ordering in either the [110] or [1$\overline{1}$0] sections. The nature of the composition modulations responsible for the diffraction contrast seen by XTEM was clearly seen in plan view. The dark field image of Fig.2, from the x = 0.285, y = 0.630 sample using g = 400, displays a series of modulations aligned nearly perpendicular (within ±5°) to the [110] direction. The strain fields associated with the composition modulations (the black-white contrast effects seen in Fig.2) vary in magnitude, but the largest of these, from regions extending up to 300-500nm in length, always had a dark field image consistent with local regions of the thin foil that are under compression, i.e. the white contrast lay on the side of +ve g (see Fig.2) (Ashby and Brown 1963).

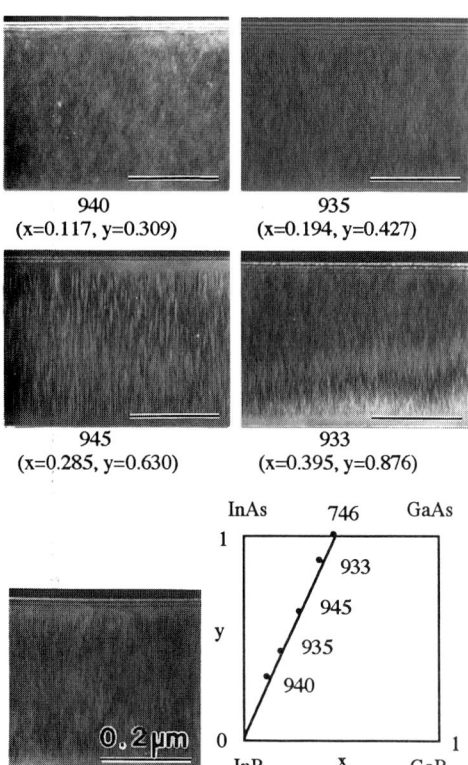

940
(x=0.117, y=0.309)

935
(x=0.194, y=0.427)

945
(x=0.285, y=0.630)

933
(x=0.395, y=0.876)

746 (x=0.467, y=1)

Figure 1: Dark field images, g = 220, of modulations in 300nm thick lattice-matched layers grown at 470°C. The compositions of the layers are shown on the lattice - matched line.

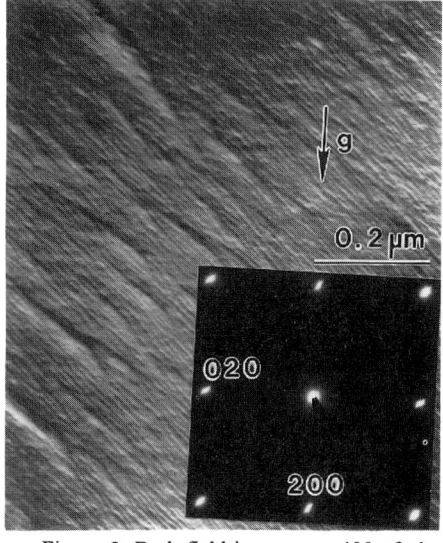

Figure 2: Dark field image, g = 400 of plan view sample (x = 0.285, y = 0.630) showing modulations and anomalous black-white strain contrast. The diffraction pattern (in the inset) shows streaking in or near to the [110] direction.

Lattice matched and compressively strained layers always grew with a planar <001> surface, but thicker layers grown in tension developed a faceted growth surface, bounded by two {hhl} facets where l/h ≈ 3-4 (Fig.3). The tendency to faceted growth was more pronounced at higher growth temperatures (490°C) and for compositions that lay well within the miscibility gap. Layers grown with strains in the range of 1-2% developed more pronounced facets than those with a strain of 0.5% (Okada 1995). At lower values of strain (0.3% or less) faceting was not observed. XTEM observations demonstrate an interesting correlation between the growth mode and the scale over which the compositional modulations develop. In the example

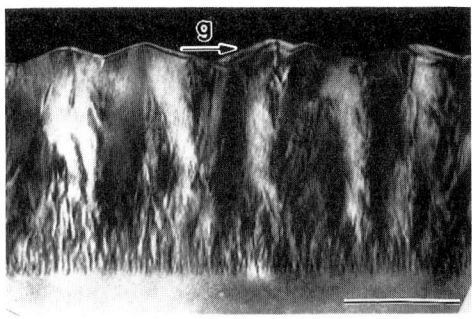

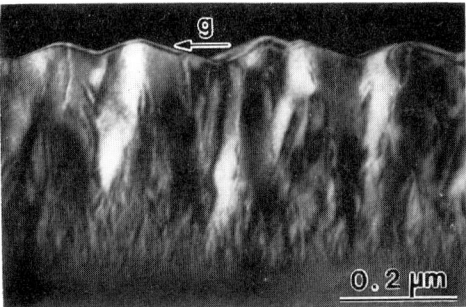

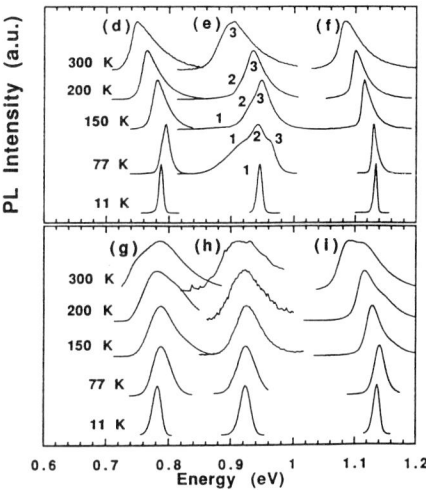

Figure 3: g = ± 220 dark field images of tensile strained InGaAs layer (x= 0.554, y = 1) showing surface faceting and black-white constrast reversal when the sense of g is reveresed.

Figure 4: Temperature dependence of PL spectra from lattice matched ((d) to (f)) and 0.5% tensile strained ((d) to (i)) InGaAsP layers. The compositions are (d) x = 0.47, y = 1, (e) x = 0.32, y = 0.69, (f) x = 0.17, y = 0.38, (g) x = 0.54, y = 1, (h) x = 0.38, y = 0.69, (i) x = 0.24, y = 0.38.

of Fig.3, the modulations at the onset of growth have a wavelength ≈ 10nm, comparable to the value found for the lattice matched layers. When the layer was about 150nm thick, the modulation wavelength (as measured by the scale of the white-black contrast reversals, Fig.3) increased to about 150nm, remaining roughly constant thereafter. Other experiments have demonstrated that the onset of faceting during growth follows a similar trend, i.e. the initial stages of growth in layers under tension are planar, but facets develop progressively as growth proceeds. We conclude that in these layers, faceting and phase separation are interrelated phenomena.

3.2 PL Measurements

The linewidths (FWHM), integrated intensities and temperature dependence of the PL spectra provide additional evidence for phase separation. The temperature dependence of three lattice-matched (d,e,f) and three tensile-strained (g,h,i) layers are shown in Fig.4. Samples (e) and (h) which lie deepest within the miscibility gap have the broadest linewidths, displaying the largest deviations from theoretical calculations of the linewidths based on thermal and alloy broadening in a random alloy (Benzaquen et al 1978; LaPierre et al 1995). Three distinct PL peaks (marked 1,2 and 3 in Fig.4) were found for sample (e). The temperature dependence of these peaks has been attributed to electron-hole recombination in InAs-rich

256

regions (peak1), regions of nominal composition (peak2) and GaP-rich regions (peak3) (LaPierre et al 1995). Since the scale of the composition modulation is much smaller than the carrier diffusion length ($\approx 1\mu m$), the carriers become trapped after diffusing into the smaller bandgap InAs-rich regions at low temperatures. This results in a decomposition induced red-shift of the PL emission compared to that expected for the average composition. As the temperature is increased, the carriers acquire enough energy to diffuse out of the InAs-rich regions, and carrier recombination in both the nominal and GaP-rich regions contribute to the spectra.

4 DISCUSSION

Both the PL and TEM results from the lattice matched samples are consistent with phase separation. The degree to which this occurs in the different samples correlates with the position of their compositions vis-a-vis the miscibility gap in the quaternary system. The strongest evidence for phase separation comes from samples grown well within the miscibility gap. However it is also clear that none of the existing theories of phase separation can account for all of our observations. When the effects of elastic strain energy associated with phase separation are included in the analysis, a growth temperature of 450°C lies above the coherent miscibility gap for composition modulations along <100> or <110> directions in the plane of the film (Glas 1987; Ipatova et al 1993; Okada 1995). Furthermore, <100> rather than <110> modulations would be predicted as the <110> direction is elastically stiffer than <100> in InGaAsP. Just as ordering in InGaAsP alloys grown at higher temperatures is controlled by surface reconstruction, with the incorporation of different sized atoms being site specific (Philips et al 1994), so it is likely that the present results are associated with surface driven phenomena.

The maximum composition fluctuations in the lattice matched layers can be estimated from the PL data for sample (e) (Fig.4). The details of the analysis, given by LaPierre et al (1995), have shown that the fluctuations correspond to variations of about 2% in the group III and V components. These small changes in composition give regions of varying lattice parameter (above and below the nominal values). This qualitatively accounts for the strain contrast seen in Figs.1 and 2, which is associated with lattice rotations at the surface of the thin foil (Treacy et al 1985). However, as Fig.2 demonstrates, the regions under local tension or compression are not uniformly distributed throughout the thin film.

The influence of a superimposed tensile strain on promoting phase separation and non-planar growth (Fig.3) may be explained by the recent analysis of Guyer and Voorhees (1995). These authors have considered the coupling between the stress field induced by a non-planar growth mode in strained layers and the redistribution of solute atoms in systems which lie within a miscibility gap. Their analysis does not specifically account for the role of the surface, which we believe to be important in the present study, but it might provide an explanation for why only tensile, and not compressively strained layers, show this instability.

REFERENCES

Ashby M F and Brown L M 1963 Phil. Mag. **8** 1649
Benzaquen R, Charbonneau S, Sawady N, Roth A P, Hobbs L and Knight G 1978 J. Appl. Phys. **49** 5944
Glas F 1987 J. Appl. Phys. **62** 3201
Guyer J E and Voorhees P W 1995 "Morphological Stability of Alloy Thin Films" Phys. Rev. Lett. in press
Ipatova I P, Malyshkin V G and Shchukin V A 1993 J. Appl. Phys. **74** 7198
LaPierre R, Okada T, Robinson B J, Thompson D A and Weatherly G C 1995 submitted to J. Cryst. Growth
Macrander A T and Lau S 1991 J. Electrochem. Soc. **138** 1147
Okada T 1995 Ph.D dissertation McMaster Univ. Canada
Philips B A, Norman A G, Seong T Y, Mahajan S, Booker G R, Skowronski M, Harbison J P and Keramidas V G 1994 J. Cryst. Growth **140** 249
Spencer B J, Voorhees P W and Davis S H 1993 J. Appl. Phys. **73** 4955
Stringfellow G B 1982 J. Electronic Mater. **11** 903
Treacy M M J, Gibson J M and Howie A 1985 Phil. Mag. A **51** 389

Inst. Phys. Conf. Ser. No 146
Paper presented at Microsc. Semicond. Mater. Conf., Oxford, 20–23 March 1995
© *1995 IOP Publishing Ltd*

Characterization of highly strained $Ga_{0.5}In_{0.5}P/InP$ interfaces

H Lakner, J Stammen, *Q Liu and *W Prost

Werkstoffe der Elektrotechnik, *Halbleitertechnik/Halbleitertechnologie, Sonderforschungsbereich 254, Gerhard-Mercator-Universität Duisburg, D 47048 Duisburg, F R Germany

ABSTRACT: Highly strained $InP/Ga_{0.5}In_{0.5}P/InP$ interfaces have been investigated by STEM, CBED and XRD. We found an asymmetric behaviour of the chemical composition across the interfaces. The lower interface - in the growth direction - is abrupt compared to the upper one which is smeared out over 3nm. Such highly strained $Ga_{0.5}In_{0.5}P$-layers can be used as wide band-gap material in InP-based HFET structures. The arising advantages are demonstrated and discussed.

1. MOTIVATION

$Ga_xIn_{1-x}P$ grown as a wide band-gap material in InP-based heterostructure field-effect transistors (HFET) promises improved electrical properties compared to other barrier materials. It offers a high Schottky barrier giving low gate leakage. As a spacer layer it offers increased band discontinuity at the channel interface. There are however problems in the growth of such structures as well, especially for high gallium-concentrations ($x \approx 0.5$) - which are neccessary for high band discontinuities - although this material is far from being lattice matched to InP. As a consequence only for very thin layers of GaInP (less than the critical thickness) pseudomorphic growth (and therefore no mismatch-induced defects due to relaxation) can be expected. The lattice mismatch of pseudomorphic $Ga_{0.5}In_{0.5}P$ on InP is very high ($\approx 3.6\%$). This may be one of the reasons why up to now there is not much work reporting on the growth of highly strained layers of GaInP on InP. In this contribution we report on the growth of such heterostructures and on the results of the characterization by scanning transmission electron microscopy (STEM), convergent beam electron diffraction (CBED) and X-ray diffraction (XRD). Additionally, examples for the incorporation of highly strained $Ga_{0.5}In_{0.5}P$-layers in InP-based HFETs will be shown and the advantages will be outlined.

2. GROWTH AND APPLIED CHARACTERIZATION TECHNIQUES

The $Ga_{0.5}In_{0.5}P/InP$ heterostructures investigated here were grown in a horizontal low pressure metalorganic vapour-phase epitaxy (MOVPE) apparatus at a total pressure of 20 mbar on exactly [100]-oriented semi-insulating InP-substrates. The growth temperature was 650°C. The precursors were trimethyl-(gallium and/or indium) and phosphine. The group III-element exchange at the interfaces was realized by a growth stop (group III-element switched off for 5s under continuous phosphine flow). In order to get information on the critical thickness we grew several samples with a nominal composition x=0.5 for all but with different nominal layer thicknesses of the $Ga_xIn_{1-x}P$ layer (0.5 nm, 1.1 nm and >1.1 nm, respectively).

Cross-sectional specimens were investigated in a field-emission STEM operated at 100keV (VG Microscopes: HB 501). Bright-field imaging was used for the characterization of structural defects. The interface abruptness was determined by Z-(atomic number) contrast imaging qualitatively. CBED performed with a subnanometer electron probe was used to measure the tetragonal distortion in the ultrathin highly strained layers. For XRD characterization we used a STADI P triple-crystal diffractometer.

3. RESULTS AND DISCUSSION

The electron microscopy investigations delivered the following results: In the heterostructures containing very thin layers of $Ga_{0.5}In_{0.5}P$ (0.5 nm and 1.1 nm nominal thickness) no mismatch-induced defects due to relaxation were observed by bright-field imaging. This is not the case for thicker layers (>1.1 nm nominal thickness), where many defects on {1 1 1} planes starting - in growth direction - at the lower GaInP/InP interface can be observed (for example see fig. 1). This is a first indication that the thin layers (0.5nm and 1.1nm thickness) are really strained, while layers with thicknesses >1.1nm tend to relax. Therefore we estimate the critical thickness for $Ga_{0.5}In_{0.5}P$ on InP to be ≈1nm.

By means of Z-contrast imaging we investigated the abruptness of the chemical composition across the interfaces. Z-contrast imaging is sensitive to the local chemical composition of heterostructures (Lakner et al 1992). Areas with high (low) mean atomic number appear bright (dark). The qualitative chemical composition profile across the

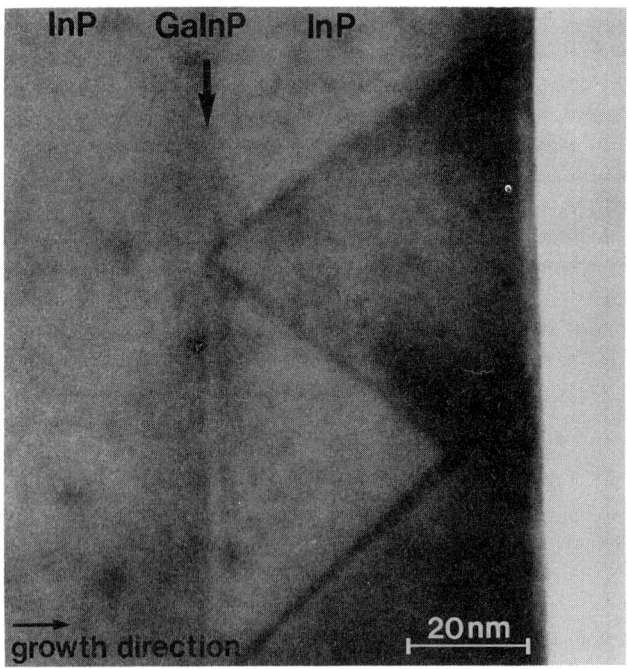

Fig.1 STEM brigth-field image of mismatch-induced defects in a relaxed $Ga_{0.5}In_{0.5}P$ /InP-heterostructure. The critical thickness for $Ga_{0.5}In_{0.5}P$ is exceeded.

interface was extracted from Z-contrast line-scans and is superimposed on the Z-contrast micrograph. In the very thin GaInP layers (nominally ≤ 1.1 nm) the two interfaces are of quite different quality (see fig. 2). The lower interface (in growth direction) is much sharper compared to the upper one, which is smeared out over 3nm. This result indicates that the upper interface is a 3nm thick graded $Ga_xIn_{1-x}P$ layer with in growth direction decreasing Ga-concentration. This behaviour can be explained by a strain-induced segregation of Ga-atoms at the upper interface.

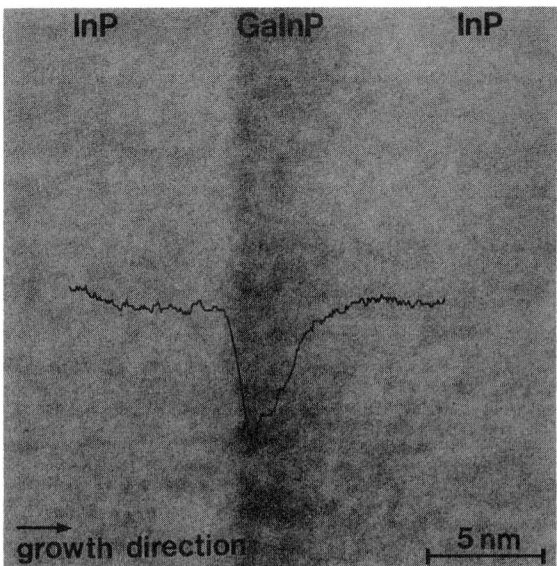

Fig. 2 STEM Z-contrast image of a highly strained $Ga_{0.5}In_{0.5}P$/InP heterostructure with a nominal thickness of 0.5nm for $Ga_{0.5}In_{0.5}P$

CBED patterns recorded directly from the thin GaInP layers using a subnanometer electron probe (Lakner et al 1995) show the typical effect of tetragonal distortion due to pseudomorphic growth of $Ga_{0.5}In_{0.5}P$ on InP. The position of the HOLZ (High Order Laue Zone) -lines in the patterns is different compared to those recorded from InP. The point of intersection of the HOLZ-lines is shifted (see arrows in fig. 3). This result proves the presence of high strain in the thin layers (≤1nm).

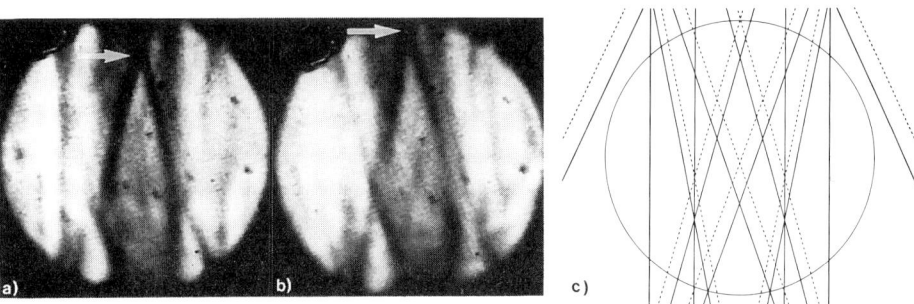

Fig. 3 CBED-patterns showing strain-sensitive HOLZ-lines: a) recorded in InP b) recorded in highly strained GaInP   c) simulations for a) ___ and b)------

Additionally we observed an interesting effect at the lower (abrupt) GaInP/InP-interface. The Kikuchi-lines (here: vertical lines in the CBED pattern shown in figure 4) recorded directly at the interface using a subnanometer electron probe exhibit a splitting of the {200}- and {400} -lines (see arrows) compared to the pattern recorded in InP. The line-splitting can be attributed to surface relaxation effects in the thinned x-sectional specimen (Chou et al 1994).

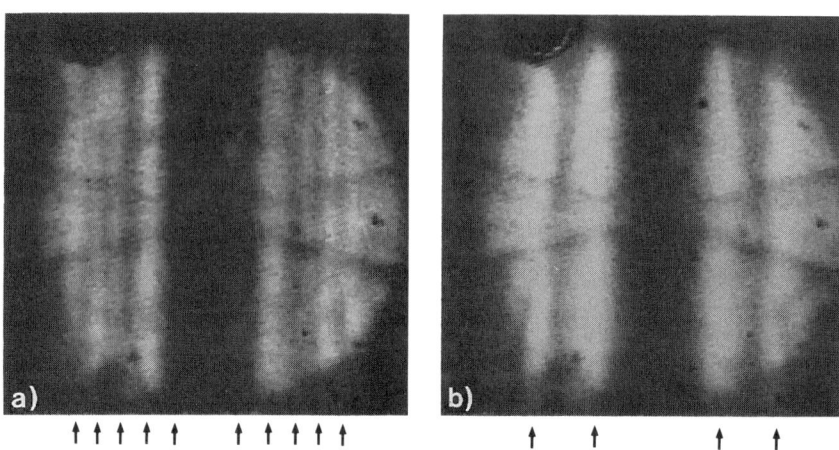

Fig. 4 CBED-patterns recorded:
   a) directly at the lower (abrupt) GaInP/InP-interface   b) in InP (for comparison)

Keeping in mind the behaviour of the chemical composition across the GaInP/InP-hetero-structure (compare figure 2) we used a simplified model of the asymmetry at the interfaces for the dynamical XRD-simulations. The lower interface was approximated to be abrupt. For 0.5nm thickness a composition of $Ga_{0.5}In_{0.5}P$ was supposed, followed by a graded $Ga_xIn_{1-x}P$-layer as sketched in figure 5.

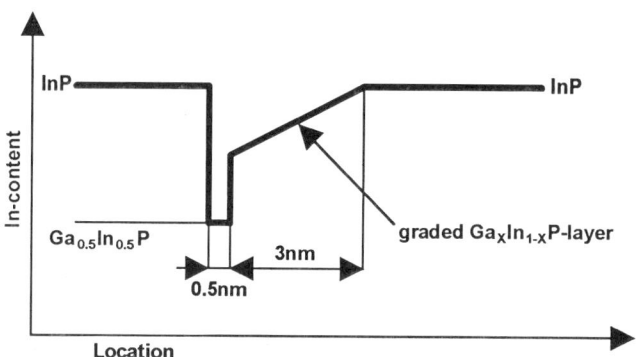

Fig. 5 Model of the chemical composition across the GaInP/InP-interfaces
   for the XRD-simulations

Recorded XRD-rocking curves and simulated rocking curves (dynamical theory) match each other only when the asymmetric behaviour of the chemical composition profile across the interfaces and the tetragonal distortion of the GaInP-layer are taken into account in the simulations (see figure 6).

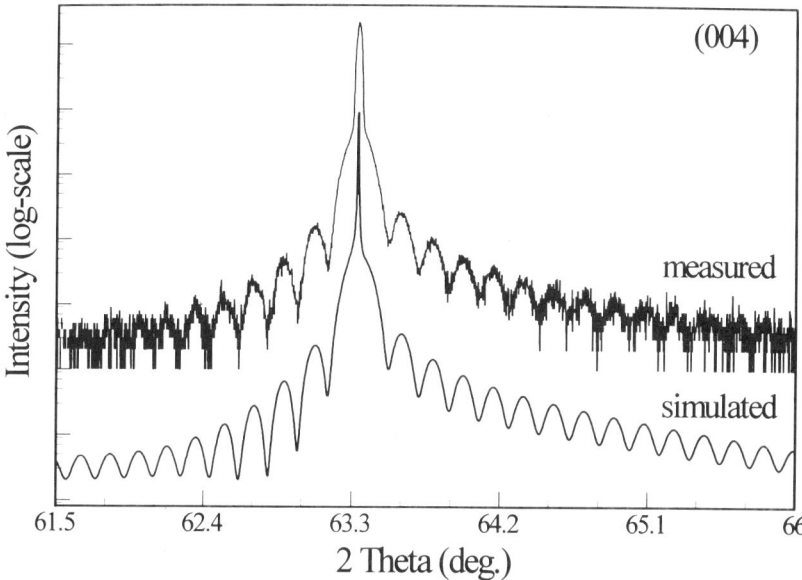

Fig. 6 XRD- {400} rocking curve of the highly strained GaInP/InP-heterostructure

The influence of the In-content x in the 0.5nm thick Ga_xIn_{1-x}P-layer on the XRD- rocking curve is simulated in figure 7. For x = 0.5 we obtained the best fit to the experimental data. The XRD-simulations and -results confirm clearly that the model with the asymmetric interfaces is a good description of the highly strained GaInP-layer embedded in InP.

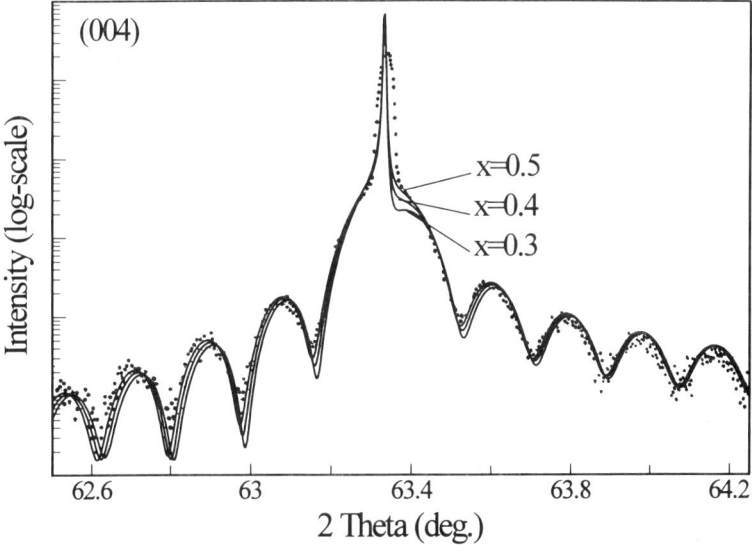

Fig. 7 Influence of different In-contents x in the 0.5nm thick GaInP layer on the simulated XRD- {400}-rocking curve

4. THIN $Ga_{0.5}In_{0.5}P$ LAYERS IN InP - BASED HFET - STRUCTURES

Thin $Ga_{0.5}In_{0.5}P$ layers have been introduced in InAlAs/InGaAs/InP-HFET - structures as wide band-gap spacer materials (see figure 8). The transport properties of the two dimensional electron gas (2DEG) in the channel should not suffer because the - in growth direction - lower interface to $Ga_{0.5}In_{0.5}P$ can be expected to be abrupt. The graded upper interface does not affect the device performance.

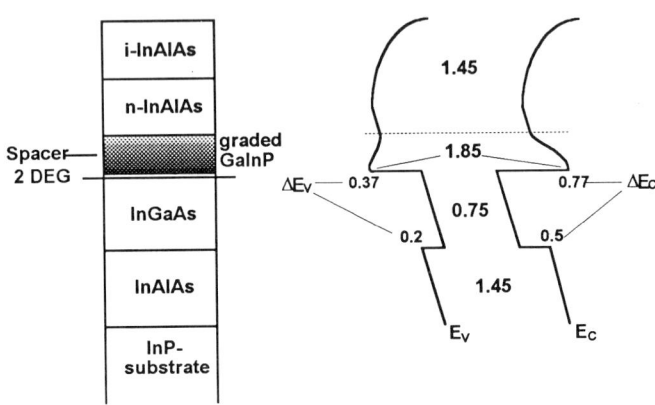

Fig. 8 Band structure of a InAlAs/InGaAs/InP-HFET with highly strained $Ga_{0.5}In_{0.5}P$ layer

This is proved by measured 2DEG transport data (room temperature) giving a sheet density of $n_s = 2.3 \; 10^{12} cm^{-2}$ and a mobility of $\mu_H = 11300 cm^2/Vs$. These data are comparable to data obtained from HFETs without $Ga_{0.5}In_{0.5}P$ layers. However, there are several advantages due to the use of the $Ga_{0.5}In_{0.5}P$ spacer layers: The increased valence band discontinuity ΔE_V yields a suppression of the hole gate leakage generated by impact ionisation. Improvements in DC and RF performance especially for high voltage applications (e.g. high breakdown voltages up to 10V) have been achieved. We measured cut-off frequencies $f_{max} > 150GHz$ for both, high drain bias and high drain current in devices with $0.7\mu m$ gate length. A detailled discussion of DC and RF performance is given by Prost et al 1995 and Scheffer et al 1995.

REFERENCES AND ACKNOWLEDGEMENTS

Chou C T, Anderson S C, Cockayne D J H, Sikorski A, Vaughan M R
    1994 Electron Microscopy Vol 1 905
Lakner H, Maywald M, Balk L J, Kubalek E 1992 Surface and Interface Analysis 19 374
Lakner H, Ungerechts S, Kubalek E 1995 in this proceedings
Prost W, Scheffer F, Liu Q, Lindner A, Lakner H, Gyuro I, Tegude F J
    1995 J Crystal Growth 146 538
Scheffer F, Lindner A, Liu Q, Heedt C, Reuter R, Prost W, Lakner H, Tegude F J
    1995 J Crystal Growth 145 326

The financial support of the Deutsche Forschungsgemeinschaft within SFB 254 is gratefully ackowledged.

Inst. Phys. Conf. Ser. No 146
Paper presented at Microsc. Semicond. Mater. Conf., Oxford, 20–23 March 1995
© *1995 IOP Publishing Ltd*

Study by transmission electron microscopy of GaInSb layers grown on (001) GaAs substrates by molecular beam epitaxy

G Aragón, M J de Castro, J J Pérez-Camacho*+, F Briones* and R García

Departamento de Ciencia de los Materiales e Ingeniería Metalúrgica y Química Inorgánica. Universidad de Cádiz. Apdo 40, Puerto Real, 11510-Cádiz. Spain.
*Instituto de Microelectrónica de Madrid. Serrano 144. 28006-Madrid. Spain.
+Present address: Optronics Ireland Research Centre. Department of Pure and Applied Physics. Trinity College Dublin. Dublin 2. Ireland.

ABSTRACT: The defect structure of GaInSb layers has been studied by Transmission Electron Microscopy. A change with In-content for the defect structure is observed. 60°-type threading dislocations appear when the In-content is less than 40%, whilst vertical-type threading dislocations appear when the In-content is greater than 40%. This defect structure suggests that the cation mixing (In/Ga ratio) seems to be an adequate candidate to be correlated to the appearance of vertical dislocations.

1.INTRODUCTION

Antimonide compounds have received increasing attention as the alternative materials for devices in the mid-infrared region of the spectrum, with a variety of applications in optical gas sensors and communication systems based on fluoride fibres. Mid-infrared emitters and detectors are required to operate efficiently at (or close to) room temperature. Heterostructure GaInAsSb/AlInAsSb lasers, lattice matched to GaSb substrate, emitting at wavelengths beyond $2\mu m$ (300K), have been reported (Caneau et al 1985, Choi and Eglash 1992).

Mismatched heterostructures have also proved to be appropiate for these purposes. High detectivity InAsSb mid-infrared photodiodes has been obtained on GaAs and Si substrates (Dobbelaere et al 1992, Dobbelaere et al 1993). Besides the application of this material system to photodetectors it is also used to fabricate light emitting diodes. The GaInSb ternary alloy provides another material available for device operation in the 2-7μm range.

Semiconductor heterostructures containing GaInSb layers will inevitably be strained since this ternary compound is not lattice-matched to any III-V substrate. Pascal-Delannoy et al (1992) reported the room temperature operation of GaInSb (40% In-content) p-i-n photodiodes with a 2% of mismatch on GaSb substrates. We have recently demonstrated (Pérez-Camacho 1994) a similar performance in GaInSb (40% In-content) p-i-n photodiodes grown on GaAs substrates despite the larger mismatch (10%). The assumption that larger lattice mismatch generates higher dislocation density seems natural. Dislocations are expected to be electrically active as recombination centres. The main consequences of dislocations in photodiodes are to generate an excess dark current (Longeway and Smith 1988) and to reduce the minority carrier lifetime, with the subsequent decrease in photosensitivity. A Transmission Electron Microscopy (TEM) study provides a deeper insight for a proper understanding of defect structure and its possible correlation with the operation of photodiodes.

2.EXPERIMENTAL

A series of $Ga_{1-x}In_xSb$ samples was grown by Molecular Beam Epitaxy (MBE) on (001) GaAs semi-insulating substrates, ranging from x=0 to x=0.4. A GaAs buffer layer was grown (at 580°C) in order to provide a smooth GaAs surface for epitaxy. The $Ga_{1-x}In_xSb$ layers were grown at 420°C under Sb_4 flux slightly higher than the minimum required for a Sb-stabilized growth.

The crystal defect structure was investigated by TEM in the two <110> cross section and planar view. These samples were thinned by mechanical polishing and Ar+-ion milling. The TEM observations was performed with a JEOL 1200-EX at accelerating voltage of 120kV.

3.RESULTS

3.1 GaSb layers grown by MBE at 420°C

The defect structure in GaSb epilayers consists of 60° threading dislocations as shown in fig. 1. The threading dislocation density was $8 \cdot 10^8$ cm^{-2} which is of the same order of magnitude as GaAs on Si epitaxies (Fang et al 1990) in spite of different lattice mismatches between GaSb on GaAs (8%) and GaAs on Si (4%).

Misfit dislocations have been observed by weak beam as shown in fig. 2. They are characterized as 90° dislocations and their spacing corresponds to a relaxed GaSb epilayer. Moreover, half-period-shifts of 90° misfit dislocations are observed (fig. 2). These shifts have been explained as the interaction with 60° misfit dislocations (Zhu and Carter 1990) and they have been observed early in GaSb epilayer growth at different temperatures and thickness (Rocher et al 1990, Kang et al 1994).

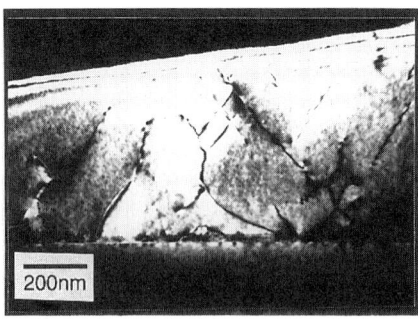

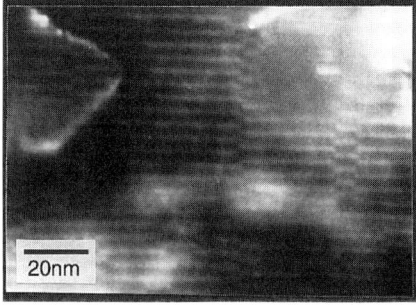

200nm

20nm

Fig. 1. XTEM dark field micrograph with {220} reflexion showing the 60° threading dislocation distribution in the GaSb epilayer.

Fig. 2. PVTEM weak beam micrograph with {220} reflexion showing 90° misfit dislocations and the half-period-shifts between them in the GaSb/GaAs interface.

3.2 Ga$_x$In$_{1-x}$Sb layers grown by MBE at 420°C (x=0.07)

Fig. 3 shows the defect structure in the epilayer of GaInSb with 7% In-content. The defect structure consists of 60° threading dislocations and is similar to the GaSb epilayer previously described.

3.3 Ga$_x$In$_{1-x}$Sb layers grown by MBE at 420°C (x=0.4)

The defect structure in GaInSb with 40% In-content is now totally different in comparison with one in GaInSb with 7% In-content. The defect structure consists mainly of a high density of threading dislocations which run nearly parallel to the growth direction as shown in fig. 4. These vertical threading dislocations are possibly edge-type dislocations (Tamura et al 1992).

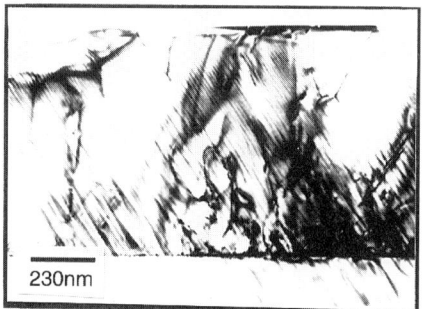 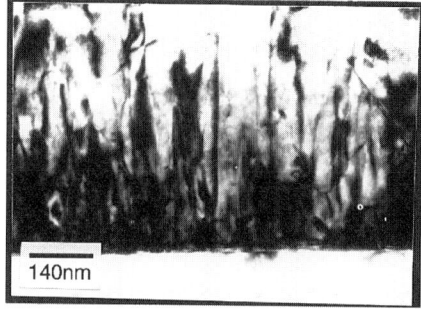

Fig. 3. XTEM bright field micrograph with {220} reflexion illustrating 60° threading dislocations in the GaInSb epilayer with 7% In-content.

Fig. 4. XTEM bright field micrograph with {220} reflexion illustrating a high density of vertical threading dislocations in the GaInSb eiplayer with 40% In-content.

4. DISCUSSION

According to our TEM study, the defect structure of GaInSb layers on GaAs(001) changes from 60°-type to edge-type threading dislocations when the In-content increases. The observed sequence of defect structure according to In-content in GaInSb on GaAs is similar to the defect structure in GaInAs on GaAs (Chang et al 1989, Tamura et al 1992). In the GaInAs/GaAs system, vertical-type threading dislocations appear for an In-content higher than 30%. Chang et al (1989) and Tamura et al (1992) suggested that the smaller the islands, the larger the number of vertical-type threading dislocations. However, the coalescence of many small islands may explain the high density of threading dislocations, but may not explain the generation of vertical-type threading dislocations. This type of threading dislocations may be generated from the coalescence of islands with different thickness where the spacing of 90° misfit dislocations is somewhat different due to the dependence of their spacing upon island thickness (Kang et al 1994). The interaction between 90° misfit dislocations may generate isolated 90° misfit dislocations which have to thread up to surface. The threading of 90° misfit dislocations is dominant where small, thin islands have many occasions of coalescence for growth conditions given by high Ga/In ratio (Ga/In approximately one) and low substrate temperature (420°C).

5.CONCLUSION

The defect structure of GaInSb layers grown on (001) GaAs substrates was studied by TEM. Vertical-type threading dislocations appear in GaInSb for high In-content. This type of theading dislocation generates from the interaction between 90°-misfit dislocations in small and thin islands.

ACKNOWLEDGMENTS

The samples were grown at the *Instituto de Microelectrónica de Madrid* and the TEM study was carried out at the Electron Microscopy Facilities of the *Universidad de Cádiz*. The present work has received financial support from the European Union ESPRIT III Project 6374 and from the *Junta de Andalucía* under group 6020.

REFERENCES

Caneau C, Srivastava A, Dentai A G, Zyskind J L and Pollack M A 1985 Electron. Lett. 21 815
Chang K H, Bhattacharya P K and Gibala R 1989 J. Appl. Phys. 66 2993
Chang S Z, Chang T C and Lee S C 1993 J. Appl. Phys. 73 4916
Choi H K and Eglash S J 1992 Appl. Phys. Lett. 61 1154
Dobbelaere W, De Boeck J, Bruynseraede C, Mertens R and Borghs G 1993 Electron. Lett. 29 890
Dobbelaere W, De Boeck J, Heremans P, Borghs G, Luyten W and Van Landuyt J 1992 Appl. Phys. Lett 60 3256
Fang S F, Adomi K, Iyer S, Morkoç H, Zabel H, Choi C and Otsuka N 1990 J. Appl. Phys. 68 R31
Kang J M, Nouaoura M, Lassbatère L and Rocher A 1994 J. Cryst. Growth 143 115
Longeway P A and Smith R T 1988 J. Cryst. Growth 89 519
Pascal-Delannoy F, Bougnot J, Allogho G G, Giani A, Gouskov L and Bougnot G 1992 Electron. Lett 28 531
Pérez-Camacho J J 1994 PhD Thesis Universidad Politécnica de Madrid
Rocher A, Da Silva F W O and Raisin C 1990 Revue Phys. Appl. 25 957
Tamura M, Hashimoto A and Nakatsugawa Y 1992 J. Appl. Phys. 72 3398
Zhu J G and Carter C B 1990 Phil. Mag. A62 319

Inst. Phys. Conf. Ser. No 146
Paper presented at Microsc. Semicond. Mater. Conf., Oxford, 20–23 March 1995

Quantitative convergent beam electron diffraction (CBED) from III-V semiconductor heterostructures and interfaces with nanoscale spatial resolution

H Lakner, S Ungerechts and E Kubalek

Werkstoffe der Elektrotechnik, Sonderforschungsbereich 254,
Gerhard-Mercator-Universität Duisburg, D 47048 Duisburg, F R Germany

ABSTRACT: We investigated cross-sectional specimens from ternary and quaternary heterostructures of $Ga_xIn_{1-x}As_yP_{1-y}$ on InP- or GaAs- substrates by CBED using subnanometer electron probes in a field-emission STEM. The quality of the obtained CBED patterns is sufficient to perform local strain measurements with 1 nm spatial resolution and with a sensitivity of $(\Delta a/a)_\perp \approx 10^{-3}$. This is proved by a CBED line-scan across an alternately strained quaternary superlattice. CBED patterns recorded at interfaces directly exhibit symmetry violations. Further simulations are necessary for a detailed quantitative understanding of CBED patterns from internal interfaces.

1. INTRODUCTION

Epitaxially grown ternary and quaternary heterostructures of $Ga_xIn_{1-x}As_yP_{1-y}$ on InP- or GaAs- substrate are increasingly used for the fabrication of e.g. optoelectronic devices. One of the key parameters for the performance of such devices is the crystalline quality and especially the amount of tetragonal distortion in strained layers on a nanometer scale. Convergent Beam Electron Diffraction (CBED) is in principle a very powerful technique for the investigation of local crystal properties. But in practice the applicability of CBED is often limited by a lack of ultimate spatial resolution and/or of sensitivity. The use of a field-emission electron gun for CBED offers two advantages: Nanometer and even subnanometer electron probes can be formed which contain enough current to perform CBED-measurements with good signal-to-noise ratio within seconds of recording time. As a consequence, high spatial resolution and high sensitivity can be achieved simultaneously. Additionally, a field-emission gun is a coherent electron source so that coherent CBED patterns can be generated. It is the intention of this contribution to demonstrate the performance of CBED in detailed crystal structure analysis on the nanometer or even subnanometer scale.

2. EXPERIMENTAL DETAILS

In our investigations we used a field-emission Scanning Transmission Electron Microscope 'STEM' (VG Microscopes: HB 501) equipped with a high-resolution pole-piece. The minimum diameter of the electron probe is approximately 0.3 nm. The CBED patterns are recorded by a single-crystal YAG-scintillator which is lens-optically coupled to a thermoelectrically cooled 512^2 pixel^2 high dynamic range (18 bit) CCD detector (EG&G: 1530-P). We investigated cross-sectional specimens of ternary and quaternary $Ga_xIn_{1-x}As_yP_{1-y}$ heterostructures on InP- or GaAs- substrate. Quantitative information from experimentally recorded CBED patterns was evaluated by comparison with dynamically simulated CBED patterns. For the simulations we used the Bloch-wave dynamical program developed by Spence and Zuo (1992). For the determination of strain we used the shift in the position of HOLZ lines (High Order Laue Zone) due to tetragonal distortion. All the CBED patterns shown here are recorded at 100 keV using a electron probe with an aperture half-angle of ≈19 mrad. With this electron probe we are able to resolve the 0.34 nm fringes in graphite by STEM bright-field imaging. Therefore we estimate the probe diameter (FWHM) to be considerably less than 1 nm. In order to have the interfaces in the cross-sectional specimens parallel to the optical axis and to avoid strong dynamical effects which exist e.g. in {1 1 0} zone axis orientation we tilted the specimens off the {1 1 0} pole towards an orientation close to the {3 2 0} zone axis. For our investigations we used specimen thicknesses in the range of 60-80 nm in order to get representative information on bulk material and to minimize the effects of surface relaxation (Chou et al 1994) or specimen preparation surface damage which create serious problems for CBED investigations in very thin specimens.

3. CBED-RESULTS USING SUBNANOMETER ELECTRON PROBES

An example for the quality of CBED patterns recorded with a subnanometer electron probe is given in figure 1. The upper left pattern is the as-recorded central CBED disc for GaAs near {3 2 0} orientation. The typical HOLZ lines are visible but appear blurred compared to the corresponding simulated pattern for an ideal detector (see lower right pattern). This can be explained by the contributions of inelastic scattering and by the effect of

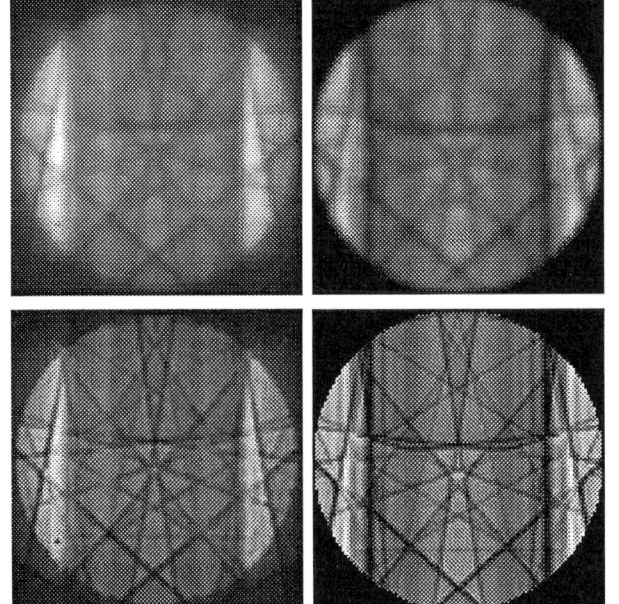

Figure 1: CBED pattern of GaAs near {3 2 0} orientation

Left: Experimental pattern (as-recorded)
Right: Simulation (point spread function of the detector is considered)

Left: Experimental pattern (high-pass filtered)
Right: Simulation for an ideal detector

the point spread function of the CBED detector used (Zuo 1994). Simple high-pass filtering applied to the experimental pattern reveals clearly the position of the HOLZ lines (see lower left pattern) and is therefore used to determine the position of the HOLZ lines in future. The simulated pattern in which the point spread function of the detector used is considered (see upper right pattern) fits much better to the experimental CBED disc. The remaining differences between experimental and simultated CBED discs can be attributed to inelastic contributions which are not considered fully in the simulations, and to the fact that neighbouring CBED discs overlap with the experimental pattern. The quality of the experimental CBED patterns obtained using subnanometer electron probes is sufficient to obtain quantitative information on strain by measuring shifts in the position of the HOLZ lines.

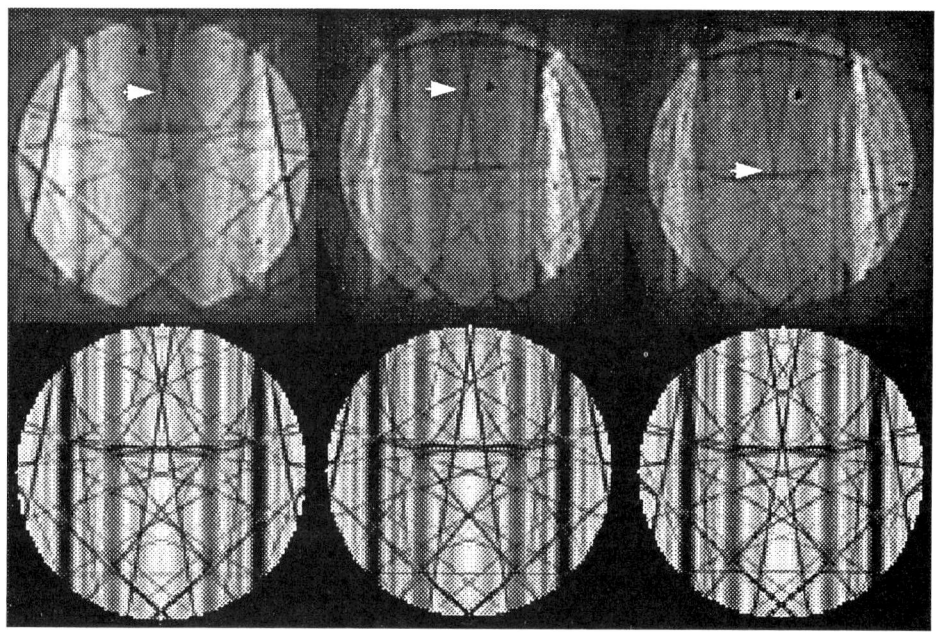

Figure 2: Experimental (top) and simulated (bottom) CBED-patterns of a alternately strained GaInAsP superlattice on InP (near {3 2 0} orientation)
Left: InP - substrate
Middle: $Ga_{0.26}In_{0.74}As_{0.8}P_{0.2}$ with tetragonal distortion +1.7%
Right: $Ga_{0.5}In_{0.5}As_{0.8}P_{0.2}$ with tetragonal distortion -1.3%

In order to determine the spatial resolution and sensitivity of the applied technique with respect to local determination of strain we investigated an alternately strained $Ga_{0.26}In_{0.74}As_{0.8}P_{0.2}$ / $Ga_{0.5}In_{0.5}As_{0.8}P_{0.2}$ superlattice grown on InP. We found no hints for strain relaxation. Figure 2 shows the CBED patterns obtained from the centers of the different layers (top row). The corresponding simulations are shown in the bottom row. The point of intersection (see arrows) of one pair of HOLZ lines is shifted into opposite directions compared to InP. The simulations with the best fit to the experimental results indicate the presence of tetragonal distortion of +1.7% and -1.3% respectively. As a next step we recorded a CBED line-scan across the alternately strained superlattice. The path of the linescan is marked in figure 3a. The electron probe position was moved in 0.66 nm steps while the 30 CBED patterns were recorded. The shifts in the HOLZ lines were converted into quantitative values for strain and are shown in figure 3b. The interfaces appear to be asymmetric and not abrupt. But the line scan shows clearly that a spatial resolution of ≈ 1 nm and a sensitivity of $(\Delta a/a)_\perp = 10^{-3}$ were achieved.

An interesting effect is exhibited by the CBED patterns recorded directly at the interfaces (the corresponding probe position is marked by arrows in figure 3a and 3b). The symmetry in the intensities of the HOLZ lines within each pattern is violated (see marking in figure 3c) but the two patterns exhibit a mirror symmetry with respect to each other. This results demonstrates clearly that the subnanometer probe CBED patterns are sensitive to interface structure and properties. For a detailed understanding it will be necessary to model the interface by corresponding multislice CBED simulations.

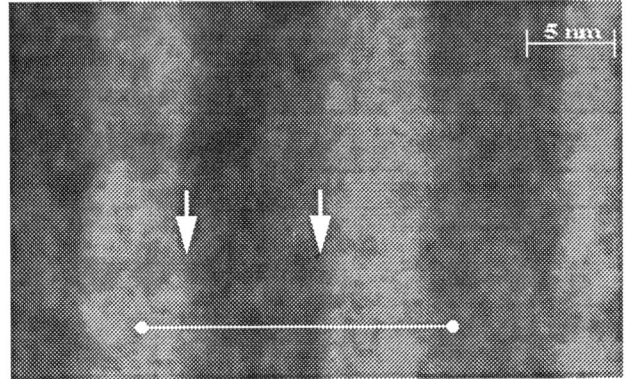

Figure 3a):
STEM micrograph of a quaternary $Ga_{0.26}In_{0.74}As_{0.8}P_{0.2}$ / $Ga_{0.5}In_{0.5}As_{0.8}P_{0.2}$ superlattice

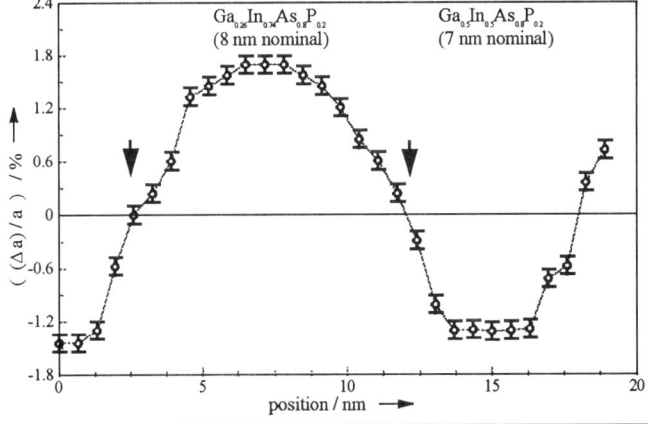

Figure 3b):
Strain profile across the heterostructure shown above.
The trace of the line-scan is marked in fig. 3a.

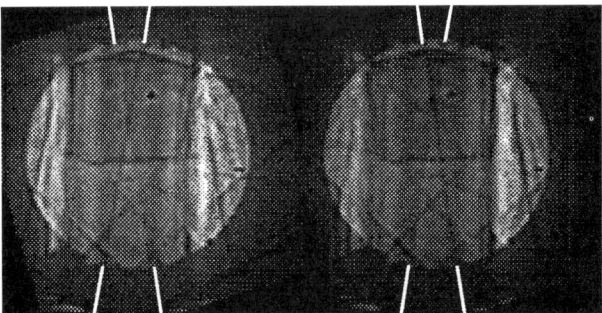

Figure 3c):
CBED patterns recorded at the interfaces (see arrows in fig 3a and 3b).
The symmetry within the patterns is violated (see marking).

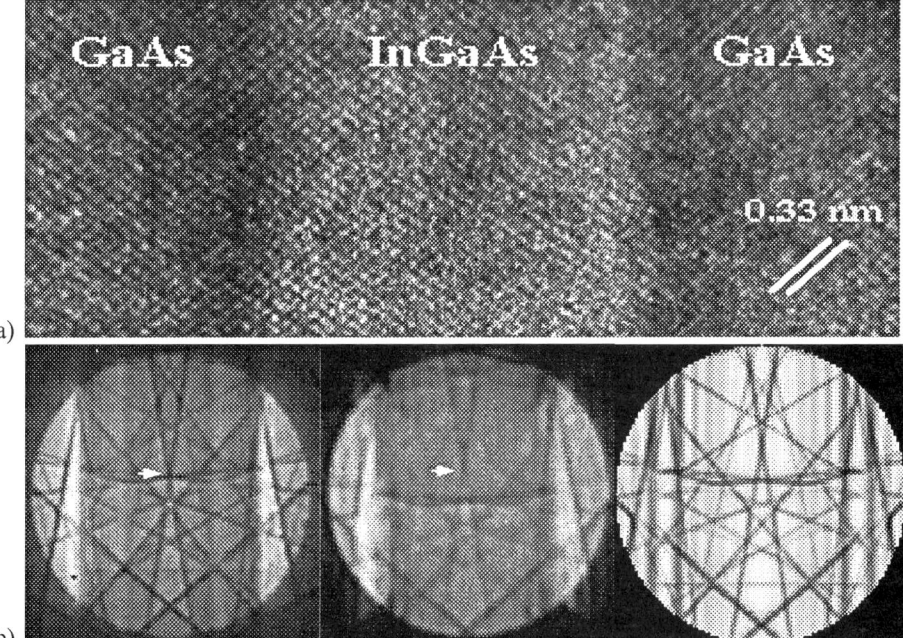

Figure 4a) STEM micrograph of a strained InGaAs/GaAs heterostructure
   b) CBED patterns from the heterostructure shown above
   Left: Experimental pattern recorded in GaAs
   Middle and right: Exp. and simulated patterns for the strained InGaAs layer

CBED patterns recorded in strained layers can also be used to determine the chemical composition of the strained layer (Pike et al 1991). An example of this is given in figure 4. The high resolution bright-field image for a {1 1 0} oriented specimen shows the strained InGaAs layer embedded in GaAs. Again the specimen was tilted close to the {3 2 0} orientation in order to perform the strain measurement. By comparison with simulations (see fig. 4b) we measured a tetragonal distortion of 1.2%. Due to the fact that no hint of relaxation is present we converted the measured tetragonal distortion into a chemical composition of $x = 0.096 \pm 0.008$ for the $In_xGa_{1-x}As$ layer. This is an accuracy which is very difficult to achieve with other analytical techniques using subnanometer electron probes. Additionally, small thickness variations in the specimens have little influence on the determination of chemical composition by CBED compared to energy dispersive x-ray spectroscopy or electron energy-loss spectroscopy.

4. DISCUSSION

The combination of a field-emission STEM and a high dynamic range (18 bit) CCD detector for CBED allows the recording of high quality diffraction patterns using nanometer and even subnanometer electron probes. The application to semiconductor heterostructures enables us to study individual ultra-thin layers of cross-sectional specimens by CBED. Dynamical simulation series for heterostructures with a systematically varied amount of tetragonal distortion and corresponding calibration curves allow a quick conversion of measured HOLZ line shifts into quantitative values for strain. Measurements in individual layers and mapping of strain with a nm-resolution and a sensitivity of $(\Delta a/a)_\perp \approx 10^{-3}$ are possible. This is demonstrated by CBED line-scans across a $Ga_xIn_{1-x}As_yP_{1-y}$ heterostructure. The achieved high spatial resolution of the CBED technique allows the characterization of interfaces with regard to local strain variations. In some cases of pseudomorphic heterostructures (no relaxation present) the results of strain mapping can be converted into chemical composition maps showing an accuracy and sensitivity which exceeds other analytical techniques like energy dispersive x-ray spectroscopy or electron energy-loss spectroscopy.

Additionally in CBED patterns recorded directly at interfaces, violations of the symmetry within the diffraction patterns can be seen. This may offer new possibilities for the characterization of internal interfaces by CBED. Unfortunately, internal interfaces can not be modeled with the Bloch-wave programs used. For a detailed understanding of CBED patterns recorded directly at interfaces it will be necessary to use multislice simulation programs for CBED which allow to model both perfect and imperfect interface structures.

The results demonstrate that CBED on the nanometer scale is a very powerful inspection tool for heterostructure characterization. It is complementary to the already available imaging and chemical analysis techniques (Lakner et al. 1992) in scanning transmission electron microscopy.

Further improvements can be expected by the application of energy-filtering in CBED which can be performed with the present STEM instrumentation according to the techniques described by Pike (1993), by Holmestad et al (1992) and by Krivanek et al (1994). As a consequence of energy-filtering, structure factor measurements will be possible even when using nanometer or subnanometer probes.

REFERENCES

Chou C T, Anderson S C, Cockayne D J H, Sikorski A, Vaughan M R
    1994 Electron Microscopy Vol 1 905
Holmestad R, Marthinsen K, Høier R 1992 Electron Microscopy Vol 1 185
Krivanek O L, Bui D N, Ray D A, Boothroyd C B, Humphreys C J
    1994 Electron Microscopy Vol 1 167
Lakner H, Maywald M, Balk L J, Kubalek E 1992 Surface and Interface Analysis 19 374
Pike W T et al 1991 J. Cryst. Growth 111 925
Pike W T 1993 Ultramicroscopy 51 117
Spence J C H, Zuo J M 1992 Electron Microdiffraction Plenum Press New York and London
Zuo J M 1994 Electron Microscopy Vol. 1 215

ACKNOWLEDGEMENTS

The authors like to thank Dr. J. M. Zuo (Arizona State University) for the provision of the Bloch-wave simulation program and for fruitful discussions.

The financial support of the Deutsche Forschungsgemeinschaft within SFB 254 is gratefully ackowledged.

Inst. Phys. Conf. Ser. No 146
Paper presented at Microsc. Semicond. Mater. Conf., Oxford, 20–23 March 1995
© *1995 IOP Publishing Ltd*

TEM method of study of (AlGa)As and (InGa)(AsP) heterostructures on profiled substrates

J Kątcki, J Ratajczak, A Maląg and M Piskorski

Institute of Electron Technology, Al. Lotników 32/46, 02-668 Warsaw, Poland

ABSTRACT: (AlGa)As and (InGa)(AsP) heterostructures epitaxially grown by liquid phase epitaxy (LPE) on V-grooved substrates of GaAs and InP respectively were studied by cross-sectional TEM. Preparation of TEM cross-sections requires gluing together two pieces of the specimen with epoxy. During ion milling the epoxy is etched away faster than the semiconductor leaving the surface of heterostructure uncovered. In the next stage of specimen thinning an ion beam etches the top layers of the heterostructure. It is of great importance to measure the distance of interfaces from the real surface of the heterostructures. In order to avoid excessive etching of the heterostructure surface and to be sure that the distance is measured accurately from the top of heterostructure, a thin layer of poly-silicon was deposited onto the surface of heterostructure. Low-temperature processing was used which did not cause any surface reaction. The thickness of poly-Si was about 600 nm. This was enough to mark the surface of heterostructure. During specimen preparation the poly-Si layer adhered well to the surface of heterostructure.

1. INTRODUCTION

Liquid phase epitaxy (LPE) is a very old technique but still very useful when an epitaxial layer is to be grown on a profiled substrate. The local crystallographic orientation of such a substrate is not the same at every point on the surface, this causes a local difference in epitaxial growth rate.

The purpose of our study was to find a reliable method of evaluating the geometry of III-V compound heterostructures grown on profiled substrates. Two heterostructures were chosen (AlGa)As/GaAs and (InGa)(AsP)/InP. Cross-sections of these heterostructures were studied in a transmission electron microscope.

Cross-sectional transmission electron microscopy is a very popular tool for study of multilayer structures. However, the investigation of regions lying close to the surface is frequently difficult. Being situated close to the epoxy layer a subsurface region of heteroepitaxial structure can be easily removed during ion milling of the cross-sectional TEM specimen. Possible erosion of the subsurface region makes geometrical measurements which start at the surface of heterostructure uncertain.

Erosion of subsurface layers of heterostructure during ion milling is even more drastic when the milled sample has a complicated surface as in case of the growth on a profiled substrate. In order to protect the subsurface layer(s) of heterostructure we deposited a thin polysilicon layer on the top surface of the heterostructure.

2. EXPERIMENTAL

As a substrate material (100)-oriented GaAs and InP wafers were chosen. Using a standard photolithographic procedure V-grooves were chemically etched in both materials. The grooves were oriented in a <110> type direction. To prepare a V-grooved substrate of GaAs a $NH_4OH:H_2O_2:H_2O = 20:7:973$ solution was used. V-shaped grooves in InP substrates were prepared by chemical etching in a $H_3PO_4:HCl = 1:3$ solution. (AlGa)As and

(InGa)(AsP) heterostructures were epitaxially grown by the LPE method on V-grooved substrates of GaAs and InP respectively. Four epitaxial layers were grown on V-grooved substrates of GaAs. They were: $Al_xGa_{1-x}As$:Sn, GaAs, $Al_xGa_{1-x}As$:Ge and GaAs:Ge. On the profiled InP substrate the following layers were grown: InP:Sn, InGaAsP, InP:Zn and InGaAsP:Zn.

Before preparing the TEM cross-sectional specimens a 600 nm thick polysilicon layer was deposited on the top surfaces of both (AlGa)As and (InGa)(AsP) heterostructures. Polysilicon was deposited by RF magnetron sputtering at a deposition rate of 0.10-0.12 nm/s. During deposition the substrate temperature was below 200°C.

To prepare cross-sectional specimens for transmission electron microscopy study a procedure originally developed in our laboratory was employed. For several years the procedure has been successfully applied to prepare specimens from processed silicon wafers. The first step of our procedure was cutting rectangular pieces from a wafer and gluing those pieces together face-to-face with epoxy. A block formed by two pieces of wafer (1) was mounted in a specially prepared aluminum rod (2) 3 mm in diameter (Fig. 1a). Each rod had a milled groove 0.6-0.8 mm thick. This allowed the block of heterostructure to be tightly mounted inside the groove using epoxy. The rod was then cut into disk-shaped slices (Fig. 1b). The thickness of each slice (3) was about 200 μm. Using this method it is possible to cut 10-15 slices from one rod. The slices were then mechanically thinned to 100 μm and dimpled to reach a thickness in the centre of a sample of approximately 20 μm. Finally the specimen was ion milled in Ar^+ ions at 5 kV. In the case of InP/InGaAsP heterostructures the specimens were cooled with liquid nitrogen during ion milling. Additionally, in order to avoid artifacts in a final stage of ion thinning an iodine ambient was applied.

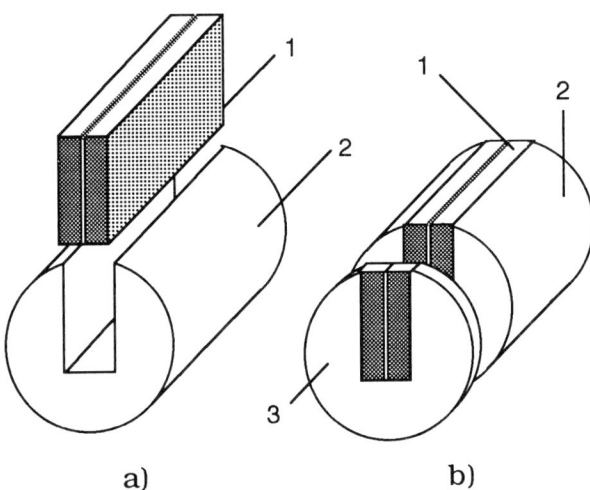

a)   b)

Fig. 1. Initial stages of specimen preparation of TEM cross-sections: a) inserting two pieces of heterostructure glued together, b) cutting specimen slices

3. RESULTS AND DISCUSSION

Surface irregularities such as edges, steps and grooves are more susceptible to damage during ion milling than the flat-surface heterostructures. The top surface of the heterostructure grown on a profiled substrate is not flat especially close to the groove. In the case of removing epoxy by ion beam milling, the protruding areas of the heterostructure surface were eroded first. To protect these surface areas the heterostructures were covered by low-temperature polysilicon.

Polysilicon is used as a gate material in the semiconductor industry. During ion milling it behaves similarly to other layers of semiconductor structures. Vanhellemont et al (1983) successfully applied a thin-layer of low-pressure chemical vapor deposited (LPCVD) polysilicon layer to protect oxidized silicon structures.

RF magnetron sputtering was carried out at a temperature below 200°C. Therefore, covering both (AlGa)As and (InGa)(AsP) heterostructures with polysilicon using this technique did not cause interfacial reactions between polysilicon and the heterostructure material. On the other hand, polysilicon adhered well to the surface and protected the heterostructure against possible erosion by the ion beam. No stresses were observed either in polysilicon or the substrate/heterostructure.

In Fig. 2 an AlGaAs layer grown at the edge of the groove is shown. The epitaxial layer gradually fills the groove causing smoothing of the surface. During the epitaxy small symmetrically positioned grooves were formed on both sides of the former groove (Fig. 2a). One of these, indicated by an arrow, is visible in Fig. 2b. The entire surface of the AlGaAs layer is covered with a polysilicon layer of equal thickness. The polysilicon layer bonded very tightly to the surface protecting it against unwanted erosion by the ion beam.

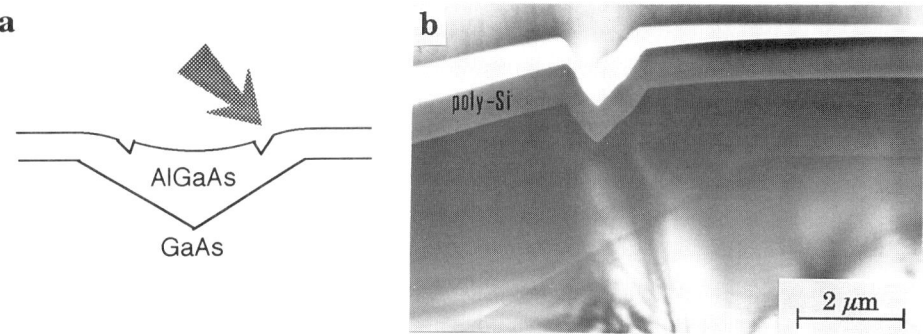

Fig. 2. An AlGaAs layer grown on a V-grooved GaAs substrate: a) schematic drawing, (b) cross-sectional TEM micrograph (the area shown is indicated with an arrow on a schematic drawing)

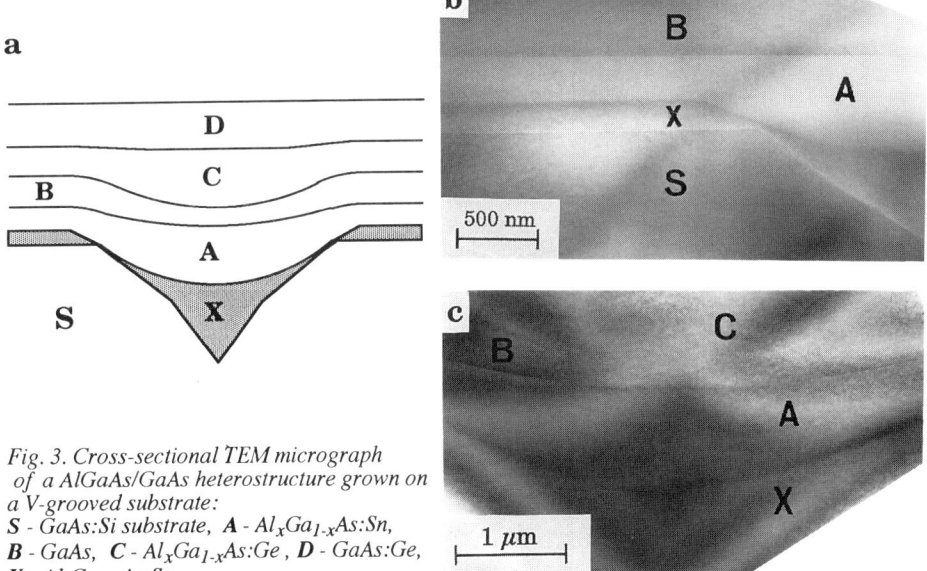

Fig. 3. Cross-sectional TEM micrograph of a AlGaAs/GaAs heterostructure grown on a V-grooved substrate:
S - GaAs:Si substrate, A - $Al_xGa_{1-x}As$:Sn, B - GaAs, C - $Al_xGa_{1-x}As$:Ge, D - GaAs:Ge, X - $Al_yGa_{1-y}As$:Sn

276

In Fig. 3a a schematic drawing of a GaAs/AlGaAs laser heterostructure is shown. Cross-sectional TEM micrographs of interesting areas of such a heterostructure are shown in Fig. 3b and 3c. Subsequent layers are visible in the micrograph. Both in Fig. 3a and 3b an additional $Al_yGa_{1-y}As$ layer (y<x) can be distinguished (denoted by **X**). This layer was formed due to the transition from melt etch to growth. Despite the fact that no chemical etching was applied, subsequent layers can be easily distinguished.

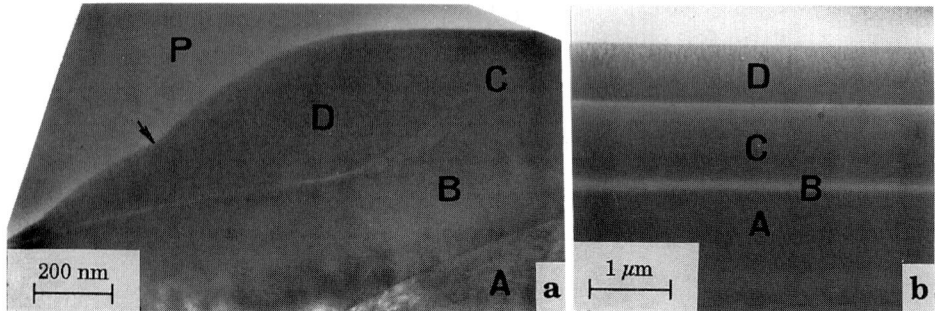

*Fig. 4. Cross-sectional TEM micrograph of InP/InGaAsP heterostructure grown on a V-grooved substrates: **S** - InP:Sn substrate, **A** - InP:Sn, **B** - InGaAsP, **C** - InP:Zn, **D** - InGaAsP:Zn, **P** - poly-Si*

A cross-sectional TEM view of an InGaAsP/InP heterostructure grown on a V-grooved substrate is shown in Fig. 4a. Under a polysilicon layer covering the V-groove edge a local perturbation of epitaxial growth can be found. During the growth, the edges of the groove were dissolved and new layers were formed at these sites. For comparison an area of the correct growth is shown in Fig. 4b. Subsequent layers are visible in the micrograph. The presence of polysilicon also allowed slight surface roughness at the groove edges (denoted by an arrow) to be revealed.

4. CONCLUSIONS

In order to study (AlGa)As and (InGa)(AsP) heterostructures cross-sectional TEM has been chosen. An original specimen preparation method has been presented. Our research proved that a polysilicon layer deposited on the surface of a heterostructures protects it against erosion by the ion beam. It is especially important when heterostructures are grown on profiled substrates. This allows easier study of subsurface layers of heterostructures.

ACKNOWLEDGMENT

This publication is based on work sponsored by the Polish Government under the project #88016.9102. The authors are very much indebted to Mr I Wójcik for polysilicon sputtering, Ms D Szczepańska for assistance in specimen preparation and Ms J Wiącek for careful preparation of micrographs.

REFERENCE

Vanhellemont J, Bender H, Claeys C, Van Landuyt J, Declerck G, Amelinckx S and Van Overstraeten 1983 Ultramicroscopy 11 pp 303-306

Inst. Phys. Conf. Ser. No 146
Paper presented at Microsc. Semicond. Mater. Conf., Oxford, 20–23 March 1995

X-ray topographic imaging and x-ray elastic measurements in the GaAs/Ge heteroepitaxial system

N Burle and B Pichaud

Laboratoire MATOP, URA CNRS No 1530, Universite Aix-Marseille III, Faculte des Sciences St Jérome, case 151, 13397 Marseille Cedex 20, France

ABSTRACT: The occurrence of two different regimes of misfit dislocation nucleation is clearly found in epitaxial system GaAs/Ge. As expected, the first one was identified as based on the development of threading dislocations from the substrate (the Matthews' model), and an hypothesis was given to explain the second one. It is shown that x-ray topography is perfectly suited to determine the mechanical state in the layer and the substrate of a low misfit system.

1. INTRODUCTION

It is well known that, in the case of heteroepitaxial growth of a film on a substrate, misfit dislocations (MD) are nucleated to relax extra stresses as soon as the elastic energy in the film exceeds the energy of MD formation. In the last few years, a lot of studies have revealed that the layers very often exhibit two stages of MD development: the first one is the mechanism suggested by Matthews et al (1976) involving threading dislocations coming from the substrate; this is now well defined and characterized. However, little is known about the other one: it can only be predicted that this second stage involves a high dislocation density so that complete relaxation can be achieved, but the mechanisms invoked in various materials seem to be different from each other.

The GaAs and Ge lattice parameters are very closed at room temperature (a_{GaAs}=5.6533Å, a_{Ge}=5.6576Å) so that, in the case of epitaxial growth of one on the other, the misfit is notably low ($7.6 \ 10^{-4}$) and allows good quality epitaxy, which implies a high critical layer thickness.

In these conditions (perfect epitaxy, thick layers), the development of misfit dislocations is very progressive, so that low MD densities can be observed and TEM observations are not convenient. Nevertheless, x-ray topography (XRT) is a large scale, non destructive, well suited technique to observe the successive steps of deformation both in the film and in the substrate.

2. EXPERIMENTAL DETAILS

GaAs layers were deposited by closed space vapour transport (Burle et al, to be published) onto (001) Ge substrates at growth rates of 0.8 or 0.4μm/min and cooling rates from 5 to 70°C/min.

These samples were studied by means of two x-ray topographic techniques, giving

278

complementary information (Fig. 1):
- X-ray transmission topography, using planes perpendicular to the interface. The images are a summation of the film and the substrate contributions, so that the whole sample is observed. A value of the radius of curvature of both the layer and the bulk, R, can be deduced from the angular displacement of the Bragg peak along the length of the sample.
- X-ray reflection topography, using planes slightly inclined relative to the interface. With a suitable wavelength, diffracted beams on (hkl) planes from the film and the substrate can be separated and their angular deviation $\Delta\varphi$ can be measured: in this way, one can obtain an image from the layer alone and the tetragonal deformation of the layer can be determined from $\Delta\varphi$.

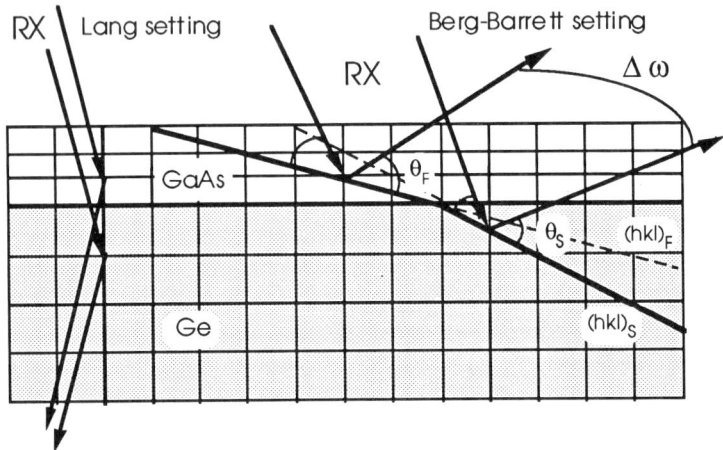

Figure 1: Epitaxial system and relative positions of reflecting plans in transmission and reflection conditions.

3. RESULTS AND DISCUSSION

Up to 40 samples with increasing layer thicknesses have been observed by XRT and may be classified in three regimes of thickness, illustrated in Fig. 2:
- regime A, below thickness t_a; no dislocations at all are developed in the interface, so that the deformation is purely elastic; $\epsilon_{xx}=\epsilon_{yy}$ in the interfacial plane = the misfit f;
- regime B, below thickness t_b; heteroepitaxial nucleation occurs following the Matthews' model; a maximal 1 % relaxation is reached;
- regime C; a second nucleation mechanism is observed giving rise to high dislocation densities.

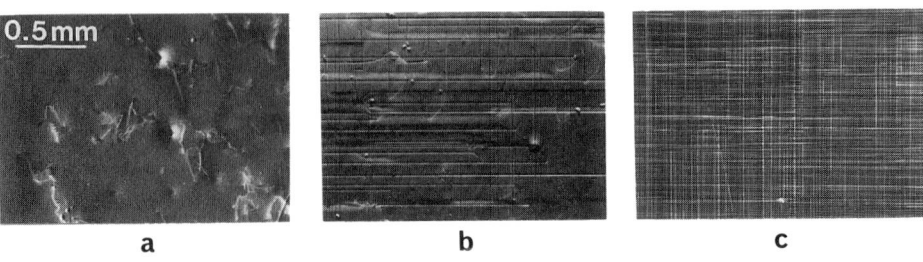

Figure 2: The three regimes observed with increasing layer thickness.

3.1 First mechanism

In case of equilibrium state, the step limit t_a would be equal to the theoretical critical thickness t_c given by the Matthews' model, ie $0.3\mu m$, which was not observed in most experimental cases. We obtained values of t_a between $0.8\mu m$ and $1.3\mu m$, depending on the cooling rate; $1.3\mu m$ is obtained with the usual "quick" cooling (70°C/min) (Burle et al 1993, and in press) while $0.8\mu m$ is related to the slowest cooling.

Between t_a and t_b, a few dislocations are nucleated from threading ones, as predicted by Matthews et al (1976). A particular shape was still noticed in a previous paper (Burle et al 1993): straight arms elongated in the interface are very often connected with arms developed in the substrate, so the defect has an hairpin shape. This unusual hairpin shape was observed in the Ge(B)/Ge system by Prokhorov et al (1993). They demonstrated, following Matthews, the existence of film critical thicknesses t_c^{II} and t_c^{III} is relative to dislocation development in the compression and tension parts of the substrate. Applying this model to our system we found that t_c^{II} and t_c^{III} are respectively equal to $0.6\mu m$ and $1.2\mu m$, in rather good agreement with our observations. This type of dislocation configuration, related to three critical thicknesses, has been rarely reported because it requires large thickness and metastable layers (no MD generation at t_c, in order to reach t_c^{III}), which implies very low misfit. Each threading dislocation can give only one interfacial line through the sample and so, due to the low density of dislocations in the Ge substrates, this step only allows 1% of the stress to be released.

3.2 Second mechanism

Beneath t_b a new nucleation process is then activated and a high density of dislocation can be obtained: at least 10^4 cm/cm^2, which is the resolution limit of XRT and corresponds to 10% relaxation. The extent of relaxation cannot be determined using XRT images, but it can be calculated from tetragonal parameter and curvature radius measurements. Until now, we have not determined whether t_b is a sharp limit but it can be estimated to be about $2.0\mu m$.

In a large number of samples we observed small segments of dislocations which seem to be sometimes connected to 60° MD (Fig. 3c). Their identification from the contrast they exhibit on x-ray topographs is not possible in detail, but in the most probable hypothesis these defects would be stacking faults (SF) extended from the interface, bounded by $a/6(\overline{1}12)$ partials (Fig. 3a). Such SF were observed by TEM in GaAs/Ge samples grown by MBE (Frigeri et al 1994) and are correlated with 60° glissile dislocations which would be MD. In X-ray topographs as well, the short segments are related to long MD lines (Fig. 3c). So, even if it is energetically unfavourable, one can imagine that this second nucleation mechanism (Fig. 3b) is due to the glide of an $a/2<110>$ dislocation given by the decomposition of the edge partial bounding stacking faults in the layer, as follows:

$$a/6(\overline{1}12) \rightarrow a/2(\overline{1}01) + a/6(21\overline{1}).$$

So even if it cannot be asserted that this decomposition is the actual 2nd mechanism implied by sample relaxation, it is a plausible candidate. In any case, whatever this mechanism could be, it might cost as much energy as the decomposition given here.

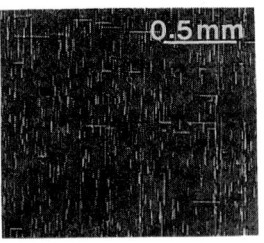 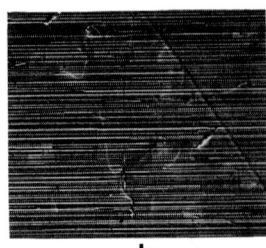 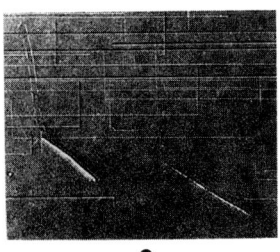

| a | b | c |

Figure 3: a) Small edge segments.   b) 2nd nucleation mechanism occurring.
c) Relation between a) and b) might be considered.

4. CONCLUSIONS

Several steps (thicknesses t_a, t_b) were observed by XRT for MD nucleation:
- below t_a, no dislocations at all are developed in the interface;
- between t_a and t_b a few dislocations are nucleated from threading ones, as predicted by Matthews et al (1976). The hairpin shape is relevant to metastability of the film which allows it to reach a critical thickness for dislocation development in the substrate, three times higher than expected from the Matthews criterion;
- beneath t_b a new nucleation process is activated and a high density of dislocations can be obtained. This second mechanism would be due to the glide of a $a/2 < 110 >$ dislocations given by the decomposition of the edge partial bounding stacking faults in the layer. Complete identification is in progress. In any case, edge defects as well as MD would be relevant to an homogeneous nucleation mechanism, which must be energetically justify.

Thus XRT is better than TEM in allowing the determination of a lot of information about the early stages of misfit dislocation nucleation in very low misfit epitaxial systems.

REFERENCES

Burle N, Pichaud B, Guelton N and Saint-Jacques R G 1993 Inst. Phys. Conf. Ser. **134**, 9, 573
Burle N, Pichaud B, Guelton N and Saint-Jacques R G (to be published in Thin Solid Films)
Frigeri C, Attolini G, Pelosi C and Longo F, 1st Conf. on Materials for Microelectronics, Barcelona, 17-19 Oct 1994 (to be published in Material Science and Technology)
Matthews J W, Blakeslee A E and Mader S 1976 Thin Solid Films **33**, 253
Prokhorov I A, Zakharov B G, Ma'shin V S and Shulpina I L 1993 J. Phys. D, Appl. Phys. **26**, A76

Inst. Phys. Conf. Ser. No 146
Paper presented at Microsc. Semicond. Mater. Conf., Oxford, 20–23 March 1995
© *1995 IOP Publishing Ltd*

Topographical, compositional, and dopant contrast from cleavage surfaces of GaAs-Al$_x$Ga$_{1-x}$As superlattices

MR Castell[1], DD Perovic[2], A Howie[3], DA Ritchie[3], C Lavoie[4] and T Tiedje[4]

1. Department of Materials, University of Oxford, Parks Road, Oxford OX1 3PH, UK.
2. Department of Metallurgy and Materials Science, University of Toronto, Toronto M5S 1A4, Canada.
3. Cavendish Laboratory, Madingley Road, Cambridge CB3 0HE, UK.
4. Department of Physics, University of British Columbia, 6224 Agricultural Road, Vancouver V6T 1Z1, Canada.

ABSTRACT: We present images of (110) cleavage surfaces of GaAs-Al$_x$Ga$_{1-x}$As superlattices obtained by scanning force microscopy (SFM) and field-emission scanning electron microscopy. Topographical information is mapped by secondary electrons (SEs) and SFM, compositional differences are imaged through SEs and backscattered electrons (BSEs), and information on dopant type is gained through SEs only. Models are presented explaining the contrast observed in each case.

1. INTRODUCTION

Semiconductor superlattices usually have periodicities of nanometre dimensions and have therefore predominantly been studied using transmission electron microscopy (TEM) and scanning transmission electron microscopy. These techniques provide information on dislocation density and type, crystal quality, nature of the superlattice interfaces, and periodicity of the materials.

With the development of immersion-lens field-emission scanning electron microscopes (FE-SEMs) which can achieve resolutions better than 1nm, it has become possible to image superlattices from cleaved cross-sections using secondary electrons (SEs) or backscattered electrons (BSEs) (Ogura 1991, Merli and Nacucchi 1993, Bleloch et al 1994, Perovic et al 1994). Scanning force microscopy (SFM) can also be used to investigate the topography of the superlattice cleavage surfaces and these results may be correlated with SEM images. In this paper experimental results are presented on contrast between layers in GaAs-Al$_x$Ga$_{1-x}$As superlattices that is due to either topography, atomic number differences or n-p type doping.

2. COMPOSITIONAL CONTRAST

It is well known that the BSE signal in the SEM is an increasing function of atomic number. BSE contrast can therefore be observed between layers in GaAs-Al$_x$Ga$_{1-x}$As superlattices as long as the probe size of the incident electron beam is sufficiently small and the molar fraction x is large enough. A BSE image of the (110) cleavage surface through a GaAs-Al$_x$Ga$_{1-x}$As heterostructure taken in a Hitachi S-4500 FE-SEM is shown in Fig.1a. The higher atomic number 20nm thick GaAs layers appear bright relative to the 20nm Al$_{0.5}$Ga$_{0.5}$As layers.

Detailed studies show that SE emission is closely related to the stopping power of the specimen (e.g. Bleloch et al 1989). Computations of the various contributions to stopping power in Al (Ashley et al 1979) show that at primary energies above 1keV about 60% of SE generation arises from excitation of the 2p and 2s shells while plasmon excitation and single electron valence excitations each contribute only around 20%. Further studies on ZnS using

100keV primaries (Pennycook and Howie 1980) also indicate that inner shell excitations dominate SE generation. Therefore, as long as the materials that are being compared do not differ in their valence configurations the high atomic number material should cause greater SE generation. This is seen in Fig.3a, taken in a Hitachi S900 FE-SEM, where the GaAs layers appear bright and the AlAs layers dark. The same superlattice that was imaged with BSEs in Fig.1a is shown imaged by SEs in Fig.1b. Here the GaAs layers again appear bright relative to the $Al_{0.5}Ga_{0.5}As$ layers, but there is a strong change in SE signal levels half way though the superlattice which is due to dopant effects, discussed later.

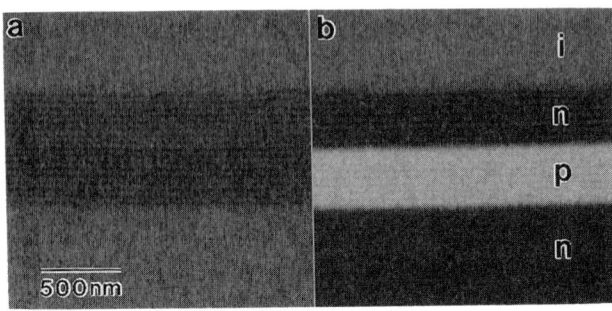

Fig.1 Identical area SEM images of a GaAs -Al_xGa_{1-x} As heterostructure imaged by BSEs at 30kV (a) and by SEs at 20kV (b). In (a) only the GaAs-$Al_{0.5}Ga_{0.5}As$ superlattice is visible. In (b) both the superlattice and doped regions can be seen.

3. DOPANT CONTRAST

In some doped materials a SE imaging effect may be observed that is related to a form of voltage contrast. SE contrast between differently doped materials has been reported by a number of researchers. Aven et al (1972) observe SE contrast between n and p doped regions of $ZnSe_xTe_{1-x}$ and Sawyer and Page (1978) have shown that SE imaging reveals contrast within SiC crystals caused by the distribution of trace impurities. In semiconductors, SE contrast may be observed between n, p, and i type material as seen in Fig.1b. The dopant concentrations are $2 \cdot 10^{18}$ carriers/cm^3 and in the micrograph the n, p, and i regions appear dark, bright and grey, respectively. As the atomic number differences between the differently doped regions are negligible as regards compositional contrast the effect must have another origin.

We explain the n-p-i contrast by examining the band structure and work functions of the doped materials. Surface states and/or surface contaminants pin the Fermi energy E_F of semiconductors within the band-gap with the result that for doped semiconductors the bands are bent within the depletion (or accumulation) region, as can be seen in Fig.2. For n-type material (Fig.2b) the doping causes E_F to be raised to the donor level in the bulk (at 0K), but at the surface the high density of states is filled to E_F, causing E_F to be pinned in the band-gap. Electrons that fill the surface states to E_F are drawn from the depletion region of the semiconductor, resulting in ionisation of atoms in the depletion region. A dipole is created, with a high negative charge density at the surface and a positive charge distribution in the depletion region. In p-type material (Fig.2a), E_F is lowered to the acceptor level in the bulk, and at the surface the bands are bent the opposite way to the n-type case, resulting in a positively charged surface layer. The vacuum levels far from the surface need to be equivalent, which causes bending of E_{vac} close to the surface.

After an SE has been generated within the accumulation region of p-type material, it will experience a force from the bulk to the surface by the Coulomb field of the positive surface charge discussed above. This field will increase the number of SEs that can escape from the surface. The escape depth for SEs is of the order of a few nanometres which is a significantly smaller value than the depth of the accumulation region which for $n_c = 10^{18}$ carriers/cm^3 in GaAs is approximately 15nm. This means that all SEs detected from the p-type regions will have been accelerated from the bulk towards the surface. As a result the SE yield of the p-type material will be increased relative to undoped regions. Conversely, as the field acts the opposite way in n-type material, the SEs in the depletion region will experience a force into the bulk and away from the surface, resulting in a lower SE yield than in the case of undoped or p-type material. When n and p doped material are in neighbouring regions, the vacuum levels of the materials are bent in the manner shown in Fig.2. So, although E_{vac} is equivalent for both types of material far away from the surface, the local work functions are different near the surface. An SE would be able to escape more easily through the type of graded barrier appropriate to p-type material than that of n-type material (Garcia 1995). This effect also enhances SE emission from p-type over n-type material.

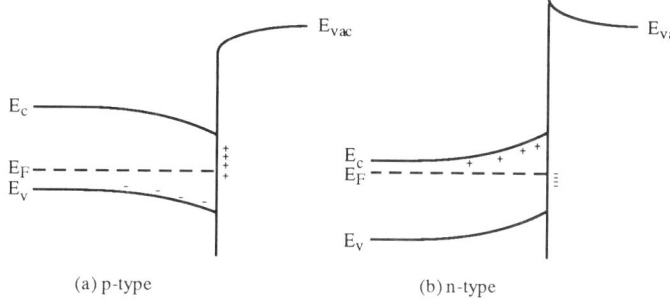

Fig.2 Band bending due to Fermi level pinning of the surface states. An SE would find it easier to escape from p-type material (a) than n-type (b).

(a) p-type   (b) n-type

4. TOPOGRAPHICAL CONTRAST

The topography of cleavage surfaces of large periodicity GaAs-AlAs superlattices was investigated with a Nanoscope II SFM, and the results compared with SE images taken at 20keV in a Hitachi S900. Figs.3a-c shows corresponding SFM and SEM images, as well as an SFM line profile, of a 200nm periodicity GaAs-AlAs superlattice. GaAs has an average atomic number of 32, compared with AlAs which has a value of 23, and this difference is large enough to cause SE contrast between the layers. The AlAs layers appear dark and the GaAs layers appear bright, as can be seen in the SE image in Fig.3a. The SFM image of the cleavage surface also shows the periodicity of the superlattice structure where the AlAs layers stand proud of the GaAs. The AlAs ridges are seen on both cleavage surfaces, so an effect like fibre pull-out in a fibre composite material is not occurring. It seems most likely that sample cleavage results in the AlAs layers being plastically deformed before fracturing, whereas the GaAs layers seem to cleave without undergoing plasticity. This results in the creation of AlAs ridges on both sides of the exposed cleavage surfaces, as shown schematically in Fig.3d. SE images of GaAs-AlAs superlattices under indentations also show that the AlAs layers deform plastically prior to fracture (Castell et al 1993). The behaviour of the GaAs-AlAs system seems analogous to that of a metal reinforced ceramic.

The probable sequence of events of the cleavage crack running through the GaAs-AlAs superlattice is as follows. When the crack reaches an AlAs layer it will not immediately be able to propagate into it as the stress field at the crack tip will not be great enough to cause fracture. This is because the fracture toughness is larger for AlAs than GaAs. As the layer interfaces are relatively weak, cracks will run perpendicular to the cleavage crack resulting in debonding at the interfaces. Initially the AlAs layer will experience tensile elastic deformation and the crack will re-nucleate in the GaAs on the far side of the AlAs layer. This situation is shown in Fig.3d. By now the AlAs layer has started to deform plastically and continues to do so until crack nucleation occurs and causes the layer to fracture. How many AlAs layers are being deformed at a given time is not known, but only one is shown in Fig.3d.

The type of cleavage mechanism discussed above is entirely different from the situation in a bulk sample of GaAs or AlAs. As the crack propagates through the superlattice there are continuous crack re-nucleation events as well as plastic deformation of the AlAs layers. This means that the fracture toughness of one of these types of superlattice should be greater than the average of its components. Size dependent effects are commonly observed in fracture mechanics and could explain why the AlAs layers behave differently from the bulk material.

The SE contrast across the layers in Fig.3a can be explained in the following way. The grey regions are GaAs, and the dark layers are AlAs, which is due to atomic number differences as discussed previously. The layer widths measured from the micrograph are 90nm and 109nm for the AlAs and GaAs layers, respectively. At the edges of the AlAs layers the image appears bright due to the increased SE yield at edges. The dark region immediately next to the bright edge is caused by a very low SE yield from the deep cavities due to delamination of the layers. The sample topography as imaged by SFM, is shown in Fig.3b. It is immediately apparent that the image has been adversely affected by the effects of shape of the tip. Where one would expect a trace of approximately the form of a square wave, the linetrace in Fig.3c shows that during scanning the SFM tip barely reaches the bottom of the troughs between the AlAs ridges. Although the SFM data cannot be viewed as an accurate representation of the true surface topography, it contains reliable information on the periodicity of the superlattice and the height of the AlAs ridges. The periodicity was measured to be 200nm, which may be compared with 199nm determined from the SE micrograph in Fig.3a. AlAs ridge heights were approximately 55nm, as can be seen in Fig.3c.

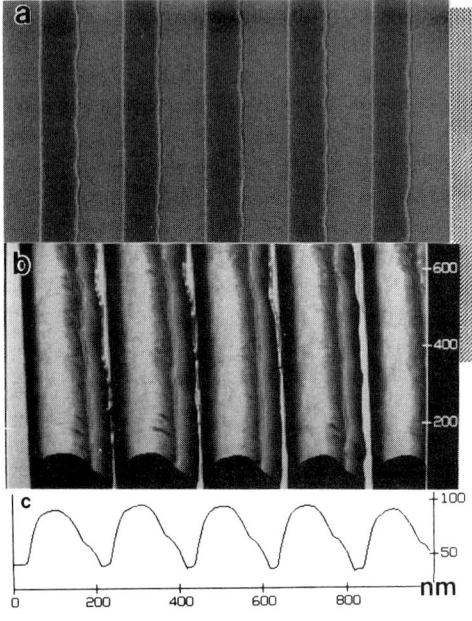

Fig.3 Creation of AlAs ridges by cleavage of a GaAs-AlAs superlattice. In the SE image (a) the GaAs appears bright and the AlAs dark. The SFM image (b), which has curved profiles due to tip effects, shows that the AlAs layers have formed 55nm high ridges as seen in the SFM linescan (c). A model describing the origin of the ridges is shown in (d).

5. CONCLUSIONS

We have shown SE, BSE and SFM images of GaAs-Al$_x$Ga$_{1-x}$As heterostructures where mapping of topography, dopant type and composition was carried out. This information is very valuable when characterising these materials and can be obtained with much less effort than would be necessary for TEM analysis. Models presented here also explain the mechanisms that give rise to dopant contrast and topographical features.

REFERENCES

Ashley JC, Tung CJ and Ritchie RH 1979 Surf. Sci. 81 409
Aven M, Devine JZ, Bolon RB and Ludwig GW 1972 J. Appl. Phys. 43 4136
Bleloch AL, Castell MR, Howie A and Walsh CA 1994 Ultramicroscopy 54 107
Bleloch AL, Howie A and Milne RH 1989 Ultramicroscopy 31 99
Castell MR, Howie A, Perovic DD, Ritchie DA, Churchill AC and Jones GAC 1993 Phil. Mag. Lett. 67 89
Garcia N 1995 private communication
Merli PG and Nacucchi M 1993 Ultramicroscopy 50 83
Ogura K 1991 JEOL News 29e 26
Pennycook SJ and Howie A 1980 Phil. Mag. A 41 809
Perovic DD, Castell MR, Howie A, Lavoie C, Tiedje T and Cole JSW 1994 Ultramicroscopy in press
Sawyer GR and Page TF 1978 J. Mater. Sci. 13 885

Inst. Phys. Conf. Ser. No 146
Paper presented at Microsc. Semicond. Mater. Conf., Oxford, 20–23 March 1995
© 1995 IOP Publishing Ltd

T E M Characterisation of GaN layers grown by MOCVD on Al₂O₃(0001)

J -L Rouvière, M Arlery, A Bourret, R Niebuhr* and K Bachem*

C.E.A. - Centre d'Etudes Nucléaires de Grenoble - Département de Recherche Fondamentale sur
la Matière Condensée - Service de Physique des Matériaux et Microstructures.
17 rue des Martyrs 38054 Grenoble Cedex 9 France
* Fraunhofer-Institut for Applied Solid State Physics Tullast. 72, 79108 Freiburg i.Br. Germany

ABSTRACT : We report TEM observations of wurtzite GaN layers grown by Metal- Organic
Chemical Vapour Deposition (MOCVD) on (0001)Al₂O₃. Apart from dislocations generally
associated with basal stacking faults, most of the observed defects are Inversion Domain
Boundaries (IDB) either on the prismatic (0,1,-1,0) plane or on the prismatic {2,-1,-1,0}
planes. High Resolution Electron Microscopy (HREM) images allow us to propose a new
atomic model for these {2,-1,-1,0} IDBs.

1. INTRODUCTION

At the beginning of 1994, Nichia (Nakamura et al 1994) commercialised a high
brightness blue emitting diode made from InGaN/GaN/AlGaN, demonstrating the potential of
these materials in optoelectronics. Since then, the activity in studying AlN/GaN layers has
increased world-wide. In this paper we present Transmission Electron Microscopy (TEM)
characterisation of GaN layers deposited by MOCVD directly on sapphire (Al₂O₃) or on a
predeposited AlN buffer layer. Part 2 of this paper summarises the experimental procedures.
Part 3 presents the different kinds of defects observed in all GaN layers. Part 4 gives more details
on the AlN buffer layer characterisation. Part 5 is devoted to the atomic structure determination
by HREM of the {2,-1,-1,0} planar defects.

2. EXPERIMENTAL DESCRIPTION

The investigated GaN layers were MOCVD deposited on (0001) Al₂O₃ substrates at a
temperature of 1000°C, with or without an AlN buffer layer. A flow of trimethylgallium (TMG)
and ammonia was cracked above the sample. The substrate was not rotated leading to significant
sample thickness differences between the upstream (100nm) and the downstream (600nm)
regions. The average full width at half maximum (FWHM) for (0004) beam of X-ray rocking
curves is about 18 minutes. X-ray diffraction, SIMS and photoluminescence measurements have
been realised on these layers (Niebuhr et al). **The present paper focuses on TEM results .**
Specimens for TEM were prepared using the standard techniques : mechanical polishing
and Argon ion milling. TEM observations were realised on a JEOL4000EX electron microscope,
specially equipped for HREM (Scherzer resolution about 0.17nm). GaN remains very stable
under the electron beam during long observations. HREM simulations and numerical analysis
were respectively realised with the EMS (Stadelmann 1987) and SEMPER softwares.

3. DEFECT CHARACTERISATION OF GAN LAYERS

The investigated GaN layers have a wurtzite structure and grow epitaxially on the (0001)
Al₂O₃ substrate, with [0001] and [2,-1,-1,0] GaN respectively parallel to [0001] and [1,0,-1,0]
Al₂O₃ (Kuwano et al 1991). The observed defects can be classified into two categories; those
that tend to annihilate during the GaN deposition and those that have a constant density
throughout the layer thickness. The former should not be disturbing for the electrical activity as
they are limited to regions near the interface and tend to disappear when more and more
material is deposited.

286

Stacking faults on basal planes are of the first type. They are generally associated with threading dislocations ($10^{10}cm^{-2}$) that come from the GaN/Al$_2$O$_3$ interface (fig. 1c). Two beam dark field images obtained with g=(0,0,0,1) do not show the stacking faults and their associated dislocations, indicating that the burgers vector of these dislocations is contained in the basal plane (fig. 1d). Stacking faults are larger in the thin upstream regions.

Planar defects (PDs) on the prismatic (0,1,-1,0) planes , which are of the second type, are the major type of defects (density around $10^{10}cm^{-2}$). They are out of contrast (fig 1c) when g=(0,1,-1,0) indicating that the "displacement" (translation or rotation) associated with this defect is contained in the (0,1,-1,0) plane. These (0,1,-1,0) PDs start at the interface and extend to the surface. HREM images show that GaN has the same crystallographic orientation on both sides of the PD(fig 2ab) and we measure no translation along the c-axis (fig 2a). These PDs appear in only one specific prismatic variant whose plane normal (0,1,-1,0) is parallel to the flow of the gases in the reactor cell (fig. 2b). They are associated in pairs. These defects are compatible with an IDB, but the exact atomic structure of the interface is not yet determined.

Planar defects (PDs) on the {2,-1,-1,0} planes with their associated variants (fig. 4ab) are occasionally observed instead of the previous (0,1,-1,0)IDBs. These defects, which belong to the second type of defect, are associated with the presence of hexagonal steps (fig. 4a) observed at the top surface of the samples with optical microscopy (Niebuhr et al). The HREM contrast when viewed exactly along the [0001] axis exhibits a mirror symmetry with respect to the defect plane. This is compatible with a translation defect or with an inversion symmetry-translation defect (IDB). However, the observed contrast variation on both sides of the defects away from the exact c-axis is only compatible with an inversion-translation defect. The reason for the

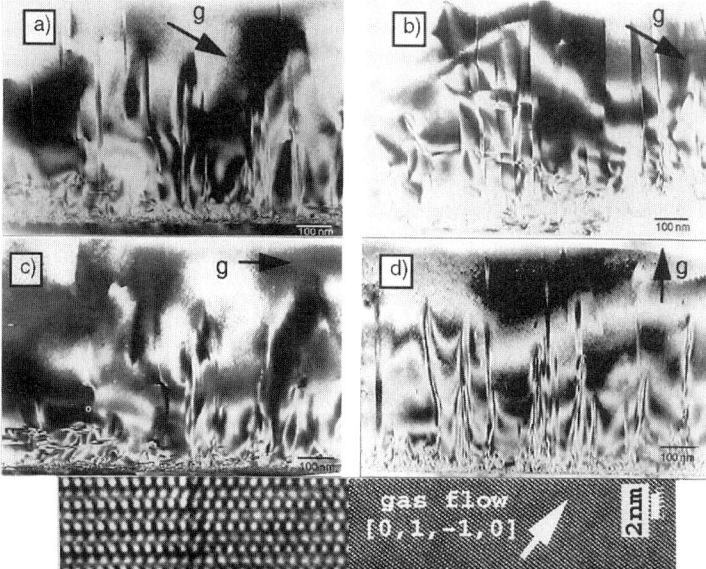

Fig 1. Dark field two beam TEM images of GaN layers taken at :
a) g=(0,1,-1,-1) GaN with AlN buffer
b) g=(0,1,-1,-1) GaN with no AlN buffer
c) g=(1,0,-1,0) GaN with AlN buffer
d) g=(0,0,0,1) GaN with AlN buffer.

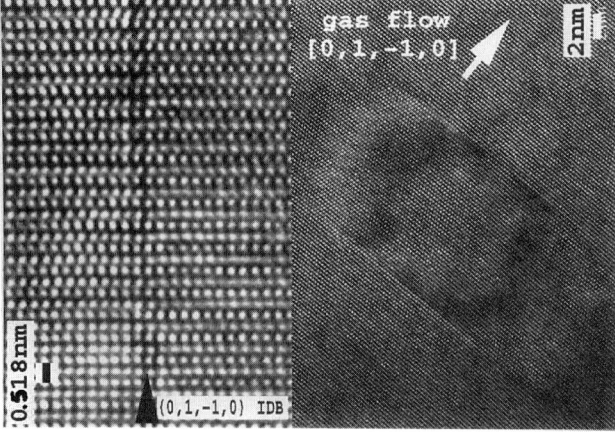

Fig 2. HREM image of the (0,1,-1,0) IDB
a) viewed along the [2,-1,-1,0] direction
b) viewed along the [0,0,0,1] direction.

presence of one or the other defect species is still unclear. The exact atomic structure of the {2,-1,-1,0} IDB interface is proposed below.

4. AlN BUFFER LAYER

We measure the AlN buffer layer thickness in cross sectional HREM images. The [0,1,-1,0] direction is particularly suited to realise this analysis, as in this direction the HREM mainly visualises the c-plane stacking. By measuring the c-interplanar distances, we can locally measure the AlN buffer thickness (around 3nm) and confirm that the brighter region near the interface corresponds to a rather uniform AlN layer (fig 3).

We detect the same kind of defects when the GaN layer is grown with or without an AlN buffer layer. The only difference seems to be that the dislocations and stacking faults on the basal planes are kept closer to the interface when a thin AlN buffer layer is introduced (fig 1ab).

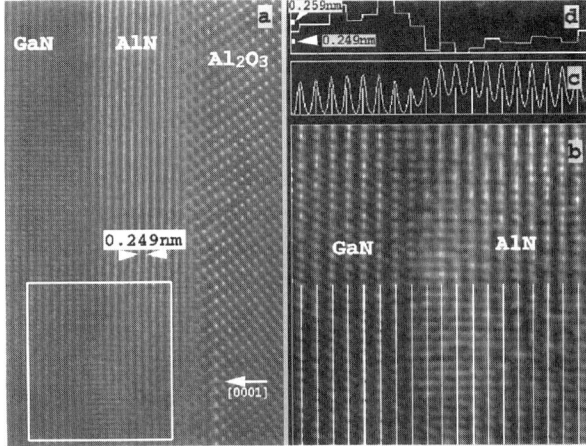

Figure 3
a) HREM image of GaN grown with an AlN buffer layer viewed along the [0,1,-1,0] direction. Only the c-planes stacking are resolved in the deposited layers
b) Image extracted from image 3a
c) Vertical projection (along [2,-1,-1,0]) of image 3b. Each maxi-mum is associated with a basal plane
d) Interplanar distances between the c-planes :
    GaN   c/2=0.2592 nm
    AlN   c/2=0.249 nm

5. ATOMIC STRUCTURE OF THE {2,-1,-1,0} PLANAR DEFECTS

{2,-1,-1,0} planar defects have already been observed in GaN materials (Amelinck 1979) either coming from the movement of Shockley partials or created during growth. However, the two proposed models are not compatible with our HREM images (fig. 4cd) that exhibit a mirror symmetry in projection on the c-plane.

Four models can be constructed with this mirror symmetry in projection. Two of them are Stacking Faults (SF), two of them are IDBs. These models (fig 4ghijk) can be built by starting from a perfect GaN crystal, cutting it into 2 parts, applying or not one inversion transformation, (whose inversion point is situated in between two adjacent Ga and N atoms in order that one specie transformes into the other one) and translating one part with respect to the other. For the different models, the respective operations are : **IDB1**, an inversion and a translation 1/2[0,1,-1,0], **IDB2**, an inversion and 1/2[0,1,-1,0]+1/2[0,0,0,1], **SF1** a translation 2/3[0,1,-1,0] + 1/2[0,0,0,1] and **SF2**, 2/3[0,1,-1,0]+1/4[0,0,0,1]. All models have an apparent contraction in projection of a/2=0.1595nm along the a-direction which is exactly what we measured on digitised HREM images. In all of them, each atom has 4 neighbours, but only in models IDB1 and SF1 the 4 neighbours are of the opposite species. As explained in part 3, we rejected the SF models as these models with only a transitional state should not introduce a contrast variation between the two sides of the defect when observed slightly away from the exact c-axis. IDB1 is thus the more likely model. An observation along the [0,1,-1,0] direction would determine exactly the structure.

The HREM image simulations do reproduce the main characteristic of the models even though no local atomic relaxations have been allowed in the simulated structure. At a defocus value of -35nm, only the tunnels of the structure are visualised and the alternation of small and large tunnels is well reproduced. At a defocus of -67nm, all the atomic columns of the structure are visualised as white spots, the only white spots that do not correspond to an atomic column are the brighter ones that represent the large tunnels of the structure. In the experimental image, this brighter spot is not so well marked. Local relaxations at the interface will certainly improve the

288

fit, but this large tunnel could also be partly occupied by atom impurities, like carbon impurities which we know are present in our samples (Niebuhr et al).

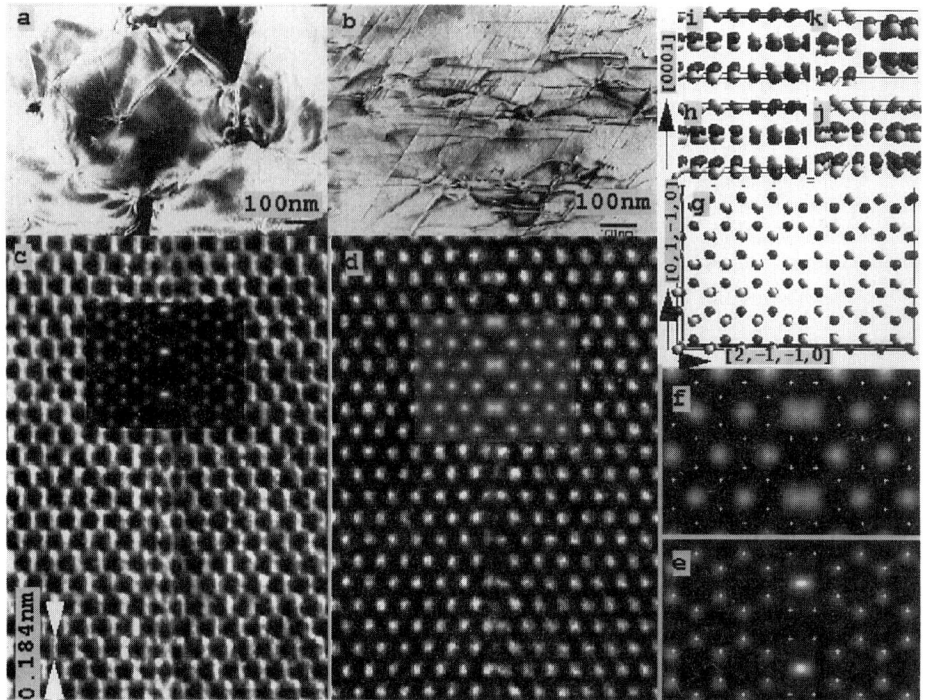

Figure 4. a) Plan view image of the GaN layer near the sapphire substrate : {2,-1,-1,0} IDBs tend to form hexagonal stars b) Same as before but in another region c)d)e)f) HREM images and simulations of {2,-1,-1,0} IDBs taken at 2 defoci : (ce) -67nm and (df) -35nm , atomic columns are drawn on the simulations. ghijk) Projections of the models, on the c-plane (g) and on the (0,1,-1,0) planes h:ID2, i:ID1, j:SF1, k:SF2

6. CONCLUSION

TEM observations have allowed us to characterise the main planar defects presents in the MOCVD grown GaN layers on sapphire. Besides dislocations and basal stacking faults, the main defects were identified as inversion domain boundaries (IDBs) on two different crystallographic planes : (0,-1,1,0) and {2,-1,-1,0} .Work is still under progress to determine the interface structure of the (0,-1,1,0) IDBs, but thanks to HREM images, we were able to propose an atomic model for the {2,-1,-1,0} IDBs. This model should be confirmed by an HREM observation along a second direction and the comparison between simulated and experimental images should be increased by relaxing the atoms at the interface with ab-initio numerical techniques. This model contains large tunnels that could easily incorporate impurities. We hope to correlate this structural characterisation with optical properties.

REFERENCES

Amelinck S 1979 Dislocations in Solids, eds Nabarro (North-Holland publ. comp.) 2 pp 67-460
Nakamura, S Mukai T and Senoh M 1994 Appl. Phys. Lett. 64 13
Kuwano N Shiraishi T Koga A Oki K Hiramatsu K Amano H Itoh K and Akasaki I 1991 J. Cryst. Growth 115 381
Niebuhr R, Bachem K, Dombrowski K, Maier M, Pletschen W and Kaufmann U submitted to J. Electr. Mat.
Stadelmann P A 1987 Ultramicroscopy 38 265

We thank Dr. U. Kaufmann for his contributions to this work.
Part of this work has been supported by the EU commission under the Brite-Euram II programme.

Inst. Phys. Conf. Ser. No 146
Paper presented at Microsc. Semicond. Mater. Conf., Oxford, 20–23 March 1995
© 1995 IOP Publishing Ltd

Microstructure of GaN epitaxially grown on hydrogen plasma cleaned 6H-SiC substrates

P Vermaut, P Ruterana and G Nouet
A Salvador*, A Botchkarev*, B Sverdlov* and H Morkoç*

Laboratoire d'Etudes et de Recherches sur les Matériaux, URA CNRS n°1317, ISMRA, 6 Bd du Maréchal Juin, 14050 Caen Cedex, France
* University of Illinois, Materials Research Laboratory and Coordinated Science Laboratory, 104 South Goodwin Avenue, Urbana, Illinois 61801, USA

ABSTRACT: The microstructure of GaN layers deposited on SiC-6H substrates with AlN buffer has been studied. A high density of threading dislocations has been found. Dislocation half-loops are observed to nucleate at the substrate buffer layer interface and seem to glide toward the surface in the prismatic planes. All the compatible Burgers vectors are observed except [0001].

1. INTRODUCTION

III-IV nitride semiconductors present a great interest for short wavelength optoelectronic applications (Morkoç et al.1994) . One of the problems in the development of these materials is the lack of good quality substrates. The more commonly used substrate is sapphire for which good results have been obtained (Amano et al. 1986). SiC substrates should be better than sapphire as they have superior lattice and thermal match, however it is not easy to remove the oxide layer.

Recently, enhancement of the substrate preparation by addition of a hydrogen plasma step (Lin et al. 1993) has allowed us to overcome this problem. This new procedure leads to good quality epilayers exhibiting sharp X-ray peaks with FWHM of 10 min and photoluminescence peaks at 3.474 eV with a FWHM of 5 meV. The aim of this paper is to study the microstructure of these layers.

2. EXPERIMENTAL PROCEDURE

SiC-6H wafers were cut 3.5°off the basal plane toward <11$\overline{2}$0>. The (0001)$_{Si}$ surface was treated using the classical method followed by a hydrogen plasma step in order to reduce the amount of oxygen-carbon bondings down to below the X-ray photoemission detection limit. The details of this procedure were reported by Lin et al. (1993). Depositions started with an AlN buffer layer which has intermediate thermal and lattice parameters between GaN and SiC. The thickness of the AlN buffer layer was 20 nm in the first sample and 50 nm in the second one. Growth is performed by electron cyclotron resonance plasma-enhanced molecular beam epitaxy at a rate of 40 nm/h with a substrate temperature between 600 and 650°C. For the second sample, alternate AlN/GaN layers of 3 nm thickness were then deposited before the final GaN layer grown at the same rate. The final thickness was 450 nm for the first sample and 1000 nm for the second one.

TEM cross section samples were thinned down to 100 μm by mechanical grinding and dimpled down to 10 μm. Electron transparency was achieved by ion milling with a LN_2 cold stage at 5 kV. A final stage at 3 kV was used to remove possible ion damage. TEM observations were carried out on a JEOL JEM 200 CX operating at 200 kV.

3. RESULTS

The 6H-SiC has a hexagonal structure with the stacking sequence ..AαBβCγAαCγBβ... Its parameters are a=0.308 and c=1.5112 nm. AlN and GaN materials have the wurtzite structure, their lattice parameters are a=0.3111, c=0.4979 nm, and a=0.3186, c=5.178 nm respectively. A general view of the microstructure of the deposited layers of samples 1 and 2 are shown on figure 1a and b respectively. In the two cases, the contrast observed in the buffer layers illustrates the high density of defects. This is striking because the lattice mismatch between AlN and SiC is only of 1.03%. New samples with the GaN layer deposited directly on SiC substrates will be studied by TEM to understand the role of the AlN buffer in the case of SiC substrates. On the other hand, no dislocation lines could be observed in the substrate, confirming their good quality. The densities of defects in the GaN layers are high and estimated to be 10^{11} cm^{-2} in sample 1 with only a 20 nm thick AlN buffer layer, and 10^{10} cm^{-2} in sample 2 with a 50 nm thick AlN buffer layer and a 6x3 nm thick AlN/GaN multilayer.

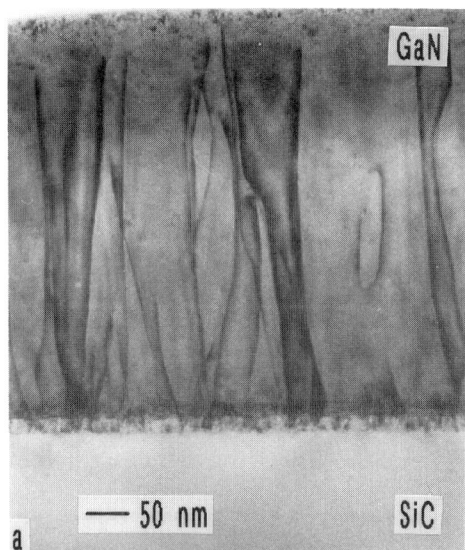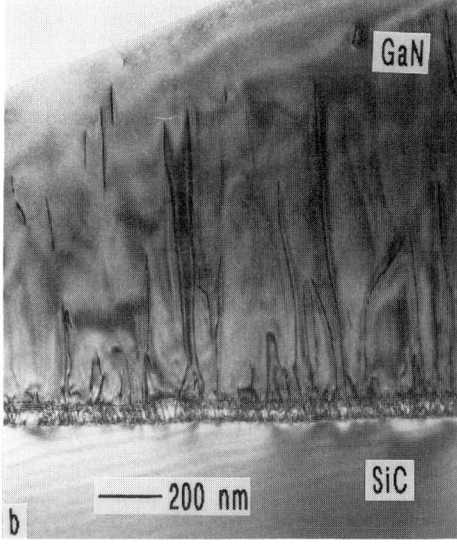

Fig. 1: Microstructure of the GaN layer, a) deposited on a 20 nm thick AlN buffer b) deposited on a 50 nm thick AlN buffer followed by a 6x3 nm thick AlN/GaN multilayer.

Numerous dislocation lines perpendicular to the interface are observed and they are seen to thread across the GaN layer. The contrast observed shows evidence of large distorsions of the lattice all over the layer due to the high density of these defects.

Since AlN and GaN materials have the wurtzite structure, the dislocation types have the following possible Burgers vectors:

$1/3<11\bar{2}0>$, $<0001>$, $1/3<11\bar{2}3>$ for the perfect dislocations, and

$1/3<10\bar{1}0>$, $1/2<0001>$, $1/6<2\bar{2}03>$ for the partial dislocations.

Most of these Burgers vectors have been found by using the $g.\underline{b}=0$ criteria, except the $<0001>$ one, confirming that the growth mode is bidimensional and not a spiral one.

The dislocation lines are mainly perpendicular to the interface, with some deviations visible. However, the dislocation half-loops generally present an angular form due to their high Peierls energy in these materials.

Numerous dislocations half-loops were observed close to the multilayer, in sample 2. Their configuration Fig.2, clearly shows that they nucleate at the SiC/AlN interface and thread across the AlN buffer and the multilayer.

A tilt experiment has been carried out in order to determine the habit planes of the dislocation half-loops (Fig.3a and b). The $[11\bar{2}0]$ zone axis is obtained at zero tilt. The dislocation half-loop arrowed A seems to be in the $(2\bar{1}\bar{1}0)$ plane. For the second one, arrowed B, the images obtained for two opposite tilt angles are inversed meaning that it is in the $(\bar{1}100)$ plane. Therefore, one can conclude that dislocation half-loops are lying in both prismatic planes.

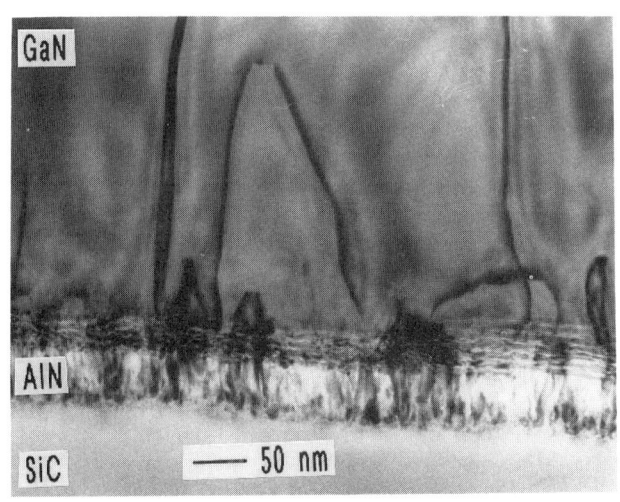

Fig.2: Dislocation half-loops are clearly observed to nucleate at the SiC/AlN interface and thread across the AlN/GaN multilayer.

4. DISCUSSION

The density of threading dislocations in the GaN layers is observed to be lower when a 50 nm thick AlN buffer layer is used with a AlN/GaN multilayer than in the case of a 20 nm thick AlN buffer layer only. This result is in agreement with that of Kuznia et al. (1993) who reported an optimized thickness of the AlN buffer layer of 50 nm in the case of sapphire substrates. On the other hand, no clear evidence has been found on the efficiency of the AlN/GaN multilayer system in decreasing the defect density as numerous dislocations are clearly observed to cross it.

In cubic materials, several models show that misfit dislocations are formed by glide of semicircular half-loops from the surface towards the interface, after the growth of a critical layer thickness. In this case the grown material is under compression. This is in good agreement with the results on the (001) InP/InGaAs system reported by Wagner et al. (1989).

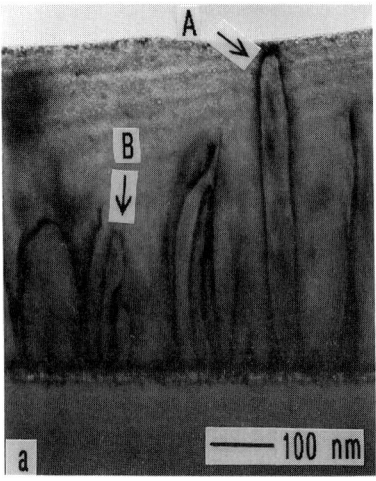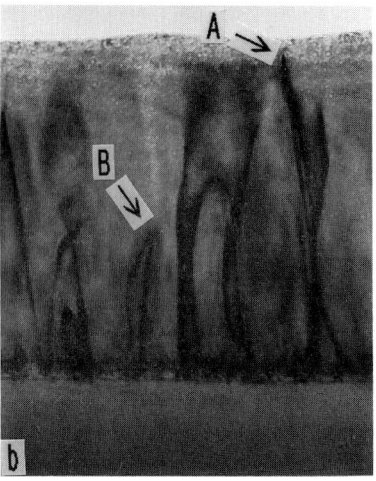

Fig. 3: Determination of the habit plane of the dislocation half-loops A
and B which are found to lie respectively in the $\{11\bar{2}0\}$ and $\{10\bar{1}0\}$
planes. The tilt axis is perpendicular to the interface. a) tilt angle of -
25° b) tilt angle of +25°.

In our case the GaN layer is also in compression, however, our observations show the
dislocations half-loops to form at the interface and then to thread across the layer toward the
surface.

Geometrical considerations point out that inclined glide planes i.e. pyramidal ones should
be activated. It has been determined that dislocation half-loops are present in the $\{11\bar{2}0\}$ and
$\{10\bar{1}0\}$ planes. This results could be explained by taking into account a possible climb
mechanism, or an activation of glide systems perpendicular to the interface.

5. CONCLUSION

The results obtained on optimized AlN buffer thickness on sapphire substrates by Kuznia
et al. seem to be in good agreement with SiC substrates. Results obtained in cubic system on
the origin of threading dislocations, are not in agreement with our observations. A HREM
study of the AlN/SiC-6H interface is now carried out in order to analyse their core structure
and probably try to understand the mechanisms underlying their formation.

REFERENCES

Amano H., Sawaki I. and Toyoda Y. 1986 Appl. Phys. Lett. 48, 353
Kuznia J N, Asif Khan M, and Olson D T 1993 J. Appl. Phys. 73, 4700
Lin M E, Strite S, Agarwal A, Salvador A, Zhou G L, Teraguchi N, Rockett A, and Morkoç H
    1994 J. Appl. Phys. 62, 702
Morkoç H, Strite S, Gao G B, Lin M E, Sverdlov B, and Burns M 1994 J. Appl. Phys. 76,
    1363
Powel R C, Lee N E, Kim Y W, and Greene J E 1993 J. Appl. Phys. 73, 189
Wagner G, Gottschalch V, Rhan H, and Paufler P 1989 Phys. Stat. Sol. (a) 113, 71

Inst. Phys. Conf. Ser. No 146
Paper presented at Microsc. Semicond. Mater. Conf., Oxford, 20–23 March 1995
© 1995 IOP Publishing Ltd

Structural characterization of $In_{1-x}Mn_xAs$ and $Ga_{1-x}Mn_xAs$ epitaxial magnetic films grown by MBE on GaAs

H Bender, A Van Esch, W Van Roy, R Oesterholt, J De Boeck and G Borghs

IMEC, Kapeldreef 75, B-3001 Leuven, Belgium

ABSTRACT: Epitaxial layers of $In_{1-x}Mn_xAs$ and $Ga_{1-x}Mn_xAs$ have been grown by molecular beam epitaxy (MBE) on (001) GaAs substrates at various temperatures and with different Mn concentrations. The structural properties of these layers were investigated by High Resolution Electron Microscopy (HREM) for the as-grown layers and for different subsequent anneal. Several types of orientation relationships between the MnAs and $In_{1-x}Mn_xAs$ or $Ga_{1-x}Mn_xAs$ films have been determined and will be discussed.

1. INTRODUCTION

Diluted magnetic semiconductors based on III-V materials can be grown epitaxially on GaAs substrates. The major material investigated so far is $In_{1-x}Mn_xAs$ (Munekata et al, 1990, Guha and Munekata 1993). On the other hand the epitaxial growth by MBE of MnAs thin films on (001) GaAs has recently been reported (Tanaka et al, 1994). Such layers are interesting materials for the study of fundamental aspects of materials science and electronic transport properties, and are potential candidates for new applications.

MnAs exists in three different phases (Okamoto 1989): α-MnAs (hexagonal with NiAs-structure, a=0.3724nm and c=0.5706nm, ferromagnetic), β-MnAs (orthorhombic with MnP-structure, a=0.5724nm, b=0.3363nm and c=0.6367, paramagnetic) and γ-MnAs (hexagonal with NiAs-structure, paramagnetic). The phase transitions $\alpha \to \beta$ and $\beta \to \gamma$ occur at 40°C and 125°C, respectively.

In this work, epitaxial layers of $In_{1-x}Mn_xAs$ and $Ga_{1-x}Mn_xAs$ grown by molecular beam epitaxy on (001) GaAs substrates are structurally characterized.

2. EXPERIMENTAL DETAILS

$In_{1-x}Mn_xAs$ layers are grown with different Mn contents (x=0.09, 0.12 and 0.15) at 290°C on an InAs buffer layer which is grown on (001) GaAs. Some samples are annealed at 450°C for 30s. The volumetric amount of MnAs precipitates and the remaining Mn in the $In_{1-x}Mn_xAs$ are estimated by saturation magnetisation measurements (Van Esch et al 1995a).

Low temperature GaAs is grown on (001) GaAs at 230-280°C. Mn is incorporated during the growth resulting in $Ga_{1-x}Mn_xAs$ with x between 0.08 and 0.15.

The layers are investigated as a function of the growth and anneal conditions by High Resolution Transmission Electron Microscopy (HREM) on $[110]_{GaAs}$ cross-sectional samples.

3. RESULTS AND DISCUSSION

3.1 $In_{1-x}Mn_xAs$

The as-grown $In_{1-x}Mn_xAs$ layers show a defected structure with different types of MnAs precipitates and extended defects. The MnAs clusters grow upon annealing and have irregular sizes and shapes. Their density as well as the extended defect density increases with Mn content of the layers. The following types of defects can be distinguished: misfit dislocations at the InAs/GaAs interface, dislocations in the $In_{1-x}Mn_xAs$ layer, twins and stacking faults in the upper part of the $In_{1-x}Mn_xAs$ layer, needle-like precipitates, and irregular oval-shaped or elongated α-MnAs precipitates. Magnetization measurements at room temperature show a clear hysteresis due to the presence of these precipitates, while the ferromagnetic behaviour totally disappears above the Curie temperature (40°C) of MnAs (Van Esch et al, 1995a).

Needle precipitates : In the samples with $x \leq 0.12$ needle-like precipitates occur (Fig. 1). They nucleate on the $In_{1-x}Mn_xAs$/InAs interface and grow in the $[001]_{InMnAs}$ direction through the lower 100-150nm of the $In_{1-x}Mn_xAs$ layer. As their widths are only a few nm, the interpretation of their HREM images is not straightforward and the lattice spacings cannot be determined with good accuracy. In no case a clear zone can be identified parallel with the $[110]_{InMnAs}$ cross-sectional direction or by tilting away from this orientation around the $[001]_{InMnAs}$ or $[1\bar{1}0]_{InMnAs}$ directions. On some defects a lattice spacing of 0.28-0.29 nm is present for planes making an angle of 60° with the $(002)_{InMnAs}$ surface plane (see along the arrows on Fig. 1). Within a single defect such planes can be observed in different regions under +60° or -60° tilts. This lattice spacing can correspond with the $(01\bar{1}1)_{\alpha MnAs}$ (0.2806nm) or $(0002)_{\alpha MnAs}$ (0.2855nm) planes of the α-phase, or with the $(200)_{\beta MnAs}$ (0.2862nm) plane of the β-phase. In view of the absence of a periodicity doubling (see below), the $(0002)_{\alpha MnAs}$ can be excluded. On other defects a spacing of 0.29-0.30nm is present parallel with the surface. This does not fit well with any spacing of either the α- or the β-phase. The presence of other Mn-As phases cannot be excluded. In conclusion, the exact structure determination of these defects needs further investigation.

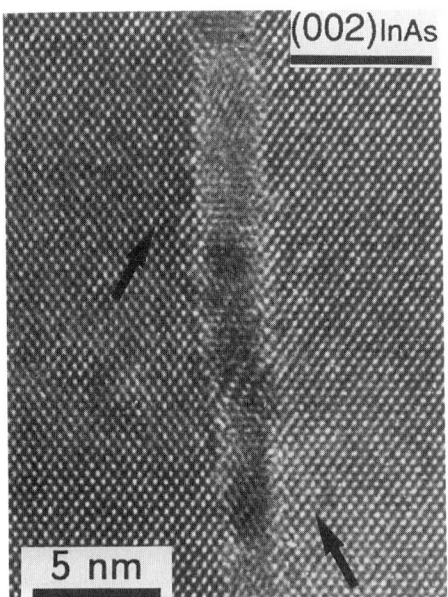

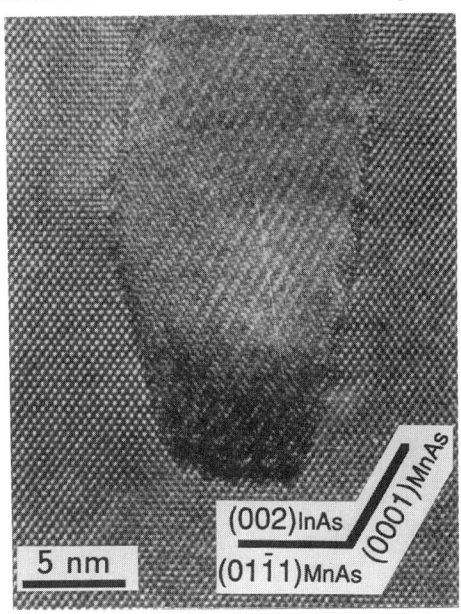

Fig. 1. Needle-shaped precipitate nucleated at the $In_{0.88}Mn_{0.12}As$ interface. Along the arrows, planes with a spacing of 0.28-0.29nm can be observed.

Fig. 2. Elongated oval-shaped precipitate in the $In_{0.88}Mn_{0.12}As$ sample with orientation Ia

Oval-shaped precipitates : The oval-shaped precipitates show Moiré fringes in the thicker specimen regions. For the anneal conditions resulting in the largest precipitates, direct imaging of the precipitate lattice structure is possible when no $In_{1-x}Mn_xAs$ is present above or below the precipitates. In this way two different precipitate zones are identified parallel with the $[110]_{InMnAs}$ cross-sectional direction. Both can be interpreted as due to hexagonal α-MnAs.

The most common zone is (Fig. 2):

$$[2\bar{1}\bar{1}0]_{MnAs} \,\text{//}\, [110]_{InMnAs}$$

with
$$(01\bar{1}1)_{MnAs} \,\text{//}\, (002)_{InMnAs} \qquad\qquad [Ia]$$

which implies that $(0001)_{MnAs}$ makes an angle of $5.8°$ with $(1\bar{1}1)_{InMnAs}$ and $60.6°$ with $(002)_{InMnAs}$. The investigation of the exact orientation of a large number of precipitates, shows that a considerable deviation from this epitaxial relationship occurs, such that the precipitates are rotated around the zone axis up to $6°$. This extreme value corresponds with a situation that the parallel planes are: $(0001)_{MnAs}\text{//}(1\bar{1}1)_{InMnAs}$ [Ib]. Assuming that the lattice parameter of InAs can be used for the $In_{1-x}Mn_xAs$ layer (ie x small), the misfits in the parallel planes are in the first case: -13.1% for $2[2\bar{1}\bar{1}0]_{MnAs}$ versus $[110]_{InAs}$ and 2.0% for $2[01\bar{1}\bar{2}]_{MnAs}$ versus $3[1\bar{1}0]_{InAs}$, while in the latter case one has -13.1% for the three $2<2\bar{1}\bar{1}0>_{MnAs}$ versus $<110>_{InAs}$. In view of these misfits, the former case is the more likely epitaxy between the MnAs and $In_{1-x}Mn_xAs$. On most HREM pictures a periodicity doubling ($0.57nm$) is observed along the $[0001]_{MnAs}$ direction. This is due to slight misalignment of the zone axis with respect to the incident electron beam (Coene et al, 1985).

A second zone of MnAs is observed for some precipitates such that:

$$[\bar{1}101]_{MnAs} \,\text{//}\, [110]_{InMnAs}$$

with
$$(1\bar{1}02)_{MnAs} \,\text{//}\, (002)_{InMnAs} \qquad\qquad [II]$$

It does not belong to an orientation variant of the first epitaxy, but is a second type. The misfits in the parallel planes are in this case : 0.5% for $[\bar{1}101]_{MnAs}$ versus $[110]_{InAs}$ and -13.1% for $2[\bar{1}\bar{1}20]_{MnAs}$ versus $[1\bar{1}0]_{InAs}$.

With the majority of the precipitates no structural HREM image could be obtained. This is due to the occurrence of the above epitaxies in all kinds of orientation variants, but could also be due to the presence of other types of epitaxy or random orientations of the MnAs precipitates such that no low index zone is close to the $[110]_{InMnAs}$ direction.

3.2 $Ga_{1-x}Mn_xAs$

The as-deposited $Ga_{1-x}Mn_xAs$ layers grow epitaxially with the zincblende structure on the GaAs substrate and are free of extended defects. Although no observable MnAs clusters are found by HREM, a weak magnetization is observed for the as-grown layers. This can be due to the occurrence of tiny MnAs clusters ($<2nm$) which cannot be revealed from the HREM images. Annealing at $600\text{-}650°C$, for all growth conditions, leads to phase separation by formation of globular α-MnAs precipitates with uniform size and distribution, while the extended defect density remains very low. The magnetization increases strongly by annealing (Van Esch et all, 1995b).

The MnAs precipitates are mostly aligned with the basal hexagonal plane of the MnAs parallel to a $(111)_{GaMnAs}$ plane within one degree, ie according to epitaxy Ib (Fig. 3). Only very few precipitates are found with orientation Ia. Hence most of the precipitates are rotated over $5.8°$ compared to the common case for the $In_{1-x}Mn_xAs$ layers. This difference can be understood by the fact that the misfits in the parallel plane are now much less : -6.9% for the three $2<2\bar{1}\bar{1}0>_{MnAs}$ versus $<011>_{GaAs}$ assuming that the GaAs lattice parameter can be used for the $Ga_{1-x}Mn_xAs$ layer.

In the $Ga_{1-x}Mn_xAs$ samples also a third epitaxy is identified:

$$[2\bar{1}\bar{1}0]_{MnAs} \,\text{//}\, [110]_{GaMnAs}$$

with
$$(01\bar{1}0)_{MnAs} \,\text{//}\, (1\bar{1}1)_{GaMnAs} \qquad\qquad [Ic]$$

Compared to epitaxy Ib, this corresponds to a $90°$ rotation around the $[2\bar{1}\bar{1}0]_{MnAs}$ zone axis. The misfit along the c-axis is large : 23.7% for $3[0001]_{MnAs}$ versus $[112]_{GaAs}$, so that this relationship is not very likely.

All the epitaxies reported here are different from the one found for MnAs grown on GaAs by MBE, in which case the $(\bar{1}100)_{MnAs}$ plane is found parallel with $(001)_{GaAs}$ (Tanaka et al, 1994). Epitaxy II corresponds to one of the XRD peaks observed by Munekata et al (1990), who found, however, no peaks corresponding to epitaxy I.

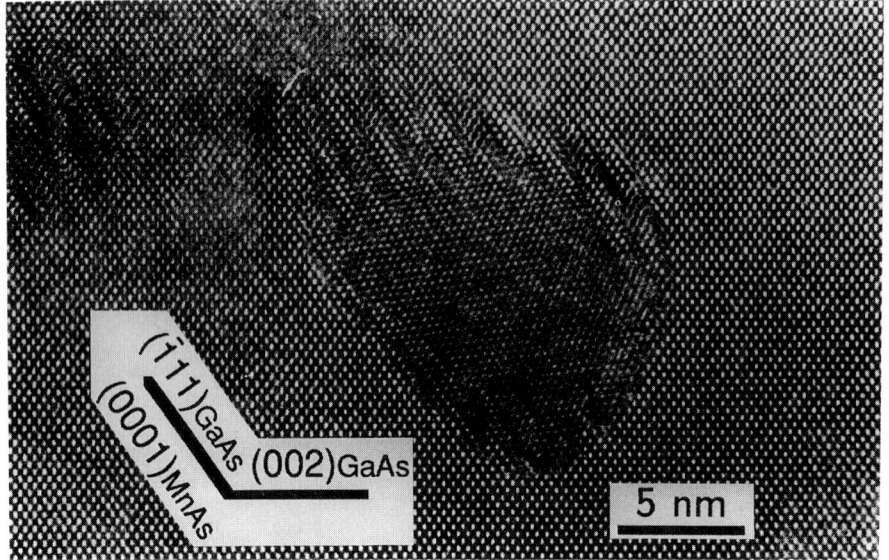

Fig. 3. Globular precipitate with orientation Ib in $Ga_{0.94}Mn_{0.06}As$ annealed at 600°C.

4. CONCLUSION

The orientation relationships of MnAs precipitates in $In_{1-x}Mn_xAs$ and $Ga_{1-x}Mn_xAs$ grown epitaxially on GaAs are discussed. Precipitates are already present in the as-grown $In_{1-x}Mn_xAs$ layers, which also contain a high density of other lattice defects. On the other hand the as-grown $Ga_{1-x}Mn_xAs$ has a high crystal perfection and contains no precipitates visible by TEM. Upon annealing a uniform distribution of MnAs precipitates forms in these layers.

This work is carried out with the TEM at the University of Antwerpen (RUCA).

REFERENCES

Coene W, Bender H, Lovey F C, Van Dyck D and Amelinckx S 1985 phys. stat. sol. (a) 87 483
Guha S, Munekata H 1993 J Appl Phys 74 2974
Munekata H, Ohno H, von Molnar S, Harwit A, Segmüller A, Chang L L 1990 J. Vac. Sci. Technol. B 8 176
Okamoto H 1989 Bulletin of Alloy Phase Diagrams 10 549
Tanaka M, Harbison J P, Sands T, Cheeks T L, Keramidas V G and Rothberg G M 1994 J. Vac. Sci. Technol. B 12 1091
Van Esch A, Van Roy W, De Boeck J, François I, Bender H, Borghs G, Van Bockstal L, Bogaerts R and Herlach F 1995a Proc Int Workshop on Semimagnetic Semiconductors, September 1994, Linz, Austria, in press.
Van Esch A, De Boeck J, Oesterholt R, François I, Bender H, Van Roy W, Borghs G, Van Bockstal L, Bogaerts R and Herlach F 1995b Proc VIIIth EuroMBE, Sierra Nevada, Spain.

Inst. Phys. Conf. Ser. No 146
Paper presented at Microsc. Semicond. Mater. Conf., Oxford, 20–23 March 1995
© 1995 IOP Publishing Ltd

TEM investigation of Ge_xSi_{1-x}/Si(111) heterostructures grown by MBE

O I Lebedev, N A Kiselev, A G Vasiliev[1], A A Orlikovsky[1]

Institute of Crystallography Russian Academy of Sciences, Leninsky pr.59,
117333 Moscow, Russia
[1] Institute of Physics and Technology Russian Academy of Sciences, Krasikov str.25A,
117218 Moscow, Russia

ABSTRACT: The aim of this work was the investigation of Ge_xSi_{1-x}/Si(111) heterostructures obtained by MBE deposition of alternating Ge and Si layers of different thicknesses in high vacuum. The samples were formed at T_s from RT to 700°C. The TEM and RBS investigation has enabled us to propose a model for the GeSi/Si heterostructure formation process during MBE. The data of mutual Ge and Si diffusion at different T_s were obtained and the temperature range of interface mixing was defined. HREM investigation of the film and interface structure made it possible to define the conditions of 2D and 3D film growth. The new superstructure $Ge_{0.25}Si_{0.75}$ was revealed.

1. INTRODUCTION

Two directions may be outlined in the course of investigation of Ge/Si structures: formation of Ge_nSi_m composite superlattices and formation of Ge_xSi_{1-x} heterostructures.

Substrate orientation and the substrate surface conditions to some extent affect the quality of epitaxial films, interfaces and the critical film thickness (Hull et al 1991). The critical thickness for different orientations of the substrate surface changes is as follows: h_c (111) > h_c (100) > h_c(110) (Jagannadham et al 1991). The inter-dislocation distance and stress relaxation energy in stable and metastable strained Ge_nSi_m layers are a function of thickness, Burgers vector and lattice misfit parameters (Jain et al 1991). The conversion from two-dimensional (2D) to three-dimensional growth (3D) in Ge_xSi_{1-x} on Si(100) occurs at thicknesses much lower than the critical one. The thickness for different "x" values of the Ge_xSi_{1-x} structure were given by Wang et al (1991). Segregation of Ge in the growing Si layer was the main reason for interface mixing of the Ge_xSi_{1-x} layers in the strained superlattice (Fukatsu et al 1991, Bowman et al 1991). The coefficients of Si-Ge interdiffusion (D) and activation energy (E) were found by Prokes et al (1985).

The paper investigates Ge_xSi_{1-x} /Si(111) heterostructures and Ge and Si layer mixing on the interface during MBE.

2. EXPERIMENTAL

A set of samples for TEM was prepared in a high vacuum electron beam co-evaporation system in a vacuum 10^{-9} Torr at film growth speed of 0.5-1 Å/s.

The thickness of alternating Si and Ge layers was increased above the critical thickness. The total thickness of the structure varied in the range 80-140 nm. The samples were obtained in the range of temperatures from RT to 700°C. Structure thickness and growth rate were controlled with a quartz micro-scale.

After-growth investigations of the structures were carried out using HREM and RBS.

The specimens for HREM were prepared using (110) and (211) cross-sections and ion thinning in Gatan 600. The specimens were investigated in a Philips EM-430ST with an accelerating voltage of 200 kV.

3. RESULTS AND DISCUSSION

HREM investigation of Ge_xSi_{1-x} /Si(111) heterostructure formation shows that at temperature of substrate of up to $T_s=300^\circ$C the films retain their initial layer structure (Fig1a). At higher temperature intensive interaction between Ge and Si begins, resulting in a uniform film formation. HREM shows the possibility of cluster formation in the film with varying Ge and Si content at temperature 400°C$<T_s<600^\circ$C (Fig.1b). At $T_s=700^\circ$C uniform epitaxial films are formed (Fig.2) with the following content $Ge_{0.3}Si_{0.7}$ (based on RBS data).

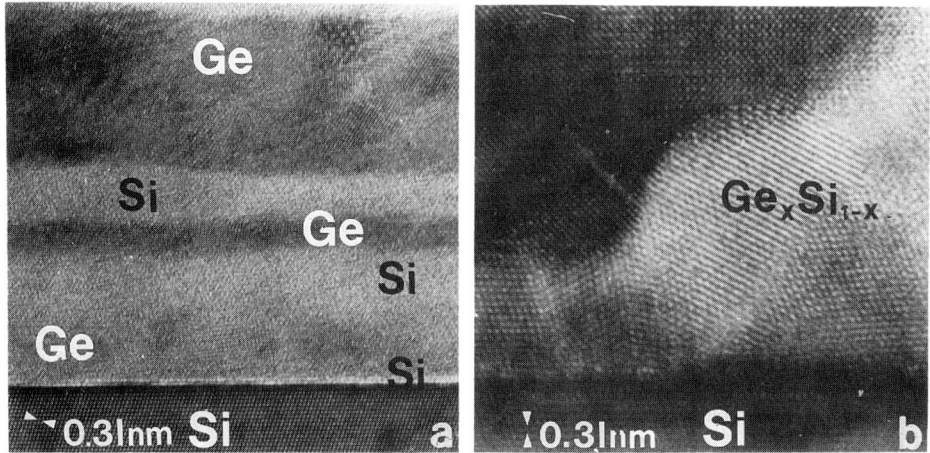

Fig.1 HREM cross-section image of the Ge_xSi_{1-x} /Si(111) film obtained at $T_s=300^\circ$C (a) and $T_s=600^\circ$C (b)

High resolution images (Fig.2) show twin formation close to the interface with twinning planes parallel to the surface. Their concentration drops considerably with growth distance from the substrate surface. Stacking faults were observed in the films lying in planes normal to the substrate. Studies of HREM images and diffraction patterns from an area of the Ge_xSi_{1-x} /Si interface indicate that the film grows in the unusual orientations: $(02\bar{2})$ Ge_xSi_{1-x} $||$ (111)Si; $[\bar{2}33]$ Ge_xSi_{1-x} $||$ $[01\bar{1}]$Si. The diffraction pattern shows that superstructure reflections are present at $1/2\{311\}$ and $1/2\{02\bar{2}\}$. Image calculation for $[\bar{2}33]$ sections (Fig.3) indicates that they are in agreement at the content $Ge_{0.25}Si_{0.75}$. The grains of superstructure were revealed at $T_s=600^\circ$C (Fig.4) The unit cell (a=10.965 Å) of this new phase where the Ge atoms replace Si atoms is shown in Fig.5.

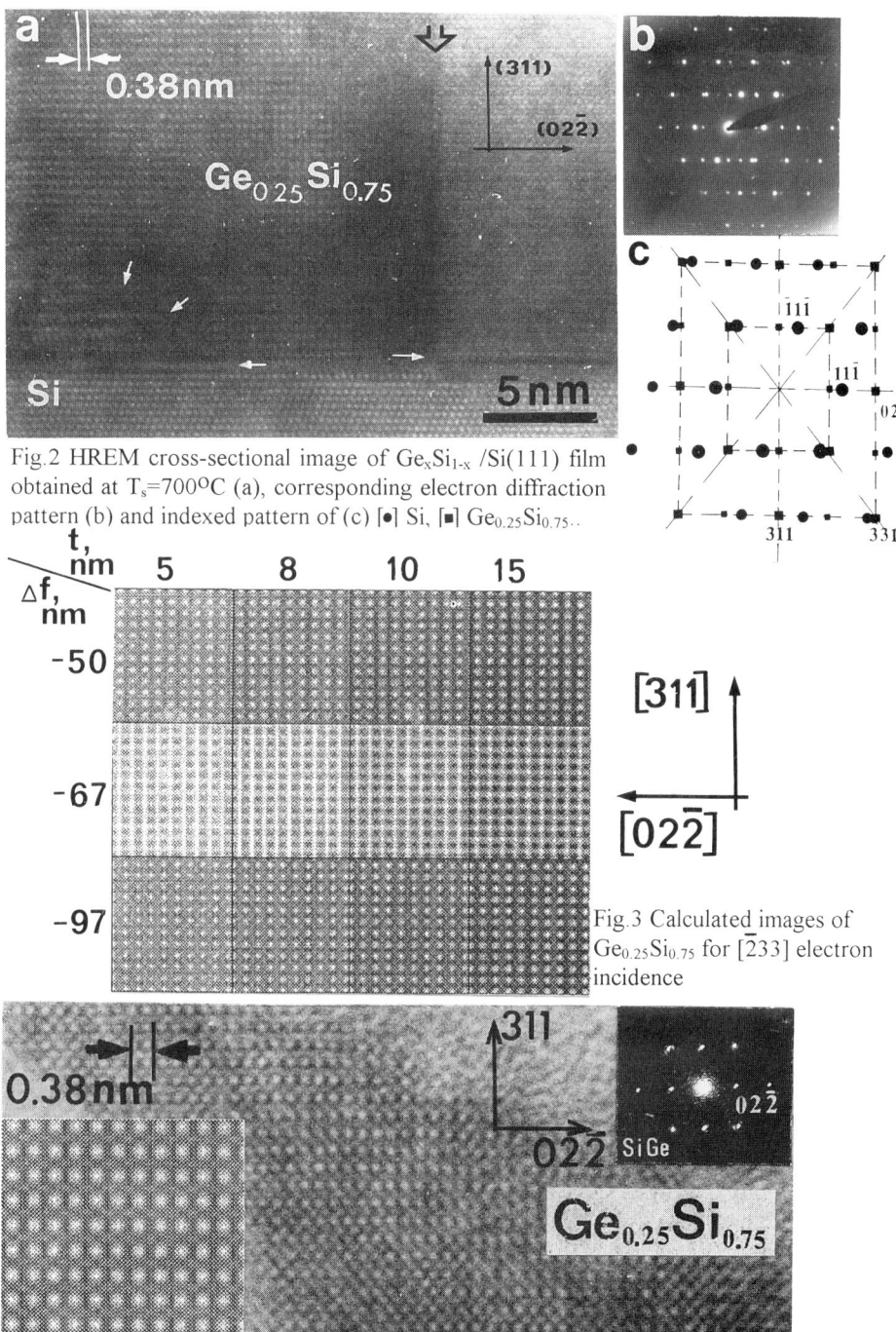

Fig.2 HREM cross-sectional image of Ge_xSi_{1-x} /Si(111) film obtained at T_s=700°C (a), corresponding electron diffraction pattern (b) and indexed pattern of (c) [•] Si, [■] $Ge_{0.25}Si_{0.75}$..

Fig.3 Calculated images of $Ge_{0.25}Si_{0.75}$ for [$\bar{2}$33] electron incidence

Fig.4 HREM image of $Ge_{0.25}Si_{0.75}$ grain of films obtained at T_s=600°C (a). An optical transform from this grain and calculated image are inset.

300

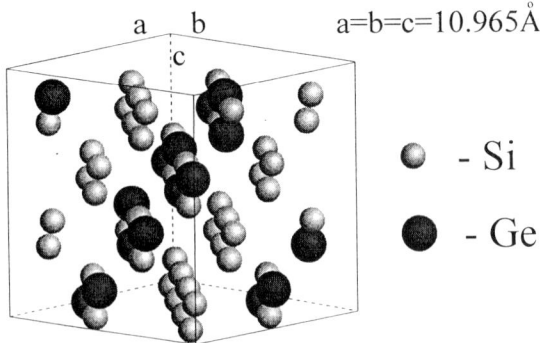

a b   a=b=c=10.965Å

c

○ - Si

● - Ge

Fig.5 Model of new unit cell of $Ge_{0.25}Si_{0.75}$

Analysing the experimental data one can conclude that at growth temperatures exceeding a certain limit (~ 400°C) complete mixing of layers is observed with the formation of a uniform structure. Evidently, diffusion processes in the film growth differ from those taking place in the bulk of a solid film and are related both to the growth kinetics and to the existence of two different growth mechanisms (Kiselev et al.,1995)

4. CONCLUSIONS

The investigation allowed one to propose a model for the Ge_xSi_{1-x} heterostructure formation process on Si(111). The data on mutual Ge and Si diffusion at different substrate temperatures were obtained and the temperature range of layer interface mixing was defined. HREM investigation of the structure of Ge_xSi_{1-x} /Si films at different substrate temperatures allows one to define the conditions of 2D and 3D film growth. It was shown that the result of interface mixing was the formation of a $Ge_{0.3}Si_{0.7}$ /Si(111) epitaxial film at T_s=700°C. Formation of a new superstructure with content $Ge_{0.25}Si_{0.75}$ was discovered at T_s=600-700°C. The model of the new $Ge_{0.25}Si_{0.75}$ unit cell was proposed.

ACKNOWLEDGEMENT

This work was carried out with the aid of grant 93-02-3523 from the Russian Foundation of Basic Science and grant 3-004 of the Russian Ministry of Sciences.

REFERENCES

Bowman R C, Adams Jr P M, Chang S J et al 1991 Mat.Res.Soc.Symp.Proc 220
Fukatsu S, Fujita K, Yaguchi H et al 1991 Mat.Res.Soc.Symp.Proc. 220
Hull R, Bean J C, Peticolas et al 1991 Mat.Res.Soc.Symp.Proc. 220 153
Jagannadham K and Narayan J 1991 Materials Science and Engineering 138 107
Jain S, Balk P, Goorsky M S et al 1991 Microelectronics Engineering 15 131
Kiselev N A, Lebedev O I, Vasiliev A L et al 1995 Vacuum 46(3) 269
Prokes S M, Spaepen F 1985 Appl. Physics Lett. 47(3)
Xun Wang, Zhon C L, Zhon T C et al 1991 Mat.Res.Soc.Symp.Proc. 220 241

Inst. Phys. Conf. Ser. No 146
Paper presented at Microsc. Semicond. Mater. Conf., Oxford, 20–23 March 1995
© *1995 IOP Publishing Ltd*

TEM characterization of $Si_{1-x-y}Ge_xC_y$ and $Si_{1-y}C_y$ layers grown with molecular beam epitaxy on (001)Si substrates

E Bugiel, S Ruvimov* and H J Osten

Institute of Semiconductor Physics, PO BOX 409, D-15204 Frankfurt (Oder), Germany
*Ioffe Institute, Russian Academy of Science, 194021 St.Petersburg, Russia

ABSTRACT: The strain in pseudomorphically grown $Si_{1-x}Ge_x$ layers can be reduced by adding a small amount of carbon without a loss in structural identity and without introducing crystalline defects. The carbon increases the stability of the SiGe layer not only by reducing the misfit strain but also by decreasing the dislocation mobility. After an annealing step a lot of small precipitates (probably β-SiC) have been observed in the $Si_{1-y}C_y$ layer. We demonstrate the possibility to incorporate carbon as highly concentrated strain-stabilized Si_nC δ-layer with a nominal thickness of 1-1.5 monolayers (ML) carbon without introducing crystalline defects.

1. INTRODUCTION

The behavior of compressive strained pseudomorphic $Si_{1-x}Ge_x$ layers on Si substrate is currently attracting considerable interest. Due to the high strain in these systems there is stability only to a certain equilibrium critical thickness. Beyond this critical thickness they are merely metastable and relax during post growth annealing. The main relaxation mechanism is the strain relief by the introduction of misfit dislocations at the interface. Another process relieving the strain in the layer is the diffusion of the Ge atoms out of the layer setting in at temperatures around 1000°C. To avoid the high strain in these heterostructures the incorporation of carbon in the layer is a promising possibility (see Osten et al 1994a). The lattice constant of carbon is smaller than that of silicon or germanium and therefore additional carbon may reduce the strain in the epilayers. Nevertheless, to study the influence of carbon it is very important to investigate the behavior of the tensile strained $Si_{1-y}C_y$ layers. Since the mismatch between carbon and silicon is about 10 times higher than the mismatch between silicon and germanium, one needs only a few percent carbon to get visible strain. Carbon has a low solubility in silicon (about 10^{-4} at.% at the melting point of silicon). During the growth processes this bulk solubility need not be considered since the "surface solubility" becomes important. In a kinetically limited growth process working under supersaturation the adatoms become rapidly buried in subsurface states, which serve as an energetic sink against further migration of atoms. This incorporation process determines the surface solubility, which can be orders of magnitude larger than the bulk solubility. Recent resonant Rutherford backscattering experiments on highly C-doped Si layers (above 1% carbon) support the picture of mainly substitutional incorporation of carbon (see Endisch et al 1995). Rücker et al (1994) investigated the possible existence of strain-stabilized highly concentrated (up to 20 at.% of carbon) Si_nC δ-layers embedded in silicon.

2. EXPERIMENTAL

The layers were grown in a three chamber molecular beam epitaxy equipment with a base pressure below 10^{-10} mbar. Silicon was evaporated from an e-beam evaporator, germanium from an effusion cell and a pyrolytic graphite-filament sublimation source has been used as a solid carbon source. P-type Si(100) substrates were prepared by an appropriate *ex situ* and an *in situ* cleaning procedure, resulting in an atomically clean and (2x1) reconstructed silicon surface. After the growth of the Si buffer layer at 600°C, the $Si_{1-x-y}Ge_xC_y$ and $Si_{1-y}C_y$ layers were grown at 500°C and a growth rate of 0.5 A/s. The amount of carbon coevaporated with the silicon was varied between 0.5 and 2% for the individual samples. The deposition process was controlled *in situ* by reflection high energy electron diffraction using 10 kV electrons under an incident angle of 0.5°. Conventional TEM has been carried out using a Philips CM 30T at 300 kV, while JEOL JEM4000EX (MPI Halle) was used for HREM studies. Both (110)-oriented cross-sections and (100)-oriented wedge-shaped samples were prepared for TEM/HREM investigations. HREM image simulations were carried out using a commercial CERIUS program package (Molecular Simulations Inc.).

3. RESULTS AND DISCUSSION

3.1. $Si_{1-x-y}Ge_xC_y$ layers

In a first set of experiments we investigated the reduction in compressive strain in SiGe layers. A sample with 20% germanium and 0.9% carbon in the upper half of the SiGe layer was grown. The misfit in the SiGeC layer should be reduced to a value equivalent to a SiGe layer with 11.5% germanium. This sample shows in plan-view TEM an unusual low density of dislocations corresponding to a degree of relaxation of about 1% or 2% (estimated according to Bugiel and Zaumseil 1993). Figure 1 shows the cross-sectional TEM image.

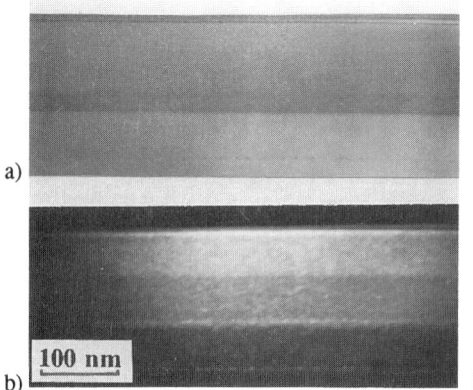 

Fig.1: XTEM micrographs of a sample with 20% germanium and 0.9% carbon in the upper half imaged under a) bright field and b) dark field conditions.

Fig.2: XTEM micrographs of a sample with 10% germanium and 1.3% carbon in the upper half imaged under a) bright field and b) dark field conditions.

No extended crystallographic defects could be found. We observed only a very weak constrast difference between the SiGe and SiGeC layers in the bright field (Fig. 1a), but a bright contrast for the SiGeC layer in the dark field image (Fig. 1b). We assume that the carbon atoms in lattice positions act as strong scattering centers, whereas local lattice deformations around the carbon atoms seem to be responsible for the observed bright contrast. In a next set of experiments we tried to grow an epitaxial SiGe layer under tensile stress on silicon. We chose a germanium concentration of 10% and a carbon concentration in the upper part of the SiGe layer of 1.6%. Plan-view TEM investigations of this sample do not show any dislocations. It seems that the presence of the carbon suppresses the injection and movement of misfit dislocations, although we could observe several possible sources for misfit dislocation generation. In the cross-sectional TEM image (Fig. 2) we find a strong dark contrast for the SiGeC layer already in the bright field image (Fig. 2a) and an increased bright contrast in the dark field image. We conclude from the non-uniform contrast that the carbon atoms are probably not uniformly distributed within the film. HREM investigations did not show any crystalline defects.

3.2. Stability of layers containing carbon

A sample with 20% Ge in the lower part and 25% Ge and 0.5% C in the upper part was prepared. In the unrelaxed state we should expect in both parts the same tetragonal distortion. After relaxation we see that the movement of the threading dislocations was hindered in the carbon containing layer. Fig.3. shows an example of the beginning generation of a dislocation dipole.

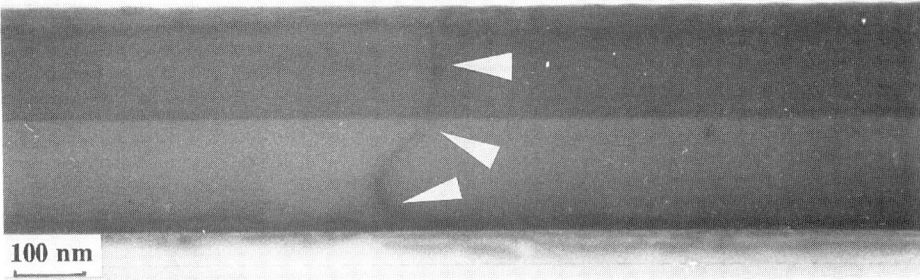

Fig. 3: XTEM micrograph of a sample with 20% germanium in the lower part and 25% germanium and 0.5% carbon in the upper part.

Also in the case of Ge layers additional carbon increases the stability and reduces the relaxation (see Osten et al 1994b). We prepared a Ge layer with about 1% additional carbon (see Fig. 4). The lattice parameter $a_{\parallel}$ can be obtained from the plan-view TEM micrograph (see Bugiel and Zaumseil 1993). Measuring the average distances of the moire fringes by performing a fast Fourier transformation (d=0.72 nm), we can calculate $\Delta a_{\parallel}/a_{Si}$ independently of any other material constant, and obtain 0.027 ± 0.003. Assuming the strain-equivalent structure of $Si_{0.1}Ge_{0.9}$. layer, we could estimate a degree of relaxation of R=(72 10)%. This independtly determined value is in excellent agreement with the XRD result. The degree of relaxation is surprisingly low. For a 30 nm thick Ge layer antimony-mediated grown on Si(001) at a temperature above 450°C, we always find a complete relaxation to the structure of bulk germanium (Osten and Klatt 1994; LeGoues,

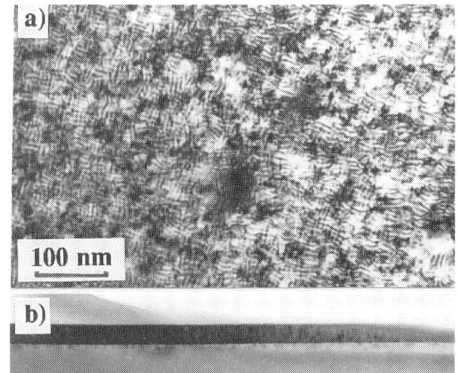

Fig. 4: Plan-view (a) and cross-sectional (b) TEM micrographs show a fully covered $Ge_{1-y}C_y$ layer.

Copel and Tromp 1990). For a 30 nm thick $Si_{0.05}Ge_{0.95}$ layer antimony-mediated grown on Si(001) at 450°C, we determined a degree of relaxation of approximately 95% (Bugiel and Zaumseil 1993).

During the relaxation of carbon rich $Si_{1-y}C_y$ alloy no typical misfit dislocation network was generated. We observed β-SiC precipitates in agreement with Powell et al (1993) and Strane et al (1994).

3.3. Si_nC δ-layers

Fig. 5 shows a HREM image of a (110)-oriented cross-section with a single carbon δ-layer embedded in the Si matrix. Based on the amount of deposited carbon (1 monolayer, estimated by SIMS), we obtained a concentration of carbon higher than 10%. Nevertheless, no extended defects were observed.

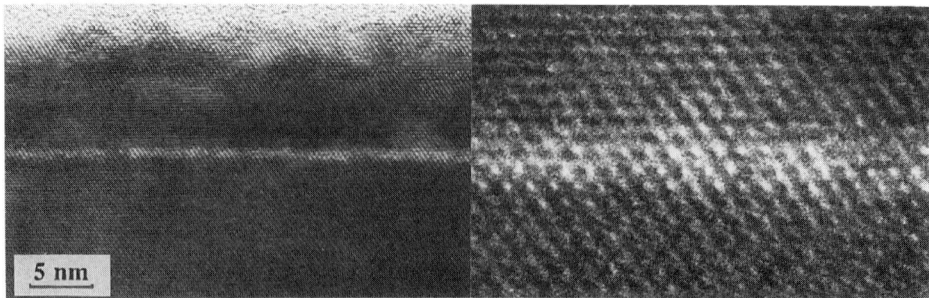

Fig. 5: HREM micrograph of single δ-layer of Si_nC. The cross-section shows the defect free structure including the cap. Inserted in a) is the 5 times magnified fragment of the carbon rich layer.

Summarizing, the Si cap layer first grows as islands resulting in the reduction of carbon-induced strain. These defect free islands can further coalesce without defect generation. Increasing the amount of deposited carbon will lead to an increase of strain and finally to defect generation. Nearly perfect structures of the predicted Si_nC δ-layers embedded in silicon have been observed. They show quasi-periodic variations in the carbon distribution and some ordering over the layer. The observed microroughness of the layer interfaces is probably the result of carbon diffusion during the formation of the Si_nC system.

4. CONCLUSIONS AND ACKNOWLEDGEMENTS

The addition of carbon into pseudomorphically grown SiGe layers opens a new perspective in strain manipulation. Using special MBE techniques it is possible to overcome thermodynamical restrictions and produce highly-concentrated Si_nC δ-layer which is energetically forbidden.

The authors are very thankful to Dr. H. Rücker, Dr. M. Methfessel (IHP-Frankfurt/Oder) and to Dr. P. Werner (MPI-Halle/Saale) for assistence with HREM.

REFERENCES

Bugiel E and Zaumseil P 1993 Appl. Phys. Lett. 62 2051

Endisch D, Osten H J, Zaumseil P and Zinke-Allmang M 1995 Nuclear Instruments and Methods B (in press)

LeGoues F K, Copel M W and Tromp R M 1990 Phys. Rev. Lett. B42 11690

Osten H J, Bugiel E and Zaumseil P 1994a Appl. Phys. Lett. 64 3440

Osten H J, Bugiel E and Zaumseil P 1994b J. Cryst. Growth 142 322

Osten H J and Klatt J Appl. Phys. Lett. 1994 65 5630

Powell A R, Eberl K, LeGoues F E, Ek B A and Iyer S S 1993 J. Vac. Sci. Technol. B 11 (3) 1064

Rücker H, Methfessel M, Bugiel E and Osten H J 1994 Phys. Rev. Lett. 72 3578

Ruvimow S, Bugiel E and Osten H J 1994 Appl. Phys. Lett. (unpublished)

Strane J W, Stein H J, Lee S R, Picraux S T, Watanabe J K and Mayer J W 1994 J. Appl. Phys. 76 (6)

Inst. Phys. Conf. Ser. No 146
Paper presented at Microsc. Semicond. Mater. Conf., Oxford, 20–23 March 1995
© 1995 IOP Publishing Ltd

Deformation microtwinning in heteroepitaxial films on offcut (001) substrates

T T Cheng, X L Wei[1], M Aindow and I P Jones

School of Metallurgy and Materials, The University of Birmingham, Edgbaston, B15 2TT, UK.
[1]Beijing Laboratory for Electron Microscopy, Chinese Academy of Sciences, Beijing, China.

ABSTRACT: Cross-sectional TEM has been used to reveal that for two different systems, MOCVD CdTe on 2°-off (001)GaAs and MBE GaAs on 4°-off (001)Si, the microtwins are distributed anisotropically with more than 90% lying on one of the {111} planes in each case. It is argued that these twins must arise by shear processes since they lie on the plane for which the Schmid factor is highest.

1. INTRODUCTION

Where microtwins arise in heteroepitaxial thin films their distribution is often highly anisotropic. For epitaxial semiconductors with the sphalerite structure, (111) films tend to contain twins only on the plane (111) parallel to the substrate surface whereas (001) films contain more twins on ($\bar{1}$11) and ($\bar{1}\bar{1}$1) than on ($\bar{1}$11) and (1$\bar{1}$1) or *vice versa*. For (111) films it is widely accepted that the twins are produced by double positioning on the advancing planar growth front. For (001) films, however, two different explanations have been proposed. Where the twins are growth twins produced by double positioning on {111} facets of initial island nuclei, the anisotropy is usually ascribed to differences in growth rates and/or facet areas for the (111)$_A$ and (111)$_B$ surfaces (e.g. Ernst and Pirouz 1989). Where the twins are deformation twins produced by the motion of partial dislocations on the {111} planes, the anisotropy is usually explained as the result of differences in the mobility of the partials on the two types of planes (e.g. Brown *et al.* 1989). Under certain circumstances other distributions can be produced in (001) deposits including microstructures where almost all of the twins lie on one of the {111} planes. In this paper, we describe a cross-sectional TEM study of two systems which exhibit such microstructures and discuss the ways in which the anisotropy could arise.

2. EXPERIMENTAL PROCEDURE

CdTe deposits were grown by MOCVD using dimethylcadmium and di-isopropyltellurium as precursors. The substrates used were Epi-ready GaAs off-cut from (001) by 2° about [1$\bar{1}$0] towards [110], prepared by heat cleaning only at 565-585°C for 30 min. in 1 atm. of H_2 in the growth chamber. Initial nucleation layers were deposited for up to 5 min. at growth temperatures of between 280°C and 320°C before heating to 370°C for deposition of the "bulk" of the CdTe film. For one control sample an intial layer was annealed at 370°C for 5 min.

GaAs deposits were grown by MBE on Si off-cut from (001) by 4° about [1$\bar{1}$0] towards [110]. The substrates were prepared by annealing at 700°C in the growth chamber to remove any surface oxide films. A layer of GaAs ≈1 μm in thickness was deposited onto the substrate at 340°C under a balanced flux of Ga and As such that one would expect two-dimentional growth to occur. The wafer was then cooled to 280°C and a capping layer of GaAs was deposited.

Cross-sectional TEM specimens were produced with two portions of each wafer rotated by 90° with respect to one another such that the specimen normal was parallel to the two <110> directions in the substrate surface. The discs were dimpled, Ar$^+$ ion beam milled to perforation at liquid nitrogen temperatures and examined in Philips CM20, JEOL 200CX and JEOL 2010 TEMs all operating at 200kV.

3. RESULTS

3.1 CdTe on GaAs

Examination of the initial layers showed that the continuous deposits are polycrystalline and consist of grains which have one of the two orientation relationships: $(001)_{CdTe}//(001)_{GaAs}$ with $[110]_{CdTe}//[110]_{GaAs}$ and $(111)_{CdTe}//(001)_{GaAs}$ with $[1\bar{1}0]_{CdTe}//[110]_{GaAs}$ which we will refer to as the (001) and (111) orientations, respectively. One example is shown in Figure 1 which is a high resolution electron microscopy (HREM) image obtained with the beam direction parallel to $[1\bar{1}0]$ from a layer grown at 280°C for 5 min. The individual (001) and (111) grains contain some growth defects and lateral twins. Examination of many cross-sectional images suggests that (111)-oriented grains may form at flat regions of the GaAs surface whereas (001)-oriented grains form where there is some roughening of the substrate surface.

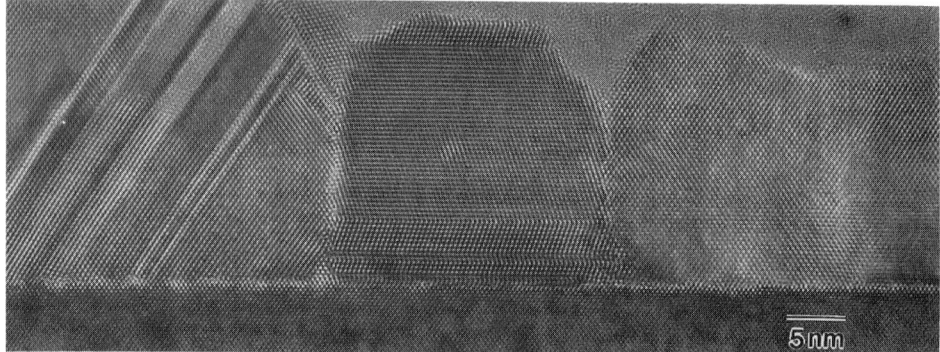

Figure 1 - HREM image obtained from an initial layer with the beam direction parallel to $[1\bar{1}0]$.

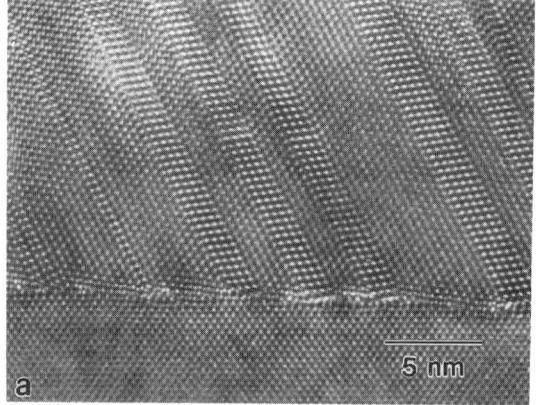

Figures 2(a) and (b) - HREM images obtained from 'bulk' CdTe/GaAs layer with beam direction parallel to $[1\bar{1}0]$ and $[110]$, respectively.

The thick "bulk" films exhibit a very different microstructure - there are no regions with (111)CdTe parallel to the substrate surface, and the deposit has the same anisotropic defect distribution as that observed previously for CdTe deposits on thermally etched GaAs substrates (Cheng *et al.* 1994) . These single crystal deposits have subgrain boundaries and 90°-type misfit dislocations along [110], i.e. perpendicular to the off-cut axis, and planar faults and an intimate mixture of roughly equal numbers of 60°-type and 90°-type dislocations along $[1\bar{1}0]$, i.e. parallel to the off-cut axis. Typical HREM images of this microstructure are shown in Figs. 2(a) & (b) - it can be seen that all of the twins, planar faults and 60°-type dislocations lie on the plane (111) producing a large rotation between the substrate and deposit lattices about the off-cut axis; the magnitude of this rotation is ≈7.2° for the sample shown. For a more detailed description of this microstructure the reader is directed to Cheng *et al.* (1994). The microstructure of the annealed control sample was very similar to that of the thick films with almost no (111)-oriented CdTe and the same distinctive distribution of defects in the (001)-oriented regions (Figs. 3(a) & (b)).

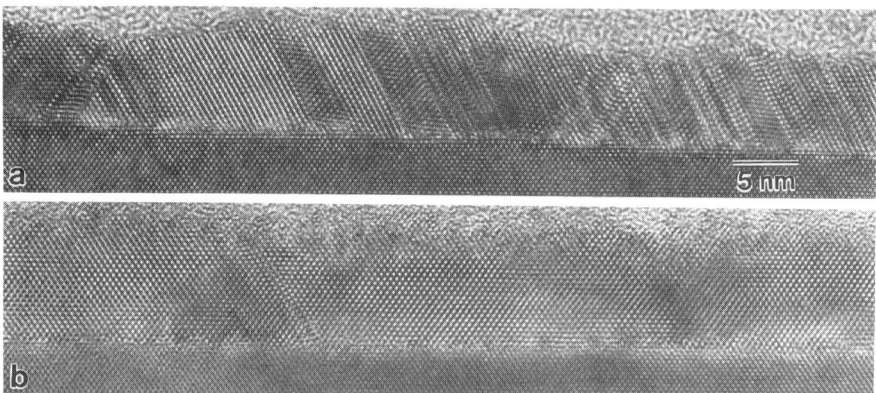

Figs 3 (a)&(b) - HREM images from the control sample with **B** // [1T0] and [110], respectively.

3.2 GaAs on Si

Analysis of HREM images obtained with the beam direction, **B**, parallel to [110] and [1T0] shows that the misfit dislocations lying in the interface are an irregularly spaced mixture of 60°- and 90°-type dislocations (e.g. Figure 4). Whilst it is difficult to determine the misfit dislocation content of the interface accurately it is clear that this is significantly less than that required for complete misfit accommodation. Diffraction contrast imaging of the defect microstructure within the GaAs film shows that there is a pronounced anisotropy in the nature and distribution of the defects. Figs 5 (a) and (b) are bright field images which show the microstructure of the film viewed along [1T0] and [110], respectively. There is a very high density of twins in the deposit on (111) but very few on (T11), (1T1) and (TT1). These twins exhibit a lenticular shape and are not present in the thinnest areas of the specimen such as that shown in Fig. 5(a). For further details see Wei and Aindow (1994).

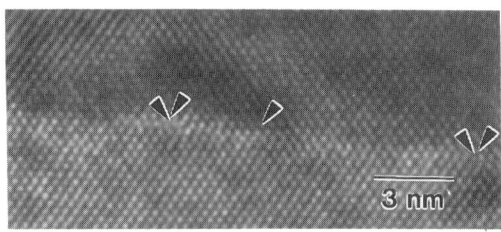

Figure 4 - HREM image with **B** // [1T0].

Figs 5 (a) & (b) - Bright-field images with **B** ≈ [1T0] and [110], respectively.

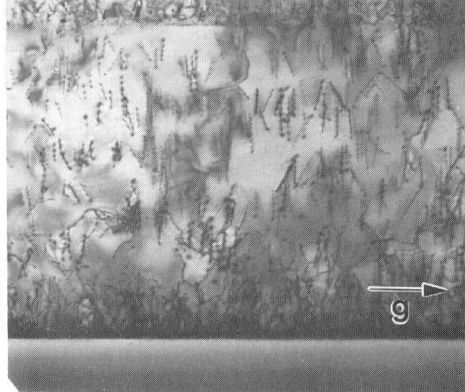

4. DISCUSSION

For the CdTe deposits the defect anisotropy is not present in the (001)-oriented regions of the initial layers but is, instead, induced when the (111)-oriented regions transform on heating to the bulk growth temperature. It has been shown by Bourret *et al.* (1993) that the interface between (111)CdTe and (001)GaAs surface is reconstructed and does not contain misfit dislocations in the usual sense. Thus when the (111)-oriented regions transform, additional defects must be introduced to accommodate the misfit and so the microtwins must arise by shear processes. In a similar manner, the GaAs deposits were grown under conditions where Frank-van der Merwe growth occurs and thus the microtwins can only arise by shear processes.

In considering the formation of twins in such deposits it is helpful to calculate the Schmid factors for the possible twinning systems. For pseudomorphic deposits grown on nominal substrates, the Schmid factors for twinning on each of the systems $[11\bar{2}](111)$, $[\bar{1}1\bar{2}](\bar{1}11)$, $[1\bar{1}\bar{2}](1\bar{1}1)$, and $[112](\bar{1}\bar{1}1)$ are equal to 0.471, which is more than twice that for twinning on any other system. For pseudomorphic growth on substrates offcut by 2° and 4° about $[\bar{1}10]$, however, the Schmid factors for the system $[11\bar{2}](111)$ are 0.482 and 0.490, respectively. In the latter case, the resolved shear stress for twinning on this system will be more than 4% higher than that on any other system. Whilst this difference is modest, its significance lies in the fact that one would expect deformation twins and stacking faults on (111) to form before those on other systems and thus any twins or faults which form subsequently on $(\bar{1}11)$, $(1\bar{1}1)$ or $(\bar{1}\bar{1}1)$ will be blocked by those on (111) and thus the observed anisotropy could develop in GaAs on Si. For CdTe on GaAs, however, the resolved shear stress for twinning on the system $[11\bar{2}](111)$ will only be $\approx 2\%$ higher than that on any other system. This small difference will, however, be enhanced by the presence of (001)-oriented regions in which the misfit is accommodated completely by 90°-type dislocations since the stresses induced when the film transforms will be inclined to the deposit surface. Thus here again the observed anisotropy could develop when twins and stacking faults form first on (111).

We note that the twin anisotropy observed here is not a characteristic of these particular materials systems: if CdTe is deposited on etched GaAs such that no (111)-oriented regions are produced, or if GaAs is deposited on Si under an imbalanced flux such that growth occurs in Volmer Weber mode, then high densities of twins form on (111) and $(\bar{1}\bar{1}1)$ and few on $(\bar{1}11)$ and $(1\bar{1}1)$ or *vice versa*. Thus microstructures in which the majority of twins lie on one plane in a (001) deposit arise when there are high levels of stress in continuous films on vicinal substrates giving anisotropic resolved shear stresses for deformation twinning processes.

5. CONCLUSIONS

It has been shown that for deposits such as MOCVD CdTe on heat-cleaned GaAs, and balanced flux MBE GaAs on Si, where there is a significant induced or residual stress in the continuous film, the effect of vicinal offcut is to give anisotropy in the shear stresses on the {111} planes which can lead to highly anisotropic distributions of microtwins.

ACKNOWLEDGEMENTS

The authors would like to thank Dr J.E. Hails who grew the CdTe on GaAs samples and Dr. Y.K. Li who grew the GaAs on Si samples. This work has been supported in part by Defence Research Agency, Malvern, Ministry of Defence, United Kingdom and in part by the Chinese Academy of Sciences and the Chinese Natural Science Foundation.

REFERENCES

Bourret A and Feuillet G 1993 Mat. Res. Soc. Symp. Proc. 295 71.
Brown P D, Russell G J and Woods J 1989 J. Appl. Phys. 66 129.
Cheng T T, Aindow M, Jones I P, Hails J E, Williams D J and Astles M G 1994 J. Crystal Growth 135 409.
Ernst F and Pirouz P 1989 J. Mater. Res. 4 834.
Wei X L and Aindow M 1994 Appl. Phys. Lett. 65 1903.

Inst. Phys. Conf. Ser. No 146
Paper presented at Microsc. Semicond. Mater. Conf., Oxford, 20–23 March 1995

Depth resolved X-ray studies of tilt distributions in CdTe/GaAs epilayers

R I Port, C D Moore, B K Tanner, K Durose and J E Hails*

Department of Physics, University of Durham, South Road, Durham, DH1 3LE, UK.
*DRA Electronics Division, St Andrew's Road, Malvern, Worcs. WR14 3PS, UK.

ABSTRACT: A systematic HRXRD study of CdTe/GaAs (001) epilayers grown under identical MOVPE conditions is presented. HRXRD was carried out using conventional and synchrotron sources. The width of the tilt distribution decreased rapidly with increasing layer thickness up to a thickness of 8μm after which the rate of change was considerably reduced. The lack of surface specific information obtained using conventional radiation sources is discussed and compared with more surface sensitive methods avaliable using synchrotron radiation.

1. INTRODUCTION

CdTe is grown as a buffer layer on GaAs substrates for the subsequent growth of cadmium mercury telluride, important properties of the buffer layer being the dislocation density at its top surface, surface morphology and the extent to which diffusion of substrate species occur through it. The high lattice mismatch (14.6%) causes a very high density of Lomer and 60° dislocations in the interface which relieve most of the misfit strain. This generates a high density of threading dislocations which participate in the relief of residual strain. Both dislocation density and residual strain decrease with increasing layer thickness (Tatsuoka et al 1991, Tatsuoka et al 1992, Durose and Tatsuoka 1993).

The object of this work is to investigate how the threading dislocation density in CdTe/GaAs varies with increasing layer thickness using conventional and synchrotron high resolution X-ray diffraction (HRXRD) techniques. The tilts imparted on a layer by the component of the Burgers vector perpendicular to the interface result in broadening of the X-ray rocking curve. In the absence of strain and lattice dilatations the dislocation density may be estimated from the full width at half maximum (FWHM) of the rocking curve using for example the formula of Gay et al (1953). However this formula has been shown to give an overestimate of dislocation density for some heteroepitaxial layers compared with etch pit density measurements (Watson 1993).

2. EXPERIMENTAL

A series of CdTe layers were grown under identical conditions by MOVPE on (001)GaAs wafers offcut by 4° towards <110>. The substrates were prepared using conventional etching techniques and the layers grown at 370°C using diisopropyl telluride and dimethylcadmium precursors in H_2 carrier gas. Thickness maps of the wafers were obtained by transmission FTIR fringe spacing measurement.

HRXRD measurements of layers in the thickness range 0.3-25 μm were made using $CoK_{\alpha 1}$ radiation. The symmetric 004 reflection was used in the nondispersive setting with InSb(001) as the reference crystal. In order to establish the influence on FWHM of continued growth, the 25 μm thick layer was repeatedly etched with 2% bromine in 1:4 methanol:ethylenediol and remeasured by HRXRD.

The broadening of the X-ray rocking curve includes contributions from lattice dilatations and tilts. Triple axis X-ray measurements of the samples have shown that broadening due to lattice dilatations is very small compared with that due to layer tilts. Therefore the FWHM of the HRXRD rocking curve of CdTe/GaAs gives a direct measure of the layer tilt distribution - and by inference the threading dislocation density (Gay et al 1953).

When a monoenergetic beam of X-ray photons of intensity I_0 passes through a medium of thickness t it suffers a decrease in intensity, according to equation 1. μ is the linear attenuation coefficient and is a strong function of wavelength as shown in Table 1.

$$I = I_o exp\left[-\mu(\lambda)t\right] \qquad\qquad \text{equation 1}$$

Synchrotron radiation has the advantage of being continous in wavelength. By selecting different wavelengths to be diffracted by the reference crystal, the depth penetration profile of the X-ray beam in the layer can be changed, thus providing a non-destructive means of investigating tilt variations in the bulk of the layer. HRXRD measurements with variable wavelength were performed at the Daresbury Laboratory Synchrotron Radiation Source. A strain relieved Si(111) crystal was used to select the wavelengths reaching the specimen. Of the wavelengths satisfying $n\lambda=2d_{111}sin\theta_B$ only the Si 333 reflection gave significant intensity from the specimen set at the Bragg angle for CdTe 004 diffraction. Therefore only this wavelength contributed to the recorded rocking curve. Since it is known that layer thickness and X-ray rocking curve width vary considerably over the area of a wafer (Tanner et al 1993), care was taken to record rocking curves from the same position on the sample after each etch step and when the incident wavelength was changed.

source		Synchrotron			$CoK_{\alpha1}$	$CuK_{\alpha1}$
Wavelength / Å	1.95	1.25	0.85	0.68	1.789	1.541
μ / μm^{-1}	0.25	0.082	0.030	0.017	0.21	0.14

Table 1: Linear absorption coefficients for CdTe at X-ray wavelengths used in this work (International Tables 1962).

3. RESULTS AND DISCUSSION

Figure 1 shows the FWHM of the HRXRD rocking curve for $CoK_{\alpha1}$ radiation for the five individual layers. Also shown in Figure 1 are the rocking curve widths obtained after repeated etching of the 25μm thick layer. There is a considerable reduction in FWHM with increasing layer thickness up to a thickness of 8μm, followed by a continued decrease at a much slower rate up to the limit of the experimental range. There is no definite minimum reached.

The rocking curve width of a 25μm layer as a function of wavelength is shown in Figure 2. The values of FWHM are those obtained from curve fitting a Gaussian line shape to the data. The more penetrating the X-rays (ie the shorter the wavelength) the greater the contribution from the region near the interface. Assuming a Gaussian distribution of tilts with mean angular position $\bar\theta$ (the angle at rocking curve maximum) and standard deviation σ_n within each monolayer, the intensity I_θ diffracted from the layer at a given incident angle θ and wavelength λ is given by equation 2. $A_{\theta\lambda}$ contains the structure factor and other diffraction terms, $N_{\bar\theta n}$ is the number of tilt domains at $\bar\theta$ in the nth layer, z is the distance of the nth monolayer from the top surface and θ_B is the Bragg angle for the chosen wavelength.

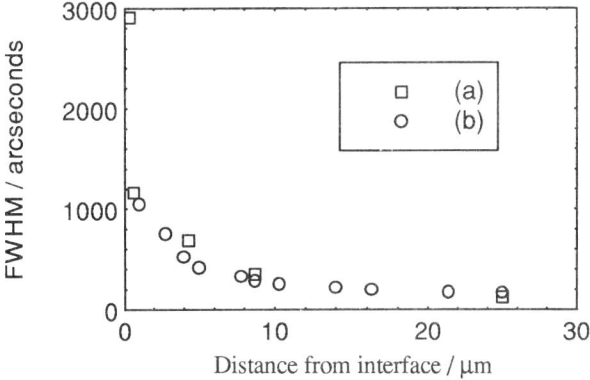

Figure 1: FWHM of X-ray rocking curves recorded using CoKα1 radiation as a function of distance from the interface for (a) five individual layers and (b) repeated etching and measurement of the 25μm thick layer.

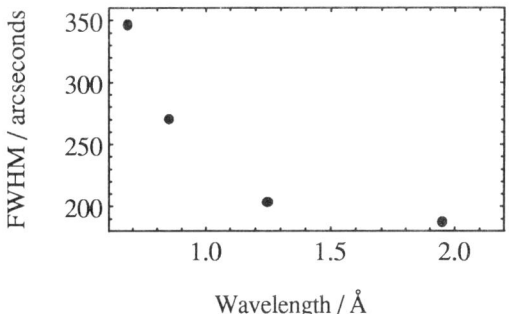

Figure 2: FWHM of X-ray rocking curves of a 25μm thick layer recorded using different wavelengths

$$I_\theta = \sum_{n=1}^{m} A_{\theta\lambda} N_{\bar{\theta}n} \exp\left[-\frac{1}{2}\left(\frac{\theta-\bar{\theta}}{\sigma_n}\right)^2\right] \exp\left[-2\mu(\lambda)z\text{cosec}\theta_B\right] \qquad \text{equation 2}$$

Equation 2 is kinematical in approach and does not take into account the reduction of intensity due to diffraction by each monolayer. The contribution from each monolayer is dependent on the wavelength of the incident photons. A longer wavelength will give a rocking curve much more representative of the surface tilt distribution than a shorter wavelength, provided an absorption edge is not crossed. In an attempt to obtain more detailed quantitative information of tilt distributions within the bulk of the layer, the 25μm layer was represented by five separate layers each of which was assumed to be represented by a Gaussian distribution of tilts. Deconvolution of these contributions was attempted using simultaneous equations derived from equation 2. The solutions of the simultaneous equations gave tilt distributions which were not self consistent and this is thought to be due to the complex nature of the term $A_{\theta\lambda}$ which is heavily dependent on wavelength and geometry.

312

In conventional source experiments $CoK_{\alpha 1}$ radiation of wavelength 1.789Å was chosen rather than the standard $CuK_{\alpha 1}$ radiation of wavelength 1.541Å due to its more surface sensitive nature. A better estimate of the width of the tilt distribution of the surface layer may be obtained for these layers using the data obtained with different wavelengths by extrapolation of the data to infinite absorption. This is best achieved by plotting FWHM against $1/\mu$ and extrapolating to $1/\mu=0$ as shown in Figure 3. The different extent to which diffraction condition may be optimised and the different polarization states of conventional and synchrotron sources means that data from these two sources is not directly comparable.

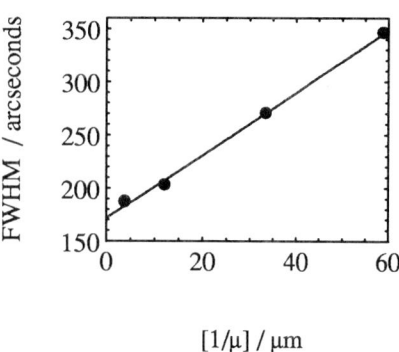

$[1/\mu] / \mu m$

Figure 3: Extrapolation to infinite absorption.of FWHM values obtained from the 25μm thick layer.

4. CONCLUSIONS

In this material HRXRD rocking curve widths represent tilt distributions in the layers. It has been found that the range of tilts, and by inference the threading dislocation density, rapidly decreases with increasing thickness up to about 8μm after which the rate of change decreases significantly. The improvement in layer quality continues to the limit of the experimental observations at 25μm. Further growth appears to improve the quality of underlying layers relative to a layer whose growth was terminated at a given thickness.

X-ray rocking curves represent a distribution of tilts sampled from within the bulk of the layer. The contribution to the rocking curve from a given depth is dependent upon the wavelength chosen, therefore the comparison of FWHM for different layers is only valid if measured at the same wavelength. Contribution to the rocking curve width from underlying, more dislocated regions explains the observation that the model due to Gay et al (1953) gives an overestimate of dislocation density in some heteroepitaxial semiconductor layers relative to etch pit density (Watson 1993).

A FWHM value representative of the surface region may be obtained by extrapolating a graph of FWHM versus $1/\mu$ to infinite absorption.

ACKNOWLEDGEMENTS

This work has been carried out with the support of the Defence Research Agency (Malvern) and the Engineering and Physical Sciences Research Council.

REFERENCES

Durose K and Tatsuoka H 1993 Inst. Phys. Conf. Ser. 134 581
Gay P, Hirsch P B and Kelly A 1953 Acta Metallurgica 1 315
Intenational Tables for X-ray Crystallography 1962 (Birmingham: Kynoch) vol. 3.
Tanner B K, Hallam T D, Funaki M and Brinkman A W 1993 Mat. Res. Soc. Symp. 280 635
Tatsuoka H, Kuwabara H, Nakanishi Y and Fujiyasu H 1991 Thin Solid Films 201 59
Tatsuoka H, Kuwabara H, Nakanishi Y and Fujiyasu H 1992 J. Cryst. Growth 117 554
Watson C C R 1993 Ph. D. Thesis University of Durham

Inst. Phys. Conf. Ser. No 146
Paper presented at Microsc. Semicond. Mater. Conf., Oxford, 20–23 March 1995
© *1995 IOP Publishing Ltd*

Effect of substrate temperature on the microstructure of CdTe thin films grown by plasma-enhanced MOCVD on (001) GaAs

R Chester, T T Cheng, M Aindow, I P Jones and D J Williams[†].

School of Metallurgy and Materials, The University of Birmingham, Edgbaston, B15 2TT.
[†]Electronics Division, DRA, St. Andrews Road, Malvern, Worcs, WR14 3PS

ABSTRACT: Cross-sectional transmission electron microscopy (TEM) has been used to reveal that CdTe thin films grown on (001)GaAs by plasma-enhanced metalorganic chemical vapour deposition (MOCVD) exhibit a distinctive polycrystalline microstructure consisting of heavily twinned columnar grains. The films are epitaxial at the interface but this breaks down with increasing thickness and the reasons for this are discussed.

1. INTRODUCTION

Cadmium Telluride (CdTe) is a useful x-ray detector material and can be used as a buffer layer for the growth of the IR detector material $Cd_xHg_{1-x}Te$. Single crystal epitaxial thin films of CdTe and $Cd_xHg_{1-x}Te$ have been grown successfully by a number of techniques, including MOCVD but for good quality films this requires substrate temperatures of 350-400°C. For layered device structures, however, abrupt heterojunctions are required and at these growth temperatures there are problems due to the interdiffusion rate of Hg and Cd.

It is well established that one of the main reasons for using elevated growth temperatures in MOCVD is to promote efficient cracking of the precursors at the substrate surface. Thus one approach to reducing the growth temperature and suppressing interdiffusion is to precrack the alkyls before they impinge upon a substrate held at some lower temperature. The first successful application of such an approach was by Lu *et al.* (1986) who grew single crystal $Cd_xHg_{1-x}Te$ onto CdTe substrates held at 225°C. Subsequent work by Benjoshis *et al.* (1990) in which a high frequency plasma discharge has been used to crack the alkyls has shown that $Cd_xHg_{1-x}Te$ can be deposited onto a buffer layer of CdTe on (001)GaAs. The films obtained were polycrystalline for temperatures less than 115°C, and single crystal for temperatures above this.

Attempts to apply this approach to systems in which the lattice misfit is higher have been less successful. Haq (1988) used ultra-violet irradiation to crack the alkyls for MOCVD of CdTe on GaAs. The films were found to exhibit an epitaxial relationship at the interface but were polycrystalline consisting of equiaxed grains ≈500Å in diameter. It was also shown that the grain size and shape depended on the Cd:Te metalorganic ratio. In this paper we present a cross-sectional TEM study of the microstructure of CdTe thin films grown onto (001)GaAs at reduced temperatures by plasma-enhanced MOCVD and discuss the way in which this microstructure might develop.

2. EXPERIMENTAL

Layers of CdTe were grown on 2°-off (001) GaAs by plasma-enhanced MOCVD using dimethylcadmium and di-isopropyltellurium as precursors. Prior to growth the substrates were prepared by etching in a 5:1:1 solution of $H_2SO_4/H_2O_2/H_2O$. The alkyls were precracked in a hydrogen plasma before impinging on the substrate which was held at 180°C. Cross-sectional TEM specimens were produced in the usual manner. The discs were dimpled, ion beam milled to perforation using Ar^+ ions at liquid nitrogen temperatures and examined in Philips CM20 and JEOL 200CX TEMs both operating at 200 kV.

314

3. RESULTS

Low magnification bright field images reveal that the thin films are polycrystalline with columnar grains perpendicular to the interface (e.g. Figure 1). The width of the grains varies from 50-200nm with *most* grains (>90%) having widths of 100-150nm. The columnar grains are heavily twinned with all of the twins in a particular grain appearing to lie on the same {111} plane and with each twin appearing to extend across the whole width of the grain. A selected area diffraction pattern obtained from an area which includes about six grains and the substrate with the beam direction, **B**, parallel to [110]GaAs is shown in Figure 2. There is insufficient space to present a full analysis of such a pattern here but it can be shown that they consist of: a [110] zone axis pattern (ZAP) from the GaAs substrate, two types of <110> ZAPs from CdTe - some of which are only misaligned slightly with respect to the substrate and others which are twin related to these, and <112> ZAPs from CdTe which are oriented with the 222-type reflections perpendicular to the substrate surface.

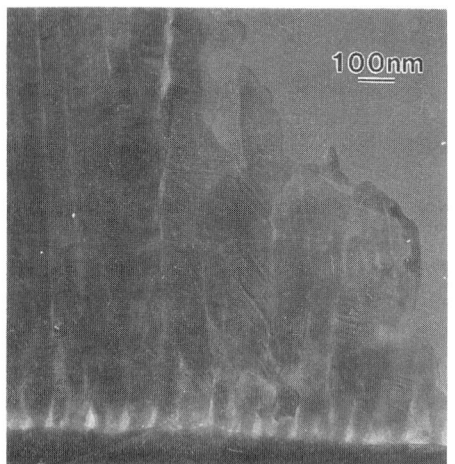

Figure 1. Bright field image of the overall microstructure with **B**//[110]GaAs.

Figure 2. SADP from a region near to the CdTe/GaAs interfacewith **B**//[110]GaAs.

Figure 3. Typical HREM image of the CdTe/GaAs interface region obtained with **B**//[110] showing the character of the misfit dislocations and the misorientation.

The presence in Figure 2 of [110] ZAPs from CdTe in which the diffraction vectors are nearly parallel to the corresponding ones in GaAs indicates that a significant proportion of the material is epitaxial and HREM images confirm this. Figure 3 is a typical HREM image obtained from the CdTe/GaAs interface with **B** // [110]GaAs. At the interface, the CdTe is highly distorted but epitaxial at all points. Significantly, there are twins and stacking faults on more than one {111} plane at the interface and there is some evidence for preferential blocking of those on one plane by those on another (e.g. at A). The misfit at the interface would appear to be accommodated by a mixture of 90°-type and 60°-type misfit dislocations. All of the 60°-type dislocations lie on the same plane as the preferred plane for the twins and this results in a rotation of the deposit with respect to the substrate of 4.5°. We should emphasise that whilst such an anisotropic microstructure was observed in all HREM images, the sense and extent of the anisotropy varies locally.

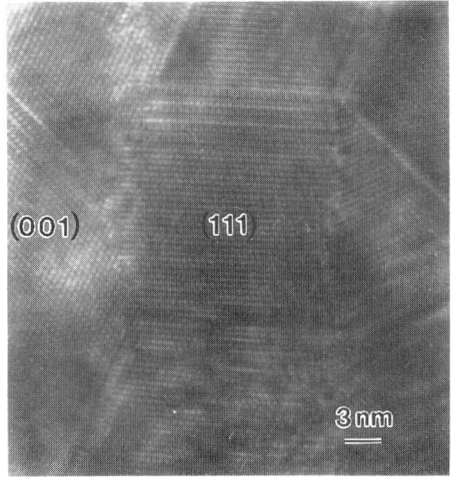

Figure 4. HREM image of a (111)-oriented region obtained with **B**//[110]GaAs

Whilst a highly twinned microstructure such as that shown in Figure 3 is consistent with the diffraction maxima observed in SADPs such as Figure 2, it cannot account for the presence of a faint [112] ZAP. HREM images such as Figure 4 which was obtained from a region further away from the interface and between columnar grains indicate how such patterns arise. At the junction of the grains there is a small pocket of material oriented with (111) parallel to the substrate surface and [1$\bar{1}$0]CdTe //[110]GaAs. For a region with the equivalent orientation, [1$\bar{1}$0]CdTe//[1$\bar{1}$0]GaAs, then an SADP with **B**//[110] would contain a [11$\bar{2}$]CdTe ZAP.

In summary, the microstructure consists of columnar subgrains each of which start out epitaxial but with different misorientations due to the numbers of 60°-type misfit dislocations. Further from the interfaces, the stacking faults and twins on one {111} plane dominate within each subgrain with occasional small pockets of (111)-oriented CdTe between them.

4. DISCUSSION

In considering the development of this distinctive columnar microstructure, we should first emphasise that under the hydrogen plasma used for this study, the cracking efficiency of the precursors is sufficiently high that any microstructural peculiarities must be related to the kinetics of nucleation and growth rather than of the gas decomposition. Furthermore, the presence of large numbers of 90°-type misfit dislocations indicates that growth is occuring in Volmer-Weber mode, as one might expect for a system with 14.6% lattice mismatch. Thus we must consider the effect of island nuclei on the final defect microstructure. It is tempting to speculate that each columnar grain corresponds to an initial supracritical island nucleus, or a group of nuclei with very similar misorientations which would require an island density of greater than 10^{10} cm^{-2}. Such high island densities are only produced under conditions where homogeneous nucleation on the terraces of the substrate is favourable, but this is consistent with the substrate temperatures used in this work since at lower temperatures the critical nucleus size will be small and the nucleation rate will be high. Moreover, this is consistent with the formation of epitaxial deposits at the interface since homogeneous nuclei tend to adopt the orientation of the substrate more readily than do heterogeneous ones (Markov and Stoyanov 1987).

We consider next the density of twins within the columnar grains. The thickness of the twins varies from 1-2 atomic layers up to around 200 atomic layers - these have very flat coherent interfaces and extend across the full width of the grains. Such parallel-sided twins are often associated with "double positioning" processes whereby atoms adopt positions

316

which correspond to errors in the stacking sequence on exposed {111} facets of island nuclei (e.g. Ernst and Pirouz (1989)). It is, however unusual to find such high densities of twins arising by such a process and we have attempted to evaluate whether this could be due to the growth temperature. Models in which the atoms adopt the "wrong positions" on a (111) facet require that such atoms either have insufficient energy, or that the deposition rate is so high that the atoms do not have sufficient time, to jump into the "correct" sites on the (111) facet. Estimates of the critical temperature for the former case yield values of 200-300°C (Chester 1995) and whilst the data used in such calculations is not sufficiently accurate, the critical temperatures do lie between those which are used normally for MOCVD of CdTe and those used for the present study.

Whereas the high density of twins can be explained on the basis of the low growth temperature, the anisotropic defect distribution is more surprising. The presence of large numbers of twins, stacking faults and 60°-type misfit dislocations on only one of the {111} planes has been observed previously for MOCVD CdTe on vicinal (001)GaAs (Cheng *et al.* 1994, 1995). In the previous study, however, the anisotropy was more uniform throughout the film and was induced when regions of the initial CdTe deposit with (111) parallel to the substrate surface transformed on heating to the bulk growth temperature. Such anisotropic microstructures are also observed in deposits such as GaAs on (001)Si where the use of vicinal substrates gives different resolved shear stresses for the introduction of defects on the four {111} planes (Wei and Aindow 1994, Cheng *et al.* 1995). If the high density of homogeneous nuclei were to coalesce to give a continuous film at an early stage of growth then there could be significant residual stress in the deposit. The effect of a rough interface would then be to give resolved shear stresses on the four {111} planes which varied from place to place and the Burgers vectors and glide planes for the 60°-type misfit dislocations introduced in response to these stresses would vary similarly. Since differences in the numbers of 60°-type misfit dislocations will alter the misorientation, it could be just such an effect which leads to the breakdown of the epitaxial deposit into subgrains. It is not possible, however to explain the microtwin distributions on this basis and thus we must assume that the favoured twin plane corresponds to the exposed growth facet in a particular subgrain and that the heavy twinning occurs by double positioning as discussed above.

5. CONCLUSIONS

The effect of reduced substrate surface temperature in the growth of (001)CdTe on (001)GaAs has been investigated by cross-sectional TEM. The deposits exhibit a heavily twinned columnar microstructure with a rough interface and an anisotropic defect distribution which varies from grain to grain. We have proposed that the reduced growth temperature favours homogeneous nucleation and that local differences in shear stresses could give variations in defect content when residual strain is relieved. Moreover, it is suggested that the heavy twinning occurs due to the low surface mobility of deposit species at these temperatures

ACKNOWLEDGEMENTS

The authors would like to thank Professors J.F. Knott and R.E. Smallman for the provision of laboratory facilities. Thanks are also due to EPSRC and DRA for financial support.

REFERENCES

Benjoshis T I, Vasilevski M I, Gurilev B V, Ershov S N, Karzhin G A, Ozerov A B and Parker T D 1990 J. Tech. Phys. (Russia) 60 (1) 160.
Cheng T T, Aindow M, Jones I P, Hails J E, Williams D J and Astles M G 1994 J. of Crystal Growth 135 409.
Cheng T T, Wei X L, Aindow M and Jones IP 1995 these proceedings.
Chester R 1995 unpublished work.
Ernst F and Pirouz P 1989 J. Mater. Res. 4 834.
Haq S 1988 PhD Thesis, University of Birmingham.
Lu P-Y, Wang C-H, Williams L M, Chu S N G and Stiles C M 1986 Appl. Phys. Lett. 49 20.
Markov and Stoyanov 1987 Contemp. Phys. 28 267.
Wei X L and Aindow M 1994 Appl. Phys. Lett 65 1903.

Inst. Phys. Conf. Ser. No 146
Paper presented at Microsc. Semicond. Mater. Conf., Oxford, 20–23 March 1995

TEM Study of (Cd,Hg)Te grown by Molecular Beam Epitaxy

E Selvig*, K Gjønnes*, A Olsen*, T Colin** and S Løvold**

*Centre for Materials Research/Department of Physics, University of Oslo, Norway
**Norwegian Defence Research Establishment. P.O. Box 25, N-2007 Kjeller, Norway

ABSTRACT: Multilayer structures of (Cd,Hg)Te have been grown by molecular beam epitaxy (MBE) and studied by transmission electron microscopy (TEM) and selected area electron diffraction (SAD). Depending on the growth conditions, the samples may contain various defects like twins, misfit dislocations and precipitates. SAD patterns of the (Cd,Hg)Te specimens show non-radial diffuse streaks along the <111>* directions from both the CdTe and (Cd,Hg)Te layers. TEM images show that the main contribution to the diffuse scattering originates near the substrate/MBE grown layer interfaces. The diffuse streaks are interpreted as due to static displacements in the {111} planes.

1. INTRODUCTION

The (Cd,Hg)Te system has important applications in the detection of infrared (IR) radiation. The defect structure plays an important role in device applications. At the University of Oslo in Norway we are studying molecular beam epitaxy grown materials of (Cd,Hg)Te by electron microscopy and diffraction. The project is carried out in collaboration with the Norwegian Defence Research Establishment at Kjeller where the samples are grown by molecular beam epitaxy (MBE) (Skauli, Colin and Løvold 1994).

The present paper describes a transmission electron microscopy (TEM) study of (Cd,Hg)Te multilayer structures grown by MBE. Various layer structures of CdTe and (Cd,Hg)Te including quantum well structures and superlattices have been grown on CdTe and (Cd,Zn)Te substrates. These substrates are known to have only a small mismatch with (Cd,Hg)Te. Depending on the growth conditions, the layers may contain various defects like twins, threading and misfit dislocations and precipitates. A tight control of the process parameters is required in the MBE machine in order to grow crystals with sufficient quality for infrared detector applications.

2. EXPERIMENTAL

The samples were grown by molecular beam epitaxy carried out in a Riber 32P MBE machine. The substrates were single crystals of CdTe or $(Cd_{0.96},Zn_{0.04})$Te. A significant improvement in the control of the substrate temperature has been achieved by improving the thermal contact between the thermocouple and the substrate holder (Skauli, Colin and Løvold 1994). Three effusion cells containing CdTe, Te and Hg respectively were used. Individual shutters in front of the cells were operated to control the transition between (Cd,Hg)Te and HgTe layers. The growth was monitored by using reflection high energy electron diffraction (RHEED). Fig. 1 gives details of the specimens discussed in the present paper. The orientations of the substrates were (111)B, (776)B and (211)B where B indicates that the top atomic layer of the substrate is Te. The (776)B orientation of the substrate surface corresponds to a misorientation relative to the (111)B plane (4° off around [$1\bar{1}0$] towards (110)). The (211)B orientation is a misoriented (111)B (19.5° off around [$1\bar{1}0$] towards (001)).

Cross section specimens were prepared from the MBE grown samples. Two pieces of the same sample were glued together by using Super epoxy. This epoxy was found to be very useful for making cross sections of such brittle materials as (Cd,Hg)Te and to avoid heating of the material and diffusion during the specimen preparation procedure. After mechanically polishing the cross sections down to 55 - 100 µm in thickness, thin specimens suitable for TEM work were prepared by ion-beam thinning in a Gatan Dual Mill model 600. All TEM specimens were studied in a JEOL 200CX transmission electron microscope. In order to prevent electron beam damage during the observations in the electron microscope, the specimens were cooled down to liquid nitrogen temperature during the TEM examination by using a Gatan cold stage specimen holder. Fig. 2 shows two examples of the investigated layer structures and their thicknesses.

Specimen	Substrate	Substrate orientation	Composition of MBE grown layer	Layer thickness (µm)
CT 5	CdTe	(111)B	CdTe	2.1
CT 9	CdTe	(111)B	CdTe	1.25
CMT 68	(Cd,Zn)Te	(776)B	$(Cd_{0.34},Hg_{0.66})Te$	5.73
CMT 142-2	(Cd,Zn)Te	(211)B	$(Cd_{0.30},Hg_{0.70})Te$	8.28
QS 91	CdTe	(111)B	$(Cd_{0.23},Hg_{0.77})Te$ layers*	3.27*

Fig. 1. Details of the cross section samples described in the present paper. * See Fig. 2a).

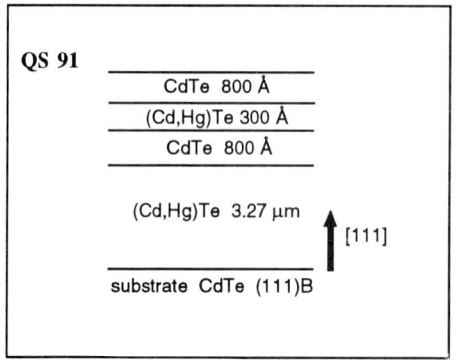
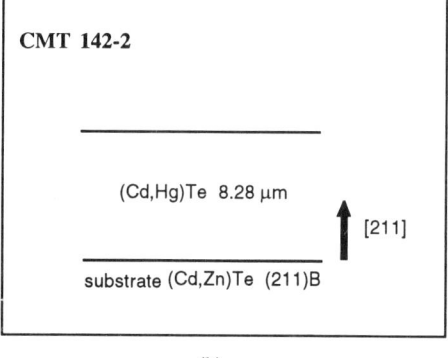

(a)   (b)

Fig. 2. Two examples of the investigated samples.

3. RESULTS

3.1 MBE grown structures

Examples of some typical multilayer structures and their defects are shown in the Figs. 3 - 5. Fig. 3 is a typical dark field (DF) image of the multilayer stucture of (Cd,Hg)Te (QS 91) showing a layer of (Cd,Hg)Te (indicated by arrows) between twinned layers of CdTe. Fig. 4 shows a bright field (BF) image of (111) twins in a CdTe layer (CT 9). Microtwins are observed near the interface. The twins become thicker as we move away from the interface. Fig. 5 shows an example of threading dislocations starting at the interface between the substrate and the MBE grown (Cd,Hg)Te layer (CMT 68) and extending up through the MBE grown layer. Fig. 6 shows misfit dislocations along the interface between the (Cd,Zn)Te substrate and the grown (Cd,Hg)Te layer (CMT 68).

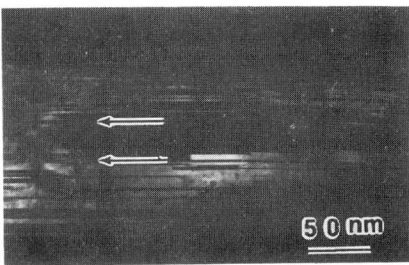

Fig. 3. DF image of the multilayer structure of (Cd,Hg)Te. Specimen: QS 91. Projection: [1$\bar{1}$0].

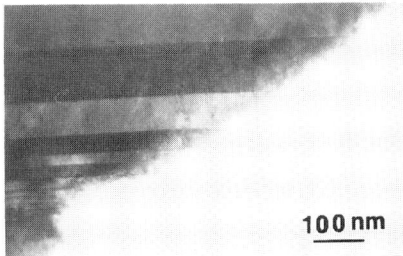

Fig. 4. BF image of the twinned CdTe layer. Specimen: CT 9. Projection: [1$\bar{1}$0].

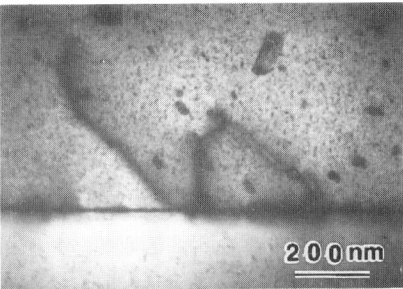

Fig. 5. DF image (11$\bar{1}$ reflection) of (Cd,Hg)Te showing threading dislocations propagating in the MBE grown layer of CMT 68. Projection: [1$\bar{1}$0].

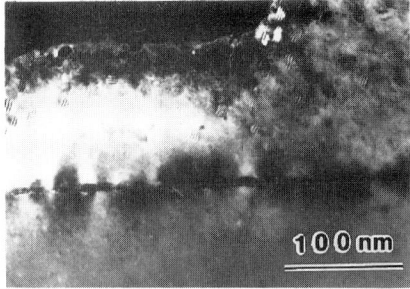

Fig. 6. DF image (11$\bar{1}$ reflection) of the grown (Cd,Hg)Te layer on the (Cd,Zn)Te substrate. Specimen: CMT 68. Projection: [1$\bar{1}$0].

3.2 Diffuse scattering

During the TEM study non-radial diffuse scattering was observed in several of the selected area electron diffraction (SAD) patterns. Examples of the diffuse scattering are shown in the Figs. 7 -10. Fig. 7 shows a SAD pattern of (Cd,Hg)Te with continuous diffuse streaks along the [111]* and the [11$\bar{1}$]* directions and running through the Bragg reflections. Note that there are no diffuse streaks in the [110]* direction and only short segments along [001]*. The diffuse streaks were found all over the QS 91 specimen, but DF imaging of the diffuse scattering shows that the main contribution to the diffuse streaks comes from the regions near the interface between the substrate and the MBE grown layer (Fig. 8).

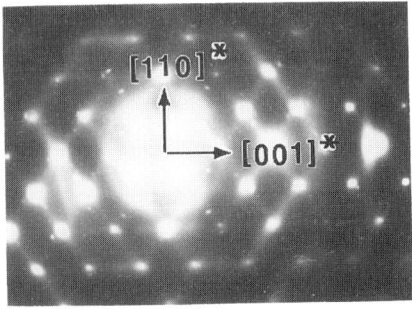

Fig. 7. SAD pattern of (Cd,Hg)Te. Specimen QS 91. Projection [1$\bar{1}$0].

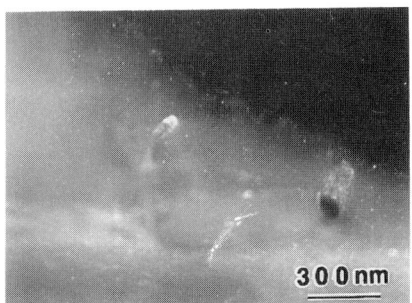

Fig. 8. DF image of QS 91 indicating disorder near the interface. Projection: [1$\bar{1}$0].

320

Fig. 9 shows a SAD pattern with non-radial diffuse streaks connecting some of the reflections. Note that the streaks are continuous along the <111>* directions (except through the origin), but they are not continuous along other directions. There are no streaks along the $[1\bar{1}0]$* direction as indicated by the arrow. It is also important to notice that the diffuse streaks along the <111>* directions become weaker and disappear before the Bragg reflections with increasing scattering angle. This may be due to the curvature of the Ewald sphere and indicates that the diffuse scattering is streaks and not diffuse planes in the reciprocal space. Fig. 10 shows a SAD pattern from a twinned (Cd,Hg)Te specimen taken near the $[12\bar{3}]$ orientation of one of the crystals. The SAD pattern shows diffuse streaks in the $[111]$* direction of this crystal, but also along the $[5\bar{1}1]$* direction. This latter diffuse scattering is due to diffuse streaks along the $[\bar{1}1\bar{1}]$* direction in the twins.

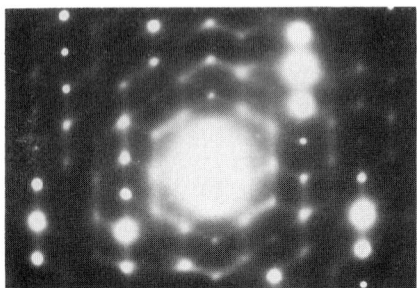 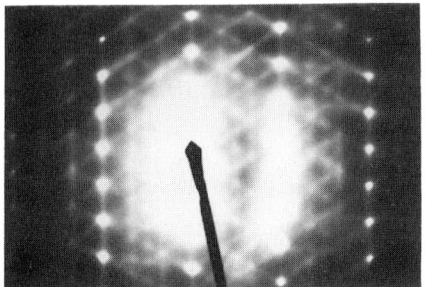

Fig. 9. SAD pattern of CdTe. Specimen orientation near the $[11\bar{2}]$ projection. Specimen: CT 5.

Fig. 10. SAD pattern from a twinned microsctructure of CMT 142-2. Projection: $[12\bar{3}]$.

3.3 Interpretation of the diffuse scattering

Characteristic diffuse non-radial streaks in electron diffraction patterns from silicon, germanium and various substances with the ZnS type structure have been noticed for many years (Honjo, Kodera and Kitamura 1964, Glas, Gors and Hénoc 1990). Such non-radial streaks may be due to thermal vibrations or static displacements of the atoms away from their equilibrium positions. Thermal vibrations in the sphalerite type structure can be described as transversal vibrations about the <110> directions. The corresponding non-radial diffuse electron scattering will then be diffuse planes with <110> as the normal direction. These diffuse planes will then appear as streaks in the diffraction patterns due to an intersection of the diffuse planes and the Ewald sphere (Honjo, Kodera and Kitamura 1964).

However, the diffuse scattering found in the present work is not in agreement with their interpretation. It may be interpreted more like the explanation of the diffuse scattering observed by Finlayson, Goodman, Olsen, Norman and Wilkins in 1984 in their work on a pre-martensitic In - Tl alloy. Firstly, the main contribution to the diffuse scattering found in the present (Cd,Hg)Te study comes from the regions near the interface between the substrate and the MBE-grown layers. Secondly, the diffuse scattering was observed at low temperatures (liquid nitrogen temperature) and consequently it is more probable that the diffuse scattering is due to static displacements than thermal vibrations. Thirdly, the strong tendency for diffuse streaks along the <111>* directions indicates that the responsible static disorder is within the {111} planes.

REFERENCES

Honjo G, Kodera S and Kitamura N 1964 J. Phys. Soc. Japan <u>19</u> 351
Finlayson T R, Goodman P, Olsen A, Norman N and Wilkins S W 1984 Acta Cryst. <u>B40</u> 555
Glas F, Gors C and Hénoc P 1990 Phil. Mag. <u>62</u> 373
Skauli T, Colin T and Løvold S 1994 J. Vac. Sci. Technol. <u>A12</u> 274

Inst. Phys. Conf. Ser. No 146
Paper presented at Microsc. Semicond. Mater. Conf., Oxford, 20–23 March 1995
© 1995 IOP Publishing Ltd

Analysis of epitaxial CuInSe$_2$ layers grown onto GaAs substrates by a hybrid magnetron sputtering technique

C A Mullan*, **C J Kiely and A Rockett**[1]

Department of Materials Science and Engineering, University of Liverpool, Liverpool L69
[1]Engineering Science Building, University of Illinois at Urbana-Champaign, Urbana, USA
*now at CEMD, McMaster University, Hamilton, Ontario, L8S 4M7, Canada

ABSTRACT: Transmission electron microscopy of epitaxial CuInSe$_2$ layers grown on GaAs(001) and (111)B substrates by a hybrid magnetron sputtering process is presented. Epitaxial layers grown on GaAs(111) substrates have three energetically degenerate orientational variants, whereas those grown on GaAs(001) exhibit unique epitaxy. The CuInSe$_2$ overlayers have a high density of stacking faults and threading dislocations which have been characterised. The interfaces between the CuInSe$_2$ and GaAs are found to be highly faceted and to contain a large number of voids which have a deleterious effect on film adhesion.

1. INTRODUCTION

CuInSe$_2$ is a direct bandgap ternary (I-III-VI$_2$) semiconductor which has been used in the polycrystalline thin film form as an absorber layer in photovoltaic devices. It has a tetragonal chalcopyrite structure ($\bar{I}42d$) with a and c lattice parameters of 5.79Å and 11.61 Å respectively. The chalcopyrite structure can be derived from the sphalerite structure by substituting Cu and In atoms onto the cation sublattice in an ordered fashion. There is considerable interest at present in growing epitaxial single crystal CuInSe$_2$ layers on GaAs substrates for infra-red detector applications. To date only two growth techniques, namely flash evaporation (Schumann et al. 1986) and reactive magnetron sputtering (Chung Yang et al. 1993) have successfully been used to achieve epitaxy in this system. In this paper we present a microstructural assessment of CuInSe$_2$ layers deposited on (001) and (111)B GaAs substrates by a hybrid magnetron sputtering process.

2. EXPERIMENTAL

The hybrid magnetron sputtering technique used in this study has been described elsewhere (Chung Yang et al. 1991). Briefly, GaAs wafers were cleaned in solvents and rinsed in de-ionised water before heating to 550°C. Copper and indium were then sputtered onto the wafer which was held at a temperature of 525°C. Selenium was simultaneously evaporated onto the surface. It was found that it was necessary to minimise exposure of the GaAs surface to Se vapour during the outgassing because Se can etch the wafer and form pits. The films were made into plan view specimens by chemical back thinning with a Cl/methanol etch. Cross section samples were thinned using 5kV Ar ions with an 18° incidence angle. A JEOL 2000FX transmission electron microscope, operating at 200kV, equipped with a LINK EDX analyser was used to study the specimens.

3. RESULTS AND DISCUSSION

3.1 CuInSe₂/GaAs(111)

The epitaxial relationship observed in this sample was $CuInSe_2[221]//GaAs[111]$ and $CuInSe_2[1\bar{1}0]//GaAs[01\bar{1}]$. Two other energetically degenerate orientational variants are possible in this system because the (112) $CuInSe_2$ plane, with m point symmetry, has been matched onto the (111)GaAs surface which has 3m point symmetry. The presence of these additional variants which can be generated by rotations of 120° and 240° about the GaAs[111] substrate normal was confirmed by Mullan (1994) from selected area diffraction experiments.

Figure 1 is a cross sectional image of the $CuInSe_2$/GaAs(111) interface, taken along the GaAs[01$\bar{1}$] direction. The epilayer is about 3μm thick and is highly defective. EDX measurements yield an epilayer composition of 25.7at%Cu, 23.9at%In and 50.4at%Se. The imbalance of Cu and In cations in the lattice leads to the formation of a population of Cu_{In} antisite defects which give the epilayer a p-type character. An obvious feature in figure 1 is that the interface regions exhibit pyramidal and needle like intrusions which can extend up to 130nm into the GaAs substrate. These features are thought to arise when the $CuInSe_2$ deposit fills in pre-existing surface pits in the substrate formed by excessive exposure to Se vapour before growth. This proposition is borne out by the observation of vertical moiré fringes with a 27-32Å periodicity in the vicinity of these pyramidal features. These correspond well to the 29 Å parallel moiré fringes expected from the beating of reflections created by the (02$\bar{4}$) $CuInSe_2$ and (2$\bar{2}$0)GaAs planes which have a mismatch of 2.3%.

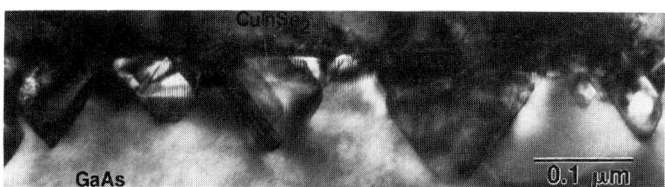

Figure 1 Axial [01$\bar{1}$] cross sectional image of the $CuInSe_2$/GaAs(111) interface.

Plan view micrographs of the epilayer, such as that in figure 2, show that it contains a very high density of stacking faults and dislocations. Systematic dark field and weak beam imaging experiments (Mullan 1994) have allowed us to deduce the crystallographic nature of some of the defects. Three sets of stacking faults on inclined (1$\bar{1}$2), ($\bar{1}$12) and (11$\bar{2}$) planes were identified, which intersect the foil surface along [$\bar{2}$04], [0$\bar{2}$4] and [2$\bar{2}$0] directions. The visibility criteria and displacement vectors, **R**, of these faults are summarised in table 1.

	Habit plane	**R**	$\bar{2}\bar{2}0.\mathbf{R}$	$\bar{2}04.\mathbf{R}$	$0\bar{2}4.\mathbf{R}$
Fault 1	(1$\bar{1}$2)	1/12[241]	-1/3(√)	0(x)	-1/3(√)
Fault 2	($\bar{1}$12)	1/6[1$\bar{1}$1]	2/3(√)	1/3(√)	1(x)
Fault 3	($\bar{1}$1$\bar{2}$)	1/12[4$\bar{2}$1]	1(x)	-1/3(√)	2/3(√)

Table 1. A summary of the behaviour of the stacking faults in figure 1
(√ implies that the fault is visible, x that it is not)

Three types of perfect edge dislocations (one of which is labelled in figure 2) were also observed in these films. These have line directions which lie in the plane of the foil and lie above the interface plane. The **g.b** visibility criteria for these defects are summarised in table 2.

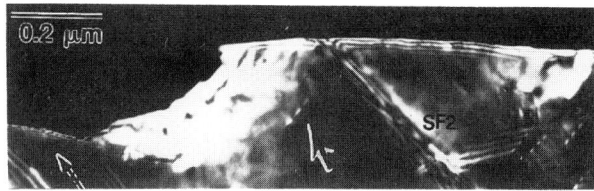

Figure 2 Plan view image of the $CuInSe_2/GaAs(111)$ epilayer ($g=0\bar{2}4$)

b	u	$2\bar{2}0.b$	$\bar{2}04.b$	$0\bar{2}4.b$	$\bar{1}14.b$
$[1\bar{1}0]$	$[\bar{1}14]$	$4(\sqrt{})$	$-2(\sqrt{})$	$2(\sqrt{})$	$0(x)$
$[20\bar{1}]$	$[1\bar{2}2]$	$4(\sqrt{})$	$-8(\sqrt{})$	$-4(\sqrt{})$	$6(\sqrt{})$
$[02\bar{1}]$	$[\bar{2}12]$	$-4(\sqrt{})$	$-4(\sqrt{})$	$0(x)$	$6(\sqrt{})$

Table 2. The invisibility criteria for the dislocations observed in the $CuInSe_2/GaAs(111)$ films ($\sqrt{}$ implies that the fault is visible, x that it is not)

3.2 $CuInSe_2/GaAs(001)$

The epitaxial relationship between the epilayer and substrate was found to be $CuInSe_2//GaAs[001]$ and $CuInSe_2[010]//GaAs[010]$. Since the GaAs (001) and $CuInSe_2(001)$ planes both have mm point symmerty, no energetically degenerate orientational variants exist. Cross-sectional micrographs of this sample again showed pyramidal and needle-like intrusions into the substrate, as well as interfacial voids. EDX analysis of the epilayer composition from such cross section specimens confirmed the previous SIMS depth profiling results of Chung Yang et al(1993), who deduced that considerable Ga diffusion across the interface occurs resulting in a graded composition $CuIn_{1-x}Ga_xSe_2$ quaternary layer. We believe that some of the interfacial voids observed in this sample, which consequently gave rise to poor epilayer adhesion, may be generated by the Kirkendall effect.

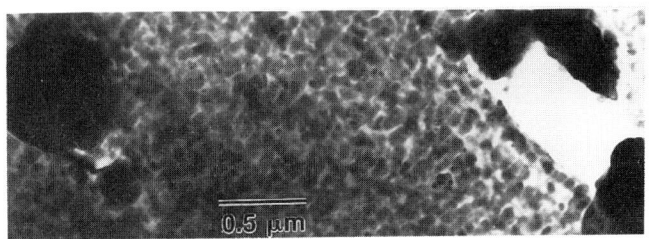

Figure 3 Axial [001] plan view image of the $CuIn_{1-x}Ga_xSe_2/GaAs(001)$ epilayer

Figure 3 shows a plan view image of this $CuIn_{1-x}Ga_xSe_2$ epilayer. Second phase particles are visible, although some have been preferentially etched away, leaving behind facetted rectangular pits. The pit edges are along [110] and [1$\bar{1}$0] directions whereas the pit sides consist of inclined {112} type planes. The approximate 3:1 aspect ratios of these pits are a simple consequence of the different surface energies associated with Se terminated and cation-terminated {112} planes. EDX analysis of the associated second phase material gave typical compositions of 10at%Cu, 4at%In, 24at%Ga, 7at%As and 55at%Se.

The (001) $CuIn_{1-x}Ga_xSe_2$ epilayers showed a high density of paired threading dislocations, some of which are labelled in figure 4. Systematic **g.b** analyses of these defects suggested that they were superdislocations (as previously observed by Hennig-Michaeli et al (1989) and Kiely et al(1991)). The diffraction contrast observed by defects I and II are consistent with them being of mixd character with pair I having **b**=1/2[110] and pair II having **b**=1/2[1$\bar{1}$0]. Simarily, pairs III and IV have **b**=1/2[201] and **b**=1/2[021] respectively.

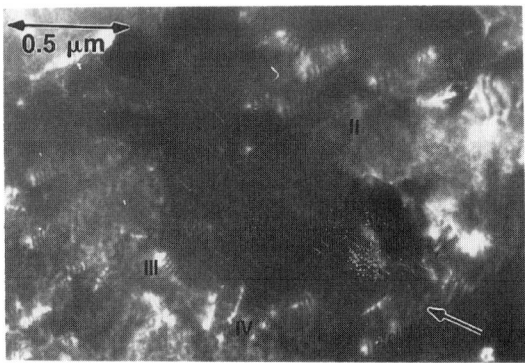

Figure 4 Plan view image of the $CuInSe_2$/GaAs(001) epilayer, **g**=040

4. CONCLUSIONS

It has been demonstrated that epitaxial $CuInSe_2$ can be grown on both GaAs(001) and (111) substrates by the hybrid magnetron sputtering technique. Problems associated with the fabrication process include the production of hightly facetted interface structures due to Se etching of the substrate prior to growth. There is also potential for elemental interdiffusion across the interface which can lead to the formation of graded composition $CuIn_{1-x}Ga_xSe_2$ layers. Finally, the layers were found to contain a multitude of stacking faults and dislocation structures which are likely to be detrimental to the opto-electronic properties of the layer.

REFERENCES

Chung Yang L, Chou LJ, Agarwal A and Rockett A 1991, Proc.22nd IEEE Photovoltaic Specialists Conference, 185
Chung Yang L, Berry G, Chou LJ, Kenshole G, Rockett A, Mullan CA and Kiely CJ 1993, Proc.24nd IEEE Photovoltaic Specialists Conference
Hennig- Michaeli C and Couderc JJ, Eur. J. Minerals 1989, 1, 295-310
Kiely CJ, Pond RC, Kenshole G and Rockett A 1991, Phil. Mag. A, 63(6), 1249-1273
Mullan CA, PhD Thesis 1994, University of Liverpool
Schumann B, Temple A and Kuhn G 1986, Solar Cells, 16, 43-63.

ACKNOWLEDGMENTS
We would like to thank EPRI, DoE and EPSRC for supporting this work.

Inst. Phys. Conf. Ser. No 146
Paper presented at Microsc. Semicond. Mater. Conf., Oxford, 20–23 March 1995
© 1995 IOP Publishing Ltd

Domain structures in CuGaSe$_2$ epitaxial films

Gin-Lern Gu, Song-Bin Lin and Bae-Heng Tseng

Institute of Materials Science and Engineering, National Sun Yat-Sen University, Kaohsiung 80424, Taiwan, ROC

ABSTRACT: Thin films of CuGaSe$_2$ are grown on GaAs substrates by molecular beam epitaxy. For the growth of the film with chalcopyrite structure on a single crystal substrate which has a zincblende structure, two kinds of domain structure will develop. They are the orientation domain and the anti-phase domain. In this work, we demonstrate that orientation domains may not occur with the use of an (001) GaAs substrate, while anti-phase domain boundaries can be eliminated by thermal annealing in a Se-beam flux. Electron diffraction and imaging techniques using diffraction contrast are performed to verify these results.

1. INTRODUCTION

I-III-VI$_2$ compound semiconductors have found applications in nonlinear optics and optoelectronic devices such as solar cells (Michalsen and Chen 1982). Among them, CuGaSe$_2$ with 1.71 eV bandgap is a candidate for fabrication of light emitting devices or use as a high-gap layer to form superlattices with CuInSe$_2$. It is essential to develop the techniques for the growth of high quality epitaxial films to realize the device applications.

Epitaxial growth of CuGaSe$_2$ films on single crystal substrates, such as GaAs, has two structural problems which need to be solved, i.e. domain structure and interfacial structure. The work on interfacial structure will be reported elsewhere. In this paper, our emphasis is on the study of domain structures of CuGaSe$_2$ epitaxial layers, in order to find ways to eliminate these domains in the films. Although papers concerning epitaxial growth of I-III-VI$_2$ compounds have been published by other groups, the present topic has not been addressed (Schumann et al 1986). Thermal annealing of the film in the Se-beam flux immediately after film deposition has been carried out for the control of defect chemistry and to reach an equilibrium state which is determined by the annealing temperature and Se overpressure (Tseng et al, in press). The annealing process does not only reduce intrinsic point defects at the equilibrium concentration but also effectively reduces the dislocation density in the epitaxial films (Tseng et al, in press). In this work, evidence will be given that the annealing process also eliminates the APB's.

2. EXPERIMENTAL PROCEDURES

Thin films of CuGaSe$_2$ were grown on GaAs substrates by an MBE technique. The background pressure of the MBE system was 4×10^{-9} torr after bakeout. The temperatures of the Cu source and Se source were kept at 1050°C and 210°C, respectively. The temperature of the Ga source was varied from 1000°C to 1050°C in order to control the Cu/Ga ratio of the films.

The crystallinity and microstructure of the films were evaluated by X-ray diffractometry (XRD) and transmission electron microscopy (TEM), respectively. TEM specimens were prepared by ion milling. A liquid-nitrogen cooled stage was used to prevent the damage caused by ion bombardment.

3. RESULTS AND DISCUSSION

3.1 Domain Structures

The crystal structure of $CuGaSe_2$ has a tetragonal unit cell with c/a ratio about 1.96. It is so-called chalcopyrite structure which is normally found in $I-III-VI_2$ and $II-IV-V_2$ compounds. For growing a film with chalcopyrite structure on a single-crytalline substrate with a zincblende structure, two kinds of domain structures will develop. One is *orientation domain* which is caused by the three variants of the alignment of c-axis on the substrate with known crystalline orientation (Tseng and Wert 1989). For the illustration purpose, Fig. 1 shows how $CuInSe_2$ unit cells can be stacked on a (001) substrate with a zincblende structure. There are three possibilities that the c-axis of chalcopyrite structure may orient perpendicularly or parallel to the substrate surface. The other is *anti-phase domain* which is caused by the three variants of the displacement of cation planes, i.e., (0, 1/2, 1/4), (1/2, 0, 1/4) and (1/2, 1/2, 0) shift, see Fig. 2.

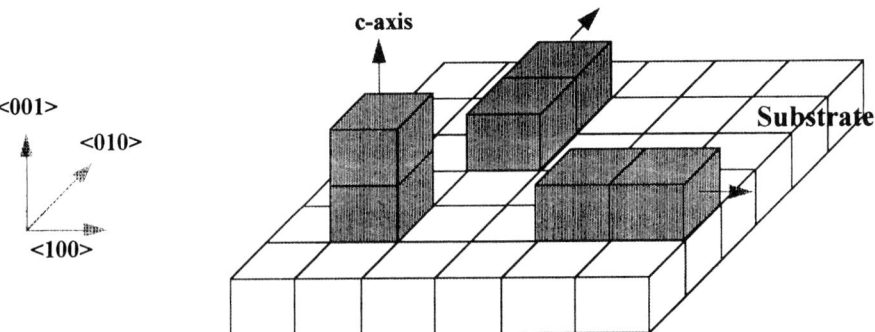

Fig. 1   Formation of orientation domains with c-axis parallel or perpendicular to [001] direction.

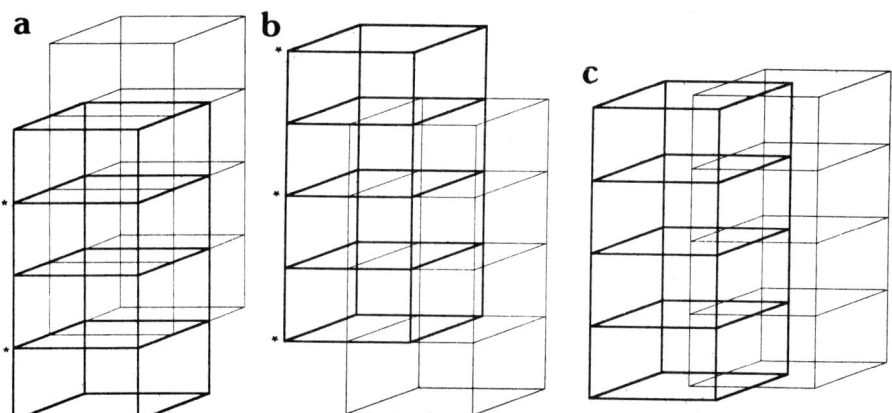

Fig. 2   Formation of anti-phase domains with the shifts in the cation planes (a) (0, 1/2, 1/4), (b) (1/2, 0, 1/4) and (c) (1/2, 1/2, 0) shifts.

3.2 Growth of Epitaxial Films Free of Orientation Domains

Epitaxial layer of CuGaSe₂, which is free of orientation domains, has been successfully grown on (001)GaAs. The structure of orientation domains can be identified by electron diffraction. The sketch of reciprocal lattice of CuGaSe₂ is shown in Fig. 3. The large dots represent the *fundamental reflections* caused by the Se sublattice while the small dots represent the *superlattice reflections* caused by the ordering of cation atoms. For the film grown on (001)GaAs substrate, if the c-axis aligned in a common [001] direction then the orientation domains do not exist. The diffraction pattern using [001] beam direction is simply a (001) pattern. Otherwise, a composite of (001), (010) and (100) patterns is observed. Figure 4a shows a typical diffraction pattern of CuInSe₂ obtained from several areas in a plane-view TEM specimens. It is a (001) pattern indicating the film is free of orientation domains.

For films grown on (111)GaAs, the orientation domains are detected. Figure 4b shows a composite pattern. An atomic model may give explanation for the growth of epitaxial film without the formation of orientation domains (Tseng et al).

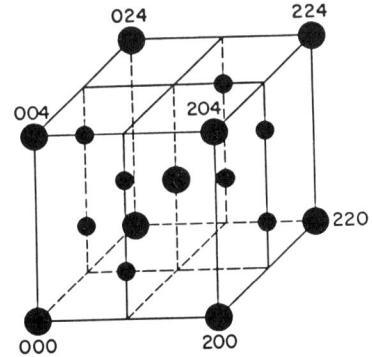

Fig. 3   Sketch of a three-dimentional reciprocal lattice of CuInSe₂.

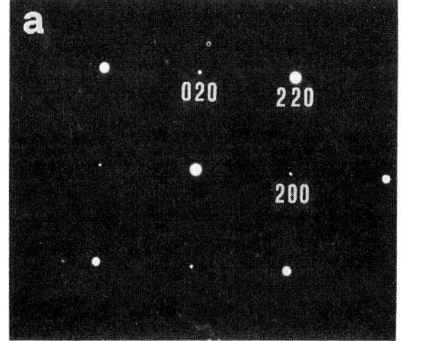

Fig. 4   Typical (001) pattern obtained from a plane-view TEM specimen prepared from the epitaxial film grown on (a) (001)GaAs substrate and (b) (111)GaAs substrate.

3.3 Elimination of Anti-Phase Domains

Thermal anneals of the films in the ambient containing Se (Se vapor or H_2Se) will completely remove the anti-phase domains in the films. The reason that the anneal is carried out in Se-containing atmosphere is to prohibit thermal decomposition of the film. This process can be conducted in the MBE chamber immediately after film deposition by exposing the film to the Se-beam flux for annealing.

The anti-phase domains in the film can be observed in the TEM using diffraction contrast to form the image. The APB's can be observed by the excitation of reflections for which the phase angle $\alpha = \pi$ (Edington 1976). A bright-field image using this technique reveals the ribbon-like APB's, see Fig. 5a. After an anneal at 400°C for 30 min, APB's are completely annihilated. A TEM image using the same diffraction condition as that of Fig. 5a does not show the APB's in the films, see Fig. 5b.

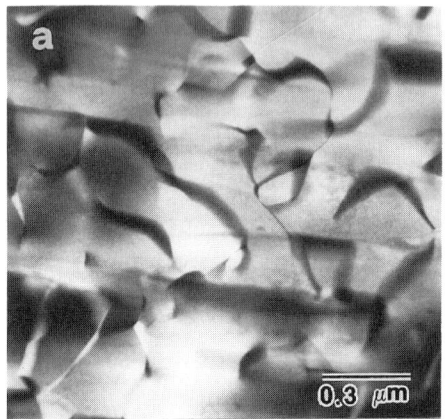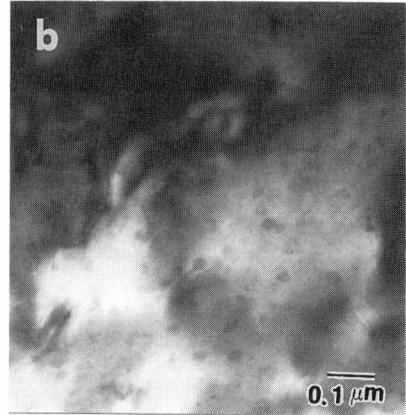

Fig. 5 Bright-field TEM micrographs show (a) ribbon-like APB's in a as-deposited film and (b) the disappearance of APB in an annealed film.

4. SUMMARY

For the growth of epitaxial films with chalcopyrite structure on the substrate with zincblende structure, both orientation domains and anti-phase domains may be developed. Domain structures should be eliminated since the domain boundaries are detrimental to the transport properties of charge carriers. In this work, we demonstrate that $CuGaSe_2$ epitaxial films free of orientation domains were successfully grown on (001)GaAs substrates by an MBE technique. Also, thermal anneal of the film in the Se-beam flux immediately after film deposition can   effectively eliminate the APB's.

REFERENCES

Edington J W 1976 Practical Electron Microscopy in Materials Science (London: Macmillian) pp 149-153
Michalsen R A and Chen W S 1982 Proc. 16th IEEE Photovoltaic Specialists Conf. p 781
Schumann B, Tempel A and Kuhn G 1986 Solar Cells 16 43
Tseng B H and Lin S B and Gu G L submitted to J. Appl. Phys.
Tseng B H and Wert C A 1989 J. Appl. Phys. 65 2254

Inst. Phys. Conf. Ser. No 146
Paper presented at Microsc. Semicond. Mater. Conf., Oxford, 20–23 March 1995

Lattice defects in non-stoichiometric CuInSe$_2$ thin films

E Thanner, O Eibl* and P Pongratz

Institut für Angewandte und Technische Physik, TU Wien, Wiedner Hauptstrasse 8-10, A-1040 Wien, Austria
*Siemens Research Laboratories, Otto-Hahn-Ring 6, D-8 München, Germany

ABSTRACT: A HREM characterisation of dislocations, twins and planar defects is given. A correlation of the size of platelike defects and their number densities for different nonstoichiometric CuInSe$_2$ films is presented, the Cu/In/Se concentrations of which have been measured locally by EDX analysis.

1. INTRODUCTION

Polycrystalline thin films of CuInSe$_2$ (CIS) are interesting because of their possible large scale photovoltaic applications. Rockett et al (1994) summarized current understanding and technology of CIS. Highly efficient devices have been achieved but detailed TEM characterization of the films is important for understanding the cell performance and its improvement (Bode et al 1993, Chen et al 1992, Kiely et al 1991). The structure of stoichiometric CIS is of chalcopyrite-type with tetragonal symmetry (c/a= 2.006) which is topologically related to cubic sphalerite type. Ordered substitution of sphalerite cations by Cu and In atoms introduces a doubling of unit cell dimensions in [001] directions. Pseudocubic Miller indexing can be used therefore. It is clear, however, that surface morphology, crystal quality and solar cell efficiency are influenced drastically by the Cu/In atomic ratio in nonstoichiometric films. The present study is concerned with a HREM-characterisation (Jeol 4000EX) of dislocations, planar defects, twins and secondary phases, which where found in these films. The results of local EDX-microanalysis (in a Jeol 2000FX with HPGe detector) of the films were also compared with nominal Cu/In-concentration ratios which have been used during film deposition.

2. EXPERIMENTAL

The polycrystalline films of CIS were deposited on Mo-coated sodalime glass substrates by DC-magnetron sputtering and/or evaporation and have been synthesized by Rapid-Thermal-Processing (RTA) of elemental Cu, In and Se. Processing details were presented elsewhere (Karg et al 1993)

3. RESULTS

Three different films will be discussed in this paper:
(a) <u>Cu-rich CIS</u> : Atomic ratios are 31.36% Cu : 24.16% In : 44.46% Se . This film has a high number density of planar defects, (1-3) 10^4 μm^{-3} ,the size of which is between 5-100 nm on (111)$_{pc}$ planes (pseudocubic indexing).

(b) <u>In-rich CIS</u> : Atomic ratios are 27.47% Cu : 25.22 % In : 47.29% Se. The number density of the platelike defects is (10.3 +/-3.4) 10^4 μm^{-3}, their size is smaller (<5nm)

330

(c) <u>Se-poor CIS</u> : Atomic ratios are 30.03 % Cu : 23.82 % In : 46.13 % Se. This film contained a high density of platelet-like precipitates on (111)pc planes, their size was between 5-10 nm and their number density was $(9.6 +/-3.2)\ 10^4\mu m^{-3}$.

These measurements have been compared with a standard CIS spectrum for which almost perfect stoichiometry was found (Atomic ratios were In/Cu = 1.09 , Se/Cu = 1.02 , In/Se = 1.07) . Standard errors were +/-5% for In/Cu and Se/Cu ratios and 2% for the In/Se ratio.The grain sizes of these films were very different:
For Cu-rich film (a) we found grains between 2µm and 3.5µm, for the In-rich film (b) we had defect sizes of 80nm-200nm <u>and</u> 500nm-800nm. and for the Se poor film (c) grain size was from 500nm to 2µm. If one compares   these data, the increasing of the grain size correlates with the increasing of Cu content. This is also true for the size of the defects, which are much larger for Cu-rich films.

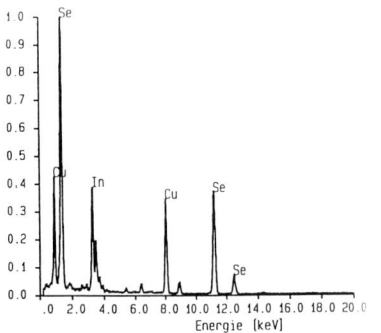 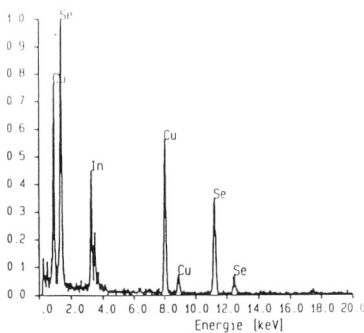

On figure 1 the EDX-spectra of CIS standard (a) and for the Cu-rich CIS film (b) are shown .

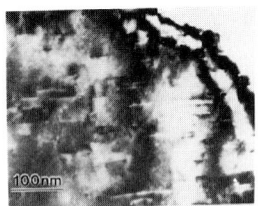 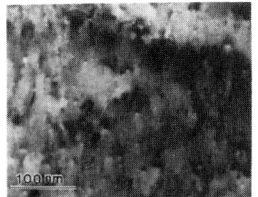 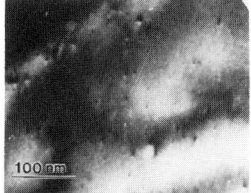

Figure 2 shows bright field images of the films (a) , (c) , (b)

All HREM micrographs have been taken at 400 kV. Comparison with calculated structure images has shown, that using a defocus of df =-45nm, dark contrast lobes correspond to projections of atomic columns (Cu+In) and Se in [110] projection.

Lattice defects in Cu-rich CIS may be seen on figure 3.

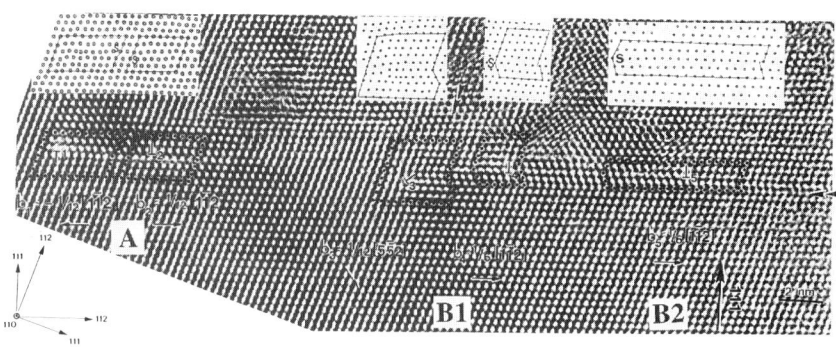

Figure 3: HREM-image of CIS (Cu-rich) in [110]-projection. The Burgers circuit is marked with black spots. On the left side (defect A) a stacking fault bounded by two Shockley-partial-dislocations with the Burgers vector a/6 [112] was identified. The defect beside it is a microtwin which shows steps in the twin-plane. The dislocation B1 on the left side border of the microtwin results from a Frank-partial dislocation and the projection of a Shockley partial-dislocation: a/3 [111] + a/ 6 [1-1 2].

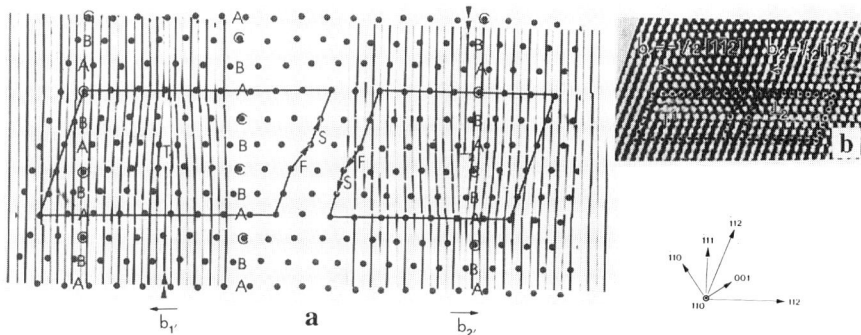

Figure 4: In the distance of a few nm there are two Shockley partial dislocations. Geometrical models of the defects have been drawn to identifiy them. The figure 4 shows the geometrical model of defect A of the HREM-image (fig.3). The added (112)-planes of the Shockley partial dislocations are visible.

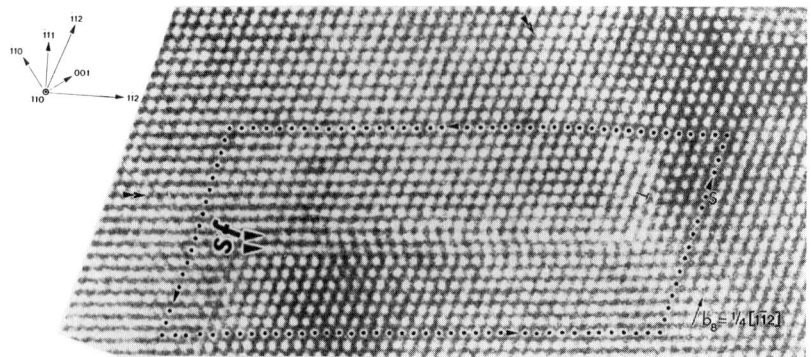

Figure 5 is a HREM image of CIS (Cu-rich) in the [110]-projection. A perfect dislocation with the component of Burgers vector in the image plane which is a/4 [-112] could be identified (A) .The Burgers circuit is marked with black spots. In the distance of a few nm a dislocation, the Burgers vector of which may be explained as the sum of a perfect dislocation and a Shockley partial, was observed. Below the HREM image the geometrical model of the defect is shown. The added planes (-111, 001, -110) are marked with broken lines.

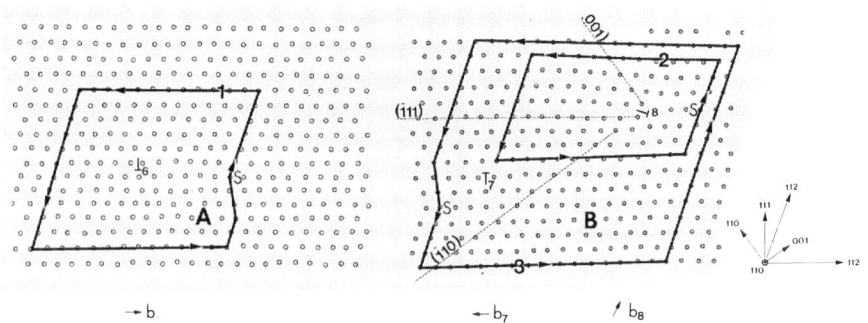

4. CONCLUSIONS

The EDX-analysis of thin CuInSe$_2$ films with different Cu/In/Se concentration ratios (Cu-rich, Se-poor, In-rich) has proved a correlation with defect properties and grain size of the fiilms. Platelike defects on the $\{111\}_{pc}$ lattice-planes with differences in size (from less than 5nm to about 100nm) were found for different Cu/In ratios. HREM of the defects has shown that Burgers vectors of perfect dislocations, Shockley and Frank type as well as microtwins were found in these films. Because of the puzzling variety of defects which were found, localized chemical analysis is shown to be essential to get better insight into the formation processes.

REFERENCES

Bode M H, Al-Jassim, Tuttle J, Albin D 1993 MRS Symp. Proc. Vol.283 pp 927
Kiely C, Pond R C, Kenshole G and Rockett A 1991 Phil Mag A63 1249
Karg F et al 1993 Proc. 23rd IEEE Photovoltaic Specialists. Conf. , Kentucky, 10-15 May
Rockett R et al 1994 Thin Solid Films 237 1

Inst. Phys. Conf. Ser. No 146
Paper presented at Microsc. Semicond. Mater. Conf., Oxford, 20–23 March 1995
© *1995 IOP Publishing Ltd*

Structural aspects of the combination of Si and $YBa_2Cu_3O_{7-x}$

A L Vasiliev[1], G Van Tendeloo[2], Yu Boikov[3], E Olsson[4], Z Ivanov[4],
T Claeson[4], N A Kiselev[1]

[1]Institute of Crystallography, 117333 Moscow, Russia
[2]EMAT, University of Antwerp (RUCA), B 2020 Antwerpen, Belgium
[3]Ioffe Physical-Technical Institute, 194021 St-Petersburg, Russia
[4]Chalmers University of Technology, S 41296 Göteborg, Sweden

ABSTRACT: The microstructure of defects and interfaces as well as interfacial reactions of the $YBa_2Cu_3O_{7-x}$ (YBCO) thin films on Si or Si on sapphire with single Y-stabilized ZrO_2 (YSZ), double CeO_2/YSZ or triple MgO/CeO_2/YSZ buffer layer has been characterized by transmission electron microscopy. The complex buffer made it possible to prevent detrimental interdiffusion and to control the orientation of YBCO layers.

1. INTRODUCTION

The combination of Si and $YBa_2Cu_3O_{7-x}$ (YBCO) thin film technologies requires the formation of a reliable buffer layer to prevent mutual interdiffusion and interfacial reactions (Mogro-Campero 1990, Fork et al 1991). The use of Y-stabilized ZrO_2 (YSZ) buffer layers on Si enables the growth of high quality YBCO layers with appropriate properties (Fork et al. 1990). Double buffer layers on Si can improve the performance of the YBCO thin films (Wecker et al 1992). The necessity of a second buffer layer stems mainly from the chemical reaction of YSZ with YBCO resulting in a $BaZrO_3$ thin layer (Tietz et al 1989, Eibl et al 1991, Hwang et al 1991). Fluorite-type oxides such as CeO_2 or Y_2O_3 are successful as second buffer layers (Bardal et al 1993, Copetti et al 1993). CeO_2 has a very good match to Si (Si-Fd3m, a=0.543nm; CeO_2-Fm3m, a=0.541 nm). The thermal expansion coefficient (TEC) value of CeO_2 is between those of YBCO and Si and is close to that of YSZ which is very convenient for the combination of CeO_2 and YSZ. Due to the differences in TEC the YBCO films on buffered Si contain cracks. The tension due to the mismatch in TEC can be released by using Si on sapphire (SOS) substrates instead. For microelectronic applications, including Josephson junction based and high-frequency devices, a specific YBCO film orientation is required. For *c*-axis oriented YBCO films different in-plane rotations can be achieved by sandwiching additional MgO seed layers (Char et al 1991).

The YBCO films were grown by laser ablation on initially buffered SOS. Polycrystalline YSZ, CeO_2/YSZ or MgO/CeO_2/YSZ buffer layers were formed on Si or SOS in-situ prior to the YBCO film deposition. Details of the film deposition and electrical characteristics were described by Boikov et al (1995). Plane-view and cross-sectional specimens were prepared and investigated in a Philips EM-430ST, a Jeol-2000FX and a Jeol-4000EX electron microscope.

2. YBCO/poly-YSZ/Si

It has been found that polycrystalline YSZ buffer layers do not prevent detrimental interdiffusion between Si substrates and YBCO films during film deposition at 750°C. The diffusion is relatively fast along the grain boundaries of the columnar structured YSZ buffer layers (fig.1) and results in the formation of Cu rich precipitates at the Si /YSZ interface

334

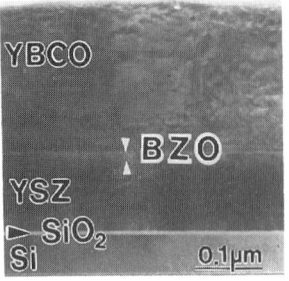

(ZrSi(Cu) Cmcm; a=0.3763nm, b=0.9944nm, c=0,3747nm) and in amorphous regions at the YBCO/BaZrO3/YSZ interface. In addition, a local surface roughness of the YSZ upper interface gives rise to significant deviation from the c-axis orientation of the YBCO film ([001] axis of YBCO parallel to the surface normal of the Si substrate). It is therefore important to maintain a smooth surface as well as an epitaxial growth of the YSZ buffer layer.

Fig.1. Cross-section image of YBCO/poly-YSZ/Si system. Thin BaZrO3 (BZO) layer between YBCO and YSZ is shown.

3. MgO/CeO2/YSZ/Si

A low-magnification image of the cross-section of the MgO/CeO2/YSZ/Si is shown in fig.2. The CeO2 layer grew epitaxially on the YSZ surface. An array of misfit dislocations

Fig.2. Cross-section image of the MgO/CeO2/YSZ/Si

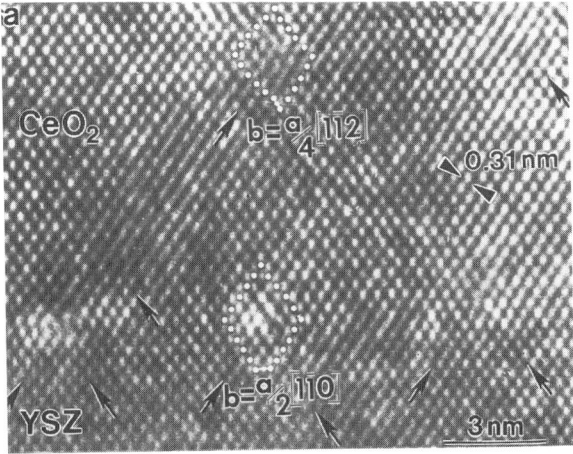

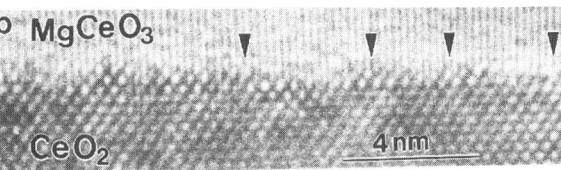

Fig.3
a) HREM image of the CeO2/YSZ interface with the 60° misfit dislocations. 90° dislocations are present in the CeO2 layer. Extra half planes are arrowed. b)HREM image of the MgCeO3/CeO2 interface. Misfit dislocations are arrowed.

having the projection of the Burgers vector on the (110) plane is a/2[1$\bar{1}$0] (referred to the YSZ unit cell) decorated the interface (fig.3a). In some areas two closely spaced undissociated dislocations with the projections of the Burgers vector equal to a/4[1$\bar{1}$2] and a/4[11$\bar{2}$] were observed. In the vicinity of the CeO2 layer 60° dislocations, located at different distances from the CeO2/YSZ interface, were found. HREM images of the interface between the top buffer layer and the CeO2 layer revealed flat areas parallel to (001) CeO2 (fig.3b) and faceted areas. In the flat part two different crystal lattices generated a set of misfit dislocations with a period of 2.6 nm which was larger than the average distance expected due to the mismatch between (200) MgO and (220) CeO2 planes (2.2nm). The faceted parts consisted of interfaces parallel to the {111} and {001} CeO2 planes. Optical diffraction (OD) and FFT from some grains of the top layer exhibited features incompatible with the MgO. The results showed that a reaction of MgO with CeO2 caused a formation of a MgCeO3 pseudo-cubic perovskite-like compound.

4. YBCO/CeO$_2$/YSZ/Si and YBCO/MgO/CeO$_2$/YSZ/Si

An image of YBCO/CeO$_2$/YSZ/Si is displayed in fig.4a. The YBCO layer was mainly oriented with the *c* -axis perpendicular to the substrate surface and parallel to the [001] axis of CeO$_2$. The facets and atomic steps on the CeO$_2$ surface do not influence the orientation of the film. In order to analyze the HREM contrast of the YBCO/CeO$_2$ interface in the experimental images a comparison of the contrast of double CuO layers in fig.6a with that at the interface was made. The CuO chain layers appeared as white dots confined between the black BaO planes. This pointed to an initial layer of BaO or CuO layer on the CeO$_2$ surface. The best fit for a flat interface is obtained when the first layer is assumed to be BaO with the stacking sequence CeO$_2$-BaO-CuO-BaO-CuO$_2$-Y... (fig 5a,b).

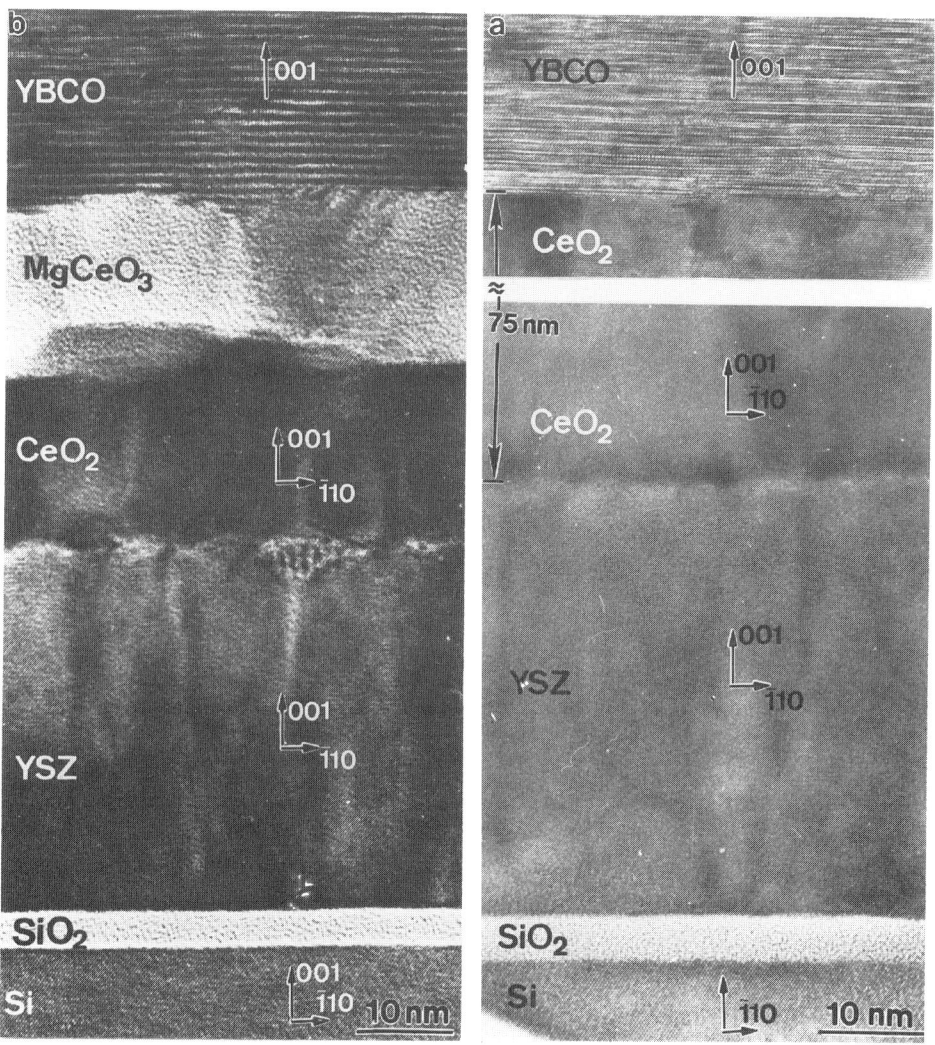

Fig.4. The cross-section images of the YBCO films on Si with the CeO$_2$/YSZ buffer -a) and MgO/CeO$_2$/YSZ buffer -b)

336

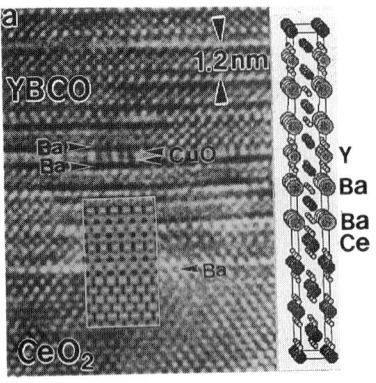

HREM image of YBCO/MgO/CeO$_2$/YSZ/Si is shown in fig.4b. The top buffer layer was polycrystalline. Its surface was uneven. The interplanar distances derived from the HREM images, FFT and OD patterns of the top buffer layer indicated that all data could be attributed to the MgCeO$_3$ structure. The orientation of the YBCO film did not follow the top buffer layer surface morphology but consisted mainly of c-axis oriented grains.

Fig.5. a)-HREM image of YBCO/CeO$_2$ interface. The simulated image (defocus Δf= -40 nm, thickness t=3.2 nm) is inserted. b)The model of stacking sequence on the YBCO/CeO$_2$ interface.

The plane-view images of the YBCO films on CeO$_2$/YSZ/Si and MgO/CeO$_2$/YSZ/Si demonstrated a rectangular arrangement of cracks mostly along {100} and {010} of YBCO. The average distance between the cracks was 1.8 µm for the first case and 1.5 µm for the second one. The YBCO films on the MgO/CeO$_2$/YSZ buffer were polycrystalline in contrast to the YBCO films on the CeO$_2$/YSZ which tended to be single crystals. The dimension of the YBCO grains on the MgO/CeO$_2$/YSZ buffer was in the range 0.2-2 µm. The average distance between the twin boundaries was slightly larger (0.04µm) on the CeO$_2$/YSZ than on the MgO/CeO$_2$/YSZ buffer (0.03µm). This difference evidently appeared due to a difference in the lattice match for the different buffers.

5. YBCO FILMS ON BUFFERED SOS

The Si layer with microtwins and dislocations was covered with a MgO/CeO$_2$/YSZ buffer. The interfaces between the YSZ, CeO$_2$ and the MgO seed layer are flat. Additional YBCO and SrTiO$_3$ layers were deposited on the complex buffer surface. The gold layer which protected the YBCO film from the wet air was deposited on the second thick YBCO film surface.

Acknowledgments
A.L.Vasiliev is indebted to the Belgian Government, Prime Minister's Office of Science Policy Programming for his fellowship at the University of Antwerp (RUCA). Financial support from the Royal Swedish Academy of Sciences, the Swedish Institute, the Swedish Natural Science Research Council, the National Board for Industrial and Technical Development and the Swedish Research Council for Engineering Sciences is gratefully acknowledged.

REFERENCES

Bardal A, Eibl O, Matthée Th, Friedl G, Wecker J 1993 J.Mater.Res. 8 2112
Boikov Yu, Ivanov Z G, Vasiliev A L, Claeson T 1995 J.Appl.Phys. 77 1654
Char K, Colclouph M S, Garrison S M, Newman N, and Zaharchuk G 1991 Appl.Phys.Lett. 59 733
Copetti C A, Schubert J, Soltner H, Zander W, Hollricher O, Buchal Ch, Schulz H, Tellman N and Klein N 1993 Appl.Phys.Lett. 63 1429
Eibl O, Hradil K, and Schmidt H 1991 Physica C 177 89
Fork D K, Fenner D B, Barton R W, Phillips J M, Connell G A N, Boyce J B, and Geballe T H 1990 Appl.Phys.Lett. 57 1161
Fork D K, Fenner D B, Barrera A, Phillips J M, Geballe T H, Connell G A N and Boyece J B 1991 IEEE Trans.Appl.Supercond. 1 67
Hwang D M, Ying Q Y, and Kwok H S 1991 Appl.Phys.Lett. 58 2429
Mogro-Campero A 1990 Supercond. Sci.Technol. 3 155
Tietz L A, Carter C B, Lathrop D K, Russek S E, Buhrman R A, and Michael J R 1989 J.Mater.Res., 4 1072
Wecker J, Matthée Th, Benher H, Friedl G and Samwer K 1992 Mater.Res.Soc.Symp. Proc. 275 1240.

Inst. Phys. Conf. Ser. No 146
Paper presented at Microsc. Semicond. Mater. Conf., Oxford, 20–23 March 1995
© *1995 IOP Publishing Ltd*

TEM and X-ray diffraction studies of III-V lattice mismatched multilayers and superlattices

G Salviati, C Ferrari, L Lazzarini, S Franchi, A Bosacchi, F Taiariol[‡], M Mazzer*, C Zanotti-Fregonara*, F Romanato** and A V Drigo**,

C.N.R.-MASPEC Institute, Via Chiavari 18/A, I-43100 Parma, Italy
‡ CSELT, Via G. Reiss Romoli 274, I-10147 Torino, Italy
* Materials Department, Imperial College, Prince Consort Road, SW7 2BP London, England
** INFM-University of Padova, Physics Department, Via Marzolo 8, I-35131 Padova, Italy

ABSTRACT: Transmission Electron Microscopy, High Resolution X-ray Diffraction, Rutherford Backscattering and Channeling, X-ray reciprocal lattice maps and low temperature Spectrally Resolved Cathodoluminescence have been employed for studying plastic relaxation and nucleation of extended defects in lattice mismatched InGaAs/GaAs multilayers and InGaAs/InP multi quantum wells grown under compressive and tensile conditions respectively. For InGaAs/GaAs compositionally graded structures, on the basis of the assumption that the elastic energy per unit interface area remains constant during the epilayer relaxation, the possibility of growing buffer structures with prefixed residual strain and composition is discussed in terms of misfit gradients inside the layers.

1. INTRODUCTION

The study of plastic relaxation and nucleation of extended defects in lattice-mismatched heterostructures is important in view, for instance, of integrated optoelectronics that demands the ability of growing buffer layers in which plastic relaxation provides the change in the lattice constant. Moreover, the study of strained layer structures is also interesting because the band structure of semiconductors is modified by the strain, that can be compressive or tensile, imposed internally by lattice mismatch.

Rutherford Backscattering (RBS) and Channeling, and High Resolution X-ray Diffraction (HRXRD) techniques have been employed for measuring the composition, the thickness and the degree of strain release of the InGaAs/GaAs samples. (110) oriented cross sectional TEM (XTEM) and (001) oriented plan view investigations were carried out for studying the dislocation nature, distribution and density inside the structures. Room temperature Panchromatic Cathodoluminescence (PCL) in the scanning electron microscope (SEM) was also employed for large area investigations of the misfit dislocation and crack distribution. X-ray diffraction maps (XDM) were carried out for assessing possible compositional gradients, strain release and extended defect location inside the structures in InGaAs/InP MQWs. Low temperature spectrally resolved cathodoluminescence (LTSCL) investigations were carried out in order to study, on a microscopic scale, the presence of possible inhomogeneities of well thickness and/or strain release in the MQWs grown under tensile conditions.

2. EXPERIMENTAL

$In_xGa_{1-x}As/GaAs$ (0.05 < x < 0.10) not intentionally doped double heterostructures were grown by Molecular Beam Epitaxy (MBE) at the CNR-ICMAT Institute in Roma. Step graded and continuosly graded buffer layers (0.11 < x < 0.51) have been grown by MBE both at CSELT and at MASPEC.

$In_xGa_{1-x}As/InP$ (0.47 < x < 0.60) test structures based on Multi Quantum Wells (MQWs) for advanced Stark effect Electrooptical Modulators (EOM) have been grown under tensile lattice mismatch by Low Pressure Metal Organic Chemical Vapour Deposition (LP-MOCVD) at CSELT.

The growth details of all the structures are reported (Ferrari et al. 1994, Antolini et al. this conference) or will be reported elsewhere (Bosacchi and Franchi 1995).

RBS Channeling measurements were carried out at the Laboratori Nazionali - Legnaro by using a high precision goniometer sample holder and $^4He^+$ beam of 2-5 MeV energy.

TEM analyses were performed at MASPEC in a 2000FX JEOL microscope working at 200 kV on samples mechano-chemically thinned and then finished by room temperature Ar ion milling. SEM-PCL studies were also carried out at MASPEC in a 360 Cambridge Stereoscan, in the emission and transmission geometries at accelerating voltages ranging between 3 and 30 KV and beam currents between 500 pA and 500 nA.

HRXRD measurements on InGaAs/GaAs heterostructures were performed at MASPEC on a double crystal diffractometer in the 117 parallel geometry corresponding to a Bragg angle θ_B of 76.64° and an asymmetry angle $\phi=\pm11.4°$ or 335 parallel geometry with $\theta_B=63.3$ and $\phi=\pm.40.32°$.

XDM around (004) and (-1-15) reciprocal lattice points were performed in CSELT-Torino using a commercial triple axis Philips diffractometer. Further, asymmetric 115 and -1-15 reflections were also employed for revealing strain release asymmetries in InGaAs/InP MQWs.

The optical quality of as grown InGaAs/InP structures was assessed in CSELT by a Fourier Transform low temperature Photoluminescence (FTPL) FTS-40 Biorad Spectrometer equipped with a liquid Nitrogen cooled Ge detector and with a Liquid Helium continuous flow cryostat. The optical investigations on InGaAs/GaAs compositionally graded buffer layers were carried out at MASPEC on a similar spectrometer.

LTSCL investigations were carried out on the same specimens at the Materials Department of the Imperial College of Science Technology and Medicine-London both at liquid Nitrogen and liquid Helium temperatures in a JEOL 840 SEM equipped with a non-commercial system based on a high sensitivity liquid Nitrogen cooled Ge detector and with a Liquid Helium continuous flow cryostat. Beam energies ranging from 2 to 30 keV and beam current values ranging from 20 pA to 10 nA were used.

3. RESULTS AND DISCUSSION

3.1 $In_xGa_{1-x}As/GaAs$ double heterostructures

Our previous results (Ferrari et al 1994) on double heterostructures with constant thickness t=500 nm and composition of the inner layer $x_1=0.05$ and of the top layer $x_2=0.10$, revealed that the top layers showed residual strain values much larger than predicted by the equilibrium theory (Matthews and Blakeslee 1974). On the contrary, despite the lower In content, the deeper InGaAs layers exhibit a much larger strain release and appeared nearly completely relaxed. This simply evidenced that the strain release of the first ternary layer depends on the total thickness of the structure that must be considered as a whole and not as made of two individual layers. The same conclusion was previously drawn by Kidd et al. (1993) on similar structures despite they found that the strain relaxation followed the Matthews and Blakeslee predictions in agreemen with previous work of Dunstan et al. 1991 on single layers.

On the basis of the above mentioned results, the residual strain of the inner layer and the dislocation propagation from the first to the second interface have been studied as a function of the top layer thickness in the structures sketched in figure 1. The structures were grown as test

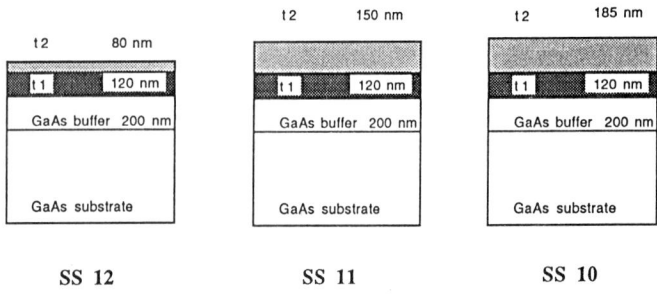

Figure 1. Sketch of the double InGaAs/GaAs heterostructures investigated.

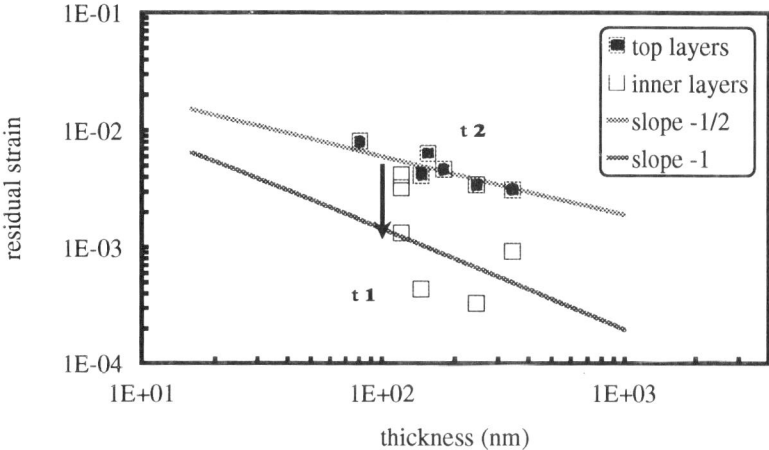

Figure 2.   Plot of the residual parallel strain vs the thickness for the structures of figure 1.

layers for investigations on step and linearly graded heterostructures. Here the inner ternary layer is grown with constant thickness (t_{fs}=120 nm) and composition (x=0.05) and the top one with variable thickness (80<t_{ss}<185) and constant composition (x=0.10). As evidenced by the vertical arrow in figure 2, the residual strain of the three inner layers, decreases by increasing the thickness of the uppermost layers. From the same plot, it can be also noticed that the residual parallel strain values of the upper layers follow the curve of -1/2 slope obtained by Drigo et al. (1989) on InGaAs/GaAs single heterostructures of similar composition and thickness and in disagreement with the equilibrium theories. This result leads to the assumption that, once the critical thickness is overcome, the elastic energy per unit interface area remains constant as suggested by Romanato (1993) (see § 3.2).

Besides misfit dislocations, both single and multiple dislocation loops extending

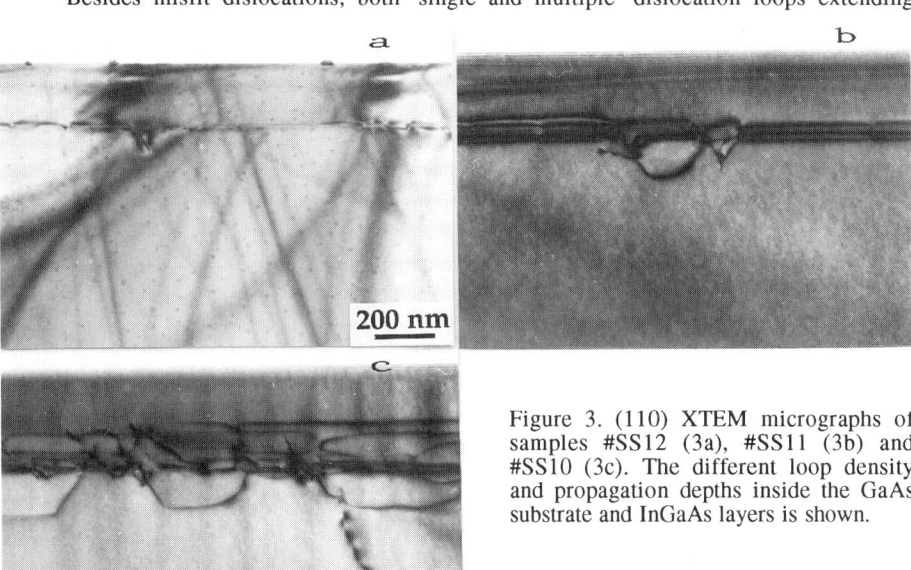

Figure 3. (110) XTEM micrographs of samples #SS12 (3a), #SS11 (3b) and #SS10 (3c). The different loop density and propagation depths inside the GaAs substrate and InGaAs layers is shown.

mainly from the deeper interface inside the GaAs substrate were also found. Such a behaviour has been observed by other groups on SiGe/Ge (LeGoues et al. (1992a) and InGaAs/GaAs superlattices (Lefebvre et al. (1991)) and graded heterostructures (Chang JCP et al. (1993), Chang KH et al. (1990), Krishnamoorthy et al. (1991)) and on InGaAs/GaAs single layers (Fitzgerald et al 1989). The presence of these loops has been correlated to the activation of Frank-Read (F-R) sources for the generation of misfit dislocations during the strain relaxation process. In SiGe/Si structures, it is suggested that this kind of strain relaxation occurs only in very pure and slowly ($f \approx 1\%/\mu m$)compositionally graded layers (from 0% to 40% of Ge), (LeGoues et al. 1992a).

A comparison among XTEM maps of the above mentioned samples (figure 3) shows that only a few single loops are revealed in the sample SS-12, that multiple dislocation loops are present in #SS-11 and that the highest density of dislocation loops both inside the substrate and the inner epilayer is found in #SS-10 (figure 3c). This demonstrates that the dimension, location and penetration of multiple dislocation loops depend on the total amount of elastic energy in the structure and on the strain release in the inner InGaAs layer. Both misfit dislocations and dislocations loops were of 60° type with Burgers vector of $a/2$ [110] type on a similar {111} glide plane.

Our observations also reveal that dislocations first nucleate and multiply   at the inner interface and that only once the dislocation density has reached the value corresponding to the minimum residual strain of the inner layer, the dislocations start crossing the first ternary layer. A maximum threading dislocation density of about $5 \ 10^5$ cm^{-2} has been estimated at the top layer of the structure #SS10 by comparing molten KOH chemical etching micrographs and plan view TEM maps.

3.2 Compositionally graded In$_x$Ga$_{1-x}$As/GaAs buffer heterostructures

A good buffer layer should prevent the propagation of threading dislocations to the active regions of devices and the onset of 3D growth. Both the aims could in principle be achieved by properly grading the In concentration profile inside, for instance, an InGaAs buffer layer; i.e. by performing an increasing profile of the misfit as a function of the buffer thickness, f(t).   Not intentionally doped linearly graded InGaAs/GaAs (0.005 < x < 0.35; 1000 < t < 2300 nm) buffer layers and almost complete structures grown as test for optical modulators, grown with different misfit profiles inside the buffer layer were studied.

Figure 4 shows a XTEM micrograph of a linearly graded InGaAs/GaAs sample of nominal composition 0.1 < x <0.3 and thickness t_{lg} =1002 nm. The majority of dislocations are distributed in about the first 600 nm of the layer thickness, leaving a consistent portion of the structure almost defect free. The threading dislocation density was estimated to be about 5-6

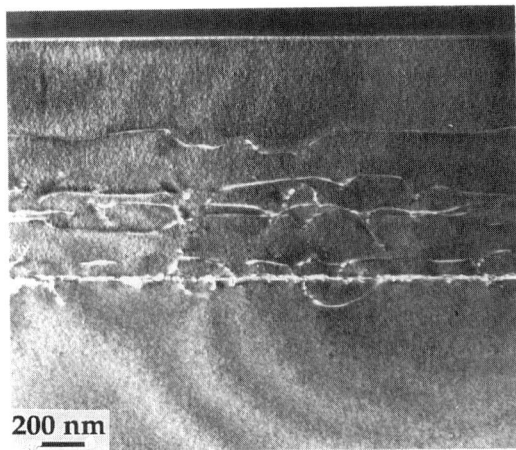

200 nm

Figure 4   (110)   XTEM micrograph   showing   the distribution of MDs in a linearly graded layer. **g-3g** (022) W.B.

10^5 cm^{-2} (Salviati et al. 1994). Similarly to the double layer structures, due to the initial infinite misfit gradient, the highest dislocation density is at the interface between substrate and graded layer; further, some loops are present in the GaAs substrate. After the critical thickness is overcome, MDs nucleate most likely from preexisting substrate dislocations at the interface and start releasing the strain. As the growth continues, the number of dislocations at the interface increases until the interface misfit is released. At this time, due to the concentration gradient, there is still elastic energy inside the structure and more dislocations are needed to release the excess of elastic energy. As the growth proceeds, new dislocations nucleate, for example as a consequence of heterogeneous nucleation or multiplication.

The strain inside the buffer layer is the major driving force that moves threading dislocations that nucleate during the growth. During the glide of a threading dislocation (TD), the interaction with a MD on an intersecting glide plane can result in an additional force that prevents a further propagation of the TD (Freund 1990). When a new threading dislocation takes place, it designs a segment of a misfit dislocation on an ideal surface where the released strain, ε_r, equals the misfit $f(t)$.

The increase of the misfit profile inside the buffer during the growth results in the increase of ε_r. As a consequence the position, t_r, of the ideal surface mentioned above, shifts away from the buffer/substrate interface according to the following relationship:

$$f(t_r) - \varepsilon_r = 0 \tag{1.}$$

Therefore, differently graded compositional profiles should influence the interaction of MDs and TDs by distributing the MDs throughout a certain thickness whose extension depends on the misfit profile inside the buffer layer (Tersoff 1993). If $n(t)$ is the MD density profile (i.e. $n(t) \, \Delta t \, L$ is the number of parallel MDs intersecting a rectangle of dimensions L and Δt perpendicular to the growth surface), the density of MDs in a layer of thickness $(t, t+\Delta t)$ must release an amount of misfit

$$\Delta f = f(t+ \Delta t) - f(t)$$

and

$$n(t) \, \Delta t = \Delta f/b_{eff} \tag{2.}$$

where b_{eff} is the effective component of the Burgers vector for releasing the strain.
This approach foresees a higher MD density profile where the misfit gradient is higher; i.e.:

$$n(t) = \begin{cases} (1/b_{eff}) \, df(t)/dt & \text{for} \quad 0 \le t \le t_r \\ 0 & \text{for} \quad t_r \le t \le t_b \end{cases} \tag{3.}$$

where t_b is the total thickness of the buffer layer.

In the layer between t_r and t_b no MDs should be expected because a residual elastic energy density remains which could be large enough to bend new or residual MDs.
Previous experimental results (Maree et al 1987, Drigo et al. 1989 and Ferrari et al. 1994) showed a dependence of the strain relaxation fitted by the following equation:

$$(f - \varepsilon_r)^2 \, t = K \tag{4.}$$

where K is a fitting parameter. It follows that equation 4 suggests that only once a critical value of the elastic energy density (proportional to $(f - \varepsilon_r)^2$) is overcome, new MDs nucleate and strain relaxation can be in some way "controlled". If we apply equation 4 to a generic misfit profile, we obtain:

$$\int_{t_r}^{t_b} (f(t) - \varepsilon_r)^2 \, dt = K \tag{5.}$$

Equation 5 together with eq. 1 allow to determine t_r and ε_r . We have applied this model to misfit profiles with different power dependences on the thickness like:
$$f(t) = A \, t^\alpha \tag{6}$$

342

with $\alpha=1/2$, 1, 2. The results of the calculations for a total buffer layer thickness $t_b = 1500$ nm and a surface In atomic fraction $x_b=0.30$, are reported in figures 5a,b,c, where n(t) and ε_r are plotted for the three values of α. The comparison of the three figures shows that the profile with $\alpha=1/2$ results in a decrease of MD concentration due to the misfit gradient decrease (according to equation 3) from the buffer layer/substrate interface to the surface, and in the lowest surface residual strain, $f(t_b)$-ε_r.

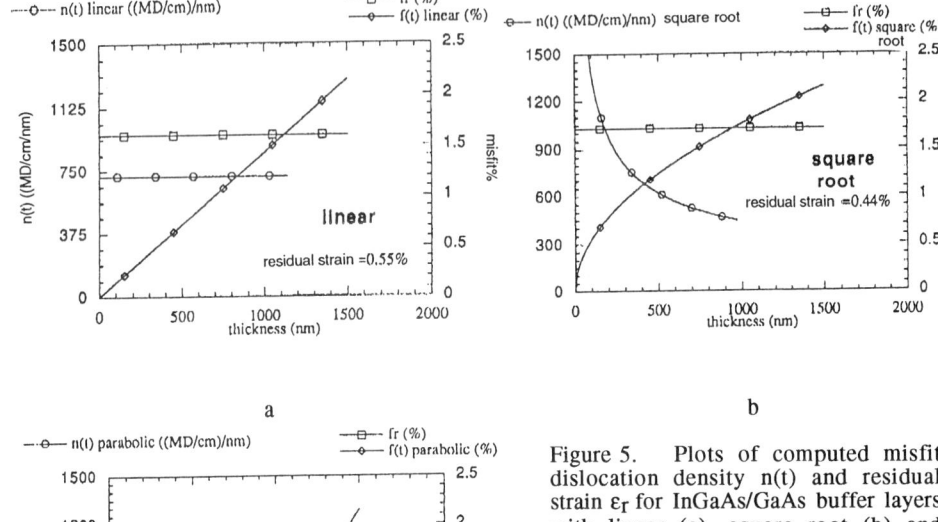

a   b

Figure 5.   Plots of computed misfit dislocation density n(t) and residual strain ε_r for InGaAs/GaAs buffer layers with linear (a), square root (b) and parabolic (c) misfit profiles (diamond). The origin of the thickness is at the interface with the GaAs substrate. ε_r open squares; n(t) circles. n(t) is expressed in number of dislocations per cm of length parallel to the surface and per nm of thickness. The calculated surface residual strain, $f(t_b)$-ε_r, is 0.55% (a), 0.44% (b) and 0.66% (c).

c

In order to check the predictions of the numerical model, the two structures of figure 6 were grown at MASPEC by conventional MBE. The only difference between the structures of figure 7a and b was due to the buffer misfit gradient that followed $t=1$ and $t^{1/2}$ misfit profiles in samples #855 and #869 respectively.
The misfit dislocation distribution inside these structures has been investigated by XTEM. Figure 8 shows two typical XTEM micrographs of the structures #855 (8a) and #869 (8b).

855 - 869

MQW/SL	barrier, GaAs, UD, 35 Å	x30
	well, InGaAs x=.50, UD, 75 Å	
	barrier, GaAs, UD, 35 Å	
	buffer, InGaAs 0.005-0.35, 2.3 μm	
	substr: GaAs	

Figure 6.   Sketch of test structures for optical modulators grown with $\alpha=1$ (sample #855) and 1/2 (#869) misfit profiles inside the buffer layers.

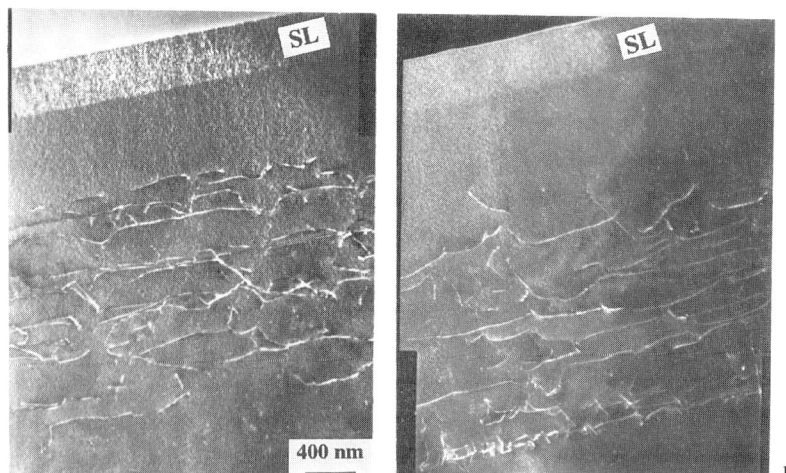

Figure 7   (110) XTEM pictures of the structures of figure 6. a) sample #855, **g**-3**g** (022) W.B; b) sample #869, **g**-3**g** (002) W.B

The calculations performed on the structures gave a MD free thickness of 670 nm and 450 nm in the sample #869 (α= 1/2) and #855 (α=1) respectively. The average experimental values obtained from XTEM maps were about 630 nm in the sample #869 and about 550 nm in the sample #855. Further, the calculations of the residual parallel strain gave 0.40% and 0.50% in the case of the sample #869 and #855 respectively.

The MD distribution inside the two specimens are only apparently similar; a much higher MD density at the InGaAs/GaAs interface can be noticed in figure 7b. The same interface can hardly be distinguished in figure 7a. This is consistent with the calculations reported in figure 5 where a different MD density is predicted for different misfit gradients.

Further, the optical and crystal quality of the SL grown on the top were quite different as assessed by PL and HRXRD investigations that revealed the sample of figure 7b to be the best. These results are consistent with the presence of MDs at the SL/buffer interface and with a poorer interface planarity of the SL #855 with respect to the sample #869 as found by XTEM investigations. As predicted by the calculations, a different residual strain at the top surfaces of the two buffer layers could be responsible for the different defect distribution and interface quality in the two SLs.

3.3 Lattice plane tilts in InGaAs/GaAs heterostructures

Single and double heterostructures were also suitable specimens for studying the influence of substrate misorientation on epilayer lattice plane tilts. The substrates used had typical misorientation angles of about 0.2 degrees as obtained by HRXRD analyses.

According to the works of Ayers et al. (1991) and Kavanagh et al. (1992), LeGoues et al (1992b) HRXRD investigations evidenced small tilts (300-400 sec. of arc) between the buffer layer and the substrate lattice in all the samples (Ferrari et al 1994). These tilt angles, observed by HRXRD after a 180° rotation along the sample surface axis with the same diffraction geometry, were due to the low angle grain boundary produced by the dislocation network at the buffer layer-substrate interface. Moreover, this tilting effect did not appear to be correlated with the asymmetry of the strain release found in several samples along the two <110> directions parallel to the interface and with the substrate misorientation.

The tilt value, α, is related to the unbalance, ($\rho^+ - \rho^-$), in the linear density of dislocations having opposite $b^{\perp}_{eff}$ component of the Burgers vector perpendicular to the interface (Mazzer et al. (1990)):

$$\alpha = b^{\perp}_{eff}\left(\rho^+ - \rho^-\right)$$

344

This result has been confirmed by X-ray diffraction maps performed in the vicinity of the 004 reciprocal lattice point as shown in figure 8.

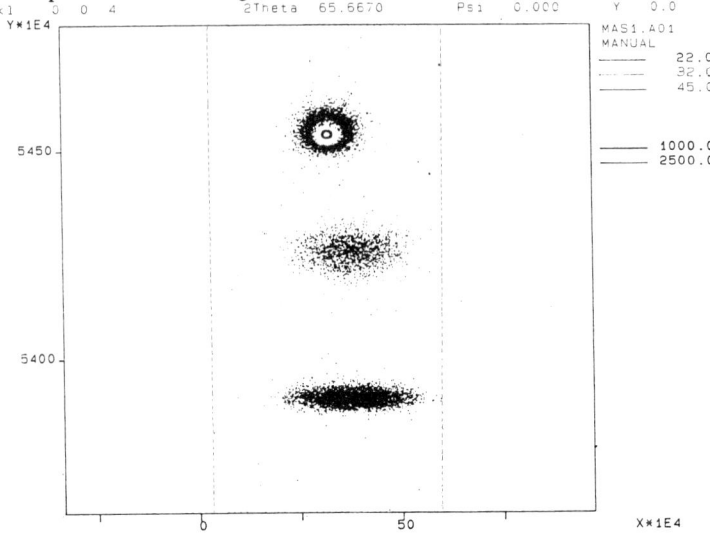

Figure 8.   XDM of sample #SS10. The misalignement between the diffraction peaks of the two InGaAs layers and the GaAs substrate is indicative of the epilayer lattice plane tilt. The larger diffuse scattering distribution around the InGaAs peaks is due to the presence of dislocations at the two interfaces.

In the frame of the study of the influence of composition on strain release (Ferrari et al. 1995), the correlation between substrate miscut and epilayer lattice plane tilts has also been studied in $In_xGa_{1-x}As/GaAs$ single layers with composition $0.083 < x < 0.324$ and constant thickness t = 250 nm. The samples were grown at MASPEC by conventional MBE on semi-insulating substrates with a substrate misorientation angle lower than 0.2 degrees. Also in this case the results, reported in Table I, do not show any correlation between miscut and epilayer lattice plane tilt. The same result has been found also by Eberl (1994) in graded InGaAs/GaAs layers.

sample	Φ (degrees)	$\theta_r-\theta_s$ (arcsec)	$\Delta\omega_s$ (degrees)	tilt angle (arcsec)
<u>767</u>	0°	-1392		
(x=0.083)	180°	-1380	+0.016°	+6
	90°	-1300		
	270°	-1417	-0.010°	-58.5
<u>765</u>	0°	-1277		
(x=0.097)	180°	-1523	-0.017°	-123
	90°	-1437		
	270°	-1449	-0.038°	-6
<u>762</u>	0°	-2040		
(x=0.131)	180°	-1268	-0.021°	+386
	90°	-1754		
	270°	-1640	-0.047°	+57
<u>759</u>	0°	-3419		
(x=0.324)	180°	-3264	-0.011°	+72
	90°	-3289		
	270°	-3282	-0.028°	+3.5

Table I   Observed epilayer lattice tilts vs substrate miscut. CuKα 004 diffraction.

The results shown here are in disagreement with respect to those of other groups (Kavanagh et al. 1992, LeGoues et al 1992b) that correlate a substrate miscut of 2° or more with epilayer lattice plane tilts. Taking into account the very high accuracy used for avoiding artefacts in our experimental measurements, as a comment we can simply say that our observations suggest that the amount of substrate misorientation influences the epilayer plane tilts probably only for values larger than those we have investigated.

3.4 In$_{1-x}$Ga$_x$As/InP MQws grown under tensile conditions

The In$_{1-x}$Ga$_x$As/InP system is characterized by a negative lattice mismatch , resulting in a tensile strain inside the epilayer for Ga concentration values x>0.47. As shown in the literature for different III-V systems (Matthews and Blakeslee 1974, Franzosi et al. 1986, Biefeld et al. 1988, Maignè et al. 1995), a tensile strain condition inside single and MQW heterostructures, in addition to dislocations, can also result in the generation of cracks. Sometimes, under particular misfit conditions, cracks develop first and instead of MDs (Antolini et al this conference). Since in principle this behaviour represents a different way to release the strain with respect to the compressive case, it is interesting, both from a fundamental and a methodological point of view, to compare the results of structural characterization on strain release performed by conventional techniques like TEM and XRXRD and less conventional ones like XDM and LTSCL.

As an example, a comparison between two partially relaxed 20 period thick In$_{0.44}$Ga$_{0.56}$As/InP MQWs with well/barrier of 107/80 Å, and 107/120 Å, misfit respectively and 100 nm thick InP cap grown as test structures for Stark effect electrooptical modulators is presented. The growth conditions and the experimental details are reported by Antolini et al., this conference, in the frame of a more complete study on samples with different well/barrier values and In concentration.

SEM pictures of the top surface of the two MQWs showed cracks mainly oriented along one of the two [110] directions (figure 9a and 9b). PCL micrographs of the same areas did not reveal any misfit dislocations (figure 9c and 9d).

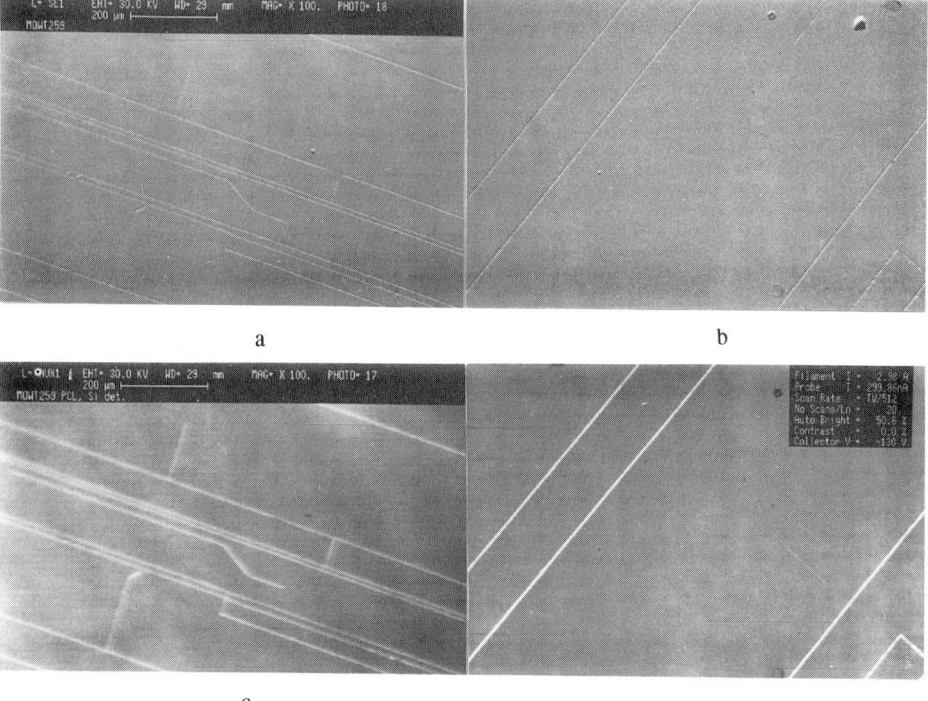

a   b

c   d

Figure 9.   a) and c) SEI and PCL images respectively of the InGaAs/InP MQW with well/barrier of 107/80 Å; b) and d) same for MQW with well/barrier of 107/120 Å

In the pictures 9c and 9d, cracks appear as white lines; this is because the PCL analyses were performed in the transmission geometry by using a Si detector. This means that the PCL signal comes mainly from the InP substrate. As a consequence, cracks should propagate from the top of the MQWs down to the InP substrate. These PCL observations have been confirmed by (110) oriented XTEM micrographs that revealed cracks propagating from the top of the MQWs into the substrate for distances larger than some micrometers.

It is worth noting that when MQWs with higher misfit (x=0.60) were studied, both cracks and MDs could be seen by SEM and PCL micrographs. The results on these MQWs indicate that in this system, cracks seem to have a lower threshold than MD generation at this misfit value. A similar result has been recently reported by Maignè et al. (1995) on InAlSb/InSb single heterostructures.

The MQWs investigated were under metastable conditions as revealed by the cracks that started propagating on the specimens surface some weeks after the growth. Later on, average shift values in the MQW peak positions were found by LTSCL with respect to the PL measurements performed on "as-grown" samples. Typical peak shifts of about 40 meV have been found.

In addition to an asymmetric propagation, also an inhomogeneous crack distribution was found on the top MQW surface; as a consequence an inhomogeneous strain release on the MQW growth plane was expected. Since PL investigations did not reveal any inhomogeneities, LTSCL investigations were carried out by taking advantage of higher lateral resolution of this technique. When LTSCL measurements were performed at magnifications lower than 1000 times, the same energy transition values of the PL analyses were found and the LTSCL data were in agreement with the calculated optical transitions for strained materials obtained by using the Daniel-Duke model (Bastard 1988). At higher magnifications (10000-15000 x) at higher energy a shoulder appeared and it was ascribed to a partial strain relaxation.

In figure 10, the peak values and the FWHM of different LTSCL spectra obtained in the same experimental conditions on different areas of the MQW of figure 9b are reported; the shifts of the peak positions have been ascribed to strain release inhomogeneities. The effect was more evident when the investigated areas were in proximity of cracks. This implies that the LTSCL technique can discriminate between different degrees of strain relaxation in the material.

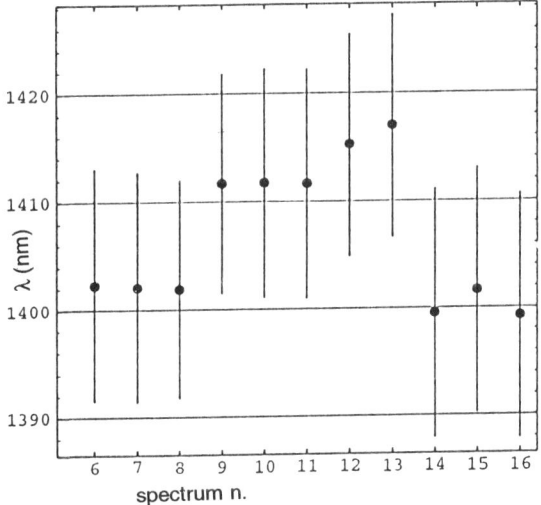

Figure 10.   Optical transition peak values of different LTSCL spectra obtained in different areas of the sample with well/barrier value of 107/120 Å. Vertical lines represent the FWHM values of the peaks. E_0=10 keV, $I_{beam} \approx$1-2 nA, Mag=10.000 x, T=78 K.

According to the crack distribution, asymmetric strain release was confirmed by HRXRD analyses performed on the same sample by using the (115) and (-1-15) asymmetric reflections both in the perpendicular and parallel geometry with respect to the cracks direction. Parallel residual strain values of $1.55 \cdot 10^{-4}$ and $5.45 \cdot 10^{-3}$ were respectively found.

When X-ray diffraction maps (Fewster 1993) of the samples of figure 9 were compared, additional information could be obtained. Two typical maps recorded by using the 004 symmetric reflection are reported in figure 11. Compositional gradients inside the wells along the growth axis, probably due to a memory effect inside the MOCVD reactor correlated to non optimized growth interruption times during the MQW growth (Salviati et al. 1993), are shown in the map of figure 11a which corresponds to the sample of figure 9b. Further, a mosaic structure on the InP peak position, correlated to the cracks inside the substrate can also be seen.

Looking at the map of the sample of figure 9a, a larger diffuse scattering distribution around the 004 reciprocal lattice point of the InP peak can be observed. This can be correlated

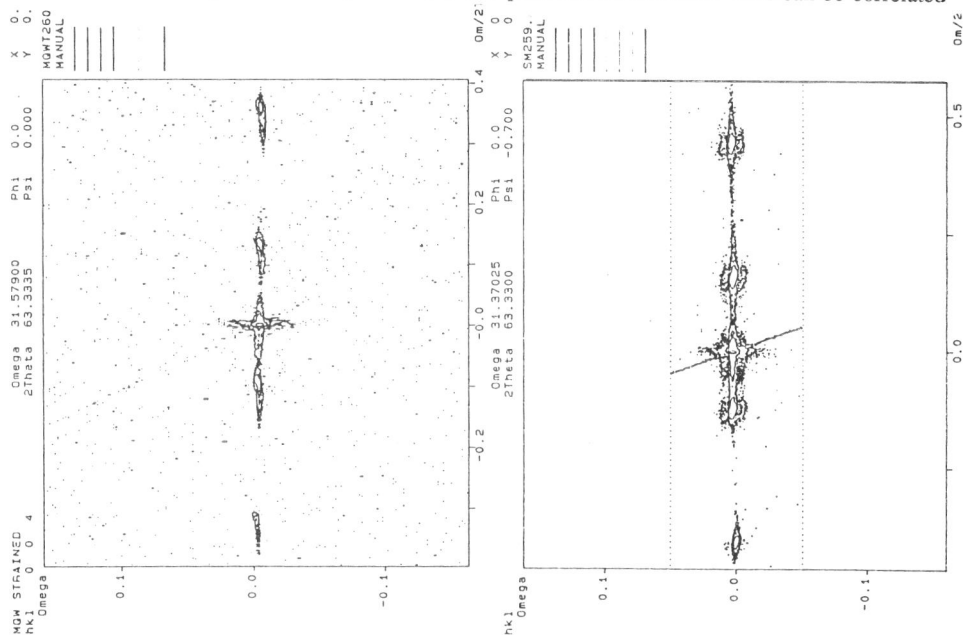

Figure 11.   Comparison between XDMs of the MQWs of figure 9b (left) and 9a (right)

to the higher crack density found in this sample as a consequence of the lower barrier thickness with respect to the other MQW. Further, SCL analyses carried out at 29 K on the same MQW revealed the presence of a shoulder on the side of the higher energies of the main peak that could be due both to strain release and/or to a well thickness wariation. As a support of this hypothesis, XTEM investigations performed on the same sample evidenced a random waving of the well interfaces.

4. CONCLUSIONS

A study of plastic relaxation and nucleation of extended defects in lattice mismatched multilayers grown under both compressive and tensile conditions has been presented. In the case of $In_xGa_{1-x}As/GaAs$ compositionally graded layers the strain release has been discussed on the basis of a numerical model based on the assumption that the elastic energy per unit interface area remains constant during the epilayer growth and after strain relaxation starts. The possibility of growing buffer layers with prefixed residual strain and composition is discussed in terms of the misfit gradient inside the layer for compositionally graded structures. Lattice plane tilts between epilayers and substrates have been found in single and double heterostructures given by the umbalance in the linear density of misfit dislocations with opposite component of the Burgers vector perpendicular to the interface. Cracks propagating from the top surface to the substrate in $In_{1-x}Ga_xAs/InP$ multi quantum wells grown under tensile strain conditions have been revealed by panchromatic cathodoluminescence and TEM investigations; no misfit dislocations were observed for anerage Ga concentrations lower than 51%. Compositional gradients along the growth axis have been found by high resolution X-ray diffraction and X-ray diffraction maps. The X-ray diffraction maps obtained by using -1 -1 5 asymmetric reflections also evidenced a mosaic structure around the reciprocal lattic epoint of the InP. Low temperature spectrally resolved cathodoluminescence investigations proved to be effective in measuring an inhomogeneous strain release on the growth plane due to the inhomogeneous crack distribution.

348

Aknowledgements
Thanks are due to Mr. M. Scaffardi for technical assistance.
The work carried out at MASPEC has been partially supported by the CNR-PS-GaAsNET.

References
Ayers JE, Ghandhi SK, Schowalter LJ 1991 J. Cryst. Growth 113 430

Antolini A, Papuzza C, Schiavini G, Soldani D, Taiariol F, Lazzarini L, Salviati G, Mazzer M, Zanotti-Fregonara C this conference

Bastard G Wave mechanics applied to semiconductor heterostructures Holsted Press NY

Biefeld RM, Hills CR, Lee SR 1988 J. Cryst. Growth 91 515

Bosacchi A and Franchi S 1995 Appl. Phys. Lett. in press

Chang KH, Bhattacharya PK, Lai R 1990 J. Appl. Phys. 67 3323

Chang JCP, Chin TP, Tu CW, Kavanagh KL 1993 Appl. Phys. Lett. 63 500

Drigo AV, Aydinly A, Carnera A, Genova F, Rigo C, Ferrari C, Franzosi P and Salviati G 1989 J. Appl. Phys. 66 1975

Dunstan DJ, Kidd P, Howard LK, Dixon RH 1991 Appl. Phys. Lett. 59: 3390

Eberl K 1994 Proc. of "Interface formation and dynamics" May 9-12 Toronto, Canada, in press on Scanning Microscopy

Ferrari C, Franzosi P, Lazzarini L, Salviati G, Berti M, Drigo AV, Mazzer M, Romanato F, Bruni MR, Simeone MG 1994 Mat. Scie. Eng. B28 510

Ferrari C, Bocchi C, Lazzarini L, Milita S, Salviati G, Bosacchi A, Franchi S 1995 in press on Appl. Phys. Lett.

Franzosi P, Salviati G, Scaffardi M, Genova F, Pellegrino S, Stano S 1988 J. Cryst. Growth 88 135

Fewster PF 1993 Semic. Science and Technology 8 1915

Fitzgerald EA, Ast DG, Kirchner PD, Pettit GD, Woodall JM 1988 J. Appl. Phys. 63 693

Freund LB 1990 J. Appl. Phys. 68 2073

Kavanagh KL, Chang JCP, Chen J, Fernandez JM, Wieder HH 1992 J. Vac. Sci. and Technol. B10 1820

Kidd P, Dunstand DJ, Grey R, David J, Fewster PF, Andrew NL, Molina SI, Kiely CJ 1993 Oxford 1993 Ints. Phys. Conf. Ser. 134 321

Krishnamoorthy V, Ribas P, Park RM 1991 Appl. Phys. Lett. 58 2000

Lefebvre A, Herbeaux C, DiPersio J 1991 Phil. Mag. A 63: 471

LeGoues FK Meyerson BS Morar JF Kirkner PD 1992 J. Appl. Phys. 71 4230

Le Goues FK, Mooney PM, Chu OJ 1992b Appl. Phys. Lett. 62 140

Lord SM, Pezeshki B, Kim SD,Harris SJ Jr 1993 J. Cryst. Growth 127: 759

Maignè P, Dharma-Wardana WC, Loockwood DJ, Webb JB (1995) J. Appl. Phys. 77(4) 1466

Maree PMJ, Barbour JC, van der Veen JF, Kavanagh KL, Bulle-Lieuwma CWT, Viegers MPA 1987 J. Appl. Phys. 62 4413

Matthews JW and Klokholm E 1972 Mat. Res. Bull. 7 213

Matthews JW, Blakeslee AE 1974 J. Cryst. Growth 27 118

Romanato F 1994 Study and Structural Characterization of Compound Semiconductors Epitaxial Layers. Ph. D. Thesis, Physics Department, University of Padova Ch. 6, 4

Salviati G, Ferrari C, Lazzarini L, Genova F, Rigo C, Schiavini GM, Taiariol F 1993 Ints. Phys. Conf. Ser. 134 471

Salviati G, Lazzarini L, Ferrari C, Franzosi P, Milita S, Romanato F, Berti M, Mazzer M, Drigo AV, Bruni MR, Simeone MG, Gambacorti N 1994 Proc. of "Interface formation and dynamics" May 9-12 Toronto, Canada, in press on Scanning Microscopy

Tersoff J. 1993 Appl. Phys. Lett 62 693

Inst. Phys. Conf. Ser. No 146
Paper presented at Microsc. Semicond. Mater. Conf., Oxford, 20–23 March 1995
© *1995 IOP Publishing Ltd*

Interfacial structure of GaSb/InAs superlattices grown by migration enhanced epitaxy

M E Twigg, B R Bennett and B V Shanabrook

Electronics Science and Technology Division, Naval Research Laboratory, Washington DC, 20375-5347, USA

ABSTRACT: Using High-Resolution Transmission Electron Microscopy to study InAs/GaSb superlattices grown by migration enhanced epitaxy, we have found that GaAs-like interfaces are rougher than InSb-like interfaces for both GaSb and InAs buffer layers grown on (100) GaAs substrates. Both GaAs-like and InSb-like interfaces are found to be rougher in superlattices grown on InAs buffer layers than in those grown on GaSb buffer layers, which agrees with studies of other semiconductor systems showing heteroepitaxial growth to result in greater surface roughness for layers under compression than for layers under tension.

1. INTRODUCTION

In order to grow the optimal strained layer $Ga_{1-x}In_xSb$/InAs superlattice (SL) for a specific infrared detector application, the thickness and composition of the layers are tailored to achieve the desired bandgap (Smith and Maihiot 1987, Miles et al. 1990). Reducing strain effects and dislocation generation in a SL, for a SL in which the InAs layers are significantly thicker than the $Ga_{1-x}In_xSb$ layers, may require growing on an InAs buffer layer instead of a GaSb buffer layer. It is not clear, however, that such a strategy would succeed. There is some indication in other systems that a surface grown under compression is rougher than a surface grown under tension (Ghaisas and Madhukar 1988, Xie et al. 1994). This is an important consideration because interface roughness may play a critical role in transport within the SLs. In InAs/GaSb SLs, GaSb layers are under compression when grown on an InAs buffer layer, whereas InAs layers are under tension when grown on a GaSb buffer layer. Therefore, one might expect an InAs/GaSb SL grown on a GaSb buffer layer to have less interface roughness than an InAs/GaSb SL grown on an InAs buffer layer. We have addressed this question by growing InAs/GaSb SLs on both GaSb and InAs buffer layers, and, using High-Resolution Transmission Electron Microscopy (HRTEM), monitoring interface roughness in each case.

2. EXPERIMENTAL

Using molecular beam epitaxy (MBE), we grew 40-period SLs with nominal structures of 8 monolayers (MLs) InAs and 12 MLs GaSb at 400^0C, the optimal growth temperature for this system (Shanabrook et al. 1992, Bennett et al. 1993). To insure well-defined interfaces, growth at the interfaces was controlled by migration enhanced epitaxy (MEE). The SLs characterized in this study were grown so that the heteroepitaxial interfaces were either InSb-like or GaAs-like. These structures were grown on a 0.1 μm GaSb buffer layer on a (100) GaSb substrate, a 1 μm GaSb buffer layer on a (100) GaAs substrate, and a 1 μm InAs buffer layer on a (100) GaAs substrate.

Using the focusing action of the objective lens, HRTEM imaging along the [001] zone axis can be tuned to accent the contribution of the {200} reflections (Twigg et al. 1994). Applying the expressions that Thoma and Cerva (1991) derived from non-linear imaging theory, we calculated the {200} contribution to the intensity at the cation site for both GaSb and InAs. For GaSb, the {200} contribution to intensity at the Ga site is strongly negative at a defocus value of 20 nm, and strongly positive at a defocus value of 60 nm. For InAs, the {200} contribution to the intensity at the In site is strongly positive at a defocus of 20 nm while strongly negative at a defocus of 60 nm. Similarly, the contribution of {200} anion contrast can be calculated from non-linear imaging theory. It should be noted that at a defocus of 60 nm, the interpretation of the image is in accord with simple intuition: atomic columns corresponding to small atomic numbers (i. e. Ga and As) appear bright; atomic columns corresponding to large atomic numbers (i. e. In and Sb) appear dark. This is analogous to the interpretation of this defocus condition by Ourmazd et al. (1990) in their study of $Al_xGa_{1-x}As/GaAs$. At a defocus of 20 nm, we have the opposite effect: atomic columns corresponding to small atomic numbers (Ga and As) appear dark; atomic columns corresponding to large atomic numbers (In and Sb) appear bright.

3. RESULTS AND DISCUSSION

HRTEM imaging experiments also reveal that some interfaces are locally abrupt, implying that interface widths are derived primarily from the presence of islands and steps rather than from interdiffusion and exchange reactions. An HRTEM image recorded from a cleaved specimen at a defocus of 60 nm, showing the initial growth of an InAs/GaSb SL on a GaSb substrate, is displayed in the left panel of Fig.1. In order to reduce surface roughness, the GaSb buffer layer was grown at 500°C, a suitable temperature for achieving a smooth GaSb surface. During an interrupt in the buffer layer growth and a lowering of the substrate temperature to 400°C, we observed an increase in the specular RHEED (reflection high-energy electron diffraction) intensity, indicating the formation of large islands on the surface. After the interrupt, the growth of the buffer layer was continued at 500°C. Following the completion of the buffer layer, the temperature was reduced to 400°C and the SL was grown.

In order to make a more quantitative assessment of the interfaces imaged by HRTEM, we developed an image processing algorithm that is sensitive to composition. Using Fourier-transformed image intensities (Thoma and Cerva, 1991), we digitally filtered the recorded image so as to allow only the spatial frequencies associated with the {200} reflection to contribute to the processed image, as shown in the right panel of Fig.1. A quantitative analysis of these data indicates that a portion of the InSb-like interface is abrupt. The claim that these interfaces are locally abrupt is also supported by Raman spectroscopy, in which we observed the strong longitudinal optical phonon peaks identified as GaAs planar vibrational modes (Shanabrook and Bennett 1994). The modes are present in SLs with nominally GaAs-like bonds and absent for nominally InSb-like bonds. These results suggest that significant numbers of exchange reactions are not occurring at the InSb-like interfaces.

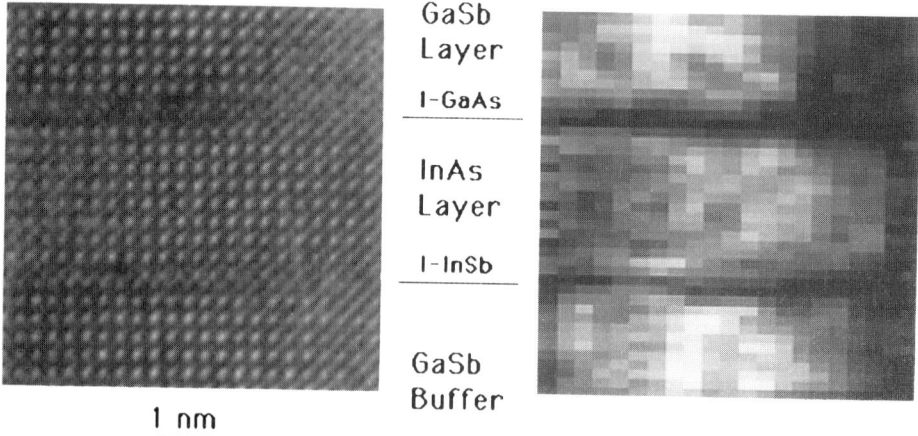

Fig.1 To the left is a HRTEM image of cleaved specimen taken from an InAs/GaSb SL grown on a GaSb buffer layer and (100) GaSb substrate. Defocus value is 60 nm. The processed image on the right, taken from the HRTEM image to the left, maps intensities associated with the {200} reflections.

Fig.1 also suggests a difference between InSb-like and GaAs-like interfaces: GaAs-like interfaces are rougher than InSb-like interfaces. The difference between InSb-like and GaAs-like interfaces in SLs grown on a GaSb buffer and GaAs substrate is also significant, as shown in Fig.2. We simulated SL images with abrupt GaAs and InSb interfaces as well as those in which the interfaces were replaced by 1 or 2 MLs of the composition $In_{0.5}Ga_{0.5}As_{0.5}Sb_{0.5}$ in order to mimic the contrast effects of 1 or 2 MLs of interface roughness. By comparing the simulations to the data, the interface roughness was found to be on the order of 1 ML for the InSb-like interface and 2 MLs for the GaAs-like interface. In Fig.3, we show images of SLs grown on InAs buffers. The interface roughness for SLs grown on an InAs buffer was found to be on the order of 2 ML for the InSb-like interface and 3 MLs for the GaAs-like interface. From these results, it seems that not only is interface roughness greater for GaAs-like interfaces than for InSb-like interfaces, but the roughness is greater for SLs grown on an InAs buffer layer than for SLs grown on a GaSb buffer layer. For the Ge_xSi_{1-x} and $In_xAl_{1-x}As$ systems, a rougher surface occurs for films under compression than under tension (Ghaisas and Madhakar, Xie et al). Such may also be true in our system, since the GaSb SL layers are under compression when grown on an InAs buffer layer, whereas the InAs SL layers are under tension when grown on a GaSb buffer layer.

InSb Bonds, GaSb Buffer   GaAs Bonds, GaSb Buffer

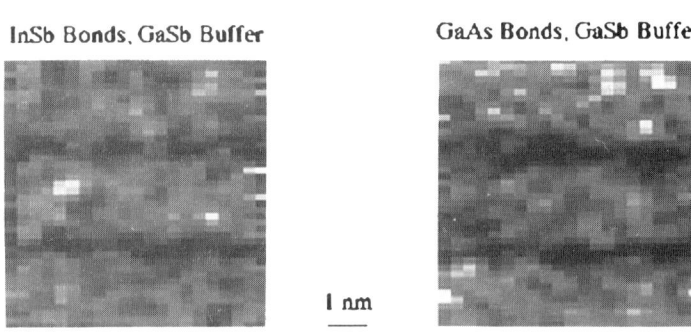

1 nm

Fig.2 Processed HRTEM images of 8 MLs InAs/12 MLs GaSb SLs grown on GaSb buffer layers on (100) GaAs substrates. Like Fig.1, the InAs layer is in the middle with a GaSb layer on top and bottom, and the growth direction is from bottom to top. The InSb-like interfaces (on the left) have a roughness of 1 ML. The GaAs-like interfaces (on the right) have roughness of 2 MLs.

InSb bonds, InAs Buffer   GaAs bonds, InAs Buffer

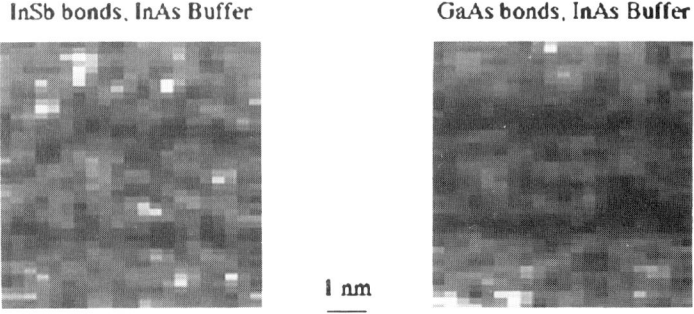

1 nm

Fig.3 Processed HRTEM images of 8 MLs InAs/12 MLs GaSb SLs grown on InAs buffer layers on GaAs substrates. Like Fig.1, the InAs layer is the middle with a GaSb layer on top and bottom, and the growth direction is from bottom to top. The InSb-like interfaces (on the left) have a roughness of 2 MLs. The GaAs-like interfaces (on the right) have roughness of 3 MLs.

352

The question of interface roughness in InAs/GaSb superlattices has also been addressed by Feenstra et al. (1994), using cross-sectional scanning tunneling microscopy. Their observations indicate that the InAs-on-GaSb interface has more intermixing than the GaSb-on-InAs interface. This was found to be the case for both Sb_2 and As_2 soaks. This observation is rationalized in terms of Sb having a lower surface free energy than As, producing exchange of Sb and As when InAs is grown on GaSb, but not for growth of GaSb on InAs. This tendency for an enhanced exchange of Sb and As, however, does not seem to be present in our SLs. This lack of exchange reactions is probably due to the deposition of a cation monolayer accompanying the soak that is part of our MEE growth procedure. This cation monolayer is meant to serve as a firewall between adjacent Sb and As layers, and thereby prevent the exchange of the two anion species.

Because anion exchange reactions are reduced by MEE, the principle mechanism for roughening would be the tendency for islanding. HRTEM micrographs show the local abruptness of interfacial regions for GaAs and InSb bonded InAs/GaSb superlattices grown on GaSb substrates. Nevertheless, this tendency for islanding appears to be arrested somewhat by the growth of InSb interfaces. The binding energy of InSb is considered the lowest of the III-V compound semiconductors (Yano et al. 1991). Therefore, when an InSb interface is formed, by soaking In-terminated InAs in an Sb flux, or when an In monolayer is deposited on Sb-terminated GaSb, the interface may flatten out.

4. CONCLUSION

In conclusion, we have used image processing of HRTEM images to analyze interface roughness in InAs/GaSb superlattices grown on both GaSb and InAs buffer layers. We have found that the smoothest interfaces are InSb-like interfaces grown on GaSb buffer layers. The roughest interfaces were GaAs-like interfaces grown on InAs buffer layers. InSb-like interfaces on InAs buffer layers and GaAs-like interfaces grown on GaSb buffer layers seemed comparable in the degree of roughness. It is our conjecture that InSb-like interfaces are smoother than GaAs-like interfaces because the surface energy of InSb is lower than that of GaAs. We also propose that the interfaces may be less rough when grown on a GaSb buffer layer (as compared to an InAs buffer layer) because the SL is grown under tension instead of compression.

5. ACKNOWLEDGEMENTS

This work was sponsored by the Office of Naval Research. We thank Larry Ardis for expert technical assistance.

REFERENCES

Bennett B R, Shanabrook B V, and Wagner R J, Davis J L, and Waterman J R 1993 Appl. Phys. Lett. 63 949.
Feenstra R M, Collins D A, Ting D Z-Y, Wang M W, and McGill T C 1994 Phys. Rev. Lett. 72 2749
Ghaisas S V and Madhukar A 1988 Appl. Phys. Lett. 53 1599
Miles R H, Chow D H, Schulman J N, and McGill T C 1990 Appl. Phys. Lett. 57 801
Ourmazd A, Baumann F H, Bode M, and Kim Y 1990 Ultramicrosc. 34 237
Shanabrook B V, Waterman J R, Davis J L, and Wagner R J 1992 Appl. Phys. Lett. 61 2338
Shanabrook B V and Bennett B R 1994 Phys. Rev. B 50 1695
Smith D L and Maihiot 1987 J. Appl. Phys. 62 2547
Thoma S and Cerva H 1991 Ultramicrosc. 38 265
Twigg M E, Bennett B R, Shanabrook B V, Waterman J R, Davis J L and Wagner R J 1994 Appl. Phys. Lett. 64 3476
Xie Y H, Gilmer G H, Roland C, Silverman P J, Buratto S K, Cheng J Y, Fitzgerald E A, Kortan A R, Schuppler S, Marcus M A, and Citrin P H 1994 Phys. Rev. Lett. 73 3006
Yano M, Yokose H, Yoshio Y, and Inoue M 1991 J. Crystal Growth, 111 609

Inst. Phys. Conf. Ser. No 146
Paper presented at Microsc. Semicond. Mater. Conf., Oxford, 20–23 March 1995
© 1995 IOP Publishing Ltd

HRTEM imaging of monolayer interfaces in MOCVD grown GaSb/InAs superlattices

I J Murgatroyd, N J Mason*, P J Walker* and G R Booker

Department of Materials, University of Oxford, Parks Road, Oxford OX1 3PH
*Clarendon Laboratory, University of Oxford, Parks Road, Oxford OX1 3PU

ABSTRACT: High resolution transmission electron microscopy has been used to obtain <100> lattice images of GaSb/InAs strained layer superlattices (SLSs) grown on (001) GaAs substrates by MOCVD. Displacement of perpendicular {200} lattice fringes reveals the presence of InSb- or GaAs-like interface layers (ILs). The ILs exhibit abrupt interfaces with the surrounding GaSb and InAs layers. Steps can be also be seen at the interfaces. SLSs grown on vicinal substrates exhibit step-bunching with sharp ILs formed in the regions between the steps.

1. INTRODUCTION

GaSb/InAs strained layer superlattices (SLSs) are promising systems for infrared detector applications. The electrical properties of this system are being investigated, as is the ability to tune the bandgap from a semiconductor to a semi-metal using a combination of quantum size effects and strain incorporated at the interfaces of the SLS (Symons *et al.* 1995). The optical and transport properties of the SLSs are expected to be very sensitive to the structure of the interfaces between the constituent GaSb and InAs layers. In order to understand the structure of these interfaces, we have grown series of (001) GaSb/InAs SLSs with different types of interfaces, i.e. where the interface between the two constituent layers is either GaAs-like or InSb-like. In this paper we apply high resolution transmission electron microscopy (HRTEM) in determining the sharpness of these two different types of interfaces.

HRTEM imaging of GaSb/InAs SLSs has been carried out on MBE grown GaAs/InSb SLSs (Twigg *et al.* 1994) where nominally 1 monolayer (ML) (3.04Å) thick InSb interface layers were observed to be roughened by a further 1 ML, while nominally 1ML thick GaAs interface layers were seen to be roughened by a further 2MLs. However, no information was reported as to the lateral roughness of the interfaces nor the effect of interface steps on the HRTEM images obtained. Spitzer *et al.* (1995) reported on the sharpness of interfaces by HRTEM in the related AlSb/InAs SLS system.

2. EXPERIMENTAL

The SLSs for this structural study were grown using metal organic chemical vapour deposition (MOCVD) (Goodings *et al.* 1989, Mason *et al.* 1991 and Booker *et al.* 1994). The substrates were semi-insulating nominally (001) GaAs wafers onto which was grown a 500Å thick coating layer of epitaxial GaSb. This was followed by a second epitaxial growth consisting of an approximately 1μm thick GaSb buffer layer, followed by the SLS of alternating InAs and GaSb layers. Growth was carried out at atmospheric pressure using a substrate temperature of 500°C.

The interface layers were grown (Booker *et al.* 1994) as shown in fig 1, to produce either GaAs-like interfaces (gas switching sequence A) or InSb-like interfaces (gas switching B). Between the growth of the GaSb and InAs layers the MOCVD reactor was flushed with H_2 for a time period, T, of 5 seconds. 4K mobilities of the GaSb/InAs SLSs grown were typically better than 100,000cm^2/Vs.

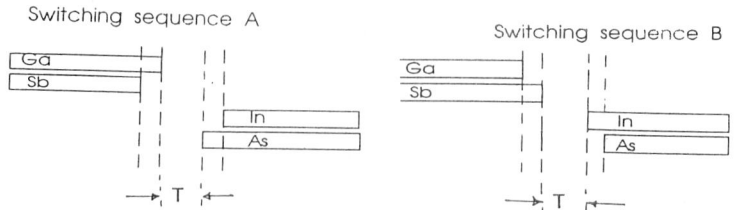

Fig 1 : MOCVD gas flow conditions used to grow SLSs with GaAs-like and InSb-like interfaces.

<100> cross-sectional HRTEM specimens were prepared of all of the GaSb/InAs SLSs using Ar$^+$ ion beam milling with the specimen cooled to liquid nitrogen temperature. A JEOL 4000EX was used to obtain lattice images, operating at 300keV to avoid damage of the HRTEM specimens by high energy electrons, which we have previously observed at 400keV (Seong 1994). The Scherzer resolution was less than 2.0Å. Lattice images were obtained using diffracted beams whose reciprocal lattice vector were less than 0.5Å^{-1}, i.e. the {220}, {200} and (000) beams. {200} and {220} fringes were obtained.

3. RESULTS AND DISCUSSION

HRTEM images of the above SLSs were obtained in which the specimen thickness and defocus were chosen so as to maximise the visibility of {200} fringes in both the GaSb and InAs layers, whilst minimising the appearance of {220} fringes and, in addition, imaging the group III atoms in one layer and the group V atoms in the other layer. As a result, the {200} fringes perpendicular to the interface in the GaSb and InAs layers are offset relatively by 1.52Å, i.e. half a monolayer (ML), thus revealing the position of the interface.

Fig 2(a) shows a lattice image of an InSb-like interface in which {200} fringes are seen in both the GaSb and InAs layers with a spacing of 3.04Å. In the region of the interface {200} fringes are not visible but {220} fringes can be seen. This region is part of neither the GaSb nor the InAs layers and therefore represents an image of an interface layer (IL). The half-ML offset of {200} fringes is most visible for a microscope defocus of approximately -420Å, but is visible over a range of defocus from approximately -350Å to -500Å. The {200} fringe offset is visible for HRTEM specimen thicknesses from approximately 50Å to greater than 200Å; the effect is not apparent in very thin crystal due to the presence of {220} fringes in both the GaSb and InAs layers. A consequence of imaging the ILs using half-ML displacements of the perpendicular {200} fringes is that the IL thicknesses have half-integer values (e.g. 1.5ML or 2.5ML or greater), as observed in this work and by Twigg *et al.* (1994).

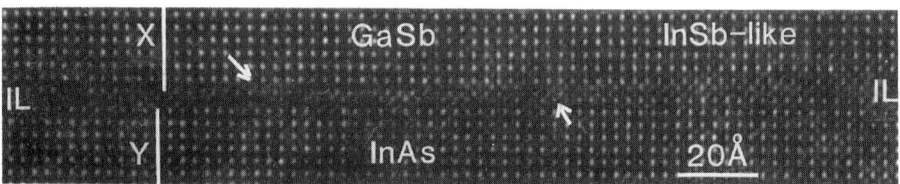

Fig 2(a) : Lattice image of GaSb/InAs interface with InSb-like interface layer (IL). (Note half-ML lateral displacement of perpendicular {200} fringes between GaSb and InAs layers indicated by lines X and Y)

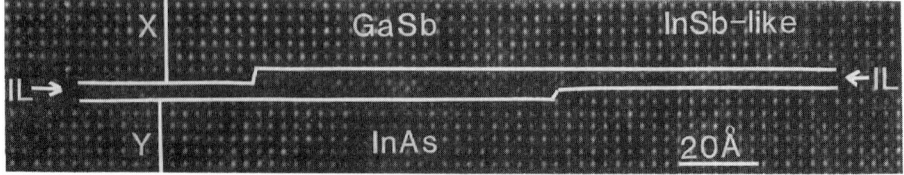

Fig 2(b) : Lattice image of fig 2(a) with addition of lines connecting final bright spots of perpendicular {200} fringes, indicating position of the InSb-like interface layer (IL).

The IL can more easily be visualised by tracing the perpendicular {200} fringes in both the GaSb and InAs layers (lines X and Y, respectively) and identifying the last bright spot which terminates the {200} fringe in the GaSb and InAs layers. Joining all such terminal spots with a line along the (002) planes identifies the limit of the GaSb and InAs layers and therefore marks the boundary of the IL. Fig 2(b) is the lattice image of fig 2(a) but with the addition of white lines to mark the position of the final bright spot of the perpendicular fringes, hence revealing the position of the InSb-like IL. Steps are seen both at the InAs/IL boundary and the IL/GaSb boundary and are clearly marked by the white lines; these same steps are also seen in fig 2(a), where they are marked with arrows, but are less easy to distinguish.

In fig 2(b), some regions of the IL are imaged to be 1.5MLs thick whereas some regions are 2.5MLs thick, with the thicker region approximately 80Å wide. This thicker region of the IL is thought to be due to a monolayer high interface step, aligned in the <110> direction, and which will have a projected width in the <100> lattice image; alternatively it may be two separate monolayer interface steps, aligned in the <100> direction, 80Å apart, running approximately parallel to the electron beam. Furthermore, because the IL is clearly imaged to change in thickness by 1ML, we may conclude that the boundaries between the IL and the adjacent InAs and GaSb layers are abrupt..

Fig 3 shows a lattice image of a SLS with GaAs-like ILs which are 2.5 to 3.5MLs thick. Firstly, it is evident that the ILs are nearly as wide as the surrounding InAs and GaSb layers themselves and this would be expected to affect the bandgap properties of such a SLS, and would indicate a limit on the minimum thickness of the GaSb and InAs layers in these SLSs. Secondly, the GaAs-like ILs of fig 3 are imaged to be thicker than the InSb-like ILs of fig 2. The GaAs-like ILs may indeed be thicker than InSb-like ILs, for instance due to interdiffusion or mixing of the IL and the surrounding layers. Alternatively, the difference in IL thickness may be due to a greater numbers of interface steps at GaAs-like rather than InSb-like ILs.

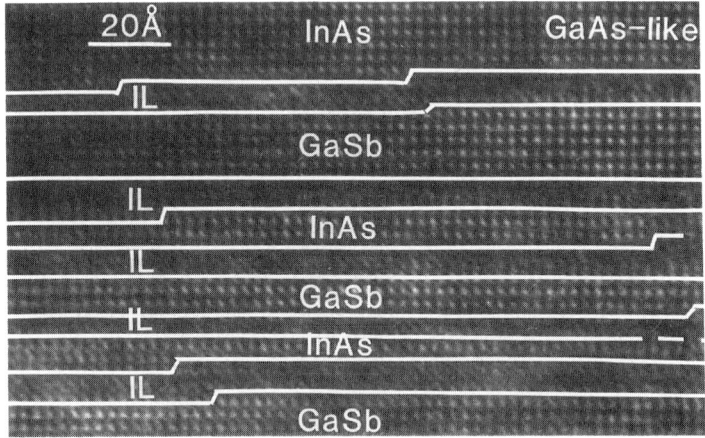

Fig 3 : Showing GaAs-like interface layers (ILs) with thicknesses of 2.5MLs or greater; the ILs are of similar thickness to GaSb and InAs layers.

The effect of interface steps on the lattice image can further be understood from the lattice image of a SLS with GaAs-like ILs which was grown on a vicinal substrate (2° away from (001) towards a <110> pole), as shown in fig 4. On average, <110> substrate steps would be expected approximately every 86Å, i.e. similar to the HRTEM specimen thickness, and so <100> lattice images of the ILs might be expected to be blurred by the presence of steps, leading to an increase in the apparent thickness of the IL. However, fig 4(a) shows the presence of sharply defined 2.5ML thick ILs in some regions, whereas in other regions the IL is imaged to be 4.5MLs or 5.5MLs thick. The occurrence of step-bunching would explain these lattice images, with several steps at one position on the interface giving an image of the IL which is blurred, while other positions on the interface have substantially no steps and give a lattice image of a sharp IL. Therefore, we may conclude that, if steps are present in the HRTEM sample, they can be clearly seen in the lattice image. Conversely, where no steps are seen, the apparent thickness of the IL in the lattice image is a reliable measure of its sharpness.

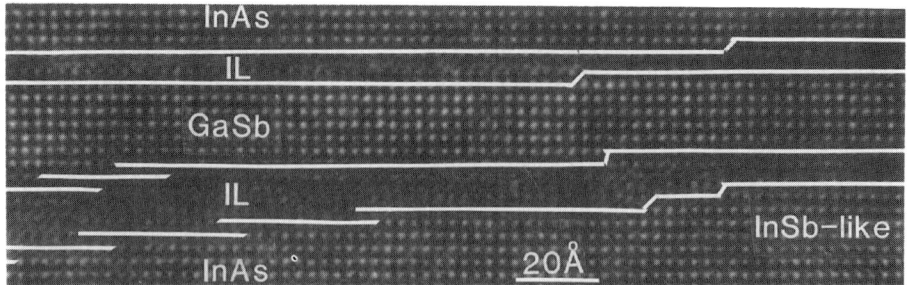

Fig 4 : Showing InSb-like interface layers (ILs) for SLS grown on a vicinal substrate (2° off [001] towards <110>). Step bunching can be seen in some areas of the lower IL with a 4.5ML thickness, whereas step-free regions of the IL are approximately 2.5ML thick.

Further work is required to investigate differences between InSb-like and GaAs-like ILs. In this work, InSb-like ILs have been imaged to be 1.5-2.5MLs thick, whereas GaAs-like ILs have been seen to be 2.5MLs thick or greater. Although these results indicate that InSb-like interfaces are more abrupt than GaAs-like interfaces, a finding confirmed by Twigg *et al.* (1994) in MBE grown GaSb/InAs SLSs, we may assume that the thickness and roughness of both InSb- and GaAs-like ILs are likely to be determined by the MOCVD growth conditions.

4. CONCLUSIONS

This is the first time that lattice images have been obtained to image the presence and shape of the InSb-like and GaAs-like interface layers (ILs) in GaSb/InAs SLSs grown by MOCVD. ILs exhibit abrupt interfaces with the adjacent GaSb and InAs layers. InSb-like ILs were imaged to be 1.5-2.5MLs thick whereas GaAs-like ILs were seen to be 2.5MLs thick or greater. Steps have been imaged in nominally on-axis SLSs. SLSs grown on vicinal substrates exhibit step-bunching but have abrupt ILs in the regions between the steps. This technique shows promise for judging interface sharpness with monolayer accuracy.

5. ACKNOWLEDGEMENTS

The authors would like to thank A G Norman and S A Lyapin for useful discussions and the EPSRC for funding.

REFERENCES
Booker G R, Klipstein P C, Lakrimi M, Lyapin S A, Mason N J, Nicholas R J, Seong T-Y, Symons D M, Vaughan T A and Walker P J 1994 J.Cryst.Growth 145 778
Goodings C, Mason N J, Walker P J and Jebb D P 1989 J.Cryst.Growth 96 13
Mason N J and Walker P J 1991 J.Cryst.Growth 107 181
Seong T-Y 1994, unpublished
Spitzer J, Höpner A, Kuball M, Cardona M, Jenichen B, Neuroth H, Brar B and Kroemer H 1995 J.Appl.Phys. 77 (2) 811
Symons D M, Lakrimi M, van de Burgt M, Vaughan T A, Nicholas R J, Mason N J and Walker P J 1995 Phys.Rev.B 51(3) 1729
Twigg M E, Bennett B R, Shanabrook B V, Waterman J R, Davies J L and Wagner R J 1994 Appl.Phys.Lett 64 (25) 3476

Inst. Phys. Conf. Ser. No 146
Paper presented at Microsc. Semicond. Mater. Conf., Oxford, 20–23 March 1995
© *1995 IOP Publishing Ltd*

Strain relaxation of lattice-mismatched In$_{0.2}$Ga$_{0.8}$As/ GaAs superlattices on GaAs(001) substrates

M Lentzen, D Gerthsen[1], A Förster[2] and K Urban

Institut für Festkörperforschung, Forschungszentrum Jülich GmbH, D-52425 Jülich, Germany
[1]Laboratorium für Elektronenmikroskopie, Universität Karlsruhe, Kaiserstraße 12,
D-76128 Karlsruhe, Germany
[2]Institut für Schicht- und Ionentechnik, Forschungszentrum Jülich GmbH,
D-52425 Jülich, Germany

ABSTRACT: A characterization of the strain relaxation of a series of In$_{0.2}$Ga$_{0.8}$As/GaAs superlattices grown on GaAs(001) substrates by molecular beam epitaxy has been performed by transmission electron microscopy. The superlattices were found to be partially relaxed with a residual strain of 0.5 %, in contrast to predictions of relaxation models minimizing the sum of the strain and the dislocation energies with respect to the strain. A detailed analysis showed that three mechanisms impeding the formation and glide of dislocations mainly contribute to the reduced relaxation.

1. INTRODUCTION

The heteroepitaxial growth of In$_{0.2}$Ga$_{0.8}$As layers on GaAs(001) substrates is associated with a lattice-parameter mismatch of 1.4 %. In the case of a single layer on a substrate plastic relaxation takes place beyond the critical thickness of the layer (Frank and van der Merwe 1949, Matthews et al. 1970, Matthews 1975, People and Bean 1985, 1986, Dodson and Tsao 1987, 1988), and is mediated by the generation and multiplication (Hagen and Strunk 1978) of misfit dislocation segments in the interface between layer and substrate. For the case of InGaAs/GaAs superlattices on GaAs(001) substrates two strain relaxation modes can be distinguished (Matthews and Blakeslee 1974). Firstly, the relaxation of the superlattice as a whole with respect to the substrate due to the bending of substrate threading dislocations and the nucleation and extension of dislocation half-loops; and secondly the relaxation of individual layers due to the extension of dislocation dipoles in "hair-pin" configurations.

The occurrence of these growth modes is strongly dependent on the individual layer thicknesses h(InGaAs) and h(GaAs) and the total thickness of the superlattice, h(SL). In the present study the influence of the individual layer thicknesses on the strain was investigated, keeping h(SL) and the indium content constant, for identical growth conditions.

2. EXPERIMENTAL TECHNIQUES

A series of In$_{0.2}$Ga$_{0.8}$As/GaAs superlattices with a total thickness of h(SL) = 800 nm was grown by molecular beam epitaxy on GaAs(001) substrates at a growth temperature of $T = 460°$C. The GaAs layer thickness was varied in the series, h(GaAs) = 2, 6, 10, 16, 20, 25, 30 nm, whereas the InGaAs layer thickness h(InGaAs) = 18 nm was kept constant in all cases. Using these layer thicknesses and an indium content of $x = 0.2$ the lattice mismatch parameter

358

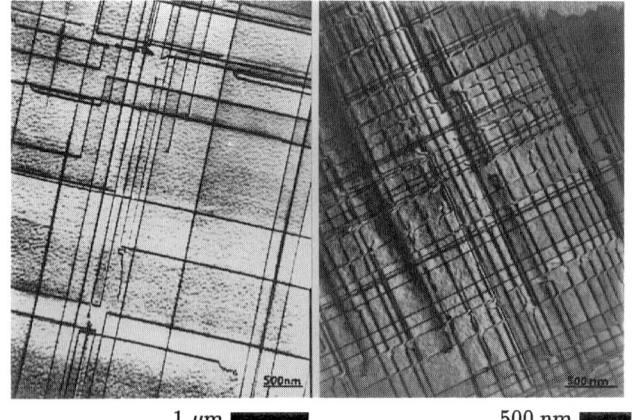

Fig. 1 Bright-field plan-view images, $\vec{g} = (220)$, of an 18 nm/16 nm $In_{0.2}Ga_{0.8}As/$ GaAs strained layer superlattice on a GaAs(001) substrate with misfit dislocation arrays, $\vec{u} = [110]$, $[\bar{1}10]$, between the individual GaAs and InGaAs layers (left) and a dense array at the superlattice-substrate interface (right).

1 μm ■■■■■■   500 nm ■■■

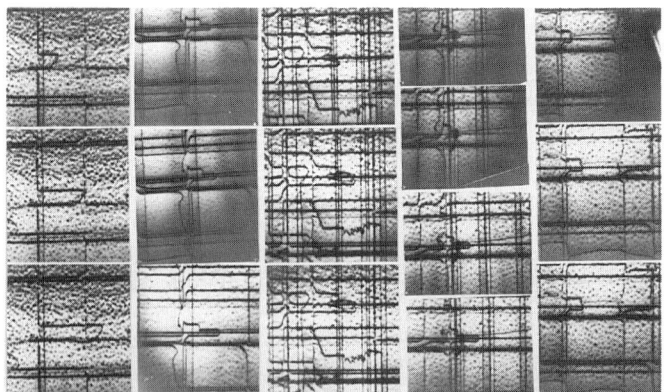

Fig. 2 Bright-field plan-view images, $\vec{g} = (200)$, of an 18 nm/25 nm $In_{0.2}Ga_{0.8}As/GaAs$ strained layer superlattice on a GaAs(001) substrate. The columns show the extension of dislocation dipoles under a small additional stress induced by the focussed electron beam (sequence from top to bottom, respectively). 1 μm ■■■■■■

between individual layers $f_L = 1.4$ % remains constant, while the mismatch parameter of the superlattice with respect to the substrate f_{SL} is "tuned" from 1.26 % to 0.50 %, respectively.

In a second step $\langle 001 \rangle$ plan-view and $\langle 110 \rangle$ cross-section samples were prepared using the standard technique of mechanical grinding, polishing and ion milling with Ar^+ ions at 4 keV and an incident angle of 18°. Finally, the samples were investigated using conventional transmission electron microscopy.

3. EXPERIMENTAL RESULTS

In all samples dislocation arrays were found located at the interfaces between the individual layers and the interface between the superlattice and substrate. The arrays consist of dislocation lines running along [110] and [1$\bar{1}$0], as shown in Fig. 1 and Fig. 3 C. Burgers vector analysis showed that the dislocations are of the glissile 60°-type.

Dislocations in adjacent superlattice interfaces are frequently found in dipole configurations. It was observed that in some cases such dipoles are terminated by small threading segments forming a hairpin configuration, as can be seen in Fig. 2. In one sample the extension of a small number of such hairpin configurations was observed under a small additional stress induced by the focussed electron beam (Fig. 2). Some misfit segments at the interface between the superlattice and the substrate bow out to form half-loops extending into the substrate (Fig. 3 A, B), or they split into a partial dislocation at the interface and a partial dislocation in the substrate leaving a stacking fault behind (Fig. 3 D).

The dislocation densities of the different arrays were measured by counting the number

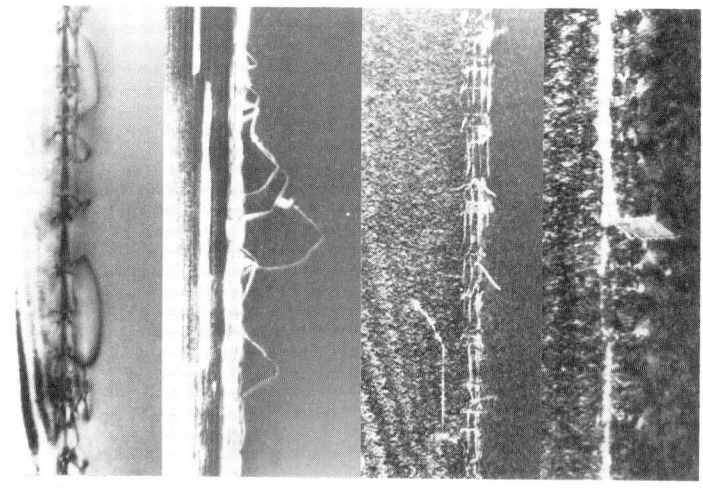

100 nm ▬▬ 200 nm ▬▬ 100 nm ▬▬ 30 nm ▬▬

(A)BF$\vec{g} = (220)$ (B)WB$\vec{g} = (400)$ (C)WB$\vec{g} = (220)$ (D)WB$\vec{g} = (220)$

Fig. 3 Cross sectional view of an 18 nm/ 6 nm (A, B) and an 18 nm / 2 nm (C, D) In$_{0.2}$Ga$_{0.8}$As/ GaAs superlattice. (A, B) Misfit dislocations bow out to form half-loops extending into the substrate. (C) Misfit dislocation array at the substrate interface and a misfit segment connected with a threading segment in the superlattice. (D) Stacking fault extending into the substrate.

of dislocation lines over distances in the order of microns. Subsequently, values for the elastic layer strain ϵ were calculated using the relation $\epsilon = f - |\vec{b_e}|/d$, with f being the lattice mismatch parameter, $\vec{b_e}$ the edge component of the Burgers vector lying in the interface plane and d the mean dislocation spacing. The corresponding values of superlattice mismatch parameter (f_{SL}) and superlattice elastic strain (ϵ_{SL}) as a function of the individual layer thickness h(GaAs) are shown in Fig. 4. For superlattices with thin GaAs layers half of the misfit strain is reduced by the dislocation arrays, whereas for superlattices with thick GaAs layers (h(GaAs) > 15 nm) the strain reduction is relatively small. In all cases a residual elastic strain of 0.5 % $\leq \epsilon \leq$ 0.6 % is observed. The density of dislocations in the arrays between individual adjacent layers is very low and provides only a small contribution to the strain relaxation.

The measured relaxation contradicts the predictions of relaxation models minimizing the sum of the strain and dislocation energies with respect to the strain, or equivalent force models balancing the dislocations line tension and forces arising from the strained layer. These models predict, in the appropriate cases, dislocation-free interfaces within the superlattice because the individual layers are sub-critical, and nearly completely relaxed superlattices with a high dislocation density at the substrate interface because the total thickness of the superlattice h(SL) is much larger than the critical thickness.

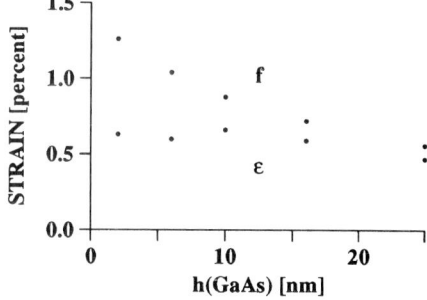

Fig. 4 Dependence of the misfit parameter f_{SL} and the residual elastic strain ϵ_{SL} on the GaAs layer thickness.

4. DISCUSSION AND CONCLUSION

In order to explain the discrepancy between experimental observation and the predictions of the relaxation models a number of possible mechanisms which are normally neglected in energy and force balance models are briefly discussed. These mechanisms concern the generation and glide of dislocations and the magnitude of the driving force.

Firstly, during the generation of new misfit segments by half-loop nucleation at the crystal

360

surface, dislocation glide is impeded by image forces. The boundary condition at the free surface, $\sigma_{ik}(\vec{r})n_k = 0$, results in an image force $\vec{F}$ with magnitude

$$F = \frac{G \sin \alpha}{4\pi d \,(1 - \nu)} \left(b_e^2 + (1 - \nu)\, b_s^2\right)$$

in the glide plane. $\sigma(\vec{r})$ denotes the stress tensor, $\vec{n}$ is the surface normal, G the shear modulus, α the angle between the substrate surface and the glide plane, d the loop radius, ν Poisson's ratio, b_e and b_s are edge and screw components of the Burgers vector, respectively. For the considered strained layer superlattices the loops must exceed a critical radius of 3.5 to 4.0 nm at a residual strain of 0.5 % to overcome the barrier imposed by the image forces.

Secondly, elastic distortion of the substrate can reduce part of the misfit strain f_{SL}, decrease the driving force for dislocation glide and therefore reduce the equilibrium density of misfit dislocations. The decrease of the elastic strain from a value of 0.5 % at the superlattice-substrate interface to a value of zero in the substrate was modelled using an exponential decrease. A fit to the experimental data resulted in an estimate of the thickness of the distorted area below the interface of about 30 to 50 nm. These values coincide roughly with the diameters of the loops observed experimentally below the substrate interfaces (Fig. 3 A, B).

Thirdly, during the relaxation of the superlattices with respect to the substrate, the glide of dislocations from the crystal surface to the superlattice-substrate interface relaxes the misfit strain in the InGaAs layers but *builds up* strain of opposite sign in the GaAs layers at the same time. The total elastic strain energy, however, still decreases during this process. With increasing degree of misfit relaxation more and more strain is built up in the GaAs layers until finally dislocation glide is effectively stopped by this barrier. The barrier itself cannot be removed by the relaxation of individual layers because the GaAs layers are still sub-critical with respect to the single-layer critical thickness. It should be noted that the occurrence of this mechanism is a special characteristic of the plastic relaxation of superlattices and has no equivalence in the case of single-layer misfit strain relaxation.

Further, two other contributions possibly leading to *partial relaxation* were considered but turned out to be small in comparison with the three mechanisms discussed above; i.e. the interaction energy corresponding to the superposition of a single dislocations strain field and the layers strain field; and the small additional force impeding dislocation glide which arises from a periodic deviation of the homogeneous layer strain when a gliding dislocation approaches an existing periodic misfit dislocation array at the substrate interface.

Finally, it is concluded that the experimentally observed defect morphology is that which is typical for the two relaxation modes in the strained-layer superlattice case. The experimentally observed dislocation density at the substrate interface, however, deviates significantly from theoretical prediction. It was shown that three mechanisms mainly contribute to the *partial relaxation:* (1) image forces at the surface opposing the nucleation of half-loops; (2) an elastic distortion of the substrate reducing part of the misfit strain, and (3) the glide of dislocations being impeded by the build-up of strain in the GaAs layers during plastic relaxation of the superlattices.

REFERENCES

Dodson B W and Tsao J Y 1987 Appl. Phys. Lett. <u>51</u> 1325; 1988 <u>52</u> 852
Frank F C and van der Merwe J H 1949 Proc. Roy. Soc. A <u>198</u> 205; <u>198</u> 216
Hagen W and Strunk H 1978 Appl. Phys. <u>17</u> 85
Matthews J W 1975 J. Vac. Sci. Technol. <u>12</u> 126
Matthews J W and Blakeslee A E 1974 J. Crystal Growth <u>27</u> 118
Matthews J W, Mader S and Light B T 1970 J. Appl. Phys. <u>41</u> 3800
People R and Bean J C 1985 Appl. Phys. Lett. <u>47</u> 322; 1986 <u>49</u> 229

Inst. Phys. Conf. Ser. No 146
Paper presented at Microsc. Semicond. Mater. Conf., Oxford, 20–23 March 1995
© *1995 IOP Publishing Ltd*

A TEM study of InGaAs/GaAs SQWs grown by MOVPE on (100) and 2° off (100) GaAs substrates

C Frigeri*, A Di Paola°, D M Ritchie°, F Longo*, A Brinciotti^, M Riva° and F Vidimari°

*CNR-MASPEC Institute, via Chiavari 18/A - 43100 Parma (Italy)
°ALCATEL-Telettra, via Trento 30 - 20059 Vimercate (Italy)
^OPTEL, Mesagne SS7 Km 7.3 - 72100 Brindisi (Italy)

ABSTRACT: InGaAs SQWs grown on vicinal (100) GaAs substrates exhibit wide macrosteps that cause marked lateral and vertical fluctuations in composition and thickness that increase with increasing layer thickness and composition. Macrosteps are absent for growth on exact (100) oriented substrates. Dislocation loops in the substrate and threading dislocations, both generated by reaction between misfit dislocations, are created more easily for growth on vicinal substrates than substrates of exact orientation.

1. INTRODUCTION

The generation and structure of misfit dislocations in InGaAs quantum wells (QW) on vicinal GaAs substrates have been widely studied by TEM (e. g., Kightley and Goodhew 1991, Chen et al 1993). However, QW layer uniformity, as regards composition and thickness, has been little studied by TEM for the case of growth on vicinal substrates, e.g. a paper has appeared recently by Hiramoto et al (1994). Layer quality has been more often investigated by photoluminescence as it is recognized that the optical properties of quantum wells are strongly affected by fluctuations in thickness and indium composition. We report here on a TEM study of InGaAs SQWs grown by MOVPE on vicinal surfaces, both with regard to layer uniformity and dislocation structure.

2. EXPERIMENTAL

$In_xGa_{1-x}As$ SQWs were grown by low pressure metalorganic vapour phase epitaxy (LP-MOVPE) at 650 °C on GaAs substrates oriented either exactly (100) or 2° off (100) towards <110>. The V/III ratio was varied between 155 and 180. A GaAs buffer and cap layer, 1 and 0.2 μm thick, respectively, were used as confinement barriers for the InGaAs well. In order to exclude run-to-run fluctuation, growth was performed on half 2" wafers of both exactly oriented and misoriented substrates. The nominal In composition, x, was varied between 0.15 and 0.22 for nominal well thicknesses, t, of 3.5 to 25 nm. Both cross sectional and plan view specimens, prepared by Ar ion milling, were investigated by bright field and (200) dark field-TEM and HREM.

3. RESULTS AND DISCUSSION

3.1 SQW homogeneity

The (200) DF micrographs of fig. 1 a) and b) compare the structure of InGaAs SQWs grown on exactly (100) and 2° off (100) oriented substrates, respectively. Growth on the exact (100) substrate produces a very homogeneous SQW (fig. 1 a) having a very sharp InGaAs-on-GaAs interface whereas the GaAs-on-InGaAs interface exhibits roughness

362

Fig. 1. (200) DF image of a 7 nm thick In$_x$Ga$_{1-x}$As SQW (x = 0.22) grown on a) (100) and b) 2° off (100) GaAs substrates. Bar = 10 nm.

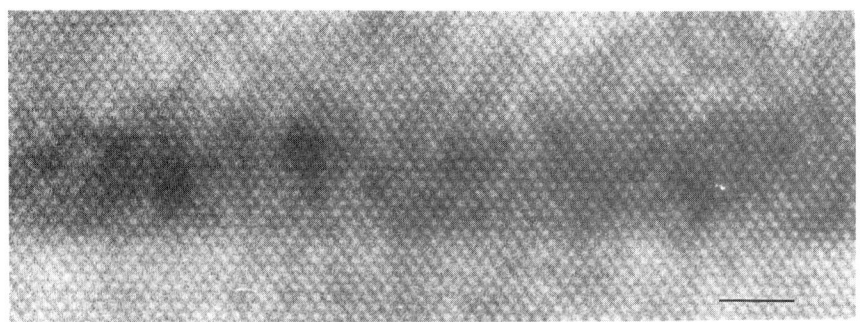

Fig. 2. HREM image of a 3.5 nm thick In$_x$Ga$_{1-x}$As SQW (x = 0.22) grown on an exactly oriented (100) GaAs substrate. Bar = 2 nm.

Fig. 3. (200) DF images of 14 nm thick In$_x$Ga$_{1-x}$As SQWs grown on GaAs substrates oriented 2° off (100) towards <110> with a) x = 0.22 and b) x = 0.17. Bar = 20 nm.

363

extending over ~3 to 5 MLs as can be seen from HREM images (fig. 2). Such interface roughness is ascribed to In segregation in the growth direction (Frigeri et al 1994). No noticeable difference has been detected between samples of differing In composition or thickness. On the other hand, the quality of layers grown on 2° off (100) GaAs substrates is markedly worse than for exactly oriented substrates for all layer thicknesses and compositions. The GaAs-on-InGaAs interface exhibits macrosteps with lateral periodicity of ~180±30 nm (arrowed in fig. 1b). (200) DF images show a lateral decrease of In content along the macrostep. A similar result was obtained by Hiramoto et al (1994) by EDX analysis. The maximum height of macrosteps in the growth direction increases for increasing x, from ~4 to ~8 nm for x ranging between 0.15 and 0.22, respectively (fig. 3), as well as for increasing well thickness. For x = 0.22 the InGaAs wells with t ≥ 14 nm exhibit a stripe-like structure when imaged in (200) DF (fig. 3a) or in HREM mode (not shown here), indicating that compositional modulation also exists along the growth axis.

The formation of macrosteps occurs by a step bunching mechanism whose occurrence requires that the lateral growth velocity of InGaAs on GaAs is smaller than that of InGaAs on itself (Cox et al 1990). Lateral compositional non-uniformity along the macrosteps may be due to enhanced tendency of group III elements to attach to step sites and, additionally, to the larger diffusion length of In with respect to Ga (Hiramoto et al 1994). The increase of macrostep height with increasing In content may be ascribed to enhanced surface migration of group III elements to reduce the surface free energy due to strain. Owing to this enhanced

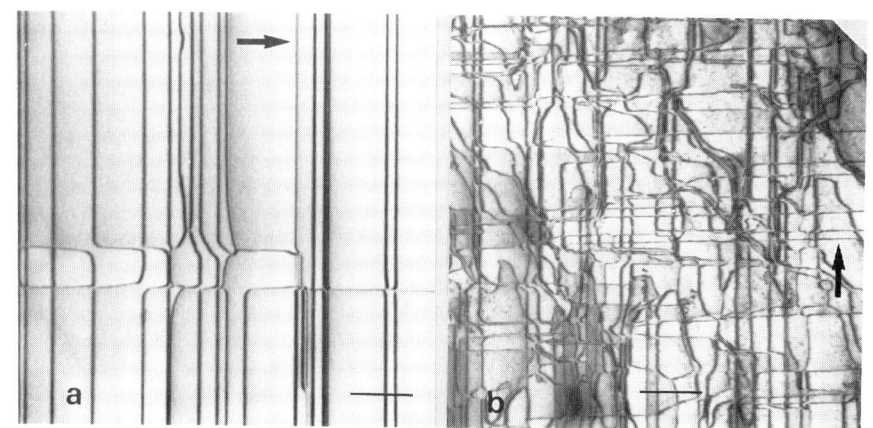

Fig. 4. TEM micrographs of SQW structures grown on a) exact (100) and b) 2° off (100) GaAs substrates. x = 0.22 and t = 25 nm in both cases. **g** = [022], bar = 0.5 μm.

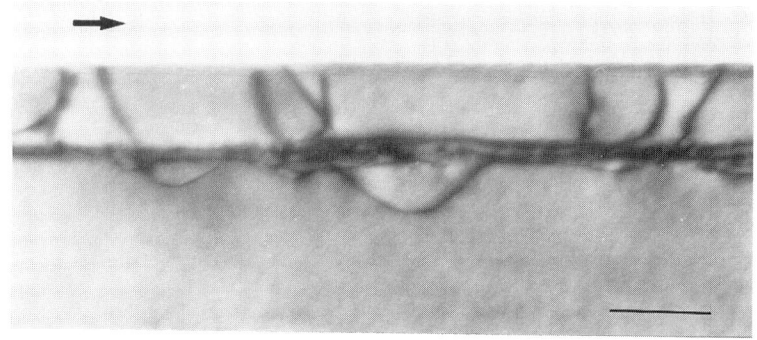

Fig. 5. Cross sectional view of an $In_xGa_{1-x}As$ SQW grown on a 2° off (100) GaAs substrate with x = 0.22 and t = 25 nm. **g** = [2$\bar{2}$0], bar = 0.2 μm.

In surface migration as the strain increases, it is also expected that the compositional difference in the growth direction between successive macrosteps is higher for high vaues of x than for low values of x. As layer thickness is increased above the first macrostep, growth proceeds via formation of another macrostep. If a sufficiently thick layer is grown (i.e. $t \geq$ 14 nm), the macrosteps formed in the initial stages of growth would tend to disappear because InGaAs grows on itself (Bhat et al 1991). Elimination of macrosteps, however, is expected to be less efficient for the highest values of x because of the larger difference in composition between successive macrosteps. This would lead to formation of a stripe-like structure in the thicker layers having a high In content ($x = 0.22$), as shown in fig. 3 a). Consequently, the nominal SQW structure behaves as a multilayer-like structure due to the In concentration modulation between the macrosteps. In such a case the step bunching effect is amplified and the macrostep height increases (Bhat et al 1991), as we observed.

3.2 Dislocation structure

For layers grown on the exactly oriented (100) substrates a crossed grid of <011> oriented misfit dislocations of the 60° type have been detected only in the layers with $x = 0.22$ and thickness $t \geq 18$ nm (fig. 4a), whereas thinner layers are defect-free. Even for the 2° off (100) oriented substrates misfit dislocations are present only for $x = 0.22$ and $t \geq 18$ nm (fig. 4 b). They have undergone, however, many reactions with one another (fig. 4 b), which is greatly favoured by the 2° offcut (Kightley and Goodhew 1991), with the generation of numerous dislocation segments oriented along directions different from <011>, which is attributed to the operation of the Hagen-Strunk (1978) mechanism. Because of these reactions, dislocation loops of the 1/2<011> type are created in the buffer layer, threading dislocations glide to the cap and new dislocations are generated at positions inside the well other than the interface plane, as shown in fig. 5. Dislocations bowing into the substrate have also been seen for layers grown on exactly (100) oriented substrate, but in much smaller number. For epilayers on exactly (100) oriented substrates, it has been suggested (Lefebvre et al 1991, LeGoues et al 1992, Chang et al 1993,) that dislocations in the substrate are introduced due to a type of Frank-Read mechanism, where the Frank-Read sources are the nodes formed by intersecting misfit dislocations. Such nodes act as pinning points. The higher density of dislocation loops in the substrate in our 2° off (100) case may be explained by the higher density of pinning points, because of the greater probability of interaction beween misfit dislocations. Pinning by interacting dislocations could also be responsible for the formation of the threading dislocations (Hull et al 1989, LeGoues et al 1992).

ACKNOWLEDGEMENTS

Thanks are due to Mr M Scaffardi for technical support. Work supported by PN-MIA.

REFERENCES

Bhat R, Koza M A, Hwang D M, Kash K, Caneau C and Nahory R E 1991 J. Crystal Growth 110 353

Chang J C P, Chin T P, Tu C W and Kavanagh K L 1993 Appl. Phys. Lett. 63 500

Chen Y, Zakharov N D, Werner P, Liliental-Weber Z, Washburn J, Klem J F and Tsao J Y 1993 Appl. Phys. Lett. 62 1536

Cox H M, Aspnes D E, Allen S J, Bastos P, Hwang D M, Mahajan S, Shahid M A and Morais P C 1990 Appl. Phys. Lett. 57 611

Frigeri C, Di Paola A, Gambacorti N, Ritchie D M, Longo F and Della Giovanna M 1994 Mater. Sci. Eng. B28 346

Hagen W and Strunk H 1978 Appl. Phys. 17 85

Hiramoto K, Tsuchiya T, Sagawa M, and Uomi K 1994 J. Crystal Growth 145 133

Hull R, Bean J C and Buescher C 1989 J. Appl.Phys. 66 5837

Kightley P and Goodhew P J 1991 Inst. Phys. Conf. Ser. 117 515

Lefebvre A, Herbeaux C, Bouillet C and Di Persio J 1991 Phil. Mag. Lett. 63 23

LeGoues F K, Meyerson B S, Morar J F and Kirchner P D 1992 J. Appl. Phys. 71 4230

Inst. Phys. Conf. Ser. No 146
Paper presented at Microsc. Semicond. Mater. Conf., Oxford, 20–23 March 1995
© *1995 IOP Publishing Ltd*

Structural and optical study of InGaAs/InP single layers and multi quantum wells grown under tensile strain condition

A Antolini, C Papuzza, G Schiavini, D Soldani, F Taiariol, [1]L Lazzarini and [1]G Salviati, [2]M Mazzer and [2]C Zanotti-Fregonara

CSELT, Via Reiss Romoli 274, 10148 Torino, Italy
[1]C.N.R.-MASPEC, Via Chiavari 18/A, Parma , 43100 Italy
[2]Imperial College of Science, Technology and Medicine, Dept. of Materials, Prince Consort Road SW7 2BP, London , UK

ABSTRACT: Structural and optical studies on single InGaAs layers and InGaAs/InP Multi Quantum Well structures have been performed. The samples were grown with different tensile strain conditions by the MOCVD technique and several conventional and less conventional structural and optical characterization techniques have been used. MQWs have shown inhomogeneous stress relaxation between the crystallographic orientations $[1\bar{1}0]$ and $[110]$ with cracks crossing the heterostructures only along the $[110]$ direction as observed with Transmission Electron Microscopy and Cathodoluminescence analysis. Reciprocal space maps have revealed this relaxation by the enlargement of the reciprocal lattice points of InP. The strained materials and the partially relaxed materials give the same Photoluminescence signal. The same behavior has been found by Spectral Cathodoluminescence in MQWs at low magnification suggesting that emission in inhomogeneous samples is mainly related to the strained areas.

1. INTRODUCTION

InGaAs/InP Multi Quantum Well (MQW) heterostructures are of great interest for devices in optical fiber communication systems. In order to realize electro-optical modulators (EOM) using the quantum confined Stark effect (QCSE) , several test MQW structures were grown. The introduction of compositional strain in $In_{1-x}Ga_xAs$ quantum wells has a strong influence on the valence band structure.It is possible to change the type of optical transitions as a function of strain. Moreover a strong increase of the QCSE shift was observed for the tensile strained samples when compared to the lattice matched samples (Härle et al. 1993). Since the Stark shift is proportional to the well thickness, strained wells have to be grown under appropriate tensile conditions in order to increase the well thickness keeping the energy transition constant .

The aim of this work is to study structural and optical properties of strained, partially and totally relaxed tensile MQW structures. To this purpose conventional and less conventional characterisation techniques were used. Structural techniques: High Resolution X-Ray, Diffraction (HRXRD), Triple Axis Diffraction (TAD), Scanning Electron Microscopy Panchromatic Cathodoluminescence (PCL), Transmission Electron Microscopy (TEM) and optical techniques: Low Temperature Photoluminescence (LTPL), Scanning Electron Microscopy, Low Temperature Spectral Cathodoluminescence (LTSCL) allowed us to evaluate structural and optical properties and hence degradation of MQW-EOM structures .

2. EXPERIMENTAL

The growth of all the investigated structures was carried out in a Low-Pressure Metal Organic Chemical Vapour Deposition (LP-MOCVD) system, with a "T" shaped reactor of rectangular cross section, working at 100 mbar and with total hydrogen flow rate of 12 l/min. The growth temperature was 560 °C. A pre -heating of 1 hour at 750 °C in hydrogen flow was performed for all the growths in order to minimize the memory effect of the reactor. A series of ternary single layers 200 nm thick with Ga composition in the range .47<x<1, grown in order to calibrate the MOCVD parameters, were also studied for assessing their optical and structural quality. All the MQW heterostructures were grown with the same optimized values of 5,0,0 and 1,1,5 seconds of growth interruption time (GIT) at the binary/ternary and ternary/binary interfaces (Salviati et al 1993). The intervals of time represent respectively the time in the flux of the growing hydride, the purge time with transport of hydrogen carrier gas, the conditioning time with the new hydride. All structures were grown on InP (100) oriented Sulfur doped substrates, with 500 nm of InP unintentionally doped as buffer layer, 20 periods of $In_{1-x}Ga_xAs$ well and InP barrier and 100 nm of InP as cap layer.

The MQWs composition and thickness, described in table A, were designed to obtain an energy transition ranging from 0.8 to 0.84 eV at room temperature.

Table A: Solid composition of the InGaAs wells, barrier and well thickness in nm, parallel and perpendicular zero-order peak mismatch and average gallium composition in the MQW structures, x_{ave}, (i.e., that corresponding to the average misfit of the MQW stack).

sample	x	well	barrier	period	$m_0//$	$m_0 \perp$	x_{ave}
MQW232	.52	13.8	6.2	20	$<$-2 10^{-4}	-2.7 10^{-3}	.50
MQW259	.56	10.7	8.0	18.7	$<$-2 10^{-4}	-4.8 10^{-3}	.52
MQW260	.56	10.7	12.	22.7	$<$2 10^{-4}	-4.4 10^{-3}	.51
MQW247	.60	13.5	6.5	19.2	-1.4 10^{-2}	-1.5 10^{-2}	.55

All the diffraction rocking curves were obtained with two Philips high resolution diffractometers (HR-XRD) using a 4 crystal monochromator. The (220) reflection of the Ge monochromator crystal was used with Cu K-a1 radiation. Spectra from the (004) symmetrical reflection and from the ($\bar{1}\bar{1}5$) and (115) asymmetrical reflections were obtained in order to evaluate the perpendicular and parallel lattice mismatch of all the heterostructures.
TEM analyses were performed in a 2000FX JEOL microscope working at 200 kV on mechano-chemically thinned samples finished by room temperature argon ion milling.
PCL studies were also carried out in a 360 Cambridge Stereoscan microscope both in the emission and transmission geometry at accelerating voltages ranging between 7 and 30 kV.
Two dimensional reciprocal space maps around (004) and ($\bar{1}\bar{1}5$) reciprocal lattice points (RELPs) were obtained using a Philips triple axis diffractometer technique (TAD).
Optical quality of the structures was assessed with Low Temperature Photoluminescence (LTPL) using a Bio-Rad FTS40 Fourier Transform spectrometer equipped with a nitrogen-cooled Germanium detector. LTSCL investigations were carried out both at liquid nitrogen and at helium temperatures in a JEOL 840 SEM, equipped with a non-commercial system based on a liquid nitrogen-cooled germanium detector.
TEM and PCL measurements were performed at MASPEC in Parma, HRXRD, TAD and LTPL were performed at CSELT in Turin. LTSCL were performed at the Materials Department of the Imperial College of Science, Technology and Medicine in London.

3. RESULTS AND DISCUSSION

3.1 InGaAs single layers

Theoretical parallel and perpendicular mismatch vs. gallium concentration for strained, partially and totally relaxed materials are shown in fig.1. It is possible to observe that the experimental points reveal that layers are strained in the composition range 0.47<x<0.55 mantaining good surface morphology, while in the range of composition 0.55<x<0.75 a partial relaxation exists and the layers show cross hatched morphology and cracks. For x>0.75 the layers are completely relaxed and their surface morphology shows evidence of three dimensional nucleation. InGaAs compressive layers of the same thickness were also grown . Experimental points of perpendicular and parallel mismatch values in the compressive part of the theoretical curve (Ga content, x<0.47), show that relaxation occurs for lower absolute values of mismatch than in the tensile strain condition. Theoretical band gaps for $In_{1-x}Ga_x$ As/InP vs. x is sketched in fig.2 with the reported LTPL experimental values. The E0 transition is related to the relaxed gap of the material and E1 and E2 are related to the degeneracy solution of $|3/2\ 3/2>$ and $|3/2\ 1/2>$ gap due to hydrostatic and tetragonal distortion of the stressed material.

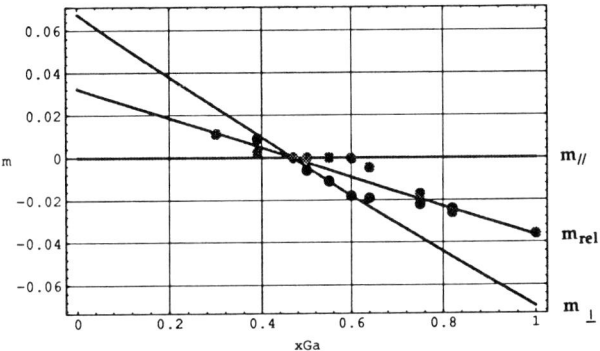

Figure 1: Experimental points, parallel and perpendicular calculated mismatch for $In_{1-x}Ga_xAs/InP$ single layers vs. x for completely strained and relaxed materials.

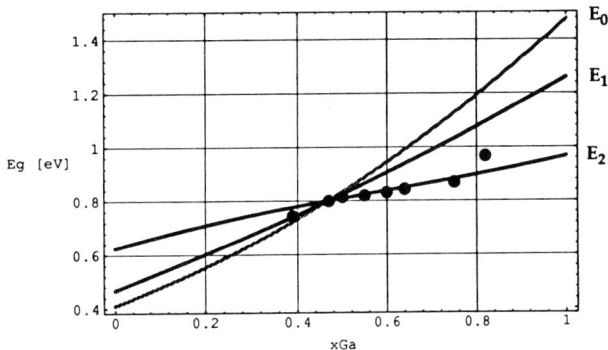

Figure 2: Calculated band gap and experimental LTPL points of $In_{1-x}Ga_xAs/InP$ layers vs. x. E0 relaxed, E1 strained $|3/2\ 3/2>$, E2 strained $|3/2\ 1/2>$ transition.

The agreement between experimental and theoretical data predicted by the elastic model theory (Wang 1990) is quite good. The LTPL experimental data always lie on the strained E2 curve of fig.2, also for the relaxed samples. This is due to the small contribution of the relaxed areas, with high dislocation or crack density, to the PL signal. The signal remains almost constant up to x=0.55 and then dramatically decreases towards zero when x=0.7.

368

Some experimental results obtained using LTSCL performed several months later at low magnification (see elsewhere) have shown no relaxation of the layers.

3.2 Tensile strained MQW

The parallel ($m_{\parallel}$) and the perpendicular ($m_{\perp}$) mismatches of the zero-order peak have been evaluated by the use of the HRXRD technique and data are reported in table A. It is noteworthy that the value of the $m_{\parallel}$ increases as the average content of gallium in the MQW increases. The MQWs investigated were in a metastable condition and cracks started propagating in the specimens some weeks after the growth. The presence of cracks along the [110] direction has been observed, using the PCL technique, in samples with a low gallium concentration, while cracks and perpendicular misfit dislocations have been dectected for higher Ga content as shown in fig.5 (Salviati et al.1995). This phenomenon could be ascribed to inhomegeneity of the relaxation energy along the two perpendicular crystalline directions which enhances the preferential formation of cracks instead of misfit dislocation (Maigné 1995). TEM analysis reveals that these cracks propagate from the top to the substrate crossing the MQW. The penetration of cracks into the substrate in tensile InGaAs layers on InP has already been observed (Franzosi et al 1988). In the reciprocal lattice maps, obtained with the TAD technique (Fewster 1993), around the reciprocal lattice points ($\overline{11}5$) an increase in the diffused X-ray intensity of the reciprocal lattice point related to InP is observed as the average composition of gallium in the MQW increases. This effect is more evident for the sample with a higher gallium content, x=0.60 (figs.3 and 4) and could be ascribed to a diffused mosaic-like structure. Moreover a compositional inhomogeneity in the growth direction, due to a memory effect inside the MOCVD reactor correlated to non optimized GIT during the growth (Salviati et al. 1993) and this is revealed by the elongation of satellite peaks of the MQW.

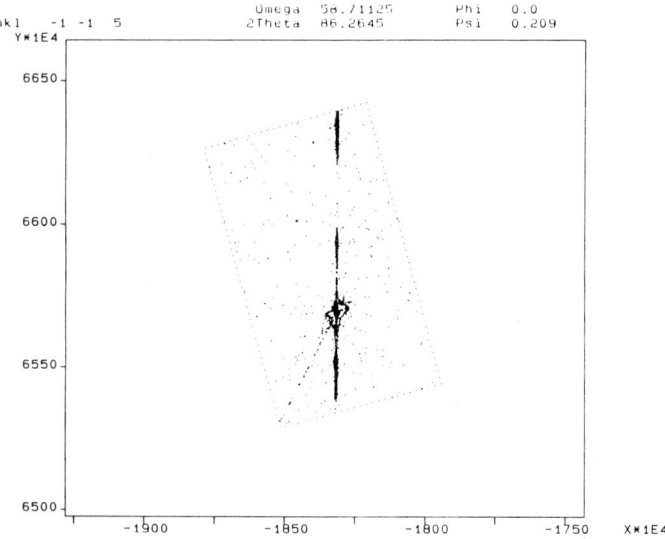

Figure 3: TAD reciprocal space map around ($\overline{11}5$) of MQW259.

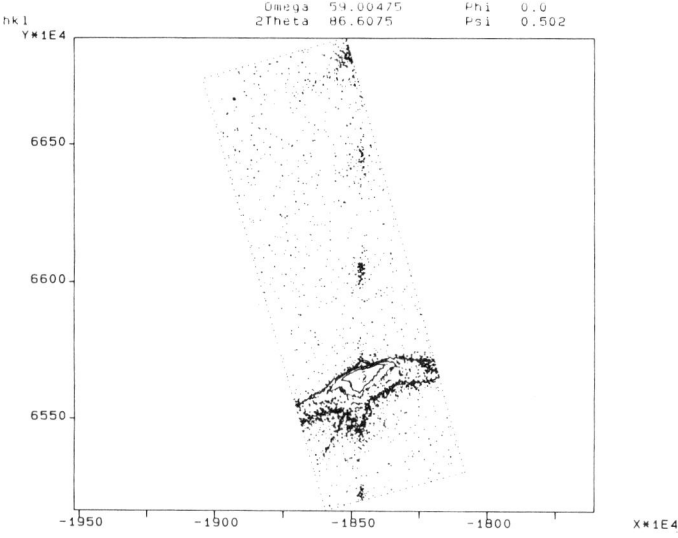

Figure 4: TAD reciprocal space map around ($\bar{1}\bar{1}5$) of MQW 247.

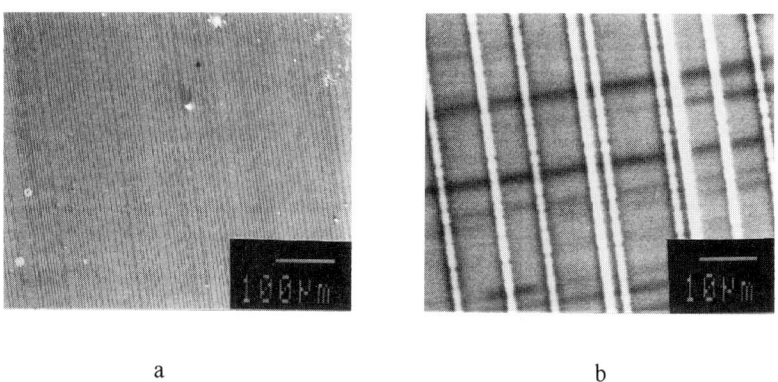

a   b

Figure 5: a) SEM image of MQW247 surface morphology b) PCL image of the same area. Cracks along [110] direction and perpendicular misfit dislocation are shown.

The LTSCL data are in agreement with the calculated optical transition for strained materials by using a simple Daniel-Duke model (G.Bastard 1988) even if shoulders are sometimes present. The presence of the shoulder at higher energy could be attributed to an optical transition with a partial layer relaxation. When small areas are irradiated (high magnification) the intensity of the shoulder increases with respect to the main peak as the beam current is increased only in the areas containing cracks. This effect is not detectable if larger areas are irradiated (low magnification) because the influence of the relaxed region becomes less important. The experimental results are reported in table B and table C. The agreement is also good for samples showing evidence of relaxation. Sample instability did not modify the value of the detectable optical transition, because the main part of the signal is related to the strained and not to the relaxed material. This phenomenon is the same as that observed in the single layers with LTPL.

370

Table B: MQW 259. LTSCL, value of the peak energy in different areas.

Sample area	Peak energy (meV)	Beam current (nA)
with cracks	0.895	5
without cracks	0.888	5

Table C: MQW259. LTSCL, peak-shoulder intensity ratio as a function of current and magnification in area containing cracks.

Beam current (nAmp)	5	2.66
Peak shoulder intensity ratio magnified x10000	0.98	2.89
Peak shoulder intensity ratio magnified x500	2.96	2.94

4. CONCLUSIONS

A study of optical and structural properties of single InGaAs layers and MQW InGaAs/InP grown under tensile strain conditions is presented. For single layers a relationship between relaxation vs Ga content for the tensile and compressive parts was obtained. Photoluminescence analysis also shows a good agreement with the elastic model for partially relaxed samples. For MQW, TEM and PCL analyses have shown cracks along the [110] direction and perpendicular misfit dislocations for average Ga content (i.e., that corresponding to the average misfit of the MQW stack) of 0.55 while only cracks exist for lower average Ga content. Reciprocal lattice maps around the ($\bar{1}\bar{1}5$) lattice points of InP reveal a mosaic-like structure and a compositional inhomogeneity in the grown direction is also observed. Spectral CL analysis shows that for low magnification the main peak is in agreement with the hypothesis that the main part of the emission is related to the strained region of the material and at higher magnification it is possible to distinguish between strained and relaxed areas.

REFERENCES

Bastard G. Wave mechanism applied to semiconductor heterostructures 1988 Halsted Press N.Y.
Fewster P.F. Semiconductor Science and Technology **8**, 1993
Franzosi P., Salviati G., Scaffardi M., Genova F., Pellegrino S., Stano S., 1988 J. Crystal Growth **88**, 135
Härle V., Bolay H., Scholz F., Tutken T., Zimmermann M., Hangleiter A., Qeisser I., 1993. EW-MOVPE V Proceedings Malmoe 2-4 June
Maigné P., Dharma-Wordana M.W.C., Lockwood O.J. and Webb J.B. 1995 J.Appl.Phys. 77 1466
Salviati G., Ferrari C., Lazzarini L., Genova F. et al. 1993 Microscopy of Semiconductor Material Proceeding Oxford 5-8 April.
Salviati G., Ferrari C., Lazzarini L., Franchi S., Bosacchi A., Mazzer M., Zanotti-Fregonara C., Taiariol F., Romanato F. and Drigo A.V. This conference.
Wang T.Y., Strigfellow G.B. 1990 J. Appl. Phys. **67** 344

Inst. Phys. Conf. Ser. No 146
Paper presented at Microsc. Semicond. Mater. Conf., Oxford, 20–23 March 1995

Fabrication and TEM characterisation of V-groove InGaAs quantum wires

P-H Jouneau, F Bobard, U Marti[1], J Robadey[1], F Filipowitz[1], D Martin[1], F Morier-Genoud[1], P C Silva[1], Y Magnenat[1] and F K Reinhart[1]

Centre for Electron Microscopy, EPFL, CH-1015 Lausanne, Switzerland
[1]Institute of Micro- and Optoelectronics, EPFL, CH-1015 Lausanne, Switzerland

ABSTRACT: GaAs and $In_xGa_{1-x}As$ quantum wire arrays grown by molecular beam epitaxy on patterned substrates are investigated by transmission electron microscopy (TEM). We show that this fabrication method allows the production of a high density array of uniform wires. In some cases, growth defects may occur, consisting mainly of dissociated dislocations lying in {111} planes and running from the substrate up to the surface. For InGaAs wires with high In content (21%), evidence of plastic strain relaxation is obtained.

1. INTRODUCTION

Interest in 1D quantum wire structures has strongly increased over the past few years. The 2D confinement of electrons and holes leads to a density of states with a narrower energy distribution compared to that of a quantum well, with the consequence of enhanced binding energy for excitons and enhanced optical nonlinearity. These structures can be used to produce semiconductor lasers with low threshold currents, narrower linewidths, and reduced temperature sensitivity (Kapon 1993). They are also very promising for optical modulators and optical switching applications. However, several criteria must be met in order to produce efficient 1D structures (Kapon 1993): (i) lateral dimensions should be in the 10 nm range to resolve 1D energy subbands (for GaAs/AlGaAs materials); (ii) low defect density in the wires is required for reducing non-radiative recombination processes; (iii) low size fluctuations are needed to restrain inhomogeneous broadening and (iv) a high density of wires is required to maintain high interaction with light. Of the various approaches attempted to fabricate such 1D structures (see, e.g., the review by Cingolani and Rinaldi 1993), one successful method is epitaxial growth on non-planar substrates, either by metal organic chemical vapour deposition (Kapon 1993) or by molecular beam epitaxy (MBE) as presented here.

2. WIRES FABRICATION

The first step in the present wire fabrication process is to produce non-planar substrates with uniform V-shaped surfaces, over an area of about 2 cm². This is achieved on GaAs (001) by a combination of holographic lithography (using the UV line of a doubled Ar⁺ laser at λ=257 nm) and etching (Marti et al 1991). In this process, a thin Si_3N_4 layer is used as an antireflective coating to eliminate interference fringes parallel to the surface. After patterning by reactive ion etching, the Si_3N_4 serves as a mask for wet chemical etching with perfect adhesion. V-groove orientation is chosen along the [110] direction, with the side facet orientations close to {$\bar{1}\bar{1}\bar{1}$}B. However, a few samples with grating lines oriented along [1$\bar{1}$0] have also been prepared for comparison. The grating periodicity is about 250 nm and the depth close to 120 nm.

The second step is MBE growth on these substrates. $In_xGa_{1-x}As$ (0≤x≤0.21) quantum wires embedded between two $(AlAs)_n(AlGaAs)_m$ superlattices (indices refer to the number of nominal monolayers) are grown under standard conditions, with substrate temperature between 550°C and 600°C, V/III ratio of 2:1 and a rotating substrate. For all samples, TEM analysis is performed at 300 kV using a Philips EM 430 ST microscope. The specimens are prepared as <110> cross sections or as <001> plan views by mechanical polishing and subsequent Ar⁺ ion milling.

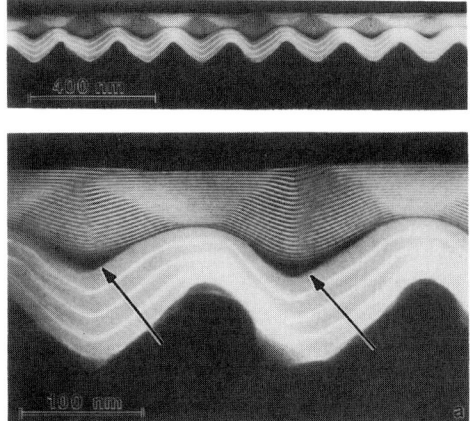

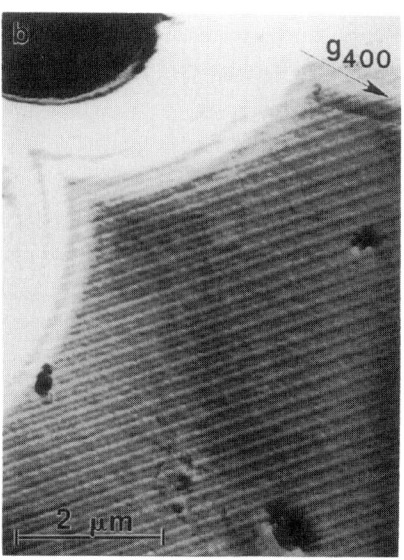

Fig. 1 GaAs wires (18 ML plane nominal thickness) embedded between $(AlAs)_4(Al_{0.33}Ga_{0.66}As)_{55}$ and $(AlAs)_4(GaAs)_8$ superlattices. (a) 200 dark-field cross-section and (b) 400 dark field plan view showing well shaped wires.

TEM observations reveal that uniform arrays of quantum wires can be formed with this growth method (Fig. 1). The wires (arrowed) are lying at the bottom of the V-grooves, with a crescent shape and a lateral size of about 30 nm. Their vertical thickness is maximum in the centre and decreases continuously towards the sides. Such a shape results from step mass transport which occurs during the growth interruption following the (In)GaAs layer deposit. Moreover, all the interfaces are obtained *in situ*, during epitaxial growth and therefore present few defects.

We also note the following features:

i) Differences of adatom diffusion lengths on the different facets of the V-grooves lead to different growth rates on these facets. The consequence is the planarization of the structure after the growth of typically 100 nm, with the appearance of (001) facets on the top of the structure. Nevertheless, this planarization is beneficial for wire uniformity. Indeed, it tends to reduce the

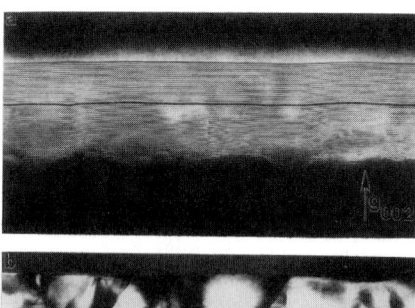

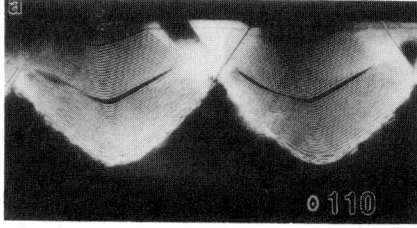

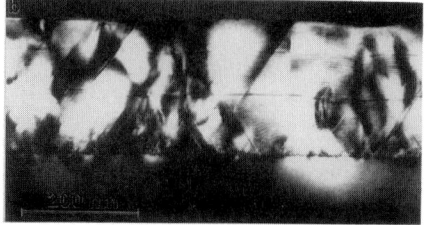

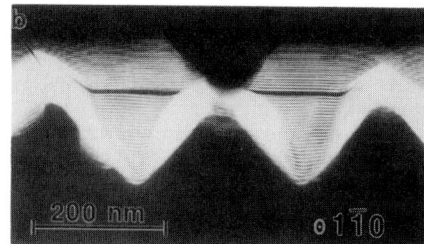

Fig. 2 Influence of oxide desorption. These two GaAs/$(AlAs)_4(GaAs)_8$ samples have been grown under the same conditions, after (a) a complete or (b) a partial oxide desorption.

Fig. 3 Influence of V-groove orientation on wire shape. These two GaAs/$(AlAs)_4(GaAs)_8$ samples have been grown under the same conditions, at the same time on the MBE machine.

impact of substrate imperfections due both to imperfect lithography and oxide desorption. This phenomenon is also seen in Fig. 2a. The surface roughness left by oxide desorption on an unpatterned reference substrate is quickly planarized by the first AlAs/GaAs superlattice.

ii) The planarization behaviour is not the same for the different compounds. Ga and In atoms are highly mobile and lead to a rapid planarization of the V-groove. On the other hand, the low mobility of Al atoms tends to preserve the shape of the grating.

iii) A strong difference is observed for samples grown under the same conditions but with grating lines oriented along [110] or [1 1 0], as shown in Fig. 3. Planarization is faster when the substrate is grooved with channels oriented along [1 1 0] direction. This is explained by the difference of surface diffusion lengths for adatoms along the two respective directions. This phenomenon, already observed by Turco et al (1990), renders the growth of well shaped wires difficult on [1 1 0] oriented V-grooves and justifies our choice of [110] oriented V-grooves.

3. GROWTH DEFECTS

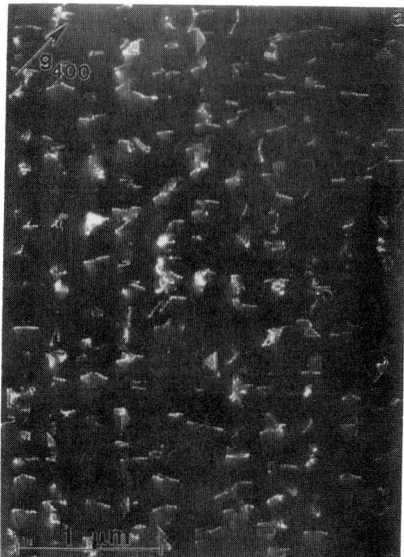

Fig. 4 $In_{0.17}Ga_{0.83}As$ wires (18 ML plane nominal thickness) embedded between two $(AlAs)_4(GaAs)_8$ superlattices. Dark field (a) 400 plan view and (b) 002 cross section.

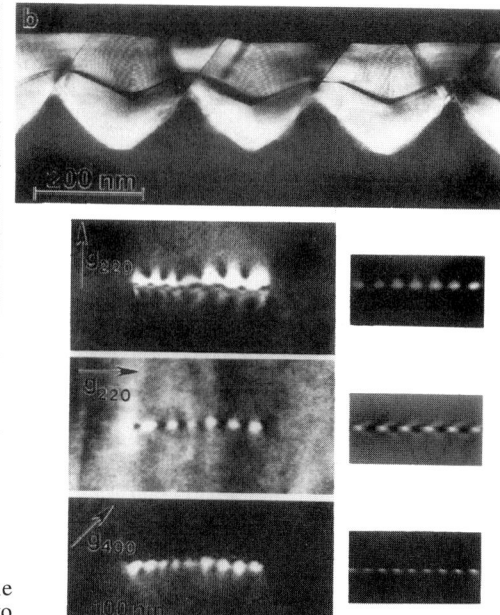

Fig. 5 A typical dissociated dislocation observed in dark field plan view for various **g**, and the corresponding simulated images.

Some defects may occur in our quantum wire structures, sometimes with high densities when growth is carried out under inadequate conditions. This is the case for the sample shown in Fig. 4. A complete description of these defects will be the subject of a forthcoming paper (Jouneau et al 1995). In cross-sectional images (Fig. 3a, Fig. 4a), we observe stacking faults running from the top of the V-grooves up to the free surface. Plan view images with various **g** conditions (Fig. 4b, Fig. 5) reveal that these stacking faults are due to dissociated dislocations running between the substrate and the surface. The more common configuration is the one displayed in Fig. 5, which shows an edge dislocation with line along [112] going down from the surface to an edge of the substrate and returning to the surface along [11$\bar{2}$]. This dislocation is dissociated with Burgers vectors given by the classical relation $a/2[1\bar{1}0] \rightarrow a/6[121] + a/6[21\bar{1}]$. The distance between the two partials is estimated to be about 5 nm from image simulations.

The mechanism leading to these defects is not yet fully understood, but it appears strongly related to two parameters:

i) Desorption of the native oxide. A thermal oxide is intentionally formed on the surface prior to loading the substrate into the growth chamber. The desorption of this oxide before growth helps to reduce contamination resulting from substrate preparation, with the consequence of smaller impurity related transitions in the luminescence spectra. The cost is an increase in the surface roughness (Fig. 2a). But if little traces of this oxide remain at the start of the growth, TEM (Fig. 2b) shows that stacking faults are easily formed in the epitaxial layer.

ii) Profile and uniformity of the grating: The etched grating profile must be smooth enough to prevent the formation of stacking faults, i.e. with facets no steeper than (111) (see, e.g., the differences between grating profiles of Fig. 1 and Fig. 4a). In the same way, an accurate grating profile is of prime importance to avoid non uniformities of the wire shape.

4. STRAIN RELAXATION

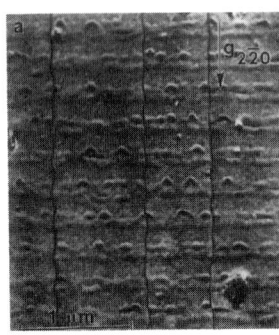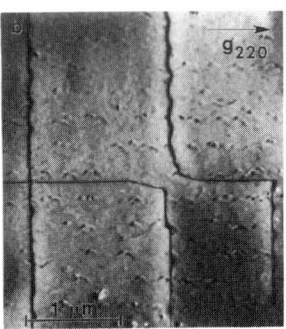

Fig. 6 Dark field plan views of an $In_{0.21}Ga_{0.79}As$ wire array (18 ML plane nominal thickness) embedded between two $(AlAs)_4(GaAs)_{53}$ superlattices.

Finally, TEM observations reveal a plastic strain relaxation of the InGaAs wire array with high indium content (21 %) (Fig. 6). 60° dislocations, running along [110] and [1$\bar{1}$0] directions, relax part of the 1.5 % lattice mismatch between $In_{0.21}Ga_{0.79}As$ and GaAs. There is, however, a difference with the case of relaxation in a quantum well. Dislocation lines appear to be straight along the wire direction, but oscillate in the normal direction as they follow the undulating grating surface. These first results also suggest a smaller critical thickness for wires than for an equivalent quantum well. However, additional experiments are required to confirm this hypothesis. There is also a need for the development of new theoretical models for interpreting the strain relaxation in quantum wire structures.

5. CONCLUSION

Uniform and high density arrays of GaAs and InGaAs quantum wires can be grown by molecular beam epitaxy on patterned substrates with V-grooves running along the [110] direction. However, our TEM experiments suggest that the fabrication process (especially substrate patterning and oxide desorption) must be precisely controlled in order to avoid growth defects in the structures. Additional photoluminescence studies by Rinaldi et al (1994) have proven that these structures exhibit clear evidence for 1D confinement, with a splitting of about 15 meV between the $n_y=1$ and $n_y=2$ excitonic levels. This work is part of the OPTIQUE priority program of the board of the Swiss Federal Institutes of Technology.

REFERENCES

Cingolani R and Rinaldi R 1993 Rivista del Nuovo Cimento 16 1
Jouneau P H et al (1995) (unpublished)
Kapon E 1993 Optoelectronics 8 429
Marti U, Proctor M, Monnard R, Martin D, Morier-Genoud F, Reinhart F K, Widmer R, and Lehmann H 1991 AIP Conf. Proc. 227 80
Rinaldi R, Ferrara M, Cingolani R, Marti U, Martin D, Morier-Genoud F, Ruterana P, and Reinhart F K 1994 Phys. Rev. B 50 11795
Turco F S, Simhony S, Kash K, Hwang D M, Ravi T S, Kapon E, and Tamargo M C 1990 J. Cryst. Growth 104 766

Inst. Phys. Conf. Ser. No 146
Paper presented at Microsc. Semicond. Mater. Conf., Oxford, 20–23 March 1995
© *1995 IOP Publishing Ltd*

Microscopy characterization of quantum wires grown on grooved substrates

A Gustafsson, B Dwir, F Reinhardt, G Biasiol, J-M Bonard and E Kapon

Institut de Micro- et Optoélectronique, Département de Physique, Ecole Polytechnique Fédérale de Lausanne, CH-1015 Lausanne, Switzerland .

ABSTRACT: We have used transmission electron microscopy in the high resolution and the dark field modes as well as atomic force microscopy to compare quantum wire structures grown by organometallic chemical vapour deposition at different pressures. The interfaces are more abrupt and develop better defined facets when growing at low pressure than at intermediate and atmospheric pressure. Growth at low pressure therefore has a potential for fabrication of smaller structures with less size variations than the conventional atmospheric pressure growth.

1. INTRODUCTION

Quantum wires (QWRs) are expected to exhibit novel and interesting optical and electrical properties related to the two-dimensional confinement of the carriers in these one-dimensional (1D) structures (Weisbuch and Vinter 1991). One of the more promising ways to produce these QWRs is to grow them by organometallic chemical vapour deposition (OMCVD) on V-grooved substrates (Kapon *et al.* 1989). The growth of a thin GaAs layer on an AlGaAs V-groove yields a crescent-shaped QWR, formed by self-ordering at the bottom of the groove. A very important feature is the shape of the barrier at the bottom of the V-groove where the QWRs are formed. The two quasi {111} planes of the side walls of the V-groove meet in a rounded corner. The radius of curvature (ρ) of this corner is given by the growth conditions for the thick AlGaAs (Kapon 1994), as shown schematically in Fig. 1. The growth of the AlGaAs barrier on the quasi {111} planes results in a sharp corner at the bottom of the groove, with a typical radius ρ_l. The steady-state value depends only on growth conditions and conventional atmospheric pressure OMCVD can reach a ρ_l down to ~8-10nm. A thin (2-4nm) GaAs layer deposited on the barrier causes a linear increase in ρ (up to ~20nm), thus forming a crescent-shaped GaAs QWR with a bottom radius ρ_l and a top radius ρ_u (Kapon *et al.* 1995). Further deposition of AlGaAs caps the QWR and reduces the corner to the steady-state radius, thus enabling the growth of a new and identical QWR on top of it. The typical radii of the QWR, ρ_l and ρ_u, determine its energy levels, whose separation increases as the QWR becomes smaller. To be able to characterize and use (e.g. in electro-optical devices) QWRs at room temperature, the 1D energy level spacing must be greater than ~25meV, thus implying radii below 10nm. Furthermore, the GaAs/AlGaAs interfaces must be abrupt since composition fluctuations at the interfaces lead to energy band broadening and to smearing of the 1D effects.

Another aspect of the AlGaAs growth on a V-grooved substrate is the formation of a

376

vertical quantum well (VQW) (Walther *et al.* 1992), which is also shown schematically in Fig. 1. This region is a vertical sheet of Ga-rich AlGaAs in the center of the V-groove, forming due to segregation of the Ga atoms during growth. Its importance comes not only from its potential use as a vertical (instead of a conventional horizontal) QW structure, but mainly from its role in enhancing carrier capture to the QWR, which contributes to increased efficiency of the QWR luminescence, e.g. in lasers. Here again, the nature and quality of the VQW interfaces are important to the role of the VQW in promoting efficient carrier capture.

One direction to take in order to reduce the QWR radii and improve its interfaces is to grow the structures at reduced pressure. Growth at low pressure (<100mbar) generally leads to sharper and more well defined interfaces of QW structures (Razeghi 1989). Here we make a comparison of growth at two different pressures in the same reactor, 150mbar and 20mbar. The 150mbar growth gives features that are comparable with growth at atmospheric pressure in terms of interface quality and QWR shape. Both structures were grown on (100) GaAs substrates with a 3.5μm-pitch grating of V-grooves, oriented along the [01$\bar{1}$] direction. A stack of several GaAs QWRs was grown in between Al$_{0.48}$Ga$_{0.52}$As barriers. Sample A was grown at 150mbar and at 750°C and the nominal GaAs thickness was 15nm. Sample B was grown at 20mbar and at 650°C and the nominal GaAs thickness was 7.5nm.

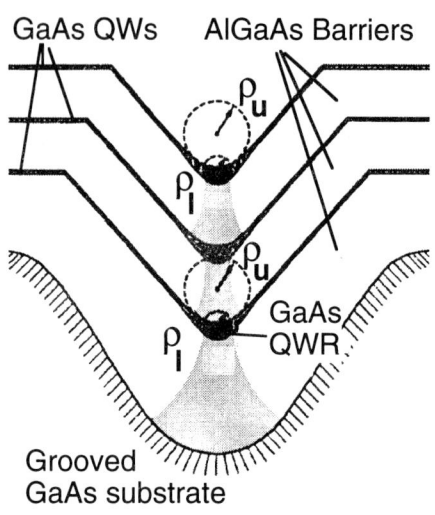

Fig. 1 A schematic drawing of the structure, explaining the radii of curvature. The formation of the VQW is also illustrated.

2. THE QUANTUM WIRES

To investigate the shape of the QWRs we have used (200) dark field (DF) imaging and for the abruptness of the interfaces we have used high resolution (HR) transmission electron microscopy (TEM). Fig. 2 shows two DF images of (a) sample A and (b) sample B. In these images, the shape of the QWRs are similar, though there is a difference of two in magnification. In these images the rounding of the lower interface is larger in sample A ($\rho_l \approx 11$nm) than in sample B ($\rho_l \approx 6$nm). There is also a difference in the upper interface. Sample A shows the typical rounding of the upper interface, normally associated with atmospheric growth, whereas sample B shows two flat facets at the upper interface. In the HR image of Fig. 3 (b) these facets can be determined to be {113} planes. Already in the DF images a difference in interface quality can be observed, with sample B showing sharper features. This is clearer in the HR images of a part of the QWRs in Fig. 3, (a) of sample A and (b) of sample B. In the images, the transition from the GaAs QWR to the AlGaAs barrier takes place over 2-4 monolayers (MLs) in sample A, whereas this is reduced to 1-2MLs in sample B. The sharper interfaces and faceted growth in sample B is interesting in itself, as it is an indication that there is a self-ordering mechanism on the atomic level, which could limit the size fluctuations of the QWRs. It is worth pointing out that the overall shape and interface abruptness of sample A is comparable to the normal atmospheric pressure OMCVD growth. For the "normal" atmospheric growth, ρ_l can be made as small as 8 - 10nm, but clearly the low pressure growth has the potential for fabricating QWRs with a significantly smaller ρ_1. The goal within reach is to fabricate QWRs with effective widths well below 10nm.

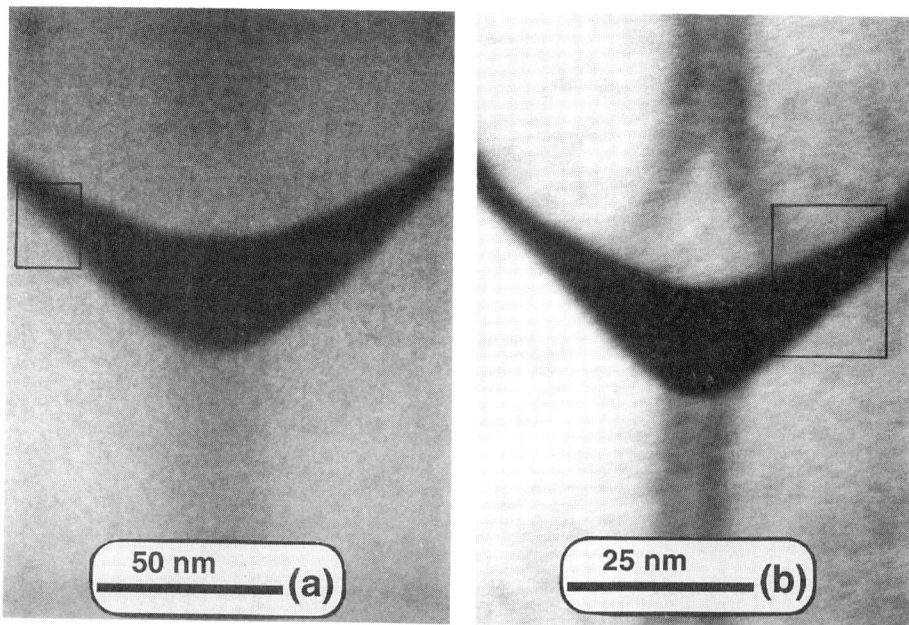

Fig. 2. (200) DF TEM images of the QWRs and VQWs of sample A (a) and sample B (b).

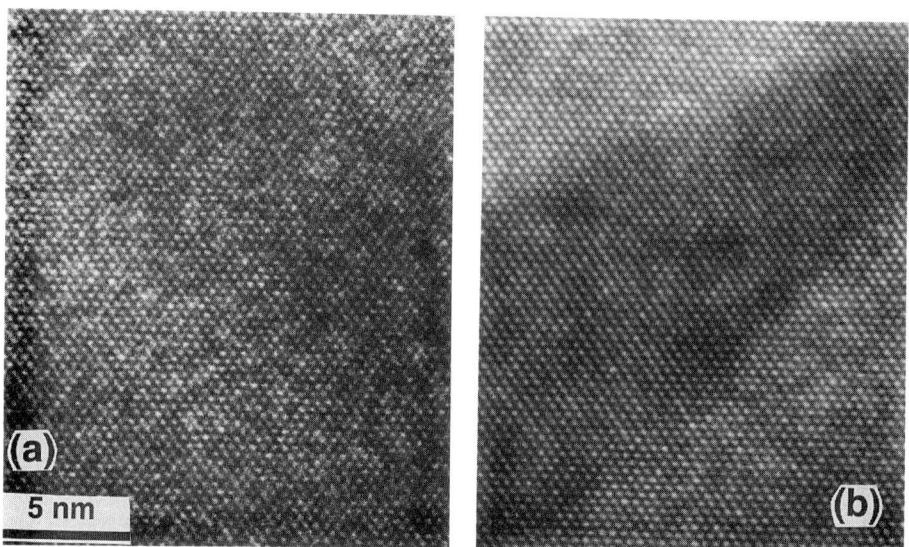

Fig. 3. HRTEM images of part of the QWR of sample A (a) and sample B (b).

3. THE VERTICAL QUANTUM WELLS

The DF images of Fig. 2 are more sensitive to compositional variations and the features of the VQWs are therefore more prominent. Sample A shows a VQW comparable with the

378

atmospheric pressure growth. It is ~30nm wide and has a constant width along its vertical axis. Sample B exhibits a double VQW structure, where each {113} facet gives rise to a branch. As the growth of the AlGaAs barrier restores the {111} side walls, these two branches are brought closely together, but still separated, forming two parallel VQWs, each <5nm wide. These features are much more well defined and smaller than the VQWs in sample A. The point where the two VQWs become parallel is of importance, as it indicates where the perturbation of the GaAs layer to the shape of the groove is restored. This indicates the thinnest AlGaAs barrier permitting growth of a second identical QWR.

4. ATOMIC FORCE MICROSCOPY

For a simpler and quicker analysis, yet with sufficiently high resolution, we have used a newly developed technique of cross-sectional atomic force microscopy (AFM). We used selective etching to translate composition differences to height differences, which we observe using the AFM. The sample is cleaved and then etched in 5% H_2O_2:H_2O, pH 7.05 $\pm$ 0.05, adjusted with NH_4OH, T=20° C. Etching for 30s gives an etch depth of ~30nm of the GaAs, whereas it does not attack the AlGaAs. The trench that is formed is therefore a replica of the GaAs layer, which can be imaged by the AFM as a height difference. This technique gives a resolution that is near that of conventional TEM, but has two main advantages: i) The sample preparation is much simpler and faster than TEM sample preparation; ii) Larger areas of the sample can be imaged (typical scan size of up to 100×100μm^2), and the scanned region can be positioned at will on an area of a few mm^2. Fig. 4 shows the AFM image corresponding to the DF TEM image of Fig. 2 (a).

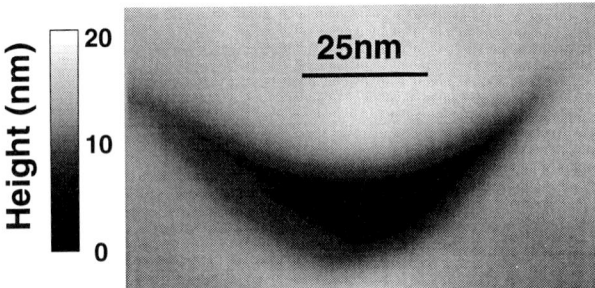

Fig. 4. An AFM image of the etched cross-section of a QWR of sample A

Acknowledgments: This project was supported by Fonds National Suisse de la recherche scientifique.

REFERENCES

Gustafsson A, Samuelson L, Hessman D, Malm J-O, Vermeire G and Demeester P 1995 J. Vac. Sci. Techn. 13 (2)

Kapon E, Hwang D M and Bhat R 1989 Phys. Rev. Lett. 63 430

Kapon E 1994 Epitaxial Microstructures ed A C Gossard in Semiconductors and Semimetals 40 259 (Boston: Academic)

Kapon E, Biasiol G, Hwang D M and Colas E 1995 to appear in Microelectronics Journal

Razeghi M 1989 The MOCVD Challenge vol. 1 (Bristol: Adam Hilger)

Walther M, Kapon E, Christen J, Hwang D M and Bhat R 1992 Appl. Phys. Lett. 60 521

Weisbuch C and Vinter B 1991 Quantum Semiconductor Structures: Fundamentals and Applications (San Diego: Academic)

Inst. Phys. Conf. Ser. No 146
Paper presented at Microsc. Semicond. Mater. Conf., Oxford, 20–23 March 1995
© 1995 IOP Publishing Ltd

Growth, Structure and Photoluminescence of SiGe/Si quantum wires and dots on patterned Si (001) substrates

Ch Dieker, A Hartmann, W Jäger*, U Bangert+, L Vescan and H Lüth

Institut für Schicht - und Ionentechnik, Forschungszentrum Jülich, D-52425 Jülich,Germany
*Institut für Festkörperforschung, Forschungszentrum Jülich, D-52425 Jülich, Germany
+Department of Pure and Applied Physics, UMIST, Manchester M60 1QD,UK

ABSTRACT: Self-organized growth of $Si/Si_{1-x}Ge_x/Si$ quantum wire and quantum dot structures on patterned Si(001) substrates during low pressure chemical vapour deposition was investigated for $0.2 \leq x \leq 0.4$ and for nominal layer thicknesses $t \leq 4nm$. Depositions were performed on V-shaped grooves bounded by {111} facets, grooves with L-shape corners bounded by {111} and {100} facets, and inverse pyramids bounded by {111} facets. Structure, strain state and composition were analysed by transmission electron microscopy and by spatially resolved energy-dispersive X-ray spectroscopy. The results were compared with photoluminescence measurements.

1. INTRODUCTION

Low-dimensional semiconductor heterostructures, such as quantum wires (QWR) and quantum dots (QD), are expected to exhibit new quantum confinement related effects. For QWR structures grown on V-groove patterned substrates in the AlGaAs/GaAs system (e.g. Kapon et al. 1989) and in the InGaAs/GaAs system (e.g. Walter et al. 1993) two-dimensional confinement was demonstrated. Recently Usami et al. (1993) and Hartmann et al. (1995a,b) reported the growth of SiGe QWR structures using V-groove patterned Si (001) substrates and also photoluminescence (PL) measurements of these structures. This paper presents investigations of the structural and optical properties of QWR and QD structures grown on patterned substrates. Their formation and correlation between structure and optical properties are discussed.

2. EXPERIMENTAL TECHNIQUES

Patterning was performed by standard optical lithography, subsequent reactive ion etching and wet chemical etching as reported by Hartmann et al. (1995b). V-shaped grooves bounded by {111} facets (V-QWR structures), grooves with L-shape corners bounded by {111} and {100} facets (L-QWR structures), and inverse pyramids bounded by {111} facets (QD structures) were fabricated on a Si (001) substrate. Low-pressure chemical vapour deposition (LPCVD) at 700°C was then used to deposit a 20 nm thick Si buffer layer, a SiGe layer of Ge concentration $0.2 \leq x \leq 0.4$ and thickness $t \leq 4nm$ capped by 1μm of Si.

Structure and strain state were investigated by transmission electron microscopy (TEM) of cross-section (CS) and plan-view (PV) samples using a JEOL 4000 FX at 400kV. Composition analysis was performed by spatially resolved EDX in a Fison VG HB601UX STEM at 100kV with a LINK detector and a Philips CM20 FEG at 200kV with an EDAX detector. PL spectra

380

were taken at 4K and 40K using a Fourier transform spectrometer (BIO RAD FTS40) and an Ar-ion laser emitting 50mW at 488nm.

3. EXPERIMENTAL RESULTS

3.1 TEM investigations

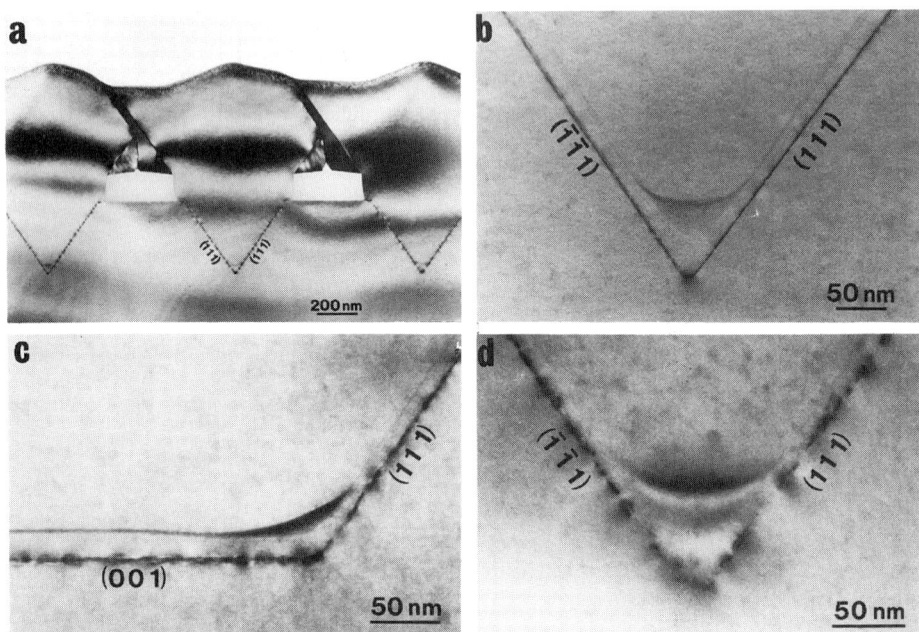

Fig.1: ⟨110⟩ cross-section bright-field images of a patterned Si (001) substrate showing (a) an overview of a V-groove QWR structure, (b) the bottom part of a V-groove, (c) QD structure grown in an inverse pyramid, (d) and a L-corner QWR structure.

Fig.1 shows cross-section TEM bright field images of Si/Si$_{1-x}$Ge$_x$/Si QWR and QD structures on a patterned Si(001) substrate with nominal Ge concentration of x = 0.3 and layer thickness of t=1.5nm. An overview of a V-groove patterned area is given in Fig.1a. Clearly visible are two SiO$_2$ etching masks (bright) and the Si cap layer. The V-shaped dark line contrasts mark the growth interface between substrate and the Si buffer layer. In addition to the two-dimensional quantum well (QWL) layers covering the {111} facets, QWR structures in ⟨110⟩ directions are formed by preferential growth of SiGe along the intersections of the facets. Fig.1b shows an en-larged cross-sectional view of the intersection. The crescent-shaped SiGe QWR structure exhibits a thickness of 20 nm, which surmounts by far the value of ~0.7 nm estimated for the QWL on {111}. Fig.1c shows a QD structure in the tip of an inverse pyramid. It is also crescent-shaped, with an estimated thickness of 25 nm. 15 nm thick L-QWR structures are formed at the intersec-tion of (001) and {111} Si buffer interfaces as shown in Fig.1d. The different thicknesses of the Si buffer layers on (001) and {111} indicate that the Si growth rate on (001) is by a factor of 2 larger than that on {111}. A similar behaviour is found for the growth of SiGe layers on these planes. Thicknesses of V-QWR and L-QWR structures are at least ten times larger than the thickness of the SiGe QWL layers.

Fig. 2 shows plan-view TEM images of a 3x3µm^2 (Fig.2a) and a 10x10µm^2 (Fig.2b) structure. The deposition of the SiGe layer was performed for a nominal Ge content x = 0.3 and a nominal thickness of 4.3 nm which is much higher than the critical limit for Stranski-Krastanov growth of 2 nm (Vescan et al. 1992). The centre part of each structure represents the (001) QWL layer and is bordered by {111} facets with a projected width of 500 nm. A dense island population is observed in the centre of the $10 \times 10 \mu m^2$ structure with an inter-island spacing of approximately 200 nm. A depletion zone free of islands of a width of 500 nm separates this region from the wire grown in the L-corners. Frequently contrast fluctuations are visible along the wire indicating inhomogeneities of the wire thickness (Fig 2b). In the $3 \times 3 \mu m^2$ structure the (001) QWL layer grows uniformly without islands. No thickness inhomogeneities are observed in the wire. Dislocations originating at the Si buffer-substrate interface are present in both structures.

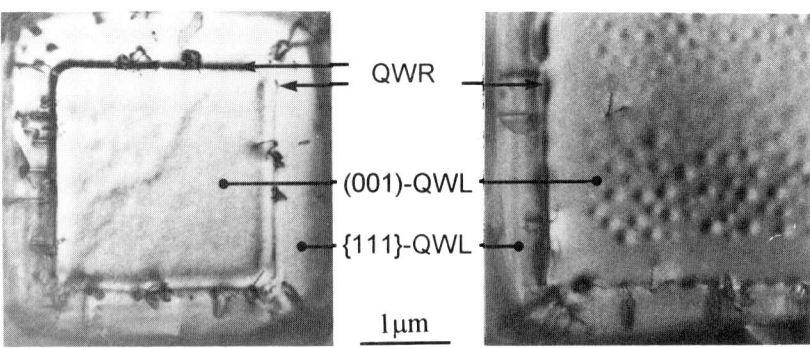

Fig.2 Plan-view bright-field TEM images of (a) $3 \times 3 \mu m^2$ and (b) $10 \times 10 \ \mu m^2$ L-QWR structures.

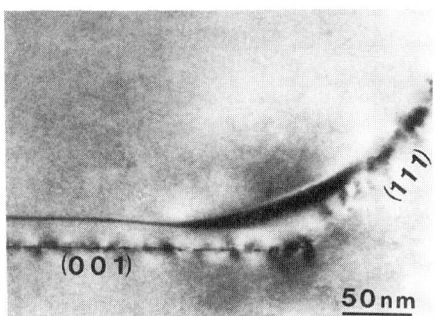
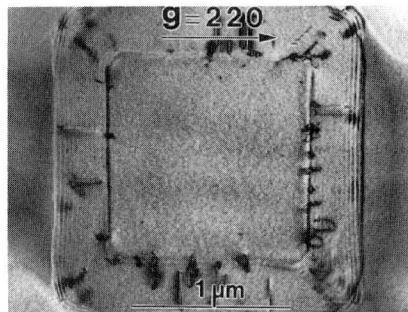

Fig.3 TEM bright-field image of a L-QWR structure. Dynamical imaging conditions **g**=(400).

Fig.4 Plan-view TEM bright-field image of a $3 \times 3 \ \mu m^2$ structure. Dynamical imaging conditions **g**=(220).

In order to estimate the extension of the strain field around L-QWR structures TEM investigations were performed for different dynamical imaging conditions (**g** = {220}, **g** = {004}). Fig.3 shows as an example the diffraction contrast of such a structure using **g** = (004). Our contrast experiments indicate that the dimensions of the strain field around the wires is of the order of 200 nm. Fig.1d shows the same wire under kinematic conditions (**g** = (004)) where the strain contribution to the contrast of the wire is strongly suppressed. The bright-field contrast of the wire appears bright in the corner and dark near the upper interface. This asymmetry indicates that there is a gradient of the Ge-content across the wire. A similar contrast behaviour is observed also for the V-QWR structures (Fig.1b).

Fig.4 shows a plan-view image of QWR structures in a $3 \times 3 \ \mu m^2$ structure sample under dynamical imaging conditions (**g**=220). The QWR structures show contrast behaviour similar to that of dislocations. For the imaging conditions applied almost complete contrast extinction occurs for the wires parallel to **g** whereas the wires perpendicular to **g** exhibit strong contrast. This indicates strong perturbation of the {220} planes parallel to the wire while the {$\overline{2}20$} lattice planes perpendicular to the wire are barely disturbed.

3.2 EDX measurements

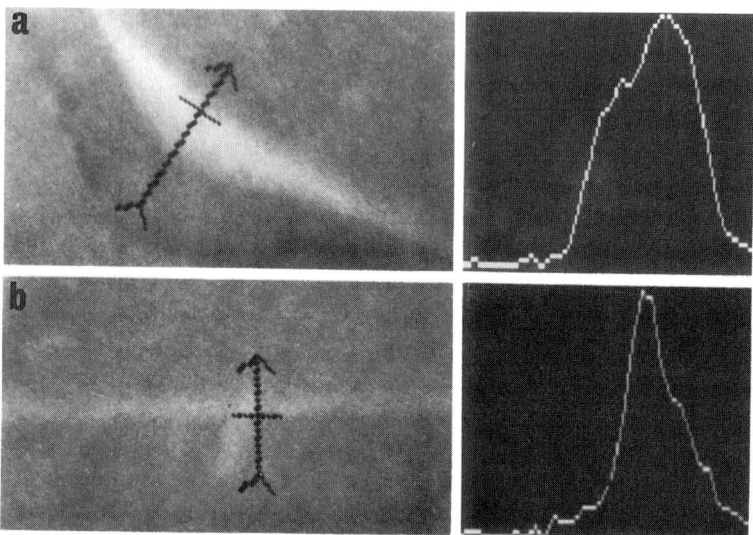

Fig.5 Large angle dark-field STEM images in [110] projection of (a) a L-QWR structure and (b) a (001) QWL layer. The corresponding composition profiles obtained from the analysis of the GeK$_\alpha$ signal are also shown (intensities not to scale, arbitrary units). Length and direction of the scans are indicated by the arrows: l = 32 nm (a), l= 16 nm (b).

X-ray point analyses and X-ray line scans have been performed for L-QWR structures grown in a 10 μm and a 3μm structure as result of a deposition of a nominally 1 nm thick Si$_{0.6}$Ge$_{0.4}$ layer (sample 863). For the quantitative determination of the Ge concentration profiles a deconvolution procedure, described by Bangert et al. (1995), has been applied in order to correct for electron beam broadening effects. Fig.5 shows large angle dark-field STEM images of a L-QWR structure (Fig 5a) and a (001) QWL layer (Fig 5b). Arrows indicate the length and direction of the EDX-linescans using the GeK$_\alpha$ signal. The profile of the L-QWR structure is clearly asymmetric, with highest Ge concentration being close to the upper interface of the wire. Results for the L-QWR structures and the QWL layers are summarized in Fig. 7, where the positions of the line scans are also sketched. The values of Ge compositions are mean values averaged from 3 to 5 scans, with an estimated error of ±10%. EDX point analysis using a Philips CM20 FEG were carried out on the same QWR samples yielding values for maximum Ge concentrations in good agreement with the STEM point analyses.

3.3 PL measurements

Fig. 6 shows the PL spectra of 3 μm and 10 μm line structures (sample 863) taken at 4 K and 40 K. These are the same structures which were investigated by EDX measurements (§3.2). The spectra taken at 4 K show at the higher energy side excitonic recombination in the Si substrate (Si). These are followed to lower energy by the no phonon (NP), transverse acoustical (TA) and transverse optical (TO) transitions of free excitons in the $Si_{0.6}Ge_{0.4}$ - (001) QWL. The blueshift of the QWL transitions observed as the width of the line structures is reduced from 10 μm to 3 μm was attributed to the outdiffusion of Ge from the QWL into the QWR during growth (Hartmann et al. 1995a,b). Assuming a constant QWL thickness of 0.9 nm the observed peak positions can be used to calculate the Ge concentrations of the QWL layer. The computed values are given in the table of Fig. 7.

As the measurement temperature is increased from 4 K to 40 K the PL spectra change considerably, QWR-related NP- and TO- PL peaks are obseved. The spectrum of the 3 μm structure is dominated by the TO- and NP- peaks of the QWR. The QWR-related peaks are also found at somewhat lower energy for the 10 μm structure, in addition to the QWL peaks, which are still visible. In this case the QWR NP peak coincides with the QWL TO peak, leaving only three distinguishable peaks in the spectrum. Taking a thickness of 10 nm for the QWR structures, as observed by TEM, and assuming a homogeneous Ge profile, the energetic positions of the PL peaks are used to calculate the Ge concentration of the QWRs resulting in the values given in Fig. 7. A more detailed PL investigation of the QWR structures will be the subject of a different publication.

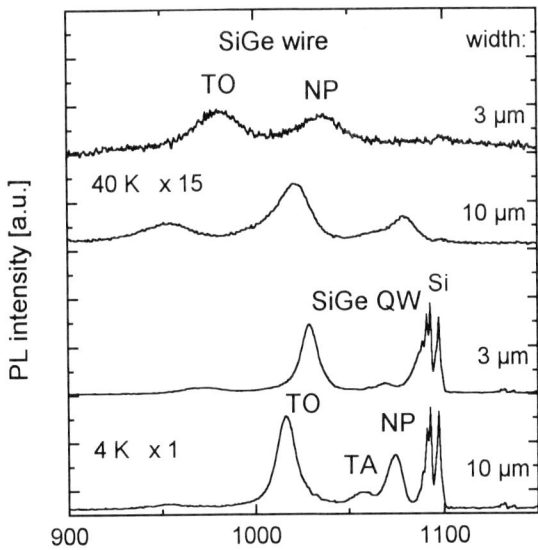

Fig 6: Photoluminescence spectra of 3μm and 10μm line structures of sample 863 taken at two different temperatures. The 40K spectra are magnified with respect to the 4K spectra by a factor of 15.

4. DISCUSSION

The presence of sharp convex corners between crystal planes enhances the local growth rate of SiGe by a factor of at least ten for all structures investigated, as compared to that observed for adjacent planes. Enhanced Ge incorporation at such corners is supplied by surface diffusion of adatoms during growth. This surface diffusion process leads to a Ge depleted zone next to the QWR which extends into the QWL layers over a distance comparable to the surface diffusion length. Direct observation of this zone is possible for depositions above a critical layer thickness for which islands growth occurs (Fig.2b). Island growth is suppressed for the smaller structure (Fig.2a). A method to determine the extent of this Ge depleted zone by a combination of PL and a surface diffusion model was introduced by Hartmann et al. (1995a,b). There as the distance between the QWR structures was decreased, the observed blueshift of the PL peaks was related to

384

the degree of overlap of the Ge depletion zones. From this it was possible to deduce a Ge surface diffusion length of 2.5±0.6 μm at 700°C.

The (001) QWL layer Ge composition profile resulting from the diffusion model is schematically shown in Fig. 7. In addition, the resulting potential wells for the holes in the valence band are sketched together with the position where quantitative STEM/EDX measurements were taken.

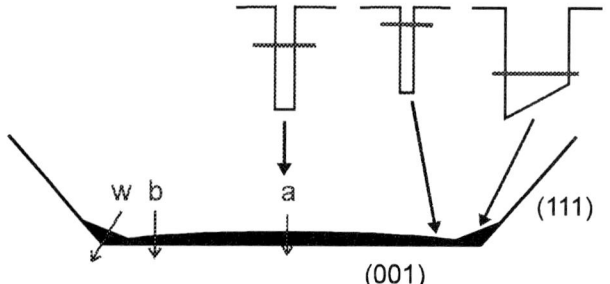

width [μm]	10	3
PL "w"	20%	16%
PL "a"	39%	35%
EDX "w"	20%	16%
EDX "a"	39%	31%
EDX "b"	35%	31%

Fig.7: Schematic illustration of a line structure and summary of measured Ge concentrations.

The temperature dependence of the PL spectra (Fig.6) can be explained as follows. At temperatures of 4K and 40K most holes are captured by the QWL layer because of its large area. At low temperatures the holes will accumulate right in the middle of the QWL ("a") because only very few holes have enough energy to overcome the potential barrier ("b") that separates the thick middle part of the QWL from the much thicker QWR. Mostly QWL-related PL signals are visible. In contrast, at higher temperatures, the holes can overcome the potential barrier ("b") and then accumulate in the deeper QWR potential ("w") resulting in strong QWR-related PL. PL values of the Ge concentration at points "a" and "w" are in good agreement with the quantitative STEM/EDX analysis (shown in table of Fig.7). Also the existence of the potential barrier at point "b" is confirmed for the 10 μm structure by a reduced Ge concentration as measured by EDX analysis. For the 3 μm structure we expect a smaller potential barrier from the diffusion model because of the strongly overlapping Ge depletion zones. This trend is also found in the EDX analysis.

The authors gratefully acknowledge experimental support by R J Keyse.

REFERENCES

Bangert U, Harvey A J, Keyse R J, Dieker C and Hartmann A this proceedings
Hartmann A, Vescan L, Dieker C and Lüth H 1995a J. Appl. Phys. 77 (5) 1959
Hartmann A, Vescan L, Dieker C and Lüth H 1995b Mat. Sci. Techn. to be publ.
Kapon E,. Hwang D M and Bhat R 1989 Phys. Rev. Lett. 63 430
Usami N, Mine T, Fukatsu S and Y.Shiraki 1993 Appl. Phys. Lett. 63 2789
Vescan L, Jäger W, Dieker C, Schmidt K, Hartmann A and Lüth H 1992 Mat. Res. Soc. Symp. Proc. Vol. 263 23
Walter M, Kapon E, Caneau C, Hwang D M and M.Schiavone 1993 Appl. Phys. Lett. 62 2170

Inst. Phys. Conf. Ser. No 146
Paper presented at Microsc. Semicond. Mater. Conf., Oxford, 20–23 March 1995
© *1995 IOP Publishing Ltd*

Quantum wire-like morphology in $In_xGa_{1-x}As$ single quantum wells: dependence on substrate misorientation and intrinsic strain

F Peiró[1,2], A Cornet[1] and J R Morante[1]

[1]Laboratori de Caracterització de Materials per a la Microelectrònica (LCMM). Dept. Física Aplicada i Electrònica, Universitat de Barcelona. Avd. Diagonal 647. E-08028 Barcelona, Spain.
[2]Serveis Científico-Tècnics, Univ. Barcelona. Lluís Solé i Sabarís, 1-3. E-08028 Barcelona, Spain.

ABSTRACT: The growth mode and strain relaxation of $In_{0.53}Ga_{0.47}As/In_yAl_{1-y}As$ heterostructures grown on exact and vicinal (100)InP surfaces have been analyzed. TEM examinations have revealed the appearance of In-rich and Al-rich regions oriented close to the {133} and {122} planes in the InAlAs buffer layer of misoriented samples. Such regions induce the formation of Ga-rich/In-rich regions inside the InGaAs single well, giving rise to quantum-wire like domains. The formation of such a lateral composition modulation is discussed on the basis of the misorientation surface steps and the mechanism of stress relaxation.

1. INTRODUCTION

In the recent years there has been much interest in devices based on carrier confinement in quantum wires and dots driven by their promising electronic and optical properties. Indeed, high carrier mobilities, reduced electron scattering and high luminescence efficiency can be achieved (Weisbuch 1993). In particular the reduction of the dimensions of the active channels in laser devices leads to a significant reduction of the threshold current for laser emission (Vurgaftman 1994). Nevertheless the restrictions imposed by the fabrication techniques, such as lithography resolution, still limit the design of the structures. Therefore, following the reports about the appearance of lateral contrast modulations induced by phase separation or lateral ordering within the as-designed-grown superlattices depending on the growth temperature T_g, interest has moved into the assessment of the formation mechanism of such spontaneous configurations, which will significantly reduce the costs of optoelectronic devices based on low dimensional structures.

2. EXPERIMENTAL DETAILS

In this work we have analyzed the quantum-wire like morphology of $In_{0.53}Ga_{0.47}As$ single quantum wells grown by MBE over $In_yAl_{1-y}As$ buffer layers on exact and vicinal (100)InP surfaces. The growth of both oriented (**A**) and misoriented samples [(**B**), (100) 4° off towards (111)] was carried out in the same run, in order to avoid additional possible contributions to their final morphology other than the misorientation itself. The general structure of the samples studied is 5 nm GaAs/50 nm $In_{0.52}Al_{0.48}As$/18 nm $In_{0.53}Ga_{0.47}As$ (SQW)/ 2μm $In_yAl_{1-y}As$ buffer layer/InP. Two sets of samples with tensile stress in the $In_yAl_{1-y}As$ buffer layer were examined, the first one (samples **1**) with $y_{In}=48\%$ and the second (samples **2**) with $y_{In}=50\%$, the latter being less mismatched with respect to the InP substrate. The growth rates were about 1μm/h for all the layers and the As_2 flux was fixed at 2.2×10^{-5} Torr. The first ≃200 nm of the buffer layer were grown at $T_g=530°$ in order to avoid As-P exchange at the interface. However, for the rest of the growth, T_g was increased to 580°C so as to favour the appearance of composition variations in InAlAs since this higher growth temperature has been found to enhance composition inhomogeneities in this material (Peiró 1993).

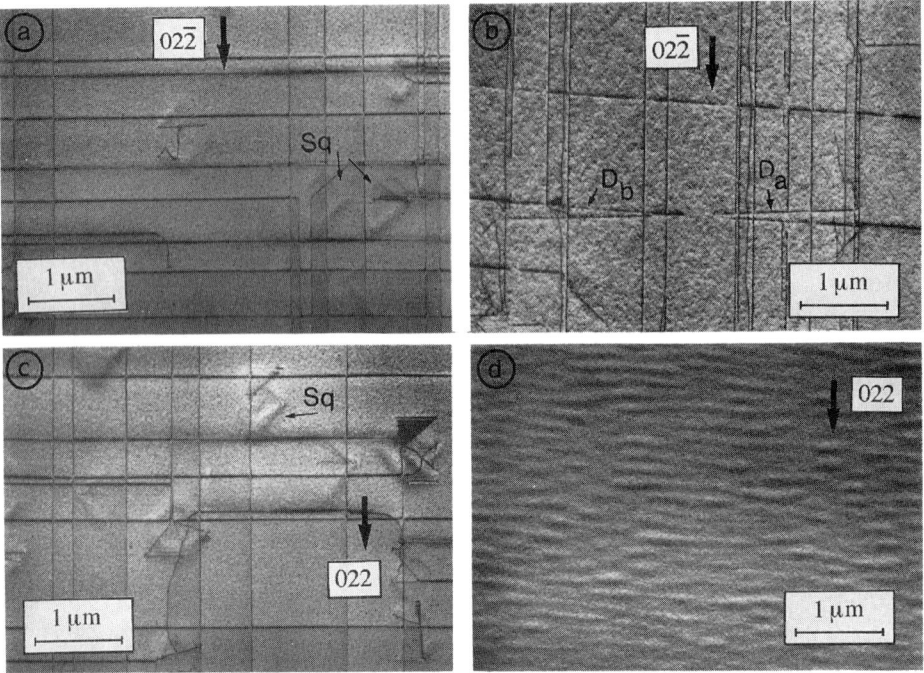

FIG. 1. Plan view images of the $In_y Al_{1-y} As/InP$ interface along the (100) zone axis: sample 1A **(a)**, 1B **(b)**, 2A **(c)**, and 2B **(d)**. The dislocation densities are lower in the misoriented samples (B). Besides, the squared shaped contrast in samples A (Sq), driven by the growth of InAlAs at high T_g are not observed in samples B.

3. RESULTS AND DISCUSSION

The state of the layer surface was first examined by Optical Normarsky interference (NI). Samples A showed some surface irregularities which were related to square shaped contrast inhomogeneities in the epilayer induced by the growth of the InAlAs buffer at high temperature (Peiró 1993). These surface inhomogeneities did not appear in the misoriented samples B. In addition, in all the cases we also observed the presence of a cross-hatched pattern due to the existence of a network of misfit dislocations at the interface, with the exception of sample 2B. These preliminary observations have been confirmed by the results of plan view TEM (PV) examination. The images 1a and 1b correspond to the buffer-substrate interface of the more mismatched samples (intrinsic mismatch $f = 2.6 \times 10^{-2}\%$), and we noticed a slight reduction on the dislocation density for the misoriented sample 1B, besides the extinction of the square shaped contrast (Sq). These dislocations are mainly 60° misfit with lines along $\langle 011 \rangle$. Moreover, in the misoriented sample 1B, we distinguish to sets of dislocations lines (D_a and D_b) at an angle of $\simeq 5°$ with the [011] direction, value which is close to the angle $\beta = 4.9°$ expected for a misorientation of 4°off towards (111), having then a good experimental check of the specified substrate tilt. In the case of the less mismatched layers ($f = 1.3 \times 10^{-2}\%$), the misfit dislocation network is present at the interface of the sample 2A grown on exact [100] (Fig. 1c) but not in the misoriented sample 2B (Fig. 1d).

Despite this reduction on the dislocation density at the InAlAs/InP interface, the main difference between the oriented and misoriented samples is evident when imaging the InGaAs well. At this position the PV images in strong two beam (g = 022) conditions show a strong anisotropic and quasiperiodic contrast modulation along the [011] direction with parallel dark bands oriented on [01$\bar{1}$] separated by a distance of $\simeq 125$-135 nm (Fig. 2a). The contrast pattern disappears when the foil is imaged with the reflection g = 02$\bar{2}$ (Fig. 2b). This anisotropic behaviour has been corroborated by cross-section (XTEM) observations, as shown in figure 3. Indeed, the contrast pattern is related to an asymmetrical rippling of the well, which presents surface undulations only when observed with g = 200 along the [01$\overline{1}$] direction (Fig. 3a), whereas the well remain flat when viewed along the [011]. Whether the sample is imaged with the reflection g = 022, a contrast modulation appears crossing the whole epilayer from the InAlAs buffer, with dark-bright bands at an angle of $\simeq$ 13-15° with respect to the growth direction (Fig. 3b). We have seen the strongest black-white contrast transition for g = 022 dark field images, for which the bands appear well correlated with the valleys of the surface rippling. For the reflection g = 111 there appear some domains whose trace analysis on stereographic projection reveal them to lie on the planes {133} and {122}. The valleys and hillocks of the well undulation correlate very well with such contrast domains (arrows in Fig. 3b). XTEM has also shown that these domains are originated at $\simeq$ 200 nm above the InP, corresponding to the change on T_g from 530°C to 580°C. As an example, Fig. 3c shows this region in the sample 1B. There we can observe the distribution of misfit dislocations at both the InAlAs/InP interface (I) and on the plane corresponding to the change in T_g (II), with dislocations loops (DL) on

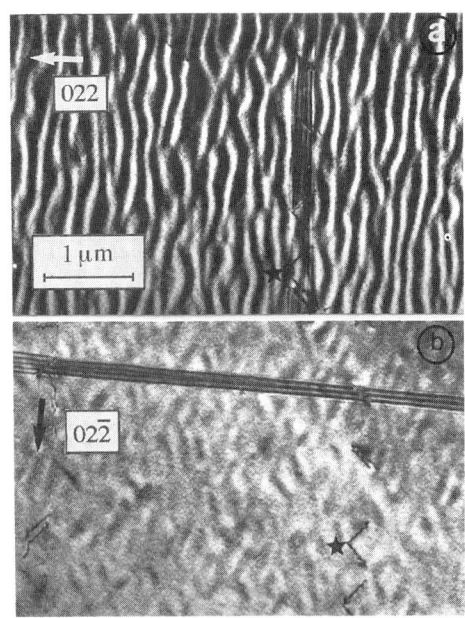

Fig. 2. (100) images showing exactly the same region of the InGaAs well, with different operating reflections: g = 022 (a) and g = 02$\overline{2}$ (b). The asteriscs help to superpose the images.

{111} linking both sets of misfit networks. These loops were also observed in PV micrographs of the oriented samples (Fig. 1). Conversely, the XTEM image of sample 2B do not exhibit dislocations at this point, although the contrast domains appear also at the "second" interface.

In view of all these results, we would like to make the following points: i) the lateral contrast modulation has been observed only in misoriented samples, hence the misorientation induced steps at the interface play an essential role on the development of such modulations; ii) the contrast domains appear just at the moment when the growth temperature has been increased to 580°C, thus, the modulation is a thermally activated process; iii) the appearance of the lateral modulation leads to a reduction (or even the elimination in the less mismatched samples) of the misfit dislocation densities at the interface, therefore the lateral modulation acts as a mechanism for strain accommodation at low mismatch. According to the model of Hasegawa et al. (1992), the single atomic height steps tend to group together developing a giant step with a facet along {133} (in our case with an uncertainty between {133} and {122} due to the error in the angle measurements). At a growth temperature of 530°C, the limited diffusion of ad-atoms leads to homogenous growth of the $In_yAl_{1-y}As$ strained layer. However, when the growth temperature is increased to 580°C, the nucleation of dislocations to relax the strain is thermally activated and the diffusion rate of ad-atoms to preferential sites is also increased. Since the sticking coefficients at the step surfaces are different for Al and In (Tsuchiya et al. (1989), this growth at high T_g may enhance the diffusion of the species leading to domains of different compositions around these giant misorientation steps. Assuming that the growth rate is higher in less strained regions, a rough growth front would be formed inducing also a wavy growth for the layers grown over the buffer layer. In our case, the In-rich regions should correspond to the hillocks of the rippling, since for $In_yAl_{1-y}As$ with tensile stress a higher molar fraction of In falls near to $y_{In} = 52\%$, which is the value for lattice match to InP. As growth proceeds, the inhomogeneous strain would strongly influence the growth of the InGaAs, giving rise to In-rich/Ga-rich domains in the InGaAs well.

388

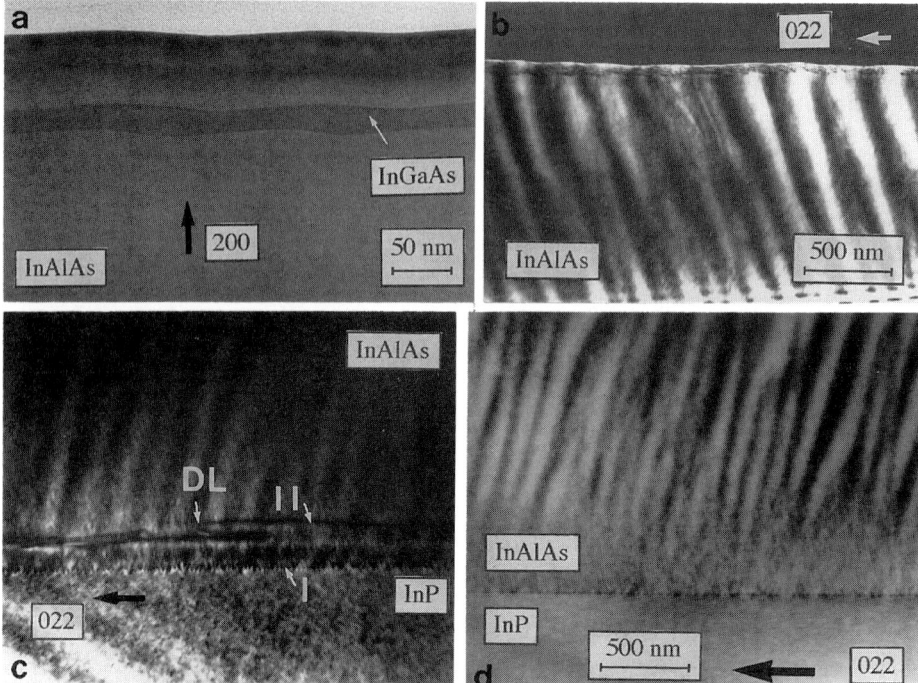

FIG. 3. (a) XTEM [01$\bar{1}$] of the sample 1B revealing the undulation of the InGaAs well. (b) With the reflection g=022 there appear some white-dark domains across the heterostructure at $\simeq 15°$ with respect to the growth direction. Detail of the location of the origin of such domains near the substrate in samples 1B (c) and 2B (d).

4. CONCLUSIONS

We have analyzed the growth mode of $In_{0.53}Ga_{0.47}As/In_yAl_{1-y}As$ heterostructures on exact (100) and misoriented 4°off towards (111) InP substrates. The misorientation steps and the growth at high T_g (580°C) induce the formation of lateral composition variations leading to In-rich and Al-rich regions oriented close to the {122} or {133} planes. This lateral modulation contributes to the accommodation of the tensile strain of the $In_yAl_{1-y}As$ buffers, as shown by the evident reduction of dislocation densities on misoriented samples with respect to the oriented ones. The propagation of such domains to the top of the structure leads to In-rich and Ga-rich regions inside the InGaAs well, developing a spontaneous quantum-wire like morphology. Accurate control of the dimensions of such spontaneous domains would be very valuable for optoelectronic applications.

ACKNOWLEDGEMENTS

This work was funded by the Spanish CICYT program MAT93-0564 and the Greek-Spanish bilateral cooperation program "Clustering and Ordering Phenomena in MBE Semiconductor Alloys".

REFERENCES

Hasegawa S, Kimura K Sato M et al. 1992 Surf. Sci. 267 5
Peiró F, Cornet A, Morante J R et al. 1993 Appl. Phys. Lett. 62, 2265
Tsuchiya M, Pettrof P M and Coldren L A 1989 Appl. Phys. Lett. 54 1690
Vurgaftman I, Singh J. 1994 Sol. Stat. Electr. 37 1263
Weisbuch C 1993 J. Cryst. Growth 127 742

Inst. Phys. Conf. Ser. No 146
Paper presented at Microsc. Semicond. Mater. Conf., Oxford, 20–23 March 1995

Convergent beam electron diffraction study of CdZnSe/ZnSe strained-layer superlattices

D Manno+, R Cingolani+, L Sorba*, L Vanzetti*, A Franciosi*

+ Dipartimento di Scienza dei Materiali - Università di Lecce (Italy)
* Laboratorio TASC - INFM - Area della Ricerca, Padriliano 99, Trieste (Italy)

ABSTRACT: Convergent beam electron diffraction (CBED) performed on plan-view samples has been considered in order to analyse $Cd_xZn_{1-x}Se/ZnSe$ strained superlattices. The composition modulation along the growth axis of multilayer materials and the strain modulation at the interfaces give rise to special features in the higher order Laue zone (HOLZ) reflections. A simple model based on the expansion-contraction of the lattice spacing in each layer along the growth direction is proposed to simulate rocking curves in the kinematical approximation.

1. INTRODUCTION

II-VI semiconductor strained layers are of great interest for the development of new electronic devices. It is well known that strain changes the band gap and the band offset, as described by Cingolani et al (1995), so it is very important to get reliable methods for the quantitative evaluation of the lattice mismatch in II-VI heterostructures.

So far much information on heterostructures has been obtained by high resolution electron microscopy or diffraction contrast techniques applied to cross-sectional specimens. Despite the near atomic resolution of these methods, the characterisation of strain is difficult due to surface relaxation effects which cause bending of the specimen surfaces, as shown by Gibson and Treacy (1984).

On the contrary, layer strain remains essentially unrelaxed in the plan-view geometry. In addition, electron rocking curves observed in convergent beam electron diffraction (CBED) patterns can give information on layer thicknesses and strain profiles of multilayer samples [for a review: Cherns et al. (1991), Gong and Schapink (1991), Xie et al. (1990)].

In this study, CBED has been used to generate rocking curves from plan-view ZnCdSe/ZnSe superlattices. Rocking curves obtained from reflections inclined with respect to the [001] growth direction, show distinct superlattice sideband peaks which are strongly sensitive to the actual layer strain.

2. EXPERIMENTAL

$Cd_xZn_{1-x}Se/ZnSe$ superlattices were grown by molecular beam epitaxy (MBE) as described previously by Sorba et al (1992). A nominally 10 periods $Cd_{0.1}Zn_{0.9}Se(3nm)/ZnSe(20nm)$ superlattice was analysed. Plan-view specimens for TEM were prepared in the following way: samples were first mechanically polished to about 40 μm and then chemically thinned. We found out that the chemical thinning provides suitable areas for TEM observations which are slightly misoriented with respect to the [001] growth direction. CBED patterns were normally obtained from dislocation free areas by a JEM 2010 transmission electron microscope at acceleration voltages of about 200 kV. Fig.1 shows the [611] zone-axis pattern. We note that the m symmetry of this zone-axis is destroyed by the presence of the superlattice which results in asymmetric sidebands in the CBED patterns (see, for example, HOLZ reflections $(2\overline{1}2 2)$ and $(\overline{2}212)$ in Fig.1),

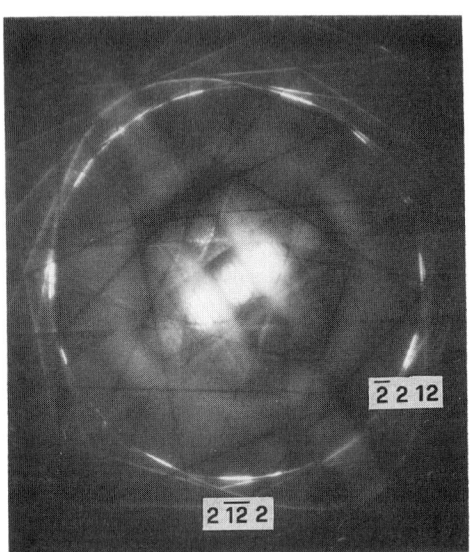

Figure 1 (a) [611] CBED zone-axis pattern

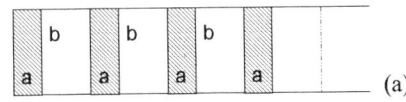

(a)

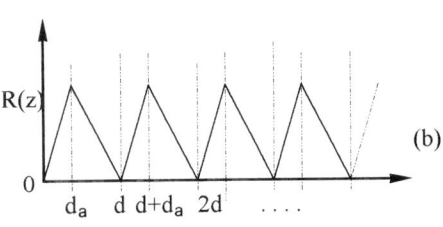

(b)

Figure 2 (a) scheme of a multilayer material, and (b) displacement vector R(z) of unit cell in superlattices.

In what follows we present a theoretical analysis of the intensity profiles of these CBED reflections, which provide quantitative information on the actual strain in the quantum wells.

3. COMPUTING DETAILS

We refer to a generic specimen $A_xB_{1-x}C/BC$ multilayer as in Fig. 2a. Let N be the number of alternating double layers ("a": layer $A_xB_{1-x}C$ + "b" layer BC), d_a and d_b the respective layer thicknesses, a_a and a_b the lattice spacings. Assuming $a_a > a_b$, layer a is compressed and layer b is expanded in the interface plane (x-y). Along the growth direction z (normal to the interface) layer a is expanded and layer b is compressed by a factor $\phi = (1 + \nu)/(1 - \nu)$, where ν is the Poisson ratio.

The layers match at the interface, so no relative displacement exists in this plane. On the contrary, in the z-direction contraction and expansion displacements occur. If we assume a linear displacement R(z) for the unit cell along the growth direction, as shown in Fig.2b, we obtain the following analytical expression for R(z):

(1)
$$\begin{cases} R_a(z) = e_a^{\perp}(z - nd) & nd < z < nd + d_a \\ R_b(z) = e_b^{\perp}(z - (n+1)d) & nd + d_a < z < (n+1)d \end{cases}$$

where $e_a^\perp$ and $e_b^\perp$ are the strains of the two layers along the z-direction. They are related to the in-plane strains $e_a^\parallel$ and $e_b^\parallel$ by:

$$(2) \qquad e_a^\perp = -\phi e_a^\parallel \;\; ; \;\; e_b^\perp = -\phi e_b^\parallel ,$$

Assuming relatively low Cd content ($0<x<0.3$) and taking into account that no relative displacement exists in the interface plane, we have: $e_a^\perp = \phi d_b \delta / d$; $e_b^\perp = \phi d_a \delta / d$ (we recall, also, that $e_a^\parallel - e_b^\parallel = \delta$ where δ is the lattice mismatch of the layers).

In the kinematical theory of electron diffraction, the intensity of a reflection **g** is given by $|\phi_g|^2$, where, in the general case proposed by Hirsch et al. (1977) $\phi_g = \int F_g \exp[-2\pi i (s\tau + \mathbf{g} \cdot \mathbf{R})] d\tau$. In our case:

$$(4) \qquad \phi_g = \sum_{n=0}^{N-1} F_a \int_{nd}^{nd'+d'_a} \exp\left\{-2\pi i \left[s\tau + g_z e_a^\perp (\tau \cos\vartheta - nd)\right]\right\} d\tau +$$

$$\sum_{n=0}^{N-1} F_b \int_{nd'+d'_a}^{(n+1)d'} \exp\left\{-2\pi i \left[s\tau + g_z e_b^\perp (\tau \cos\vartheta - (n+1)d)\right]\right\} d\tau$$

$d' = d / \cos\vartheta$ and $d'_a = d'_a / \cos\vartheta$; ϑ is the angle between [001] and the chosen zone axis.

From eq.(4) we can calculate $|\phi_g|^2$, by taking into account the continuity conditions at interfaces and neglecting the phase factor, eq. (4) become:

$$(5) \qquad |\phi_g|^2 = \frac{\sin^2 \pi s N d}{\sin^2 \pi s d} \left[F_a^2 \frac{\sin^2 \alpha d'_a}{\alpha^2} + F_b^2 \frac{\sin^2 \beta d'_b}{\beta^2} + 2 F_a F_b \frac{\sin \alpha d'_a \sin \beta d'_b}{\alpha \beta} \cdot \cos \pi s d' \right]$$

were N: number of periods, $\alpha = \pi (s + g_z e_a^\perp \cos\vartheta)$ and $\beta = \pi (s + g_z e_b^\perp \cos\vartheta)$

The scattering factor F_a and F_b were determined by a linear interpolation between the scattering factors of materials BC and AC calculated by the EMS program, worked out by Stadelmann (1987), as a function of x. Linear interpolations have also been used in order to determine the elastic constants and strains.

4. RESULTS AND DISCUSSION

The calculated intensity profiles of the HOLZ reflections $(2\,\overline{1}2\,2)$ and $(\overline{2}2\,12)$ of Fig.1 are shown in Fig.3. As we have already noted, the reflections are not related by mirror symmetry, the asymmetry being reproduced by the simulated profiles (compare Fig.3a and 3b). In addition, the intensity profile of the $(\overline{2}2\,12)$ reflection is shifted toward the positive s values (Fig.3b) with respect to the $(2\,\overline{1}2\,2)$ one. This is consistent with the experimental findings: the sidebands in the $(\overline{2}2\,12)$ reflection are inside the HOLZ ring, whereas, the $(2\,\overline{1}2\,2)$ ones are near by the HOLZ ring.

We stress that the shift of the sidebands and the intensity profile are a function of x and, due to the scalar product $\mathbf{g} \cdot \mathbf{R}$, these effects are more evident in reflections with high ℓ values. This behaviour has been also evidenced in Fig. 3 where the intensity profiles have been determined for different values of x. The strong sensitivity of the rocking curve to the actual layer strain is quite evident for the $(\overline{2}\,2\,12)$ HOLZ reflection. By comparing the intensity profile and the actual sidebands in $(\overline{2}\,2\,12)$ reflection we can conclude that the x

392

content ranges between 0.10 and 0.12, and the strains of the well $Cd_xZn_{1-x}Se$ amount to $e_a^{\parallel} = (-7.0 \pm 0.7)10^{-2}$; $e_a^{\perp} = (1.5 \pm 0.2)10^{-2}$, this latter value is only a factor two larger than the one obtained from the analysis of the optical spectroscopy data reported by Cingolani et al. (1995).

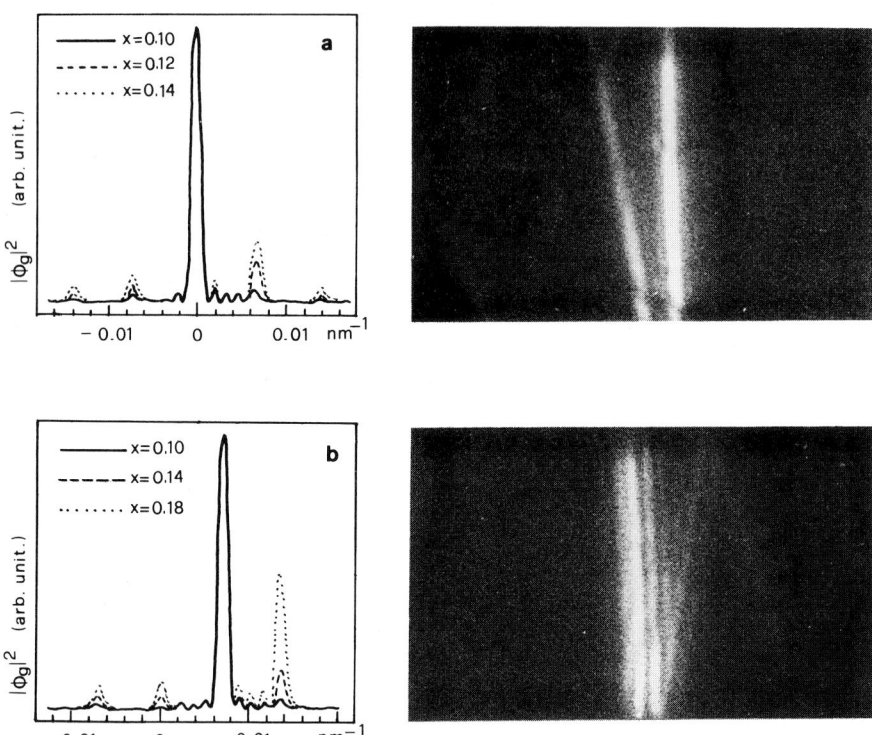

Figure 3 - Calculated intensity profiles of $(2\,\overline{12}\,2)$ (a) and $(\overline{2}\,2\,12)$ (b) HOLZ reflections together with the corresponding enlarged reflection.

REFERENCES

Cherns J A, Touaita R, Preston A R, Rossouw C J and Houghton D C 1991 Phil. Mag. A 64 397

Cingolani R, Prete P, Greco D, Giugno P V, Lomascolo M, Rinaldi R, Calcagnile L, Vanzetti L, Sorba L and Franciosi A 1995 Phys. Rev. B to be published

Gibson J M and Treacy M M J 1984 Ultramicroscopy 14 345

Gong H, and Schapink FW 1991 Ultramicroscopy 35 171

Hirsch P, Howie A, Nicholson RB, Pasheley DW, Whelan MJ 1977 Electron Microscopy of Thin Crystals (Robert E Krieger Publishing Company, Malabar, Florida,)

Sorba L, Bratina G, Antonini A, Franciosi A, Tapfer L, Migliori A, Merli P G 1992 Phys. Rev. B 46 6834

Stadelmann P 1987 Ultramicroscopy 21 131

Xie Q H, Fung K K, York P K, Fernandez G E, Eades J A and Coleman J J 1990 Appl. Phys. Lett. 57 1978

Inst. Phys. Conf. Ser. No 146
Paper presented at Microsc. Semicond. Mater. Conf., Oxford, 20–23 March 1995
© 1995 IOP Publishing Ltd

X-ray characterisation of quantum well device structures

L Hart[1], C Roberts[1], M Ghisoni[2†], J M Roberts[3], J J Harris[3], G Parry[2] and M Hopkinson[4]

[1]Interdisciplinary Research Centre for Semiconductor Materials, Imperial College, Prince Consort Road, London SW7 2BZ
[2]Interdisciplinary Research Centre for Semiconductor Materials, Department of Engineering Science, University of Oxford, Parks Road, Oxford OX1 3PJ
[3]Interdisciplinary Research Centre for Semiconductor Materials, Department of Electronic and Electrical Engineering, University College London, Torrington Place, London WC1E 7JE
[4]Department of Electronic and Electrical Engineering, University of Sheffield, Mappin Street, Sheffield S1 3JD

ABSTRACT: A number of InGaAs quantum well structures has been grown using molecular beam epitaxy on GaAs and InP in order to study their electrical and optical properties. This includes single quantum wells, stepped quantum wells and multiple quantum well structures, grown on both (001) and (111) substrates. High-resolution x-ray diffractometry can be used as a rapid, non-destructive technique to determine the composition, well width and crystal quality, and to identify any growth problems.

1. INTRODUCTION

InGaAs quantum well (QW) structures have a wide range of applications in electronic and optoelectronic devices such as power FETs (field effect transistors), optical modulators and lasers. Growth on (111) substrates can also be used to produce piezoelectric devices. Control of the molecular beam epitaxy (MBE) growth process and consequently device quality requires a rapid, non-destructive technique for the accurate determination of composition, layer thickness and crystal quality. High-resolution x-ray diffraction is such a technique, which can be used to characterise quantum well structures both before and after processing into devices (provided there is a large enough area of material, at least 1 mm^2). It is then possible to determine the indium concentration, well width and degree of strain relaxation, if any, to a high degree of accuracy. Growth problems such as surface segregation, which leads to graded interfaces and In incorporation in the barriers, can also be identified.

More detailed information on strain-depth profiles can be obtained by comparing the experimental x-ray data with simulations calculated using dynamical diffraction theory. Device structures often have a large number of layers, and their effects can often only be separated by simulation. This can also enable characterisation of quantum wells down to less than a monolayer thick (Fewster, 1993). With the use of a triple-axis diffractometer, reciprocal space mapping can also be performed to enable direct measurement of strain, relaxation and tilt in the layers.

---

† Present address: Department of Optoelectronics and Electrical Measurements, Chalmers University of Technology, S-41296 Göteborg, Sweden

2. X-RAY DIFFRACTION TECHNIQUES

X-ray diffraction measurements were carried out using a Philips high-resolution x-ray diffractometer with a 4-reflection Ge 220 monochromator and a 2-reflection Ge 220 analyser using Cu $K\alpha_1$ radiation.

Rocking curves were measured without using the analyser crystal, but with a slit in front of the detector, giving an angular acceptance of ~0.3°. The sample was scanned through the Bragg angle, with the detector rotated at twice the rate, in an ω-2ω scan (where ω and 2ω are the angles of the sample and detector relative to the incident beam). Rocking curves taken around the symmetric 004 or 333 Bragg reflections were used to determine the lattice parameters in the [001] or [111] growth directions.

Reciprocal space mapping also enabled direct measurement of residual strain and tilt in the layers. A reciprocal space map is a plot of the diffracted x-ray intensity in reciprocal space. This is obtained by using the analyser crystal to reduce the angular acceptance to 12 arc seconds, and measuring a series of ω-2ω scans, each with the sample angle, ω slightly offset. An ω-2ω scan measures a range of lattice spacings for a single lattice tilt angle, while an ω scan measures a range of lattice tilts. When a reciprocal space map has been plotted, direct determination of the lattice parameters perpendicular and parallel to the surface planes is possible. Layer tilting and mosaic spread can also be obtained directly. Each data point in a rocking curve, however, will have components of both strain and tilt, which are more difficult to separate (Fewster, 1993).

Modelling and simulation using dynamical diffraction theory was performed assuming Vegard's law to be valid, giving a linear variation in strain with composition. Simulations based on dynamical diffraction theory (Fewster and Curling, 1987) were used to determine the thicknesses and compositions of the quantum wells, and also to give some indication of the thickness uniformity and interfacial abruptness.

3. SINGLE QUANTUM WELLS

The single quantum well samples were grown by MBE on (001) GaAs substrates. Sample 1(a) is an InGaAs QW (nominally 15% In, 80Å wide) grown at 510°C with a 200Å GaAs capping layer. Experimental and simulated 004 rocking curves are shown in Fig.1(a). The sharp, intense substrate peak and the broad, weak InGaAs layer peak can be seen, as well as a series of interference fringes due to the GaAs capping layer. From the simulation, the In content and well

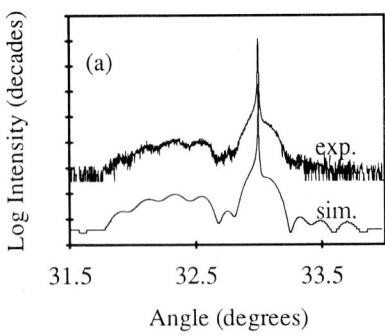
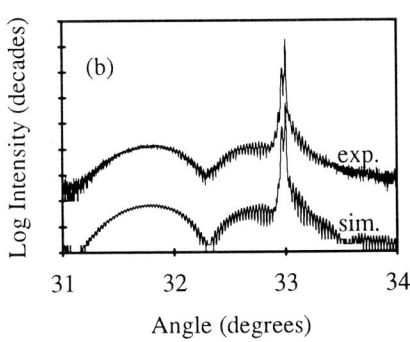

Fig. 1.   Experimental and simulated 004 rocking curves for (a) InGaAs single QW with thin GaAs cap and (b) InGaAs single QW with thick AlGaAs barriers.

width were found to be 16% and 90Å respectively. The fit between the experimental and simulated curves is good, although in the simulation, the first two fringes to the low angle side of the substrate peak are slightly low in intensity compared with the experiment, indicating possible grading of the interfaces. Some surface segregation was thought to have occurred.

Sample 1(b) is an InGaAs QW (nominally 25% In, 85Å wide) with thin layers of GaAs (one Si δ-doped) and thick barriers of AlGaAs (nominally 33% Al, 0.2μm) on either side, grown for transport measurements. The growth temperatures were 420°C for the InGaAs and 620°C for the GaAs. The experimental and simulated 004 rocking curves are shown in Fig.1(b). The long- and short-period fringes relate to the thin InGaAs and thick AlGaAs layers respectively. By simulation, it was possible to determine the InGaAs composition and well width (26% In, 90Å), and the AlGaAs composition and barrier width (30% Al, 0.17μm). These were in good agreement with the specified values.

4. STEPPED QUANTUM WELLS

The stepped quantum well samples were grown by MBE at 450°C on (001) InP substrates, in order to achieve higher In contents. Sample 2(a) contains a well with three stepped compositions of (nominally) 25%, 50% and 75% indium, each 33Å wide, with AlInAs barriers lattice-matched to the InP substrate. The stepped well was designed to be strain-balanced, having equal and opposite tensile and compressive strains in the 25% and 75% In layers, separated by an almost lattice-matched 50% In layer. The 25% and 75% In contents were produced by digital alloying (periodically interrupting the flux of In/Ga). The 25% In layer was Si δ-doped. A high-resolution 004 rocking curve is shown in Fig. 2(a). The sharp, intense peaks are from the InP substrate and, at slightly higher angle, the AlInAs barriers. The weak peaks on either side of the AlInAs peak are due to the strain-balanced 25% (at higher angle) and 75% (at lower angle) In layers. By performing a series of simulations, it was possible to determine the three different compositions (25%, 51% and 80% In), and the average step thickness was found to be 38Å. Satellites due to the digital alloying periodicity were also observed, Fig. 2(a).

In sample 2(b) the three InGaAs compositions were grown in three wells separated by 100Å AlInAs, with three repeats to produce greater intensity in the x-ray analysis. Using the digital alloying technique, 33Å of 25% In was produced by 11 repeats of GaAs/InGaAs/GaAs

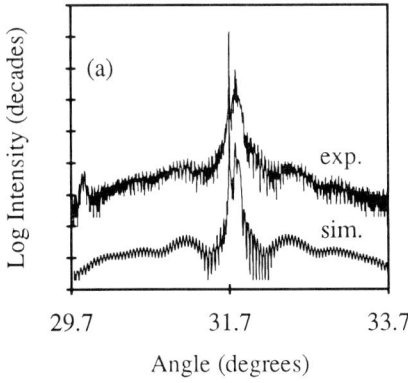
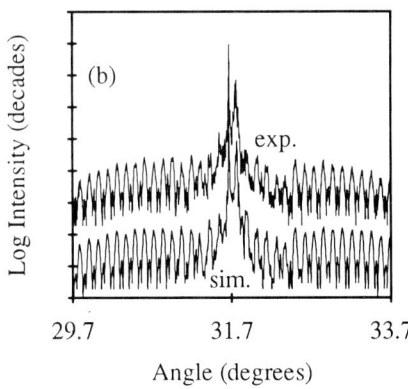

Fig. 2.   Experimental and simulated 004 rocking curves for (a) InGaAs stepped QW grown on InP and (b) three separate InGaAs QWs repeated three times.

(1Å each) and 33Å of 75% In was produced by 11 repeats of InAs/InGaAs/InAs (1Å each). The experimental 004 rocking curve in Fig.2(b) shows satellite peaks from a 440Å periodicity. Simulations showed that the wells were slightly thicker than specified (38Å). Because of the increased intensity in Fig.2(b) compared to Fig. 2(a), it was possible to achieve a higher degree of accuracy in the determination of the three different compositions (19%, 52% and 83% In).

5. MULTIPLE QUANTUM WELLS

A 10-period InGaAs (nominally 15% In) multiple quantum well (MQW) structure was grown by MBE at 540°C on a (111)B GaAs substrate misorientated 2° towards <112>. The InGaAs wells and GaAs barriers were 45Å and 60Å wide. An experimental 333 rocking curve, showing the average (zero-order) layer peak and 2 orders of satellites due to the MQW periodicity, is shown in Fig.3. The fringes between the satellites relate to the total MQW thickness. Two simulations are given, (a) with 11% In in the wells, and (b) with 9.8% In in the wells and an additional 1% In incorporation in the barriers, which gives a better fit to the experimental data. In both simulations, the wells and barriers are 49Å and 60Å wide. The effect of incorporating In in the barriers, and so decreasing the strain modulation, is to reduce the satellite intensities. Interface grading would also reduce the intensities of the higher satellite orders, but in order to affect the first-order satellite intensities as observed, a decrease in the strain modulation is required. Reciprocal space maps of the 333 reflection showed that this sample was not relaxed (see below). Surface segregation is thought to have occurred, resulting in graded interfaces and In in the barriers.

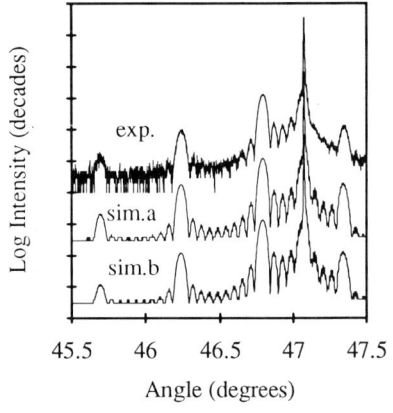

Fig. 3. Experimental 333 rocking curve for InGaAs MQW grown on (111)B GaAs and simulations, (a) 11% In wells, and (b) 9.8% In wells and 1% In barriers.

A series of 10-period InGaAs (5-20% In, 100Å wide) multiple quantum wells with AlAs barriers (100Å wide) to provide increased confinement in the wells, was grown by MBE on (001) GaAs substrates. Layers of AlGaAs (30% Al) and GaAs on either side of the MQW formed a p-i-n structure (Ghisoni et al 1994). The InGaAs, GaAs and AlGaAs were grown at 520°C, 580°C and 630°C respectively. The samples were characterised by high-resolution x-ray diffractometry both before and after processing into 400 × 400 μm² mesas using standard photolithographic techniques. The x-ray beam size was 0.5 × 3 mm², and in the processed samples, a row of three mesas was measured simultaneously.

Figs. 4(a) and (b) show 004 reciprocal space maps of samples with 9 and 12% indium (Hart et al 1995). The substrate, AlGaAs layer and 0th, 1st and 2nd order satellites from the MQW are shown. In Fig. 4(a), the satellite peaks due to the MQW periodicity are narrow and sharp, and interference fringes relating to the total MQW thickness can be seen along the surface normal (the diagonal streak is due to the analyser crystal). In Fig.4(b), the satellite peaks have become much broader and the fringes have disappeared. This indicates disruption of the interfaces by misfit dislocations (Kidd et al 1993). The dislocations introduce microscopic tilting or mosaic spread, which broadens the peaks (Hart and Fewster, 1993). It was presumed that the misfit dislocations were introduced immediately below the MQW region, so that the InGaAs wells and AlAs barriers remained strained with respect to each other, but partially relaxed with respect to the substrate. Relaxation, indicated by a shift towards the substrate peak and broadening of the AlGaAs layer peak was also observed, Fig. 4(b). There may have been misfit dislocations at the interface immediately above the MQW region as well as below.

For the partially relaxed layer, the average lattice parameter of the MQW was determined from the position of the zero-order layer peak in the 004 and 115 reflections, and the average In composition in the wells was then calculated assuming that the thicknesses of the wells and barriers were equal (shown by simulation in the unrelaxed layer). The degree of relaxation in the 12% In sample was found to be 18%. The x-ray results showed that there were no significant differences in the average composition and relaxation before and after device processing.

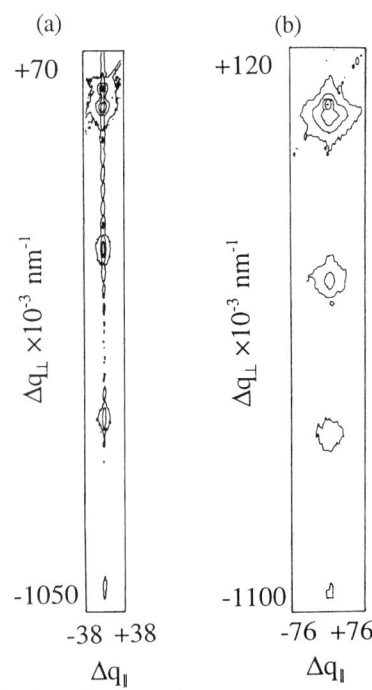

Fig. 4. Reciprocal space maps near 004 reflection for InGaAs MQWs grown on (001) GaAs, (a) 9% In and (b) 12% In wells.

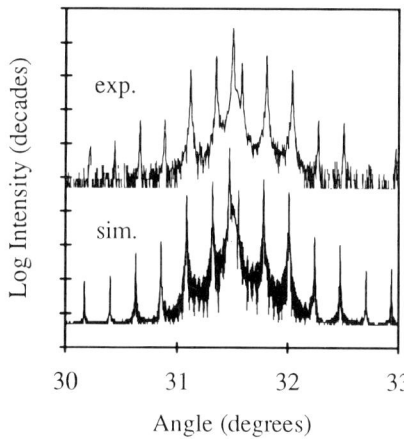

Fig. 5. Experimental and simulated 004 rocking curves for InGaAs MQW grown on (001) InP. The simulation includes 10% As incorporation in the InP barriers.

A 30-period InGaAs (nominally 47% In, 110Å wide) MQW with InP barriers was grown at 480°C by MBE on an (001) InP substrate. Experimental and simulated 004 rocking curves are shown in Fig.5. Increased satellite intensities in the experimental data compared to initial simulations led to the conclusion that there was As incorporation in the InP barriers. This would result in the wells being under tensile strain and the barriers being slightly compressive, so increasing the strain modulation while keeping the average strain constant. This is the opposite effect to that shown in Fig. 3, where the satellite intensities are reduced because of a decrease in strain modulation (caused by In in the GaAs barriers). The simulation shown in Fig. 5 was calculated using a model with InGaAs wells (45% In) and InAsP barriers (10% As), both 112Å wide.

6. CONCLUSIONS

High-resolution x-ray diffractometry has been used to determine accurately the indium content and well width in a number of InGaAs quantum well structures, including single, stepped and multiple quantum wells grown on both (001) and (111) substrates. It was possible to characterise samples both before and after processing into devices. The composition and thickness and the interfacial abruptness were determined by comparing with simulations. It was also possible to determine whether incorporation of In/As in the GaAs/InP barriers had occurred. This resulted in a change in the MQW satellite intensities, depending on whether the strain modulation had increased or decreased.

Reciprocal space mapping enabled direct measurement of strain and microscopic tilting due to the presence of misfit dislocations. The degree of strain relaxation was also determined, which was important for device considerations.

ACKNOWLEDGEMENTS

The authors would like to thank R. Grey, M.S. Skolnick and P. Kidd for the (111)B MQW sample.

REFERENCES

Fewster P 1993 Semicond. Sci. Technol. 8 1915
Fewster P and Curling C J 1987 J. Appl. Phys. 62 4154
Ghisoni M, Parry G, Hart L, Roberts C and Stavrinou P N 1994 Appl. Phys. Lett. 65 3323
Hart L, Ghisoni M, Stavrinou P N, Roberts C and Parry G 1995 Mat. Sci. Technol. 11 in press
Hart L and Fewster P 1993 Inst. Phys. Conf. Ser. 134 569
Kidd P, Fewster P F, Andrew N L and Dunstan D J 1993 Inst. Phys. Conf. Ser. 134 585

Inst. Phys. Conf. Ser. No 146
Paper presented at Microsc. Semicond. Mater. Conf., Oxford, 20–23 March 1995

Characterization of bulk as-grown and annealed III-Vs by photo-etching and complementary methods

J L Weyher

Research Centre Jülich (IFF-KFA), Postfach 1913, D-52428 Jülich, Germany
Present address: Fraunhofer-Institut (FhG-IAF), Tullastrasse 72, 79108 Freiburg, Germany

ABSTRACT: Characteristic features of photo-etching (DSL method: $CrO_3/HF/H_2O$ etchants used with light to study defects in III-Vs) are described. The role of illumination in the ultra-high sensitivity of the method for revealing crystallographic and chemical inhomogeneities is discussed and demonstrated. In S.I. undoped GaAs all structural features such as decoration and matrix precipitates and As-related recombinative defects formed in as-grown and annealed material are made visible by DSL etching. Photo-etching, dark field infra-red and conventional transmission optical microscopy were used to fully control the processing of the TEM specimens containing individual or clustered microdefects. Clear recognition by the DSL etching of small (up to 60 nm) and large (up to 3 μm) dislocation loops in heavily Si-doped GaAs is illustrated. By the combined use of photo- and dark etching (projective etching) it is demonstrated that the dislocations move not only in {111} but also in {110}, {100}, {320} and {311}) glide planes.

1. INTRODUCTION

Structural etching has become a routine method for evaluating the density and the distribution of dislocations and is often used in studying complex defects in compound semiconductors, including interaction between crystallographic and chemical inhomogeneities. To this category belong etching processes controlled by surface kinetics, which are influenced by compositional and structural imperfections. The classical, "orthodox" etching method was developed and is still used for revealing dislocations in the form of etch pits. Both the theory and the practice of this method is well recognized and described in detail for different materials including semiconductors by e.g. Miller and Rozgonyi (1980), Heimann (1982) and Sangwal (1987). Two characteristic examples of the defect-selective preferential etching of compound semiconductors are presented in Fig. 1. Both the etch pits formed on GaAs in the popular molten KOH, Fig. 1a, and on InP in the recently introduced BCA solutions (HBr-$K_2Cr_2O_7$-H_2O etching system, Weyher et al 1994a), Fig. 1b, are elongated indicating unequivocally the [110] direction. Additional information, important for the considerations in § 2, which can be obtained from the shift of the bottom-point of the crystallographic etch pits of Fig. 1a, is the direction of each dislocation line. Using the classification given by Takenaka et al (1978) for the dislocations in epitaxial layers, the pits from Fig. 1a indicate the presence in the bulk of GaAs of dislocations with the line-direction along <001>, <011> and <112>, marked A, B and C, respectively.

The second group of etch features formed on dislocations during selective etching covers hillocks or ridges, as shown in Fig. 2. The shape of such hillocks depends on the position of dislocations with respect to the surface and on the etching parameters (time and composition of the etching medium). In addition, the use of the light increases the selectivity of etching on the defect-sites, leading to the formation of complex, non-crystallographic mounds or valleys around the dislocation-related tips, Fig. 2b,c. Though there is as yet no well-established theory describing the formation of such complex etch features, some principles are already becoming clear. This refers to the DS(L) etching system containing HF-CrO_3 aqueous solutions which has been thoroughly studied for GaAs regarding aspects of the kinetics and the electrochemistry of the surface reactions (van de Ven et al 1986 and Notten et

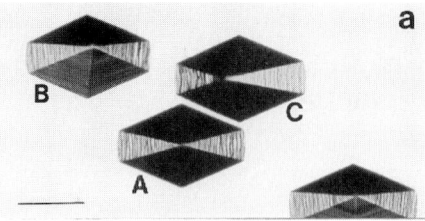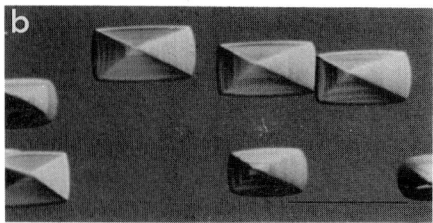

Fig. 1. Dislocation-related pits formed on the (001) surface of (a) GaAs during etching in molten KOH and (b) InP in HBr-H$_2$O solution. Markers represent 20 μm, DIC images.

al 1991). From these studies it follows that the surface coverage of GaAs by the passivating film, Θ, containing chromium complexes constitutes the main factor which contributes to the final morphological characteristics of the etch features. The thickness Θ depends on: (i) the concentration of the electroactive species in the etching medium (CrO$_3$ and HF which directly control the formation of the passivating film and its dissolution, respectively); (ii) the intensity of illumination (the photo-generated holes are used to rupture the Ga-As bonds) and (iii) the degree of deformation of the GaAs lattice (increased deformation e.g. in the strain field of dislocation reduces the Ga-As bonds strength and consequently facilitates local higher passivation). As a result, when a DS solution of the same composition is used, etching in the dark results in the formation of hillocks representing the deformation of the lattice around dislocations (Fig. 2a). During the light-assisted DSL etching, apart from the sharp tip formed due to the effective recombination on the dislocation itself, all factors which influence the availability of photo-generated carriers (electrically active impurities and the related band bending, cathodic protection phenomenon) take over, leading to the formation of the complex etch features around dislocations (Fig. 2b,c). Note that the size and shape of the features formed on dislocations introduced by indentation is similar after DS and DSL etching because they are undecorated and have no recognizable interaction zones (Weyher et al 1994b). Schematic representation of the characteristic etch features formed on dislocations during DS and DSL etching of different types of GaAs is given in Table 1. The detailed experimental data for this summary can be found in several earlier publications: Weyher and Giling (1985a), Weyher and van de Ven (1986), Frigeri and Weyher (1989), Weyher at al (1990), Visser at al (1990) and Weyher (1994). Also in InP hillocks or complex features are formed during DSL etching (Weyher and Giling 1985b, Fornari et al 1993, Weyher et al 1994a and Frigeri et al 1994), however more experimental work is needed for the conclusive description of the characteristic etch patterns.

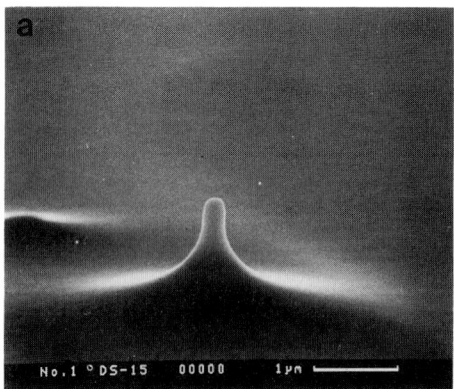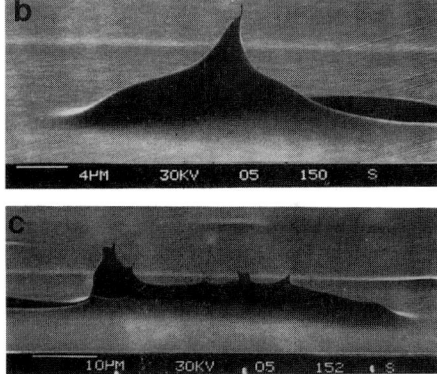

Fig. 2. Morphology of the etch figures formed on dislocation outcrops during dark DS etching (a) and photo-etching of individual (b) or clustered (c) dislocations in Si-doped (001) GaAs. SEM images: ((a) courtesy of R. Dian, (b, c) courtesy of C. Frigeri), tilt 80°.

In the following sections some examples will be shown of the exploration of the characteristic features of DSL and DS etching in studying specific defects in GaAs.

Table 1.
Schematic representation of the profiles of DS/DSL etch features on dislocations in GaAs

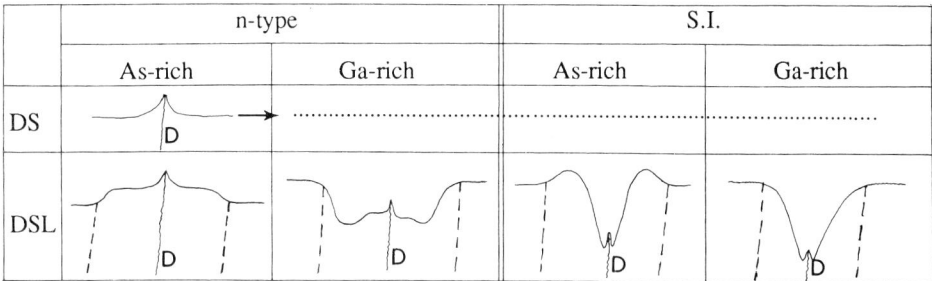

D: dislocation; ------: extent of electrically active zone with the properties different in comparison with those of the dislocation-free matrix

2. PROJECTIVE ETCHING

It has been shown that due to the ultra-high sensitivity of DSL etching in the presence of electrically active point defects it is possible to reveal the traces of dislocations that move under the thermal stress arising during post-growth cooling (G-S dislocations: Weyher and Giling 1985, Weyher and van de Ven 1986). Based on the detailed studies by photo-etching, high spatial resolution PL and EBIC it was concluded that most probably $AsGa$ defects are the by-product of the interaction of the moving dislocation with the point defects present in the host GaAs lattice (Weyher et al 1992). These defects account for the recombination of the photo-generated carriers and the consequent formation of the etch ridges, named T (traces) of the moved dislocations, Fig. 3. Making use of the difference in the mechanisms of the DS and DSL etching, as discussed in the §1, the concept of projective etching was introduced (Weyher 1984) which is presented in Fig. 3a for the clarity of the following consideration. It consists of shallow (submicron) photo-etching (1), deep DS etching (2), repeated shallow photo-etching (3) and, if necessery, preferential etching in molten KOH (4). During step 1 all elements of the moved dislocation are visualised: the starting point of the dislocation A, the trace T and the final position of the dislocation at B (Figs 3a and 3b-1). If the glide plane of the moving dislocation is for instance (111), during step 2 the intersection of this plane with the (001) surface that is actually being etched is continuously moving. The final position of this intersection, represented by the trace A'B' (Fig. 3a) is, however, almost not visible (see Fig. 3b-2), while all features from Fig. 3b-1 are clearly seen due to the memory effect (Stirland and Ogden 1973). Note that the points A, T and B are projected on the new GaAs surface, therefore they are labelled A_p, T_p and B_p in Fig. 3b-2. The final position of the intersection of the whole G-S defect can now be made visible by the second shallow photo-etching (step 3), see A', T' and B' in Fig. 3b-3. This procedure once more visualizes the basic difference between the mechanisms of the DS and DSL etching: the trace of the moved dislocation is clearly seen only after photo-etching due to its recombinative property for photo-generated carriers (Weyher et al 1992a), while strain-sensitive dark etching is unable to reveal such electrically active defects.

After the stage 3 of the projective etching it is possible to determine the angle α between the glide plane and the (001) oriented surface of the sample i.e. to define the glide plane (tg α=p/x, where p is the depth of the DS etching step and x is the distance between the projected and the final intersections of the glide plane with the DSL etched surfaces (see Fig. 3a)). In addition, step 4 can now be employed (etching in molten KOH) in order to find out the direction of the dislocation line from the shift of the bottom point of the etch pit. Fig. 3b-4 shows a fragment of the G-S defect after all 4 stages of etching. The glide plane of this dislocation, which was found to be {111} and the etch pit of the c-type (Takenaka et al 1978,

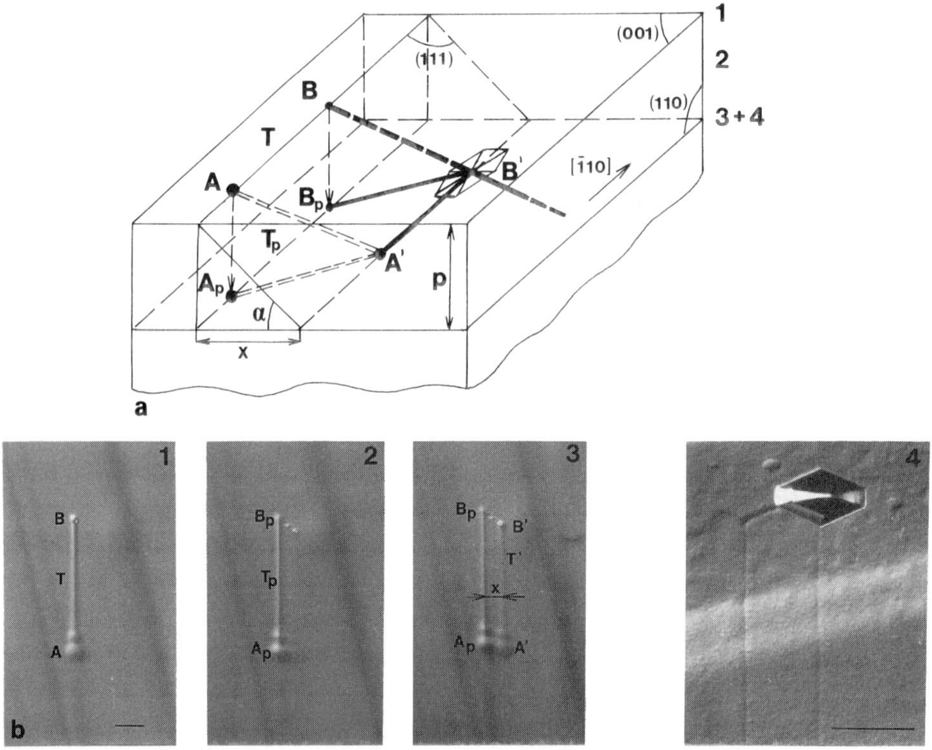

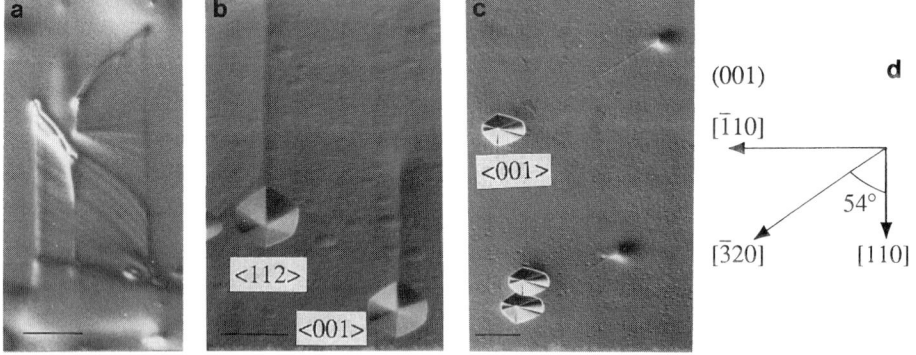

Fig. 3. (a) Schematic representation of projective etching. (b 1-4) DIC images of the subsequent steps: (1) shallow DSL, (2) deep DS, (3) shallow DSL and (4) KOH etching. Markers represent 20 μm.

Fig. 4. G-S dislocations on the (001) Surfaces of GaAs after: (a) shallow DSL etching, (b-c) 4-stage projective etching. (d) Scheme of the coordinate system. Markers represent 20 μm.

Weyher 1994, see also Fig. 1a), indicate the <112> type direction of the dislocation line. Using these 3 and 4 stage procedures of the projective etching different glide planes and directions of the dislocation lines have been found; some representative examples are shown in Fig. 4. The glide event in the (001) plane, as recorded in Fig. 4a, has been only seldomly observed after DSL etching because of the small chance of having a defect just in the vicinity of the photo-etched (001) surface. However, it was reported by Qin and Roberts (1989) that after indentation the {001}<110> slip system operates in GaAs.

The slip of dislocations in the (111) plane, which is inclined vs the (001)-oriented etched surface under 54,7°, gives two intersections after projective etching (e.g. Fig. 3b-4). On the contrary, when moving dislocations are in (100), (110) and (320) planes, which are perpendicular to the (001) surface, only one trace is formed, as can be seen in Figs 4b-c. It should be remarked, that except for the {320} glide plane, all other glide planes reported here were already predicted by Hornstra (1958). In practice it is usually assumed that the {111}<110> slip system is the one that operates in GaAs. After examination of many Si-doped GaAs crystals containing a low dislocation density it can be concluded, however, that movement of defects occurs often in the other systems, particularly in the {110}<110> and {320}<320> systems. Recently Sonnenberg and Altmann (1994) have shown by photo-etching of the (110)-oriented wafers cut from the VGF GaAs crystals that some dislocations were gliding in the (113) plane.

Using projective etching it is also possible to visualize changes of the glide planes e.g. a dislocation initially moving in the {111} plane and in the <110> direction passes over to the {100}<100> system and then "returns" to one of the {111}<110> systems (Weyher 1994). Such a change in system may occur when the dislocation line is exactly in the <011> direction, which is the only common direction for both the {111} and the {100} planes. More surprisingly a "non-crystallographic" change is sometimes observed, that suggests a remarkable involvement of the interaction of the moving dislocation with point-defects. Two characteristic examples of such complex G-S defects are shown in Fig. 5.

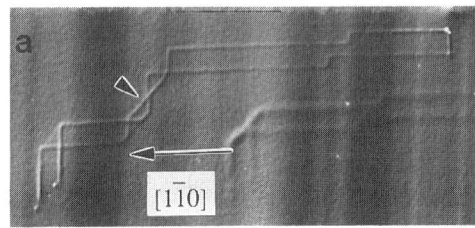 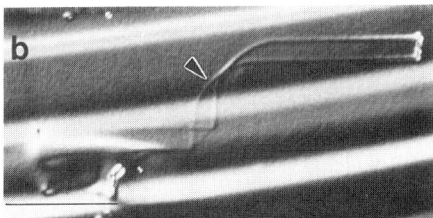

Fig. 5. Complex slip of dislocations in GaAs revealed on the (001) surface by projective etching (steps 1-3). The arrows indicate slip in: (a) (100) plane and (b) "non-crystallographic" continuous transition between the slip in two different {111}<110> systems. Markers represent 100 μm.

3. STRUCTURE OF S.I. ANNEALED GaAs

It has been well documented that defect-selective etching and photo-etching are very effective in studying properties of the annealed GaAs (Stirland 1990, Stirland 1991 and Weyher et al 1992b). However it seems worth pointing out the difference between etching with and without purposely used illumination. Fig. 6 summarizes results of the DS (etching in the dark box) and DSL etching of S.I. undoped annealed GaAs. In order to enable quantitative comparison, the samples were cut from the same GaAs wafer and DS/DSL etched to the same etch depth p=2 μm. Apart from the remarkable increase of the uniformity parameter (U as defined by Stirland et al 1988: $U_{DS}=0,05$ and $U_{DSL}=10$ from Figs 6c and 6d, respectively) and image contrast after DSL etching (compare Figs 6a-6b), the main advantage of this method is its ultra-high sensitivity to the variations of the electrical properties of the material across the dislocation cell structure. From the etch depth profiling it is clear that the formation of decoration and matrix precipitates (DPs and MPs, respectively) induces remarkably higher etch rates. Previous studies (Molva et al 1990, Brozel at al 1991, Weyher et al 1992b and Weyher 1994) showed that the presence of the spherical or spheroid-shaped "bubbles" of matrix precipitates is always accompanied by the increased photoluminescence and cathodo-luminescence yield. Such a decreased concentration of the non-radiative defects is equivalent to the decreased density of recombinative defects for the photo-generated carriers. As a result

404

the photo-etching rate increases in the cell interiors, being diversified across the MPs "bubble": the highest etch rate is always recorded at the shell of the spheroid where the size of the MPs is the largest (Fig. 6d, Weyher et al 1994c). It was therefore reasonable to conclude that the species involved in the formation of microdefects have recombinative properties: As_i, V_{As} and V_{Ga} were considered to be the most probable candidates (Molva et al 1990, Weyher et al 1992b, Brozel et al 1991). Recent TEM studies revealed the presence of both voids (Stirland 1991, Williams et al 1991) and As precipitates, sometimes coupled with voids (Stirland et al 1990, Weyher et al 1994c, Frigeri and Weyher 1995). Despite these findings the relative contribution of interstitial- or vacancy-type defects in the formation of matrix inclusions still remains speculative.

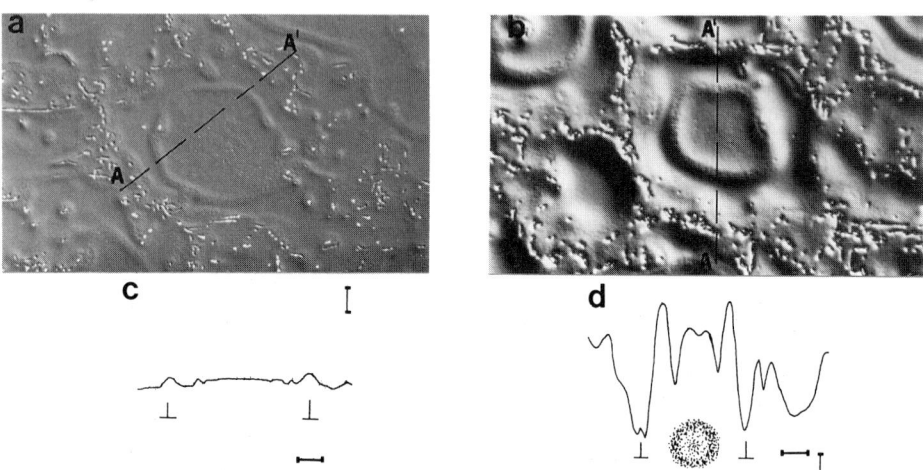

Fig. 6. Structural features in annealed GaAs after DS (a) and DSL (b) etching, followed by surface profiling along the A-A' lines (c and d, respectively). Symbol   indicates position of dislocation cell walls and the inset in (d) distribution of MPs in the cell interior. Markers represent: horizontal 100 μm, vertical 25 nm (c) and 50 nm (d).

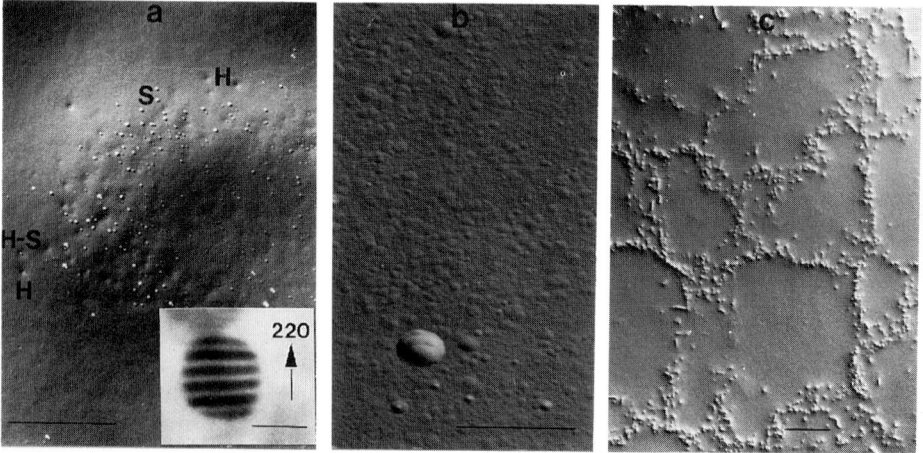

Fig. 7. (a) Fragment of a spheroid of MPs inside dislocation cell revealed by shallow photo-etching in the ingot-annealed GaAs. Inset: dark-field TEM image of the As precipitate which shows 2,83 nm Moiré fringes (courtesy of C. Frigeri). (b) Microstructure roughness at cell centre revealed by deep DS etching. (c) Homogeneous structure revealed by shallow DSL etching in MWA GaAs. Markers represent: (a-b) 50 μm, inset: 10 nm; (c) 100 μm. DIC images.

Fig. 7a shows a magnified fragment of the spherical "bubble" of the precipitates in the matrix revealed by shallow photo-etching and a TEM image of the precipitate identified as a hexagonal As inclusion as an inset to this Figure (the spacing D of the Moiré fringes from this inset is 2,83 nm, d_1 for 220 GaAs is 0,1999 nm which gives $d_2=0,1867$ nm and fits reasonably well to the As $d_{110}=0,1879$ nm). From the detailed inspection of the photo-etched annealed GaAs (Fig. 7a) it follows that before the precipitates are etched away to form S-pits, relatively large (as compared to the size of the precipitates) etch hillocks are formed (H in Fig. 7a). Note that some hillocks have already the S-pits on the top (H-S). Using sequential photo-etching and photography it has been established that each hillock formed during the first shallow etching is transformed into an S-pit. Evidently the hillocks represent the zones around the As precipitates enriched in the recombinative defects. These zones have dimensions about two orders of magnitude larger than the size of the precipitates themselves. Such detailed analysis is not possible when deep AB or DS etching is employed. In Fig. 7b is shown a fragment of the zone with MPs after DS etching to a depth of 20 μm i.e. equivalent to the AB etching for 10 minutes. The details of the structure are lost because of the memory effect leading to the overlapping of the etch features from the whole volume of the material removed. A high degree of homogeneity of the undoped GaAs, which can be achieved by the multiple-temperature wafer annealing (MWA, Mori et al 1990), can be recognised and quantified by shallow photo-etching. The characteristic features of the material after such thermal treatment i.e. homogeneous distribution of the recombinative centres (low DIC contrast and low uniformity parameter) as well as the absence of the S-pits both on the dislocation-related hillocks and in the matrix are seen in Fig. 7c. This confirms effective surface losses of the excessive As after dissolution of DPs and MPs during MWA. Such a structure is, however, observed only in the near-surface layer (50 μm, Mori el al 1990) while in the bulk of the wafers the As precipitates are still present and the higher U-parameter values are recorded (Weyher 1995).

4. DIRECT CALIBRATION OF ETCHING-RELATED SURFACE FEATURES WITH TEM AND OPTICAL TRANSMISSION METHODS

One of the main difficulties in the application of the TEM method to study defects in low defect density GaAs, e.g. VGF or VB-grown material, is to have the defect(s) of interest within the TEM specimen. This difficulty becomes even more important when the defects are distributed non-uniformly on a macro-scale. It was shown that the use of shallow photo-etching allows us to prepare good TEM specimens containing fragments of extended defects such as dislocation cell walls (Wurzinger et al 1991, Frigeri et al 1993) as well as clustered microdefects in the matrix (Frigeri et al 1991a, Weyher et al 1994c). Detailed recent studies have demonstrated that due to the combined use of IR microscopy and DSL etching it is possible to analyse by TEM a selected individual microdefect located in the bulk of the large GaAs samples (Weyher et al 1994d). Fig. 8 demonstrates this approach together with some additional practical details resulting from the procedure employed. After selecting the interesting defect (e.g. a row of DPs, Fig. 8a) by the dark-field IR method and bringing it to the surface by controlled mechano-chemical polishing, the sample is photo-etched in order to reveal one (or more) of the DPs in the form of an S-pit (S in Fig. 9b). The subsequent back-side thinning with the polishing solution ($H_2SO_4:H_2O_2:H_2O=5:1:1$) results in the formation of the elongated S-pits on the DPs, as shown in Fig. 8c. The use of visible light in the transmission mode serves as a test method to control the presence of a precipitate in the thin foil (P in Fig. 8d). The corresponding TEM image in Fig. 8e shows all elements recorded by the surface and transmission optical means: S-pit formed during the DSL etching (S), dislocation and the DPs (D, P, respectively) and S-pits formed during "5:1:1" thinning ($S_1...S_4$). It has also been unequivocally established by calibration with molten KOH, that the longer axis of the "5:1:1"- and DSL-related S-pits are parallel to the $[\bar{1}10]$ direction. (Note, that pits S and $S_1...S_4$ are on the opposite sides of the specimen i.e. their longer axes are in fact parallel). Using this method numerous thin foils containing DPs and MPs in LEC and VGF GaAs were prepared and analysed in the TEM; the results will be discussed elsewhere (Weyher et al, in preparation).

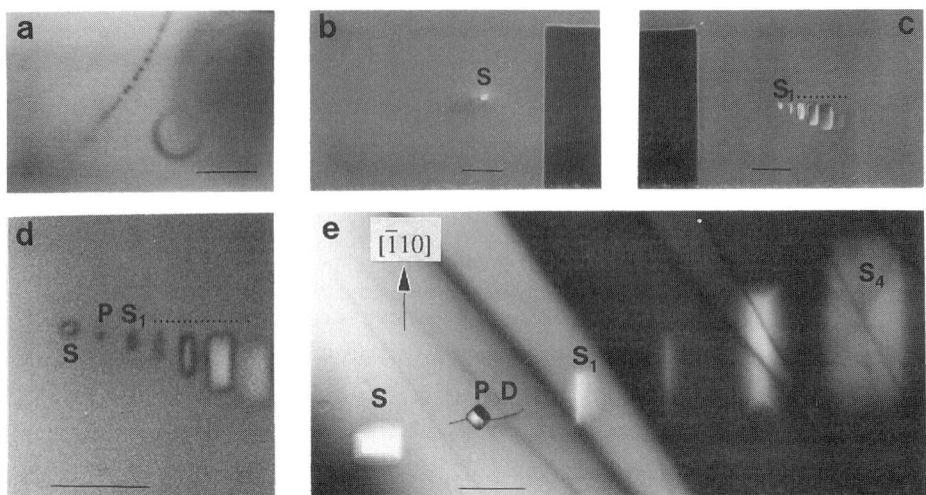

Fig. 8. (a) DF IR image of DPs along a single dislocation inside a VB undoped GaAs wafer. (b-c) DIC surface images of decorated dislocation: (b) after DSL etching and (c) after "5:1:1" thinning of the same sample from the back-side. (d) Transmission optical image (visible light) of the thin foil which brings together all etch features from Figs (b-c) and additionally reveals a precipitate P situated inside the foil. (e) Corresponding to (d) TEM image. Markers represent:

(a) 20 μm; (b-d) 10 μm; (e) 2 μm.

5. REVEALING OF MICRODEFECTS IN HEAVILY DOPED GaAs

The microdefects resulting from the intentional heavy doping or implantation of undoped GaAs are the reason for the micro-roughness during photo-etching. Apart from the As matrix precipitates which after exposing to the surface give rise to the formation of S-pits, as shown in § 3, the microdefects in the form of dislocation loops were supposed to produce

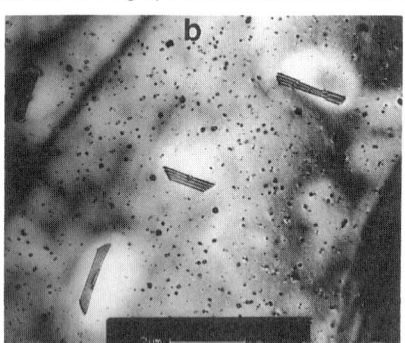

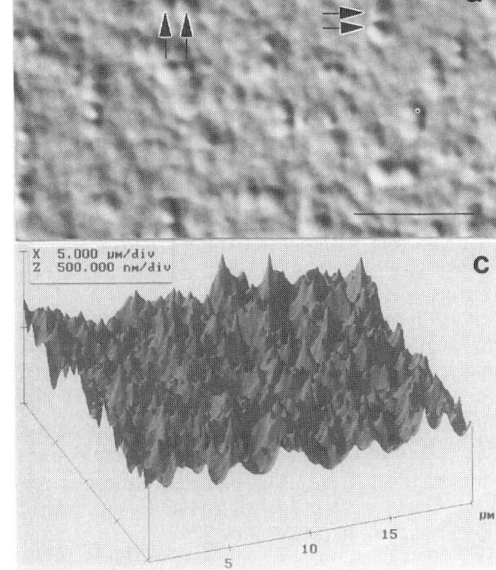

Fig. 9. Heavily Si-doped VB GaAs: (a) DIC surface image after shallow DSL etching; (b) TEM image from the specimen prepared in the area of the same density of surface features as in (a); (c) 3D tapping-mode AFM image (courtesy of S. Müller) of surface features formed during DSL etching. Marker represents

10 μm.

small etch hillocks when DSL etching was performed (Frigeri et al 1991a, Frigeri et al 1991b). However, the very high density of micro-defects in these materials and consequently the high density of surface features allowed only speculative judgment of their character (hillocks-pits). By the combined use of the three complementary methods i.e. DSL, TEM and AFM (atomic force microscopy) the presence of sub-micron hillocks formed during the DSL etching on the dislocation loops has been confirmed and is shown in Fig. 9. In Fig. 9a, the optical DIC

image of the photo-etched GaAs doped with Si to 2×10^{19} cm^{-3} level is presented. The arrows point out the position of two of several pairs of features having 2-3 μm spacing and being arranged along the orthogonal <110> directions. Fig. 9b shows the TEM image obtained from the same material. From the detailed TEM analysis (Schober at al 1994) it follows that the two types of dislocation loops seen in this Figure are: (i) small, below 60 nm, perfect loops in the

{110} planes (b=1/2<110>) and (ii) large, up to 3 μm, faulted loops in the {111} planes (b=1/3<111>). From both the orientation, size and density of the large loops, as found by the TEM analysis, it becomes clear that the pairs of features in Fig. 9a are formed on these large defects, while the overal background microroughness corresponds to the small, high density micro-loops. The expected nature of the features formed on the loops during photo-etching (i.e. etch hillocks due to the recombination on dislocations of photo-generated carriers) has been confirmed by AFM surface study, as demonstrated in Fig. 9c.

6. CONCLUDING REMARKS

(i) The use of light during etching of GaAs in DS solutions (the range of compositions with the kinetically controlled action) makes this method ultra-sensitive to structural and chemical inhomogeneities.
(ii) due to the calibration with other physical methods, particularly with TEM, DS photo-etching has become a quick and reliable tool in studying the density, the distribution and the nature of defects in III-Vs.
(iii) Using characteristic features of dark- and photo-etching it has been demonstrated that dislocations in GaAs can move in different glide planes including: {111}, {110}, {100}, {311} and {320}.
(iv) Shallow photo-etching in combination with different optical methods constitutes a useful tool in preparing TEM samples from materials containing a very low density of defects.

ACKNOWLEDGMENTS

I would like to thank C. Frigeri, K. Sonnenberg, W. Jantz and N. Herres for the stimulating discussions, and T. Schober for sharing with me his extensive knowledge of TEM analysis.

REFERENCES

Brozel M R, Breivik L, Stirland D J, Williams G M and Cullis A G 1991 Appl. Surf. Sci. 50 475
Fornari R, Frigeri C, Weyher J L, Krawczyk S K, Krafft F and Mignoni G 1993 Proc. 7th Conf. on Semi-insulating III-V Mat., eds C J Miner, W. Ford and E. Weber (Bristol: IOP) pp 39-44
Frigeri C and Weyher J L 1989 J. Appl. Phys. 65 4646
Frigeri C, Weyher J L and Gall P 1991a Inst. Phys. Conf. Ser. No 117 (Bristol: IOP) pp 353-6
Frigeri C, Weyher J L and De Potter M 1991b Appl. Surf. Sci. 50 115
Frigeri C, Weyher J L and Alt H Ch 1993 Phys. Stat. Sol. (a) 138 657
Frigeri C, Ferrari C, Fornari R, Weyher J L, Longo F and Guadalupi G M 1994 Mat. Sci. Eng. B28 120
Frigeri C and Weyher J L 1995 (in preparation)
Hornstra J 1958 J. Phys. Chem. Solids 5 129
Heimann R B 1982 in: Crystals: Growth, Properties and Applications 8, ed J Grabmaier

(Berlin: Springer-Verlag) pp 173-224

Miller D C and Rozgonyi G A 1980 in: Handbook on Semiconductors Vol. 3, ed S B Keller (Amsterdam: Elsevier) pp 218-46

Molva E, Bunod Ph, Chabli A, Lombardot A, Dubois S and Bertin F 1990 J. Crystal Growth 103 91

Mori M, Kano G, Inoue T, Shimakura H, Yamamoto H and Oda O 1990 Proc. 6th Conf. on Semi-insulating III-V Mat., eds A G Milnes and C J Miner (Bristol: IOP) pp 155-60

Notten P H L, van den Meerakker J E A M and Kelly J J 1991 Etching of III-V Semiconductors: an Electrochemical Approach (Oxford: Elsevier)

Qin C-D and Roberts S G 1989 Inst. Phys. Conf. Ser. No 104 (Bristol: IOP) pp 321-6

Sangwal K 1987 Etching of Crystals (Amsterdam: Elsevier)

Schober T, Rucky A, Jager W, Weyher J L and Sonnenberg K 1994 (Unpublished results)

Sonnenberg K and Altmann A 1994 Berichte des Forschungszentrums Jülich 2939

Stirland D J and Ogden R 1973 Phys. Stat. Sol. (a) 17 K1

Stirland D J, Warwick C A and Brown G T 1988 Proc. 5th Conf. on Semi-insulating III-V Mat., eds G Grossmann and L Ledebo (Bristol: IOP) pp 93-8

Stirland D J 1990 in: Defect Control in Semicoductors, ed K. Sumino (Amsterdam: Elsevier) pp 783-94

Stirland D J, Gall P, Brozel M R, Breivik L, Williams G M, Cullis A G and Fillard J P 1990 Inst. Phys. Conf. Ser. No 112 (Bristol: IOP) pp 55-60

Stirland D J 1991 Inst. Phys. Conf. Ser. No 117 (Bristol: IOP) pp 327-36

Takenaka T, Hayashi H, Murata K and Inoguchi T 1978 Japan. J. Appl. Phys. 17 1145

van de Ven J, Weyher J L, van den Meerakker J E A M and Kelly J J 1986 J. Electrochem. Soc. 133 799

Visser E P, van der Wel P J, Weyher J L and Giling L J 1990 J. Appl. Phys. 68 4242

Weyher J L 1984 Proc. 1984 Meeting on Caratterizzazione Anal.-Strutturale di Materiali per l'Elettronica eds S. Carra, C. Ghezzi, P.G. Merli and C. Paorici (Parma: Tecnografica) pp 1-33

Weyher J L and Giling L J 1985a Defect Recognition and Image Processing in III-V Compounds, ed J.P. Fillard (Amsterdam: Elsevier) pp 63-71

Weyher J L and Giling L J 1985b J. Appl. Phys. 58 219

Weyher J L and van de Ven J 1986 J. Crystal Growth 78 191

Weyher J L, Frigeri C and van der Wel P J 1990 J. Crystal Growth 103 46

Weyher J L, van der Wel P J and Frigeri C 1992a Semicond. Sci. Technol. 7 A294

Weyher J L, Gall P, Le Si Dang, Fillard J P, Bonnafé J, Rüfer H, Baumgartner M and Löhnert K 1992b Semicond. Sci. Technol. 7 A45

Weyher J L, Fornari R, Görög T, Kelly J J and Erné B 1994a J. Crystal Growth 141 57

Weyher J L, Frigeri C, Schohe K, Krawczyk S K, Wosinski T and Fornari R 1994b Inst. Phys. Conf. Ser. No 135 (Bristol: IOP) pp 225-8

Weyher J L, Frigeri C, Gall P and Kremer R 1994c Proc. 8th Conf. on Semi-insulating III-V Mat., ed M. Godlewski (Singapore: World Scientific) pp 163-6

Weyher J L, Sonnenberg K and Schober T 1994d Proc. 8th Conf. on Semi-insulating III-V Mat., ed M. Godlewski (Singapore: World Scientific) pp 105-10

Weyher J L 1994 in: Handbook on Semiconductors Vol. 3 ed S. Mahajan (Amsterdam: Elsevier) pp 995-1031

Weyher J L 1995 unpublished results

Williams G M, Cullis A G and Stirland D J 1991 Appl. Phys. Lett. 59 2585

Wurzinger P, Oppolzer H, Pongratz P and Skalicky P 1991 J. Crystal Growth 110 769

Inst. Phys. Conf. Ser. No 146
Paper presented at Microsc. Semicond. Mater. Conf., Oxford, 20–23 March 1995
© *1995 IOP Publishing Ltd*

TEM study of GaAs/InP heterostructures fabricated by wafer bonding

G Patriarche°, F Jeannès, F Glas and J L Oudar

France Telecom, Centre National d'Etudes des Télécommunications, Paris B, Laboratoire de Bagneux, 196 avenue Henti Ravéra, BP 107, 92225 BAGNEUX Cedex, France
°Institut des Matériaux de Nantes, 2 rue de la Houssinière, 44072 NANTES, France

ABSTRACT: We present a TEM study of mismatched heterostructures realised not by epitaxy, but by direct wafer bonding. We consider both the bonding of two single homoepitaxial layers and of complex structures. Threading dislocations are totally eliminated, but different interface and defect morphologies are observed and interpreted. The influence of the annealing temperature is assessed and we show that TEM allows the determination of optimal bonding conditions.

1. INTRODUCTION

When epitaxy is used to fabricate structures with strongly mismatched semiconductors, such as GaAs and InP, large densities of threading dislocations are always observed in the whole thickness of the sample. These residual dislocations originate from the network of misfit dislocations which appears at the interfaces between the mismatched materials in order to accomodate their different lattice parameters. Whatever the stacking of layers which may be introduced in the structure to try to prevent the propagation of the threading dislocations, the density of the latter hardly decreases below 10^7 cm^{-2}. Such a density is often incompatible with the fabrication of optoelectronic components. However, some components require the use of such mismatched heterostructures.

Recently, a new technique allowing the fabrication of such structures by direct bonding, without using epitaxy, was developed (Liau and Mull 1990, Wada et al 1993). In this technique, the two misfitting wafers are kept in contact and annealed. Atomic bonds reconstruct across the interface to produce a single structure. This technique generates no threading dislocations (Lo et al 1991, 1993).

By using this technique, we have realised (Jeannès et al) a bistable optical switch operating at 1.55 μm, which includes an active double heterostructure InP/InGaAsP/InP bonded to a GaAs/AlAs superlattice acting as lower mirror (for details, see Fig. 4). Here, the heterostructure and the mirror are mismatched whereas the individual layers constituting these two parts are lattice-matched. A multilayer coating is subsequently deposited to act as an upper mirror after selectively removing (thanks to an etch-stop layer) the InP substrate on which the InP-based heterostructure was epitaxially deposited. Using the same technique, Dudley et al (1992) had realised a vertical cavity emitting laser.

We first studied the bonding of an homoepitaxial layer of GaAs to an homoepitaxial layer of InP in order to find the best operating conditions, and then the structure described above. Our TEM results reveal how several parameters influence the apparition of defects during the bonding process.

2. EXPERIMENTAL

2.1 Bonding conditions

We always performed bonding between the surfaces of an InP and a GaAs layers fabricated by metalorganic chemical vapour deposition. In the first series of samples, these two layers are homoepitaxial, and in the second series they are heteroepitaxial. The operating mode is the following. The two layers are first deoxidised in a diluted HF bath. They are placed in a graphite holder so that two of their <110> directions coincide to within less than 2°, without taking into account the difference between [110] and [1$\bar{1}$0]. The furnace is first purged under nitrogen, and then a hydrogen flux is established. The temperature is ramped during 10 minutes, then kept at 650°C and decreased for 20 minutes down to 300°C; the holder is then extracted from the furnace. Good contact between the two surfaces is provided by the dilatation of the holder. The effective pressure at the bonding interface is of the order of 1 kg.cm^{-2}. The bonded areas are about 6x6 mm^2.

2.2 TEM specimen preparation and observations

All specimens were prepared for cross-sectional examination. Different techniques were used for the two types of specimens. For the GaAs/InP structures, we thinned the specimens mechano-chemically by using a solution of bromine in methanol. Usually this technique cannot be applied for cross-sectional specimens, because of the presence of resin at the interface; this is obviously not the case here. For the mirror/heterostructure device, thinning was performed after testing the device. Since the InP substrate has then been removed, the interface of interest lies close to the stack surface. We then have to resort to an argon/iodine ion beam thinning. To prevent the formation of a step at the interface during this operation, the structures had to be mechanically prethinned down to 10 μm. The specimens were examined at 200 keV in a Philips CM20 microscope.

3. EXPERIMENTAL RESULTS

3.1 Bonding of the homoepitaxial GaAs and InP layers

The annealing time of these structures was varied between 25 and 35 minutes, without producing any noticeable change in the interface morphology. There is no void left at the bonded interface. However, irregularly distributed pyramidal inclusions with sizes up to more than 1 μm have appeared (Fig. 1). The bases of these pyramids, which extend in GaAs as well as in InP, lie in the interface plane; their other faces are {111} planes. Often, two of these pyramids, one in each binary, have two nearly aligned faces and share only a narrow strip of their bases. Sometimes, the pyramids are only partially filled. EDX microanalysis shows that the pyramids extending into InP are constituted by an $In_xGa_{1-x}P$ alloy with x close to 0.5, whereas those extending into GaAs are made of $In_xGa_{1-x}As$ with x close to 0.5. These compositions are uniform in each pyramid and independent of its size.

HREM images of the bonded area away from the inclusions show the presence of a misfit dislocation network localized in the interface (Fig. 2). At variance with the study of Okuno et al (1995), no amorphous interfacial material is ever observed. However, there is always an interdiffusion zone of variable width (around 10 biatomic monolayers) extending on either side of the network plane. HREM observation of the pyramids reveals the absence of any dislocations in their bases, which confirms the close lattice-match between the pyramid and the binary on which it rests revealed by EDX. Of course, mismatch dislocations are present in the {111} facets. Except these misfit networks, there are no extended defects in the heterostructure, and in particular no threading dislocations.

3.2 Bonding of the complex heterostructures

The annealing time of these samples was varied between 18 and 30 minutes. The interface morphology differs from the previous one. We occasionally find filled truncated pyramidal inclusions, extending only in the 100 nm thick upper GaAs layer and over the same depth in the opposite InP upper layer. Irregularly shaped voids, up to 100 nm wide, are present in the bonding interface (Fig. 3 a). In addition, non-bonded areas, up to several μm wide, are found. These areas are sometimes filled with an amorphous phase. Their density, which depends much on the annealing time, decreases with the latter to vanish at 30 minutes (Fig. 3 b); it also seems to depend on the uniformity of the pressure applied. The well-bonded part of the GaAs/InP interface is similar to the interface between the

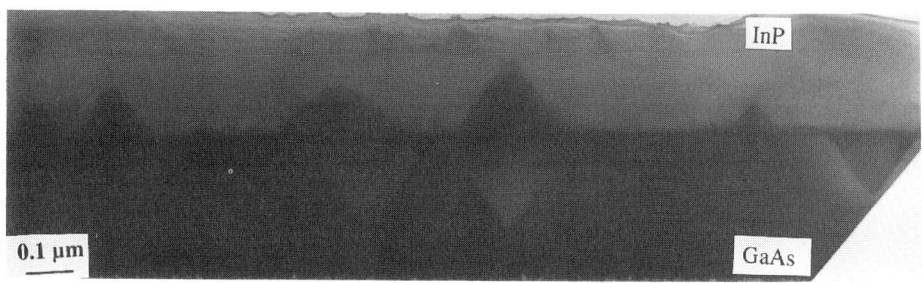

Fig. 1: TEM bright field image of the bonded interface between homoepitaxial GaAs and InP layers.

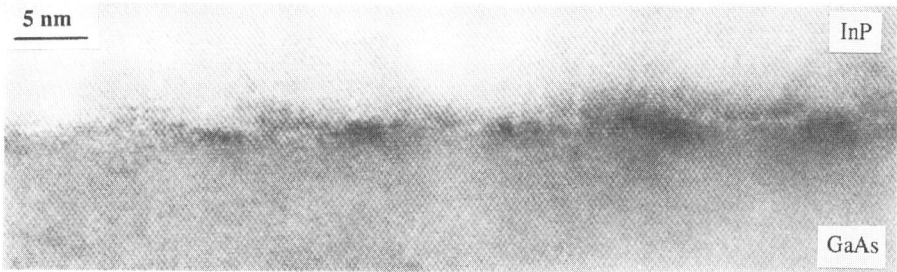

Fig. 2: HREM image of the bonded interface between the mismatched heteroepitaxial structures.

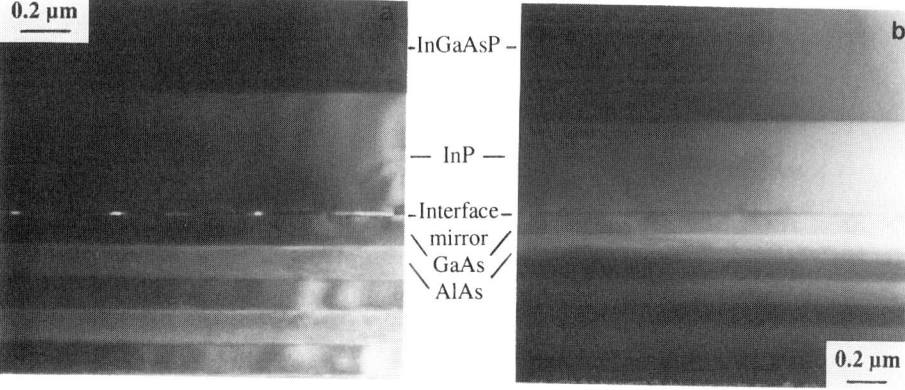

Fig. 3: TEM images of the bonded heteroepitaxial structures: (a) with interfacial voids (220 dark field micrograph); (b) without interfacial voids (bright field micrograph).

homoepitaxial layers (section 3.1). The same interdiffusion zone is observed, but the misfit dislocation network is less planar, the dislocations being distributed over a depth of several nm.

4. DISCUSSION

From section 3, the morphology of the bonded interface and the structural defects present depend much upon the nature of the structures involved.

In the case of the homoepitaxial GaAs and InP layers, the pyramidal inclusions are most likely formed in two steps. During the heating and annealing stages, the surfaces are locally etched in the areas of imperfect contact which necessarily exist in the $1 \ cm^2$ interface. Two processes may be considered, either a direct thermal etch, or a dissolution of the solids by drops of gallium or indium formed following preferential loss of the group V elements by the materials. In both cases, some group III-rich liquid is formed, which, during the cooling stage, redeposits two solids. Each of these solids is lattice-matched to the substrate lying under its (001) facet: this shows that growth proceeds from the bonding plane (which is confirmed by the occasionnal observation of pyramids with hollow apices) and that the underlying substrate imposes its own lattice parameter to the local overgrowth; this is an instance of the latching effect common in liquid phase epitaxy (Stringfellow 1972). These solids are near ternaries, although there are large ranges of quaternary alloys lattice-matched respectively to GaAs and InP. Moreover, the etching/regrowth process does not lead to significant losses of material. To explain these surprising facts, we propose that the liquid is formed from the simultaneous dissolution of approximately equal quantities of GaAs and InP (which is confirmed by the frequent observation of coupled pyramids of similar volumes). At 650°C, the liquid bath formed must be mainly constituted of group III elements in which the liberated group V elements can only be very partially incorporated: they must subsist as gases, with only trapped arsenic on the GaAs side and phosphorus on the InP side. During regrowth, the only solid which can be formed on each side is then the particular ternary which is indeed observed.

Such pyramidal inclusions are nearly absent when the complex heterostructures are bonded. We find a low density of what could be partially filled truncated pyramids, whose lateral facets are not well defined crystallographical planes. On the GaAs side, the truncation happens at the first GaAs/AlAs interface, and somewhat surprisingly, a symmetrical truncation occurs on the other side in the midst of the InP layer. However, the most important fact is the existence of non-bonded areas whose density depends sensitively on the annealing temperature. It is likely that the interface is first built up in the areas of good contact. The non uniform thickness of the interdiffusion zone also indicates that the bonding does not start everywhere at the same time, but probably proceeds laterally from the first bonded zones (patches of residual oxide may locally slow down the process). To fill the gaps which initially exist between the two surfaces, surface diffusion of the species must occur. The large remaining cavities, which were probably not present at the start, might be the modified residue of some initials gaps and the counterpart of this transport of matter. Finally, the difference of morphology with the homoepitaxial cases is probably due to a lesser planarity of the initial GaAs surface, which is the upper surface of a several μm thick GaAs/AlAs superlattice. Nevertheless, TEM has allowed us to determine optimal bonding conditions where neither pyramids nor voids appear.

REFERENCES

Dudley J J, Ishikawa M, Babic D I, Miller B I, Mirin R, Jiang W B, Bowers J E and Hu E L 1992 Appl. Phys. Lett. 61 3095
Jeannès F, Oudar J L, Patriarche G, Glas F, Azoulay R and Ougazzaden A, in preparation
Liau Z L and Mull D E 1990 Appl. Phys. Lett. 56 737
Lo Y H, Bhat R, Hwang D M, Chua C and Lin C H 1993 Appl. Phys. Lett. 62 1038
Lo Y H, Bhat R, Hwang D M, Koza M A and Lee T P 1991 Appl. Phys. Lett. 58 1961
Okuno Y, Uomi K, Aoki M, Taniwatari T, Suzuki M and Kondow M 1995 Appl. Phys. Lett. 66 451
Stringfellow G B 1972 J. Appl. Phys. 43 3455
Wada H, Ogawa Y and Kamijoh T 1993 Appl. Phys. Lett. 62 738

Inst. Phys. Conf. Ser. No 146
Paper presented at Microsc. Semicond. Mater. Conf., Oxford, 20–23 March 1995
© 1995 IOP Publishing Ltd

Zinc diffusion into Gallium Arsenide:
influence of diffusion source composition and temperature

A Rucki, W Jäger, K Urban, H - G Hettwer[*], N A Stolwijk[*], H Mehrer[*]

Institut für Festkörperforschung, Forschungszentrum Jülich, D-52425 Jülich Germany
[*]Institut für Metallforschung, Universität Münster, D-48149 Münster Germany

ABSTRACT: Zn diffusion and the generation and evolution of diffusion-induced defects in GaAs single crystals were studied at temperatures of 906 °C and 1050 °C using closed ampoules and different diffusion sources containing Zn, As and Ga. The diffusion-induced microstructure was characterized by analytical transmission electron microscopy of cross-section specimens and compared with Zn concentration profiles obtained by the spreading-resistance technique and by electron microprobe measurements. The results show the strong influence of the ambient conditions and defects on Zn diffusion into GaAs single crystals.

1. INTRODUCTION

Diffusion of Zn to high concentrations into GaAs is associated with the formation of dislocations and Ga-precipitates inside voids. The formation of these defects is connected with the incorporation of interstitial Zn onto Ga sublattice sites (Luysberg et al. 1989, 1992, Tan et al. 1991). Previous experiments at 906 °C with different source compositions showed that during Zn indiffusion, a Ga-rich crystal is always first produced, but that the progressing Zn incorporation in the near surface region depends on the presence or absence of As in the diffusion source (Jäger et al. 1993). In this paper we compare results of TEM investigations of the Zn diffusion-induced defect structures in semi-insulating GaAs with Zn penetration profiles obtained by electron microprobe analysis. The investigations concentrate on the influence of diffusion-source composition at different temperatures.

2. EXPERIMENTAL

Zn was diffused at 906 °C and 1050 °C for diffusion times between 60 min and 435 min into semi-insulating GaAs wafers with a dislocation density below 10^5 cm^{-2} using sealed quartz ampoules. In order to study the effects of different Zn_1 and As_4 partial pressures different diffusion-source materials were chosen containing either no or low weight fractions of As (As-poor conditions: Zn_{100}, $Zn_{15}As_{15}Ga_{70}$) or substantial weight fractions of As (As-rich conditions: $Zn_{50}As_{50}$, $Zn_{30}As_{70}$). The total amount of Zn in the ampoule volume of typically 6-8 cm^{-3} was chosen large enough (5-10 mg) to avoid source-depletion effects during the experiments. The Zn concentration profiles were determined using the spreading-resistance technique (SRT) and electron microprobe analysis (EMP). EMP measures the total concentration of Zn, C_{Zn}, whereas SRT monitors the electrically active substitutional Zn acceptor atom concentration but is also sensitive to precipitation and electrical compensation effects (Hettwer et al. 1991). Defects in the Zn diffused regions were studied using transmission electron microscopy (TEM) at 400 kV in a JEOL 4000FX microscope and spatially resolved energy dispersive X-ray spectrometry (EDX).

3. RESULTS

Anomalous Zn diffusion profiles are observed for all conditions. Fig. 1a shows box-shaped Zn concentration profiles as measured by EMP for As-poor conditions at 906 °C. The maximum Zn concentrations and the penetration depth at comparable diffusion times are considerably higher for the Zn_{100} than for the $Zn_{15}As_{15}Ga_{70}$ source. Fig. 1b shows profiles for As-poor and As-rich conditions at 1050 °C. A box-shaped profile is obtained for the Zn_{100} source, and kink-and-tail profiles characterized by two plateaus and two steps are observed for the As-rich conditions. The maximum Zn concentration at the front plateau is higher for the $Zn_{50}As_{50}$ source (8×10^{19} cm^{-3}) than for the $Zn_{30}As_{70}$ source (1.3×10^{20} cm^{-3}). For all profiles the maximum Zn concentrations close to the surface amount to 6×10^{20} cm^{-3}, whereas the penetration depth increases with increasing weight fraction of Zn in the source. For both temperatures the Zn concentration values close to the surface are larger by a factor of up to 4 than the values determined by SRT (not shown).

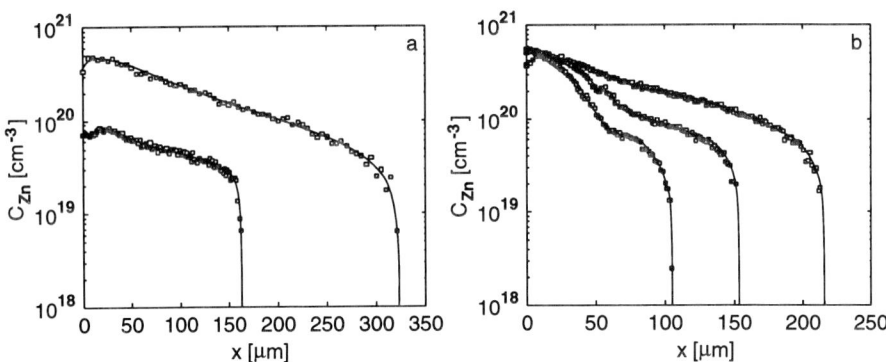

Fig. 1 Zn diffusion profiles obtained by EMP (□): (a) 906 °C: 435 min ($Zn_{15}As_{15}Ga_{70}$), 330 min (Zn_{100}) (left to right), and (b) 1050 °C, 60 min, $Zn_{30}As_{70}$, $Zn_{50}As_{50}$, Zn_{100} (left to right). Estimated respective values of (P_{Zn_1}, P_{As_4}, given in atm) are (a) (0.11, 3.9×10^{-6}) (0.43, 8.0×10^{-5}) (b) (0.46, 0.49) (0.60, 0.15) (0.80, 6.7×10^{-3}) (left to right). Solid lines are drawn to guide the eye.

Structural investigations showed that extended defects have formed throughout the Zn diffused regions. Dislocation loops, dislocations, and polyhedral precipitates (30-50 nm) with or without a small void volume are present in the diffusion front regions, independent of the diffusion conditions. The loops are identified to be perfect interstitial-ype loops, and EDX analyses showed that the precipitates essentially consist of Ga. The defect structures in the near-surface regions consist of dislocations and void/precipitate structures, which show significant differences for the various diffusion conditions. For As-poor conditions the precipitate fraction remains substantial throughout the Zn-diffused zone. Using EDX, it was found that under these conditions the precipitates contain also Zn, whose fraction increases with decreasing distance to the surface. For the Zn_{100} source at both temperatures the precipitate structure coarsens during the diffusion process, i.e., larger precipitates inside voids with a tetrahedral shape consisting predominantly of Zn are observed closer to the surface (Fig. 2). Similarly, Zn dissolution in the Ga-precipitates (Fig. 3c) is detected for As-rich conditions at 1050 °C, becoming substantial for Zn concentrations above $5x10^{19}cm^{-3}$. However, significantly different void/precipitate structures are observed during further diffusion. In a narrow zone (≤ 7 μm) close to the profiles' kink position the void volume of the precipitates increases, connected with formation of dislocation loops of vacancy and interstitial type (Fig. 3b). In an adjacent zone (width ≈ 5 μm) polyhedral voids and a dislocation network are observed (Fig. 3a). For both As-rich conditions neither Ga-Zn-precipitates nor voids are present in the near-surface region. Crystal regions below the surface (depths smaller than 18 μm) contain Zn_3As_2 precipitates.

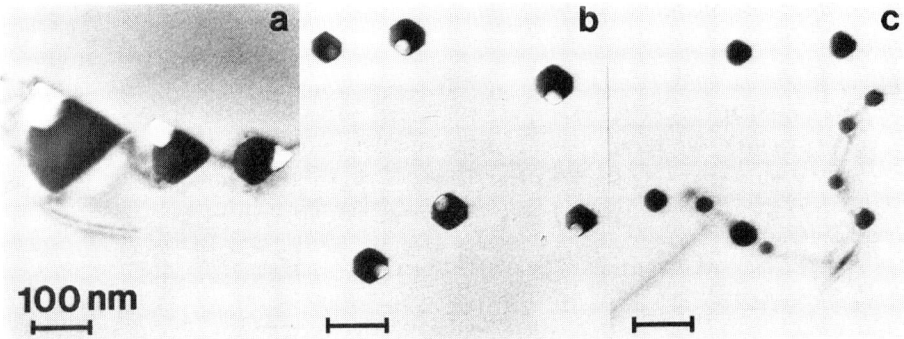

Fig. 2 Coarsening of the void/Ga-Zn precipitate pairs for As-poor conditions at 906 °C. Distance to surface increasing from left to right. Bright field TEM micrographs.

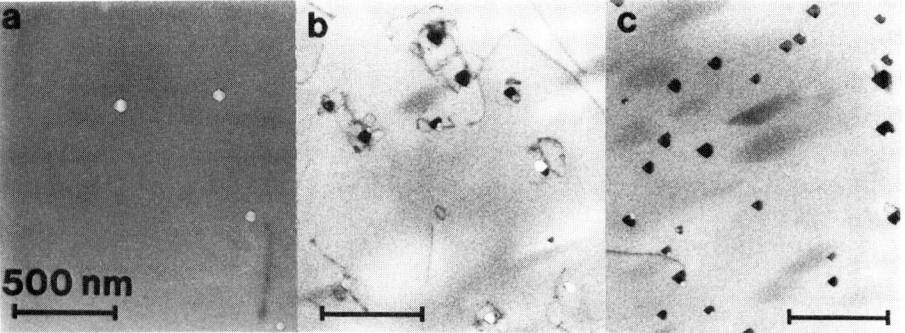

Fig. 3 Voids in the kink region (a), depletion of Ga-Zn precipitates and loop formation in the transition near the profile kink region (b) and Ga-Zn precipitates inside voids in the front region (c); As-rich condition at 1050 °C. Bright field TEM micrographs.

4. DISCUSSION AND CONCLUSIONS

It has been recognized earlier that the defect formation during Zn indiffusion at 906°C can be understood as result of supersaturations of Ga interstitials, Ga_i , generated at the diffusion front by Zn incorporation into the Ga sublattice via the kick-out mechanism (Luysberg et al. 1992). Independently of the conditions, the Ga_i supersaturation decays by

416

formation of interstitial dislocation loops and dislocation networks. Loop growth and climb of dislocations lead to emission of As vacancies, V_{As}, which condense to form voids with Ga precipitates. Since Ga is liquid at the diffusion temperatures a Ga-rich crystal region in contact with a Ga melt inside voids forms as result of the Zn diffusion. These defects are mediators in establishing local equilibria of the involved point defects. Our results confirm this view also for Zn diffusion at 1050 °C.

During further diffusion, incorporation of Zn into the Ga sublattice behind the diffusion front is accompanied by increasing dissolution of Zn into the Ga-precipitates. The extent to which Ga in the precipitate is replaced by Zn increases with both Zn_1 partial pressure and diffusion time and is further influenced by the presence of As in the ampoule. For comparison the Zn_1 and As_4 partial pressures as estimated following the scheme of Shih et al. (1968) are given in the captions of Fig. 1.

For As-poor conditions Zn-Ga precipitates are observed throughout the diffused zone, and the Zn concentration inside precipitates increases continuously towards the surface. The presence of Zn-Ga precipitates inside voids in regions with high Zn concentrations and its temporal evolution during diffusion indicates that the liquid in the void tends to attain a composition close to that of the respective melt for the diffusion source composition applied. Accordingly the local Ga_i concentrations will be affected, their values being kept closed to the equilibrium values by climb of dislocations. Indeed, box-shaped concentration profiles as observed under such conditions (Fig.1) have been modelled assuming local equilibrium for Ga interstitials (Tan et al. 1991).

For As-rich conditions Zn-Ga precipitates are only present in the front regions. The presence of void/precipitate pairs with enlarged void fractions in the transition regions of the profiles and the formation of voids (Fig.3) indicate that a different equilibrium is established which leads to the depletion of Ga precipitates. In accordance with this are the observations for As-rich conditions at 906 °C (Jäger et al. 1993). The profile shapes are similar to those observed at 1050°C (Fig.1), in the front region the maximum Zn concentration is only 2×10^{19}cm^{-3}, and no significant dissolution of Zn into the Ga-precipitates has been detected by EDX. However, at 906 °C voids are present in a substantial portion of the near-surface region. At both temperatures higher As_4 partial pressures in the ampoule (Fig. 1, captions) apparently make the substrate surface a more effective sink for Ga thus enhancing the Ga outdiffusion and creating GaAs crystal regions with compositions closer to an As-rich crystal in equilibrium with the vapour phase. As diffusion proceeds further incorporation of Zn atoms occurs until the solubility of the As-rich crystal is reached. The rate at which this situation is reached depends on the ratio of the Zn_1-to-As_4 partial pressures and hence leads to the two-step profile shapes typically observed for As-rich conditions.

REFERENCES

Casey H C 1973 Atomic Diffusion in Semiconductors Ch.7 (London:Plenum Press) 351
Hettwer H G, Lerch W, Lentfort B, Stolwijk N A and Mehrer H 1991 Appl. Surf. Sci. 50 470
Jäger W, Rucki A, Urban K, Hettwer H-G, Stolwijk N A , Mehrer H and Tan T Y 1993
  J. Appl. Phys. 74 4409
Luysberg M, Jäger W, Urban K, Schänzer M, Stolwijk N A and Mehrer H 1989 Mat.
  Res. Soc. Symp. 163 659 and 1992 Mat. Sci. Eng. B13 137
Shih K K, Allen J W and Pearson G L 1968 J. Phys. Chem. Solids 29 367
Tan T Y, Gösele U and Yu S 1991 Crit. Rev. Sol. St. Mat. Sci. 17 47

Inst. Phys. Conf. Ser. No 146
Paper presented at Microsc. Semicond. Mater. Conf., Oxford, 20–23 March 1995
© *1995 IOP Publishing Ltd*

417

Zn diffusion and defect generation in undoped and Fe-doped InP

D Wittorf[1], A Rucki[1], W Jäger[1], K Urban[1], H-G Hettwer[2], N A Stolwijk[2], H Mehrer[2], H Holzbrecher[3] and U Breuer[3]

[1]Institut für Festkörperforschung, Forschungszentrum Jülich, D-52425 Jülich, Germany
[2]Institut für Metallforschung, Universität Münster, D-48149 Münster, Germany
[3]Zentralinstitut für Chemische Analysen, Forschungszentrum Jülich, D-52425 Jülich, Germany

ABSTRACT: Zn-Diffusion into undoped and Fe-doped InP single crystals was carried out in closed quartz ampoules at 700 °C for 80 min using diffusion sources consisting of different amounts of Zn and P. Analytical transmission electron microscopy was used to investigate the diffusion-induced defects. The observations indicate that diffusion of Zn into InP involves a kick-out reaction. In Fe-doped InP substitutional Fe on In sublattice sites can be replaced by Zn leading to redistribution and precipitation of Fe.

1. INTRODUCTION

Zn diffusion into InP is applied for p-doping of optoelectronic components for communication technology. Controlled fabrication of devices by doping with high Zn concentrations requires a detailed understanding of Zn diffusion behaviour and defect generation in InP. It is widely accepted that diffusion of Zn in InP occurs at high temperatures via an interstitial-substitutional exchange mechanism involving intrinsic point defects. Substitutional Zn on In sublattice sites acts as a shallow acceptor, substitutional Fe acts as a deep acceptor. Two models were suggested to account for the exchange between indiffusing Zn and group III atoms in III-V compound semiconductors: A dissociative model is suggested for the Zn diffusion into InP involving In vacancies (Tuck 1988) and a kick-out model is suggested for the Zn diffusion into GaAs whereby interstitial Ga atoms are produced (Gösele and Morehead 1981). Previous investigations by transmission electron microscopy (TEM) of InP single crystals after Zn diffusion showed that dislocations and precipitates are formed in the Zn-diffused crystal region (Dixon et al. 1993). Zn profiles with a plateau region at the diffusion front were observed after Zn diffusion from a Zn-doped SiO₂ film into Fe-doped InP substrates at 520 and 600 °C (Bauer et al. 1991) indicating a redistribution of Fe.

In this contribution we report the dependence of the diffusion source composition on the Zn diffusion behaviour and compare the effects of Zn diffusion into homogenously Fe-doped InP with Zn diffusion into nominally undoped InP substrates. The diffusion-induced defects were investigated by analytical TEM of cross-sectional samples and compared with Zn concentration profiles measured by electron microprobe analysis (EMP) and secondary ion mass spectrometry (SIMS) on the same sample material.

418

2. EXPERIMENTAL TECHNIQUES

Zn was diffused into nominally undoped and semi-insulating Fe-doped InP single crystals ($C_{Fe} = 1.2 \times 10^{17}\,cm^{-3}$) with dislocation etch pit densities lower than $10^4\,cm^{-2}$ (substrate orientation is < 111 > for undoped and < 100 > for Fe-doped InP). Diffusion anneals were carried out in closed Ar-flushed quartz ampoules (ampoule volume $\approx$ 8-9 cm^3) at 700 °C for 80 min.

Diffusion sources of pure Zn or with compositions of Zn and P in weight ratios of 5 : 1 (Zn^5P^1) or of 1 : 1 (Zn^1P^1) with total amounts ranging from 3 to 7 milligrams of Zn were used. Total amounts of Zn in the diffusion sources are much higher than those which are dissolved in the InP crystal after diffusion. Subsequent to the diffusion anneal, the ampoules were quenched in water to room temperature. Depth profiles of the total Zn concentration, $C_{Zn}(x)$, were measured by SIMS and by EMP (Hettwer et al. 1991). SIMS analyses used CsZn$^+$ secondary ions in a scanning mode (area 200 × 200 μm) providing two-dimensional plots of the Zn distribution. Depth profiles were obtained by integration of the signal across a window (lateral extension 10 μm, depth extension 3 μm). EMP used Zn-K$_\alpha$ X-rays having a detection limit of about $5 \times 10^{18}\,cm^{-3}$, as is indicated by the scattering of our data. SIMS measurements have a much lower detection limit of about $5 \times 10^{16}\,cm^{-3}$.

The defects were studied in cross-sectional samples with a JEOL 4000FX operating at 200 and 400 kV. TEM samples were thinned by iodine-assisted argon ion milling at 2.5 keV and 0.15 mA. Compositions of precipitates were determined by energy-dispersive X-ray spectroscopy (EDX) by analysing Zn-K$_\alpha$, P-K$_\alpha$, In-L$_\alpha$, and Fe-K$_\alpha$ lines.

3. RESULTS

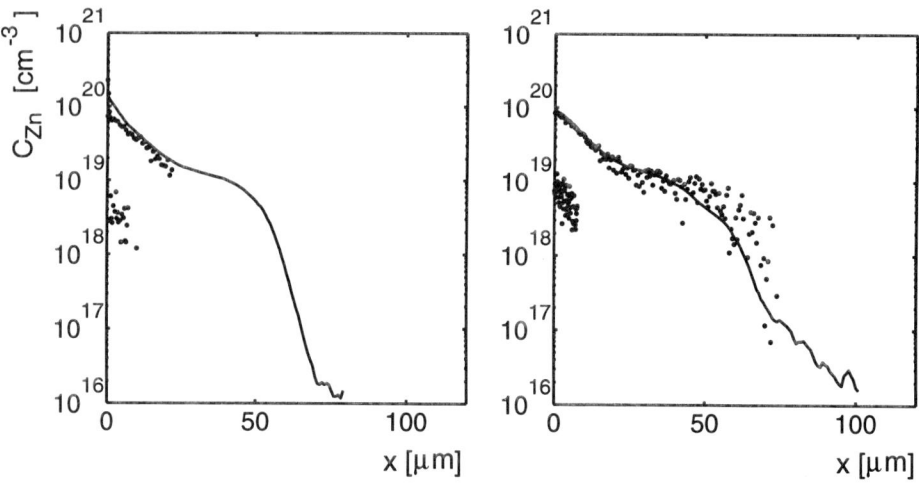

Fig. 1 : Zn concentration profiles measured by EMP (dots) and by SIMS (solid lines) after diffusion at 700 °C for 80 min into undoped InP (left) and into Fe-doped InP (right). Shown is the dependence on the diffusion source composition: Zn^5P^1 sources with Zn concentrations at the surface of $C_{Zn}^S \approx 10^{20}\,cm^{-3}$, Zn^1P^1 sources with $C_{Zn}^S \approx 10^{19}\,cm^{-3}$.

Figure 1 shows Zn concentration profiles after diffusion at 700 °C for 80 min into undoped and Fe-doped InP using Zn^5P^1 or Zn^1P^1 diffusion sources. For both crystals high Zn concentrations at the surfaces of about $C_{Zn}^S = 1 \times 10^{20}\,cm^{-3}$ result from the use of Zn^5P^1 sources. After a steep decrease in the near-surface regions both profiles show a plateau region at depths between 20 and 50 μm. For diffusion into Fe-doped InP substrates the penetration depth of Zn is higher and the profile does not drop down at the diffusion front as steeply as in the case of undoped InP substrates. Zn concentrations determined by EMP correspond well with those determined by SIMS above the EMP detection limit, particularly for Fe-doped InP substrates. The small deviation of the Zn concentration at the surface between both measurements on undoped InP can be attributed to laterally inhomogenous Zn concentrations of diffused samples.

Diffusion with Zn^1P^1 sources of higher P weight fraction leads to lower Zn concentrations at the surface ($C^S_{Zn} \approx 10^{19}\,cm^{-3}$) and much lower penetration depths (Fig. 1). Using this source for longer diffusion times (900 min, not shown here) leads to a higher penetration depth of Zn in Fe-doped as compared to undoped InP. For Fe-doped InP this is connected with a

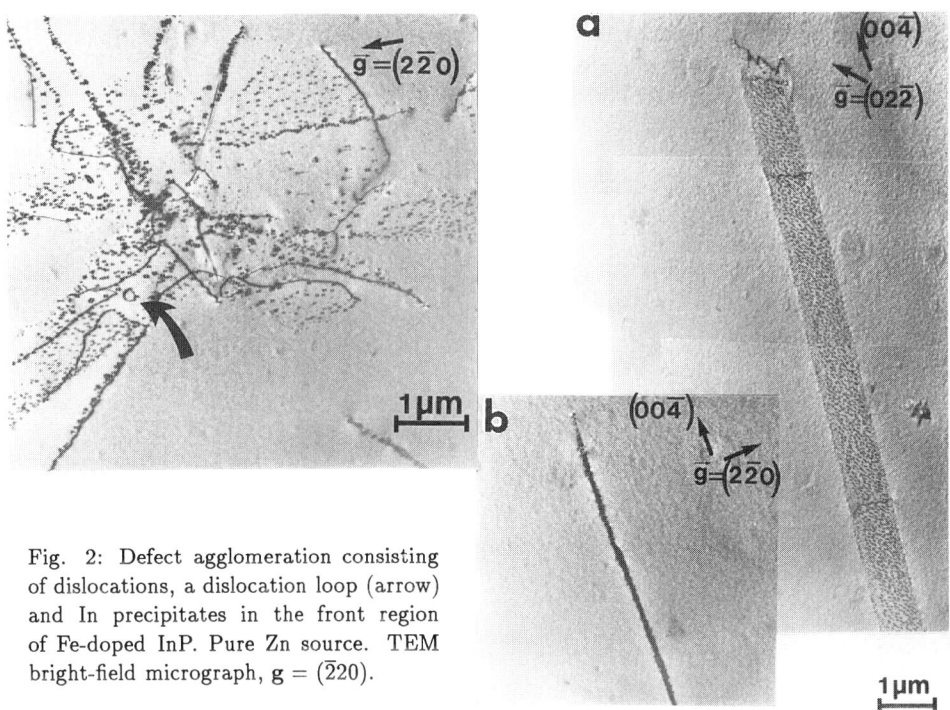

Fig. 2: Defect agglomeration consisting of dislocations, a dislocation loop (arrow) and In precipitates in the front region of Fe-doped InP. Pure Zn source. TEM bright-field micrograph, $\mathbf{g} = (\bar{2}20)$.

Fig. 3: Planar arrangement of precipitates and dislocations on a $(1\bar{1}0)$ plane (a) in projection (b) shown egde-on. Fe-doped InP, pure Zn source.

Zn concentration plateau ($C_{Zn} \approx 10^{17}\,cm^{-3}$) extending to a depth of about 80 μm, as compared to a penetration depth of 40 μm for undoped InP. The level of this Zn concentration plateau corresponds to that of the Fe concentration before diffusion.

Zn diffusion with strong Zn sources (Zn, Zn^5P^1) leads, for both undoped and Fe-doped InP substrates, to defect formation in the diffused crystal regions, except for a near-surface region extending to a depth of about 10 μm. The defects observed are dislocation loops, dislocations and In precipitates inside voids. Typical of the diffusion-induced defect structure is the

420

inhomogenous spatial arrangement of defects in the form of agglomerations and large planar arrays of precipitates.

In Fig. 2 an agglomeration of defects in the front region is depicted. It consists of dislocations, a dislocation loop and In precipitates inside voids which are depicted as black spot contrast at this magnification. From the center of such agglomerations large planar arrays of precipitates emanate which are bounded by dislocations. These planar arrays are observed frequently on {110} planes. Figure 3(a) shows such a planar array of precipitates on the (1$\bar{1}$0) plane in projection. In Fig. 3(b) this array is shown edge-on indicating the planar nature of such precipitate arrangements. Contrast analyses of dislocation loops were performed in

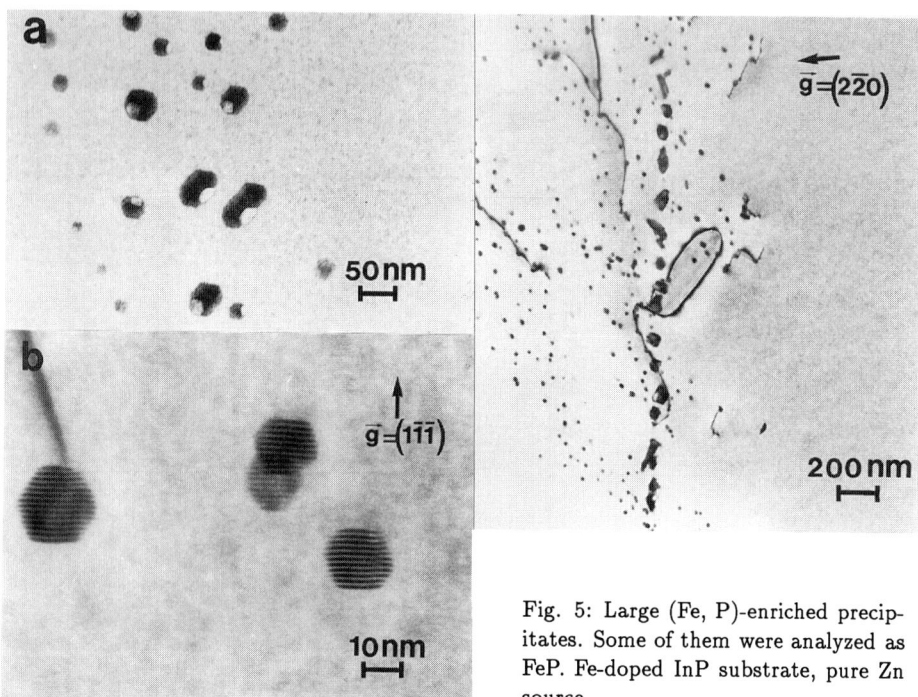

Fig. 5: Large (Fe, P)-enriched precipitates. Some of them were analyzed as FeP. Fe-doped InP substrate, pure Zn source.

Fig. 4: (a) Facetted In precipitates inside voids. Kinematical imaging conditions. (b) Moiré contrast of In precipitates. This contrast is only obtained with imaging vectors g = {111}. Undoped InP substrate, Zn^{5}P^1 source.

different depths by means of the "inside-outside" contrast method (Föll and Wilkens 1975). All loops analyzed are of interstitial-type. Most of the dislocation loops are perfect with {110} habit planes and Burgers vectors of the $\frac{a}{2}$ < 110 > − type. Only few of the analyzed dislocation loops have {111} habit planes and $\frac{a}{3}$ < 111 > − Burgers vectors. These loops show stacking fault contrast. These results indicate that the loops are composed of layers containing In and P atoms at the stoichiometry of the crystal.

The facetted shapes and the crystalline nature of In precipitates inside voids is shown in Fig. 4. The void fraction of the In precipitates is shown in Fig. 4(a) under kinematical imaging conditions. The composition of the precipitates was analyzed by EDX and the structure by selected area diffraction and bright-field imaging. Fig. 4(b) shows moiré fringe contrast of precipitates which indicates their crystalline nature. The EDX analyses yield in all cases considerable enrichment of In in precipitates. These results are compatible with the existence of crystalline In precipitates inside voids without a detectable Zn fraction.

Figure 5 shows a row of large precipitates in spatial arrangement with small In precipitates, a dislocation loop and dislocations in the case of Fe-doped InP and a pure Zn source. The composition of these large precipitates is determined by EDX to be enriched both in Fe and in P. Selected area diffraction patterns obtained from areas with such precipitates are compatible with the presence of FeP.

No defects are present in the whole bulk crystal after diffusion from Zn^1P^1 sources. For long diffusion times (900 min) large (Zn, P)-rich surface precipitates are observed by scanning electron microscopy and also by EDX analysis in TEM investigations of cross-sectional samples. The EDX measurements indicate the formation of crystalline Zn_3P_2 surface precipitates.

4. DISCUSSION AND CONCLUSIONS

From the results presented we conclude that Zn concentration profiles (Fig. 1) are strongly influenced by the composition of the diffusion source governing the generation of extended defects.

Under diffusion conditions imposed by strong Zn sources (Zn, Zn^5P^1) defects are generated during Zn indiffusion in the InP bulk crystal. The formation of dislocation loops and In precipitates inside voids during Zn indiffusion can be described by a model which is based on the kick-out mechanism (Wittorf et al. 1995). Interstitially indiffusing Zn atoms substitute In matrix atoms in a kick-out reaction. This can lead to an increase of the In interstitial concentration and a supersaturation of In interstitials.

As a consequence, interstitial dislocation loops are formed whereby P interstitials are provided by the emission of P vacancies during loop growth which themselves aggregate and form voids containing In. Earlier TEM studies of Zn diffusion-induced defects in GaAs have led to a similar model for the generation of interstitial dislocation loops and Ga precipitates inside voids (Luysberg et al. 1989, Jäger et al. 1993).

From the observation of planar arrangements of precipitates inside voids on {110} planes surrounded by dislocations we conclude that growth of dislocation loops and climb of dislocation segments occurs very effectively during diffusion thus leading to the formation of these arrays. This is also indicated by the observation of further dislocation loops inside such planar arrangements of precipitates (Wittorf et al. 1995). Generally in III-V compound semiconductors point defects of both sublattices are involved in dislocation climb processes, thus allowing the crystal to reach a local point defect equilibrium (Petroff and Kimmerling 1976, Luysberg et al. 1989).

The plateau regions observed for Zn^5P^1 sources (Fig. 1, $20\,\mu m \leq x \leq 50\,\mu m$) probably have their origin in the reduction of In self-interstitial concentrations by the presence of defects. The incorporation of Zn on In sublattice sites via the kick-out mechanism increases the In interstitial concentration. An annhilation of In interstitials at extended defects acting as sinks would therefore increase the number of Zn atoms becoming incorporated on substitutional sites, thus leading to the observed plateau region.

No defects were present in the near-surface region ($x \leq 10\,\mu m$) under diffusion conditions with Zn or Zn^5P^1 sources and in the whole crystal using Zn^1P^1 diffusion sources. The surface obviously acts as effective sink for In interstitials and therefore a supersaturation leading to defect generation does not build up in the near-surface region. Under diffusion conditions with sufficient P vapour pressure (Zn^1P^1 source) the formation of a Zn_3P_2 phase is observed (section 3). The formation of Zn-rich phases obviously reduces the Zn partial pressure in the quartz ampoule thus leading to lower Zn concentrations at the surface and therefore to In interstitial concentrations in the bulk crystal which are below the critical value for defect formation.

Effects of substrate doping with Fe are seen in the Zn diffusion-induced precipitation of Fe (Fig. 5) and in higher penetration depths in comparison to undoped substrates under otherwise identical diffusion conditions (section 3). Obviously during Zn diffusion, redistribution of Fe atoms leads to the formation of Fe-enriched precipitates and to an accelerated Zn incorporation into the crystal lattice. We suggest as a likely explanation for these phenomena a kick-out reaction between indiffusing interstitial Zn and substitutional Fe atoms on In sublattice sites.

In conclusion, our results show that defect formation is induced by diffusion sources consisting of pure Zn and by sources consisting of Zn and small amounts of P. The defect formation occurs both in undoped and in Fe-doped InP single crystals. It can be understood by indiffusion of interstitial Zn followed by an interstitial-substitutional exchange via a kick-out reaction which leads to a supersaturation of the In interstitial concentration. As a consequence, dislocation loops and In precipitates inside voids are formed. For diffusion sources consisting of equal weight fractions of Zn and P the diffused crystal regions remain free of defects. In this case (Zn, P)-rich surface precipitates form reducing obviously the Zn partial pressure in the quartz ampoule to values below the critical value for defect formation. The formation of FeP precipitates and the observation of higher Zn penetration depths in Fe-doped substrates provide evidence that interstitial-substitutional exchange of Zn can also be indicated by a kick-out reaction which involves Fe on In sublattice sites instead of In.

ACKNOWLEDGEMENTS

Helpful discussions with Prof. Dr. U. Gösele (MPI Halle, Germany) and Prof. Dr. T.Y. Tan (Duke University Durham, U.S.A.) are gratefully acknowledged.

REFERENCES

Bauer J G, Treichler R, Hillmer T, Müller J and Ebbinghaus G 1991
  Appl. Surf. Sci. 50 138
Dixon R-H, Jäger W, Rucki A, Urban K, Hettwer H-G, Stolwijk N and Mehrer H 1993
  Inst. Phys. Conf. Ser. No. 134 539
Föll H and Wilkens M 1975 Phys. Status Solidi (A) 31 519
Gösele U and Morehead F 1981 J. Appl. Phys. 52 4617
Hettwer H-G, Lerch W, Lentfort B, Stolwijk N A and Mehrer H 1989
  Res. Soc. Symp. 163 659
Jäger W, Rucki A, Urban K, Hettwer H-G, Stolwijk N A, Mehrer H and Tan T Y 1993
  J. Appl. Phys. 74 4409
Luysberg M, Jäger W, Urban K, Schänzer M, Stolwijk N A and Mehrer H 1989
  Mat. Res. Soc. Symp. 163 659 and 1992 Mater. Sci. Eng. B13 137
Petroff P M, Kimmerling L C 1976 Appl. Phys. Lett. 29 461
Tuck B 1988 Atomic Diffusion in III-V Semiconductors (Bristol: Adam Hilger)
Wittorf D, Rucki A, Jäger W, Dixon R H, Urban K, Hettwer H-G, Stolwijk N A and
  Mehrer H 1995 J. Appl. Phys. 77 2843

Inst. Phys. Conf. Ser. No 146
Paper presented at Microsc. Semicond. Mater. Conf., Oxford, 20–23 March 1995
© 1995 IOP Publishing Ltd

Zn_3P_2 precipitates in dislocation-free LEC InP heavily doped with Zn

C Frigeri*, R Fornari*, J L Weyher°, G M Guadalupi^ and F Longo*

*CNR-MASPEC Institute, via Chiavari 18/A - 43100 Parma (Italy)
°IFF, Forschungszentrum Jülich - 52428 Jülich (Germany)
^TEMAV, via delle Industrie 39 - 30175 Porto Marghera (Italy)

ABSTRACT: Zn_3P_2 precipitates are identified by SAD in dislocation-free LEC InP heavily doped to the saturation hole density of $3.0 \cdot 10^{18}$ cm^{-3} by adding $1 \cdot 10^{19}$ Zn at/cm^3. The calculated density of Zn atoms forming the precipitates shows that precipitation is not the sole mechanism responsible for the saturation of the hole density. The features of the transition from dislocation-free to dislocated crystal observed by slightly decreasing the doping level are related to the LEC growth conditions.

1. INTRODUCTION

A method for achieving dislocation-free bulk InP is given by the solution hardening, i. e. by heavy doping with specific dopants so as to harden the host lattice against generation and propagation of dislocations. Heavy doping has, however, several drawbacks related to the solubility limit of the dopant and the destination of the excess dopant atoms. Among the dopants, Zn is a very effective p-type impurity to obtain dislocation-free InP. The behaviour of Zn in bulk InP has been widely studied by diffusing Zn into InP substrates from external sources (Kin Man Yu et at 1993, Dixon et al 1993, Chan et al 1991, Hooper et al 1974). It has been reported that high doping of Zn-diffused InP produces precipitates, generally of the Zn_3P_2 type (Chan et al 1991, Eger et al 1987). Less attention has been paid to Zn-doped InP grown from LEC melts, especially as regards the identification of the precipitates present for hole concentration p >~3-5·10^{18} cm^{-3}, which is the saturation hole density that cannot be overcome whatever the amount of Zn added to InP (Kin Man Yu et at 1993, Chan et al 1991). A TEM study of the precipitates in LEC InP heavily doped with Zn is presented here.

2. EXPERIMENTAL

The two (100) LEC InP crystals examined had a hole density p of $3.0 \cdot 10^{18}$ and $2.6 \cdot 10^{18}$ cm^{-3}, respectively, obtained by doping with elemental Zn. The samples were investigated by TEM, DSL photoetching (Weyher and Van de Ven 1986) and BCA etching, which is based on the HBr-$K_2Cr_2O_7$-H_2O system (Weyher et al 1994). TEM observations in the two beam diffraction contrast mode were performed on plan view specimens thinned to electron transparency by bombardment with Ar ion beams at liquid nitrogen temperature.

3. RESULTS AND DISCUSSION

Etching by BCA showed that all the wafer surface of the most heavily doped sample (p = $3.0 \cdot 10^{18}$ cm^{-3}) exhibits a very high density of homogeneously distributed shallow pits (S-pits) (fig. 1 a). The S-pits have no crystallographic symmetry and are thus due to microdefects (Weyher et al 1994). No dislocation-related etch pit was detected by either

424

BCA or DSL photoetching. TEM confirms that the crystal contains precipitates (fig. 1 b) in a density of $\sim 7 \cdot 10^9$ cm^{-3}. The size of the precipitates is $\sim 170 \pm 30$ nm. The crystal turned out to be dislocation-free even by TEM. The absence of dislocations can be ascribed to the strengthening of the InP matrix against the generation of dislocations because of the very high density of precipitates.

Fig. 2 is the typical selected area diffraction pattern from a precipitate. The precipitates are identified as Zn_3P_2 since the experimental lattice spacings fit several lattice spacings of Zn_3P_2 as given by Pearson (1967) (Table I). Zn_3P_2 belongs to the tetragonal system, space group $P4_2/mmc$, and has lattice constants a = 0.8113 nm and c = 1.147 nm (Pearson 1967).

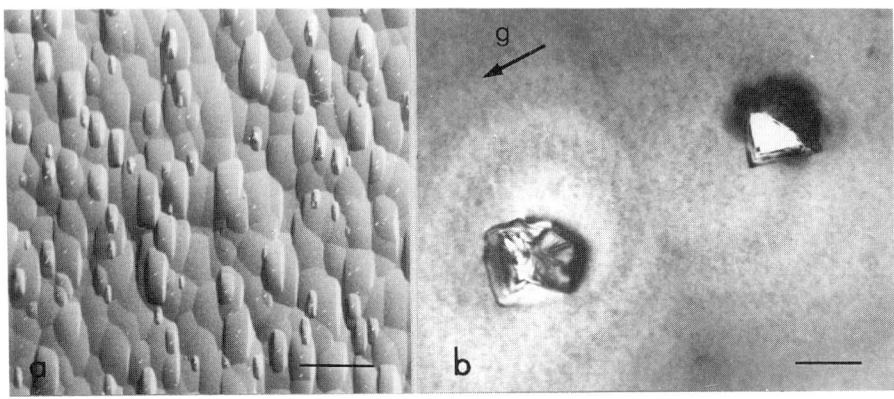

Fig. 1. InP crystal doped to p = $3.0 \cdot 10^{18}$ cm^{-3}. a) Optical microscopy image of a wafer submitted to BCA etching. Bar = 25 μm. b) TEM micrograph of two large precipitates, **g** = [022], bar = 0.2 μm.

TABLE I - Comparison between the experimental lattice spacings of the precipitates and some lattice spacings of Zn_3P_2.

d_{exp} (nm)	d_{Zn3P2} (nm)	hkl
0.3464	0.3463	103
0.3335	0.3325	202
0.2810	0.2868	004
0.2633	0.2648	301
0.2603	0.2582	310
0.2088	0.2041	400
0.1981	0.1980	140
0.1805	0.1825	240

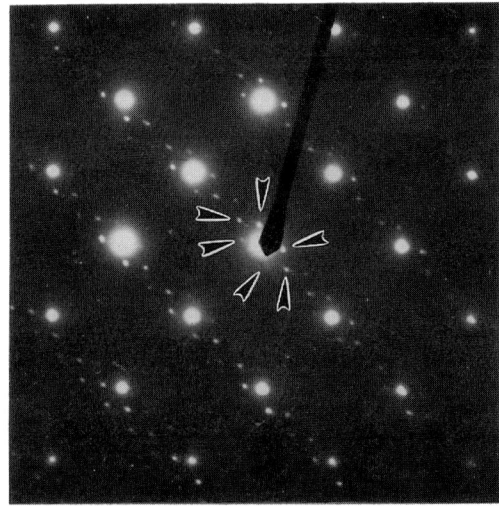

Fig. 2. Electron diffraction pattern from a precipitate like those shown in fig. 1 b).

Several diffraction spots of fig. 2 do not belong to the diffraction pattern of Zn_3P_2 or InP but are due to double diffraction, i. e. diffraction by the precipitate of the beams already diffracted by the InP matrix, like the arrowed ones around the primary beam. The bigger spots with (100) zone axis symmetry are due to the InP matrix. No clear orientation relationship between the Zn_3P_2 precipitates and the InP matrix could be detected, which is contrary to the case of Zn-diffused InP where the Zn_3P_2 precipitates often have the (220) planes parallel to the (200) planes of InP with the c axis parallel to the InP [100] direction (Nakahara et al 1984, Eger et al 1987). Such orientation relationships are due to the fact that in the case of diffusion experiments Zn_3P_2 has a strong tendency to grow epitaxially on the substrate. Such orientation constriction is not necessarily effective for precipitation during growth from the melt. Even Schlossmacher et at (1992) did not report on any orientation relationship between the As_2Zn_3 and As_2Zn precipitates they found in LEC GaAs and the matrix. The existence of Zn_3P_2 as a stable compound in the phase diagram of the ternary In-Zn-P system has been shown by several authors (Hooper et al 1974, Eger et al 1987). The Zn_3P_2 precipitates have a quite irregular shape and for nearly 90% of them the precipitate volume is partially empty.

The density of Zn atoms forming the Zn_3P_2 precipitates can be calculated like this. The number of molecules, n, in the Zn_3P_2 unit cell is given by n=DNV/M where D is the density of the Zn_3P_2 cell, N Avogadro's number, V the cell volume and M the molecular weight (MacLachlan 1957). By using D = 4.5 cm^{-3} (Pearson 1967), n = 7.93 (approximated to 8), i. e. there are 23.8 (approximated to 24) Zn atoms per unit cell or $3.15 \cdot 10^{22}$ Zn atoms/cm^3 in the cell. The density of the zinc atoms in the precipitates, [Zn], is then given by [Zn] = $3n\rho V_p/V$, where ρ and V_p are the average density and volume of the precipitates, respectively. By assuming that the precipitates have on the average a cubic shape and are half filled, one gets [Zn] = $5.4 \cdot 10^{17}$ cm^{-3} with an error of at least 50-60 % (mostly due to uncertainty in the precipitate shape and in ρ). As the amount of Zn contained in the more heavily doped crystal is 500 ppm, as measured by glow discharge mass spectroscopy, corresponding to $\sim 10^{19}$ at/cm^3, [Zn] is only 5.5 % of the total Zn concentration in the crystal. This indicates that precipitation is not the sole mechanism responsible for the saturation of the free holes to $3 \cdot 10^{18}$ cm^{-3}. Three other possible mechanisms can account for such saturation (Chan et al 1991), namely: a) compensation of Zn acceptors by donor-like native defects, b) compensation of Zn acceptors by Zn interstitials which are known to act as donors and c) formation of neutral complexes between Zn atoms and P vacancies. By our techniques it is not possible to establish which one of these mechanisms is more effective.

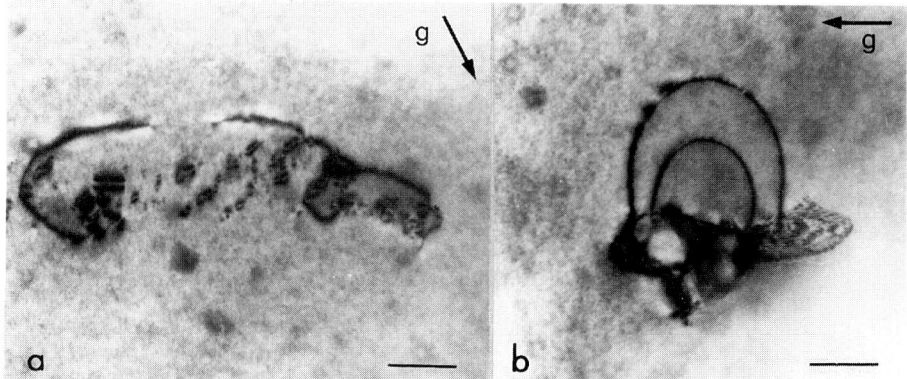

Fig. 3. TEM micrographs of typical defects in the rim region of the InP crystal doped to p = $2.6 \cdot 10^{18}$ cm^{-3}. a) Dislocation with associated microdefects and b) dislocations punched out from a Zn_3P_2 precipitate. $\mathbf{g}$ = [022], bar = 0.2 µm.

426

The low value of [Zn] also indicates that the formation of the Zn_3P_2 precipitates cannot be caused by exceeding the solubility limit of Zn in InP, which is significantly higher than the saturation hole density (Chan al al 1991). A similar conclusion was also drawn in the case of Zn-As phase precipitates in LEC GaAs (Schlossmacher et al 1992). It is likely that the formation of the Zn_3P_2 precipitates is governed by the supersaturation of P interstitials rather than of Zn interstitials. This hypothesis can be supported by the theoretical finding of Chan et al (1991) according to which the density of P interstitials increases with increasing p-type doping level.

When p is decreased to $2.6 \cdot 10^{18}$ cm^{-3} both chemical etching and TEM show that in the rim part of the InP wafers the Zn_3P_2 precipitates almost disappear whereas dislocations start to appear (fig. 3). Some dislocations have been seen to punch out from the few Zn_3P_2 precipitates remained in the rim region (fig. 3 b). Only the centre part of the wafers is dislocation-free but still contains Zn_3P_2 precipitates in a density as high as $2 \cdot 10^9$ cm^{-3}. The strong reduction of the density of the Zn_3P_2 precipitates in the rim part of the wafers has to be ascribed to the reduction of both Zn and P interstitials, as a consequence of the reduced doping level, as well as to a temperature effect. In fact, beside a sufficient high density of Zn and P interstitials, the formation of Zn_3P_2 precipitates also requires that such interstitials are mobile enough to cluster together in a sufficient number for precipitation, so that the temperature might be a key parameter when the density of available interstitials decreases. The absence of precipitates in the wafer rim could also be due, therefore, to the fact that in this part of the crystal the interstitials are less mobile and have not enough time to cluster together because the rim part of the crystal cools faster than the centre part which, instead, remains at higher temperature for a longer time.

Because of the reduced density of the Zn_3P_2 precipitates, their strengthening effect is also expected to be significantly reduced so that dislocations are more easily introduced, especially in the rim regions where the thermal stresses are higher, as detected. A quite large number of microdefects have always been seen to be associated with all dislocations (fig. 3). They are either precipitates or dislocation loops, as shown by TEM, and have been generated by the climbing of the associated dislocation. The formation mechanism of these microdefects is discussed elsewhere (Frigeri et al 1994).

ACKNOWLEDGEMENTS

Thanks are due to Mrs G Mignoni for skilfull technical help. Work supported by PN-MIA (OPTEL).

REFERENCES

Chan L Y, Kin Man Yu, Ben-Tzur M, Haller E E, Jaklevic J M, Walukiewicz W and Hanson C M 1991 J. Appl. Phys. 69 2998

Dixon R H, Jäger W, Rucki A, Urban K, Hettwer H-G, Stolwijk N A and Mehrer H. 1993 Inst. Phys. Conf. Ser. 134 539

Eger D, Springthorpe A J, Margittai A, Shepherd F R, Bruce R A and Smith G M 1987 J. Electron. Mater. 16 163

Frigeri C, Ferrari R, Fornari R, Weyher J L, Longo F and Guadalupi G M 1994 Mater. Sci. Eng. B28 120

Hooper A, Tuck B and Baker A J 1974 Solid-State Electron. 17 531

Kin Man Yu, Walukiewicz W, Chan L Y, Leon R, Haller E E, Jaklevic J M and Hanson C M 1993 J. Appl. Phys. 74 345

MacLachlan D, Jr 1957 X-Ray Crystal Structure (New York: McGraw-Hill) ch. 4

Nakahara S, Gallagher P K, Felder E C and Lawry R B 1984 Solid-State Electron. 27 557

Pearson W B 1967 A Handbook of Lattice Spacings and Structures of Metals and Alloys (Oxford: Pergamon) vol 2

Schlossmacher P, Otte M, Urban K and Rüfer H 1992 J. Crystal Growth 121 671

Weyher J L, Fornari R, Görög T, Kelly J J and Erné B 1994 J. Crystal Growth 141 57

Weyher J L and Van de Ven P 1986 J. Crystal Growth 78 191

Inst. Phys. Conf. Ser. No 146
Paper presented at Microsc. Semicond. Mater. Conf., Oxford, 20–23 March 1995
© 1995 IOP Publishing Ltd

Enhancement of secondary electron emission induced by defects in Si-implanted GaAs

F Iwase, T Matsuda, H Maruya and K Sekine

Yokohama R & D Laboratories, The Furukawa Electric Co Ltd, 2-4-3 Okano, Nishi-ku Yokohama 220, Japan

ABSTRACT: Si-implanted GaAs wafers were observed by scanning electron microscopy. In the case of isothermal annealing at 750°C and 950°C for 30 min, the secondary electron emission was enhanced for the samples implanted at 150keV with dose equal to $3 \times 10^{14} \text{cm}^{-2}$. From the assessment of the defects in these samples by transmission electron microscopy, secondary ion mass spectrometry and photoluminescence measurements, it is proposed that the enhancement may be associated with the presence of Si_{As} and $(V_{Ga} + Si_{Ga})$.

1. INTRODUCTION

Scanning electron microscopy (SEM) is usually used to determine surface morphology, sometimes to obtain electrical information in semiconductors, eg. the EBIC technique (Mishima 1988) which shows induced current in semiconductors by primary electron irradiation.

On the other hand, the defects in semiconductors have important roles in electrical activation, especially in ion-implanted GaAs. Ion implantation of Si into GaAs substrates is a common technique to prepare the n-type channel in GaAs devices. This technique needs two processes, ion implantation and annealing. Through these processes, complicated reactions occur between defects and these are often elucidated by transmission electron microscopy (TEM) and photoluminescence (PL) measurements. The former technique is powerful in microscopic analysis (Hull 1993), though it is unable to assess point defects. The latter technique is a useful method to assess point defects in semiconductors, although their microscopic distribution is not revealed (Bindal 1989). The positron annihilation method is effective to detect point defects and their distribution (Simpson 1992) but is not straight forward to apply.

In our present work (Iwase 1994), we have found that the presence of Si_{As} and another defect is likely to reduce the activation efficiency in Si-implanted GaAs and enhance the secondary electron emission. In the present work, we have assessed the details of the defects by TEM, secondary ion mass spectrometry (SIMS) and PL methods, and studied the relationship between the enhancement of the secondary electron emission and the defects present.

2. EXPERIMENTAL TECHNIQUES

2.1 Samples

Undoped semi-insulating, liquid encapsulated Czochralski (LEC)-grown (100) GaAs wafers and Si-doped LEC GaAs (100) wafers (n-doped GaAs) were used. Si^+ ions were implanted at 150keV with a dose equal to $3 \times 10^{14} \text{cm}^{-2}$: the Si concentration and the carrier concentration of the n-doped GaAs wafer were $1.5 \times 10^{18} \text{cm}^{-3}$ and $2 \times 10^{17} \text{cm}^{-3}$, respectively. The samples were coated with $0.1 \mu m$ SiO_2 and annealed at between 750°C and 950°C for 30 min in an N_2 ambient. To assess the influence of the presence of Si on the electron emission and to eliminate damage from ion-implantation, we also used undoped semi-insulating GaAs wafers. All samples were cleaved in air to observe cross-sectional SEM and TEM images.

2.2 HR-SEM Observation and PL Measurement

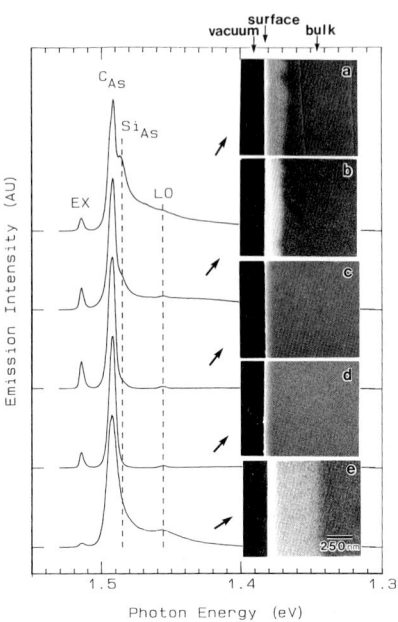

Fig. 1. Cross-sectional SEM micrographs and PL spectra of the samples with the doses equal to $3 \times 10^{14} \mathrm{cm}^{-2}$ annealed at (a) 750, (b) 800, (c) 850, (d) 900 and (e) 950°C for 30min.

SEM observations were carried out with an ultrahigh resolution SEM (UHR-SEM; Hitachi S-900) that has a field emission electron gun and an electromagnetic lens in which samples are set. The accelerating voltage was equal to 3keV with incident angle of 90°. We also measured PL spectra on the samples at 4K using the 514.5nm line of an Ar^+ laser. The PL spectra were obtained from the front surface. Before annealing, SEM images showed slight contrast between the implanted and non-implanted regions. Figure 1 shows the SEM micrographs and PL spectra. The results were obtained under the same experimental conditions, and SI substrates were always used for comparison. The samples that show the enhancement of the secondary electron emission (a,b,e) also have the shoulder on the lower energy side of the C_{As} peak. The PL intensity is normalized to the C_{As} peak in order to compare the intensity of the shoulder. The shoulder originated from Si_{As} (Ashen 1975). The undoped semi-insulating sample that included C_{As} annealed at 750 and 950°C for 30 min did not show the enhancement of the electron emission, though an n-doped sample annealed at 750 and 950°c for 30 min which included Si_{As} showed the emission enhancement.

It may be thought that the contrast could be a form of Voltage Contrast. The surface regions of the Si-implanted samples are semi-conductors after annealing though the substrates are always semi-insulating. If the potential is higher in the surface region than in the bulk region (in the case of $\delta < 1$: δ is the secondary yield), the secondary electrons escaping from the bulk region have a larger energy compared with that of the surface region. In the present case, we used an in-lens HR-SEM, so that the electrons with smaller energy would be easily detected. We also carried out another experiment using C-implanted and annealed samples. For these samples, the surface regions are semi-conductors, although they did not show the emission enhancement. If the enhancement is due to Voltage Contrast, all the samples would show it in the surface region. However, some samples

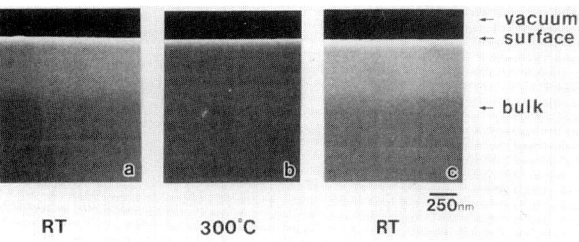

Fig. 2. The temperature dependency of SEM micrographs on the sample was doses equal to $3 \times 10^{14} \mathrm{cm}^{-2}$ annealed at 950°C for 30min. Observation temperature is (a) RT, (b) 300°C and (c) RT (after cooling).

including those with Si_{As} showed the enhancement while other samples without Si_{As} did not show the enhancement. From these considerations, we conclude that the enhancement may have been associated with the presence of Si_{As}, while C_{As} possibly plays an additional role.

Next, we observed the SEM images of the ion-implanted sample annealed at 950°C for 30 min during heating between room-temperature (RT) and 300°C. Figure 2 shows the SEM micrographs.

The contrast between the surface region and the bulk region disappears at 300°C. The SEM images are the same before and after annealing From these experiments, we concluded that the enhancement may have been associated with the presence of Si_{As} and the defects that cause the formation of deep levels.

2.3 TEM Observation and SIMS Measurement

We made TEM (Hitachi H-9000) observations to assess the defects resulting from the ion-implantation and annealing processes and SIMS (ATOMIKA ADIDA-3000) measurements to determine the distribution of Si. Figure 3 shows the cross-sectional TEM micrographs and Si distribution profiles. The ion-implanted sample (without annealing) shows dislocation loops, <10nm in diameter, in the region extending from the surface down to 250nm. The 750°C annealed sample shows voids in the surface region (-150nm) and dislocation loops in the deeper region (-250nm). The 850°C annealed sample also shows voids and dislocation loops. The size of the dislocation loops is larger, though the voids are similar to those in the 750°C sample. The Si distribution profile of the 850°C sample differs from the as-implanted sample and the 750°C sample. No voids are seen in the surface region on the 950°C annealed sample, although dislocations are present. Si has diffused significantly from point B to the surface region and the bulk region.

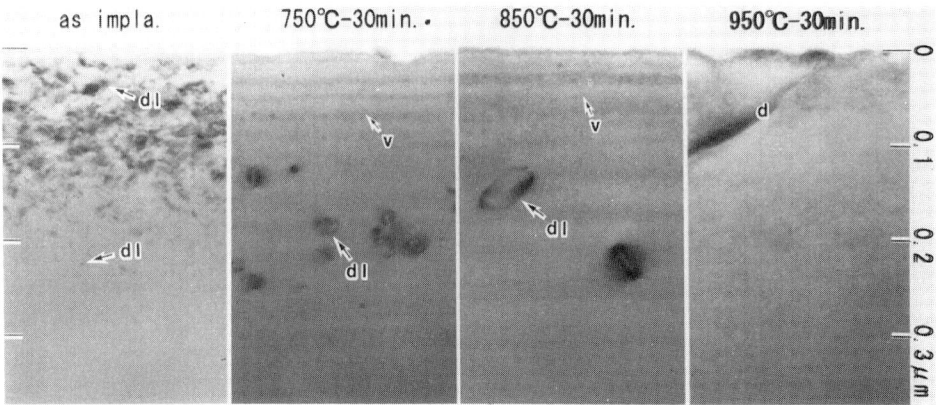

Fig. 3. Cross-sectional TEM micrographs of the samples with doses equal to $3x10^{14}cm^{-2}$. The samples are as-implanted and annealed at 750, 850, and 950°C.

3. RESULTS AND DISCUSSION

Comparing SEM images (Fig. 1) with TEM images (Fig. 3), it is not clear that the enhancement of the secondary electron emission does not necessarily occur in the region where the voids or dislocation loops exist. The ion-implanted sample is thought to include V_{Ga}, V_{As}, Ga_i (due to recoil processes), As_i, Si_i and some Si_{Ga} and Si_{As}. For the 750°C sample, the vacancies collect together to form voids in the surface region (150nm) and the dislocation loops increase in size to 10-20nm diameter in the deeper region. In this deeper region (from point A to point C), it is likely that Si_i in the dislocation loops are released by annealing and react with V_{Ga} that diffuse faster than V_{As} (Chiang 1975) and make Si_{Ga}, which increases carrier concentration (activation efficiency; 4%). The reaction between Si_i and V_{As} is also thought to occur. We think that the former reaction occurs more prominently in the 800°C sample in the deeper region. The enhancement region decreases compared with the 750°C sample and the activation efficiency increases (7%). The Si impurity of the 800°C sample diffused a little (compared with the 750°C sample) though not so much as for the 850°C sample (not shown): we suggest that the activation of the implanted Si occurs from the deeper region. In these cases, as Ga diffuses into the SiO_2 film during the annealing process (Kazuhara 1989), V_{Ga} is thought to diffuse from the surface. For the 850°C sample, Si diffuses from point B to the surface region and the bulk region. The Si_i diffusing from the deeper region reacts with V_{Ga}, introduced from the surface, to give Si_{Ga}

(activation efficiency; 14%). On the other hand, the enhancement of the secondary electron emission in the SEM micrographs disappears. For the 950°C sample, Si diffuses greatly and voids are not seen. Chen et al (1989) reported that voids form early in the annealing process, eg. 20s from the onset of RTA at 1050°C. In this case, the voids are thought to be present early in the annealing process and then decompose into individual vacancies. These vacancies may react with Si_i to make more Si_{As} because PL results show the presence of Si_{As}. The enhancement of the secondary electron emission is evident in the

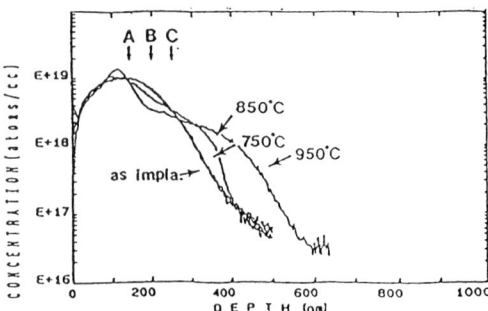

Fig. 4. Si distribution of the samples with doses equal to $3 \times 10^{14} cm^{-2}$. The samples are as-implanted and annealed at 750, 850, and 950°C.

surface region and the activation efficiency is almost zero. Chen et al (1990) also reported that the voids may compensate the electron concentration and lead to Si diffusion anomalies. For the 850°C sample, the activation efficiency reaches a maximum, though the voids are still present. As the voids disappear, it is likely that the pinned Si starts to be released and reacts with V_{As}. In this sample, V_{Ga} diffuses greatly from the surface and Si_{As}, V_{Ga} and Si_{Ga} are thought to be present in the surface region. Furthermore, the defect $(V_{Ga} + Si_{Ga})$ is known to cause the formation of a deep level (Ky 1991). The temperature dependence of the secondary electron emission appears to be correlated with the presence of this defect. From the assessment of these measurements, we propose that the enhancement of the secondary electron emission is likely to be associated with the presence of the defects $Si_{As}, V_{Ga}, + Si_{Ga}$ in the 750°C, 800°C and 950°C samples.

4. CONCLUSIONS

We consider the relationship between the enhancement of the secondary electron emission and the presence of lattice defects. From the defect assessment, we have concluded that the enhancement of the secondary electron emission may be associated with the presence of Si_{As} and $V_{Ga} + Si_{Ga}$.

To determine how the dopant activation mechanism works, we need to carry out further experiments under different annealing time regimes, because the samples used in this work were annealed at different temperatures. Using cross-sectional SEM observations, we consider that it may then be possible to assess the microscopic activation process in Si^+ ion-implanted GaAs.

We would like to thank Dr Nakamura for valuable discussions and for his help with PL measurements, and H Kuwabara for his help in ion implantation.

REFERENCES

Ashen D J, Dean P J, Hurle D T J, Mullin J B and White A M 1975 J. Phys. Chem. Solids. 36 1975
Bindal A, Wang K L, Chang S J and Kallel M A 1989 J. Appl. Phys. 65 1246
Chen S, Lee S T, Braustein G and Tan T Y 1989 Appl. Phys. Lett. 55 1194
Chen S, Lee S T, Braustein G, Ko K Y and Tan T Y 1990 J. Appl. Phys. 29 L1950
Chiang S Y and Pearson G L 1975 J. Appl. Phys. 46 2986
Hull R, Bahnck D, Stevie F A, Koszi L A and Chu N G 1993 Appl. Phys. Lett. 62 3408
Iwase F, Nakamura Y and Furuya S 1994 Appl. Phys. Lett. 64 1404
Kuzuhara M, Nozaki T and Kamejima T 1989 J. Appl. Phys. 66 5833
Ky N H, Pavesi L, Araujo D, Ganiere J D and Reinhart F K 1991 J. Appl. Phys. 69 7585
Mishima O, Eva K, Tanaka J and Yamaoka S 1988 Appl. Phys. Lett. 53 962
Simpson P J and Schultz P J 1992 J. Appl. Phys. 72 1799

Inst. Phys. Conf. Ser. No 146
Paper presented at Microsc. Semicond. Mater. Conf., Oxford, 20–23 March 1995

Point defect interactions in doped II-VI compounds under ion and electron beam irradiation

YY Loginov, PD Brown* and CJ Humphreys*

Department of Physics, Krasnoyarsk University, Svobodnii Prospect 79, 660062 Russia;
*Department of Materials Science and Metallurgy, University of Cambridge, Pembroke Street, Cambridge, CB2 3QZ, UK.

ABSTRACT: The chemistry of decomposition of doped II-VI compounds is examined under ion beam and electron beam irradiation. Assignment of moiré fringes exhibited by displaced particles against possible reaction products indicates preferential anion expulsion from n-type material and accelerated oxidation.

1. INTRODUCTION

The study of electron or ion beam induced point defect interactions within a material can potentially reveal much information about the material itself. These processes introduce the concept of 'material as a self probe,' and need to be characterised in detail if methodologies for the localised control of point defects within a semiconductor are to be developed [Loginov et al, 1994]. Information may be obtained from the study of the nature, rate, mechanistics and chemistry of material decomposition, with control external to the foil being mediated by the incident beam flux, energy and profile, coupled with temperature and time of annealing. Conversely, control within the material may be mediated by sinks (i.e. the sample foil surface, defects and interfaces), localised strain with associated bond directionality effects and competing chemical energies, and drifting under induced electric fields. In this context, the material may be described with reference to its stacking fault energy (SFE) and Fermi level, dopant concentration and native defect microstructure, while the nature of the interaction with the irradiating electron or ion beam must be understood. Additionally, point defect lifetimes and diffusivities become important. Given the large number of parameters, care must be taken to ensure that the operative processes are fully understood, and that artefacts from the thin foil preparation process do not hinder investigation of the effects of electron beam irradiation. We present information relating to the chemistry of decomposition of doped ZnS and CdTe under Ar^+, I^+ and electron beam irradiation through examination of moiré fringes exhibited by displaced reaction products.

2. EXPERIMENTAL

Bulk II-VI compounds were prepared in thin foil form by sequential mechanical polishing and argon ion milling at 5kV, $20\mu A$ and 15° with liquid nitrogen cooling until electron transparent. Improved sample foils were obtained following iodine reactive ion sputtering [Chew and Cullis, 1987] for 5 to 10min at 3kV, $10\mu A$ and 15° at room temperature. II-VI compounds with very low SFE which exhibited an artefact defect structure after this process were then given a brief chemical polish as appropriate. Samples were examined and irradiated in a Jeol 4000EX-II electron microscope operated at 400keV ($j=4x10^{19}e.cm^{-2}s^{-1}$).

3. RESULTS AND DISCUSSION

Figs. 1a,b are <110> HREM images taken from Ar^+ milled ZnS:Ga confirming the presence of dislocations loops, typically <40nm in size and density >$10^{11}cm^{-2}$, on {111} with b=1/6<211>, and native 60° dislocations dissociated into 90° and 30° partials in a manner

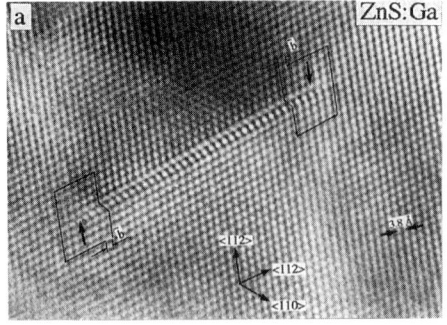

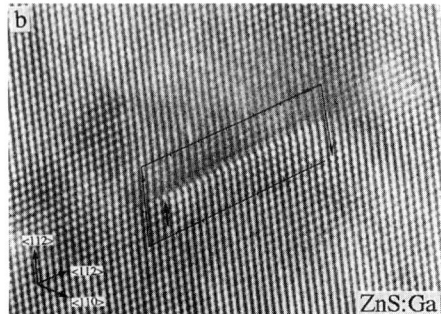

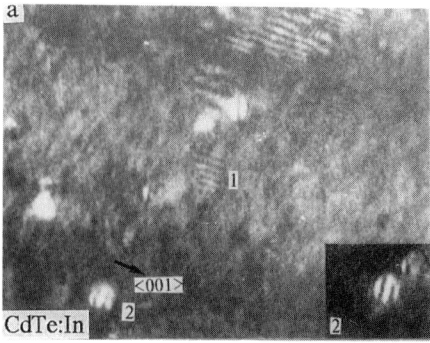

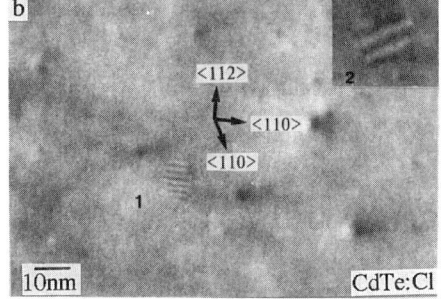

Fig 1 5kV Ar⁺ milled ZnS:Ga showing (a) dislocation loop on {111} and (b) dissociated 60° dislocation.

Fig. 2 (a) 5kV Ar⁺ milled CdTe:In and (b) 3kV I⁺ milled CdTe:Cl showing moiré fringes from displaced reaction products.

similar to that found for GaAs [Gerthsen et al, 1989]. Indeed, this microstructure is generally found for all low SFE II-VI compounds prepared by argon ion milling. The remnant structure present in Ar⁺ milled In-doped CdTe additionally exhibits moiré fringes from displaced material (Fig. 2a) and feature 1 was attributed to tetragonal InTe, hexagonal In_3Te_4 or hexagonal CdTe (which are predicted to show comparable moiré spacings for this orientation and hence cannot be distinguished), with evidence for tetragonal InTe being obtained from feature 2. Remnant structures were present even following I⁺ milling with fringes in Cl-doped CdTe, for example, being attributed to the hexagonal phase of CdTe (Fig. 2b).

Fig. 3a shows a bright field image of Ga-doped ZnS following I⁺ milling. However, as shown by the associated weak beam image (Fig. 3b), a high density of small dislocation loops are still remnant within the sample foil. In addition, striations due to the polytype structure of this compound are more readily apparent. Annealing of this foil at 450°C for 10min caused little change in the foil microstructure, while continued annealing at 520°C for a further 6 min induced loop growth, with continued loop growth ($\approx$55nm, $4x10^{10}cm^{-2}$) and dislocation segment formation following a further 5min anneal at 600°C.

The ZnS:Ga foil shown in Fig. 4a was given a brief chemical polish following Ar⁺ and I⁺ milling to remove the remnant artefact structure, and irradiated under the imaging electron beam for 5 min. The beam was centred to the right of this image and damage is delineated by the beam circumference. Video recordings of the evolution of this defect microstructure demonstrate that point defect clusters once formed migrate as a whole and in an irregular fashion. Fig. 4b is a weak beam image of the same region after 25min irradiation. The resultant distribution of dislocation loops (2.5 to 45nm, $1.4x10^{11}cm^{-2}$) may be seen, while the associated high resolution image (Fig. 5) demonstrates the presence of voids and displaced particles exhibiting moiré fringe contrast, which were attributed to ZnO_2.

Dislocation loops are also produced within CdTe under 400keV irradiation. HREM imaging demonstrates these to be situated on {112} and {111} with b= a/2<110> and

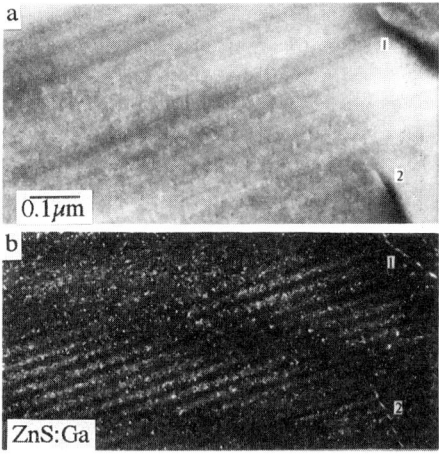

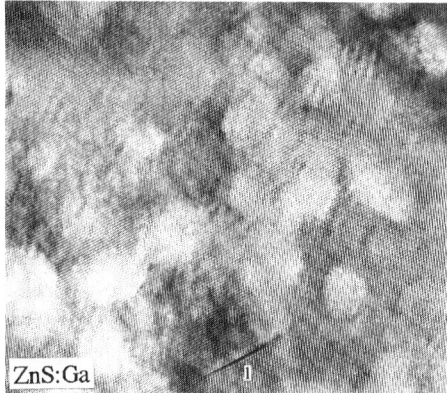

Fig. 3 3kV I+ milled ZnS:Ga (a) bright
field and (b) weak beam (g=220)

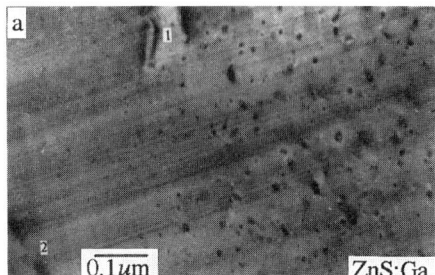

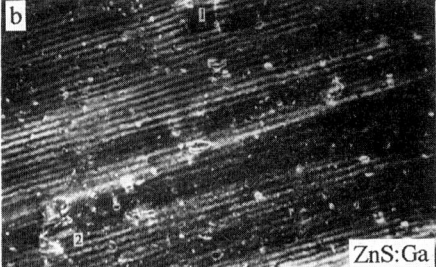

Fig. 4 400keV electron beam irradiated
ZnS:Ga after (a) 5min (g=220, bright
field) and (b) 25min (g=220, weak beam).

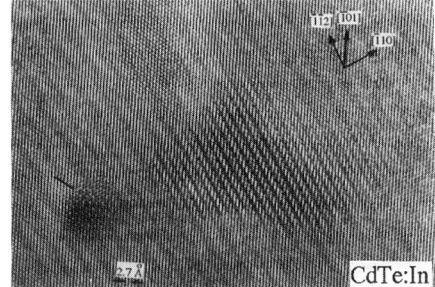

Fig. 5 <110> HREM image of 400keV
electron beam irradiated ZnS:Ga showing
voids and moiré fringes attributed to
ZnO_2.

Fig. 6 <111> HREM image of 400keV
electron beam irradiated CdTe:In showing
moiré fringes attributed to CdO.

a/3<111> respectively. In addition, platelets exhibiting moiré contrast are commonly
produced. Fig. 6 shows an HREM image (<111> projection) of CdTe:In following 400keV
electron beam irradiation for 20min. This displaced particle was attributed to cubic CdO.
Features attributed to InTe and CdTe were similarly identified.

Annealing of Ar+ and I+ induced loops promoted loop growth and interaction in CdTe and
ZnS, while subsequent electron beam irradiation caused loop shrinkage and eventual
removal (prior to the introduction of electron beam induced damage), though whether this is
due to absorption of vacancies or to glide of loops out of the sample foil requires
clarification.

Below the threshold incident electron energy for direct displacement damage, ionisation
damage results in atomic displacements with the creation of Frenkel defect pairs. For
100keV irradiated II-VI compounds the damage typically evolves to form voids decorated
by
metallic precipitates, and the nature and extent of the damage was found to follow the trend
ZnS>ZnSe≈CdS>CdTe>ZnTe>(Hg,Cd)Te≈(Hg,Zn)Te [Loginov et al, 1991]. The same
trend of damage was also found for ion beam milled material which suggests that ionisation

CdTe undoped	$10^{10}\Omega$cm	CdO, and cubic, hexagonal and tetragonal CdTe
CdTe:P	p-type, 80 and 800Ωcm	tetragonal Cd_3P_2
CdTe:In	n-type, 10Ωcm	CdO or InTe in the first instance, followed by possible phases of In_2Te_3, In_3Te_4 and CdTe
CdTe:Cu	p-type, 80Ωcm	Cu_2Te, $CuTe_x$ and CdO_2
CdTe:Cl	p-type, $4 \times 10^4 \Omega$cm	hexagonal and tetragonal CdTe
ZnS:Ga	n-type, 500ppm	ZnO_2, cubic $ZnGa_2O_4$

Table 1 Decomposition products of 400keV irradiated II-VI compounds

damage is similarly important during ion beam milling of II-VI compounds. 400keV electrons are sufficient to cause direct displacement damage in these II-VI compounds (as is ion beam milling), but sub-threshold multiple ionisation effects appear to be dominant.

The decomposition products produced by 400keV electron beam irradiated doped CdTe and ZnS are summarise in Table 1. The main point being that differences in decomposition behaviour are evident with dopant type. The tendency for the initial formation of oxides of Cd or Zn indicates the preferential removal of the anion leading to accelerated oxidation [Lu and Smith, 1988; Thangaraj and Wessels, 1990]. No evidence for Cl-based compounds is found and this is not surprising given the volatility of chlorine. Irradiated doped CdTe shows a higher density of displaced phases showing moiré contrast than undoped material, indicating that dopant species are indeed influential on the process of secondary defect formation.

It is also noted that differences with dopant type are also manifest during the process of in-situ annealing under the electron beam. The critical temperatures for loop formation within undoped CdS, ZnSe and ZnS are found to be 300, 350 and 400°C respectively (c.f. Yoshiie et al, 1983). Whereas doped CdS, ZnSe and ZnS all show such effects at substantially lower temperatures, i.e. 200, 250 and 300°C. The higher proportion of heterogeneous nucleation sites within doped material probably reflects enhanced gettering of interstitials and vacancies at these centres.

The long term aim is to develop methodologies for in situ degradation of device structures, while improved knowledge of point defect interactions, in view of the differential mobilities of interstitials and vacancies, may facilitate local type conversion.

Acknowledgements

With thanks to A W Brinkman, K Durose and N Thompson of Durham University for provision of the material used in this study. Also E O'Keefe of GEC-Marconi Infrared is thanked for provision of Cu-doped CdTe. YYL wishes to acknowledge the Royal Society for financial support, and PDB wishes to thank SERC for support under contract No. GR/J37966.

REFERENCES

Chew N G and Cullis A G, 1987, Ultramicroscopy 23 175
Gerthsen D, Ponce F A and Anderson G B, 1989 Phil Mag. A59 1045
Loginov Y Y, Brown P D, Thompson N and Durose K, 1991 J. Crystal Growth 117 682
Loginov Y Y, Brown P D and Humphreys C J, to be published in the Materials Research Society conference proceedings', Symposium Y: Microstructure of Irradiated Materials, Boston, Fall 1994
Lu P and Smith D J 1988 phys. stat. sol. (a) 107 681
Thangaraj N and Wessels B W 1990 J. Appl. Phys. 67 1535
Yoshiie T, Iwanaga H, Shibata N, Suzuki K, Ichihara M and Takeuchi S, 1983, Phil. Mag. A47 315

Inst. Phys. Conf. Ser. No 146
Paper presented at Microsc. Semicond. Mater. Conf., Oxford, 20–23 March 1995
© 1995 IOP Publishing Ltd

Ion - implantation and annealing of 6H-SiC

J Heindl, H P Strunk, A Heft*, T Bachmann*, E Glaser*, E Wendler* and W Wesch*

Universität Erlangen-Nürnberg, Institut für Werkstoffwissenschaften - Mikrocharakterisierung, Cauerstraße 6, D-91058 Erlangen, Germany
*Friedrich-Schiller-Universität Jena, Institut für Festkörperphysik, Max-Wien-Platz 1, D-07743 Jena, Germany

ABSTRACT: Single crystals of 6H-SiC were implanted with 300 keV Sb^+-ions along the c-axis and subsequently annealed at various temperatures. Cross-sectional transmission electron microscopy shows a bilayer structure. After implantation below 500K a double layer structure is observed consisting of an amorphous surface layer and a damaged defect band below it. After implantation above 500K, a monocrystalline surface layer and a defect band exists. During annealing above 1100K, the amorphous layer crystallizes, but even at 2000K not as a perfect single crystal.

1. INTRODUCTION

Silicon carbide is an interesting material for the fabrication of high power and high temperature devices. Doping of SiC by diffusion requires unacceptably high temperatures (Davies et al 1988). A promising alternative is ion-implantation in conjunction with subsequent annealing. Up to now this process has not been completely understood (see e.g. McHargue and Williams 1993, Pezold et al 1993). This paper presents results obtained from Sb^+ implantation in 6H-SiC by systematic variations of the implantation and annealing temperatures.

2. EXPERIMENTAL

The damage produced by 300 keV Sb^+- ions in single crystal 6H-SiC parallel to the c-axis and their annealing behaviour has been analysed by cross-sectional transmission electron microscopy (XTEM) and by Rutherford backscattering spectroscopy (RBS). The Sb^+-ions were implanted with a dose of $1 \times 10^{15} cm^{-2}$ at different temperatures between 80K and 1273K without annealing, respectively with a dose of $3 \times 10^{14} cm^{-2}$ at 300K followed by annealing in a double graphite strip heater under Ar atmosphere at temperatures between 873K and 2000K for a time between 90s to 30s. The XTEM specimens were mechanically polished down to 10 μm and etched by Ar^+-ions to electron transparency under liquid nitrogen cooling. Details concerning the RBS measurements are published elsewhere (Wesch et al 1995).

3. RESULTS AND DISCUSSIONS

3.1. As-implanted SiC irradiated between 80K and 1273K

A bilayer structure is formed consisting of a heavily disturbed but crystalline defect band deep in the bulk and - depending on the implantation temperature - an amorphous or monocrystalline surface layer on the top (see fig. 1). Up to 500K the 150 nm thick surface layer is amorphous and followed by a 60 nm thick defect band. Beyond 500K the surface layer is monocrystalline and about 70 nm thick while the thickness of the defect band increases from 60 nm at 673K to 90 nm at 1273K (fig.

436

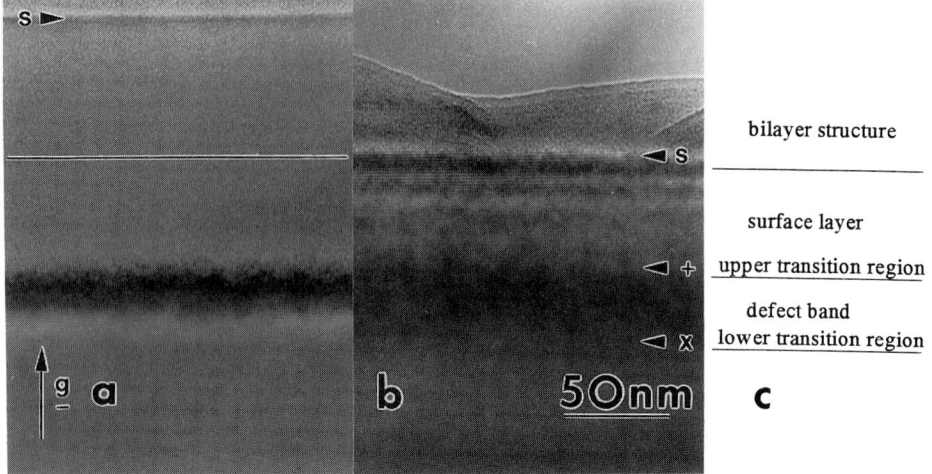

Fig. 1: SiC as-implanted

    a) at 300 K: an amorphous surface layer exists (s to x); the monocrystalline areas are recognizable from the surface-parallel black-grey stripe contrasts (6H-superlattice fringes), the original surface is indicated by a line; two-beam bright field; diffraction vector g =00.12;

    b) at 673K: the surface layer is monocrystalline; multi-beam imaging along [11.0];

    c) scheme of the bilayer structure; (s: actual surface; x, + see fig. 2); XTEM.

2). Due to volume changes during the treatments, absolute depth distributions should be measured with reference to a fixed marker. We use the position of the lower transition region (see fig. 1c) as a marker, since its position remains, as verified with different experiments (RBS), almost constant with the original surface. Due to volume increase of a-SiC the observed surface (s in fig. 1) is shifted upwards by approximately 80 nm with reference to the original surface. By assumption of a constant mass of the layer this shift is synonymous with a decrease in density (Heera et al 1994, McHargue and Williams

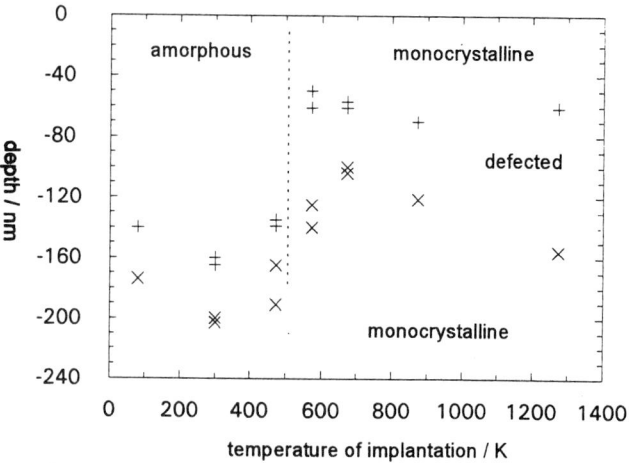

Fig. 2: Depth distribution of implantation damage, as measured from the actual surface versus temperature
+: upper transition region
x: lower transition region

1993). This thickness difference of the amorphous and crystalline surface layers cannot be detected by RBS. The modified outdiffusion model (Morehead and Crowder 1970, Haynes and Holland 1991) gives a temperature of 700K at which amorphisation is prevented (Wesch et al 1995). According to this model no defects should exist above this temperature, however, we observe a defect band still to be present. This defect band consists of a dense dislocation network and its position is in good agreement with the region of the maximum energy deposition due to the Sb-beam.

3.2. Annealing behaviour dependence of the implantation damage

Specimens are implanted at 300K and then annealed between 873K and 2000K. Before annealing all these specimens showed an amorphous surface layer followed by a defect band. Up to 1100K only a slight increase in the density of the amorphous layer can be observed (see fig. 3), above 1100K this layer crystallizes. This crystallized surface layer can be devided into two sublayers: a topmost granular and a columnar layer (see fig. 4b bright and dark regions, respectively). The granular layer consists of statistically orientated three-dimensional grains with an average diameter of 12 nm. The columnar layer consists of columns with c-axis orientation parallel to the c-axis of the bulk material and an average diameter of 9 nm. The transition region of these two layers (marked with o in fig. 3) is the deeper in the material the higher the applied temperature of annealing is. We

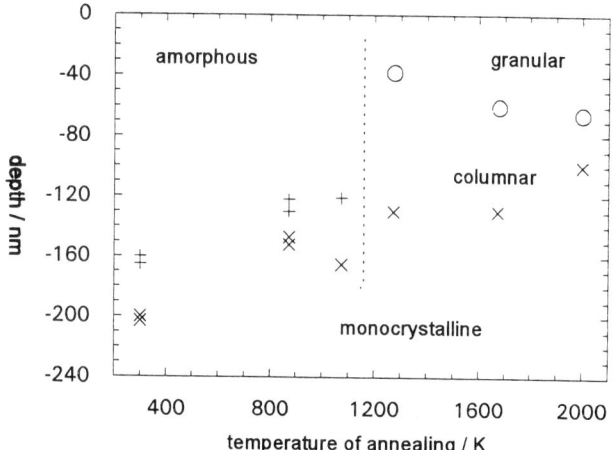

Fig. 3: Distribution of implantation damage after
annealing versus temperature
below 1100K: +: upper transition region
x: lower transition region
above 1100K: x: lower transition region
o: transition region between granular
and columnar structure

propose two competing growth mechanisms: a statistical nucleation of grains within the amorphous layer or at its surface which generates the granular layer and an epitaxial growth from the defect band which generates the columns. The higher the temperature of annealing, the more the granular growth dominates over the columnar growth. The decrease in the thickness of the whole bilayer at 2000K is caused by epitaxial regrowth of the defect band. At 2000K annealing the uppermost 10 nm are perfectly crystallized.

4. SUMMARY AND CONCLUSIONS

After implantation of 300keV Sb^+ into 6H-SiC in the dose range $3 \times 10^{14} cm^{-2}$ to 1×10^{15} cm^{-2} a bilayer is formed consisting of a defect band and an amorphous (implantation temperature < 500K), or respectively monocrystalline (implantation temperature > 500K) surface layer. After annealing at 1100K first indications of recrystallisation occur. However, even after annealing at 2000K the material is not

438

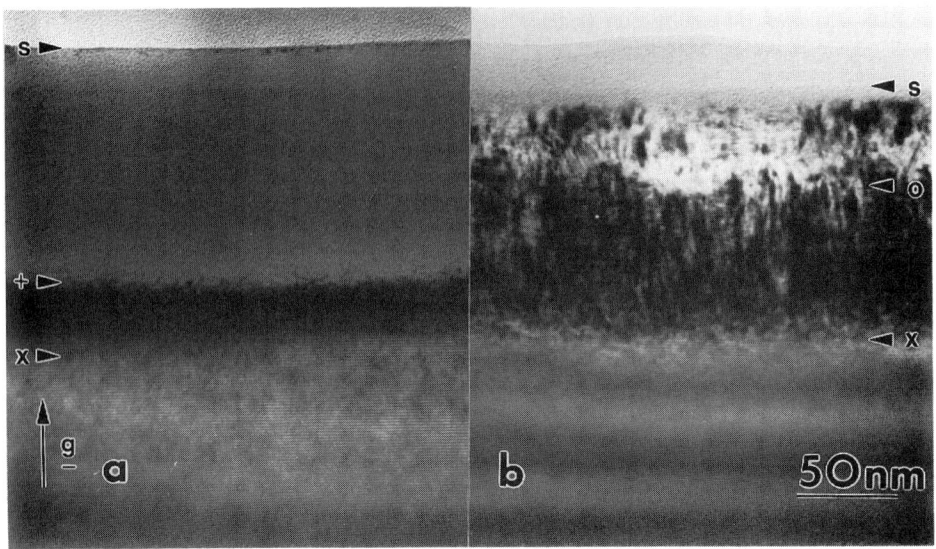

Fig. 4: SiC implanted at 300K and annealed at
  a) 873K: the amorphous surface layer is not crystallized;
  b) 1273K: the surface layer is crystallized in a granular (bright contrast, s to o) and a columnar structure (dark contrast, o to x);
  two-beam bright field (s: actual surface ; x, +, o see fig 3). XTEM

completely epitaxially crystallized. The morphology of the strongly disturbed band is independent of the implanting and annealing temperatures.

ACKNOWLEDGEMENT

This work was supported by the Deutsche Forschungsgesellschaft Sonderforschungsbereich 292

REFERENCES

Davis R F, Sitar Z, Williams B E, Kong H S, Kim H J, Palmour J W, Edmond J A, Ryu J, Glass J T, Carter C H 1988 Mat. Sci. and Engineering B1 77
Haynes T E, Holland O W 1991 Appl. Phys. Lett. 59 452
Heera V, Stoemenos J, Kögler R, Skorupa W 1994 submitted to J. Appl. Phys.
McHargue C J, Williams J M 1993 Nucl. Instr. and Methods B80/81 889
Morehead F F, Crowder B L 1970 Rad. Eff. 6 27
Pezold J, Kalnin A A, Moskwina D R, Salveyev W D, 1993 Nucl. Instr. and Methods B80/81 943
Wesch W, Heft A, Heindl J, Strunk H P, Bachmann T, Glaser E, Wendler E 1995 Nucl. Instr. and Methods to be published

Inst. Phys. Conf. Ser. No 146
Paper presented at Microsc. Semicond. Mater. Conf., Oxford, 20–23 March 1995
© *1995 IOP Publishing Ltd*

Structure transformation of 6H-SiC during room and high temperature ion implantation

A A Sitnikova, A A Suvorova and A V Suvorov*

Ioffe Physico-Technical Institute, 26 Polytechnicheskaya St., St. Petersburg 194021,Russia
*Cree Research Inc., Durham, NC,27713, USA

ABSTRACT: Defect structure formed in 6H-SiC after Al^+ implantation at room (RT) and high (HT) temperature and structure transformation after annealing were studied using TEM. The separation of the amorphous layer from the crystalline part of 6H-SiC by the transition region was observed despite the temperature during implantation. The annealing leads to the differences in the defect structure of the recrystallized layer in the cases of the RT and HT implantation. The recrystallization is accompanied by the accumulation of the impurity at the interblock regions in the case of the RT implantation.

1. INTRODUCTION

Silicon carbide is an attractive material for electronic devices. Ion implantation has been used successfully in SiC technology. In this paper we present a TEM study of 6H-SiC defect structure produced during room and high (1700 K) temperature implantation and its transformation after annealing.

2. EXPERIMENTAL

6H-SiC n-type epitaxial films were implanted with 90 keV Al^+ ions to a dose of 5×10^{16} cm^{-2} at room temperature or at high (1700 K) temperature (RT and HT samples). The annealing was performed at 2100 K for 5s.

The samples for TEM study were prepared by mechanical thinning followed by Ar^+ ion milling. A PHILIPS EM 420 TEM operating at 100 kV was used to examine the samples. Both plan-view and cross-sectional samples were observed.

3. RESULTS

3.1 Room temperature implantation

An amorphous layer 110 nm thick was produced by RT implantation as shown in Fig.1. The amorphous layer is separated from the crystalline part of the 6H-SiC by a transition region, which appears to be the crystal saturated by defects. This transition region is 100 nm thick. Annealing at 2100 K resulted in the microstructure seen in Fig.2. The propagation of defect region deep into the crystal up to 1 μm is observed. The recrystallized layer contains stacking faults and polytype inclusions. The vacancy nature of stacking faults was identified by contrast analysis. Plan-view TEM study (Fig.3) showed that the sample, after annealing, has the block structure of the near-

440

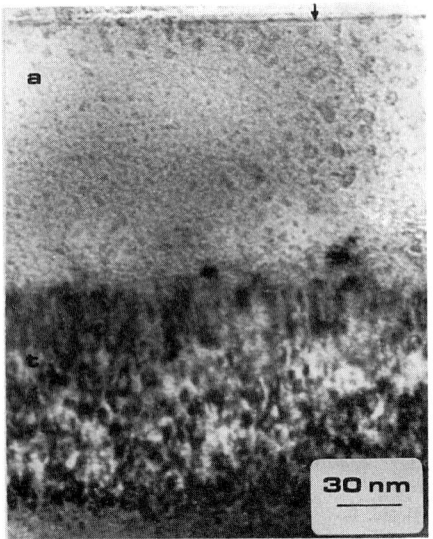

Fig.1 Cross-sectional TEM micrograph of
as-implanted RT sample
(a- amorphous region;
t- transition region).

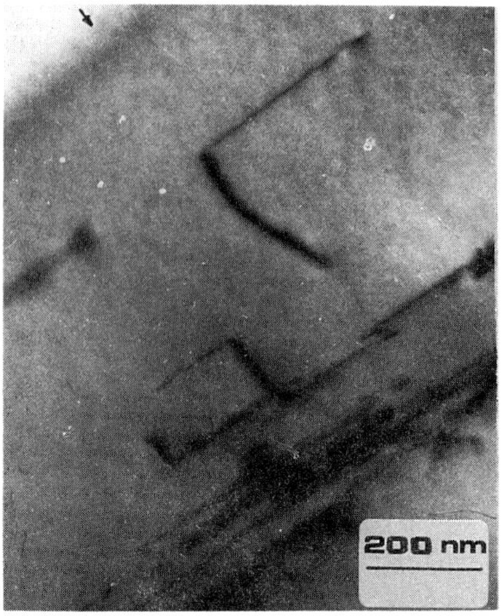

Fig.2 Cross-sectional TEM micrograph of
annealed RT sample

surface layers. The selected area diffraction pattern obtained from the interblock region showed the presence of impurity inclusions. The dislocations are generated at the interblock boundaries.

3.2 High temperature implantation

The sample implanted at 1700 K has an amorphous layer 55 nm thick and a transition region 100 nm thick as shown in Fig.4. Annealing at 2100 K resulted in the formation of a distinct implanted layer 110 nm thick (Fig.5). Plan-view TEM micrographs of the annealed sample are shown in Fig.6. The network of dislocations is observed. The investigation of stereo pairs showed that the network of dislocations apparently localizes at the boundary from which the recrystallization starts.

Fig.3 Plan-view TEM micrograph of annealed RT sample. The impurity inclusions at the interblock boundaries are shown by the arrows.

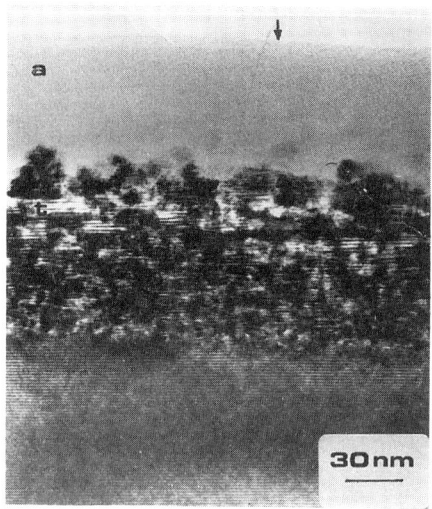 

Fig.4 Cross-sectional TEM micrograph of
as-implanted HT sample
(a- amorphous region;
t- transition region)

Fig.5 Cross-sectional TEM micrograph of
annealed HT sample

4 DISCUSSION

The results of TEM investigations show that the damage caused by ion implantation affects substantially the material behavior during annealing.

Computer simulations using TRIM90 (Ziegler 1985), a Monte Carlo ion implantation simulator, were performed. The TRIM simulations were used to determine the Al concentration profile and the depth range in which damage occurs during Al implantation. The values of the displacement energy and lattice binding energy were taken from work of Suvorov et al (1989). The maximum depth of Al profile and damage distribution obtained using TRIM are 100 nm and 80 nm, respectively. The comparison of TRIM and TEM results obtained for RT sample shows the difference in the depths. This difference can be explained by the increasing of implanted layer depth due to the smaller density of amorphous SiC.

The annealing of the RT sample leads to a substantial structure transformation. As   was reported by Sitnikova et al (1993), vacancy type defects formed in SiC under neutron irradiation become mobile and form clusters and then dislocation loops in the temperature region 1900-2300K. The presence of vacancy type stacking faults in RT sample after annealing is a confirmation of this fact.

The accumulation of Al at the regions of high defect concentration (interblock boundaries) during post-implant annealing in the RT sample has been found to occur   as   was obtained by Suvorov et al (1992).

In the HT sample according to the work of Suvorov et al (1993) the part of the amorphous layer we observed is graphitized. The dislocations that relieve the lattice mismatch are formed in the layer from interface during annealing.

It should be mentioned that the depth ratio depends on the perfection of the sample structure and implantation process conditions.

442

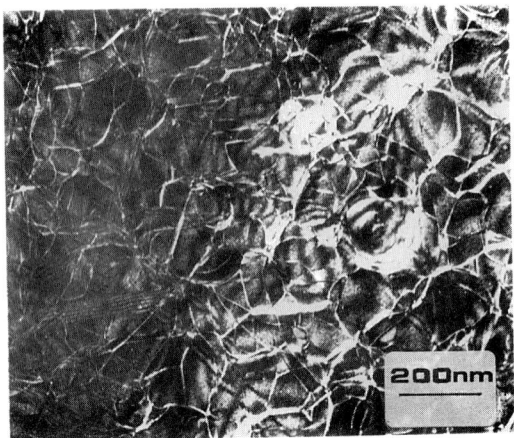

Fig6. Plan-view TEM micrograph of annealed HT sample.

5. CONCLUSION

Study of 6H-SiC after high-dose aluminum implantation was carried out. The results of our investigation are summarized below.
- The implantation leads to the separation of the amorphous layer from the crystalline part of 6H-SiC by the transition region despite the implant temperature.
- The difference between the calculated and experimental depth of the implanted layer of the RT sample could be explained by the implanted layer depth increasing due to the smaller density of the amorphous SiC.
- The recrystallization is accompanied by the accumulation of the impurity at the regions of high defect concentration (interblock regions) in the case of the RT implantation.
- The annealing leads to differences in the defect structure in recrystallized layer in the case of the RT and HT implantation. In the case of RT implantation the annealing results in the defect region propagating deep into the crystal, while in the case of the HT implantation such propagation of the defect region is not observed.

REFERENCES

Sitnikova A A, Mokhov E N, Radovanova A A 1993 Phys.St. Sol. A135 k45
Suvorov A V, Burdelle K K, Chechenin N G and Makarov V N 1989 Phys.St. Sol. A112 707
Suvorov A V, Chechenin N G, Burdelle K K, A X Kastilio-Vitloch 1992 Nucl.Instr.and Meth. B65 341
Suvorov A V, Makarov V N, Plotkin D A 1993 Proc.5th Conf. on Silicon Carbide and Related Materials (Washington) pp 545-547
Ziegler J F, Biersack J P and Littmark U 1985 The Stopping and Range of Ions in Solids (New York: Permagon)

Inst. Phys. Conf. Ser. No 146
Paper presented at Microsc. Semicond. Mater. Conf., Oxford, 20–23 March 1995
© *1995 IOP Publishing Ltd*

Investigations of $Cd_{1-x}Zn_xTe_{1-y}Se_y$ solid solutions by cathodoluminescence and X-ray standing wave methods

N N Mikheev[1], **E A Sozontov**[1], **M A Stepovich**[2] and **V I Petrov**[3]

[1] The Institute of Crystallography of the Russian Academy of Sciences, Kaluga Branch, Akademicheskaya St. 2, Kaluga 248640, Russia

[2] Moscow Bauman State Technical University, Kaluga Branch, Bazhenov St. 4, Kaluga 248600, Russia

[3] Moscow Lomonosov State University, Department of Physics, Moscow 119899, Russia

ABSTRACT: The electrophysical parameters of $CdZnTeSe$ were measured by cathodoluminescence and correlated with the atomic positions determined by the X-ray standing wave technique. A connection between the position and occupancy of sites and the composition of the solid solutions was also observed.

1. INTRODUCTION

Single crystal compounds of $CdTe$ are used as substrates for $Hg_{1-x}Cd_xTe$ optoelectronic devices. Cathodoluminescence (CL) in Scanning Electron Microscope and X – Ray Standing Wave (XRSW) methods are noncontact and nondestructive methods for investigating semiconductor composition and properties.

2. SPECIMENS

Undoped $Cd_{1-x}Zn_xTe_{1-y}Se_y$ single crystals with $x = 0.0096$ and $y = 0.036$ were obtained by the vertical gradient freeze method. Bulk crystals were cut into wafers and given a standard surface preparation, which included chemical polishing. The specimens were plates 1 – 2.5 mm thick and 10 – 20 mm on each side. The wafers were (111) oriented to within ±10 arc minutes.

3. CATHODOLUMINESCENCE INVESTIGATIONS

The CL intensity I_{CL} dependence on the electron beam energy E_0 was first proposed by Wittry and Kyser (1967) for the measuring of electrophysical parameters of semiconductors. In this work we used a new mathematical model, developed by us, which describes the dependences $I_{CL}(E_0)$ more correctly than before. The main principles of this model were discussed in the work of Mikheev et al (1992). The model used allows measurement of some electrophysical parameters of direct gap semiconductors, such as the diffusion length of minority carriers L, the thickness of the surface layer depleted with majority carriers l_s, the spectral behaviour of the absorption coefficient $\alpha(\lambda)$ and the reduced surface recombination rate of excess carriers S.

The CL of solid solutions is calculated for the case of normal incidence of the electron beam on the plane surface of a semi-infinite semiconductor with a large relative refractive index. For a wide beam and a low level of injection, the intensity of monochromatic CL from the sample is given by the expression:

$$I_{ol} \propto (1+\delta) \int_{l_s}^{\infty} \int_{0}^{\infty} \Delta p(z,z_0) exp(-\alpha z) dz_0 dz \qquad (1)$$

Here $\Delta p(z,z_0)$ is the distribution function of the minority charge carriers diffusing from a plane source which is at a depth z_0 and parallel to the sample surface and $\delta = \delta(E_0)$ is a coefficient characterizing an influence of the surface on CL - see Mikheev et al (1992).

The electrophysical parameters of semiconductors are obtained by comparison. Some results of CL investigations are presented in Fig.1 - Fig.3 and discussed below.

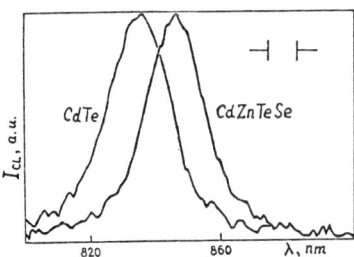

Fig.1. CL spectra of single crystals investigated at 300K: *CdTe* and *CdZnTeSe*.

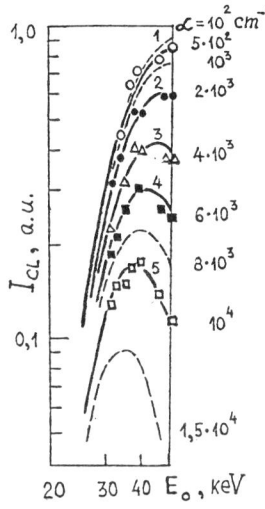

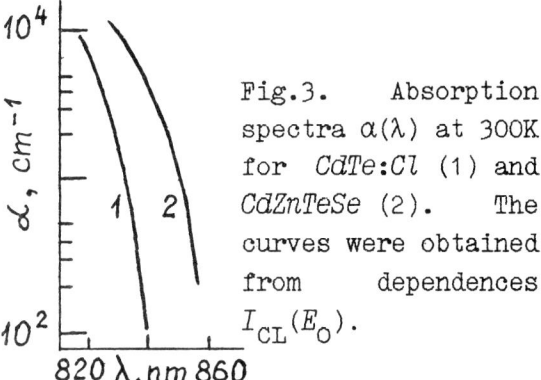

Fig.2. Dependence of I_{CL} on E_O for CdZnTeSe at λ: 855 (1), 847 (2), 840 (3), 835 (4) and 830 nm (5) at 300K.

Fig.3. Absorption spectra α(λ) at 300K for CdTe:Cl (1) and CdZnTeSe (2). The curves were obtained from dependences $I_{CL}(E_O)$.

From optimum fitted curves (solid lines in Fig.2) we obtained for $Cd_{1-x}Zn_xTe_{1-y}Se_y$ the values: l_s = (2.0 ± 0.5) μm, L_p = (0.5 ± 0.2) μm and S > 50 (i.e. in fact $S \longrightarrow \infty$). It should be noted that for CdTe:Cl L_p = (2.0 ± 0.4) μm.

4. X - RAY STANDING WAVES INVESTIGATIONS

The XRSW technique, which has been intensively developed in recent years, is widely used for the detecting of atoms positions of a single crystalline matrix. This method, which relies on dynamical Bragg diffraction, has been primarily used with perfect single crystals. It has also been shown to be highly effective for real crystalline structural analysis of the atomic distribution in multicomponent garnet crystals. The main principles and some practical using of XRSW method were discussed in the review of Kovalchuk and Kohn (1986).

The experiments were carried out using X-ray synchrotron radiation with an incident photon energy of E_γ = 13 keV and ΔE = 12 eV. This was chosen as the optimal energy for this experiment, since it would excite both the L fluorescence from the Cd and Te atoms and K fluorescence from the Zn and Se atoms. Some results of XRSW method investigations of $Cd_{1-x}Zn_xTe_{1-y}Se_y$ crystals are presented in Fig.4 and Fig.5 and discussed below.

446

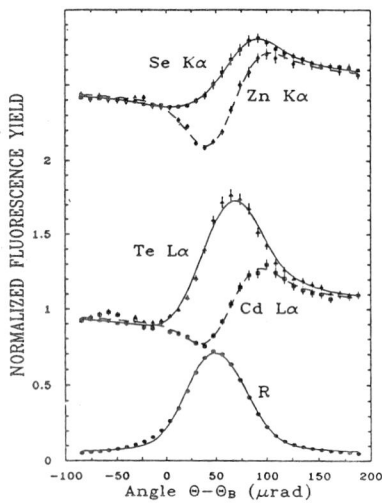

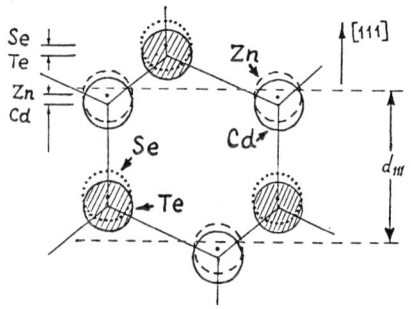

Fig.4. Theoretical (solid curves) and experimental results of XRSW method usage for *CdZnTeSe*.

Fig.5. Schematic diagram of *CdZnTeSe* crystalline lattice.

It is established that Zn and Se atoms occupy cation and anion sublattices, respectively, and their positions are Cd and Te positions in sublattices by about 1% d_{111} (Fig.5). So, by using the XRSW method we have obtained direct, detailed information on the crystalline structure of these sumples.

5. CONCLUSION

The results of our investigations showed: 1) the decrease of the band-gap energy by adding Zn and Se atoms (about 0.03 eV for the given composition); 2) some worsening of ordering of the crystalline structure of the solid solutions when the concentration of Zn and Se atoms is higher, which shows itself directly in the registered XRSW displacements of the atoms and indirectly in sharp decrease of the value of diffusion length of minority carriers, obtained from CL measurements.

REFERENCES

Kovalchuk M V and Kohn V G 1986 Sov.Phys.Usp. <u>29</u> 426
Mikheev N N, Petrov V I and Stepovich M A 1992 Bulletin of the Russian Academy of Sciences, Physics <u>56</u> 426
Wittry D B and Kyser D F 1967 J Appl.Phys. <u>38</u> 375

Inst. Phys. Conf. Ser. No 146
Paper presented at Microsc. Semicond. Mater. Conf., Oxford, 20–23 March 1995
© *1995 IOP Publishing Ltd*

Ordered double Bi layers in the structure of $Ba_mBi_{m+n}O_y$ series

V I Nikolaichik, L A Klinkova* and I I Khodos

Institute of Microelectronics Technology and High Purity Materials, RAS,
142432 Chernogolovka, Russia
*Institute for Solid State Physics, RAS, 142432 Chernogolovka, Russia

ABSTRACT: Samples of $Ba_m Bi_{m+n}O_y$ with the ratio Ba/Bi=0.5-0.8 were prepared both by slow cooling from the melting temperature and by two procedures of quenching followed by annealing in oxygen atmosphere. The slowly cooled samples consisted of $BaBiO_3$ and $BaBi_3O_y$ compounds. Quenched samples consisted either of alternating striped domains about 10 nm in width, formed by phases with different Ba/Bi ratios or of large regions of uniform composition, with a superstructure formed by double Bi layers parallel to {001} planes. The type of structure depends on the cooling procedure and on the Ba/Bi ratio. Satellite reflections corresponding to the superstructure in $BaBi_3O_y$ and $Ba_7Bi_9O_y$ are observed in commensurate positions whereas in the case of the samples consisting of domains, satellites lay in incommensurate positions.

1. INTRODUCTION

A layered structure of metal-oxide planes, characteristic of all intensively studied copper-oxide-based high-T_c compounds, is considered to be a very important element in the possible mechanism of superconductivity. On the other hand, another class of superconducting compounds, bismuth-oxide-based crystals, e.g. $Ba_{1-x}K_xBiO_3$ compounds, regarded as a substitution solution of Ba by K, is believed to lack the layered structure (Pei et al. 1990). This essential difference between the crystal structures strongly disagrees with the presence of some general properties of these two classes of oxides.

Superconductors Ba-K-Bi-O can be prepared by potassium intercalation from the melt of semiconducting double oxides of the $Ba_mBi_{m+n}O_y$ series. It was shown (Klinkova et al. 1993) that the insertion of K ions into the matrices of these oxides is a reversible process. This may be taken as evidence that K evidently does not directly substitute for Ba in the oxide lattice but rather fills in vacant sites. The structure of the compound formed should inherit some features of the double oxide structure and this is studied in this work.

2. EXPERIMENTAL

Samples with the ratio of atomic Ba/Bi concentration in the range from 0.5 to 0.8 in the reacting mixture of components were examined. Samples were prepared by three methods: 1) slow furnace cooling from the melting point $t_1=1100^\circ C$ to room temperature for periods of several hours to 8 days, 2) quenching from the temperature t_1 and 3) slow furnace cooling from t_1 to $t_2=850^\circ C$ and quenching from t_2. All quenched samples were annealed in oxygen at $450^\circ C$ for 2 hours. Crystal fragments obtained by crushing the samples were mounted on a copper grid and examined by a JEM-2000FX electron microscope, using TEM, SAD and EDX techniques.

3. RESULTS AND DISCUSSION

3.1. Slowly cooled samples

The EDX analysis revealed a pronounced disproportionation of the chemical composition in the samples. The SAD patterns and EDX analysis detected only two phases with the Ba/Bi ratio about 1.0 and 0.33, respectively (Fig.1). The diffraction pattern in Fig.1a does not exhibit any satellite reflections and corresponds to the monoclinic perovskite phase $BaBiO_3$, in which the Bi ions in two valency states (+3 and +5) have approximately equal concentrations.

The SAD pattern in Fig.1b corresponds to a compound $BaBi_3O_y$, previously unknown. The pattern exhibits a) strong basic spots indicating that the basic crystallographic unit cell is a bcc (or pseudo-bcc) cell with the parameter approximately equal to that of the $BaBiO_3$ pseudo-cubic perovskite cell; b) very weak basic spots of the 100 type; c) satellite spots $q_1=1/4g_{001}$ and $q_2=1/4g_{110}$ in commensurate positions around the strong and weak basic spots correspondingly. The related chemical analysis has shown that all Bi ions in $BaBi_3O_y$ are in valence state +3.

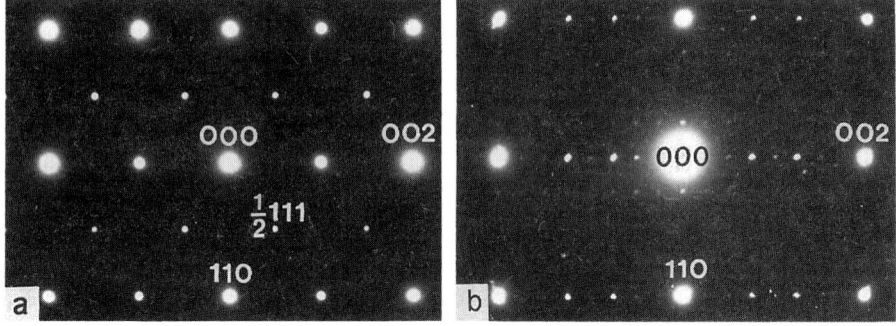

Fig.1. SAD patterns of slowly cooled sample. (a) $BaBiO_3$ phase; (b) $BaBi_3O_y$ phase.

3.2. Quenched samples

3.2.1. A striped domain structure

Quenching from the melt followed by annealing in an oxygen atmosphere is a widely used technique for the preparation of high-temperature superconductors, $Ba_{1-x}K_xBiO_3$ included. The technique allows one to obtain metastable superconducting phases that are impossible to synthesize using slow-cooling procedure. TEM studies showed that the structure of the samples of all compositions, quenched according to (2), and samples with the composition Ba/Bi=0.5-0.7, quenched according to (3) are inhomogeneous and consist of plate-like microdomains with an average width of about 10 nm (Fig.2). This small size provides an explanation for the fact that the domains were not detected by X-ray and EDX

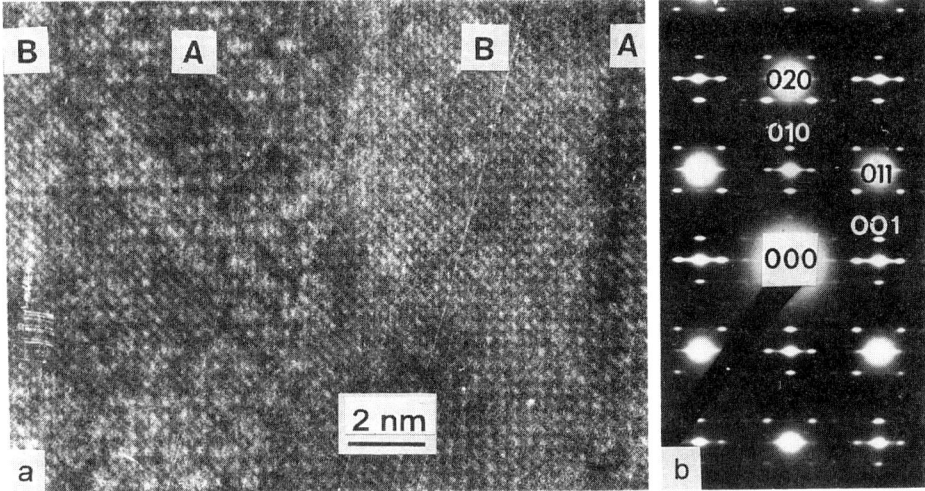

Fig.2. The domain structure of the quenched samples. (a) TEM image and (b) SAD pattern.

analyses of the bulk material. Fig.2a shows a mutual alternation of domains of two types, those having a structure modulation along the <001> and <110> directions (domains A) and without it (domains B). Domains B are apparently due to the compound $BaBiO_3$ with the perovskite lattice. The positions of the satellite reflections in Fig.2b are very similar to their position in Fig.1b. This allows us to conclude that the structure of the modulated domains closely resembles that of $BaBi_3O_y$.

However, Figs.1b and 2b exhibit some essentially different features. The satellite reflections in Fig.2b appear in incommensurate positions, whereas Fig.1b shows commensurate modulation. The modulation wavelengths q_1 and q_2 around the weak and strong basic spots, respectively, are in the ranges $q_1=(0.21-0.24)g_{001}$ and $q_2=(0.21-0.24)g_{110}$. A correlation is apparent between the q_1 and q_2 values, viz., the smaller (or the greater) q_1, the smaller (the greater) q_2 is.

3.2.2. A layered structure of the $Ba_7Bi_9O_y$ compound

The diffraction patterns of some particles in the samples with a composition $Ba/Bi=0.75-0.8$ quenched according to (3) differ radically from the pattern in Fig.2b. Fig.3a presents a SAD pattern of a particle in the sample with the composition $Ba/Bi=0.75$. The specific features of this pattern are (i) The most intensive satellite reflections aligned unidirectionally are in commensurate positions at distances of $1/4g_{001}$ and $1/8g_{001}$ from the strong and weak basic spots, respectively; (ii) Satellites along the <110> directions are not observed. Fig.3b,c show the lattice image of this sample and the schemes of the {001} type planes alternation. These figures demonstrate a superlattice of plane pairs formed by atoms of the same kind. A regular alternation of double Bi layers in 3.5 and 4.5 interplane distances of the {001} type is mainly observed. Only small local areas in the specimen lack this superlattice. This regular alternation of Bi double planes separated by the perovskite blocks corresponds to the $Ba_7Bi_9O_y$ compound. The observed superstructure agrees with the proposal about specific ordered arrangement of Ba and Bi planes and Bi^{3+} and Bi^{5+}-ions (Klinkova 1994).

450

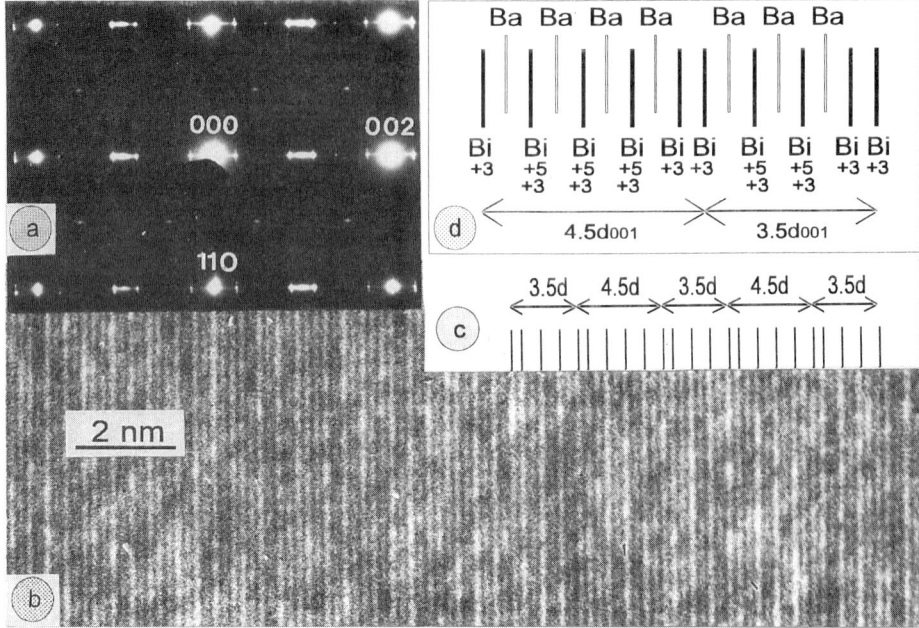

Fig.3. a) SAD pattern, b) the lattice plane image, c) the scheme of the (001) Bi planes positions and d) the scheme of one period of Ba and Bi planes alternation of the quenched from the melting point Ba/Bi=0.75 oxide.

4. CONCLUSIONS

The established dependence of synthesized oxide material on the cooling rate, the domain character of the structure of rapidly cooled samples, the different elemental content of the alternative domains, as well as the presense of some compositions giving rise to a homogeneous structure, not broken down into domains, suggest that a complex phase diagram exists. The periodicity in the positioning of the Bi-depleted or Bi- enriched domains in quenched samples suggests that the stripe-like structure is formed during the spinodal decomposition of oxides which are only stable within certain temperature ranges. As a result, transient concentration states on the way to the final stable state are formed. The evidence for this is provided by the deviation of the supercell satellites, to a greater or lesser extent, from the commensurate 1/4-positions which are characteristic of the final stable compound $BaBi_3O_y$ (Fig.1b). Knowledge of the phase diagram would help to obtain single-phase high-T_c superconducting compounds with a narrow region of transition into their superconducting state.

REFERENCES

Klinkova L A, Filatova M V, Volegova O A 1993 Superconductivity: Physics, Chemistry, Technology **6** 1917 (in Russian)

Klinkova L A 1994 Superconductivity: Physics, Chemistry, Technology **7** 418 (in Russian)

Pei S, Jorgensen J D, Dabrowski B, Hinks D G, Richards D R, Mitchell A W 1990 Phys. Rev.B **41** 4126

Inst. Phys. Conf. Ser. No 146
Paper presented at Microsc. Semicond. Mater. Conf., Oxford, 20–23 March 1995

{311} defects in ion-implanted Si: the cause of transient diffusion, and a mechanism for dislocation formation

DJ Eaglesham, PA Stolk, J-Y Cheng, H-J Gossmann, TE Haynes and JM Poate

AT&T Bell Laboratories, 600 Mountain Avenue, Murray Hill, NJ 07974, USA

ABSTRACT: Ion implantation is used at several critical stages of Si integrated circuit manufacturing. We show how {311} defects arising after implantation are responsible for both enhanced dopant diffusion during annealing, and stable dislocations post-anneal. We observe {311} defects in the earliest stages of an anneal. They subsequently undergo rapid Ostwald ripening and evaporation. At low implant doses evaporation dominates, and we can quantitatively relate the interstitials emitted from these defects to the transient enhancement in diffusivity of dopants such as B and P. At higher doses Ostwald ripening is significant, and we observe the defects to undergo a series of unfaulting reactions to form both Frank loops and perfect dislocations. We demonstrate our ability to control both diffusion and dislocation formation by the addition of small amounts of carbon impurity.

1. INTRODUCTION

Ion implantation is the predominant technique used in production to form electrical junctions for CMOS devices in Si. A typical processing sequence shows between 12 and 18 implant steps during the "front-end" half of the process. These implants may include deep gettering, isolation (beneath the device), threshold adjustment (near the junction), guard rings, and heavy implants into poly-Si. The most critical implants are those near the source and drain, and our experiments focus on typical implant doses and energies for these steps. There is usually a source-drain implant which is extremely high dose and low energy ($>10^{15}$cm^{-2}, $\approx$50keV), and serves the purpose of establishing a low-resistance contact to a lightly-doped region of the transistor adjacent to the gate. In most manufacturing processes, this critical junction uses a much lighter implant ($\approx$10^{13}cm^{-2}). It should be noted that although ion implantation has been very extensively studied in TEM in the past, most studies have concentrated on high implant energies, and doses much larger than that used for the critical junction.

Two issues are of primary concern: first, that defects from the implant should not extend into the active junction; and second, that the position of the implanted dopants after annealing should be well-controlled (and, preferably, predictable). This does not preclude the existence of residual (post-anneal) defects in the very heavily-doped region at the S/D, provided these defects are well removed from the depletion regions. Consequently, many manufacturers deliberately employ a strategy that leaves extended defects in this region. The precise conditions for excluding defects from the active rgion have been defined by intensive experimentation. Because of our lack of understanding, extension to new regimes (such as high-energy ion implants) is obviously problematic. A more serious issue is the control and prediction of the diffusion of implanted dopants during the anneal required to activate the implant. It has long been established that ion damage causes enhanced diffusion of dopants such as B and P. These impurities are interstitialcy diffusers, meaning (rather vaguely) that their diffusivity is enhanced in the presence of supersaturations of Si self-interstitials (e.g. during the oxidation of the Si surface). Thus transient enhanced diffusion (TED) of implanted B is attributable to elevated levels of interstitials in the implant region. Several problems exist with our knowledge of this phenomenon. Notably, the most widely accepted values for the interstitial diffusivity would allow all point defects to diffuse to the surface well before B motion was possible. Empirical solutions have been developed which use a slow Si_i diffusivity, but even these need to invoke some store for interstitials. Here, we will identify the source of interstitials driving TED, and use diffusion measurements to explain the slow-moving interstitial. In addition, we will show the mechanism by which dislocations evolve from ion damage, and establish the regimes for extended defect stability.

2. THE SOURCE OF THE INTERSTITIALS

As a model implant for a device, we used a 5×10^{13} cm^{-2} 40keV Si implant. This removes possible complications from the effects of the implanted species. (Comparison between Si, B and P implants suggests that this plays some role for very high doses ($\approx 10^{19}$cm^{-3})). Figure 1 shows the microstructure seen after extremely short anneals (rapid thermal anneal RTA at 800°C, 5s). There is a high density of extended defects which resemble the "rod-like defects" well-known to occur in electron-irradiated Si, as well as a variety of other conditions (Davidson and Booker, 1970; Salisbury and Loretto, 1979; Bourret, 1987). Cross-sections show that these defects are concentrated at 500-800Å below the surface, coinciding with the projected range of a 40keV Si implant. High resolution confirms the {311} habit plane expected for these defects (Figure 2).

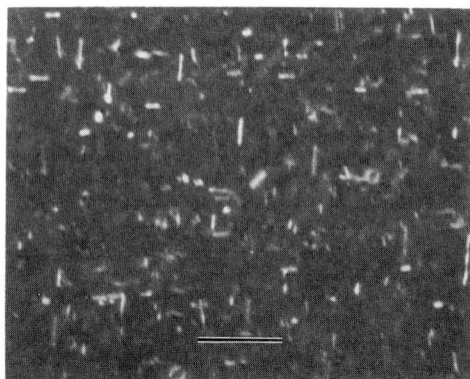

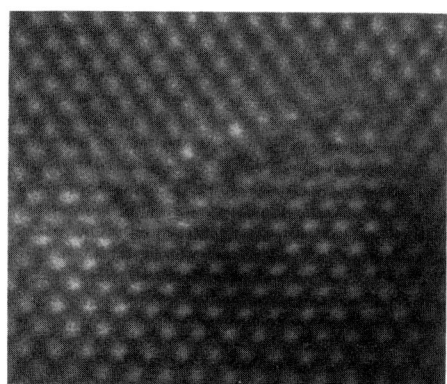

Figure 1 Weak-beam image of rod-like defects in implanted Si

Figure 2 HREM showing {311} habit

{311} defects in Si have previously been widely reported under conditions of B or metal implantation, and oxide precipitation as well as electron irradiation. Through their extensive history they have been variously identified as B precipitates, or coesite (platelets of hexagonal SiO_2), but a consensus has emerged in the electron microscopy community that they are agglomerates of Si self-interstitials, with a local structure resembling hexagonal Si (Bourret, 1987). A structure which incorporates Si_i with 4-fold coordination of Si everywhere was proposed some time ago (Tan et al., 1980; Tan, 1981), which can be regarded as locally hexagonal. The structure determined from HREM studies of He-implanted Si (Takeda, 1991; Takeda and Kohyama, 1993) differs from this in incorporating several additional structural units.

3. INTERSTITIAL EMISSION FROM {311} DEFECTS

Given that {311} defects consist of self-interstitial agglomerates, we can monitor the number of interstitials stored using defect counting. We measure defect widths from HREM cross-sections, and lengths and densities from plan-view weak-beam images. [Defect statistics are obtained from particle-counting of scanned micrographs in the NIH "Image" software. Manual measurements confirm the accuracy of this routine]. Figure 3 shows the evaporation of {311} defects at different temperatures: all curves use the same $5 \times 10^{13} \mathrm{cm}^{-2}$ Si implant to form the initial damage. The evaporation rate is strongly temperature-dependent, with an activation energy of $3.6 \pm 0.1 \mathrm{eV}$. The characteristic evaporation time for these defects has been compared directly with measurements of transient enhanced diffusion (Stolk et al., 1994; Stolk et al., 1995). The close agreement between the time to evaporate defects and the duration of the transient confirms that {311} defects are the source of interstitials causing TED. Moreover, we can now directly measure the number of interstitials injected to cause this enhancement.

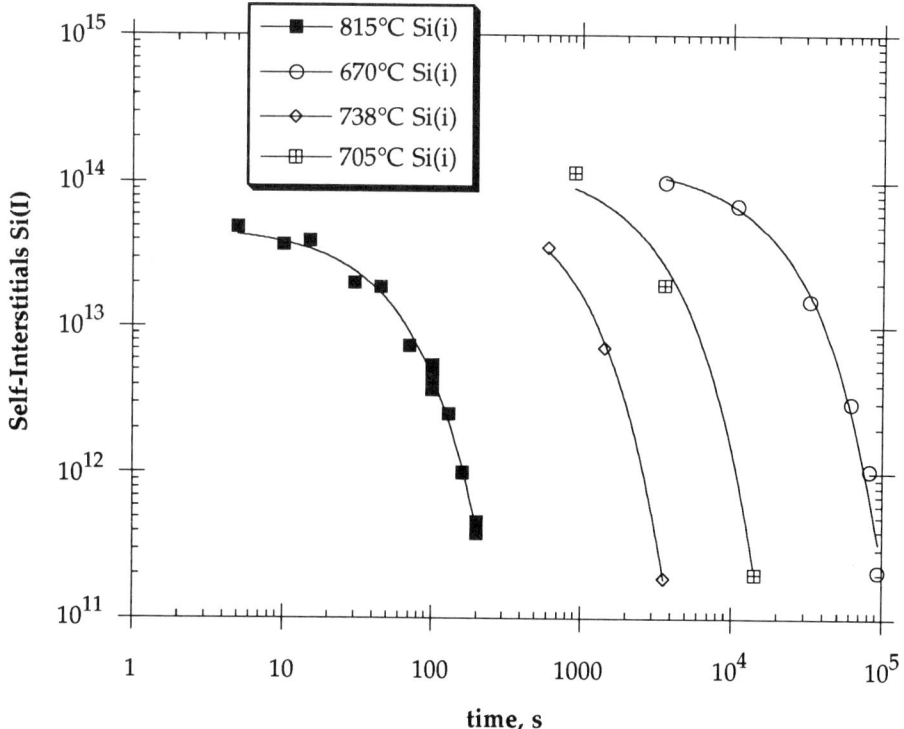

Figure 3 Evaporation of interstitials from {311} defects as a function of T

The total number of interstitials observed correlates closely with the "plus one" model. This hand-waving approximation for the post-implant point defects suggests that each implanted ion forms a cascade of ≈ 1000 Frenkel-pair formation events, and comes to rest in an interstitial position. Hence each implanted ion injects about 1000 vacancies and 1000 interstitials, plus one. On annealing, all Frenkel pairs annihilate, leaving the "plus one" interstitial. This picture has never been taken very seriously, despite agreement with empirical models for diffusion (Pinto, et al., 1992), and some support from experiments using pre-existing dislocations as point defect "detectors" (Listebarger et al., 1993). In Figure 4 we provide the first precise test of the plus-one model, with the measured interstitial content of an implanted sample plotted as a function of implant dose. The close agreement suggests that Frenkel pair annihilation is extremely effective at removing the vast majority of ion damage. Deviations from "plus one" arise from subtleties in the behaviour of point defects: plus 1.4, for instance, could be attributed to a small number (0.4 per ≈1000) of vacancies reaching the surface.

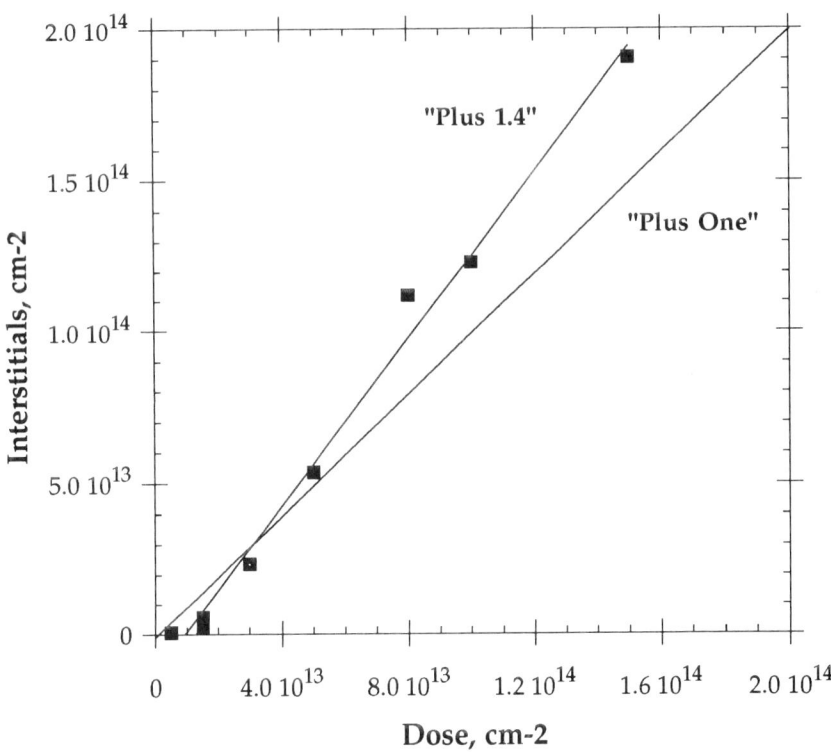

Figure 4 Testing the "plus one" model for interstitial introduction. Dose-dependence of the number of interstitials in {311} defects.

The large number of interstitials emitted from defects impacts our understanding of a key problem in Si diffusion. This is the diffusivity of the Si self-interstitial, whose reported values vary across more the 8 orders of magnitude at the processing temperature(Taylor et al., 1989). The large concentration of interstitials, along with measurements of diffusion implicating traps (Stolk et al., 1995), suggest that all measurements of diffusion in Si are dominated by trap limited diffusion of interstitials, and that traps are present at high concentrations. This in turn suggests C as a possible trap (Stolk et al., 1995). A consistent picture is finally beginning to emerge where trap-limited diffusion gives rise to the whole range of observed behaviour.

4. DISLOCATION FORMATION FROM {311} DEFECTS

The observation that most damage annihilates, leaving only excess interstitials, poses new problem. It is well known that high-dose implants lead to dislocation formation. So, how do dislocations form from the relatively small point defect excess arising from the implant? On increasing the dose, we enter a regime where both {311} defects and dislocations are observed. Fig 5 shows the microstructure resulting from a 1.5x10^{14}cm^{-2} 145keV Si implant annealed at 900°C, 15 minutes. The defects observed are predominantly 1/3[111] Frank loops frequently arranged in linear chains, suggesting that the loops may form from the ≈1μm long rod-like defects seen at lower doses and shorter anneals. Remnants of a {311} defect are observed connecting the Frank loop chain in Fig. 5. A plausible "unfaulting" reaction could convert the burgers vector at the {311} defect to a Frank loop (1/21<116> + 1/21<111>=1/21<777>). Despite the fact that this would require the habit plane of the defect to twist onto {111}, the observation does strongly suggest that Frank loops are forming from rod-like defects. (Previous unfaulting reactions of {311} defects during electron irradiation involved formation of perfect 1/2<110> dislocations (Salisbury and Loretto, 1979)). We expect Frank loops to further unfault into perfect dislocations, giving a dislocation formation sequence of {311}⇒Frank⇒perfect.

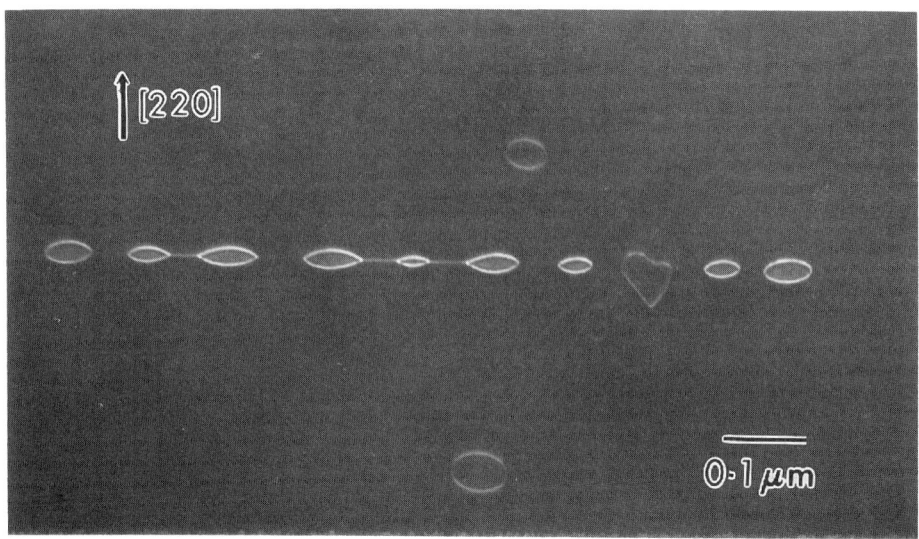

Figure 5 Frank loops forming from a {311} defect

Why do {311} defects dominate interstitial agglomeration in Si, when radiation damage in most materials (metals etc.) involves loop formation? Primarily because of the high energy of dislocations in semiconductors. The calculated energy of {311} defects is low, ≈0.5-0.9eV/interstitial (Takeda and Kohyama, 1993), consistent with tight binding of interstitials to these defects. This binding should lead to stability of these defects with respect to Frank loops for sufficiently small size. Model calculations (Figure 6) suggest that for clusters up to 100 atoms, {311} defects are more stable than the corresponding Frank loop. The system then gets trapped into a metastable situation where {311} defects grow far beyond the size where Frank loops are more stable.

5. CONCLUSIONS

In summary, we have demonstrated that interstitial evaporation from {311} defects is responsible for TED in Si. Post-implantation, Frenkel pair annihilation dominates the initial anneal, and subsequent behaviour is dictated by the small excess of interstitials, about 1.4i per implanted ion. These interstitials rapidly agglomerate into {311} defects because of the smaller activation energy for formation. Evaporation of these defects correlates perfectly with the observed diffusion transient. The diffusion is trap-limited, with the large concentration of traps implicating carbon. At high doses, the {311} defects can also give rise to stable Frank loops, and this reaction seems to be responsible for most extended defects seen after prolonged high-temperature anneals.

456

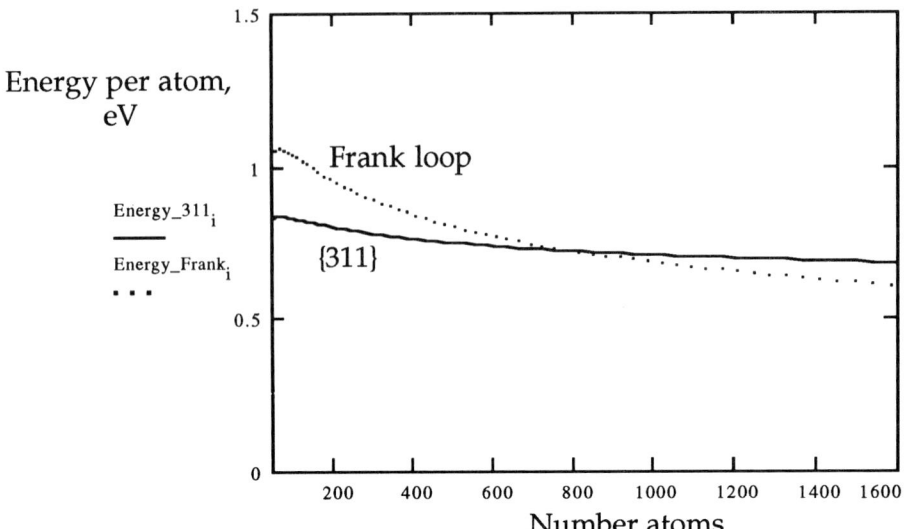

Figure 6 Calculated energies of Frank loops and {311} defects as a function of defect size. Neither the dislocation core parameter nor the {311} energy are known with any degree of certainty, but the curve does suggest why nucleation of {311} defects dominates over the more stable dislocations

REFERENCES

Bourret, A. (1987). *Institute of Physics Conference Series* **87**: 39.
Davidson, S. M. and G. R. Booker (1970). *Radiation effects* **6**: 33.
Listebarger, J. K., K. S. Jones and J. A. Slinkman (1993). *Journal of Applied Physics* **73**(10): 4815.
Pinto, M. R., D. M. Boulin, C. S. Rafferty, R. K. Smith, W. M. Coughran, I. C. Kizilyalli and M. J. Thoma (1992). Proceedings International Electronic Devices Meeting, (San Francisco), p. 923.
Salisbury, I. G. and M. H. Loretto (1979). *Philosophical Magazine* **A 39**(3): 317-323.
Stolk, P. A., H.-J. Gossmann, D. J. Eaglesham, D. C. Jacobson and J. M. Poate (1995). *Applied Physics Letters* **66**(5): 568-570.
Stolk, P. A., H.-J. Gossmann, D. J. Eaglesham and J. M. Poate (1994). 10th International Conference on Ion Implantation Technology, Catania,
Takeda, S. (1991). *Japanese Journal of Applied Physics* **30**: L639.
Takeda, S. and M. Kohyama (1993). *Institute of Physics Conference Series* **134**: 33.
Tan, T. Y. (1981). *Philosophical Magazine* **44**(1): 101-125.
Tan, T. Y., H. Föll and W. Krakow (1980). *Applied Physics Letters* **37**: 1102.
Taylor, W., B. P. R. Marioton, T. Y. Tan and U. Gosele (1989). *Radiation Effects and Defects in Solids* **111/112**(1-2): 131-150.

Inst. Phys. Conf. Ser. No 146
Paper presented at *Microsc. Semicond. Mater. Conf.*, Oxford, 20–23 March 1995
© 1995 IOP Publishing Ltd

Epitaxial regrowth of As-doped polysilicon at 850°C induced by fluorine implantation and a pre-anneal

C D Marsh, N E Moiseiwitsch*, G R Booker and P Ashburn*

Dept of Materials, University of Oxford, Parks Rd, Oxford OX1 3PH, UK.
* Dept of Electronics & Computer Science, University of Southampton, Southampton SO9 5NH, UK.

ABSTRACT: The role of fluorine and the combined role of fluorine and a pre-anneal in the break-up of the native oxide layer and the regrowth of the polysilicon (poly-Si) emitter of bipolar transistors at an emitter drive-in (anneal) temperature of 850°C is investigated by TEM. The results show that when fluorine is present and a pre-anneal is used, substantial regrowth (60%) of the poly-Si layer can be achieved after annealing at 850°C for 30 minutes and near complete regrowth (>92%) can be achieved after annealing for 120 minutes. This is the first time that regrowth has been achieved at a temperature as low as 850°C.

1. INTRODUCTION

Arsenic (As) doped polycrystalline silicon (poly-Si) is used for the fabrication of emitters for high speed bipolar transistors. The poly-Si is deposited onto single crystal Si and as a result a thin oxide layer is present at the poly-Si/single crystal Si interface (Wolstenholme *et al*, 1987). The presence of the oxide has the advantage of increased transistor gain, but the disadvantage of higher emitter resistance (Wolstenholme *et al*, 1987). The processing conditions which lead to the lowest emitter resistance and the most uniform transistor gains are those which result in the complete break-up of the oxide layer and the complete regrowth of the poly-Si during the emitter drive-in.

Hoyt *et al* (1987) showed that with a high As concentration ($5 \times 10^{20}/cm^3$), complete epitaxial regrowth occurs for an emitter drive-in of 1150°C for 20s. Recently Williams and Ashburn (1992) showed that complete epitaxial regrowth occurs for an emitter drive-in of 1100°C for 10s if a pre-anneal of 1000°C for 10 mins is used and, together with Wu *et al* (1994), that the break-up of the oxide and the regrowth of the poly-Si can occur more rapidly if fluorine (F) is present. As device dimensions decrease it is important to reduce the use of high temperature processes. In the present paper we use TEM to investigate, for the two oxide thicknesses that result from using HF and RCA pre-cleans, the role of F and the combined role of F and a pre-anneal on the break-up of the oxide layer and the regrowth of the poly-Si at the much lower temperature of 850°C.

2. EXPERIMENTAL

Cz Si (100) wafers, p-type resistivity 5-35Ωcm, were given either a HF or RCA pre-clean which resulted in a surface oxide layer of 4Å or 14Å respectively. This was immediately followed by a low pressure chemical vapour deposition (LPCVD), at 610°C, of 400nm of undoped poly-Si. Some of the poly-Si layers were implanted with F ($1 \times 10^{16}/cm^2$, 50keV) and then some of these samples and some without F were given a "pre-anneal" at 1000°C for 10 mins. All the poly-Si layers were then implanted with As ($1 \times 10^{16}/cm^2$, 70keV), capped with 600nm LPCVD of oxide to prevent subsequent As loss and annealed (equivalent to an emitter drive-in) at 850°C for times between 30 and 480 mins. After the anneal the As concentration was $2 \times 10^{20}/cm^3$. The samples were examined in a Philips CM20 TEM. The samples not given a pre-anneal are called SD and those given a pre-anneal are called AD.

458

3. RESULTS AND DISCUSSION

Table 1 summarises the TEM results for the regrowth of the samples without F after the anneal at 850°C for 480 mins. In the SD samples (i.e. no pre-anneal) there was no break-up of the oxide and hence no regrowth of the poly-Si (fig. 1). In the AD samples (i.e. given a pre-anneal) the oxide layer had started to break-up but only ≤1% of the poly-Si had regrown (table 1 and fig. 1).

F dose (/cm^2)	Sample type	Wafer pre-clean	Anneal (mins)	Oxide break-up	Regrowth pinned to oxide*	Regrowth to surface*	Amount of regrowth (%)
none	SD	HF	480	✗	-	-	0
"	"	RCA	"	✗	-	-	0
"	AD	HF	"	✓	✓	✗	0.5
"	"	RCA	"	✓	✓	✗	1

Table 1 : Summary of the regrowth of the samples not containing fluorine (*in all or part of sample).

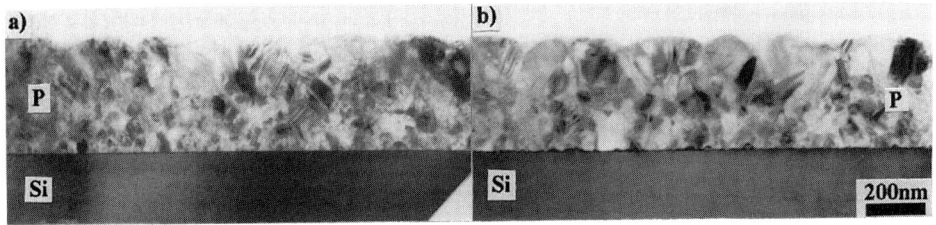

Fig. 1 : HF samples without fluorine after annealing for 480 minutes a) SD and b) AD (P = poly-Si).

Table 2 summaries the TEM results for the regrowth of the samples implanted with F after the anneal at 850°C as a function of the anneal time. In the SD samples (HF and RCA) there is no oxide break-up after 30 mins but after 120 mins the oxide layer has partially broken-up, enough for 2-3% of the poly-Si to regrow to single crystal Si (fig. 2). This regrowth is pinned to the oxide layer (fig. 2). After 240 mins the oxide layer is completely broken up and 45-60% of the poly-Si layer had regrown. This regrowth is not pinned to the original oxide layer. After annealing for 480 mins 60-70% of the poly-Si had regrown, although the regrowth has not reached the surface of the sample (fig. 3).

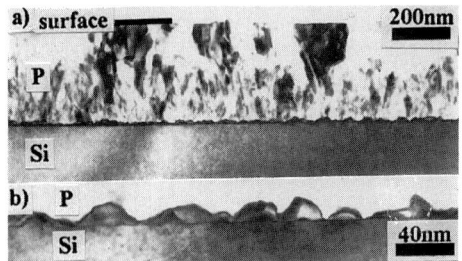

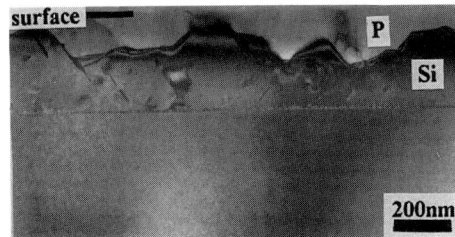

Fig. 2 : SD RCA sample annealed for 120 mins a) overall structure and b) regrowth pinned to the oxide. P = poly-Si.

Fig. 3 : SD HF sample annealed for 480 mins, showing 70% of the poly-Si layer has regrown. P = poly-Si.

In the AD samples (HF and RCA) the oxide layer has completely broken up after annealing for only 30 mins and significant regrowth of the poly-Si (50-60%) has occurred (fig. 4). After annealing for 120 and 240 mins >92% and near complete (97-98%) regrowth of the poly-Si has occurred respectively (fig. 4). In these samples small regions of poly-Si remain at the surface (fig. 4). The similar amount of regrowth, in both the HF and RCA samples, in those annealed for 240 mins and 480 mins (table 2) suggests that the last bit of poly-Si is difficult to regrow to single crystal Si.

F dose (/cm²)	Sample type	Wafer pre-clean	Anneal (mins)	Oxide break-up	Regrowth pinned to oxide*	Regrowth to surface*	Amount of regrowth (%)
10^{16}	SD	HF	30	✗	-	-	0
"	"	"	120	✓	✓	✗	3
"	"	"	240	✓	✗	✗	45
"	"	"	480	✓	✗	✗	70
"	"	RCA	30	✗	-	✗	0
"	"	"	60	✓	✓	✗	0.1
"	"	"	120	✓	✓	✗	2
"	"	"	240	✓	✗	✗	60
"	"	"	480	✓	✗	✗	60
"	AD	HF	30	✓	✗	✗	60
"	"	"	60	✓	✗	✓	75
"	"	"	120	✓	✗	✓	92
"	"	"	240	✓	✗	✓	97
"	"	"	480	✓	✗	✓	98
"	"	RCA	30	✓	✗	✗	60
"	"	"	120	✓	✗	✗	75
"	"	"	240	✓	✗	✓	98
"	"	"	480	✓	✗	✓	98

Table 2 : Summary of the regrowth of the samples containing fluorine (*in all or part of sample).

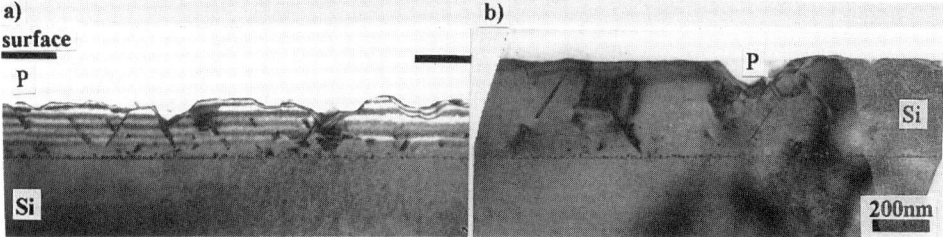

Fig. 4 : AD HF samples after annealing for a) 30 minutes, showing substantial regrowth (60%) of the poly-Si layer and b) 120 minutes, showing nearly complete regrowth (>92%) of the layer (P = poly-Si).

In both the SD and AD samples the interface between the epitaxially regrown single crystal Si and the poly-Si is irregular and stacking faults are present in the single crystal Si. In the samples where the oxide layer is sufficiently broken up that the regrowth was not pinned to the oxide layer, oxide particles are present at the depth of the original oxide layer. These oxide particles have typical dimensions of 8nm in the HF samples and 20nm in the RCA samples (fig. 5). The larger size of the oxide particles in the RCA samples (fig. 5) arises from the thicker oxide layer that results from the RCA pre-clean compared to the HF pre-clean. Electrical measurements on these samples showed that the regrowth of the poly-Si produced a decrease in the sheet resistance (Moiseiwitsch et al, 1995).

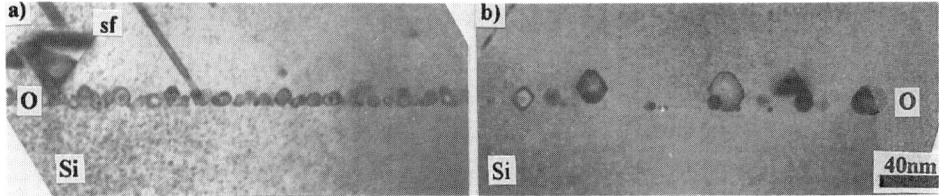

Fig. 5 : Oxide particles (O) present at the depth of the original oxide layer in samples where the oxide layer had completly broken up a) HF sample and b) RCA sample (sf = stacking faults).

The absence of any regrowth of the poly-Si after annealing in the SD sample without F (fig. 1) shows that even with a high As concentration ($10^{20}/cm^3$), an emitter drive-in (anneal) at 850°C for 480 mins is insufficient to regrow the poly-Si. The regrowth of the poly-Si in the SD samples containing F, (fig. 3) compared to the absence of regrowth in the SD samples without F (fig. 1), clearly demonstrates the importance of the F in breaking up the oxide and enabling regrowth of the poly-Si to take place.

The negligible regrowth of poly-Si in the AD samples without F (fig. 1 and table 1) suggests that the pre-anneal without F does not significantly enable regrowth to occur. This is confirmed in fig. 6. The regrowth in the AD samples containing F (fig. 4) compared to the negligible regrowth in those without F (fig. 1) further demonstrates the importance of the F in breaking up the oxide layer and enabling regrowth of the poly-Si to take place. The faster regrowth in the AD samples containing F compared to the SD samples with F demonstrates that the combination of a pre-anneal with F plays an important role in the break-up the oxide and the regrowth of the poly-Si. This is because during the pre-anneal the F starts to break up the oxide layer and some regrowth of the poly-Si then takes place (fig. 6).

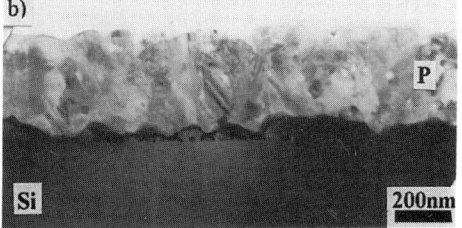

Fig. 6 : RCA samples after the 1000°C for 10 mins pre-anneal a) without F and b) with F. P = poly-Si.

The results also clearly demonstrate that regrowth can be achieved for both HF and RCA pre-cleans. The thicker oxide in the RCA samples might be expected to take longer to break-up and hence the poly-Si to take longer to regrow. Table 2 shows that for the same set of samples (SD or AD) there is negligible difference in the regrowth between the HF and RCA pre-cleaned samples (only in the AD samples annealed for 120 mins was a significant difference observed between the HF and RCA samples (table 2)). The similar speed of regrowth suggests that the presence of the F is a more dominant factor in the regrowth of the poly-Si than the difference in oxide thickness that result from the two pre-cleans.

4. CONCLUSIONS

When F is present the break-up of the interfacial oxide layer and substantial epitaxial regrowth (70%) of the poly-Si layer, for both HF and RCA pre-cleaned wafers, can be achieved during an emitter drive-in (anneal) at 850°C. Combined with the use of a pre-anneal near complete regrowth (>92%) can be achieved at 850°C. This is the first time that such epitaxial regrowth has been achieved at a temperature as low as 850°C. This reduction in the required thermal budget is potentially very important for the fabrication of high speed bipolar and BiCMOS circuits.

5. ACKNOWLEDGEMENTS

This authors would like to thank the EPSRC for financial support and the staff of the University of Southampton clean room for fabrication work.

REFERENCES

Hoyt J L, Crabbe, Gibbons J F and Pease R, 1987 Appl. Phys. Lett. 50 751.
Moiseiwitsch N E, Marsh C D, Ashburn P and Booker G R, 1995 Appl. Phys. Lett. 10/4/95 in press.
Williams J D and Ashburn P, 1992 J. Appl. Phys. 72 3169.
Wolstenholme G, Jorgensen J, Ashburn P and Booker G R, 1987 J. Appl. Phys. 61 225.
Wu S L, Lee C L, Lei T F, Chen C F, Chen L J, Ho K and Ling Y, 1994 IEEE Elec. Dev. Lett. 15 120.

Inst. Phys. Conf. Ser. No 146
Paper presented at Microsc. Semicond. Mater. Conf., Oxford, 20–23 March 1995
© *1995 IOP Publishing Ltd*

Structural and optical characterisation of Er-implanted silicon

M Q Huda, T Taskin, J H Evans, A R Peaker, P Liu*, J P Zhang*, G Curello* R J Wilson*, Z H Jafri*, S S Rao* and P L F Hemment*

Centre for Electronic Materials, UMIST, PO Box 88, Manchester, M60 1QD, UK
*Department of Electrical Engineering and Electronics, University of Surrey, Guildford, GU2 5XH, UK

ABSTRACT: As a preliminary to realising an Er-doped Si LED emitting at 1.54μm we have characterised silicon implanted with erbium over the dose range 5×10^{13} cm^{-2} to 1×10^{15} cm^{-2}, co-implanted with oxygen or fluorine, and re-grown by a two stage anneal procedure. For the highest dose, damage accumulation results in a fully amorphous layer extending up to the surface, and transmission electron microscopy reveals that, after annealing, threading dislocations extend to the surface in these layers. Extended defects detected by TEM are shown to be optically active, and degrade the erbium-related luminescence.

1. INTRODUCTION

Currently there is considerable interest in fabricating silicon light emitting structures with emission wavelengths suitable for communications technologies. The erbium-doped silicon system, where the light output is due to an intra-f-shell atomic transition in the erbium atom and is relatively independent of the semiconductor host, has received particular attention (Franzò et al 1994, Efeoglu et al 1993). It is known that the optical activity of erbium in a semiconducting matrix is enhanced when a co-implant of a light element such as oxygen, carbon or fluorine is carried out (Favennec et al 1990, Michel et al 1991). However, it has been calculated that, if the erbium concentration is kept below the solid solubility limit in silicon ($\sim 10^{18}$ cm^{-3}), the overall emitted power from an LED, assuming that all the erbium is in the optically active state (Er^{3+}), is only in the order of microwatts (Xie et al 1991). Consequently, effort has recently been devoted to studies of the implantation of erbium into silicon and subsequent regrowth using procedures that result in erbium peak volume concentrations well above the solid solubility limit (Custer et al 1994).

We report here a study of erbium-implanted silicon that has been re-grown by solid phase epitaxy. The doses of implanted erbium gave peak volume concentrations above 10^{18} cm^{-3}, and F$^+$ or O$^+$ co-implants were also carried out to enhance the luminescence efficiency. Rutherford Backscattering (RBS) and Transmission Electron Microscopy (TEM) have been used to characterise the material after re-growth and we have correlated these results with the emission spectra of the silicon:erbium system, as measured by photoluminescence (PL). Two issues are addressed in this paper. The first is the effect on the regrowth and subsequent optical efficiency of the degree of amorphisation resulting from the erbium implant, and the second is the microstructural effect of co-implanting a lighter atom.

2. EXPERIMENTAL

The substrates used in the experiments were p-type (100) Czochralski-grown (CZ) silicon wafers with resistivities in the range 10 - 20 Ωcm. Wafers were implanted with   Er$^+$ only, Er$^+$ plus O$^+$, or Er$^+$ plus F$^+$. Table I lists the implant conditions for each sample studied. The F$^+$ ion energy was calculated using the TRIM simulation program to achieve a concentration depth profile to match the erbium implant profile.   Multiple O$^+$ implants resulted in a flat oxygen profile up to approximately twice the range of the erbium. The F^+ implant gave a volume concentration more than twice that of the erbium, and the O^+ implants gave a volume concentration of either two or ten times that of the erbium. Epitaxial regrowth, amorphous layer thickness and the erbium profiles in the silicon were studied by random and channelling 1.5MeV He^+ RBS with a detector resolution of 14keV and a backscattering angle of $160°$.

Sample No	Er$^+$ implant			O$^+$ implant	F$^+$ implant	
	Energy (MeV)	Dose (cm^{-2})	Peak conc. (cm^{-3})	Peak conc. (cm^{-3})	Energy (keV)	Dose (cm^{-2})
1	1	5×10^{13}	5.0×10^{18}	1×10^{19}	--	--
2	1	1×10^{14}	1.0×10^{19}	1×10^{20}	--	--
3	2	1×10^{15}	3.5×10^{19}	--	300	1×10^{16}

TABLE I. Sample Details

After implantation, all wafers were regrown by solid phase epitaxy (SPE). A two stage annealing process was employed: the amorphous layers were first regrown during a 15 minute anneal (samples 1 and 2) or a 180 minute anneal (sample 3) in flowing nitrogen at $600°C$. This was followed by an anneal at $900°C$ for 90 seconds (samples 1 and 2) and 30 minutes (sample 3) to optically activate the erbium. A two stage anneal procedure is necessary as the erbium concentration incorporated during SPE is anneal-temperature dependent, with the limit reducing as the anneal temperature is increased (Polman et al 1993). The second anneal leads to local re-ordering and is believed to result in the formation of erbium/impurity complexes (Er/O and Er/F in this case), in which it is thought that the erbium has the required trivalent state to ensure optical activity.

The microstructures of the samples were observed by cross-sectional transmission electron microscopy (XTEM). The PL measurements were carried out at 14K or 80K and excitation was by an argon ion laser tuned to 514nm. The signal was collected by a grating spectrometer and detected by a cooled germanium detector. Signal processing was by conventional lock-in techniques.

3. RESULTS AND DISCUSSION

Fig. 1 shows XTEM from sample 2 after the first $600°C$ anneal. The erbium dose used for this wafer was not sufficient to amorphise the layer up to the surface, and at approximately 0.2 μm from the surface an interface between two dislocated regions is observed. Higher resolution XTEM (not illustrated) showed the lower dislocated region to be a twinned structure. We suggest that, because full amorphisation was not achieved, re-growth during this anneal proceeded from both the upper and lower amorphous/crystalline (a/c) interfaces. During SPE regrowth both regrowth fronts move (towards each other in this case) resulting in redistribution and trapping of Er in the regrown layer (Polman et al 1993, Custer et al 1994) while some erbium will be pushed ahead as the fronts sweep through the layer. There is an apparent interface in this

Surface

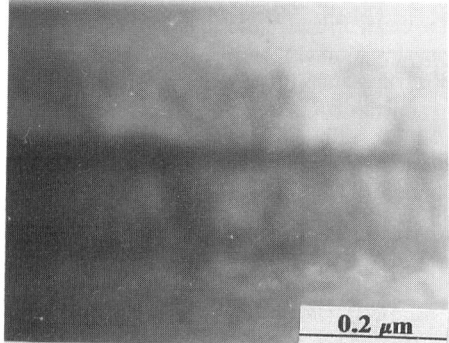

0.2 μm

Figure 1.XTEM of sample 2, Si implanted with Er^+ and O^+, but not fully amorphised, after a $600°C$ anneal.

sample associated with extended defects where the two growth fronts meet.

Figure 2 shows the PL from sample 2 after the first anneal (curve A) and after the 900° anneal (curve B). The spectrum recorded after the first anneal shows defect related PL at energies below 1eV. After the 900°C anneal emission due to erbium is observed at 0.805eV but defect-related features are still apparent in the spectrum.

XTEM was also carried out on sample 1 (not shown), and indicates that the crystalline quality of the annealed layer in sample 1 is better than that of sample 2. The intensity of the Er-related PL from sample 1 is greater, despite the lower Er concentration, and the lineshape is sharper with some crystal-field splitting induced features. The reduction of the Er related PL in sample 2 is therefore ascribed to the increased dislocation density in this sample, and the increased number of competing recombination routes that this will introduce in this layer.

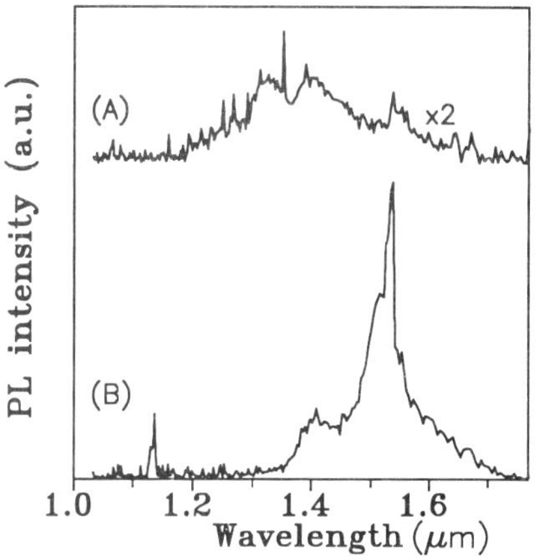

Figure 2. PL from sample 2, Si:Er:O, after 600°C anneal (A) and after 900°C anneal (B). The measurement temperature was 14K.

Surface

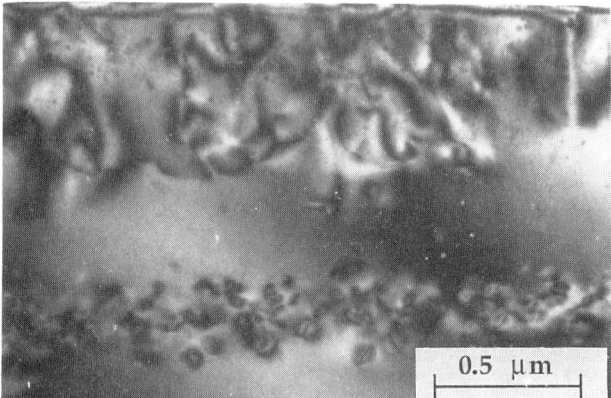

Figure 3. XTEM from Si implanted with Er$^+$ with a dose of 1x10^{15}cm^{-2} at 2MeV which is amorphised up to the surface, after regrowth at 600°C.

The existence of an amorphous layer extending up to the surface in sample 3 was confirmed by TEM before annealing (not shown). Fig. 3 is the XTEM micrograph after SPE regrowth of a sample identical to sample 3 but without a F$^+$ co-implant. A band of EOR defects can be seen at the original a/c boundary. Above this band and originating at a depth that corresponds to the peak of the erbium concentration there are entanglements of hairpin dislocations that extend up to the surface.

Fig. 4 shows the XTEM micrograph of sample 3. The same dislocation structure is evident, consisting of EOR and hairpin dislocations, but it is evident that the density of the hairpin dislocation is lower in the F$^+$ implanted sample, and RBS experiments confirm that there is a lower density of scattering centres in this sample. As the F atoms are light and have a small atomic diameter, we speculate that they are effective in relieving stress caused by the incorporation of the large erbium atoms in the Si lattice, and, thus, facilitate the annealing out of lattice defects.

RBS measurements show that in the sample implanted with erbium alone, the Er atoms have redistributed to form a broad distribution centred at a depth of about 430nm. However, very little redistribution of Er has occurred in sample 3, co-implanted with F$^+$. It is suggested that this is due to formation of Er-F complexes, as discussed by Ren et al (1994). In these samples, RBS also indicates that the majority of the erbium is not on substitutional sites.

Fig. 5 shows the PL from sample 3 with and without a F$^+$ co-implant. Erbium related emission

464

is centred at $1.537\mu m$. A large intensity enhancement is achieved by co-implanting with F^+, and this is consistent with the observation that this co-implant results in better crystalline quality than implanting with erbium alone.

4. CONCLUSIONS

By comparing TEM and PL from regrown erbium implanted silicon samples which exhibit either partial amorphisation, or full amorphisation up to the surface, we deduce that this is a crucial issue in the optical efficiency of such layers. Partial amorphisation results in a highly defective regrown layer, which gives rise to extended defect emission in the PL spectra. Therefore these dislocations act as competing recombination routes. Sample 1, implanted with half the erbium dose, and therefore containing fewer extended defects after regrowth, produced more intense Er-related emission, supporting the suggestion that incomplete regrowth results in competing recombination routes in the layer.

Full amorphisation results in hairpin dislocations after SPE which extend to the surface. This dislocation density is reduced, and a reduced erbium re-distribution during SPE is detected, when a co-implant of F^+ is carried out. This results in improved optical efficiency. Therefore, we conclude that an implantation procedure that guarantees full amorphisation is desirable, such as a silicon post-amorphisation step after all other implants have been carried out.

Surface

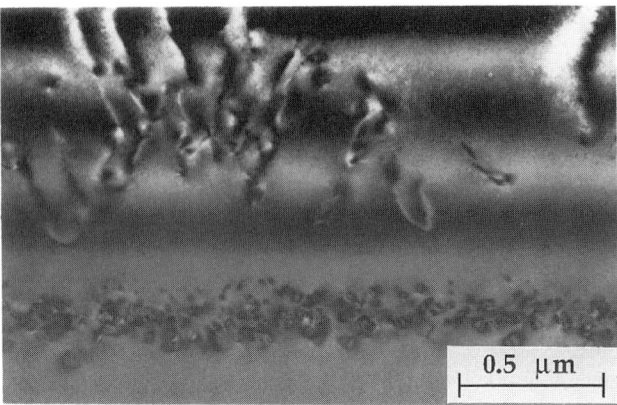

0.5 μm

Figure 4. XTEM of sample 3, Si implanted with Er^+ and F^+, after regrowth at $600°C$.

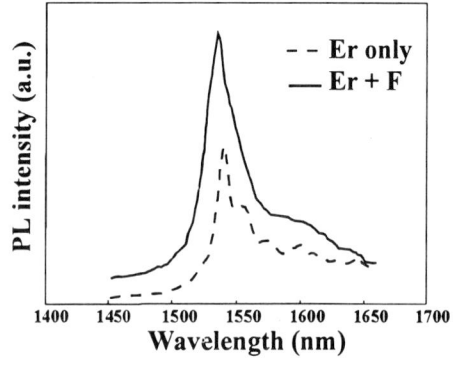

Figure 5. PL from sample 3, Si implanted with Er^+ (A), and Er^+ plus F^+ (B). The measurement temperature was 80K.

REFERENCES

Custer J S, Polman A and van Pinxteren H M, J Appl Phys 1994 75 2809

Efeoglu H, Evans J H, Jackman T E, Hamilton B, Houghton D C, Langer J M, Peaker A R, Perovic D, Poole I, Ravel N, Hemment P and Chan C W 1993 Semicond Sci Technol 8 236

Franzò G, Priolo F, Coffa S, Polman A and Carnera A 1994 Appl Phys Lett 64 2235

Favennec P N, L'Haridon H, Moutonnet D, Salvi M and Gauneau M 1990 Jap J Appl Phys 29 L524

Michel J, Benton J L, Ferrante R F, Jacobson D C, Eaglesham D J, Fitzgerald E A, Xie Y-H, Poate J M and Kimerling L C 1991 J Appl Phys 70 2672

Polman A, Custer J S, Snoeks E and van den Hoven G N 1993 Appl Phys Lett 62 507

Ren F Y G, Michel J, Jacobson D C, Poate J M and Kimerling L C 1994 Mat Res Soc Symp Proc 316 493

Xie Y-H, Fitzgerald E A and Mii Y J 1991 J Appl Phys 70 3223

Inst. Phys. Conf. Ser. No 146
Paper presented at Microsc. Semicond. Mater. Conf., Oxford, 20–23 March 1995

465

High resolution transmission electron microscopy investigation of ion implanted silicon after irradiation with ultrashort laser pulses

A Koch, M Seibt, G Böhne and W Schröter

IV.Physikalisches Institut der Universität Göttingen and Sonderforschungsbereich 345, Bunsenstr.13-15, D-37073 Göttingen, Federal Republic of Germany

ABSTRACT: The amorphous-crystalline interface of Ge-implanted (001) Si has been studied by HRTEM after irradiation with 100 fs laser pulses. The micrographs show early stages of the crystallization process upon annealing by the pulse. The gaussian beam profile leads to distinct regions of interfacial morphology. In regions of low incident laser fluences we find microtwins on {111} planes at the interface. With increasing fluences the twins are behind and at the advanced interface. At high fluences no twins are found and interface roughness is reduced by a factor of about three.

1. INTRODUCTION

Rapid transient melting and resolidification of semiconductors can be achieved by irradiation with short laser pulses. In ion implanted semiconductors, laser annealing is well established in the nanosecond pulse regime for melting and epitaxial regrowth of the amorphous layer and removing radiation damage due to implantation. The epitaxial regrowth can persist up to a critical growth rate v_c which is higher for (001) Si than for (111)Si. Amorphous layers or defective zones containing twins are formed at velocities above v_c (Cullis et al 1984).

For 25 ps pulses melt-in velocities of 800 m/s (von der Linde et al 1988) and superheatings of several hundred degrees (Fabrizius et al 1986) can be obtained. In crystalline Si irradiated with 15 ps ultraviolet laser pulses the regrowth velocity was 25 m/s and the undercooling amounted to about 700 K (Bucksbaum and Bokor 1984).There is a basic difference between the melting process induced by femtosecond pulses and laser pulses in the ns and ps pulse regime. For pulse durations in the order of tens of ps or longer the liquid-solid phase transition occurs during the pulse (Liu et al. 1982). For irradiations with 100 fs pulses the energy of the pulse is absorbed well before the phase transition can developed. Applying 100 fs pulses a precursor phase to the melt has been observed for Si (Tom et al. 1988) and GaAs (Schröder et al. 1990, Sokolowski-Tinten et al. 1991, Saeta et al., 1991) In this work we studied the amorphous-crystalline(a/c) interface of Ge-implanted Si by HRTEM after irradiation with single ultrashort (100 fs) laser pulses.

2. EXPERIMENTAL

Samples of p-type FZ (001) Si (5-10 Ωcm) were implanted with Ge (energy : 70 keV, dose 6×10^{14}cm^{-2}) followed by a BF$_2$-implantation (energy : 30 keV, dose

: 6×10^{13}cm^{-2}) resulting in an overall amorphous layer thickness of 75 nm. 100 fs laser pulses(λ= 620 nm) were generated by a colliding pulse modelocked laser and amplified with a seven pass dye laser pumped by a copper vapour laser. Single laser pulses with a gaussian fluence distribution (F = $F_0 \times$exp(-r^2/σ^2) with F_0= 0.29J/cm^2 and σ = 6 μm) are focussed on a undisturbed region of the (001) Si surface.

For cross-sectional TEM studies samples were prepared by mechanical polishing and argon ion milling. Electron micrographs were obtained at 120kV in a Philips 420ST with a point resolution of 0.3 nm.

3. RESULTS

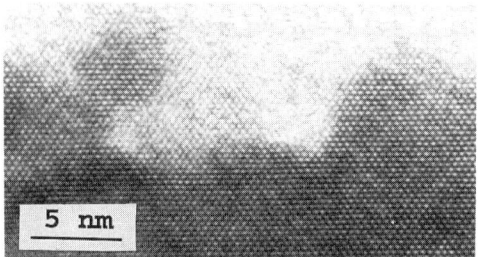

Fig. 1 : $\langle 110 \rangle$ lattice image of the as-implanted a/c interface showing small crystalline islands in the amorphous layer and amorphous pockets in the crystalline substrate. (The following pictures are shown in the same projection.)

Due to the gaussian beam profile distinct regions of interfacial morphology can be identified. In as-implanted parts crystalline islands lying inside the amorphous layer and amorphous pockets in the crystalline substrate are observed (see Fig. 1, and also Seibt et al. 1993). In regions where the incident laser fluence has dropped to 1/e of the maximum fluence we find microtwins on {111} planes (size of the smallest twin is about 2 nm) at the interface (Fig.2a) and at the crystalline islands(Fig. 2b). With increasing fluence the microtwins appear behind and at the advanced interface (Fig. 3). In the centre of the beam profile (see Fig.4) microtwins are absent and the interface roughness is decreased by a factor of three compared to non-irradiated regions. The maximum thickness of the regrown crystalline layer due to the irradiation is about 20 nm.

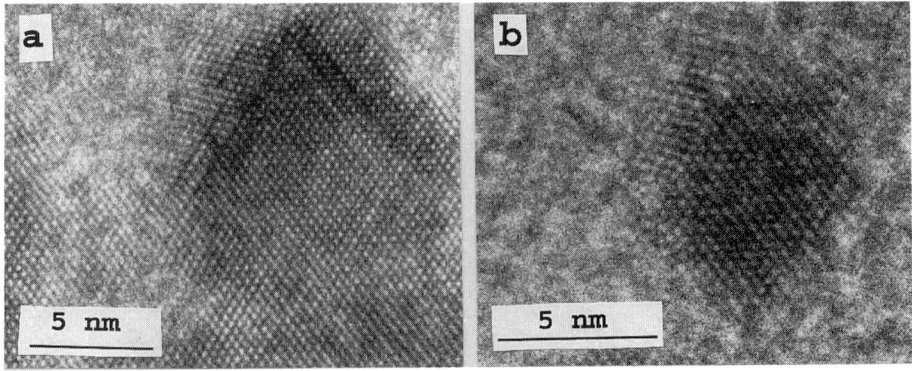

Fig. 2 : Microtwins on {111} planes a) at the interface and b) at detached micro-crystallites in a region of low incident laser fluence.

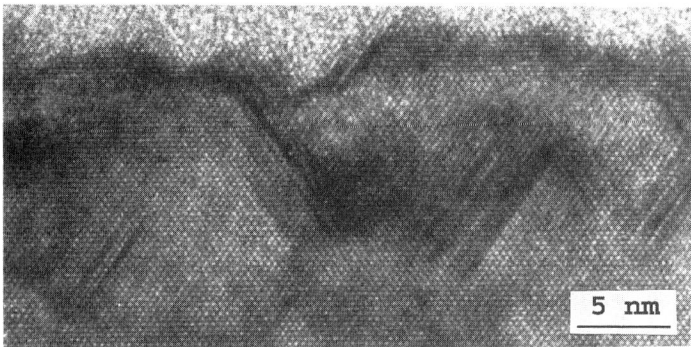

Fig. 3 : With increasing fluence the advanced crystallization front has left microtwins behind the interface.

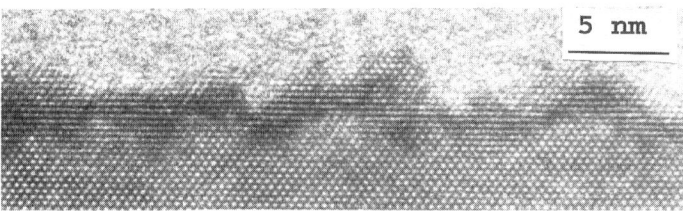

Fig. 4 : At high fluences the interfacial layer is smoothed without any microtwins observed

4. DISCUSSION

We have found early stages of the crystallization process in Ge-implanted Si upon 100 fs laser annealing accompanied by a decrease of the a/c interface roughness. For nanosecond pulsed laser melting of both crystalline and amorphous ion-implanted silicon Cullis et al. (1984) have demonstrated the breakdown of epitaxial growth when the resolidification velocity exceeds a critical value v_c which depends on substrate orientation. The breakdown results either in the formation of an amorphous overlayer or in the formation of an defective zone containing twins. During the first stages of the resolidification process the velocity of the liquid/solid interface increases due to increasing undercooling of the melt leading then to twinned or amorphous layer. Drosd and Washburn (1982) have shown that twins can only be formed on {111} planes.

After 100 fs laser pulses we find that crystallization breaks down after the epitaxial regrowth of only a few monolayers in regions of low fluence resulting in the formation of microtwins on islands. The formation of twins at the detached microcrystallites in the amorphous layer provides evidence that the liquid-solid interface velocity exceeds v_c at the *onset* of crystallization. The pyramidal islands (see Fig. 2) can be understood in terms of orientation depended critical growth rate. Thus the interface is terminated by the slower growing {111} planes compared to {110} or {100} planes. Due to the different melting points of crystalline and amorphous Si it is possible to melt an amorphous layer completely without melting the underlying crystalline substrate resulting in an undercooled liquid. We believe that this situation occurs in the region where twins have been found at the detached microcrystallites.

At maximum fluence the dwell time of the melt is enhanced. We therefore expect a lower resolidification velocity in accordance with the observed epitaxial regrowth

468

and decreasing of interface roughness in the centre of the laser beam.

To check that melting takes place we investigated evaporated amorphous silicon layers with a thickness of 300 nm on a (001) Si substrate under identical irradiation conditions. Similar to the results of Narayan and White (1984) we observe, as a function of depth, distinct regions of both fine and large polycrystals up to a depth of 200 nm in the centre of the laser beam. The thickness of the polycrystalline layer is lower than the amorphous ion-implanted layer only in the outer part of the beam profile. Therefore we conclude that the observed microtwins grow into a highly undercooled melt. In Fig. 5 we depict a model how microtwins can appear behind the advanced interface. The picture shows the development of the as-implanted a/c interface with time assuming (a) orientation depended critical rates for epitaxial regrowth (Cullis et al 1984) and (b) isotopic velocity of liquid-solid interface. For simplification the interface is terminated only by $\{111\}$ and $\{100\}$ planes rendering amorphous pockets and protruding crystalline material similar to Fig. 1. With increasing time the critical velocity v_c is exceeded first on $\{111\}$ planes resulting in the formation of twins. At the same time the amorphous pockets are filled. At later stages the interface is built up essentially by $\{100\}$ surfaces which can still regrow epitaxially due to the higher v_c value. Further investigations will be necessary to understand under which conditions this process takes place.

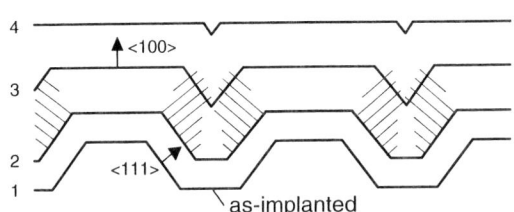

Fig. 5 : Model for the temporal development of the a/c interface leading to the formation of buried twins. Hatched areas correspond to twinned crystals. Numbers denote the growth front at different times.

REFERENCES

Bucksbaum P H and Bokor J 1984 Phys. Rev. Lett. <u>53</u> 182
Cullis A G, Chew N G, Webber H C and Smith D S 1984 J. Cryst. Growth <u>68</u> 624
Drosd R and Washburn J 1982 J. Appl. Phys. <u>53</u> 397
Fabrizius N, Hermes P, von der Linde D, Pospieszczyk and Stritzker B 1986
    Solid State Commun. <u>58</u> 239
Liu J M , Kurz H and Bloembergen N 1982 Appl. Phys. Lett. <u>41</u> 643
Narayan J and White C W 1984 Appl. Phys. Lett. <u>44</u> 35
Saeta P, Wang J K, Siegal Y, Bloembergen N and Mazur E 1991
    Phys. Rev. Lett <u>67</u> 1023
Schröder T, Rudolph W, Govorkov S V and Shumay I L 1990 Appl. Phys. A <u>51</u> 49
Seibt M, Imschweiler J, Hefner H-A 1993 Inst Conf Ser No <u>134</u> 137
Sokolowski-Tinten K, Schulz H, Bialkowski J and von der Linde D 1991
    Appl. Phys. A <u>53</u> 227
Tom H W K, Aumiller G D and Brito-Cruz C H 1988 Phys. Rev. Lett. <u>60</u> 1438
Von der Linde D, Danielzik B, Sokolowski-Tinten K and Harten P 1988
    Springer Series in Chemical Physics ed. by Schäfer F P, Vol.48: Ultrafast Phenomena VI, pp420-424 ed. by Yoshihara T, Harris C B and Shionoya S

Inst. Phys. Conf. Ser. No 146
Paper presented at Microsc. Semicond. Mater. Conf., Oxford, 20–23 March 1995
© 1995 IOP Publishing Ltd

Surface roughness of Si_xGe_{1-x} layers obtained by high dose Ge implants in Si after oxidation treatments

V Raineri, K Kyllesbech Larsen, F La Via, S Lombardo and F Iacona

CNR-IMETEM-Stradale Primosole 50, I-95121 Catania, Italy

ABSTRACT: Si_xGe_{1-x} layers were obtained by high dose Ge implants in (100) silicon substrates and solid phase epitaxy. Wet cleaning treatments (H_2SO_4/H_2O_2) based on silicon surface oxidation greatly increase surface roughness. The thin oxide layer formed on the Si_xGe_{1-x} layer is essentially composed by SiO_2 while no GeO_2 has been detected. Furthermore, Rutherford backscattering spectrometry reveals Ge segregation at SiO_2/Si_xGe_{1-x} interface. These effects are compared with those produced by the growth of a thicker (15 nm) thermal oxide. The observed phenomena suggest that great care must be used in Si_xGe_{1-x} surface cleaning treatments.

1. INTRODUCTION

The potential of heterojunction devices has long been recognised, and many efforts have driven in order to develop a technology comparable with Very Large Scale Integration (VLSI). Si_xGe_{1-x} layers can be grown epitaxially on silicon and then they can be used to realise heterojunction devices joining their advantage with standard silicon technology (Iyer 1989, Pfiester 1991 and Selvakerman 1991).

Heterojunction bipolar transistors (HBT) with cut off frequencies exceeding 100 MHz with a Si_xGe_{1-x} base have already been fabricated (Crabbé 1993). However, the demonstration of the compatibility of Si_xGe_{1-x} technology with standard procedures used in silicon device fabrication is still in progress. In particular during HBT fabrication thin oxide layers are normally grown on the base while surface cleaning and chemical treatments are performed before the polycrystalline silicon emitter formation. Metal Oxide Semiconductor (MOS) device characteristics can be improved by realising the channel in Si_xGe_{1-x}. In this case an oxide layer should be formed directly on Si_xGe_{1-x}. The oxidation of Si_xGe_{1-x} has been investigated in the past showing that Ge segregates at Si_xGe_{1-x}/SiO_2 interface while thermal silicon oxide grows (Srivatsa 1989).

Cleaning procedures are based on chemical oxidation of silicon forming very few nanometres of SiO_2 on the surface and then HF dipping (Heyns 1993). This procedure can induce an increase in the surface roughness (Heyns 1993). We have found a great increase in the surface roughness if similar cleaning treatments are performed on Si_xGe_{1-x}. This effect has been investigated by using several analytical techniques to relate surface roughness to Ge segregation.

2. EXPERIMENTAL

70 keV Ge ions were implanted in (100) silicon wafers at doses of 3×10^{15} and 3×10^{16}/cm^2. In both cases a surface amorphous layer is produced. Wafers were then annealed at temperatures above 700°C, by using a rapid thermal processor (RTP) under N_2 flow in order to regrow epitaxially the amorphous layer. Transmission Electron Microscopy and High Resolution Transmission Electron Microscopy (HRTEM) were employed to characterise the quality of Si_xGe_{1-x} layers.

A very thin thermal oxide layer (15 nm) was grown on some wafers at 920°C for 8 minutes under O_2 + HCl flux while the up and down ramps were performed in N_2 flow.

The effects of the H_2SO_4/H_2O_2 cleaning procedure on the Si_xGe_{1-x} surface were studied by using a mixture of 4:1 at a temperature of 140°C. In order to enhance the effects we used longer dipping times (30 min instead of the usual 10 min). After this step some wafers were analysed immediately, while others have been dipped for 20 s in 2 % HF and then measured.

The surface roughness was determined by Atomic Force Microscopy (AFM) in tapping mode by using a Digital Instrument Nanoscope III equipment while the chemical composition was obtained by Rutherford Backscattering Spectrometry (RBS) of 2.0 MeV He^+ ions in random and channelling configuration and X-ray Photoelectron Spectroscopy (XPS).

a)

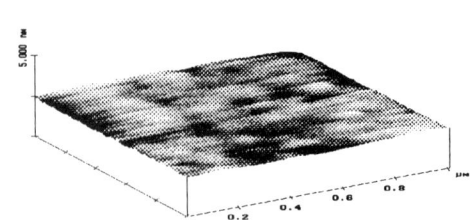

b)

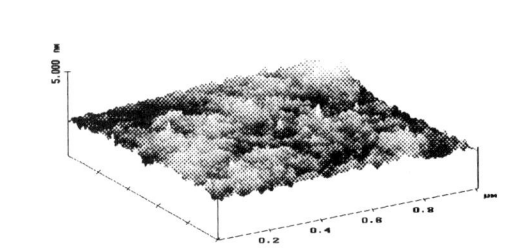

Fig. 1 a) AFM image of an unimplanted wafer after a H_2SO_4/H_2O_2 treatment. b) AFM image on a wafer implanted with 70 keV 3×10^{16}Ge/cm^2 and treated with H_2SO_4/H_2O_2.

3. RESULTS AND DISCUSSION

In Fig. 1a the AFM image of a virgin silicon wafer after a H_2SO_4/H_2O_2 step is reported in a 3D representation. The surface roughness R was calculated as:

$$R = \frac{1}{L_xL_y}\int_0^{L_x}\int_0^{L_y}|f(x,y)|dxdy \qquad (1)$$

with Lx and Ly dimensions of the surface and f(x,y) surface relative to the centre plane.

Wafer description	R [nm]	R after H_2SO_4/H_2O_2 [nm]	R after HF dip [nm]
Virgin	0.045	0.038	0.046
70 keV 3×10^{15}/cm^2	0.063	0.081	0.105
70 keV 3×10^{16}/cm^2	0.056	0.150	0.174
virgin + 1100°C 10 s	0.050	0.090	0.105
70 keV 3×10^{15}/cm^2 + 1100°C 10 s	0.094	0.211	0.263
70 keV 3×10^{16}/cm^2 + 1100°C 10 s	0.140	0.210	0.268
	R [nm]	R after thermal oxidation [nm]	R after HF dip [nm]
virgin + 1100°C 10 s	0.050	0.054	0.073
70 keV 3×10^{15}/cm^2 + 1100°C 10 s	0.094	0.221	0.554
70 keV 3×10^{16}/cm^2 + 1100°C 10 s	0.140	0.438	0.874

Table 1 - Roughness of wafers as determined by AFM measurements after different treatments

In the case of Fig. 1a the roughness was measured to be 0.038 nm. It is compared with the surface of a wafer implanted with 70 keV 3×10^{16} Ge/cm^2 after the same chemical treatment (Fig. 1b). The increase in roughness (from 0.038 to 0.150 nm) is evident.

All the roughness values for the most representative wafers are summarised in table 1. The roughness increases greatly in all the samples where Ge is present. Moreover, the roughness increase during the chemical treatment seems related to the amount of Ge present at the surface. In the same table the roughness data for a thermal oxidation are also reported for comparison. The oxide layer formed in this case is 15 nm, at least 10 times larger than produced by H_2SO_4/H_2O_2 treatments, and it has been determined by RBS measurements. The roughness produced by the growth of 15 nm of oxide is greater than that produced by the H_2SO_4/H_2O_2 treatment. The reason for this difference is difficult to be understood by only AFM measurements. By XPS measurements the growing layer gives in all cases SiO_2, while no GeO_2 was detected. The RBS spectra of a sample implanted with 70 keV 3×10^{16}Ge/cm^2 and treated with H_2SO_4/H_2O_2 is reported in Fig. 2a. The spectra indicates Ge segregated at the Si_xGe_{1-x}/SiO_2 interface. For comparison in Fig. 2b the spectra obtained for a wafer implanted with 3×10^{16}Ge/cm^2 and thermal oxidised is reported. A larger amount of Ge is segregated at the interface.

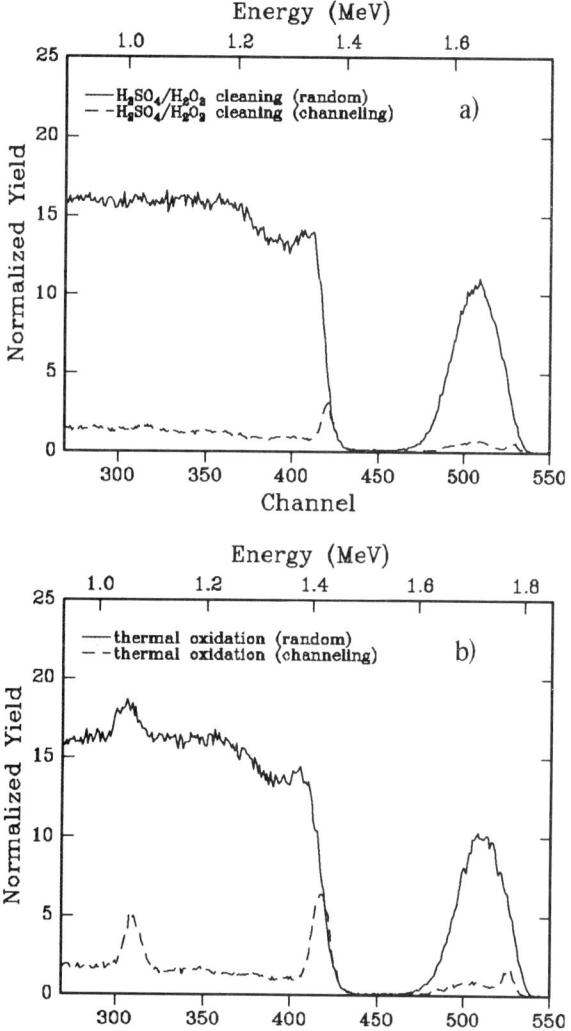

After oxidation treatments, both by thermal or chemical processes, Ge segregates at the interface SiO_2/Si_xGe_{1-x}. This effect is confirmed by all the measurements performed. The amount of Ge segregated at the interface depends on its surface concentration. In the case of the two implants reported (3×10^{15}

Fig. 3 a) RBS spectra of a sample implanted with 70 keV 3×10^{16}Ge/cm^2 and treated with H_2SO_4/H_2O_2. b) RBS spectra of a sample implanted 70 keV 3×10^{16} Ge/cm^2 and thermal oxidised.

and 3×10^{16}Ge/cm^2) a surface concentration of 1×10^{20} Ge/cm^3 and 1×10^{21} Ge/cm^3 was determined, respectively. Then, a larger effect is expected for the higher implanted dose, in agreement with the roughness data. The roughness increase is due to the specular covering by SiO_2 of the Si surface where Ge is segregating but not forming a uniform layer. After an HF dip roughness increases further. As shown in Tab. 1 this effect is probably due to the fact that SiO_2 removal exposes Ge precipitates.

During device fabrication many cleaning processes are performed (after photolithography, implantation,...) so that more than one cycle is a common procedure. If subsequent cyclesof H_2SO_4/H_2O_2 10 min immersion and HF dip are performed sequentially the roughness increasesmore and more and a value of 0.3 nm was reached on a wafer implanted with 70 keV $3\times10^{16}Ge/cm^2$ after 5 cycles.

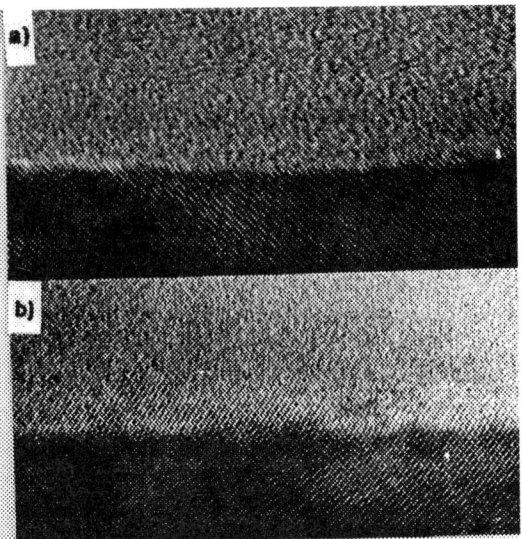

The Ge segregated at the Si_xGe_{1-x} /SiO_2 interface was observed by HRTEM. The HRTEM images of the wafers implanted at 70 keV with $3\times10^{16}Ge/cm^2$ after treatment by H_2SO_4/H_2O_2 and after a thermal oxidation are reported in Figs 3a and b, respectively. The Si_xGe_{1-x}/SiO_2 interface is quite rough, as already established by AFM measurements and crystalline objects are present at the interface. Considering the previous measurements we can identify these objects as crystalline Ge segregated at the Si surface.

Fig. 3 HRTEM of the Si_xGe_{1-x} /SiO_2 interface of a wafer implanted with 70 keV $3\times10^{16}Ge/cm^2$ and a) treated with H_2SO_4/H_2O_2 . b) thermal oxidised.

Ge is not distributed in a uniform layer but it is agglomerated in epitaxial islands. Their dimensions depend on the thickness of the SiO_2 grown. During oxidation Si at the Si_xGe_{1-x} /SiO_2 interface moves through the SiO_2 layer while Ge segregates at the interface. During low temperature processes (chemical treatments) the diffusion of the species is negligible and only an oxide of 1.4 nm can be grown. Then, the amount of Ge segregated is low forming smaller islands.

4. CONCLUSIONS

The roughness measured on as-implanted and annealed samples was comparable to the one measured on virgin silicon. Treatments based on Si_xGe_{1-x} surface oxidation and etching increase the surface roughness and result in Ge segregation at the Si_xGe_{1-x} /SiO_2 interface. The same phenomenon is observed for thermal oxidation, enhanced by the fact that much thicker oxide layers are formed. Most of the Ge segregated at the interface remains also after an HF dip producing a further increase of the roughness. Multiple oxidation processes increase surface roughness more and more damaging the Si_xGe_{1-x} surface. The observed roughness can be detrimental if subsequent processes (for example emitter formation or oxide layer to form MOS) are performed on the wafer.

REFERENCES

Crabbé E, Meyerson B S, Stork J M C and Horama D 1993 Tech. Dig. IEDM 93 83

Heyns M M, Verhaverbek S, Meuris M, Mertens P W, Schmidt H, Kubata M, Philipossian A, Dillenbeck K, Grof D, Schnegge A and de Blank R 1993 Proc. Material Research Society symp. vol. 315, eds G S Higashi, E A Irene and T Ohmi (Pittsburgh: MRS) p.35-45

Iyer S S, Patton G L, Sark J M C, Meyerson B S and Horama D L 1989 IEEE Trans. Electron Devices ED-36 2043.

Pfiester J R and Alvis J R 1991 IEEE Electron Device Letters 12 441

Selvakerman C R and Hecht B 1991 IEEE Electron Device Letters 12 12

Srivatsa A R, Sharon S, Holland O W and Narayan J 1989 J. Appl. Phys. 65 (1989) 4028

Inst. Phys. Conf. Ser. No 146
Paper presented at Microsc. Semicond. Mater. Conf., Oxford, 20–23 March 1995
© 1995 IOP Publishing Ltd

Structural defects in ion beam synthesised semiconducting FeSi$_2$

Z Yang, G Shao[1], K P Homewood, M S Finney, M A Harry and K J Reeson

Department of Electrical and Electronic Engineering, University of Surrey, Guildford, Surrey GU2 5XH
[1]Department of Materials Science and Engineering, University of Surrey, Guildford, Surrey GU2 5XH

ABSTRACT: Both as-grown and annealed ion beam synthesised (IBS) semiconducting FeSi$_2$ (β) have been studied by transmission electron microscopy. In the as-implanted sample, there is a high density of defects with a high level of lattice distortion. A continuous polycrystalline β-layer is formed during subsequent thermal annealing. A streaking contrast normal to [200]$_\beta$ was found within the β grains. This contrast is caused by the interfaces between coexistent β ordered domains, which are 90° rotated to one another around [200]$_\beta$. The mechanism for the formation of ordered domain boundaries is discussed.

1. INTRODUCTION

Semiconducting β phase FeSi$_2$ (space group Cmca) has been receiving growing interest in recent years due to its compatibility with silicon technology and its reported direct band gap which would open a number of potential applications in areas such as silicon-based optoelectronic components (Bost et al 1985). β layers can be fabricated on Si substrates using different techniques including ion beam synthesis (IBS) (Mantl et al 1993). Structural studies have shown a buried IBS β layer with preferred orientation relationships between the β grains and the Si substrate (Gerthsen et al 1992, Tavares et al 1993, Yang et al 1995). A striking feature of our IBS β layers is the existence of internal streaking contrast within the β grains. In this paper, the mechanism to form the internal streaking contrast will be reported.

2. EXPERIMENTAL

A dose of 4 ×10^{17} cm^{-2} Fe$^+$ was implanted at 200 keV into a n-type single crystal Si(100) wafer with a resistivity of 10-20 Ωcm. During the implantation, the wafer was heated to a temperature of 350°C, using the power of the incident ion beam. After the implantation, the sample was annealed in a halogen lamp annealing system with a nitrogen atmosphere at 900°C for 18 hours. Transmission electron microscopy (TEM) studies were carried out on a JEOL 2000-fx electron microscope.

3. RESULTS AND DISCUSSION

Figs. 1a and 1b show a cross sectional TEM (XTEM) image and a large angle convergent beam electron diffraction (LACBED) pattern respectively for the as-implanted β sample. A high density of defects is presented in the Si substrate in the vicinity of the β layer. Moire fringes are observed in the layer, as shown by arrows in Fig. 1b, indicating the existence of embedded small Si crystals. The LACBED pattern reveals a high level of lattice distortion at/near the β/Si interface due to the damage induced strain.

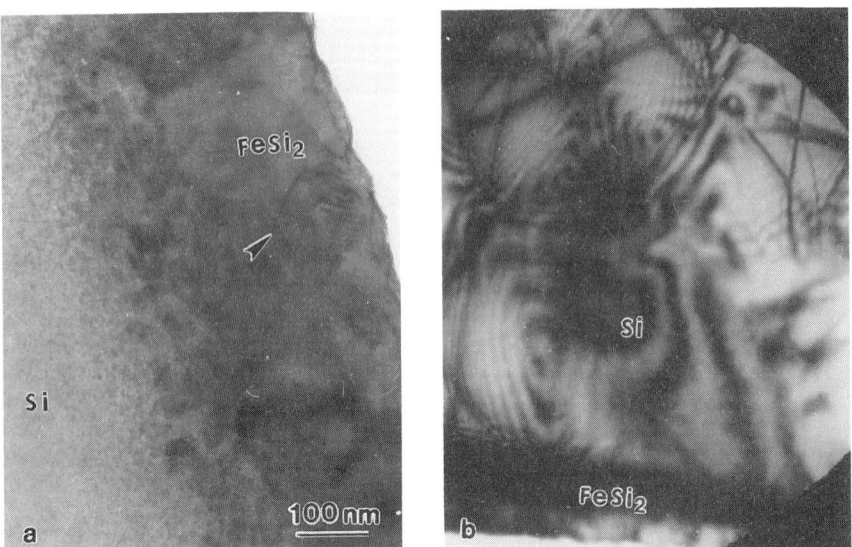

Figure 1: a XTEM image with arrows indicating Moire fringes (a) and a LACBED pattern (b) for as-implanted β-FeSi₂ sample.

Both plan-view and XTEM images for the annealed sample are shown in Figs. 2a and 2b, with superimposed $[012]_\beta$ selected area diffraction (SAD) pattern. It is obvious that thermal annealing removed the implantation damage and led to a well defined continuous layer of coarsened β grains. A streaking contrast is seen clearly when the electron beam (**B**) is perpendicular to $[200]_\beta$ and vaguely when tilting the specimen off the **B**⊥[200] condition. This contrast is due to the presence of planar defects which leads to elongated diffraction spots along the (200) reciprocal vector in the $[012]_\beta$.

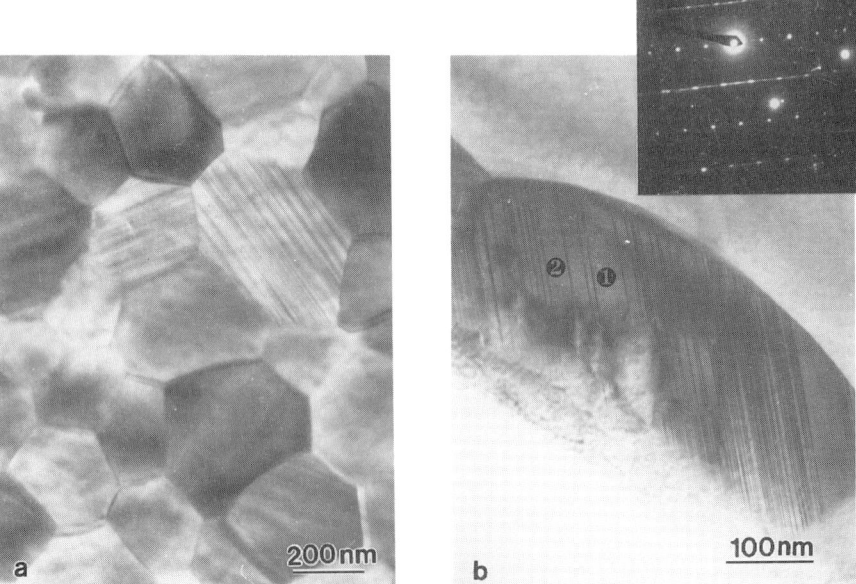

Figure 2: A plan-view TEM image (a) and a XTEM image with superimposed SADP $<01\bar{2}>_\beta$ (b) for annealed β sample.

Figs. 3a-3c are <111>$_\beta$ microdiffraction patterns (MDP) taken from the positions 1, 2 and 3 in Fig. 2b, respectively. It is noted that with a larger electron probe size, some reflections occur at forbidden positions, the equivalent (10$\bar{1}$) positions in Fig. 3a. However, these extra spots could be eliminated from the MDP when the illumination region was free of the streaking contrast. Figs. 3b and 3c show MDPs taken from domains "1" and "2", respectively, using a much smaller probe size. The pattern shown in Fig. 3a is, therefore, clearly a result of the superimposition of Figs. 3b and 3c. The key to Fig. 3a is shown in Fig. 3d. The extra spots in Figure 3a other than those shown in Figure 3d were caused by the double diffraction effect since they could be eliminated by a tilting experiment.

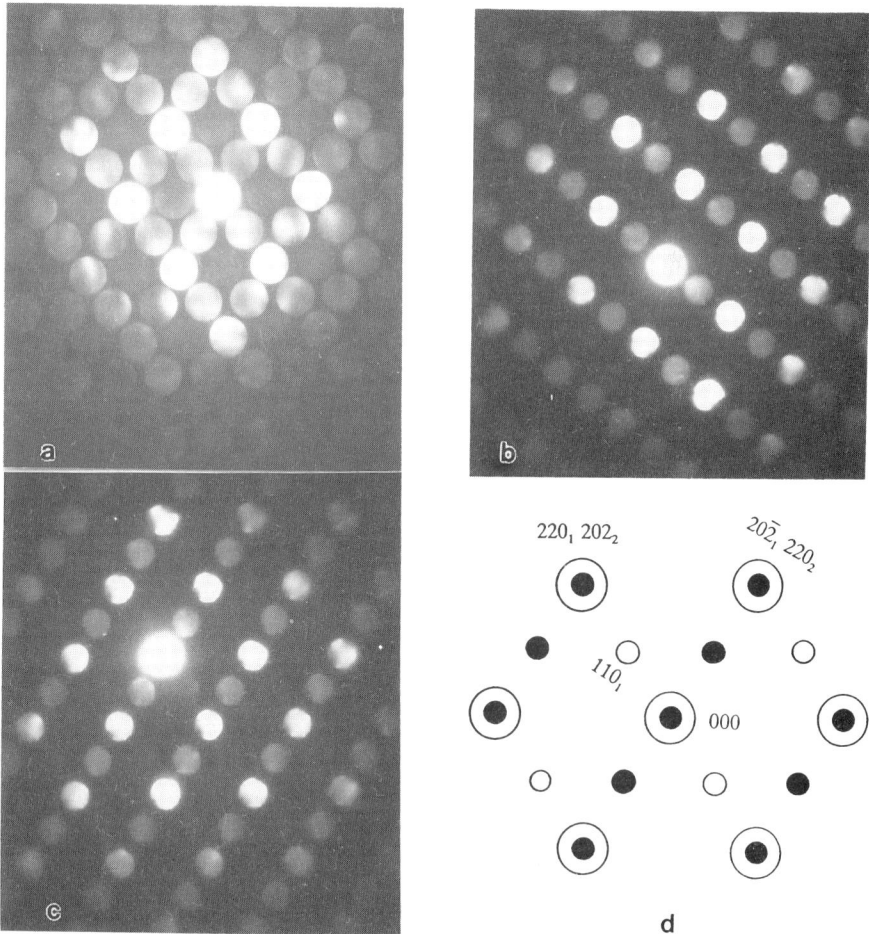

Figure 3: MDP [1$\bar{1}$1]$_\beta$ from the grain in Fig. 2b using an electron probe size of 40nm (a), corresponding MDPs from domain "1" (b) and "2" (c) using a probe size of a few nm. (d) is the key for (a), where open and filled circles represent reflections from domain "1" and "2", respectively.

The crystallographic relationship between domains "1" and "2" is shown in Fig. 4, where domain "2" is rotated 90° around the a-axis of domain "1". The coexistence of such two domains then allows the presence of reflections at both allowed (110)$_1$ and forbidden (101)$_1$ positions (Fig. 3a). In fact, the equivalent position of (110)$_1$ corresponds to (101)$_2$, and (101)$_1$ will be at the same position of (1$\bar{1}$0)$_2$, as shown in Fig. 3d. The trace lines in Fig. 2b is clearly attributed to the boundaries between such 90° related domains which can be defined as **order domains** (ODs), since the actual difference between them lies in the difference in the arrays of the component elements of the β phase.

476

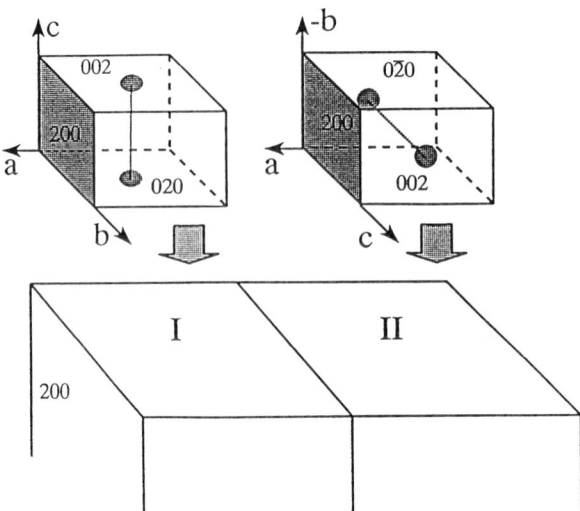

Figure 4: Crystallographic relationship between domain "1" and "2".

The formation of the order domain boundaries (ODBs) in the IBS β-FeSi$_2$ layers can be explained in terms of nucleation and growth. Since a lot of impurity atoms and clusters which are built up at the initial stage of the implantation provide preferential nucleation sites, a large number of precipitate nuclei distribute randomly in the silicon substrate at the initial stage of precipitation. The algebraic similarity between the b- and c- axes of the β phase could permit the nucleation of neighbouring order domains which are 90° to one another around [200]$_\beta$. As (200)$_\beta$ is the most compact lattice plane for the β phase, the growth rate along [200] will be lower than other directions. The preferential growth along the directions perpendicular to [200]$_\beta$ will lead to plate-shaped crystals. The impingements of these β plates will then form the (200)$_\beta$ order domain boundaries via a "growing-in" mechanism. It is worth pointing out that the interfacial energy of such order domain boundaries is expected to be fairly low since the difference between b- and c- of the β phase is very small.

4. CONCLUSION

In the as-grown IBS β-FeSi$_2$ sample, a high density of implantation damage existed and fine Si shreds were embedded in the β precipitates. A continuous buried layer with larger β grains formed after subsequent annealing. Order domains which are 90° to one another around [200]$_\beta$ result in parallel streaking contrast within β grains. The interface between adjacent order domains is (200)$_\beta$. The mechanism for the formation of ODBs is attributed to the impingements of separately nucleated growing silicide nuclei during the process of ion implantation and subsequent thermal annealing.

REFERENCES

Bost, M. C. and Mahan, J. E., J. Appl. Phys., 58 (1985) 2696.
Gerthsen, D., Radermacher, K., Dieker C. and Mantl, S., J. Appl. Phys., 71 (1992) 3788.
Mantl, S., Nuclear Instruments and Methods in Physics Research B 80/81 (1993) 895.
Tavares, J., Bender, H., Lauwers, A., Maex K. and Rossum, M. V., Inst. Phys. Conf. Ser., 134 (1993) 181.
Yang, Z., Homewood, K. P., Finney, M. S., Harry, M. A. and Reeson, K. J., J. Mater. Sci. Lett. (1995), in press.

Inst. Phys. Conf. Ser. No 146
Paper presented at Microsc. Semicond. Mater. Conf., Oxford, 20–23 March 1995
© 1995 IOP Publishing Ltd

TEM characterization of β-SiC synthesized by high dose carbon ion implantation into silicon

A Romano-Rodríguez[1], A Pérez-Rodríguez[1], C Serre[1], L Calvo-Barrio[1], JR Morante[1], R Kögler[2] and W Skorupa[2]

[1]EME, Dept. Física Aplicada i Electrònica, Universitat de Barcelona, Avda. Diagonal 645-647, E-08028 Barcelona, Spain
[2]Forschungszentrum Rossendorf e.V., Institut für Ionenstrahlphysik und Materialforschung, Postfach 510119, D-01314 Dresden, Germany

ABSTRACT: A buried layer containing nanometer-sized β-SiC precipitates has been directly synthesized by high dose carbon ion implantation into silicon at 500°C. The precipitates obtained are almost perfectly aligned to the silicon substrate and are nearly free of defects. After implantation, the crystallinity of SiC is improved by an annealing step (1150°C). Although the precipitates are aligned with the silicon lattice, they present incoherent interfaces to it, which gives rise to a small misfit.

1. INTRODUCTION

SiC is a large band-gap semiconductor, which shows very interesting electronic, chemical and mechanical properties, that make it suitable for high temperature and power electronic applications, as well as for sensors operating in aggressive environments. From the different SiC polytypes, the so-called 3C-SiC or β-SiC (cubic), is the one that presents the most interesting electronic properties. A drawback is that this phase cannot be grown as bulk material, being usually deposited by chemical vapour deposition (CVD), molecular beam epitaxy (MBE) or sputtering.

β-SiC can also be obtained by ion implantation. Implanting high doses of carbon ions into silicon at room temperature leads to the formation of amorphous SiC_x compounds, which regrow into polycrystalline β-SiC after annealing. Implanting at high temperatures, however, directly forms β-SiC precipitates and the synthesized phase shows special orientation relationship to the silicon substrate. So, different authors have observed the synthesis of β-SiC by implanting at temperatures higher than 850°C (Chayahara et al (1993), Martin et al (1990), Nejim et al (1994)). More recently, Simon et al (1995) report the formation of crystalline SiC by implanting at temperatures higher than 400°C. However, this is deduced from Infrared spectroscopy, and no direct observation of the precipitates and their relationship with the Si lattice is presented. De Veirman (1990) presents some TEM results from implantation at 500°C, however for a lower implantation dose.

In this work, a detailed microscopic analysis by TEM and XRD of SiC precipitates formed by carbon ion implantation into Si at 500°C is presented. The results obtained confirm the direct synthesis of the β-SiC phase already at this implantation temperature, being the precipitates nearly free of defects and almost perfectly aligned to the Si substrate. The high crystalline quality of the precipitates obtained gives interest in this technology for the formation of a buried continuous β-SiC layer.

478

2. EXPERIMENTAL

2 inch (001) B doped device grade Czochralski Si wafers (16-24 Ωcm) were carbon implanted at an energy of 300 keV and doses ranging from 1 to 5×10^{17} cm^{-2}. The implantations were performed at an angle of 7° to minimize channelling effects and the substrate was kept during implantation at a temperature of 500°C. After implantation, part of the wafers were annealed in nitrogen ambient at 1150°C for 10 hours.

The samples were characterized by Transmission Electron Microscopy (TEM), both in plan-view and in cross-section, and by X-Ray Diffraction (XRD). Correlation with Fourier Transform InfraRed (FTIR) spectroscopy results was also made.

3. RESULTS AND DISCUSSION

3.1 Transmission electron microscopy

Fig. 1 shows cross-section TEM images of the samples implanted at the dose of 5×10^{17} cm^{-2} (a) as-implanted and (b) annealed. The main feature is that, already after implantation, a 500 nm thick buried layer is formed, which contains a very high density of spherical precipitates, of about 7-10 nm in size. The electron diffraction pattern shown in fig. 2a is obtained in cross-section from the buried area. The pattern confirms that the precipitates are monocrystalline β-SiC, with an almost perfect alignment related to the silicon substrate. The relationship between the two lattices is as follows:

$$[001]_{Si} || [001]_{\beta\text{-SiC}}$$
$$(110)_{Si} || (110)_{\beta\text{-SiC}}$$

Proceeding towards the surface, the density of precipitates diminishes and a zone denuded of precipitates, of about 70 nm, is seen at the surface. Starting at the surface down to a depth of about 180 nm, the presence of {113}-type defects (also called rod-like defects), nucleated during the implantation, is shown. Below the buried layer, first a zone denuded of defects is visible, while below again {113}-type defects are formed. This defects, which are an agglomeration os silicon self-interstitials, have also been observed below and, sometimes, above the buried layer synthesized for SOI structures by oxygen and/or nitrogen ion implantation (De Veirman et al 1991). Plan-view TEM images from this sample show that the upper {113}-type are not uniformly distributed throughout the sample but form a network-like distribution, with bands along <110> directions. This result can clearly be seen in fig. 3, in which also some β-SiC precipitates present in the upper silicon layer are visible.

Upon annealing some changes occur in the structure, but, in spite of the annealing temperature and time, the buried layer remains apparently unchanged. The main features observed are the disappearance of the {113}-type defects, as they are only stable up to temperatures of about 700°C, and the formation of dislocations at their place, but mainly below the buried layer. Electron diffraction of the buried zone of this sample shows, again, the presence of β-SiC spots with the same orientation

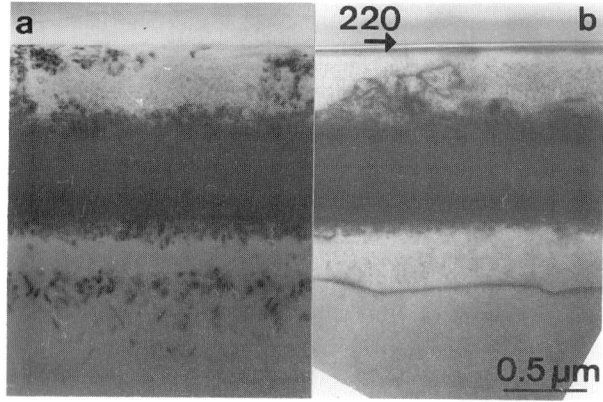

Fig. 1: Cross-section TEM image of the samples implanted at a dose of 5×10^{17} cm^{-2}, a) as-implanted and b) annealed at 1150°C.

relationship to the Si lattice, although a strong increase in their intensity is evident, indicating an improvement in the crystallinity. The very good alignment between the two materials is also verified by the presence of a Moiré fringe pattern in the <111> directions. However, some small misalignments are visible in some of the precipitates.

Regarding the β-SiC precipitates, it is important to point out that the images obtained by conventional TEM do not show any strain contrast around them. Furthermore, HREM images show that the precipitates are surrounded by an amorphous region, as shown in fig. 4. From this result it is possible to conclude that the β-SiC precipitates are well aligned with the silicon lattice, but present incoherent interfaces to it and, thus, show no strain effects.

For the lower implantation doses, the obtained TEM results are similar to those of the higher implantation dose. A variation in the precipitates' density is visible, as a consequence of the lower amount of carbon introduced and, thus, to the lower carbon concentration at the implantation peak. Furthermore, in the as-implanted sample the self-interstitial defects above the buried layer do not appear.

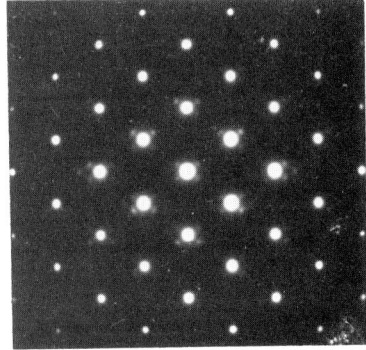

Fig. 2: Electron diffraction pattern of the buried zone of the sample as-implanted, showing the presence of Si and β-SiC spots.

3.2 X-Ray Diffraction

θ-2θ scans, performed on the samples implanted at the highest dose show the presence of the (004) Si peak, together with the (002) β-SiC peak, thus confirming the preferential orientation of the precipitates with the lattice. For the lower dose no peaks corresponding to the SiC are visible, probably due to the low concentration of SiC. For the as-implanted sample the (002) β-SiC peak intensity is quite low, while it strongly increases for the annealed one, which corroborates the increase of the crystallinity with annealing, already shown by TEM. In the annealed sample ω-scans are performed around the (002) and around four of the {220} peaks of β-SiC, in order to estimate the degree of misalignment. After correcting for instrumental broadening, the FWHM of the diffraction peaks is of about 2.4°. This might be interpreted as the precipitates being oriented to the silicon within an angle of about $\pm 1.2°$. Chayahara et al (1993) and Martin et al (1990), who produced samples implanting at temperatures in the range of 850-880°C without post-implantation annealing, report a FWHM of 4.8° (the former) and a misalignment of 3.5° (the latter, probably this value corresponding to the FWHM).

3.3 Discussion

As can be deduced from the results, in spite of the relatively low implantation temperature, compared to the conditions used by other authors (Martin et al (1990), Chayahara et al (1993), Nejim et al (1994)), the β-SiC precipitates are directly formed in the buried layer and they are almost perfectly aligned with the silicon lattice. Concerning the orientation relationships, they are also reported by De Veirman (1990) and Martin et al (1991), although the latter observe a large density of twins. These defects are only rarely

Fig. 3: Plan view image showing the network-like distribution of the {113}-type defects in the as-implanted sample.

present in our samples. Very important is the absence of strain effects around the precipitates, which show that no epitaxial relation exists between the Si and the β-SiC, but that the precipitates grow with their own lattice parameter.

After an annealing step the samples improve their crystallinity, which is indicated by an increase in the intensities of the β-SiC diffraction spots in the TEM and of the (002) β-SiC peak in the XRD spectra. The improvement of the crystallinity is further confirmed by FTIR absorbance spectra, which show a strong decrease of the FWHM of the Si-C absorption peak, centred at 796 cm^{-1} (Serre et al (1995)).

As shown by the low FWHM value (2.4°) of the ω-scans, the β-SiC precipitates are almost perfectly aligned with the Si lattice. Comparison with the results by other authors is difficult, as the experimental set-up differs from one instrument to another. However, in principle, the values obtained in this work are better than those reported by Martin et al (1990) and by Chayahara et al (1993). The highest crystalline quality of the precipitates synthesized in this work is also demonstrated by the comparative intensity of the Si and SiC spots in the electron diffraction patterns.

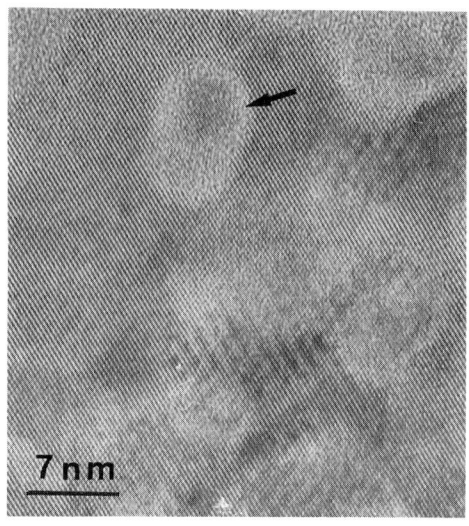

Fig. 4: HREM image showing the β-SiC precipitates at the buried layer. Moiré fringes show the presence of the precipitates and the amorphous interface surrounding one precipitate is indicated (arrow).

4. CONCLUSIONS

Nanometer-sized β-SiC precipitates have been directly formed by high dose carbon ion implantation into silicon substrates at a temperature of 500°C. The precipitates are almost perfectly aligned with the silicon substrate within about $\pm 1.2°$, which, to our knowledge, is the best result reported up to now. The precipitates are coherent with the substrate, but show incoherent interfaces to it. For this reason the mismatch between the substrate and the precipitates is small (0.33%) and no strain effects are observed.

REFERENCES

Chayahara A, Kiuchi M, Kinomura, A, Mokuno Y, Horino Y and Fujii K 1993 Jpn. J. Appl. Phys 32 Part 2 L1286

De Veirman 1990 Ph. D. Thesis (University of Antwerpen)

De Veirman A, Van Landuyt J and Skorupa W 1991 Phil. Mag. A 64 513.

Martin P, Daudin B, Dupuy M, Ermolieff A, Olivier M, Papon AM and Rolland G 1990 J. Appl. Phys. 67 2908.

Nejim A, Hemment PLF and Stoemenos J, Proc. 6th Int. Symp. on SOI Technology and Devices, ed. S Cristoloveanu (ed.), (The Electrochemical Society Proceedings vol. 94-11 (1994), pp. 167-172

Serre C, Pérez-Rodríguez A, Romano-Rodríguez A, Morante JR, Kögler R and Skorupa W 1995 J. Appl. Phys. 78 (in press)

Simon L, Mesli A, Grob JJ, Heiser T and Balladore JL 1995 Mater. Res. Soc. Symp. Proc. (in press)

Inst. Phys. Conf. Ser. No 146
Paper presented at Microsc. Semicond. Mater. Conf., Oxford, 20–23 March 1995
© 1995 IOP Publishing Ltd

Formation of cavities in Si and their chemisorption of metals

D M Follstaedt and S M Myers

Sandia National Laboratories, Albuquerque, New Mexico USA 87185-1056

ABSTRACT: Nanometer-size cavities formed in Si by He$^+$ implantation and annealing are examined with cross-section TEM. During annealing at 700°C or above, He degasses from the specimens, leaving UHV cavities with reactive Si bonds on their walls. Cavity microstructures have been characterized in detail for an implanted fluence of 1×10^{17} He/cm^2: Cavity volume remains approximately constant (0.75 lattice sites/He) for anneals from 700 to ~1000°C, while surface area (3 to 7 times the wafer area) decreases with temperature as the cavities coarsen. The cavities are found to getter up to ~1 monolayer of Cu or Au from solution in Si and a second-phase is not found, thus identifying the trapping mechanism as chemisorption on the cavity walls.

1. INTRODUCTION

Bubbles are formed when semiconductors are implanted with He (Cullis et al 1978, Follstaedt et al 1992, Griffioen et al 1987). Since He is slightly permeable in Si the bubbles can be degassed by vacuum annealing, leaving nanometer-size empty cavities with surfaces as clean as those prepared in ultra-high vacuum. The cavities have been exploited to study the binding of H on their surfaces (Myers et al 1993), faceting and free energies of crystallographic planes (Cullis et al 1978, Follstaedt et al 1993, Eaglesham et al 1993), trapping of transition metal impurities (Follstaedt et al 1995, Myers et al 1995), and trapping of charge carriers at the dangling surface bonds (Seager et al 1994). In these studies, well characterized cavities with microstructures that can be manipulated are very useful; the cavities must also be relatively stable. Here we characterize cavity volumes and surface areas in detail using cross-section transmission electron microscopy (TEM). When Cu and Au are gettered to cavities from their equilibrium phases with Si, they saturate at amounts consistent with ~1 monolayer (ML) of solute on the cavity walls. Furthermore, TEM examination of cavities with trapped Cu show no precipitated phases. These findings lead us to conclude that the trapping is due to chemisorption of the metals on cavity walls.

Cavities were formed by ion implanting 1×10^{17} He/cm^2 at 30 keV into (001) float-zone Si. Cross-section TEM specimens were prepared using mechanical polishing and ion milling, as discussed elsewhere (Follstaedt et al 1992). Imaging was done with a Philips CM20T operating at 200 keV and capable of 0.27 nm resolution. To observe cavities in the midst of lattice damage from the implantation, specimens were tilted several degrees from the [110] orientation to reduce diffraction contrast from the damage, and cavities were imaged using underfocussing to highlight their edges with Fresnel contrast. In general, a greater underfocus (~1.5 μm) was required for optimum contrast with small cavities (~1 nm), lower magnifications (≤ 100 kx), or thick areas with high projected cavity densities, while less underfocus (~600 nm) was required when fewer, larger cavities (>~20 nm) were imaged, at higher magnification (≥ 200 kx), or in thin areas. Cavities were identified by their contrast change on going from underfocus (light inside/dark outside) to overfocus (dark inside/light outside). Cavities were counted and sized by their largest diameter, and their average area and volume were obtained from these distributions. The density of cavities per wafer area was obtained by measuring the specimen thickness using contours from two-beam diffraction conditions or Kikuchi lines in convergent-beam diffraction patterns (Kelly et al 1971).

482

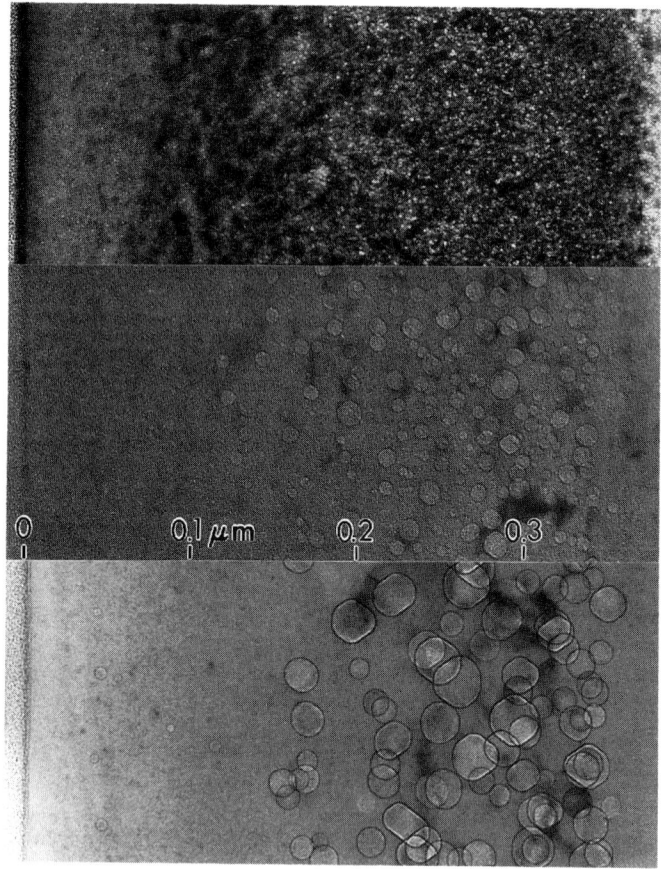

Figure 1. Bright-field, cross-section TEM micrographs of Si implanted with $1x10^{17}$ He/cm^2 at 30 keV.

a) As implanted; imaged at -1 μm (underfocus).

b) Annealed 1/2 hour at 700°C; imaged at -1.2 μm.

c) Annealed 1 hour at 900°C; imaged at -600 nm.

2. CAVITY MICROSTRUCTURE EVOLUTION

Figure 1a) shows the microstructure formed by implanting at room temperature with no anneal. A high density of 1-2 nm bubbles is found at depths of ~0.05 to 0.4 μm; the density is highest at the He projected range, 0.3 μm, where a peak concentration of 9 at.% He is predicted (Ziegler 1990). The dark contrast is a residual effect of the dense lattice damage, although individual dislocations were not resolved. Upon annealing the implanted Si in vacuum for 1/2 hour at 700°C, the cavities enlarge to an average diameter of 8.5 nm and a well defined layer is found from 0.15 to 0.35 μm, as in Fig. 1b). The overlayer is almost cavity-free. After 1 hour at 900°C larger (20 nm) cavities form, as in Fig. 1c). A few isolated dislocations can be seen with residual contrast in the images of annealed cavity layers. Numerous dislocations are found under two-beam conditions with low-index reflections. Annealed cavities often show facets, predominantly {111}, although many appear round. Some cavities are elongated along <111> directions and probably formed by coalescence. Detailed counting and sizing were done for three anneals as summarized in Table I. Cavity diameter increases with temperature and cavity density decreases, while volume remains approximately constant. The microstructure thermally evolves but the cavity layer persists, even after processing at 1180°C (Medernach et al 1995). The volume corresponds to 0.75 Si lattice sites per implanted He atom, suggesting that the He atomic density in bubbles approached that of a solid. It is notable that the 0.2 μm-wide cavity layer has a porosity of ~7 %. The internal cavity surface area is also high: 3 to 7 times the wafer area.

Table I. Summary of Cavity Microstructures and Surface Trapping Sites
(001) Si implanted with $1 \times 10^{17} He/cm^2$ at 30 keV and subsequently annealed as indicated.

Anneal:	½ hr. 700°C	1 hr. 800°C	8 hr. 800°C	1 hr. 900°C	½ hr. 1000°C
Avg. Diameter (nm)	8.5	13.3	16	20	28
[a] Areal Density of Cavities (cm^{-2})	2.9×10^{12}		6×10^{11}	3.1×10^{11}	
Cavity Volume per Wafer Area (nm^3/nm^2)	12.4		16	15	
Cavity Area to Wafer Area Ratio	6.7		4.5	3.7	
[b] Calculated Trap Sites per Wafer Area (cm^{-2})	4.7×10^{15}		3.2×10^{15}	2.6×10^{15}	
Trapped Solute per Wafer Area $(at./cm^2)$	4.8×10^{15} Cu		3.3×10^{15} Au	2.3×10^{15} Cu	

[a] $\pm 15\%$ uncertainty in cavity density and related quantities.

[b] Cavity/Wafer Area Ratio multiplied by an areal density of 7 sites/nm^2.

3. GETTERING METAL IMPURITIES

We have investigated gettering of transition metals (Cu, Au, Ni, Co and Fe) to cavity layers and found them to be trapped strongly, with energies ≥ 1.5 eV with respect to solution (Follstaedt et al 1995, Myers et al 1995). The metals were introduced by ion implantation, annealed to form equilibrium phases, and their buildup at a separated cavity layer monitored with ion backscattering. Here we examine Cu and Au, which bind more strongly to cavities than the equilibrium phases and saturate the cavity sites. In a typical experiment, Cu_3Si was formed on one side of a wafer and cavities on the other; the Cu was gettered from the silicide across the wafer to the cavities during annealing at 600°C. The amount of trapped Cu increased with time and then saturated; 4.7×10^{15} Cu/cm^2 were found for cavities formed by annealing 1/2 hour at 700°C (Myers et al 1994).

The ratio of cavity area to the implanted wafer area can be used to predict the number of cavity-wall trapping sites per wafer area, assuming one metal atom per Si atom. The Si surface density differs for each facet plane; we use 7 Si bonds/nm^2, which is appropriate for unreconstructed {111} planes as well as 1x2 reconstructed {100} planes. The predictions are given in Table I along with the saturated amount of trapped Cu and Au found using cavities formed by annealing as indicated at the top of the table. Both metals saturate at levels consistent with 1 ML coverage of the cavity walls. The rather exact agreements are fortuitous given the 15% uncertainty in the area ratio and the use of a single surface density. Nonetheless, such agreement with two elements in several experiments with differing cavity areas is compeling support for chemisorption binding on cavity walls. Saturation at such low levels appears inconsistent with precipitation, which would probably continue until much of the implanted Cu (1×10^{17} Cu/cm^2) was gettered to the cavities.

Cavities saturated with Cu were examined with TEM to look for precipitation directly. No evidence of second phases was observed in bright-field while tilting between high-symmetry directions nor in diffraction patterns. Moreover, only Si lattice spacings are seen in high-resolution (lattice) images, as in Fig. 2. Such images exhibit a contrast change between the interior and exterior of the cavity, but no indications of a second phase on the cavity walls are seen. The absence of Cu precipitation is readily understood, since silicide is already present and there is no thermodynamic driving force for additional precipitation. In some images, a single {111} atomic facet plane appears dark, which could be due to either Cu on the surface or surface reconstruction.

484

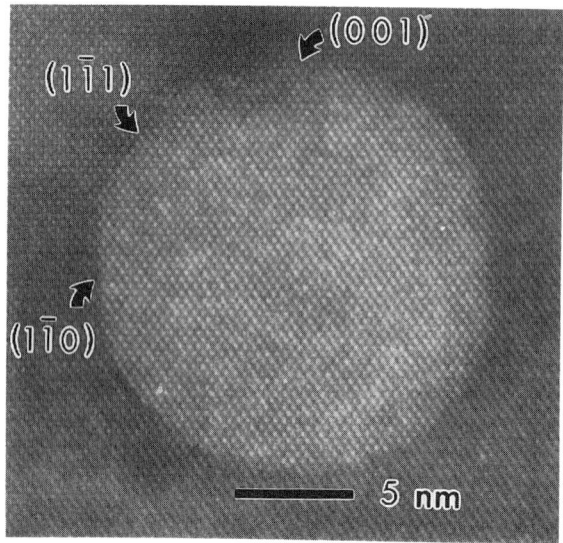

Figure 2. High-resolution (lattice) image of a cavity from a layer in Si saturated with Cu. The cavity was formed by annealing 1 hour at 900°C and then gettering Cu for 26 hours at 600°C. Prominent facets are indicated.

In Fig. 2, flat {111} and {110} facets are readily identified. The cavity is also truncated in the (001) direction, but the surface rounds toward adjacent {111} planes, as also seen in pure Si (Follstaedt 1993). The images indicate that Cu binds on all three types of facets.

The binding energy of Cu (2.2 eV) to cavities has been used to model its gettering (Myers et al, 1994), and under representative technological conditions, the residual solution concentration could be reduce by several orders of magnitude more than achievable by precipitation-based gettering. Lower levels are achievable because the cavity gettering mechanism relies upon solute trapping which operates even at low concentrations, whereas precipitation can reduce concentrations only to the impurity solubility. Cavities appear viable as front-side gettering centers in future microelectronics devices, provided other detailed considerations are met (Medernach et al 1995).

This work supported by Division of Materials Sciences, Office of Basic Energy Sciences of the United States Department of Energy under contract DE-AC04-94AL85000.

REFERENCES

Cullis A G, Seidel T E and Meek R L 1978 J. Appl. Phys. 49 5188.

Eaglesham D J, White A E, Feldman L, Moriya N and Jacobson D C 1993 Phys. Rev. Lett. 70 1643.

Follstaedt D M, Myers S M, Wampler W R and Stein H J 1992 Proceedings of the Electron Microscopy Society of America (San Francisco Press) Part I, 334-335.

Follstaedt D M 1993 Appl. Phys. Lett. 62, 1116.

Follstaedt D M, Myers S M, Petersen G and Medernach J 1995 J. Electron. Mater., in press.

Griffioen C C, Evans J H, de Jong P C and Van Veen A 1987 Nucl. Inst. Meth. B27 417.

Kelly P M, Jostsons A, Blake R G and Napier J G 1975 Phys. Stat. Sol a31 771.

Medernach J W, Hill T, Myers S M and Headley T J 1995 to be published.

Myers S M, Follstaedt D M, Stein H J and Wampler W R 1993 Phys. Rev. B47 13,380.

Myers S M, Follstaedt D M, Bishop D M and Medernach J W Semiconductor Silicon 1994, eds H Huff, W Bergholz and K Sumino (Electrochem. Soc, Pennington, NJ) PV94-10 808.

Myers S M, Follstaedt D M, Petersen G A, Seager C H, Stein H W and Wampler W R 1995 Nucl. Inst. Meth. B in press.

Seager C H, Myers S, Anderson R, Warren W and Follstaedt D 1994 Phys. Rev. B50 2458.

Ziegler J F 1990 private communication.

Inst. Phys. Conf. Ser. No 146
Paper presented at Microsc. Semicond. Mater. Conf., Oxford, 20–23 March 1995
© *1995 IOP Publishing Ltd*

485

Effect of D-defects and oxygen precipitates on oxide integrity

J-G Park[1], **A Romanowski**[1], **K-C Cho**[2], **C-S Lee**[2], **R Falster**[3] and **G A Rozgonyi**[1]

1. Department of Materials Science and Engineering, North Carolina State University, Raleigh NC 27695-7916.
2. Analysis Technology Department, Samsung Electronics, San #24, Nongseo-Lee, Kihung-Eup, Youngin-Gun, Kyungki-Do, Korea.
3. MEMC, Via Gherzi 31, 28100 Novara NO, Italy

ABSTRACT: Morphology observations of oxide breakdown sites via X-TEM clearly differentiate oxide breakdown mechanisms due to D-defects and oxygen precipitates. The D-defect related oxide breakdown is associated with both local oxide thinning and shear stresses around the defect site; whereas, that due to oxygen precipitates is associated with precipitate induced shear stress. The D-defect induced shear stresses enhance a local oxide tunneling current which increases with oxide thickness. The D-defect induced MOS/EBIC contrast is strongly dependent on the applied bias and the polarity of bias.

1. INTRODUCTION

In DRAM fabrication, the thin gate oxide integrity (GOI) is a key factor in achieving high yield and reliability of devices. Recently, much effort has been made to investigate the impact of as-grown and process induced crystal defects on GOI, particularly D-defects(called flow pattern defects, Yamagishi et al 1992), oxygen precipitates(Satoh et al 1992), and IR laser scattering tomography defects(Umeno et al 1993). Investigating the BV_{ox} dependence on t_{ox} is very useful in identifying the origin of oxide breakdown, since D-defects and oxygen precipitates exhibit different oxide breakdown voltage(BV_{ox}) dependence on oxide thickness(t_{ox}). As shown Fig.1, for t_{ox}, from 200 to 2000Å, BV_{ox} due to the presence of D-defects decreases with increasing t_{ox} up to $\approx$1000Å, and then recovers with further increase in t_{ox} (Park et al 1994). On the other hand, oxide breakdown degradation due to the presence of oxygen precipitates attenuates with increasing t_{ox} above 200Å, see Fig.2. One-to-one correlation of D-defects and oxygen precipitates with oxide breakdown sites via MOS/EBIC, FIB, and XTEM enable us to propose separate oxide breakdown mechanisms. In addition, the D-defect related MOS/EBIC imaging mechanism is discussed.

2. EXPERIMENTAL PROCEDURES

In order to isolate individual defect induced oxide breakdown sites, BV_{ox} was measured for which the leakage current reached -25μA/capacitor. The detailed microscopic morphology of defect induced oxide breakdown sites was determined through i) isolating the defect position via MOS/EBIC imaging, ii) marking the defect site with FIB, iii) preparing the exact site as a XTEM foil using the FIB, and iv) investigating the structure of defects in TEM. Similarly, the morphology of the D-defect induced gate oxide breakdown site in DRAM devices was observed through i) isolating the defect site via Emission Microscope for Multi Inspection (EMMI), ii) marking and vertical etching of the defect site using FIB, and iii) X-SEM observation at the defect. To understand the D-defect induced MOS/EBIC imaging, the D-defect induced leakage current during slow line scanning of the electron beam was investigated at a beam current of 100pA and accelerating voltage of 15kV as a function of bias voltage and polarity using an x-y recorder.

3. EXPERIMENTAL RESULTS

Figure 3 is a representative XTEM micrograph obtained from a BV_{ox} failure site induced by a D-defect. All D-defect related TEM observations exhibit three important morphological characteristics: 1) the D-defect is a cavity of 0.25 to 0.50µm diameter located at the Si/SiO_2 interface, 2) the oxide layer which has grown around the D-defect exhibits a large thickness variation wherein the normal oxide thickness of ≈ 100nm is reduced to a minimum of ≈ 20nm, and 3) the aluminum layer of the MOS capacitor is a convex cap over the D-defect and it's locally thinned oxide, which suggests that the D-defect initially creates a convex surface thickness variation with an aspect ratio of ≈ 32 % (height/width $= \approx 0.16$ µm/ ≈ 0.5 µm) in a bare wafer. Figure 4 is a representative SEM micrograph of D-defect induced oxide breakdown sites in an actual DRAM device. D-defects are typically observed as surface etch pits in gate oxide regions. Comparing Figs.3 and 4, it is evident that the morphology of the D-defect cavity in an MOS capacitor is correlated with a surface etch pit created after various DRAM etching and oxidation processes. Figure 5 is a representative XTEM micrograph for an oxygen precipitate induced oxide breakdown. It shows that polyhedral oxygen precipitates are incorporated into the oxide layer without a noticeable oxide perturbation. This is fundamentally different from the morphology of the D-defect induced oxide breakdown site.

As shown in Fig.6, a local D-defect induced tunneling current yields EBIC contrast at the defect site during electron beam scanning. The dependence of EBIC imaging contrast on bias polarity is shown in Figs. 6(a) and (b) where a change in bias polarity from accumulation to depletion/inversion results in the defect contrast changing from dark to bright, while the background contrast remains essentially unchanged. This behavior is indicative of a defect site spatially located at the Si/SiO_2 interface, or in the oxide itself, but not in the bulk(Kirk et al 1994). In addition, the D-defect induced EBIC imaging contrast is strongly dependent on oxide thickness of MOS devices and the polarity of applied bias. As shown in Figs.7(a) and (b), the D-defect induced EBIC contrast for t_{ox} of ≈ 105Å can only be observed under depletion/transition bias, and the contrast intensity is relatively weak, i.e., ≈ 4.5%. and independent of electric field. For t_{ox} of ≈ 191Å the EBIC contrast decreases rapidly with increasing electric field. For t_{ox} of ≈ 500Å the maximum EBIC contrast of ≈ 200% is observed at electric field of ≈ 0.1MV/cm, and then it decreases with increasing electric field. For t_{ox} of ≈ 1000Å the EBIC contrast intensity is weak and uniform over a wide range of applied electric field.

4. DISCUSSION

There are two oxide breakdown mechanisms related to the presence of D-defects; namely, a local oxide thinning and local oxidation induced shear stresses around the defect site. According to XTEM observations of D-defect related oxide breakdown sites, these defects are cavities of 0.25 to 0.5µm in extent, see Fig.3. The oxide layer grown around the D-defect site is subjected to a large thickness variation with the normal t_{ox} of ≈ 1000Å being reduced to a minimum of ≈ 200Å. The D-defect related oxide breakdown occurs at the edge of the D-defect site, which is ≈ 400Å in t_{ox}, although the oxide above the defect site shows a minimum thickness of ≈ 200Å. The BV_{ox} of ≈ 12MV/cm is applied to the oxide at the edge of the D-defect, point **a** in Fig.3, but the BV_{ox} of ≈ 4.8MV/cm is only applied to the oxide above the defect site, point **c**. The extent of local oxide thinning around the defect site increases for

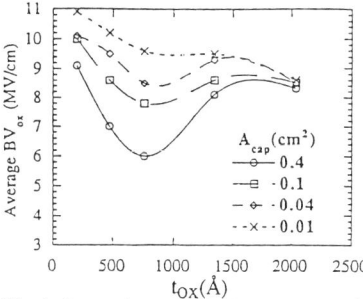

Fig.1. Dependence of the D-defect induced BV_{OX} on t_{OX} for wafer grown with a crystal pull rate of rate1.4mm/min.

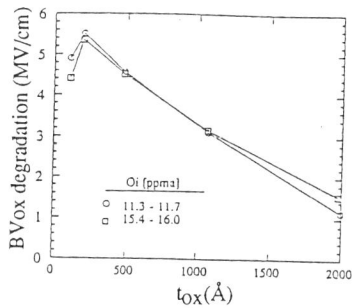

Fig.2. BV_{OX} degradation due to only presence of oxygen precipitates as a function t_{OX}.

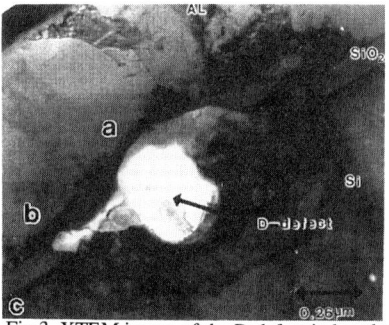

Fig.3. XTEM image of the D-defect induced oxide breakdown site (4.06MV/cm).

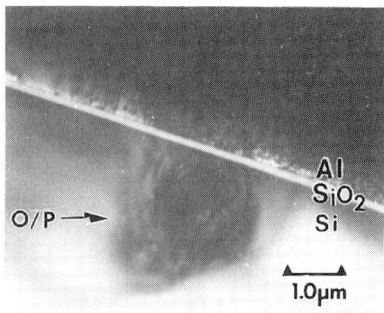

Fig.4. XTEM image of oxygen precipitate(O/P) induced oxide breakdown site (3MV/cm).

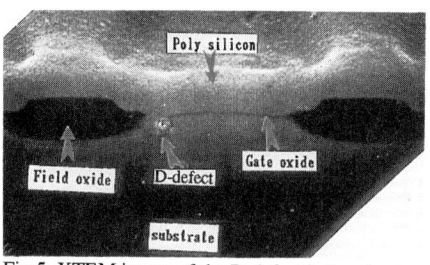

Fig.5. XTEM image of the D-defect induced gate oxide breakdown in a DRAM device.

(a)   (b)

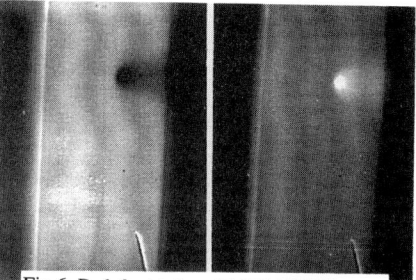

Fig.6. D-defect induced MOS/EBIC image. (a) accumulation and (b) depletion/transition bias.

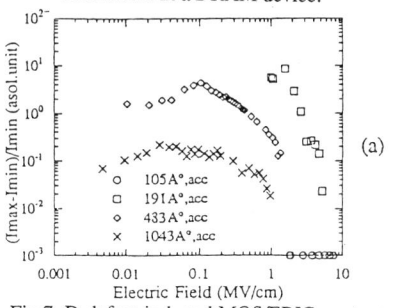
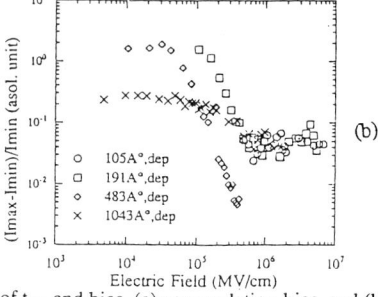

Fig.7. D-defect induced MOS/EBIC contrast as a function of t_{OX} and bias. (a) accumulation bias and (b) depletion/transition.

thicker oxides since the oxide growth above a D-defect is limited by the finite silicon thickness capping a defect. In addition to local oxide thinning, a stress induced oxide breakdown is also operative when D-defects are in close proximity to the Si/SiO$_2$. As shown in Fig.8, the D-defect induced shear stresses increase with oxide thickness prior to an oxide collapses. As a result, the BV$_{ox}$ decreases with increasing t$_{ox}$ due to the presence of local oxide thinning and oxidation induced resolved shear stress, as shown in Fig.1. At t$_{ox}$ of $\approx$ 1000Å, the resolved shear stresses at t$_{ox}$ of $\approx$ 1000A$^\circ$ exceed the critical value of $\approx$ 16GNm^{-2}, resulting in an oxide rupture at the edge of the D-defect site, the point a in Fig.3. Recall that the shear stress required to break oxide bonds is $\approx$ 16GNm^{-2}. The occurrence of an oxide rupture advances the oxide layer grown around the D-defect to be collapsed for t$_{ox}$ above $\approx$ 1000Å. Thereafter, the overall oxide thickness begins to equalize since the collapsed region, which no longer contains the D-defect, oxidizes faster than the surrounding previously oxidized host crystal areas. As a result, the thickening oxide layer becomes progressively more resistant to field induced breakdown, and the BV$_{ox}$ increases for t$_{ox}$ above $\approx$ 1000Å, as shown in Fig.1. In MOS/EBIC observations, a local D-defect induced tunneling current increases with t$_{ox}$, as shown in Fig.10. For a given applied bias, a local current increase in a MOS/EBIC observation originates from enhancing impact ionization in a locally damaged oxide layer caused by the D-defect induced resolved shear stresses. As expected, the locally D-defect induced tunneling current increases with the D-defect induced resolved shear stress, see Fig.11. It is evident that the D-defect induced stresses increase with t$_{ox}$.

The formation of oxygen precipitates in the Si matrix during thermal treatment develops compressive strains due to either a growth residual strain or a cooling strain. Both types of oxygen precipitate induced stresses degrade oxide integrity due to the presence of stress induced oxide breakdown. Based on TEM observations of oxygen precipitate related oxide breakdown sites in Fig.4, the remaining volume of oxygen precipitates existing the Si/SiO$_2$ interface dramatically decreases with increasing oxide thickness. As a result, the remaining oxygen precipitate induced shear stress decreases with increasing oxide thickness, as shown in Fig.9. Thus, the extent of the oxide integrity degradation due to the presence of oxygen precipitates themselves attenuates with increasing t$_{ox}$, as shown in Fig.2.

In MOS/EBIC observation, electron beam scanning passing through the oxide creates electron-hole pairs in the oxide by breaking Si-O bonds, which produces a tunneling current in the oxide. To isolate the D-defect induced oxide breakdown site, all MOS devices were pre-stressed by applying the total leakage current of -25μA/cm^2. Since oxide growth around the D-defect site induces a local shear stresses, many of the Si-O bonds near the Si/SiO$_2$ interface are weak, thereby reducing the impact ionization energy. As a result, the EBIC current at the D-defect site locally increases, as shown in Fig.6. To understand the detailed current conduction in the oxide, three carrier transport processes were reviewed, as shown in Table.1.

Table.1. Conduction Processes in Insulators(Sze 1991 and Ma et al 1989)

Process	Expression
Tunnel emission	$J \propto E^2 \exp[-\dfrac{2\sqrt{2m^*}(q\phi_B)^{\frac{2}{3}}}{3\pi qhE}]$
Space-charge-limited	$J = \dfrac{8\varepsilon_i \mu V^2}{9d^3}$
Hopping transport or trap mediated valence band hole conduction	$J \propto E \exp(-E_{ac}/kT)$

ϕ_B barrier height, E= electric field, ε_i=insulator dynamic permittivity, m^*=effective mass, d=oxide thickness, h=plank constant, V= voltage and E_{ac}=activation energy of conduction carrier.

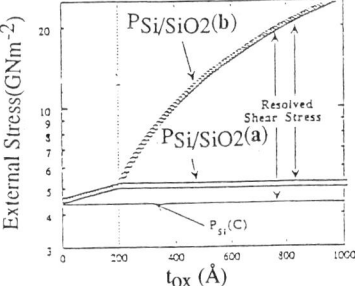

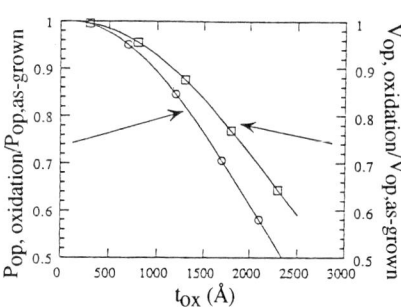

Fig.8. Simulated D-defect induced shear stresses as a function of t_{ox}.

Fig.9. Simulated oxygen precipitate induced shear stress as a function of t_{ox}.

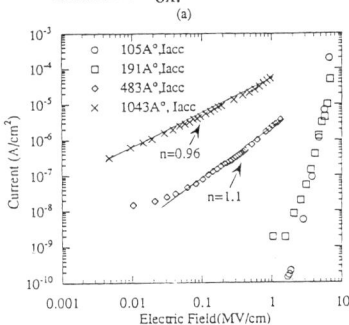

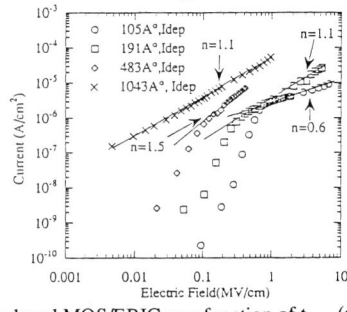

Fig.10. log I vs. log E characteristics for the D-defect induced MOS/EBIC as a function of t_{ox}. (a) accumulation bias and (b) depletion/transition bias.

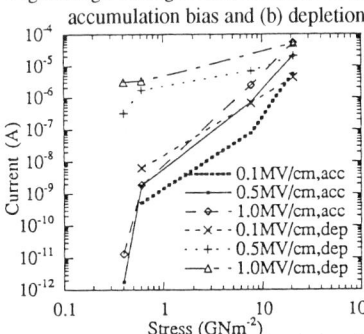

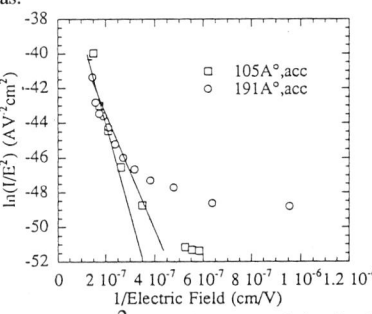

Fig.11. Correlation of the D-defect induced MOS/EBIC with the D-defect induced shear stress where acc is accumulation and dep is depletion /transition bias.

Fig.12. $\ln(1/E^2)$ vs. 1/E characteristics for the D-defect induced MOS/EBIC for t_{ox} of 105Å under accumulation bias. Inserts show tunneling emission.

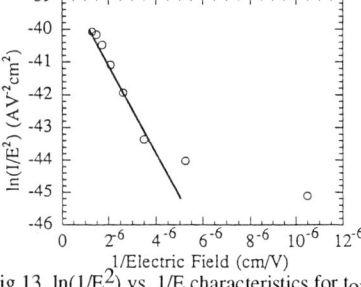

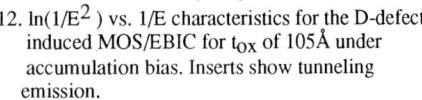

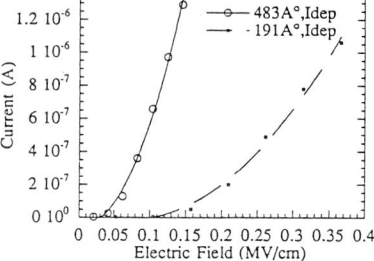

Fig.13. $\ln(1/E^2)$ vs. 1/E characteristics for t_{ox} of 105Å under depletion/transition bias. Insert shows tunneling emission.

Fig.14. I vs. E characteristics for the D-defect induced MOS/EBIC for t_{ox} of 105 and 191Å under depletion/transition. Inserts show space-charge conduction including hopping transport.

For MOS devices with t_{ox} of 105 and 191Å under an accumulation bias, the current conduction is dominated by tunnel emission, as shown Fig.12. For these oxide thickness the resolved shear stresses are less than 1GNm^2. Therefore, the oxide sustains a very high electric field across the oxide, corresponding to tunneling emission. For thicker oxides, the D-defect induced shear stress rapidly increases with t_{ox}, allowing current conduction through the oxide at lower the electric fields. Still higher D-defect induced shear stresses induce oxide damage, creating many oxide traps. The current conduction is then dominated by hopping transport or trap-mediated valence band hole conduction, which produces a log Current log vs. Voltage slope of one. As expected, MOS devices with t_{ox} of 483 and 1043Å under an accumulation bias exhibits a slope of log Current vs. log Voltage close to one, see Fig.10(a). Comparing Figs.10(a) and 10(b), it is evident that depletion/inversion results in higher D-defect induced EBIC current than accumulation bias, due to space-charge-limited conduction from carriers injected into the insulator to compensate the SCR charge under depletion. For MOS devices with t_{ox} of 105, 191, and 483Å under a depletion/transition bias, two different current conduction mechanism obtains, as shown in Fig.9(b). At a relatively high electric field, they seem to follow hopping transport or trap-mediated valance band hole conduction. However, at a lower electric fields, the tunneling emission is dominant for the t_{ox} of 105Å, see Fig.13, while the space-charge-limited conduction including hopping transport is dominant for oxide thickness of 191 and 483Å, see Fig.14. Otherwise, for a MOS device with the t_{ox} of 1043Å, the dominant current conduction is hopping transport or trap-mediated valence band hole conduction since the oxide growth of 1043Å induces very high D-defect induced shear stress above 15GNm^2.

5. CONCLUSIONS

The morphology and the BV_{ox} dependence on t_{ox} indicate the defect nature. D-defects are vacancy related defects while oxygen precipitates are interstitial related defects. The D-defect induced EBIC current rapidly increases with oxide thickness due to the presence of the D-defect induced shear stresses. The correlation of the D-defect induced EBIC current with the D-defect induced stress indicates that the D-defect induced MOS/EBIC contrast follows tunneling current microscopy (Lau et al 1993) contrast.

REFERENCES

Kirk H, Radzimski Z, Buczkowski A and Rozgonyi G A 1994 IEEE Trans. Electron Devices **41**, 6.
Lau W S, Chan D S H, Phang J C H, Chow K W, Chow K S, Pey K S, Lim Y P and Chonquist B 1993 Appl. Phys. Lett. **63**, 2240
Ma T P and Dressendorfer P V 1989 Ionizing Radiation Effects in MOS Devices and Circuits (New York: Wiley) pp 87-192
Park J G, Ushio S, Takeno H, Cho K-C, Kim J-K and Rozgonyi G A 1994 Diagnostic Techniques for Semiconductor Material and Devices eds D K Schroder, J L Benton and P Rai-Chudhury (Pennington: ECS) **94-33**, 57
Satoh Y, Murakami Y, Furuya H, Shingyouji T 1993 Appl. Phys. Lett **64**, 303.
Sze S M Physics of Semiconductor Devices 2nd edition, p 402.
Umeno J, Sudamitsu S, Murakami H, Hourai H, Sumita S, and Shigematsu T 1993 Jpn. J. Appl. Phys. **32**, L699
Yamagishi H, Fusegasa I, Fujimaki N and Katayama M 1992 Semicond. Sci. Tech. **7**, A135

Inst. Phys. Conf. Ser. No 146
Paper presented at Microsc. Semicond. Mater. Conf., Oxford, 20–23 March 1995
© *1995 IOP Publishing Ltd*

High performance oxide on SiGe and the interface

Siu-Wai Chan[1], L Zhao[1], C Chen[1], P W Li[2] and E Yang[2]
[1] Dept. of Chemical Engineering, Materials Science and Mining Engineering,
[2] Dept. of Electrical Engineering, Columbia University, New York, NY 10027

ABSTRACT: High resolution transmission electron microscopy was applied to study the interface of $Si_{0.85}Ge_{0.15}$ and its oxide grown by using an electron-cyclotron-resonance (ECR) oxygen-plasma system. The interface is atomically abrupt and nanoscopically smooth, while the oxide layer is uniform in thickness within 5%. There are no pinholes observed. Unlike other oxidation methods, neither the precipitation of Ge nor the massive segregation of Ge are observed. These findings are consistent with the x-ray photo-emission spectroscopy results, electrical measurements and the high quality pMOSFET devices which have been made using the ECR oxide.

1. INTRODUCTION

A good oxide or native dielectric is a key element for making Metal-Oxide-Semiconductor devices (e.g. MOSFETs, CMOS, CCD) and circuits (e.g. DRAM, BICMOS, A/D converters). Many devices which could be benefitted by the advantages of $Si_{1-x}Ge_x$ alloyed semiconductor over Si are impeded by the quality of the oxide which can be grown on $Si_{1-x}Ge_x$. Previous investigations by LeGrouse et al (1989) and Liou et al (1991) have shown the segregation of the Ge to the thermal oxide/ $Si_{1-x}Ge_x$ interface when the oxide was grown by a standard thermal oxidation process similar to those used for the normal Si device processing. Silicon is more prompt to oxidation than Ge. During the oxidation of a SiGe alloy, Si is preferentially oxidized over Ge. The precipitation of the leftover Ge to the interfaces has caused the interfaces to be unsuitable for device applications. Various methods of SiGe oxidation such as rapid thermal oxidation by Nayak et al (1992), high pressure oxidation by Piane et al (1991), and low energy ion beam oxidation by Vancauwenberhe et al (1991) were proposed but none of these have produced sufficiently high quality oxides. The best of these results from rapid thermal oxidations but only gives a minimum fixed charge density Q_f above 5×10^{11} /cm^2 and a lowest interface state density D_{it} above 10^{12}/cm^2-eV. Clearly a better oxidation method is needed for device quality interfacial electrical properties.

Oxide layers on $Si_{1-x}Ge_x$ were grown in oxygen plasma generated using an electron-cyclotron-resonance (ECR) microwave source by Li et al (1992). The electrical properties of the ECR-grown SiGe oxide layers were characterized by current-voltage (I-V) as well as high frequency and quasi-static capacitance-voltage (C-V) measurements. These properties are compared with those of the conventional properties of the Si/SiO_2. From the C-V characteristics of aluminum-gate MOS capacitors indicate the fixed charge and the interface state densities were lower than earlier literature results of oxides on SiGe alloys. The performance of the p-MOS field effect transistors (FET's) devices with one micron Al-gate fabricated with the ECR oxide was high and showed high saturation transconductance and channel hole mobility both at 300K and 77K. The feasibility of making high speed, low power devices with the ECR oxides for SiGe/Si system has been demonstrated as comparable to the traditional devices made with Si/SiO$_2$. The important parameters as well as interface properties for device applications are listed in Table I for oxide layers grown by ECR oxidation, rapid thermal oxidation of SiGe alloys and traditional Si/SiO$_2$ interfaces. In particular, the formation of stoichiometric SiGe oxides by ECR plasma was verified through x-ray photo-emission spectroscopy (XPS) study by Li et al (1992) and the SiGe pMOSFET's made with ECR gate oxides were shown to have a high performance by Li et al (1994). Such outstanding characteristics of the ECR oxide have motivated us to study the structural aspect of the ECR oxide/SiGe interface.

Table 1: The important interface properties for device applications for oxide layers grown by ECR oxidation, and rapid thermal oxidation of SiGe alloys, compared to those of the traditional Si/SiO$_2$ interfaces.

Types of oxides	Fixed Charge Density, Q_f, (10^{11}cm^{-2})	Interface state density, D_{it} (10^{11}cm^{-2} eV^{-1})	Breakdown Field (MV/cm)	Channel Mobility at 77K (cm/s-V)
ECR at 15°C	-1.1	0.9	8-10	530
rapid thermal oxide[1]	>5	>10	N/A (Leaky)	-
SiO$_2$/Si	0.1-0.5	0.1-0.5	>10	400

2. EXPERIMENTAL

A layer of 10nm thick oxide was grown at 15°C by ECR oxygen plasma on 100nm thick MBE Si$_{0.85}$Ge$_{0.15}$ which was on (001) Si with a 100nm thick buffer MBE Si. The ERC oxygen plasma was generated under a 250 Watt microwave power with oxygen at 2 mTorr. The oxide layer was capped with a layer (50nm-100nm thick) of evaporated polycrystalline Al. (110) cross sectional samples of the ECR oxide on Si$_{0.85}$Ge$_{0.15}$ with oxide grown were made and studied with a Jeol 4000FX transmission electron microscope (TEM) with a point-to-point resolution of 0.22nm.

3. RESULTS AND DISCUSSIONS

The oxide layer was observed (see Fig. 1) to be continuous, uniform in thickness to within 5% and without pinholes along a few micrometers of the oxide

---

[1]   Nayak et al (1992)

493

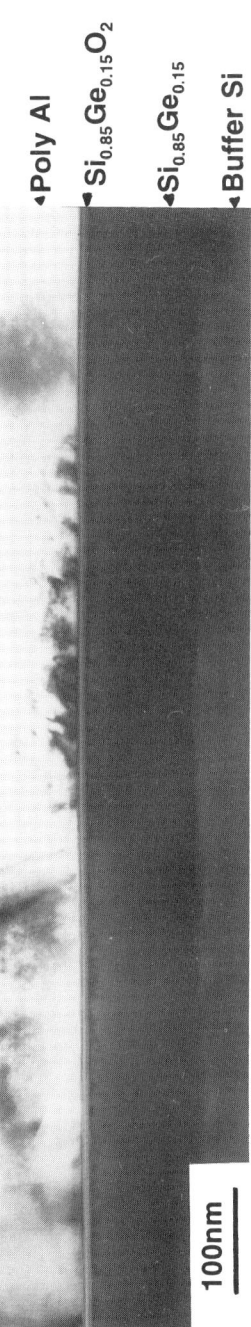

Figure 1: Low magnification TEM micrograph showing the extent of the oxide layer and its uniformity.

Figure 2: High resolution TEM micrograph showing the nanoscale smoothness of the oxide/SiGe interface.

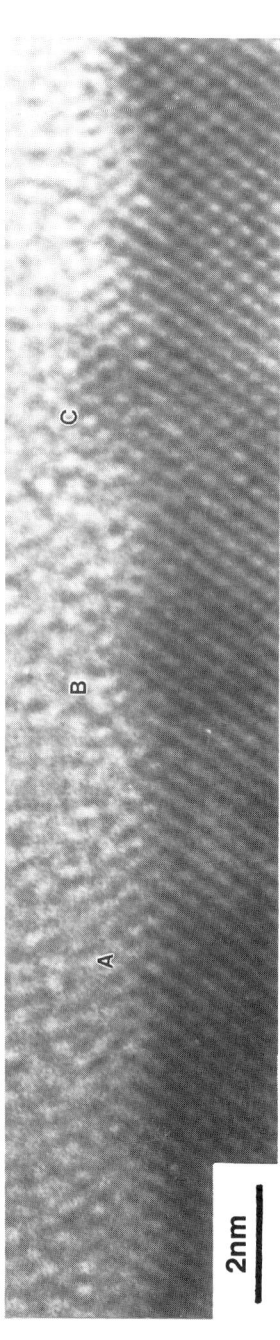

Figure 3: HRTEM micrograph showing the truncated octahedral, cuboidal protrusions (A, B & C) at the interface.

layer under TEM observation. This microscopy result is consistent with early electrical measurements by Li et al (1994) where high breakdown voltages (8-10x 10^6 V/cm) were observed. No precipitation of Ge to the oxide/SiGe interface was observed in the high resolution micrographs (see Figs. 2 and 3). This is in agreement with the previous results of XPS data of the oxide interface by Li et al (1992) where no distinct peaks corresponding to pure Ge were observed in XPS spectra.

The interface between the ECR oxide and the $Si_{0.85}Ge_{0.15}$ is atomically abrupt with no distinct areas or features that can be associated with massive Ge segregation. If there were local segregation of Ge, it would be at a very low level because no distinct areas at the interface can be characterized to have a higher electron absorption characteristic of a higher Z element segregation. The roughness of the interface is kept to a very low level. Some half-octahedrals (see areas A, B and C marked in Figure 3) bounded by 4 {111} planes as well as some truncated semi-octahedral, cuboidal protrusions with {111}, {110} and {001} terminating surfaces of $Si_{0.85}Ge_{0.15}$ were observed. The remarkable coincidence is that all these protrusions are more or less of the same height, which keeps the interface smooth down to the nanometric scale. Such smooth interface is an important structural characteristic for the superb electrical properties of the oxide layer.

In order to have a low interface state density at the oxide/SiGe interface, it is important to have a low density of dangling bonds originating from the underlying semiconductor SiGe layer. Presently the TEM technique cannot observe such fine details of the interface.

4. CONCLUSIONS

Our TEM observations agree with previous XPS, C-V, I-V results of the ECR oxide on $Si_{0.85}Ge_{0.15}$. The interface is nanoscopically smooth and the oxide layer is uniform in thickness with no pinholes observed. Neither the precipitation of Ge nor the massive segregation of Ge were observed. These findings are consistent with the electrical measurements and the high quality pMOSFET devices which have been made from the use of the ECR oxide.

REFERENCES

LeGrouse F. K., Rosenberg R, and Meyerson B. S., (1989) Appl. Phys. Lett. 54, 644

Li P.W., Liou H.K., Yang E.S., Iyer S. S., Smith III T. P., and Lu Z., (1992) Appl. Phys. Lett. 60, 3265

Li P.W., Yang E. S., Yang Y. F., Chu J., and Meyerson B. S., (1994) IEEE Electron Devices Lett. 15, 402,

Liou H. K., Mei P., Gennser U., and Yang E. S., (1991) Appl. Phys. Lett. 59, 1200

Nayak D. K., Kamjo K., Park J. S., Woo J.C. S. and Wang K.L., (1992) IEEE trans. Electron Devices, ED-39, 56

Piane D. C., Caragianis C., and Schwartzman A.F., (1991) J. Appl. Phys., 70, 5076

Vancauwenberhe O., Hellman O.C., Herbots N., and Tan W. J., (1991) Appl. Phys. Lett., 59, 2031

Inst. Phys. Conf. Ser. No 146
Paper presented at Microsc. Semicond. Mater. Conf., Oxford, 20–23 March 1995
© 1995 IOP Publishing Ltd

TEM and HREM study of microstructure of FZ-silicon crystal grown at high varying rate

L M Sorokin, J L Hutchison*, K Scheerschmidt**, G N Mosina and N B Ponomariova

Ioffe Physical-Technical Institute, Russian Academy of Sciences, 194021 St.Petersburg, Russia
*Department of Materials, University of Oxford, Parks Road, Oxford OX1 3PH, UK
**Max-Planck Institute for Microstructure Physics, Weinberg 2, 06120 Halle/Saale, Germany

ABSTRACT: Two types of microdefects (MDs) have been found in as-grown FZ Si. They resemble dislocation loops of very small size and the nanoparticles of a square platelet form. The latter are stable after 950°C - 20h treatment. The focus dependence of HREM contrast of nanoparticles has been studied. The cell-like structure of Si grown at about 9-10mm/min has been revealed. Specific feature of images related to this microstructure are studied, and a possible mechanism of its formation is discussed.

1. INTRODUCTION

The D-type MDs having the smallest size and the highest concentration in FZ Si grown at 5-8 mm/min rate are the most difficult to study as compared to the A- and B-type MDs according to the well known classification of Veselovskaya et al.(1977) . First, they do not cause additional reflections in microdiffraction patterns, either because of their small size or due their non-crystalline nature. Second, the size of an MD is usually less than the total thickness of the specimen under study. Thus a problem in revealing MDs arises, since the image contrast is formed not by the contrast from the MD itself only, but also by the contrast from the matrix above and beneath it. Disk-like defects (DLDs) were also revealed by HREM in FZ silicon grown at ~ 10 mm/min rate (Sorokin et al. 1993). Such defects have not been found in the crystals with lower growth rates examined earlier.The contrast features between DLDs and the DLD-matrix boundary may have an alternative explanation than that given by Sorokin et al (1993).

2. AIMS OF THE INVESTIGATION

The aims of this investigation were: a) to obtain new information on the nature of D-MDs in the FZ-Si crystals and on the influence of high temperature annealing on these MDs; b) to explain the contrast features at the DLD-matrix boundary in crystals grown at ~ 9-10mm/min rate, and c) to suggest a possible DLD formation mechanism.

3. EXPERIMENTAL

A p-type FZ Si crystal (~ 30mm in diameter and ~ 12cm in length) was grown in vacuum at ~ 5-10mm/min varying rate. The concentration of oxygen and carbon was less

496

than 1×10^{16} cm^{-3} and about 1.6×10^{16} cm^{-3} respectively. For a selective etching pattern the crystal was cut parallel to the [111] growth axis in the (112) plane. XRT, XRD, TEM and HREM were used in the investigation. In addition, a specimen cut from another crystal grown at 4mm/min rate was used, all other factors being the same. The TEMs used were JEOL 7A and 4000EX(II) microscopes.

4.RESULTS

Both the selective etching and XRT for the part of the crystal with growth rate Vgr $\sim$10mm/min show a high density of dislocations ($\sim 10^5$ cm^{-2}) which suggests that this part of the crystal had been subjected to the high thermal stress during cooling. On the selective etching pattern outside the dislocation area (Vgr $\sim$5-9mm/min) the etch pits are seen arranged along the growth striations.

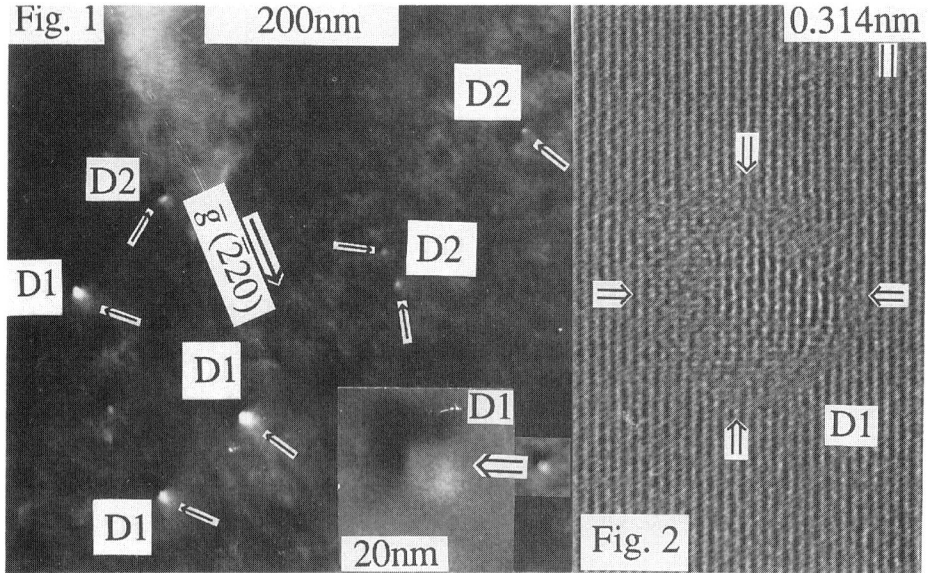

Fig.1 (112) TEM image; Vgr $\sim$4mm/min, showing two types (D1,D2) of MD with BW contrast. Inset: enlarged platelet-like nanoparticle (D1) far from extinction contour (s $\neq$ 0). Fig.2 (111) Si, (Vgr $\sim$4mm/min) viewed along <112>. HREM image of a D1 defect situated inside matrix. Note irregularity on periphery.

For the specimen from the crystal with Vgr $\sim$4mm/min TEM investigation showed two types of small MDs having black-white (BW) contrast (Fig.1). Some of them having a broken up line of no contrast (LNC) are larger than others (with straight LNC) resembling very small dislocations loops. *Not straight* LNC for image of defects with BW contrast indicates a nonspherical shape for the particle. But for a two beam approximation for s = 0 the particle shape can not be seen clearly. However it can be revealed rather clearly far from an extinction contour at s $\neq$ 0 (inset in Fig.1). It should be noted that the nanoparticles are present in all the crystals grown at Vgr $\sim$4-8mm/min rates. They are stable at high temperature annealing (950°C for 20h): their sizes and shape do not change.

The HREM image of a square shaped particle similar to the inset in Fig.1 is presented in Fig.2. Analysis of this image showed the following contrast features: inward and outward bending of lattice planes; in the area of particle projection the number of lattice fringes is one fewer than in the corresponding area in the matrix; termination and shifting of fringes. In Fig. 2 the lattice fringe distances inside the particle projection is on average the same as in the matrix, i.e. 0.314 nm. The decrease of the number of fringes with respect to matrix appears through irregularities at the periphery of the particle. The average value of fringe distance through the whole particle projection with periphery is 0.336 nm. The lattice fringe image features observed correlate in most cases with the calculated data of Scheerschmidt et al.(1989) obtained for various imaging conditions.

The images of particles situated inside the matrix depend essentially on defocus (Fig.3), irregularities of lattice plane images being shifted from the centre of particle projection. One of the main conclusions should be done from studying lattice fringes contrast: one may not judge the sign of matrix deformation from the appearance of lattice fringe bending.

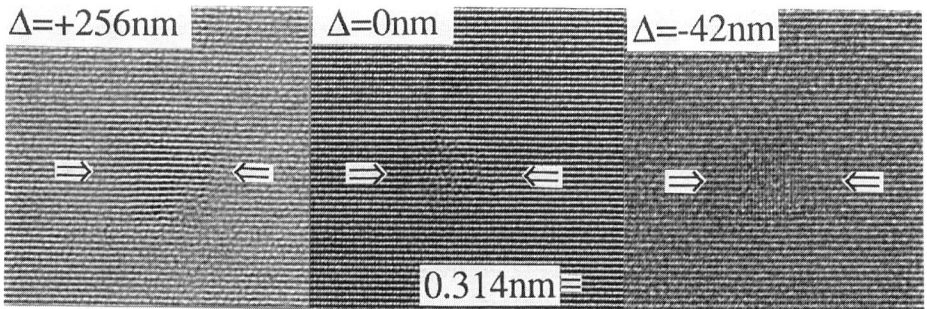

Fig.3 (111) Si (Vgr $\sim$4mm/min) viewed along $<112>$, showing the variation with defocus of lattice image contrast for particle inside matrix.

As mentioned, the DLDs are the typical microstructural features of FZ Si grown at Vgr $\sim$ 10mm/min. It was interesting to discover whether they also appeared in the region grown at lower rate (Vgr $\sim$9mm/min). It turned out that the specimens cut from the region corresponding to this lower growth rate showed a cell-like structure. The cells appear as areas of the same size as the DLDs, but without sharp boundaries (Fig.4a). So they resemble the DLDs at an earlier stage of their formation. In the region of cellular structure Moire-like fringes are seen. In this area the diffuseness of every fourth atomic row is evident. The atomic column projection in the distorted rows appears as enlarged, white diffuse spots (Fig.4b). This feature could be caused by a small misorientation of neighbouring cells (by about 1°). The continuity of atomic planes is reached by gradual bending without disturbance of the coherence between the cells. This bending leads to changes in lattice distances and to local displacement of atoms, giving the overall appearance of Moire fringes. These new observations allow us to suggest an alternative mechanism for the microstructure formation.

It was shown by Muller (1988) that during growth the probability of thermal fluctuations at the crystallization front increases at high growth rate. This results in increasing concentration of regions with a higher temperature than the average for the growth front. Thus the capture of melt droplets by the growing crystal is possible. The droplets become trapped inside the solid matrix and cannot crystallize freely. The stresses arising at solidified drop-matrix boundary may cause various kinds of deformation. For example, when the torsion deformation component is predominant, the rotation of one crystal part about another will be present. A multi-slice simulation of such an image is shown in Fig. 5b.

498

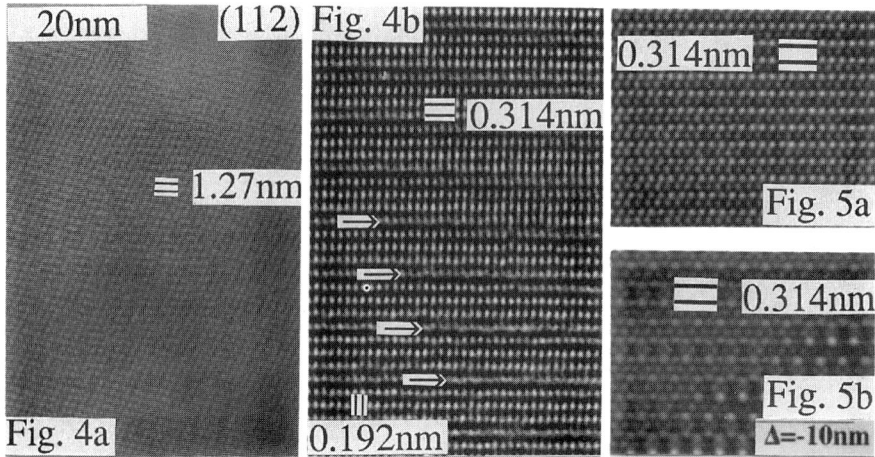

Fig.4 Si(111) (V ~ 9mm/min), viewed along < 112 >. a) cell-like structure image with fringes resembling Moire pattern; b) enlarged area with Moire fringes.
Fig.5 Si(111) (V ~ 10mm/min), viewed < 112 >. Compare experimental image (a) with the simulated image (b) of two (112) rotated lattices. See text for details.

The simulations were performed (15 beams contributing in the image) with the following parameters: accelerating voltage = 400 kV, Cs = 0.9mm, thickness range 7.98 - 15.3 nm, surfaces inclined by 19° to beam direction. This image is obtained by superposition of two Si slabs, rotated by 2° around < 112 > with respect to one another and projected along < 112 >. This is in reasonable agreement with the HREM image in Fig. 5a. Using this simulated image the appearance of the extra planes in images of areas between neighbouring DLDs for the crystal region with Vgr ~ 10mm/min can be explained rather well. Further calculations to extend this analysis are now in progress.

5. CONCLUSION

MDs of two types were found in the FZ Si (Vgr ~ 4-8mm/min): a) MDs of non-dislocation nature - nanoparticles having the shape of square platelets, b) MDs resembling dislocations loops of very small size. For crystal with Vgr ~ 9mm/min a cellular structure was revealed which resembled the DLDs at an early stage of their formation.

A mechanism for the formation of the DLDs and cell structure was suggested. Some features of the HREM images related with above mentioned structures were discussed on the basis of torsion one part about another, or misorientation of two neighbouring areas without disturbing the overall coherence.

REFERENCES

Muller G 1988 Convection and Inhomogeneities in Crystal Growth from the Melt 12.
    Crystals, Properties and Applications, ed. H C Freyhardt (Berlin:Springer-Verlag) 143-51
Scheerschmidt K, Hillebrand R and Heydenreich J 1989 Phys.St.Sol.a 116 123
Sorokin L M, J L Hutchison, Ponomariova N B and Fal'kevich E S 1993 Inst.Phys.Conf.
    Ser.No.134: sect.3 107
Veselovskaya N V, Sheikhet E G, Neimark K N and Fal'kevich E S 1977 in "Growth and
    doping of semiconductor crystals and films" publ. by Academy of Sciences, Moscow, 2 284

Inst. Phys. Conf. Ser. No 146
Paper presented at Microsc. Semicond. Mater. Conf., Oxford, 20–23 March 1995
© 1995 IOP Publishing Ltd

Surface plasticity of silicon at low temperature

Y G Shreter[1,2], Y T Rebane[1,2], D V Tarhin[1], S A Khorev[1], D Cherns[2] and J W Steeds[2]

[1] A F Ioffe Physico-Technical Institute, Russian Academy of Sciences, St Petersburg, Russia
[2] H H Wills Physics Laboratory, University of Bristol, Tyndall Avenue, Bristol BS8 1TL, UK

ABSTRACT: A new type of surface defect has been observed around scratches made at 77 K or 300 K and at high loads, 80 g, on a diamond stylus. The defects are responsible for surface plasticity (long-range residual stress) around scratches and form a characteristic surface contrast when samples are etched. The patterns can be locally described by a director which is usually oriented along the scratches. It was found that the defects can move under applied stress and disappear after annealing at 600° C for 1 hour. It is suggested that the defects are dense clouds of vacancies.

1. INTRODUCTION

Scratching and indentation of silicon at room temperature gives rise to a large variety of complicated and interesting physical phenomena.

Eremenko and Nikitenko (1972) have found a high density of immobile dislocations in the close vicinity of the indentor. Gridneva et al (1972) have discovered that the semiconductor silicon transforms into a metallic state as a result of indentation. Hill and Rowcliffe (1974) attributed dislocations observed around indentations to a mechanism that they described as a crystal block slip. Clark et al (1988) found that single-crystal silicon was converted to a conducting amorphous state directly under an indentor. Pirouz et al (1990) investigated the generation of hexagonal silicon and martensitic transformations under indentation at T > 400° C. Cahn (1992) has reviewed problems related to the metallic state of silicon under indentation.

Stickler and Booker (1963) found dislocation networks and cracking around scratches. Renninger (1972) discovered long-range strain fields around scratches, which he attributed to microplastic deformation of the brittle material. Sunada (1974) observed small crystal misorientations which he assigned to incomplete healing of microcracks in the vicinity of scratches. Badrick et al (1977) found that the long-range residual strain was an in-plane-compression and associated it with a system of cracks. Directional effects in the appearance of scratches in <112> directions have been investigated in detailed by Puttick et al (1980). Recently Minowa and Sumino (1992) have discovered stress-induced amorphisation of a silicon crystal by surface scratching at room temperature.

In this paper we have found unusual etching patterns around scratches. The patterns are attributed to an ensemble of vacancies which group into linear surface defects (LS-defects). These defects have been investigated by optical microscopy, SEM and TEM methods combined with a very sensitive preferential etching technique.

2. EXPERIMENTAL

Scratches have been made on an oxidised surface of FZ n-Si at room temperature by a 120°-cone diamond stylus with loads of 80 g. The radius of the curvature of the diamond point used was about 2 μm. The silicon used in these experiments was 2000 Ω cm dislocation-free single crystal. The thickness of the wafers was about 300 μm. Prior to scratching the surfaces {100} and {111}, were mechanically and then chemically polished and oxidised at 1050 C for 40 min. After scratching the samples were etched in "Secco" etch for 15 s. SEM-micrographs were obtained on a JEOL 6400; TEM-micrographs were obtained using a Philips EM 430 operating at 250 kV.

3. RESULTS

3.1 Pattern around a single scratch

Fig 1. shows an optical micrograph of etching patterns around the scratches made at room temperature on {100} surfaces, after removal of the oxide layer from a part of the crystal surface. The typical pattern widths are about 30 μm and there are defect-free zones of about 5-10 μm in width around the scratch-furrows.

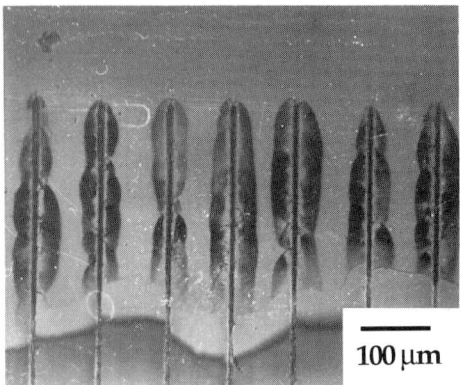 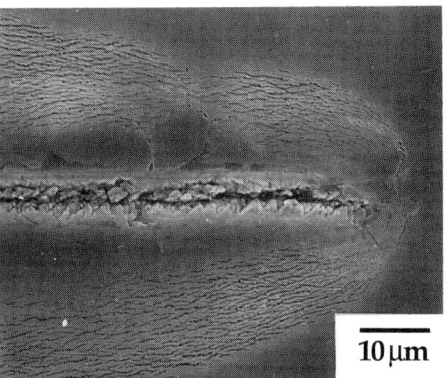

Fig. 1.   Fig. 2.

SEM investigation showed that the patterns consist of the LS-defects and that some peculiarities occurred in the patterns near the chevron cracks and the scratch tips (Fig 2.). Sometimes, the LS-defects seem to be associated with microcracks or, alternatively, the microcracks may be formed from agglomerations of the LS-defects (Fig.3).

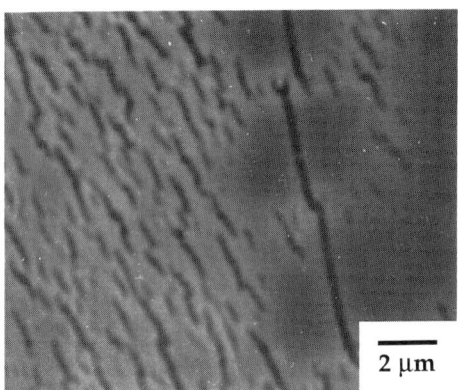 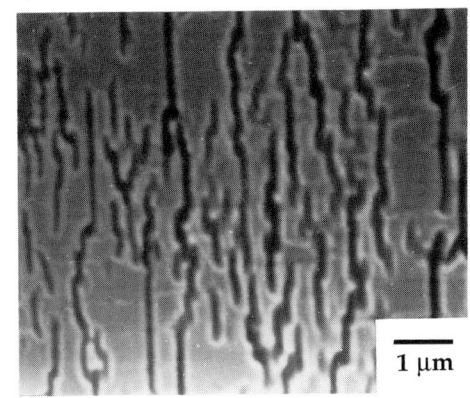

Fig. 3.   Fig. 4.

The LS-defects can be also created by scratching at 77 K but are absent in scratched foils subsequently annealed at 600 C for 1 hour.

SEM investigation at higher magnification showed that the etching pattern consisted of segments with length 1-2 μm and width of 0.1 μm (Fig.4). The LS-defect orientation can be characterised locally by a director (bi-directional vector). The director was mostly parallel to the scratch line but it changed orientation near the scratch tip (Fig.2).

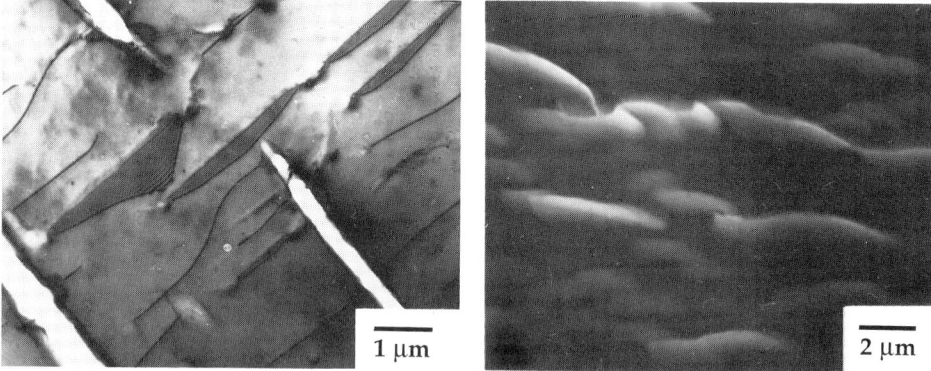

Fig. 5.   Fig. 6.

Fig.5 shows a TEM micrograph from an area near the scratch on an etched sample with deliberately introduced oxidation induced stacking faults (OISFs) and 90°-dislocations (details of the method are described by Lilikov et al (1992)). It can be seen that the LS-defects are much more sensitive to the etching then OISFs and dislocations.

Fig.6 shows a TEM micrograph from an area near the scratch in a sample without deliberately introduced OISFs and dislocations. The front surface of the sample has not been preferentially etched but, for the sample thinning, a non-preferential etch from back surface was used. It can be seen from this picture that the LS-defects were localised near the surface. In contrast to the SEM surface etching patterns, the LS-defects appear as wide trenches of about 1-2 μm in width which is an order of magnitude large than that observed by SEM.

3.2 Patterns around the scratch crossings.

Fig.7 shows an optical micrograph of a preferentially etched sample with scratches. It can be seen that the individual etching patterns do not simply superimpose on each other at the crossing points.

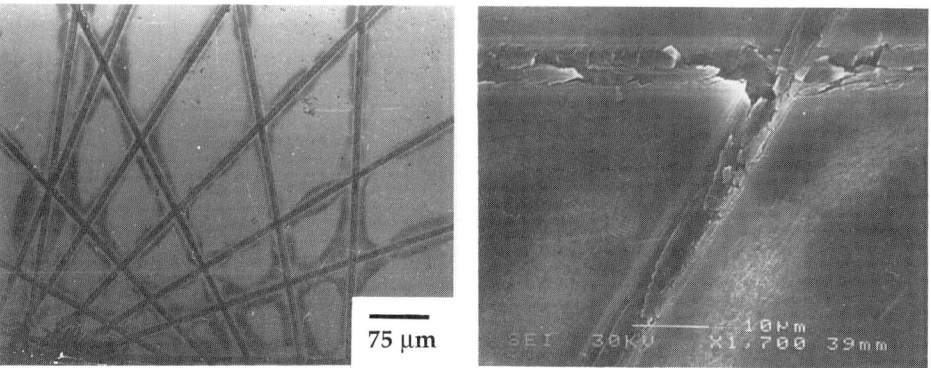

Fig. 7.   Fig. 8.

Higher resolution SEM micrographs of scratch intersections (Figs.8,9) show that there is an interaction of the patterns from two scratches. This interaction may lead to a continuous change in the

502

direction of the LS-defects from an alignment along one of the scratches to an alignment along the other scratch. It can be seen that the behaviour is different for acute and obtuse angles (Fig.8). Sometimes, for angles close to a right angle, both types of behaviour are present and this give rise to a disclination in the LS-defect field (Fig.9).

4. DISCUSSION

TEM observation (Fig.6) shows that LS-defects are not related to Hertz-type surface microcracks or dislocations. An investigation by CBED showed that the crystal was perfect inside the LS-defects and therefore they were not regions of local amorphisation.

There are two main features which have to be explained.

i) The etching rate of LS-defects was much higher than that of dislocations or stacking faults (Fig.5) and therefore LS-defects should have a high concentration of dangling bonds.

ii) LS-defects can move along the crystal surface as can be deduced from Figs. 2,7 and 8. This motion did not produce detectable damage of the crystal lattice.

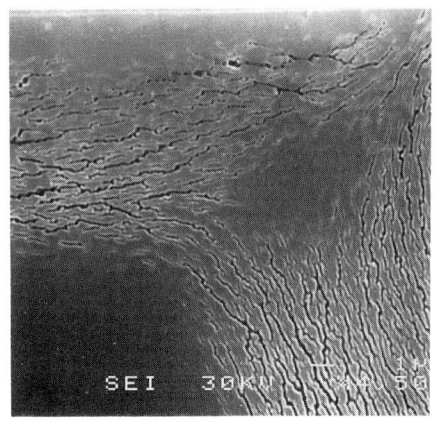

Fig. 9.

It is difficult to imagine an extended defect that possesses these properties.

We suggest that the LS-defects are dense clouds of vacancies. The vacancies might be generated directly from the surface in response the high compressive stress during the process of scratching. The SEM-pattern is dictated by the pattern of surface steps where the vacancy formation energy is lower. The TEM-patterns show extended regions, revealed by etching, into which the vacancies have been injected from the surface. This model is compatible with the mobility of the LS-defects. At very high vacancy densities the LS-defects can develop into microcracks or Griffith's flaws by coalescence (Fig.3). Thus, the LS-defects may be important for an adequate description of the phenomena of strength and fatigue of brittle materials.

ACKNOWLEDGEMENTS

We would like to thank L.Farmer, S.Wright and I.I. Khodos for help with TEM measurements, and the EPSRC for financial support.

REFERENCES

Badrick A S T, Eldeghaidy F, Puttick K E and M A Shahid 1977 J. Appl. Phys. 10 195
Cahn R W 1992 Nature, 357 645
Clark D R, Kroll M C, Kirchnet P D and Cook R F 1988 Phys. Rev. Lett. 60 2156
Eremenko V G and Nikitenko V I 1972 Phys. Stat. Sol. a1 4 317
Gridneva I V, Milman Y V and Trefilov V I 1972 Phys. Stat. Sol. 14 177
Hill M J and Rowcliff DJ 1974 J.Mat.Science, 9 1569
Lelikov Y S, Rebane Y T, Ruvimov S S, Sitnikova A A, Tarhin D V and Shreter Y G 1992
    Sov. Phys. Solid State, 34 804
Minova K and Sumino K 1992 Phys. Rev. Letters, 69 320
Puttick K E and Hosseini M M 1980 J.Phys.D 13 875
Renninger M 1972 J.Appl.Cryst. 5 163
Stickler R and Booker G R 1963 Phil. Mag. 8 859
Sunada J 1974 Jap. J. of Appl. Phys. 13 1944

Inst. Phys. Conf. Ser. No 146
Paper presented at Microsc. Semicond. Mater. Conf., Oxford, 20–23 March 1995
© 1995 IOP Publishing Ltd

Crystal growth characterization of polycrystalline silicon films obtained by hot-wire chemical vapour deposition

M C Polo[1], F Peiró[1,2], J Cifre[1], J Bertomeu[1], J Puigdollers[1] and J Andreu[1]

[1] Dept. Física Aplicada i Electrònica, Universitat de Barcelona, Av. Diagonal, 647, E-08028 Barcelona, Spain
[2] Serveis Científico-Tècnics, Universitat de Barcelona, Lluís Solé i Sabarís, 1-3, E-08028 Barcelona, Spain

ABSTRACT: Polycrystalline silicon (poly-Si) films were obtained at moderate temperatures (280-500°C) from a mixture of silane and hydrogen in a hot wire CVD reactor. SEM and TEM results revealed a columnar growth of poly-Si grains with a preferential orientation of the crystals perpendicular to the substrate surface along the [110] direction. Plan view examinations along the [110] axis revealed a needled shape of the crystals (0.3-1 μm) with the largest axis randomly distributed on the plane. The high quality of the polycrystalline samples obtained makes the hot-wire technique very promising.

1. INTRODUCTION

Polycrystalline silicon films have a great interest owing to their wide range of applications in large area optoelectronic and photovoltaic devices, such as thin film transistors for active displays and solar cells, respectively.

Several methods have been used to obtain poly-Si films, such as low pressure chemical vapour deposition (LPCVD) (Harbeke 1983), crystallization by rapid thermal process (RTP) (Bonnel 1991) and solid phase crystallization (SPC) (Matsuyama 1990). However, in these techniques high temperatures are needed either during the film growth or in post-annealing processes, which represent a drawback to obtain poly-Si films on glass substrates.

This work deals with the growth of poly-Si films by the hot wire chemical vapour deposition (HWCVD) method using a mixture of silane and hydrogen. By this technique poly-Si films with good electronic properties can be obtained at low growth temperatures (<500°C) and high deposition rates, also avoiding the need of post-annealing treatments (Matsumura 1993). The aim of this paper is then to asses the usefulness of the HWCVD method for the growth of poly-Si films from the scanning (SEM) and transmission electron microscopy (TEM) characterizations of the crystalline quality of these films. The results of other complementary characterization techniques, X-ray diffraction (XRD), Raman spectroscopy and Secondary Ion Mass spectroscopy, will be also commented on.

2. EXPERIMENTAL

The films were grown on fused silica substrates in a hot-wire reactor from a mixture of 10 % silane - 90 % hydrogen. The reaction gases were activated by a tungsten filament, 1 mm thick and 150 mm long, that covered the whole sample surface homogeneously. The temperature of the filament was set to 1600 °C as measured by an optical pyrometer. The substrate temperatures were varied in the range 280-500 °C by placing the sample holder at different distances from the hot tungsten wire. The substrate temperature was measured by a calibrated thermocouple attached to the

504

substrate holder. In all experiments the total pressure was 21 Pa and the silane and hydrogen flows were 2 and 18 sccm, respectively.

The films were morphological and structural characterized by SEM, TEM, XRD and Raman Spectroscopy. For the TEM and TED analysis plan view and cross-section specimens were prepared. SIMS was used for the study of tungsten sample contamination during growth.

3. RESULTS

The study of the films showed the possibility of growing large columnar crystalline structures at moderate temperatures and high deposition rates and also the influence on these structures of the substrate temperature. No presence of tungsten contamination in the films was observed by SIMS.

The thicknesses of the deposited films were between 2 and 8 μm which correspond to values of growth rate as high as 40 Å/s. The increasing of the deposition temperature produced a decrease in the growth rate but even at the lowest substrate temperature (280 °C) the growth rate was very high (about 10 Å/s).

Small differences have been observed in the crystallinity of the films obtained. The XRD spectra of the samples showed the characteristic peaks of crystalline structures. The relative intensities of (111) (220) (311) (331) (422) diffraction peaks in the XRD spectra were similar. Moreover, when comparing these spectra with those of randomly oriented silicon, a (220) preferential orientation perpendicular to the substrate was observed in all samples. The crystalline character of the silicon films was also corroborated by Raman spectroscopy by the presence of an intense peak characteristic of poly-silicon centered al 520 cm^{-1} whereas no evidence of amorphous silicon was found.

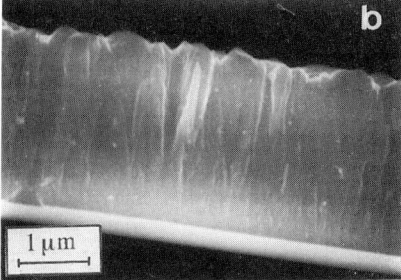

Fig. 1. SEM micrographs of surface (a) and cross section (b) for a sample deposited at 330 °C.

The morphology of the poly-Si samples can be observed in the SEM micrographs of Fig. 1. Plan views for all samples showed a structure formed by grains (Fig. 1a). The sample obtained at the highest temperature (500 °C) presented the largest grains with sizes ranging from 0.5 to 1 μm. The size of the grains decreased with the sample-filament distance. This effect could be attributed not only to differences in substrate temperature but also to the different thicknesses of the samples because of the columnar growth structure observed in the cross-sectional views (Fig. 1b).

A more detailed characterization of the poly-crystalline microstructure was carried out by TEM and TED analyses. Cross-sectional views (Fig. 2a) confirmed the columnar crystal growth already observed by SEM. The extension of the twin of the crystal shown in Fig. 2a allows us to outline the size of the crystalline columns which is larger than 1.3 μm. TED patterns of the cross section of the samples obtained from a selected area of 1μm diameter (inset in Fig. 2a) consisted of regular spots, whereas the diffraction rings, typical of polycrystalline material with small crystals, were not seen. Such diffraction patterns showing single-zone-axis reflections ([-111] zone axis) therefore corroborated the growth of films with large silicon crystals at moderate deposition temperatures (< 500 °C).

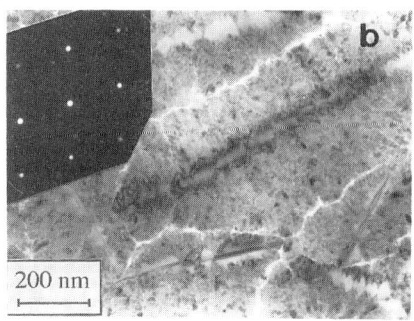

Fig. 2. TEM bright-field micrographs of silicon grains viewed along the [-111] direction in a XTEM view (a) and along the [110] direction (b) of a sample grown at 400 °C. The insets show the diffraction patterns.

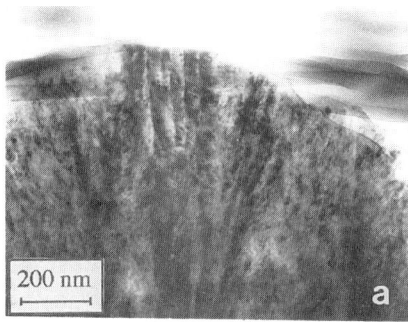
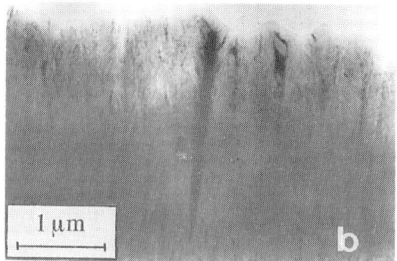
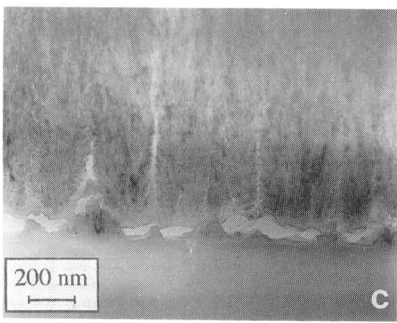

Fig. 3. Cross-section view of the crystal along all the film thickness.

In order to find out the growth direction of the crystals, a plan-view TEM study was performed (Fig. 2b). Needle shaped crystals ranging from 0.3 to 1 μm in length were found in the middle of the grains seen in the SEM images (corresponding to the white areas of Fig. 1a) and they were surrounded by a nanocrystalline material. Nanodiffraction patterns of most of the tested crystals showed a [110] orientation (inset in Fig. 2b) in agreement with the results of XRD. Moreover, nanodiffraction patterns obtained along the length of the crystals always showed the same zone axis [110], which confirmed monocrystallinity of the grains.

Figure 3 shows that the monocrystal nuclei of the grains extended from the interface (Fig. 3c) across all the thickness of the poly-Si films (Fig. 3b) up to the top of the film (Fig. 3a).

Figure 4 illustrates the effect of the temperature on the film morphology. An increase of the crystal sizes was observed when increasing the deposition temperature. In these images it is also observable that the poly-Si grains are not perfectly compact. They are surrounded by very porous grain boundaries.

4. CONCLUSIONS

Morphological and structural studies performed with XRD, Raman spectroscopy, SEM, TEM and TED showed that poly-Si films could be obtained in a hot-wire CVD reactor when silane diluted in hydrogen was used as gas precursor. A columnar crystalline structure with 0.3-1 μm was observed with preferential orientation in [110] direction. In all polycrystalline samples, crystals columns extended from the substrates to the top of the films. The growth rate of poly-Si films ranged from 10 Å/s to 30 Å/s.

506

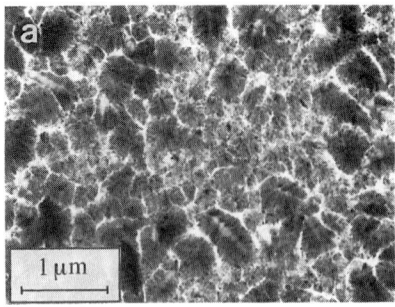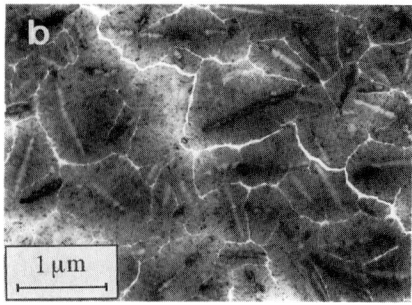

Fig. 4. TEM bright-field images of poly-Si samples deposited at 280 °C (a) and 400 °C (b).

These features associated with the high deposition rates and the scaling up possibilities make hot-wire CVD a very promising method for large thin film photovoltaic device development, for opto-electronic applications. Finally, we would like to point out the relevance of the presented results for the application of the obtained material for the fabrication of low cost, large area, poly-crystalline solar panels.

ACKNOWLEDGEMENTS

This work was supported by the DGICYT of the Spanish Government under programmes PB89-0236 and MAT93-0703-C03-02.

REFERENCES

Bonnel M, Duhamel N, Guendouz M, Haji L, Loisel B and Ruault P 1991 Jpn. J. Appl. Phys. 30 L1924

Harbeke G, Kransbaur L, Steigmeier E F, Widmer A C, Kappert H F and Neugebauer 1983 Appl. Phys. Lett. 42 249

Matsumura H, Hosoda Y and Furukawa S 1993 Mat. Res. Symp. Proc. 283 623

Matsuyama T, Wakisaka K, Kameda M, Tanaka M, Matsuoka T, Tsuda S, Nakano S, Kishi Y and Kuwano Y 1990 Jpn. J. Appl. Phys. 29 2327

Inst. Phys. Conf. Ser. No 146
Paper presented at Microsc. Semicond. Mater. Conf., Oxford, 20–23 March 1995
© 1995 IOP Publishing Ltd

Growth of microcrystallites in thin silicon films prepared by PECVD

M Luysberg, P Hapke, F Finger and R Carius

Institut für Schicht- und Ionentechnik, Forschungszentrum Jülich, 52425 Jülich, Germany

ABSTRACT: Microcrystalline silicon prepared by plasma enhanced chemical vapour deposition consists of variable volume fractions of amorphous phase, grain boundaries and crystalline grains. To study the influence of the microstructure on the electrical transport properties a set of samples is investigated, in the as-deposited state and after annealing, by transmission electron microscopy and Raman spectroscopy. The changes of the microstructure due to different growth conditions and upon annealing and its relation to the conductivity give evidence for percolation processes dominating the electrical transport.

1. INTRODUCTION

Large area deposition by plasma enhanced chemical vapour deposition (PECVD) at low temperatures of 200°C makes microcrystalline Si (μc-Si) a promising material for large area electronic devices such as flat panel displays (Kanicki 1992). In addition the material has potential as a nucleation seed for thin crystalline silicon. μc-Si is a composite material of amorphous phase, grain boundaries and crystalline grains with dimensions from a few to a few hundreds of nanometers. The microstructure of these films strongly influences their electronic and optical properties (Hapke 1993, 1995). To investigate the electrical transport properties as a function of the microstructure we characterized the structure of μc-Si films grown under different conditions. In addition, the initial microstructure was modified by thermal annealing.

2. EXPERIMENTAL

The μc-Si:H samples were grown on glass substrates using PECVD at 70 MHz at a substrate temperature of 180 °C, up to a film thickness of 300 nm. The dilution of silane in hydrogen in the gas phase was varied from 1.5% for samples B to 4.5% for sample A, which were both doped with phosphine in the gas phase (doping level 1.5%). For the third sample C prepared at 1.5% silane in hydrogen diborane was used as doping gas. The samples were annealed at temperatures up to 700 °C under high vacuum conditions for 20 minutes. The microstructure was characterized by TEM using a JEOL 2000EX microscope operated at 200 kV and a JEOL 4000EX microscope operated at 400 kV. The TEM specimens were prepared as cross sections, which allows the observation of depth dependent structural modifications. Raman spectra were recorded with a double monochromator coupled with a cooled photomultiplier in the photon counting mode using the 514.5 nm line of an Ar-Laser for excitation. The conductivity was measured in the temperature range between 80 K and 300 K in a coplanar gap geometry.

3. RESULTS AND DISCUSSION

Fig. 1 shows the Raman spectra of sample A at different annealing stages. In the as deposited state only a very weak signal of the crystalline TO-mode of Si centered at 520 cm^{-1} can be observed be-

508

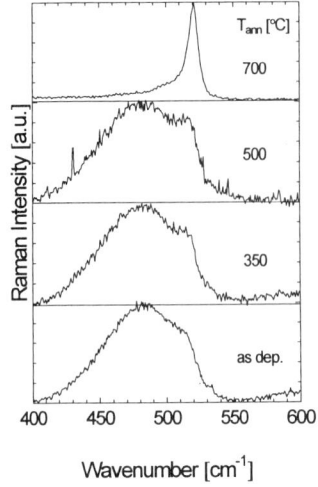

Fig. 1: Raman spectra of sample A at various annealing stages.

sides the dominating signal of the TO-mode of amorphous Si at 480 cm⁻¹. Annealing at low temperatures does not induce a distinct change of the spectra indicating that the amorphous volume fraction remains nearly constant upon annealing up to 500°C. A significant change of the spectrum is observed after annealing at a temperature of 700°C, where only the signal of the crystalline TO-mode is found.

Although there is no significant crystalline contribution detectable in the Raman spectrum of sample A in the as deposited state, a high density of small crystallites can be observed in dark field images obtained by TEM (Fig. 2a). The individual crystallites, which are of sphere-like shape, have typical sizes from 3 nm to 10 nm. In the corresponding diffraction patterns diffuse rings arising from amorphous silicon and rings of reflections indicating a random distribution of crystallographic orientations of the Si-crystallites are superposed. The crystalline grains are embedded in amorphous Si as is observed in high-resolution TEM images, where no structural defects within the crystallites could be detected. From the high-resolution images a crystalline volume fraction of about 20% is estimated. Annealing at 700°C induces the growth of the crystallites to a typical size of 5 nm to 40 nm (Fig 2b). In few cases sizes up to 200 nm could also be observed. Some of the crystallites reveal a Moiré-fringe contrast. The fringes of irregular width are caused by twinning, as could be deduced in several cases from additional spots arising in the diffraction patterns characteristic for twinning. Additionally performed high-resolution studies revealed also the existence of twinned crystallites. From the superposition of several dark field images an amorphous volume fraction smaller than 10% is estimated which is beyond the resolution limit of this technique. The structural modification is confirmed by the Raman measurements (Fig. 1).

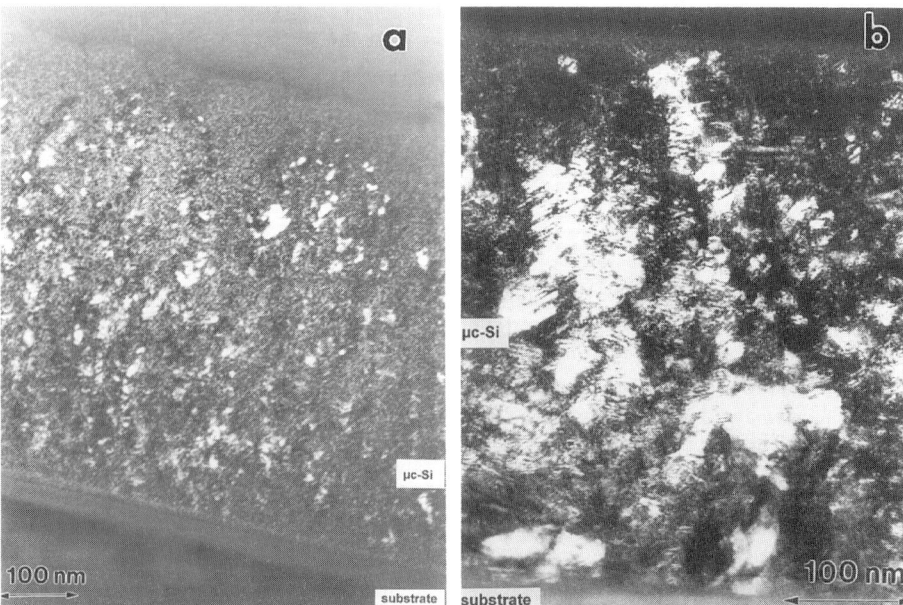

Fig. 2: Dark field images of sample A (n-type) in the as deposited state (a) and after annealing at 700°C (b)

From the structural changes observed in sample A upon annealing it can be deduced that the crystallites in the as deposited state act as nuclei for the growth of larger crystals at temperatures above the crystallisation temperature. Thereby a high density of twins ist formed. This growth mode was also observed by Batstone (1993) who examined the crystallization of silicon from an amorphous film by in-situ TEM. As in our samples they observed the growth of heavily twinned crystallites, which grow until they attach adjacent grains. Thus the grain size is determined by the initial density of the nuclei.

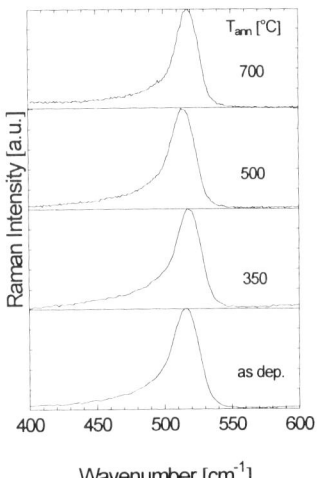

Fig. 3: Raman spectra of sample B at various annealing stages.

Fig.4: Dark-field image of sample B in the as deposited state.

Preparing samples with a lower silane to hydrogen ratio of 1.5% results in the growth of films of higher crystalline volume fraction. Fig. 3 shows as an example the Raman spectra of sample B. The spectrum of the as deposited state is clearly dominated by the crystalline signal at 520 cm^{-1}. Furthermore, no significant changes are observed after annealing at different temperatures indicating that the crystalline volume fraction remains unchanged. Indeed, the TEM investigations of this sample in the as deposited state and after annealing at 700°C show no significant changes of the structural properties. Fig. 4 shows a typical dark field image of sample B in the as deposited state. A columnar structure parallel to the growth direction is observed. The columns, which consist of small crystallites (diameter: 10 nm) of the same crystallographic orientation with respect to the electron beam, are up to 70 nm in width and up to 150 nm in length. Some of the columns increase in diameter from the glass substrate to the surface indicating a competive growth of crystalline regions starting from different nuclei. The dark contrasts within one column correspond either to crystalline regions out of contrast or to amorphous regions. The superposition of several dark field images and high-resolution studies points to a small amorphous volume fraction (<10%).

In contrast to sample B, sample C is doped with diborane in the gas phase, which is known to hinder the crystalline growth (Prasad 1991). In the Raman spectra, dominated by the crystalline signal, only a small contribution of the amorphous phase was found, which decreases continuously with increasing annealing temperature. The TEM studies revealed no significant structural differences of sample C compared to sample B. This indicates that the growth mode is not severely influenced by doping. However, a strong influence of the silane to hydrogen ratio is observed. At a large ratio (sample A) small crystallites are embedded in an amorphous phase, whereas at small ratios (samples B and C) a columnar structure within the highly crystalline films is observed.

In Fig. 5 an overview of the temperature dependent conductivity is shown in the as-deposited state and after annealing at 700°C. For all samples the shape of the curves is the same at all annealing stages. Thus the temperature dependence of the conductivity is in all cases governed by the same dominant transport mechanism. In addition, a strong deviation from a singly activated behaviour is observed. For annealing up to 500°C the conductivity of sample A is by two orders of mag-

510

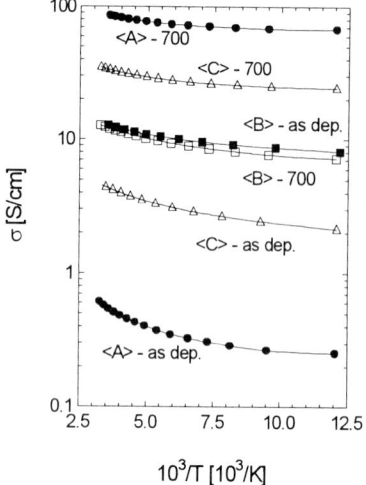

Fig. 5: Dark conductivity of all samples in the as-deposited state and at an annealing temperature of 700°C.

nitude larger than in highly doped amorphous Si despite the an amorphous volume fraction of 80%. This can consistently interpreted by assuming percolation of carriers through the crystallites. Indeed, a percolation threshold of 18% crystalline volume fraction was reported by Tsu (1982), who examined the conductivity in Si-F alloys. This is in line with the theoretical prediction of a percolation threshold below 20% (Scher 1979). After annealing at 700°C the conductivity of sample A increases by two orders of magnitude. It appears as if additional transport paths of the same type are opened up. In sample B the temperature dependent conductivity remains unchanged over the whole range of annealing temperatures in agreement with the small changes of the microstructure. However, compared to sample A, which is also n-type, sample B shows a considerably lower conductivity after annealing at 700°C, which can be attributed to the slightly smaller grain sizes observed for sample B. For sample C (p-type) the conductivity increases continuously with increasing annealing temperature in accordance with a decreasing amorphous volume fraction with the temperature treatment.

In conclusion it can be stated that dependent on the dilution of silane in hydrogen the growth mode changes. In the case of a small silane to hydrogen ratio high crystalline volume fractions and columnar structures are observed, while at larger ratios small crystallites embedded in amorphous tissue were found. The structural and electrical data imply even for samples with small crystalline volume fractions that the transport is dominated by percolation paths through the crystallites.

ACKNOWLEDGEMENT

We thank the Institut de Microtechnique, Neuchâtel (Switzerland), for providing samples. This work was supported by the Bundesministerium für Forschung und Technologie (Germany).

REFERENCES

Batstone J L 1993 Phil. Mag. A 67 51

Hapke P, Finger F, Carius R, Wagner H, Prasad K and Flückiger R 1993 J. Non-Cryst. Solids 164-166 981

Hapke P, Luysberg M, Carius R, Finger F and Wagner H 1995 Semiconductor processing and characterization with lasers 1994, eds M Brieger, M Klose, T Dietrich, H W Schock and J H Werner (Aedermannsdorf Switzerland: Transtech) pp 249-254

Kanicki J 1992 Amorphous and microcrystalline semiconductor devices (Boston: Artech House)

Prasad K, Kroll U, Finger F, Shah A, Dorier J L, Howling A, Baumann J and Schubert M 1991 Mat. Res. Soc. Symp. Proc. 219 383

Scher H and Zallen R 1979 J. Chem. Phys. 53 3759

Tsu R, Gonzalez-Hernandez J, Chao S S, Lee S C and Tanaka K 1982 Appl. Phys. Lett. 40 534

Veprek S 1972 J. Chem. Phys. 57 952

Inst. Phys. Conf. Ser. No 146
Paper presented at Microsc. Semicond. Mater. Conf., Oxford, 20–23 March 1995
511

Dynamic observation of electrochemical etching in silicon

Frances M Ross and Peter C Searson*

National Center for Electron Microscopy, Lawrence Berkeley National Laboratory, 1 Cyclotron Road, Berkeley, CA 94720, USA
*Department of Materials Science and Engineering, The Johns Hopkins University, Baltimore, MD 21218, USA

ABSTRACT: We have designed and constructed a TEM specimen holder in order to observe the process of pore formation in silicon. The holder incorporates electrical feedthroughs and a sealed reservoir for the electrolyte and accepts lithographically patterned silicon specimens. We describe the system and present preliminary, *ex situ* observations of the etching process.

1. INTRODUCTION

Porous semiconductors represent a relatively new class of materials formed by selective etching of a single or polycrystalline substrate. Although porous silicon has received considerable attention due to its novel optical properties (Canham 1990), porous layers can be formed in other semiconductors such as GaAs and GaP (Krumme and Straumanis 1967, Faktor et al. 1975, Chase and Holt 1972). These materials are characterised by very high surface area and electrical, optical and chemical properties that may differ considerably from bulk. The pore morphology can be controlled by adjusting the processing conditions and the dopant concentration. A number of novel structures can be fabricated using selective etching. For example, self-supporting membranes can be made by growing pores through a wafer (Searson 1991), films with modulated pore structure can be fabricated by varying the potential during growth (Munder et al. 1993), composite structures can be prepared by depositing a second phase into the pores and silicon-on-insulator can be formed by oxidising a buried porous layer (Tsao et al. 1989).

In all applications of porous silicon the ability to grow nanostructures controllably is critical. We are attempting to gain a better understanding of the formation mechanism of porous silicon by allowing pore formation to occur in real time within the microscope. We have designed and constructed a specimen holder for a JEOL 200CX TEM featuring a sealed reservoir which can be filled with the desired electrolyte (such as HF). Electrical feedthroughs are provided so that an anodization potential can be applied while in the microscope. The specimen itself requires a complex design to allow pore propagation laterally through an electron transparent region while avoiding the release of HF. We find that pores propagate at speeds of up to 100 nm s^{-1} on application of a 2-3V potential. Our aim in observing the dynamics of the process in real time (for example as a function of voltage) is to test different models of pore formation and to control the morphology of the pores produced. In this paper we will describe the issues we have encountered in setting up these experiments and present *ex situ* observations of pore formation. We ultimately hope to apply the experience gained to the study of different electrochemical processes and we will therefore discuss the extension of these techniques to other systems.

2. ANODIC ETCHING OF SILICON

Electrochemical etching of silicon is carried out by applying a potential of several volts between silicon and the HF electrolyte. Current densities are 1-100 mA cm^{-2}, silicon goes into solution as H_2SiF_6 and a network of unetched silicon is left behind. H_2 gas is a product of the reaction. The structures formed have been documented in considerable detail. In p-type silicon, the etching process leaves an interconnected network of filaments with typical dimensions 3-5 nm

(Cullis and Canham 1991, Smith and Collins 1992, Ross et al. 1995a). In n-type silicon, the pores can range in size from nanometers to microns and tend to grow in well defined crystallographic directions with a variable degree of branching (Theunissen 1972, Chuang et al. 1989, Smith and Collins 1992, Lehmann 1990, 1993, Searson et al. 1992).

The interfacial reactions which occur are reasonably well understood (see for example Smith and Collins 1992) and will not be discussed further here. However, knowledge of the overall electrochemical reaction does not give any information about the spatial distribution of the current during etching. It was in order to make a detailed study of models proposed to explain the morphology found in n-type silicon (Smith et al. 1988, Beale et al. 1985a, b, Erlebacher et al. 1994), including the formation of side branches and the reaction of the system to changes in the applied potential, that we decided to examine the pore formation process in real time.

3. DESIGN OF AN *IN SITU* ELECTROCHEMICAL SYSTEM

In order to observe electrochemical etching both the specimen holder and the specimen must be designed within fairly severe constraints. Etching must proceed in a well defined geometry within an electron transparent region while the specimen remains in electrical contact with the electrolyte, and this must all take place within the microscope vacuum. The specimen holder (figure 1) features an enclosed 3x4x6mm reservoir made of machinable ceramic in which the electrolyte is sealed. Electrical contact to the electrolyte is made by a platinum screw which extends into the reservoir and also acts as a means to fill the reservoir. The specimen is sealed into the other end of the reservoir using HF-resistant epoxy, leaving an area of about 1mm^2 in contact with the electrolyte. The reservoir and specimen can be removed from the holder for testing.

The specimen itself requires a complex design to avoid the release of HF during etching (figure 2). The initial substrate was a lightly doped SIMOX wafer on which MBE was used to add a lightly doped barrier layer, a heavily doped "active" layer and a further barrier layer. This sequence was chosen to maximise the selectivity of the etching, since the order of reactivity for a given applied potential is $n^+>p^+>p^->n^-$ (Smith and Collins 1992). The purpose of the buried oxide layer was twofold, firstly to act as a barrier to current flow within the bulk of the specimen and secondly to act as an etch stop for the HF/HNO$_3$/HAc etchant which was used to form electron transparent windows by etching from the back surface. Before inserting the specimen into the reservoir, black wax was painted around the edges of the specimen to minimise current leakage, and the top surface electrical contact was made with Ag paint.

Pore growth was carried out at a constant voltage from 0.5-3V using a Fischione power supply. A typical current was 0.01mA, corresponding to 1000mA cm^{-2} over the "active" cross sectional area (but see below).

Figure 1. Electrochemical biasing specimen holder. A second (sectioned) reservoir is placed on the holder for illustration.

4. *EX SITU* RESULTS

Figure 3 shows the results of etching *ex situ* (i.e. at 1 atm.). The pores have propagated 250µm from the initial window to the electron transparent region. Cross sectional SEM confirms that etching occurs in the n$^+$ layer, so the aspect ratio of these pores is 250µm/400nm ~ 600. The pore front is very straight so all pores are propagating at the same speed. Rather than forming the well separated, highly branched pores which have been observed in n-type silicon, the etching process has removed most of the material, leaving behind walls less than 50nm thick. Although pore morphology is known to be a sensitive function of the etching conditions, the morphology observed here is unusual and may be related to the confinement of the pores into a small cross sectional area. The striking light and dark bands in the image were created artificially by modulating the voltage. This technique allows us to mark the pore front at selected times and suggests some exciting possibilities related to the creation of density superlattices. Our films generally show small scale modulations which we believe are due to fluctuations in the current.

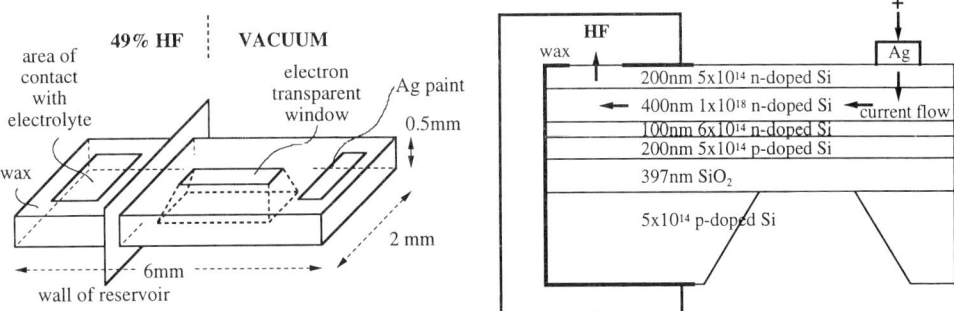

Figure 2. Schematic diagrams showing the overall specimen geometry and details of the layers.

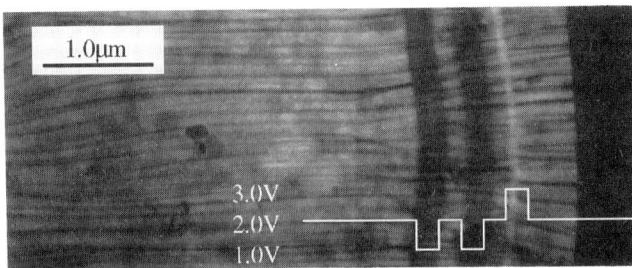

Figure 3. Bright field, plan view image of the pore front after etching *ex situ* in the (100) direction for 32 min. at 2.0V and 0.016mA. Inset is the time-varying voltage which was applied to create the bands of varying porosity. This enables us to calculate etch rate as a function of applied voltage.

We now consider some specimen geometry related issues which have arisen from these *ex situ* experiments. We were initially concerned that pores would not propagate far enough to reach the electron transparent window. Figure 3 shows that this is not a problem, and we have propagated pores as far as 1500μm. Furthermore, I-V curves from specimens etched in the reservoir are identical to those from specimens etched in a large beaker of HF. This suggests that exhaustion of the HF by saturation or slow diffusion of etching products is not limiting the reaction.

Vertical propagation is a problem if it allows the escape of electrolyte. Cross sectional SEM shows that pores do grow vertically, and we see a distinctive cellular morphology in plan view

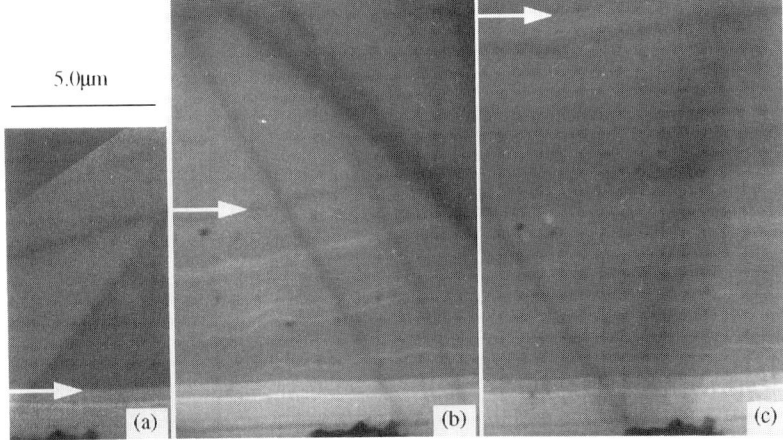

Figure 4. (a) A pore front after etching for 62 min. at an average of 0.013 mA total current. (b) After a further 10 min. 30 s. at 0.007 mA. (c) After a further 10 min. 15s. at 0.010 mA. Between (a) and (b) the pore front has advanced irregularly, probably because the specimen was left overnight in between. However (b) and (c) were carried out within an hour and the pore front at (c) has retained its profile at (b).

when this happens. This actually simplifies the specimen design since it is not necessary to pattern contact windows through the barrier layer to the active layer. However, it is a potential problem in experiments under vacuum. Barrier layers 200 nm thick appear to provide a compromise between containing the HF (*ex situ*) and still allowing reasonable electron transmission. It is possible that replacement of the top layer with a thinner Si_xN_y layer will improve image quality further.

At present, we have not observed pore growth in real time within the microscope, and are working on a reliable vacuum seal between the specimen and the reservoir. However, in figure 4 we show an interesting series of images taken after successive ten minute periods of etching, again *ex situ*. The pore front has propagated at speeds of 10.8 and 10.2 nm s^{-1}, corresponding to a current density of 4.2 mA cm^{-2} (ssuming a porosity of 50%).

5. CONCLUSIONS AND FUTURE DIRECTIONS

We have demonstrated that it is possible to carry out electrochemical etching inside a buried layer in the electron transparent region of a silicon specimen. After further vacuum testing we hope to allow this process to take place within the microscope. We are very excited by the possibility of examining the etching reaction in real time and at moderately high resolution. Our initial objectives are to study the porosity, morphology and etch rate as a function of voltage and doping level, and to observe the process of pore branching. As the voltage increases the system enters the electropolishing régime where material is removed isotropically. We expect to find changes in morphology as this transition point is approached. Another area of interest is the behaviour of the etch rate when the voltage is removed, which could indicate the presence of certain species at the silicon/HF interface. Finally, the flow of current along dislocations can be studied by using partially relaxed Ge_xSi_{1-x}/Si substrates with misfit dislocations at the interfaces.

There are analogous systems for which this type of *in situ* experimentation would prove valuable, such as n-type GaAs, where the pores run in <111>a directions and have triangular or hexagonal cross sections and interesting branching geometry (Ross et al. 1995b). Modifications of the holder and specimen design can extend these experiments further and we are presently fabricating specimens in which we can study the dynamical behaviour of ionic conductors and ferroelectric thin films under an applied field.

We gratefully acknowledge John C. Bean and J. Alex Liddle of AT&T Bell Laboratories, NJ for MBE growth and scanning electron microscopy, respectively, Andrew Wittkower of SOITEC/USA, MA, for kindly providing the SIMOX wafers, and M. L. Tech, CA, for construction of the specimen holder. This work was supported by the Director, Office of Energy Research, Office of Basic Energy Sciences, Materials Science Division, U. S. Department of Energy under contract DE AC-03-76SF00098, and by the National Science Foundation under grant DMR-9202645.

REFERENCES

Beale M I J, Chew N G, Uren M J, Cullis A G and Benjamin J D 1985a, Appl. Phys. Lett. **46**, 86
Beale M I J, Benjamin J D, Uren M J, Chew N G and Cullis A G 1985b, J. Cryst. Growth **73**, 622
Canham L T 1990, Appl. Phys. Lett. **57**, 1046
Chase B D and Holt D B 1972, J. Electrochem. Soc. **119**, 314
Chuang S-F, Collins S D and Smith R L 1989, Appl. Phys. Lett. **55**, 154 and 675
Cullis A G and Canham L T 1991, Nature **353**, 335
Erlebacher J, Sieradzki K and Searson P C 1994, J. Appl. Phys. **76**, 182
Faktor M M, Fiddyment D G and Taylor M R 1975, J. Electrochem. Soc. **122**, 1566
Krumme J-P and Straumanis M E 1967, Trans. Met. Soc. AIME **239**, 396
Lehmann V, J. Electrochem. Soc. 1993 **140**, 2836
Lehmann J-P and H Föll 1990, J. Electrochem. Soc. **137**, 653
Munder H, Berger M, Frohnhoff S, Thonissen M and Luth H 1993, J. Luminescence **57**, 5
Ross F M, Natarajan A, Oskam G and Searson P C 1995a, in preparation
Ross F M, Oskam G, Searson P C, Macaulay J M and Liddle J A 1995b, sub. to Phil. Mag.
Searson P C 1991, Appl. Phys. Lett. **59**, 832
Searson P C, Macaulay J M and Ross F M 1992, J. Appl. Phys. **72**, 253
Smith R L, Chuang S-F and Collins S D, J. Electronic Materials **17** 1988, 533
Smith R L and Collins S D 1992, J. Appl. Phys. **71**, R1
Theunissen M J J 1972, J. Electrochem. Soc. **119**, 351
Tsao S S, Guilinger T R, Kelly M J and Clews P J 1989, J. Electrochem. Soc. **136**, 586.

Inst. Phys. Conf. Ser. No 146
Paper presented at Microsc. Semicond. Mater. Conf., Oxford, 20–23 March 1995
© *1995 IOP Publishing Ltd*

Microstructure of critical point dried porous silicon

G Wakefield, J L Hutchison and P J Dobson*

Department of Materials, University of Oxford, Parks Road, Oxford OX1 3PH
*Department of Engineering Science, University of Oxford, Parks Road, Oxford OX13 3PJ

ABSTRACT: Highly luminescent porous silicon may be produced by anodisation of a silicon wafer in an HF solution. Porous layers of depth greater then a few microns are susceptible to crack formation and, in some cases, disintegration during air drying. Critical point drying overcomes this problem, and produces uniform layers.

Analysis of cross-sections of critical point dried porous silicon at high resolution show that, although larger scale cracking is avoided, there is still misalignment of the quantum dots, and stresses in the quantum wires, that make up the silicon structures. It is suggested that Van der Waals forces play a role in this misalignment.

1. INTRODUCTION

Although the formation of a porous film produced by anodic dissolution of silicon in an HF solution has been known for over thirty five years (Turner 1958), the recent discovery of efficient luminescence from such films (Canhan1990) has generated a great deal of interest. It is proposed that this effect is due to quantum confinement within silicon quantum wires or dots (Voos 1990, Kanemitsu 1992).

Transmission Electron Microscopy (TEM) studies of such porous layers show that they are indeed composed of small (< 5nm) silicon particles (Nishida 1992, Cole 1992), or arrays of silicon wires (Cullis 1991).

A major problem in the formation of porous silicon layers is the fact that the films are prone to disintegration as a result of surface tension forces if they are dried in air. This problem can be avoided if the films are subject to critical point drying (CPD) techniques (Bartlett 1975), and this has indeed been used on porous silicon (Canham 1994, Frohnhoff 1995), producing films of greater thickness with higher porosity.

In this paper we consider the microstructure of such CPD porous silicon films in cross-section, and show that surface tension effects are responsible for large scale cracking, but not the smaller scale misalignment of the quantum structures.

2. EXPERIMENTAL

Porous silicon is formed in a single cell with a platinum cathode parallel to a single crystal silicon wafer (P-type ρ=5 Ωcm). The wafer is backed with aluminium and baked at 500°C for fifteen minutes in a nitrogen atmosphere in order to provide an Ohmic contact. The cell fluid is HF:H$_2$O:C$_2$H$_5$OH=1:1:2 (Ethanoic HF), and the wafer is anodised at a current density of 60 mAcm^{-2}, for one minute in the case of air dried specimens, and for five minutes in the case of CPD porous silicon. Following anodisation, the samples are leached for up to two hours in the cell fluid in the dark. The porous silicon layers are a golden colour, and exhibit photoluminescence in the orange with a broad peak centred on 600nm.

The thicker porous layers are stored under acetone and placed in a Polaron Critical Point Drier, where the acetone is gradually replaced by flushing with liquid CO$_2$ over a period of about a day. The drier is then heated to 40°C sending the CO$_2$ above its critical point, the gas is slowly released at constant temperature, the drier is cooled and the sample removed.

TEM cross-sectional samples are prepared by mechanical polishing followed by low voltage ion milling at grazing incidence which, as well as the use of a retarding potential on the specimen stage, reduces any artefacts which may be present. Analysis of a scraped sample shows that the features observed in the cross-section are as expected, and are not altered by the specimen preparation procedure.

3. RESULTS
3.1 Scanning Electron Microscopy Images of Porous Silicon

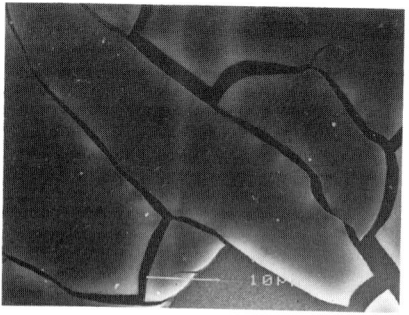

Figure 1 Air dried porous silicon

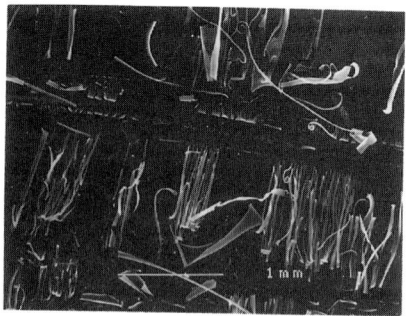

Figure 2 Disintegration under air drying

Air dried porous silicon appears cracked under the SEM. The extent of this cracking is determined by both the thickness of the layer, and the size of the silicon microstructures. This is in turn determined by the anodisation time and the length of the post-anodisation etch.

A one minute anodisation followed by a one hour etch results in a structure such as that shown in figure 1. The majority of the coating is intact, however cracking is visible within the porous silicon layer. Obviously this is undesirable for potential electroluminescence applications as it provides a short circuit between the silicon substrate and the contact layer. Additional etching time results in higher porosity, and complete porous silicon disintegration, as in figure 2. In both these cases the layer thickness is approximately one micron.

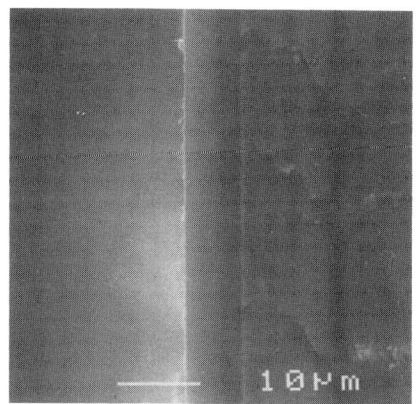

Critical point drying allows thicker layers to be produced without cracking. In figure 3 a cleaved cross-section of CPD porous silicon can be seen. This sample has had a two hour post-anodisation etch and has a thickness of 6.5 microns. No cracking is observed within the structure.

Figure 3 Cross-section of CPD porous silicon

3.2 Transmission Electron Microscopy of CPD Porous Silicon

A cross-section of CPD porous silicon is shown in figure 4. A network of interconnecting silicon structures can be seen, surrounded by amorphous material. The majority of these structures appear spherical, although some have a wire-like appearance. There is little evidence of and directionality and no evidence of a porosity gradient. There is quite a large range of crystal size, with features ranging from ≈25nm down to <4nm.

Figure 5 shows a number of these quantum dots in high resolution TEM. A large 15nm grain is visible, surrounded by a number of smaller crystals of sizes down to ≈2 nm. This size of crystal will exhibit quantum confinement effects sufficient to allow visible luminescence from the porous silicon. These quantum dots are misaligned with respect to each

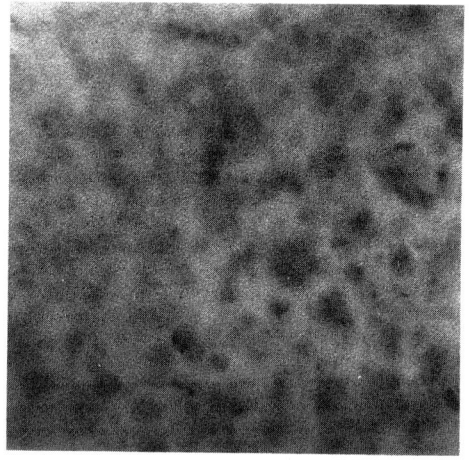

50nm

Figure 4 Cross-section of CPD porous silicon

other. As surface tension forces are not a factor in this experiment, a possible source of the misalignments are Van der Waals forces acting between the hydrogen passivated pore walls (Gruning 1995).

Longer wire-like structures are also visible in the cross-section, as in figure 6, in which a wire of width ≈2-3 nm can be seen. The atomic planes, {111} 3.1 Å, are rotated by a few degrees over the length of the wire. This is another example of the stress occurring within the porous layer.

518

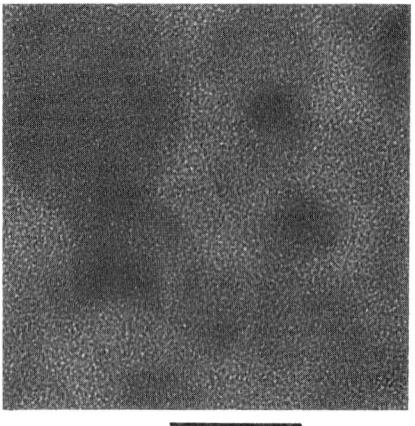

10nm
Figure 5 Quantum dot structures

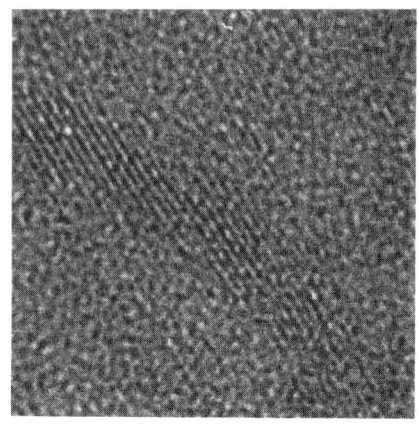

20Å
Figure 6 Wire-like structure

4. CONCLUSIONS

It has been shown that critical point drying is a useful technique that is useful in producing high porosity porous silicon films with a thickness greater then that available with air drying. The high surface tension forces experienced by the film during air drying are sufficient to crack and, in some cases, completely destroy the films.

Films that have undergone critical point drying do not exhibit large scale cracking, but do show a misalignment of the quantum dot structures within the porous silicon. Although these films are principally composed of dots, a number of wire-like structures are visible and these also exhibit stress effects. It is suggested that anisotropic stresses produced by Van der Waals forces, acting at varying distances within the highly porous structure between passivated pore walls, are responsible for these effects.

REFERENCES

Bartlett AA 1975 Scanning Electron Microscopy 305
Canham LT 1990 App.Phys.Lett. 57 1046
Canham LT, Cullis AG, Pickering C, Dosser OD, Cox TI and Lynch TP 1994 Nature 368 133
Cole MW, Harvey JF, Lux RA, Eckart DW and Tsc R 1992 Appl.Phys.Lett. 60 2800
Cullis AG and Canham LT 1991 Nature 353 335
Frohnhoff S, Arens-Fischer R, Heinrich T, Fricke J, Arntzen M and Theiss W 1995 Thin Solid Films 225 115
Gruning U and Yelon A 1995 Thin Solid Films 225 135
Kanemitsu Y, Suzuki K, Uto H, Masumoto Y, Masumoto T, Kyushin S, Higuchi K and Masumoto H 1992 Appl.Phys.Lett. 61 2446
Nisida A, Nakagawa K, Kakibayashi H and Shimada T 1992 Jpn.J.Appl.Phys part2 31 L1219
Turner DR 1958 J.Electrochem.Soc. 105 402
Voos M, Uzan P, Delalande C, Bastard G and Halimaoui A 1992 Appl.Phys.Lett. 61 1213

Inst. Phys. Conf. Ser. No 146
Paper presented at Microsc. Semicond. Mater. Conf., Oxford, 20–23 March 1995
© *1995 IOP Publishing Ltd*

Chemical composition of luminescent porous silicon layers

S Gardelis[a], U Bangert[a], B Hamilton[a], R F Pettifer[b], D A Hill[b] and R Keyse[c]

[a] Department of Physics, University of Manchester Institute of Science and Technology, Manchester M60 1QD, England
[b] Department of Physics, University of Warwick, Coventry CV4 7AL, England
[c] Department of Material Science and Engineering, University of Liverpool, L69 3BX, England

ABSTRACT: In this study we have used high resolution electron energy loss spectroscopy performed in a scanning transmission electron microscope to probe the chemical nature of porous silicon layers grown in electrolytes of different hydrofluoric acid concentration. Dramatic shifts of the plasmon peaks and of the Si L-edges between bulk Si and SiO_2 indicate that in porous silicon layers some sort of silicon suboxide is present depending on the preparation conditions of the layers. The presence of this phase might affect the luminescence properties of porous silicon.

1. INTRODUCTION

Since the report of Canham (1990) that porous silicon emits efficiently in the visible even at room temperature, many groups around the world have reproduced this observation and tried to explain the origin of the effect, the motivation being the potential application of porous silicon in optoelectronics. In order to elucidate the origin of the light emission in porous silicon one must investigate the chemical composition of this material. Chemical information was obtained from parallel electron energy loss spectroscopy (PEELS) carried out in a scanning transmission electron microscope. With this method we examined the L-edges of silicon, oxygen and carbon as well as the plasmons in porous silicon, in regions as small as the electron beam size (3 Å). The porous layers were grown electrochemically in electrolytes of different hydrofluoric acid concentration. All the layers investigated here emit light. In this paper we demonstrate that porous silicon is not chemically uniform and consists of crystalline regions (ie small crystallites of size less than 50 Å) and amorphous-like regions which are between the crystalline regions or surround them, the amorphous region being silicon oxide (Gardelis et al 1995). We also demonstrate that the amount of the oxidation of porous layers depends on the growth conditions, in this case on the hydrofluoric acid concentration in the electrolyte used for the anodic reaction. The presence of the amorphous like regions might affect the luminescence from porous silicon.

2. EXPERIMENTAL

Single crystal p-type silicon substrate of resistivity 0.06 Ωcm was used for the formation of the porous silicon layers in this study. Aluminium was evaporated at the back of the substrate to form an ohmic contact. Subsequent sintering of the aluminium contacts was performed at 450° C, in order to improve the ohmic contacts and ensure a uniform anodic current distribution. Samples were cut in squares of 1 cm² from the original wafers and each of them was mounted on a metal contact incorporated in a teflon plate. The edges of the samples were coated with lacomite to prevent current flow from the aluminium to the hydrofluoric acid solution, ensuring a uniform anodic reaction on the sample surface. Silicon anodic dissolution was performed under dark conditions with continuous stirring of the electrolyte, which was a mixture of 7 volumes of aqueous hydrofluoric acid solution and 3 volumes of ethanol. The use of ethanoic solutions results in more uniform porous layers, since ethanol reduces considerably the size of the gas bubbles formed at the interface between the electrolyte and the surface of the semiconductor and removes the bubbles more easily from the surface. Three different hydrofluoric acid concentrations were used to grow the porous silicon layers, 48%,20% and 10% by weight. All samples were grown at a current density of 40 mA/cm^2 to a thickness of about 2 μm. The porous layers had porosities 50%, 80% and 90% respectively. These samples are labelled in the text as sample #1, #2 and #3, respectively. All samples immediately after their growth were rinsed in ethanol and blown dry with nitrogen. They were cleaved in the <110> direction. Each sample was kept in an airtight box for about 6 hours prior to its insertion in the UHV chamber of the microscope. High resolution PEELS was carried out in a scanning transmission electron microscope, model HB601UX of VG MICROSCOPES. The accelerating voltage of the electrons in the beam was 100 KV. The convergence angle of the beam was 14.2 mrad. The beam current was 50 pA. The beam size was 3 Å. A parallel detector GATAN Model 666 was used to detect the electrons of different energies. The collection angle was 6.8 mrad. The resolution of the detection was 0.1 eV/ channel.

3. RESULTS AND DISCUSSION

All the PEELS results presented in this study were obtained by sampling regions of the porous layers which appeared to be either crystalline or amorphous in the high resolution scanning transmission electron images which we could obtain simultaneously with the PEELS measurements.For example, Fig.1a shows two silicon L-edge spectra of two different regions of sample #2. A PEELS spectrum from region B, which appears to be crystalline in the HRSTEM micrograph of Fig.1b, shows a silicon-like L-edge with an onset at 99 eV. There is also a small contribution of oxidised silicon as we can observe from the feature peaking at 108 eV. In the PEELS spectrum of region B where no lattice fringes were observed the onset of the L-edge of silicon shifted by $\sim$ 3 eV which does not coincide with the onset of the L-edge of silicon in SiO$_2$ (105 eV, Batson 1993). Moreover strong and clear peaks appeared at 108 and 115 eV which are features observed in the near edge region of the EELS spectrum of SiO$_2$. These two facts lead us to conclude that a silicon suboxide is present in region A.

In another set of experiments performed on sample #3 we moved the beam from inside the layer at a point where the substrate/ porous layer interface can be seen (60 nm away from the edge) to its very edge. Fig.2a shows the near edge structure of three points of this set of measurements; A situated 60 nm away from the edge, B situated 50 nm away from the edge and C situated at the edge. Fig. 2b shows the low loss region of the PEELS spectra obtained from these points. Fig. 2c shows the K-edge of the oxygen detected at these points. It can be clearly observed that there is a gradual oxidation of the porous layer from inside to the edge. Three facts support this evidence as we moved the beam from point A to point C: (a) The plasmon peak shifts from 17 to 22 eV and also broadens. (b) The half maximum of the Si L-edge shifts from 100.5 to 106 eV. This shift is accompanied by an increase of the peaks at 108 and 115 eV and a gradual shift of the L$_1$ edge to higher energies. (c) The shift of the plasmon peak and that of the Si L-edge are accompanied by a gradual increase of the oxygen in the sampling volume. It turns out that points B and C have features of both Si and SiO$_2$. This suggests that an intermediate phase, ie a silicon suboxide, is present at these points.

Finally we compared the low loss and the near Si L-edge PEELS spectra of the three samples with those obtained from bulk silicon. Fig. 3a shows the plasmon peaks and Fig. 3b shows the near Si L-edge structure PEELS spectra obtained from the samples. These graphs are representative of each sample and if there are differences in different regions within the same layer, these are minor compared to the differences between the samples. From the shift of the plasmon peaks and the shift of the Si L-edge it turns out that as the concentration of the hydrofluoric acid in the electrolyte decreases, oxidation of porous silicon becomes more important. For example, sample #1 is similar to bulk silicon. Samples #2 and #3 have features of both Si and SiO$_2$ in their spectra. This is in agreement with previous measurements (Gardelis 1993). It also agrees with results obtained from the study of the x-ray absorption near edge structure (XANES) of the K-edge of the same samples. These measurements show that oxidation of the porous layers increases with increasing

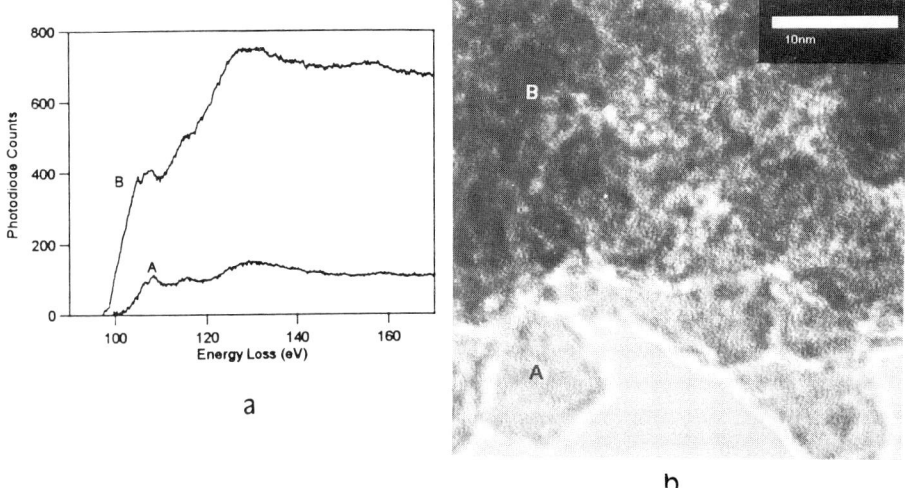

Fig.1 a) Near Si L-edge structure of an amorphous- like region A and of a crystalline region B in sample #2. b) High resolution STEM micrograph showing the regions A and B.

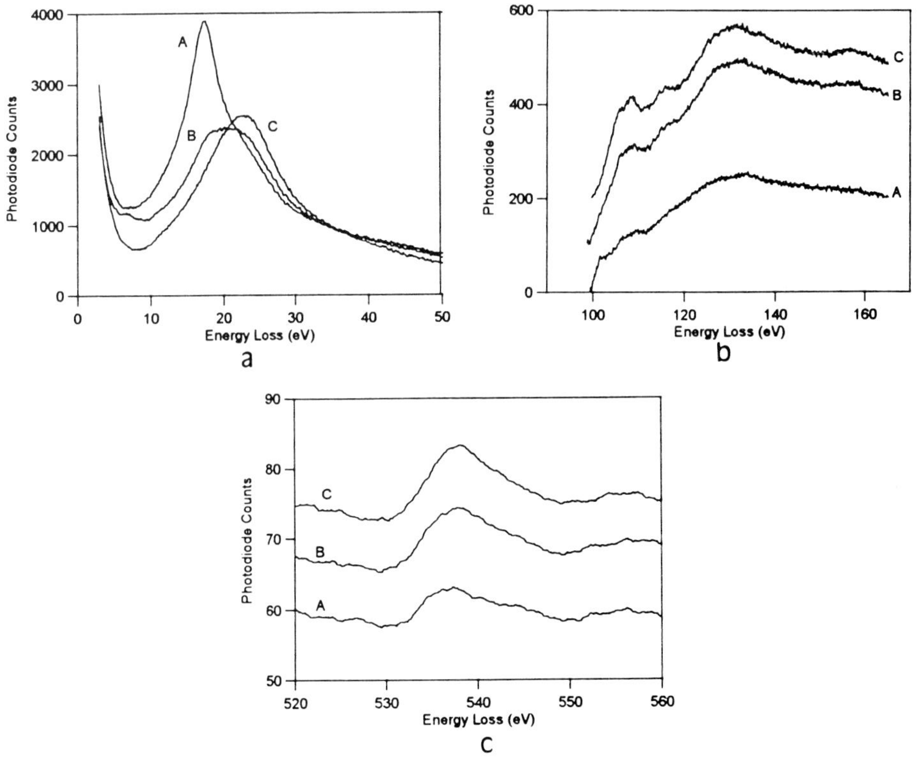

Fig. 2 a) Low loss ,b) near Si L-edge structure and c) oxygen K-edge PEELS spectra obtained from three different points of sample #3.

522

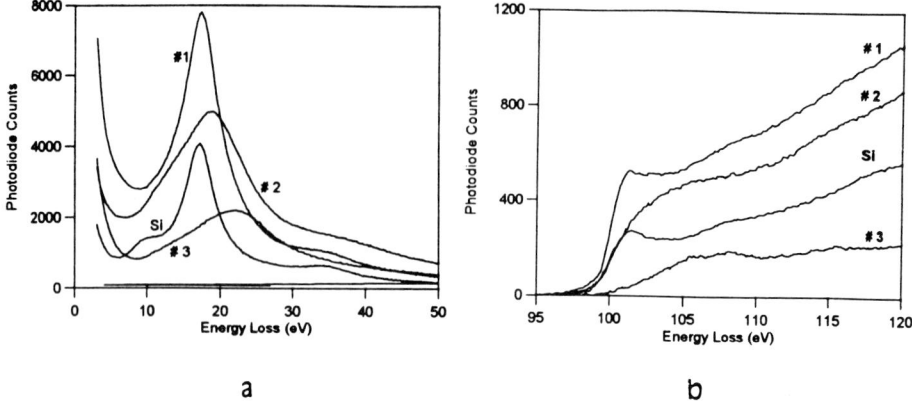

Fig. 3 a) Low loss and b) near Si L-edge structure PEELS spectra of silicon and the three porous silicon layers.

dilution of the aqueous hydrofluoric acid solution used for their preparation. These results will be published elsewhere. Given that the vacuum during the measurements was 10^{-9} mbars, it is very unlikely that the observed oxidation occured under the electron beam. On the contrary under these conditions we would expect ejection of light atoms such as oxygen. Therefore the observed oxidation of the layers is due to their preparation.

From this study it can be concluded that porous silicon is a non uniform material at the atomic scale. Only techniques such as high resolution PEELS can be used to reveal the chemical microstructure of this material which is important in order to understand the effect of the light emission from this material. The presence of some form of SiO_x might affect the characteristics of the luminescence.

ACKNOWLEDGEMENTS

The authors would like to acknowledge the support of the Engineering and Physical Sciences Research Council in this work.

REFERENCES

Batson P E 1993 Phys. Rev. B 47 6898
Canham L T 1990 Appl. Phys. Lett. 57 1046
Gardelis S 1993 PhD Thesis UMIST Manchester England
Gardelis S, Bangert U, Harvey A J and Hamilton B 1995 to be published in J. Electr. Soc.

Inst. Phys. Conf. Ser. No 146
Paper presented at Microsc. Semicond. Mater. Conf., Oxford, 20–23 March 1995
© *1995 IOP Publishing Ltd*

Defects and microstructure of CVD diamond as observed with the TEM

P Wurzinger, M Joksch and P Pongratz

Inst. f. Angewandte und Technische Physik, Technical University of Vienna, Wiedner Hauptstr. 8-10/137, A-1040 Vienna, Austria

ABSTRACT: CVD diamond layers with various film textures have been investigated with the TEM. Intersecting microtwin lamellae originating from point imperfections at growing {111} facets were observed in layers with a texture near to ⟨100⟩. Domains with predominant twin lamella orientation were found for film textures between ⟨111⟩ and ⟨110⟩. Some samples with this film texture and a layer grown epitaxially on {100} silicon were virtually free of microtwins. The nucleation and growth mechanisms of the twin lamellae with respect to the growth conditions are discussed.

1. INTRODUCTION

The concept of evolutionary growth (Van der Drift 1967) has recently been used successfully to explain the development of film textures during CVD (chemical vapour deposition) diamond growth on non-diamond substrates (Wild et al 1993): The texture is controlled by the parameter $\alpha = (v_{100}/v_{111}) \cdot \sqrt{3}$ which describes the ratio of the growth rates on the {100} and {111} facets. It varies from ⟨111⟩ for $\alpha=1$ through ⟨110⟩ for $\alpha=1.5$ to ⟨100⟩ for $\alpha=3$. Strongly fibre textured films can thus be grown from randomly oriented nuclei while films consisting entirely of nearly heteroepitaxial grains can be deposited following an oriented nucleation process (Jiang et al 1993, Hessmer et al 1995). α is only a function of the relative facet growth rates. Therefore, the same value of α can be obtained by different settings of the deposition parameters, e.g. growth temperature, total gas pressure, and gas composition (Wild et al 1994).

Previous TEM studies on CVD diamond films have revealed regions with a high density of microtwin lamellae (Williams and Glass 1989, Zhu et al 1989, Wurzinger et al 1993). The location of these regions within the grains depends on the growth texture (Clausing et al 1989, Joksch et al 1994) and there is some evidence that they are directly connected to growth on the {111} crystal facets (Wang et al 1994). The present study once more focuses on these microtwin lamellae, especially on their formation and development during the layer growth and on their dependence on the film texture.

2. EXPERIMENTAL

Several diamond layers grown with different film textures from randomly oriented nuclei and one sample grown heteroepitaxially on {100} silicon were investigated. All samples had been grown by microwave assisted plasma CVD, the detailed process parameters will be published elsewhere (Joksch et al 1995, Hessmer et al 1995). The layers were removed from the substrate and prepared as plan-view TEM specimens: They were ion milled mainly from the substrate side in order to study the regions near to the layer surface which are influenced by the growth process rather than by the nucleation. The TEM investigations were performed using a JEOL 200CX TEM operated at 200 kV. The layer texture was determined by measuring the orientation of at least 20 grains with respect to the layer surface. Bright and dark field as well as weak beam imaging were performed to image and analyze the defects.

3. RESULTS AND DISCUSSION

Figure 1a shows a defect configuration which is typical for CVD diamond films with a texture close to ⟨100⟩ (Joksch et al 1994, Wang et al 1994): A defect free column (C) exists in the interior of the grains while agglomerations of microtwin lamellae (T) can be observed near to the grain boundaries. According to Wang et al (1994) the defect free column develops by growth on the {100} crystal facet while the twin lamellae origin from growth on the {111} facets. Growth on the {111} facets is directed towards the grain boundaries (Fig. 1b). Therefore, the twin lamellae expanding towards the grain boundaries in Fig. 1a give evidence that the twin growth starts from single points (labelled S in Fig. 1a). The defect configurations originating from these points are actually not single twin lamellae,

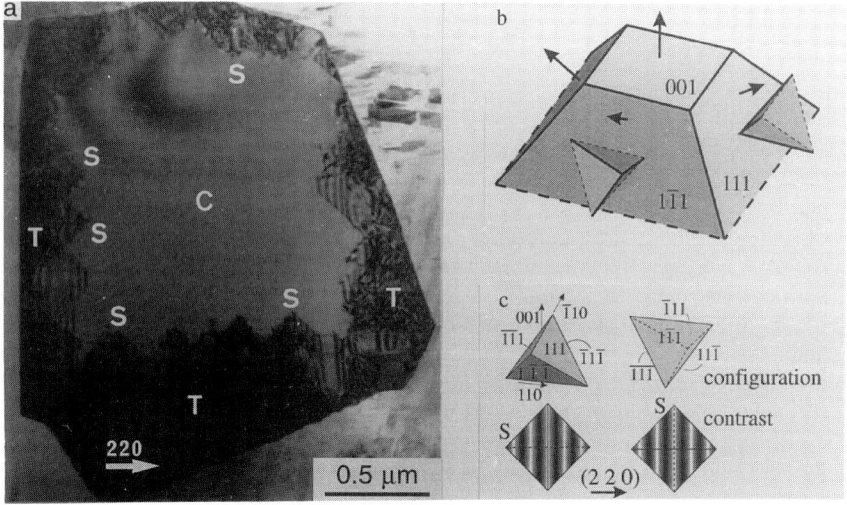

Figure 1: *Growth of microtwin lamellae in ⟨100⟩ textured layers.*
a): Bright field image taken under two-beam conditions with (220) strongly excited near to the [001] zone axis; C: defect free column; T: Regions containing agglomerations of microtwin lamellae; S: Starting points of twin growth
b): Schematic representation of microtwin tetrahedra growth on {111} crystal facets: The tetrahedra nucleate with one corner at an imperfection at the growing surface and widen during subsequent growth. Arrows indicate the growth directions of the respective facets, the indices of the facet planes are given
c): Bright field contrast of tetrahedra formed by stacking faults or microtwin lamellae for g=(220) strongly excited: Shear vectors R=n/3 {111}+r ({111} perpendicular to the respective twin plane (cf. given indices); r: arbitrary lattice vector), therefore g·R≠n only for the (1 1 1) and (1 1 $\bar{1}$) planes → only these planes cause the contrast; the contrast is similar for both orientations of the tetrahedron; S: Starting point of growth corresponding to (a) and (b)

but tetrahedra formed by microtwin lamellae (Fig. 1b; referred to as 'twin tetrahedra' further on) as can be demonstrated from the contrast in Fig. 1a: The contrast of twin lamellae parallel to ($1\bar{1}1$) or ($\bar{T}11$) vanishes under the applied diffraction conditions because the dot product of the diffraction vector and the possible shear vectors is always an integer (Hirsch et al 1967). The strong contrast of the configurations starting at the points S can, therefore, only be caused by twin lamellae parallel to (111) or ($11\bar{T}$) (cf. Fig. 1c). Because of the crystal symmetry the configurations starting from the ($1\bar{T}1$) facet must be isomorphic (though rotated by 90°) with the configurations starting from (111). Therefore, the contrast in Fig. 1a gives evidence for the presence of twin lamellae parallel to all three {111} matrix planes which are inclined to the growth facet. These sets of twin lamellae form incomplete tetrahedra with the fourth plane being the {111} growth facet (Fig. 1b). At the intersections of the twins with this facet re-entrant grooves should form which may favour and drive the further growth of the lamellae (Angus et al 1992).

The simultaneous growth of twin lamellae as open twin tetrahedra is consistent with two main features of the twin agglomerations in CVD diamond observed so far: (i) When that kind of twin tetrahedron is cut perpendicular to a ⟨110⟩ direction, the two twin lamellae perpendicular to the cut will appear as a v which encloses an acute angle and opens towards the growth direction. This is exactly what Shechtman et al (1993) reported from HREM images taken along ⟨110⟩ zone axes. (ii) The twin agglomerations contain intersecting twin lamellae parallel to more than two {111} matrix planes (Wurzinger et al 1993).

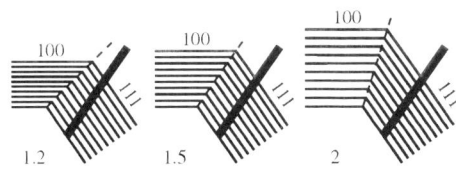

Figure 2: *Progress of adjacent {111} and {100} facets during growth with different parameters α (given below the figures). Thick solid line: Twin lamella nucleated on the {111} facet; Broken line: Advance of the facet intersection;*

As the growth of twin tetrahedra is connected to the {111} facets, it also depends on the development of these facets during the layer growth. Figure 2 represents schematically the progress of adjacent {100} and {111} facets during growth with different α parameters (cf. sec. 1) and a twin lamella nucleated on the {111} facet: The intersection of the twin lamella with the crystal surface will always stay on the {111} facet for α≥1.5, while it eventually will change to the {100} facet for α<1.5.

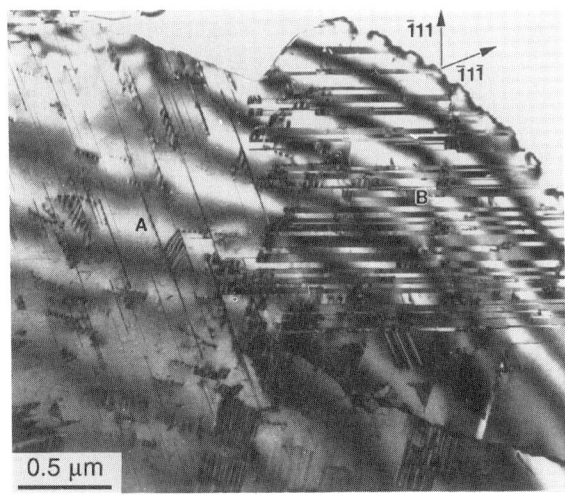

0.5 μm

Figure 3:
Twin lamella domains in a CVD diamond layer with film texture between ⟨111⟩ and ⟨110⟩; bright field image near to the [110] zone axis
A,B: Domains with different predominant orientation of the twin lamellae

Therefore, the tetrahedral growth of twin lamellae presented above for a film texture near to $\langle 100 \rangle$ $(2.5 < \alpha \leq 3)$ need not necessarily apply to layers grown with $\alpha < 1.5$. In fact, as can be seen from Fig. 3, the configuration of twin lamellae in such layers is completely different from the intersecting twin lamellae described so far: There are domains in which one type of twin lamellae clearly dominates. The single grains contain more than one domain with different lamella orientations (A, B in Fig. 3). The development of these twin lamella domains is not completely understood, yet. The facet growth competition represented by Fig. 2 certainly is involved with it. However, twin nucleation at {100} facets also may play an important role as is suggested by geometrical analyses for the growth of extended twins (Tamor and Everson 1994, Wild et al 1994).

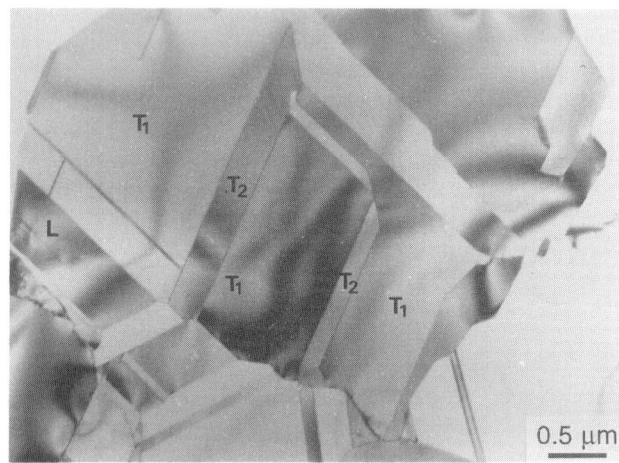

Figure 4:
Bright field survey of a CVD diamond layer containing very few microtwin lamellae (L). The film textur is near to $\langle 110 \rangle$ (image taken without tilting the specimen).
T_1, T_2: *Examples for $\Sigma 3$ twins*

Most of the CVD diamond samples investigated so far in our laboratory contained microtwin lamellae in one of the two configurations described above. A few samples, however, were virtually free of such defects. Figure 4 gives an example of that. The film texture of this layer is near to $\langle 110 \rangle$. The grains contain many extended twins (T_1, T_2 in Fig. 4) but only a very limited number of twin lamellae (L in Fig. 4). The large number of extended twins is not surprising because extended twins, once nucleated, are not overgrown but grow together with the grains for $\alpha \approx 1.5$ (film texture near <110>), as has been shown by Tamor and Everson (1994) and Wild et al (1994). Both the size of the twins and the absence of microtwin lamellae, however, indicate that the nucleation rate of twins is small.

Another example for a virtually twin-free CVD diamond film is a part of a layer where heteroepitaxial growth (with $\alpha \approx 3$) on {100} silicon could be achieved (Fig. 5). This specimen has been thinned only from the interface side of the layer, therefore the original layer surface formed by {100} facets is visible in the micrograph. All grains of this specimen are well aligned to the substrate. The mean angular deviation of the crystal axes is as low as 5.6°. The grains are separated by two types of low angle grain boundaries:

(i) Curved boundaries (D in Fig. 5) consisting of parallel dislocations which are generally of the edge type. They are essentially tilt boundaries with the tilt axis parallel to the dislocation lines. As macroscopic grooves or steps do not exist at these boundaries, they are considered as being formed by the fusion of neighbouring grains. The dislocations which grew along with the crystal compensate for the tilt misalignment between the grains. The elastic stress which can be recognized from the bending contours in the vicinity of these boundaries is probably caused by small values of twist misalignment.

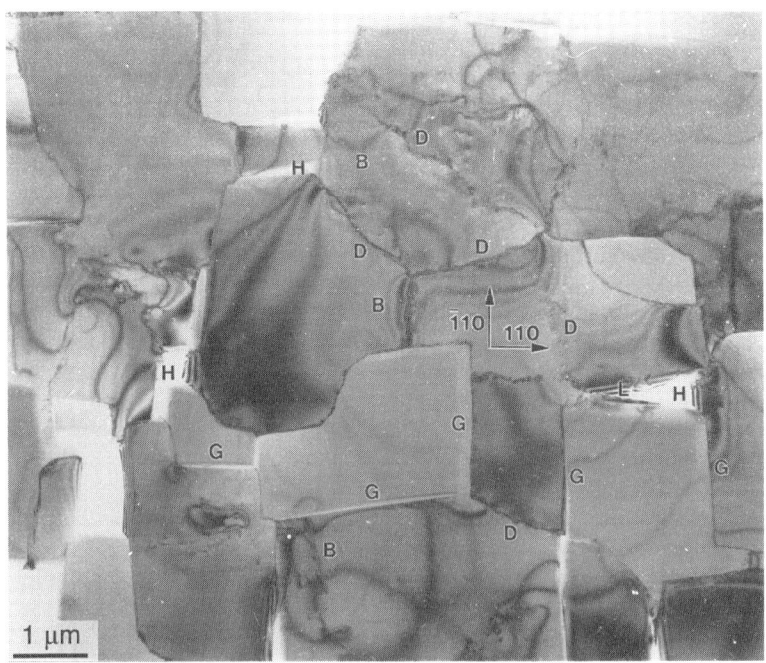

Figure 5: *Surface region of a diamond layer part grown heteroepitaxially on {100} silicon;
Bright-field survey parallel to the surface normal; the arrows indicate the crystal directions of the
respective grain; G: Grain boundaries connected with grooves or steps at the layer surface; H: Holes
formed by ion milling where grooves deeper than the actual thickness of the TEM specimen existed;
D: Low angle grain boundaries consisting of parallel dislocations; B: Bending contours indicating
the presence of residual elastic stress*

(ii) The second type of grain boundaries is connected with steps and grooves at the layer
surface (G in Fig. 5). Holes (H in Fig. 5) in between the grains are vestiges of such
boundaries where the grooves were originally deeper than the actual TEM specimen
thickness. These grain boundaries are generally flat and parallel to {110} lattice planes. They
are of the general tilt-twist type. A transformation to parallel dislocations, therefore, has not
been possible. The sidewalls of the grooves and steps associated with these boundaries are
residues of the {111} growth facets. Only very few twin lamellae (L in Fig. 5) can be
observed at these facets, most of them are completely free of microtwins.

Figure 1 clearly shows that the nearly complete absence of microtwin lamellae in Fig. 5
can not be attributed to ⟨100⟩ textured growth. However, a specific layer texture (as ⟨100⟩)
can be obtained by different sets of deposition parameters (Wild et al 1994). The variation
of these parameters, of course, might influence the formation probability of growth
imperfections on the {111} facets from which the twin lamellae can nucleate. In fact, Fig. 5
gives evidence that a set of parameters for ⟨100⟩ textured growth exists which reduces this
formation probability sufficiently to avoid microtwin nucleation on the small area of residual
{111} facets. The existence of similar parameter sets for other growth textures is indicated
by Fig. 4. Therefore, the film texture (or the parameter α) determines the possible
configurations of microtwin lamellae, as has been shown above, but the density of the
lamellae is controlled by other factors which may be optimized independently.

4. CONCLUSIONS

The possible configurations of microtwin lamellae in CVD diamond layers depend on the film texture. Domains of parallel lamellae may form in layers with a texture between ⟨111⟩ and ⟨110⟩ while intersecting lamellae can be observed in layers with a texture near to ⟨100⟩. The latter nucleate at point imperfections on the {111} facets and start to grow together with these facets as tetrahedra formed by twin lamellae. The density of microtwins is not a function of the textured growth but is connected to other growth parameters which influence the formation probability of growth imperfections from which the twins can originate. TEM observations of the microtwin lamellae may, therefore, be used to determine optimum growth parameters.

ACKNOWLEDGEMENTS

The authors are indebted to P.K. Bachmann (Philips Research Laboratories Aachen) and to R. Hessmer and M. Schreck (University of Augsburg) for providing the diamond samples for this research. We also gratefully acknowledge financial support of the present work by the Austrian Science Foundation FWF (project S5903) carried out under the auspices of the trinational "D-A-CH" cooperation of Germany, Austria, and Switzerland on the "Synthesis of Superhard Materials".

REFERENCES

Angus J C, Sunkara M, Sahaida S R and Glass J T 1992 J. Mater. Res. 7 3001

Clausing R E, Heatherly L, More K L and Begun G M 1989 Surf. Coat. Technol. 39/40 199

Hessmer R, Schreck M, Geier S, Rauschenbach B and Stritzker B 1995 to be published in Diamond Relat. Mater. 4

Hirsch P B, Howie A, Nicholson R B, Pashley D W and Whelan M J 1967 Electron Microscopy of Thin Crystals (London: Butterworths)

Jiang X, Klages C-P, Rösler M, Zachai R, Hartweg M and Füsser H-J 1993 Appl.Phys.A 57 483

Joksch M, Wurzinger P, Pongratz P, Haubner R and Lux B 1994 Diamond Relat. Mater. 3 681

Joksch M, Bachmann P, Wurzinger P and Pongratz P 1995 to be published

Shechtman D, Hutchison J L, Robins L H, Farabaugh E N and Feldman A 1993 J. Mater. Res. 8 473

Tamor M A and Everson M P 1994 J. Mater. Res. 9 1839

Van der Drift A 1967 Philips Res. Repts. 22 267

Wang Z L, Bentley J, Clausing R E, Heatherly L and Horton L L 1994 J. Mater. Res. 9 1552

Wild C, Koidl P, Müller-Sebert W, Walcher H, Kohl R, Herres N, Locher R, Samlenski R and Brenn R 1993 Diamond Relat. Mater. 2 158

Wild C, Kohl R, Herres N, Müller-Sebert W and Koidl P 1994 Diamond Relat. Mater. 3 373

Williams B E and Glass J T 1989 J. Mater. Res. 4 373

Wurzinger P, Joksch M and Pongratz P 1993 Inst. Phys. Conf. Ser. 134 157

Zhu W, Badzian A R and Messier R 1989 J. Mater. Res. 4 659

Inst. Phys. Conf. Ser. No 146
Paper presented at Microsc. Semicond. Mater. Conf., Oxford, 20–23 March 1995
© *1995 IOP Publishing Ltd*

Structure characterization of boron doped CVD diamond films

F Peiró[1,2], M C Polo[1], J Cifre[1], G Sánchez[1] and J Esteve[1]

[1] Dept. Física Aplicada i Electrònica, Univ. Barcelona, Av. Diagonal 647, 08028 Barcelona, Spain
[2] Serveis Científico-Tècnics, Univ. Barcelona, Lluís Solé i Sabarís 1-3, 08028 Barcelona, Spain

ABSTRACT: Boron doped diamond films were obtained by using two deposition techniques: hot filament chemical vapour deposition (HFCVD) and microwave plasma chemical vapour deposition (MWCVD). In both systems, films were deposited from mixtures of methane and hydrogen and variable amounts of trimethylboron $B(CH_3)_3$ (TMB) as the boron gaseous source. The effects of boron in the growth process and in the film structure were investigated. TEM and SEM showed that high boron contents in the films induced well-defined morphological effects.

1. INTRODUCTION

The potential of diamond as a semiconductor material has promoted many studies on the doping of CVD diamond films during their growth. Methods of p-doping of diamond are based on the introduction of boron compounds in the gas mixture (CH_4 and H_2) used in the deposition process. The use of gaseous sources over the solid (Mort 1989) and liquid (Cifre 1992) ones allows an accurate control of the boron concentration in the films by changing the ratio of the B-compound in the gas mixture. The most commonly reported gaseous boron source in diamond doping is diborane (B_2H_6) (Okano 1989), a highly poisonous compound. We have obtained boron doped diamond films by using trimethylboron $B(CH_3)_3$ (TMB)(Polo 1994), a novel gaseous boron source. TMB is easily mixed with most carrier gases at precise concentrations, has fairly good chemical stability and has a lower toxicity than diborane gas. The aim of this work is to study not only the effects of boron incorporation in the microstructure of doped diamond films but also the influence of the growth technique. Structural characterization of the films was carried out by SEM and TEM analyses. Raman spectroscopy and X-ray diffraction results are also discussed.

2. EXPERIMENTAL

Two series of doped diamond films were grown on Si substrates, with different amounts of TMB in the gas mixture, by both hot filament CVD and microwave plasma CVD, which are the most commonly used growth techniques. The reaction gases were a mixture of CH_4 and H_2 and the boron source was trimethylboron diluted in helium at 0.5 vol.%. In the hot filament system, reaction gases were activated by a folded tungsten filament 1 mm thick and 150 mm long heated at 2000 °C. Before deposition the filament was heated in a methane atmosphere until totally carburised. Substrates were placed on a graphite holder 2 cm below the filament. The microwave reactor was of the waveguide type. The gases were activated in the microwave plasma and the substrates were placed in a graphite holder at the centre of the waveguide. In both cases, diamond films were grown onto polished silicon substrates scratched with $1\mu m$ diamond powder and cleaned in a methanol ultrasonic bath. Two sets of doped samples were obtained at standard deposition parameters. The methane-to-hydrogen concentration ratio was fixed at 1%, the total gas pressure at 4500 Pa and the substrate temperature at 920 °C. The TMB-helium volume concentration in the reaction gas was in the range of 0%-4.5%.

530

The incorporation of boron in the films was analyzed by secondary ion mass spectrometry (SIMS) and wavelength dispersive electron-probe X-ray microanalysis (WDS). SIMS depth profiles showed a uniform level of boron through the whole diamond film and an increase in boron content with increasing TMB concentration in the gas. WDS measurements allowed us to quantify the atomic content of boron from the area ratio of C K_α and B K_α peaks. For the same TMB concentrations in the reaction gas, boron incorporation in the films was more efficient when deposited in the HFCVD system but it saturated at 3 at.% for 0.03 vol.% TMB. However, in the MWCVD samples the boron content steadily increased up to a concentration of 9 at.% for a 0.024 vol.% of TMB in the gas, without showing saturation. Thus, we can conclude that in both deposition systems the boron incorporation efficiency is very high.

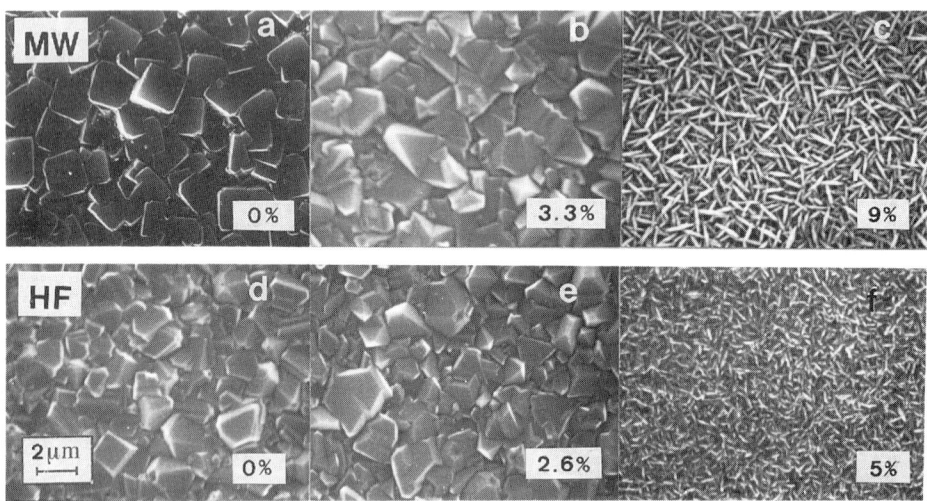

Fig. 1. SEM surface images showing the evolution of crystal structure with increasing the boron concentration in the diamond films obtained by MWCVD (a), (b), (c) and HFCVD (d), (e), (f).

3. RESULTS

Cross section SEM analyses showed that thick and compact films were obtained in the whole range of TMB concentrations, with a clear columnar structure. From layer thickness measurements, we found that the boron incorporation notably affected the growth rate of the doped MWCVD films. The highest growth rate, about 1 μm/h, corresponded to the undoped one and it decreased to 0.25 μm/h when increasing the boron content up to 3 at.%. However, the growth rate of the doped HFCVD samples did not change appreciably for similar boron levels, only decreasing from 0.3 to 0.1 μm/h. Surface images of MWCVD (Fig. 1a) and HFCVD (Fig. 1d) undoped samples showed different morphologies. The surface of the former consisted mainly of squared {100} faces whereas the surface of the latter showed multiple randomly orientated crystals. It was found that MWCVD samples lost the {100} texture when doped even for boron contents as low as 0.1 at.% and the grain size decreased as the boron concentration raised up to 3 at.% (Fig. 1b). In HFCVD samples, no appreciable changes in surface morphology or grain size are observed when doping the films (Figs. 2d and 2e). For all samples boron concentrations over 3 at.% induced a drastic diminution of crystal size and for the highest boron contents (above 5 at.%) films surface did not show the crystalline morphology characteristic of diamond (Figs. 2c and 2f).

These results were corroborated by TEM analysis. Plan-view bright field images and the correspondent TED patterns of diamond films grown by MWCVD containing different amounts of boron are shown in Fig. 2. The undoped MWCVD films were composed of a large amount of cubic

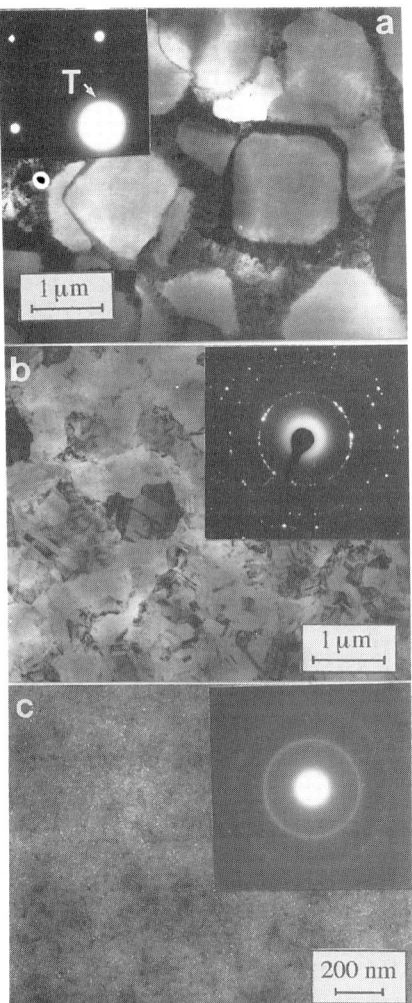

Fig. 3. Diamond film with a content of boron of 0.35 at.% obtained by HFCVD.

Fig. 2. BF TEM images of MWCVD films revealing the change of the crystallinity with increasing boron concentration. (a) undoped film, (b) 1 at%. and c) 9 at.% in boron.

{100} faceted crystals (Fig. 2a). Although the growth direction for these crystals was mainly [100], they did not show a preferential orientation with respect to the [010] and [001] directions, as it is also seen in Figs. 1a and 2a. These square-shaped crystals (with sizes between 2-3 μm) did not contain planar defects, but they were surrounded by very defective small grains. As it was found from the SEM results, boron doping led to a loss of the [100] morphology. In figure 2, we can observe the polycrystalline nature of a 1 at.% doped film. Crystals were found to be randomly orientated, most frequently in the [111], [110] and [311] directions in agreement with the XRD spectra. Despite the fact that the intensity ratios of the XRD peaks did not appreciably change, the size of the crystals was monotonically reduced when the boron content in the films increased. We had thus the smallest ones in the heaviest doped sample (9 at.% in Fig. 2c), which presented nearly an amorphous structure containing only nanocrystals.

Diamond films grown by HFCVD showed different features. Indeed, the undoped film did not present such a strong [100] morphology and also exhibited a lower crystal size (between 0.5-1.5μm) than the MWCVD films. Moreover, the grain size remained nearly constant up to boron contents of 3 at.% (Fig. 3). Similarly to MWCVD films, for relatively high boron concentrations the material became rather amorphous. The amorphization of the films for boron contents of 9 at.% (MWCVD) and 5 at.% (HFCVD) was confirmed by XRD and Raman analysis. X-ray diffraction peaks related to the diamond phase were not observed, and in Raman spectra, the narrow diamond peak at 1332 cm^{-1} disappeared giving rise to a broad band centred around 1550 cm^{-1} typical of disordered graphitic phases.

Besides the decrease in crystal size when boron was added, which was attributed to an effect of boron which enhanced nuclei formation during the first steps of growth, another significant result can be noticed when comparing the MWCVD and HFCVD deposition techniques. Whereas in doped HFCVD samples no preferential orientation was observed (Fig. 4a), in doped HFCVD films (Fig. 4d)

532

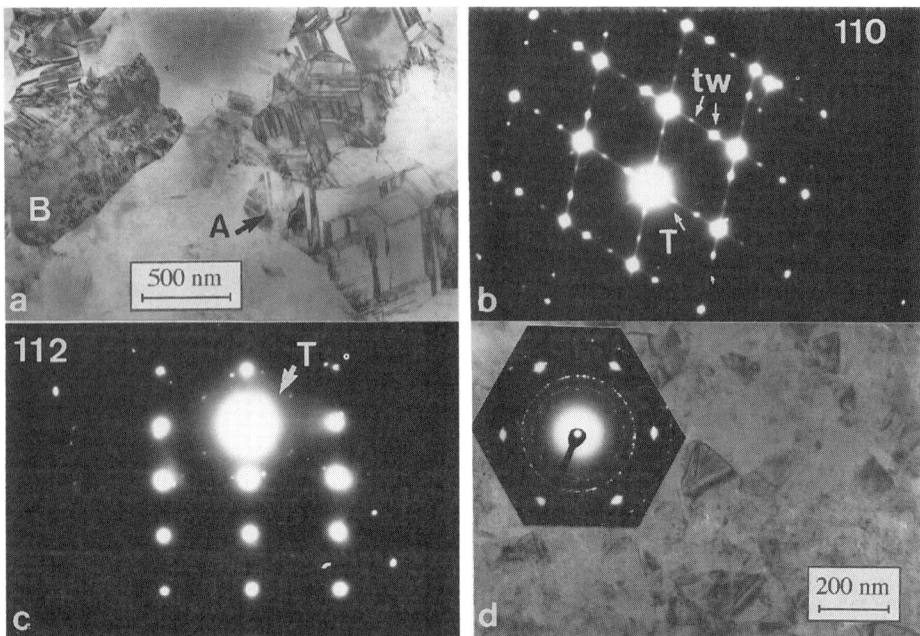

Fig. 4. Diamond films with a boron concentration of $\simeq 3$ at%. (a) Sample obtained by MWCVD; we notice the coexistence of grains (A and B) with orientations [110] (b) and [112] (c) respectively, revealing the presence of twins (T) and threading dislocations (TD). (d) HFCVD film, with preferential growth along [111].

the growth direction was found to be mainly [111] (Fig. 4e). These images also revealed the nature of the defects inside the diamond grains. Hence, for the [110] orientation (grain A in 4a, with corresponding TED in 4b), a faulted structure with multiple twins (T) in {111} planes was observed, coexisting with a relatively high density of threading dislocations (TD) clearly observable in other crystal orientations as [112] (Grain B in Fig. 4a with its TED in 4c).

The correlation of the boron content with the conductivity of the films (Polo 1994), suggested an incorporation of the dopant in substitutional sites for concentrations up to 1at.%, and interstitially for higher doping levels.

4. CONCLUSIONS

Good quality p-doped diamond films were obtained by both HFCVD and MWCVD deposition methods. Low doping levels improved the quality of the films by decreasing the amount of graphitic phases, but higher concentrations induced a decrease of the crystal sizes and even an amorphization of the films for the highest contents of boron analyzed. The deposition technique also affected the growth direction of the crystals, giving rise to a strong [111] orientation in doped HFCVD films.

ACKNOWLEDGEMENTS
This work has been financially supported by CICYT of the Spanish Government under Contract MAT93-C03-02.

REFERENCES
Cifre J, López F, Morenza J L and Esteve J 1992 Diamond Relat. Mater. 1 500
Mort J, Kuhman D, Machhonkin M, Jansen F, Okamura K et al. 1989 Appl. Phys. Lett. 55 1121
Okano K, Naruki H, Akiba Y, Kurosu T, Iida M, Hirose Y et al. 1989 Jpn. J. Appl. Phys. 28 1066
Polo M C, Cifre J and Esteve J 1994 Vacuum 45 1013

Inst. Phys. Conf. Ser. No 146
Paper presented at Microsc. Semicond. Mater. Conf., Oxford, 20–23 March 1995
© *1995 IOP Publishing Ltd*

HREM characterization of ion beam synthesized ternary silicides in (111) silicon

J Tavares[1,2], H Bender[1], M F Wu[3,4], A Vantomme[3], G Langouche[3], C Lin[5]

[1]IMEC, Kapeldreef 75, B-3001 Leuven, Belgium
[2]LSI, Universidade de São Paulo, Brasil
[3]Instituut voor Kern- en Stralingsfysika, University of Leuven, B-3001 Leuven, Belgium
[4]Department of Technical Physics, Peking University, Beijing, China
[5]Shanghai Institute of Metallurgy, Chinese Academy of Sciences, China

ABSTRACT: Buried ternary NiFe- and CoFe-silicides are synthesized by sequential equal-dose implantations of both metals. TEM analysis shows strong evidence for the formation of a metastable $Ni_{0.5}Fe_{0.5}Si_2$ and $Co_{0.5}Fe_{0.5}Si_2$ phase during the implantation. Annealing of these structures results in the formation of a continuous buried layer in which almost full phase separation occurs in case of the CoFe-silicide, while only the onset of phase separation is found for the NiFe-silicide.

1. INTRODUCTION

The fundamental properties of the binary silicides $CoSi_2$ and $NiSi_2$ have been extensively studied in the past because these materials are potential candidates for applications in microelectronic devices. Both silicides have the cubic fluorite structure and can be grown epitaxially on silicon substrates. $FeSi_2$ has a more complex phase diagram and occurs in several phases : the high temperature metallic tetragonal α-$FeSi_2$-phase (stable above 960°C), the semiconducting orthorhombic β-$FeSi_2$, and the metastable cubic γ-$FeSi_2$ which has the fluorite structure. With the aim to improve the epitaxial quality and to modify the band gap of the semiconducting β-$FeSi_2$, mixed Co-Fe and Ni-Fe silicides are explored.

In this paper the results will be discussed of a TEM investigation of the structure of $Ni_{0.5}Fe_{0.5}Si_2$ and $Co_{0.5}Fe_{0.5}Si_2$ layers prepared by Ion Beam Synthesis (IBS) in (111) oriented silicon substrates. The occurring phases and their epitaxial relationships will be discussed.

2. EXPERIMENTAL DETAILS

The samples are consecutively implanted with equal doses of $0.6\times10^{17}cm^{-2}$ for each metal species with Fe^+ and Ni^+ or with Co^+ and Fe^+ at an energy of 90 keV, while the silicon substrate temperature is kept at 350°C. Different subsequent single and two-step anneal sequences are considered.

The material is analyzed in cross-section by combination of conventional and high resolution electron microscopy and electron diffraction after implantation and after the different annealings.

3. RESULTS AND DISCUSSION

Investigation of the as-implanted samples by cross-sectional high resolution electron mi-

534

croscopy (HREM) reveals a high density of precipitates with a size of 5-15 nm in a 90 nm wide region centered at a depth of about 90 nm (Fig. 1a). The upper 50 nm of the silicon shows less precipitates but has a high density of twins and stacking faults parallel with the surface. The precipitate distribution corresponds to the Gaussian-like metal distribution as observed by Auger (AES) depth profiling and Rutherford Backscattering Spectrometry (RBS). The HREM images and diffraction patterns are consistent with a fluorite structure for the precipitates. No other phases as e.g. α- or β-FeSi$_2$ are found. As NiSi$_2$ (or CoSi$_2$) and γ-FeSi$_2$ precipitates will show similar images and have almost equal contrast compared to the silicon matrix and cannot be distinguished by TEM from their only slightly different lattice parameters, we cannot uniquely decide from the TEM observations whether a mixture is present of these binary phases or whether a ternary Ni$_{0.5}$Fe$_{0.5}$Si$_2$ (or Co$_{0.5}$Fe$_{0.5}$Si$_2$) phase with fluorite structure has formed. However, Mössbauer measurements on these samples give no indication for the presence of γ-FeSi$_2$ (Vantomme et al. 1995b). Also the EXAFS and XRD observations of Tan et al. (1993) on as-implanted samples prepared with 2×10^{17}cm^{-2} Co and 1.5×10^{17}cm^{-2} Fe at 150 keV are in favour of the formation of a ternary silicide. Our TEM results show that the precipitates occur predominantly in B-type orientation, i.e. twinned with respect to the silicon matrix. The B-type precipitates give rise to the characteristic extra spots along the $\{111\}_{Si}$ directions in the selected area electron diffraction patterns. No reflections due to other phases are observed.

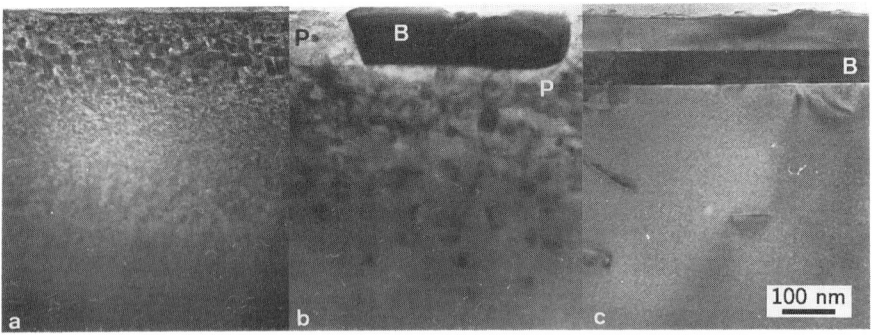

Fig. 1 : Cross-section TEM images of the structure of the Ni-Fe implanted sample : a) as-implanted, b) annealed at 1030°C for 15 s in an N$_2$ ambient, and c) annealed in vacuum at 1000°C for 30 min. Precipitates are marked 'P'.

Annealing the Ni-Fe implanted sample at 1030°C for 15 s in an N$_2$ ambient, results in the formation of large epitaxial silicide grains with B-oriented fluorite structure and which reach the surface (Fig. 1b). Between and below these large grains, small precipitates occur. A more interesting configuration is obtained by annealing at 1000°C for 30 min in vacuum. This treatment results in a nearly continuous buried silicide layer with abrupt planar interfaces (Fig. 1c). The buried layer has a thickness of approximately 50 nm with a top silicon layer of 70 nm. Precipitates outside the silicide layer are only rarely observed. Dislocations occur until a depth of ~500 nm. These observations agree with the AES and RBS measurements on these samples which show that the initially Gaussian-like metal distribution coalesces to a buried layer with sharp interfaces during the vacuum anneal (Vantomme et al. 1995a).

In the buried layer two different structures can be distinguished (Fig. 2). For the Ni-Fe samples, at least 90% of the layer shows a high resolution image which is due to the B-type orientation of a fluorite structure (Fig. 2,3). The twinning occurs almost exclusively on the $(11\bar{1})_{Si}$ plane parallel with the wafer surface. The rest of the buried layer is characterized by two types of high resolution images and related diffraction patterns which can both be interpreted as due to an α-silicide structure. For the vacuum annealed Co-Fe samples both phases occur with nearly equal volume. Both structures run through the whole buried layer. In case of the CoFe they occur in lamellae with a thickness of 5 to 40 nm. Due to this small width, it has not been possible to obtain exclusive evidence of the composition of both materials by energy dispersive X-ray analysis, but it could be established that the composition of the buried layer is inhomogeneous with Co and Fe rich regions.

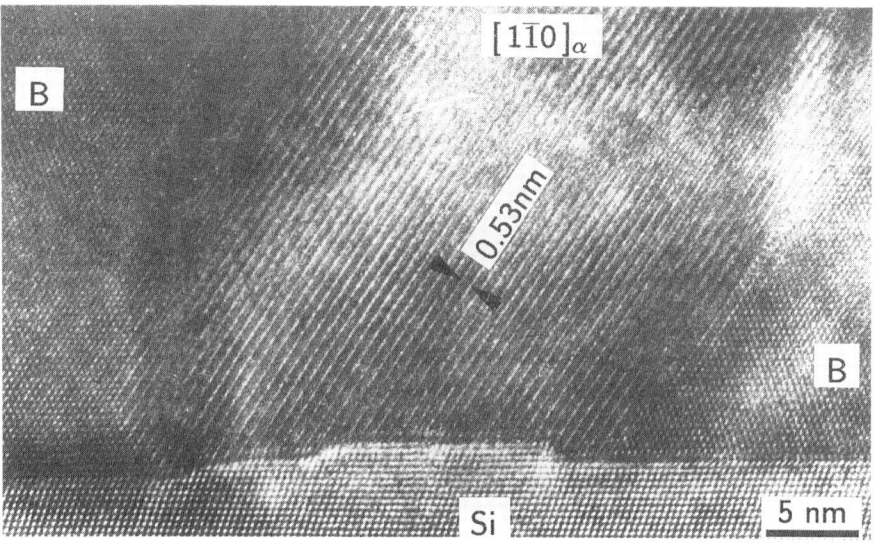

Fig. 2 : Cross-section HREM image of the lower interface of the buried layer with the silicon matrix for the vacuum annealed Fe-Ni-sample showing the B-oriented $Ni_xFe_{1-x}Si_2$ ($x \sim 0.5$) and $[1\bar{1}0]_\alpha$ oriented α-$Fe_yNi_{1-y}Si_2$ ($y > 0.8$).

As the implanted doses of Ni and Fe are equal, it can be concluded that these phases are respectively B-$Ni_xFe_{1-x}Si_2$ with x close to 0.5 and α-$Fe_yNi_{1-y}Si_2$ with y unknown. A probable lower limit for y can be estimated as $y=0.8$ from the observations of Panknin et al. (1993) that up to 20% Co can be incorporated in β- and probably also in α-$FeSi_2$. On the other hand, in case of the Co-Fe samples, the volume of both phases is roughly the same so that one should have B-$Co_xFe_{1-x}Si_2$ and α-$Fe_xCo_{1-x}Si_2$ both with x close to 1. This implies that almost full phase separation occurs during the anneal of the ternary $Co_{0.5}Fe_{0.5}Si_2$, while only the onset of phase separation is found for the $Ni_{0.5}Fe_{0.5}Si_2$. Also XRD measurements confirm that much less α-phase material is present in the vacuum annealed NiFe- than in the CoFe-sample. The reason for the higher stability of the ternary $Ni_{0.5}Fe_{0.5}Si_2$ is not clear for the moment and the temperature dependence of this phenomenon has to be investigated in more detail. Tan et al. (1993) also reported phase separation for Co-Fe silicides into $CoSi_2$ and β-$FeSi_2$ for annealings at 700°C. In (100) silicon Hunt et al. (1994) found phase separation for Co-Fe silicides in a two layer structure consisting of an epitaxial α-$FeSi_2$ and an aligned A-type $CoSi_2$ after annealing at 1000°C. All these observations indicate that the ternary cubic silicide is a metastable phase.

Two types of high resolution images are obtained for the α-phase. The first one (Fig. 2) shows a periodicity of 0.53 nm for the planes parallel with the $(100)_B$ planes of the twinned phase. The image and diffraction pattern correspond to the $[1\bar{1}0]$ zone of the tetragonal α-phase such that : $[1\bar{1}0]_\alpha$ // $[011]_B$ (i.e. also // $[011]_{Si}$), with $(001)_\alpha$ // $(100)_B$ and $(112)_\alpha$ // $(11\bar{1})_B$. The second type of images corresponds to the $[02\bar{1}]$ zone of the α-silicide such that : $[02\bar{1}]_\alpha$ // $[011]_B$ (i.e. also // $[011]_{Si}$), with $(100)_\alpha$ // $(100)_B$ and $(112)_\alpha$ // $(11\bar{1})_B$. Both zones are due to different orientation variants of the same epitaxial relationship between the α-phase structure and the B-oriented phase, i.e. : $(001)_\alpha$ // $(100)_B$ with $[100]_\alpha$ // $[010]_B$ and

536

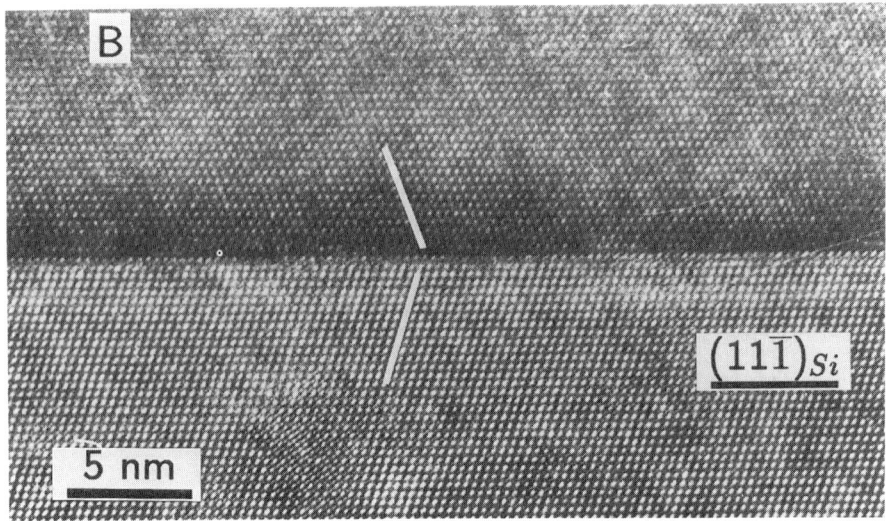

Fig. 3 : HREM image of the lower interface of the buried layer with the silicon matrix for the vacuum annealed Fe-Ni-sample showing the B-oriented $Ni_xFe_{1-x}Si_2$ ($x \sim 0.5$).

$[010]_\alpha$ // $[001]_B$. The epitaxy corresponds with an α-phase structure aligned with respect to the B-phase. Hence both are twinned with respect to the silicon matrix.

4. CONCLUSIONS

It is shown that by IBS in (111) silicon, ternary $Ni_{0.5}Fe_{0.5}Si_2$ and $Co_{0.5}Fe_{0.5}Si_2$ silicide precipitates with the fluorite structure (mainly B-orientation) are formed during equal-dose implantation at 350°C. This phase turns out to be metastable, leading at 1000°C to the onset of phase separation in case of the Ni-Fe silicide and almost full phase separation in case of Co-Fe silicide.

JT acknowledges the financial support of the Conselho Nacional de Desenvolvimento Científico e Tecnológico (CNPq) from Brasil. AV is postdoctoral researcher NFWO (National Fund for Scientific Research, Belgium). This work is carried out with the TEM at the University of Antwerpen (RUCA).

REFERENCES

Hunt T D, Sealy B J, Hanebeck J, Reeson K J, Homewood K P, Gwilliam R M, Meekison C D and Booker G R 1993 Mat. Res. Soc. Symp. Proc. <u>279</u> 893

Panknin D, Wieser E, Skorupa W, Vohse H and Albrecht J 1993 Nucl. Instr. Methods in Phys. Res B <u>74</u> 213

Tan Z, Namavar F, Heald S M and Budnick J I 1993 Appl. Phys. Lett. <u>63</u> 791

Tavares J, Bender H, Wu M F, Vantomme A, Langouche G, Lin C 1995 Appl. Phys. Lett. submitted

Vantomme A, Wu M F, Degroote S, Dekoster J, Langouche G, Tavares J and Bender H 1995a Proc. Conf. on Ion Implantation Technology, Catania, Italy, eds Rimini E, Ferla G, Priolo F and Coffa S, in press

Vantomme A, Wu M F, Langouche G, Tavares J and Bender H 1995b Proc. Ion Beam Modification of Materials (IBMM95)

Inst. Phys. Conf. Ser. No 146
Paper presented at Microsc. Semicond. Mater. Conf., Oxford, 20–23 March 1995
© *1995 IOP Publishing Ltd*

Gadolinium silicide films grown on (111) Si

C Vannuffel

CEA-DTA-LETI-DOPT-SCPM, 17 rue des Martyrs, 38054 Grenoble cedex 9, France

ABSTRACT: The microstructure of layers of co-deposited Gd and Si is reported. Strong and complex ordering of vacancies occurs in the Si sublattice. Annealing has a great influence upon the structural quality (crystal defects) and the ordering of the layer.

1. INTRODUCTION

Rare-earth silicides on Si are attractive for their Schottky-barrier heights, which are amongst the lowest known on n-type Si substrates. They are in the range 0.3-0.4eV (eg Tu et al 1981, Norde et al 1981) and specifically around 0.37-0.38eV for gadolinium silicide (Suu et al 1986, Tu et al 1981). The value found for our layers is 0.35eV (Pescher et al 1995a). For comparison, the value for Al is around 0.6eV. Therefore, these silicides may be useful as contact materials for integrated circuits in silicon.

2. PREPARATION AND STRUCTURE OF THE SILICIDE

Layers were grown on (111)Si substrates by co-evaporation in a UHV environment as described elsewhere (Lollman et al 1993) roughly with a 1Å/min growth rate for Gd and a Si/Gd ratio of 1.7. They were then annealed: sample A (20 nm thick) at 640°C for 20 min and 720°C for 15 min; sample B (10 nm thick) at 450°C for 40 min. The silicide stoichiometry is $GdSi_{1.7}$ (Pescher et al 1995b)

Cross sectional and plan view samples were observed at 200 kV with an Akashi 002B TEM fitted with a UHR pole piece. In order to avoid oxidation, care was taken to store the samples - covered with wax - under dry vacuum using silica gel. Samples were polished down to $30\mu m$ and then were ion milled immediately before observations.

The Gd silicide structure is hexagonal Gd_3Si_5: it can be considered a Si deficient $GdSi_2$ and, therefore, also written as $GdSi_{2-x}$ or $GdSi_{1.7}$. The structure of hexagonal $GdSi_2$ is a stacking of Gd planes and Si planes alternately along the [0001] axis. The projection on (0001) is given in Fig. 1: Gd atoms are represented by large grey circles and Si by small open circles. The parameters are a=3.88Å and c=4.17Å (e.g. Knapp and Picraux 1986).

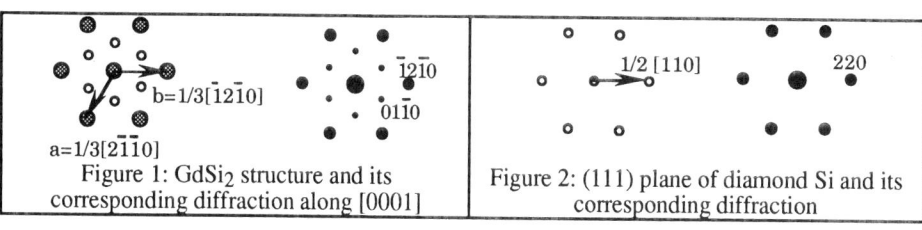

Figure 1: $GdSi_2$ structure and its corresponding diffraction along [0001]

Figure 2: (111) plane of diamond Si and its corresponding diffraction

Side 1 of the hexagon, formed by the Si atoms (Fig. 2) in {111} planes of the diamond structure, is 3.84Å. Therefore, good epitaxy is expected to occur on (111)Si which has the same symmetry as the basal plane of the silicide hexagonal structure and parameters are similar in this plane ($\Delta l/l = 1\%$).

3. MICROSTRUCTURE OF THE SILICIDE

Plan views and corresponding diffraction patterns from samples A and B are displayed in Figs. 3 and 4, respectively. Reflections are well defined spots and form a perfect 0001 zone axis diffraction pattern revealing the single-crystal morphology of the layer. The silicide is epitaxial upon the substrate; $11\bar{2}0$ GdSi$_{1.7}$ and 220 Si reflections are overlapping. The epitaxial relationship is then $<11\bar{2}0>$GdSi$_{1.7}$//$<110>$Si and [0001]GdSi$_{1.7}$//[111]Si. This has been reported for other rare-earth (including Y) silicides (e.g. Lee et al 1992, Knapp and Picraux 1986, Kaatz et al 1993, Arnaud d'Avitaya et al 1989).

For wide angle reflections, it is possible to distinguish the Si reflections from those of the silicide, which allows us to verify that the value of the $\Delta l/l$ ratio (see above) is indeed roughly 1%. With a $<110>$ Si cross-section (not shown) lattice parameters have been determined: a=3.9Å, c=4.2Å, which are consistent with the hexagonal structure.

The silicide has stacking faults lying in the {$1\bar{1}00$} planes. These defects produce the diffuse streaking in the $<1\bar{1}00>$* directions. The density is much larger for sample B than for sample A showing the improvement generated by the annealing process. Other defects not shown here are pits whose density increases with the annealing temperature.

4. ORDERING

Diffraction patterns also reveal extra reflections from the plain AlB$_2$ structure attributed to the ordering of silicon vacancies in GdSi$_{2-x}$. Similar diffraction spots have been observed for other rare-earth silicides (e.g. Lee et al 1992, Kaatz et al 1993, Arnaud d'Avitaya et al 1989).

The ordering occurs in domains revealed by HREM images. These domains are of relatively small size, unfortunately too small to obtain diffraction patterns from single domains. Microbeam diffraction is also not suitable, since it operates under convergent beam conditions and provides overlapping reflection discs.

Most of the ordering is revealed in stripes parallel to the three {$11\bar{2}0$} symmetrically equivalent planes (Fig. 3). It does not seem that one is represented more than the others. Roughly, these stripes are 20 to 40Å wide and produce the extra reflections in the $<11\bar{2}0>$* directions. Actually, there is a modulation on the width of these stripes which streaks the extra reflections in the $<11\bar{2}0>$* directions. Furthermore, the atomic structure inside the stripes is quite complex (Fig. 5).

Two types of extra spots with well defined positions occur at $1/2<1\bar{1}00>$* and $1/3<11\bar{2}0>$* implying the doubling and tripling of periodicities in the corresponding real space directions. Domains with such periodicities have been observed. Firstly, planes arranged with a doubling of the periodicity along one $<1\bar{1}00>$ are shown on Fig. 6; three types of planes exist producing all reflections at $1/2<1\bar{1}00>$*. Secondly a domain with large hexagons - with a side of 6.7Å - (Fig. 7) produce a tripling of the periodicity along all three $<11\bar{2}0>$ directions. However, few of both types of domains are present over the probed area. This is consistent with the relatively weak intensity of the associated reflections.

Layers are mostly composed of ordered domains; actually there is no evidence of non ordered domains. These domains may be separated by planar defects (say of the AlB$_2$ structure) but this is not necessarily so (see Fig.3). The sharpening of diffraction patterns suggests that the ordering may increase with the annealing temperature.

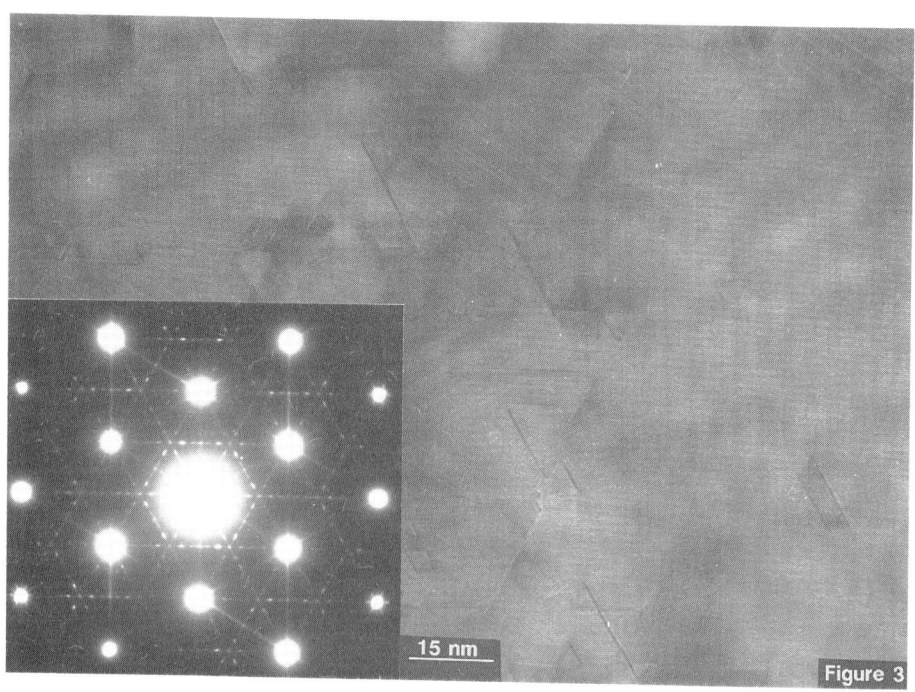

15 nm

Figure 3

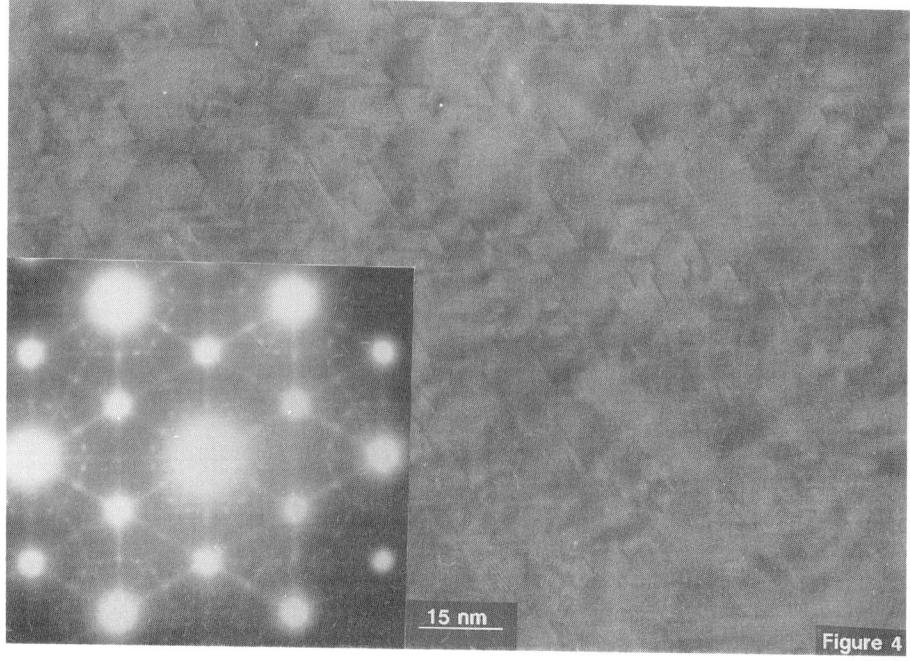

15 nm

Figure 4

540

Figure 5

Figure 6

Figure 7

5. CONCLUSIONS

The Gd-silicide layers grown by co-deposition are single crystalline and epitaxially oriented on the Si substrate. Though planar defects are present in rather high density, their numbers are dramatically decreased through annealing.

Strong ordering of Si vacancies occurs in the silicide. It is rather complex and exists in small domains which seem to be present throughout the layer. This ordering increases with the annealing temperature, but there is still work to be done to fully describe the ordering.

6. ACKNOWLEDGEMENTS

Nguyen Tan T A, Veuillen J-Y from CNRS-LEPES and Pescher C are gratefully acknowledged for providing the layers.

REFERENCES

Arnaud d'Avitaya F, Perio A, Oberlin J-C, Campidelli Y and Chroboczek J A 1989 Appl. Phys. Lett. 54, 2198-200

Kaatz F H, Graham W R and Van der Spiegel J 1993 Appl. Phys Lett. 62, 1748-50

Knapp J A and S.T. Picraux S T 1986 Appl. Phys. Lett. 48, 466-8

Lee T L, Chen L J and Chen F R 1992 J. Appl. Phys. 71, 3307-12

Lollman D B B, Nguyen Tan T A, Veuillen J-Y, Muret P, Lefki K, Brunel M and Dupuy C 1993 Appl. Surf. Sci. 65/66, 704-711

Norde N, De Sousa Pires J, d'Heurle F, Pesavento F, Petersson S and Tove P A 1981 Appl. Phys. Lett. 38, 865-7

Pescher C, Ermolieff, Veuillen J, Nguyen Tan T and Brunel M 1995b accepted for publication in Solid State Communications

Pescher C, Pierre J, Ermolieff A and Vannuffel C 1995a submitted to Thin Solid Films

Suu H V, Paszti F, Mezey G, Petö G, Manuaba A, Fried M and Gyulai J 1986 J. Appl.Phys. 59, 3537-9

Tu K N, Thompson R D and Tsaur Y 1981 Appl. Phys. Lett. 38, 626-8

Inst. Phys. Conf. Ser. No 146
Paper presented at Microsc. Semicond. Mater. Conf., Oxford, 20–23 March 1995
© *1995 IOP Publishing Ltd*

541

Structural characterization of thin films with the use of RHEED azimuthal plots

Z Mitura[1,2], **P Mazurek**[2], **K Paprocki**[2], **P Mikołajczak**[2], **P A Maksym**[1] and **J L Beeby**[1]

[1]Department of Physics and Astronomy, Leicester University, University Road, Leicester LE1 7RH, U.K.
[2]Institute of Physics, Marie Curie-Skłodowska University, pl. M.Curie-Skłodowskiej 1, 20-031 Lublin, Poland

ABSTRACT: A quantitative method of characterizing ultra thin films based on a precise analysis of RHEED azimuthal plots is presented. An interpretation of experimental data collected for dysprosium silicide layers grown on a Si(111) substrate is carried out. Detailed agreement for experimental azimuthal plots and theoretical ones obtained from a numerical solution of the Schrödinger equation is demonstrated.

1. INTRODUCTION

Researchers who deal with preparation of crystalline materials must carry out at some stage of the work structural characterization of their samples. We can discern two different important cases: (1) characterization of unknown structures (2) verification of whether samples are grown with the structures expected. Substantially different treatments are necessary for these two cases. In the first case one must deal with all details extremely precisely using as many experimental techniques as possible. In the second case taking care of fine details of the structural arrangement is usually less important while it is essential to apply a method of structural characterization which is convenient and fast. Researchers dealing with preparation of 3D (bulk) structures prefer to use standard X-ray diffraction to verify their samples. It seems that the best choice to realise the same aim with respect to 2D structures is to use Reflection High Energy Electron Diffraction (RHEED).

The grazing geometry of RHEED measurements makes this technique very suitable for use in UHV chambers designed for epitaxial growth of 2D structures. It is important that the intensity of the diffracted electrons is very sensitive to any changes of position of atoms at surfaces. Usually, RHEED oscillations, in which the intensity of the specularly reflected electron beam is observed while depositing the material, are recorded to select proper growth conditions and a RHEED pattern at a screen is observed to get some structural information on a sample. Interpretation of experimental data is mostly carried out within kinematic diffraction theory. Finally, from the measurements just

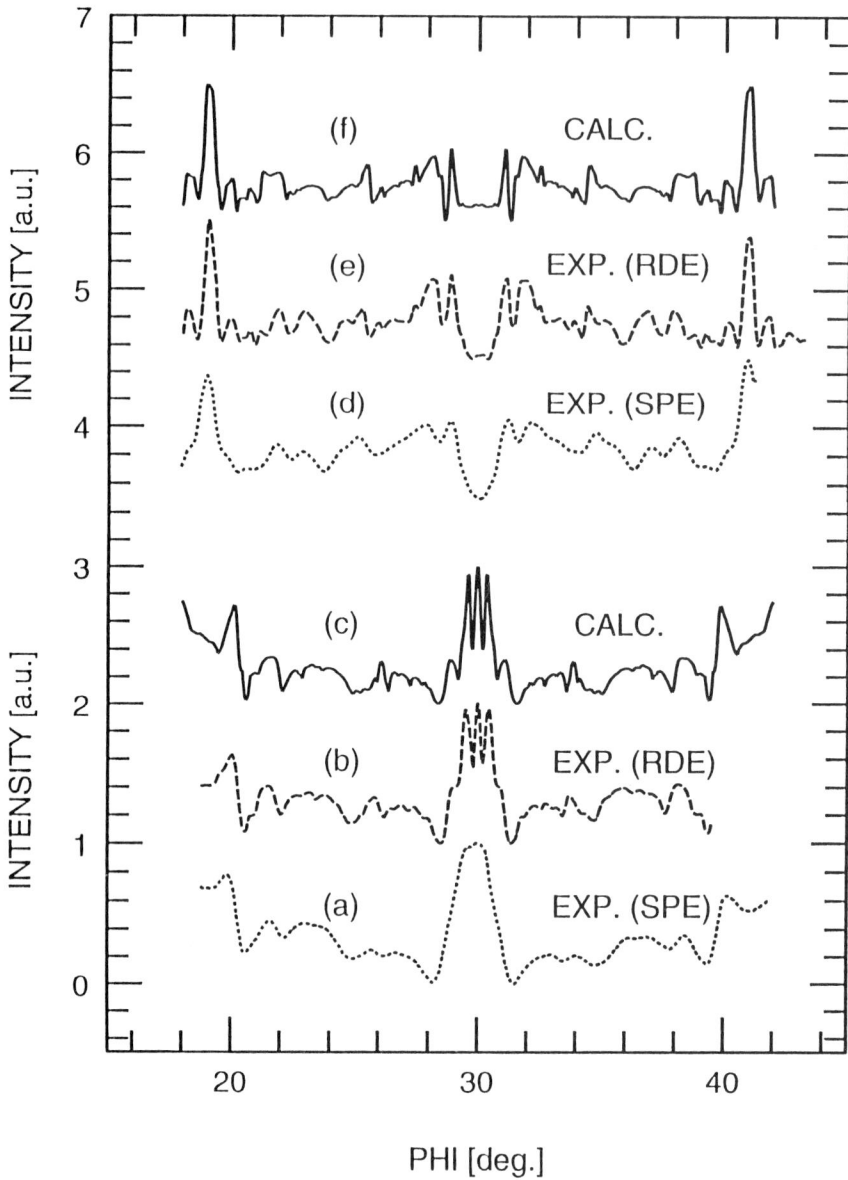

Figure 1. Experimental (a,b,d,e) and calculated (c,f) azimuthal plots for a $DySi_{2-x}$ layer on a Si(111) substrate. Results are shown for the two different procedures for sample preparation: Solid State Epitaxy (SPE) (a) and Reactive Deposition Epitaxy (RDE) (b). Plots a,b,c refer to the 0004 Bragg reflection, plots d,e,f to 0006. The electron primary beam energy is 18 keV. The value of the azimuthal angle of 30 deg. corresponds to the azimuth $< 11\bar{2} >$ of a Si(111) substrate. The best fits shown (c,f) are for a Si subplane separation of 0.5 Å.

mentioned, it is possible to get qualitative information (for example rough estimates of lattice constants) which can be useful in some cases. However in many other cases further, more detailed information (like details of reconstruction) is required.

In this paper we present a quantitative method of structural characterization of thin layers based on an analysis of RHEED azimuthal plots. In this method the specular beam intensity is measured while rotating a sample around the axis perpendicular to the surface. Next, the experimental data are analyzed using numerical solutions of the Schrödinger equation for scattered waves. We apply the method to characterize dysprosium silicide layers.

2. EXPERIMENTAL PROCEDURE AND RESULTS

Two methods of preparation of dysprosium silicide samples (about 150 Å thick) were used: solid state epitaxy (SPE) and reactive deposition epitaxy (RDE). In both cases the substrates were Si(111) wafers. During SPE, in the first step a thin template layer (about 10-20 Å thick) of Dy is deposited on the substrate held at room temperature. In the second step the sample is annealed to about 600-800° C to improve its crystallographic quality. The sequence is repeated several times to get finally a thickness of $DySi_{2-x}$ of about 150 Å. During RDE, a flux of Dy atoms is sent directly onto a hot substrate (held at about 600° C). Comparison of RHEED patterns observed at the screen and azimuthal plots (Fig. 1a,b, d,e) allows us to deduce that RDE leads to samples with better surface atomic arrangements than SPE.

We analyzed experimental data using a two dimensional Bloch wave approach introduced by Maksym and Beeby (1981), and Ichimiya (1983). Detailed information about this approach can be found, for example, in a review article of Beeby (1988).

In carrying out the analysis we applied a model of the $DySi_{2-x}$ structure introduced by Baptist et al (1990). In this model (Fig.2) it is assumed that the structure below the surface resembles bulk $DySi_{2-x}$ with alternate planes of Dy and Si. At the surface the structure is terminated with Si atoms which, however, do not form a single plane (as in the bulk) but form two subplanes separated by some distance. Actually Baptist et al (1990) introduced their model for YSi_{2-x} and determined the value of the Si subplane separation at the surface to be 0.8 Å. We treated the value of the separation as a parameter and found that for the case of our $DySi_{2-x}$ layers the best fits were obtained for 0.5 Å (Fig.2 and Fig.3).

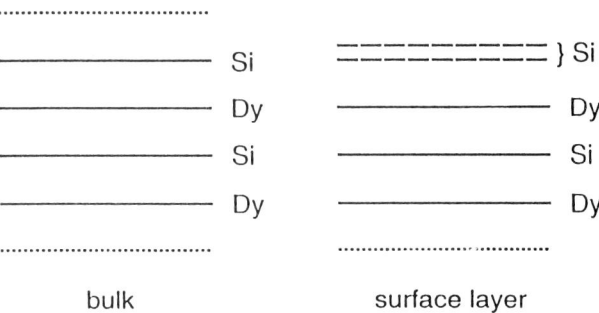

Figure 2. Side view of the structure of $DySi_{2-x}$.

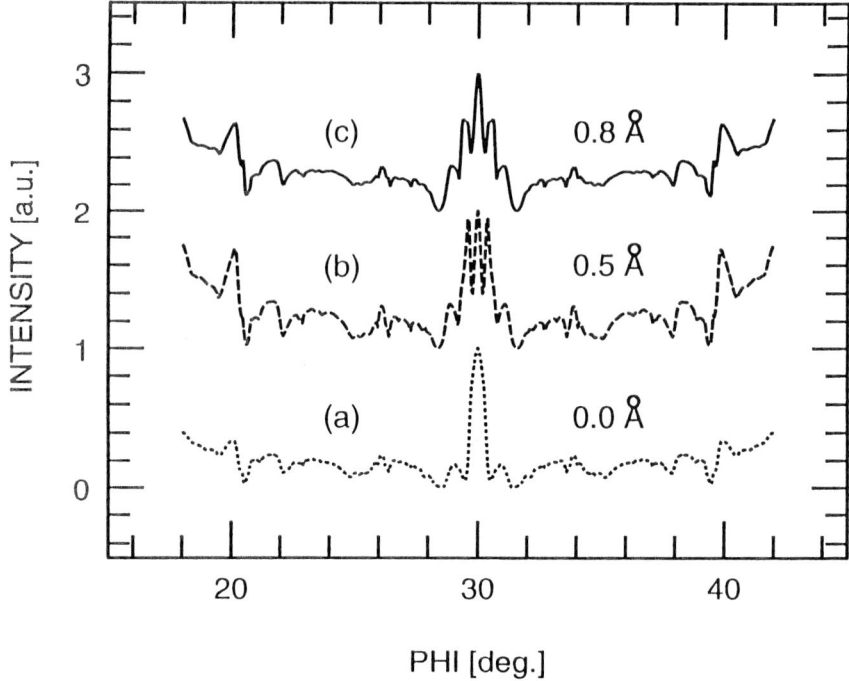

Figure 3. Calculated azimuthal plots for $DySi_{2-x}$ for different values of the Si subplane separation at the surface: (a) 0.0 Å, (b) 0.5 Å, (c) 0.8 Å.

3. CONCLUSION

In conclusion, we have demonstrated a method of structural characterization of thin layers based on an analysis of RHEED azimuthal plots. The method seems to be especially useful if one is interested in a rapid check of the degree of order in an arrangement of atoms at the surface. It also allows one to verify some details of models of reconstruction at the surface.

ACKNOWLEDGMENTS

Financial support from the U.K. Engineering and Physical Sciences Research Council (Grant No. Gr/J23020) and the Polish Scientific Research Committee (Grant No. 2 0346 91 01) is acknowledged.

REFERENCES

Baptist R, Ferrer S, Grenet G and Poon H C 1990 Phys. Rev. Lett. 64 311
Beeby J L 1988   RHEED and Reflection Electron Imaging of Surfaces, eds P K Larsen and P J Dobson (New York: Plenum), pp 29-42
Ichimiya A 1983 Jpn. J. Appl. Phys. 22 176
Maksym P A and Beeby J L 1981 Surface Sci. 110 423

Inst. Phys. Conf. Ser. No 146
Paper presented at Microsc. Semicond. Mater. Conf., Oxford, 20–23 March 1995
© *1995 IOP Publishing Ltd*

Metastable silicide formation at grain boundaries in gold/silicon multilayers

M Seibt, S Buschbaum and U Gnauert

IV.Physikalisches Institut der Universität Göttingen and Sonderforschungsbereich 345, Bunsenstr.13-15, D-37073 Göttingen, Federal Republic of Germany

ABSTRACT: We have studied the formation of metastable Au_3Si in silicon-gold multilayers. High-resolution transmission electron microscopy and X-ray diffraction show that silicide nucleation takes place at grain boundaries in the gold films followed by silicide growth perpendicular to the substrate normal. It is argued that the observed growth mode results from grain boundary and interface diffusion being the dominant paths of interdiffusion.

1. INTRODUCTION

The formation of silicides in metal-silicon thin film diffusion couples and multilayers is of fundamental and technological importance. Relaxation to thermal equilibrium of such systems often involves the formation of metastable phases during early stages of the reaction. Although there has been considerable theoretical and experimental effort, basic rules and criteria of phase selection under large driving forces ($\gg kT$) are still a matter of discussion.

This paper is concerned with phase formation in the Au:Si system, which is a simple eutectic, i.e. no gold silicides exist in thermal equilibrium (Okamoto and Massalski (1983)). However, metastable silicides have been produced by various techniques including laser annealing (von Allmen et al 1980), fast quenching from the melt (Anantharaman and Suryanarayana (1971)), ion-implantation (Tsaur and Mayer (1981)), precipitation from supersaturated solutions of Au in crystalline Si (Baumann and Schröter 1991) and solid-state reaction in multilayers (Hultman et al (1987)).

We report results obtained in multilayers consisting of polycrystalline gold (c-Au) and amorphous silicon (a-Si). Transmission electron microscopy investigations of different stages of the reaction show that silicide nucleation occurs at grain boundaries of the c-Au films followed by two-dimensional growth perpendicular to the substrate normal. Our results provide evidence, that phase selection criteria have to take into account the microstructure of the films, especially under conditions where interface and grain boundary diffusion is the dominant path of atomic transport.

2. EXPERIMENTAL

Multilayers consisting of crystalline Au (c-Au) and amorphous silicon (a-Si) with thicknesses in the range of 2.2-7.8nm and 4-20nm, respectively, have been prepared by evaporation at background pressures of 10^{-9}mbar on (001) Si substrates cooled by liquid nitrogen. Differential scanning calorimetry (DSC), Rutherford backscattering (RBS) and x-ray diffraction (XRD) have been used to characterize solid-state reactions occuring below the eutectic temperature of 363°C .

For TEM investigations special structures consisting of a single 5.6nm thick c-Au layer between two 4.5nmn thick a-Si layers on Si substrates have been used. In order to investigate different stages of silicide formation, they were heated at a rate of 10K/min to a certain temperature and subsequently quenched, thus allowing a direct comparison with DSC measurements. Plan-view samples have been chemically etched from the back using mixtures of HF/HNO_3. This procedure was essential to avoid silicide formation, which resulted from the preparation of cross-section specimen involving ion beam thinning at liquid nitrogen temperature. TEM micrographs were obtained at 200kV in a Philips CM200-UT-FEG.

3. RESULTS

3.1 Reactions in the Au-Si System

A detailed characterization of solid-state reactions occuring in the Au:Si system is given elsewhere (Gnauert (1994), Gnauert et al (1995)). Briefly, three exothermic reactions occur below the eutectic temperature of 363°C , i.e. Au_3Si formation at about 100°C followed by metal-induced crystallization of a-Si and finally the decomposition into c-Au and c-Si, which represents thermal equilibrium (compare DSC spectrum in Fig.1). From combined RBS- and cross-section TEM studies the composition is deduced as Au_3Si , which has a formation enthalpy of (1.6 ± 0.3)kJ/mole as measured by DSC. X-ray diffraction and TEM show that the silicide remains stable during Si crystallization, but that the layered structure is destroyed in agreement with *in-situ* TEM studies of the Ag:Ge system (Konno and Sinclair (1993)).

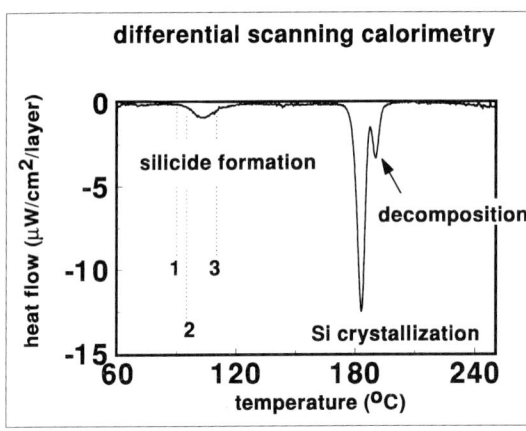

Fig.1: Differential scanning calorimetry experiment at 10K/min; three exothermic reactions are obtained, i.e. Au_3Si formation at about 100°C , (metal-induced) crystallization of the remaining a-Si at 180°C and decomposition into c-Si and c-Au at 190°C . Samples quenched from temperatures indicated by the numbers 1,2 and 3 were investigated by TEM (compare figures 2a, 2b and 2c, respectively)

3.2 Au-silicide formation

We now describe structural investigations during silicide formation. As-prepared multilayers are strongly textured with Au{111} parallel to the growth direction. Heating up to about 80°C leads to recrystallization of the c-Au involving twinning and the destruction of the growth texture. HRTEM micrographs obtained in plan-view (Fig.2a) reveal grain boundaries mostly of high-angle type; no evidence for silicide formation is observed in this stage.

In-situ XRD allows to determine the thickness d_{Au} of the Au films during silicide formation by measuring the width of the Au(111) diffraction peak. Our data show, that d_{Au} stays essentially *constant* during the reaction, which excludes one-dimensional silicide growth perpendicular to the a-Si/c-Au interfaces. In fact,

HRTEM reveals Au$_3$Si formation at grain boundaries in the gold films in samples quenched from 95°C . Fig.2b shows such a silicide grain formed at a triple point of Au grain boundaries, which appear to be prefered nucleation sites. Further silicide fromation then occurs by *lateral* growth along a-Si/c-Au interfaces resulting in columnar Au$_3$Si and c-Au grains existing side by side as revealed by the plan-view micrograph in Fig.2c.

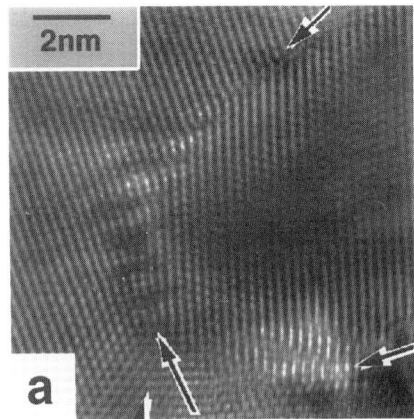
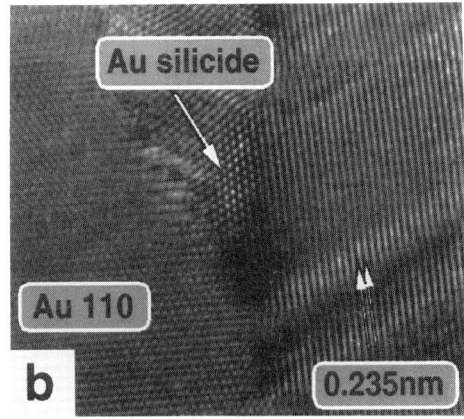
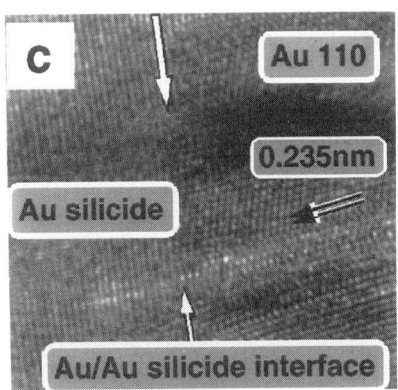

Fig.2: HRTEM micrographs of different stages of Au$_3$Si formation; (a) reaction stopped at 80°C : grain boundaries (arrows) in the gold film showing no silicide; (b) reaction stopped at 95°C : nucleation of Au silicide at triple points of grain boundaries in the Au film; (c) reaction stopped at 110°C : arrows show interphase boundary between Au$_3$Si (left hand part) and unreacted Au (right hand part).

4. DISCUSSION

Fig.3 schematically shows the process of Au$_3$Si formation. The most prominent features are (i) silicide nucleation at grain boundaries in the Au films followed by (ii) lateral growth along c-Au/a-Si interfaces. In addition to our HRTEM observations *in-situ* XRD measurements and reaction kinetics obtained from DSC are consistently described by this growth mode. In order to discuss our experimental observations, we have to consider the role of interfaces and grain boundaries as possible diffusion paths and nucleation sites. Recently, Coffey and Barmak (1994) have pointed out that diffusion along grain boundaries and interphase boundaries can considerably contribute to atomic transport in thin film reactions. The authors use irreversible thermodynamics to calculate time dependent chemical potentials of the reacting species in grain and interphase boundaries. They argue that the phase selected is not necessar-

548

ily that with the lowest free energy but that whose free energy curve lies below that of the grain or interphase boundary.

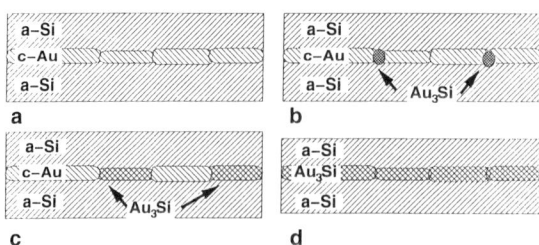

Fig.3: Silicide formation at grain boundaries: (a) as-grown state consisting of polycrystalline Au and a-Si; (b) nucleation of Au_3Si at grain boundaries of the Au; (c) silicide growth parallel to c-Au/a-Si - interfaces; (d) coalescence of Au silicide grains.

In the present case, Au_3Si formation starts at temperatures of about 80°C . At this low temperature bulk diffusion of Si in c-Au or Au in a-Si are slow (Coffa et al (1992)). However, Pasa et al (1991) attribute SiO_2- formation at $\simeq100$°C on top of Au films to silicon diffusion along grain boundaries in the Au. Furthermore, our observation of recrystallization in the Au films preceeding silicide formation inicates that Au atoms are mobile in grain boundaries even below 80°C . Hence, grain boundaries in the Au may be regarded as the dominant diffusion path in the system.

Heterogeneous compound nucleation at interfaces or grain boundaries has been discussed recently in the framework of classical nucleation theory (Bormann 1994). Our HRTEM investigations provide evidence that the product phase (Au_3Si) grows epitaxially at least on one of the adjacent grains of the parent phase (Au) indicating that grain boundary energies contribute to a reduction of nucleation barriers.

In conclusion, we have shown that Au_3Si formation in Au-Si multilayers proceeds by nucleation at grain boundaries in the Au films followed by lateral growth of the silicide. More generally, our results demonstrate that phase selection criteria have to take into account the microstructure of the films, especially under conditions where interface and grain boundary diffusion is the dominant path of atomic transport.

REFERENCES

Anantharaman T R and Suryanarayana C (1971) J Mat Sci 6 1111
Coffa S, Poate J M, Jacobson D C, Frank W, and Gustin W 1992
    Phys. Rev. B 45 8355
Coffey K R and Barmak K 1994 Mat. Res. Soc. Proc. 343 193
Baumann F H, Schröter W 1991 Phys. Rev. B 43 6510
Bené R W 1987 J. Appl. Phys. 61 1826
Bormann R 1994 Mat Res Soc Proc 343 169
Gnauert U 1994, Thesis, University of Göttingen
Gnauert U, Buschbaum S, Seibt M, and Schröter W 1995, to be published
Gösele U and Tu K N 1989 J. Appl. Phys. 66 2619
Hultman L, Robertson A, Hentzell H T G, Engström I and Psaras P A 1987
    J. Appl. Phys. 62 3647
Konno T J and Sinclair R 1993, Inst. Phys. Conf. Ser. 134 173
Okamoto H and Massalski T B 1983 Bulletin of Alloy Phase Diagrams 4 190
Pasa A A, Paes H R, Losch W 1991 J. Non-Cryst. Sol. 137&138 1087
Tsaur B Y and Mayer J W 1981 Phil. Mag. A43 345
von Allmen M, Lau S S, Mäenpää, and Tsaur B Y 1980 Appl. Phys. Lett. 36 207

Inst. Phys. Conf. Ser. No 146
Paper presented at Microsc. Semicond. Mater. Conf., Oxford, 20–23 March 1995
© 1995 IOP Publishing Ltd

549

Epitaxial Al on compound semiconductors: structural and electrical studies

S J Pilkington, M Missous and *U Bangert

Centre For Electronic Materials, U.M.I.S.T., PO Box 88, Manchester, M60 1QD, UK.
*Dept. Of Pure and Applied Physics, U.M.I.S.T., PO Box 88, Manchester, M60 1QD, UK.

ABSTRACT: High Resolution Transmission Electron Microscopy has revealed the existence of an AlGaAs phase at the interface of epitaxial Al/GaAs grown by MBE after a 500 °C anneal. The presence of this AlGaAs phase correlates well with the increase in the barrier height reported previously upon similarly annealed Al/GaAs Schottky diodes.

Electrical and structural studies of intimate $Al/In_{0.53}Al_{0.47}As$ Schottky diodes grown by MBE have shown that, although the Al does not appear to be a continuous layer (it is single crystal (110) with typically 0.15 μm sized islands of (100)), thermionic emission of carriers over the barrier is the dominant current transport mechanism.

1. INTRODUCTION

Metal-semiconductor contacts are the building blocks of many electronic devices and their integrity (electrical and structural) is probably the most important parameter controlling the performance of the semiconductor devices. We report here on intimate 'in situ' grown single crystal Aluminium on GaAs and InAlAs using Molecular Beam Epitaxy. Such contacts lead to highly ideal and thermally well behaved metal-semiconductor interfaces.

2. EXPERIMENTAL DETAILS

The Molecular Beam Epitaxy growth of the InAlAs or GaAs layers was performed on 2" epi-ready wafers InP (100) S-doped or GaAs (100) Si doped in a VG V90H reactor. The layers were grown at 500 °C (580 °C for GaAs) under exact dynamic stoichiometry conditions and were intentionally doped to 1 x 10^{16} cm^{-3}. Immediately after semiconductor growth, the substrate was allowed to cool to room temperature before deposition of the Al upon a perfectly reconstructed (3x1) InAlAs or (2x4) GaAs surface. The Al deposition, assessed using Reflection High Energy Electron Diffraction (RHEED), was started at a rate of 1 Å/min to follow the initial growth dynamics and it was clear that the growth modes were almost identical ie a Stranski-Krastanov mode with an initial ad-layer of Aluminium and subsequent island formations. In the case of Al/GaAs, 3 Al orientations co-existed until ≈ 400 Å, after which (100) dominated and subsequently the film was predominately (100). In the case of Al/InAlAs, the 3 Al orientations co-existed to a greater thickness, ≈ 1000 Å, after which (110) dominated. The Al layers were ≈ 1000 Å thick in both systems. After growth, the InAlAs layers were assessed by Double Crystal X-ray diffraction which revealed an In composition of 0.534, $\Delta a/a$ of 7.7 x 10^{-4} and with a full width at half maximum of 80 arc sec.

Plan view and cross-sectional TEM samples were prepared by chemical etching from the substrate side and by conventional Ar-milling respectively. The samples were viewed in a Philips EM430 microscope operating at 300 kV.

3 EPITAXIAL Al/GaAs SCHOTTKY DIODES

3.1 Current-Voltage Relationship

Current-voltage (I-V) characterisation of Al/GaAs Schottky contacts grown by MBE have been extensively studied by Missous et al (1986a, 1986b) and they found that the current transport mechanism was dominated by the thermionic emission of carriers over the barrier. This highly ideal behaviour (n=1.01) indicates an intimate interface of good perfection between the Al and the GaAs. Annealing studies up to 500 °C, revealed that the thermionic emission of carriers still dominated, but the barrier height increased from 0.78 eV to 0.84 eV.

3.2 TEM Analysis

When deposited by Molecular Beam Epitaxy upon freshly grown GaAs (100), aluminium deposits as (100) and as two different (110) phases. Fig 1 shows a cross sectional lattice image of the interface between the GaAs and the Al in the predominant (100) orientation (85% in this case).

The conditions for the growth of any one of the above orientations and the proportions in which different orientations co-exist in the same epitaxial film have been reported by various authors (Beanland and Kiely 1993, Kiely and Cherns 1989, Landgren et al 1982, Missous et al 1986b, Petroff et al 1982). The thermal instability of the [110] orientation has been demonstrated by Missous et al (1986a) and was discussed by Landgren et al (1982). Short rapid thermal anneals at 500 °C result in the complete disappearance of the (110) islands and recrystallization to (100).

HREM observations lead us to believe that the recrystallization is connected with or might even be triggered by diffusion of As into the Al-film, leading to AlAs or AlGaAs phases within the Al (Bangert et al 1995). Fig 2 shows a lattice image of a sample annealed at 500 °C for 90 s, the imaging conditions are such that the Al lattice image is not visible. The interface region between the Al and GaAs appears diffuse, with a broad band of approximate thickness between 60 and 100 Å (marked)   seen to protrude into the Al-layer. At places within this band (encircled) the (110) sphalerite lattice can be seen. This indicates interdiffusion occurring at this temperature, most likely being As out-diffusion into the Al. Extended AlGaAs phases near the interface would explain the increase in the Schottky barrier height.

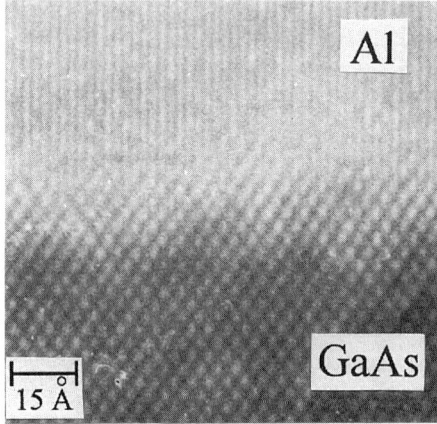
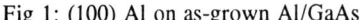
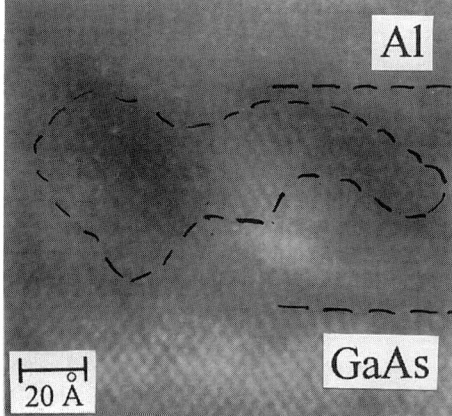

Fig 1: (100) Al on as-grown Al/GaAs

Fig 2: Al/GaAs annealed at 500 °C.
Note the broad diffused region (dashed lines)
and sphalerite phases (encircled).

4 EPITAXIAL Al/In$_{0.53}$Al$_{0.47}$As SCHOTTKY DIODES

4.1 Current-Voltage Characterisation

Typical I-V characteristics of intimate Al/InAlAs diodes at various temperatures, are shown in fig 3. The corresponding log I / [1 - exp(-qV/kT)] versus voltage plots, which were linear at all biases up to 0.25 V, indicating the applicability of the thermionic emission theory, with the extracted ideality factors being less than 1.06. This is in complete disagreement with other authors (Clark et al 1994, Luo et al 1993) who either found that, intimate evaporated Al contacts to MBE grown In$_{0.52}$Al$_{0.48}$As Schottky diodes were not well characterised by thermionic emission (n = 1.63), or that there was a significant number of traps in the MBE grown InAlAs that manifested as a defect-assisted tunnelling current in the forward I-V characteristics of intimate evaporated Au or Cu Schottky diodes.

Close agreement was found between the theoretical reverse I-V characteristics calculated due to image force lowering of the barrier height, and the observed reverse I-V characteristics. This shows that the reverse characteristics are well explained by the image force alone, indicating the highly intimate near ideal nature of the interface.

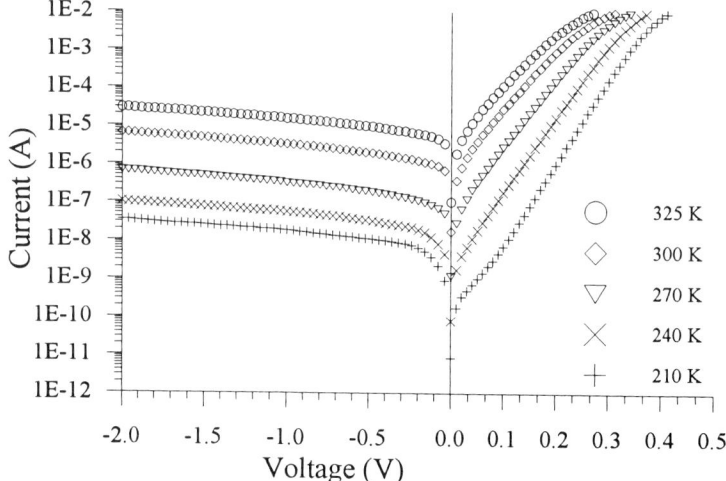

Fig 3: Typical forward and reverse I-V characteristics at several temperatures.

Linear Arrhenius plots of log I/T^2 versus 1/T were obtained, indicating the dominance of thermionic emission over the temperature range 250-400 K. The zero bias barrier height obtained was 0.555 eV and the Richardson constant was 39 A.cm^{-2}.K^{-2}, which compare very favourably with values obtained by Gueissaz et al (1992) upon MBE grown intimate Al/In$_{0.52}$Al$_{0.48}$As.

4.2 TEM Analysis

Fig 4 shows a large area diffraction pattern of the Al film. From this it is clear that the Al film is predominately a (110) single crystal. Fig 5 shows a diffraction contrast image of the Al film, and although the film appears to be interrupted, islands within the (110) single crystal are obvious. These islands are typically ≈ 0.15 μm and are orientated (100), and some (110)R islands. This situation is the complete reverse of the case for epitaxial Al/GaAs, where the Al films are (100) with (110) and some (110)R islands. The single crystal appears to be distorted, leading to a slight change in the lattice orientations between adjacent areas within the (110) structure. This is in complete agreement with RHEED, which showed facetting of the islands.

552

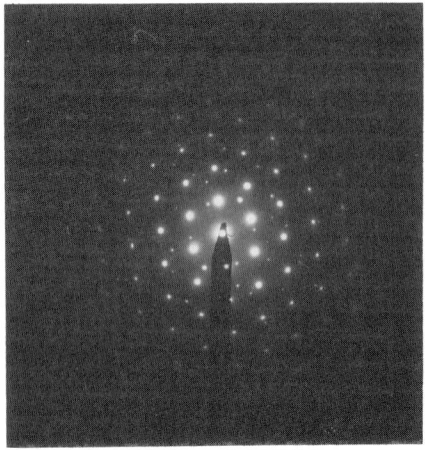

Fig 4: Large area diffraction pattern of Al film, showing single crystal (110).

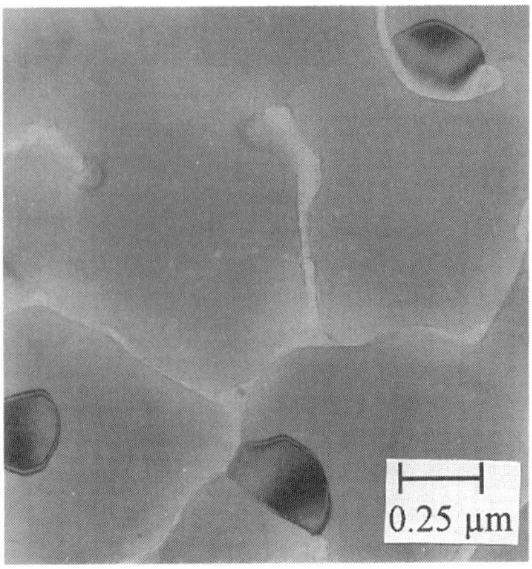

Fig 5: TEM Micrograph showing (100) islands in the majority single crystal (110).

5. CONCLUSIONS

In the case of Al/GaAs, the appearance of the AlGaAs phase at the interface correlates well with the increase in barrier height since it is well known that Al makes higher barriers with AlGaAs (Missous et al (1990)), and even though anticipated for a long time, this is the first time that such direct evidence has been shown.

In the case of the Al/InAlAs, although the Al is single crystal ((110) with (100) islands), the apparently interrupted Al film has not hindered the formation of thermionic emission dominated Schottky barriers grown by MBE. The transport properties of metal/InAlAs Schottky diodes depend primarily on two main parameters :

(i) The quality of the InAlAs semiconductor,
(ii) The integrity of the metal/InAlAs interface.

The perception that metal/InAlAs Schottky diodes grown by MBE are a poor second to similar structures made using MOCVD, because of the low growth temperatures involved, are proven not to be the case here. However, we also demonstrate that the intimate nature of the contact is vital in establishing a near perfect diode, especially for such highly reactive surfaces as InAlAs.

REFERENCES

Bangert U, Tang B and Missous M 1995 J. Cryst. Growth., in print.
Beanland R and Kiely C J 1993 Int. Sci. 1 99
Clark S A, Wilks S P, Morris J I, Woolf D A and Williams R H 1994 J. Appl. Phys., 75 2481
Gueissaz F, Galihanou M, Houdre R and Ilegems M 1992 Appl. Phys. Lett., 60 1099
Kiely C J and Cherns D 1989 Phil. Mag. A 59 1
Landgren G, Ludeke R and Serrano C 1982 J. Cryst. Growth 60 393
Luo J K, Thomas H, Clark S A and Williams R H 1993 J. Appl. Phys., 74 6726
Missous M, E H Rhoderick and K E Singer 1986a J. Appl. Phys. 59 3189
Missous M, E H Rhoderick and K E Singer 1986b J. Appl. Phys. 60 2439
Missous M, W S Truscott and K E Singer 1990 J. Appl. Phys. 68 2239
Petroff P M, Feldman L C, Cho A Y and Williams R S 1981 J. Appl. Phys. 52 7317

Inst. Phys. Conf. Ser. No 146
Paper presented at Microsc. Semicond. Mater. Conf., Oxford, 20–23 March 1995
© *1995 IOP Publishing Ltd*

TEM characterization of ion beam mixed Au/GaAs contacts

B Pécz, G Radnóczi and E Jároli*

Research Institute for Technical Physics of HAS, H-1325 Budapest, P.O. Box 76
*KFKI Research Institute for Materials Science, H-1525 Budapest, 114, P.O. Box 49

ABSTRACT: Au(40 nm)/GaAs(100) samples were treated by ion beam mixing i.e. by Xe ion implantation through the gold layer. The energy of the ions was in the range of 300-700 keV, while low (10^{14} ions/cm^2) and high (10^{16} ions/cm^2) doses have been applied, respectively. The structure of the implanted specimens has been studied using cross sectional transmission electron microscopy (TEM). Some of the specimens have been annealed at 400°C resulting in the crystallization of the amorphous regions while GaAs grains appeared in the top covering layer as a consequence of gold diffusion into the GaAs substrate.

1. INTRODUCTION

GaAs is a frequently used semiconducting material and ion beam mixing has been applied to different contact layers on GaAs e.g. on Au/GaAs (Palmstrom et al 1984 and Barcz et al 1987), AuGe/GaAs (Bhattacharya et al 1985) and AuGeNi/GaAs (Zhao Jie et al, 1989) systems. Usually the investigators reported a good surface morphology (Palmstrom et al 1984 and Zhao Jie et al, 1989) in contrast to the conventional heat treatment. The defect structure caused by ion implantation has a great influence on the solid phase reactions taking place between gold and GaAs as we have shown in our earlier work (Pécz et al 1992). In this paper we report on the defect structure of Xe implanted Au/GaAs samples studied by cross sectional transmission electron microscopy.

2. EXPERIMENT

A 40 nm thick gold layer was deposited onto (100) GaAs slices and these samples were treated by ion beam mixing i.e. by ion implantation through the gold layer. High energy Xe ions have been used for that purpose. The energy of the ions was in the range of 300-700 keV, while low (10^{14} ions/cm^2) and high (10^{16} ions/cm^2) doses have been applied respectively. The structure of the implanted specimens have been studied using cross sectional transmission electron microscopy (TEM). Some of the specimens have been annealed at 400°C and the effect of annealing has been studied as well.

Transparent specimens have been prepared for TEM investigations by ion beam milling following the method of Barna (1992).

2.1 Defect structure of as-implanted specimens, low dose case

The bombardment of Xe ions caused the amorphization of the GaAs beneath the gold layer even at the lowest energy. In the case of 300 keV, low dose implantation, the thickness of the amorphous GaAs region is about 50 nm (Fig.1.a), while the range of the ions in GaAs is 100 nm at least. The gold layer is still continuous and uniform in thickness. Increasing the energy of the ions to 400 kV increases the thickness of the amorphous GaAs region to 85 nm (Fig.1.b). On increasing the energy of the implantation an interesting phenomena has been observed: more and more crystalline GaAs grains appeared in the amorphous region of GaAs. In Fig.1.c crystalline GaAs grains can be clearly seen close to the gold layer. This phenomenon leads to the structure observed in our earlier work (Pécz et al 1992) as implantation of 700 keV Xe ions resulted in the formation of a polycrystalline GaAs region beneath the gold layer (Fig. 1.d).

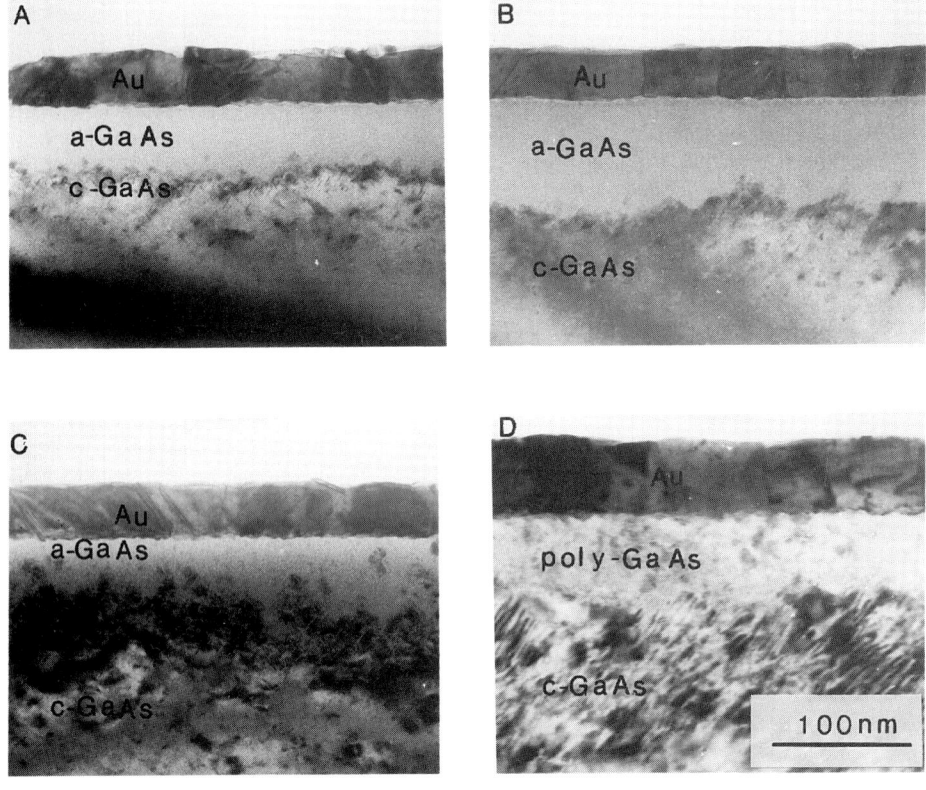

Fig.1. Cross sectional images of Xe implanted Au/GaAs specimens at low dose. a: 300 keV, b: 400 keV, c: 600 keV, d: 700 keV. (The specimen in Fig.1.d has been prepared for our earlier experiments and the gold layer was slightly thicker.)

The defects at the amorphous/crystalline GaAs interface have been studied by high resolution TEM and found to be stacking faults. In the low dose case the gold layer has not been destroyed even at the highest energy.

2.2 High dose case

In the case of the higher dose applied (10^{16} ions/cm^2) the bombardment destroyed the continuity of the gold layer (even at the lowest energy) leaving behind large metallic grains. The amorphous/crystalline interface looks like a wavy line (Fig.2) due to the shadowing effect of the large metallic grains. The defects at the amorphous/crystalline GaAs interface are typically stacking faults (Fig.3) and are similar to the low dose case. In all of the high dose cases there are no crystalline GaAs grains in the amorphous GaAs region near to the amorphous/crystalline interface.

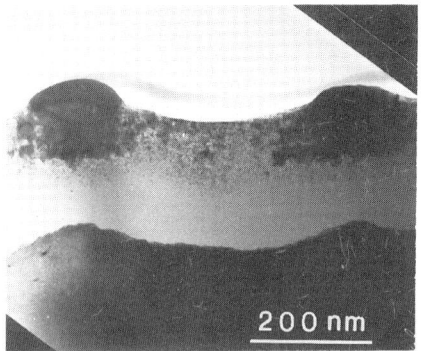

Fig. 2: Cross section of Au/GaAs specimen implanted at 400 keV, at high dose.

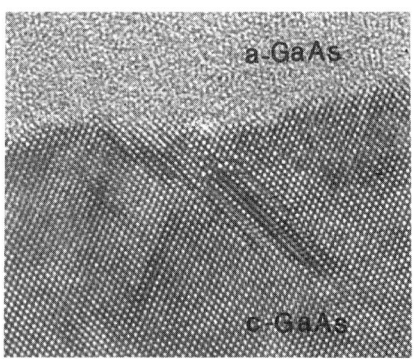

Fig. 3: High resolution image of the amorphous/crystalline GaAs interface. (600 keV, high dose)

2.3 Effect of annealing

Annealing of the specimens resulted in the crystallization of the amorphous region. A typical cross section of annealed specimen is shown in Fig. 4. (The structure of the same specimen before annealing can be seen in Fig.1.b.) Large metallic grains grew into GaAs, while the grains of top layer show two quite different contrasts. Taking EDS (Energy Dispersive System) spectra we concluded that the brighter grains (see for example grain marked A) are GaAs grains, while the darker ones (see for example grain labeled B) are gold grains. The latter grains probably also contain some gallium in solid solution. Typical size of the GaAs grains found in the top layer is 50 nm thick and 500 nm long. The EDS spectra of the appropriate grains are shown in Fig.5.

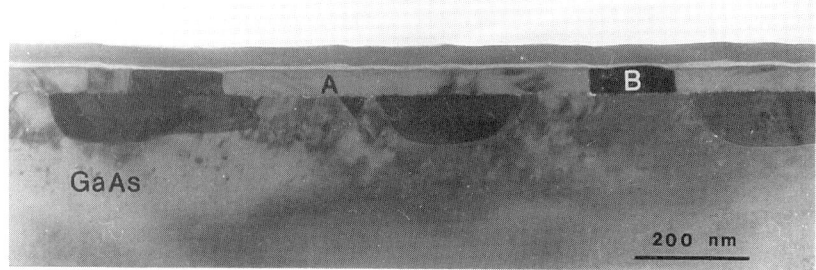

Fig. 4: Cross section of the Au/GaAs specimen implanted at 400 keV, low dose and annealed at 400°C for 10 minutes.

These results are in good agreement with our earlier findings (Pécz et al 1992) i.e. the whole amount of gold diffused through the polycrystalline layer which had been formed after implantation at 700 keV (Fig.1.d) during annealing. In that case the whole top layer covering the specimen was a polycrystalline GaAs layer.

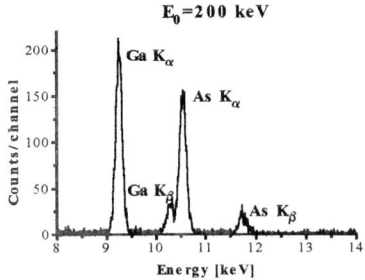 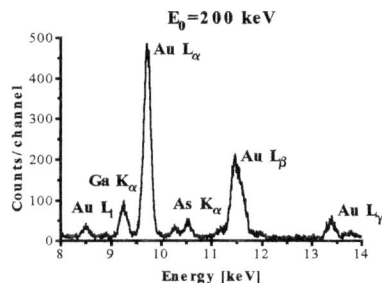

Fig. 5 EDS spectra of the grains of the Au/GaAs specimen implanted at 400 keV, low dose and annealed at 400ºC for 10 minutes. a: grain marked A in Fig.4, b: grain marked B in Fig.4.

3. CONCLUSION

As the energy of the Xe ions increases more and more crystalline grains are found within the amorphized region of GaAs. The recrystallization of those grains occurred during the implantation by a dynamic process of amorphization caused by the incoming Xe ions and of recrystallization due to the local temperature. At the highest energy (700 keV, low dose case) these processes resulted in the formation of a polycrystalline GaAs region beneath the gold layer having very low grain size.

Annealing of the specimens resulted in the crystallization of the amorphous region. GaAs grains appeared in the top covering layer as a consequence of gold diffusion into the GaAs substrate. In addition at all energies of implantation large, flat GaAs grains have been observed embedding smaller Au(Ga) grains.

4. ACKNOWLEDGMENT

The authors are grateful to Dr. J. Lábár for his valuable help at the EDS measurements. This work was supported by the OTKA project No. F 007612.

REFERENCES

Barcz A J, Domanski M, Jagielski J and Kaminska E 1987 E Nucl. Instr. and Meth. B., 19/20, 773
Barna Á 1992 Mat. Res. Soc. Symp. Proc. Ser. 254 3,
Jie Zhao and Thompson D A 1989 Vacuum 39 303
Palmstrom C J, Kavanagh K L, Hollis M J, Mukherjee S D and Mayer J W Mat. 1984 Res. Soc. Symp. Proc., 37, 473
Pécz B, Radnóczi G, Horváth Zs J, Barna P B, Jároli E and Gyulai J 1992 J. Appl. Phys. 71 3408

Inst. Phys. Conf. Ser. No 146
Paper presented at Microsc. Semicond. Mater. Conf., Oxford, 20–23 March 1995

TEM study of annealed behaviour of Au/InP contacts

J Kątcki, E Mizera*, A Piotrowska and J Ratajczak

Institute of Electron Technology, Al. Lotników 32/46, 02-668 Warsaw, Poland

* Institute of Physics, Polish Academy of Sciences, Al. Lotników 32/46, 02-668 Warsaw, Poland

ABSTRACT: In order to study interfacial reactions between InP and gold two contact structures were prepared. The first was a gold layer deposited onto the surface of InP. The thickness of the gold layer was 90 nm. In the second sample a 200 nm thick SiO_2 layer was deposited on a gold layer on an InP surface forming an InP/Au/SiO_2 structure. Selected InP/Au and InP/Au/SiO_2 structures were annealed at temperatures of 380°C and 420°C for 3 min. Cross-sectional transmission electron microscopy was used to study the formation of new phases in the Au/InP contacts during annealing. In order to identify new phases selected area electron diffraction in the transmission electron microscope was used. The effect of the SiO_2 layer on the InP/Au contact structure was determined.

1. INTRODUCTION

Gold and gold based alloys are the most commonly used materials for ohmic contacts to high speed and optoelectronics devices. Good quality ohmic contacts are of the great importance to these devices. For a better understanding of the interfacial reactions taking place at the interface of Au/InP system the material was annealed. The formation of new phases during annealing of the Au/InP ohmic contact at 320-360°C was discussed by Piotrowska et al (1981). The behaviour of this contact at 375°C and 400°C was investigated by Pécz et al (1991).

2. EXPERIMENTAL

(100) Sn-doped InP wafers ($n=10^{17}cm^{-3}$) were used as a substrate material. Prior to deposition of the contact material the substrates were cleaned by sequential boiling in trichloroethylene, acetone and methanol, deoxidized in 5% HCl solution, etched in HCl-CH_3COOH-H_2O_2, rinsed in de-ionised water and blown with dry N_2. Au metallization was deposited by vacuum evaporation at a pressure of $5*10^{-7}$ Torr. The thickness of a gold layer was 90 nm. A capping layer of SiO_2 200 nm thick was deposited by radio-frequency magnetron sputtering with a power density of 7 W/cm^2, at a pressure of $4*10^{-3}$ Torr and with a background pressure better than $1*10^{-6}$ Torr. Heat treatments were carried out by furnace annealing under flowing H_2 for 3 min. For our study we did not remove the capping layer. Usually it is removed by wet etching in buffered HF after annealing.

In order to study the ohmic contacts in a transmission electron microscope cross-sectional specimens were prepared. Two rectangularly shaped pieces were glued together face-to-face with epoxy. From this "sandwich" 200 μm thick slices were cut. The slices were mounted into aluminium disks with a diameter of 3 mm. To fix a slice in the aluminum disk epoxy was used. Such a sample was then mechanically thinned to 120 μm and dimpled until the thickness of the central part of the sample reached 20 μm. After grinding in a dimpler the sample was ion milled in Ar$^+$ ions at 5 keV. The angle between the ion beam and the specimen surface was 3-5°. To avoid heating the specimen the holder was cooled by liquid nitrogen during the entire ion milling process. In the final stage of ion milling iodine was introduced into the specimen chamber. The iodine ambient was used to avoid formation of In dots on the surface of the sample. The specimens were studied in transmission electron microscopes JEM2000EX (Institute of Physics, PAS, Warsaw) and JEM4000EX (Max-Planck Institut für Metallforshung, Stuttgart).

3. RESULTS AND DISCUSSION

During annealing of an InP/Au contact phosphorus atoms tend to diffuse out of the surface. In order to slow down the interfacial reactions taking place in the InP/Au contact an SiO_2 capping layer was deposited on top of the contact layer.

A cross-sectional TEM view of the as-deposited InP/Au contact is shown in Fig. 1. The local thickness of the gold layer depends on the grain size. At those sites where grains were smaller than 50 nm the thickness of the gold layer was 90 nm. However, if the local grain size exceeded 50 nm the thickness of the layer was equal to the grain size. This resulted in the thickness of the contact layer being locally less than 90 nm.

Fig. 1. Cross-sectional TEM micrograph of an as-deposited contact structure of InP/Au /SiO$_2$

Annealing of the InP/Au contacts at the temperature of 380°C for 3 min. caused the formation of Au_9In_4 in the contact layer and in the substrate (Fig. 2a). This phenomenon was reported by Pécz et al (1993). Due to the migration of indium atoms into the gold layer and the solid state reaction of gold with indium a cubic Au_9In_4 phase was formed. In the dark-field micrograph (Fig. 2b) it is clearly seen that the formation of the Au_9In_4 phase was not limited to the contact layer. Extended grains of the Au_9In_4 phase were also formed in InP directly under the contact layer. In other micrographs not presented here the grains of Au_9In_4 phase were also observed to be embedded in an Au_2P_3 matrix (similar to that shown in Fig 3 and 4). They can also be found below the Au_2P_3 grains.

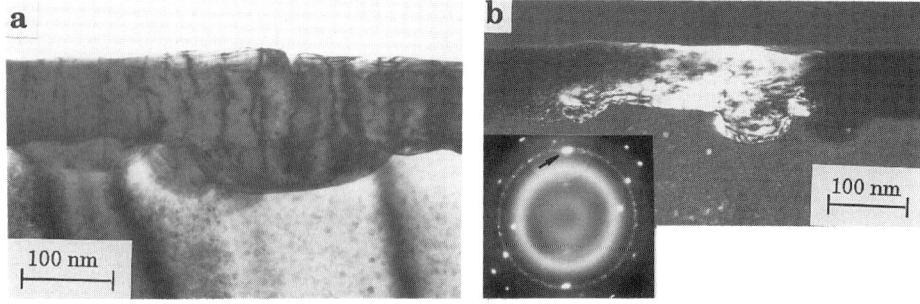

Fig. 2. Cross-sectional TEM micrographs of a InP/Au contact annealed at 380°C

Deposition of a 200 nm thick SiO_2 capping layer onto the gold layer resulted in no Au_9In_4 being observed in the contact layer. After 3 minutes annealing at a temperature of 380°C the gold layer remained unchanged. Directly under the contact layer big grains of monoclinic Au_2P_3 were formed. Au_9In_4 grains were observed to be embedded in bigger grains of Au_2P_3 (Fig. 3a). Horizontally extended grains of Au_9In_4 were also observed at the edge of Au_2P_3 grains on the InP side. Grains of Au_9In_4 embedded in an Au_2P_3 matrix are smaller and more regularly shaped than those lying at the bottom edge of grains of the Au_2P_3. A dark field micrograph shown in Fig. 3b confirms that the grains embedded in the Au_2P_3 matrix are of the Au_9In_4 phase. To take this micrograph one of Au_9In_4 spots (indicated by an arrow) was selected in the diffraction pattern. This caused bright contrast to appear in one of the Au_9In_4 grains.

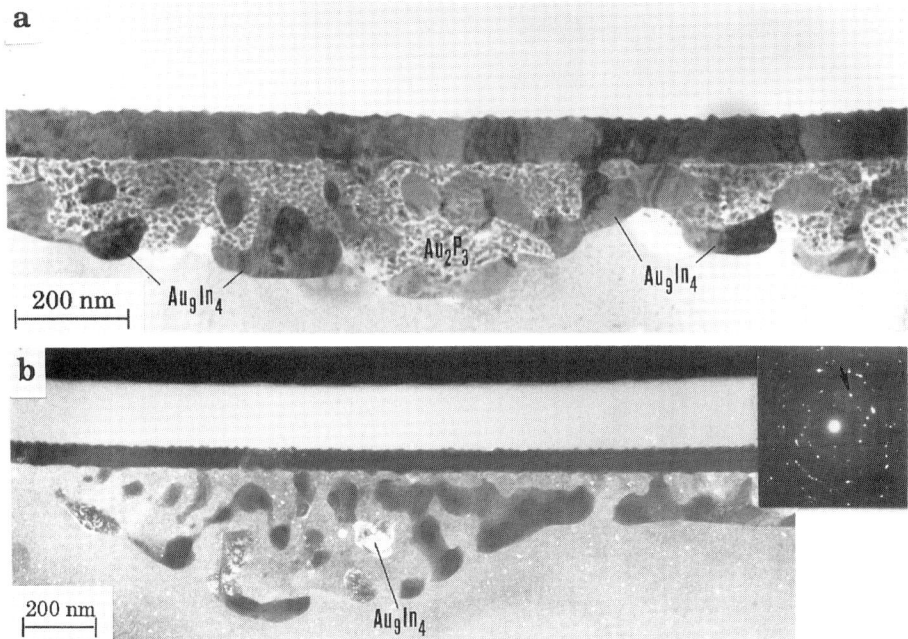

Fig. 3. Cross-sectional TEM micrographs of the InP/Au/SiO_2 contact structure annealed at 380°C

Fig. 4. Cross-sectional TEM micrographs of the InP/Au contact annealed at 420°C

In Fig. 4 a cross-sectional view of the InP/Au contact annealed at a temperature of 420°C for 3 min. is shown. The structure is similar to that seen after annealing at 380°C but bigger grains of the Au_9In_4 phase tended to touch the contact layer. In these samples the depth of grains of the Au_2P_3 phase was smaller than after annealing at 380°C. On the other hand, there were more small grains of the Au_9In_4 which almost completely occupied the Au_2P_3 grains.

Annealing of the $InP/Au/SiO_2$ contact structure at a temperature of 420°C for 3 min. resulted in the grains of the Au_9In_4 phase situated at the bottom edge of the Au_2P_3 matrix becoming bigger and there were almost no big grains of the Au_9In_4 phase in the Au_2P_3 matrix.

a   **b**

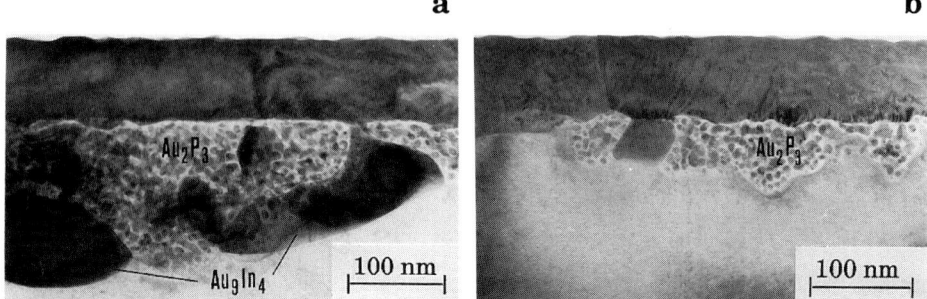

Fig. 5. Cross-sectional TEM micrographs of the $InP/Au/SiO_2$ contact structure annealed at 420°C

4. CONCLUSIONS

Annealing the InP/Au contacts causes the formation of a new phase both in the contact layer and below it. In a contact structure without a capping layer annealing at temperatures higher than 380°C causes complete replacement of gold with the Au_9In_4 phase. Applying the SiO_2 capping layer to the InP/Au contacts resulted in gold still being detected within the contact layer. An application of the SiO_2 capping layer did not influence the depth of penetration of the new phases.

ACKNOWLEDGMENT

Investigations were partially performed on a 400keV JEM4000EX at the Max-Planck Institut für Metallforschung. J Kątcki wishes to express his gratitude to Prof W Frank and Dr F Phillipp for making these studies possible. The Authors are very much indebted to E Kamińska for collaboration in material processing, D Szczepańska for assistance in specimen preparation and J Wiącek for careful preparation of micrographs.

REFERENCES

Piotrowska A, Auvray P, Guivarc'h A, Pelous G and Henoc P 1981 J. Appl. Phys. 52 pp 5112-5117.
Pécz B, Veresegyházy R, Randóczi, Barna A, Mojzes I, Geszti O and Vincze Gy 1991 J. Appl. Phys. 70 pp 332-336.
Pécz B, Randóczi G and Barna P B 1993 Microscopy of Semiconducting Materials 1993 (IOP Conf Ser No 134) p 203-206

Inst. Phys. Conf. Ser. No 146
Paper presented at Microsc. Semicond. Mater. Conf., Oxford, 20–23 March 1995
© *1995 IOP Publishing Ltd*

Granular superconducting contacts to buried two-dimensional electron gases

D A Williams, A M Marsh[1], T D Moore[1] and W M Stobbs[2]

Hitachi Cambridge Laboratory, Cavendish Laboratory, Madingley Road,
Cambridge CB3 0HE, U.K.
[1]Microelectronics Research Centre, Cavendish Laboratory, University of Cambridge,
Madingley Road, Cambridge CB3 0HE, U.K.
[2]Department of Materials Science and Metallurgy, University of Cambridge, New Museums
Site, Pembroke Street, Cambridge CB2 3QZ, U.K.

ABSTRACT: Superconducting contacts to GaAs:AlGaAs two-dimensional electron gases have been characterised by scanning and transmission electron microscopy. The contacts were formed by the rapid thermal annealing of a tin:chrome:gold multilayer in an electron beam system, which results in junctions which can be very transmissive and show a large degree of Andreev reflection. They also have a relatively high critical temperature and critical magnetic field. The contacts were found to be granular to a depth well below the original position of the heterojunction, and to contain a variety of alloy phases and voiding.

1. INTRODUCTION

The observation by Kastalsky (1991) of a supercurrent carried by ballistic electrons in a high-mobility electron gas between superconducting contacts has led to many investigations of such structures, and a renewed interest in the mechanism of Andreev reflection. This is the process whereby two electrons from a normal material can pair and enter the ground state of an adjacent superconductor, even if there is a tunnel barrier between the two materials. [See for example the review by Beenakker (1995)].

Many experiments involving superconducting contacts to high-mobility electrons have used InAs based semiconductors, as the high-mobility region is at the surface, where an elemental superconductor can be directly deposited in a relatively simple fabrication procedure. However, the much higher electron mobilities available in the GaAs:AlGaAs heterojunction make this an attractive system, despite the considerable difficulty of making a superconducting contact to the electron channel, which is typically 100 nm below the surface. A supercurrent mediated by ballistic transport has been observed by Marsh et al. (1994a) in a GaAs:AlGaAs heterostructure with diffused indium contacts, and tin-based contacts have been used by Marsh et al. (1994b), Lenssen et al. (1994) and Gao et al. (1995) to study Andreev reflection and related electron transport phenomena.

Here we describe the analysis of contacts formed using tin:chromium:gold multilayers which have been used for experiments on Andreev reflection and similar phenomena.

2. CONTACT FABRICATION

The contacts were formed on modulation-doped GaAs:AlGaAs heterostructures grown by molecular beam epitaxy, and with mobilities in the range 10^5-$2 \ 10^6$ cm^{2}v^{-1}s^{-1}. The two-dimensional electron gas (2DEG) was 80nm below the surface. A metal layer of 250nm Sn, 100nm Cr and 60nm Au was deposited by thermal evaporation on the pre-cleaned substrate. In samples to be used for electrical measurements, the metal was deposited in a pattern defined by electron-beam lithography and lift-off, and unpatterned samples were also made for analysis by transmission electron microscopy (TEM). High-resolution scanning electron microscopy (SEM) was used to ensure that both types of sample had a similar morphology.

After deposition, the material was sintered in an electron-beam rapid thermal annealing system (McMahon et al. 1985). The thermal cycle used was a very rapid heat-up followed by exponential cooling, with a peak temperature of approximately 630 °C for 10ms. The samples were heated from the back, with proximity capping to minimise arsenic out-diffusion during the anneal in vacuum.

The purpose of the multilayer structure was that the tin should act as the superconductor and also dope the semiconductor, and the chromium should act as a refractory layer to hold the tin in place during annealing. If a refractory layer is not used, tin will flow across the semiconductor surface, potentially shorting contacts which can be as close as 1 μm. The gold layer was added to minimise oxidation of the chromium, and ensure good electrical contact to further layers of metallisation.

The electrical characteristics of the contacts were measured at temperatures from 50mK to room temperature, with and without applied magnetic field. The contacts have been found to have a spread of critical temperatures, with a significant number having a T_C of 7.2K. This is markedly higher than that of bulk elemental tin (3.4K), and such contacts also show Andreev reflection in applied magnetic fields greater than 3T (Marsh et al. 1994b).

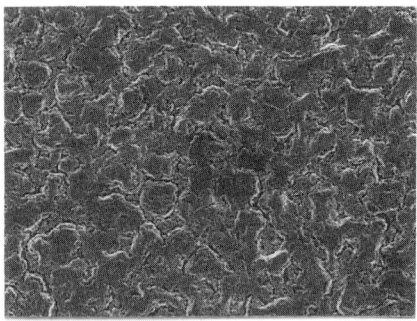 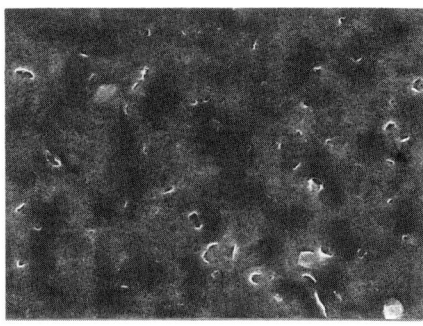

[a] _____ 3 μm [b]

Figure 1: *Planar SEM micrographs of the contact metallisation before and after annealing. (a) shows the original roughness due to the granular tin growth. (b) The surface after annealing is grossly deformed, with surface ripples relating to underlying voids.*

3. ANALYSIS

The samples were analysed using cross-sectional and planar high-resolution SEM, and cross-sectional TEM. The morphology before and after annealing is shown in Fig. 1. The as-deposited metal layer can be seen to be uneven, the main cause of which is the granular growth of the tin layer. After annealing, the contact structure is grossly uneven, and the appearance has changed from gold to silver. Cross-sectional TEM shows the material to be largely polycrystalline to a depth of around 250nm, well below the original position of the 2DEG. The chromium layer, rather than entirely remaining in place as a refractory cap, has alloyed with the other materials present as has, to some degree, the gold. However, a comparison between the edges of patterned contacts formed from tin, and those made using the multilayer structure, showed that there was very little lateral flow if the multilayer structure were used, wheras pure tin flowed several microns across the surface.

The alloy phases going downwards from the surface are found to retain partial orientation relationships with the substrate, and are occasionally but not usually faceted. Voiding is seen in many places, both in the surface metal layer and in the substrate, and can be seen in cross section in Figs 2 and 3, and as holes and bubbles in plan view in Fig. 1b. A rounded trapezoidal grain as seen in Fig. 2b and at position D in Fig. 3b is common, and is reminiscent of the 'spikes' seen in annealed Au:Ge:Ni ohmic contacts to GaAs. The voiding is believed to be due to the formation of gaseous arsenic during the anneal, a process which is also seen in Au:Ge:Ni contact formation by rapid thermal annealing.

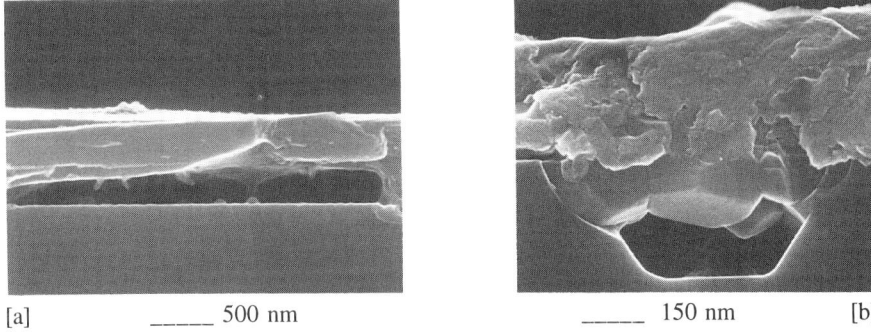

[a] _____ 500 nm _____ 150 nm [b]

Figure 2: *Cross-sectional scanning electron micrographs of annealed contacts. (a) The metallisation is seen to have reacted with the substrate in many places to a depth of 200-300 nm, with considerable voiding. (b) A typical rounded trapezoidal inclusion.*

Compositional analysis shows that there are many alloy phases present, and a large proportion of the alloy present in the downward reaction product is CrAs based. Fig. 3 shows evidence of small localised particles in a region of the substrate GaAs which shows changes in local lattice parameter and As depletion. The edge of this region of altered lattice parameter is sometimes decorated with a reaction product, as can be seen at point E in Fig. 3a. There was evidence of Ga up-diffusion, and the Au was predominantly near the surface, although some had diffused downwards.

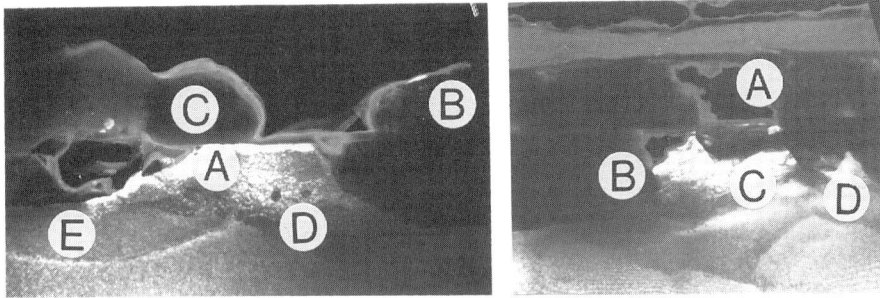

[a] _____ 200 nm [b]

Figure 3: *Cross-sectional TEM of annealed contacts: (a) A is near the original surface of the semiconductor. B is a metallic grain near the surface. C is a void in the surface metallisation. D indicates two precipitates in a region of changed GaAs lattice parameter. E indicates the end of this region, and is decorated with small precipitates. (b) A and B are voids. The substrate at C is heavily deformed. D shows a typical rounded trapezoidal inclusion as seen in Fig. 2b.*

4. DISCUSSION

It has not yet been possible to determine which of the compounds present is the dominant superconducting phase, but it would seem likely to be one of the tin-based intermetallic alloys. The high critical magnetic field is due to the granularity of the contact material, as has been observed by Beamish et al. (1990) for granular tin films.

564

The contact alloying has destroyed the semiconductor to a depth well below the original position of the heterojunction, and contact to the remaining 2DEG through which electron transport is measured must be lateral. The measured high probability of Andreev reflection in some of these junctions is almost certainly due to the combination of local disorder and an interface barrier, as modelled by Marmorkos et al. (1993). This contrasts with the system used by Verwerft et al. (1995), where a titanium diffusion barrier was used between the tin and the semiconductor substrate; high resolution electron microscopy showed that there was then no alloying, and electron focusing experiments showed that a two-dimensional electron gas was present underneath the contact, therefore contact to the 2DEG was vertical. A similar approach was used by Lenssen et al. (1994).

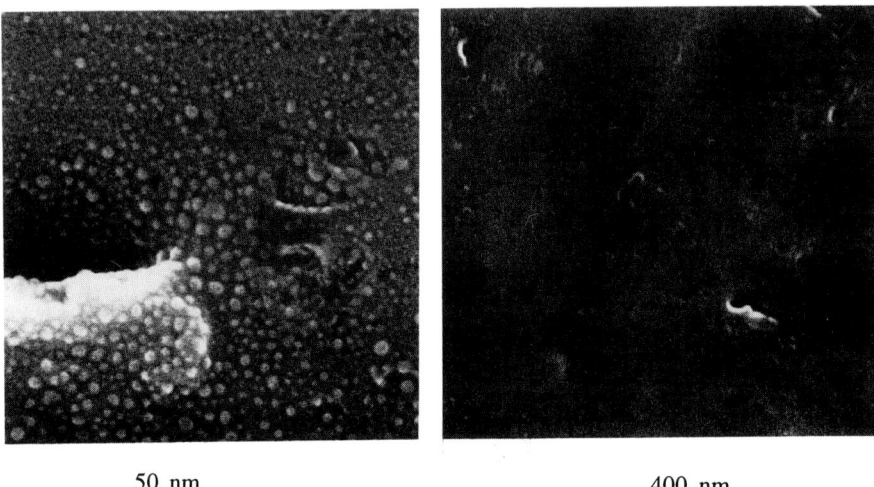

_____ 50 nm   _____ 400 nm

Figure 4: Planar SEM at two different magnifications showing the various inhomogeneities of structure in the contacts.

REFERENCES

Beamish J R, Patterson B M and Unruh K M 1990 Proc. Mat. Res. Soc. <u>195</u> pp 385-390
Beenakker C W J 1995 To be published in: Mesoscopic Quantum Physics, eds. E Akkermans G Montambaux and J-L Pichard (Amsterdam: North Holland)
Verwerft M, Gao R F, De Hosson J Th M and Klapwijk T M 1994 Proc. Int. Conf. Electron Microscopy p681
Kastalsky A, Kleinsasser A W, Greene L H, Bhat R, Milliken F P and Harbison J P 1991 Phys.Rev.Lett. <u>67</u>(21) pp 3026-3029
Lenssen K-M H, Westerling L A, Harmans C J P M, Mooij J E, Leys M R, van der Vleuten W and Wolter J H 1994 Surface Science <u>305</u> pp 476-479
Marmorkos I K, Beenakker C W J and Jalabert R A 1993 Phys.Rev.B <u>48</u>(4) pp 2811-2814
Marsh A M, Williams D A and Ahmed H 1994a Phys.Rev.B <u>50</u>(11) pp 8118-8121
Marsh A M, Williams D A and Ahmed H 1994b Physica B <u>203</u> pp 307-309
McMahon R A, Hasko D G and Ahmed H 1985 Rev. Sci. Instrum. <u>56</u>(6) pp 1257-1261

Inst. Phys. Conf. Ser. No 146
Paper presented at Microsc. Semicond. Mater. Conf., Oxford, 20–23 March 1995
© *1995 IOP Publishing Ltd*

Electron microscopy applications to semiconductor devices

B Cunningham, T W Joseph, L Gignac and A Domenicucci

IBM East Fishkill, 1580 Route 52, Hopewell Junction, NY 12533, USA

ABSTRACT: Electron microscopy is rapidly becoming one of the major analytical tools in semiconductor device analysis. Although scanning electron microscopy (SEM) has been firmly established in the semiconductor industry for many years, the application of transmission electron microscopy (TEM) has, until recently, been limited. This paper will discuss the factors responsible for the increases in the applications of both SEM and TEM to semiconductor devices, and give examples showing the scope and diversity of present day analysis. A brief discussion on the future role of electron microscopy to device analysis will also be given.

1. INTRODUCTION

In the last 5-10 years there has been a large increase in the application of electron microscopy to semiconductor device problems. This increase is due to several factors. Higher density chips and decreasing device dimensions have led to structural components which are frequently in the nanometer range. Electron microscopes are ideally suited for imaging such components, making them invaluable for routine constructional analysis. Also, as device dimensions decrease, the structure and composition of thin interfacial layers become increasingly important to device performance. The ability to examine these interfaces in actual devices is extremely important and TEM is proving to be uniquely suited to this task. Crystal defects, such as dislocations, are major yield detractors in many semiconductor technologies. Determining the defect location and origin are critical to yield enhancement, and the analysis of crystal defects in devices is an area in which electron microscopy excels.

Although the major factor responsible for the increase in electron microscopy applications to semiconductor devices is the decrease in device dimensions and the associated consequences, two more areas have been contributing factors. The first of these is the development of sample preparation techniques. Most analyses of semiconductor devices require fast turnaround and high throughput. Historically, sample preparation for electron microscopy, in particular TEM, has been laboriously slow. Over the last several years however sample preparation techniques have been refined to the point where the preparation of device samples is routinely accomplished in a matter of hours (see e.g. Anderson et al., 1989). Whether the requirements of the semiconductor industry have driven the need for improved sample preparation techniques or whether the improved sample preparation techniques have resulted in increased usage of electron microscopes is a point open to debate. Nevertheless, it is safe to state that the semiconductor industry and the microscopy community have both benefited from these developments.

The other factor which has contributed to the increase in electron microscopy usage is the development of the electron microscopes themselves. In the case of SEM, improvements have kept pace with the size of manufactured integrated circuits. Ease of use and reliability have made the SEM a common inspection tool, to the point where they are now incorporated into the clean room environment of manufacturing lines. One of the most recent developments, the in-lens field emission microscope, allows high resolution images to be recorded at low accelerating voltages. This capability is important for imaging defects in thin layers when beam penetration is a concern. The development of TEM has allowed for easy imaging at lattice resolution, reliable operation and microanalysis capabilities. The recent advances in field emission TEM have dramatically increased the analytical capabilities of TEM.

The following examples show the types of applications where electron microscopy is used in semiconductor device development. The emphasis has been placed on the types of applications, rather than in-depth analyses of specific problems. Many problems encountered in semiconductor devices are seldom solved by the application of one analytical technique. Electron microscopy is often complemented by other techniques. For example, prior to examination of defects in devices, the location of the defects has to be determined. In such cases, EBIC, OBIC and emission microscopy are often used prior to electron microscopy. Where relevant, a description of these complementary techniques will be given.

2. APPLICATIONS

2.1 Constructional Analysis

A major application of electron microscopy is in the area of constructional analysis. Simply stated, constructional analysis is the determination of device dimensions and the arrangement of the various layers and elements comprising the device. Figure 1 shows an example of the type of routine analysis performed by SEM. The source and drain regions of a field effect transistor (FET) have been delineated with a junction etch. The advantage of this type of analysis is that large areas incorporating many devices can be easily scanned in a short period of time (Curling et al., 1989). The resolution of the in-lens SEM is demonstrated in Figure 2, which shows the sidewall of an emitter contact in a polysilicon emitter bipolar transistor. The application of TEM in constructional analysis is best demonstrated in the examination of the thin dielectric films used for gate dielectrics in FET technologies and for trench liners in dynamic random-access memory (DRAM) storage capacitors (Oppolzer et al. 1987, Song et al. 1992). Figure 3 shows a thin oxide layer of a gate dielectric as it passes over shallow trench isolation. Not only can the thickness of the oxide be quickly and accurately determined, but the changes in shape and thickness at the gate edges can be assessed. No other analytical technique has the spacial resolution required to image components with nm dimensions, and the importance of this capability to semiconductor device development cannot be overemphasized.

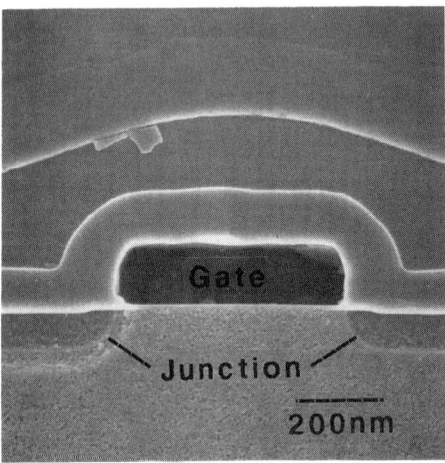

Figure 1. SEM micrograph of an FET. The source and drain regions have been delineated with a junction etch.

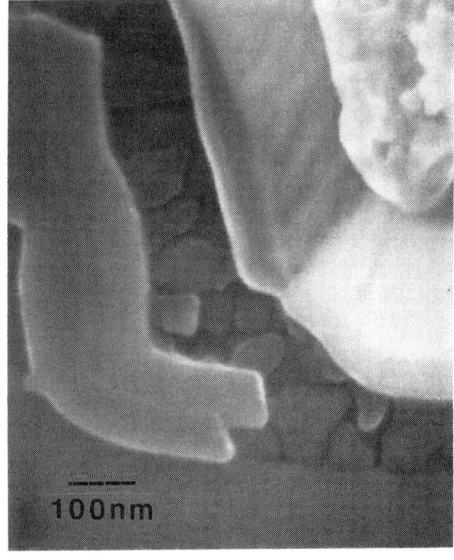

Figure 2. High resolution SEM micrograph of the emitter sidewall of a bipolar transistor.

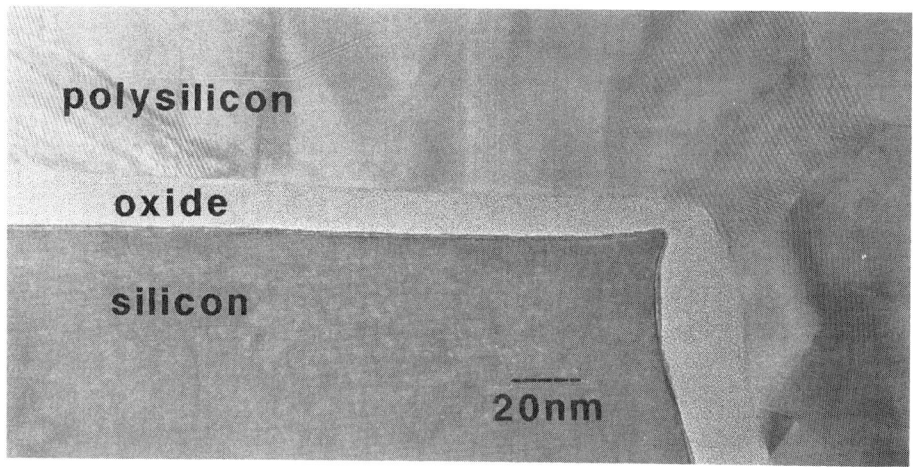

Figure 3. Cross-sectional TEM image of a thin gate oxide, showing thickness variations at different locations.

As stated previously, one of the factors responsible for the increase in TEM analysis of devices is the decrease in device dimensions. An example of the trend towards larger scale integration is in DRAM development. Figure 4 shows plan view micrographs from 4Mb and 256Mb DRAM chips with deep trench storage capacitors. Figure 4(a) shows a single cell in a 4Mb DRAM. Figure 4(b) was taken at the same magnification, and demonstrates how the cell size decreases as device density increases. These micrographs show cells from memory devices which are both commercially available (4Mb) and under development (256Mb). Figure 5(a) shows a low magnification cross-sectional image of two cells from a 256Mb DRAM. Each cell contains many different elements. The value of this type of analysis is demonstrated in Figure 5(b). The arrowed region at the top of the trench capacitor shows the formation of a bird's beak around the collar oxide. It is well known that an oxide bird's beak can create a region of high stress in the adjacent silicon, making it a potential site for dislocation generation. The number of applications of electron microscopy in constructional analysis, and the benefits of such analysis, are too numerous to mention. However, with the trend towards decreasing device dimensions in many technologies, this type of analysis is likely to used more and more frequently.

2.2 Thin Films and Interfaces

As device dimensions decrease, there is a corresponding increase in the contribution of thin films and interfaces to device performance. Figure 6 shows a contact between a W stud and W line. Ti is deposited between the tungsten as an adhesion layer. The contact shown in Figure 6 had a resistance of 4 ohm, twice the specification value. The interfacial layers marked in Figure 6 were identified by small probe EELS in a field emission STEM. Contacts with a nominal resistance of 2 ohm showed only a thin Ti layer between the two W layers. In order to understand the cause of the increased resistance, we first have to understand the processing steps required to make the contact. The via onto the W line is opened up by reactive ion etching (RIE). This process step leaves a polymeric film on the W which is subsequently removed in an oxygen ash. The ashing however creates a thin layer of oxide on top of the W line. After ashing, Ti is deposited, followed by the W stud deposition. The Ti reduces the tungsten oxide, leaving an oxygen rich Ti layer between the two W layers. In the contact shown in Figure 6, the oxide layer formed during the ashing step was too thick to be fully reduced by the Ti layer. A WO(x) layer was left in the contact, producing a contact with a higher than normal resistance. Identification of the layers indicated which processing step was responsible for the high contact resistance. Examination of the asher showed that the temperature calibration was incorrect, and that the asher was in fact operating at a higher than

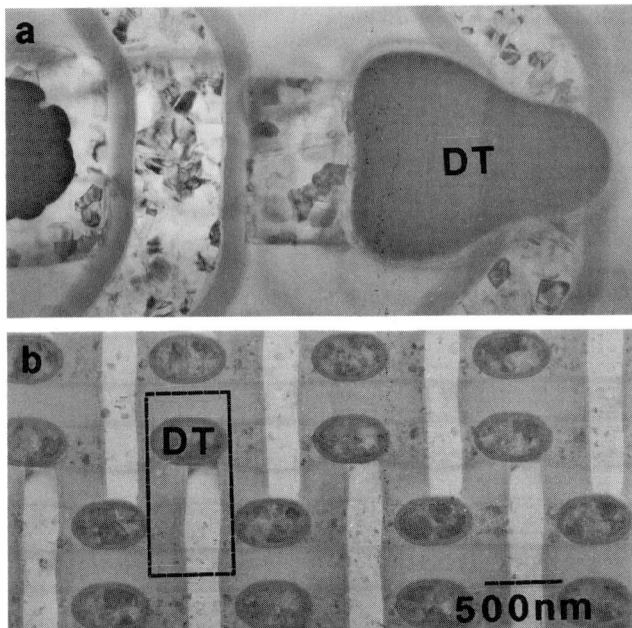

Figure 4. Plan-view TEM images of a) one cell from a 4Mb DRAM chip, and b) the equivalent area from a 256Mb DRAM chip, with a single cell outlined. The deep trench storage capacitor in each is labelled DT.

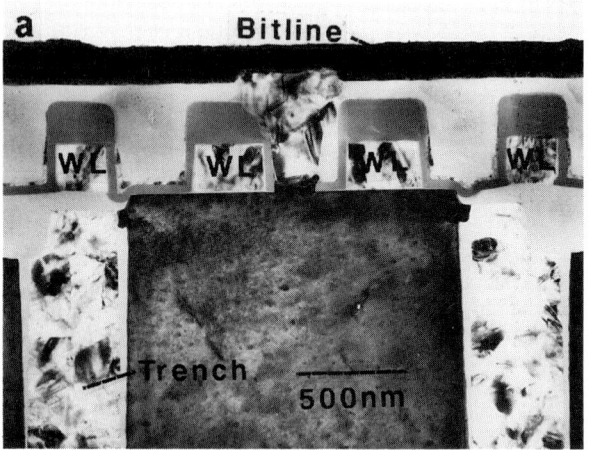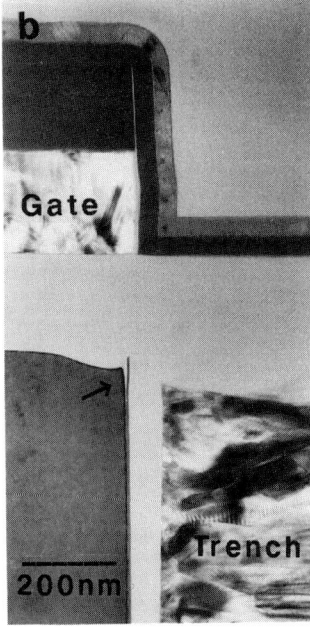

Figure 5. Cross-sectional TEM images of a 256Mb DRAM. a) A low magnification image showing the layout of two cells. The word lines are labelled WL. The arrowed region in b) shows the formation of an oxide bird's beak at the top of the trench collar.

normal temperature, thereby growing a thicker tungsten oxide than expected. This example shows the importance of being able to quickly identify interfacial films in small contacts. Although other analytical techniques can be used for similar analysis, only TEM has the spatial resolution required to analyze small contacts in actual devices. Another point to note is the resistance value of the contact. Although the contact resistance was out of spec, the value of 4ohm/contact is not abnormally high. The value of 2ohm/contact was specified because of the requirements for total stud resistance. Since the number of studs on a chip increases with increasing integration, the resistance of individual contacts must be reduced if the total resistance remains the same. In less dense technologies, the value of 4ohm/contact might not have been a problem, thereby demonstrating the relative importance of the interfaces as a function of the level of integration.

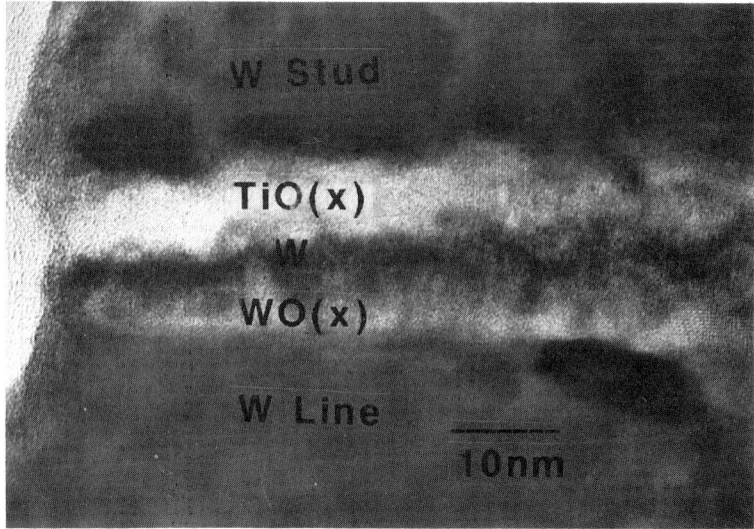

Figure 6. Cross-sectional TEM micrograph showing the interfacial layers in a W/Ti/W contact with high resistance.

The next example demonstrates a rather unique attribute of TEM in studies of semi-conductor devices. For many years an often cited disadvantage of TEM has been that only a very small volume of material is analyzed. It has been debated, in areas such as metallurgy, that the small volume sampled by TEM might not be representative of the bulk material. Microscopists in the semiconductor industry can now make the unique claim that the volumes they are examining are not only representative of the devices under study, but that in many cases the device dimensions are such that the entire device can easily be imaged in the TEM. Along with the structural information obtained by TEM, contacts to the devices can provide complete electrical characterization.

Fuses are commonly used in semiconductor technologies for chip repair. Redundancy is built into many chips to allow for a certain density of cell fails. Typically, memory chips are fabricated in "blocks" of cells. A 64k SRAM chip may consist of 8x8k blocks, with an additional 8k block so that a defective block in the original 8x8k array can be replaced. During wafer test, the defective block can be replaced by the redundant 8k block by "blowing" fuses on the chip. Figure 7(a) shows a SiCr fuse which was designed to be blown by an electrical pulse. The fuse material is actually a silicon rich $CrSi_2$ phase, the reasons for which are unimportant to the present discussion. The fuse is several μm long, approximately 1μm wide at the center of the fuse, with a thickness of about 50nm. The fuse is dog-boned in shape with metal contacts at each end. It is fabricated on upper level metal layers, and is sandwiched between two layers of SiO_2 which are several microns thick. It is relatively easy to prepare a sample for the TEM which contains the entire volume of the fuse. Figures 7(a) and 7(b) show an unblown fuse and a fuse after a 4V voltage pulse has been applied respectively. Prior to the application of any voltage pulse the fuse has a microstructure typical of a crystallized $CrSi_2$ thin film. After the application of the voltage pulse, the microstructure of the fuse changes dramatically. Silicon has piled up at one end of the fuse forming a polysilicon region, and areas of Cr_3Si and Cr_5Si_3 are present, Figure 7(c). In addition, a crack has formed in the fuse, and the fuse material has delaminated from the over and underlying SiO_2 layers. Obviously a considerable amount of "material" movement has occurred during the voltage pulse, the duration of which was only several milliseconds. By studying fuses with different applied voltages and pulse times it was possible to determine a mechanism for the fuse blow. During the application of the voltage pulse, material moves in the fuse due to electromigration and thermal gradients. In addition, the voltage pulse increases the fuse temperature and causes phase formation. A complete description of the voltage blown fuse mechanism is outside the scope of the present paper, however the material presented here again shows how TEM can be applied to semiconductor device problems. As with previous examples, the dimensions of the fuse are such that examination of its structure would, at the very least, be extremely difficult by any other analytical technique.

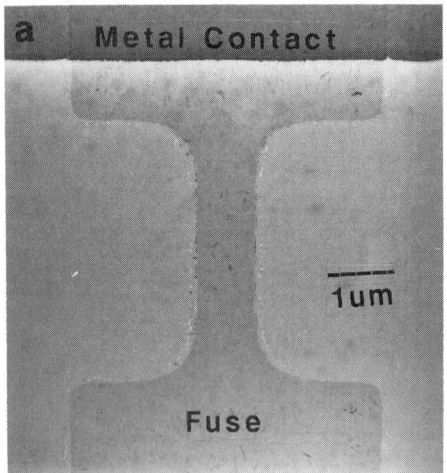
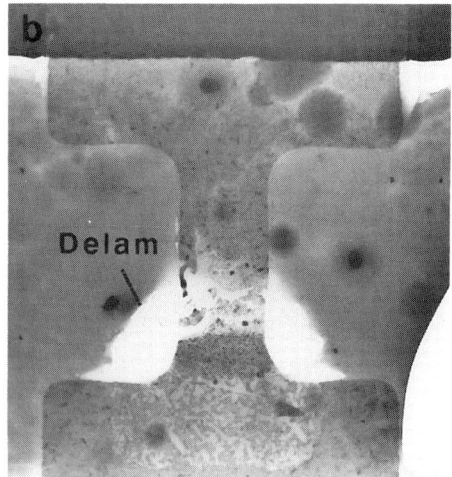

Figure 7. Plan-view TEM micrographs of SiCr fuses. a) as deposited fuse b) a fuse after the application of a 4V voltage pulse c) higher magnification image of the "blown" portion of the fuse shown in b).

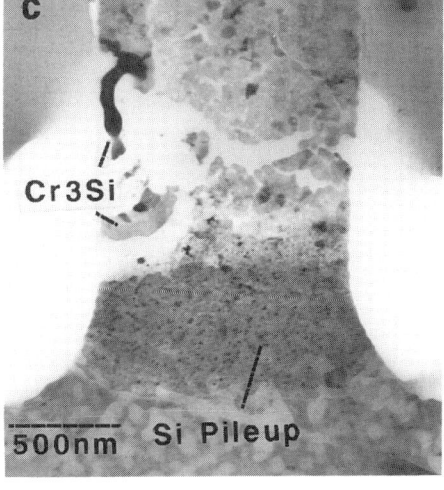

2.3 Defect Analysis

Defects, such as dislocations, can be created during silicon wafer processing by many steps, or combination of steps. During ion implantation the silicon lattice is damaged, which in turn can lead to the formation of dislocations during subsequent annealing steps. Processing steps such as oxidations can produce stresses in the silicon which can also generate dislocations and, in some cases, stacking faults. A combinations of processes, such as overlapping an ion implant with a region of high stress, can generate dislocations, when neither process by itself will generate dislocations. Figure 8 shows three regions, an implanted region, a region of high stress from a deep trench isolation structure, and a region where the ion implant overlaps the stress field of the trench structure. No defects were generated by the trench isolation, and only small residual dislocation loops are visible in the ion implanted region. However, when the ion implant was close to the trench structure, a large number of glissile dislocations was generated.

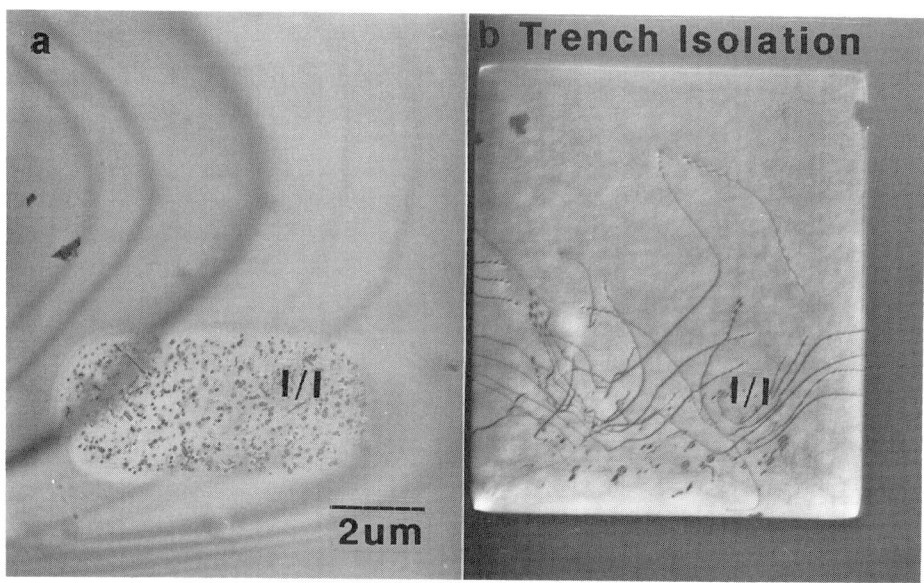

Figure 8. Images showing the effect of implanting close to a trench structure. a) residual dislocation loops from an ion implant, and b) trench isolation with the same ion implant adjacent to one side of the trench structure.

Overlapping effects in semiconductor processing are also a consequence of larger scale device integration. As device dimensions shrink, it becomes harder to keep the individual units in a semiconductor cell separated. A glance at Figure 4 will quickly demonstrate this point. An additional effect of decreasing device size and increasing packing density is that the defect limited yield decreases for a fixed defect density. A dislocation density which may be tolerable in one technology may give zero yield for a higher density technology.

Gross defect problems which impact large numbers of cells in a particular technology are relatively easy to examine. Figure 9 shows a region from a 256Mb DRAM chip which contains a high density of dislocations. The density is such that examination of any region on the chip reveals the problem. Process changes can be made and their effects on the defect density assessed quickly by TEM. When the defect density decreases to a level where random examination of chips is unlikely to "find" the defects, other localization techniques have to be used. In the case of DRAMs, a bit map can be generated, and the locations of single cell fails mapped. Figure 10 shows an example of a single cell fail in a 4Mb DRAM. The exact site was located in the TEM after the bit map had shown which cell had failed.

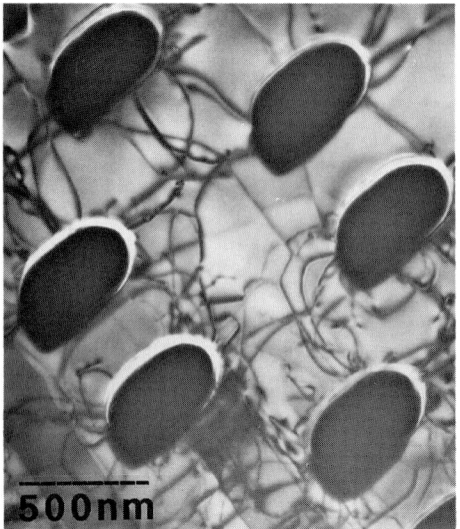

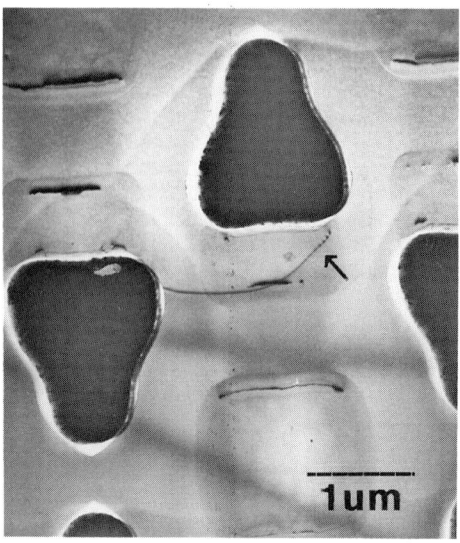

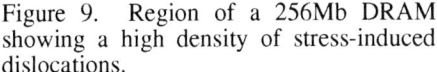

500nm

1um

Figure 9. Region of a 256Mb DRAM showing a high density of stress-induced dislocations.

Figure 10. Single cell fail in a 4 Mb DRAM. Image shows that the fail was caused by a dislocation.

For application-specific integrated circuits (ASIC), microprocessors, analog components etc., it is not always possible to locate the defective component without many hours of electrical probing. Even then, for large devices, electrical isolation techniques can only pinpoint the failing component, and not the location within the component. Techniques such as electron beam induced current (EBIC) and optical beam induced current (OBIC) have been use for many years to locate defective devices on integrated circuits (e.g. Joseph et al., 1992). Both of these techniques require scanning a probe, electron or laser, across the sample to create carriers in the silicon. When these carriers cross a junction, they generate a current which can be measured externally. Whenever the carriers recombine at a defect the EBIC or OBIC signal is enhanced. Typically, a current amplifier is used and the image is formed by mapping the signal intensity to beam position. Several features make OBIC easier to use than EBIC. The laser can penetrate thick layers of oxide and nitride, whereas the electron beam cannot, and no vacuum is needed for OBIC. An additional technique for defect location, which has been refined in recent years, is light emission microscopy (Kolzer et al., 1992). The light emission system consists of an optical microscope, sample stage with provision for contacting the device, and a sensitive detection system such as a photocathode image intensifier or CCD camera. Electrical contact is made through either wire bonds to an open top carrier or by probes. Power is applied either by a simple power supply or by a parametric tester so that current flows through the device. Recombination radiation arises when electrons and holes recombine in a forward biased junction, generating photons which are collected by the detection system. This emission mechanism can be used to locate junctions which are present due to defects such as collector-emitter shorts, or gate oxide defects such as gate to silicon shorts. Avalanche breakdown from impact ionization at a reverse biased junction is also an efficient producer of photons. This type of radiation can be used to identify areas within a diode which are responsible for high reverse current leakage.

Figure 11 shows a gate oxide defect localized by OBIC, and the corresponding area imaged with in-lens field emission SEM after deprocessing and gate removal. The image shows that the defect is a pinhole in the gate oxide. This example also highlights the advantage of the field emission source and its low voltage capability. At higher voltages the defect is not visible due to a strong signal from the underlying silicon.

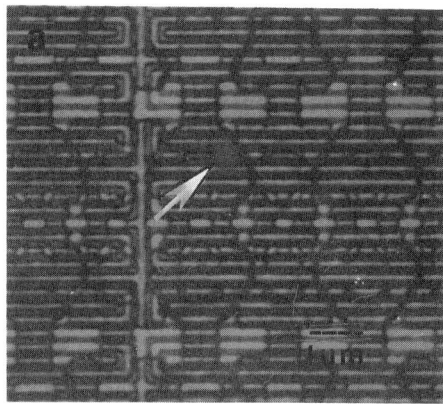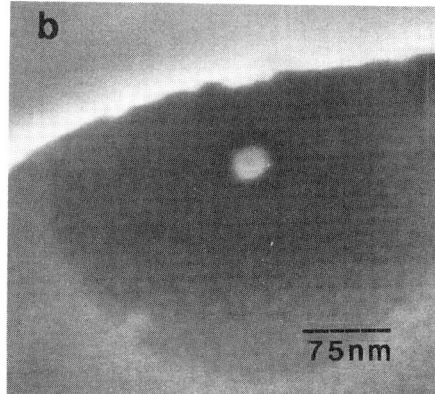

Figure 11. a) OBIC image showing location of leakage in a gate array, and b) defect in gate oxide imaged in SEM.

Another example showing the benefits of defect localization and electron microscopy analysis is given in Figure 12. Light emission microscopy was used to locate the failing region in a bipolar transistor array and TEM was subsequently used to examine the same location. A dislocation was found beneath the emitter polysilicon, which had caused a collector to emitter short. Since the transistor array was large, it is unlikely that the dislocation would have been found by random inspection.

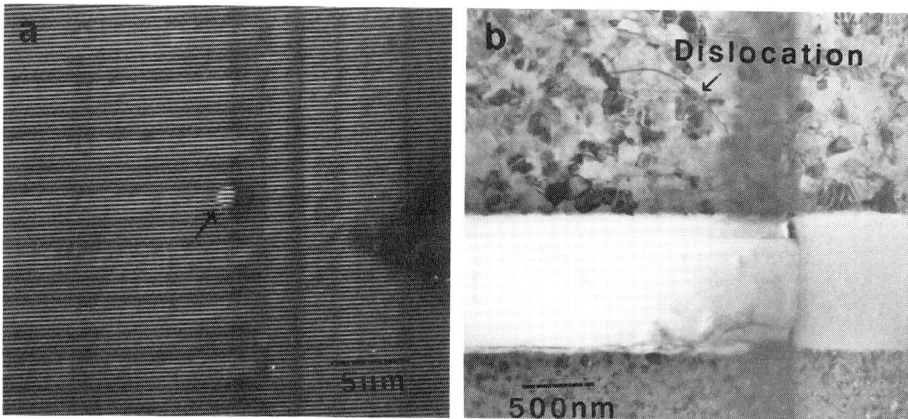

Figure 12. a) light emission "hot spot" in a bipolar array, and b) TEM image taken at hot spot location showing a dislocation.

2.4 Field Emission TEM

The recent advances in field emission TEM's have made them powerful analytical tools in device analysis. The small probes and high beam currents available in the probes make them ideal for the analysis of the thin layers and interfaces encountered in many of todays semiconductor devices. The thin layers shown in Figure 6 are one example of the application of field emission TEM to semiconductor devices. Another example is shown in Figure 13. A thin oxide layer approximately 5nm in width has been imaged in the X ray map using the oxygen K line. The ability to perform elemental mapping at high magnifications using small probes offers great potential for detecting variations in the composition of interfacial layers.

574

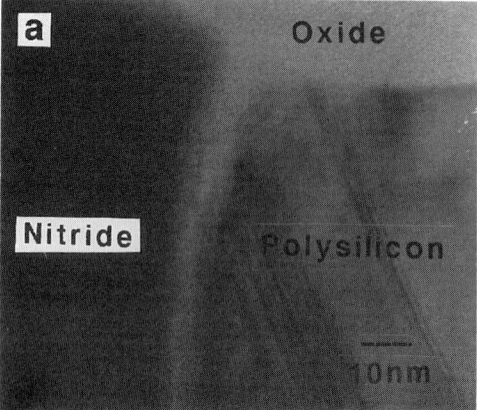 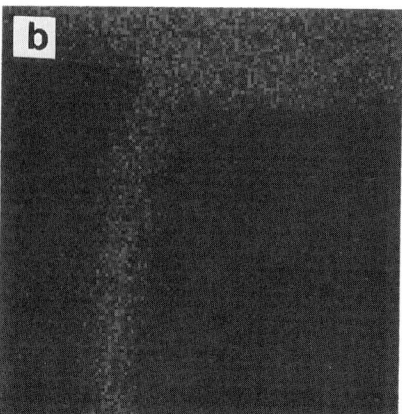

Figure 13. a) STEM image of a thin oxide layer between nitride and polysilicon, and b) X ray dot map of the same area using the oxygen K line.

3. DISCUSSION

The preceding examples clearly show that electron microscopy is a valuable tool in semiconductor device development. SEM and TEM have applications in almost every aspect of device manufacturing, from development to yield enhancement, and from thin film analysis to bulk defect analysis. The usage of both SEM and TEM is likely to continue to grow as it has in the last few years. In particular, it is likely that in a few years TEM will be used for constructional analysis to a similar extent as SEM is today. The developments in field emission TEM will almost certainly have an impact on device analysis. Tighter requirements on device and product performance will drive reductions in process variations and a lower tolerance for defects. These requirements, together with the reduction in device dimensions, indicate that electron microscopy analysis will continue to play a major role in semiconductor device development.

ACKNOWLEDGEMENTS

The authors wish to acknowledge L Palmer and J Miller for technical assistance, and P Batson for his EELS data and many discussions. The data for Figure 13 were taken on a JEOL 2010F with the special help of M Kawasaki and K Ibe of JEOL Ltd.

REFERENCES

Anderson R, Klepeis S, Benedict J, Vandygrift W G and Orndorff M 1989 Inst. Phys. Conf. Ser. **100**, 491
Curling C J, Hokke R and Reader A H 1989 Inst. Phys. Conf. Ser. **100**, 531
Joseph T, Berry A L and Bossmann B 1992 ISTFA Microelectronic Proceedings
Kolzer J, Boit C, Dallmann A, Deboy G, Otto J and Weinmann D 1992 J. Appl. Phys. **71**, 11
Oppolzer H, Cerva H, Fruth C, Huber V and Schild S 1987 Inst Phys. Conf. Ser. **87**, 433
Song M, Hashimote H, Yokota Y, Matsukawa T, Ajika N and Oguh I 1992 J. Electron Microsc. **41**, 337

Inst. Phys. Conf. Ser. No 146
Paper presented at Microsc. Semicond. Mater. Conf., Oxford, 20–23 March 1995
© *1995 IOP Publishing Ltd*

Sub-nanometric elemental analysis on ULSI devices with FEG-TEM

Kyung-ho Park, S Hashimoto, M Kawasaki* and K Ibe*

Advanced Materials and Processes, Texas Instruments Tsukuba R & D Center Ltd.,
17 Miyukigaoka, Tsukuba, 305 JAPAN
* Electron Optics Division, JEOL Ltd., 3-1-2 Musahino, Akisima, Tokyo, JAPAN

ABSTRACT: An FEG-TEM analysis was applied to detect the dopant segregation at the interface between a poly-crystalline silicon gate (poly-Si) and a gate silicon oxide (SiO_2), as well as at the grain boundary in the poly-Si of the device equivalent 64M DRAM. Phosphorus segregation profiling at the interface between Si and SiO_2, across and along a grain boundary was obtained. Observations indicate that phosphorus segregation exceeds 3 % at the grain boundary, whereas 2 nm away from the grain boundary, segregation was below the detectable limit for the EDX analysis.

1. INTRODUCTION

Due to the continuous shrinkage of the cell dimension and dopant aggregation, in heavily-doped silicon, only a fraction of the total dopant concentration is electrically active. For thermal cycles used for VLSI fabrication, the segregation process of the dopant elements playsa critical role in determining the active doping levels and thus device electrical characteristics (Baikei 1993). Although the dopant segregation phenomenon at a grain boundary in poly-crystalline silicon has been proven by the analytical method for a sample which contained a relatively high concentration level of dopant (2×10^{20} As atoms per cm^3, which is about 0.4 at%)(Grovenor 1984), few studies have been undertaken to prove a similar phenomenon as well as the dopant segregation at the layered interface between the poly-Si gate and the gate SiO_2 on the dopant level of a poly-Si gate in ULSI. The quantitative analysis of the dopant segregation for such kind of sample requires ultra-high spatial resolution and high sensitivityon the sub-nanometre scale (Maiti 1993).

Transmission electron microscopy (TEM) techniques have proved to be indispensable in the analysis of materials structure and chemistry from the microscopic to atomic scale. A major advance in this regard comes from the recent availability of field-emission gun (FEG) TEM instrumentation where both structural and chemical information can be obtained from a sub-nanometre region. In this work, we applied the FEG-TEM instrumentation for the analysis of the device equivalent 64M DRAM where both structural and chemical information can be obtained from sub-nanometer regions by both EDX and EELS methods (Hosokawa 1993).

It is well known that grain boundary chemistry can have a profound influence, either detrimental or beneficial, on the properties of polycrystalline materials. Accordingly, there has been a tremendous amount of theoretical and experimental work on the study of interfacial segregation under equilibrium or non-equilibrium conditions. In the former, the soluble impurity concentration depends sensitively on the number of available atom sites (i.e. boundary structure) in the interface and thus the impurity enrichment extends over only a few atomic diameters, for example the effective grain boundary width. Alternatively, under non-equilibrium conditions, as a result of point defect assisted impurity diffusion for example, the segregate can extend over several microns on either side of the boundary plane(Perovic 1993).

2. EXPERIMENTAL PROCEDURE and RESULTS

The microscope was operated at 200 kV with Schottky emission source capable of probe sizes as small as 0.5 nm with relatively high probe currents. The objective lens was of the analytical type with a spherical aberration coefficient of 1.0 mm which is giving a point resolution of 0.19 nm and an information limit of 0.18 nm. The smallest probe size attainable is 0.4 nm (FWHM). The chemical analysis is using the ultra thin window EDX detector with 25 degree take-off angle, 0.13 sr solid angle and an energy resolution of 148 eV. The data acquisition time was set for the 30 seconds live time, enough time for getting a significant peak, even ten seconds would be realistic for the data aquistition. The beam drift was checked by setting the beam on several different pointsof the O/N/O sample for three minutes, watching any change of peak appearance on EDX spectra but could not find any indication of beam drift.

The samplesare thin films and the fully processed TEG equivalent 64M DRAM. For both samples; substrates are p-type(100), samples are first oxidized at 1123 K to form the 12 nm gate oxide, then an in-situ doping amorphous silicon is deposited at 798 K. The thickness of the poly-Si after the crystallization was 250 nm for the thin film samples and 140 nm for the fully processed TEG samples. The crystallization are done while the following over layer oxide deposition process at 1073 K for 3 hours as sample named <Standard(STD)>, on the other hand, some samples are lamp annealed at 1273 K for 10 Sec. for the crystallization, named as <RTA(Lamp Anneal)>. All TEG sample received the STD thermal process for the gate integration. The TEG sample was analyzed from two direction, namely the direction vertical to the Poly-Si gate line, as <Pattern-X(STD)>, and along to the gate line as <pattern-Y (STD)>. The cross-sectional TEM specimen was prepared by an ordinary ion-milling method.

Figure 1 shows a <011> cross-sectional high resolution lattice image at the interface between the gate poly-Si and the oxide for the sample of <Standard(STD)> and <Pattern-Y(STD)>. The inset micrograph shows the point of the cross-sectional structure of the device. Using the HREM image, a series of EDX profiles were obtained using a probe size of 0.5 nm. Circle points in the micrograph are indicating the point of the elemental analysis by the EDX. An example of a raw EDX spectrum (30 sec acquisition) using the 0.5 nm probe size taken from the region of the points of the interface and inside the poly-Si gate 2 nm away from the interface is shown in Fig.2. The phosphorus peak appears in the spectrums. This clearly appears on the elemental mapping of Si-K, O-K and P-K according with the STEM image as shown

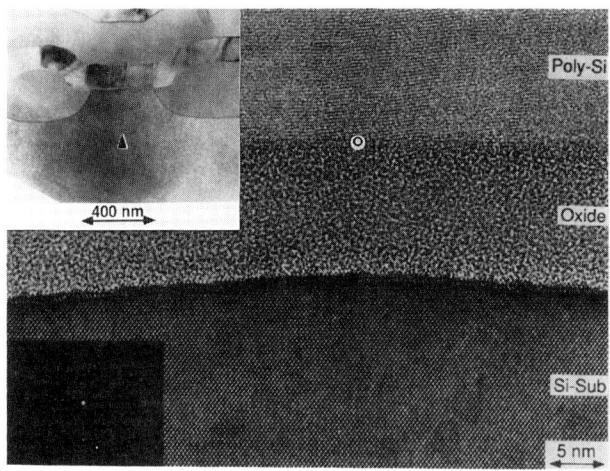

in Fig. 3. Elemental concentrations were obtained following a standardless k-factor analysis. However, since background subtractions were not performed prior to quantification, absolute elemental concentrations cannot be determined. The large peak-to-background ratio of the P-Ka signal allowed for semiquantitative profiling across the interface, are shown in Fig.4.

Fig. 1. Lattice imaging at the interface of the gate poly-Si, showing the points of EDX analysis.

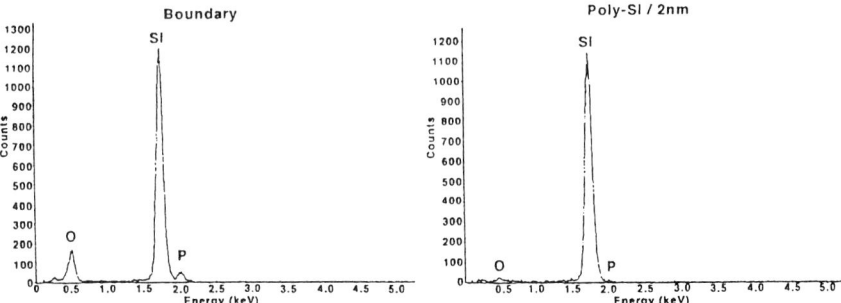

Fig. 2. EDX spectrum these are get at the point of the "boundary 1" and side points.

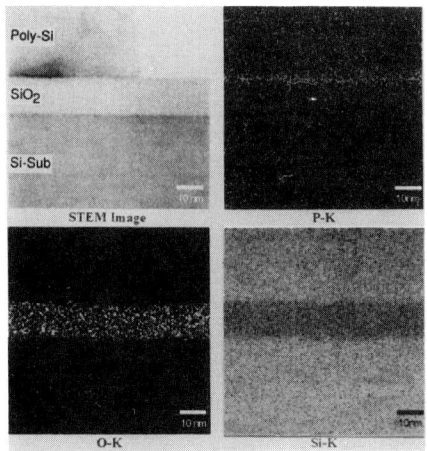

Fig. 3. Elemental mapping for Si, O and P.

It is showing results from a profile obtained with a 0.5 nm probe following beam shifting at 0.5 nm intervals perpendicular to the interface for both STD and RTA samples as well as Pattern-X and -Y. It is observed that the phosphorus segregation exceeds 3 % at the grain boundary whereas it is not detected at the area only 2 nm away from the interface, although the concentration condition are strongly dependent on the gate annealing process, and the morphology of the gate line. Nevertheless, the profiles clearly indicate the presence of P enrichment at the interface.

The semi-quantitative profile of phosphorus along the grain boundary is plotted in Fig.5, where the profile is continued to the gate silicon oxide. It is also observed the

578

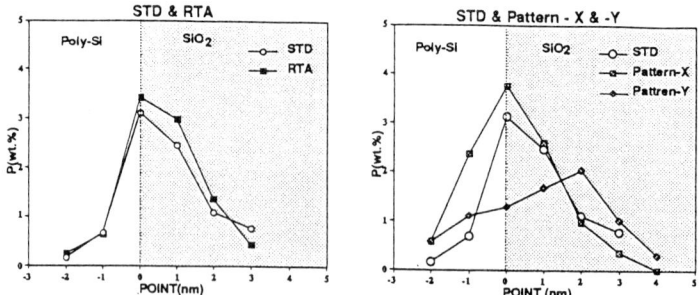

Fig. 4. Semi-quantitative EDX profile of phosphorus across the interface.

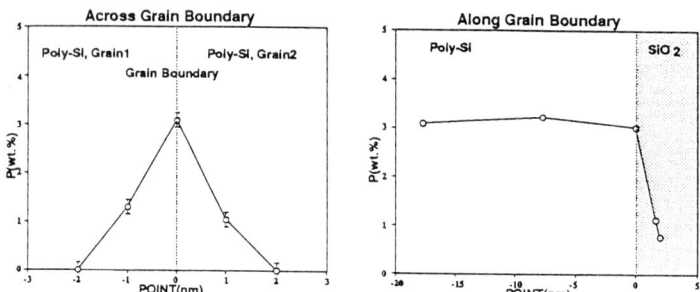

Fig. 5. Semi-quantitative EDX profile of phosphorus at the grain boundary.

phosphorus is accumulated in the silicon oxide near the grain boundary on the interface between Si and SiO_2. Although the boundary misorientation relative to the electron beam makes this kind of analysis very difficult, it is one of the key issues to understand the breakdown phenomena of the gate oxide. The FWHM of the profile reflects the current distribution in the probe coupled with a finite degree of beam broadening. The asymmetry of the profile is most likely to be due to a slight tilt of the interface relative to the electron beam. In light of the sub-nanometre scale phosphorus profile, it is clear that the phosphorus is present as an equilibrium segregant, localized within regions as large as a few nm.

3. SUMMARY

In conclusion, FEG-TEM analysis was applied to detect the dopant segregation at the interface between a poly-Si gate and a gate silicon oxide, as well as at the grain boundary in a poly-Si of the device equivalent 64M DRAM. Phosphorus segregation profiling at the interface between Si and SiO_2, across and along a grain boundary was obtained. Observations indicate the presence of Phosphorus enrichment at the interface. The phosphorus segregation exceeds 3 % at the interface and the grain boundary, whereas 2 nm away from the interface, segregation was below the detectable limit for the EDX analysis, although the concentration conditions are strongly dependent on the gate annealing process, and the morphology of the gate line.

REFERENCES
Baikie, I.D. and Bruggink, G.H. 1993 MRS symposium Proc. 309, 35.
Grovenor, C.R.M., et.al. 1984 Phil. Mag. A, 50, 409.
Hosokawa, F., et.al. 1993 Electron Microscopy and Analysis, Inst. phys. conf.Ser. 138, 531.
Maiti, B. et. al. 1993 MRS symposium Proc. 309, 3.
Perovic, V., et.al. 1993 J. of Nucler Materials, 199, 102.

Inst. Phys. Conf. Ser. No 146
Paper presented at Microsc. Semicond. Mater. Conf., Oxford, 20–23 March 1995

Electron diffraction contrast imaging as a tool for nano-range strain analysis and application to a semiconductor laser structure

K G F Janssens[1,2], J Vanhellemont[2], H E Maes[2], O Van der Biest[1] and R Hull[3]

[1]Department of Metallurgy and Materials Engineering (MTM), KULeuven, de Croylaan 2, B-3001 Leuven, Belgium
[2]IMEC, Kapeldreef 75, B-3001 Leuven, Belgium
[3]Department of Materials Science, Univ of Virginia, Charlottesville, VA 22903-2442, USA

ABSTRACT: One of the critical challenges when developing new deep submicron semiconductor technologies is understanding and controlling localized strain in devices. Electron Diffraction Contrast Imaging (EDCI) can provide a way to study strain with nanometre resolution. Interpretation of EDCI experiments requires substantial computer simulation, due to the non-straightforward correlation between EDCI observations and the strain field of the specimen. In this contribution, the EDCI technique is described including recently developed software to solve the EDCI simulation problem for arbitrary strain fields. A semiconductor laser structure is used as a case study.

1. INTRODUCTION

In the recent past strain has shown to be one of the important factors in mastering semiconductor technology. The impact of strain can be divided into two main areas. The first area of importance is *processing,* where strain influences e.g. growth characteristics, defect generation and diffusion processes. The second area of importance is impact of strain on *device characteristics during operation.* Strain influences e.g. electrical and optical properties of the composing materials and is one of the factors in understanding reliability of a device.

In the present contribution a technique is presented which is suited to *re-emerge* as a tool for characterization of localized strain fields on a micrometer to nanometer scale. The word re-emerge is used because more than 30 years have passed since electron diffraction contrast imaging (EDCI) was used for the first time by Ashby and Brown in 1963 to analyse the local strain field in and around small coherent precipitates. In later stages the same technique was used to identify straight dislocations by simulating the EDCI contrast resulting from the strain field of a dislocation and comparing it with experimental observations by Head et al. (1973). Since then the technique was developed further by a small number of researchers, most of whom programmed their own dedicated algorithms to solve the problem of EDCI image simulation for the particular problem they were studying at that time. Programming dedicated algorithms was necessary because the computation power was still limited.

2. SIMULATING EDCI IMAGES: SIMCON

2.1 Electron Diffraction Contrast Imaging

In view of the fact that computers have become much faster during the past decade, it was decided to develop an algorithm which can solve the EDCI image simulation problem for local strain fields of arbitrary geometry (Janssens et al., 1992). An overview of the EDCI procedure can be found in figure 1. On the left the experiment is depicted. Standard but state of the art TEM is used to record a *picture* of the strain field present in the thin film specimen. The information in the picture which relates to the strain in the specimen is the *contrast* in the EDCI image. The intensity of a particular pixel in the EDCI image depends on the orientation of the crystal lattice planes through which the electrons pass. If a localized strain field exists in the specimen the lattice plane orientation will locally change, and thus a contrast will be generated in the image. The right half of figure 1 depicts the simulation part of EDCI. Because of the dynamic aspects of high energy electron diffraction, the correlation between EDCI image contrast and the strain field in the specimen is not straightforward. The only way to overcome this is to simulate the EDCI process and compare the experimentally taken image with the simulated one. Based on this comparison the strain field in the specimen can be characterized.

The procedure of image simulation comprises two stages. In the first stage the strain field is modelled. For this purpose analytically formulated models as well as finite element formulated ones can be implemented. In the second stage the strain field model is used as input for EDCI simulation. Iteration of both stages is necessary to obtain a fitted image from which statements can be made regarding the characteristics of the strain field. The resolution of EDCI is limited to >1nm because of approximations made in the theory of dynamic electron diffraction.

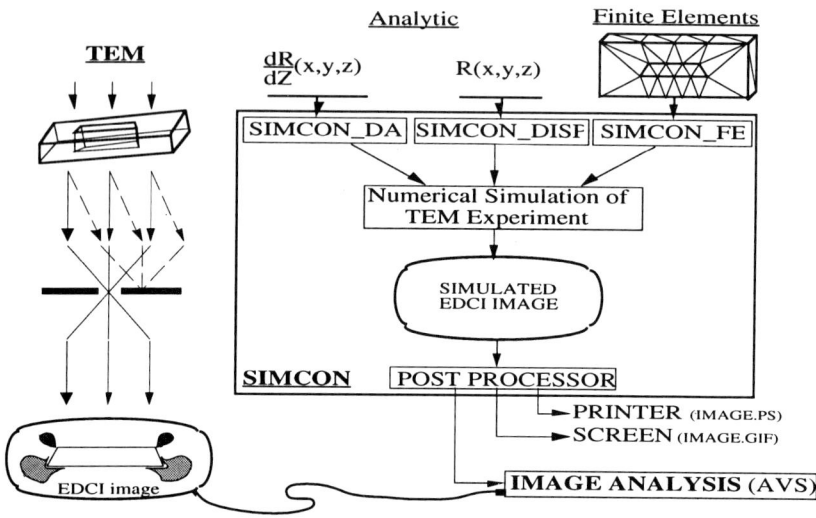

Figure 1:
Electron Diffraction Contrast Imaging (EDCI): Experiment (left) and Interpretation (right)

2.2 Functionality of SIMCON

The front end of SIMCON is an easy-to-use interface which gives the user access to several modular units. Each of these modules is closely related to a step in the simulation of an EDCI observation. The simulation of such an EDCI view is a three step process:
1. The first step is to come up with a mathematical model for the displacement field. In the most simple case such a model would exist of a set of analytic expressions for the strain field (i.e. the displacement field). In more complex cases one may want to use finite element modelling. The latter will also allow to take into account the fact that the strain field is relaxed at both TEM specimen surfaces, which becomes important when studying TEM cross sections.
2. The second step is simulation of the process of dynamic electron diffraction. Detailed description of the algorithms behind this step is beyond the purpose of this contribution. The user interface allows to enter different boundary conditions, such as diffraction parameters, image resolution setup, coordinate system related data and file logistics information.
3. Step three of the simulation process is analysis of the simulated images and comparison of the images with the experimentally recorded ones, and interpretation of the strain field.

3. STRAIN CHARACTERIZATION IN A SEMICONDUCTOR LASER STRUCTURE

Strain in semiconductor laser structures is known to influence the degradation as well as the performance of the devices (see Ueda (1988) and Hashimoto et al.(1993)). In the case study below it is demonstrated that the strain field in the structure can be characterized in a semi-quantitative manner. Further experiments necessary to fully quantify the analysis are given in the end of this paragraph.

3.1 Experiment

The semiconductor laser structures are thinned for examination in the transmission electron microscope using mechanical grinding followed by focused ion beam sputtering as is described by Hull et al. (1993). A schematic of the cross section can be found in figure 2a.

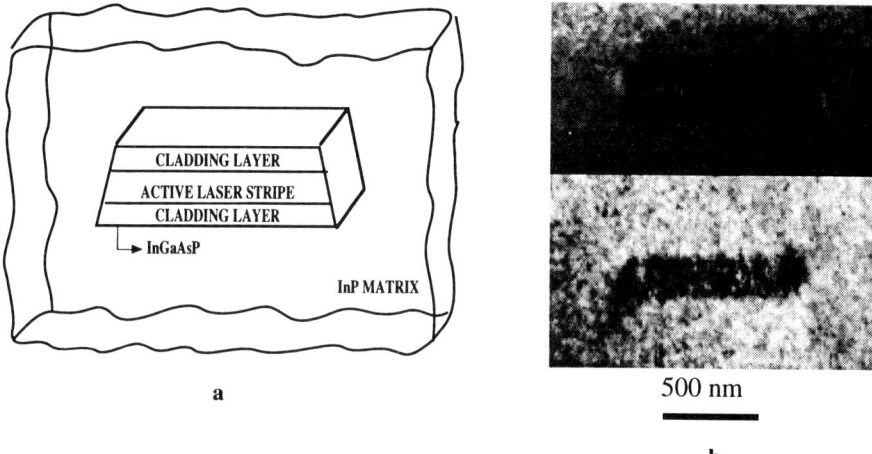

a

500 nm

b

Figure 2:
a) Cross section through the semiconductor laser structure. b) Experimentally obtained 200 kV EDCI observations of a cross section of the semiconductor laser structure, in both images Bragg deviation is about zero, diffraction vector [400] pointing upwards (left view) and [0$\bar{2}$2] pointing to the right (right view). Specimen thickness is 360nm.

Next 200 kV TEM was used to obtain EDCI images from these structures. In figure 2b two examples of such figures are shown. Absolute crystallographic indexation leads to the crystal vector [100] pointing upwards, [0$\bar{1}$1] pointing to the right and [011] normal to the image plane and pointing towards the reader.

3.2 Simulation

The strain field is modelled using the finite element method. The approach used was to cool the structure from processing temperature to room temperature, thereby introducing the thermal strain component in the model. In addition the difference in thermal expansion coefficients was enlarged to account for misfit strain between matrix and embedded layer. EDCI observations simulated based on this model are shown in figure 3.

Figure 3:
Simulated EDCI observations for the same diffraction conditions as in figure 2. See text for details.

In figure 3 the image on the left (diffraction vector [400]) is obtained when specimen surface relaxation is not taken into account. The image in the middle and the one on the right is obtained when specimen surface relaxation is taken into account (diffraction vector [400] and

[2$\bar{2}$0] respectively). Obviously this relaxation is the main source of contrast under these diffraction conditions.

The question arizes whether relaxation at both specimen surfaces is of such importance that strain values at the middle plane of the TEM specimen are much different from strain values under plane strain conditions, as is the case in the bulk situation (assuming infinite long laser stripes). Detailed analysis shows that for the particular boundary conditions in this case study, critical TEM specimen thickness is about 100nm, i.e. specimens should be thicker than 100nm for valid strain analysis. An indication of this can be found in figure 4: due to surface relaxation the magnitude of the shear stress τ_{xy} increases near the surfaces. The figure shows that the relaxation effect proliferates into the specimen to about 50nm at each side.

< Top Specimen Surface

Specimen thickness 500 nm

< Bottom Specimen Surface

Figure 4:
Strain relaxation at the TEM specimen surfaces. The magnitude of τ_{xy} is shown as a *view-through* (ray traced) the sample, looking in the direction bottom-to-top in figure 2a. Brighter grey level is equivalent to higher shear stress.

4. CONCLUSIONS

A technique has been presented to characterize localized strain with nanometer resolution. Interpretation of EDCI observations resulting from specimens with arbitrary strain fields requires extensive computer simulation, for which new software is presented. It has been shown that three dimensional analysis of the strain field (e.g. using finite element calculations) is necessary to explain the contrast in the EDCI observations of cross sections through electronic structures. It has also been shown that, for the boundary conditions in the case study, strain analysis is valid and may be quantified for TEM specimens thicker than 100nm.

ACKNOWLEDGEMENT

Koenraad G F Janssens is indebted to the Flemish IWT Foundation for his Fellowship. AT&T Bell Labs is acknowledge for provision of the samples.

M F Ashby and L M Brown, Phil. Mag 8(1963)1083.
J Hashimoto et al., IEEE J. Quantum Electronics 29(1993)1863.
A K Head et al., Computed Electron Micrographs and Defect Identification, North Holland (1973).
R Hull et al., Appl. Phys. Lett. 62(1993)3408.
K G F Janssens et al., Ultramicroscopy 45(1992)323.
O Ueda, J. Electrochem. Soc. 135(1988)11C.

Inst. Phys. Conf. Ser. No 146
Paper presented at Microsc. Semicond. Mater. Conf., Oxford, 20–23 March 1995
© 1995 IOP Publishing Ltd

Combined focused ion beam/electron microscopy investigation of laser diodes

A Dietzel,† A Jakubowicz,† and R F Broom‡

† IBM Research Division, Zurich Research Laboratory, 8803 Rüschlikon, Switzerland

‡ Dept of Materials Science and Metallurgy, Cambridge University, Pembroke Street, Cambridge CB2 3QZ, England

ABSTRACT: We used local focused ion beam (FIB) milling to expose cross-sectional facets and to fabricate thin membranes, with submicron precision, for SEM/EBIC and TEM investigations of strained InGaAs/AlGaAs quantum well ridge-type laser diodes. We analyzed the laser structure and mirror coating by conventional and high-resolution TEM, and studied intentionally generated defects in overstressed lasers. For this we applied EBIC imaging to select the defective areas for local FIB milling.

1. INTRODUCTION

In recent years, the focused ion beam (FIB) technique has been applied successfully to the failure analysis of microelectronic devices, particularly VLSI circuits (Nikawa 1991). This technique allows cross sections to be prepared quickly and with a high positioning accuracy of about 100 nm. Recently, its application was extended to the preparation of thin membranes for TEM investigations (Young *et al* 1990). The feasibility of locating the membranes precisely, and the absence of sputter-rate selectivity effects (due to a beam incidence angle of $\sim 0°$) make this technique enormously superior to conventional preparation techniques. As electron microscopy plays an essential role in the investigation of laser diodes, the application of FIB was a logical consequence (Hull *et al* 1993).

We applied FIB, SEM/EBIC and TEM to analyze the structure and mirror coating of 980-nm strained InGaAs/AlGaAs quantum well (QW) ridge-type laser diodes. We also studied lasers purposely damaged, using EBIC to select defective areas for local FIB milling.

2. INVESTIGATION OF LASER STRUCTURE

FIB milling. We used the preparation procedures illustrated in Fig. 1, which are similar to those used by Young *et al* (1990). Prior to milling, an ion-beam-assisted tungsten or platinum deposition over the object of interest is performed. This minimizes the ion damage, absorbs rounding of the upper edge and reduces shadow effects from metallization grains. In order to reduce the volumes to be sputtered, some of the laser chips were lapped until they formed slices about 100 μm thick. Trenches were cut inwards from two opposite chip/slice edges up to a few microns away from either side of the feature of interest using a high beam intensity (rough milling). Two parallel facets forming the boundaries of a thin membrane were then fabricated with reduced beam intensity (fine milling). We used the weakest available beam to inspect the membranes at 60° tilting. The membrane thickness, estimated by the Kossel–Moellenstedt fringe method, was typically about 200 nm.

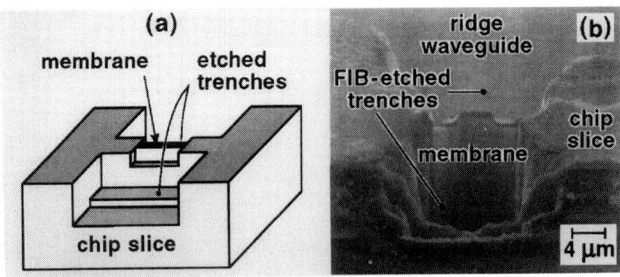
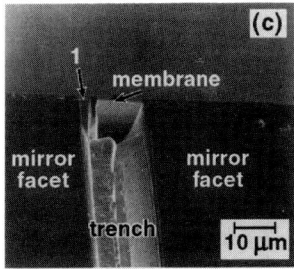

Figure 1. FIB preparation technique: (a) schematic of membrane preparation; (b) SEM view of a thin membrane prepared by FIB; (c) mirror cross section in the QW plane. Initially the mirror coating was milled off, as indicated by arrow 1, to locate the QW.

Mirror coatings. TEM examination of multilayer coatings at the ends of the laser cavity yields information on the thickness and uniformity of the layers, and, importantly, details of the interface between dielectric and semiconductor, where strain and crystal defects can lead to degradation of the laser's output. For the mirror cross section, the FIB was aligned perpendicular to the 90% reflecting mirror surface, and a trench was cut on either side of the plane of the active region to form a membrane enclosing the InGaAs QW beneath the ridge, as shown in the SEM photograph Fig. 1c. A bright-field TEM image of such a cross section is shown in Fig. 2. It illustrates two unique features of the FIB technique: First, owing to the very shallow milling angle ($\leq 1°$), the thickness of the membrane is very uniform despite the large differences in milling rates of the layers, 6:1 for GaAs:TiO$_2$. Second, the membrane is precisely located at the QW beneath the ridge. Neither of these conditions can be met with conventional lapping and ion milling. Other multilayer dielectric combinations, such as Al$_2$O$_3$/Si/... or Si$_3$N$_4$/Si/... yield similar well-resolved results.

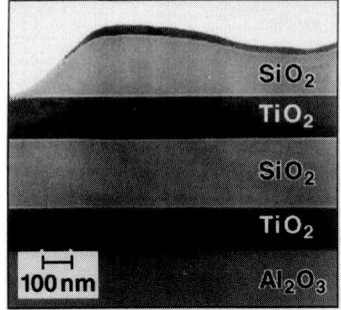

Figure 2. Bright-field TEM image of a multilayer dielectric mirror.

High-resolution imaging. To obtain sufficiently thin regions for lattice fringe imaging, wedge-shaped membranes were etched as shown in Fig. 3a. Figure 3b is a bright-field TEM image of such a membrane showing the laser's vertical structure within the ridge waveguide. As for the preparation damage, one can see two regions. There is an amorphous region at the wedge's edge, which, where strain is present, extends towards the thicker part of the wedge (around the strained QW, see arrow 2, and also below the metallization visible at higher magnification). The amorphous region is adjacent to a disturbed crystalline region (arrow 1), which appears only in the ternary epitaxial layers, but not in the GaAs (arrow 3). From the width of the amorphous and disturbed regions, the surface damage is found to penetrate less than 10 nm into the material (using EBIC, we also saw a conversion from n- to p-type in a thin layer adjacent to the treated surface within the Si-doped lower cladding). Figure 3c shows a high-resolution image of a region around the QW, far enough from the wedge's edge, where the image is dominated by the nondisturbed crystal. The lattice contrast is weaker than in conventional samples owing to background 'noise' from damaged surfaces.

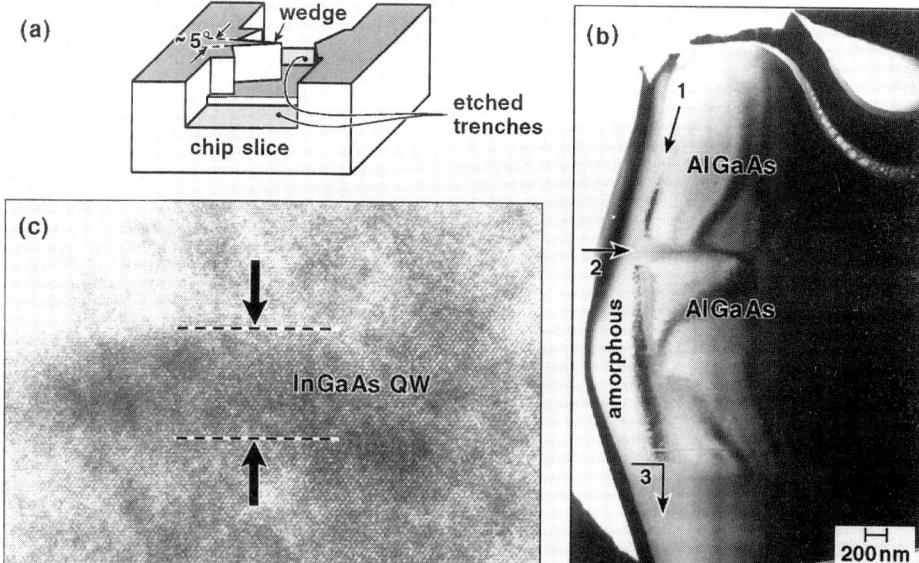

Figure 3. High-resolution TEM using FIB preparation. (a) Schematic of the wedge preparation. (b) Wedge-shaped membrane in bright-field TEM image shows the vertical structure of the waveguide, the amorphization areas, and lattice distortion (arrow 1). (c) Lattice fringe image of the QW region.

3. STUDIES OF DAMAGE IN LASERS

In order to be able to demonstrate the above method for studying defects in laser diodes, we have intentionally damaged laser by severe overstress. We then localized the damaged regions by EBIC-imaging the lasers' p-n junction plane. Figure 4(a) shows such an image for the case of a device that was damaged. The reduced EBIC signal area (RESA), indicated by the arrow, represents a region modified by laser overstress. Figures 4(b) and (c) show cross-sectional TEM images of the region indicated by the arrow in (a), Dislocation loops and small strain fields are visible. These defects are clearly related to the failure, since none are visible in fresh lasers. The defects are confined to the active region, and we observed no preferential growth orientation. From their overall character we concluded that they grew within the active region.

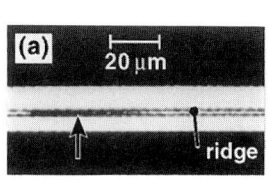
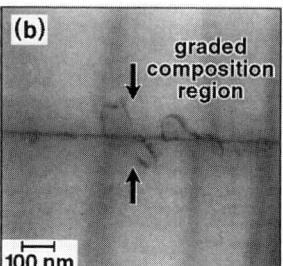
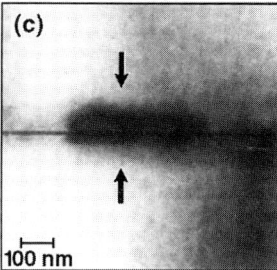

Figure 4. (a) Top view EBIC image of the ridge waveguide of an overstress-damaged laser. (b) TEM bright-field image of small dislocation loops and (c) of strain fields around the QW in the region of a RESA in an overstressed laser.

586

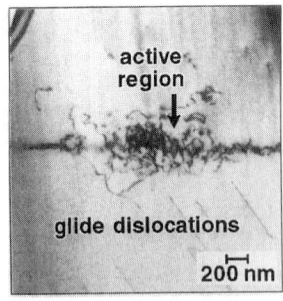

Figure 5. TEM bright-field image of a cross section through a RESA adjacent to the mirror after catastrophic facet damage. It shows a tangle of dislocations around the QW and extended glide dislocations on {111} glide planes.

Figure 5 illustrates the case of catastrophic facet damage due to package-induced failure (PIF, Sharps 1994) (the TEM membrane was etched 1 μm behind the mirror facet). In contrast to the previous overstress damage, the dislocation tangle extends far beyond the active region. In addition, glide dislocations on {111} planes are present. This result and the fact that no damage on the facet is visible in the SEM is in contrast to the conventional catastrophic mirror damage (COMD). Figures 4 and 5 evidently indicate different microscopic damage mechanisms.

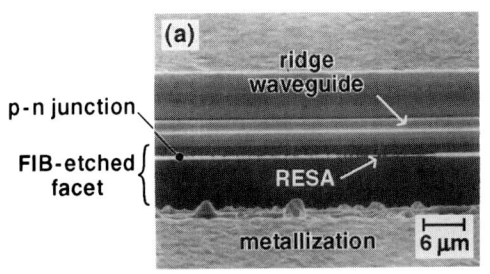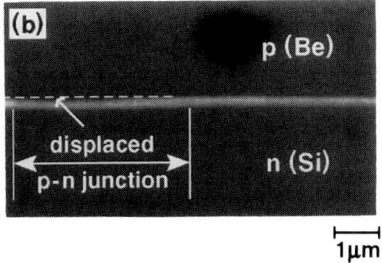

Figure 6. Cross-sectional SEM/EBIC images (a) of a RESA in an overstress-damaged laser and (b) of a downward shift of the *p-n* junction.

Figure 6 shows two EBIC images of cross sections along the waveguide, prepared by FIB in devices that were damaged in overstress. In Fig. 6a the defective region appears dark owing to loss of EBIC signal. The advantage of cross-sectional EBIC imaging is that lower accelerating voltages can be used, which improves the lateral resolution. In addition, artifacts caused by the device geometry/topography can be eliminated. Figure 6b shows a shift of the *p-n* junction towards the lower *n*-type cladding. We attributed this shift to Be diffusion caused by severe overstress (Be is the *p*-type dopant of the upper cladding).

We gratefully acknowledge the valuable contribution of John F Walker, FEI Europe Ltd, for fabricating a number of the samples used in this study.

References

Hull R *et al* 1993 *Inst. Phys. Conf. Ser.* **134** 259
Nikawa K 1991 *J. Vac. Sci. Technol.* **9**(5) 2566
Sharps J A 1994 *Proc. LEOS '94 Conf., Boston, 1994* (IEEE) Vol 2, p 35
Young R J *et al* 1990 *Mat. Res. Symp. Proc.* **199** (Pittsburgh: Materials Research Society) p 205

Inst. Phys. Conf. Ser. No 146
Paper presented at Microsc. Semicond. Mater. Conf., Oxford, 20–23 March 1995
© *1995 IOP Publishing Ltd*

Heating and damage of InGaAs/GaAs/AlGaAs laser facets

I Rechenberg[1], A Höpner[2], J Maege[1], A Klein[1], G Beister[1] and M Weyers[1]

[1] Ferdinand-Braun-Institut für Höchstfrequenztechnik Berlin,
   Rudower Chaussee 5, D - 12489 Berlin, Germany
[2] Max-Planck-Institut für Festkörperforschung, D - 70569 Stuttgart, Germany

ABSTRACT: The temperature rise of coated laser facets during operation has been investigated by means of cathodoluminescence (CL). The peak shift of the signals from confinement and cladding layers indicates heating of the front facet by up to 300 K for ridge waveguide (RW) laser diodes driven at an output power of 150 mW. Contrast changes in transmission electron microscopy (TEM) images indicate that the high temperature in the laser spot on the facet can result in interdiffusion of aluminium and indium.

1. INTRODUCTION

InGaAs/GaAs/AlGaAs strained quantum well (QW) laser diodes emitting at 980 nm have been widely studied due to their suitability for pumping of Erbium^{3+}- doped fiber amplifiers in fiber communication systems. In AlGaAs-based laser diodes optical damage of the laser facet is one of the most important failure mechanisms. Although lifetimes of more than 10^5 h under operating conditions can be extrapolated from accelerated lifetime tests of our lasers, sudden failure after stable operation over extended periods still presents a problem. This sudden failure is caused by catastrophic optical degradation (COD) due to facet heating by nonradiative recombination processes of photo- and current induced carriers (Fukuda et al 1994, Eliseev 1995).

We have investigated the temperature increase on facets of 980 nm laser diodes by means of spectrally resolved CL in dependence on operation current. Laser facets on which damage has been observed by SEM were investigated by TEM.

2. EXPERIMENTAL

Separate confinement heterostructure (SCH) laser diodes were prepared on (100) n-type GaAs substrates by MOVPE. The structure consists of a 7 nm $In_xGa_{1-x}As$ QW (x = 0.18) sandwiched between GaAs spacer layers and $Al_yGa_{1-y}As$ confinement and cladding layers (sample type A: double QW, with 10 nm barrier and 15 nm spacer layers, 500 nm $Al_{0.24}Ga_{0.76}As$, 1000 nm $Al_{0.29}Ga_{0.71}As$; sample type B: single QW, 22 nm spacer layers, 150 nm $Al_{0.2}Ga_{0.8}As$, 1200 nm $Al_{0.5}Ga_{0.5}As$). Doping concentrations were 1×10^{17} cm^{-3} for confinement and 5×10^{17} cm^{-3} for cladding layers (dopant elements were zinc for p-type and silicon for n-type) (Bugge et al 1994). The laser facets were coated with antireflecting films on one side and highly reflective dielectric mirrors (Al_2O_3/TiO_2) on the other side.

The laser diode was operated in a SEM (Fig. 1). The submount with soldered laser

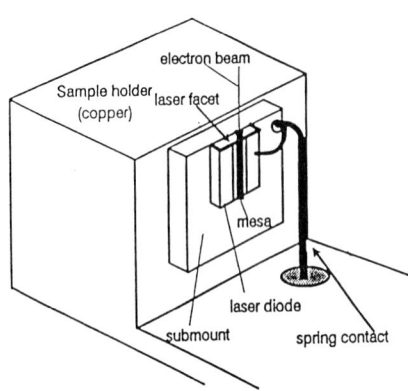

diode was attached to the cooled sample holder (90 K). The electron beam hits perpendicular to the laser facet at an accelerating voltage of 25 kV and a sample current of 100 nA. The CL spectra are recorded by a combination of plane mirror, spheric and aspheric lenses together with a grating monochromator. Laser emission is suppressed by a short pass filter (> 800 nm). CL spectra from confinement and cladding layers were recorded in dependence on laser operation current. From the peak shift for different operation currents J in relation to the peak position at J = 0 the temperature increase was evaluated by means of the Varshni equation (Adachi 1993).

Fig. 1 Schematic of laser diode mounted on SEM sample holder

Laser diodes showing facet damage after stress tests at 40 °C were prepared for TEM investigations by mechanical preparation (Klein et al 1994) followed by focussed ion beam thinning. The TEM investigations of the facet region were carried out in two microscopes operating at 400 and 1000 kV.

3. RESULTS

3.1 Facet heating

Fig. 2 shows spectra taken on a laser diode of type A. Curve 1 is the CL spectrum of the cladding and confinement layers without laser operation (driving current J = 0). Curve 2 shows the spectrum obtained with the laser diode switched on (J = 150 mA). When the laser diode is switched on but the electron beam switched off spectrum 3 is recorded. The peak around 680 nm stems from electroluminescence (EL) from the confinement layers. This contribution has to be subtracted to yield the CL spectrum under laser operation (curve 4). The peak shift between curves 1 and 4 gives the temperature increase in the confinement ($\Delta\lambda_2$) and cladding layers ($\Delta\lambda_1$). Fig. 3 shows the temperature increase with increasing operation current. At high current (230 mA) a temperature increase of up to 300 K in the confinement layers together with a strong temperature gradient to the cladding layers is observed.

3.2 Facet damage

Laser diodes showing a sudden increase in the current necessary to obtain a fixed output power in lifetime tests at higher temperature were investigated by different SEM modes. Especially images produced with backscattered electrons show a strong (bright) contrast in the laser light emission region. In TEM images of this region noticeable changes in layer contrast were observed. Fig. 4 shows the TEM image of the facet region of a damaged laser of type B. The contrast between the GaAs spacer layers adjacent to the InGaAs QW and the AlGaAs confinement layers disappears indicating interdiffusion of aluminium.

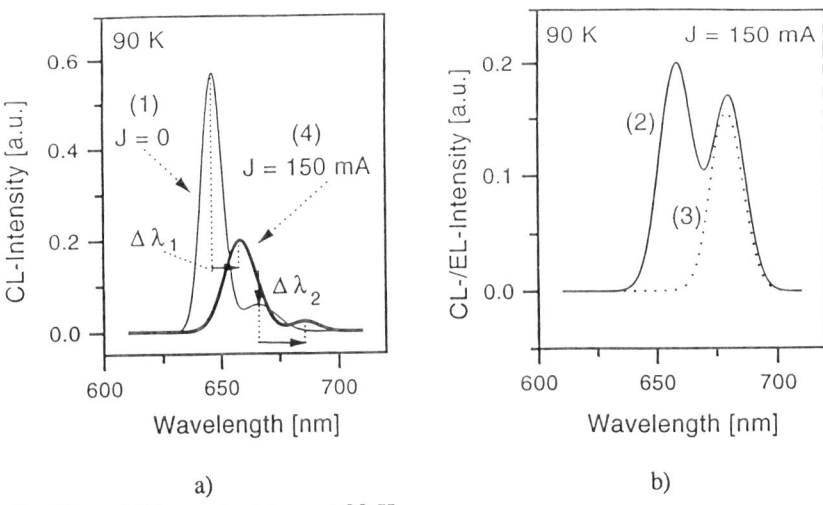

a)   b)

Fig. 2   CL and EL spectra taken at 90 K
a) curve 1: CL spectrum with laser diode switched off (J = 0);
   curve 4: CL spectrum with laser on (J = 150 mA; EL (curve 3) subtracted);
b) curve 2: full CL spectrum (J = 150 mA);
   curve 3: EL spectrum (J = 150 mA), electron beam switched off.

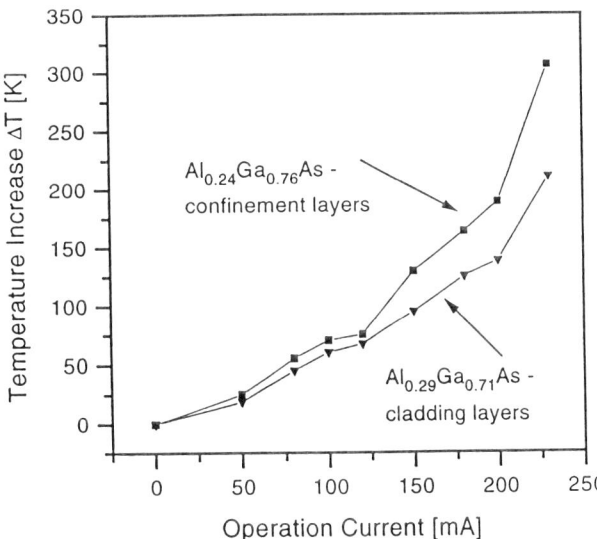

Fig. 3   Temperature increase in dependence on the operation current of the laser diode

The position of the InGaAs QW appears to be displaced towards the p-side by about 5 nm. The bright lines of lenticular form in Fig. 4 resemble the near field intensity distribution calculated for this structure. Here a temperature gradient exists which causes a high stress field. Dislocation contrast as a result of this stress field is observed especially in the two corners of the lens. In this region also the contrast of the QW vanishes indicating interdiffusion of indium.

4. CONCLUSIONS

From the peak shift of the CL spectra of the waveguide and cladding layers in a SCH laser diode the temperature increase on the laser facet in dependence on operation current can be determined. The lateral resolution of CL also allows an estimation of the temperature distribution on the facet. This CL method is suited to study heating behaviour and the effectiveness of facet coating processes. The facet damage caused by the heating during laser operation includes both interdiffusion of aluminium and indium (early stage of COD) and defect generation (later stage of COD) in the laser light emission region.

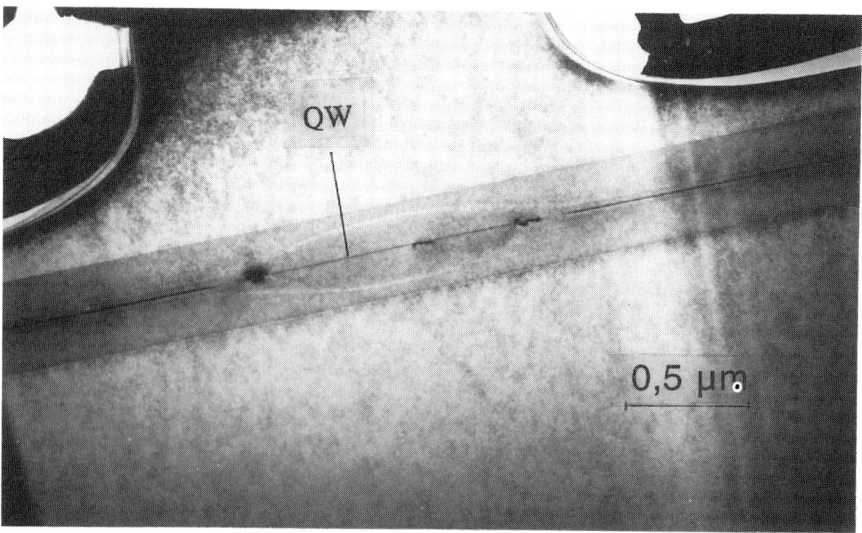

Fig. 4 TEM image of damaged laser facet region. The bright lines indicate the light emission region.

REFERENCES

Adachi S 1993 Properties of Aluminium Gallium Arsenide,
    emis DATAREVIEWS SERIES No. 7,Chapt. 4.6, 79
Bugge F, Beister G, Erbert G, Gramlich S, Rechenberg I, Treptow H and Weyers M
    1994 J. Cryst. Growth 145 907
Eliseev P G 1995 to be published
Fukuda M, Okayasu M, Temmyo J and   Nakano J
    1994 IEEE J. Quantum Electron. 30 471
Klein A, Rechenberg I, Richter U and Gesatzke W
    1994 to be published Proc. EDO Saarbrücken

Inst. Phys. Conf. Ser. No 146
Paper presented at Microsc. Semicond. Mater. Conf., Oxford, 20–23 March 1995

Characterization of integrated GaInAsP laser/modulators grown by selective and non-planar epitaxy

R E Mallard, N Puetz, C J Miner, J Zorzi, D Adams, M Cleroux, G Hillier, and R Moore

Bell-Northern Research, P O Box 3511, Station C, Ottawa, Ontario, Canada, K1Y 4H7

ABSTRACT: We report on the characterisation of a GaInAsP integrated laser/phase modulator structure grown by selective Metal Organic Chemical Vapour Deposition (MOCVD). Energy dispersive X-ray analysis in the transmission electron microscope shows that selectively grown quaternary material is rich in P and In by approximately 7 atomic percent in regions within 1μm of the edge of the mask. The compositional variation is sufficiently large as to induce the generation of interfacial misfit dislocations. This spatial distribution of defects correlates with a diminution of intensity and increase in wavelength of spatially resolved photoluminescence. Selective MOCVD growth is compared to a growth process in which the dielectric mask is removed prior to the second stage growth, which offers improved control of the defect density at the laser/modulator interface.

1. INTRODUCTION

The cost efficient fabrication of present and next generation prototypes of III-V semiconductor optoelectronic circuits requires the continuous development of epitaxial growth techniques for the deposition of increasingly laterally complex structures. The trend in device development is towards higher levels of integration, for which differing device structures must be deposited on adjacent discrete regions of the same substrate. The selective Metal Organic Chemical Vapour Deposition (MOCVD) process typically involves growth through a patterned dielectric layer on the surface of the substrate, which serves as a mask for selective area etching and also as a means of locally preventing growth during multiple epitaxial growth stages. This paper describes a transmission electron microscope (TEM) and spatially-resolved Photoluminescence (sPL) investigation into the microstructure and composition of an integrated GaInAsP/InP laser/phase modulator which is fabricated by selective MOCVD.

One particular requirement for efficient light coupling and reliable operation of such a device is that the junction region between the laser and modulator be free from extended defects. However, the gas flow dynamics and surface kinetics which are of critical importance in the MOCVD growth process are perturbed by the presence of the dielectric mask. This can result in the formation of regions of chemical inhomogeneity and strain in the epitaxially deposited material. Gibbon et al (1993) have described the defect structure and compositional variations in GaInAs which was selectively grown on InP. They report an enrichment of In in the ternary alloy, which diminishes with increasing distance from the mask edge, to the point at which the material becomes Ga rich, before eventually tending towards the unperturbed, InP lattice matched composition. These compositional variations are ascribed to the differences in lateral gas phase diffusion rates of the Ga and In containing precursors over the masked regions of the substrate, as well as to their respective reaction rates in exposed regions of the masked substrate. The lateral distance over which these variations can occur is on the order of hundreds of μm, and depend strongly on the size, orientation and areal fraction of the dielectric masked areas, as well the MOCVD growth conditions employed.

The effects are much less pronounced when GaInAsP selective overgrowth is performed. The same work also investigated the compositional variations of selectively grown GaInAsP. An enrichment of In immediately adjacent to the mask was again observed, but to a much smaller degree than was the case in the ternary growth. In addition, the group V compositional profiles are uniform to within approximately 1μm of the mask edge. Thompson et al (1992) have used sPL to characterise selectively grown GaInAsP, and found that the photoluminescence of the material is uniform to within less than 2μm of the mask edge. The differences in the behaviour between the ternary and quaternary systems suggests that the presence of P in the MOCVD reactor changes the growth dynamics in such a way that the excursion in composition of the deposited material from the nominal lattice matched value is smaller than that which is obtained if ternary material is grown.

2. EXPERIMENTAL

Integrated laser/phase modulator test structures were grown by selective MOCVD. Briefly, this structure was fabricated by a first growth of a distributed feedback multiple quantum well loss-coupled laser which consists of a double quantum well of GaInAs and a 4 period GaInAsP multiple quantum well. This is followed by dielectric masking and wet chemical etching, which is used to define the shape of the laser regions on the device. The mask consists of a series of 0.5mm wide lines of 30nm thick SiO_2, which are separated by 1mm. The modulator, which consists of an InP buffer layer and a 20 period $Ga_{.4}In_{.6}As_{.85}P_{.15}$ ($\lambda=1.51\mu m$)/InP superlattice, with nominal thicknesses of 11nm and 9nm respectively, is then deposited during a second growth onto the masked substrate. This structure is compared to that of another wafer, having an identical etched laser substructure, and regrown in the same second run, but having the oxide mask removed prior to overgrowth. Epitaxial deposition was carried out in a low pressure MOCVD reactor of inverted horizontal geometry. The wafers were rotated during growth, to achieve improved uniformity in film thickness and composition (typically ±1.5% thickness variation and ±1.5nm photoluminescence variation). The operating pressure was 150mTorr and the growth temperature was 610°C. Growth rates for InP and $Ga_{.4}In_{.6}As_{.85}P_{.15}$ were 1.65μm/h and 2.4μm/h respectively. Prior to the regrowth, the patterned wafer surfaces were treated in a solution of 1 HBr:4 HNO_3:400 H_2O, in order to remove any process damage and surface contamination. Photoluminescence analysis of the samples was performed using a Waterloo Scientific SPM200. SPL maps were generated using a YLF laser ($\lambda=1046nm$) focused to a 20μm diameter spot, which is rastered across the sample. Maps of intensity at constant wavelength (λ), peak λ, and λ at the upper half-maximum of the PL peak ($\lambda_{1/2\ max}$) were acquired. The latter is a more reliable indicator of the lowest bandgap transition in cases where there is some degree of interfacial roughness. The interfacial region between the laser and selectively grown modulator was studied in the JEOL 2000FX transmission electron microscope (TEM). The selectively grown modulator material was analysed by energy dispersive X-ray analysis (EDX) in the TEM with a 20nm diameter electron probe.

3. RESULTS AND DISCUSSION

A threefold increase in thickness of the selectively grown modulator material is observed at the edges of the oxide mask. This local growth rate enhancement is due to the arrival of an excess flux of reactant gases from over the masked regions. The degree of growth rate enhancement decreases with increasing distance from the mask edge, and a 10% enhancement persists as far as 1000μm from the mask edge. The variations in epilayer thickness are accompanied by variations in the epilayer composition and strain. Misfit dislocations are observed in selectively grown material when the strain level due to compositional perturbation surpasses a critical value (Mallard et al 1993). Plan view TEM analysis reveals the presence of a dense tangle of predominantly 60° dislocations originating at the mask edge. The dislocations propagate up to 50μm from the mask edge, before they thread upwards and terminate at the surface of the epilayers. The areal density of the dislocations drops rapidly with distance from the mask edge.

These same features are observed in the TEM cross section of the selectively grown laser/modulator in fig. 1a. Near the oxide mask, the modulator quaternary and InP layer thicknesses are 35nm and 30nm respectively, illustrating the large amount of growth rate enhancement which occurs. A high density of dislocations is observed at the edges of the modulator, particularly in the regions where the growth is non-planar. Misfit dislocations are predominantly observed at the superlattice/buffer layer interface, although a number of dislocations also lie on GaInAsP/InP interfaces within the modulator superlattice, and other dislocation segments thread through the epilayers and terminate at the wafer surface. In contrast, there is little or no growth rate enhancement of the modulator in the unmasked sample shown in fig. 1b. Misfit dislocations due to the growth perturbation at the step edge are not observed. The presence of the defects at the base of the etched laser sidewall is related to surface preparation, which has not been optimised for the growth conditions employed.

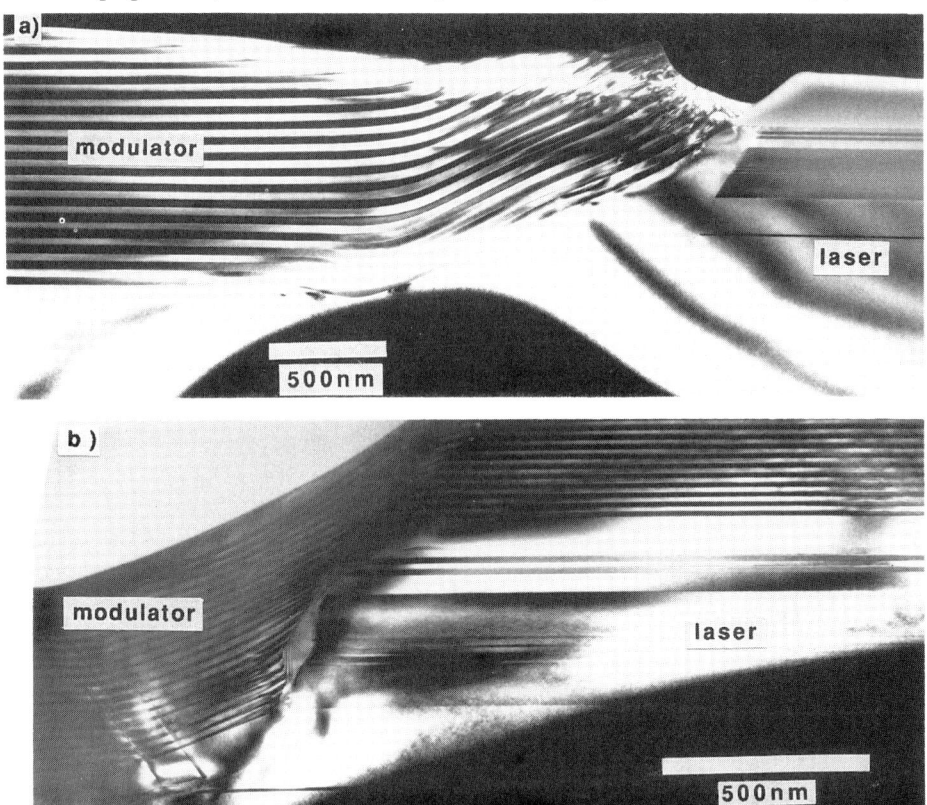

Fig. 1. (200) dark field TEM micrographs of (a) selectively grown laser/modulator interfacial region and (b) equivalent structure with the oxide mask removed prior to overgrowth.

The EDX linescans in fig. 2a show that the GaInAsP wells in the selectively grown modulator are rich in In and P by 7 atomic % within 1μm of the edge of the oxide mask. Fig. 2b shows the corresponding variation in the InP composition, which is negligible because the InP growth rate enhancement depends only upon the diffusion rate of one species (In) across the mask. SPL analysis shows that there is a gradual increase in peak luminescence wavelength of approximately 20nm from the edge to the centre of the selectively grown modulator material over a lateral scale of 500μm. This increase is accompanied by a drop in luminescence intensity. In contrast, the PL wavelength at $\lambda_{1/2 \text{ max}}$ exhibits a decrease in wavelength of 20-25nm within 200μm from the mask edge. This apparent discrepancy is due to the variation in PL peak width, which is large at the edge of the masked regions, and drops with increasing distance from the mask edge. We attribute the change in peak width to local variations in defect density. As indicated by the presence of misfit dislocations as far as 50μm from the mask edge, it is likely that the compositional perturbations due to the selective growth process continue beyond the lateral field of our EDX measurements. However, these

594

composition variations may not necessarily induce a change in luminescence wavelength, because the combined effects of enrichment in In and P move the bandgap of the GaInAsP roughly along an iso-energy contour. The increase in $\lambda_{1/2 \text{ max}}$ near the mask edge is on the order of that expected due to a decrease in quantum confinement caused by the observed increase in quantum well thickness. No sPL variations are observed in the case of the sample which had the oxide mask removed prior to overgrowth.

Although the measured variations in modulator composition do not induce a large change in luminescence wavelength, they do induce considerable lattice strain. The EDX data indicate that the lattice mismatch in the quaternary at the edge of the oxide masks is on the order of 0.8% (compressive). Lattice relaxation of individual quaternary layers via misfit dislocations is energetically favourable at a critical thickness equal to approximately that of the GaInAsP layers observed at the edges of the oxide mask. However, for the 20 period superlattice in the present study, a more stringent relaxation criterion is for the superlattice as a whole, relative to the InP substrate. The modulator thickness of 1200nm at the mask edge is easily sufficient to induce relaxation of the superlattice relative to the substrate, and to account for our observation of misfit dislocations at the superlattice/buffer layer interface.

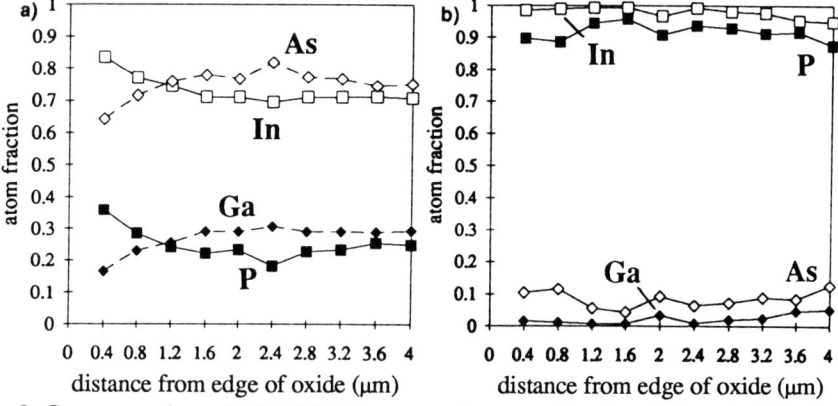

Fig. 2. Convergent beam TEM-EDX linescans of compositional variations in (a) quaternary, and (b) InP selectively grown modulator material as a function of distance from the oxide mask edge. The measurements are from the first modulator period above the InP buffer layer.

5. CONCLUSIONS

EDX analysis of selectively grown GaInAsP reveals a small enrichment of both In and P within 1μm of the edges of the oxide mask. SPL measurements do not reveal a strong compositionally dependent wavelength variation, but nevertheless indicate an increase in luminescence wavelength with increasing proximity to the mask edge, which is consistent with the observed degree of growth rate enhancement in that region. The lattice misfit resulting from the compositional changes in the selectively grown modulator material is sufficient to induce lattice relaxation through the introduction of misfit dislocations. The growth of an equivalent structure using non-planar epitaxy, where the oxide mask is removed prior to overgrowth appears to offer improved control over composition and luminescence properties. Using the non-planar growth technique, we have successfully fabricated integrated laser/modulators with high coupling efficiencies.

REFERENCES

Gibbon M, Stagg J P, Cureton C G, Thrush E J, Jones C J, Mallard R E, Pritchard R E, Collis N and Chew A 1993 Semicon Sci Tech **8** 998
Mallard R E, Thrush E J, Gibbon M A and Booker G R 1993 Materials Sci Eng **B20** 48
Bassignana I C, Miner C J and Puetz N 1989 J Appl Phys **65** 4299
Thompson J, Wood A K, Carr N, Charles P M, Moseley A J, Pritchard R, Hamilton B, Chew A, Sykes D E and Seong T-Y 1992 J Cryst Growth **124** 227

Inst. Phys. Conf. Ser. No 146
Paper presented at Microsc. Semicond. Mater. Conf., Oxford, 20–23 March 1995
© 1995 IOP Publishing Ltd

Electron holography of p-n junctions

B G Frost, D C Joy, L F Allard and E Völkl

HTML, Oak Ridge National Laboratory, Oak Ridge, TN 37831-6064, USA   and
Walters Life Science Bldg, University of Tennessee, Knoxville, TN 37996-0810, USA

ABSTRACT: The phase of an electron wave modulated by the electric field in and around p-n junctions is investigated by electron holography. Hologram fringes indicate positively and negatively charged areas in a sample by the direction of the fringe bending. Different reconstructions of phase distributions allow us to analyze the potential distribution and to display the electric field as well. Since the phase distribution also depends on the alignment of the microscope it is difficult to evaluate the precise voltage at a junction.

1. INTRODUCTION

Semiconductor devices have been investigated at $100\,kV$ by Frabboni et al (1985) using image plane off-axis electron holography. Utilizing a Hitachi HF-2000 field emission electron microscope at $200\,kV$ we can coherently image even thicker samples. With the objective lens switched off we detect for the first time the potential distribution and electric field lines both inside and outside a p-n junction in silicon.

2. PRINCIPLES OF ELECTRON HOLOGRAPHY

A plane electron wave $\exp(i\vec{k}\vec{r})$ modulated by a sample is given by

$$o(\vec{r}) = a(\vec{r})e^{i\varphi(\vec{r})}. \tag{1}$$

where $a(\vec{r})$ and $\varphi(\vec{r})$ is the amplitude distribution and the phase distribution respectively.

When taking a hologram the specimen covers only one half of the object plane, while the other half remains empty to provide a plane reference wave (Fig.1). Superimposing the object wave on the reference wave using a charged electron biprism and imaging onto the image plane results in an interference pattern, a hologram. Its intensity distribution is given by

$$I(\vec{r}) = 1 + a^2(\vec{r}) + 2a(\vec{r})\cos[2\pi\vec{q_c}\vec{r} + \varphi(\vec{r})], \tag{2}$$

The carrier frequency q_c is defined by the spacing s of the hologram fringes to $q_c = 1/s$. Amplitude modulation of the image wave causes a contrast variation of the interference fringes of the hologram, and phase modulation causes a corresponding fringe displacement. When taking holograms following the setup shown in Fig.1 the fringes are bent away from an electrically positive sample, and are bent towards a negative sample (Fig.2).

The numerical reconstruction of the object wave from a hologram is staightforward: Evaluating the Fourier transform of the hologram represented by eq.(2) results in

$$
\begin{aligned}
FT(I(\vec{r})) \ = \ & \delta(q) + FT(a^2(\vec{r})) \\
& + FT(a(\vec{r})\exp[i\varphi(\vec{r})]) \otimes \delta(\vec{q_c} + \vec{q}) \\
& + FT(a(\vec{r})\exp[-i\varphi(\vec{r})]) \otimes \delta(\vec{q_c} - \vec{q})
\end{aligned}
\tag{3}
$$

where the first row shows the zero order of diffraction plus the autocorrelation and the second and third row show the two sidebands representing the Fourier transform of the image wave. If the carrier frequency q_c is at least three times as big as the highest spatial frequency of the object then the distance of the sidebands to the autocorrelation is sufficient to separate one sideband from the rest of the diffraction image without loss of information on the structure of the object. After separation and centering of one sideband the object wave is obtained in its amplitude and its phase by an inverse Fourier transformation.

In our laboratory the numerical reconstruction of the holograms aquired by a 1024 by 1024 slow-scan CCD-camera is performed by the software "HoloWorks" (Gatan Inc.) on a Macintosh Quadra 950 computer fitted with an array processor.

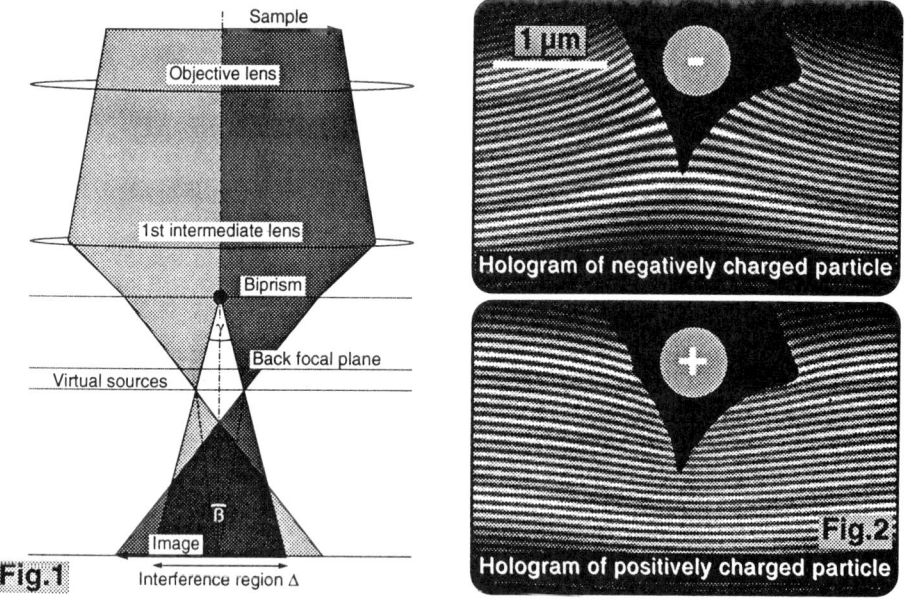

Fig.1: Setup for taking image plane off-axis electron holograms. The Möllenstedt-type biprism is between the first and second intermediate lenses.

Fig.2: Hologram fringes are bent towards a negatively charged particle (top) and away from a positively charged particle (bottom).

3. PHASE SHIFT CAUSED BY ELECTRIC POTENTIALS

The phase shift $\delta\varphi$ of an electron wave caused by the electric potential V of the sample and by a magnetic vector potential $\vec{A}$ was evaluated by Aharonov and Bohm (1959) and is given by

$$\delta\varphi = \frac{e}{\hbar}\int(V dt - \vec{A}(\vec{r})d\vec{r}) \tag{3}$$

where e is the electron charge, h is Planck's constant and the path of integration goes over any closed circuit in space-time. Neglecting any magnetic influence the phase is modulated only by the electric potential which comprises the mean inner potential V_{in} of the sample and the potential due to electric fields e.g. charging of the sample in the electron beam or fields in and around p-n junctions. The phase shift caused by the mean inner potential is given by

$$\delta\varphi(x,y) = \frac{e}{\hbar}\frac{1}{v}\int_{z_1(x,y)}^{z_2(x,y)} V_{in}(x,y,z')dz' \tag{4}$$

where $z_1(x,y) - z_2(x,y)$ is the thickenss of the sample at (x,y). Though it is in principial possible to separate the phase caused by the inner potential from the phase caused by the field distribution, as shown by Gajdardziska-Josifovska and McCartney (1993), it is still difficult to evaluate the correct number of electric charges from a hologram because in our experience the phase strongly depends on the alignment of the microscope (to be published).

The measured phase lines are contour lines of the potential projected onto the image plane. In a first approximation the electric field lines are perpendicular to the phase lines. However, inside the sample strong thickness variations spoil the orthogonality and the electric field lines become more difficult to plot.

3.1. Phase Shift Caused by p-n Junctions

Following Ashcroft and Mermin (1976) the potential V_{pn} of a p-n junction is given by

$$V_{pn}(x) = \begin{cases} \phi(\infty) - c_1(x - d_n)^2, & d_n > x > 0 \\ \phi(-\infty) + c_2(x + d_p)^2, & 0 > x > -d_p \end{cases}$$

where $x = -d_p$ and $x = d_n$ are the boundaries of the depletion layer (arbitrary units) and c_1 and c_2 are constants. Assuming a uniform thickness s of the sample the phase distribution caused by this potential is given by $e/\hbar \cdot s \cdot V_{pn}$. The line scan in Fig. 3 (right), which was simulated using this phase distribution, is in good agreement with the line scan across the measurement, shown in Fig. 3 (left).

4. RESULTS

The phase of the numerically reconstructed complex object wave (eq.(1)) is defined only modulo 2π. This means that the relation of the 256 gray levels used for display is single valued only as long as the phase differences are $\delta\varphi < 2\pi$. At bigger phase shifts between two object points the gray level between the related image points in the reconstructed phase distribution jumps from black to white forming a line especially easy

598

to follow with the eyes. Since the phase difference between two neighbouring lines of constant phase is constant, they are contour lines of the wave front.

It is possible to vary the phase difference between the contour lines. A smaller phase difference increases the number of contour lines, which is advantageous when plotting electric field lines because they are perpendicular to the phase lines. A phase display where the 256 gray levels are single-valued over the total measured phase range does not show contour lines. However, in this case it is easier to interpret the phase along a line scan.

Fig.3 shows a band on the left side in which the phase is displayed without contour lines. The measured line scan was taken in this part. The phase distribution with contour lines suggests the electric field lines which are schematically displayed perpendicular to the phase lines.

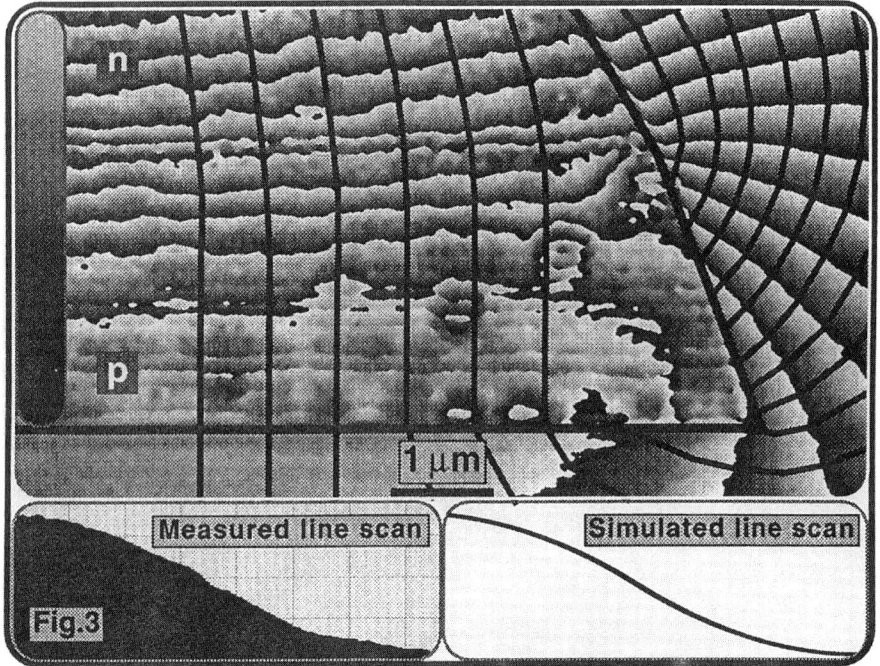

Fig.3: Measured phase distribution of a p-n junction. Since the thickness of the sample is nearly uniform the phase inside represents the potential distribution of the p-n junction. The electric field lines are perpendicular to the phase lines. The line scan across the measured distribution is in good agreement to the simulation.

REFERENCES

Aharonov Y, Bohm D 1959 Phys. Rev. 115 485
Ashcroft N W and Mermin N D 1976 Solid State Physics (Saunders College Philadelphia) p596
Frabboni S et al 1985 Phys. Rev. Lett. 55 2196
Gajdardziska-Josifovska M and McCartney M R 1994 Ultramicroscopy 53 291

We thank B.Cunningham/IBM for the semiconductor sample.

Inst. Phys. Conf. Ser. No 146
Paper presented at Microsc. Semicond. Mater. Conf., Oxford, 20–23 March 1995
© *1995 IOP Publishing Ltd*

599

Quantitative SEM-DVC imaging of semiconductor devices

S Mil'shtein, D Kharas and C Lee

Electrical Engineering Dept., University of Massachusetts, Lowell MA 01854, USA

ABSTRACT: Quantitative measurements of potential distribution across silicon diodes, solar cells, GaAs MESFETs and quantum wells were obtained using Dark Voltage Contrast (DVC) a new SEM microscopy technique. The device is imaged twice, first with no external excitation then again under bias or illumination. Subtracting both digitized images, and calibrating the resultant frame according to the applied excitation conditions, allows us to measure potential distribution. Simultaneous use of Light and Electron Beams (LEBEAMS) and video taping allows us to study the dynamic behavior of devices under changing bias or light conditions.

1. INTRODUCTION

After the first (Mil'shtein et al, 1993, 1994) quantitative field distribution measurements of various semiconductor devices were obtained with the new Dark Voltage Contrast (DVC) method, it became apparent that this technique serves the best in the area of design and testing of new semiconductor devices. The current study enlightens important details in methodology of the DVC technique and presents new results obtained recently in testing of Field Effect Transistors (FETs) and photo-detectors etc. The SEM static voltage contrast technique developed in the early sixties (Smith, 1956) was later developed into dynamic contrast testing. This became known as stroboscopy (Plows and Nixon 1970) testing for Very Large Scale Integrated (VLSI) circuitry. The quantitative DVC method does not employ special hardware attachments to the SEM, such as electron spectrometers (Fleming and Ward, 1970).

Contrast in electron microscopy is formed by both surface morphology and the potential beneath the surface. Although contrast itself is a relative value, measured in percent, the change in contrast with changing bias conditions is linear for a small range of voltages. Thus the measure of the change in contrast can be used to measure the voltage drop across a device. The appearance of the word "Dark" in the title of our method reflects the uniqueness of handling the so called gray line of an image, and the specifics of our calibration method, it also implies that the contrast does not change sign with increased external excitation. The important technical difference of our technique versus other contrast measurements is the necessity to image a few square micrometers of a semiconductor device surface, requiring high magnification microscopy. DVC consists of recording first a digitized image of a semiconductor device; diode transistor etc. with no bias. While maintaining the same microscope conditions namely, not changing the position of the SEM controls, and disregarding slight degradation (if any) of the image

600

due to the applied bias, the same area of the device is imaged under bias (or other excitation) conditions. Subtraction of the two images reveals the contrast change due to the change in potential from unbiased to biased, it also removes much of the morphological features of the surface that remain constant in both the biased and unbiased images. An image is calibrated by locating an area of known potential on the image, and assigning this potential to the contrast in that area. The distribution of the potential will follow the distribution of contrast. Once the contrast is calibrated by the applied bias at known equipotential locations within the device, the Dark Voltage Contrast technique quantitatively defines the internal electrical potential.

2. EXPERIMENT

To illustrate the quantitative DVC measurement let us first discuss testing of Zener diodes which present a nonlinear, well defined distribution of electrostatic potential across the abrupt p-n junction. The Zener diodes were produced by Allegro Microsystems Company of Massachusetts with doping at the n-side $N_d = 10^{17}$ cm^{-3} and $N_a = 10^{19}$ cm^{-3} at the p-side. The heavy doped p-side was diffused through a square opening in the oxide window to create a four sided vertical, non planar p-n junction. Therefore cleavage of the chip would open only one part of the p-n junction. Zener diodes were reverse biased under the SEM electron beam. The applied voltage created a voltage drop across the p-n junction. The voltage drop across the bulk of the device due to the to the small reverse current, was insignificant. The image of an unbiased diode was subtracted from the image of a biased device. The contrast across the p-n junction was calibrated to the bias applied at the terminals. Fig. 1 shows the potential distribution across two Zener p-n junctions, Zener 18 is reverse biased at 3.0 V and Zener 19 is reverse biased at 3.30 V. The superposition of a local potential with the applied bias creates a spike of the voltage above the bias of the device, in Zener 19. From the I-V characteristics we determine that Zener 18 has a break down voltage at V_{br}=-6.4V and for Zener 19 V_{br}=-5.9 V. One can correlate the break down information with the potential distribution of the device. The first derivative of the potential distribution would give a field profile while a second derivative would demonstrate a fluctuation of the doping profile. Both plots are excluded due to space limitations of this paper.

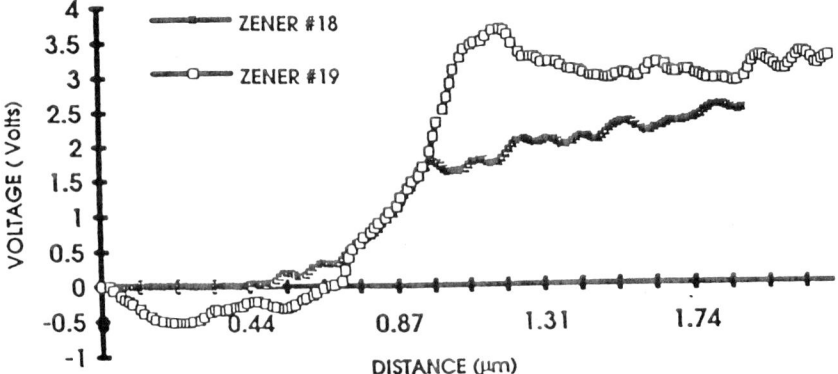

Fig. 1 Potential distribution across Zener 18 and 19, rev biased at 3.0V and 3.3 V.

To understand the influence of the applied bias we examined an unbiased p-n junction of a solar cell, where the change of a built-in potential was created by illumination through a fiber optic cable. While the cell was illuminated it was simultaneously imaged by the SEM using the Light and Electron Beams (LEBEAMS) testing technique (Mil'shtein

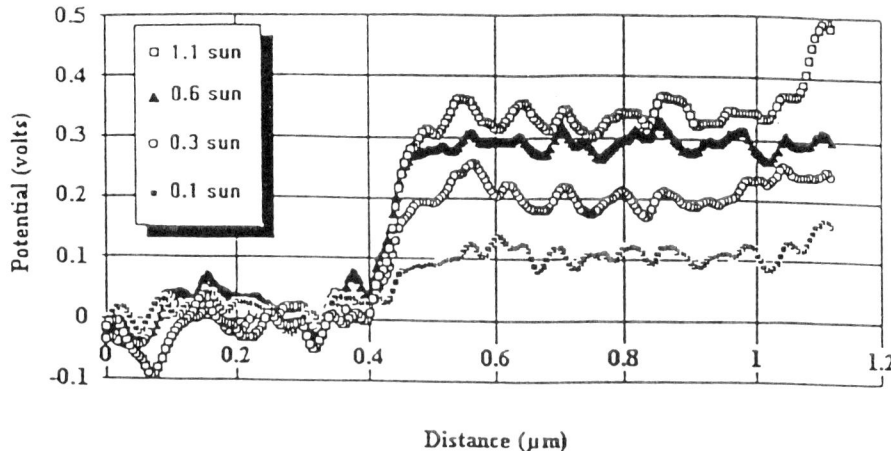

Fig. 2 Potential Distribution across a solar cell with increasing illumination.

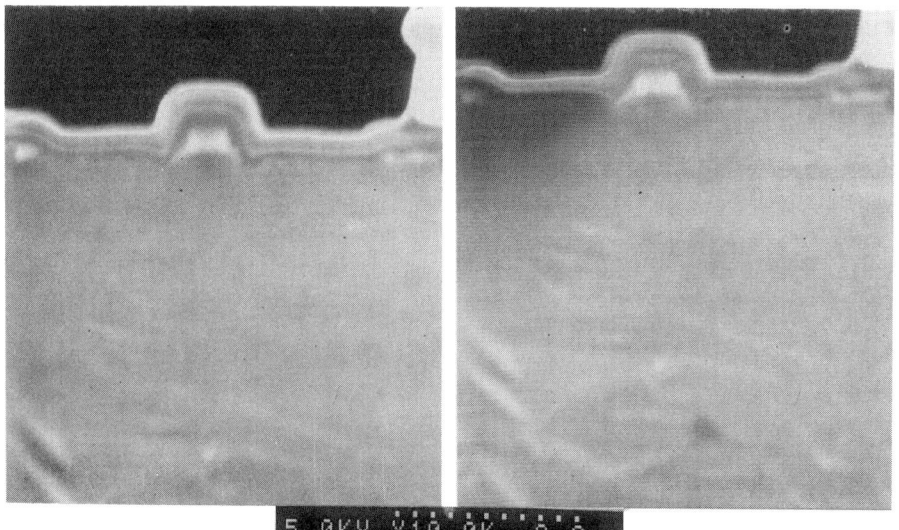

Fig. 3A   Fig. 3B

Fig. 3A Micrograph of an unbiased MESFET, the drain appears at left, gate is in the middle and the source is on the right.

Fig. 3B Micrograph of a Biased MESFET with drain at 5V and gate at 3.5V, one can see the dark area beneath the drain.

et al, 1995). The solar cell illumination was increased from 0.2 sun to one sun conditions. The DVC measurements were performed and the appearance of power at the device terminals was recorded. A digitized image of a solar cell junction under no illumination was subtracted from digitized images taken at various levels of illumination. The voltages which the solar cell produced were used to calibrate each image. Fig. 2 presents one-dimensional DVC scans across the p-n junction of the solar cell under illumination.

Built-in potential in illuminated solar cells measured by DVC experiments came very close to those defined by the I-V characteristics. The DVC measurements of Zener diodes sometimes measured depletion regions of similar size to those measured by C-V profiling, but significant deviations were also recorded. We will discuss these discrepancies in the conclusion.

The MESFET was tested to obtain the dynamic distribution of potential along the transistor channel. Along with MOSFETs these transistors present the first three-terminal devices tested by the quantitative DVC method. Figure 3A and 3B show micrographs of an unbiased and biased MESFET. With the 10 thousand times magnification one can see the drain, gate, source and the area of the channel. Another interesting feature is the broken metalization on the drain contact in the upper left of the micrograph. The I-V characteristics and the plot of potential distribution shows an area of increased resistance corresponding to the broken drain metalization, this presents a problem for impurity profiling. Figure 3B shows the area of contrast associated with biased drain and gate contacts. We used the drain to source voltage drop for calibration of the two dimensional plot of potential distribution. Figure 4A presents a two-dimensional distribution of potential along the MESFETs channel for $V_d=5$ V and $V_g=0$ V. Figure 4B presents the potential distribution across the same MESFET at pinch-off conditions $V_d=5$ V, $V_g=3.5$ V.

Finally we tested a laser heterostructure to study the potential distribution in a quantum well. The structure is shown as an insert to Figure 5. Above a n^+ GaAs substrate, a n^+ GaAs epi layer, and a 1.5µm layer of $Al_{0.2}Ga_{0.8}As$ donor doped to $N_d= 5x10^{17}$ cm^{-3}, a 0.2µm GaAs undoped layer was grown. Then a 10-20 nm layer of undoped $In_{0.2}Ga_{0.8}As$ was deposited followed by a second 0.2µm layer of undoped GaAs and 1.5µm layer of p-$Al_{0.2}Ga_{0.8}As$ acceptor doped to and $N_a =10^{17}$cm^{-3}. Completing the structure are a 30 nm layer of p^+ GaAs doped to 10^{19}cm^{-3} and a 20 nm p^+ $In_{0.2}Ga_{0.8}As$ contact layer doped to 10^{19}cm^{-3}. Examining this vertical structure one finds that a quantum well is formed by the two 0.2µm layers with the $In_{0.2}Ga_{0.8}As$ layer in between. One clearly sees that an applied voltage would drop primarily over the quantum well area since all three layers are highly resistive. Figure 5 presents a potential distribution across the quantum well structure with a 0.8V bias applied in both the forward and reverse directions. The DVC scan begins in the middle of the p-$In_{0.2}Ga_{0.8}As$ layer and ends in the n-$In_{0.2}Ga_{0.8}As$ layer. The potential distribution does not appear as a square box potential but rather as a smooth potential profile. The central minimum on the graph represents the middle of the $In_{0.2}Ga_{0.8}As$ undoped layer. The width of the peaks, shrinks with changes in the direction of the applied bias, but the height in both cases exceeds the 0.8V applied to the device.

3. CONCLUSION

Various semiconductor devices were tested using the quantitative Dark Voltage Contrast (DVC) method. Subtraction of two digitized images with and without bias produces a resultant image which contains contrast changes generated by the applied voltages. To justify the quantitative linkage between contrast and distribution of potential

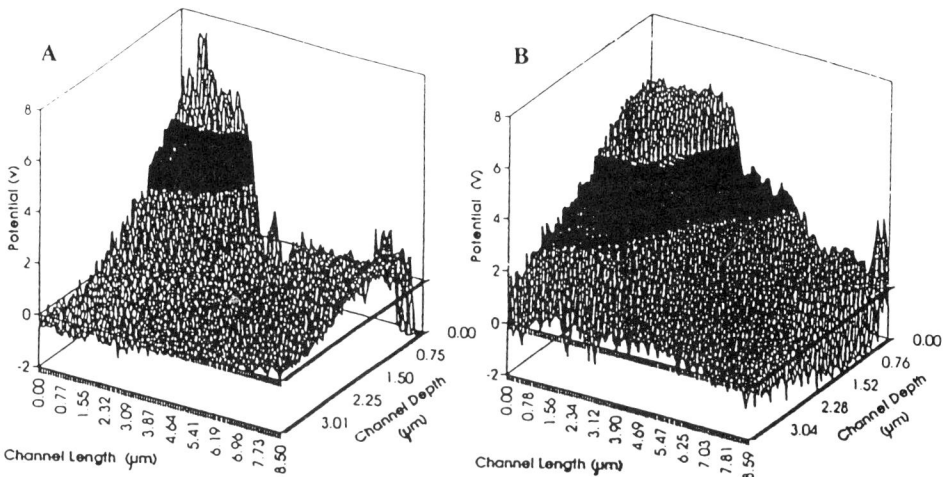

Fig. 4A and 4B Potential distribution across a MESFET, Drain appears at left,
Gate at the middle and Source at right. A) $V_d=5$ V and $V_g=0$ V,
B) $V_d=5$ V, $V_g=3.5$ V.

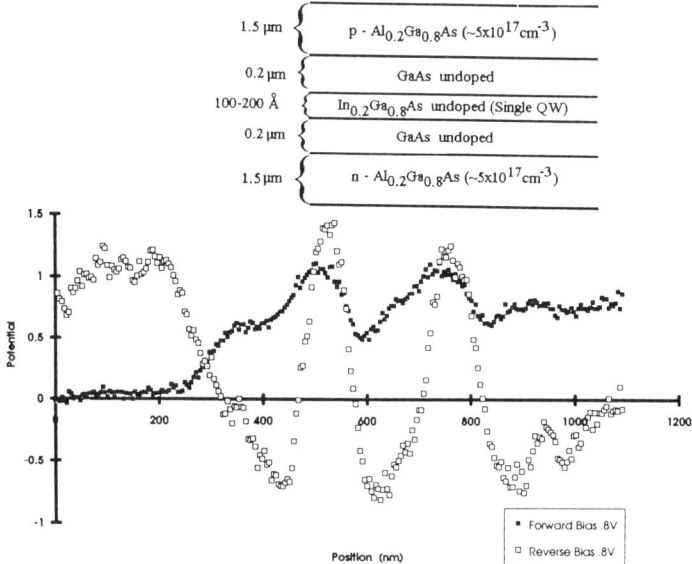

Fig. 5 Potential Distribution across a GaAs quantum Well.

in the device the issues of morphological contrast, and Fermi potentials should be discussed. Micrographs of Zener diodes and DVC scans showed irregularities of the potential distribution which can cause early breakdown as observed in some manufactured Zeners. DVC testing of solar cells showed that low efficiency solar cells usually have weaker electric fields in the junction area than higher efficiency cells. Testing of MESFETs produced the first two dimensional scans of potential distribution along the channel. Examination of DVC plots of the MESFET revealed a drain resistance increase due to irregularities in the metalization. All the experiments described above show that DVC issensitive to the position of Quasi Fermi Levels (QFLs) in localized areas of the device. Change of QFLs in a p-n junction or a transistor changes the local work function for secondary electrons . Experiments with illuminated solar cells are rather convincing, in this case terminal voltages are not applied but a voltage distribution is generated by the light. Light introduced internal changes of Fermi energy are quantitatively related to the contrast readout, by using the values of generated voltage to calibrate the contrast. The width of the depletion region, and built-in potential measured by DVC was compared with built-in potential defined by I-V characteristics.

In the case of Zener diodes, when a few external volts are applied the linearity remains in place as expected. Morphological contrast is difficult to completely eliminate due to the electric field induced changes in the surface (surface states or defects trapping the charge at the surface). Non active features that are present in both biased and non biased images can be subtracted out. Attention should be paid to occasional disagreements between DVC measurements of the depletion region width W of a p-n junction and C-V profiling results. The DVC measurements of QFLs are strongly influenced by the closeness to the cleaved surface. Band bending at places where the p-n junction crosses the surface is different for p and n materials and depends on the density of surface states and unfortunately on the cleanness of the surface.

The profiling of the potential across the single quantum well demonstrates feasibility of DVC characterization of nanoscale devices. However reconstruction of the potential barriers of the quantum well is complicated by several factors. We tend to believe that DVC measurements produce a profile of Quasi Fermi energies. However the energy band discontinuity, presence of unintentional doping and the unknown positions of the peaks of the electron wave functions makes it very difficult to produce a well justified energy diagram of the tested quantum well. The detailed reconstruction of these energy diagrams and closeness to the surface are the subject of our current study.

In summary the DVC method is a direct quantitative measurement of profiles of Fermi energies of free electrons or profiling of potential and fields across various devices. Videotaping of a DVC test gives information about dynamic behavior of all described above devices.

REFERENCES

Fleming J and Ward E, Scanning Electron Microscopy 11 TRI. p465-470, (1970)
Mil'shtein S, Iatrou S, Kharas D, Bell R, and Sandstrom D , MRS 283, p921, (1993)
Mil'shtein S, Kharas D , Iatrou S , MRS, 305,p637, (1994)
Mil'shtein S,Bakker B,Iatrou S ,Kharas D ,and Bell R , Mater Sci Forum 174,p325,(1995)
Plows S and Nixon W , J Sci Instrum. 1 595-600 (1970)
Smith K C A , PhD dissertation Cambridge Univ (1956)

Inst. Phys. Conf. Ser. No 146
Paper presented at Microsc. Semicond. Mater. Conf., Oxford, 20–23 March 1995
© 1995 IOP Publishing Ltd

Application of secondary electron imaging to dopant profiling of semiconductors

D Venables and D M Maher

Department of Materials Science and Engineering, North Carolina State University, Raleigh, North Carolina 27695-7907

ABSTRACT: Doping-dependent contrast in secondary electron images of p/n junctions in silicon was observed and characterized. Secondary electron contrast profiles of in-situ boron doped epitaxial silicon structures showed an excellent correlation with spreading resistance and SIMS profiles. The magnitude of the contrast scaled with carrier concentration and the precision in locating the abrupt p+/n junctions was approximately 10 nm. The sensitivity of the technique extends to a carrier concentration of ~10^{16} cm^{-3} under optimum conditions. This phenomenon is employed to extract two-dimensional dopant profiles of suitable test structures.

1. INTRODUCTION

Experimental information concerning the depth distribution of dopants in semiconductor devices has traditionally been of paramount importance for device modeling and process simulation. As the minimum feature sizes of devices continue to decrease, it has become critical to obtain information about lateral, as well as vertical dopant distributions. To date, microscopy based techniques have centered on junction delineation by means of preferential etching of p or n regions with subsequent examination by SEM or TEM (Subrahmanyan 1992). Recently, it has been demonstrated that doped regions in semiconductors can give rise to contrast in secondary electron images without aid from a chemical delineation process (Farrow et al. 1991, Perovic et al. 1994, 1995, Venables and Maher 1994, 1995). In this paper, we investigate the effect of doping and carrier concentration and the abruptness of the p/n junction on this "electronic" contrast. In addition, we show that this contrast can be use to extract two-dimensional dopant profiles.

2. EXPERIMENTAL PROCEDURES

In-situ boron doped silicon epitaxial layers, ~100 nm thick, were grown on 0.7 Ω-cm, n-type, <100> oriented silicon wafers in an ultra-high vacuum rapid thermal chemical vapor deposition system. These layers were capped with ~150 nm of unintentionally doped silicon. Patterned and blanket p+/n junctions were fabricated in 0.4 Ω-cm, n-type, <100> oriented silicon wafers using an ion implanted polysilicon overlayer as a diffusion source. The 300 nm thick polysilicon layer was implanted with boron at 15 kV to a dose of 1×10^{16} cm^{-2} (Zhang 1992). The samples were then capped with a 30 nm deposited oxide to reduce out-diffusion during subsequent annealing. Rapid thermal anneals were performed at 950 to 1100°C for 10 sec to diffuse the dopant into the substrate.

Cleaved cross-sections of these test structures were examined in field-emission SEMs at an accelerating voltage of 1 kV. Previous work (Venables and Maher 1994, 1995) has shown that the measured contrast levels are maximized at low accelerating voltages (~1 kV) and when a through-the-lens secondary electron detector configuration is employed. The images presented here were obtained on an Hitachi S-4500 SEM with a through-the-lens detector and on a JEOL 6400F SEM which uses the conventional detector configuration below the objective lens. The contrast between p and n regions was easily obscured or altered by beam-induced contamination. Therefore, quantitative analysis was always performed on fresh areas of the sample. Wherever possible, secondary electron images were acquired in a digital format by video frame store. Digital densitometer traces were used to obtain intensity profiles from these images. Results are presented in terms of image contrast, defined as the difference between the brightness values of

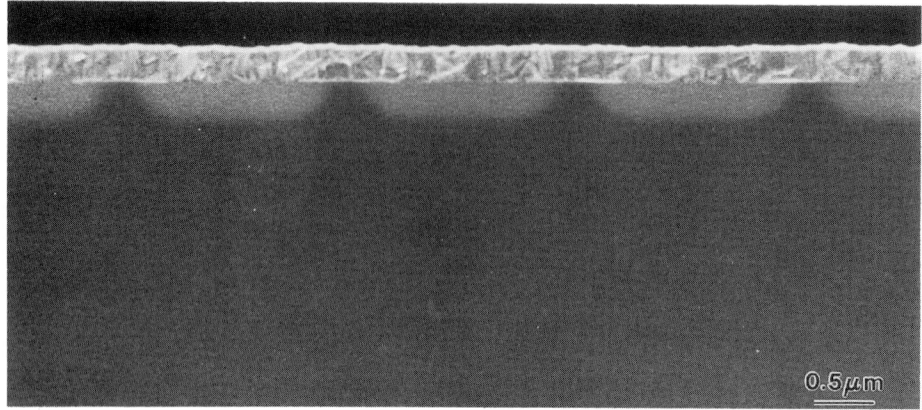

Fig. 1: Secondary electron image of a cleaved cross-section of boron doped p$^+$ diffusion wells in a n-type silicon substrate showing doping-dependent contrast.

the doped and substrate regions, normalized to the number of brightness levels (256) available on the digital system. It should be noted that image contrast defined in this manner is different from physical contrast. These secondary electron (SE) contrast profiles were compared to secondary ion mass spectroscopy (SIMS) depth profiles of the atomic concentration of the selected dopant species and to spreading resistance probe (SRP) profiles of the carrier concentration. SIMS analysis was performed on a Perkin-Elmer 6300 quadrapole SIMS with a 7 kV, 250 nA Cs$^+$ primary beam at a sputter angle of 30 or 60° or with a 7 kV, 500 nA O$^+$ beam at a sputter angle of 60°. Quantification of the SIMS depth profiles was obtained with suitable boron implant standards.

3. RESULTS AND DISCUSSION

Fig. 1 is a secondary electron image obtained with a through-the lens detector. The image is from a cleaved cross-section of a patterned, boron doped test structure annealed at 1050°C, 10 sec. The boron doped p-type diffusion wells are significantly brighter than the n-type substrate and the potential of this technique for direct, qualitative imaging of doped layers in semiconductors is self-evident.

To assess the relationship between the SE contrast and the doping concentration, in-situ boron doped epitaxial layers with doping levels ranging from 4×10^{16} cm^{-3} to 1×10^{19} cm^{-3} were imaged with a conventional detector configuration. The SE contrast, SIMS and SRP profiles for the sample with the highest doping level are compared in Fig. 2a. The SE contrast profile shows a fairly low level of contrast because a through-the-lens detector was not used. Nevertheless, the contrast profile mimics the SRP and SIMS profiles quite well. The contrast profile locates the abrupt p$^+$/n junction at the interface with the substrate with an accuracy of ~10 nm. In Fig. 2b, the relationship between SE contrast and boron concentration for the epitaxial layers is shown. The SE contrast scales with doping concentration, although not in a linear manner (note that one scale is logarithmic and the other is linear). The contrast at 4×10^{16} cm^{-3} is essentially zero, suggesting that the sensitivity to dopant level is about 10^{17} cm^{-3} when a conventional detector configuration is used.

SE contrast, SIMS and SRP profiles across a p$^+$/n junction obtained by boron in-diffusion from polycrystalline silicon are shown in Fig. 3. In this case, however, the images were acquired with a through-the-lens detector. The contrast levels are much higher than in the previous example and the contrast and SRP profiles show a good correlation. However, the SIMS profile is slightly deeper, showing a bias in the direction of sputtering. In addition, comparing the SE contrast and SRP profiles suggests a sensitivity of ~10^{16} cm^{-3} in carrier concentration when a through-the-lens detector is employed.

These results suggest that the spatial resolution and sensitivity exhibited by doping-dependent contrast are sufficient for two-dimensional profiling of p$^+$/n junctions. To assess this possibility, an image processing procedure was developed for producing iso-contrast contours. The procedure consists of converting the original digitally acquired image (with a relatively arbitrary brightness scale) to a "contrast" image. To achieve this end, an image consisting

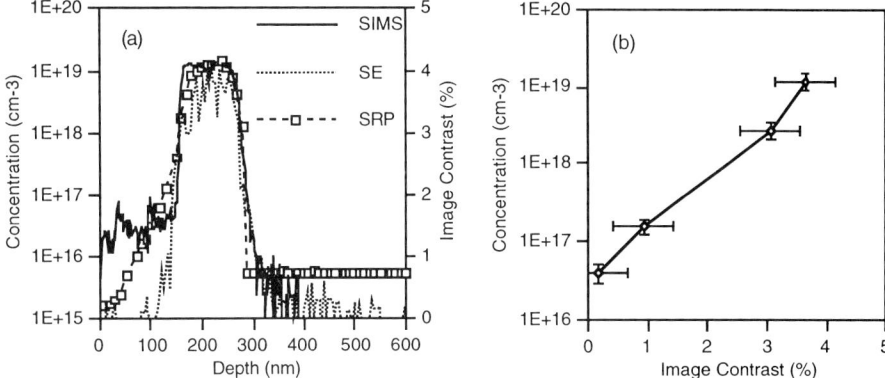

Fig. 2: In-situ boron doped epitaxial layers (a) comparison of SIMS, SE contrast and SRP profiles for highest doping concentration and (b) relationship between SE contrast and boron concentration for four doping levels.

entirely of the average substrate brightness level was created. This background image was subtracted from the original image and the result was normalized to produce a contrast image. The value of each pixel in this contrast image thus represents the contrast level with respect to the substrate. Iso-contrast contours were produced by creating a binary image, where the threshold between black and white pixels was set at the desired contrast level. The final contours are simply the border between the black and white regions of these binary images. A median filter was used to reduce noise in the images. More aggressive filtering schemes, such as pixel averaging (smoothing) or erosion and dilation operations were avoided since they shifted the contours noticeably. An example of the iso-contrast contours resulting from the application of this method to a p$^+$/n diffusion well that was imaged with a through-the-lens detector is shown in Fig. 4. An empirical correlation between the contrast levels in this image and the SIMS depth profile from the corresponding blanket structure was performed. Based on this correlation, the 5, 15 and 25% iso-contrast contours shown in Fig. 4 correspond to boron doping concentrations of approximately 3×10^{17}, 2×10^{19}, and 1×10^{20} cm^{-3}. Thus, it is possible to use this technique to produce quantitative two-dimensional dopant profiles.

To exploit the phenomenon of doping-induced contrast for two-dimensional profiling of p/n junctions it is necessary to work under a set of optimum observation conditions (Venables and Maher 1995). These instrumental requirements include high resolution operation at low accelerating voltages, a through-the-lens detector (or other energy-filtered detector configuration) and a clean vacuum system. In addition, the ability to acquire and store images digitally is important for obtaining quantitative intensity profiles from the SE images.

The application of this doping-induced contrast phenomenon to the problem of two-dimensional dopant profiling has several advantages. The technique appears to provide adequate sensitivity to the carrier concentration ($\sim10^{16}$ to 10^{17}cm^{-3}), at least for p$^+$/n junctions. In addition, modern field-emission scanning electron microscopes have inherently high spatial resolution, even at low accelerating voltages. Sample preparation consists simply of cleaving the wafer to expose a cross-section and transferring the specimen to the microscope. Visual, qualitative two-dimensional maps are then immediately available from the image. Iso-contrast contours can then be extracted within minutes, provided the images are acquired digitally. Converting these intensity or contrast profiles into concentration profiles requires calibration against a one-

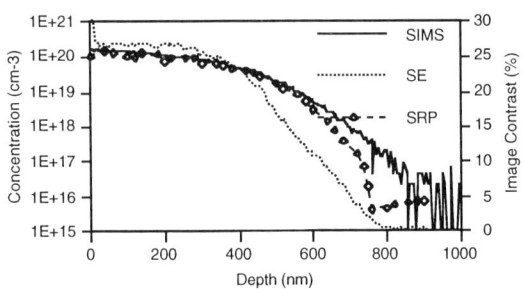

Fig. 3: Comparison of SIMS, SE contrast and SRP profiles for p$^+$/n junction obtained by solid-source diffusion.

608

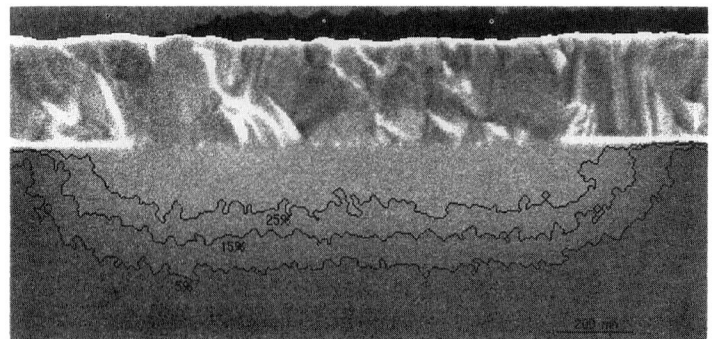

<u>Fig. 4</u>: Two-dimensional map of iso-contrast contours at 5%, 15% and 25% contrast levels, superimposed on the original digital image. The contrast contours correspond to doping levels of $\sim 3\times10^{17}$, 2×10^{19}, and 1×10^{20} cm^{-3}.

dimensional profiling technique. To alleviate this difficulty, it may be possible to image a well characterized "standard" sample at the same time, and under the same experimental conditions as the unknown sample. Under these circumstances, the contrast level in each sample should correspond to the same doping concentration if the substrate doping levels are not significantly different.

An important disadvantage of the technique is the possible confusion introduced by artifacts, topographical features and other contrast features that are typical of a cleaved surface. In particular, dielectric layers tend to image with high contrast because of charging under the electron beam. It may be possible to reduce this effect by selectively etching the dielectric layer before imaging the sample. A second, and rather severe, disadvantage is that, thus far, quantitative results have only been obtained for p$^+$/n structures.

4. SUMMARY AND CONCLUSION

Secondary electron contrast profiles correlate well with SRP and SIMS depth profiles for p$^+$/n junctions. The contrast level scales non-linearly with dopant level and a sensitivity of $\sim10^{16}$ to 10^{17}cm^{-3} in carrier concentration is demonstrated. The utility of the phenomenon for simple, qualitative two-dimensional images of doped regions in the p$^+$/n case is demonstrated. In addition, an empirical technique for extracting quantitative two-dimensional profiles which relies on calibration of the profiles with a suitable one-dimensional technique is shown. However, a fully quantitative 2-D profiling method based on the phenomenon requires a more detailed understanding of the contrast mechanism.

5. ACKNOWLEDGMENTS

The contributions of R J Gordon, M A Hernandez and T W Reilly of Nissei Sangyo America, Ltd., R Brennan of Solecon Laboratories, Inc., D C Joy of the University of Tennessee and M Ray of the Microelectronics Center of North Carolina are gratefully acknowledged. This work has been supported by the NSF Engineering Research Centers Program through the Center for Advanced Electronic Materials Processing (Grant CDR 8721505).

REFERENCES

Farrow R C, Maher D M, Ellington M B, Katz A, and Weir B E 1991 Scanning 13 I-46
Perovic D D, Castell M R, Howie A, Lavoie C, Tiedje T and Cole J S W 1994 Proc. 13th Int.
    Cong. on Elec. Microsc., eds J Jouffrey and C Colliex (France: Les Editions de Physique) 91
Perovic D D, Castell M R, Howie A, Lavoie C, Tiedje T and Cole J S W 1995 Ultramic. in press
Subrahmanyan R 1992 J. Vac. Sci. Technol. B10 358
Venables D and Maher D M 1994 Proc. 52nd Ann. Meeting Microsc. Soc. Amer., eds G W
    Bailey and A J Garratt-Reed (San Francisco: San Francisco Press) 1024
Venables D and Maher D M 1995 Proc. 3rd Int. Workshop on Ultra-Shallow Doping Profiles in
    Semiconductors, eds J Ehrstein, R. Mathur and G McGuire in press
Zhang B 1992 MS Thesis North Carolina State University

Inst. Phys. Conf. Ser. No 146
Paper presented at Microsc. Semicond. Mater. Conf., Oxford, 20–23 March 1995

SEM imaging of contrast arising from different doping concentrations in semiconductors

C P Sealy, M R Castell, A J Wilkinson and P R Wilshaw

Department of Materials, University of Oxford, Parks Road, Oxford OX1 3PH.

ABSTRACT: A technique for the direct imaging of 2-dimensional doping profiles in layered semiconductor structures is reported. The SE signal of standard and field-emission SEMs is used to produce images in which contrast between *n*- and *p*-doped layers is visible. The contrast arising from the differently doped layers is electronic in origin. The contrast is small, 1-5%, and the spatial resolution is found to be better than 0.1 μm for the samples imaged in this work.

1. INTRODUCTION

Semiconductor devices consisting of *n*- and *p*-type layers on doped or undoped substrates are widely used for many purposes. The advent of growth technologies, such as Molecular Beam Epitaxy (MBE) and Metal Organic Vapour Phase Epitaxy (MOVPE), has enabled the fabrication of structures with layers on the nanometre scale with improved and often novel electronic and optoelectronic properties. The fabrication of devices on such scales has demanded a corresponding improvement in characterisation techniques.

Existing techniques, such as secondary-ion mass spectrometry, spreading resistance and capacitance-voltage profiling, are employed to analyse the doping profiles of semiconductor structures. However, these techniques have limited spatial resolution and do not readily provide 2-dimensional doping concentration information. Microscopy techniques, such as scanning tunnelling microscopy, conventional and scanning transmission electron microscopy, are becoming more popular as their potential for higher spatial resolution and concentration sensitivity is realised (Osburn and McGuire 1992, Subrahmanyan 1994). Dopant effects may be studied indirectly using scanning electron microscopy (SEM) in conjunction with thinning and/or staining of specimens, or related techniques such as electron-beam induced current and specimen biased voltage contrast measurements. More recently it has been reported by Bleloch et al (1994) and Perovic et al (1994a) that semiconductor structures can be directly imaged in secondary electron (SE) or backscattered electron (BSE) mode (Merli and Nacucchi 1993).

This paper reports on a direct SE imaging technique for semiconductor structures which produces full 2-dimensional doping concentration profiles in cross-section. Homogeneous silicon (Si) structures containing doped layers and indium phosphide (InP) based laser devices are imaged using this technique. In addition, quantitative contrast measurements as a function of accelerating voltage are presented.

2. EXPERIMENTAL TECHNIQUES

In this work semiconductor structures containing differently doped layers based on Si and InP were investigated. InP-based etched-mesa laser (gratefully received from Robert Hull while at AT&T Bell Labs) and Si-based *p-n-i-n-p* layered structures (gratefully received from Lew Reynolds of AT&T Bell Labs) were fabricated using low pressure MOVPE. The laser structure consists of an

InGaAsP active layer lattice-matched to n-InP (001) substrate. The blocking layers, which are regrown over the etched mesa, consist of one 2000 nm thick p-layer doped with $1-2\times10^{18}$ cm^{-3} Zn, and two 800 nm thick n-layers doped with $1-2\times10^{18}$ cm^{-3} Si separated by an 800 nm thick i-(or semi-insulating) layer doped with 5×10^{17} cm^{-3} Fe. A 2000 nm epitaxial p-layer doped with $1-2\times10^{18}$ cm^{-3} Zn is grown on top of the blocking layer sequence. Finally, a capping layer of highly p-doped InGaAs is present. The Si structure consists of 300 nm thick p- and n-layers doped with 10^{18} cm^{-3} B and As respectively.

The samples were prepared simply by cleaving in air to give a smooth cross-sectional surface. The freshly cleaved Si structures were dipped in 1% HF for 5 minutes, according to the method described by Takahagi et al (1988) and Burrows et al (1988). This serves to saturate the dangling bonds at the surface with hydrogen, while removing less than 1 nm of Si. A standard LaB$_6$ JEOL JSM 6300 and a field-emission gun JEOL JSM 840F SEM were used to image the samples.

3. RESULTS

The technique is sensitive to changes in doping concentration and produces images in which n- and p-type material appears dark and bright respectively. A combination of SE and BSE imaging allows for the position of the active layer and the variations in doping around it to be accurately mapped. Dopant concentrations of 10^{17} - 10^{18} cm^{-3} have been successfully imaged in Si- and InP-based structures.

In Fig. 1(a) the blocking layer sequence of an InP-based etched mesa type laser structure is imaged in cross-section using the SE signal. The contrast between the differently doped n, p and i (semi-insulating) blocking layers is clearly visible. The p-layers appear bright, the n-layers dark, and the i-layers intermediate between the two levels. Similarly, in Fig. 1(b) the active region of the same laser structure is imaged in SE mode. The mesa region containing the InGaAsP active layer and the adjacent blocking layers can clearly be seen. Fig. 1(c) shows the same region using the BSE signal. In this image the active and capping layers are visible due to compositional contrast mechanisms, but none of the other structure can be seen. Although this implies that there is a contribution from atomic contrast in the SE images of the active layer, the contrast from the blocking layers must be electronic in origin (see below). The spatial resolution of the images shown in Fig. 1(a, b) is estimated to be between 0.05 - 0.1 μm.

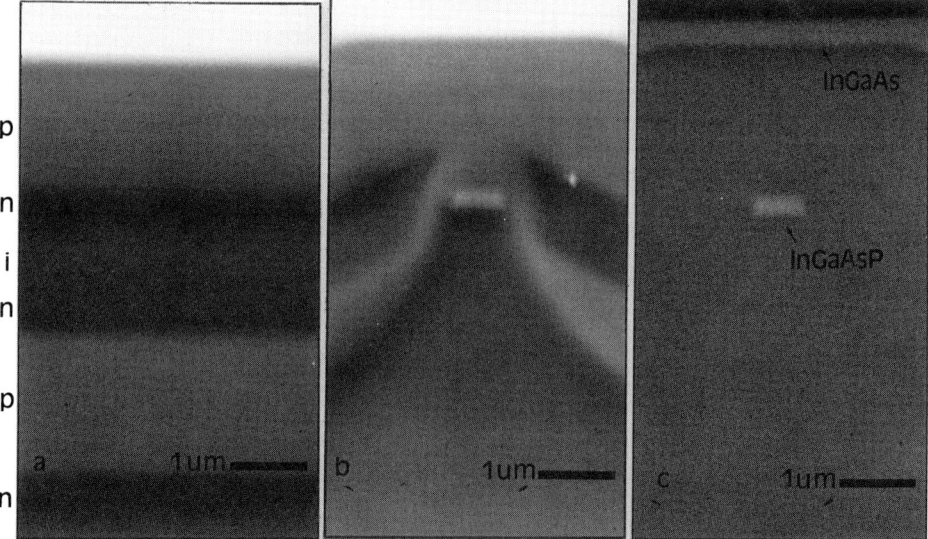

Fig. 1(a) SE image of InP-based structure blocking layers (at 10 kV); (b) SE image of InP-based structure active layer (at 10 kV); (c) BSE image of same region (at 10 kV).

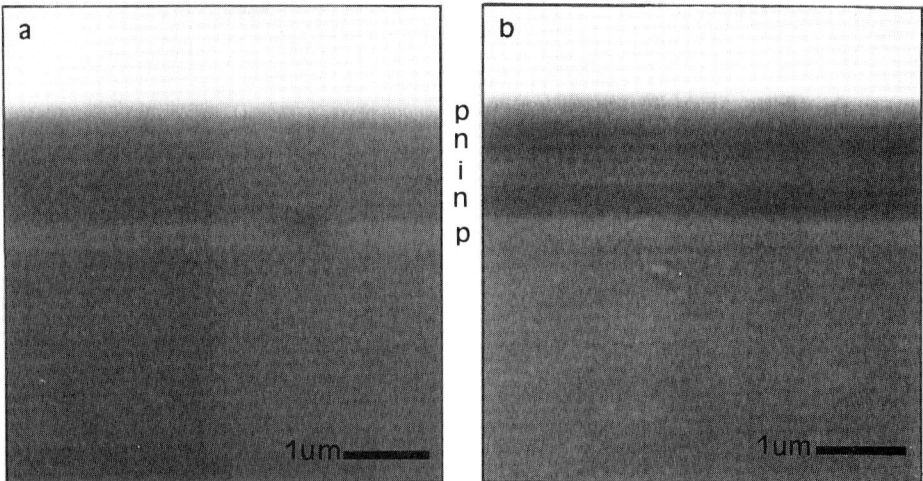

Fig. 2(a) SE image of Si structure before HF dip (at 2 kV); (b) SE image of Si structure after HF dip (at 2 kV).

Fig. 2(a, b) shows the Si structure before and after dipping in dilute HF. The comparison of these images, which were taken under identical conditions, clearly shows that the contrast is markedly stronger after this treatment. This is the first time that doping contrast in Si has been imaged using the present technique.

The contrast between the *n*- and *p*-layers in the InP structure was measured quantitatively and was found to be 3-5% for accelerating voltages in the range 1-10 kV. The contrast measurements, shown in Fig. 3(a, b), demonstrate that at 7 kV the accuracy of the doping profiling near the surface is reduced compared to 1 kV because of beam breakthrough. The contrast level in the Si structure was observed to be lower than in the InP structure, and was estimated to be 1-2% at 2 kV.

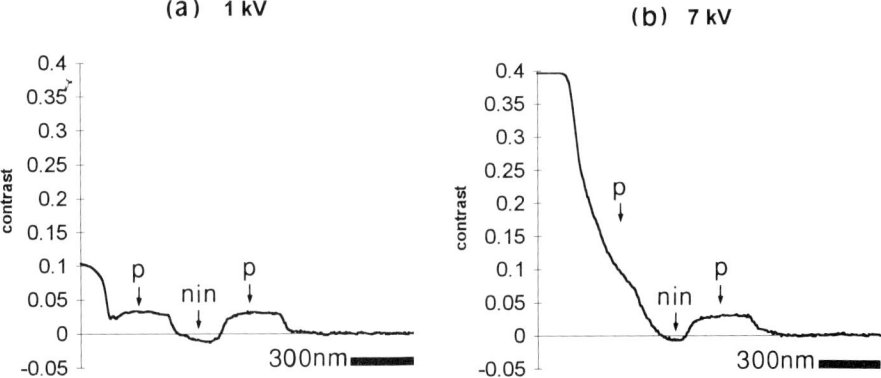

Fig. 3 Contrast between *p*- and *n*-layers in InP structure at (a) 1 kV and (b) 7 kV.

4. DISCUSSION

It has been shown here that SE contrast may arise from differently doped layers. Since the dopant concentrations are ~1 dopant atom in ~50000, the contrast cannot arise from conventional compositional contrast mechanisms. The origin of doping contrast must be explained electronically in terms of the band structure of doped semiconductors.

Previous work on contrast between n- and p-doped regions was carried out by Aven et al (1972) and Sawyer and Page (1978), but no comprehensive explanation of the effect was offered. The most recent work on the subject by Castell (1995) and Perovic et al (1994b) describes a model in terms of the surface depletion region electric field and the differences in local work function.

In this work, it has been reported that the doping concentrations in semiconductor structures and devices can be mapped in 2-dimensions using SE imaging in conjunction with BSE imaging. The spatial resolution for the samples imaged here has been found to be between 0.05 - 0.1 μm. It is expected that the resolution will depend upon the spatial extent of band bending in the surface region. The contrast measurements displayed here demonstrate that low accelerating voltages are necessary to avoid edge effects, such as beam breakthrough, and enable the imaging of the surface region of structures. It has been found that HF treatment of Si structures is necessary to image clearly the differently doped layers.

5. CONCLUSION

We have reported on a direct 2-dimensional imaging technique using standard or field-emission SEMs to observe contrast between n-, p- and i-doped layers in various semiconductor structures. InP laser devices and doped Si layered structures were imaged in this way. The technique is sensitive to doping concentrations of $\sim 10^{18}$ cm^{-3} and has spatial resolution of better than 0.1 μm. The contrast mechanism is electronic in origin and arises from differences in band structure between n- and p-type material, but further work is necessary for a more complete explanation of the contrast mechanism. The technique is swift and easy to use, sample preparation is minimal, and a qualitative 2-dimensional image of differently doped regions is produced very rapidly. The advantages of this technique make it potentially useful for many technological applications.

ACKNOWLEDGEMENTS

C P Sealy would like to thank Robert Hull, Lew Reynolds, Dick Lum and AT&T Bell Labs for their continued support and provision of specimens, and also the EPSRC for their support. P R Wilshaw gratefully acknowledges the support of the Royal Society.

REFERENCES

Aven M, Devine J Z, Bolon R B and Ludwig G W 1972 J.Appl. Phys. 43 (10) 4136
Bleloch A L, Castell M R, Howie A and Walsh C A 1994 Ultramicroscopy 54 107
Burrows V A, Chabal Y J, Higashi G S, Raghavachari K and Christman S B 1988 Appl. Phys. Lett. 53 (11) 998
Castell M R 1995 PhD Thesis, University of Cambridge
Merli P G and Nacucchi M 1993 Ultramicroscopy 50 83
Osburn C M and McGuire G E 1992 J. Vac. Sci. Technol. B 10 (1) 286
Perovic D D, Castell M R, Howie A, Lavoie C, Tiedje T and Cole J S W 1994a ICEM 13-PARIS 91
Perovic D D, Castell M R, Howie A, Lavoie C, Tiedje T and Cole J S W 1994b Ultramicroscopy (in press)
Sawyer G R and Page T F 1978 J. Mater. Sci. 13 885
Subrahmanyan R 1994 J. Vac. Sci. Technol. B 12 (1) 163
Takahagi T, Nagai I, Ishitani A, Kuroda H and Nagasawa Y 1988 J. Appl. Phys. 64 (7) 3516

Inst. Phys. Conf. Ser. No 146
Paper presented at Microsc. Semicond. Mater. Conf., Oxford, 20–23 March 1995   613

Observation of strong transmission electron microscope contrast from doped layers in InP-based structures.

R Hull , M Moore, D Bahnck*, M Geva**, R F Karlicek**, F A Stevie+ and J F Walker++

University of Virginia, Department of Materials Science, Charlottesville, VA 22903, USA
* AT&T Bell Laboratories, 600 Mountain Avenue, Murray Hill, NJ 09794, USA
**AT&T Bell Laboratories, 9999 Hamilton Blvd., Breinigsville, PA 18031, USA
+ IBM East Fishkill Facility, 1580 Rte. 52, Hopewell Junction, NY 12533
++FEI Europe Inc., Brookfield Business Center, Cottenham, Cambridge CB4 4PS

ABSTRACT: We have observed strong contrast (~20-50%) between p-doped (~ 10^{18} cm^{-3} Zn), n-doped (~ 10^{18} cm^{-3} Si) and intrinsic (~ 10^{17} cm^{-3} Fe) InP layers in transmission electron microscopy (TEM) of semiconductor laser diode structures, following sample preparation by Ga^{+} focused ion beam (FIB) sputtering. The operative contrast mechanism is differential "absorption contrast", i.e. high angle scattering out of the instrumental image collection system. The critical question is what physical mechanism could produce such large amounts of differential scattering from the dopant atom concentrations ~ 0.001%.

1. INTRODUCTION

In Figure 1, we show a typical TEM observation of doped InP layers in a CMBH (Capped Mesa Buried Heterostructure) semiconductor laser diode structure. Strong contrast is observed between the light n-doped layers (~ 10^{18} cm^{-3} Si) and substrate (~ 10^{18} cm^{-3} S) and the dark p-doped (~ 10^{18} cm^{-3} Zn) and intrinsic (~ 10^{17} cm^{-3} Fe) InP layers. This contrast is only observed in samples which have been prepared for TEM by Ga^{+} focused ion beam (FIB) sputtering. Such images represent enormous compositional sensitivity - a capability at which TEM does not generally excel. If we define compositional sensitivity by the ratio, $\Delta I/\Delta Z$, where ΔI is the intensity difference between two image elements with an atomic number change ΔZ, then in the present work $\Delta I/\Delta Z > 10^4$. More typical TEM values for $\Delta I/\Delta Z$ might be in the range 10^{-2} - 10^{-1} (e.g. a change in atomic number of 10 between two layers will produce contrast levels between 10 and 100%).

2. EXPERIMENTAL DETAILS

Samples of the active regions of CMBH laser structures are prepared for TEM by 25-30 keV Ga^{+} FIB sputtering (FEI 610 and 200 instruments) as described in Hull et al (1993). A brief synopsis of the process is provided by Figure 2. A site of specific interest in the sample (in this case a section of the active laser stripe, A, and surrounding region) is positioned within a thin membrane, M..M, by FIB sputtering of two trenches, T..T, either side of the remaining membrane. The membrane then becomes the electron transparent sample required for TEM imaging. Membrane thicknesses in the range 100-300 nm may be routinely fabricated. Secondary electron images are generated during Ga^{+} sputtering to monitor trench positions and dimensions. Beam damage is minimized on the final membrane surface by use of low Ga^{+} currents (~10 pA) for the final polish, and by the fact that the beam is incident at

614

almost perfect grazing incidence on the sample surface.

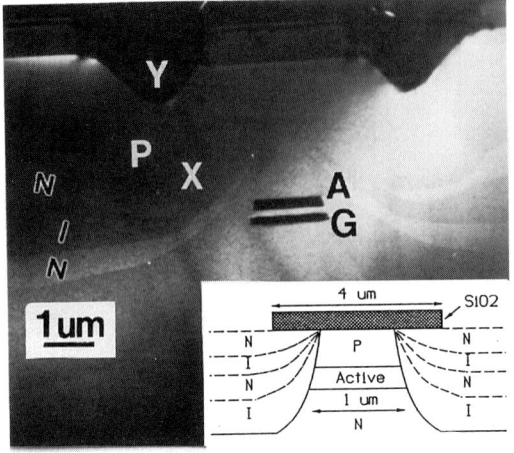

Figure 1: FIB-TEM image of current blocking layers in an InP CMBH laser structure. Layers marked "P", "I" and "N" refer to p-type (Zn-doped), intrinsic (Fe-doped) and n-type (Si-doped) InP layers respectively. A is the InGaAsP active region of the laser and G is part of an InGaAsP grating. Inset shows a schematic illustration of a sidewall growth scheme for InP current blocking layers during fabrication of CMBH laser structures, with some approximate dimensions. Solid lines denote the etched mesa structure, upon which the doped blocking layers (dashed lines) are grown. On top of the mesa is an SiO$_2$ mask.

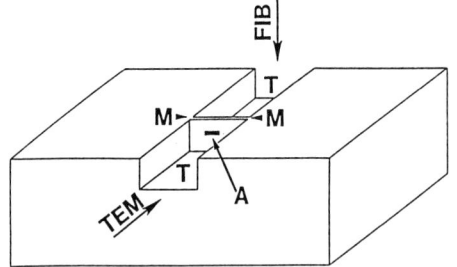

Figure 2: Schematic illustration of the FIB fabrication process for TEM samples

Electron microscopy is performed at 200 kV on a JEOL 2000FX. The CMBH laser structures are grown by metalorganic vapor phase deposition (MOCVD) at temperatures in the range 600-650°C. The laser fabrication process is complex, involving several deposition and etching stages. The critical stage relevant to this work is illustrated in the inset to Figure 1. Following a planar MOCVD deposition, laser active stripes are defined by etching around an array of oxide stripes on the wafer surface, thereby defining a linear array of etched mesas. Doped "current blocking" layers are then grown over the array of capped mesas. These blocking layers provide reverse-biased p-n junctions with respect to the forward-biased junctions in the active region, thereby confining current flow to the active region. The current blocking layer growth is a highly complex process, involving MOCVD deposition over a highly non-planar structure.

3. EXPERIMENTAL OBSERVATIONS

The salient experimental observations relating to this dopant contrast are:

(a) We have quantified the magnitude of this contrast by careful measurement of electron intensities (Hull et al 1995): observed contrast levels between n- and p-doped materials are as high as 20-30% with the sample at room temperature, and 50% with the sample at liquid nitrogen temperature.
(b) The contrast is only observed after sample preparation with a Ga$^+$ focused ion beam (25-30 keV, at grazing incidence), and not after standard Ar$^+$ ion milling (2-5 keV Ar$^+$ at incidence angles 12-18°).
(c) Although, the contrast is visible immediately upon observation in the TEM, it increases in intensity under 200 kV electron irradiation, with a time constant of order 20 minutes.
(d) Measured contrast increases with decreasing sample temperature from 80 to 300 K.

(e) Although the magnitude of the contrast does vary with the precise diffraction conditions in the TEM, the sign of the contrast (i.e. n-type light and p-type dark) does not vary under <u>any</u> experimental conditions, e.g. g, s, bright field or dark field (see Figure 3). The contrast is still visible if an aperture is put around both beams in a two-beam condition. The sign of the contrast is independent of objective aperture size for radii ~ 0.5 - 20 mr, and the sign of the contrast does not reverse if one scans out as far as 50 mr into reciprocal space with a ~ 1 mr aperture. There is no observable deflection of extinction contours across p-n interfaces. <u>These observations are consistent only with "absorption" contrast in the TEM, i.e. differential loss of the electrons from the final image by high angle scattering out of the image collection system</u>

(f) To date, we have observed this effect only in InP. In this material it is entirely reproducible over the range of dopants we have studied (Zn, Si, Fe, S). Preliminary experiments in other systems (Si:B, Si:Sb, GaAs:C, GaAs:Si) have not revealed this contrast to date.

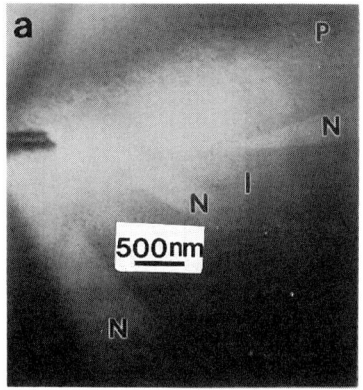
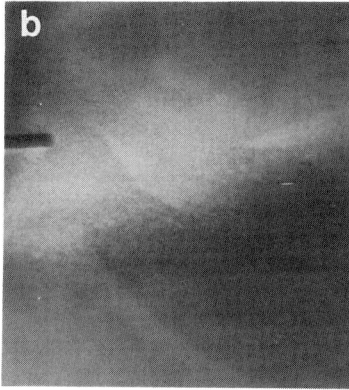

Figure 3: Blocking layers imaged under (a) 200 BF and (b) 200 DF conditions, showing absence of contrast reversal.

4. PREVIOUS WORK

The salient question then remains: what is the physical origin of the observed 20-30% difference in high angle scattering from the change in only ~ 1 part in 10^5 atoms between p- and n- type materials? Voss et al (1980) measured the changes in structure factors for doping levels of ~ 10^{19} cm^{-3} of P and As in Si, and determined maximum differences of ~ 3% for F_{111}. This corresponds to an order of magnitude weaker signal at an order of magnitude greater doping than the present work. Perovic et al (1991) studied TEM contrast from Si:B and Si:As. For the Si:As case, they found only very weak contrast at levels > 10^{19} cm^{-3}, consistent with structure factor contrast. For Si:B they observed contrast levels comparable to the present work, but at doping concentrations ~ two orders of magnitude higher. They explained this relatively high contrast on the basis of elastic displacements around the Si:B site (the match of covalent radii of B to Si is far worse than As to Si), and modeled it using the concept of a "static Debye-Waller factor" introduced by Hall et al (1966). Although this static Debye-Waller factor concept is qualitatively compatible with our present observations, quantitatively, we would expect to need at least two orders of magnitude greater doping to produce the observed contrast. Note also that in the present work, the covalent radii of dopant atoms and host atoms match more closely than Si:B. In summary, existing proposed mechanisms can qualitatively, but not quantitatively explain the present data.

5. PHYSICAL ORIGIN OF CONTRAST

So what is causing this contrast? From the dependence of the observed contrast upon irradiation (Ga ion and electron), and the operative TEM contrast mechanism, we believe it likely that the contrast is associated with point defect creation. We then need to be able to provide a plausible explanation for differential contrast arising from different point defect densities in p- and n-type materials. One

616

possible mechanism is to consider the creation of charged point defects under irradiation (bonding in InP is approximately 50% ionic and 50% covalent). Equilibrium charged point defect densities under irradiation will depend upon the energy difference of the charged defect state to the Fermi level; this will necessarily vary from p- to n- type material as the band edges move. Differential charged point defect densities would then produce differential high angle scattering between p- and n- type material. Note, however, that when the charged point defect density becomes sufficiently high, it will pin the Fermi level, and this differential mechanism saturates. InP is also near a critical ionicity-covalency boundary where point defect behavior is complex (Phillips 1984), and point defects may significantly delocalize in the lattice. Thus, each point defect would have a "magnifying factor" by disturbing many lattice sites around it. This is analogous to the enhanced contrast in Si:B, but does not in the present case arise from strain.

In summary, we have determined the operative TEM scattering mechanism producing the observed dopant contrast. We are still trying to determine the precise physical mechanism which gives rise to the highly anomalous differential scattering between p- and n- type materials.

6. APPLICATIONS

The ability to image such low dopant concentrations with the high spatial resolution inherent to TEM affords several important fundamental and technological opportunities. Dopant "mapping" can provide substantial insight into laser device fabrication, operation and degradation. For example, in Figure 1, there is evidence of the complex blocking layer growth geometry resulting in incomplete blocking of the second n-type layer (marked at X in the image), and for a highly non-planar morphology developing under the oxide cap which propagates during subsequent growth to the semiconductor-contact interface (marked Y in the image), where it causes local defect generation, enhanced metal diffusion and local current concentrations. We are also developing use of this dopant contrast as strain-free "marker layers" for the study of growth and diffusion over complex, non-planar surfaces. For example, in Figure 4, we show thin n-type and p-type marker layers grown over an etched mesa. Such an image represents a map of both growth (from bilayer thicknesses) and diffusion (from ratio of p- to n-layer thicknesses) over the complex surface. For example, Figure 4 is grown under conditions which result in significant Zn diffusion, resulting in thicker p-layers than n-layers.

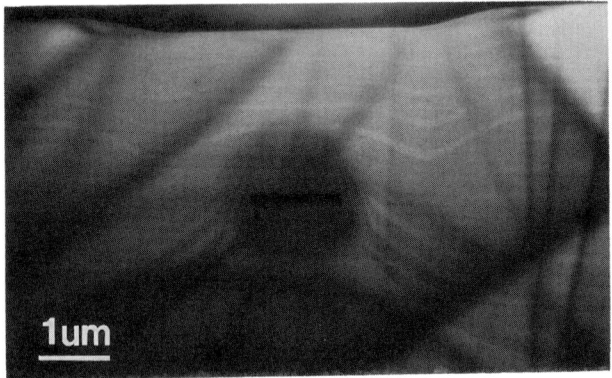

Figure 4: FIB-TEM image of a structure where a superlattice of n-doped (2×10^{18} cm^{-3} Si, light layers) and p-doped (2×10^{18} cm^{-3} Zn, dark layers) has been grown over an etched CMBH laser mesa.

1um

REFERENCES

Hall C R, Hirsch P B and Booker G R 1966 Phil. Mag. **14**, 979
Hull R, Bahnck D, Stevie F A, Koszi L A and Chu S N G 1993 Appl. Phys. Lett. **62**, 3408
Hull R, Stevie F A and Bahnck D 1995, Appl. Phys. Lett. **66**, 341
Perovic D D, Weatherly G C, Egerton R F, Houghton D C and Jackman T E 1991 Phil. Mag. A**63** 757
Phillips J C 1984, Phys. Rev. B**29**, 5683
Voss R, Lempfuhl G and Smith P J 1980, Z. Naturf. A **35**, 973

Inst. Phys. Conf. Ser. No 146
Paper presented at Microsc. Semicond. Mater. Conf., Oxford, 20–23 March 1995
© 1995 IOP Publishing Ltd

Microstructural investigation of $Si_{1-x}Ge_x$/Si(001) MODFET structures grown by gas-source MBE

W J Lee, A E Staton-Bevan, J D Russell, J M Fernández[1], J Zhang[1], and B A Joyce[1]

Department of Materials, Imperial College, University of London, London SW7 2BP.
[1]Interdisciplinary Research Centre for Semiconductor Materials, Blackett Laboratory, University of London, London SW7 2BZ, U.K.

ABSTRACT: $Si_{1-x}Ge_x$/Si MODFET structures, incorporating $Si_{1-x}Ge_x$ graded buffer layers, (x = .05 to .35), have been grown by Gas-Source Molecular Beam Epitaxy on (001) substrates. Relaxation of the buffer layers and the influence of the graded buffer layer growth temperature have been studied by TEM and EBIC-SEM. The majority of the misfit dislocations were confined within graded buffer layers and were of 60° type. Structures having graded buffer layers grown at 615°C showed undulation of the device layers and threading dislocation densities of up to $2 \times 10^7 cm^{-2}$. However, lower buffer layer growth temperatures of 565°C or 515°C produced structures with more planar device layers and threading dislocation densities $\leq 4 \times 10^6 cm^{-2}$, showing potential for MODFET device applications.

1. INTRODUCTION

$Si_{1-x}Ge_x$ alloy heterostructures and superlattices have generated considerable interest because of their novel electronic and optical properties. These materials have shown very promising performance, for applications such as Heterojunction Bipolar Transistors (HBTs), Modulation Doped Field Effect Transistors (MODFETs) and Optical Detectors (ODs).

For Ge concentrations of less than 15%, $Si_{1-x}Ge_x$/Si(001) epilayers can reach thicknesses of up to 1μm by pseudomorphic growth, although the lattice mismatch between Si and Ge is relatively large (4.17%). However, for these low Ge concentrations, fully strained layers show very small bandgap off-sets between Si and the $Si_{1-x}Ge_x$ alloy. The wide range of bandgap engineering required for potential devices includes the need for higher Ge-content epilayers. It therefore is necessary to obtain fully relaxed $Si_{1-x}Ge_x$ layers and to minimise defect densities in the device structure for the entire range of Ge composition (0<x<1). Recently it has been reported that $Si_{1-x}Ge_x$ graded buffer layers on Si(001) substrates produce a reduction in the threading dislocation density in the relaxed epilayer, above the critical thickness. These studies have utilised the growth techniques of Ultra High Vacuum Chemical Vapour Deposition (UHV-CVD) [LeGoues *et al*(1991)], Rapid Thermal CVD (RTCVD), and Solid-Source Molecular Beam Epitaxy (SS-MBE) [Fitzgerald *et al* (1991)]. However, few growth and characterisation studies of $Si_{1-x}Ge_x$/Si device structures grown by the Gas-Source MBE technique (GS-MBE) have been published. GS-MBE has the advantages of a continuous gas supply and source operation at room temperature. This paper reports a TEM and EBIC-SEM investigation of strain relaxation in GS-MBE $Si_{1-x}Ge_x$/Si(001) MODFETs, having linearly-graded buffer layer structures.

2. EXPERIMENTAL

$Si_{1-x}Ge_x$/Si(001) MODFET device structures were grown by GS-MBE. Three inch diameter Si(001) wafers were chemically etched and the silicon oxides were removed in the chamber by heating. The hydrides, disilane (Si_2H_6) and germane (GeH_4) were used as source materials. Diluted arsine (0.1% AsH_3+Ar) was also supplied as an n-type dopant. The average growth rate was approximately 1.5 nm/min. Figure 1 shows a schematic diagram of the MODFET structure. A stack of three buffer layers, consisting of a Si buffer layer, a $Si_{1-x}Ge_x$ linearly-graded buffer layer and a $Si_{0.7}Ge_{0.3}$ relaxed buffer layer, was incorporated below the MODFET device layers. The compositions and growth temperatures of the buffer layers are shown in Table 1. Standard procedures of TEM specimen preparation were used involving Ar+ ion-milling at 4-5 kV with the beam angle in the range 8-18°. The electron microscope used was a 200 kV Jeol-FX2000. A sample holder with double tilting facilities was used for defect analysis.

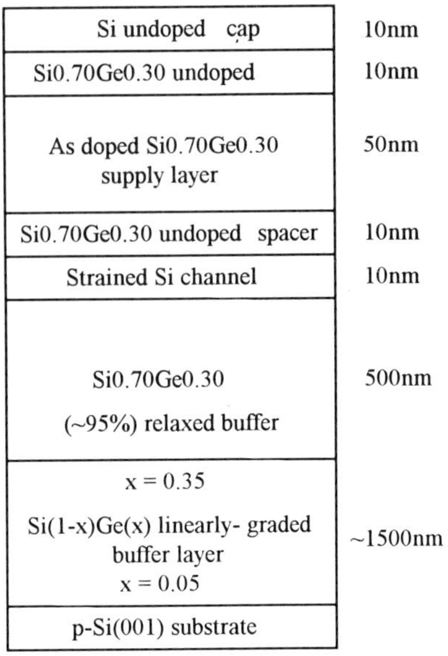

Si undoped cap	10nm
Si0.70Ge0.30 undoped	10nm
As doped Si0.70Ge0.30 supply layer	50nm
Si0.70Ge0.30 undoped spacer	10nm
Strained Si channel	10nm
Si0.70Ge0.30 (~95%) relaxed buffer	500nm
x = 0.35 Si(1-x)Ge(x) linearly- graded buffer layer x = 0.05	~1500nm
p-Si(001) substrate	

Fig.1 Diagram of MODFET structure.

For EBIC-SEM studies, samples were mounted on TO5 headers. Connection was made between the substrate and the surface layer, in order to use the in-built *pn*-junction of the MODFET structure as the charge collecting barrier. Low accelerating voltages 2-5 kV were used to probe predominantly the surface layers.

Table 1 Characteristics of the buffer layers of the $Si_{1-x}Ge_y$/Si(001) MODFET Structures.

Specimen No.	Growth temperature (°C)	Grading rate in linearly- graded buffer (% Ge/μm)	Ge content in the relaxed buffer layer (%)
BF180	Tg = 615	22	30
BF181	Tg = 565	23	30
BF173	Tg = 515	23	30

3. RESULTS AND DISCUSSION

3.1 TEM investigations

Figure 2 shows cross-sectional TEM micrographs of MODFET structures having three different graded buffer growth temperatures. The layers showing lighter contrast are the Si channels. The upper MODFET device layers and the relaxed $Si_{0.7}Ge_{0.3}$ buffer layers were found to be relatively free of dislocations. However, the linearly-graded buffer layers contained large numbers of dislocations, lying on {111} glide planes. These were predominantly of the 60° type and plan view studies showed them to lie preferentially along <110> directions.

Interaction had occurred between dislocations on intersecting {111} planes. The dislocations lying on {111} planes which are normal to the plane of the micrograph may be seen to lie on localised bands of neighbouring slip planes. The majority of the misfit dislocations were confined within the graded buffer layers, although in some places dislocations penetrated into the substrate to depths of up to 10 microns. Very occasionally, dislocations at the upper surface of the graded buffer layer penetrated into the relaxed buffer layer.

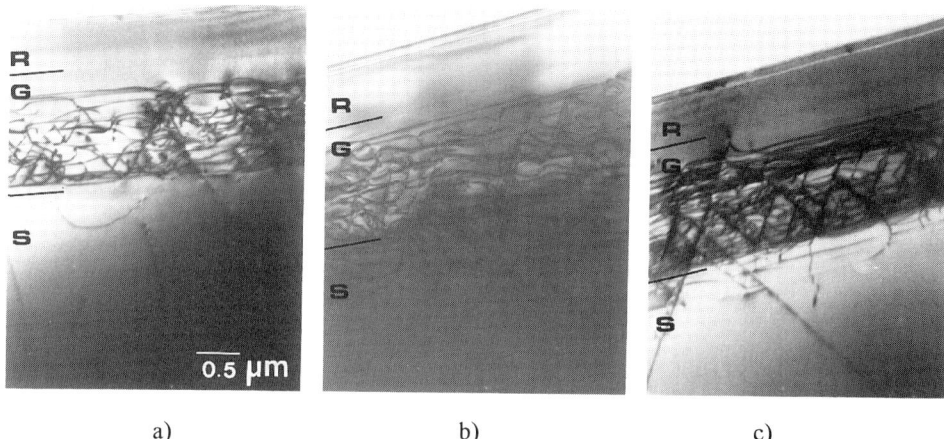

a)	b)	c)

Fig. 2 Cross-sectional TEM images of $Si_{1-x}Ge_x$/Si MODFET structures having linearly-graded buffer layer growth temperatures of a) Tg = 615°C, b) Tg = 565°C, c) Tg = 515°C (R: relaxed buffer, G: linearly-graded buffer, S: Si(001) substrate)

The growth temperatures for the linearly-graded buffer layers for the MODFET structures shown in Figures 2a, 2b and 2c were Tg = 615°C, Tg = 565°C, and Tg = 515°C, respectively. No significant differences could be distinguished between the dislocation structures, in the buffer layers for the three samples, using TEM. However, for the highest Tg = 615°C, an undulation was observed in the upper device layers (Fig. 2a). For the lower growth temperatures, Tg = 565°C and Tg = 515°C, the device layers were more planar and therefore more suitable for MODFET applications. Previous studies of surface undulations, or "ripples", on Si-Ge epilayers have been carried out, for example by Cullis *et al* 1992 (LPCVD) and Dutartre *et al* 1994a (RTCVD). Using TEM diffraction contrast, Cullis *et al* have identified oscillatory strain variations associated with the surface ripples. It is generally agreed that the driving force for ripple formation is partial elastic strain relaxation resulting from an expansion of the lattice at the ripple crests and a contraction of the lattice at the ripple troughs. An increased tendency for ripple formation with increasing growth temperature is reported for all growth techniques and is the result of surface diffusion kinetics.

3.2 EBIC-SEM studies

Figure 3 shows plan-view EBIC-SEM micrographs of MODFET structures. The faint cross-hatch dark pattern along orthogonal <110> directions is associated with enhanced carrier recombination due to bunches of misfit dislocations in the linearly-graded buffer layer region. The image in Figure 3a, for the graded buffer layer growth temperature Tg = 615°C, shows

620

strong, dark line contrast, for example at DL. This arises from inclined dislocations that exist within ~ 200 nm of the surface, in the device layers. The dark spot contrast, for example at DS, arises from threading dislocations. The other samples show very low numbers of dark line and dark spot defects, indicating that the lower growth temperatures are desirable for device applications. Estimates of dislocation densities in the MODFET layers as a function of growth temperature gave values of ~ 4×10^6-2×10^7 cm^{-2} (Tg = 615°C), ~1×10^6-4×10^6 cm^{-2} (Tg = 565 °C), and ~1×10^6-4×10^6 cm^{-2} (Tg = 515°C). This assumes that, because of the low resolution of the EBIC technique (~1μm), one dark line represents an unresolved cluster of 10 dislocations on average. For comparison, Dutartre et al (1994b) and Fitzgerald et al (1992) have reported values of 10^5-10^7 cm^{-2} for relaxed $Si_{1-x}Ge_x$ structures, having 0-32% Ge graded buffer layers (20-50% Ge/μm grading rates) and ~1.1μm $Si_{0.7}Ge_{0.3}$ relaxed layers. These were grown by RTCVD (Tg = 820°C) and SS-MBE (Tg = 700-900°C), respectively.

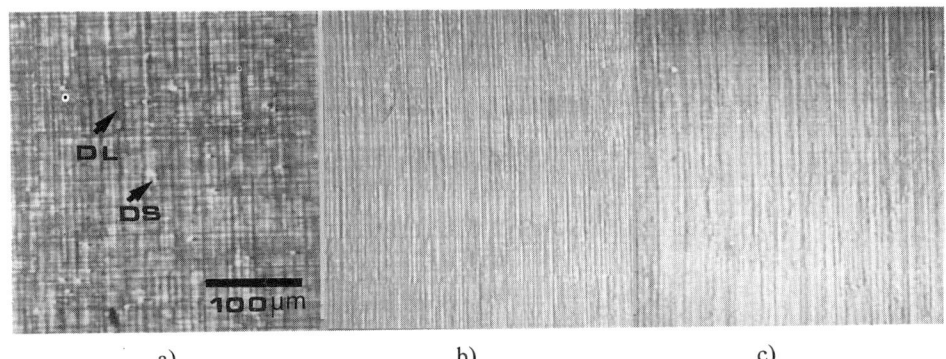

a)   b)   c)

Fig. 3 Plan-view EBIC micrographs of $Si_{1-x}Ge_x$/ Si MODFET structures as a function of graded buffer layer growth temperature. a) Tg = 615°C, b) Tg = 565°C, c) Tg = 515°C

4. CONCLUSIONS

$Si_{1-x}Ge_x$ / Si MODFET structures, on relaxed $Si_{1-x}Ge_x$ linearly-graded buffer layers, have been grown on Si (001) substrates by the Gas-Source MBE technique. The growth temperature of the graded buffer layer has been found to influence both the planarity of the device layers and the threading dislocation density. A growth temperature of 615°C resulted in undulations in the device layers and threading dislocation densities of up to 2×10^7 cm^{-2}. However, lower growth temperatures of 565°C and 515°C produced more planar epilayers and dislocation densities ≤ 4×10^6 cm^{-2}, indicating potential for MODFET applications.

REFERENCES

Cullis A G, Robbins D G, Pidduck A J and Smith P W 1992, J. Cryst. Growth 123 333
Dutartre D, Warren P, Chollet F, Gisbert F, Bérenguer M and Bérbezier I 1994a, J. Cryst. Growth 142 78
Dutartre D, Warren P, Provenier F, Chollet F, and Pério A 1994b, J. Vac. Technol. A12 1009
Fitzgerald E A, Xie Y-H, Green M L, Brasen D, Kortan A R, Michel J, Mii Y-J, and Weir B E 1991, Appl. Phys. Lett. 59 811
Fitzgerald E A, Xie Y-H, Monroe D, Silvernan P J, Kuo J M, Kortan A R, Thiel F A and Weir B E 1992, J. Vac. Sci. Technol. B10 1807
LeGoues F K, Meyerson B S, and Morar J F 1991, Phys. Rev. Lett. 66 2903

Inst. Phys. Conf. Ser. No 146
Paper presented at Microsc. Semicond. Mater. Conf., Oxford, 20–23 March 1995
© *1995 IOP Publishing Ltd*

Quantitative analysis of compound semiconductor solar cells by Electron Beam Induced Current

C Hardingham and D B Holt[#]

EEV, Waterhouse Lane, Chelmsford, Essex CM1 2QU.
Imperial College of Science, Technology & Medicine, Prince Consort Road, London SW7 2BP

ABSTRACT: EBIC measurements have been made on a variety of single crystal, compound semiconductor, space solar cells. Plots of the EBIC gain against beam voltage of homoepitaxial GaAs, heteroepitaxial GaAs/Ge and InP solar cells are presented. Direct correlation between the photovoltaic performance of the cells and the EBIC gain measurements is demonstrated. Quantifiable deterioration of the diffusion lengths, determined by applying Monte Carlo techniques, is evident after irradiation of cells with high energy protons to simulate the environment in space.

1. INTRODUCTION

This paper presents the use of Electron Beam Induced Current (EBIC) techniques in a SEM, to study compound semiconductor solar cells designed for application in space.

As with any other semiconductor device type, the performance of a solar cell is dependant on both the device design and the material properties. Since a solar cell is designed to efficiently collect carriers which are photogenerated by insolation, the device is ideally suited to study by Electron Beam Induced Current, which also relies on collection of carriers, in this case generated by the primary electron beam. Non-destructive EBIC techniques can readily be used to investigate the material properties of working production devices.

The charge collection efficiency (CCE) at any primary beam voltage can be derived from the EBIC gain (collected current/beam current). If the spatial distribution of carrier generation is known, the CCE (or EBIC gain) can be modelled. Adjustment of material parameters used in the model, in particular minority carrier diffusion length, allows a fit of the measured CCE to the model for a variety of beam voltages, and hence evaluation of these material parameters.

2. SOLAR CELLS FOR USE IN SPACE

Ever since the earliest artificial satellites in the 1950s, photovoltaic solar cells have been their main source of power. The first solar cells used in space were made from crystalline silicon, which remains the most widely used material to this day. However, although the properties of silicon are well understood, and it is relatively cheap resulting in a reliable, cost effective power source, its optical characteristics are not ideal for this application. Its band-gap of 1.1eV is somewhat lower than the optimum of 1.4 eV for the AM0 space solar spectrum; that of GaAs is 1.43eV (Hovel, 1975). Also, unlike GaAs and InP, for instance, silicon has an indirect band-gap, leading to a low absorption coefficient: 50-200μm of active device thickness is required, compared to 5-10μm for GaAs (Hardingham et al 1993). The larger volume of active material makes a silicon cell more liable to damage from high energy particle radiation. Radiation-induced performance degradation of the solar arrys is a major constraint in Satellite operational life.

Increasing interest has therefore been shown, over the last 2 decades, in the use of alternative materials. In particular GaAs solar cells, grown either homoepitaxially, or on a Ge substrate, are now finding widespread application (Datum and Billets, 1993), where their higher unit cost is offset by better radiation resistance, and balance of system savings due to their higher beginning of life (BOL) efficiencies.

High efficiency solar cells require high quality material, in order to minimise loss mechanisms such as non-radiative recombination. In particular, good minority carrier diffusion lengths are required in the active

622

region of the device, in order to maximise the collection of photo-generated carriers. Knowledge and control of the material properties are thus crucial to effective manufacture.

EBIC has been used previously to determine the diffusion lengths in GaAs, and InP, solar cells (see for instance, Chung et al, 1988, Hakimzadeh and Bailey, 1992). However these studies have all involved EBIC scans across the junction, and thus require a cleaved sample. The techniques are thus not applicable as non-destructive tests on real devices.

3. SOLAR CELL STRUCTURES

The structure of an IMLPE (Infinite Melt Liquid Phase Epitaxy) GaAs solar cell is shown schematically in Fig. 1, and described in detail by Cross et al (1989). It comprises a GaAs homojunction; an optically transparent $Al_xGa_{1-x}As$ window layer on top of the p-type emitter acts to reduce the interface recombination velocity of the top face of the emitter layer. A SiN antireflective layer maximises optical transmission to the

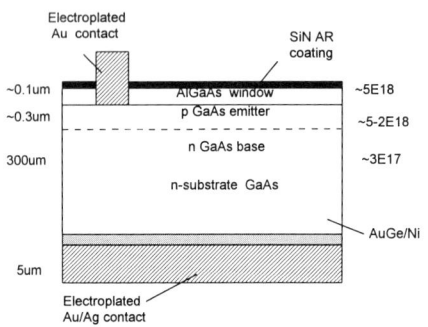

Fig. 1: Schematic section through an IMLPE GaAs solar cell structure

Fig. 2: Schematic section through a MOCVD GaAs/Ge solar cell structure.

active part of the structure. The contacts are fine grain electroplated Au and Ag, made directly onto the highly doped p-emitter and with an evaporated Au/Ge/Ni contact layer to the n-substrate.

The MOCVD GaAs and GaAs/Ge solar cells comprise a similar p-n GaAs homojunction, however, in this case the antireflective coating is $TiOx/Al_2O_3$. The n-type base layer is grown on a highly-doped n-type buffer. The buffer layer is grown either onto 300μm thick <100> GaAs substrates misorientated a few degrees towards <111>, or on a nucleation layer on a 200μm Ge substrate, misorientated a few degrees off <100>.

The InP structure, shown in Fig 3, is an n-p shallow homojunction. The structure is described fully by Hardingham et al (1991).

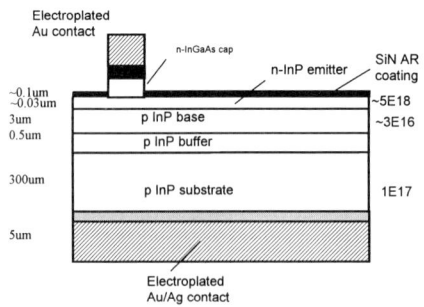
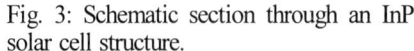
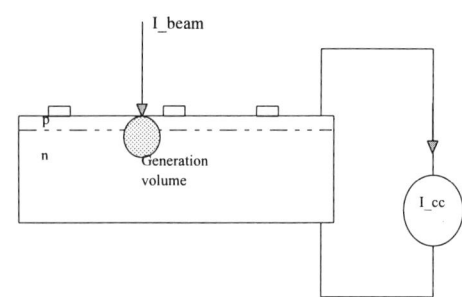

Fig. 3: Schematic section through an InP solar cell structure.

Fig. 4: Schematic of EBIC measurement arrangement.

4. EXPERIMENTAL TECHNIQUES

The solar cells to be studied were mounted, complete, using pressure contacts to front and rear faces, on an adapted specimen stub prior to loading into the Jeol 840A Microscope. The configuration is shown in Fig. 4. The technique is thus non-destructive, which is useful when working with real operational devices. The contacts were made, via a head amplifier, to a low input impedance Matelect Current amplifier, from which the Electron Beam Induced Current (I_{cc}), could be read. Use of the Matelect amplifier to detect photogenerated current in the samples during loading assisted in ensuring good electrical contact.

The beam current (I_{beam}) was measured by insertion of a Faraday cup into the beam, and measuring the collected current with the Matelect.

5. RESULTS AND DISCUSSION

5.1 IMLPE GaAs Solar Cells

The variation of EBIC gain with beam voltage for a typical GaAs solar cell is shown in Fig. 5 (triangles). The form of the curve is explained as follows: At very low beam voltages (<about 3kV), the Grün range of the incident electron beam is very low, so the generation volume only extends through the SiN antireflection layer into the AlGaAs window layer, and not the active part of the device - so no collection is observed. In the range 5-15kV the generation volumes lie close to the p-n junction, so almost all charge is generated within a diffusion length of the junction and is therefore collected. Since the charge generated scales linearly with the incident electron energy (or Vacc), the curve rises linearly in this portion of the curve. At higher beam voltages, however, the gain does not rise so fast, and even starts to tail off, since progressively more of the charge is generated deep within the base, where there is a significant chance of recombination, or even in the buffer or substrate, from which there is a very low probability of collection.

Fig. 5 also shows the variation of EBIC gain with beam voltage for a IMLPE GaAs solar cell after irradiation with 10^{12} 1 MeV protons (squares). At low beam voltages the gains are similar. However, at higher beam voltages the gain in the irradiated cell starts to tail off, indicating a higher level of recombination ie shorter diffusion length in the base of the cell. The analogous plot of the variation in quantum efficiency (QE) of the cell across the optical spectrum is shown in Fig 6. This indicates a lower QE across the whole spectrum for the irradiated cell. Since at short wavelengths the optical absorption of the GaAs is strong, a poor QE indicates poor collection from near the surface of the device. The long wavelength QE, although including a contribution from carriers generated deeper in the device, is still dominated by the emitter response. Modelling of the QE plots thus provides more accurate information about the emitter of the cell, whereas modelling of the EBIC gain provides better data for the base of the case.

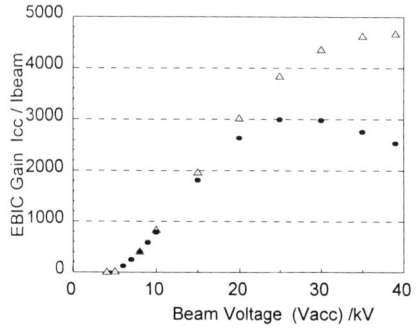

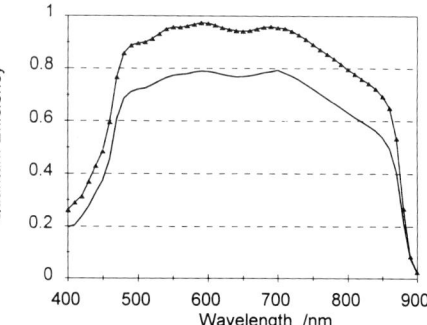

Fig 5: EBIC gain for an unirradiated IMLPE cell (▲), & after 10^{12} 1 MeV H$^+$ (●).

Fig. 6: Spectral QE of IMLPE GaAs Solar cells, unirradiated (top), & after 10^{12} 1 MeV H$^+$(lower).

5.2 Proton Irradiated MOCVD GaAs and GaAs/Ge cells.

Fig. 7 shows the effect of high energy proton irradiation (10MeV in this case), on MOCVD GaAs cells, both homo- and heteroepitaxial. The response of the cell after low levels of radiation is comparable to the unirradiated IMLPE cells; in fact, it is slightly better, which is indicative that the epitaxial base of the MOCVD cell is better quality that the bulk base of the IMLPE cell. However, after a high radiation dose, the response (particularly at higher beam voltages) is degraded, which is evidence that the lattice is being damaged by the proton bombardment.

The response of a heteroepitaxial GaAs cell after 10^{12} 10 MeV protons is also shown in Fig 7. In this case, the irradiation affects the cell at much lower beam voltages, which is indicative that the material is being affected higher in the device. (At lower doses, the response is very similar to the homo-epitaxial cell for beam voltages up to 20 kV). It is possible that the radiation is "unpinning" misfit dislocations which were initially tied to the hetero-interface.

5.3 InP cells

Fig 8 shows the variation of EBIC gain with beam voltage, for an InP shallow homojunction cell, together with a modelled response with various values of the base (electron) diffusion length. The sensitivity of the response to the diffusion length is clear: the fit demonstrates very good quality material, which is consistent with the >18% photovoltaic conversion efficiency achieved with this device.

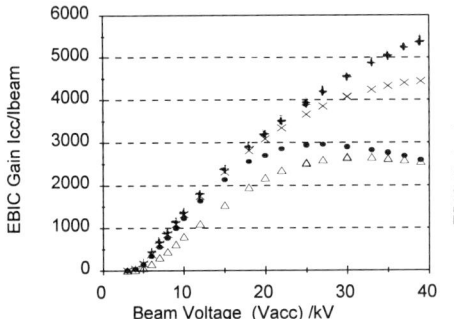

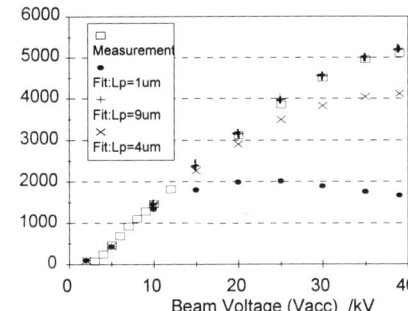

Fig 7: Irradiated MOCVD cells: GaAs @ 10^{10}, 10^{11} & 10^{12} (+, X, •), & GaAs/Ge @ 10^{12} ($\triangle$)

Fig 8: EBIC gain vs Vacc for InP cell (squares), modelled with various electron diffusion lengths.

6. CONCLUSIONS

Planar EBIC measurements have been made on a variety of III-V space solar cells. Qualitative variation in response, consistent with photovoltaic performance has been demonstrated, and a non-destructive, quantitative measurement of diffusion lengths in production devices has been demonstrated.

REFERENCES

Chung M A, Meier D L, Szedon J R and Bartko J 1988 Proc 20th IEEE Photovoltaics Specialists Conference, Las Vegas (New York: IEEE) pp 924-929

Cross T A, Burrage J, Hardingham C and Potts A 1989 Proc European Space Power Conference, Madrid (Noordwijk: European Space Agency publication SP294) pp 525-530

Datum G C and Billets S A 1991 Proc 22nd IEEE Photovoltaics Specialists Conference, Las Vegas (New York: IEEE) pp 1422-1428

Hakimzadeh R and Bailey S 1992 Proc XII Space Photovoltaics Research and Technology Conference, Cleveland (Cleveland: NASA publication CP-3210) pp 64-77

Hardingham C, Burrage J, McLeod S, Cross T A, Pearsall N, Forbes I and Winckler J 1991 Proc 2nd European Space Power Conference, Florence (European Space Agency publication SP-320) pp 543-546

Hardingham C, Huggins C, Cross T, Gray A, Mullaney K and Kitchen C 1993 Proc 23rd IEEE Photovoltaics Specialists Conference, Kissimmee (New York: IEEE) pp 1399-1403

Hovel H J 1975 Solar Cells (London: Academic Press) p75

Inst. Phys. Conf. Ser. No 146
Paper presented at Microsc. Semicond. Mater. Conf., Oxford, 20–23 March 1995

Transmission electron microscopy of CVD copper for metallisation in microelectronics

L Weaver, R Siemsen* and M Sayer*

Dept of Materials and Metallurgical Engineering, Queen's University, Kingston, Ontario, Canada K7L 3N6
* Dept of Physics, Queen's University, Kingston, Ontario, Canada K7L 3N6

ABSTRACT: Patterned aluminium thin films, deposited by either sputtering or PVD have been the most widely used interconnect structures in the fabrication of silicon devices. However, as device dimensions approach deep sub-micron sizes, the reliability of the interconnect in terms of electromigration and also the interconnect resistance become increasingly important. Therefore, lower resistance metals are being investigated to replace the present aluminium based system. Copper thin films are attractive as ULSI conductor material as the higher melting point of copper removes the maximum temperature limit imposed for low melting point aluminium thus facilitating further processing of the multilevel microelectronic devices. CVD copper deposition from an organo-metallic precursor offers the highly conformal films required to fill vias and trenches in multilevel metal interconnect architectures. These films have been examined by TEM, AFM and AES and their microstructure related to the electrical properties.

1. INTRODUCTION

Copper films have been successfully deposited by a number of methods (Gelatos et al 1993a, Kaloyeros et al 1990 and Pai et al 1989). CVD is the most promising because of the ability to deposit highly conformal films. This is important in the deep submicron regime where the aspect ratio of vias, for example, is expected to be greater than 2. Copper will also support high current densities for improved electromigration resistance. CVD is a relatively simple process, an organometallic copper precursor contacts a hot wafer and decomposes into copper and volatile byproducts which are pumped away. The precursor used in this study is Cu(1) hexafluoroaceylacetone trimethylvinylsilane (Norman et al 1991), one of the Cu^{+1} (hfac) group. The conversion efficiency of these precursors is high with deposition rates of 1000 nm/minute reported (Jain et al 1991). The deposition of selective copper is achieved by a disproportionation reaction catalysed by electron transfer (Jain 1991 and Cohen 1992). The growth characteristics of copper on conducting substrates show a transition from a discontinuous collection of copper nuclei to a continuous copper film (Gelatos 1993b). A minimum film thickness of 300 nm is required for low(<10 μohm-cm) resistivity because of the porous nature of thinner films.

2. EXPERIMENTAL

Copper films of 100 nm to 2 μm in thickness were deposited by low pressure chemical vapour deposition (LPCVD) in a single wafer, rotating disk reactor. The copper source was Cu(1) hexafluoroaceylacetone trimethylvinylsilane, known commercially as Cupra Select (Norman et al

1991). The substrates were p-type , 150mm <100> silicon wafers with a 1.2μm blanket covering of SiO₂. The SiO₂ was then coated with 1μm of aluminium, annealed at 400°C in argon for 30 minutes in the reaction chamber, or 1 μm of titanium nitride. Copper depositions were carried out for 30 minutes using a process gas flow of 4 sccm of nitrogen and a dilution gas flow of 8-10 sccm of hydrogen. The chamber pressure was maintained at 1 Torr throughout. The substrate temperature was maintained at 275 °C with a rotation speed of 50 rpm.

The resistivity of the as-deposited films were calculated from four-point probe measurements. Auger spectroscopy (AES) was carried out to determine impurity concentrations. Transmission electron microscopy (TEM) plan-view (PV) samples were prepared using a modified lift-off technique (Weaver 1995a) and cross-section (XS) samples by the aperture grid method (Weaver 1995b). These samples were examined using a Philips CM20. An Nanoscope III atomic force microscope (AFM) was used to examine the growth mechanism of the copper film.

3. RESULTS AND DISCUSSION

The precursor is composed of two key components, Cu^{+1} hexafluoroacetylacetonate (Cu^{+1}hfac) and trimethylvinylsilane (TMVS). The TMVS stabilises the copper^{+1} oxidation state while the complex is in liquid or gaseous form at moderate temperature conditions. Under CVD conditions and /or elevated temperatures (>130°C) the TVMS can be made to dissociate from the complex thereby generating unstable Cu^{+1} (hfac) species which then disproportionate to yield copper metal and volatile Cu^{+2}(hfac). Initially, chemisorption of the precursor occurs on the substrate and dissociates into Cu^{+1} (hfac) and TMVS. The TMVS, being more volatile, desorbs leaving the Cu^{+1} (hfac) behind. Two of these species react to form copper metal and Cu^{+2} (hfac)₂ which is volatile. The breakdown sequence is shown below and the structure in figure 1.

$$2Cu^{+1}(hfac)TMVS_{(g)} \longrightarrow Cu^0_{(s)} + Cu^{2+}(hfac)_{2(g)} + 2TMVS_{(g)}$$

Conductors rather than insulators (or semiconductors) appear to catalyse this reaction, probably relating to the need for electron transfer to occur. In this way, selective copper deposition can be achieved onto metallic surfaces. Since it is necessary to provide a copper impervious barrier layer such as tantalum, titanium nitride or tungsten between the copper and the silicon substrate (Chamberlain 1982 and Wang 1994) this selective deposition permits the direct growth of copper onto these surfaces.

Aluminium and titanium nitride were used as a conducting substrates to study the growth mechanism. The nucleation rate is determined by the precursor availability and once nucleation occurs, growth is favoured. The initial copper deposition is in the form of islands. These islands then act as growth centres from which the initial copper spheres enlarge and grow hemispherically The TEM plan-view sample, figure 2a, clearly demonstrates the non-continuous nature of the film while the cross-section, figure 2b, shows the growth of existing nuclei. This is also reflected in the AFM images of the same sample, figure 2c. The resistivity of this discontinuous films is so high as to be non - conducting when measured with a four-point probe.

After deposition for a period of 30 minutes, the copper islands merge into a continuous film. This film, shown in figure 2d, contains many growth twins and a high

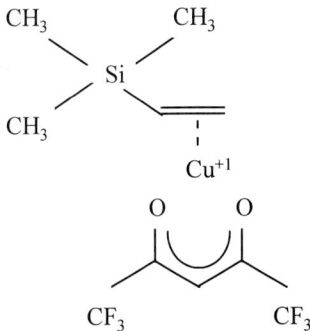

Figure 1: The structure of the precursor

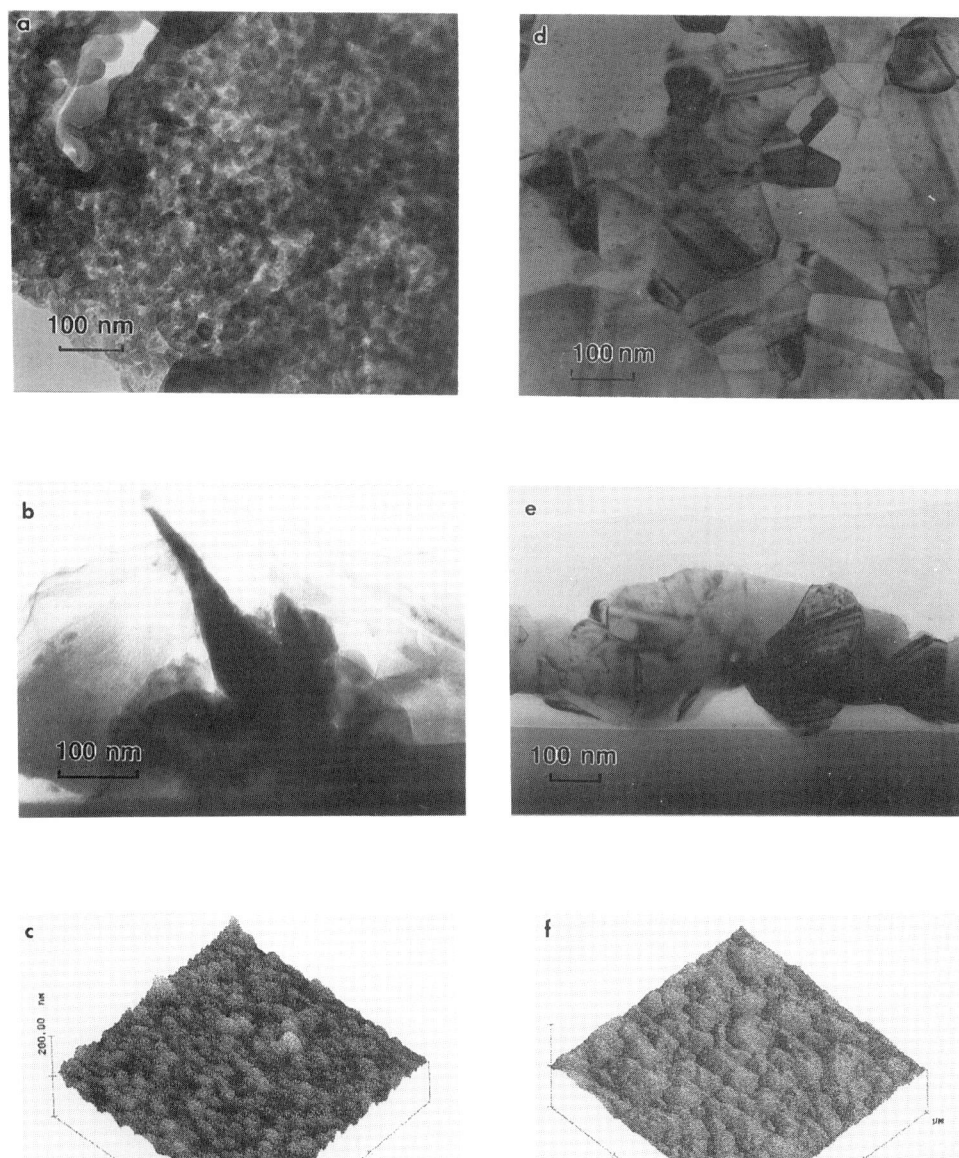

Figure 2: Initial copper deposition is in the form of islands (a) shown in plan-view (b) in cross-section showing the growth of a single nucleus and (c) AFM image .
After a longer deposition a continuous film is formed and the morphology can be see in (d) plan-view (e) cross-section (f) AFM image.

density of slip dislocations suggesting possible plastic deformation during film growth (Nakahara 1988). The stress for this film was measured at 43 MPa and the resistivity was 6.7 μohm-cm. The roughness of this film can be see in the TEM cross-section, figure 2e, and is reflected in the AFM image, figure 2f. Auger electron spectroscopy was carried out and verified that no detectable contaminants were present in the film. Contamination of CVD deposited films by carbon from organometallic precursors is detrimental to the resistivity of the deposited films.

4. CONCLUSIONS

The resistivity of CVD copper films is directly related to microstructure. The growth mechanism can be examined by electron microscopy, or , more easily by atomic force microscopy. The film roughness at the thicknesses required to achieve usable resistivities may present challenges to planarisation in the semiconductor industry.

ACKNOWLEDGMENTS

Dr Prasad Gadgill carried out the initial depositions. AES was provided by Carl Watkins of Surface Science Western at the University of Western Ontario, Canada. Dr J Szpunar of McGill University, Montreal, performed the AFM.

REFERENCES

Chamberlain M B 1982 Thin Solid Films 91 155
Cohen S, Liehr M and Kasi S 1992 Appl. Phys. Lett. 60 p50
Jain A S, Chi K-M, Kodas T T, Hampden-Smith M J, Farr J D and Paffett M F 1991 Chem.   Mater. 3 p995
Gelatos A V, Marsh R, Kottke M and Mogab C J 1993a Appl. Phys. Lett. 63 p2842
Gelatos A V, Poon S, Marsh R, Thompson M and Mogab C J 1993 Proc. of the 1993b Symp. on VLSI Technology Kyoto Japan p123
Kaloyeros A, Feng A, Garhart J, Brokks K C, Chosh S K, Saxena A N and Luers F 1990   J. Electron. Mater. 19 p271
Nakahara S 1988 Acta Metall.36 (7) p1669
Norman J A T, Muratore B A, Dyer P N, Roberts D A and Hochberg A K 1991 Proc. VMIC Conf. June 11-12, pp123-129
Pai P-L, Paunovic M and Ting C H 1989 Electrochem. Soc. Extended Abstracts 88-2 p362
Wang S-Q 1994 MRS Bulletin Vol XIX 8 pp 31-408.
Weaver L 1995a Microscopy Research and Techniques (in press)
Weaver L 1995b Microscopy Research and Techniques (in press)

Inst. Phys. Conf. Ser. No 146
Paper presented at Microsc. Semicond. Mater. Conf., Oxford, 20–23 March 1995
© *1995 IOP Publishing Ltd*

Focused ion beam sample preparation for TEM

John F Walker, J C Reiner* and C Solenthaler*

FEI Europe Ltd., Brookfield Business Centre, Cottenham, Cambs., CB4 4PS.
*ETH Zentrum, ETZ H78, CH-8092 Zurich, Switzerland.

Abstract: Microfocused ion beams have been used to prepare thin (<100nm) membranes for transmission electron microscopy. This technique is described with particular reference to two specific cases where, previously, difficulties have been found. The first describes mirrors of silicon and aluminium oxide multilayers on gallium arsenide lasers. Focused ion beam techniques were used to thin each layer equally, where conventional methods removed much of the structure before the remainder became electron transparent. The second describes the locating and imaging of a 100nm diameter defect in a 1μm square gate region. Techniques are described for reliably producing thin sections of many semiconductor materials. A quantitative analysis of these effects from TEM images and diffraction patterns is given, showing that for a direct, normal incidence beam, a 40nm deep region is observed. For glancing incidence the damaged region is under 5nm.

1. INTRODUCTION

Transmission electron microscopy (TEM) is a technique for producing extremely high resolution images of thin sections in many materials. While the resolution performance of such systems is virtually unmatched, limitations in the applicability of the technique arise from the difficulty in preparing the materials sufficiently thin for electron transmission. There are cases where TEM observation is extremely difficult or even impossible using the conventional thinning techniques of grinding and broad beam ion thinning. Here we identify two cases where conventional techniques have failed to provide a sample and describe how the use of a focused ion beam (FIB) has allowed the samples to be prepared. The FIB system used in these experiments was an FEI 200 system equipped with a 30kV liquid metal (gallium) ion source capable of better than 20nm resolution at low currents. This is a fully computer controlled system and features iodine enhanced etch and platinum deposition capabilities.

2. SAMPLE PREPARATION TECHNIQUE

Following the techniques developed by Kirk (1989) and Young (1990) and subsequently applied by Hull (1993 and 1995), small pieces of the material containing the region of interest are initially cleaved and ground to appropriate dimensions for mounting on a TEM grid. Typically, a 2-3mm square piece of semiconductor would be cleaved from the wafer and then ground down from two sides such that the feature of interest lies approximately in the middle of a 50-100μm thick sliver. A copper TEM grid with a hole or slot in the centre is prepared by cutting out a vee from one edge to allow access to the top of the sample for the focused ion beam. The sample is attached to the grid using a conductive adhesive with the feature of interest as central as possible (fig. 1). This is mounted onto the stage of the FIB system and when the pressure reaches 10^{-5}mbar or less, FIB preparation can begin. The ion beam is used at low currents in a similar way to an SEM to allow the feature of interest to be identified and centred. Normally, and particularly if a feature is close to the surface, a layer of metal is deposited to protect and planarise the surface. Platinum is selectively deposited over the feature of interest by ion beam dissociation of a platinum containing organo-matallic gas, typically laying a 1μm thick film in less than 6 minutes. High beam currents in the 5-15nA range are used to mill trenches on either side of the feature (fig. 2). These would typically be 10-20μm wide by 5-10μm deep. The dimensions must be sufficient to allow a reasonable degree of tilting in the TEM without shadowing the beam. This process usually takes from 30 minutes to an hour but can be reduced if an enhanced etch gas is directed onto the milling area. After the initial trenches have been milled, a membrane 1-2 μm thick is left; this must be thinned in successive steps, progressively reducing the ion beam current each time. The final cut would typically be done at 100pA and would remove very thin slices to leave a membrane 100-150nm thick and up to 10μm wide and 5μm deep. The specimen can then be transferred to the TEM.

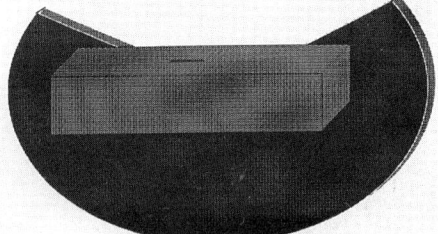

Figure 1. The sample is mounted on a modified TEM grid with conductive adhesive. The feature of interest should be approximately central.

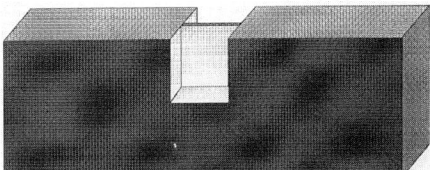

Figure 2. Trenches are milled on either side of the feature of interest.

3. SEMICONDUCTOR LASER MIRRORS

There is a substantial and continuing interest in semiconductor lasers and their degradation modes (see Ueda, 1988). One particular area of interest is in the integrity of the mirror structures which define the ends of the optical cavity. The mirrors are multilayer dielectric stacks of low and high refractive index materials whose dimensions control the reflectivity. The high photon densities can lead to catastrophic degradation which can be minimised by using refractory materials with low absorption coefficients. TEM can be used to study the microstructural integrity of the mirrors; however in many cases the very different hardnesses of the materials make preparation impossible. One case in question involves mirrors of silicon and aluminium oxide. The aluminium oxide is considerably harder and less reactive than the silicon. During conventional thinning processes, the silicon is removed before the aluminium oxide becomes sufficiently thin to be electron transparent. The mirror structure shown in figure 3 was prepared by FIB methods and shows excellent structural integrity and shows no adverse effects from the milling procedure except at the edges where the materials become very thin.

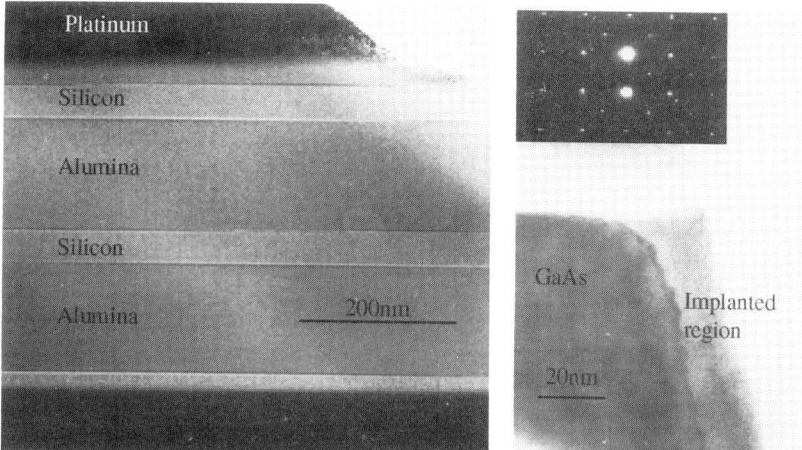

Figure 3. TEM image of the laser mirror structure. Inset is a diffraction pattern taken from the GaAs substrate.

The final membrane was tapered by milling at an angle on one side. This was done to achieve two effects; firstly to be certain that there would be a region thin enough for electron transparency, and secondly to investigate any damage caused by the ion beam on the milled faces. It is interesting to note that at these edges, where differential milling effects would be expected to be most easily seen, little or no excessive milling of one material over the other is seen.

While the motivation for the use of FIB to prepare this sample was its ability to accurately and cleanly thin the mirror area, the gallium arsenide of the substrate is interesting to look at as it gives a clear idea of any artefacts introduced by the preparation. Diffraction patterns from all regions of the membrane show many diffraction orders with little or no rings. Indeed, measurements on the thinnest edge show damage penetrating to less than 4nm, presumably from the lateral straggle of the ion beam. This is similar and certainly no worse than conventional preparation techniques.

4. SITE SPECIFIC PREPARATION

Integrated circuit technology has been inexorably moving towards faster, denser and, in particular, smaller structures than previously thought possible. Where process analysis engineers have until recently been able to rely exclusively on SEM technology to resolve processing problems, feature size, such as the sub-10nm oxide-nitride-oxide layers in trench capacitors, now dictates that higher resolution techniques such as TEM be used. The present experiment examines the failure of a gate oxide in a field-effect transistor. The structure of the device is shown in figure 4. The gate is polycrystalline silicon on a 17nm thick oxide. The gate area is 1μm by 1μm. If the gate voltage exceeds a threshold, about 17V, the oxide breaks down and passes a high current which can melt the silicon. The expected dimension of this failure is less than 100nm in diameter and conventional grinding and polishing techniques could easily fail to find the exact site.

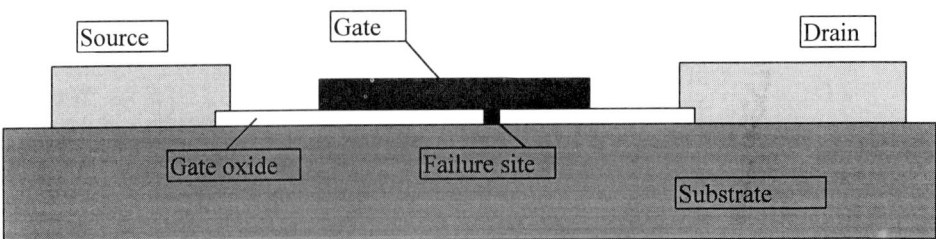

Figure 4. Diagram showing a cross-section of the device showing the expected position of the break in the oxide layer.

As in all TEM sample preparation, a small piece of the wafer containing the failed device was mounted on a TEM grid. The sample and grid were mounted in the FIB and the gate area identified from surface features. Trenches were cut to provide two channels, one on either side of the site of the transistor. Under normal circumstances, the membrane would not be viewed with the ion beam as this would introduce some additional ion beam damage. In this case, as the defect site cannot be precisely identified from surface features, some imaging of this face must be done. The dose was reduced to a minimum, using a 3pA beam and one 300nS per pixel scan. First the gate structure was identified. Successive slices were removed from the membrane, reducing the thickness by about 50nm each time.

Figure 5(a). The initial trenches are ion beam milled on either side of the transistor.

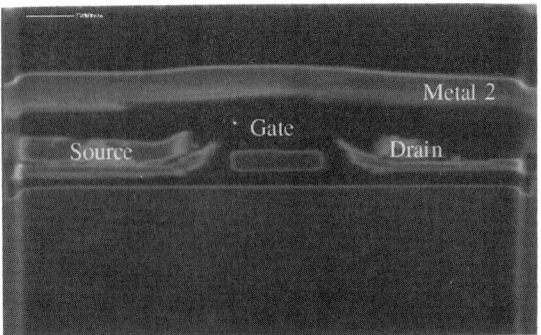

Figure 5(b). Further milling on one side begins to reveal parts of the transistor structure.

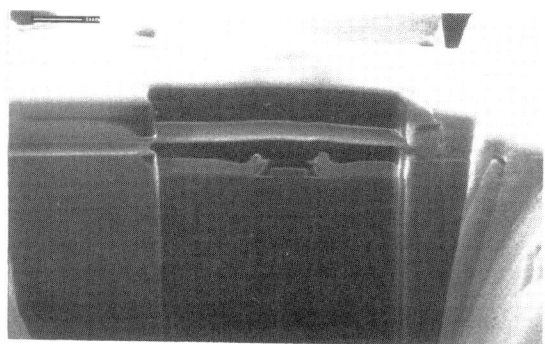

Figure 5(c). The transistor structure is clearly visible. Care must be taken to search for the defect in the oxide layer. Milling continues slice by slice.

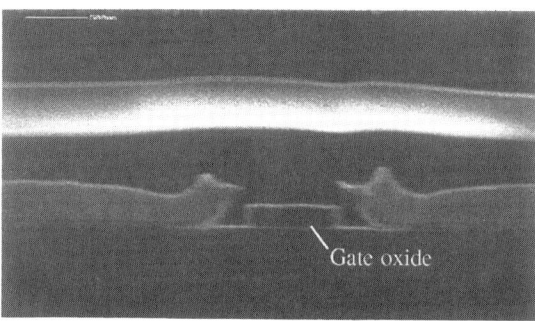

Figure 5(d). Further milling and imaging in 50nm steps continues to show an intact oxide.

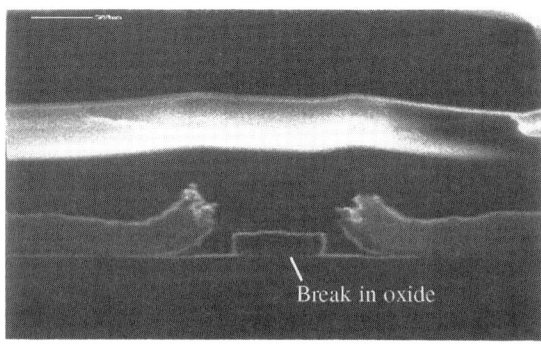

Figure 5(e). Failure found as a break in the oxide layer. Milling now continues from the rear to leave a thin membrane for TEM imaging.

After each slice was removed, with the milling beam parallel to the membrane surface, the sample was tilted to 45° and the face imaged as described above and searched for any signs of the problem (Fig. 5). When the defect appeared, as seen by a discontinuity in the oxide layer, milling of that face was stopped and started on the opposite face, which continued without imaging until the final membrane was under 100nm thick. The sample was then ready for TEM imaging.

Figure 6 shows the TEM image taken of the structure. The gate is clearly polycrystalline and this is confirmed by the diffraction pattern taken from this region. In addition, the substrate is single crystal which is again confirmed by the diffraction pattern, though the sample tilt was not sufficient to reach the exact, on-axis diffraction condition . The defect itself is a small ball of fine grain polycrystalline silicon centred on the break in the amorphous gate oxide.

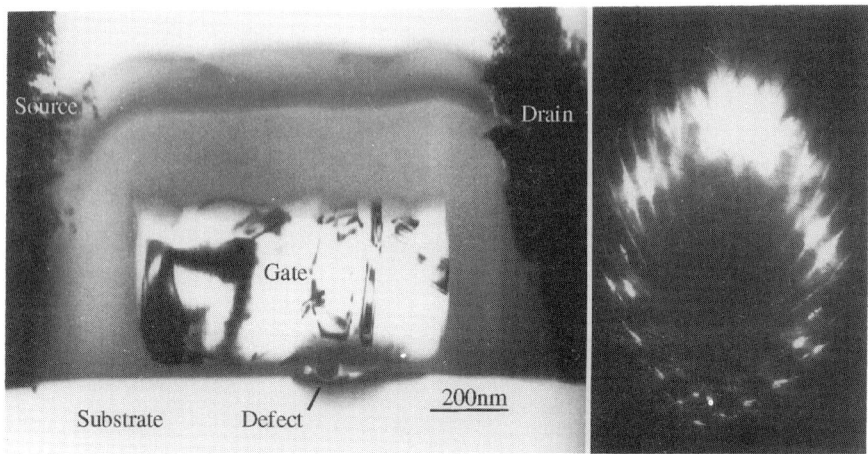

Figure 6. TEM micrograph of the structure with a diffraction image of the substrate.

5. CONCLUSIONS

A focused ion beam system has been used to prepare TEM specimens which have previously been found impossible to produce. In one case, the mixture of materials made equal thinning rates impossible by traditional techniques. FIB technology provided the precision milling required to thin the material only in those areas where the beam was directed and hence allowing uniform thickness. In the second case, the need to precisely locate the feature of interest, which otherwise cannot be seen if buried deep within the structure, was allowed by the system's high resolution imaging capability. It is likely that as the technology progresses, FIB will be used more to resolve these and other issues.

6. ACKNOWLEDGEMENT

We would like to express our gratitude to R Broom of the Materials Science Department of Cambridge University, England for his TEM work and useful discussions about TEM and semiconductor lasers.

REFERENCES

Hull R, Banck D, Stevie F A, Koszi L A, Chu S N G, 1993, Appl. Phys. Lett. **62**, 3408
Hull R, Stevie F A, Banck D, 1995, Appl. Phys. Lett., **66** (3), 1
Kirk E C G, Williams D A, Ahmed H, 1989, Inst. Phys. Conf. Ser. **100**, 501
Ueda O, 1988, J. Electrochem. Soc. **135**, 11C
Young R, Kirk E C G, Williams D A, Ahmed H, 1990, Microelectron. Eng. **11**, 409

Inst. Phys. Conf. Ser. No 146
Paper presented at Microsc. Semicond. Mater. Conf., Oxford, 20–23 March 1995
© 1995 IOP Publishing Ltd

Investigation of high electric fields in semiconductors by Franz-Keldysh effect microscopy

K Berwick and M R Brozel

Centre for Electronic Materials and Solid State Electronics Group, Department of Electrical Engineering and Electronics, University of Manchester Institute of Science and Technology, Sackville Street, Manchester, M60 1QD, UK

ABSTRACT: We have developed a system for viewing regions of high fields within semiconductors. The system utilises the near bandedge absorption resulting from the Frans-Keldysh effect. It is shown that the system is useful for imaging regions of high electric field within GaAs particle detectors and can be used to investigate breakdown effects at high fields.

1. INTRODUCTION

There is, at present, considerable interest in developing nuclear particle detectors fabricated from semiconductors other than silicon. Although Si detectors have great sensitivity they have limited stopping power and suffer from irradiation damage after moderate doses of neutrons. Gallium Arsenide has been considered for some time as an alternative to silicon because of its greater stopping power (due to the larger masses of Ga and As compared to Si) and good resistance to radiation damage (Beaumont et al, 1990, 1991). GaAs also benefits from being able to be supplied in a high resistivity form (semi-insulating, SI) which suggests that large volumes can be employed, this advantage being a result of a high volume of "intrinsic" material being available for electron-hole production without the need to apply large voltages to deplete the material. Unfortunately, detectors made from SI GaAs have generally yielded disappointing results (McGregor et al, 1992). Thus the fraction of charge produced in the GaAs by the ionizing particle that is detected in the outside electronic circuit, the charge collection efficiency (CCE), rarely achieves 50% for alpha particles. The sensitivity of the detector as a function of applied bias is also unexpected and it appears that a sensitive volume whose penetration into the material is <u>linearly</u> proportional to the applied bias is present (Berwick et al, 1994a).

In this paper we report a technique for directly viewing the magnitude and penetration of this high field region. This method is based on optical absorption via the Franz-Keldysh effect. We also demonstrate the use of this technique to image the creation and motion of high field domains which are a characteristic of high field breakdown in this material at low temperatures.

2. METHOD

The microscopic technique to be presented here uses the increase in near bandedge absorption that results from the application of a high electric field to a semiconductor. Such a field results in severe band bending with penetration of conduction and valence band states into the bandgap due to the finite potential barrier present at the bandedges. This effect, the Franz-Keldysh Effect, results in a small increase in band to band absorption for energies of a few meV below Eg (Pankove, 1971). It can be shown that the dependence of the field-induced shift in wavenumber $\Delta\omega$ on the field, E, is of the form $\Delta\omega \propto E^{3/2}$ (Keldysh, 1958, Franz, 1958).

Measurement of the absorption in this energy range can be used to derive the electric field. If regions of different electric field in a structure are expected, absorption microscopy at wavelengths corresponding to energies in this range, gives a mapping of regions where the electric field is high.

In this study we are interested in simple, parallel sided structures consisting of a slab of {100} SI GaAs (thickness ~450 microns) on to the surfaces of which have been deposited one Schottky and one Ohmic contact. The Schottky contact is formed by evaporating Au onto the previously cleaned surface; the Ohmic contact is either indium painted on the surface at ~180°C or gold-germanium evaporated under a vacuum of better than 10^{-7} Torr and then alloyed at 450°C for 30 seconds. Similar results are obtained with either contact type. These structures are, in fact, similar to those used for particle detector testing (Berwick et al, 1994b). For the present microscopical measurements they are cleaved along two orthogonal sets of parallel (110) directions to produce the structure sketched in figure 1.

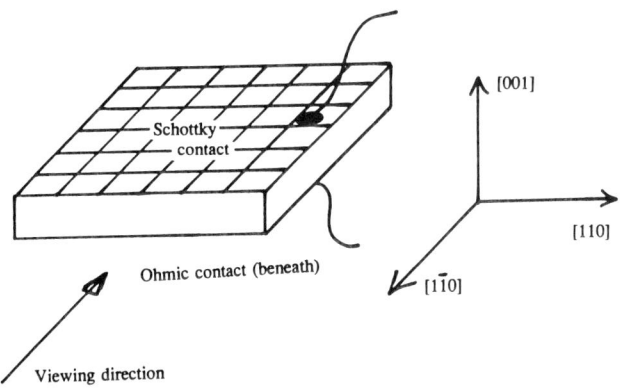

Figure 1. The orientation of the samples

The experimental set up for Franz-Keldysh microscopy is shown in Figure 2.

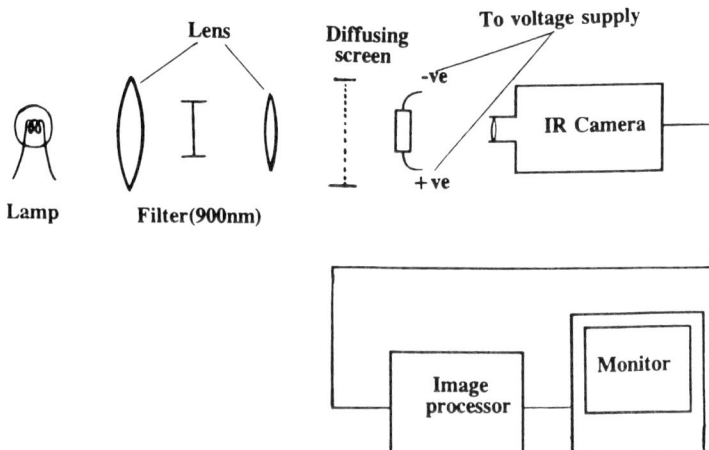

Figure 2. High electric field microscopy based on the Franz-Keldysh Effect.

The illumination is produced by a 100W incandescent lamp via an interference filter. For room temperature measurements, we have found that a narrow bandpass (bandwidth $\simeq$ 15nm) filter centred at 900nm is suitable (Il'inskii et al, 1992, Moss, 1961). It is useful to be able to vary the temperature of the sample by placing it in a variable temperature cryostat and consequently the variation of Eg with temperature must be considered. At a sample temperature of 80K, Eg has increased from 1.42eV to 1.51eV and a filter of ~840nm must be used. In this wavelength range, we find that an uncooled Si CCD camera is sufficiently sensitive, especially if it is used in conjunction with an image processor to enhance the image. We use a Hamamatsu "Argus 10" Real Time Image Processor for this purpose.

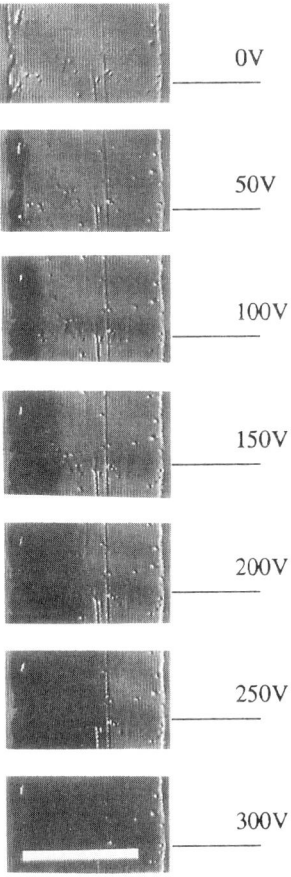

0V

50V

100V

150V

200V

250V

300V

Figure 3. Penetration of the high electric field region at a function of applied voltage. The width of the dark region corresponds to the material under a high electric field whilst the darkness is increases with the magnitude of the electric field.

3. STATIC ELECTRIC FIELD MEASUREMENTS

Figure 3 is a transmission image through a thickness of 800 microns for varying applied bias. The Schottky contact is to the left of these images and is biased negatively with respect to the Ohmic contact. This is reverse bias, corresponding to the operating condition of a detector for room temperature application. As can be seen, a dark, more absorbing region is seen below the Schottky contact. The extra absorption is due to the Franz-Keldysh effect and represents the penetration of a region of high electric field. The residual area is no more absorbing than if no bias were applied, corresponding to a region of low electric field. The dark region is of near constant opacity indicating that the electric field is relatively uniform and it does not increase in darkness as the bias is increased. In other words, the magnitude of the high field is barely affected by the applied bias but its penetration increases as the bias voltage is increases. These observations, confirmed by a direct voltage probing technique (Berwick et al, 1994b), partially explain the known variation of CCE with applied voltage. The variation of electric field with position at different applied bias voltage, as derived from the direct voltage probe but in agreement with the present measurements, is shown in figure 4. Not shown on this figure is the effect when sufficient bias is applied so that the high electric field penetrates the entire thickness of the device.

4. TEMPORAL VARIATIONS OF ELECTRIC FIELD

It has been known for some time that the application of a moderate voltage to contacts on SI GaAs can result in a current that flows in the form of pulses, the period of these being of the order of a few Hz at room temperature (Johnson et al, 1990, Sacks and Milnes, 1971). The origin of these can be investigated using Franz-Keldysh imaging. Although faint and, therefore, not reproduced in this paper, "waves" are often seen in the high field regions moving with a velocity of less than 1 mm s^{-1} especially at applied voltages greater than 50 volts. Their wavelength is a few tens of microns. The image indicates that these waves are moving regions of high electric field

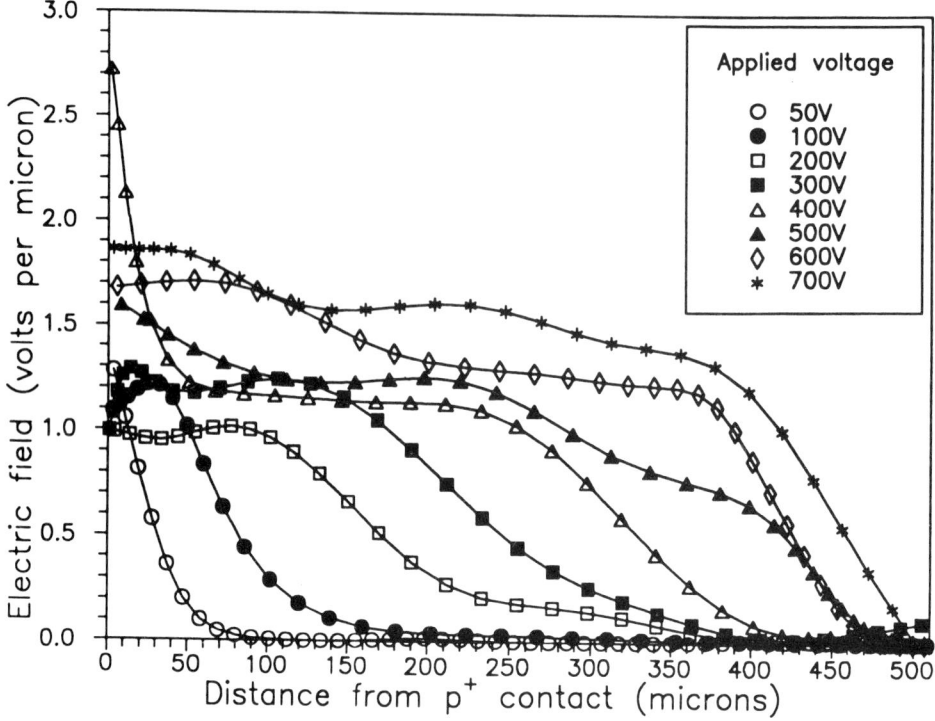

Figure 4. Quantitative estimates of the electric field penetration as a function of applied voltage. Note that the electric field barely exceeds 1.5V per micron at any applied bias.

similar to the domains that are representative of electron motion in materials like GaAs that show the Transferred Electron Effect. Indeed, the electric field strength in the high field regions ($\sim 1.0 \times 10^6$ V.m^{-1}) exceeds the critical field for this effect ($\sim 3 \times 10^5$ V.m^{-1}). However, these waves are not always seen and the criteria for observing them are unknown.

5. BREAKDOWN EFFECTS

At room temperature one effect of increasing the applied bias past the point where the high field region fully penetrates the GaAs is a marked increase in current. At this stage the sample becomes uniformly dark, apparently because of heating and the Franz-Keldysh Effect, both of which cause a bandgap energy reduction. On cooling the sample below $\sim 200K$ the leakage current and ohmic-heating become negligible and a new breakdown phenomenon is observed. The creation of a deeply penetrating region of electric field of $\sim 1 \times 10^6$V.m^{-1} is not found much below room temperature. At these low temperatures the high field region appears to be constrained to a shallow region (tens of microns) below the Schottky barrier and is, for a given bias voltage, of greater magnitude. Also a high electric field region appears below the "Ohmic" contact which becomes blocking. Apparently, it is the creation of this new high field region that results in this breakdown effect. Figure 5 shows this effect. Dark "droplets" form at the Ohmic contact and move at velocities ~ 1mm.s^{-1} to the Schottky contact. Not all these droplets reach the Schottky electrode but simply disappear in the GaAs bulk. The droplets are produced at greater numbers and move with increasing velocity as the applied bias is increased. This type of breakdown is associated with pulsing of the current in the external circuit.

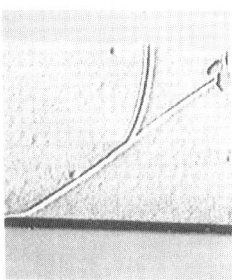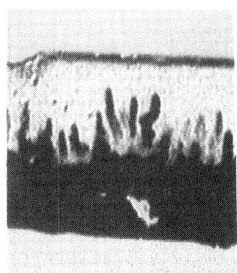

Figure 5. The formation of high electric field domains at a temperature of 85K using a filter of 850nm. From left to right, the applied bias voltages are 100V, 700V and 800V. The Schottky contact at a negative potential is at the top of the images.

The droplets represent regions of increased electric field and must result from the creation of charge dipoles. The motion towards the negative terminal demonstrates that each has a net positive charge and the low velocity demonstrates that the mobility is very low. We note that the positive charge excludes the possibility that this is due to the transferred electron effect.

6. DISCUSSION

Observations of high electric field regions in SI GaAs nuclear particle detectors by the Franz-Keldysh effect explain in part the bias dependence of CCE that is almost universally found (Beaumont et al, 1993). We are not yet able to explain the unexpected penetration of a near-constant electric field into the bulk but it has been suggested that this results from field-enhanced capture of electrons by EL2 deep donor centres (McGregor et al, 1994). EL2 defects, accepted by many workers to be the arsenic antisite, As_{Ga} defect, has an ionization level at Eg-0.8eV and gives rise to the semi-insulating nature of SI GaAs by compensating residual shallow acceptors such as carbon (Baraff, 1992). It follows that the presence of EL2 centres is necessary for this material, even if they are involved in these unwanted effects. The role of enhanced electron capture by EL2 centres is currently under investigation in our laboratory.

In general, the velocity of domains would be expected to be equal to the saturated carrier drift velocity which is $\sim 1 \times 10^7 cm.s^{-1}$ for both electrons and holes, many orders of magnitude greater than our observations of the effects recorded in sections 4 and 5. On the other hand, SI GaAs is a relaxation semiconductor and the motion of carriers in high field domains is expected to be greatly impeded by trapping and detrapping effects. High concentrations of EL2 deep donors ($\simeq 1 \times 10^{16} cm^{-3}$) may result in the negatively charged domain motion in section 4. However, only low concentrations of hole traps are thought to be present and the droplet (domain) motion seen under low temperature breakdown conditions is less easy to explain. We have observed that these droplets are initiated in a somewhat random fashion at the Ohmic contact and, although there may be a tendency for them to be created where the surface is roughened, we have no evidence to suggest that their creation is associated with the proximity of dislocations. Thus, although dislocations are present at densities from 10^4 to 10^5 cm^{-2} in this material (Stirland, 1991), we see no evidence that they represent weaknesses in the crystal where electrical breakdown is initiated. This, at least, is encouraging for the development of these high voltage devices.

7. CONCLUSIONS

We have employed absorption by the Franz-Keldysh effect to image regions of high electric field in nuclear particle detectors fabricated from SI GaAs. Static and dynamic effects have been observed which partly explain the characteristics of these devices. We believe that this technique can be used to probe other devices which operate at elevated voltages and should be applicable to many semiconductors other than GaAs.

ACKNOWLEDGEMENTS

The authors are pleased to acknowledge Miss J Gilmore for the careful preparation of this manuscript. One of us (KB) acknowledges the award of a research grant from EPSRC.

REFERENCES

Baraff G A, 1992, Proc. Semi-Ins. III-V Mats. (Bristol: IOP Publishing) pp 11-18

Beaumont S P, Bertin R, d'Auria S and members of the RD8 collaboration, 1990, European Committee for Future Accelerators: Workshop on Large Hadron Collider, Geneva (CERN) p 244

Beaumont S P, Bertin R, Zichichi and members of the RD8 collaboration, 1991, Proc. Symp. Detector Res. and Dev. for SSC (Fortworth, Texas, USA) p 169

Beaumont S P, Bertin R, Booth C N and members of the RD8 collaboration, 1993, IEEE Trans. Nucl. Sci. 40, p 1225

Berwick K, Brozel M R, Buttar C M, Cowperthwaite M, Sellin P and Hou Y, 1994a, Mats. Sci. Eng. B28, p 485

Berwick K, Brozel M R, Buttar C M, Cowperthwaite M and Hou Y, 1994b, Inst. Phys. Conf. Ser. 135, (IOP Publishing: Bristol), p 305

Franz W, 1958, Z. Naturforsch 13, p 484

Il'inskii A V, Kutsenko A B and Stepanova, 1992, Sov. Phys. Semicond. 26, p 399

Johnson D A, Puechner R A and Maracas G N, J. Appl. Phys, 67, p 300

Keldysh L V, 1958, Sov. Phys JETP, 34, p 788

Moss T S, 1961, J. Appl. Phys. 32, p 2136

McGregor D S, Knoll G F, Eisen Y and Brake R, 1992, IEEE Trans. Nucl. Sci. 39, p 1226

McGregor D S, Rojeski R A, Knoll G F, Terry F L Jr, East J and Eisen Y, 1994, Nucl. Inst. Meth. in Phys. Res. A 343, p 527

Pankove J I, 1971, Optical Processes in Semiconductors, (Prentice-Hall: New York)

Sacks H K and Milnes A G, 1971, Int. J. Electronics, 30, p 49

Stirland D J, 1991, Inst. Phys. Conf. Ser. 117, p 327

An XPS study of the effects of semiconductor processing treatments used to make InP optoelectronic devices

R L Van Meirhaeghe, L Goubert, L Fiermans, W H Laflère, F Cardon, P De Dobbelaere*, P Van Daele*

Laboratorium voor Kristallografie en Studie van de Vaste Stof, Krijgslaan 281, B-9000 GENT, Belgium
*Intec, St. Pietersnieuwstraat 41, B-9000 GENT, Belgium

ABSTRACT : A study is presented on the effects of various important semiconductor processing treatments used to manufacture optoelectronic components containing InP. Using X-ray photoelectron spectroscopy, it was found that plasma etching of SiO_x by $CF_4:O_2$ left InF_3 at the surface. Reactive ion etching caused oxidation of the InP epilayers as well as passivation of Zn-acceptors by hydrogen.

1. EXPERIMENTAL

All samples investigated consist of a p^+-doped InP substrate. A p-InP epilayer was grown on each substrate using metal organic chemical vapour deposition (MOCVD). This growth was carried out after etching the substrate in a $H_2SO_4:H_2O_2:H_2O$ solution. Such epilayer covered substrates are defined as reference samples. On these standards, several technologically important processing treatments used in optoelectronics were applied. One such process is plasma enhanced chemical vapour deposition (PECVD) of SiO_x by using $SiH_4:N_2O$. Another treatment consists of applying a photoresist cover. Afterwards, patterning of these layers occurred. For the photolithography, KOH was used. The SiO_x-layer was plasma etched in $CF_4:O_2$. For reactive ion etching (RIE) of InP itself, a $CH_4:H_2$ mixture was applied. On some samples, the photoresist was stripped using an O_2 plasma. A review of the different samples and their associated processing is given in fig. 1 on page 2. In that figure, it is shown that sample 2 was given an additional dip in concentrated H_2SO_4 to remove the native oxide layer on InP.

The XPS-measurements were carried out using a PHI 5500 apparatus. The X-ray source was Al (K_α). Firstly, a survey over a large binding energy range (0-1400 eV) was made, followed by detailed peak measuring in the appropriate energy windows for In, P, O, F and C. For fitting, Gaussian-Lorentzian forms were used and a linear background. Spin-orbit splitting energy differences of In and P were used as input. After fitting, the values obtained for the peak positions, the peak widths and the spin-orbit branching ratio were all very acceptable. The fits done for several elements (In, P, O) on the same sample were found to be consistent. This was also the case for fits of the same element on different samples.

642

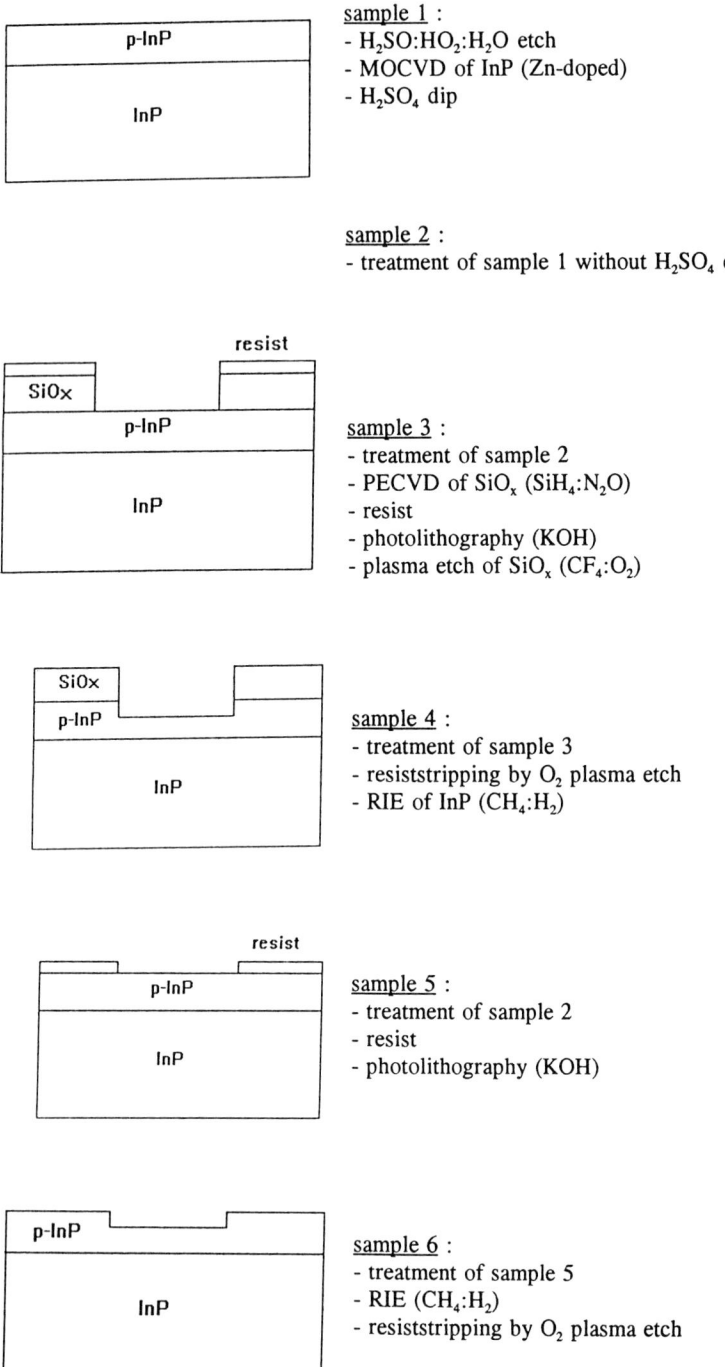

sample 1 :
- H₂SO:HO₂:H₂O etch
- MOCVD of InP (Zn-doped)
- H₂SO₄ dip

sample 2 :
- treatment of sample 1 without H₂SO₄ dip

sample 3 :
- treatment of sample 2
- PECVD of SiOₓ (SiH₄:N₂O)
- resist
- photolithography (KOH)
- plasma etch of SiOₓ (CF₄:O₂)

sample 4 :
- treatment of sample 3
- resiststripping by O₂ plasma etch
- RIE of InP (CH₄:H₂)

sample 5 :
- treatment of sample 2
- resist
- photolithography (KOH)

sample 6 :
- treatment of sample 5
- RIE (CH₄:H₂)
- resiststripping by O₂ plasma etch

Fig. 1 :Review of the different samples and their associated processing

2. RESULTS

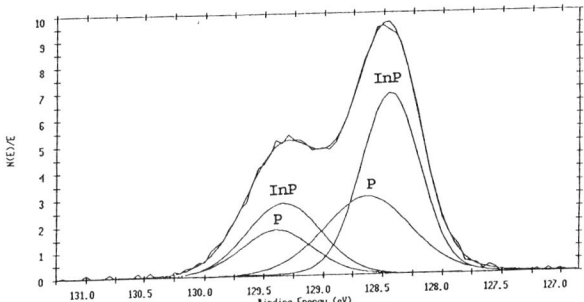

Fig. 2 : P 2p spectrum and fitting for sample 1

In fig. 2, the P 2p spectrum of sample 1 is given. In the region 132 to 134 eV (not shown on the figure), no peak could be detected (Lee et al. 1989), pointing to the removal of the native oxidized layer by the H_2SO_4 dip. This is corroborated by the fitting of the In-peak of sample 1 where no oxides or phosphates are involved. The fitting of fig. 2 consists of two peaks (and their associated spin-orbit partners) : one due to P bound to In and one due to elemental P. This elemental P is produced by the H_2SO_4 dip. An analogous situation is known for GaAs where elemental As is found (Van de Walle et al. 1993).

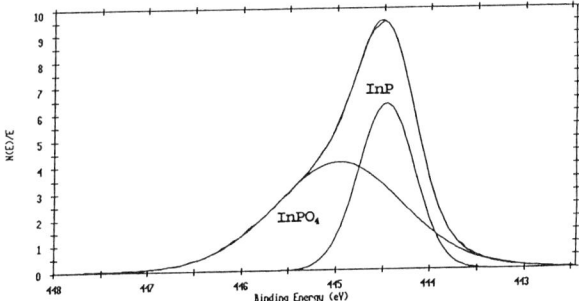

Fig. 3 : In3d 5/2 peak and fitting for sample 2.

In fig. 3, the In 3d 5/2 peak is plotted for sample 2. It could be fitted by 2 components, one from InP and the other from $InPO_4$ (Lee et al. 1989). This sample did not receive the H_2SO_4 dip so that the native oxidized layer is still present and is proven here to be $InPO_4$. This is confirmed by the P-spectrum (not shown) which contains a broad peak at 133 eV due to $InPO_4$. The fitting of the P 2p peak showed that only a small quantity of elemental P is present here.

In fig. 4 the In 3d 5/2 peak for sample 3 is fitted. Four components have to be introduced. The most important is due to InF_3. This shows that the plasma stripping of the SiO_x layer by a $CF_4:O_2$ mixture leaves an InF_3 residu on the epilayer surface. In the case of GaAs etched in CHF_3 (Williston et al. 1992), galliumfluorides were found up to a depth of 1.5 nm. These fluorides could be removed by annealing above 305 °C. In our case, ARXPS (angle resolved XPS) measurements are planned to investigate the depth of the F-penetration and to reveal if a dual layer structure of fluorides and phosphates/oxides is present. Fig. 4 also shows the presence of $InPO_4$ (increased in comparison with sample 2) and even In_2O_3. The latter two compounds are due to the presence of O_2 in the stripping environment.

644

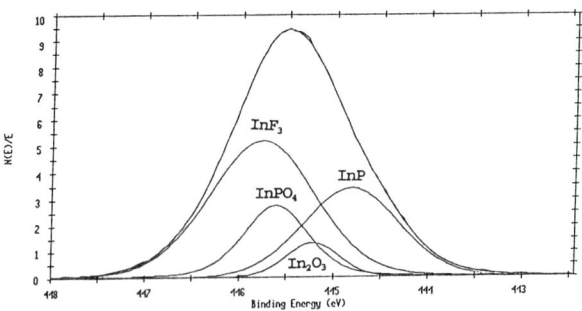

Fig. 4 : In 3d 5/2 peak and fitting for sample 3.

In fig. 5, the O 1s spectrum of sample 4 is shown. Next to the large contribution of $InPO_4$, also In_2O_3 and a small quantity of $InPO_3$ are found. These components are due to the resiststripping in an O_2 plasma etch. The fitting of the In 3d peaks (not shown here) learned us that the InF_3 component decreased but the $InPO_4$ and In_2O_3 contributions increased with respect to those displayed in fig. 4. Consequently, the resiststripping and/or RIE remove some of the InF_3. Considerable binding energy shifts were observed for this sample and we had to use a flood gun to restore the original peak positions. We assume that passivation of the Zn-acceptors (Pearton et al. 1992) due to hydrogen originating from the CH_4:H_2 RIE gas, produces an insulating region over some depth into the semiconductor. This region is charged during the XPS-measurements.

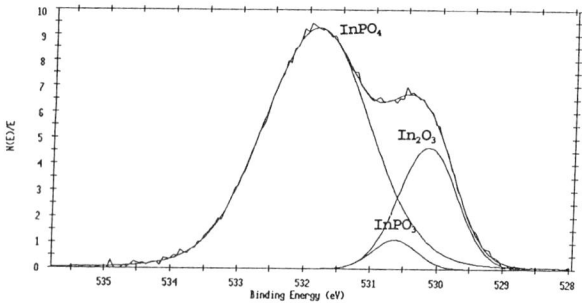

Fig. 5 : O 1s spectrum and fittings for sample 4

For sample 5, XPS results analogous to those of sample 2 were found. This shows that the photolithographic process does not cause changes   of or residus on the epilayer.

The XPS-results on sample 6 showed a large $InPO_4$ contribution and the presence of In_2O_3 and $InPO_3$. All these oxidized species are a result of the last processing step, namely the resist stripping by the O_2 plasma etch. The large binding energy shifts, already found for sample 4 were also present here. This supports our assumption about the passivating effect of the RIE.

REFERENCES

Lee Y S, Anderson W A 1989 Journ. Appl. Phys. 65 4051-6

Pearton S J, Corbett J W, Stavola M 1992 Hydrogen in Crystalline Semiconductors, Springer, Berlin, p 137

Van de Walle R, Van Meirhaeghe R L, Laflère W H, Cardon F 1993 Journ. Appl. Phys. 74 1885-9

Williston L R , Bello I, Lau W M 1992 J. Vac. Sci. Technol. A 10 1365-70

Inst. Phys. Conf. Ser. No 146
Paper presented at Microsc. Semicond. Mater. Conf., Oxford, 20–23 March 1995
© *1995 IOP Publishing Ltd*

Growth mode transitions on GaAs(110) surface

D M Holmes, J L Sudijono, J G Belk, J H Neave, T S Jones and B A Joyce

Dept. of Chemistry and Semiconductor Materials IRC, Imperial College, London, SW7 2AY, UK

ABSTRACT: STM has been used to study the MBE growth of GaAs(110). Filled-state images show that growth proceeds in a distinct layer-by-layer fashion, consistent with the observation of RHEED intensity oscillations. The two-dimensional islands take on a triangular structure with edges running along the $[\bar{1}10]$, $[1\bar{1}2]$ and $[1\bar{1}3]$ directions. The existence of a bilayer growth mode on this surface under certain growth conditions has also been confirmed. The bilayer steps grow from nucleation at the upper step-edge of a triangular monolayer island.

1. INTRODUCTION

The understanding of growth mode transitions on GaAs surfaces prepared by molecular beam epitaxy (MBE), using reflection high energy electron diffraction (RHEED), has been instrumental in defining good growth conditions for nanostructure fabrication (Joyce 1994). The existence of oscillatory behaviour of the specular RHEED beam has generally been accepted as a consequence of a layer-by-layer growth mode. From an atomistic perspective, however, the question remains as to what is oscillating on the surface. In principle, the surface morphology of a growing film is controlled by kinetic processes, such as chemisorption, nucleation, and diffusion. Understanding these processes significantly enhances the quality of the interface structure. As such, the availibity of scanning tunnelling microscopy (STM) as a real-space imaging tool has enabled the surface morphology to be probed at the atomic scale.

GaAs(110) provides an ideal surface for STM studies for several reasons. Firstly, the surface is the natural cleavage plane for GaAs. Secondly, the surface has an equal number of Ga and As atoms such that no surface reconstruction occurs.

The interest in MBE-grown GaAs(110) thin films has recently been rekindled by the RHEED observation of the transition from monolayer-by-monolayer to bilayer-by-bilayer growth (Fawcett 1993). This transition is inferred from the existence of specular RHEED oscillations with twice the period of the original growth of monolayer steps. In a systematic study using RHEED, Fawcett et al (1994) have shown the importance of growth conditions in obtaining a good surface morphology. To better understand how nucleations and growth propagation progress, STM has been used to generate real-space images of the growing film. The aim of this work is to understand the nature of the growth mode, shown by the RHEED specular intensity oscillations, in terms of nucleation and coalescence of islands on the surface. A unique feature of this work is the "quench and look" capability of our combined ultra-high vacuum (UHV) MBE-STM machine. This allows us to "freeze" the growth fronts and "look" at the morphology with an *in situ* STM. Details of the quenching procedures are described elsewhere (Holmes 1995).

2. EXPERIMENT

The experiments were carried out in our MBE-STM growth chamber with a base pressure of better than 1×10^{-10} mbar. Epi-ready, on-axis n^+ Si-doped GaAs(110) wafers were mounted with indium onto molybdenum plates and transferred to UHV via a fast entry lock chamber. After initial outgassing at 300°C, the native oxide layer was removed by heating the substrate to 630°C under an As_2 pressure. The oxide removal was monitored by the RHEED specular intensity in the [$\bar{1}$10] azimuth. A buffer layer of approximately 8ML was deposited at a substrate temperature of 500°C with a growth rate of 10ML/minute and an As_2 pressure of 2.7×10^{-6} mbar, corresponding to an As_2/Ga ratio of 2.8. The surface was then annealed at 540°C for several minutes in order to produce smooth surfaces with large terraces. These conditions were routinely used to provide a starting surface on which sub-monolayer coverage growth studies could be performed.

Each deposition of 0.3ML and 0.6ML of GaAs(110) was carried out on a new starting surface as described above. The deposition was performed at temperatures varying from 420°C to 520°C with a growth rate of 6 and 12ML/minute under the same arsenic conditions as the buffer layer preparation. The freshly grown samples were subsequently imaged by STM in a constant current mode at room temperature.

3. RESULTS AND DISCUSSION

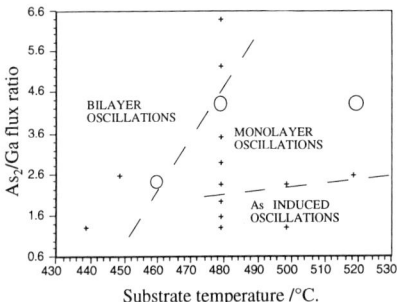

Fig. 1. GaAs(110) growth regimes as a function of As_2/Ga flux ratio and substrate temperature. The crosses refer to conditions whereby STM data are acquired, while open circles correspond to STM data discussed in this paper.

A plot of the growth conditions used to observe monolayer and bilayer RHEED oscillations is shown in Fig.1. Only STM results from the three open circles are discussed in this paper.

At the onset of growth there is a drop in the RHEED specular intensity. STM examination shows that a high density of islands are present on large terraces of the smooth starting surface. A large scale mosaic of an STM image of the surface after deposition of 0.3ML of GaAs at 520°C shown in Fig.2a. The striking feature of these randomly distributed islands is the triangular shape and preferred orientation of the apex in the [001] direction. The average island separation is 400Å and a typical dimension of the triangular island is 150Å and 200Å in the [$\bar{1}$10] and [001] directions, respectively. Some islands can grow large enough that they begin to coalesce while preserving their shape.

Fig.2. (a) GaAs surface after 0.3ML coverage at 520°C. An even distribution of islands is present. (b) Surface after 0.6ML deposition. Coalescence is prominent.

As deposition continues the islands begin to grow and eventually coalesce. Fig.2b shows an STM image of the surface after 0.6ML deposition of GaAs. The coalescence has reached a state where the overall island shape has almost lost its original triangular form. Some of the islands have different height. Close inspection reveals that the height is exactly twice the normal value of monolayer-high islands. Fig.3 shows a high-resolution STM image of such a bilayer island. The linescans A and B clearly illustrate the dimensions of a typical island. In addition, a riser step is often seen at the base of the triangle indicating that growth of the second layer starts at the apex or sides of the triangle and propagates in the $[00\bar{1}]$ direction. The sides of the triangle have been identified as those of $[1\bar{1}2]$ and $[1\bar{1}3]$ types.

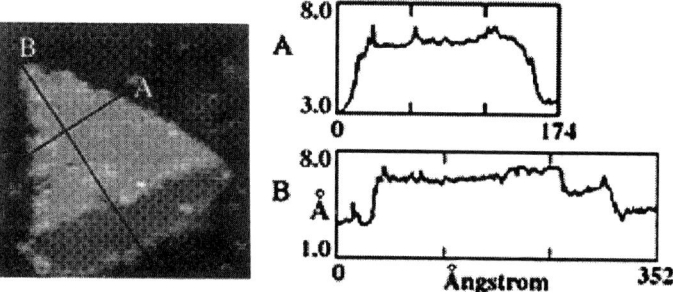

Fig.3. A high-resolution STM image of a bilayer island.

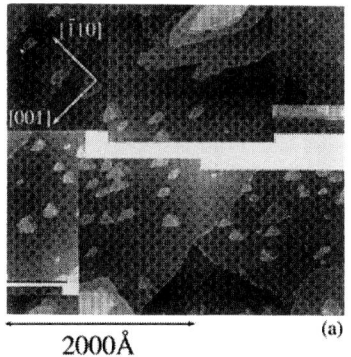

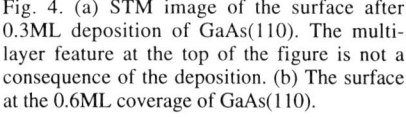

Fig. 4. (a) STM image of the surface after 0.3ML deposition of GaAs(110). The multi-layer feature at the top of the figure is not a consequence of the deposition. (b) The surface at the 0.6ML coverage of GaAs(110).

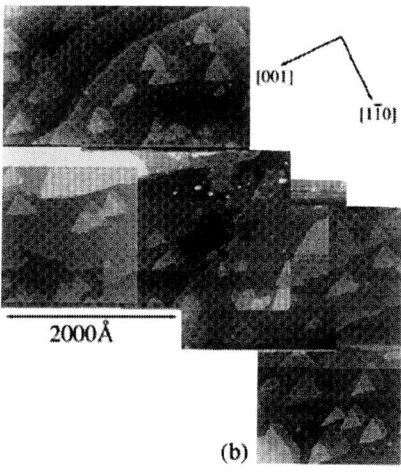

In contrast, STM images of the surface after 0.3ML deposition at 480°C (Fig.4a) show that islands on the terraces are predominantly of bilayer height. It is also apparent on comparison with Fig.2a that the island density is, on the average, lower than that with monolayer islands. This suggests that freshly deposited adatoms prefer nucleating on top of existing monolayer high islands rather than form new islands on the terraces. Further deposition to 0.6ML results in a surface as shown in Fig.4b. The bilayer islands have grown larger while maintaining the presence of the riser steps at the base. A small number of islands are seen to have merged to form larger islands. Interestingly, the triangular shape of the islands are still preserved during coalescence. This supports the fact that the

648

growing facet is the base of the triangle (the [$\bar{1}$10] steps in the [00$\bar{1}$] direction), as slowest moving steps remain at the end of growth.

As the GaAs coverage increases, the nucleation and coalesecence continue in time. In particular, as the coalescing islands become bigger, nucleations on top of the bilayer islands are very likely. However, the STM image shown in Fig.5 indicates that third layer nucleations rarely occur. While a large number of bilayer islands are present mostly in the upper left corner of the image, some islands are seen to have merged giving rise to many apparent third layer islands in the middle part of the image. Significantly, no third layer is present until the first layer has fully coalesced with neigbouring islands. As such, no "trilayer" island formation is possible under these growth conditions.

Fig.5 40ML growth of GaAs at 460°C with growth rate of 12ML/min.

This observation is probably related to the nature of step-edge barriers. The presence of bilayer islands at very low coverage indicates preferential nucleation on top of monolayer islands, despite the small catchment area, rather than formation of new monolayer islands on the lower terrace. It appears that, at this temperature, there is a significant net flow of adatoms from the lower to upper terraces. However, third layer formation is energetically unfavourable either due to insufficient temperature or a modified step-edge barrier. Conversely, at higher temperatures, the absence of bilayer islands suggests a more typical net downward flow of adatoms. A detailed study of this apparent inversion of the step-edge barriers will be presented elsewhere.

4. CONCLUSIONS

Consistent with previous RHEED observations, we have been able to confirm the existence of the bilayer growth mode on GaAs(110) for low temperature growth conditions. In the sub-monolayer deposition triangular shaped island are formed, characterised by [$\bar{1}$10] and [1$\bar{1}$2]/[1$\bar{1}$3] steps. The stability of these bilayer islands is discussed in terms of step-edge energy barriers.

REFERENCES

Joyce B A, Vvedensky D D, Foxon C T 1994 Handbook on Semiconductors (Amsterdam: Elsevier Science) pp276-368
Fawcett P N, Neave J H, Zhang J, Joyce B A 1993 Surface Science 296 67
Fawcett P N, Neave J H, Zhang J, Joyce B A 1994 J. Vac. Sci. Technol. A. 12 1201
Holmes D M, Belk, J G, Sudijono J L, Neave J H, Jones T S, Joyce B A 1995 Surface Science 341 133

Inst. Phys. Conf. Ser. No 146
Paper presented at Microsc. Semicond. Mater. Conf., Oxford, 20–23 March 1995
© 1995 IOP Publishing Ltd

The epitaxial growth of compound semiconductors observed by atomic force microscopy

I H Wilson, J B Xu and C C Hsu

Materials Technology Research Centre, The Chinese University of Hong Kong, Sha Tin, N.T., Hong Kong.

ABSTRACT: Results will be reported on atomic force microscopy (AFM) of features of homoepitaxial growth of GaAs, InP, and strained layer growth of $Ga_{1-x}In_xAs$ on GaAs by metalorganic vapour phase epitaxy (MOVPE) and homoepitaxial growth of GaAs by molecular beam epitaxy (MBE). Growth was found to be by the classical step-flow mode. Defects were seen to act as strong persistent step sources. We observed spiral growth originating at screw dislocations and ring-pattern growth, probably originating at the intersection of stacking faults with the surface. The same features were seen for MBE growth of GaAs but step growth was irregular in particular directions due to surface reconstruction. In the case of strained layer growth we observed the transition from two to three dimensional growth and the generation of misfit dislocations.

1. INTRODUCTION

Epitaxial growth is an essential first step in preparing compound semiconductor layers for the fabrication of devices such as heterojunction lasers and transistors, photodetectors and high electron mobility transistors. Metalorganic vapour phase epitaxy (MOVPE) is emerging as the most commercially attractive method and has some advantages in terms of the quality of the films over other methods such as Molecular Beam Epitaxy (MBE). Epitaxial growth has much similarity to bulk crystal growth as described by the classical Burton, Cabrera, Frank (1951) (BCF) theory. On a singular surface (a low index facet) one expects the nucleation of two dimensional (2D) islands that eventually coalesce to apparently give monolayer growth as evidenced by reflection high-energy electron diffraction (RHEED) of MBE growth (Neave et al 1985). However commercially available wafers are always slightly misoriented and when heated to the growth temperature low index surface steps will form due to mass transport to minimise surface energy (Hsu et al 1993, 1994) In this case at least two modes of growth can take place. In high supersaturation where the diffusion length is less than half the terrace width 2D nucleation will still dominate. In the case of low supersaturation the step edge will be the major nucleation site and growth will be by step-flow (Nishinaga & Cho 1988). Real crystals are not perfect and defects such as dislocations can form monatomic and submonatomic steps on the growing surface and so act as dominant and persistent step sources. In this paper we describe ex-situ atomic force microscopy (AFM) of surface morphology before and after growth by MOVPE of InP on InP, GaAs on GaAs, $Ga_{1-x}In_xAs$ on GaAs and MBE of GaAs on GaAs.

2. EXPERIMENT

An in-house Aixtron MOVPE reactor with the safe sources trimethyl-gallium/indium and tertiarybutyl-arsine/phosphine was used. Growth temperatures were 600 to 650 °C, pressure 200 mbar, III/V ratio 1:20 (InP) or 1:30 (GaAs), typical deposition rate 0.5 nm/s. MBE samples were supplied by TG Andersson and JV Thordson (Chalmers University, Sweden) grown in a Varian GEN II starting at 550 °C then raising to 700 °C to obtain step-flow growth. AFM was performed in air using a Digital Instruments Nanoscope III in the constant force mode with standard cantilevers at a loading force of approximately 10 nN. Commercial nominally (100) epi-ready wafers (usually semi-insulating) were used with no surface treatment before or after growth.

650

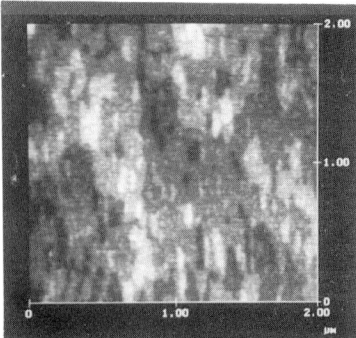

Fig. 1: AFM image of the surface morphology of
a GaAs epitaxial film grown by MBE at
580 °C under the 2D nucleation mode.
(Gray scale 2 nm)

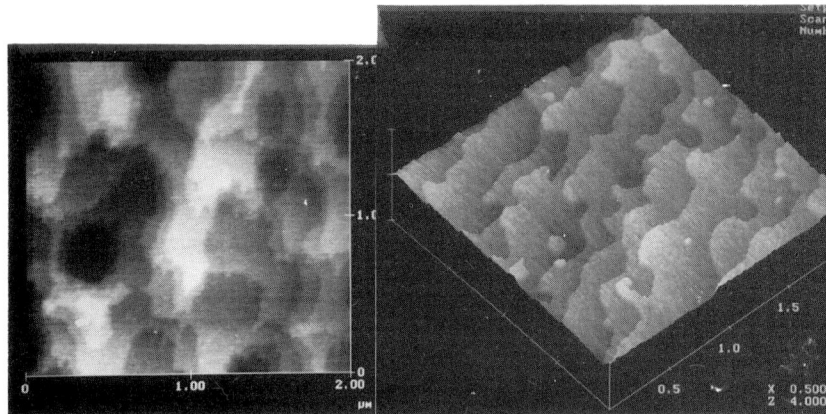

Fig. 2: Monoatomic steps formed
on the vicinal surface of InP,
nominally (100), annealed at
625 °C (gray scale 4 nm).

Fig. 3: Monoatomic steps formed on the vicinal surface
of GaAs, nominally (100), annealed at 750 °C
in a TBA/hydrogen ambient.

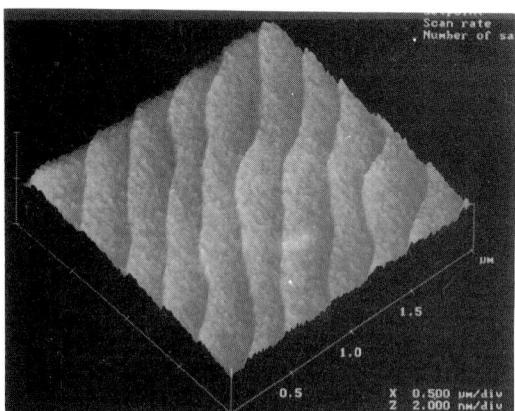

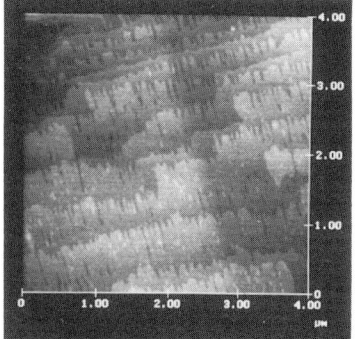

Fig. 4: GaAs surface after 600 nm of MOVPE
growth showing morphologicaly
stable step-flow growth in the
[011] direction.

Fig. 5: GaAs surface after MBE growth in
the step-flow mode at 700 °C.
Step direction [01$\bar{1}$],
(gray scale 4 nm).

3. 2D AND STEP-FLOW GROWTH

RHEED oscillations appear to indicate monolayer growth after MBE growth in the 2D nucleation mode. Figure 1 shows an AFM image of a GaAs surface grown in the 2D mode at 580 °C. Our observations are similar to those reported earlier by Johnson et al (1994), the surface is rough due to nucleation of 2D islands on incomplete surfaces due to the small (< 100nm) diffusion length. At least 4 levels of monolayer steps can be seen in Fig. 1.

The surface of nominal (100) wafers (supplied with misorientation 0.04 to 0.1°) annealed at the growth temperature are shown in Figs. 2 (InP) and 3 (GaAs). The as-received undulating surface has formed monatomic steps (0.28 nm high) in order to minimise total surface energy by forming low index facets. The steps are irregular as we have allowed insufficient time to minimise the kink density. For the GaAs example (Fig. 3) the mean terrace width is 280 nm equivalent to 0.06° misorientation. Our AFM observations indicate that growth on these surfaces by MOVPE at 625 °C (InP) or 650 °C (GaAs) is by step-flow. Fig. 4 shows the surface of GaAs after growth of a 600 nm thick epitaxial film. Regular steps have formed of similar terrace width to those on the vicinal substrate. The steps are straight and are therefore, in the terminology of crystal growth, morphologically stable (Bales & Zangwill 1990). InP surfaces after similar growth appear the same, terrace widths of up to 450 nm have been observed indicating a diffusion length in excess of 225 nm. On the larger terraces some incipient 2D nucleation is seen.

Step-flow growth can be achieved in MBE by raising the temperature, effectively lowering the supersaturation, resulting in an increase in diffusion length. In Fig. 5 is shown a terraced GaAs surface formed after growth at 700 °C in an increased As_4 flux. In this case there were no RHEED oscillations. The terrace width in this example is as much as 1μm and some 2D nucleation can be seen. The surface misorientation is in the [011] ('B') direction. The steps are rough as arsenic dimers form rows in the B direction as part of surface reconstruction. This imposes anisotropy so that [011] ('A') steps are straight whilst B steps are rough.

Step-flow growth can also be achieved in strained-layer growth of $Ga_{1-x}In_xAs$ on GaAs by MOVPE if care is taken to suppress the transition to 3D growth by controlling composition, growth temperature and film thickness. AFM images are shown in Figs. 6 (x = 0.2, grown at 650 °C, 10 nm thick) and 7 (x = 0.5, 600 °C, 5 nm thick). The layers are coherently strained. The steps in Fig. 6 are irregular due to strain effects and incipient onset of the transition from step-flow to 2D growth. In the case of Fig. 7 one can see the coexistence of 2D and step-flow growth. When GaAs is grown on top of these layers the straight morphologically stable steps are recovered.

4. DEFECTS AS PERSISTENT STEP SOURCES

Surface misorientation is not the only source of steps. Real crystals are not perfect and steps arising from the interception of dislocations with the surface can act as powerful step sources. A pure screw dislocation has a Burgers vector normal to the surface of height equal to the magnitude of the Burgers vector at the core falling to zero outside the strain field of the dislocation. This makes the whole crystal a continuous non-Euclidean surface and growth is a continuing extension of this surface. The step-flow growth front wraps itself around the dislocation core throwing out an ever widening spiral. In Fig. 8 is shown one such spiral formed during MOVPE growth of a 600 nm thick InP layer. The spiral is superimposed on misorientation steps which merge perfectly with the spiral steps demonstrating their monatomic nature. A close-up of the centre of on such spiral is shown in Fig. 9. The oval shape reflects the reciprocal Wulff polar diagram (Herring 1951) and illustrates a slight anisotropy (1.07) in the edge free energy of [001] to [011]. Fig. 10 shows a similar spiral in MOVPE grown GaAs. The shape here is rectangular with corners in the A and B directions. We have also observed spiral growth during MBE of GaAs as shown in Fig. 11. The anisotropy arises from the fact that the rate limiting step is the arrival of Ga in an excess of As and every arrival at the A step forms 3 Ga-As bonds whilst at the B step 2 bonds are formed for each Ga atom arriving. Thus the A steps grow faster. This is compounded by the surface reconstruction of arsenic dimers. This makes growth in the B direction even more difficult and irregular, but the spiral still forms. Indeed growth of this type may well be the source of the 'oval defect' seen by many workers.

We have also observed a new type of persistent step source. This is illustrated in Fig. 12

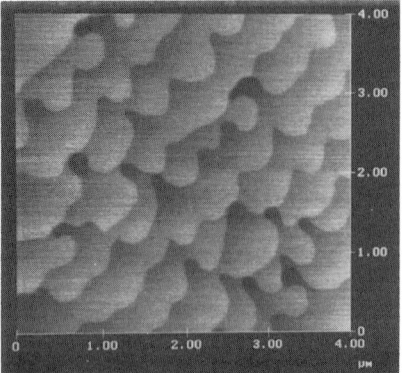

Fig. 6: Surface of a 10 nm thick $In_{0.2}Ga_{0.8}As$ on GaAs MOVPE layer grown at 650 °C (gray scale 4 nm).

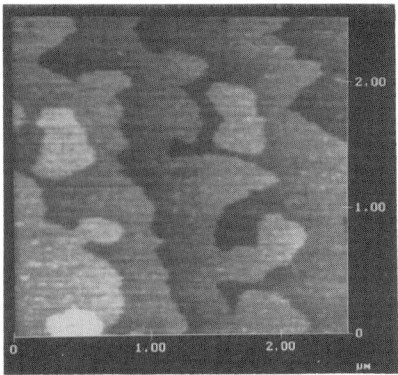

Fig. 7: A 5 nm thick $In_{0.5}Ga_{0.5}As$ grown on GaAs by MOVPE at 600°C. (gray scale 4 nm).

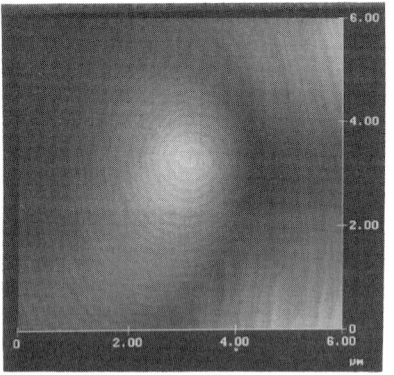

Fig. 8: Spiral step generated by a screw dislocation during MOVPE growth of InP (gray scale 14 nm)

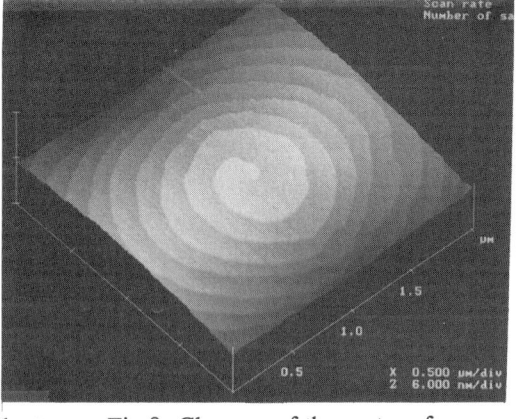

Fig 9: Close-up of the centre of an InP growth spiral showing surface energy anisotropy.

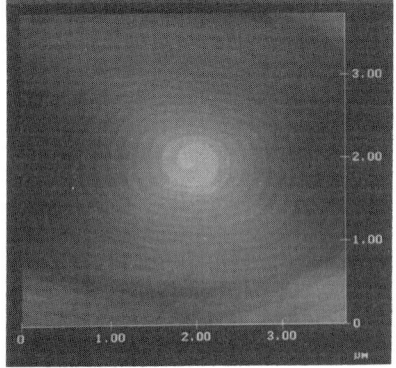

Fig. 10: Spiral step generated by a screw dislocation duringMOVPE growth of GaAs (gray scale 10 nm).

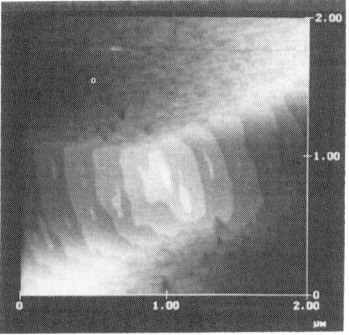

Fig. 11: Spiral step generated during MBE growth of GaAs (gray scale 4 nm).

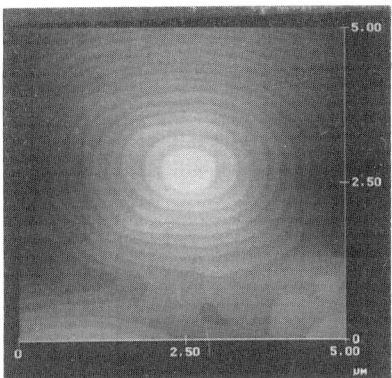

Fig. 12: A step source forms a concentric ring pattern. MOVPE of GaAs on GaAs (gray scale 8 nm).

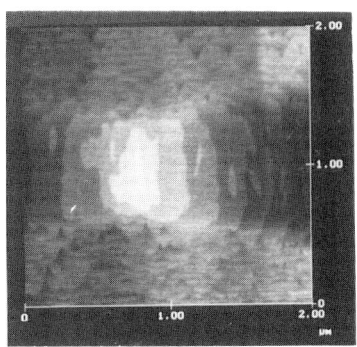

Fig. 13: A step source forms a concentric ring pattern. MBE of GaAs on GaAs (gray scale 4 nm).

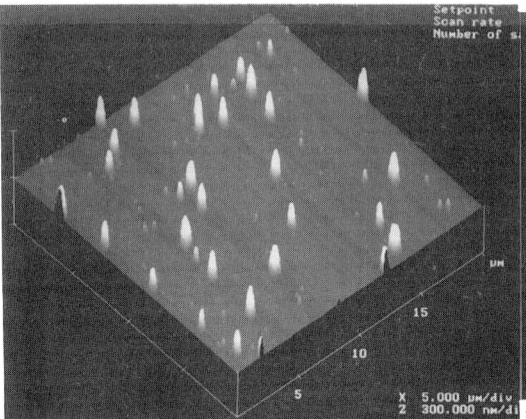

Fig. 14: Surface of a 5 nm thick $In_{0.5}Ga_{0.5}As$ on GaAs layer showing 3D growth. Growth temperature 650°C.

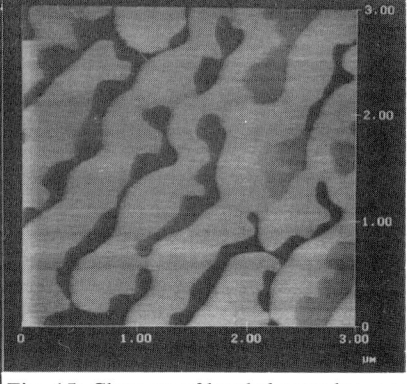

Fig. 15: Close-up of banded areas between islands in Fig. 14 showing 2 nm plateaus aligned with vicinal steps (gray scale 10 nm).

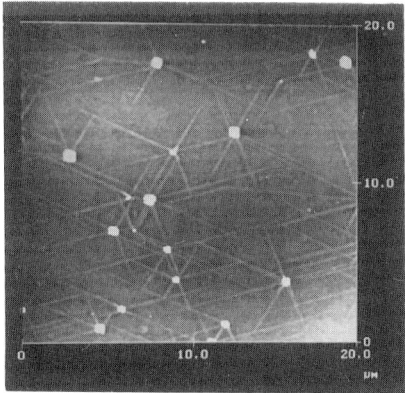

Fig. 16: Surface of a 7.5 nm thick $In_{0.5}Ga_{0.5}As$ on GaAs layer showing 3D growth. Growth temperature 600°C.

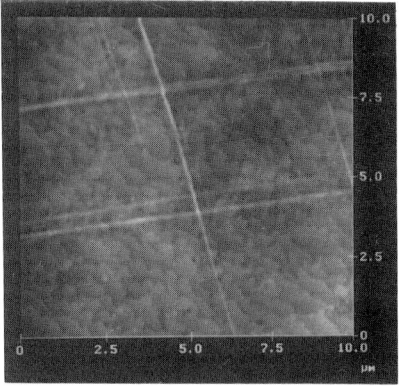

Fig.17: Close-up of the cross-hatched area in Fig.16 showing monolayer steps (gray scale 10 nm).

654

(MOVPE growth) and Fig. 13 (MBE growth) for GaAs. In this case closed rings are thrown out from a central flat topped ring. It has been pointed out by Ming (1993) that a stacking fault emerging from the surface will form a step of 1/3 monolayer height. This will act as a nucleus for surface adatoms and grow a 2D island with a step 2/3 height and a full step that moves away. Then the process repeats. This seems a most likely mechanism as we can eliminate the possibility of two screw dislocations of opposite sign as we do not see a full step that would exist between them on the centre ring. One other possible source (Frank 1981) is the strain field of an edge dislocation but it is difficult to see why this would be of sufficient magnitude to make it energetically favourable to nucleate a 2D island in the case of GaAs epitaxy.

5. STRAIN AND THE 2D TO 3D GROWTH TRANSITION

As shown in Figs. 6 and 7 coherently strained layers can be grown by step-flow and 2D growth. In Fig. 14 is illustrated the morphology of a film grown under the same conditions as Fig. 6 but with x (the In composition) increased from 0.2 to 0.5. Even though the film is half the thickness 3D growth dominates. The large islands are superimposed on a background of stripes of 2nm high plateaus with the same pattern as the vicinal steps. This is shown in detail in Fig.15. We propose that the stripes are composed of InGaAs grown preferentially on the step edges due to strain induced differences in chemical potential. Reducing the growth temperature to 600 °C restores the growth mode to 2D and step-flow as shown in Fig. 7. However, as shown in Fig. 16 if the film thickness is increased from 5 to 7.5 nm under the same conditions 3D islands form. Also seen are ridges emanating from these islands in [011] and [001] directions, forming a cross-hatched pattern typical of misfit dislocations on the surface. In Fig. 17 one can see the ridges superimposed on monolayer steps. It is possible that the ridges form from preferential growth of InGaAs on steps created by the misfit dislocations.

6. CONCLUSIONS

We have demonstrated that 2D and step-flow epitaxial growth of compound semiconductors are quite general phenomena. The mode of growth is dependent on the relationship between diffusion length and terrace width. Steps result from annealing of misoriented surfaces or from dislocations. The latter act as powerful step sources. Growth in the step-flow mode for MBE is much less regular than for MOVPE due to surface reconstruction. In the case of strained layers abnormal growth on step edges and misfit dislocation lines occur in the regime of 3D growth. The parameters for 2D growth of coherently strained layers have been determined.

REFERENCES

Bales G S and Zangwill A 1990 Phys. Rev. B **41**, 5500
Burton W K, Cabrera N and Frank F C 1951 Philos. Trans. R. Soc. **243**, 299
Frank F C 1981 J. Crystal Growth **51**, 367
Herring C 1951 Phys. Rev. **82**, 87
Hsu C C, Wong T K S and Wilson IH 1993 Appl. Phys. Lett. **63**, 1839
Hsu C C, Xu J B and Wilson I H 1994 Appl. Phys. Lett. **65**, 1394
Johnson M D, Sudijono J, Hunt A W and Orr B G 1994 Appl. Phys. Lett. **64**, 484
Ming N B 1993 J. Crystal Growth **128**, 104
Neave J H, Dodson P J, Joyce B A and Zhang J 1985 Appl. Phys. Lett. **47**, 100
Nishinaga T and Cho K I 1988 Jpn. J. Appl. Phys. **27**, L12

ACKNOWLEDGEMENT

This work was made possible by funding from the Research Grants Committee of the Hong Kong University Grants Committee.

Inst. Phys. Conf. Ser. No 146
Paper presented at Microsc. Semicond. Mater. Conf., Oxford, 20–23 March 1995

Nanoscopic detection of the thermal conductivity of compound semiconductor materials by enhanced scanning thermal microscopy

L J Balk, M Maywald and R J Pylkki*

Sonderforschungsbereich 254, Bergische Universität Gesamthochschule Wuppertal,
Lehrstuhl für Elektronik, Fuhlrottstrasse 10, D-42097 Wuppertal, Germany
*TopoMetrix, 5403 Betsy Ross Drive, Santa Clara, CA 95054-1162, USA

ABSTRACT: In this work, applications of nanoscopic thermal experiments have been demonstrated. These include investigations of diode contact areas as well as the localisation of doped material. Furthermore, an example of depth profiling is given.

1. INTRODUCTION

In addition to other techniques (Williams and Wickramasinghe 1986, Majumdar et al 1993), scanning thermal microscopy (SThM) has already been proved to enable imaging of thermal conductivity variations with high lateral resolution even in unfavourable samples such as diamond films (Maywald et al 1994a). Here, the thermal conductivity is measured by a determination of the heat flow from the scanning tip into the sample.

To improve the sensitivity of the system, it was changed from a static mode to a dynamic mode of operation: the enhanced SThM (ESThM). ESThM makes the adaption of a lock-in-amplifier feasible and thereby allows the detection of both magnitude and phase of the thermal signal even at low signal-to-noise ratios.

2. EXPERIMENTAL DETAILS

SThM is a recently developed (Dinwiddie et al 1994) scanning force microscope (SFM) based technique, which detects specimen thermal features. Using the conventional setup in constant temperature mode, information is gained by detecting the heat flux from a self-heating tip into the sample. For this purpose, a thermal feedback loop is adapted to maintain a predetermined tip temperature constant.

To increase the sensitivity in general and to enable phase-sensitive experiments to be carried out, a lock-in-amplifier (LIA) was incorporated into the system. As the LIA input must be an AC signal, the conventional static mode has to be modified into a dynamic mode. Here, two different ways of generating an AC signal are proposed.

First, a slight z-amplitude modulation of the cantilever position was applied, resulting in a variation of the effective contact area. Thereby, the detected heat flux into the sample is varied at the modulation frequency generating a lock-in input signal. In the following paragraphs, this mode will be referred to as the height modulation mode (HMM) - see Fig. 1a. Besides the HMM, a temperature modulation mode (TMM) was developed. In the case of TMM, the temperature variation is obtained by applying an additional AC voltage to the thermal feedback loop (Fig. 1b).

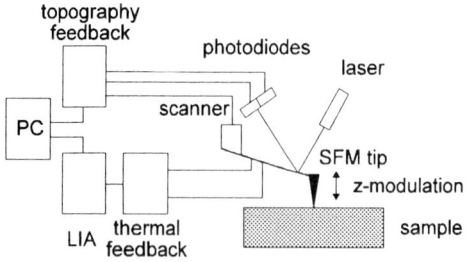

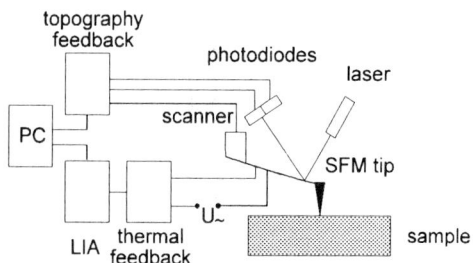

Fig.1a Experimental setup for ESThM
in height modulation mode (HMM)

Fig.1b Experimental setup for ESThM
in temperature modulation mode (TMM)

3. EXPERIMENTAL RESULTS AND DISCUSSION

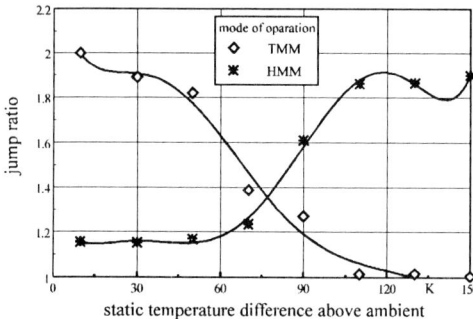

Fig.2 Contrast of the TMM and HMM versus
the static temperature

To obtain information about the general contrast versus temperature and modulation amplitude, line profiles have been carried out on a test grid consisting of Si and SiO_2 patterns. As a criterion for contrast magnitude the jump ratio was taken into account. Fig. 2a illustrates contrast curves for the TMM and HMM for different static temperatures. The data have been acquired at 32kHz modulation frequency and 5K and 25nm modulation amplitude respectively. As expected, the TMM reaches its highest values for low static temperatures, as the AC contribution of the signal increases. An opposing situation occurs for the HMM which achieves maximum contrast for high static temperatures. This again is due to a higher AC contribution. The jump ratio for the static SThM is not plotted here, because its values range between 1.00 and 1.02 therefore not being visible in this diagram. Regarding thermal near field phenomena theory (Xu et al.1994), the HMM is more sensitive to the modulation amplitude than the TMM, as it varies by z^3 (TMM~T^2). This can be observed regarding the slope of the fitted polynom which is higher for the HMM than for the TMM case.

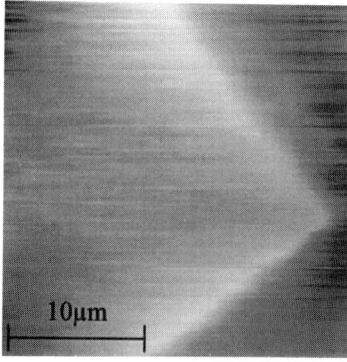

Fig.3a Topographical image of a
rectangular diode structure

Fig.3b Conventional SThM image
corresponding to figure 3a

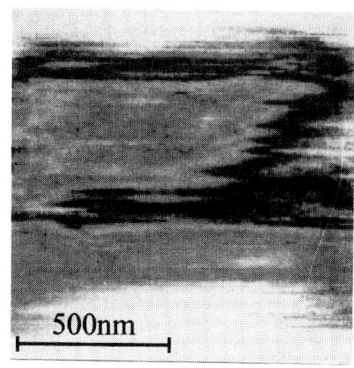

Fig.3c HMM ESThM amplitude image
corresponding to figure 3a

Fig.3d HMM ESThM amplitude image
at an enlarged magnification

Fig. 3a-3d demonstrate a first example for the increase of sensitivity using the ESThM. Here the contact areas of a GaAs based tunneling diode have been investigated at a scan range of 24µm. The topography is illustrated in fig.3a. The overall height in this image is 460nm. Fig.3b is the corresponding SThM image which has been obtained simultaneously. As can be deduced from this image the contact area appears at an almost homogeneous thermal level bordered by a high signal at the structure edge. The signal increase at the vicinity of the structure edge is likely   to be due to an enlarged effective tip contact area causing a higher heat flux. A second scan has been performed at the same location by HMM at 32kHz modulation frequency and 40nm modulation amplitude. The resulting thermal signal is visible in fig.3c. Beside the bright rim at the structure edge a vertical line in the very left of the image is visible which was not observed in the static mode. As the semiconductor layers itself are homogeneous, these features are assumed to be due to compositional variations of the contact materials (Au/Ni/Ge).

An image was obtained at a higher magnification (1.5µm scan range) to determine the lateral resolution (fig.3d) by HMM (s. marked area fig.3c). Here the resolution was achieved by measuring the width of the dark rim which varies between 28 and 150nm. Therefore the lateral resolution is assumed to be better than 30nm in this case.

Theoretical studies of pulsed heat sources acting on a semiconductor reveal the possibility to measure the thermal conductivity of subsurface structures by detecting the surface temperature (Opsal, Rosencwaig 1982). Additionally the evaluated depth can be controlled by adjusting the modulation frequency and thereby the thermal diffusion length.

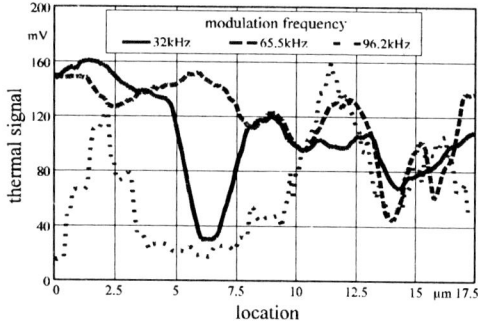

Fig.4 HMM amplitude line profiles obtained at
different modulation frequencies

Fig.4 illustrates line profiles at different modulation frequencies performed on the contact area visible in fig3. Comparing the line profiles at each location for different modulation frequencies gives information about the relative change in thermal conductivity at different specimen depth.

658

substrate s.i.GaAs
20nm undoped GaAs
100nm undoped AlAs
50nm n-doped 10^{18}cm^{-3} GaAs
100nm undoped GaAs
50nm n-doped 10^{17}cm^{-3} GaAs
100nm undoped GaAs
50nm n-doped 10^{16}cm^{-3} GaAs
undoped GaAs

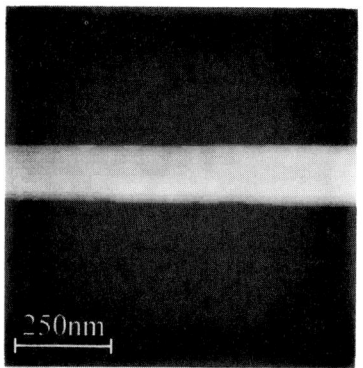

Fig.5a schematic layer structure   Fig.5b TMM amplitude image
of the investigated cross section

The determination and localization of doped layers is mostly carried out by electrical experiments (Maywald et al. 1994b). As doped layers imply a different thermal conductivity than undoped material this could be another field of application for thermal microscopy. Here a cross section of a GaAs layer structure has been investigated. The schematic layer structure is illustrated in fig.5a. The TMM experiment (32kHz, 10K) obtained on the cross section of this specimen is shown in fig.5b. The observed line is located at the position of the highest doped layer. The measured feature width is 175nm which does not correspond to the nominally value but may be due to an increased tip contact convoluting with the layer width.

4. CONCLUSIONS

It has been shown that the enhanced SThM modes imply a strong contrast increase compared to the static SThM. Both the TMM and the HMM are superior to the static mode. The presented applications showed compositional variations on diode contacts and the ability to localize doped material within a GaAs cross section. As is well known from other SFM applications tip shape and contact area are crucial factors limiting the lateral resolution. In this case a lateral resolution of 30nm could be achieved.

ACKNOWLEDGEMENTS

The authors like to thank A.Lindner from the department of solid state electronics from Duisburg University who supplied the cross section.. Also many thanks to Dr. A. Förster from the institute of Thin Film and Ion Technology (ISI) of the Forschungszentrum Jülich who supplied the diodes.

REFERENCES

Dinwiddie R B, Pylkki R J, West P E 1994 Thermal Conductivity 22 (Lancaster, Basel: Technomic Publishing Co), 668-677
Majumdar A, Carrejo J P, Lai J 1993 Appl.Phys.Lett. 62, 2501-2503
Maywald M, Pylkki R J, Balk L J 1994a Scanning Microscopy 8 (2),181-188
Maywald M, Stephan R E, Balk L J 1994b Microelectronic Engineering 24, 99-106
Opsal J, Rosencwaig A 1982 J.Appl.Phys. 53 (6), 4240-4246
Williams C C, Wickramasinghe H K 1986 Appl.Phys.Lett. 49, 1587-1589
Xu J B, et al. 1994 J.Appl.Phys. 76 (11), 7209-7216

Inst. Phys. Conf. Ser. No 146
Paper presented at Microsc. Semicond. Mater. Conf., Oxford, 20–23 March 1995
© *1995 IOP Publishing Ltd*

Nano-field-effect-microscopy of electrical inhomogeneities on InGaAs surfaces

P M Koschinski, G B M Fiege, and L J Balk

Sonderforschungsbereich 254, Bergische Universität-Gesamthochschule Wuppertal, Lehrstuhl für Elektronik, Fuhlrottstr. 10, D-42097 Wuppertal

ABSTRACT: The imaging of electrical surface inhomogeneities of a p-doped InGaAs layer is performed by means of a scanning force microscope with a biased tip acting as an extremely small electrode opposite to the semiconductor surface. The resulting locally induced band bending underneath the tip depends on tip bias and on the electronic properties of the surface. Variation of the band bending due to changes of the electronic properties modulates a current through the layer allowing spatially resolved determination of electrical inhomogeneities of the surface by measuring the current change.

1. INTRODUCTION

The analysis of electrical properties of semiconductor surfaces is a prerequisite for a subsequently application of the semiconducting materials to device fabrication. The electrical properties of semiconductor surfaces are mainly determined by surface states present, i.e. by energy levels distributed over the band gap at the surface (Many et al. 1971). The surface states can act as recombination centers reducing the amount of mobile charge carriers in a device (Kalingamudali et al. 1994a), and they can behave as traps for charge carriers leading to an undesired surface charge (Otaredian 1993).

Several characterization techniques are developed giving access to surface state properties in terms of surface state density, and energetic position or distribution within the band gap. Optical measurement techniques like the photoconductive decay technique (Derhacobian et al. 1994b) utilize the generation of electron-hole-pairs by light excitation and allow the determination of, e.g. carrier life times or in the case of the photoluminscence technique the determination of surface charge densities (Mettler 1976). Other measurement techniques use the field effect as a characterization tool, where an electrode is placed at the surface and where an electrical field is introduced into the material enabling, e.g. the determination of the frequency dependence of surface states effects (Graffeuil et al. 1986). However, all these techniques provide only integral values of the surface parameters. A locally, high spatially resolved determination of surface properties on the sub-micron scale is not possible with these techniques.

A measurement technique possessing both features, an extremely high spatial resolution and sensitivity for electrical inhomogeneities of surfaces is the nano-field-effect microscopy. In nano-field-effect microscopy a scanning force microscope (SFM) (Meyer and Heinzelmann 1992) is used in the non-contact mode (Sarid 1991) with a voltage applied between sample and conductive tip. The principle of the new measurement technique is to alter the electrical situation at the surface by introducing a local electrical field in the semiconductor layer, and to monitor the change of a current flow through the layer perpendicular to the field. The current through the layer is initiated by an applied voltage to ohmic contacts at the surface. Due to the non-contact-mode of the SFM, in which the tip-surface distance is kept constant during the measurement, neither charge carriers are impressed into the semiconductor by direct contact between the biased tip and the surface, nor by tunneling processes or field emission processes because of the great distance between tip and surface. Therefore a minimum interaction between sample and tip can be maintained.

2. PHYSICAL DETAILS

2.1 Non-contact mode of SFM

The non-contact-mode of a SFM is based on the attractive near field forces on a tip located some nanometers above a sample surface. The tip-sample distance dependence of these forces can be used to guide the tip in a fixed distance d across the surface (Fig. 2a). The tip is mounted on a cantilever and the whole system tip-cantilever, hereafter referred to as cantilever for convenience, is able to oscillate with a specific resonance frequency ω_0 The cantilever is excited to oscillate at the fixed frequency ω_0 with an

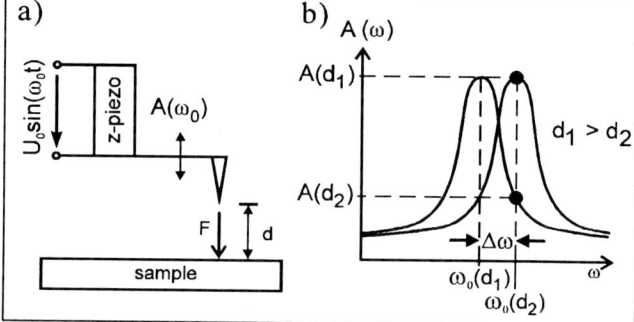

Fig. 2: SFM non-contact-mode basics

amplitude $A(\omega_0)$ at a tip-surface distance d_1 far away from the surface. In close proximity to the surface (distance $d_2 < d_1$) additional attractive forces F with a magnitude depending on the tip-surface separation shift the resonance frequency of the cantilever by $\Delta\omega_0$, leading to an reduced amplitude $A(d_2)$ of the oscillation (Fig. 2b). The amplitude of the cantilever oscillation is therefore a measure for the tip-sample distance.

2.2 The nano-field-effect

The tip is biased with a mandatory ac-component and a superimposed optional dc-component. The capacitor like set-up introduces an electrical field into the semiconductor, which is mainly localized under the tip due to the extremely small tip diameter of some ten nanometers. The alignment of the penetration depth of the field into the semiconductor is possible by varying the magnitude of the dc-component. The ac-component modulates the electrical potential at the surface at a given frequency and with specific magnitude. It is therefore possible to create locally a band bending, i.e. a depletion or enhancement region at the semiconductor surface underneath the tip with a depth variation according to the voltage between tip and surface. The variation of the band bending depends not solely on the applied potentials but on the surface states, too. By varying the surface potential surface states can become charged or uncharged, which leads to additional surface charges modulating the band bending as well as the external voltages. In order to detect this modulation the change in the current flow through the layer is measured by a lock-in amplifier.

3. EXPERIMENTAL CONDITIONS

The experimental set-up used for nano-field-effect microscopy is shown in Fig. 1. The microscope used in the measurements is a commercially available SFM (TopoMetrix TMX 2000 Explorer). The tip of the microscope can be scanned across the surface by a piezo scanner, which is driven by a personal computer controlled electronic control unit. In order to operate the SFM in the non-contact-mode a dc-voltage and a superimposed sine wave voltage are applied to the z-piezo element. The tip is excited to oscillate above the surface by the resulting periodic elongation of the z-piezo element, and the amplitude of the oscillations are monitored by a laser-photo diode array deflection detection system. The output signal of this system is demodulated by a lock-in amplifier, which provides a

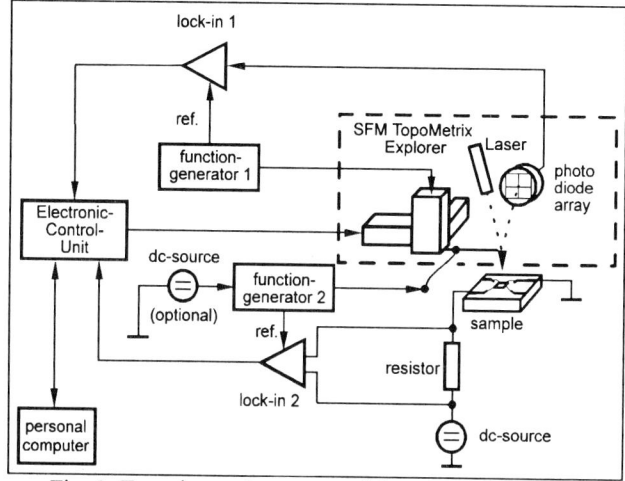

Fig. 1: Experimental set-up used for nano-field-effect microscopy

voltage proportional to the magnitude of the tip oscillation. Since the magnitude of the oscillation is a measure for the tip-surface separation, this signal is used as input signal of the feed back loop of the microscope, which maintains the tip-surface separation by adjusting the z-piezo elongation with the applied dc-voltage. The change of the dc-voltage thus reflects the topography of the sample.

The sample investigated consists of a layer structure of a 200 nm thick p-doped MBE-grown InGaAs layer (doping density $N_A = 5.0 \cdot 10^{18} cm^{-3}$), layer width 100 µm, on an n-doped InGaAs substrate (doping density $N_D = 5.0 \cdot 10^{16} cm^{-3}$). The p-layer is equipped with two ohmic contacts separated by 60 µm, to which a voltage of 18 mV is applied. A change in the current flow through the layer due to a change of the electrical surface properties can be measured as a voltage drop over a resistor of 1kΩ. This voltage is used as an input signal of a lock-in amplifier. Both signals, the topography signal and the nano-field-effect signal can be acquired simultaneously by the electronic control unit and can be displayed as color coded pictures on a computer screen.

4. MEASUREMENT RESULTS

For the measurements a dc-component of 15V and an ac-component of 4V peak to peak are applied to the tip. Fig. 3 shows the image of the sample surface (topography) obtained from the non-contact-measurement. The picture reveals a fairly rough surface for an epitaxially grown layer. The difference in height between the dark and the bright areas is approx. 32 nm. Fig. 4 depicts the change in the current flow through the layer for the same sample area as shown in Fig. 3, where bright areas correspond with great changes in the current flow (3.3 µA), and dark areas correspond with little changes (2.92 µA). It can be seen that significant electrical inhomogeneities of the electronic surface properties are present. These inhomogeneities are not randomly distributed over the surface, they are located in form

662

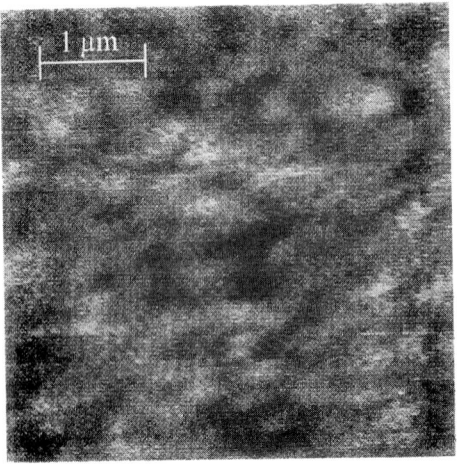 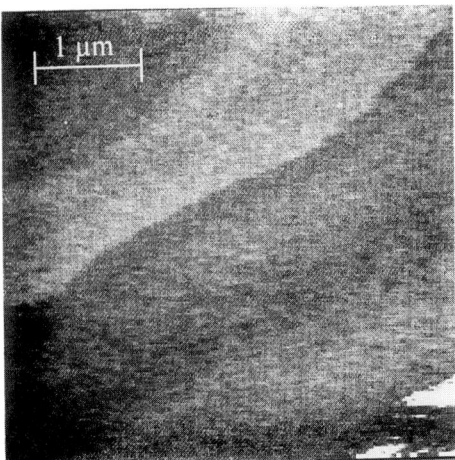

Fig. 3: Topography micrograph of the sample surface   Fig. 4: Nano-field-effect micrograph of the sample surface

of bands with various widths. This is a hint to a non-uniform distribution of surface states or to a local variation of the energic position of the states within the band gap.

5. CONCLUSIONS

Nano-field-effect microscopy allows the imaging of electrical inhomogeneities of semi-conductor surfaces with an extremely high spatial resolution of 100 nm. This characterization technique does not only provide information about the distribution of surface inhomogeneities but yields supplemental spatially resolved information by varying the potentials applied to the tip. An application of an ac-voltage with variable frequency allows the determination of the frequency response of surface states, the alignment of the dc-component allows the determination of the surface state level within the band gap by spectroscopic measurements. By changing the polarity of the dc-component the recombination behaviour of the charge carriers at the semiconductor surface can be studied.

6. ACKNOWLEDGEMENTS

The authors like to thank A. Wiersch from the solid state electronics department of the Gerhard-Mercator-University of Duisburg for supplying the semiconductor sample, and TopoMetrix GmbH, Germany for technical assistance.

REFERENCES

Derhacobian N et al. 1994b Appl. Phys. 76 (8) 4663
Graffeuil J et al. 1986 Solid-State Electronics 29 No. 10 1087
Kalingamudali S R D et al. 1994a Solid-State Electronics 37 No. 12 1977
Many A et al. 1971 Semiconductor Surfaces (Amsterdam: North Holland publishing)
Mettler K, 1977 Appl. Phys 12 p 75
Meyer E and Heinzelmann H 1992 Scanning tunneling Microscopy II, eds R Wiesendanger and H-J Güntherodt (New York:Springer) pp 99-146
Otaredian T. 1993 Solid-State Electronics 36 No. 6 905
Sarid D 1991 Scanning Force Microscopy (New York: Oxford Press)

Inst. Phys. Conf. Ser. No 146
Paper presented at Microsc. Semicond. Mater. Conf., Oxford, 20–23 March 1995
© *1995 IOP Publishing Ltd*

Resolution and sensitivity of the lateral and vertical localization of buried layers in Gallium Arsenide by contact current measurements

M Maywald, V Wittpahl and L J Balk

Sonderforschungsbereich 254, Bergische Universität Gesamthochschule Wuppertal, Lehrstuhl für Elektronik, Fuhlrottstr. 10, D-42097 Wuppertal

ABSTRACT: Doping profiles have been evaluated on a nanometer scale using an SFM based application: the contact current measurement. Additional 2-dimensional calculations based on the analytical semiconductor equations have been carried out to allow a better insight toward carrier transport mechanisms. These calculations also reveal the ability of this technique to localize subsurface structures. To compare experiments and simulations line profiles have been performed on vertical n-doped GaAs layers.

1. INTRODUCTION

The determination of doping concentrations of semiconductor layers with respect to the localization of defects is a problem which is difficult to solve, especially when reaching a nanometer scale. Recently, there have been approaches to determine doping profiles based on a scanning force microscope (SFM). These include capacitive measurements (Huang, Williams 1994), a so-called scanning resistance microscopy (SRM) (Shafai et al. 1994), and contact current measurements (CCM) (Maywald et al. 1994, Balk,Maywald 1994) Whereas the first technique has been applied for point measurements only up till now, SRM and CCM are able to obtain two-dimensional doping profiles. Additionally CCM allows a simultaneous measurement of the current signal and the topography, thereby making experiments at a constant applied load feasible.

Although contrast mechanisms often seem to be straightforward when investigating epitaxial layer structures, it is difficult to obtain quantitative results. As the parameters which contribute to the signal cannot be separated experimentally, additional simulations have been developed to allow a better insight towards carrier flow and contrast generation. In this work, the aspects of sensitivity and lateral resolution due to buried inhomogeneities have been studied at differently doped, vertical Gallium Arsenide layers by two-dimensional numerical calculations. Additionally experiments have been carried out for comparison with calculated values.

2. EXPERIMENTAL SETUP

The CCM determines the variations of the electronic properties of the sample by evaluating locally injected carriers. For this purpose a metallized SFM tip is scanned in contact mode over the sample surface and a potential is applied between the tip and a second fixed electrode (s. Fig. 1). The presence of doping faults and layer interfaces will cause a modification of the carrier trajectories, resulting in measurable signal variations. The current signal and the topography are obtained simultaneously, and thereby can be easily correlated. Although both AC and DC voltage sources can be applied in principle, only DC experiments

664

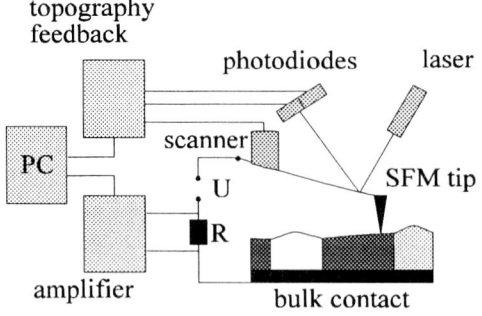

topography
feedback

photodiodes   laser

scanner

PC

U   SFM tip

R

amplifier   bulk contact

Fig.1 experimental setup for CCM

have been performed in this case in order to make a comparison of experiments and numerical calculations easier.

3. SIMULATIONS

To calculate the contact current, the simulator "MINIMOS 5.1" (Selberherr), which solves the analytic semiconductor equations (Poisson (1), continuity (2a,2b)) was adapted to our geometry. Equation (1) is the Poisson equation where λ denotes a geometric factor, Ψ is the electrostatic potential, n and p represent the carrier concentrations for electrons and holes respectively. C denotes a constant term due to doping. $D_{n,p}$ and $\mu_{n,p}$ represent the diffusivity constants and the mobility of electron and holes within the continuity equations for electrons (2a) and holes (2b).

$$\lambda^2 \, div \, grad \, \Psi - (n - p - C) = 0 \qquad (1)$$

$$div(D_n \cdot grad \, n - \mu_n \cdot n \cdot grad \, \Psi) - R(\Psi, n, p) = \frac{\partial n}{\partial t} \qquad (2a)$$

$$div(-D_p \cdot grad \, p - \mu_p \cdot p \cdot grad \, \Psi) - R(\Psi, n, p) = \frac{\partial p}{\partial t} \qquad (2b)$$

For a comparable interpretation of both experiments and simulations, the obtained current density was integrated and recalculated. Moreover the tip was assumed to form an ideal Schottky barrier and the contact on the bottom side of the sample to be of ohmic nature.

4. RESULTS AND DISCUSSION

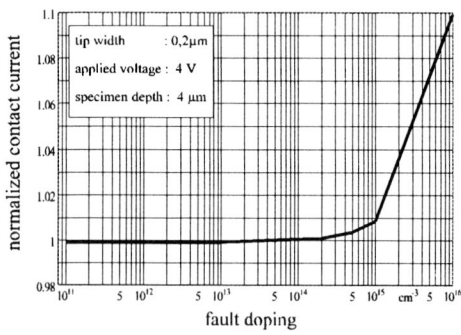

Fig.2 normalized contact current versus fault doping concentration

The first aspect of evaluation was the general sensitivity of the CCM signal towards inhomogeneity recognition. For this purpose we investigated a vertical structure consisting of an n-doped layer of low doping (10^{14} cm^{-3}, 2μm width), embedded by two higher n-doped layers (10^{16} cm^{-3}). A doping inhomogeneity of 300nm^2 size was placed in the middle of the lower doped layer at a depth of 1μm. The fault doping was changed from 10^{11}cm^{-3} to 10^{16}cm^{-3}. Fig. 2 shows the resulting signal that was normalized to the undisturbed case and plotted versus the doping concentration. As can be deduced from fig.2, the fault can be recognized disregarding its doping.

Besides the evaluation of the sensitivity in general, special interest was focused on the depth analysis of inhomogeneities. In fig.3 the normalized contact current is plotted versus the fault depth for a fault size of 500nm^2, a specimen depth of 4μm, and an applied voltage of 4V. The doping profile is analogue to the pervious case. Fig. 3 demonstrates a rapid signal decrease for a low fault depth which reaches an almost steady state close to the bulk contact. The explanation for this effect is a general broadening of the majority carrier channel, due to

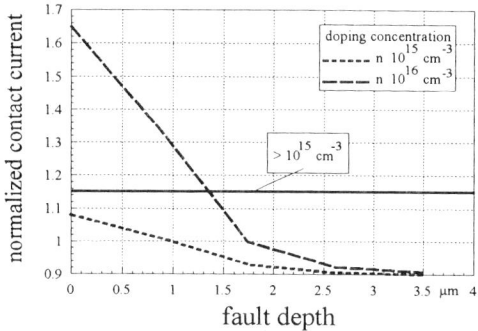

Fig.3 Normalized contact current versus fault depth

the arrangement of a point contact (tip) on top of the surface, and an area contact (bulk) on the bottom side of the sample. For this reason an inhomogeneity located near the specimen surface will be penetrated by the entire current, and therefore will cause a larger signal variation than an inhomogeneity placed close to the bulk contact. Another interesting aspect is the determination of the fault doping concentration. As both the doping concentration and the geometrical size of the fault contribute to the signal, a determination of the doping concentration is possible only when the geometry is known. In this case, a doping concentration $<10^{16}$ cm^{-3} can be excluded for values greater than 1.15 which represents a situation in which the embedded layer is completely n-doped with 10^{15}cm^{-3}. Normalized values below 1 can occur, because even in the undisturbed case the current channel will broaden, and the carriers will penetrate the embedding higher doped layers at a certain critical depth. The presence of an inhomogeneity within the lower doped layer will deflect the majority carriers and therefore change the critical depth.

In this context, line profiles of 40 points each have been calculated to evaluate a

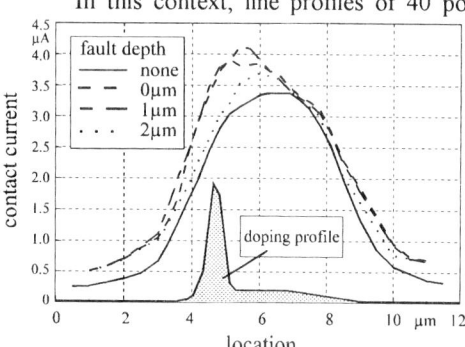

Fig.4 line profiles at different fault depths

possible lateral localization of buried faults. Fig. 4 illustrates the doping profile for a fault of 10^{16}cm^{-3} and 500nm^2 size located at the specimen surface at a transition from 10^{14}cm^{-3} to 10^{15}cm^{-3}. Line profiles have been calculated for 0µm, 1µm, and 2µm fault depths respectively. The resulting contact current was plotted versus tip position (tip size 200nm) and is shown in fig.4. Two important results can be obtained: Firstly, the presence of a fault changes the CCM signal not only close to the inhomogeneity, but also changes the signal amplitude in general. Secondly the lateral position of the fault can easily be detected even under unfavourable conditions. As already mentioned, in addition to the general channel broadening, a carrier deflection occurs due to internal fields caused by doped areas. A deflection such as this can lead to an influence of the inhomogeneity which is not only limited to its lateral extension. Fig.5 illustrates this effect by showing the transversal majority

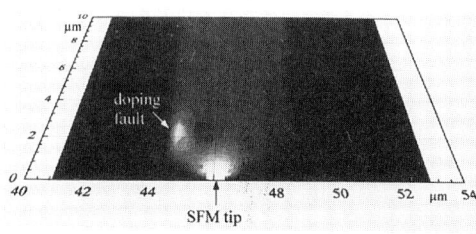

Fig.5 transversal current density

current for a fault depth of 2µm. The illustrated transversal current density corresponds to a tip position of 46µm in fig.4. To visualize the effect the contrast is enhanced in this image.

Lastly, the question of reliability of down scaling from micrometer to nanometer is investigated. The first example is a line profile of a layer interface of an HBT structure (s.fig.6a). Here the transition from an n-doped layer (10^{17})to a highly n-doped ($2x10^{18}$) was evaluated for an applied voltage of 2V. The second example consists of line profiles carried

666

out over a single layer (10^{18}cm^{-3}) at 5V applied voltage (s.fig. 6b). The calculations were performed on a micron scale and rescaled to nanometers to be comparable to the measurements. Relative tip, layer, and specimen sizes were the same in both cases. In the first example (HBT) the tip size was small compared to the layer thicknesses (1/30) whereas in the second example tip and layer widths are of the same size. Therefore as expected, the interface could be imaged quite well during the first experiment. Regarding the signal magnitudes, it can be deduced that the calculated relative jump height values differ by <10% from the measured. This is because during the calculations, ideal Schottky and ohmic contacts were assumed. These assumptions are simplifyed when working under ambient air conditions, because signal variations throughout the measurements due to contaminants or a non ideal ohmic contact at the bottom side, which will effect the signal contrast, were not taken into account in the simulations. Nevertheless it is demonstrated that in this case the major contribution of the CCM signal can be described using the analytic semiconductor equations.

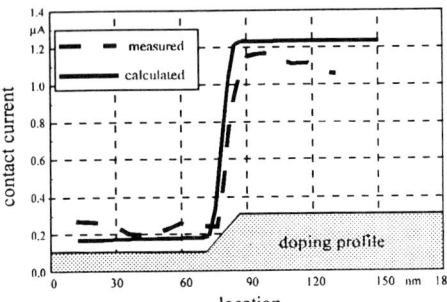 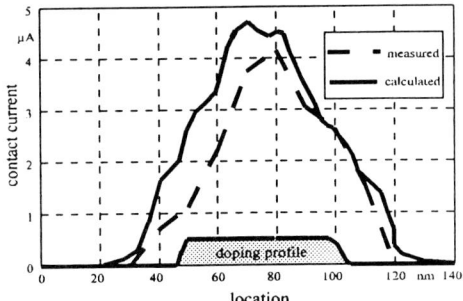

Fig.6a measured and calculated line profiles of the emitter interface of an HBT structure

Fig.6b measured and calculated line profiles of a single layer

5. CONCLUSIONS

Measurements and calculations have been performed on differently doped vertical GaAs layer structures. It is demonstrated that doping faults down to 1% layer size can be detected within a vertical structure. As the CCM signal implies a sensitivity to the depth of an inhomogeneity, a localization is possible if the faults are placed within 20% of the specimen depth beneath the surface. The exact determination of the doping concentration of a buried structure is only possible when the geometry is known, otherwise, only qualitative results can be obtained. It was shown that it is valid to use the analytic semiconductor equations for this geometry even on a nanometer scale by comparable line profiles. The difference between the performed experiments and the calculations are assumed to be due to non ideal surface contact behavior.

ACKNOWLEGDEMENTS

The authors like to thank A.Lindner from the department of solid state electronics from Duisburg University who supplied the specimen. Also many thanks to P.Koschinski for many helpful discussions.

REFERENCES

Balk L J, Maywald M, 1994 Material Science and Engineering B24, 203-208
Huang Y, Williams C C,1994 J.Vac.Sci. Technol. B12 (1), 369-372
Maywald M, Stephan R E, Balk L J, 1994 Microelectronic Engineering 24, 99-106
Selberherr, MINIMOS 5.1, Institut für Mikroelektronik, TU Vienna, Austria
Shafai C, et al., 1994 J.Vac.Sci. Technol. B12(1),378-382

Inst. Phys. Conf. Ser. No 146
Paper presented at Microsc. Semicond. Mater. Conf., Oxford, 20–23 March 1995
© *1995 IOP Publishing Ltd*

Scanning probe acoustic microscopy

M Maywald, G Brockt and L J Balk

Sonderforschungsbereich 254, Bergische Universität Gesamthochschule Wuppertal,
Lehrstuhl für Elektronik, Fuhlrottstrasse 10, D-42097 Wuppertal, Germany

ABSTRACT: It is demonstrated that purely acoustic features can be imaged on a nanometer scale. In contrast to force modulation techniques more relevant information can be obtained than just topographical enhancement. The observed contrast is assumed to be due to propagating contributions as well as to generating mechanisms.

1. INTRODUCTION

There are different techniques to monitor acoustic or elastic specimen features applying different excitation sources. Along with the conventional acoustic microscopy (Wickramasinghe 1979), there are other methods based on a laser or electron beam as a source. Whereas the contrast mechanisms are straightforward when applying acoustic microscopy, understanding of signal generation and contrasts becomes difficult using laser based techniques (Li et al. 1992) or the scanning electron acoustic microscopy (Kaufmann et al. 1994), because thermal and mechanical excitation takes place at the same time. Therefore, a separation of mechanical and thermal features can only be obtained for preferential cases.

Further attempts of detecting purely mechanical specimen features on a nanometer scale have been performed using the force modulation technique (Radmacher et al. 1993). In a scanning force microscope a sample is vibrated in z direction at a frequency of several kHz up to a few MHz (Rabe, Arnold 1994). Although it is possible to detect the mechanical elasticity of the specimen surface by this technique, it suffers from a strong topographical contribution to the detected signal. Especially on rough specimens an analysis of elasticity features is almost impossible due to the signal being overlaid by topographical edge enhancement. To overcome this problem a new technique: scanning probe acoustic microscopy (SPAM) is introduced in this paper.

2. EXPERIMENTAL SETUP

In contrast to the force modulation technique, SPAM excites mechanical strain by a vibrating tip within a scanning force microscope in contact mode. The vibrations are simply generated by modulating the z-piezo. The excited vibrations are detected by a shielded piezoelectric transducer fixed to the bottom side of the sample

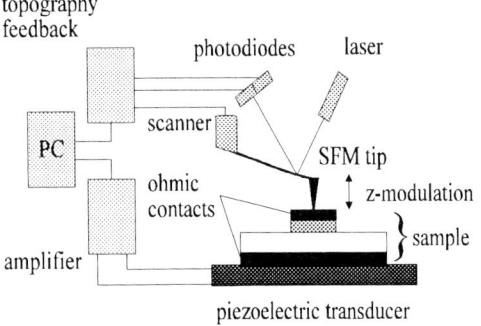

Fig.1 experimental setup for SPAM

(s.fig.1). The transducer delivers an AC voltage at the modulation frequency which is applied to a lock-in-amplifier being a direct measure of the acoustic signal generated. The amplifier output voltage is transmitted to the computer via an analogue-digital-converter. Both the detection of the acoustic signal and the topography are obtained simultaneously. This offers an easy way of correlating topographic and acoustic features.

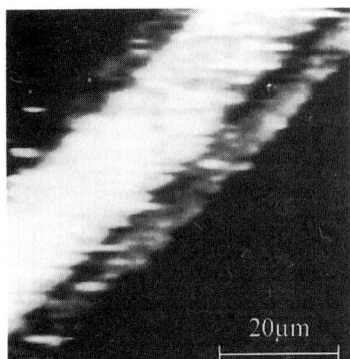

Fig.2a topography image
of a rectangular structure

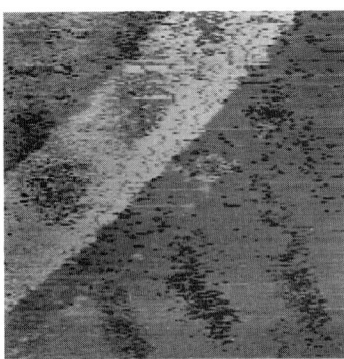

Fig.2b force modulation phase
image at 30kHz

Fig.2c SPAM phase image
at 35kHz

3. RESULTS AND DISCUSSION

With SPAM GaAs based tunnelling diodes were investigated. The diodes were equipped with ohmic contacts on both sides (s.fig.1), and etched afterwards to form rectangular structures.

Firstly, a comparison between the force modulation method and SPAM was performed. To carry out force modulation experiments the setup illustrated in fig. 1 was changed in the following way: an AC voltage applied at the piezoelectric transducer generating a vibration at a certain modulation frequency. The lock-in-amplifier was connected to the photodiode output, and the amplifier output signal was sent to the computer. Fig.2a shows the topography image of the investigated area. The bright area corresponds to a high topographic level and hence, to an ohmic contact. The overall height in this image is 850nm. Figures 2b and 2c are the corresponding force modulation and SPAM phase images obtained at 30kHz, and 35kHz respectively. Obviously, the topographic structures are enhanced throughout fig.2b. The nature of diagonal features, visible throughout the images, seem to be scan artefacts like stick-slip effects, as they do vary with the scan parameters. The corresponding SPAM phase image, obtained at 35kHz, is demonstrated in fig.2c. Within the SPAM image the rectangular structure can be observed. Furthermore a dark line, which is totally missing in the topography and in the force modulation image and therefore assumed to be subsurface, occurs. The appearance of features similar to this vary strongly with the applied modulation frequency but remained stable for each frequency throughout several scans with different parameters. Also, similar features have been observed by other techniques such as SEAM, and are assumed to be due to attenuation of the propagating wave.

The next figure demonstrates an example of another contrast mechanism. Figure 3a is the topography image of a part of a diode structure. The

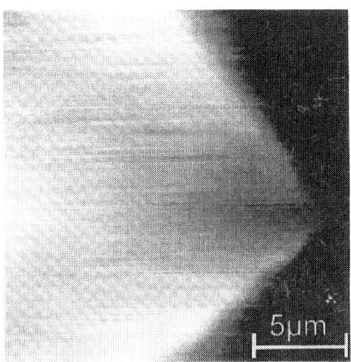

Fig.3a topography image
of a diode structure

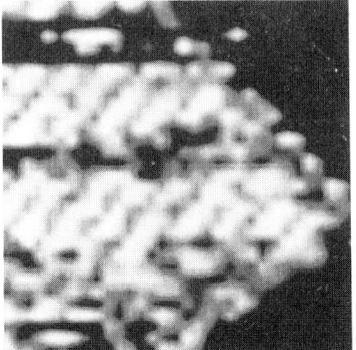

Fig.3b SPAM amplitude image
obtained at 35kHz

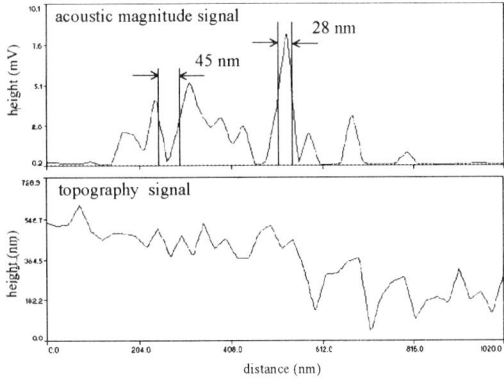

Fig.4 line profiles at a scan range of 1020nm

bright areas correspond again to a high topographical level and hence to the contact area. The overall height in this case was 460nm. The corresponding acoustic image demonstrates that a high signal magnitude was achieved only on the contact area, whereas the signal decreases rapidly outside the contact. It is also important to note that the high magnitude signal is limited very sharply to the contact area. The contrast mechanism leading to this result is not correlated to propagation, but to the signal generation and therefore to the actual surface. As the contact area itself is a continuous area, a point loading force acting on it will have an effect on the whole area and not just at the tip radius. Therefore a large acoustic signal will be generated within the contact area. To confirm this, the ohmic contact (Au/Ni/Ge) was peeled off and the experiment was repeated. A rapid SPAM signal decrease after peeling was observed.

Within this contact area, experiments were carried out at a higher magnification. Fig. 4 shows acoustic and topography line scans at a scan area of approximately 1μm. The obtained lateral resolution was determined by the FWHM criterion. As can be deduced from this measurement the lateral resolution is in the range of 50nm.

Besides the well and sharply distinguishable diodes, some structures were observed where the acoustic signal was not limited to the contact area only (s. fig.5). Fig.5a illustrates the topography of an almost rectangular structure. The overall height difference throughout the image was 450nm. In contrast to fig.3b, the corresponding acoustic amplitude (fig.5b) oviously demonstrates that an extended area supplying a high acoustic signal is present and was not detected in the topography signal. Following the thoughts of the previous example (fig3b) of a continuous area, the contrast could be generated in an equal manner. Therefore, it can be concluded that even in this case a continuous area is present. It may be caused by a misadjustment during the etching process. Additional scanning thermal microscopy (SThM) (Dinwiddie et al. 1994) confirmed this assumption (fig.5c). Bright areas related to high thermal conductivity indicate compositional variations within the contact. Further experiments to analyze this area and to confirm this assumption must be performed.

670

4. CONCLUSIONS

In this work it was demonstrated that SPAM offers new prospects of imaging and localization of purely acoustic features. Examples were shown for the evaluation of surface and subsurface features. It was illustrated that SPAM, in contrast to the force modulation technique reveals more information than just topographical enhancement. Additionally performed thermal experiments give an example for the separation of acoustic and thermal features. This may be very helpful in obtaining a better understanding of signal contrast generation of other techniques like SEAM. The typical lateral resolution that could be achieved was in the range of 50nm. Although not any contrast mechanism is understood right now, SPAM seems to be a good means to investigate acoustic properties of semiconductors on a nanometer scale.

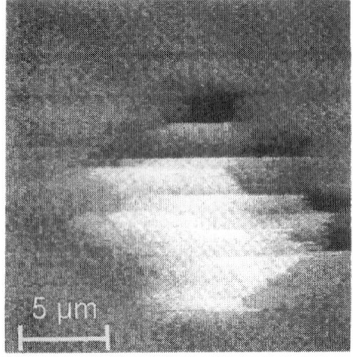

Fig.5a topography image

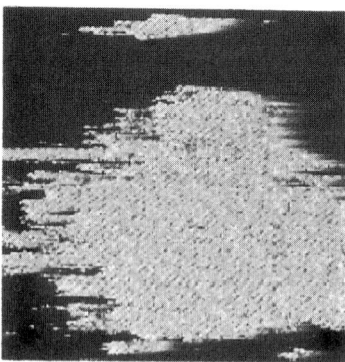

Fig. 5b SPAM magnitude image

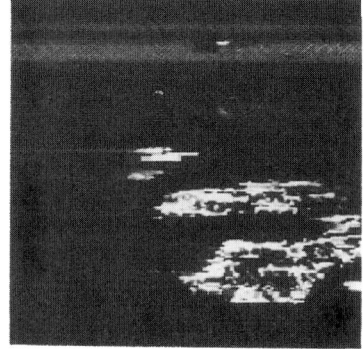

Fig5c SThM image

ACKNOWLEGDEMENTS

The authors like to acknowlegde Dr. A. Förster from the institute of Thin Film and Ion Technology (ISI) of the Forschungszentrum Jülich, Germany who supplied the specimen.

REFERENCES

Dinwiddie et al. 1994 Thermal Conductivity 22 (Lancaster,Basel: Technomic Publishing Co), 668-677

Kaufmann K, et al. 1994 J.Phys.D: Appl. Phys. 27, 2401-2413

Li G et al 1992 Springer Series in Optical Science 69, 384-386

Rabe U, Arnold W 1994 Ann.Physik 3, 589-598

Radmacher M et al. 1993 Biophys. J. Vol.64, 735-742

Wickramasinghe HK 1979 J.Appl.Phys. 50(2), 664-670

Inst. Phys. Conf. Ser. No 146
Paper presented at Microsc. Semicond. Mater. Conf., Oxford, 20–23 March 1995

Scanning tunnelling microscopy at atomic resolution: quantitative information from corrugation heights

A R H Clarke and J B Pethica

Department of Materials, University of Oxford, Parks Road, Oxford, OX1 3PH, United Kingdom.

ABSTRACT: Here we report experimental studies of Cu(100), atomically resolved by Scanning Tunnelling Microscopy (STM).We have systematically determined corrugation heights as a function of tip-surface separation and use these data to deduce a tunnel barrier using a method previously unreported. Our results show that the variation of the corrugation with separation implies that the tunnel barrier can be much lower than the workfunction. At very small separations, the corrugation height actually decreases with decreasing separation. This, we suggest, is due to a mechanical interaction between tip atom and the surface.

1. INTRODUCTION

Most STM images are interpreted in terms of the Tersoff and Hamann model (1985), in which the tip traces out contours of constant charge density above the surface. These contours are given by a combination of atomic position, the local density of states at the imaging energy, and the local electron decay length, κ, above the surface. Despite the directness of the model, in practice very few quantitative STM models of experiments have been presented. The primary reason for this is the ill defined nature of the tip and its geometry, which have a direct effect on the corrugations observed. For semiconductors there is the additional difficulty of providing accurate calculations of the local density of states close to band edges. Even such a basic parameter as the decay length above the surface, or the tunnel barrier height, is not always well understood. This relative weakness of quantitative modelling is not of serious concern where the STM image is used solely to determine lateral atomic geometry, where surface electronic structure is not complex, as for metals. However, particularly for semiconductors, which show strong effects of surface and probably tip electronic structure in the images, a better understanding of the processes involved in STM imaging is desirable. This paper presents a study of apparent tunnel barriers, using a new method. We conclude that forces between tip and surface may produce image effects additional to those in the Tersoff and Hamann model.

The normal method to measure the barrier height, ϕ, is to oscillate the tip surface separation, s, and measure the resulting changes in tunnel current, I. Previous experimentally measured values, Schuster et al. (1992) for the barrier height above clean metal surfaces using this method are in the range 2.5-4eV and are close to their respective large separation workfunction limits. However in this modulation method, it is necessary to oscillate the tip at a relatively high frequency (faster than the feedback loop response). At such a frequency it can difficult to accurately calibrate the real change in tip surface gap in terms of piezo voltage due to resonances and phase shifts. In the method described in this paper, the barrier height is deduced by measuring the rate of decay of the image corrugation as a function of tunnel current. The barrier is deduced via the parameter $\partial \mathrm{Ln}(\Delta)/\partial \mathrm{Ln}(\mathrm{I})$. A series of topographic images are taken in quick succession using the same tip. Each image is of the same area of the surface, but taken at a different tunnel current while the potential bias is kept at a constant value. Systematic measurements of corrugation heights are then made along the principal directions in each image. These data are used to make a plot $\mathrm{Ln}(\Delta)$:$\mathrm{Ln}(\mathrm{I})$ from which $\partial \mathrm{Ln}(\Delta)/\partial \mathrm{Ln}(\mathrm{I})$ can be obtained. An important feature of the method is that no distance calibration other than the well established value (e.g. from step heights) for the normal mode of the microscope is needed.

2. THEORY

Following Tersoff and Hamann (1985) the peak to trough corrugation height, Δ, of the corrugations above the surface at an average distance, s, which is large compared with the corrugation height is:

$$\Delta = \frac{C}{2\kappa} \exp(-\beta s) \qquad (1)$$

where $\beta = 2(\kappa^2 + \frac{G^2}{4})^{1/2} - 2\kappa$, $\kappa = \frac{2\pi(2m\phi)^{1/2}}{h}$, $G = \frac{2\pi}{a}$ and C is a constant.

The tunnel current varies with separation as (one dimensional model, Simmons (1963))

$$I = \frac{I_0 V \phi^{1/2}}{s} \exp(-2\kappa s) \qquad (2)$$

where V is the applied bias voltage and I_0 is a constant. If the prefactors in equations (1) and (2) are treated as constants, it is easy to show that

$$\phi = \frac{\hbar^2 G^2}{8m_e X}, \text{ where } X = \left(\frac{\partial Ln(\Delta)}{\partial Ln(I)}\right)_V \left[\left(\frac{\partial Ln(\Delta)}{\partial Ln(I)}\right)_V + 2\right] \qquad (3)$$

Inclusion of the prefactors increases the complexity of the expressions, and significantly reduces the calculated value of barrier height. In addition, the decay constant κ may itself be a function of tip surface separation, as discussed by Coombs et al. (1988) and Lang (1988), which further complicates the modelling. We describe these effects in more detail elsewhere, Clarke et al. (1995). However, they still leave $\partial Ln(\Delta)/\partial Ln(I)$ as a measure of barrier, and Δ still increases monotonically with I.

Note that the absolute values of Δ and I are not important in obtaining the final result, only their variation over a sequence of images. It is not necessary to know the tip geometry. The method avoids the need for an accurate current-distance calibration, which can be difficult to obtain experimentally. In addition, the effect of using corrugations from different surface G-vectors can be used to cross-check the results.

3. EXPERIMENTAL

The experiments in this paper were performed using an Ultra High Vacuum (UHV) compatible fast scanning STM which has been described elsewhere by Lægsgaard et al. (1988). The vacuum chamber in which all the experiments were performed is equipped with surface analysis techniques including Low Energy Electron Diffraction, Auger Electron Spectroscopy and facilities for sample sputtering and annealing. All the STM images taken in this work were recorded using the constant current mode of operation. The tip used was a tungsten single crystal oriented along [100]. A majority of images in this work were taken with a negative sample bias relative to the tip which was held at a virtual ground, although atomic resolution images were routinely taken at both polarities. The Cu(100) crystal used in this work was prepared using cycles of 15 minute spells of Neon ion bombardment (3kV, 15mA) followed by annealing of the crystal to 530°C for 15 minutes.

The first test was on Cu(100), a symmetric surface, which has two equivalent $<1\overline{1}0>$ and two $<001>$ directions along which to investigate corrugation heights. A metallic surface was chosen because metals do not usually exhibit strong voltage dependent effects in the image; they have a fairly uniform density of states near the Fermi level.

Imaging conditions used in this work were in the range I ~ 1-20nA and V ~ ±5-20mV. A typical STM image of the surface is shown in fig 1, while figure 2 shows a data set from an experimental run for corrugations measured along $<1\overline{1}0>$. Considerable care was taken to ensure no tip changes occurred within a data set. In figure 2, there are two main regions worthy of note, a linear region at low values of tunnel current which is used to deduce $\partial Ln(\Delta)/\partial Ln(I)_V$ and a roll-off region at higher tunnel currents typically characterised in these experiments by a tunnel gap resistance of 4MΩ, where the corrugation does not increase by as much as is expected and eventually starts to decrease compared with points taken at slightly lower currents. Similar qualitative behaviour was observed for data measured along <001>. Doyen et al. (1992) have observed a similar corrugation roll off in experiments on the 2x direction of the clean Au(110) (1x2) reconstruction. In that study, a maximum in the corrugation heights was seen at a tunnel gap resistance of 5.5MΩ, which is extremely close to the value reported here.

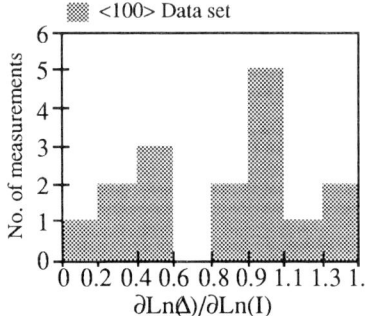

<100> Data set

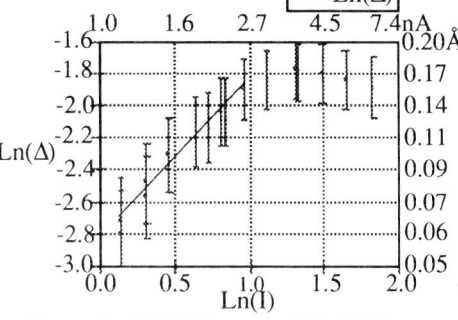

Figure 1: A STM image of Cu(100). Tunnelling parameters I=2.5nA V=10mV.

Figure 2: A Ln(Δ):Ln(I) plot for a data set analysed along <110> as described in text.

Figure 3: A frequency histogram showing the distribution of values of $(\partial Ln(\Delta)/\partial Ln(I))$ obtained in this series of experiments along <001>.

In total 16 data sets were analysed in detail. Within a single set there were small standard deviations (< 10% of their mean measured value). Between data sets there were significant variations, due without doubt to differing tip structures. All showed the linear variation of Ln(Δ) with Ln(I) at lower currents. The values of these lower current slopes were plotted in a histogram for all the data sets. Figure 3 shows that a bimodal distribution of values of $\partial Ln(\Delta)/\partial Ln(I)_v$ was observed, with a large peak in the distribution at $\partial Ln(\Delta)/\partial Ln(I)_v = 1$ and a smaller peak at $\partial Ln(\Delta)/\partial Ln(I)_v = 0.5$. It was noted that in the former instance, the data was characterised by generally lower tunnel currents (~ 2-5nA) and lower corrugation heights (~ 0.02-0.05Å along <1$\bar{1}$0> and 0.07-0.12Å along <001>), while in the latter case, the data was in general taken at higher tunnel currents (~ 6-12nA) and exhibited larger corrugation heights (~ 0.05-0.15 along <1$\bar{1}$0> and 0.1-0.2Å along <001>). We conclude from this and the histogram that two general types of tip are present.

4. DISCUSSION

Simple application of equation (3) to the data measured along <1$\bar{1}$0> yields an apparent mean tunnel barrier of 4.6eV for $\partial Ln(\Delta)/\partial Ln(I) = 0.5$, while $\partial Ln(\Delta)/\partial Ln(I) = 1$ gives a barrier height of 1.9eV. Thus the data reported here is similar to other reported values in the literature. However, the true mean barrier is significantly smaller when the approximations in (3) are taken into account. The bimodal nature of Figure 3 suggests that two general types of tip are present. For high apparent barriers the corrugations are small, suggesting a larger radius tip. The geometry might more closely resemble one-dimensional, with the corrugation being a small perturbation. Along with the lower currents, this suggests that tip and surface are further apart. In contrast, the lower apparent barriers relate to high corrugations; the tip may comprise a rather exposed single atom. Significant tip and surface interaction may be more likely.

674

The most marked feature of the data in Fig. 2 is the roll-off of corrugation with increasing current. This is not predicted by the Tersoff and Hamman model, and is a clear indication of tip-surface interaction. Indeed the 'low barrier' tips described above also show the earliest corrugation roll-off. Following Ciraci et al. (1990), we have investigated the forces between the tip and the surface using a full Molecular Dynamics (MD) simulation. In particular, we have studied the relationship between the change in tunnel gap width measured at long range (this is how displacements are measured in STM) and the change in the shortest vertical distance between the tip and surface for different tip positions above the surface, Clarke et al. (1995).

At small gap widths (~ 3.5Å), the tip apex atom feels a site dependent force due to the presence of the surface. This force is larger over a hollow site in the surface than over an atomic site for two reasons. Firstly, as a result of the condition of constant current, the tip is closer to the surface over the hollow by an amount equal to the true corrugation height and secondly the forces on the tip are larger as it is in a position of higher effective co-ordination. During a tip scan, in moving from an on top position to a hollow site, the tip apex must move a distance equal to the true corrugation height closer to the surface. Part of this displacement comes as a result of the larger force seen by the tip over the hollow site. The z-piezo element has to make up the remainder of the displacement which is now less than the true corrugation height. Hence an apparently reduced corrugation height is recorded in the topographic image. At a gap width of 3.3Å, a corrugation reduction of 0.14Å is predicted in this work. This is a little larger than the reduction observed in figure 1. However the MD simulation performed in this work was for a tip over a hollow site and better agreement between theory and experiment is to be expected by repeating this calculation for a tip over a bridge site.

Doyen et al. (1992) argue that the roll-off seen in experiments on Au(110) can be explained as being due to a short range tip sample interaction which increases the local charge density at the Fermi level due to the formation of bonding and antibonding states. Over the hollow site, the tip experiences a higher co-ordination and is subject to a stronger chemical bond. The result of this interaction leads to a reduction in the measured corrugation height.

The implications of this study are that from a study of the corrugation above a metallic surface, a point of significant tip-surface interaction can be established which can be practically exploited for example in the use of STM as an atomic writing device. Zeppenfeld et al. (1992) have shown the feasibility of the use of STM for use as an atomic manipulation device.

5. CONCLUSION

In this paper, a study has been made of the corrugation in the charge density above a clean metallic surface, Cu(100) using STM. The corrugations exhibit the expected exponential increase with decreasing gap width down to fairly small separations. At smaller separations still typically characterised by a tunnel gap resistance of 4MΩ, the corrugations exhibit a "roll off" and do not increase by as much as expected and may even decrease. Two general types of STM tip can be discerned from the data. Using a simple force calculation, the roll-off phenomenon is explained as being due to the enhanced mechanical interaction of the surface with the tip when it is midway between a pair of surface atoms compared with when it is directly over a single atom. The implications of these data are that they provide a measure of the interaction between the tip and sample which in turn is relevant to the use of STM as an atomic writing device. They also suggest that tip specific interactions might be used to differentiate between chemical species.

Clarke A R H and Pethica J B, 1995 in preparation
Ciraci S, Baratoff A. and Batra I P, 1990 Phys. Rev. B **42** 7618
Coombs J H, Welland M E and Pethica J B, 1988 Surface Science **198** L353
Doyen G, Drakova D, Barth J V, Behm R J and Ertl G, 1993 Phys. Rev. B **48** 1738
Lægsgaard E, Besenbacher F, Mortensen K and Stensgaard I 1988 J. Micros. **152** 663
Lang N D 1988 Phys. Rev. B.1988 **37** 10395
Tersoff J and Hamann D R, 1985 Phys. Rev. B **31** 805
Schuster R, Barth J V, Winterllin J, Behm R J and Ertl G, 1992 Ultramicroscopy **42-44** 533
Simmons J G, 1963 J. Appl. Physics **34** 1793
Zeppenfeld P, Lutz C P and Eigler D M, 1992 Ultramicroscopy **42-44** 128

Inst. Phys. Conf. Ser. No 146
Paper presented at Microsc. Semicond. Mater. Conf., Oxford, 20–23 March 1995
© *1995 IOP Publishing Ltd*

SEM characterization of multilayers

V V Aristov, E I Rau and E B Yakimov

Institute of Microelectronics Technology, Russian Academy of Sciences, Chernogolovka, 142432 Russia

ABSTRACT: Applications of SEM "apparatus" tomography methods for nondestructive characterization of multilayer planar structures are discussed. It is shown that the energy dispersive BSE mode allows us to obtain the thicknesses and depths of layers with thicknesses as small as 10nm. The possibilities of reconstruction of depth distributions of electrical and optical properties with submicron resolution by in situ differential EBIC and CL are demonstrated.

1. INTRODUCTION

Multilayer structures now have a lot of technical applications that stimulate the development of methods for layer characterization. The spatial resolution of characterization techniques used for this purpose should be high, because the thicknesses of layers to be studied can achieve values of about 1-10nm. Of course, the thicknesses of such thin films can be measured by transmission electron microscopy or by scanning tunnelling microscopy, but both techniques need special sample preparation which is rather complex, especially in the case when the lateral dimensions of structures under study are also small. The methods of Scanning Electron Microscopy (SEM) seem to be more suitable for these purposes because they usually do not need special sample preparation and can be used for nondestructive characterization. The lateral resolution is high enough but the depth resolution of conventional SEM methods does not allow characterization of planar structures with nanometre-thick layers.

To characterize multilayer structures it is necessary not only to measure the thickness of different layers but also the electrical and optical properties of each layer. The SEM methods can be successfully used for this purpose but obtaining the required depth resolution is the problem limiting the field of their application. Recently, it was shown (Donolato 1989, Zaitsev and Samsonovich 1990) that tomography methods not only allow reconstruction of the three-dimensional distribution of physical properties but also improve the spatial resolution. For such reconstruction it is necessary to solve the inverse problem, which in nearly all cases can only be solved by the regularization technique (Tichonov and Arsenin 1977) that needs very high measurement precision. Therefore, in many cases it is useful to develop "apparatus" tomography methods in which layer-by-layer sectioning is achieved using a specially designed setup. These methods allow us to reconstruct the distribution of properties to be measured in one direction only. Fortunately, for the characterization of most microelectronic

676

structures it is necessary to have high depth resolution and lateral resolution typically of about a few micrometres.

In the present paper the "apparatus" approach to realization of SEM layer-by-layer tomography, in which the reconstruction of internal structure and physical properties depth distribution is achieved by means of specially designed setup, has been discussed. It is shown that two in situ differential methods: the Electron Beam Induced Current (EBIC) with depletion region modulation and   Cathodoluminescence (CL) with electron beam energy and current modulation, can be used for the reconstruction of depth distribution of electrical and optical properties. The possibilities of internal structure reconstruction using the measurements in Backscattering Electron (BSE) mode have been demonstrated. It has been shown that such methods can achieve the depth resolution in the nanometre range and simultaneously have the lateral resolution usual for SEM methods.

2. BACKSCATTERING ELECTRON MODE

The possibility of layer-by-layer tomography of multilayer structures in the backscattering electron (BSE) mode has been demonstrated by Wells (1971) and by Aristov et al (1988) using the measurements of BSE coefficient dependence on the primary electron energy E_b. The layers with densities different from that of the matrix lead to an appearance of some peculiarities in such dependences which, in principal, allow the thickness of layers and their depth to be obtained (Aristov et al 1993). If the structure under study consists of layers with a different density, the more dense layers give peaks in the BSE coefficient dependence upon E_b the number of which is equal to the number of such layers. It should also be noted that for layers containing heavier elements it has been shown that there is a range of E_b for which the signal from the deeper layer is larger than that from the shallower one. By optimizing the energy of primary electrons, any specific layer in the multilayer structures can be inspected. The experiments carried out allow the estimation of the minimal thickness which can be measured by this method. It was found to be about 10 nm for layers with rather different densities (Al-Cu-Si structure) and about 100 nm for those with close densities (Al-SiO$_2$-Si structure).

An alternative method for the internal structure reconstruction from the measurements in the BSE mode is to use spectroscopy of BSE energy (Reimer et al 1991, Dryomova et al 1993). Indeed the electrons reflected from the near surface layer have not practically lost their

Fig.1. BSE energy spectra obtained on the bulk pure Al and Cu (Curves 1 and 2, respectively), on the structure Si-Cu-Al formed by thermal evaporation of 60 nm Cu layer and of 270 nm Al layer on the Si substrate (Curve 3) and on the structure Cu-Al-Cu consisting of 270 nm Al layer and 75 nm Cu layer evaporated on the Cu substrate (Curve 4).

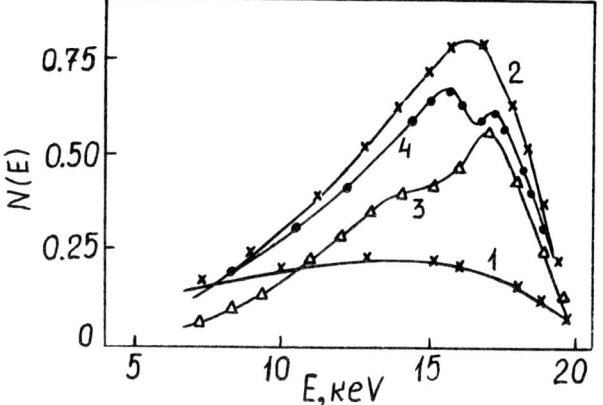

energy, while the energy of backscattered electrons reflected from the deeper layers decreases with an increase of layer depth. Therefore measurements of BSE spectrum allow one to obtain information about the depth and thickness of layers with different chemical compositions.

For such measurements, it is necessary to design the BSE energy spectrometer well adapted to the SEM with high energy resolution. For these purpose we used the energy analyzer based on toroidal deflector developed by Dryomova et al (1995) which has an energy resolution of about 100 eV at BSE energy 10 keV. The energy spectra of BSE obtained on the test structures at E_b =20 keV using this spectrometer are presented in Fig.1. It is easy to see that the spectra obtained on multilayer structures have some peculiarities in the high-energy part that allows the evaluation of the depth and thickness of heterogeneous layers. For measurements of the parameters mentioned, it is necessary to solve the inverse problem by using a computer tomography technique

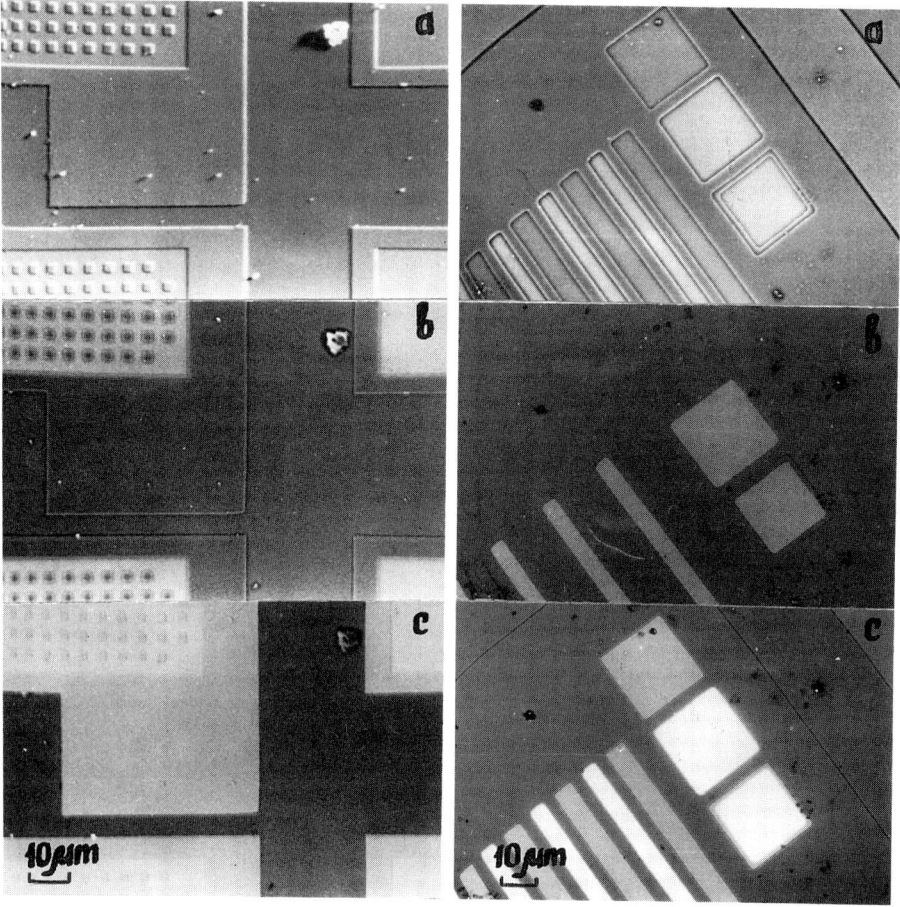

Fig.2. Image of integrated circuit fragment obtained in the SE (a) and BSE (b,c) modes. The energies of detected BSE are 23 keV (b) and 15 keV (c), E_b=25 keV.

Fig.3. Image of Si locally doped by phosphorus obtained in the SE (a) and BSE (b,c) modes. The energies of detected BSE are 14 keV (b) and 10 keV (c), E_b=15 keV.

It should be noted that the detection of BSE having the energy within a narrow energy window allowed the layer-by-layer inspection of multilayer structures and the separation of signals from layers situated at different depths. The possibilities of such "apparatus" layer-by-layer tomography are illustrated by Fig.2 and Fig.3. In Fig.2 the images of integrated circuit fragment in the secondary electron (SE) (a) and BSE (b,c) modes are presented. From the SE image the integral information about the structure under study can only be obtained, whilst the BSE energy filtration allows the separation of images of Al contact layer (b) or that of SiO_2 film (c). Thus, the use of energy dispersive detection of BSE allows to reconstruct nondestructively the topology of multilayer structure and to check the quality of different layers even for layers with close densities such as Si-SiO_2-Al structures. The images of Si locally doped by phosphorus are presented in Fig.3. The phosphorus concentration is $5 \times 10^{18} cm^{-3}$ and the depth of doping is 1 and 2.2 µm. It is seen that this method reveals doped regions and separates regions with different doping depth; the shallow layer is shown in Fig.3(b) and the deeper one in Fig.3(c).

3. METHODS FOR ELECTRICAL AND OPTICAL PROPERTIES RECONSTRUCTION

To investigate the dopant and diffusion length depth distribution and the depth distribution of radiative recombination centers, two in situ differential methods were developed; electron beam induced current (EBIC) with depletion region width modulation and cathodoluminescence (CL) with energy and current modulation have been recently proposed by Kononchuk et al (1991) and by Kireev and Razgonov (1989), respectively. The depth resolution of both methods was shown to be much better than the electron range and is determined by modulation parameters, measurement precision and characteristics of structures under study (Yakimov 1994).

3.1 Modulated EBIC

The collected current in the EBIC mode I_c changes as $\exp(-s/L)$, where $s = z - W$ is the distance between the collected junction and the generation point, z is the depth of electron-hole pair generation, W is the depletion region (DR) width and L is the diffusion length. Therefore to obtain the diffusion length the dependence of I_c on s should be measured. Usually for planar structures with collected junction perpendicular to the e-beam s is varied by changing the generation depth which can be done, for example, by a variation of primary electron energy. But s can be changed also by varying of DR width. If the DR width W is modulated by applying a small ac voltage, the measurements of first harmonic of signal selected by a lock-in amplifier allow the diffusion length in the region adjacent to the DR to be determined (Kononchuk et al 1991). In this modulated EBIC method, the signal measured is proportional to the first derivative of collected current dI_c/dW and, as shown by Kononchuk et al (1991), the diffusion length at depth $z = W$ can be obtained as

$$L^2(W) = [Q^2(W) + Q'(W)]^{-1} \qquad (1)$$

where $Q(W) = -[dI_c(W)/dW]/[I_c(W) - \int_0^w h(z)dz]$, $h(z)$ is the depth dependent excess carrier generation function. It should be stressed that in the method discussed $dI_c(W)/dW$ and $I_c(W)$ values can be measured independently, therefore $Q(W)$ can be easily calculated using the experimental data. From (1) it is easy to see that $L(z)$ distribution can be reconstructed on the base of such measurements by changing of DR width (by increasing the applied bias). This method allows to reconstruct diffusion length depth distribution with resolution determined by measurement precision, by modulation amplitude and by doping level and can achieve values much smaller than the electron range R. The results of reconstruction of L distribution in dry etched Si and GaAs by the modulated EBIC were presented by Kononchuk and Yakimov (1993). The depth resolution under diffusion length reconstruction in GaAs was demonstrated to be better than 100 nm at E_b=25 keV. The lateral resolution is the same as for conventional EBIC and depending on E_b can be changed from 0.1 to a few micrometres.

As shown by Kononchuk and Yakimov (1992) this technique can be used also for local measurements of W at any applied bias and for mapping of the dopant and diffusion length distribution separately. For this purpose it is necessary to measure $dI_c(W)/dW$ and $I_c(W)$ values at two different beam energies The depth resolution of the method discussed can be comparable with that of widely used C-V method but with the higher lateral resolution.

Under multilayer structure, characterization the modulated EBIC method allows the measurements of diffusion lengths in thin films even in the case when their thickness is smaller than diffusion length value. It is possible also to reconstruct depth distribution of dopants that can be very useful under failure analysis of structures with small lateral dimensions.

3.2. Modulated CL

In the modulated cathodoluminescence mode electron energy E_b and electron beam current I_b are modulated with frequency ω as $E_b = E_0(1 + a_1 \sin \omega t)$ and $I_b = I_0(1 + a_2 \sin \omega t)$, respectively. Under the assumption that minority carrier diffusion, non-radiative surface recombination and light absorption inside the sample can be neglected, and that the sample consists of layers infinite in x and y directions and its characteristics are constant inside every layer, the signal detected at wavelength λ can be presented as (Kireev et al 1993)

$$S_{\lambda,k\omega} = I_0 E_0^{1-\alpha} a^k [\sum_{i=1}^{n} A_i(\lambda) \int_{d_{i-1}}^{d_i} f_{k\omega}(z/R_0, a)dz] \quad (2)$$

where $k=0$ for the conventional unmodulated CL and $k=j$ for j-th harmonics of CL signal, α is the coefficient in the dependence $R \sim E_b^\alpha (\alpha = 1.75)$, $A_i(\lambda)$ is the constant depending on the parameters of i-th layer, d_i is the depth of interface between i-th and $(i+1)$-th layers, n is the number of layers, $a=a_1/a_2$, R_0 is the electron range at $E_b=E_0$, $f_{k\omega}(z/R,a) = R h(z,E_b)$ for $k=0$. The expressions for $f_{k\omega}(z/R,a)$ for the first and second harmonics of the modulated CL signal were given by Kireev et al (1993, 1994). It is very important to stress that while $f_{k\omega}(z/R,a)$ is positive at all z, $f_\omega(z/R,a)$ is equal to 0 at one point and $f_{2\omega}(z/R,a)$ is equal to 0 at two points from the range $0 \leq z \leq R$. Therefore it is possible (Kireev et al 1993) to choose

680

the **a** and E_b values such that the signals on the first harmonics from any layer of a two-layer structure or the signals on the second harmonics from any two layers of a three-layer structures can be made equal to 0.

In order to measure the CL spectrum from any layer of a two- or three-layer structure it is possible to use ccmputer simulation of signals to choose the condition when the signals from other layers are equal to 0. If at least one line from the spectra associated with any specific layer is known, the conditions which allow the elimination of the signal from this layer can be obtained experimentally. The influence of charge carrier diffusion and surface recombination, as shown by Kireev et al (1993) can also be taken into account.

4. CONCLUSION

The possibilities of SEM methods for multilayer structure characterization have been discussed. It is shown that the "apparatus" tomography approach is very promising for measuring the thicknesses of thin layers and for reconstruction of the electrical and optical properties distributed in them. Such methods allowed us to achieve depth resolution of about 10 nm and lateral resolution in the micrometre range, that is very suitable for the characterization of microelectronics structures.

ACKNOWLEDGEMENTS

This work was partially supported by Russian Foundation of Fundamental Research (Grant No 93-02-2293)

REFERENCES

Aristov V V, Dryomova N N, Zaitsev S I, Kazmiruk V V, Ushakov N G and Firsova A A 1988 Doklady AN SSSR 301 611 (In Russian)
Aristov V V, Dryomova N N, Kireev V A, Razgonov I I and Yakimov E B 1993 Acta Phys. Polon.A 83 81
Donolato C 1989 Inst.Phys.Conf.Ser. No100 715
Dryomova N N, Drokin A P, Zaitsev S I, Rau E I and Yakimov E B 1993 Izv.RAN 57 No8 9 (In Russian)
Dryomova N N, Rau E I and Robinson V N E 1995 Pribory i Technika Experimenta No1 87 (In Russian)
Kireev V A and Razgonov I I 1989 Zh.Technich.Phiz. 59 180 (In Russian)
Kireev V A, Razgonov I I and Yakimov E B 1993 Scanning 15 31
Kireev V A, Razgonov I I and Yakimov E B 1994 Mater.Sci.Engineer. B24 121
Kononchuk O V, Ushakov N G, Yakimov E B and Zaitsev S I 1991 J.Phys. IV, Coll.C6 1 C6-51
Kononchuk O V, Yakimov E B 1992 Semicond.Sci.Technol. 7 A171
Kononchuk O V, Yakimov E.B 1993 Solid State Phenomena, eds. H G Grimmeiss, M Kittler and H Richter (Switzerland: Scitec Public.Ltd) 32&33 pp 99-104
Reimer L, Bongeler R, Kassens et al 1991 Scanning 13 381
Tichonov A N and Arsenin V V 1977 Solution of Ill-Posed Problems (Washington Winston/Wiley)
Wells O 1971 Appl.Phys.Lett. 19 232
Yakimov E 1994 Mater.Sci.Engineer. B24 23
Zaitsev S I and Samsonovich A V 1990 Izv.AN SSSR, ser.Fiz. 54, 247 (In Russian)

Inst. Phys. Conf. Ser. No 146
Paper presented at Microsc. Semicond. Mater. Conf., Oxford, 20–23 March 1995

Applications of scanning infra-red microscopy to bulk semiconductors

GR Booker, Z Laczik and P Török[*]

University of Oxford, Department of Materials, Parks Road, Oxford OX1 3PH,

ABSTRACT The scanning infrared microscope (SIRM) for the imaging of inhomogeneities present in bulk semiconductor specimens is described in its transmission (T) and reflection confocal (RC) modes. Factors that affect the performance of the SIRM and the image contrasts that arise are described. These include the electric energy density distribution in the probe focused into the specimen and its variation with lens numerical aperture and focusing depth, the scattering of light by spherical and platelike particles and its variation with particle size and scattering angle, and the effect of these factors on the SIRM lateral and axial resolutions and the image signal intensity. Calculated results agree well with experimental results. The transmission polarising (TP) mode for imaging undecorated dislocations and determining dislocation Burgers vectors is described.

1. INTRODUCTION

The scanning infra-red microscope (SIRM) for the imaging of inhomogeneities present in bulk semiconductor specimens has been developed in the Materials Department, Oxford University during the last ten years. Our first SIRM (Kidd et al 1987) was a transmission instrument in which light from a 1.15 μm wavelength laser was focused into a semiconductor wafer using a 0.6 NA objective lens to give a probe ~1 to 2 μm across (~lateral resolution). The transmitted light received by a Ge photodiode was used to obtain the signal and the specimen was mechanically XY scanned perpendicular to the optical z axis to produce the image. The contrast arose from, for example, individual precipitate particles as they moved through the probe causing light to be scattered or absorbed (Fig. 1.1a). The depth of field was ~30 μm (~depth resolution). Most of the particles investigated were smaller than the lateral resolution, the individual particles then appearing as circular dark spots ~1 to 2 μm across in the bright-field (BF) image on the monitor screen. Dark-field (DF) images were obtained by placing the detector at off-axis positions. The system possessed high sensitivity enabling particles down to ~50 nm across to be imaged. In addition to the standard transmission (T) mode, this SIRM was used (Laczik et al 1991) in the transmission confocal (TC) mode to improve the resolution and in the transmission polarising (TP) mode to reveal local strains in the specimens.

Our present SIRM (Török et al 1993, Laczik and Booker 1995a) uses a 100 mW, 1.3 μm wavelength semiconductor laser and either a standard 0.6 NA lens or a high-performance infrared 0.85 NA lens with adjustable spherical aberration correction. High precision XY and XZ mechanical scanning together with direct computer controlled data acquisition, storage and display are used. This microscope mostly operates in the reflection confocal (RC) mode (Fig. 1.1b) with a lateral resolution of ~0.7μm and an axial resolution of ~5 μm. It can also operate in either the TC or RC modes, without or with either differential phase contrast (DPC) (Török et al 1993) or differential interference contrast (DIC) (Török 1994).

---

[*] Now at Multi-Imaging Centre, Univ. of Cambridge, Downing Street, Cambridge CB2 3DY, UK

682

The SIRM can be used to examine inhomogeneities within wafers up to ~1 mm thick and ~150 mm in diameter. The examinations are non-destructive and non-contacting and are performed in the laboratory environment. Individual particles can be imaged and three-dimensional particle number densities and distributions determined. Individual dislocations can be imaged because of contrast arising from either precipitate particle decoration or the strain field of the dislocation. Inhomogeneities in a wide range of bulk GaAs, InP, CdTe and Si specimens have been investigated and the structural results were correlated with results obtained by other assessment methods, e.g. etching and transmission electron microscopy (TEM), the material growth methods and processing conditions, and measured optical and electrical properties. Some of these SIRM procedures and results have recently been reviewed (Booker et al 1992, Laczik and Booker 1995a).

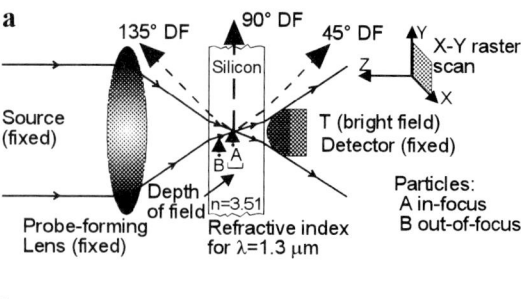

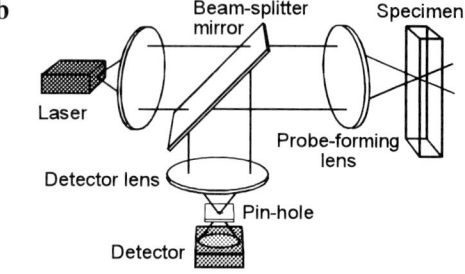

Fig. 1.1 Principle of SIRM in a) transmission (T) mode and b) reflection confocal (RC) mode.

The aim of the present paper is to give a brief description of some of the factors that are important regarding the performance of the SIRM and the image contrasts that arise, particularly from precipitate particles, and to compare calculated and experimental results. Some of the results have recently been published and other results will later be published in more detail.

2. SPECIMEN SPHERICAL ABERRATION

When the SIRM is focused at a small depth in a semiconductor specimen, the

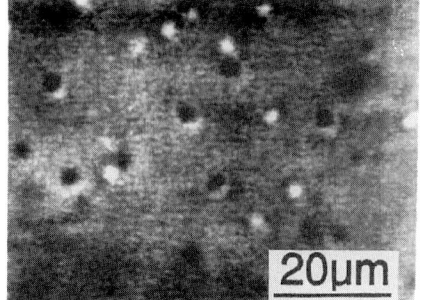

Fig. 2.1 XY T image of precipitate particles in S-doped LEC InP specimen.

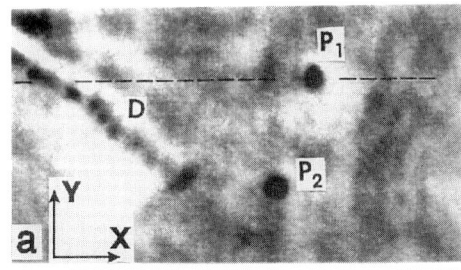

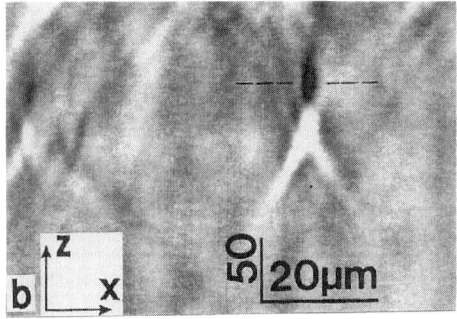

Fig. 2.2 Same region of S-doped LEC InP specimen showing precipitate particles, a) XY T image, b) XZ T image.

optical system is diffraction limited and the focused probe is small and regularly shaped. When the SIRM is focused deeper into the specimen, the specimen produces spherical aberration and the focused probe becomes larger and less regularly shaped. For transmission (T) images of individual particles, the image can change significantly, while for reflection confocal (RC) images, changes also occur but are less pronounced.

Fig. 2.1 is an XY T image, obtained using a 0.60 NA lens, of precipitate particles deep within a LEC InP specimen (Jin et al 1993). Most of the spots are circular, poorly defined and of approximately the same size, but some are dark, some are dark/bright and others are bright. Fig. 2.2a is a similar XY image showing two particles (P_1 and P_2) and a particle decorated dislocation (D). Fig. 2.2b is an XZ image corresponding to the section shown dashed in Fig. 2.2a and the complex contrast in Fig. 2.2b corresponds to particle P_1. Clearly, in this case for an XY image, if the particle is located at the depth marked in the XZ image, the particle will appear dark (as in Fig. 2.2a), if slightly deeper it will appear dark/bright, and if deeper still

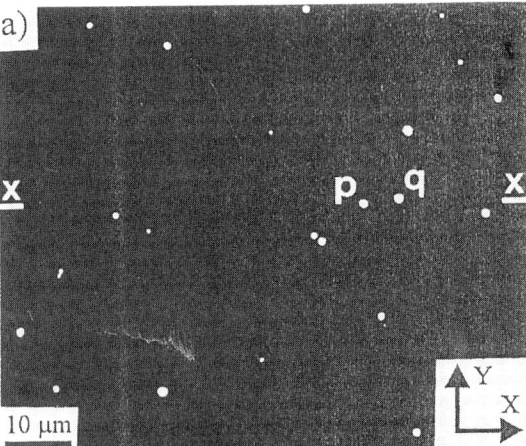

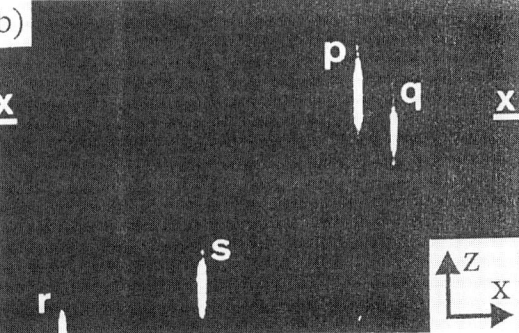

Fig. 2.3 Same regions of annealed Cz Si specimen showing oxide precipitate particles, a) XY RC image, b) XZ RC image.

it will appear bright. The complex particle contrast in these T images is mainly due to the spherical aberration arising from focusing into the specimen.

Fig. 2.3a is an XY RC image, obtained using a 0.85 NA lens, of oxide particles deep within a Cz Si specimen (Török 1994). All of the spots are circular, well defined and bright, but some are large and others are small. Fig. 2.3b is an XZ RC image corresponding to the section x-x in Fig. 2.3a. The elongated bright 'spots' p and q of the XZ image correspond to the circular bright spots p and q of the XY image. The XY image of Fig. 2.3a corresponds to the section x-x in the XZ image of Fig. 2.3b. Clearly, in this case for an XY image, if a particle is located at a depth corresponding to the middle region of an elongated spot in an XZ image, the particle will appear as a circular bright large spot (as for p and q in Fig. 2.3a). Conversely, if it is located at a depth corresponding to a 'tail' of an elongated spot, it will appear as a circular bright small spot. The more regular RC images compared with the T images is mainly due to the confocal effect (see below).

2. FOCUSED PROBE

When imaging a particle in a semiconductor specimen using the SIRM, for particles smaller than the probe, the size and shape of the particle image depend on the size and shape of the probe, i.e. on the electric energy density distribution in the probe. To calculate this

684

energy distribution we have used a high-aperture vectorial diffraction theory and applied it to a system in which a collimated and plane polarised electromagnetic wave is incident on a lens that has no spherical aberration. The lens focuses the light through air into an isotropic and homogeneous semiconductor specimen with its planar surface oriented perpendicular to the optical axis.

A vectorial theory was previously developed by Wolf (1959) and Richards and Wolf (1959) for the focusing of light by a lens without spherical aberration into a single medium. These workers treated the converging spherical wave as a coherent superposition of plane waves and derived integral formulae for the electromagnetic field to describe these waves. In our work (Török et al 1995a) we used the Wolf integral formulae to derive the electric and magnetic fields just before the air/semiconductor interface. The fields were 'traversed' across the interface using the Fresnel refraction law and then used as the boundary conditions for a second set of integral formulae. In this way mathematically rigorous equations satisfying the homogeneous wave equation were derived for the field within the specimen which could be computed to give numerical results (Török et al 1995b). Some of the results obtained for focusing light of wavelength 1.3 μm into a Si specimen are as follows.

For a lens with 0.3 NA and a focusing depth of 5 μm, the xz calculated electric energy density distribution in the region of the probe is shown in Fig. 3.1a The contours correspond to energy densities equally spaced on a log scale. The horizontal scale corresponds to distances parallel to the optical axis z, with z = 0 representing the paraxial focal plane. The vertical scale corresponds to lateral distances x, with x = 0 corresponding to the optical axis.

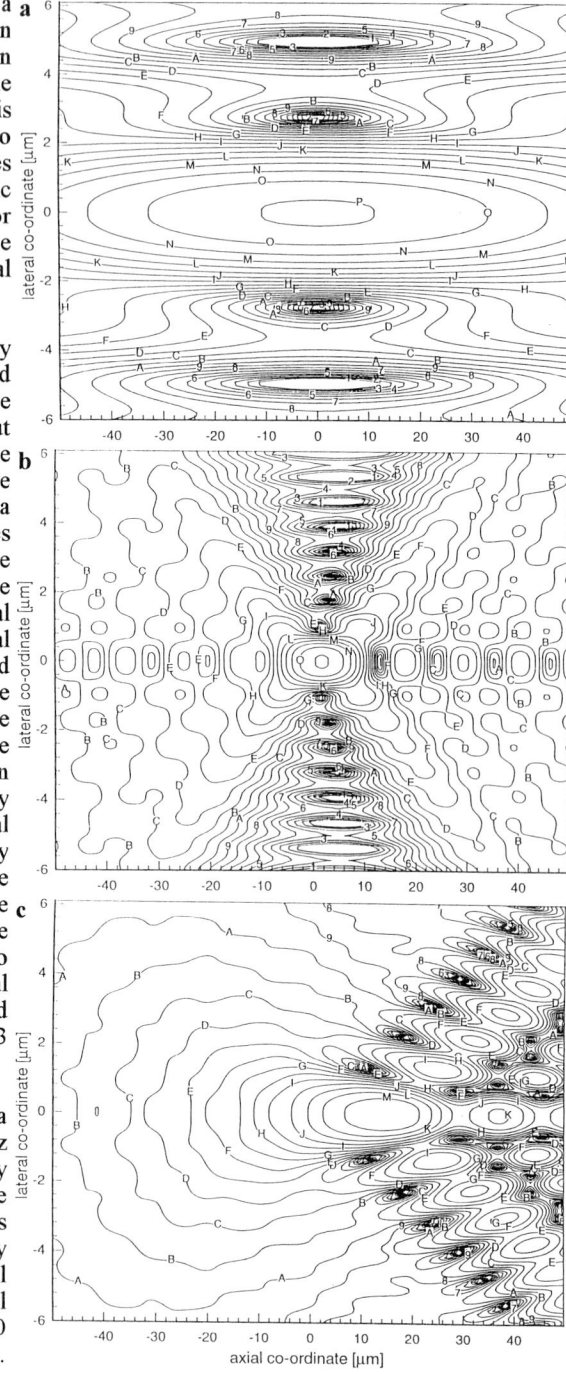

Fig. 3.1 Calculated electric energy densities in xz plane for light focused into a Si wafer. a) 0.3 NA, depth 5 μm, b) 0.9 NA, depth 5 μm, c) 0.9 NA, 80μm.

Probe-forming lens NA	Focus depth [μm]	Focused probe			
		Lateral size [μm]	Axial size [μm]	Max. energy density [a.u.]	Shift [μm]
0.3	5	2.4	84	1.3	0
0.9	5	1.0	10	64	+ 2
0.9	80	1.2	20	23	+ 22

Table 1   Probe parameters calculated from energy density distributions for $\lambda_{Si} = 1.3$ μm.

There is a large central energy density maximum which is longer in the z direction than the x direction, and there are secondary lateral maxima (and minima). The energy density distribution is symmetrical about both $x = 0$ and $z = 0$. The distribution arises because of diffraction effects, any spherical aberration due to the specimen for a 0.3 NA and a focusing depth of 5 μm being negligible. The central maximum corresponds to the probe, which has lateral and axial widths of 2.4 and 84 μm respectively (FWHMs). The maximum energy density in the probe is low, 1.3 in arbitrary units (Table 1).

For a lens with 0.9 NA and a focusing depth of 5 μm, the energy density distribution is as in Fig. 3.1b. The probe is now smaller with lateral and depth widths of 1.0 and 10 μm respectively and the maximum energy density in the probe is larger, 64 a.u. There are many secondary maxima, both axial and lateral, although the energy densities in the secondary maxima are mostly small, e.g. in the first secondary axial maxima, 2.0 a.u.. The energy density distribution is slightly asymmetrical about $z = 0$, the maximum energy density being shifted + 2 μm along the z axis and the secondary lateral maxima also being shifted. These asymmetries are due to a small amount of spherical aberration from the specimen.

For a lens with 0.9 NA and a focusing depth of 80 μm, the energy density distribution is as in Fig. 3.1c. The probe is now larger than it was when this lens was used with a focusing depth of 5 μm, the lateral and depth widths being 1.2 and 20 μm respectively, and the maximum energy density in the probe is smaller, 23 a.u. There are many secondary maxima and the energy densities in these maxima can be large, e.g. in the first secondary axial maxima, 7.1 a.u. The energy density distribution is highly asymmetrical, the maximum energy density being shifted + 22 μm along the z axis and the secondary maxima being more pronounced and progressively shifted to give V-shaped lines of intensity on the + z side, i.e. deeper into the specimen. These asymmetries are due to a large amount of specimen spherical aberration and these trends continue as the focusing depth is further increased.

The xz energy distributions in Fig. 3 were calculated for the polarisation direction of the incident plane wave along the x direction. Similar xz energy calculations with the polarisation parallel to the y direction gave slightly different results, i.e. there was a small asymmetry in the xy plane. The lateral width of the probe was ~10% smaller in the direction perpendicular to the polarisation direction.

For an XZ T image of a small particle, the size and shape of the image would be expected to depend on the intensity distribution in the probe, which corresponds approximately to the square of the energy density distribution. The lateral and axial resolutions should then correspond to the lateral and axial probe sizes respectively in intensity distribution diagrams obtained from energy density diagrams such as those of Fig. 3 and our experimental results show this to be approximately the case. For the XZ T image of a particle present in InP in Fig. 2.2b, the pair of weak tails on the side deeper in the specimen is related to the presence of V-shaped lines of intensity such as those in the calculated energy density distribution of Fig. 3.1c.

For an XZ RC image of a particle, the size and shape of the image will be different from the corresponding T image because of the selective effect of the RC pin-hole/detector on the signal collected. This causes a relative increase in the stronger intensities and a relative decrease in the weaker intensities in the images. The lateral and axial resolutions are then better (smaller) than the lateral and axial probe sizes calculated from intensity distribution diagrams for the corresponding non-confocal system, and our experimental results show this to be the case (section 5). For the XZ RC image of particles in Si in Fig. 2.3b, there are no pairs of weak tails and this is due mainly to the effect of the pin-hole/detector.

4. LIGHT SCATTERING

The contrast that occurs in a SIRM image of a particle in a semiconductor depends on the light scattered by the particle. Such scattering depends on the incident light wavelength and polarisation and whether the light is a plane or spherical wave; the particle size, shape, orientation and material; and the semiconductor material. Most scattering calculations hitherto performed have been for particles of higher refractive index than the surrounding material. For oxide particles in Si, the opposite case occurs ($\lambda = 1.3$ μm, $n_{Si} = 3.5$, $n_{ox} = 1.45$).

A rigorous theory for the scattering of light by particles requires an analytical solution of Maxwell's equations and this has so far only been possible for relatively simple geometries, e.g. a plane wave and spherical particles (Mie 1908). We have applied (Török 1994, Laczik 1995) the Mie theory to spherical oxide particles in Si for incident light of wavelength 1.3 μm (in air). Fig. 4.1 shows for unpolarised incident illumination the calculated scattered light intensity as a function of scattering angle from 0 to 180° for particles of different sizes. For a radius of 25 nm, the intensity is low and relatively independent of scattering angle. As the radius increases to 250 nm, the intensity increases rapidly becoming $\sim10^5$X larger in the forward direction (0°) and $\sim10^3$X larger in the backward direction (180°).

For an SIRM operating in the reflection mode using a lens with 0.85 NA (in air) and a Si specimen, the scattered light is collected by the same lens over an approximate scattering angle of 166 to 180° (in Si). By using the data of Fig. 4.1. and integrating over this solid angle, the curve T of Fig. 4.2 is obtained which shows how the amount of scattered light received by the lens (Ω) increases as the particle radius (r) increases. The plotting of this curve on log-log scales shows that the slope progressively changes, with Ω proportional to r^n, where n = 6, 3, 2 and 1.2 in the regions of the curve corresponding to r = 25, 70, 150 and 250 nm respectively. The n = 6 dependence corresponds to the Rayleigh scattering relationship for particles small compared with the light wavelength, and the progressive decrease in n as r increases corresponds to a deviation from this relationship as r becomes comparable with λ.

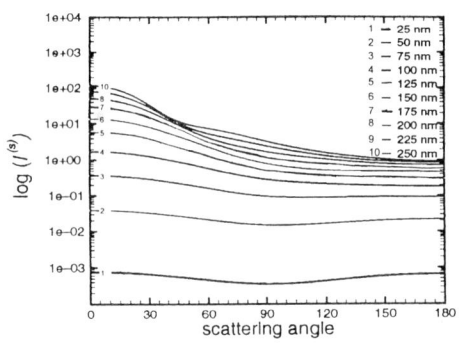

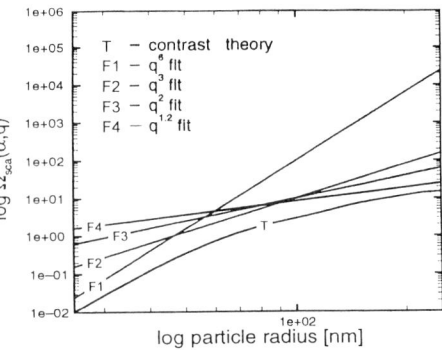

Fig. 4.1 Scattered light intensity calculated by Mie theory for individual spherical oxide particles in Si as a function of scattering angle and particle size.

Fig. 4.2 Calculated scattered light intensity from individual spherical oxide particle in Si for 0.85 NA detector lens using reflection (R) mode. Results based on data of Fig. 4.1.

In order to determine how the scattered light depends on the size, shape and orientation of particles which are not spherical, we have used the discrete dipole approximation (DDA) method as developed by Draine and Flateau (1994). The particle is replaced by an array of point-like electric dipoles and the collective radiation field of the dipoles when excited by an incident plane wave is calculated, taking into account the interactions between the dipoles.

We initially applied the DDA method to spherical oxide particles in Si (Laczik 1995) and the results were closely the same as our Mie theory results, justifying the use of the DDA

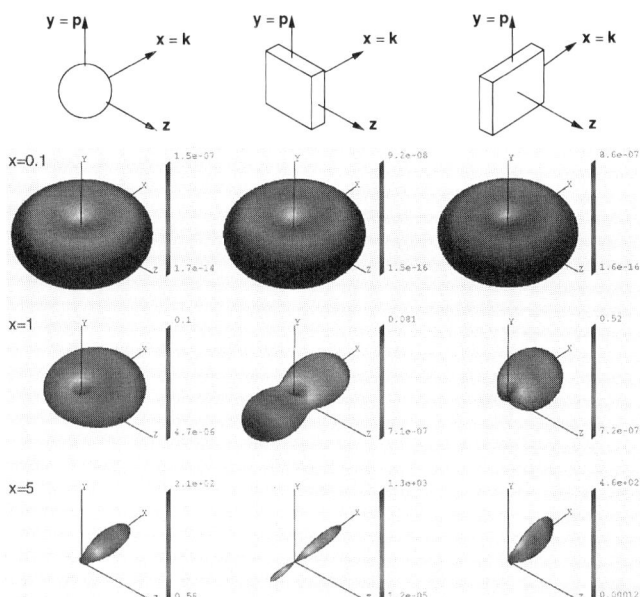

Fig. 4.3 3-D distribution diagrams showing scattered light intensity calculated by the discrete dipole approximation (DDA) method for individual oxide particles in Si. Columns - different particles shapes, rows - different normalised particle sizes. Plane wave incident illumination with wavelength 1.3 μm (in air).

method for oxide particles in Si. When we applied the DDA method to non-spherical particles, it was convenient to refer the results to a normalised particle size x, where $x = 2\pi r_e/\lambda$, r_e is the radius of a sphere with the same volume as the particle and λ is the light wavelength (in Si).

Calculated 3-D scattering intensity diagrams (Laczik and Booker 1995 a and b) are shown in Fig. 4.3. where the scales are normalised to the maximum intensity in each diagram. The incident plane wave is along the x axis (following the general convention used for such scattering diagrams) and is linearly polarised along the y axis. Rows 1, 2 and 3 correspond to $x = 0.1$, 1 and 5 respectively, which for oxide particles in Si and a wavelength of 1.3 μm (in air) correspond to particle radii r_e of 6, 60 and 300 nm respectively. Column 1 is for a sphere, column 2 for a square plate with 1:10 aspect ratio (thickness/width) and edges parallel to the y and z axes (face-on), and column 3 for a similar plate with edges parallel to the x and y axes (edge-on).

The results show that for small particles ($x = 0.1$), the 3-D scattering diagrams are very similar for particles of all shapes and exhibit the classical dipole radiation pattern. For medium particles ($x = 1$), the diagrams are similar in shape to that of the small particles but are slightly asymmetrical in the incident light direction (x axis), and the asymmetry is different for the sphere, face-on plate and edge-on plate. For large particles ($x = 5$), the diagrams are significantly different from the corresponding diagrams for medium particles. The diagrams are highly asymmetrical in the x axis direction and the asymmetry is markedly different for the sphere, face-on plate and edge-on plate.

These results show that the scattered intensity for spherical particles decreases rapidly as the particle radius decreases below 30 nm. The minimum size particle that can be imaged will depend mainly on the laser power and the background noise. For our RC SIRM it may be possible to simultaneously image and distinguish particles in the radius range 25 to 250 nm because the detected signal should then correspond to a range of $\sim 10^3 X$, and the dynamic range of our detector system is $\sim 10^3 X$.

For spherical particles, the ratio I_{180}/I_0, where I_{180} is the backscattered intensity ($180°$) and I_0 is the forward scattered intensity ($0°$), decreases progressively from 1 to ~ 0.1 as the particle radius r increases from 6 to 300 nm (Fig. 4.3). The ratio I_{90y}/I_{90x}, where I_{90y} is the $90°$ scattered intensity parallel to the polarisation direction and I_{90x} is the $90°$ scattered intensity perpendicular to the polarisation direction, increases progressively from $\sim 10^{-5}$ to ~ 0.1 over the same particle radius range. Hence, for spherical particles, measurement of these ratios could enable the particle size to the deduced.

For plates, the scattering diagrams for the larger particles are significantly different from those for spheres, and there are differences between face-on and edge-on plates. The use of such diagrams could lead to a method for deducing the size and orientation of such plates. Experimental measurements of scattered light from, for example, clouds of dust particles in air have enabled such data to be deduced for spherical and non-spherical particles but the use of the SIRM for similar measurements from particles in semiconductors could enable the data to be deduced for individual particles.

In our calculations of scattering we used models appropriate for plane wave illumination, rather then spherical wave illumination as in the SIRM. However, although the semi-angles for NAs of 0.3, 0.6 and 0.9 in air (17.5, 36.9 and $64.2°$ respectively) are large, the corresponding angles in Si (4.9, 9.9 and $14.9°$ respectively) are small due to the large refractive index of Si (3.5). These small angles, which define the cone of illumination of the particles being imaged in Si and for the RC mode also define the cone of scattered light accepted to give the image, are generally considered to correspond to the paraxial case. Hence, any errors that may arise for scattered intensities from in-focus particles by assuming a plane wave model are probably small. However, for the RC mode, before calculated scattering diagrams such as those of Fig. 4.3 can be used quantitatively to interpret the images, other features would need to be included in the calculations, e.g. the selective effect of the pin-hole/detector.

5. RC MODE IMAGES OF PARTICLES

The RC mode has high performance because it operates in the confocal mode, the same high quality lens is used to focus the light into the specimen and to receive the scattered light from the specimen, i.e. the illumination and detection system are precisely self-aligned, and it is reasonably easy to position the pin-hole. The T mode has poorer resolution than the RC mode. The TC mode can be used but in practice it is not easy to set up. A second lens needs to be placed after the specimen to focus the transmitted light onto the detector and there can be difficulties in aligning the optical axes of the two lenses and positioning the pin-hole. We recently reported experimental resolutions for the T and TC modes (Laczik et al 1991) and the results were in good agreement with calculated data.

For the RC mode, the decrease in performance that occurs when focusing deep into a specimen because of the spherical aberration arising from the specimen can be completely overcome by using a lens with a spherical aberration 'correction' collar. High performance lenses designed to be used, for example, with light of 1.3 μm wavelength and Si specimens up to 500 μm thick are now available. Fig. 5.1 shows XZ RC images (Török 1994) of an individual oxide particle at a depth of 300 μm in a Si specimen using a lens of 0.85 NA. For Figs 5.1a to f, the correction collar was set at 0, 100, 200, 300, 400 and 500 μm respectively, these corresponding to spherical aberration at the particle equivalent to depths of +300, +200, +100, 0, −100 and −200 μm respectively, where + indicates undercorrection and −

indicates overcorrection. On going from a correction of 0 to 300 μm, the lateral and axial widths of the particle image progressively decrease, and on going from 300 to 500 μm, these widths progressively increase. This sequence illustrates the effectiveness of the correction collar. The lateral and axial widths of the particle image at the optimum setting of 300 μm, as obtained from line traces, are 0.7 and 5.0 μm respectively. These correspond to the lateral and axial resolutions of the RC mode and are as expected smaller than the calculated lateral and axial probe sizes of 1.0 and 10 μm respectively for closely these conditions (Table 1, row 2) obtained from energy density distribution diagrams.

For the RC mode when focusing into the specimen, some of the incident light is back reflected by the specimen surface into the lens and can produce a bright background to the bright particle image. This background is initially apparent at a focusing depth of ~100 μm and increases as the focus depth decreases to 0 μm, i.e. when the focus is at the specimen surface. When examining large oxide particles in Si (high particle signal) in the surface region, the particles are generally observed, but when examining small oxide particles (low particle signal), the particles may not be seen because of the obscuring effect of this background.

We have overcome this difficulty by using the RC mode either in dark field (DF) (Laczik et al 1995), or with differential phase contrast (DPC) (Török et al 1993) or differential interference contrast (DIC) (Török 1994). All three of these procedures completely eliminate the surface reflected light from the image. The DF procedure is simple to use but there is a small loss in resolution because the effective NAs of the illumination and detection are decreased. The DPC and DIC procedures require additional components to be added to the optical system but there is no loss in resolution. Hence using these procedures, small and large particles can be imaged all the way to the specimen surface. This enables, for example, the depths of surface denuded zones (SDZs) present in Cz Si wafers which have been heat treated for internal oxide gettering to be directly measured from XZ RC images (Fig. 5.2). Such zones are free from oxide particles and extend from the surface to a depth in the range typically 5 to 50 μm.

Particle number densities are generally determined from

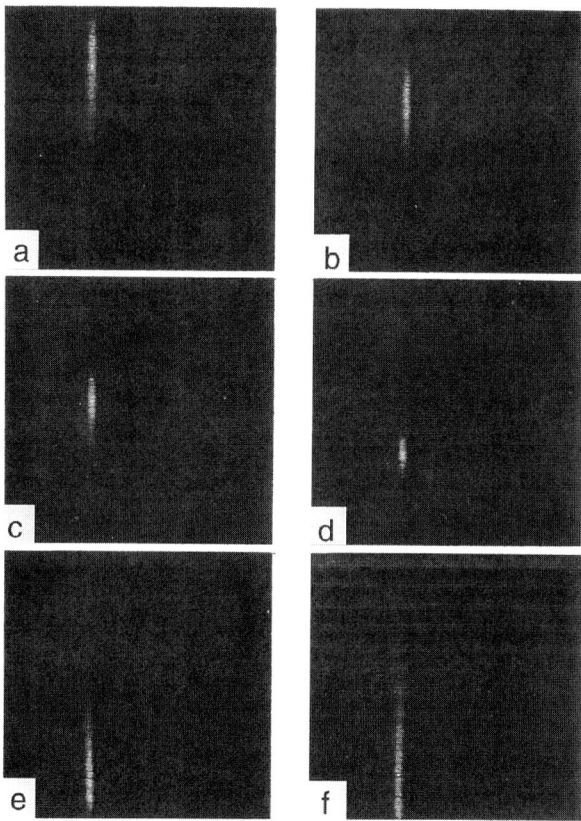

Fig. 5.1 XZ RC images from the same oxide particle located 300 μm below the surface of a Si specimen obtained using a 0.85 NA lens with a spherical aberration corrector. Images a) to f) correspond to spherical aberration from the specimen corresponding to equivalent focusing depths without correction of + 300, + 200, + 100, 0, − 100 and − 200 μm respectively. Each image frame is 40 μm horizontal by 50 μm vertical.

690

XY SIRM images by counting the number of in-focus particles per unit area in a particular image and dividing by the depth of focus, i.e. the axial resolution. Alternatively, a focal series is performed, the number of particles per unit area is counted as individual particles come in and go out of focus, and this number is divided by the depth over which the focal series is performed. In general, for low particle densities, low magnifications and a focal series are used, and for high particle densities, high magnifications and a single image are used. For the RC mode, the lowest number density that can be readily measured is $\sim 10^6$ cm^{-3} and the highest number density is $\sim 10^{11}$ cm^{-3}.

With regard to the size range of individual particles that can be imaged, particles larger than the resolution appear with their correct sizes and shapes and there is in general no upper size limit. Particles smaller than the resolution give circular spots in XY images and elongated spots in XZ images, with the spot sizes corresponding to the SIRM lateral and axial resolutions respectively. The sizes of such particles can only be deduced from the particle image contrast and two possible ways of doing this are as follows.

First, the particle contrast can be measured using plane polarised incident light and varying the imaging conditions, e.g. the direction of the detected light, and this may provide information about the individual particle size, shape and orientation (section 4). Procedures for achieving this are presently being investigated. Second, approximate particle sizes can be deduced by using an experimentally determined calibration curve of SIRM particle contrast against TEM particle size for the particular type of specimen being investigated. The SIRM images are obtained from bulk specimens by a standard procedure and the specimens are subsequently thinned for TEM examination. Experience has shown that it is extremely difficult using TEM to locate specific particles that were previously imaged by SIRM, although it can be done (Weyher 1995). Consequently, in practice a suitable series of bulk specimens is obtained and mean values for SIRM particle contrast and TEM particle size are determined for each specimen, and a calibration curve is constructed.

We recently made such an investigation (Török et al 1995c) for oxide particles in Si by examining a series of wafers that had been differently heat treated with the aim of producing particles with increasing sizes in successive wafers. The project was only partly successful because although in the initial wafers individual isolated particles occurred and progressively increased in size, in the subsequent wafers small clusters of particles occurred and the individual particles were difficult to resolve with the SIRM. Nevertheless, the SIRM and TEM results were consistent with one another and there were several new findings. For example, the TEM results showed that the oxide particles in all the wafers were mostly plates and provided detailed information concerning their sizes and shapes. The

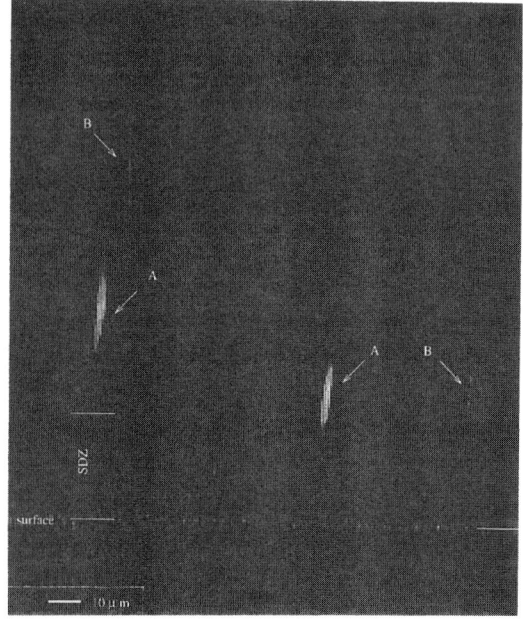

Fig. 5.2 XZ RC images of oxide particles in Si using dark-field (DF) obtained by inserting a half-stop. A - in-focus particles, B - out-of-focus particles. Surface reflected light almost completely eliminated from image. Surface denuded zone (SDZ) depth can be measured.

combined SIRM/TEM results showed that individual plates down to ~100 x 100 x 10 nm in size could be imaged using the SIRM in the RC mode, this size being equivalent on a volume basis to a spherical particle ~30 nm in radius (r_c). It is possible that smaller individual particles could have been imaged.

6. POLARISED LIGHT MODE

The SIRM can be used in the transmission polarised (TP) mode by inserting in the optical system a polariser before, and an analyser after, the specimen. With crossed polariser and analyser the image reveals local variations in specimen strain with good resolution and high sensitivity. The contrast arises because the elastic strain produces anisotropy in the matrix and this changes the amount of light transmitted by the analyser.

Fig. 6.1a is an XY T image (Jin et al 1993) of dislocations in an (001) LEC InP wafer obtained using a 0.6 NA lens and parallel polariser and analyser, the resulting image then corresponding to the standard T mode. The dislocation contrast arises mainly from light scattered by small precipitate particles decorating the dislocations, the individual particles mostly being unresolved. Fig. 6.1b shows the same specimen area with crossed polariser and analyser. The dislocations seen in Fig. 6.1a are present in this image but in addition another series of dislocations (arrowed) is revealed. The contrast for the latter arises mainly from strains associated with these dislocations, any particle decoration being insufficient to give contrast in Fig. 6.1a. The contrast for each of the arrowed dislocations in Fig. 6.1b is similar suggesting that they have the same Burger's vector. The reason why these dislocations are not decorated could be because either the atomic core structure for this dislocation type does not have suitable precipitate nucleation sites or these dislocations glided in the ingot when it was cooling down and there was insufficient time for precipitation to occur.

Fig. 6.2a is an XY T image (Laczik et al 1991) of two dislocations viewed end-on in a (001) LEC In-doped GaAs slab with parallel polariser and analyser. The contrast arises from light scattered by precipitate particles decorating the dislocations. Fig. 6.2b shows the same specimen area with crossed polariser and analyser and polarising direction parallel to <100>. Both dislocations exhibit the same

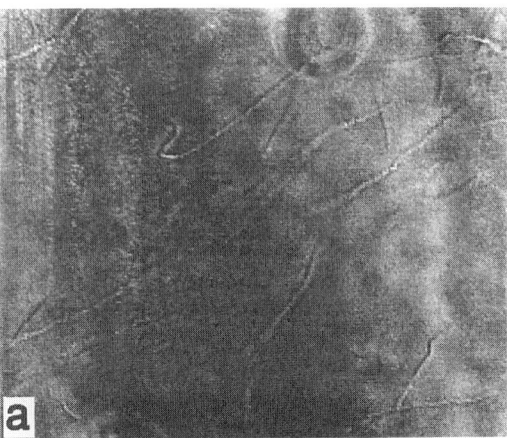
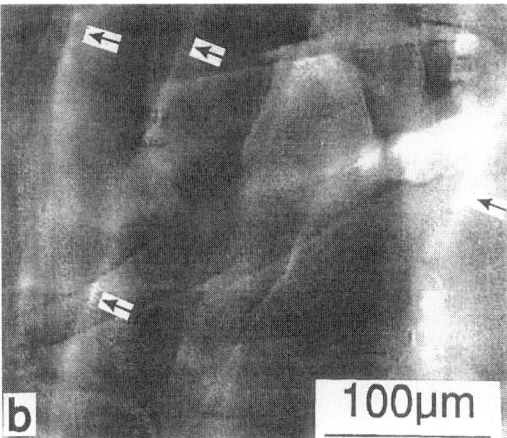

Fig. 6.1 XY T images of dislocations in S-doped InP specimen. Polariser and analyser are a) parallel and b) crossed. Additional dislocations in b) (arrowed) are revealed by crystal anisotropy due to dislocation strain fields.

692

pronounced two-lobe contrast extending up to ~50 μm from the dislocations. Fig. 6.2c shows the same area with crossed polariser and analyser and polarising direction parallel to <110>. Both dislocations exhibit the same pronounced four-lobe contrast. Comparison of these contrasts with those calculated (Booyens and Basson 1980) for edge dislocations viewed end-on with crossed polariser and analyser (left column of Fig. 6.2) enables the Burgers vector for the two dislocations to be deduced as a[100] type. In this way ~50 such dislocations were analysed and the results showed that both a<100> and a/2<110> Burgers vector types occurred, and for each type the two possible directions and two possible senses were present. The two dislocation types correlated with two different particle decoration behaviours as observed in XZ T images.

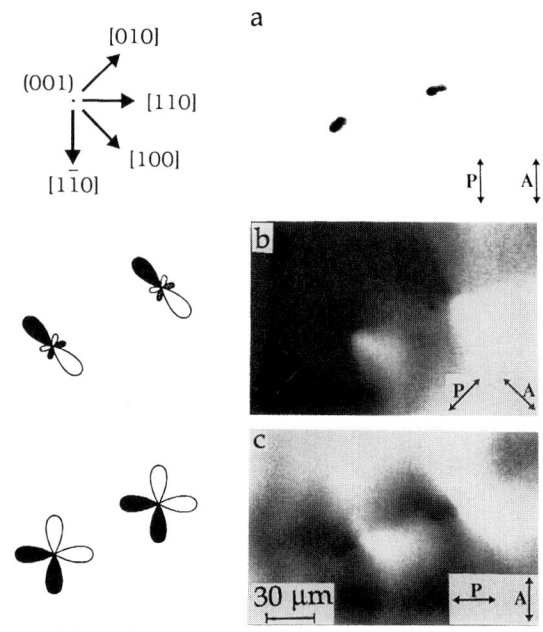

Fig. 6.2 XY T images of two end-on dislocations in In-doped LEC GaAs specimen. Polariser and analyser are a) parallel, b) <100> crossed, c) <110> crossed. Lobe contrasts in b) and c) are due to dislocation strain fields and identify Burgers vectors as a[100] type.

ACKNOWLEDGEMENTS

The authors wish to thank the Engineering and Physical Sciences Research Council (EPSRC) UK for support.

REFERENCES

Booker GR, Laczik Z and Kidd 1992 Sem. Sci. Technol. 7 A110.
Booyens H and Basson JH 1980 J. Appl. Phys. 51 4368 and 4375.
Draine BT and Flateau PJ 1994 J. Opt. Soc. Amer. A11 1491.
Jin NY, Booker GR and Grant IJ 1993 Mat. Sci. and Eng. B20 94.
Kidd P, Booker GR and Stirland DJ 1987 Inst. Phys. Conf. Ser. No.87 275.
Laczik Z, Török P, Booker GR and Falster R 1991 Inst. Phys. Conf. Ser. No. 117 785.
Laczik Z, Török P and Booker GR 1995 Inst. Phys. Conf. Ser. (in press).
Laczik Z and Booker GR 1995a Electrochem. Soc. Symp. Ser. (in press).
Laczik Z and Booker GR 1995b Appl. Phys. Lett (submitted).
Laczik Z 1995 Appl. Opt. (submitted).
Mie G 1908 Ann. Physik 25 377.
Richards B and Wolf E 1959 Proc. Roy. Soc. London Ser. A 253 358.
Török P, Booker GR, Laczik Z and Falster R 1993 Inst. Phys. Conf. Ser. No. 134 771.
Török P 1994 D.Phil. thesis Oxford University.
Török P, Varga P, Laczik Z and Booker GR 1995a J. Opt. Soc. Amer. 12 325.
Török P, Varga P and Booker GR 1995b J. Opt. Soc. Amer. (in press).
Török P, Pécz B, Laczik Z, Booker GR, Radmóczi G and Falster R 1995c Inst. Phys. Conf. Ser. (in press).
Weyher J 1995 Inst. Phys. Conf. Ser. (in press).
Wolf E 1959 Proc. Roy. Soc. London Ser. A 253 349.

Inst. Phys. Conf. Ser. No 146
Paper presented at Microsc. Semicond. Mater. Conf., Oxford, 20–23 March 1995
© *1995 IOP Publishing Ltd*

EBIC and DLTS investigations of charged dislocations in Si

O V Kononchuk, E B Yakimov and N A Yarykin

Institute of Microelectronics Technology Russian Academy of Sciences
Chernogolovka Moscow district 142432 Russia

ABSTRACT: The influence of the dislocation electrostatic barrier near a charged dislocation in Si on the ability to reconstruct the spatial distribution of dislocation-related deep level centres from the EBIC and DLTS measurements is discussed. It is shown that the DLTS method is more suitable for the study of the centres located at the periphery and beyond Read's cylinder. The EBIC technique seems to be more promising for the investigation of the centres in/near the dislocation core but adequate theory has to be developed for correct interpretation of the results.

1. INTRODUCTION

The electrical properties of dislocations in crystalline Si have been intensively investigated since the fifty's but up to now even the main features of their nature are not clear. A wide variety of methods (Hall effect, EPR, PL, DLTS, EBIC etc.) (Alexander and Teichler 1991) were applied to the investigation of dislocations. Unfortunately, it is often very difficult to compare the results of different methods due to a great number of deformation related centres. A set of these centres in each particular sample is strongly dependent on crystal impurity content, temperature and duration of deformation and subsequent annealing, dislocation velocity and the distance covered by dislocations during deformation (Eremenko et al 1977, Bondarenko et al 1980, Kononchuk et al 1994). Therefore for the correct comparison of the results obtained by different methods it is necessary to use similar deformation conditions and to control the impurity contents.

In the present paper the results of Electron Beam Induced Current (EBIC) and Deep Level Transient Spectroscopy (DLTS) investigations of electrical properties of dislocations introduced into Si single crystals by plastic deformation at temperatures lower that 700°C are discussed. There is a huge amount of experimental data on such dislocations obtained by different techniques. Low temperature dislocations are known to introduce a number of deep level centres and thought to be charged and surrounded by an electrostatic barrier. EBIC and DLTS are both sensitive to deep levels and believed to be able to reconstruct the energy spectrum and spatial distribution of dislocation related centres. The problems arising during the interpretation of EBIC and DLTS results obtained on charged dislocations in Si are discussed.

2. INVESTIGATIONS OF DISLOCATIONS BY EBIC

Dislocations are considered to enhance the recombination of excess minority carriers. Two most important questions arise when investigating the charged dislocations by the EBIC.

They are "How does the dislocation barrier influence the recombination via dislocation?" and "How does electron beam excitation affect the dislocation electrostatic barrier?". An attempt to answer the first question was made by Wilshaw and coworkers (1987, 1989). For charged dislocations they predict a logarithmic decrease of the contrast with injection level and a linear increase with temperature. In some cases this model allows quantitative parameters of the dislocation to be obtained such as barrier height or linear density of dislocation related centres. Indeed, dependences similar to those theoretically predicted were experimentally observed (Wilshaw and Fell 1989, Bondarenko and Yakimov 1990). However, the value of the contrast as well as the region of near logarithmic dependence differ significantly for dislocations in various crystals and even for dislocations in the same crystal. In the frame of the model this means that these dislocations have quite different equilibrium barrier. In our opinion, this seems to be unlikely and is inconsistent with barrier determination using a metal microprobe (Eremenko et al 1975). As follows from Wilshaw's model, the higher dislocation barrier the lower excitation level would change its value. An analysis of experimental results revealed (Bondarenko and Yakimov 1990) the opposite relation, i.e. the higher dislocation EBIC contrast (the higher barrier) the more stable its value under an e-beam excitation. Besides, all quantitative conclusions of Wilshaw's model are strongly based on the assumption that a single level only contributed to barrier formation, while at least five centres were found recently by Bondarenko et al (1993a) to contribute significantly to the dislocation IRBIC contrast.

In general, the contrast dependences observed do not unambiguously prove the existence of the barrier under e-beam excitation. It should be pointed out that the decrease of the EBIC contrast with beam current and increase with temperature rise could be described by the Shockley-Read-Hall theory. However, this holds true only for limited ranges of injection level and temperature and can hardly explain logarithmic dependence of the contrast over three orders of beam current (Wilshaw and Fell 1989).

Unfortunately, up to now there have been no direct experiments which could define to what extent the initial barrier drops under EBIC conditions. For instance, simple estimation shows that even the lowest beam current used in EBIC mode must decrease the barrier of 0.3-0.4 eV derived from metal microprobe measurements (Eremenko et al 1975). Besides, recent experiments on IRBIC contrast of $60°$ dislocations (Bondarenko et al 1993b) have shown that some dislocation related centres are located far enough from the dislocation. These centres can affect the EBIC contrast also and therefore Wilshaw's model should be modified to account for different recombination channels.

3. INVESTIGATION OF DISLOCATION CENTRES BY DLTS

The principal problem arose during interpretation of DLTS data obtained in plastically deformed Si is associated with the spatial distribution of the centres observed. The DLTS spectra of dislocated Si usually exhibit four peaks in n-type crystals (Omling et al 1985) and three at least peaks in p-type samples (Kveder et al 1982). The most pronounced peculiarity of these peaks is slow, near logarithmic dependence of their heights on the filling pulse duration (Kveder et al 1982, Omling et al 1985, Yarykin and Feklisova 1988). Since the original work by Kveder et al (1982), such dependence is usually associated with the influence of the electrostatic barrier formed due to charging of closely located centres and serves as an indicator of the dislocation related nature of the centres (although, the dependence for clusters of point-like defects is expected to be practically the same).

If one assumes that all the centres observed are located in/near the dislocation core then the question arises "How is it possible to reveal shallow levels after formation of the electrostatic barrier due to filling of deeper ones?" The most simple answer is that different centres are located at dislocations of different types (screw and edge, split and/or reconstructed or not, etc.). However, no changes in the relative intensity of DLTS peaks were observed in our measurements of the crystals with quite different dislocation structures, such as a low density of predominantly parallel 60°-dislocations and a high density of dislocations produced by uniaxial compression up to strains of a few percent. Another possible explanation is the limited number of centres of each type, so that even complete filling of deeper levels does not prevent the capture of carriers on shallow ones. However, this assumption does not seem to be able to explain the logarithmic dependence of captured charge on the filling pulse duration, since filling of the most shallow level takes place under the practically constant electrostatic barrier formed by deeper levels. This has to result only in a significant decrease of the capture cross section but not in a logarithmic dependence.

To overcome these difficulties the model of "extended dislocation" was proposed (Koveshnikov et al 1991). In this model the centres revealed by DLTS are considered to be distributed around the dislocation at the periphery of the electrostatic barrier formed by charge captured on deeper levels. In this case, the electrostatic interaction between the centres under study can be neglected due to their low local concentration and they all can be revealed by DLTS as usual point defects. The dependence on filling pulse duration in this case is determined by the spatial distribution of the centres inside the barrier. As shown by Koveshnikov et al (1991), the experimentally observed near logarithmic dependence means a roughly exponential decrease of the centre concentration with distance from the dislocation. However the constant ratio of DLTS peak amplitudes over more than two order of crystal doping level (Feklisova and Yarykin, unpublished) gives rise to some doubts regarding the adequacy of the model.

Note finally that recent results obtained by the IRBIC and QIRBIC methods (Bondarenko et al 1993) show that some of the centres contributing to the DLTS signal are really more or less concentrated near the dislocation. This means that the whole DLTS spectra measured can not be ascribed to any clusters homogeneously distributed over the crystal volume.

4. CONCLUSIONS

Thus the above consideration has highlighted some problems in characterization of dislocation electrical activity by EBIC and DLTS techniques. The most important of them are:

(i) the influence of the dislocation electrostatic barrier on the processes of capture and recombination of majority and minority charge carriers

(ii) determination of concentration of dislocation-related centres and reconstruction of their spatial distribution.

The whole set of DLTS and EBIC data available appears to argue for a wide enough distribution of some dislocation centres. The extrinsic nature of these centres is indisputable but to clarify their origin the spatial distribution and its dependence on deformation conditions and thermal treatments should be investigated.

There is a great deal of evidence that a high density of electrically active centres exists close to the dislocation core but at least some of them are located outside the dislocation core far enough from dislocation. Some kinds of these centres (the deepest ones) located in the dislocation core or very close to it are believed to determine the value of dislocation charge. The energy position and concentration of these "inner" centres determine the electrostatic barrier height but its stability under the e-beam or optical excitation and temperature dependence are determined by all centres located near dislocations. The barrier prevents filling of "inner" centres with majority carriers and hence their DLTS study. Therefore the parameters of deep "inner" centres such as their concentration and corresponding energy level can not be obtained by this technique. The EBIC seems to be more promising for this purpose because the dislocation barrier height can be controlled by changing the excitation level and temperature. But, for obtaining the parameters of dislocation-related centres from the EBIC results, adequate theory has to be developed for correct interpretation of these results. Such theory can be based on the Wilshaw's model but should take into account the existence of a few centers some of which are located outside the dislocation core. Besides, the complementary methods for reconstruction of spatial distribution of dislocation-related centres and electrostatic potential inside the dislocation space charge cylinder should be developed.

ACKNOWLEDGEMENTS

This work was partially supported by the International Science Foundation (project MRZ000). We would like to thank Dr. I.Bondarenko for stimulating discussions.

REFERENCES

Alexander H and Teichler H 1991 Mater.Sci.Technol. 4 249

Bondarenko I E, Eremenko V G, Nikitenko V I and Yakimov E B 1980 Phys. Stat. Sol.(a) 60 341

Bondarenko I E and Yakimov E B 1990 Phys.Stat.Sol.(a) 122 121

Bondarenko I, Castaldini A, Cavallini A 1993a Phys.Stat.Sol. (a) 137 411

Bondarenko I, Castaldini A, Cavallini A 1993b Proc. DRIP V (Santader, Spain)

Eremenko V G, Nikitenko V I and Yakimov E B 1975 Zh.Eksper.Teor.Fiz. 69 990 (In Russian) [Sov Phys. JETP. 1976 42 503]

Eremenko V G, Nikitenko V I, Yakimov E B 1977 Zh.Eksp.Teor.Fiz. 73 1129 (In Russian) [Sov. Phys.JETP 1978 48 598]

Kittler M and Seifert W 1993 Phys.Stat.Sol.(a) 138 687

Kononchuk O V, Nikitenko V I, Orlov V I and Yakimov E B 1994 Phys.Stat.Sol.(a) 143 K5

Koveshnikov S V, Feklisova O V, Yakimov E B.and Yarykin N A 1991 Phys.Stat.Sol.(a) 127 67

Kveder V V, Osipyan Yu A, Schroter W and Zoth G 1982 Phys.Stat.Sol.(a) 72 701

Omling P, Weber E R, Montelius L, Alexander H and Michel J 1985 Phys. Rev. B32 6571

Wilshaw P R and Booker G R 1987 Izv. AN SSSR, ser.Fiz. 51 1582 (In Russian)

Wilshaw P R and Fell T S 1989 Inst.Phys.Conf.Ser., No 104 85

Yarykin N A and Feklisova O V 1988 Defects in crystals, ed E Mizera (Singapore: World Sintific) pp 366-70

Inst. Phys. Conf. Ser. No 146
Paper presented at Microsc. Semicond. Mater. Conf., Oxford, 20–23 March 1995

Application of Monte Carlo simulation to EBIC-contrast modelling

H Mohr and D J Dunstan

Department of Physics, University of Surrey, Guildford, GU2 5XH, UK

ABSTRACT: A Monte Carlo electron trajectory simulation has been applied to the EBIC study of multi-layer structures with different carrier formation energies in the layers. The computed energy dissipation volumes were used in EBIC calculations for the evaluation of parameters of the specimen used and for lattice defects. Simulations of defect contrast are compared with EBIC signals from a strained multi-layer structure for semiconductor lasers. The simulated depth dose function shows structure details of 50nm thick layers. Line scans over misfit dislocations at various interfaces and with gettering volumes are simulated.

1. INTRODUCTION

The analysis of EBIC images from multilayer structures presents the problem of a far more complicated shape of the carrier generation volume than in homogeneous material. The form of the generation volume can be found with a Monte Carlo simulation and can subsequently be used to calculate line scans over dislocations. We applied the Monte Carlo simulation and EBIC to strained layer heterostructures with a buried pn-junction. The simulation takes into account different diffusion lengths in the p- and n-type region and a gettering volume around dislocations, which is modelled as a cylinder with a higher diffusion length. Using this model, we obtain a higher EBIC gain at the dislocation and can compare this with our experimental findings of white lines instead of dark lines for misfit dislocations in EBIC micrographs.

2. EXPERIMENTAL AND MONTE CARLO SIMULATION

A computer program written in Turbo Pascal simulates the electron beam interaction with the material and the generation of electron-hole pairs. The trajectory simulation for the beam electrons must take account of the different material parameters for each layer through which the beam electron travels, such as scattering cross section, mean free path length and carrier generation energy.

The program calculates very easily the amount of deposited energy by the electron beam in the material per unit depth, the depth dose function. Fig.1 shows the depth dose function for a beam energy of 20keV and the layer structure of the simulated III-V strained layer heterostructure. Every interface appears as a discontinuity. Each segment of the curve follows the Everhart and Hoff (1971) polynomial expression for the depth dose function for bulk material. We used a laser structure consisting of an n-type InP substrate and InGaAsP waveguides and undoped active region with four strained quantum wells, the cap consists of p-type InGaAs and InP layers.

Fig.2 shows the EBIC gain of the sample (= number of electron-hole pairs created by one beam electron and collected by the pn-junction) for different beam energies and the Monte Carlo calculation with the diffusion length as a fit parameter.

Micrographs and line scans have been taken with a Cambridge S250 and S100 SEM. Fig.4 shows a micrograph of the sample. The appearance of a bright contrast for misfit dislocations is most unusual. Threading dislocations appear as dark spots. Some of the misfit dislocations can be followed for about 11mm in the sample, while others are only a few microns long. Usually the end of a misfit dislocation is marked by a black threading dislocation.

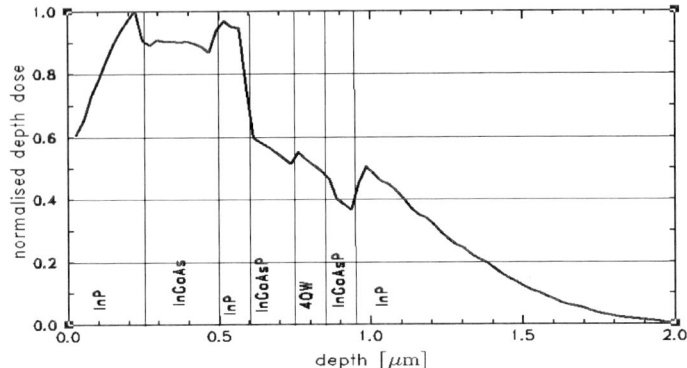

Fig.1 Depth dose function for a beam energy of 20keV and sample structure.

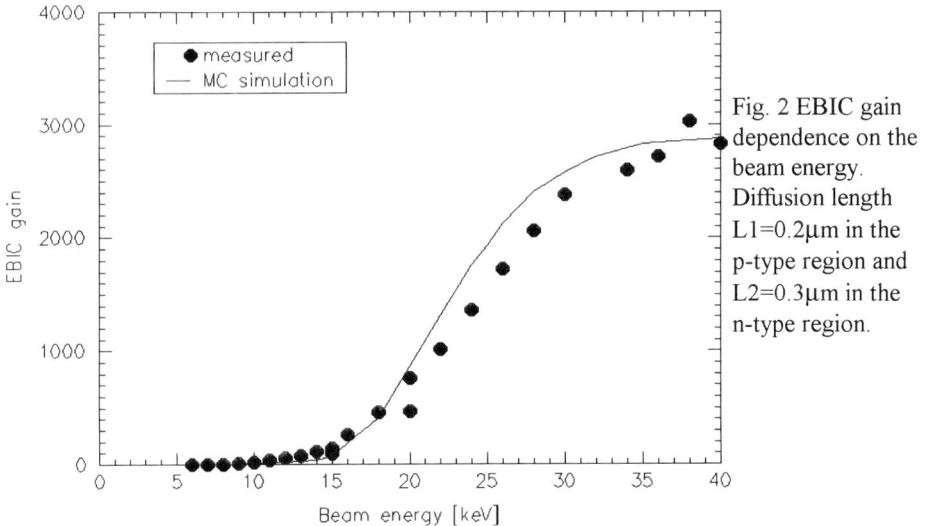

Fig. 2 EBIC gain dependence on the beam energy. Diffusion length L1=0.2μm in the p-type region and L2=0.3μm in the n-type region.

3. LINE SCAN SIMULATION

In order to simulate a line scan over a defect, the effect of recombination on the generation distribution needs to be calculated. The most accurate and flexible way of simulating the process, is by computing the problem in three dimensions. This allows the simulation of various dislocation configurations. The carrier density of each cell of the simulated generation volume is reduced by the amount which is lost due to recombination at the dislocation. For this approach we treated each cell as a point source and calculated the amount of carriers which reach one point of the dislocation by diffusion and the carrier reduction for this cell due to recombination at the defect. This procedure is repeated for all points along the dislocation and for all elements of the generation volume. This method takes account of the different diffusion lengths for the various layers between volume element position and defect position. Finally, the EBIC gain is calculated as the integral over the carrier density of all volume elements times the loss due to diffusion to the depletion region. This procedure is repeated for a sequence in which the dislocation is moved across the generation volume in its full lateral extension. The shape of such a line scan is shown in Fig.5. A similar approach would be the calculation using the Greens function method by Donolato (1978/79).

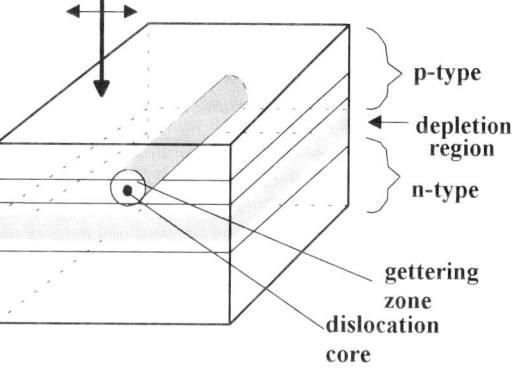

Fig. 3 Sample configuration with dislocation position

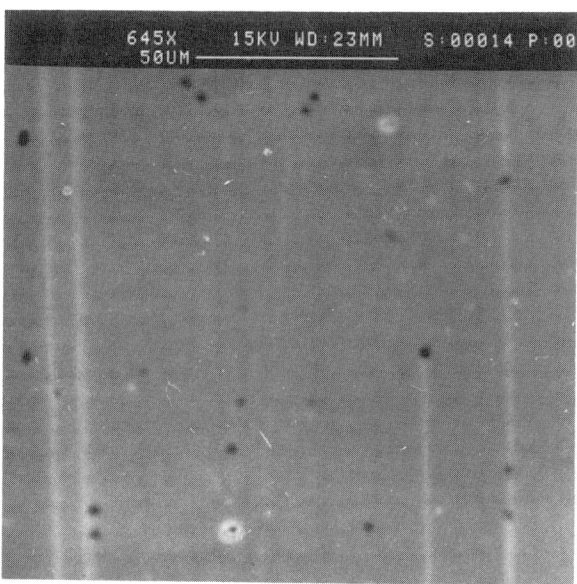

Fig. 4 EBIC micrograph of a bundle of misfit dislocations with gettering area. The misfit dislocations give a bright contrast along one <110> direction, black spots indicate threading dislocations.

The calculation can be repeated for various dislocation depths and defect strengths, as well as for different diffusion lengthsinside the gettering cylinder around the dislocation and various radii of the cylinder.

4. DISCUSSION

The simulation can explain the appearance of white lines for misfit dislocations, if a gettering volume with a higher diffusion length around the dislocation is assumed. By treating the dislocation as a recombination centre without gettering volume we obtain the normal dark contrast.

The number of simulated beam electrons is not a very crucial parameter, it is sufficient to cancel out large fluctuations in the EBIC gain calculation. This can be achieved with less then 50,000 electrons. More important is the volume element resolution for the generation volume. The resolution limits how accurately the layer structure is simulated and therefore the effect of the layer structure on the generation distribution. This can be seen in Fig.1 of the depth dose function, where 100nm thin layers show a significant effect on the generation volume. We used a vertical resolution of 25nm at a beam energy of 20keV. On the other hand, the calculation of the generation distribution needs to be done only once and is then used for all line scan simulations for this energy. Therefore, it is advantageous to run the simulation for a large number of beam electrons, about 200,000. Only if an altered beam energy is required, is a complete new simulation needed.

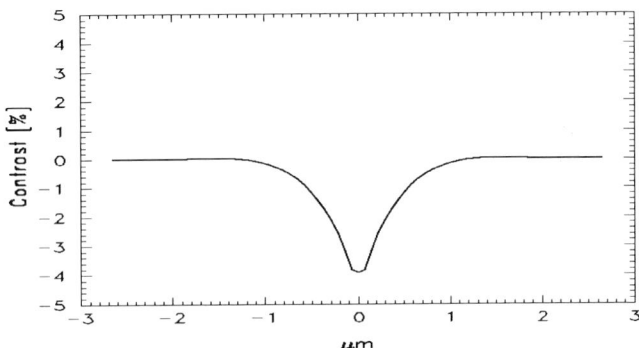

Fig. 5 Simulated line scan over a misfit dislocation with gettering volume.

Our simulation confirms the usefulness of Monte Carlo simulation for EBIC in accordance with other work by Joy (1986) and Czyzewski and Joy (1990). It can give a better understanding of the origin of the EBIC signal in heterostructures.

REFERENCES

Czyzewski Z and Joy D C 1990 Scanning 12 5
Donolato C 1978/79 Optik 52 19
Donolato C 1979 Appl.Phys.Lett. 34 80
Everhart T E and Hoff P H 1971 J.Appl.Phys. 42 5837
Joy D C 1986 Journal of Microscopy 143 233

Inst. Phys. Conf. Ser. No 146
Paper presented at Microsc. Semicond. Mater. Conf., Oxford, 20–23 March 1995

701

STEBIC of Si/Si$_{1-x}$Ge$_x$/Si and high voltage REBIC of CdTe

P D Brown and C J Humphreys

Department of Materials Science and Metallurgy, University of Cambridge, Pembroke Street, Cambridge, CB2 3QZ, UK.

ABSTRACT: Scanning Transmission Electron Beam Induced Conductivity (STEBIC) imaging of misfit dislocations in a Si/SiGe/Si heterostructure is demonstrated using a Jeol 2000FX TEMSCAN. High voltage remote contact EBIC (REBIC) images of doped CdTe indicate that sub-micron resolution of electrical activity is readily achievable.

1. BACKGROUND

The electronic properties of extended defects can only be fully understood if electrical measurements are made with high spatial resolution. STEBIC imaging of an electron transparent foil allows both electrical and structural properties of defects to be simultaneously observed, while improved resolution is afforded over the SEM/EBIC technique due to minimisation of beam spreading effects. The constraint of minority carrier diffusion length is also removed and resolution depends on the incident probe size and the width of the electron-hole pair generation zone [Laval et al, 1988]. The trade off is low electrical signals due to the small generation volume and surface recombination effects, in addition to the practicality of contacting and handling thin foils. The STEBIC technique was originally developed in the late 1970s but has received little attention in recent years. Sparrow and Valdre [1977], Fathy et al [1979] and Pennycook [1981] used 500keV STEM to profile the electrical activity within Si transistor structures to a resolution of 0.3μm. Petroff et al [1980] also used STEBIC in a STEM to examine dislocation core structures in (Ga,Al)(As,P), thereby providing the first evidence that non-radiative recombination processes at dislocations are related to jogs and kink sites. Improved technology and the wider availability of TEMSCAN instruments suggests STEBIC requires re-evaluation, particularly in view of the availability of electron sources with increased brightness to increase the generated signal, and the possibility of preparing tailored devices structures incorporating buried p-n junctions to maximise charge collection within a thin foil. Two research groups have attempted EBIC using TEMSCAN instruments in recent years. Cabanel and Lavel [1990] have successfully used 100keV STEBIC to correlate the localisation of electrical activity with structural defects in polycrystalline Si with regard to carbon which traps oxygen and impurities, to a spatial resolution of 0.2μm within a 0.6μm thick sample foil. This work demonstrated asymmetric profiles across boundaries indicating that segregation zones act as diffusion barriers and the electrical activity in this case was attributed to impurities rather than dangling bonds. High Voltage-EBIC (200keV) has been used by Perreault et al [1993] to characterise polycrystalline Si for solar cell applications. The increased penetration of electrons at higher voltages more appropriately profiles this materials response to light (since electrons have penetration depths of $\approx$3μm and $\approx$180μm at 20keV and 200keV respectively).

We extend the applicability of STEBIC to examine Si/SiGe/Si heterostructures. Our long term interest is to determine quantitatively the extent to which transition metal impurities influence the electrical activity of misfit dislocations in MBE grown material. As a precursor to this work we simply demonstrate that meaningful STEBIC images can be acquired from this materials system for varying sample geometries and contacts.

The variant of REBIC may also be used to reveal local electric fields in a material and is

702

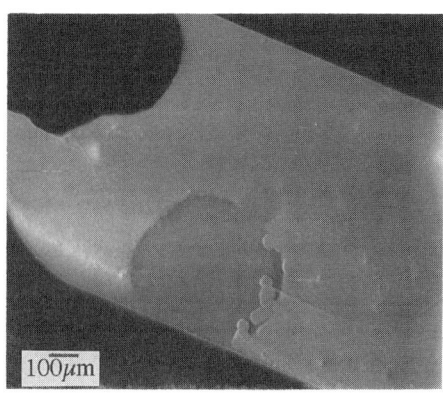

Fig. 1 STEBIC apparatus attached to a Jeol 2000FX TEMSCAN. (1) LN_2 electrical contact holder, (2) head amplifier, (3) EBIC amplifier and (4) framestore.

Fig. 2(a) Si/$Si_{0.96}Ge_{0.04}$/Si imaged in plan view using 200keV STEBIC following LN_2 cooling.

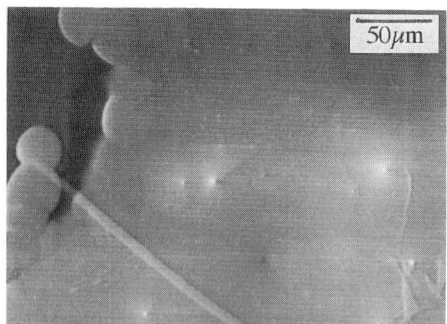

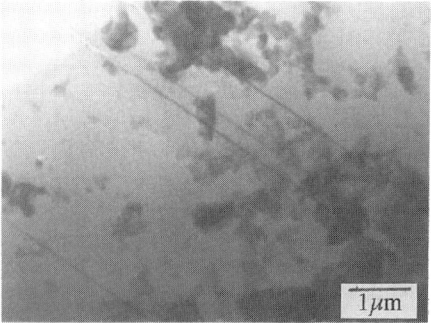

Fig. 2(b) Enlarged STEBIC image of 2(a) and (c) TEM image of edge of foil giving direct correlation of misfit dislocations with the linear feature in STEBIC (g=220).

experimentally direct in that two ohmic contacts are simply applied to opposite ends of a sample, one of which is grounded. A signal is formed as a result of separation of electron hole pairs created by the electron beam in the electric field of a charged defect. Charge separating defects within (Hg,Cd)Te have been mapped to a spatial resolution of $1\mu m$ in SEM and the technique is sensitive to fields produced by inclusions, damage, strain, p-n junctions, doping variations and possibly compositional variations [Bubulac and Tennant, 1988]. We show improved resolution of doped CdTe using High Voltage REBIC.

2. EXPERIMENTAL

Nominally Si $(0.4\mu m)$ / $Si_{0.96}Ge_{0.04}$ $(0.8\mu m)$ / Si $(0.8\mu m)$ on a Si n^{++} substrate was grown by MBE at 550°C. The SiGe epilayer and Si buffer were doped p-type using $10^{16}cm^{-3}$ Boron, while the Si cap was doped $5\times10^{16}cm^{-3}$ with Boron. A rectifying contact to the top surface was made by evaporating Al prior to preparing an electron transparent thin foil in plan view by sequential mechanical polishing and argon ion milling of the substrate. An ohmic contact to this lower surface was made using InGa eutectic. Samples of bulk CdTe grown from the vapour phase were doped with P and In as previously described [Loginov et al, 1991]. Thin foils were prepared by sequential mechanical polishing, argon ion milling and iodine reactive ion sputtering. Two ohmic contacts were made by applying $AuCl_3$ which evaporated to leave an Au contact. Connection to the external circuit was made for both STEBIC and REBIC

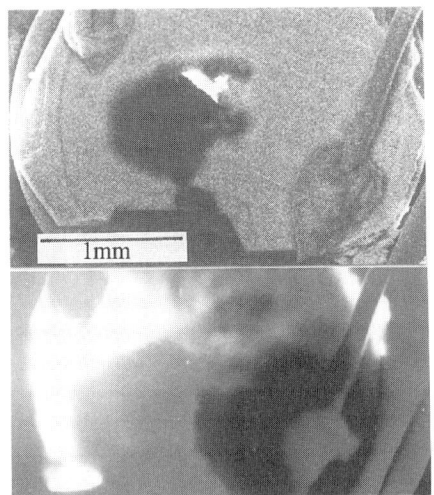

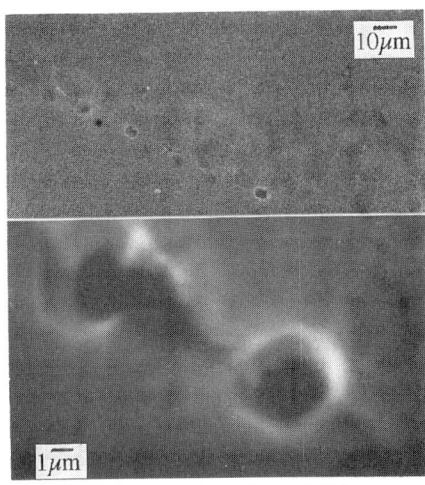

Fig. 3 Low magnification images of P-doped CdTe (a) SEI and (b) 200keV REBIC

Fig. 4 (a) low magnifications SEI image and (b) high magnification 200keV REBIC image from feature in P-doped CdTe.

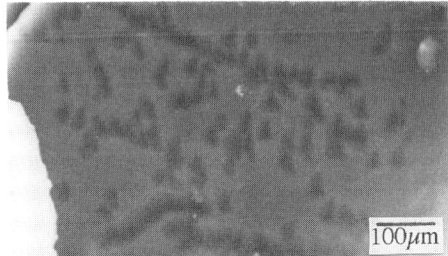

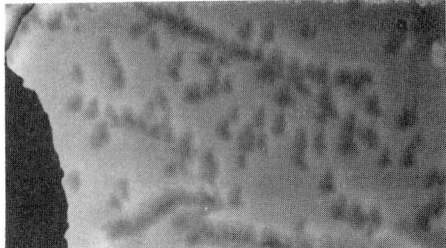

Fig. 5(a) 200keV REBIC and (b) topographic images of etch pits in bulk In-doped CdTe.

samples using wire affixed with silver paste. Images of electrical activity were acquired using a Jeol 2000FX electron microscope operated between 20 and 200keV. Samples were supported in an Oxford Instruments liquid nitrogen cooled, single tilt, electrical contact stage. Since electrical contacts are likely to be non-ideal, cooling is required for quantitative work to ensure that depletion region widths (i.e. minority carrier diffusion lengths) remain constant. The incident beam current at 200keV in TEM mode was variable up to 140nA, while in TEMSCAN mode this was reduced by an order of magnitude (as determined by observation of one of the REBIC contacts, giving a direct current reading) as the beam was rastered across the sample. Average EBIC signals were typically 1nA in REBIC (for a CdTe foil too thick for electron transmission) and 100pA in STEBIC (for an electron transparent Si/SiGe/Si foil). Signal amplification was performed using a Matelect ISM5 EBIC low noise current amplifier. Control of the scan-rate of the TEMSCAN using a frame store allowed the beam to be rastered at a rate compatible with the low bandwidth constraint of the amplifier. The experimental set up is shown in Fig. 1.

3. RESULTS AND DISCUSSION

The low magnification STEBIC image of Fig. 2 illustrates the experimental arrangement used to profile the electrical activity within this pseudomorphic $Si/Si_{0.96}Ge_{0.04}/Si$ sample. Liquid nitrogen cooling was required to enhance the STEBIC contrast from this sample foil. The top contact appears black (top left) for this geometry, as does the black outline of the supporting 3mm slot grid. Contrast to the right of the hole is enlarged in Fig. 2b with a TEM image obtained from a short section of the main linear feature being shown in Fig. 2c. The linear feature in STEBIC (of 4μm width) directly correlates with four dislocations running along the line of this feature, which are presumed to be misfit dislocations lying in the plane of the two SiGe/Si and Si/SiGe interfaces, and distributed as a band of three and a single dislocation 4μm apart. The darker, upper line of the EBIC band being associated with the group of three dislocations, while the lower, weaker band correlates with the single dislocation. Since this image is indicative of the first stage of epilayer relaxation, the potential of this technique immediately becomes evident for the purpose of locating and identifying dislocation sources within low x content $Si_{1-x}Ge_x$ epitaxial material. The bright, point like contrast features are possibly due to threading dislocations from the substrate which have not bowed over into the plane of the interface, although no direct correlation was obtained. The artefacts in the TEM image are due to the evaporated Al contact which does not contribute to the STEBIC image. Although it was difficult to obtain decent TEM images of these dislocations due to the raised height of the sample in the objective lens, contrast remained with a g=220 reflection both parallel and perpendicular to their line, which is consistent with them being of the 60° type as one might expect.

Low magnification secondary electron and High Voltage REBIC images of a P-doped CdTe sample, known to contain precipitates [Loginov et al 1991], are shown in Figs 3a and 3b. Strong black-white contrast delineates the contact regions and reflects the non-ideality of the contacts and the short separation between them for the material resistivity used in this instance. A REBIC image exhibiting contrast on the scale of <1μm is shown in Fig. 4b. One can speculate as to the nature of the contrast produced since the foil in this region was too thick to allow imaging of the associated structural image. However, the suggestion from the secondary electron image (Fig. 4a) is that some boundary has been delineated by the sample preparation process and this appears to be decorated with features showing hexagonal symmetry in REBIC, and so these features are tentatively attributed to precipitates. Such REBIC images were retained down to 80keV, while little variation in contrast was retained at 60keV with a <1nA generated signal. Upon returning to 200keV the original image had greatly deteriorated in both generated signal and contrast which suggests that enhanced beam damage while imaging at lower voltages may have occurred. Subsequent cooling with liquid nitrogen facilitated temporary recovery of an image comparable to that originally obtained.

High Voltage REBIC and topographic images of In-doped CdTe are shown in Figs. 5a and 5b respectively. The etch pit structure delineated in the sample by the preparation process is evident in each. It is not known if the contrast in the REBIC image arises either from the defect associated with the etch feature, or is associated with the surface roughness in the vicinity of these features.

In summary, STEBIC may be applied to the study of MBE grown Si/SiGe/Si structures following deposition of a Schottky contact, and facilitates direct correlation with structural defects introduced during the first stage of epilayer relaxation. High Voltage REBIC may be used to obtain <1μm resolution of contrast features within doped CdTe.

Acknowledgements With thanks to EHC Parker and R Kubiak of Warwick University for supplying SiGe/Si material; to N Thompson, K Durose and AW Brinkman of Durham University for providing the CdTe used during this study; to YY Loginov for assistance with sample preparation, and to A Wojcik of Matelect for useful discussions. PDB wishes to acknowledge SERC for support under contract No. GR/J37966.

References

Bubulac LO and Tennant W E, 1988, Appl. Phys. Lett 52 1255
Cabanel C and Lavel JY, 1990, J. Appl. Phys. 67 1425
Fathy D, Sparrow TG and Valdre U, 1979, J. Microsc. 118 263
Lavel JY, Pinet-Berger MH. and Cabanel C, 1988, J. de Phys. Colloque C5 521
Loginov YY, Brown PD, Thompson N, Alnajjar AA, Brinkman AW and Woods J, 1991, J. Crystal Growth 117 259
Petroff PM, Logan RA and Savage A, 1980, Phys. Rev. Lett. 44 287
Pennycook SJ, 1981, Ultramicroscopy 7 99
Perreault GC, Hyland SL and Ast DG, Solar Energy Materials and Solar Cells, 1993, 30 309
Sparrow TG and Valdre U, 1977, Phil. Mag. 36 1517

Inst. Phys. Conf. Ser. No 146
Paper presented at Microsc. Semicond. Mater. Conf., Oxford, 20–23 March 1995
© *1995 IOP Publishing Ltd*

P-N junction localization in semiconductor laser structures

A L Tóth

Research Institute for Technical Physics of the Hungarian Academy of Sciences
1047 Budapest, Fóti út 56.

ABSTRACT: A measuring and correction procedure is proposed to decrease a systematic error of SEM-EBIC p-n junction localization to a level compatible with the high precision of the underlying line profile characterization of the GaAlAs/GaAs/GaAlAs heterostructure.

1. INTRODUCTION

Semiconductor lasers are multilayer heterojunction devices, where the composition changes across the layers according to the required characteristics, such as confinement of charge carriers and radiation in the active layer. The aim of the work described below is the localization of the p-n junction position in the active layer with a high precision, using a correction procedure against the systematic error due to different minority carrier diffusion lengths in the structure.

2. EXPERIMENTAL

2.1. Excitation: the choice of signals

To minimize the possible errors from successive line scans, simultaneous localization of the GaAs active layer and the p/n junction is needed, which excludes the use of etching method and voltage contrast, respectively, in the high resolution secondary electron (SE) imaging mode. That is why the backscattered electron (BE) and the electron beam induced conductivity (EBIC) signals were used to localize the layers and the junction, in spite of their relatively poorer lateral resolution.

The measurement can be carried out in line profiling mode on a smooth cleaved surface, where the inflexion points of the BE profile indicate the hetero-interface, and the EBIC maximum shows the p-n junction position. These points can be obtained easily from derivative curves, measured by modulating the beam position and detecting by lock-in amplifier, or by numerical derivation of the digital line distributions.

2.2. Detection: the substage

To achieve the necessary high resolution, low beam current I_0 and low energy E_0 had to be used with a short working distance. Consequently the signals were small, and the detection had to be optimized by incorporating the EBIC contacts, the laser chip holder and the BE detectors into one compact substage. Mechanical stability, low-noise EBIC detection and fast, large solid angle BE detection were reached using small, low capacitance diodes in an optimized distance from the sample. Integrating the cosine BE distribution function over the area covered by the detectors, the measured intensity is

$$\mathbf{I} = \sin(\mathrm{atn}(\mathbf{a}+\mathbf{a_0})/\mathbf{d})) - \sin(\mathrm{atn}(\mathbf{a_0}/\mathbf{d}))$$

where $\mathbf{a}$ is the width of the detectors, $\mathbf{a_0}$ is the distance between the detectors, and $\mathbf{d}$ is the detector-sample distance. The optimum $\mathbf{d}$ distance in our case ($\mathbf{a}=\mathbf{a_0}=2\text{mm}$) is 2.6 mm, where the sensitivity of the measurement is better than that of the commercial large area BE detectors.

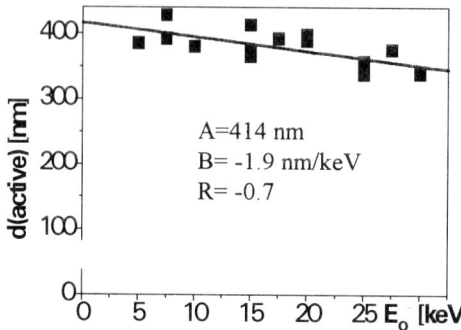

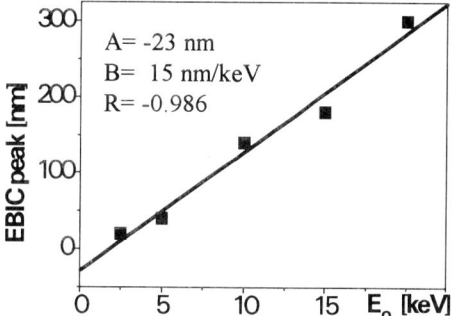

Fig.1. The widening of the apparent active layer width measured from BE profiles

Fig.2. Computed EBIC peak shift as a function of beam energy for $L_1/L_2=5$

As a result the BE line scans with 100 nm resolution can be measured, with a beam current of 50-100 pA and energy of 5-30 keV. The measured systematic error of active layer width measurement, i.e. the deviation of the distance of the inflexion points from the metallurgical layer thickness as a function of the beam energy can be determined from Fig.1.

3. SIMULATION RESULTS AND MEASUREMENT OF EBIC PROFILES

The p-n junction location coincides with the EBIC maximum only in the case of symmetric EBIC profiles, where the minority carrier diffusion lengths are equal on both sides of the junction. The peak shift as a function of diffusion lengths has been analysed by Tóth (1987), and a simple semilogarithmic relationship has been found between the ratio of diffusion lengths on the two sides of the junction and the shift of the EBIC maximum from the actual p-n junction position. However, the use of this relationship is difficult, as the diffusion length measurement is problematic over the short distances characteristic of the laser diodes.

As the information volume of EBIC measurement, according to the Donolato theory, is determined by the size of the excited volume rather than the diffusion lengths, it seems to be worthy to calculate the peak shift as a function of beam energy using the Monte Carlo (MC) simulation routine of Joy and Pimentel (1985). EBIC line profiles were simulated with $L_1/L_2=5$ ratio. The shift of the zero crossing in the first derivative of simulated curves as a function of E_0 beam energy is shown in Fig.2.

The linear dependence of EBIC peak shift on E_0 obtained by MC simulation can be confirmed by measurements. Two different methods have been used to obtain the derivative line profile.

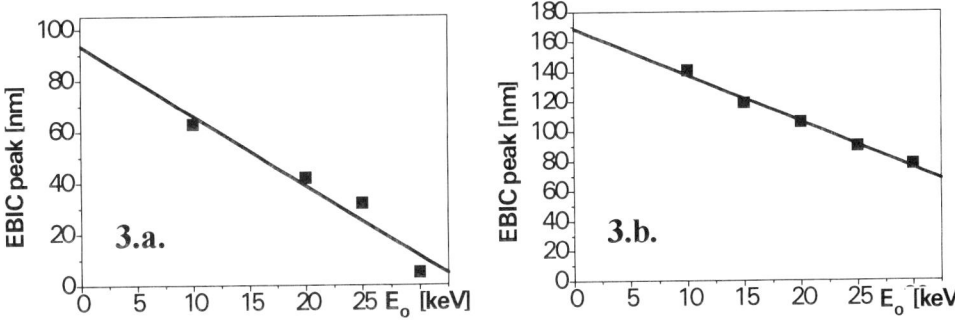

Fig.3. Measured EBIC peak position as a function of beam energy on different samples with different techniques

The numerical derivation (Fig.3.a) of digital line profiles is very easy, but its serious drawback is the high sensitivity to noise. The noise can be reduced by increasing the beam current, or by digital smoothing of the curve, but in each case at the cost of poorer resolution. The derivative signal (Fig.3.b.) can be measured by phase sensitive detection of the EBIC signal excited by a position modulated beam. The signal-to-noise ratio becomes better, but in this case the modulation amplitude decreases the resolution, and the experimental setup is more complicated. Each method has its advantages, and our practice shows that the right choice depends on the size and physics of the device under test.

4. CONCLUSION

As the E_0 dependence of active layer width obtained from BE profile is much weaker than that of the position of EBIC maximum, the heterojunction can be used as reference point of p-n junction positioning. Plotting together the results of the BE and EBIC measurements, the following combined graphs can be constructed (Figures 4a and 4b).

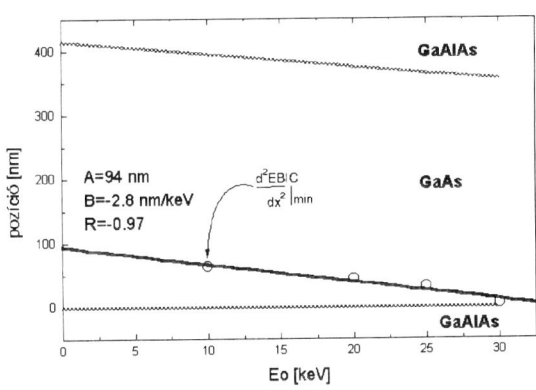

Fig.4.a. Result of a combined EBIC-BE line profile measurement using numerical derivation

708

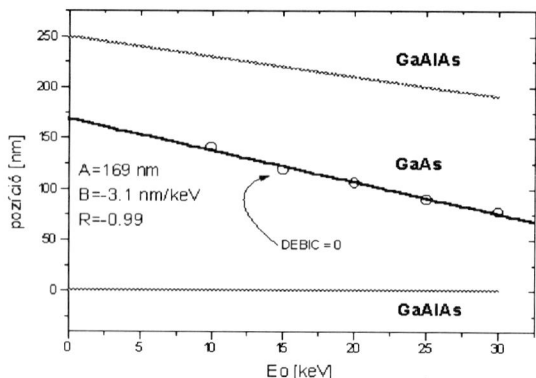

Fig.4.b. Result of a combined EBIC-BE line profiling using phase sensitive derivation

Summarizing the results it can be seen, that the simultaneous backscattered electron and EBIC line profile measurement, together with the proposed methods of signal detection, evaluation and correction can be used for the localization of p-n junctions in heterolaser structures of submicrometer size. The measurements verified the results of the Monte Carlo simulations, that the systematic error of p-n junction localization decreases to zero with decreasing beam energy, and the results of the high resolution EBIC measurements can be corrected with a precision and accuracy of 20 nm.

ACKNOWLEDGEMENTS

The author is grateful to D. C. Joy for kindly providing the core of the MC simulation code. This work was supported, in part, by the (Hungarian) National Scientific Research Fund (OTKA) through Grant T 7615.

REFERENCES

Tóth A L 1987 Inst. Phys. Conf. Ser. **87** 659
Joy D C and Pimentel C A 1985 Inst. Phys. Conf. Ser. **76** 355

Inst. Phys. Conf. Ser. No 146
Paper presented at Microsc. Semicond. Mater. Conf., Oxford, 20–23 March 1995
© *1995 IOP Publishing Ltd*

SEM/EBIC observations of CdTe/CdS thin film solar cells

S A Galloway and K Durose

Department of Physics, University of Durham, South Road, Durham DH1 3LE, UK

ABSTRACT: This work reports on preliminary studies using the SEM/EBIC technique into CdTe/CdS thin film solar cells. Three types of cells are studied which are distinguished by their post deposition processing (1) untreated, (2) 400°C heat-treatment in air and (3) exposure to $CdCl_2$ followed by a similar heat-treatment. The cells are studied in plan-view and in cross-section. Topographical features associated with the CdTe surface strongly influence EBIC plan view images and this influence can be removed by polishing. Combined plan-view and cross-section results reveal inhomogeneities in both the efficiency of, and in the position of the collecting junction with respect to the metallurgical CdS/CdTe junction.

1. INTRODUCTION

Thin film polycrystalline CdTe/CdS heterojunction technology is one of the primary candidates for the fabrication of stable, high efficiency, cost effective, large area photovoltaic solar conversion units (Chu and Chu 1993). The theoretical maximum conversion efficiency for CdS/CdTe heterojunctions reported in a review by Brinkman (1993) has recently been calculated to be ~29%. In such devices a thin layer of n-type CdS acts as a transparent conducting window to the incident radiation and the resulting photocurrent consists predominantly of electrons generated in, and collected by the junction depletion region in the p-type CdTe absorber. The most popular device structure at present uses a glass substrate coated with transparent conducting oxide upon which CdS (~0.2μm) and CdTe (several μm) are deposited. The thin film structure is feasible because of the high optical absorption coefficient of CdTe.

Substantial progress has recently been made in thin film technology especially with regard to post-deposition processing. Typical post-deposition steps of exposure to $CdCl_2$ and annealing in air have been found to play a crucial role in improving CdS/CdTe device performance (Ringel et al 1991) with the resultant conversion efficiencies increasing by up to two fold. However there is a lack of fundamental understanding of the role of grain boundaries and of the importance of the quality of the interfaces on a microscopic scale. Bulk experimental techniques (e.g. J-V-T, spectral response) are commonly employed when studying photovoltaic devices but these tend to be of limited value when studying polycrystalline samples in that they only give an average picture of the material and little information concerning inhomogeneity. For example, in bulk measurements the recorded signals may be dominated by strong signals from small areas whereas little information is determined concerning the remaining areas which present weak signals.

In the preliminary work reported here the spatial resolution of an SEM and in particular the EBIC technique is used to study the effects of post-deposition treatments on the inhomogeneity of CdTe/CdS devices with the aim of furthering understanding on a microscopic level those factors which influence the device efficiency.

2. EXPERIMENTAL

The heterojunctions were manufactured and supplied by Antec GmbH. The device structure comprises indium tin oxide (250nm) deposited on glass followed by tin oxide : strontium (50nm) and CdS (100nm) deposited at 150°C by evaporation and CdTe (5-8μm) at 530°C deposited by close space sublimation. Three different types of samples were studied and these were untreated, heat-treated only at 400°C in air for 30 minutes, and heat-treated in the same conditions after dipping in CdCl$_2$/methanol solution. Contacts were made to the CdTe using evaporated Au.

The SEM/EBIC investigations were carried out in a JEOL JSM 848 with a Matelect ISM5 EBIC amplifier. As electron beam carrier injection cannot mimic the intended operation of the cells with irradiation through the glass, EBIC signals were obtained by studying the cells either in cross-section, or by irradiating the junction (plan-view) through the back contact (Au/CdTe). For certain samples the CdTe surface roughness was removed by polishing with 0.25μm diamond paste. The resultant surface damage was removed by polishing specimens in a solution of 0.5% Br$_2$ in equal parts of methanol and ethylene glycol. Following this small areas of Au was evaporated to a thickness of ~15nm onto the CdTe. An "α-step" profiler was used where required to measure film thickness and surface roughness. Cross-section specimens were made simply by scoring the edge of the glass substrate and breaking the sample with the thin films in tension.

3. RESULTS AND DISCUSSION

The surface roughness of the polycrystalline CdTe layers was found to influence the plan-view EBIC contrast to such an extent that grain boundaries were often associated with greater EBIC signal than grain interiors. This evident in fig. 1 which shows an EBIC image of the untreated cell where the CdTe layer is ~8μm thick. The reason for this change in contrast from that which would normally be expected at the grain boundaries is due to increased generation and perhaps more important, due to a reduced distance to the collecting junction when the beam is located at the cusps between grains. Surface topography (identified by SE contrast) occasionally influenced the EBIC signal obtained away from the grain boundaries and this is also evident in fig. 1.

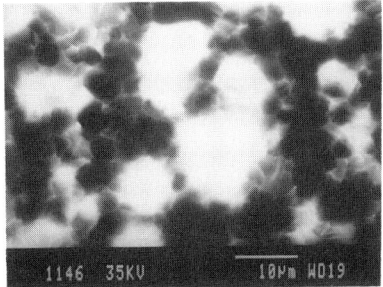

Fig. 1 Plan view EBIC image of untreated cell in as received state.

It is clear that in order to obtain plan view EBIC where surface topography does not play an important role in forming EBIC contrast it is necessary to remove surface features by polishing. Hence all three specimens were polished as described above leaving a CdTe layer thickness of ~2μm. Plan-view EBIC images of the untreated and CdCl$_2$/heat-treated cells are shown in figures 2 and 3 respectively. The grain boundaries are no longer associated with increased EBIC thus proving the unusual contrast behaviour of the unpolished cells to be an artefact associated with surface topography.

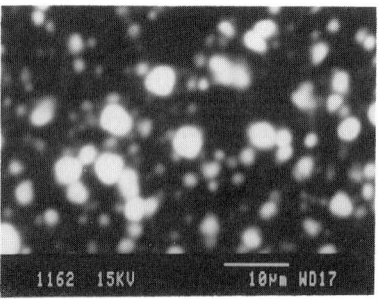

Fig. 2 Plan view EBIC image of untreated cell after polishing.

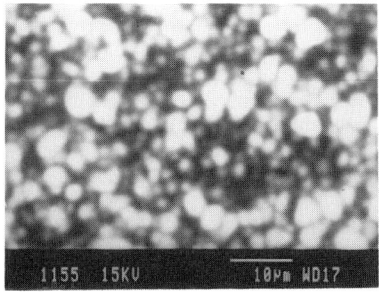

Fig. 3 Plan view EBIC image of CdCl$_2$/heat-treated cell after polishing.

The plan-view images show great differences in the strength of the EBIC signal between different regions which have areas larger than single grains. Figure 4 shows three line scans from the untreated specimen taken at 15, 25 and 35kV with the beam current adjusted accordingly to give constant power. The contrast between typical bright and dark regions falls from over 90% to ~65% as the accelerating voltage is increased from 15 to 35kV. However before such data can be interpreted it is worth noting that plan-view EBIC contrast may be due to variations in four different parameters; (1) the minority carrier diffusion

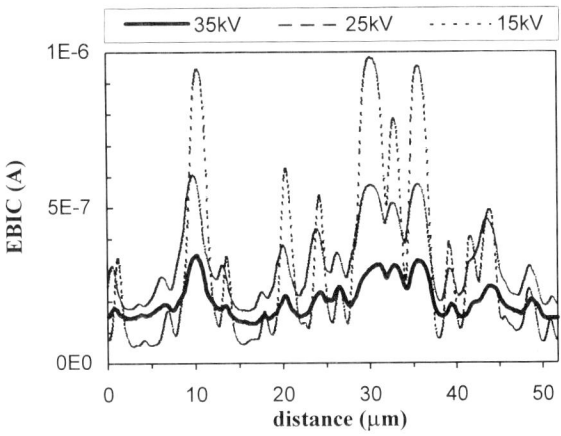

Fig. 4 EBIC signal from untreated cell after polishing as a function of accelerating voltage with constant beam power.

length of the CdTe, (2) the width of the CdTe depletion region, (3) the position of the collecting junction and (4) recombination at the interface. The voltage dependence of the plan view EBIC signal is therefore difficult to interpret without knowledge of each of these parameters.

Each specimen type (in the unpolished state) was studied in cross-section. Back scattered (BS) and secondary electron (SE) analysis were used to determine the position of the metallurgical junction between the CdS and the CdTe. The fractured surfaces of the heterojunctions showed evidence of the granular structure the films. However there was no evident link between the fractured surface topography and the EBIC signal. The shape of the cross-section EBIC signal is of considerable interest as well as the magnitude of its peak. To a first approximation the position of the EBIC peak corresponds to that of the electrical junction where the band bending and hence electric field is greatest (Matson et al 1986) whilst the magnitude reflects the junction collection efficiency. In addition, it is well known that for regions away from the depletion region electric field the shape of the EBIC profile can be used, also to a first approximation, to give the minority carrier diffusion length (Oakes et al 1977). Both of these approximations assume that the thin film sampled by fracturing behaves as bulk material with no flaws and no surface states.

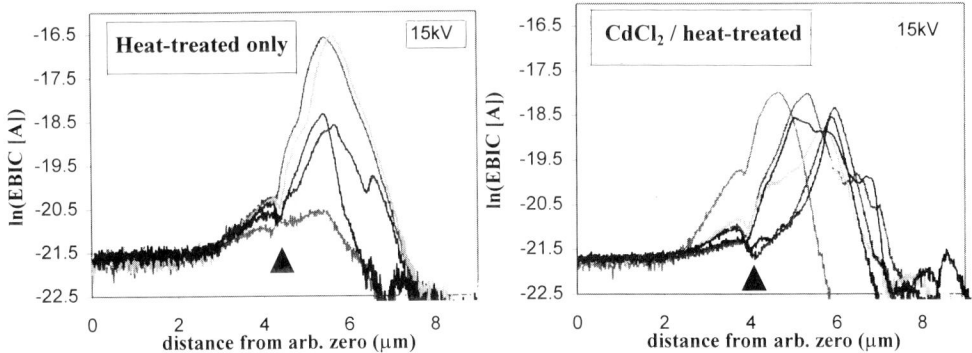

Fig. 5 Typical cross section EBIC scans from heat-treated only and CdCl$_2$/heat-treated cells. The arrow denotes the position of the metallurgical CdS/CdTe junction.

The cross-section EBIC signal was found to vary greatly in magnitude as the point at which the beam traversed was altered. The peak signal was usually in the CdTe thus suggesting homojunction behaviour (Mitchell et al 1977). The peak signal also fluctuated in position normal to the metallurgical junction with a spatial periodicity parallel to the junction of $\sim 10\mu m$. Some of these results are illustrated in fig. 5 in which the magnitude of the EBIC signal is plotted on a logarithmic scale for typical cross-section scans from the heat-treated only and $CdCl_2$/heat-treated cells. The position of the cross section EBIC peak in the heat-treated only sample is relatively steady with respect to the metallurgical junction when compared to the other two specimens.

It is not known whether the complicated nature of many of the scans (e.g. multiple peaks) are caused solely by irregularities in the fractured surface. McClure et al (1990) suggests such irregularities may be partly explained by grain boundary recombination. For EBIC profiles which showed few irregularities and where the logarithm of the EBIC signal was linear with distance, the minority carrier diffusion length was typically 0.2, 0.3 and $0.2\mu m$ for the untreated, heat-treated only and $CdCl_2$/heat-treated samples respectively.

4. CONCLUSIONS

The following conclusions are drawn from the data presented in this work:
(1) Surface roughness of the CdTe leads to artefacts in the plan-view EBIC contrast which can be eliminated by polishing. However interpretation of variable voltage plan-view EBIC is very difficult as it is not clear from such data whether contrast arises from variations in the minority carrier diffusion length, position of the depletion region or collecting junction, or in the interface recombination velocity.
(2) Cross-section studies show the collecting junction to be a homojunction in the CdTe. Fluctuations occur in the position of the collecting junction relative to the metallurgical CdTe/CdS junction and these fluctuations are a function of the post-deposition treatment. The collection junction is most uniformly positioned in the sample which was heat-treated only.
(3) Although there is some variation in the CdTe diffusion length when different regions of the device are scanned, the greatest variation is in the magnitude of the EBIC peak signal. The minority carrier diffusion length shows a slight improvement in the heat-treated only cell but no improvement in the $CdCl_2$ heat-treated cell.
(4) The $CdCl_2$/heat-treatment appears to improve the homogeneity of the junction. This is clear from plan-view EBIC results and although more data is required, also appears to be corroborated by cross-section EBIC results.
(5) The large variation in the magnitude of the cross-section EBIC peak together with the substantial EBIC contrast at high accelerating voltages from polished plan view specimens indicates that interface recombination plays an important role in hindering the efficient collection of minority carriers and this occurs in large areas of the devices.

ACKNOWLEDGEMENTS

The authors would like to thank Antec GmbH for the provision of the devices and to Drs A W Brinkman and S Oktik for useful discussions.

REFERENCES

Brinkman A W 1993 in Properties of Narrow Gap Cadmium based Compounds Ed. Capper P, EMIS Data Reviews Series No. 10 pp591-597
Chu T L and Chu S S 1993 Prog. in Photovoltaics: Res. & Appl. 1 31
Matson R J, Noufi R, Ahrenkiel R K, Powell R C and Cahen D 1986 Solar Cells 16 495
McClure J C, Chung C J and Singh V P 1990 Solid State Comm. 75 3 171
Mitchell K W, Fahrenbruch A L and Bube R H 1977 J. Appl. Phys. 48 10 4365
Oakes J J, Greenfield I G and Partain L D 1977 J. Appl. Phys. 48 6 2548
Ringel S A, Smith A W, MacDougal M H and Rohatgi A 1991 J. Appl. Phys. 70 2 88

Inst. Phys. Conf. Ser. No 146
Paper presented at Microsc. Semicond. Mater. Conf., Oxford, 20–23 March 1995
© 1995 IOP Publishing Ltd

Structure and properties of APBs in GaAs/Ge

D B Holt, M Mazzer, C Zanotti-Fregonara, C Hardingham[+], G Salviati*, L Lazzarini* and L Nasi*

Materials Department, Imperial College of Science and Technology, London SW7 2BP, U.K.
[+]EEV Ltd.Waterhouse Lane, Chelmsford, Essex CM1 2QU, U.K.
*MASPEC-CNR Institute, Via Chiavari 18/A,43100-I Parma, Italy

ABSTRACT: Antiphase Boundaries (APBs) in GaAs grown epitaxially on (100) Ge have been studied both in GaAs solar cells and in as-grown material. Scanning electron microscope EBIC (electron beam induced current) and CL (cathodoluminescence) have been used to image and to record APB contrast linescans. Values of surface recombination velocity and of minority carrier diffusion length derived by applying the Donolato phenomenological analysis to the EBIC contrast are reported. TEM studies of APB structure are also reported.

1. INTRODUCTION

It was originally suggested that APBs occur in semiconducting compounds with the sphalerite structure (Holt 1969) on the basis of experimental observations including one on the growth morphology of some of the first epitaxial films of GaAs on (100) Ge (Bobb et al 1966). Little was published in the field until the mid-1980s (e.g. Petroff 1986 and Kroemer 1987) when GaAs on (100) Si and Ge became of intense technological interest. APBs in these materials have been studied especially by TEM (e.g. Gowers 1984, Georgakilas et al 1993) but also occasionally by EBIC and/or CL (e.g. Chu et al 1988, Nauka et al 1990) although the published EBIC and CL images were not of high resolution. Some of us have recently been involved in TEM studies of APBs in as grown GaAs/Ge (Li et al 1994) and others in EBIC and CL studies of APBs in GaAs/Ge solar cells (Hardingham 1994) and we report here initial results of studies combining the two techniques.

2. EXPERIMENTAL

GaAs/Ge was grown by atmospheric pressure MOVPE (Li et al) and kindly supplied by Prof. Giling of the Research Institute of Materials, University of Nijmegen, The Netherlands. Solar cells were grown by a low pressure MOVPE process at EEV Ltd. Samples were thinned first mechanochemically and then by room temperature Ar ion milling for transmission electron microscopy study at MASPEC in a 2000FX JEOL microscope operating at 200 kV. EBIC and CL studies were carried out at Imperial College on a JEOL JSM-840A fitted with a Matelect ISM-5 system for EBIC etc and a spectroscopic CL system with a North Coast Ge detector and phase lock amplifier for work in the infrared (Napchan et al 1993).

714

The equations of the Donolato (1985) phenomenological theory of EBIC contrast for area defects were put into a Mathematica program and run on a Sun workstation to obtain values of the APB interface recombination velocity and the minority carrier diffusion lengths in the neighbouring grains.

3. RESULTS

3.1 APBs in GaAs/Ge Solar Cells

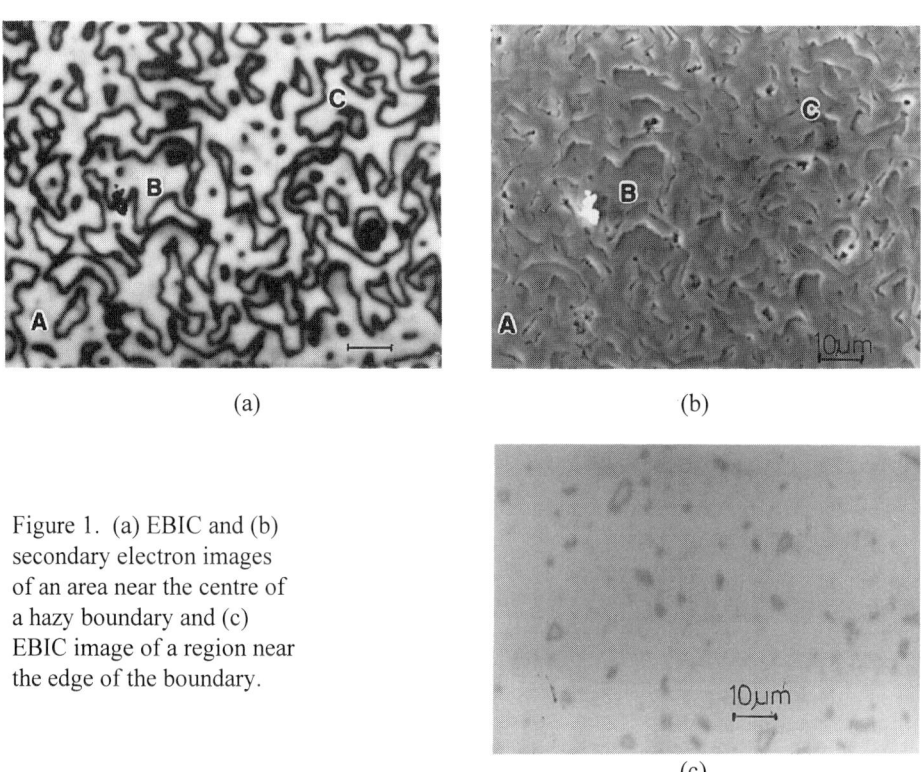

(a)   (b)

Figure 1. (a) EBIC and (b) secondary electron images of an area near the centre of a hazy boundary and (c) EBIC image of a region near the edge of the boundary.

(c)

APBs in the solar cells occur in a hazy boundary, visible to the naked eye, near the periphery of the 50 mm wafers. EBIC shows that in the middle of the boundary the material consists of an interlocking, jig-saw puzzle like structure of APDs (antiphase domains) (Figure 1a) which run right up through the cell structure to give rise to surface topography (Figure 1b) which is responsible for the diffuse reflection of light, producing the visible surface haziness. From the middle of the hazy boundary to its edges (a distance of about a mm to both sides) the APDs became smaller and less numerous (Figure 1c) so the material eventually becomes single phase. Across this APB-rich boundary region the GaAs changes from one domain polarity to the other as was shown by etching with molten KOH. (The KOH etch pits at dislocations appear as elongated hexagons and in regions of different 'phase' (polarity) their orientations are orthogonal Li et al, 1994).

EBIC linescan profiles (LSPs) were recorded by scanning at right angles to the APBs (Figure 2). Quantitative APB LSPs were recorded via the Matelect software. Since the domains are small areas, APBs often appear in pairs on opposite sides of a domain and are not fully resolved but in such cases the outer halves of the APB LSPs could be analysed and such a pair appear in Table 1 as 1L and 2R.

Figure 2. EBIC micrograph plus the superimposed straight white line along which the EBIC 'y-modulation' linescan trace, containing four dips which are the profiles of the 'four' APBs, was recorded. The form of the leftmost APB contrast trace shows that it consisted of two neighbouring APBs of comparable contrast.

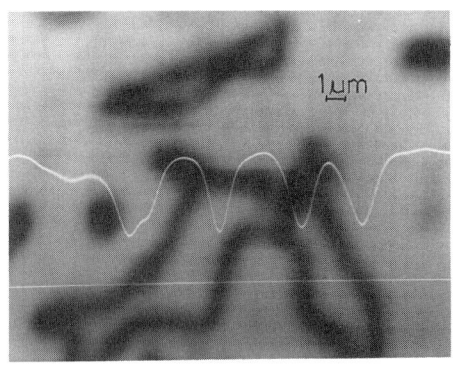

The phenomenological theory of the EBIC dark contrast of defects due to Donolato (1978/79, 1985) was applied i.e. APBs were treated as volumes of thickness h in which the minority carrier lifetime is reduced from the bulk value τ_0 to τ'. Values obtained for the recombination parameter, $s = v_s/D$ (where $v_s = h/\tau'$ is the surface recombination velocity of the APB and D is the diffusion coefficient for GaAs) and of the minority carrier diffusion length, L, in the neighbouring grains are listed in Table 1.

Table 1 Diffusion lengths and recombination parameters of APBs in GaAs/Ge solar cell 3929-3 measured at 15 keV. (I_b is the electron beam current, C is the contrast.)

APB, Left or right	I_b in nA	s in μm^{-1}	L in μm	C in %
1L	0.510	26.5	0.536	26
2R	"	61.1	0.674	46.2
A	10.73	19.6	0.482	35
B	"	19.8	0.577	37.5
C	10.70	24.9	0.719	35.3
D	10.69	7.43	0.710	27.2
E	10.68	18.4	0.821	37.3

It can be seen at A and B in Figure 3 that APBs sometimes change EBIC contrast markedly. There are also at C and D what appear to be small included APDs embedded in the APBs bounding larger domains but this cannot be (Figure 3b). One possible explanation is that either the top or the bottom boundaries of the small regions at C and D are not single APBs but unresolved pairs of close, parallel APBs like the leftmost boundary in Figure 2.

Viewing the APBs in the solar cells (and in the as-grown material) using a Si photodiode placed a short distance above, to obtain ECL (emission CL) gave images showing only rounded brighter and darker areas similar to those previously published (Chu et al 1988, Nauka et al 1990). Using the spectroscopic CL system with the Ge detector, however, it proved possible to obtain CL images (e.g. Figure 4). comparable to the EBIC images.

716

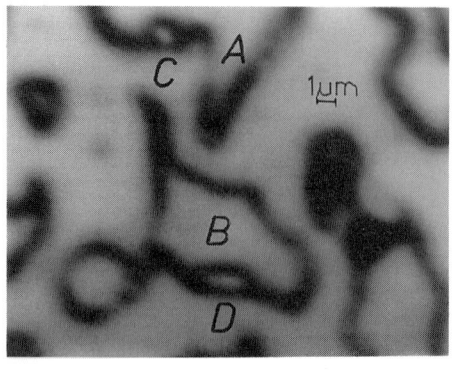 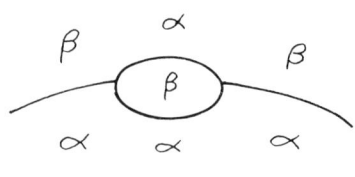

(a)   (b)

Figure 3. EBIC micrograph. On either side of the positions marked A and B the APBs change contrast from strong to weak. At C and D there are what seem to be small APDs embedded in the boundary of larger ones. However these are not possible structures (b) since the reversal of polarity across such arrangements of APBs would be mutually inconsistent. Consider the polarities derived from that (α) for the bottom APD to that for the top one at the left and right compared with that up through the middle.

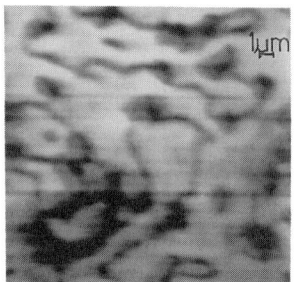

Figure 4. (a) Monochromatic CL image of solar cell 3929-3 taken with the spectroscopic detection system set for the 832.4 nm emission peak. Compare the EBIC images of Figures 1 - 3.

3.2 APBs in As-grown Material

Figure 5. CL micrograph of an area of GaAs/Ge material known to contain APBs. The bright blobs are of the size and shape to be expected for APDs. The straight dark lines are misfit dislocations (MD) or groups of closely spaced MDs.

Attempts at CL imaging of APBs produced not relatively sharp images like those in EBIC but fuzzy brighter areas surrounded by less luminescent regions (Figure 5a) as reported also by Chu et al. In this Figure a number of misfit dislocations can also just be seen running from upper left to lower right across the lower left corner of the picture. This microstructure corresponds closely to that seen in TEM pictures of the same material.

4. DISCUSSION

Both EBIC and CL defect dark contrast can be successfully accounted for by the phenomenological theory (Donolato 1978/79, Lohnert and Kubalek 1983). As both simply treat defects as volumes of enhanced recombination the two should give equal contrast except that CL contrast is depth dependent due to optical self-absorption (Pasemann and Hergert 1986, Jakubowicz 1986). For planar defects running vertically through the whole thickness of the GaAs like the APBs reported here, the self-absorption of CL will have little effect. Hence it was to be expected that the EBIC (Figure 3) and CL (Figure 4) contrast should be similar but hitherto it has not been possible to image APBs in CL adequately. The reason the solar cells give better CL images than any obtained hitherto is believed to be due to the combination of the relatively high sensitivity detection system used here plus the fact that the solar cell CL emission was unusually strong. This may be partly due to the material being of unusually high luminescence efficiency. However, another important factor is the presence of an anti-reflection coating on the cells. This will increase the critical angle for total internal reflection so the cone of rays emitted from an interior point that escape through the surface to be detected, becomes of much larger semi-angle. That is, the antireflection coating increases the fraction of the CL that is emitted.

The APBs in EBIC images of these cells were found to give strong contrast (26 - 46%). Their excellent visibility was due to their high surface recombination velocities together with the solar cell structure. This is designed to give efficient photovoltaic-effect power generation and so does also for the electron voltaic effect which gives rise to the EBIC signals used here. This made possible what appear to be the first quantitative EBIC studies of APBs. The contrast values and the consequent interface recombination velocites are large. The APB CL contrast levels were comparable to those seen in EBIC so reasonable images could be obtained. It has been found that the polarity of GaAs epitaxially deposited on Ge in a lower temperature range is the reverse of that obtained for a higher range (Li et al 1994). A possible explanation for the occurrence of the hazy boundary of APBs in the solar cells therefore would be the occurrence of a radial temperature gradient across the substrate wafers. The APBs would then occur in a transitional band along the critical temperature isothermal.

APBs are interfaces across which all the bonds are wrong (Ga - Ga or As - As) bonds. It has also been suggested that APBs in epitaxial GaAs are preferred sites for substrate atoms (Si in their case, Ge in ours) to deposit (Chu et al 1988). Such segregation of Si atoms to APBs was consistent with the preferential etching of the APBs that they observed as well as their dark cathodolumin-escence (CL) contrast. However, the higher energy (weaker bonding) and elastic strain of the APBs in the absence of impurity segregation would produce similar etching and contrast effects. The results above show that APBs can have large interface recombination velocities, whether the effect is intrinsic or extrinsic (impurity dependent).

REFERENCES

Bobb L C, Holloway H, Maxwell K H and Zimmerman E 1966 Appl. Phys.**37** 4687
Chu S N G, Nakahara S, Pearton S J and Vernon S M 1988 J. Appl. Phys. **64**, 2981

Donolato C 1978/79 Optik **52** 19

Donolato C 1985 in Polycrystalline Semiconductors (G Harbeke, Editor) (Berlin: Springer Verlag) pp. 138 - 154

Georgakilas A, Stoemenos J, Tsagaraki K, Komninou Ph, Flevaris N, Panayotatos P and Christou A 1993 J. Mater. Res. **8** 1908

Gowers J P 1984 Appl. Phys. **A34** 231

Hardingham C 1994 M.Phil. Transfer Report, Imperial College

Holt D B 1969 J. Phys. Chem. Solids **30** 1297

Jakubowicz A 1986 J. Appl. Phys. **59** 2205

Kroemer H 1987 J. Cryst. Growth **81** 193

Li Y, Lazzarini L, Giling L J and Salviati G (1994) J. Appl. Phys. **76** 5748

Lohnert K and Kubalek E 1983 in Microscopy of Semiconducting Materials 1983. Conf. Series No. 67. (London: Inst. Phys.) pp. 303 - 314

Napchan E, O'Neill D and Zanotti-Fregonara C L M in Microsc. Semicond. Mater. 1993. Conf. Series No. 134 (Inst. Phys.: Bristol) pp. 693 - 698

Nauka K, Feid G A and Liliental-Weber S 1990 Appl. Phys. Lett. **56** 376

Pasemann L. and Hergert W 1986 Ultramicroscopy **19** 15

Petroff P M 1986 J. Vac. Sci. Technol.**B4** 874

Inst. Phys. Conf. Ser. No 146
Paper presented at Microsc. Semicond. Mater. Conf., Oxford, 20–23 March 1995
© *1995 IOP Publishing Ltd*

Cathodoluminescence imaging and spectroscopy of individual impurity atoms in quantum well structures

A Gustafsson, L Samuelson* and A Petersson*

Institut de Micro- et Optoélectronique, Département de Physique, Ecole Polytechnique
Fédérale de Lausanne, CH-1015 Lausanne, Switzerland
* Department of Solid State Physics, Lund University, Box 118, S-221 00 Lund, Sweden.

ABSTRACT: In GaAs/AlGaAs quantum wells (QWs) we have identified two
luminescence peaks: the intrinsic QW peak and the peak of recombination via single carbon
acceptors. Using cathodoluminescence we have located single impurities and imaged their
distribution. We have also recorded spectra, where nearly all the recombination occurs via
a single acceptor.

1. INTRODUCTION

Impurities in semiconductors are generally treated as species of identical and averaged
properties. This is a necessity, as even in the case of low doping levels (10^{15} cm^{-3}) the density
of impurities is high enough for most semiconductor structures to contain several impurities.
In most cases there is no possibility to distinguish the effect of single impurities. There is,
however, a fundamental interest in trying to isolate single impurities and to study the influence
of them. One aspect of the single impurities is their influence on the properties of low
dimensional structures. An example is, when the size of a quantum dot (QD) is such that each
QD is expected to contain, on average, one impurity. This means that the individual QDs can
contain 0, 1 or 2 impurities, and this can determine the optical as well as the electrical
properties of the QDs. This in turn can seriously influence the performance of devices, based
on low-dimensional structures. There have been a few reports of the influence of single
impurities, mainly on the electrical properties. Telegraph noise in a small MOS structure was
related to charging and de-charging of a single impurity (Ralls *et al.*, 1984). A single donor
atom affected the resonant tunneling via a QD (Nixon and Davies, 1990). Scanning tunneling
microscopy has recently been able to locate single arsenic anti-sites (Feenstra *et al.*, 1993) and
single silicon donors (Zheng *et al.*, 1994) in GaAs. In a recent article we have shown that
single impurities can be accessed by cathodoluminescence (CL), using a special sample design
(Samuelson and Gustafsson, 1995).

2. SAMPLE DESIGN

To gain access to single impurities by CL we have studied stripes of QW material, with
separation between impurities larger than the resolution (>0.5 um). These QWs have a
thickness where the effect of the impurity can be distinguished from thickness variations. Our
aim is to study single carbon acceptors (C_{As}) in a GaAs QW. C_{As} is chosen as it is the
background impurity in metalorganic vapour phase epitaxy (MOVPE) grown GaAs, i.e. it has
the lowest possible doping level. In a normal planar (100) epitaxial layer, the incorporation of

the C_{As} is ~10^{15} cm^{-3} for GaAs. This means an inter-impurity distance (D_{imp}) of ~0.1 μm. In a 5 nm thick GaAs QW, D_{imp} is ~0.5 μm and the energy separation between the free exciton and the free-to-bound transition via the C_{As} is 20-25 meV (Masselink *et al.*, 1986). At the same time the effect of the separation in energy introduced by a difference in thickness of 1 monolayer in this QW is ~5 meV. Hence, the effect of a thickness variation is easily separated form the C_{As}. Reducing the width of the QW, with stripe widths less than the 0.5 μm of D_{imp} in the QW, will further increase their spacing beyond the 0.5 μm, as discussed above.

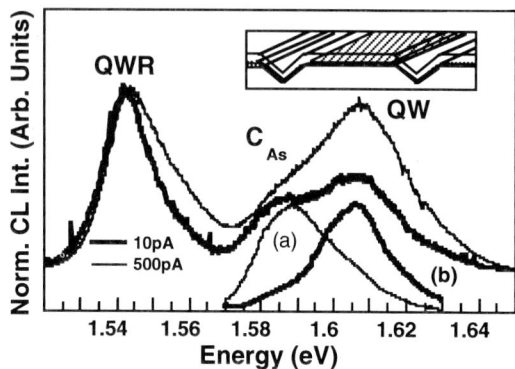

Fig. 1. Averaged spectra recorded over 5×5 μm^2 at 10 pA and at 500 pA. Spot mode spectra, recorded at (a) the site of a C_{As} and (b) in between two C_{As} sites.

These QW stripes have been realized by MOVPE growth on non-planar substrates (Kapon *et al.*, 1989) with a sub-micron period grating of V-grooves. The growth conditions used here (Vermeire *et al.*, 1992) develop two sets of planes: (100) planes in between the grooves and {111} planes on the side walls of the groove. The growth rate of GaAs is lower on the {111} planes than on the (100) planes, which means that the spectral position of the {111} QW will be on the high energy side of the planar (100) QW emission and will not interfere with the identification of the peaks of interest. At the bottom of the groove, a significantly thicker quantum wire (QWR) is formed, with a spectral position that is significantly lower in energy than the other three emissions, again no interference with the (100) QW and the C_{As} emission. Furthermore, each of the QW stripes will be surrounded by an effective diffusion barrier in the form of a high energy {111} QW on the side walls and a low energy QWR, effectively cutting communication between the QW stripes. The samples we have used in this study have either 450 nm (sample D) or 350 nm (sample B) wide (100) QW stripes of GaAs with a 5 nm thickness and surrounded by Al$_{0.35}$Ga$_{0.65}$As. The top barrier is ~20 nm thick and uncapped. All dimensions are determined from TEM images.

3. RESULTS

The CL investigations were performed at either 3 or 5 keV to ensure a high spatial resolution. The probe current measured at the sample was 30 pA unless otherwise stated. The sample temperature was 25 K throughout all experiments. Typical spectra of sample D is shown in Fig. 1 for two different probe currents. The spectra were obtained with the beam scanning over a 5×5 μm^2 area. Using a "normal" CL probe current of 500 pA, there are two main emission peaks in the spectrum. The peak of the QWR at 1.54 eV and the peak of the intrinsic QW emission at 1.61 eV, dominated by the (100) QW rather than the {111} QW. On reducing the probe current below 100 pA, a shoulder appears on the low energy side of the QW peak. At 10 pA this shoulder is well defined and positioned at ~20 meV lower energy. We attribute the low energy shoulder to free-to-bound emission, involving a C_{As}.

Fig. 2 (a) was obtained detecting the main QW peak and Fig 2. (b) detecting the low energy shoulder. These two images have one feature in common: the dark lines, separated by ~0.75 μm. These lines corresponds to the positions of the QWRs, and in Fig. 3 (a) of the

QWR emission this contrast is reversed. There is a high degree of complementary behaviour in the two images of Fig. 2: a bright spot in (a) corresponds to a dark spot in (b) and vice versa. Furthermore, we have recorded spectra in spot mode, with the spot centred on a bright spot (Fig. 1(a)) and a dark spot (Fig.1(b)) in Fig 2(b). Fig. 1(a) exhibits an almost pure C_{As} emission peak (from a single C_{As}!), whereas Fig. 1(b) only has the intrinsic QW emission peak. Sample B shows the same behaviour with a low-energy shoulder in the spectra at low excitation densities. Furthermore, the imaging also results in spots that are slightly more separated along the QW stripes, which is to be expected since the QW has the same thickness, but the QW stripe is narrower in this sample. In both these samples, the density of low-energy spots is ~35 per 5×5 μm^2, or 2.5 per μm^2 of QW area, or ~5*10^{14} cm^{-3}.

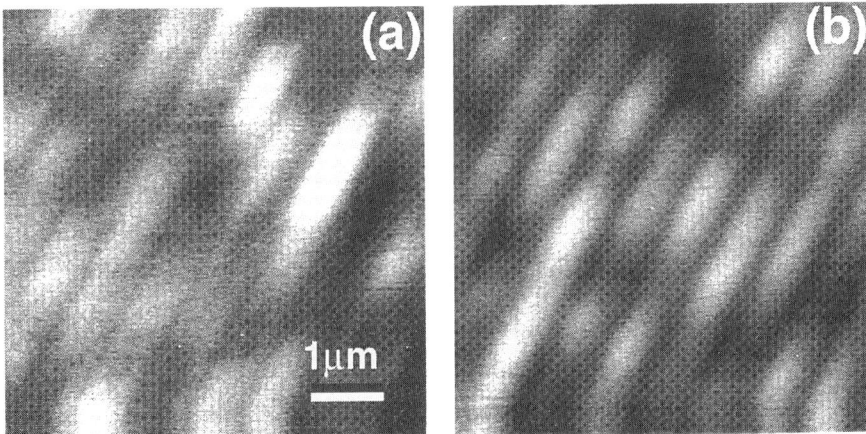

Fig. 2. CL images using the intrinsic QW emission (a) and the C_{As} emission (b)

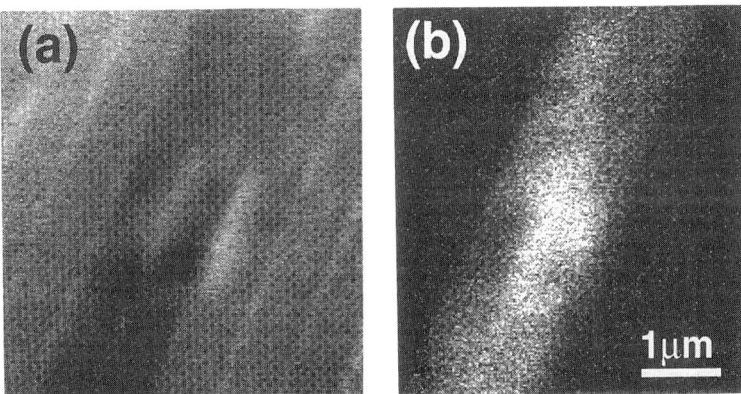

Fig. 3. CL images using the intrinsic QWR emission (a) and shifted ~20 meV down in energy (b).

To further confirm the interpretation of the low-energy peak as the emission involving a C_{As}, we have looked for the same phenomenon in the QWR as well. Fig. 3 (a) is the image of the QWR emission and Fig. 3 (b) is the image with the detection set 20 meV below the main QWR peak. It is worth pointing out that this image is recorded at 3 keV and 10 pA. Fig 3 (b) exhibits one bright spot, which shows the same complementary behaviour with the main QWR emission as Fig. 2. The shape of this bright spot is elongated along the QWR, surrounded by a region of less intensity, limited by the two adjacent QWRs. We interpret the

722

shape as the region of diffusion that feeds the single C_{As} in the QWR. The feeding along the QWR (1D exciton diffusion) is   more efficient than from the side QW and the barrier. The spectra obtained in spot mode are not as clear as the ones of the QW. In the bright areas the QWR peak still dominates, but there is a weak signal at lower energy, which only occurs in these regions, indicating that the C_{As} emission is much more sensitive to saturation in the QWR than in the QW. In fact reducing the probe current to 1 pA, which is the limit of our system, does not bring out the C_{As} emission much further. This is caused by the much higher excitation density in the QWR than in the QW.

4. DISCUSSION

The doping level in structures grown on non-planar substrates is difficult to measure. The nearest we can come is to compare with thick GaAs layers grown on planar substrates. This results in a carbon incorporation of $\sim 10^{15}$ cm^{-3}, which is close to the $5*10^{14}$ cm^{-3} we get from the density of spots in the images. Our interpretation that each spot corresponds to a single C_{As} rather than clusters is based on the fact that the charge of one C_{As} would repel the incorporation of a second C_{As} in the immediate vicinity during growth. This causes the single C_{As} to have a relatively even distribution. Furthermore, an alternative explanation would be to attribute the spots to thickness fluctuations (Herman *et al.*, 1991). We can rule out this effect as this would correspond to local variations in thickness of >1 nm, which is unlikely. There is however a slight thickness variation, mainly between adjacent QW stripes rather than along the stripes (Gustafsson *et al.*, 1995). Another point that warrants a comment is the fact that the acceptor state is a very well defined recombination path, yet the emission peak is relatively broad. This broadening could be due to the electron temperature being far higher than the lattice temperature (Grundmann *et al.*, 1994), where the recombination takes place between a thermalized hole at an acceptor and electrons with a range of energies. In summary, we have identified the emission via single C_{As} in a GaAs QW. We have imaged the distribution of, and recorded spectra from, single C_{As} in the QW. Finally, we have also imaged single C_{As} in a QWR.

We are grateful to G. Vermeire and P. Demeester for supplying the samples and to J.-O. Malm for help with the TEM investigations. This work was supported by grants from NFR, TFR and NUTEK and was performed within the "Nanometer structure consortium" in Lund.

REFERENCES

Feenstra R M, Woodall J M and Pettit G D 1993 Phys. Rev. Lett. 71 1176
Grundmann M, Christen J, Joschko M, Stier O, Bimberg D and Kapon E 1994 Semicond. Sci. Technol. 9, 1939
Gustafsson A, Samuelson L, Hessman D, Malm J-O, Vermeire G and Demeester P 1995 J. Vac. Sci. Techn. 13 (2)
Herman M A, Bimberg D and Christen J 1991 J. Appl. Phys. 70 R1
Kapon E, Hwang D M and Baht R 1989 Phys. Rev. Lett. 63 430
Nixon J A and Davies J H 1990 Phys. Rev. B 41 7929
Masselink W T, Chang Y-C, Morkoç H, Reynolds D C, Litton C W, Bajaj K K and Yu P W 1986 Solid-state Electronics 29 205
Ralls K S, Skocpol W J, Jackel L D, Howard R E, Fetter L A, Epworth R W and Tennant 1984 Phys. Rev. Lett. 52 228
Samuelson L and Gustafsson A 1995 Phys. Rev. Lett. 74 2395
Vermeire G, Yu Z Q, Vermaerke F, Buydens L, Van Daele P and Demeester P 1992 J. Crystal Growth 124 513
Zengh J F, Liu X, Newman N, Weber E R, Ogletree D G and Salmeron 1994 Phys. Rev. Lett. 72 1490

Inst. Phys. Conf. Ser. No 146
Paper presented at Microsc. Semicond. Mater. Conf., Oxford, 20–23 March 1995

The influence of hydrogen, oxygen and transition metal contamination on the optical and electrical activity of dislocations in bulk Si and SiGe

V Higgs and M Kittler[*]

Department of Physics, Kings College London, Strand WC2R 2LS, UK
* Institute Für Hableiterphysik, Walter Korsing-Str., D-153230 Frankfurt (Oder), Germany

ABSTRACT: Combined cathodoluminescence (CL) and EBIC experiments have been used to characterize dislocations in bulk Czochralski (CZ) Si and misfit dislocations in SiGe/Si epilayers after hydrogen plasma treatment. EBIC measurements showed that 90% of the shallow states associated with misfit dislocations were unaffected by hydrogen treatment whereas deeper mid-gap levels are readily passivated. Dislocations produced by plastic deformation (700^0C, 15 minutes) exhibit dislocation luminescence, after increasing the deformation time the luminescence is quenched. EBIC measurements showed that the dislocations exhibited deep level recombination. It is suggested that the oxygen-dislocation interaction produces predominately deep level states that quench the radiative recombination.

1. INTRODUCTION

The influence of transition metal impurities on the electrical and optical activity of dislocations in Si and SiGe has been studied in detail using a combination of both cathodoluminescence (CL) imaging and EBIC (Higgs and Kittler 1993). This type of study has demonstrated how transition metals can dramatically effect the recombination properties of dislocations. It was also found that the dislocation luminescence (D-bands) could not be observed in the absence of transition metal impurities (Higgs et al 1990). However, the D-bands were also not detected when non-radiative recombination was dominant due to metal related precipitates. On inspection of the spectral position of the D-bands (0.8-1.0 eV) with the band gap it is clear that these levels are relatively shallow states. These levels are in competition with other competing pathways, when the dislocation-impurity interaction introduces deep levels into the band gap, non-radiative recombination is more dominant. Dislocations can be detected more reliably using CL indirectly by measuring the dark line contrast which correlates with recombination measured by EBIC (Higgs and Kittler 1993). Detailed variable temperature EBIC contrast measurements make it possible to differentiate between recombination controlled by deep centres or recombination by shallow levels using Shockley-Read-Hall recombination statistics (Kittler and Seifert 1994). The interaction of oxygen and hydrogen with dislocations is well documented in Si but the combined interaction with transition metals impurities has not received attention. It is well known that hydrogen can passivate grain boundaries, dislocations and metal related point defects (Pearton et al 1987) and also may be incorporated during epitaxial growth. Oxygen is known to be effective in locking dislocation movement and can form dislocation donors (Yonenega and Sumino 1985). This investigation presents preliminary results concerning the interaction of hydrogen and oxygen with dislocations in bulk czochralski (CZ) Si and SiGe/Si epilayers using both CL and EBIC.

2. Experimental

CL measurements were made at T≈5 K using the CL mode of a Jeol 35 C SEM. Custom designed optics were used to collect the luminescence and direct the luminescence into a Bruker IFS 66 Fourier transform spectrometer or focus the luminescence on a Ge photodiode through a narrow band pass filter. A range of excitation conditions was used including beam energies between 5-35 keV and beam currents of 0.1 nA- 0.1 μA. EBIC measurements were performed in a Cambridge Stereoscan S 360 SEM using a Matelect amplifier and images were obtained using a Kontron image processing system. CZ Si samples were selected which contained different bulk oxygen levels ($[O]=10^{17}$-10^{19} cm^{-3}) and FZ samples were prepared as control samples to check the possibility of contamination occurring during deformation. Dislocation sources were nucleated by scratching the surface parallel to the [011] with a diamond on which a constant load was applied (0.15 N). Then the sample was deformed elastically at room temperature by cantilever bending and annealed under stress at 700^0C or 800^0C for different times (15 minutes-16 hours). SiGe alloy layers grown using both low pressure chemical vapour deposition (LPCVD) and rapid thermal chemical vapour deposition (RTCVD) were characterized by CL. These epilayers contained different oxygen levels ($[O]=10^{18}$-10^{19} cm^{-3}) as determined by secondary ion mass spectrometry (SIMS) and different densities of misfit dislocations. In addition, a selection of SiGe/Si samples grown by high temperature CVD were examined before and after intentional transition metal contamination. Hydrogen treatments were carried out in the temperature range 250-500^{0}C in a hydrogen plasma (13.56 MHz, 5-10 Torr). All the samples were surface cleaned (RCA) and placed downstream in the plasma.

3. Results and Discussion

The CL spectra recorded from the CZ Si samples ($[O]=10^{17}$-10^{19} cm^{-3}) deformed at 700^0C for 15 minutes contained the characteristic D-band features (D1-D4). As the deformation time increased (> 2 hours) the D-band luminescence features were quenched. The quenching of the D-band features was more rapid for the CZ Si samples containing more oxygen. No D-band features were observed in the CL spectra recorded from the samples that were deformed at T= 800^0C for a short deformation time (15 minutes). CL spectra recorded from the FZ control samples deformed under the same conditions contained no D-band features. D-band features were only observed in the FZ deformed samples after intentional transition metal contamination. Therefore we believe that no transition metal contamination is occurring during deformation, however we cannot eliminate the possibility that there is some trace metallic impurities already present in the CZ samples. It is not clear if the D-band features in CZ Si are produced by residual transition metal impurities already present in the starting material (as is the case in FZ Si) or that during deformation the interaction of the dislocations with oxygen or non-equilibrium point defects is a more significant factor. From our previous work with FZ Si it has been shown that the production of the D-bands is independent of type of transition metal used. Subsequent contamination of the CZ deformed samples (700^0C for 15 minutes) with Cu, Fe, or Ni increases the D-bands intensity. In the CZ samples where the D-band luminescence is quenched, transition metal contamination produced a marked increase in the non-radiative recombination processes (as measured by EBIC). We suggest that the role of interstitial transition metal atoms (in the production of the D-bands in FZ Si), is to react with the strained bonds in the dilated area of the dislocation core and kick out an interstitial atom which could become trapped in another part of the dislocation core. The trapped interstitial atom could produce the centres responsible for the D-band features, both shallow levels (from strained bonds) and deep levels due to uncoordinated dangling bonds. Similarly, in CZ Si the oxygen interstitials interact with the dislocations during deformation, and would lie in the dilated part of the dislocation, and like transition metal impurities kick out an interstitial. Both oxygen and transition metal impurities would aggregate in the dilated region, and could lead to the production of non-radiative states in the bands-gap.

Infrared absorption measurements showed that there was no detectable change in the bulk oxygen levels for all the deformation treatments carried out, TEM investigations did not reveal any changes in the dislocation structure or precipitation. In the samples where the D-band features were quenched EBIC measurements showed the dislocations exhibited deep level recombination. We believe that the effects we are observing are due to atomically dispersed oxygen atoms interacting with the dislocation core. These dislocation-oxygen interactions produce predominately deep level states and non-radiative processes dominate the recombination process and quench the radiative processes.

CL spectra were recorded from all the SiGe/Si epilayers grown by RTCVD or LPCVD methods. The majority of the epilayers contained intrinsic bound exciton recombination from the SiGe epilayer, however there were samples where no luminescence was observed. These samples contained the highest concentration of oxygen as measured by SIMS ($[O]=10^{19}$ cm^{-3}) and by measuring the intensity of luminescence centres created after electron irradiation characteristic of carbon-oxygen complexes. The SiGe layers that showed intrinsic luminescence features contained much lower oxygen levels ($<[O]=10^{18}$ cm^{-3}). CL dark line images were recorded from epilayer samples which contained intrinsic luminescence and those that showed no luminescence. CL line scans were made at fixed intervals along an individual dislocation and from 10 dislocations in the same sample. The CL contrast was determined using the standard definition $C_{CL} = (I_{CL}-I_{CLD})/ I_{CL}$, where $I_{CL}=$ is the CL intensity away from the defect and I_{CLD} is the CL intensity at the defect. The misfit dislocation contrast varied between $C_{CL}\approx10\text{-}30$ % for the layers where the intrinsic luminescence could be observed and was much higher, $C_{CL}\approx65\text{-}85$ %, in the layers exhibiting no intrinsic luminescence. These results show that the non-radiative recombination tends to increase with increasing oxygen content of the epilayer. Fig. 1 shows the CL dark line image of misfit dislocations in a SiGe epilayer sample containing high oxygen. EBIC measurements show that these samples contain mid-band gap levels controlling the recombination process.

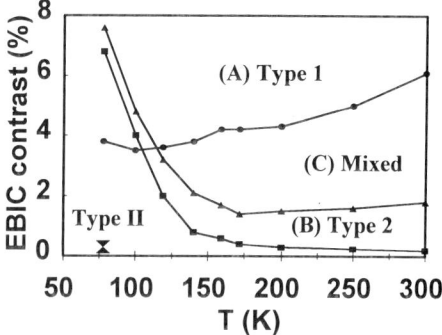

Fig. 1. CL image of misfit dislocations in SiGe/Si epilayers

Fig. 2. Temperature dependence of EBIC recombination contrast

To understand the interaction of hydrogen with misfit dislocations we selected a variety of SiGe/Si epilayer samples grown by high temperature CVD that where well characterized by EBIC and CL so we could examine the same dislocation before and after hydrogen treatment. Fig. 2 summarises the different behaviour observed from copper contaminated misfit dislocations and clean dislocations. When the activity of the dislocations is controlled by deep levels connected to impurity decoration the dislocations exhibit contrast throughout the temperature range, with contrast proportional to $T^{-0.5}$ (curve A in Fig. 2). Dislocations that show this behaviour we denote type 1. Dislocations which exhibit no room temperature (below the detection limit) contrast and show a marked increase in contrast when the temperature decreases will be denoted type 2 (curve B in Fig 2). These dislocations are controlled by shallow centres, which could be related to the elastic strain field of the dislocation or to point defects (vacancies, interstitials, and metallic impurities). Also dislocations can show both types of behaviour type 1 and type 2, and are denoted mixed character (curve C in Fig. 2). A sample was selected that contained misfit dislocations exhibiting both types of dislocation character, the dislocations were visible throughout the temperature range (T=80-300 K). After hydrogen treatment the dislocations contrast cannot be observed at T=300 K. However, on cooling the sample (T= 80 K) the dislocations can be observed. This result shows that the deep levels (type 1) are passivated but the shallow levels remain. It is not clear from our measurements, if the deep levels are passivated and the shallow levels are unaffected or that the deep levels are transformed into shallow states. A limited number of SiGe/Si samples contained dislocations which have only very small contrast at T=80 K, also no radiative recombination was observed from these layers. This type of behaviour has been denoted type II (see Fig. 2). It has been suggested that dislocations exhibiting this type of behaviour have are exhibiting the true dislocation character (intrinsic) whereas those showing other behaviour are extrinsic properties of the dislocation. However, because these epilayers have been grown in an hydrogen atmosphere it is possible that the type II dislocations are passivated. To investigate this we took samples that contained dislocations showing type II behaviour and first Cu contaminated, now the dislocations exhibited type 2 behaviour. After hydrogen treatment about 10% of the dislocations show a reduction in EBIC contrast, however they were not transformed back to type II. Figs. 3a and 3b show EBIC micrographs of misfit dislocations exhibiting type 2 behaviour (after Cu contamination) before and after hydrogen treatment. The dislocation marked with arrows in Fig. 3a shows a reduced recombination activity after hydrogen treatment (Fig 3b). The micrograph shown in Fig. 3b was not recorded at the same position as shown

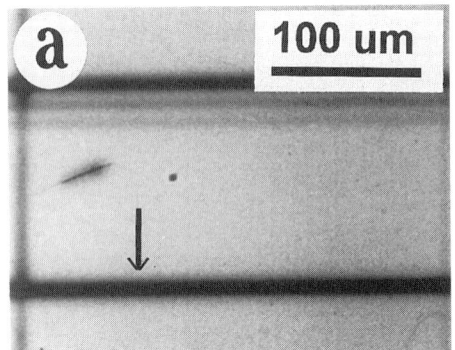 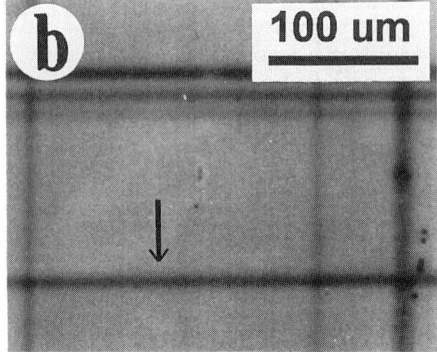

Fig. 3. EBIC micrographs of misfit dislocations a) before, b) and after hydrogen treatment.

in Fig. 3a, but they both contain the same parallel dislocation as indicated in Fig. 3a . Additional hydrogen treatments at different temperatures (T=200-400^{0}C) revealed a similar behaviour, that the majority of dislocations with type 2 character were unaffected by hydrogen. The resistance of the misfit dislocations to passivation maybe due to the nature of the shallow levels present or the role of existing impurities inhibiting passivation. Shallow levels arising from the long range strain field could not passivated by hydrogen, whereas vacancies, Si interstitials and other dangling bond defects can be passivated . Kinks and locally strained bonds in the dislocation core are expected to have shallow levels however it is not known if hydrogen can passivate these states. Post hydrogen annealing treatments showed that the original dislocation activity could be restored by annealing at 600^{0}C. Annealing the SiGe/Si sample (up to T= 900^{0}C) containing dislocations exhibiting type II behaviour had no effect on the recombination properties. This suggests that the dislocations exhibiting type II character are not hydrogen passivated. However, it must be noted that the epilayers (containing misfit dislocations) investigated in this study were grown in flowing hydrogen, and there maybe a much stronger interaction of dislocations with molecular hydrogen.

CL spectra were recorded from the SiGe/Si samples before and after hydrogen treatment. Fig. 4a shows the CL spectrum recorded from the same section as used in the EBIC measurements after Cu contamination containing dislocations with type 2 character. The spectrum contains the D-band features and the bound exciton Si substrate features. After hydrogen treatment the D1 and D2 features disappeared and there was a reduction in the intensity of D3 and D4 features (see Fig. 4b). The same effect is observed after hydrogen treatment at lower temperatures (T=200-280^{0}C). Depassivation annealing experiments showed that the D2, D3 and D4 peaks were restored to their original intensity at 400^{0}C, whereas the D1 band only reached its original intensity at 600^{0}C. Previous PL measurements and CL experiments have shown that the D1 and D2 features exhibit similar behaviour and also D3 and D4. The depassivation experiments reflect the hydrogen bonding strength to the defect centres, and these results suggest that the D1 and D2 features have different origin and D1 is not a phonon side band of D2 as has been previously suggested. The centres responsible for D1 and D2 are readily passivated indicating that these centres have unsaturated bonds, infrared measurements are underway to try to identify the Si-H bonding interaction in these dislocation related centres. As discussed previously, the origin of the D-bands is still unclear, however, preliminary experiments carried out by the authors have shown that the D1 and D2 band features can be produced without any extended defects being present, and the production of these centres depends on the complex chemistry of hydrogen, oxygen and Cu. We suggest that the D1 and D2 features may be related to Si-interstitial clusters trapped in the bulk of the deformed samples (for example at carbon) and also trapped in the dislocation strain field, which is consistent with our previously reported CL measurements. In contrast, D3 and D4 may arise from Si interstitials interacting at the dislocation core, producing new reconstruction sites or reacting with the unreconstructed parts of the core (for example at jogs).

SiGe/Si samples were further contaminated with higher concentrations of Cu, these samples were more resistant to hydrogen passivation, that is the D1 and D2 features were slightly reduced in intensity and the D3 and D4 features were unaffected. This effect could be due trapping of the hydrogen atoms at the impurity atoms before they can diffuse to the dislocation related centres.

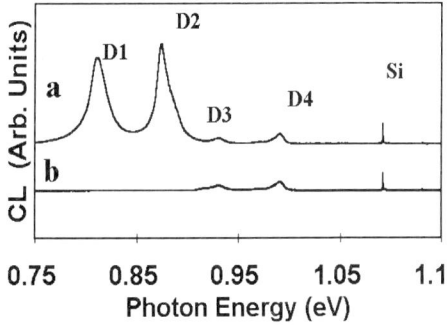 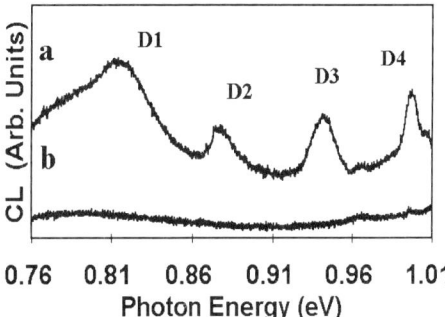

Fig. 4 CL spectra of misfit dislocations a) before, and b) after H-passivation .

Fig. 5 CL spectra of dislocations deformed in flowing a) nitrogen or b) deuterium.

To Explore the interaction of dislocations with hydrogen CZ samples were deformed in a flowing hydrogen (deuterium) gas stream at T=800^{0}C. Fig. 5a shows the CL spectrum recorded from the CZ control sample under standard conditions, very weak D-band features can be observed. Fig. 5b shows the adjacent section of the same sample deformed in hydrogen under the same conditions. No D-band features could be observed from this particular section, yet preferential defect etching revealed that there was a moderate density of dislocations present in the deformed sample ($\approx$ $10^6 cm^{-2}$). However other sections that were deformed in hydrogen did show D-band features in the as-deformed state, it was not clear why some sections deformed under the same conditions showed D-band features. Previous deformation experiments using hydrogen have shown that there is an increase in the solubility of hydrogen attributed to the presence of deformation induced defects (Kisielowski-Kemmerich and Beyer 1989). However, they found in some cases, dislocations still had significant activity even though a high concentration of hydrogen was present. They suggested that a large fraction of hydrogen trapped in the deformed material was in the molecular form. Since molecular hydrogen is electrically inactive, and it is thought that passivation requires dissociation of the hydrogen molecules, hydrogen in this form will not affect the electrical activity of the dislocations. Therefore, it is possible that in our deformed samples some proportion of the hydrogen is present in the molecular form as well as dissociated hydrogen. The CL results show that under certain conditions deformation in hydrogen can readily passivate the luminescence related defect levels. Dissociation of hydrogen may be enhanced by the presence of dislocations, increasing dissociation during surface adsorption or during pipe diffusion along the dislocation. Depassivation experiments revealed that after low temperature annealing (T=400^{0}C) the D-band features were observed from the hydrogen deformed samples. These results raise an important issue, there has been some suggestion that the effects observed in CL and EBIC were due to hydrogen passivation of the dislocation states. Our previous work (Higgs et al 1990) has shown that high temperature annealing (T>700^{0}C) had no appreciable effect on the optical and electrical activity of dislocations in Si and SiGe, therefore we suggest that hydrogen passivation is not playing a significant role in our samples. These results again suggest that when dislocations are produced in an impurity free environment they have no optical and electrical activity as measured by EBIC and CL. This is supported by theoretical studies about dislocation related levels (Marklund 1979), that suggest that the dislocation core is reconstructed without dangling bonds and therefore the electronic effects observed from dislocations are likely to be due to impurities or intrinsic impurities like jogs, vacancies or interstitials. More experiments are required to fully understand the combined interaction of impurities to elucidate the recombination processes.

4. Conclusions

The results presented here show that the interaction of dislocations with oxygen or hydrogen can modify the recombination properties of dislocations in bulk Si and SiGe epilayers. We suggest that dislocations which show no electrical or optical activity exhibit the true intrinsic properties.

ACKNOWLEDGEMENTS

We would like to thank our collaborators, Dr. G. Mariani (CNRS), Mr. R. Apetz (Inststitüt Für Ionentechnik) and Prof. G. A. Rozgonyi (North Carolina State University). We would also like to thank Prof. E. C. Lightowlers (King's College London) and Dr. R. Jones (University of Exeter) for stimulating discussions.

REFERENCES
Higgs V Lightowlers EC and Kightley P 1990 Mat. Res. Sci. Soc. Proc. 163 57
Higgs V Kittler M 1993 Inst. Phys. Conf. Ser. No. 134 703
Kisielowski-Kemmerich C and Beyer W 1989 J. Appl. phys. Vol. 66 553
Kittler M Seifert W 1994 Mat. Sci. Eng B24 78
Marklund S Phys. Stat. Solidi B92 83
Pearton SJ, Corbett JW and Shi TS 1987 Appl. Phys. A 43 153
Yonenega I and Sumino K, 1985 Dislocations in Solids Yamada Science Foundation
   Tokyo University Press 385

Inst. Phys. Conf. Ser. No 146
Paper presented at Microsc. Semicond. Mater. Conf., Oxford, 20–23 March 1995
© 1995 IOP Publishing Ltd

Cathodoluminescence and TEM investigations of stress relaxation mechanisms in $Ga_xIn_{1-x}P$/InP heterostructures

F Cléton, B Sieber, A Lefebvre, A Bensaada[1], R A Masut[1], J M Bonard[2], J D Ganière[2] and M Ambri[3]

Laboratoire de Structure et Propriétés de l'Etat Solide, URA CNRS 234, Bât. C6, Université des Sciences et Technologies de Lille, 59655 Villeneuve d'Ascq Cédex, France
[1] Groupe de Recherche en Physique et Technologie des Couches Minces, Université de Montréal, C.P. 6128, Montréal, Québec H3C 3A7, Canada
[2] I.M.O., Ecole Polytechnique Fédérale de Lausanne, CH 1015 Lausanne, Switzerland
[3] DPM, URA CNRS 172, Université de Lyon, 69622 Villeurbanne Cédex, France

ABSTRACT: Two tensile-strained $Ga_xIn_{1-x}P$ / InP (001) heterostructures (with x = 6.5% and x = 11.2%) have been studied by TEM and cathodoluminescence. In both structures stacking faults and twins, located in the epilayer, have been found preferably along the [$\bar{1}$10] direction whereas perfect misfit dislocations have been observed at the interface in the [110] and [$\bar{1}$10] directions. The two types of structures differ about the nature of the misfit dislocations and about the thickness of the twins and this is thought to be due to the different stress relaxation amounts in these structures.

1. INTRODUCTION

Up to now, the relaxation mechanisms of tensile strain in semiconducting heterostructures have not been much investigated. Previous studies dealing with the InGaAs/InP and InGaP/GaAs systems have nevertheless established that the relaxation of epilayers under tensile strain proceeds via the emission of perfect and partial dislocations (Marée et al 1987, Wagner et al 1989). In this work, the extended defects of partially relaxed biaxial tensile strain GaInP/InP structures have been identified by plan-view transmission electron microscopy (TEM). We have also used the cathodoluminescence (CL) technique to detect large scale relaxation heterogeneities as well as to measure locally the relaxation efficiency.

2. EXPERIMENTAL DETAILS

The $Ga_xIn_{1-x}P$ epilayers were grown at 640°C on (001)-oriented sulphur-doped (n = 10^{19} cm^{-3}) InP substrates using a computer-controlled cold-wall horizontal low-pressure MOCVD reactor. Prior to the GaInP deposition, an InP buffer layer was grown directly on the pre-heated surface in order to improve the crystallographic and electrical quality of the GaInP/InP interface. The GaInP and InP epilayers were doped with residual silicon impurities (n = 3.10^{14} cm^{-3}) (Bensaada et al 1992). In table I are listed the main parameters of the studied specimens :

TABLE I. Composition, lattice mismatch, relaxation level and epilayer thickness of the samples.

Samples	x (%)	$\Delta a/a$ (%)	R[$\bar{1}$10] (%)	R[110] (%)	GaInP Epilayer Thickness (μm)	InP buffer Epilayer Thickness (nm)
CE 84S	11.2	-0.80	13	43	0.55	25
CF 88S	6.5	-0.47	1.5	1.5	0.40	25

High-Resolution X-Ray Diffraction (HRXRD) and low-temperature photoluminescence were used to determine the epilayer compositions and their amount of relaxation. We have first recorded plan-view CL images on a Cambridge Stereoscan 250 MK3 scanning electron microscope (SEM), equipped with a home-made annular silicon photodiode system located above the specimen; this allowed us to obtain CL images at magnifications down to x60. On the other hand, CL spectra were recorded at 77K on a Cambridge S-360 SEM fitted with an Oxford cathodoluminescence system and a Jobin-Yvon HR250 monochromator. An accelerating voltage of 5 keV (electron penetration depth of 0.16 μm) allowed us to obtain reliable information from the uppermost layer. Thin foils have been prepared by chemical thinning in a bromide-methanol solution for plan-view TEM observations which were carried out with a Jeol 200 CX electron microscope operated at 200 keV.

3. NATURE AND DISTRIBUTION OF CRYSTALLOGRAPHIC DEFECTS

The tensile-strained GaInP epilayers we have studied present a number of common characteristics such as a higher density of extended defects lying along the [$\bar{1}$10] direction (Cléton et al 1993). However, as the two specimens exhibit amounts of relaxation ranging from 1.5% to 43% and lattice-mismatches ranging from -0.47% to -0.80%, differences are expected in the processes of relaxation.

3.1 Defects in the highly-mismatched CE84S-sample:

According to TEM observations, two types of structural defects are found: planar defects lying in the {111} planes generally extend throughout the epilayer (Fig.1) whereas misfit dislocations are confined in the InGaP/InP interface (Fig.2).

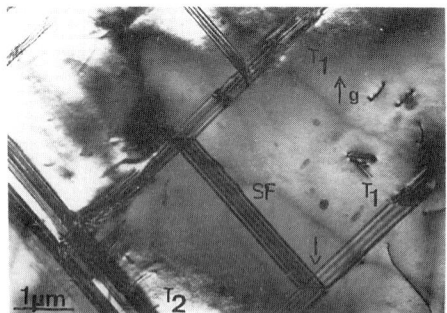
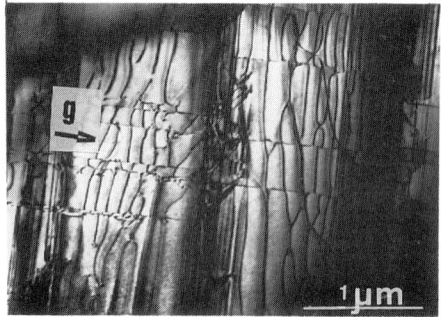

Fig.1: CE84S sample ($\Delta a/a = -0.8\%$). Microtwins and stacking faults in the epilayer; Bright-Field. $\mathbf{g} = 004$. The stair-rod dislocation resulting from a cross-slip event is arrowed.

Fig.2: Dislocation network at the interface of the CE 84S sample. Bright-field. $\mathbf{g} = 220$.

The defects parallel to [$\bar{1}$10] in Fig. 1 are T1 twins lying in the (111) and (11$\bar{1}$) planes and extending from the free surface to the InGaP/InP interface. Most defects parallel to [110] are short single intrinsic stacking faults parallel to the ($\bar{1}$11) and (1$\bar{1}$1) planes (labelled SF). A few twins are also observed in these planes (labelled T2). The partial dislocations bounding the short single stacking faults are systematically found in adjacent T1 twins. These stacking faults are thus thought to be the result of double cross-slip of partial dislocations in the T1 twins. This is illustrated in Fig. 3 in the case of a partial twinning dislocation with Burgers vector 1/6 [11$\bar{2}$] gliding into the (111) plane. The following reaction is needed for further cross-slip into the (1$\bar{1}$1) plane :

$$1/6 \, [11\bar{2}] \rightarrow 1/6 \, [\bar{1}0\bar{1}] + 1/6 \, [21\bar{1}]$$
$$(\delta C) \qquad (\delta\alpha) \qquad (\alpha C) .... (\text{Thomson notation})$$

The trailing 1/6 [$\bar{1}0\bar{1}$] stair-rod partial remains in the T$_1$ twin whereas the leading 1/6 [21$\bar{1}$] Shockley partial can glide into the (1$\bar{1}$1) cross- slip plane. A single stacking fault is created in its wake as long as it meets another T$_1$ twin and cross-slips in turn. A "ladder-like" configuration similar to those observed in Fig.1 is then obtained. In cathodoluminescence observations, these defects are imaged as highly contrasted Dark Line Defects (DLDs). Fig.4 shows that this "ladder-like" configuration is observed throughout the sample. This can be considered as a first explanation of the high relaxation amounts found by HRXRD.

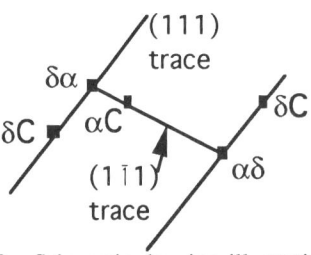

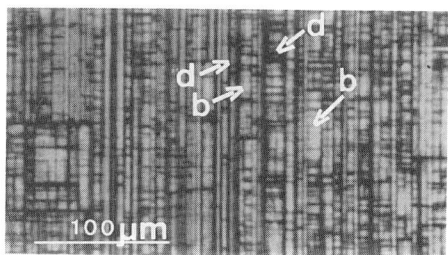

Figure 3 : Schematic drawing illustrating the mechanism of double cross-slip which results in the formation of "ladder-like" configurations. δC, $\delta \alpha$ and αC are the Burgers vectors of the partial dislocations.

Fig.4: Cathodoluminescence picture of the CE 84S sample ($\Delta a/a = -0.8\%$) obtained with the silicon photodiode device. A few bright and dark areas are labelled b and d.

The second type of defects are not connected with the first one. As shown in Fig.2, the misfit dislocations are either 1/2 [110] edge dislocations parallel to the [$\bar{1}$10] direction or 60° dislocations parallel to the [110] direction. This type of configuration is similar to those observed in GaAs/Si interfaces (Zhu and Carter 1990) and GaSb/GaAs interfaces (Kang et al 1994): the 60° dislocations interact with the edge dislocations and cause displacements of these edge dislocations. This dislocation configuration is thought to be the result of a three-dimensional growth mode for the InGaP epilayer. The dark and bright diffuse areas in CL images (Fig.4) are thought to be due to these misfit dislocations: the darker the CL contrast, the higher the density of misfit dislocations (these dislocations cannot be individually imaged by CL because their density is too high even in the brightest areas).

3.2 Defects in the low-mismatched CF88S-sample:

Stacking faults and twins are also found in this sample. As expected from the lower lattice mismatch and relaxation levels (table I), the density of these defects is much lower than in the case of the CE84S sample, even though the "ladder-like" structures can be still observed (Fig.5). It is also worth pointing out that TEM observations show that the mean twin thickness is lower in the CF88S sample, which is also consistent with the lower relaxation amount found in this sample.

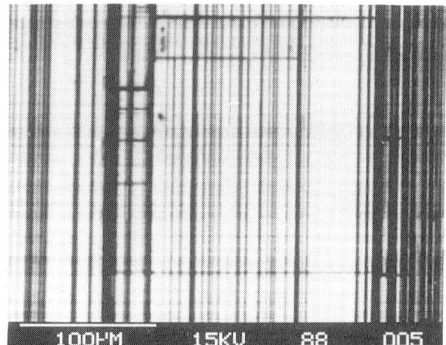

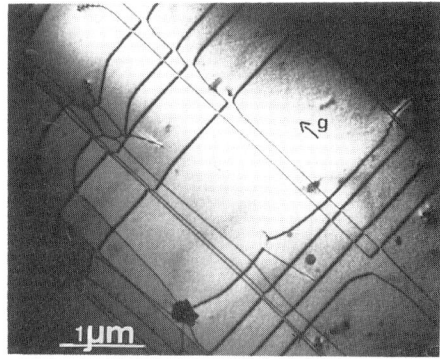

Fig.5: Cathodoluminescence picture of the CF88S sample. Notice the heterogeneous distribution of the DLDs.

Fig.6: Dislocations network at the interface of the CF 88S sample. Bright field. **g** = 220.

The second type of defects is found to be an orthogonal array of 60° misfit dislocations lying in the (001) interface plane and parallel to the [110] and [$\bar{1}$10] directions (Fig. 6). This type of dislocation configuration has been commonly observed in other systems (see for instance Herbeaux et al for the InGaAs/GaAs system and Wagner et al for the InGaAs/InP system); it is characteristic of a two-dimensional growth mode for the epilayer.

4. EVALUATION OF LOCAL STRESS RELAXATION FROM BAND-TO-BAND RADIATIVE RECOMBINATIONS IN GAINP/INP HETEROSTRUCTURES

Valence band splitting arises as a result of the stress stored in the GaInP epilayer. Radiative transitions between the conduction band and the light- and heavy-hole valence bands are then detectable, and their energetic   position is connected with the strain level as expected from the theory of Pikus and Bir:

$$\Delta E_{hh} = [-2a \frac{c_{11}-c_{12}}{c_{11}} + b \frac{c_{11}+2c_{12}}{c_{11}}] \varepsilon$$

$$\Delta E_{lh} = [-2a \frac{c_{11}-c_{12}}{c_{11}} - b \frac{c_{11}+2c_{12}}{c_{11}}] \varepsilon - (2b^2/\Delta) [\frac{c_{11}+2c_{12}}{c_{11}}]^2 \varepsilon^2$$

a is the hydrostatic deformation potential, b is the shear deformation potential for tetragonal symmetry, c_{11} and c_{12} are the elastic stiffness constants, Δ is the spin-orbit splitting between the split-off and the light/heavy holes in the unstrained bulk, and ε is the strain level (Bensaada et al 1992). Local relaxation values R are then determined with the following relation: $\varepsilon = (1-R) \Delta a/a$.

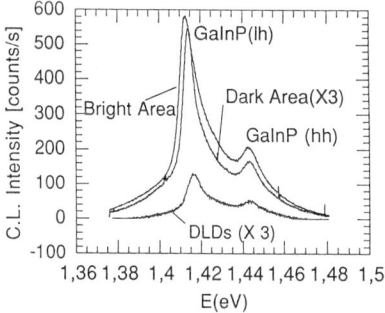

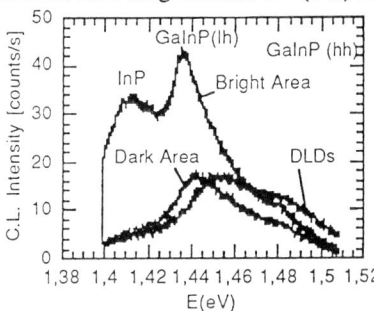

Fig.7: Typical CL spectra of the CF88S sample ($\Delta a/a = - 0.47\%$).

Fig.8: Typical CL spectra of the CE84S sample ($\Delta a/a = - 0.8\%$).

As far as the CF88S-sample is concerned, the experimental positions of the peaks correspond to high residual strain in the epilayer. R values equal to 0%, 2%, and 6% have been measured for bright areas, dark areas and DLDs respectively. The InP peak is thus hidden by the GaInP (lh) peak.

In the CE84S-sample, all the areas are notably relaxed. We have measured R values equal to 22%, 31%, and 45% for bright areas, dark areas and DLDs, respectively. These relaxation amounts are consistent with the densities of perfect interfacial dislocations and with the twin thicknesses observed in the corresponding areas.

Spatial averages of the R values all over the samples are in agreement with the macroscopic values found by HRXRD. Furthermore, CL measurements allowed us to assess the spatial heterogeneity of relaxation as well as the local radiative recombination properties.

ACKNOWLEDGEMENT

We would like to thank C. Vanmansart for technical assistance.

REFERENCES

Bensaada A, Chennouf A, Cochrane RW, Leonelli R, Cova P and Masut RA 1992 J. Appl. Phys. 71, 1737
Cléton F, Sieber B, Bensaada A, Isnard L and Masut RA 1993 Inst. Phys. Conf. Ser. No 134, 655
Herbeaux C, Di Persio J and Lefebvre A 1989 Appl. Phys. Lett. 54, 1004
Kang JM, Nouaoura M, Lassabatère L and Rocher A 1994 J. Cryst. Growth 143, 115
Marée PMJ, Barbour JC, Van der Veen JF, Kavanagh KL, Bulle-Lieuwma CWT and Viegers MPA 1987 J. Appl. Phys. 62, 4413
Pikus GE and Bir GL 1959 Sov. Phys. Sol. St. 1, 136
Wagner G, Gottschalch V, Rhan H and Paufler P 1989 phys. Stat. Sol. (a) 112, 519
Zhu JG and Carter CB 1990 Philos. Mag. 62, 319

Inst. Phys. Conf. Ser. No 146
Paper presented at Microsc. Semicond. Mater. Conf., Oxford, 20–23 March 1995
© 1995 IOP Publishing Ltd

Interface roughness of quantum wells - a scanning electron microscopy and cathodoluminescence study

U Jahn, J Menniger, R Hey, S H Kwok and K Fujiwara*

Paul-Drude-Institut für Festkörperelektronik, Hausvogteiplatz 5-7, 10117 Berlin, Germany
* Kyushu Institute of Technology, Tobata, Kitakyushu 804, Japan

ABSTRACT: The relationship between the lateral cathodoluminescence (CL) intensity variation and the interface roughness distribution is investigated in GaAs-AlGaAs single quantum wells (SQW). For roughness length scales (L_R) smaller than the spatial resolution of CL, the pattern of CL micrographs does not change significantly with L_R, but reflects the averaging procedure of CL. However, the depth of the CL intensity modulation contains information about the roughness length scale. QWs grown on vicinal GaAs surfaces contain a component of interface roughness on a large length scale, which correlates with the surface topography.

1. INTRODUCTION

The unique possibilities for imaging inhomogeneities, e.g., thickness variations in semiconductor heterostructures by combination of scanning methods and luminescence spectroscopy arise from the high spectral sensitivity of layer parameters such as thickness. Therefore, excitons can serve as probe for interface roughness of quantum wells in cathodoluminescence (CL) investigations (Christen et al 1990). If the length scale of interface roughness (L_R) is larger than the spatial resolution of CL ($L_{av} \approx 1\mu m$), CL micrographs recorded at detection energies, which correspond to different QW thicknesses, give a direct mapping of the interface structure (Bimberg et al 1987). However, this is valid only in some special cases. Usually, L_R is expected to be smaller than L_{av}. Therefore, the actually observed CL pattern results from a multiple averaging process over L_{av} including averaging due to relative motion of the electron and hole of the exciton, the scattering volume of the incident electron beam and the exciton diffusion. Consequently, in many cases a characteristic pattern with μm large irregular bright and dark regions is found in CL images of QWs, which is often attributed to the existence of large monolayer flat islands (e.g. Stützler et al 1988, Nilsson et al 1990).

In the present work, we show that even if the CL pattern is similar for different QWs, the value of the lateral CL intensity variation ($\delta I_{CL}/I_{CL}$) contains information about the length scale of interface roughness. Moreover, we discuss a reason for the occurence of long range roughness in QWs grown on vicinal (100) GaAs substrates.

2. EXPERIMENTAL

GaAs-$Al_xGa_{1-x}As$ single QWs (SQW) were grown under similar conditions by molecular beam epitaxy (MBE). The GaAs well layer is chosen to be rather thin (about 3.5nm in each case) to get a high sensitivity of the exciton transition energy with varying well thickness. In order to investigate the response of different interface roughness distribution on CL imaging we chose QWs, which differ in growth interruption time (GI) or in miscut angle of the substrate. The specific parameters of selected examples are listed in Table 1.

After growth, the QWs were analysed by photoluminescence (PL) at 4K. CL investigations were performed in a Zeiss scanning electron microscope DSM 962 equipped with an Oxford mono-CL and cooling stage system. CL spectra, spectrally resolved CL micrographs and CL line profiles were recorded at 5K. A grating monochromator and a cooled photomultiplier were used in conjunction with conventional photon counting technique to disperse and detect the CL signal, respectively.

sample	substrate miscut	GI (s)	x	PL n	PL: FWHM (for n=1)	L_{CL} (μm)	$\delta I_{CL} / I_{CL}$
I*	2° (111)A	30	0.3	1	5.5meV	1.5	0.06
II	0°	30	0.3	2		1.7	0.08
III	0°	120	0.17	4		1.8	0.26
IV*	2° (111)B	30	0.3	1	22meV	1.7	0.74

Table 1. Growth parameters and results derived from PL spectra and spectrally resolved CL line profiles of GaAs-$Al_x Ga_{1-x}$ As SQWs, where GI, n, FWHM, L_{CL} and $\delta I_{CL}/I_{CL}$ are the growth interruption time at both QW interfaces, number of single lines in the QW PL spectra, full width at half maximum in the case of one Gaussian shaped PL line, mean spacing between dark and bright regions in spectrally resolved CL images and mean lateral CL intensity variation, respectively.
(* simultaneously grown)

3. RESULTS

Results, which are derived from PL spectra and spectrally resolved CL line profiles of four SQWs are listed in Table 1. The PL (and also the CL) spectra consist in the case of sample I and IV of one line, which differ considerably in width as usually observed, when QWs are grown on (111)A and (111)B tilted substrates (e.g. Kanamoto et al 1989). The spectra of samples II and III consist of 2 and 4 lines, respectively. These single lines represent QW regions larger than the Bohr exciton radius (R_B), which differ in thickness by 1 monolayer (ML splitting).

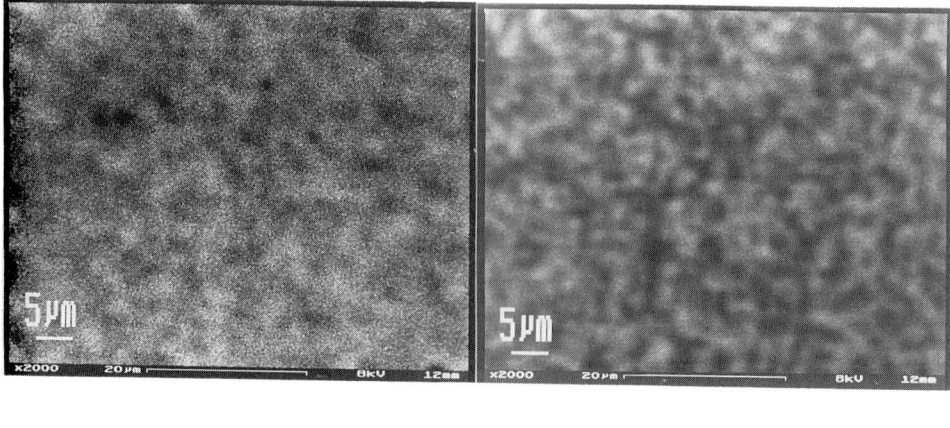

(a)   (b)

Fig. 1. Spectrally resolved CL micrographs of the samples II (a) and IV (b) of Table 1 recorded at 5K. The detection energy is chosen to correspond to the spectral position of the high energy line and the high energy flank of the Gaussian shaped PL spectrum for sample II and IV, respectively.

Fig. 1 shows CL micrographs for a fixed detection energy of the samples II (a) and IV (b). They exhibit irregular bright/dark patterns, which are also observed for the samples I and III. On the contrary to our expectation of very different interface roughness distributions related to the chosen growth conditions, the spacings between the bright and dark regions (L_{CL}) do not vary essentially among the samples (cf. Table 1). But as can be seen from the last column of Table 1, $\delta I_{CL}/I_{CL}$ varies considerably. From the statistical point of view, the CL pattern (L_{CL}) should mainly reflect the averaging procedure according to the spatial resolution of the CL method if $L_R \ll L_{av}$. However, $\delta I_{CL}/I_{CL}$ is expected to vary with $(A_{av}/A_R)^{-1/2}$, where A_{av} is the averaging area and A_R is the area of the interface islands (Runge et al 1995). Hence $\delta I_{CL}/I_{CL}$ is proportional to L_R/L_{av} and therefore contains information about the interface length scale. The nearly constant value of L_{CL} indicates that L_R is small compared to L_{av} in all four samples. Based on $\delta I_{CL}/I_{CL}$ of Table 1, L_R is approximated to be 60, 80, 260 and 740nm, when we suppose an averaging area of $A_{av} = 1\mu m^2$. The obtained tendency between the samples is consistent with the expected roughness structure related to the chosen growth conditions and to the PL and spatially resolved CL spectra, which indicate an increasing roughness length scale from I to IV. But such values have to be discussed critically, because the CL contrast is not only determined by statistical considerations but also by exciton dynamics between growth islands (Jahn et al 1995). Therefore, L_R of sample I is overestimated. It should be however smaller than $R_B \approx 15$nm because of the missing ML splitting. For sample IV we have to suppose at least two components of interface roughness, one on a nm scale (missing ML splitting) and one on a scale of the order of L_{av} (largest value of $\delta I_{CL}/I_{CL}$).

According to the fact that $\delta I_{CL}/I_{CL}$ approaches 1 in this sample, the CL pattern in Fig. 1 (b) begins to reflect the real structure of the large scale roughness. In order to demonstrate that in this case we are able to resolve QW regions, which differ in well thickness, we probed the lateral distribution of the exciton transition energy by recording CL spectra with the fixed electron beam at different positions on sample IV (spatially resolved CL spectra). The CL spectra in Fig. 2 are measured at positions with a separation of $0.5\mu m$. The shape of the spectra changes continuously from one position to the other resulting in a spectral shift between position 0 and 2.5 of nearly 12meV, which corresponds to a QW thickness variation of about 1ML.

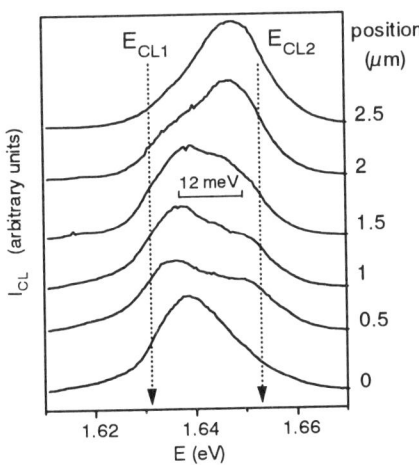

Fig. 2. CL spectra of sample IV recorded with a fixed electron beam at different excitation positions. The temperature amounted to 5K.

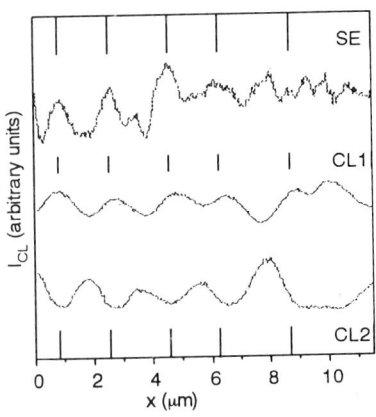

Fig. 3. SE and CL line profiles of sample IV recorded at 5K. The detection energies of the CL line profiles are marked in Fig. 2.

This fluctuation of the exciton transition energy is averaged by conventional PL and contributes to the spectral broadening compared with the simultanously grown QW No. I.

The origin of the large scale roughness of sample IV and with it the difference to sample I becomes visible by comparing the surface morphology of both specimens (not shown here). While the surface is rather smooth in case I, it exhibits a remarkable topography with mounds of several 10nm height in the case of IV. The spacing between the hills amounts to few μm and correlates with the lateral CL intensity distribution. A direct comparison between surface structure and CL pattern has been shown earlier by Menniger et al (1995) for a multiple QW structure. This comparison is also possible for the present sample IV and is given in Fig. 3. Secondary electron (SE) and CL line profiles recorded along the same line on the sample are shown. The two CL profiles differ in detection energy, which are marked as E_{CL1} and E_{CL2} in Fig. 2. They are chosen to represent low energy (wider) QW regions on the one hand and high energy (narrower) QW regions on the other hand. It is clearly seen that the CL profile CL1 correlates with the main features of the SE profile. Due to averaging effects, maxima of the SE profile closer than 1μm are not resolved any more in the CL profile (cf. positions larger than 7μm). The profile CL2 exhibits a completely complementary behavior. Since wider QW regions emit at lower photon energies than narrower ones and the maxima of the SE profile correspond to the top of the surface mounds, we conclude that the mounds contain the wider QW regions. Therefore, a MBE growth mechanism, leading to a wavy surface topography, can cause long range roughness of heterointerfaces. This example demonstrates the close relationship between surface topography and homogeneity of luminescence properties in thin heterostructures.

4. CONCLUSION

We have investigated GaAs-AlGaAs single quantum wells with different interface roughness distribution related to molecular beam epitaxy with varying growth interruption times and the choice of GaAs substrates with different miscuts. It is shown that even if the length scale of interface roughness is much smaller than the spatial resolution of the CL method, a CL pattern is observed with spacings between bright and dark regions, which are determined by the averaging procedure due to lateral resolution on the order of 1μm. Therefore, a CL pattern with structures on a μm scale can be usually expected. However, the modulation depth of the lateral CL intensity distribution varies strongly from sample to sample and contains obviously the desired information about the interface roughness length scale. For the QW grown on a substrate with an miscut angle of 2° towards (111)B we found two components of interface roughness, one on a μm scale, which correlates with the surface morphology, and one on a nm scale, indicating a fractal like interface structure.

REFERENCES

Bimberg D, Christen J, Fukunaga T, Nakashina H, Mars D E and Miller J N 1987 J. Vac. Sci. Technol. B 5 1191
Christen J and Bimberg D 1990 Phys. Rev. B 42 7213
Jahn U, Fujiwara K, Menniger J and Grahn H T 1995 J. Appl. Phys. 77 396
Kanamoto K, Fujiwara K, Tokuda Y, Tsukada N, Ishii M and Nakayama T 1989 Journ. of Crystal Growth 95 273
Nilsson S, Gustavsson A and Samuelson L 1990 Appl. Phys. Lett. 57 878
Menniger J, Kostial H, Jahn U, Hey R, Grahn H T 1995 Appl. Phys. Lett. 66 1 May
Runge E, Menniger J, Jahn U and Hey R 1995 submitted to Phys. Rev. B
Stützler F R, Fujieda S, Mizuda M and Ishida K 1988 Appl. Phys. Lett. 53 1923

Inst. Phys. Conf. Ser. No 146
Paper presented at Microsc. Semicond. Mater. Conf., Oxford, 20–23 March 1995
© 1995 IOP Publishing Ltd

Cathodoluminescence of InGaAs on patterned GaAs substrates

C E Norman [a], A R Pratt [b], R L Williams [b], M R Fahy [a], A Marinopoulou [a] and F Chatenoud [c]

[a] IRC for Semiconductor Materials, Imperial College of Science, Technology and Medicine, Prince Consort Road, London SW7 2BZ, UK.
[b] Physics Department, Imperial College of Science, Technology and Medicine, Prince Consort Road, London SW7 2BZ, UK.
[c] Institute for Microstructural Sciences and Solid State Optoelectronics Consortium, National Research Council, Ottawa, Ontario, Canada, K1A OR6

ABSTRACT: InGaAs single quantum wells (SQW) grown by MBE on patterned GaAs substrates have been characterized by cathodoluminescence (CL). Migration of indium off the pattern sidewalls during growth has been monitored as a function of the pattern size, shape, and the growth conditions. Implications for the use of the technique to fabricate closely spaced, integrated opto-electronic devices are discussed.

1. INTRODUCTION

The monolithic integration of QW-based optoelectronic components such as laser/modulator/waveguide structures is desirable as it offers minimal coupling losses between the individual component regions. Such integration requires selective area modification of the QW band gap within the individual regions. The passive waveguide section, for example, requires a much wider band gap than the modulator section, which in turn requires a slightly larger band gap than the laser section. To achieve such area selectivity using only one growth step has the advantage that it avoids complicated regrowth procedures and also circumvents post growth processing such as impurity induced or impurity free disordering, thereby avoiding the introduction of impurity atoms or point defects in significant quantities. One method of obtaining the requisite selective area band gap control during MBE growth is to use the migration of indium off A-type (predominantly gallium terminated) sidewalls of features etched into GaAs substrates. This technique was developed at IBM Zurich [Arent *et al* 1989] and subsequently used to demonstrate a working laser/modulator/waveguide combination based on a single InGaAs/GaAs QW [Brovelli *et al* 1993]. We have further refined the technique by demonstrating that the indium migration itself may be controlled by carefully choosing the pattern dimensions and the MBE growth conditions, particularly, the arsenic flux [Norman *et al* 1994] and the substrate temperature [Norman *et al* 1995]. We have also demonstrated the applicability of the technique to multilayer structures in which the *number* of active QWs can be varied from region to region, to achieve (for example) a single QW laser integrated with a three QW modulator [Pratt *et al* 1994].

Hitherto, much of the work in this field has been performed on ridge structures which, when aligned along the [1$\bar{1}$0] direction on (001) GaAs, can be wet etched to produce significant surface areas of A-type sidewall. It is, however, technologically important to assess the limits of the technique with regard to factors such as minimum/maximum feature size, feature shape and the proximity of other pattern features. To elucidate this problem we have grown nominally 8nm $In_{0.2}Ga_{0.8}As$/GaAs SQW structures under a wide range of MBE growth conditions, on a large variety of substrate patterns, such as constant width ridges, tapering width ridges, dots and hoops, plus ridges in variable proximity, and at certain angular deviations from the [1$\bar{1}$0]. All substrates in the present work were etched to a depth of 2μm with sidewall angles of 31°. To avoid confusion, features are referred to here by their masked widths, for instance, a (masked) 20μm ridge signifies a ridge defined by a 20μm photoresist stripe, but which would be reduced by subsequent etching and growth to a width of 11.75μm.

Previous work involving photoluminescence excitation and cross-sectional transmission electron microscopy measurements has confirmed that it is primarily the change in indium composition which accounts for the spectral differences between the emission from narrow ridges and unpatterned areas, since accompanying changes in the SQW thickness are very small [Norman *et al* 1994]. To spatially map the emission wavelength of the SQWs on a μm scale we use CL in the scanning electron microscope. The CL equipment used in this work comprises an Oxford Instruments Ltd. MonoCL system and CF302TC helium cryostat attached to a Leica Cambridge Ltd. S-440 SEM with an up-graded turbo + diaphragm vacuum pump system to minimize hydrocarbon contamination of the specimen.

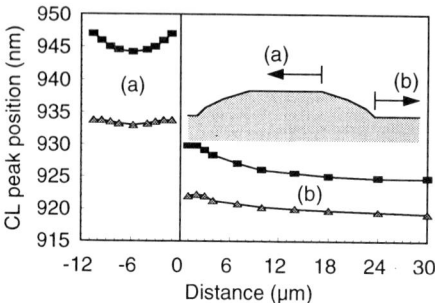

Figure 1.
The 6K SQW CL peak position as a function of distance across a (masked) 20μm ridge and the adjacent 100μm groove at 500°C (triangles) and 460°C (squares).

2. RESULTS

Figure 1 shows the variation of the 6K SQW CL peak position as a function of (a) distance across the top of a (masked) 20μm ridge, and (b) distance away from the foot of the sidewall, for two growth temperatures. The peak positions are taken from single spectra, and the profiles shown in figure 1 are repeatable along the length of the ridge. After etching, the ridge top is 11.5μm wide and, as signified by the higher emission wavelength with respect to the adjacent groove regions, appreciable pile-up of indium on the top of the ridge has occurred. The SQW grown at 460°C contains a higher concentration of indium by virtue of the lower indium desorption rate during MBE growth at lower temperatures (the same III-V fluxes were used for both growth runs), which is why the emission wavelengths are uniformly higher for the 460°C specimen. It is apparent that at the lower temperature, the indium arriving on the top of the ridge from the sidewalls has not redistributed as evenly as it has at the higher temperature, although the centre of the ridge has still received significant excess indium (approximately 2.5% more than the groove regions, according to theoretical calculations).

Figure 2 shows the 6K SQW CL emission wavelength from the centres of (masked) 20μm ridges, as a function of their misalignment with respect to the $[1\bar{1}0]$ direction (0° signifies parallel to the $[1\bar{1}0]$). It is apparent from the emission wavelengths that, at angles of $\leq 30°$, significant migration of indium onto the top of the ridges is taking place, whereas at angles $> 60°$, very little pile-up of indium is observed. The apparent maximum at 30° misorientation is somewhat misleading: at angles of 15° and 30° extra facets develop at the sidewall/ridge top intersection. These extra facets serve to reduce the width of the ridge tops from 11.5μm to 9μm ±0.5μm, thereby reducing the area into which the excess indium is incorporated by around 20%. The facetting behaviour of the ridges during MBE growth is a complex function

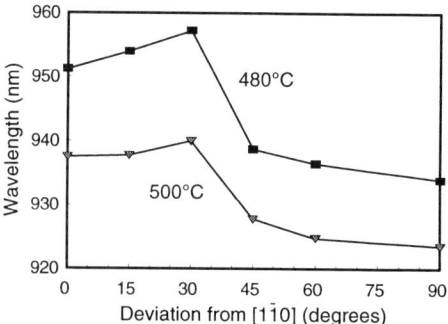

Figure 2.
The position of the 6K SQW CL peak from (masked) 20μm ridges, as a function of their misorientation away from the $[1\bar{1}0]$ direction.

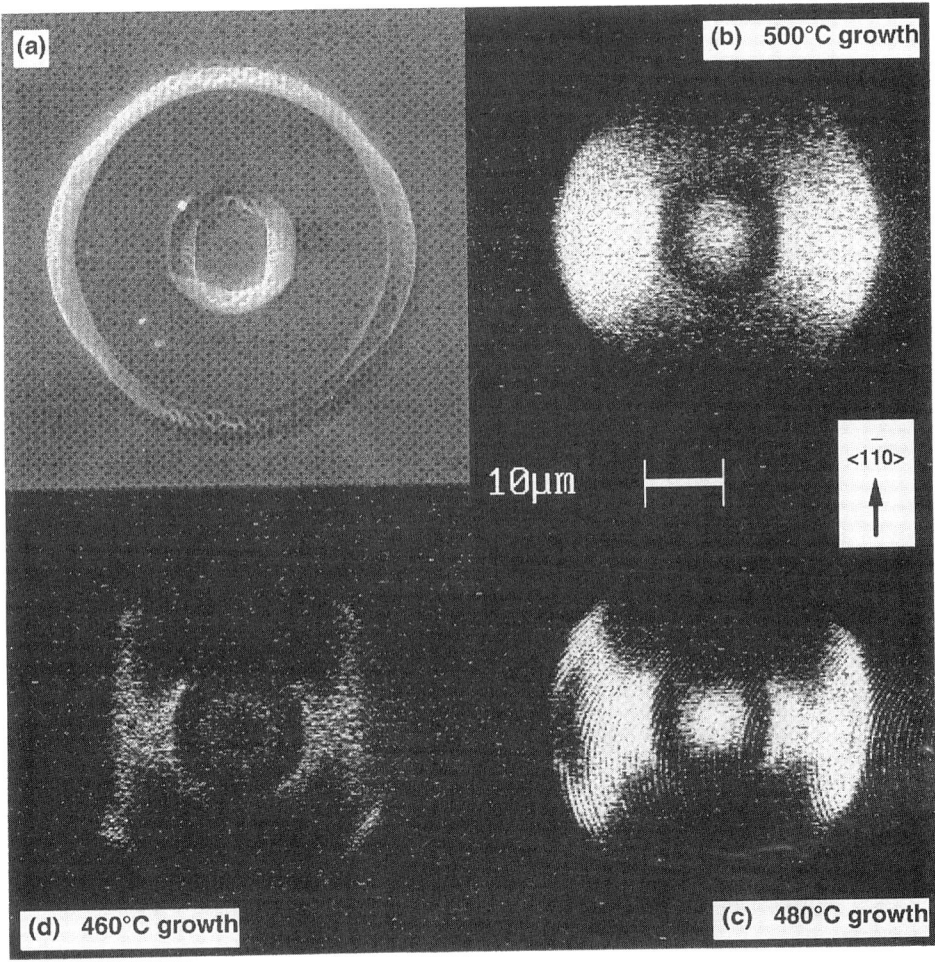

Figure 3.
Four micrographs of 50µm diameter ring structures. (a) secondary electron mode, (b) monochromatic CL (930.5nm) of 500°C SQW, (c) monochromatic CL (945nm) of 480°C SQW, (d) monochromatic CL (947nm) of 460°C SQW. CL imaging conditions: Temperature = 6K, beam energy = 10keV, beam current = 2nA, spectral bandpass = 2.5nm.

of the pattern geometry and the growth conditions, and is beyond the scope of the present work. The important observation is that significant amounts of indium still migrate off the sidewalls, even at relatively large deviations from the [1$\bar{1}$0] direction.

The results in figures 1 and 2 are clearly illustrated in figure 3 (a), (b), (c) and (d). Figure 3(a) is a secondary electron image of a 50µm diameter ring. Some facet formation during growth is evident. Monochromatic CL images recorded using the SQW peak wavelength observed from the centre of the sections of the ring aligned parallel to the [1$\bar{1}$0] direction are shown, for three different growth temperatures: (b) 500°C, (c) 480°C and (d) 460°C. The bright regions in figures 3(b),(c) and (d) therefore correspond to regions of increased indium concentration. It is apparent that as the growth temperature is reduced, the distribution of the indium around the ridge top becomes less uniform, in agreement with the results in figure 1. From figures 3(b) and (c) it is also apparent that the region of higher indium concentration extends around the ring to where the sidewalls are misaligned by some tens of degrees with respect to the [1$\bar{1}$0] direction, in agreement with the results in figure 2.

An important factor which must be considered if the technique is to be commercially viable is the effect of adjacent patterns on one another, which would limit the eventual packing density of devices. Figure 4 shows the 6K SQW CL emission wavelength from a series of proximal (masked) 10μm ridges (squares) and their adjacent grooves (triangles) as a function of the groove width. The growth temperature is 500°C. It is apparent that the ridge tops (which after etching and growth are 1.75μm wide) are unaffected by the proximity of other ridges, with the emission wavelength not varying by more than 0.25nm. The grooves do change their emission, but in an accountable way: narrower grooves accumulate more indium and shift to higher wavelengths.

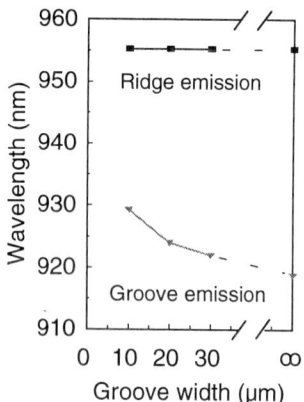

Figure 4.
The 6K SQW CL peak emission wavelength of the centres of ridges and grooves, as a function of the width of the separating grooves.

3. DISCUSSION AND CONCLUSIONS

A major problem with performing a truly systematic study of indium migration is that there are a large number of interdependent variables involved in MBE growth on patterned substrates. For instance, changing the growth temperature will not only change the indium desorption rate, but will also change factors such as indium segregation to the growing surface, and thence the SQW profile, which is not considered in this work. Furthermore, since we have previously shown that the migration process is a strong function of the growth temperature, even small temperature inhomogeneities across the sample must be avoided for results to be repeatable. Whilst a lower growth temperature has been shown to encourage indium migration off non-(001) sidewalls [Norman et al 1995], if the temperature is too low, the indium which arrives on the (001) plane is not sufficiently mobile to redistribute itself evenly over features with dimensions greater than a few μm. It is possible that this drawback may be overcome by growing at lower arsenic fluxes and/or lower overall growth rates. The absence of any proximity effects on the ridge tops after growth at 500°C is an interesting result, suggesting that once an indium atom has migrated off the sidewall, be it up onto the ridge or down into the groove, it is either not readily re-admitted to the sidewall, or it has a very low probability of crossing the entire 5μm of sidewall as part of a concentration gradient-driven process. The latter explanation might appear flawed because (i) the fact that indium leaves the sidewall in the first place shows it to be capable of movement on the sidewall, and (ii) it is then redistributed over distances of tens of μm on the (001) surface. It is possible that the facetting which develops on the sidewalls during growth is the governing factor, with the facet intersections, or even certain facets themselves, acting as kinetic barriers to surface diffusion (Arent et al 1989). Work is in progress to investigate this matter.

REFERENCES

Arent DJ, Nilsson S, Galeuchet YD, Meier HP and Walter W, Appl. Phys. Lett., 55, 2611-2613. (1989).
Brovelli LR, Germann R, Reithmaier JP, Jaeckel H, Meier HP and Melchior H, IEEE Photonics Technol. Lett., 5, 896-899. (1993).
Norman CE, Fahy MR, Marinopoulou A, Pratt AR, Williams RL and Chatenoud F, Mater. Sci and Eng B28, 299-301, (1994).
Norman CE, Pratt AR, Fahy MR, Williams RL;, Marinopoulou A and Chatenoud F, Proc. Photonics West '95, San Jose, Ca. Feb '95. (SPIE Proc. Dept. In press)
Pratt AR, Williams RL, Norman CE, Fahy MR, Marinopoulou A and Chatenoud F, Appl. Phys. Lett., 65, 1009-1011. (1994).

Inst. Phys. Conf. Ser. No 146
Paper presented at Microsc. Semicond. Mater. Conf., Oxford, 20–23 March 1995
© *1995 IOP Publishing Ltd*

Cathodoluminescence study of misfit dislocations in InGaAs/GaAs multi-quantum-wells

M Mazzer, F Ghiraldo, C Zanotti-Fregonara, D B Holt, K W J Barnham*, J Barnes*, R Grey** and J S Roberts**

Depts of Materials and Physics*, Imperial College of Science and Technology, London SW7 2AZ, G.B.
**EPSRC III-V Facility, University of Sheffield, S1 3JD, G.B.

ABSTRACT: We studied the distribution and the density of misfit dislocations in InGaAs/GaAs multi quantum well (MQW) solar cells using monochromatic cathodoluminescence (CL) imaging at 35K. Analysis of dislocation CL contrast combined with strain field calculations show that the strain generated by misfit dislocations introduces a significant modulation in the bandgap of the MQW.

1. INTRODUCTION

Generation of misfit dislocations and strain-relaxation in III-V semiconductor heterostructures is still not fully understood. All attempts to work out even a general phenomenological model have, so far, been inconclusive. This is because of the complexity of the mechanisms responsible for the generation and multiplication of misfit dislocations. Drigo et al (1989) have shown that Matthews and Blakeslee's model (Matthews 1979) is correct in predicting the critical thickness T_{1c} (i.e. the threshold for generation of misfit dislocations) and that strain relaxation above T_{1c} is almost negligible until a second critical thickness T_{2c} is reached. Strained layers whose thickness, t, exceed T_{2c} relax and the residual strain becomes a function of $1/\sqrt{t}$. On a plot of strain versus thickness the most interesting region to investigate is the one between the two critical values ($T_{1c}<T<T_{2c}$). This is a wide region within which most of the strained structures used for device application are actually situated.

2. EXPERIMENTAL

The samples studied in this paper were designed and fabricated as quantum-well solar cell devices (see Barnes et al 1994). Each sample consisted of a strained $In_xGa_{1-x}As$ MQW sandwiched between a p^+-AlGaAs/GaAs double heterostructure and an n/n^+ - GaAs double layer. The samples were grown by atmospheric pressure Metallo-Organic Vapour Phase Epitaxy and by Molecular Beam Epitaxy on (001) oriented GaAs substrates. Different combinations of In composition, x, and well and barrier thickness were studied as shown in table I. Analysis was carried out at Imperial College using a JEOL JSM-840A SEM and a spectroscopic cathodoluminescence system. The latter was fitted with a North Coast (EO-817L) Ge detector and a phase-lock amplifier (Napchan et al 1993). The samples were looked at in plan view (p+ layers on top) at liquid nitrogen and liquid helium temperatures.

3. RESULTS AND DISCUSSION

A series of plan-view CL micrographs were collected for each sample at different beam energies, E_b. Dark lines parallel to <110> directions were observed in all samples for $E_b > 6$-7 keV. Two different distributions of dark lines were found depending on whether E_b was below or above 20-25 keV. Examples are shown in Figs. 1a and 1b. The dark lines visible at low beam energies were due to misfit dislocations (MDs) located at the interface between the MQW and the GaAs cap layer (800 nm from the surface) while the others were MDs situated at the bottom interface of the MQW (~1.6 μm from the surface). This was confirmed by the results of Monte Carlo simulations (Holt et al 1994)

742

which gave a value of about 1.8 μm for the maximum depth of the e-h pair generation volume at E_b=20 keV.

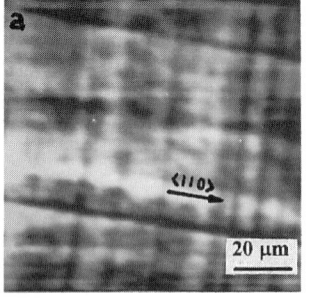 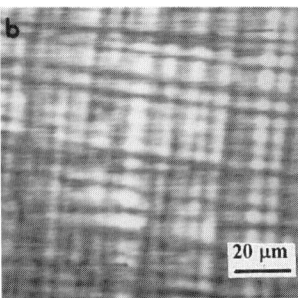

Fig.1 CL micrographs showing misfit dislocations at the top of the MQW (a) and at the bottom of the MQW (b). The lines are oriented along [110] and [1$\bar{1}$0] directions.

Sample	Dark line density		Well/Barrier thickness	Indium fraction	Number of wells	Wavelength at peak	fwhh
	$E_b \leq 15$ keV	$E_b \geq 25$ keV					
	d/μm	d/μm	(T_w/T_b) nm	(X_w)	n_w	nm	nm
QT510a	0.12 ± 0.07	~ 0.2-0.3	10/50	0.20	15	917.6†	4.3†
QT510b	0.29 ± 0.08	≥ 0.3	10/50	0.20	23	910.0	10.0
QT459b	0.02 ± 0.01	0.07 ± 0.09	11/50	0.15	10	914.9	5.8
QT459c	0.22 ± 0.07	≥ 0.3	11/30	0.15	15	916.0	5.4
QT723b	0.06 ± 0.02	0.11 ± 0.03	8/50	0.26	15	971.6	15.9
QT723c	0.09 ± 0.04	0.20 ± 0.05	8/50	0.26	15	976.0†	12.8†
RMB972	0.00	0.04 ± 0.01	4/30	0.20	15	876.3†	5.4†
RMB971	0.09 ± 0.02	~ 0.2-0.3	8/30	0.20	17	928.5	5.5
RMB1027	0.11 ± 0.03	~ 0.2-0.3	8/50	0.20	15	934.4	5.0
RMB1028	0.23 ± 0.06	≥ 0.3	8/30	0.20	23	939.8	5.4
RMB1029	0.08 ± 0.03	~ 0.2-0.3	8/50	0.20	15	933.6	4.7

Table 1. Dark line densities (i.e. average line spacing along [110] and [1$\bar{1}$0] directions), material parameters, observed emission wavelengths and full widths at half maximum (fwhh). The observations were made at 77K (except as marked †=liquid helium).

Table 1 shows the dark line densities obtained from CL micrographs of the various specimens. Since the spatial resolution of CL in our experimental conditions was of the order of 1 μm, what appears to be a single line in the micrograph is actually the effect of a number of closely spaced misfit dislocations. In fact, it is well known from TEM evidence (Salviati et al 1993) that misfit dislocations in InGaAs/GaAs tend to gather in bands as a consequence of heterogeneous nucleation and multiplication processes. This was the case in our sample as the following discussion suggests.

The overall thickness of the GaAs (top) layers, grown above the strained MQW, was about 800 nm in all samples. In order to exceed Matthews and Blakeslee's critical thickness, that is to have T_{1c}<800 nm, a strain relaxation of about 2×10^{-4} in the MQW is sufficient. The relaxation curve, on the other hand, is exceeded by the GaAs top layer when the lattice mismatch (due to the equal and opposite relaxation in the MQW) is around 2×10^{-3}. If we assume that the MQW relaxes by generating 60 degree misfit dislocations, the corresponding values for the dislocation densities at the bottom interface are expected to be between 1 and 10 lines/μm respectively. In almost all of our samples we could detect between 10^{-2} and 10^{-1} dark lines per μm at the topmost interface, this meant that the GaAs cap layer had exceeded the Matthews and Blakeslee critical thickness and we expect at least 1 dislocation per μm to be present at the bottom interface of the MQW. According to table 1 the dark line densities measured at the bottom interface (i.e. at $E_b \geq 25$keV) were of the order of 0.3 μm^{-1} at most, so we expect each dark line to be the result of an average of at least 5-10 misfit dislocations.

The strain field variation along a plane parallel to the dislocation network has an amplitude which is essentially proportional to the (average) sum of the Burgers vectors within the dislocation groups and decays rapidly with the distance from the network. This is true everywhere apart from regions which are very close to the dislocations compared with the average distance between the lines in a group.

We calculated the total strain field due to a planar distribution of groups of parallel 60 degree misfit dislocations having the same Burgers vector and equally spaced in the plane. If y is the distance from the plane, x is the in-plane coordinate perpendicular to the misfit lines, Δ is the dislocation spacing and b is the Burgers vector of a single dislocation, the strain tensor is given by a constant term, which is responsible for the average strain relaxation in the crystal, plus a periodic function of x oscillating around zero. In order to obtain the order of magnitude of the energy gap modulation due to the oscillating term we considered the trace of the tensor, that is the hydrostatic strain. The periodic part is given by:

$$\delta\varepsilon^{hy} = \frac{N_{db}\, b}{\Delta}\; \frac{1+e^Y Cos(X) - \sqrt{2}e^Y Sin(X)}{4(1+e^{2Y}+2e^Y Cos(X))} \tag{1}$$

where $X = 2\pi x/\Delta$, and $Y= 2\pi y/\Delta$ and N_{db} is the number of dislocations per band. The strain field, $\delta\varepsilon^{hy}$ obtained by taking $N_{db}=10$ and a typical value of $\Delta=3\mu m$ is plotted in fig.2 for different values of the distance, y.

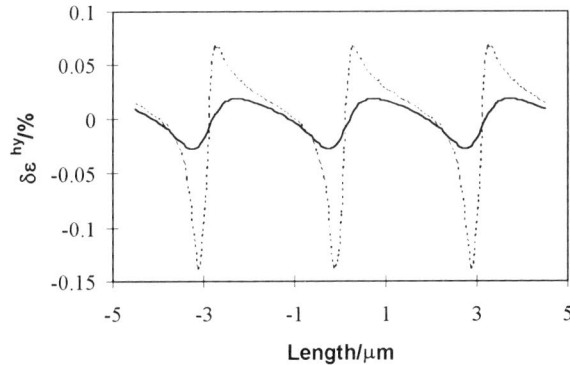

Figure 2. Hydrostatic strain field for $\Delta = 3\ \mu m$ for a band of 10 dislocations separated by 10 nm and at a distance (y) of 100 nm (dotted line) and 500 nm (solid line) from the plane containing the dislocations.

Since the total hydrostatic deformation potential for GaAs is of the order of 14 meV (Blacha 1984), the maximum variation of the band-gap associated with the strain fluctuations in fig.2 is of the order of 4 meV at y=500 nm and 20 meV at y=100 nm. This does not consider the effect of the tetragonal distortion (associated with the deformation potential b) which is responsible for the separation between the light hole and the heavy hole energy levels and decays more slowly with increasing y. These results suggest that, even at low beam energy ($E_b \sim 10$ keV), the CL intensity from the MQW at a few meV from the main peak should exhibit a detectable spatial modulation associated with the distribution of dislocations at the lower interface through their strain field.

Figure 3 shows three CL micrographs of the same area of the sample QT510a taken at E_b = 10 keV. The pancromatic image, 3(a), shows a few dark lines arising from misfit dislocations at the topmost interface. The same lines are visible also in figures 3(b) and 3(c), taken 4 meV above and below the maximum of the spectral emission respectively. These figures, however, contain additional features whose contrast has roughly the same magnitude as the original dark lines. Moreover, all and only the additional lines have complementary contrast in fig. 3(b) and 3(c). The energy range within which the contrast inversion was observed is too small to be explained by recombination events like those giving rise to the D-bands observed in SiGe (Higgs 1992) or, more recently, in InGaAs (Raisanen et al 1994), namely impurity segregation at the dislocation core or point defects left behind by the dislocation moving at the growth temperature. The D-bands are located a few tenths of an eV from the band edge, that is an energy interval between one and two orders of magnitude larger than those involved in our experiment.

The micrographs shown in figure 3, were part of a set of several images collected at different wavelengths around the emission peak. The complete sequence (not shown) showed that the contrast changed continuously, and that dark lines became bright lines passing through a situation of no contrast. We observed the same behaviour in all our samples at liquid helium temperature.

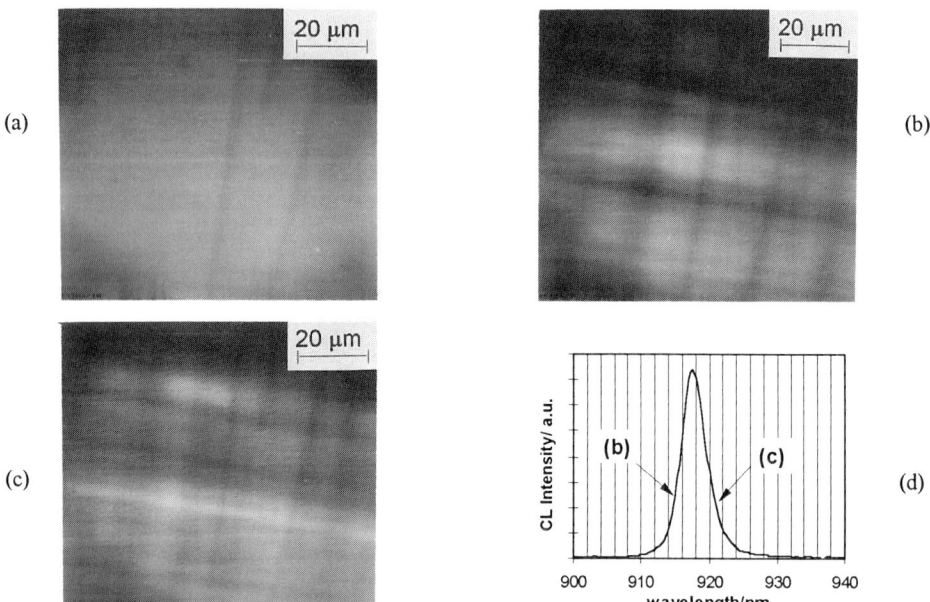

Figure 3. CL micrographs of the same area of sample QT510a taken at a temperature T=35 K and with a beam energy, $E_b = 10$ keV. (a) Pancromatic, (b) Monocromatic at λ=914.8 nm, (c) Monocromatic at λ=920.8 nm, (d) CL spectrum.

4. CONCLUSIONS

Our results have shown that besides detecting the effect of non-radiative recombination at the core of dislocations, CL is sensitive to the the strain field generated by the dislocations. The strain-induced modulation of the band-gap is controlled by the distribution of misfit dislocations. This provides the possibility to use spectroscopic CL analysis for a fully quantitative study of misfit dislocations in semiconductor heterostructures.

REFERENCES

Barnes J, Ali T, Barnham K J, Nelson J, Tsui E, Proc. 12th European Photovoltaic and Solar Energy Conf. 1994 (H S Stevens, Editor) pp 1374-1377

Blacha A, 1984 Phys. Status Solidi B126 11

Drigo A V, Aydinly A, Carnera A, Genova F, Moro L, Ferrari C, Franzosi P, Salviati G 1989 J.Appl. Phys. 66 1975

Higgs V, Lightowlers E C, Tajbakhsh S, Wright P J 1992 J. Appl. Phys. 61 1087

Holt D B, Napchan E 1994 Scanning 16 pp 78-86

Matthews J W 1979 in Dislocations in Solids (F Nabarro, Editor) (Amsterdam:North Holland)

Raisanen A, Brillson L J, Goldman R S, Kavanagh K L, Wieder H H 1994 Appl. Phys. Lett. 64 3572

Salviati G, Ferrari C, Lazzarini L, Norman C E, Bruni M R, Simeoni M G, Martelli F 1993 J.Electrochem. Soc. 140 2422

Napchan E, O'Neill D and Zanotti-Fregonara C L M 1993 in Microsc. Semicond. Mater. Conf. Series 134 (Bristol:Inst. Phys.) pp. 693 - 698

Inst. Phys. Conf. Ser. No 146
Paper presented at Microsc. Semicond. Mater. Conf., Oxford, 20–23 March 1995
© *1995 IOP Publishing Ltd*

Band-to-band recombination in N^+ InP substrate: evidence of photon recycling

F Cléton[1], B Sieber[1], L Isnart[2], R Masut[2], J-M Bonard[3] and J-D Ganière[3]

1 Laboratoire de Structure et Propriétés de l'Etat Solide, URA CNRS 234, Bâtiment C6, Université des Sciences et Technologies de Lille, 59655 Villeneuve d'Ascq Cédex, France.
2 Groupe de Recherche en Physique et Technologies de Couches Minces, Ecole Polytechnique, C.P. 6079, Montréal (Québec) H3C 3A7, Canada.
3 Institut de Micro Optoélectronique, Ecole Polytechnique Fédérale de Lausanne, Switzerland.

ABSTRACT: Quantitative nondispersive cathodoluminescence (CL) measurements and dispersive CL experiments are performed at room temperature on n^+-InP substrates as a function of the accelerating voltage (E_0). The results are explained in terms of band-to-band recombination. The red shift of the CL spectrum observed when increasing E_0, as well as the large value of the measured diffusion length with respect to the radiative one are interpreted on the basis of a high efficiency of the photon recycling process.

1. INTRODUCTION

The knowledge of the diffusion-recombination mechanisms in a homo- or heterostructure requires a detailed characterisation of the substrate. As a matter of fact, depending of its doping level with respect of that of the epilayer, the recombination properties of the substrate can play a major role in the radiative recombination properties of the structure. The work reported in this paper is the first step of a detailed characterisation of the recombination of excess carriers in n-InP/n^+-InP homojunctions. We detail here the results obtained on the substrate used in such junctions.

2. EXPERIMENTAL

A (001)-oriented InP layer was grown by low-pressure MOCVD at 630°C to a thickess of 2.5 µm on a (001)-oriented LEC InP substrate doped with sulfur to about 10^{19}/cm^3. The InP layer was n-type doped to 3.10^{14}/cm^3 by residual impurities. The polychromatic luminescence (all wavelengths) was collected in a Cambridge Stereoscan 250 MK3 scanning electron microscope (SEM), by an ellipsoïdal mirror located above the specimen, and was detected by a GaAs photocathode. The CL spectra were recorded on a Cambridge S-360 SEM also equipped with an Oxford Instruments CL system. The CL signal was focused on an optical fiber bundle connected with the entrance slit of a Jobin-Yvon HR250 monochromator, which provided a spectral resolution of 1 meV. The luminescence was detected with a Si-CCD camera. Both types of experiments were performed at 300 K, at a constant beam power of 4.10^{-5} W, in the range 5 or 10 keV to 40 keV. The luminescence was recorded either from the InP layer of directly from the substrate through a hole in the epilayer.

3. CATHODOLUMINESCENCE SPECTRA

The room-temperature luminescence in InP is associated to band-to-band (B-B) recombination (Kim and Streetman 1991). Fig. 1 shows that the peak position of the CL spectrum recorded at 10 keV shifts from 1.35 eV to 1.42 eV when the electron beam moves from the epilayer to the substrate. The 1.35 eV position corresponds to that usually found in intrinsic InP materials (Pavesi et al 1991), and the CL band shape from the epilayer is typical of undoped semiconductors (Bebb and Williams 1972).

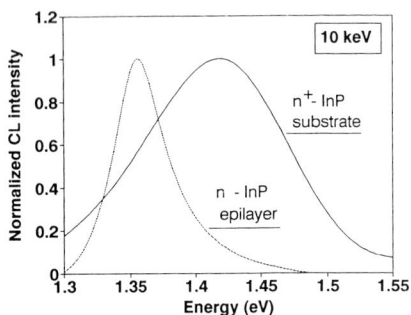 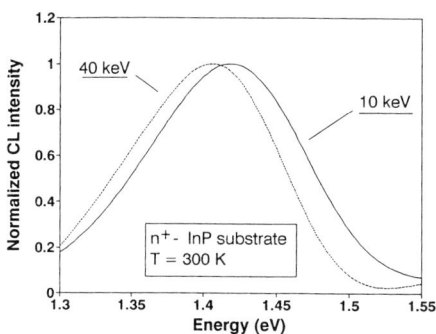

Fig.1: CL spectra recorded at 10 keV on the n-type epilayer and n⁺-substrate.

Fig.2: CL spectra recorded on n⁺InP. The CL peak is red-shifted by 12 meV by photon recycling when E₀ increases from 10 to 40 keV.

The high value of the substrate CL peak, with respect to undoped InP associated with the large full-width-half-maximum (FWHM) of its CL band gives evidence that the concentration of electrons in the substrate markedly exceeds degeneracy (Bugajski and Lewandowski 1985) which occurs at $5.10^{17}/cm^3$. In degenerate materials, the energy position E_{CL} of the CL peak can be related to the doping level by (Bugajski and Lewandowski 1985):

$$E_{CL} = E_g - E_c + E_F = E_{g0} - E_{exch} - E_c + E_F \qquad (1)$$

where E_{g0} and E_g are the energy band gaps of the nondegenerate and degenerate materials; E_{exch} is the exchange (electron-electron) interaction which produces band gap shrinkage, E_c is the Coulomb (electron -impurities) interaction and E_F is the Fermi energy in the case of non parabolic bands. Their dependence on the doping level n_0 are:

$$E_{exch} = 2.25 \ 10^{-8} \ (n_0)^{1/3} \qquad (2)$$

$$E_c = (\frac{\pi^{4/3}}{3^{1/3}} \frac{\hbar^2}{}) \ (\frac{1}{m_c})[1-1.660(3\pi^2)^{2/3}(\frac{\hbar^2}{2m_c}) \ \frac{n_o^{2/3}}{E_g}] \ n_o^{2/3} \quad (3)$$

$$E_F = 4.733 \ 10^{-14} \ (n_o)^{2/3} \ [1-2.922 \ 10^{-14} \ n_o^{2/3}] \quad (4)$$

We find a n_o value of $1.7 \ 10^{19}/cm^3$ and a reduced band gap E_g of 1.16 eV.

When E_o increases from 10 to 40 keV, the CL peak is red-shifted by 12 meV (Fig.2) and the FWHM decreases from 131 meV to 123 meV. As demonstrated in the following, the red-shift can be associated to an increase, with E_o, of the self-absorption of photons, the so-called photon recycling process (PR). The PR process, by producing an extra generation of electron–hole pairs in the material (Ahrenkiel et al 1992), can strongly modify its radiative properties. Such an interplay of generation-recombination events of electron–hole pairs induces an increase of the diffusion length and a reduced luminescence efficiency (Badescu and Landsberg 1993). The PR efficiency depends on a large number of parameters such as the surface recombination velocity, the doping level, the diffusion length L and the absorption coefficient α.

The influence of the PR process on the luminescence band comes from the variation of the absorption coefficient α with the energy $\hbar\omega$ of the emitted radiation. A larger absorption of the radiations of higher energies leads to a stronger decrease of their luminescence efficiency as compared to that of the radiations of lower energies. We simulate the resulting red-shift of the CL peak, already observed in GaAs (Ahrenkiel et al 1992), by using the correction proposed by Holt (1974):

$$I_{CL}(\hbar\omega) = (1-R) \ I_{CLo}(\hbar\omega) \ \frac{1 - \exp(-\alpha(\hbar\omega)t)}{\alpha(\hbar\omega)t} \quad (5)$$

where $I_{CL}(\hbar\omega)$ is the experimentally observed CL intensity and I_{CLo} is the CL intensity before self-absorption. t is the thickness of the material involved in the PR process; we have taken it equal 5 microns which smaller than the penetration depth of the electron beam at 40 keV. The $I_{CL}(\hbar\omega)$ curve in Fig.3 has been simulated from the 10 keV $I_{CLo}(\hbar\omega)$ curve. The $\alpha(\hbar\omega)$ function of the n^+-substrate has been derived from the 10 keV CL band by using the van Roosbroeck-Shockley relation (Bebb and Williams 1972):

$$\alpha(\hbar\omega) = I_{CLo}(\hbar\omega) \ \frac{h^3c^2}{8\pi n^2(\hbar\omega)^2} \ [\exp(\frac{\hbar\omega - \Delta E_f}{kT}) - 1] \quad (6)$$

where ΔE_f is the seperation of the quasi-Fermi levels of electrons and holes. The value of α at 1.42 eV was taken equal to 700 cm^{-1} i.e. equal to that experimentally determined by Bugajski and Lewandowski (1985) in a InP specimen doped to $10^{19}/cm^3$. The red-shift calculated with expression (5) is in agreement with that found experimentally. The FWHM of the I_{CL} calculated curve, equal to 117 meV, is also close to the experimental one.

A first important result of this work is that, due to the PR process, the doping level of the n^+-materials measured from the CL peak position depends on the value of E_o. As a matter of fact, when E_o increases, the doping level measured experimentally decreases, due to the shift of the CL peak and of the

748

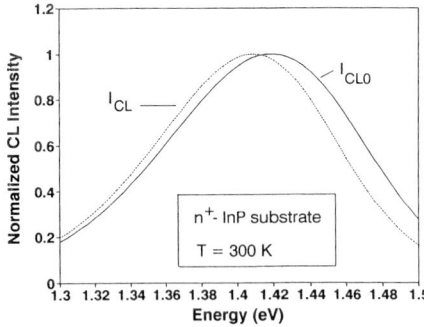

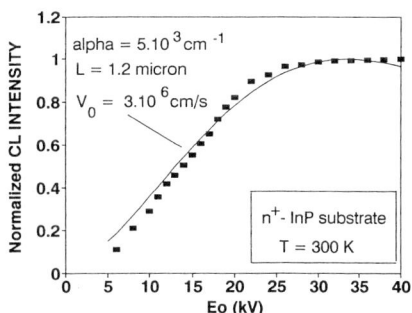

Fig.3: Calculated I_{CL} curve from the 10 keV I_{CL0}. Photon Recycling gives a red-shift of 12 meV, as measured experimentally.

Fig. 4: CL intensity versus E_0 recorded on n$^+$InP (■). The best fit is given for L=1.2 µm and $V_0 = 3.10^6$cm/s.

decrease of the FWHM. The influence of photon recycling is minimised by choosing a low E_0 value (10 keV in our experiments). This is particularly the case when the surface recombination velocity is high (Badescu and Landsberg 1993).

4. QUANTITATIVE CL MEASUREMENTS

4.1. CL intensity

The evaluation of the diffusion length and of the surface recombination velocity is made by fitting the polychromatic experimental $I_{CL}=f(E_0)$ curve. The theoretical curve is calculated by modelling the E_0 variation of the CL intensity ($I_{CL}(E_0)$) produced by carrier injection in the InP substrate. I_{CL} corresponds to the number of photons emitted per unit time; in the case of an homogeneous specimen, it is given by:

$$I_{CL} = (1-R) \int_0^\infty A(z) \, \eta \, \frac{\Delta p(z)}{\tau} \, dz \qquad (7)$$

The correction parameters of the CL system have been neglected in relation (7). $\Delta p(z)$, the minority carrier density in the substrate will be calculated in section 4.2. The factor (1-R) is the transmission coefficient through the surface of the specimen. We take the reflexion coefficient R as the ratio $(n-1)^2/(n+1)^2$, where n is the refraction index of the InP epilayer, $i.e.$ we make R independent on the angle of incidence. This relation is valid to a first approximation over $0 < \Theta < \Theta_c$ for nonnormal incidence and unpolarized light (Yacobi and Holt 1990). Θ_c is the critical angle for total internal reflexion at the free surface. A(z) accounts for optical losses in the specimen by absorption of photons, following the relation:

$$A(z) = \frac{1}{2} \int_0^{\Theta_c} \exp \left(- \frac{\alpha \, z}{\cos \Theta} \right) \sin \Theta \, d\Theta \qquad (8)$$

α, the optical absorption coefficient, is taken at the band edge equal to 5.10^3 cm^{-1}, as a result of the Burstein-Moss effect in degenerate materials. τ is the total minority carrier lifetime which accounts for both radiative and non-radiative electron-hole pair recombination $1/\tau = 1/\tau_r + 1/\tau_{nr}$. The luminescence efficiency η in relation (7) is given by the ratio τ/τ_r.

In the case of band-to-band (B-B) radiative recombination, in the standard low injection conditions ($\Delta p(z) << n_o$, n_o the free majority carrier density) the radiative lifetime τ_r is given by $\tau_r = 1/(Bn_o)$ where B is the radiative rate constant ($cm^3 s^{-1}$). Expression (7) of the CL intensity can be rewritten:

$$I_{CL} = B (1-R) n_o \int_0^\infty A(z) \Delta p(z) dz \qquad (9)$$

The experimental and calculated CL curves are normalized with respect to their maximum. The determination of the experimental parameters is made by fitting both the E_o position of the maximum and the slope(s) of the curves.

4.2. Minority carrier densities

The minority carrier density, $\Delta p(z)$, is calculated by assuming a diffusive motion of excess carriers and a Shockley-Read-Hall recombination (SRH). Thus, $\Delta p(z)$ obeys the steady-state continuity equation

$$D \frac{d^2 \Delta p(z)}{dz^2} - \frac{\Delta p(z)}{\tau} = - g(z) \qquad (10)$$

$g(z)$ is the depth-dose function (Akamatsu et al 1989). The solution $\Delta p(z)$ can be written in the form:

$$\Delta p(z) = A \exp(-z/L) + A' \exp(z/L) + \int_{e_{i-1}}^{e_i} \chi(z,z') \frac{g(z')}{D} dz' \qquad (11)$$

where $\chi(z,z')$ is the Green function of (10):

$$\chi(z,z') = \frac{L}{2} \exp(-\frac{|z-z'|}{L}) \qquad (12)$$

L and D are respectively the diffusion lengths and diffusion coefficients. The constants A and A' in expression (11) are determined by the boundary conditions:

$$D \frac{\Delta p(z=0)}{dz} = V_o \Delta p(z=0) \qquad and \qquad \Delta p(z= \infty) = 0 \qquad (13)$$

V_o is the surface recombination velocity. The analytical expression of A' is easily calculated and is not detailed here. A is equal to zero.

4.3. Electrical parameters

The $I_{CL}(E_0)$ calculations have been made with an average value of α of 5.10^3 cm^{-1}. The fit of a typical curve is displayed in Fig. 4.

The large value of V_0 could result from the presence of recombination states induced by substrate preparation and exposure to contaminants during growth. The radiative diffusion length can be estimated when the value of B, the radiative constant is known. In first approximation, we assume that B is independent of doping, and we use the Lasher and Stern expression (1964) which is valid for non-degenerate materials:

$$B = \frac{e^2\ n\ h\ (2\pi)^{1/2}}{12\ \pi\ \varepsilon_0\ c^3\ k^{3/2}} \times \frac{E_g^2\ (E_g + \Delta)}{E_g + 2\Delta/3} \times T^{-3/2}$$

$$* \left\{ \frac{[m_{hh}/(m_c+m_{hh})]^{3/2} + [m_{lh}/(m_c+m_{lh})]^{3/2}}{m_c(m_{hh}^{3/2} + m_{lh}^{3/2})} \right\} \tag{14}$$

E_g is the energy bandgap, n the refractive index; e, h, ε_0 and k have their usual meaning and T is the temperature. m_c, m_{hh} and m_{lh} are respectively the effective masses of electrons, of heavy and light holes. Δ is the split-orbit splitting between the split-off and light/heavy holes in an unstrained material. At 300 K, m_c=0.077m_0, m_{lh}=0.089m_0, m_{hh}=0.72m_0 (Bugajski and Lewandowski 1985), E_g=1.347 eV and Δ=0.108 eV gives a B_{InP} value of 9.5 10^{-11} cm^{3}s^{-1}. B would be smaller in n$^+$-InP since it decreases with doping level (Casey and Stern 1976). Thus, the minimum value of L_{rad} is 0.4 μm for a hole mobility of 100 cm^2/Vs (Iseler 1991). Without taking into account the PR process, a value of L_{rad} three times higher than that found experimentally in the n$^+$-substrate, would correspond to a B value nine times smaller than that calculated previously. Therefore, it seems more reasonnable to invoke, coherently with the results of the spectroscopic analysis, that photon recycling plays a major role in the radiative recombination properties of the n$^+$-substrate.

REFERENCES

Ahrenkiel RK 1991 in *Properties of Indium Phosphide*, INSPEC, 77

Ahrenkiel RK, Keyes BM, Lush GB, Melloch MR, Lundstrom MS and MacMillan HF 1992 J. Vac. Sci. Technol. A 10 990

Akamatsu B, Henoc P and Martins R.B. 1989 J. Microsc. Spectrosc. Electron. 14 12a

Bebb HB and Williams EW 1972 in *Semiconductors and Semi-metals 8*, 'Transport and Optical Phenomena', Eds. Willardson RK and Beer AC, Academic Press.

Bugajski M and Lewandowski W 1985 J. Appl. Phys. 57 521

Casey HC and Stern F 1976 J. Appl. Phys. 47 631

Holt DB 1974 in *Quantitative Scanning Electron Microscopy* Eds Holt DB, Muir MD, Grant PR and Boswarva IM, Acdemic Press

Iseler GW 1991 in *Properties of Indium Phosphide*, INSPEC, 25

Kim TS and Streetman BG 1991 in *Properties of Indium Phosphide*, INSPEC, 165

Lasher G and Stern F 1964 Phys. Rev. 133 A553

Pavesi L, Piazza F, Rudra A, Carlin JF and Ilegems M 1991 Phys. Rev. B 44 9052

Inst. Phys. Conf. Ser. No 146
Paper presented at Microsc. Semicond. Mater. Conf., Oxford, 20–23 March 1995

Cathodoluminescence studies of MOCVD grown, e-beam pumped, II-VI laser structures

G M Williams, A G Cullis, B Cockayne, P J Wright and P C Smith

Defence Research Agency, St Andrews Road, Malvern, Worcs. WR14 3PS, UK

ABSTRACT: Single and multiple quantum well structures composed of CdSe wells and ZnSe barriers have been grown onto GaAs (001) substrates, using Metal Organic Chemical Vapour Deposition, for the production of laser chips which will emit in the green / blue-green part of the electromagnetic spectrum. A purpose built electron-beam excitation system has been developed with the ability to excite luminescence in the sample and monitor its wavelength and intensity from the edge of the laser cavity. Low temperature (4K) scanning cathodoluminescence has been used to optimise the materials growth and structural parameters for laser applications. This paper presents results from this study which has allowed the determination of the optimum thickness and growth parameters for the CdSe well and the ZnSe buffer layers. Atomic scale non-uniformities of the quantum well thickness, on a lateral scale of 10 to 20μm, have been observed using monochromatic cathodoluminescence imaging. Evidence of sample lasing is presented.

1. INTRODUCTION

The II-VI semiconducting compounds have for many years been regarded as promising materials for the production of light emitting devices in the visible region of the electromagnetic spectrum. Laser structures and light emitting diodes can be stimulated by three methods: photo-pumping generally with nitrogen lasers; electron-beam pumping using an electron gun and focusing optics; and P-N junction diode pumping - as a consequence of recently improved doping control.

Our research is directed towards visible light emitting devices produced from CdSe/ZnSe based structures, in particular it is hoped to develop quantum well lasers pumped using an electron beam. Electrically pumped ZnSe based lasers on GaAs emitting at ~490nm have been reported fairly recently (Haase et al 1991) and with improved doping and defect control these devices will prove an attractive proposition for applications such as data storage. Electron-beam pumped lasers are, however, intrinsically simpler devices which eliminate problems such as contacting to the material. Possible applications for electron-beam pumped lasers include large area projection TV (Basov 1981) where high powers are required, together with holographic data readout (Kozlovskii et al 1980) and scanning optical microscopy (Bogdankevich et al 1980). Fast continuous tunability of the output wavelength is also a potential advantage with compositionally graded structures and the ability to selectively excite different depths of the structure by varying the electron beam voltage.

In this study we have used low temperature (4K) scanning cathodoluminescence (CL) to examine the effects of various growth parameter changes on the optical uniformity of single CdSe/ZnSe quantum well structures. Evidence of lasing from these structures, under electron beam excitation, will be presented.

2. EXPERIMENTAL METHODS

The CdSe quantum well / ZnSe buffer layer structures were deposited on (001) semi-insulating GaAs substrates at temperatures in the range 275°C to 500°C. The growth technique of Metal Organic Chemical Vapour Deposition used the reagent dimethylzinc-trithylamine adduct, dimethyl cadmium and hydrogen sulphide with purified hydrogen acting as the carrier gas. More details of the growth can be found elsewhere (Parbrook et al 1990).

The scanning CL system used to study these layers is described in detail elsewhere (Williams et al 1991). In brief, however, it is based on a Cambridge Stereoscan 150 Mk2 SEM with an LaB_6 electron source. The samples were cooled to 4K and the luminescence collected by an Oxford Instruments MonoCL 2 system. The electron-beam excitation system is a purpose built novel unit which can produce beams at voltages up to 30kV and with currents up to 1mA. The laser sample can be cooled to any temperature down to approximately 20K.

3. RESULTS AND DISCUSSION

3.1 Effects of growth parameter changes

A series of layers, each with the same structure (1μm ZnSe buffer layer, a CdSe quantum well and a 0.5μm ZnSe cap), were grown on GaAs substrates. In each case the CdSe quantum well was 0.3nm thick and the only parameter varied from run to run was the substrate temperature. It is clear from Fig. 1, which plots the quantum well CL peak intensity against growth temperature, that the optimum material is grown around 350°C. It is believed that this optimum regime is a result of balancing the improved surface mobility gained at higher temperatures against the generation of point defects, which increase as the growth temperature is increased.

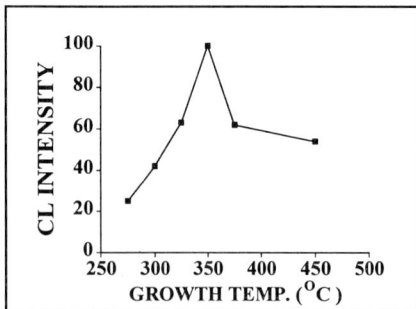

Fig. 1. CL peak intensity vrs growth temperature for a series of identical CdSe/ZnSe quantum well structures

A further series of layers were examined all of which were grown at this optimum temperature. Each of the structures had a different thickness CdSe quantum well, the thinnest being 0.3nm and the thickest 1.2nm. The data plotted in the graphs in Fig 2(a) and (b) demonstrate that the quantum well CL peak intensity decreases as the CdSe thickness increases and that simultaneously the peak FWHM increases. These effects are, we believe, caused by an increase in the number of defects as the quantum well approaches the critical thickness for CdSe on ZnSe.

Growth at lower temperatures (ie <350°C) not only resulted in a poorer CL efficiency from the CdSe quantum well but also the observation of more than one quantum well CL peak. Figure 3 shows a typical spectrum from a single quantum well structure grown at 275°C. It is obvious that the quantum well emission is present at discrete wavelengths of 495nm and 512nm. The two monochromatic images also shown in Fig. 3 are both from the same area of material. The diagonal dark band is due to a scratch on the layer and serves as a location reference. It is clear, from comparing the two images, in particular areas 'X', 'Y' and 'Z', that the two emissions originate from discrete regions $\sim 10\mu$m in dimension. This effect is thought to be due to the growth of two groups of CdSe islands, the groups being differentiated by thickness. We did not observe similar islands in layers grown at higher temperatures and hence conclude that the thickness variation results from the reduced surface mobility of the growth species at the lower temperatures. The thicknesses of the two regions were calculated to differ by one monolayer (O'Donnell et al 1992).

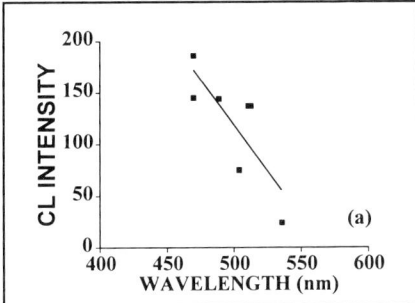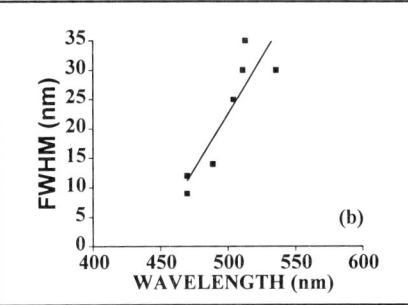

Fig. 2. A series of single CdSe quantum well structures were grown under identical conditions. Each structure had a different thickness well. (a) shows a decrease in the CL intensity with increasing well thickness and (b) an increase in the peaks full width at half maximum with increasing thickness.

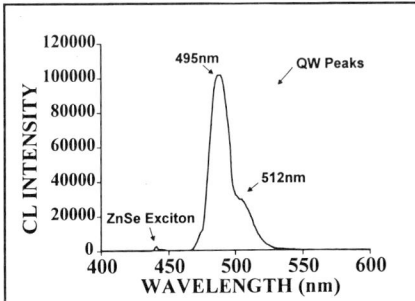

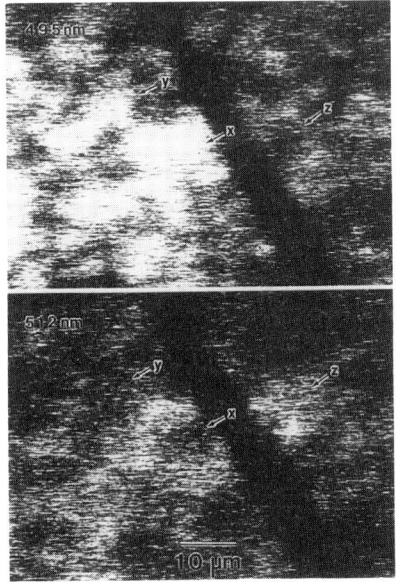

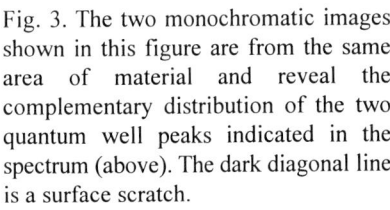

Fig. 3. The two monochromatic images shown in this figure are from the same area of material and reveal the complementary distribution of the two quantum well peaks indicated in the spectrum (above). The dark diagonal line is a surface scratch.

3.2 Lasing observations

Samples grown under optimum conditions were cleaved into 1mm wide 3mm long test specimens for electron beam pumping experiments. No coatings were applied and the cleaved faces were relied upon to form the cavity for laser oscillation. The samples were cooled to 20K and excited using a continuous electron beam of 20kV. The beam current was varied to increase the input power into the sample and the output luminescence intensity was monitored by acquiring a spectrum for each increment in power. Despite the non-optimisation of these samples we observed laser action from a large number of them as demonstrated by the data shown in Fig. 4 below. A threshold curve for a specimen containing a single CdSe quantum well ~0.3nm thick is shown in Fig. 4(a). It can also be seen from Fig. 4(b) that as the luminescence intensity of the sample increased the line width narrowed. The luminescence was also found to have a high degree of optical polarisation, which is an indication of a coherent beam, and taken together the data confirm that lasing action occurred.

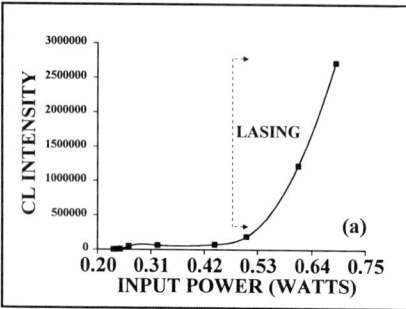 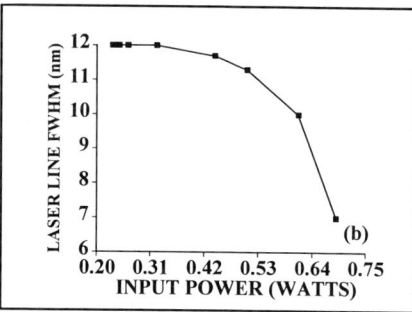

Fig. 4. The threshold curve for a single CdSe quantum well laser structure is shown in (a) and in (b) we see the way in which the line width narrows above threshold.

4. CONCLUSIONS

In this paper we have demonstrated some of the applications of low temperature scanning cathodoluminescence as used to optimise II-VI based epitaxial laser structures. Variations in the quantum well thickness on an atomic scale have been deduced. This work, together with other CL data, reveal how critical the growth parameters can be in the production of material of high optical quality which can subsequently be used to produce working laser devices.

REFERENCES

Basov N G 1981 in: Proc. Intern Conf. on Lasers p.3
Bogdankevich O D et al., 1980 IEEE J. Quantum Electron QE-16 129
Haase M A, Qiu J, DePuydt J M and Cheng H 1991 Appl. Phys. Lett 59 1272
Kozlovskii V I et al., 1980 Soviet J. Quantum Electron. 10 917
O'Donnell K P and Henderson B 1992 J Luminescence 52 133
Parbrook P J, Wright P J, Cockayne B, Cullis A G, Henderson B and O'Donnell K P 1990 J Crystal Growth 106 503-509
Williams G M, Cullis A G, Sotomayor-Torres C M, Thoms S, Beaumont S P, Stanley C R, Lootens D and Van Daele P 1991 Inst. Conf. Ser. 117 695

Inst. Phys. Conf. Ser. No 146
Paper presented at Microsc. Semicond. Mater. Conf., Oxford, 20–23 March 1995
© 1995 IOP Publishing Ltd

Cathodoluminescence and transmission electron microscopy investigation of ZnCdSe/ZnSe electron-beam-pumped laser structures

J-M Bonard, J-D Ganière, L Vanzetti[1], L Sorba[1], A Franciosi[1], D Hervé[2] and E Molva[2]

Institut de Micro- et Optoélectronique, Département de Physique, Ecole Polytechnique Fédérale de Lausanne, CH-1015 Lausanne, Switzerland

[1] Laboratorio TASC-INFM, Area di Ricerca, Padriciano 99, I-34012 Trieste, Italy

[2] LETI (CEA-Technologies Avancées) Département Optronique, 17 rue des Martyrs, F-38054 Grenoble Cédex 9, France

ABSTRACT: We present results obtained both with scanning electron microscopy (SEM) in the cathodoluminescence (CL) mode and transmission electron microscopy (TEM) on ZnCdSe/ZnSe electron-beam-pumped laser structures. This investigation aims at establishing a relationship between different parameters measurable in electron microscopy and the lasing activity of the structures.

1. INTRODUCTION

The realisation of coherent and incoherent emitters based on wide-gap II-VI semiconductors has been achieved only recently, although the first studies of these materials date back to the early 1960s. Among the main obstacles to the development of viable devices are the difficulty of growing II-VI layers of sufficient crystalline quality on the available III-V substrates, self-compensation and/or dopant migration during p-type doping of II-VI materials, and the high specific contact resistance to p-type II-VI epilayers. The last two problems can be circumvented - at least in principle - by using a high voltage (5-15kV) electron beam for the injection, instead of the traditional p-n junction structure. Compact electron-beam-pumped ZnCdSe/ZnSe blue and blue-green lasers operating up to 220K were recently demonstrated (Hervé et al 1995). The devices use graded-index, separate confinement heterostructures (GRINSCH) in combination with a lithographically patterned microtip cathode, acting as a miniaturized electron source for injection.

2. EXPERIMENTAL DETAILS

All laser structures are grown by molecular beam epitaxy on $In_{0.01}Ga_{0.99}As$ (001) wafers. $In_{0.04}Ga_{0.96}As$ and ZnSe buffer layers are grown sequentially on the substrate prior to the fabrication of the GRINSCH. The GRINSCH itself includes a $Zn_{1-x}Cd_xSe$ quantum well (QW) embedded between two 500nm thick $Zn_{1-x}Cd_xSe$ graded layers, where x varies continuously from x=0.05 (at well boundaries) to x=0 (at the surface and at the buffer). Typical well thickness examined are in the 50-100Å range, typical Cd concentration in the well are in the 0.15<x<0.25 range.

Plan-views, as well as cross-sections, are prepared of all structures by mechanical thinning down to a thickness of 50µm followed by 5keV-Ar$^+$ ion bombardment in a Gatan PIPS (Precision Ion Polishing System). All TEM observations are carried out on a Philips EM430ST at 300keV.

The CL measurements are done on a Cambridge S-360 SEM, with a modified stage (Oxford Instruments 302) allowing observations down to a temperature of 10K. The sample's CL is collected by an ellipsoidal mirror and focused either on a Si-photodiode for polychromatic imaging and measurements or through an optical fiber to a Jobin-Yvon HR250 monochromator equipped with a Si-CCD camera for spectral acquisition. The theoretical spectral resolution is better than 1Å with 0.2mm entrance slits and the 1200 lines/mm holographic grating blazed at 500nm.

3. TRANSMISSION ELECTRON MICROSCOPY

The TEM observation of cross-sections of the four examined structures (see the table in section 4.2 for details) shows a number of surprising features, as we can see on Fig. 1.

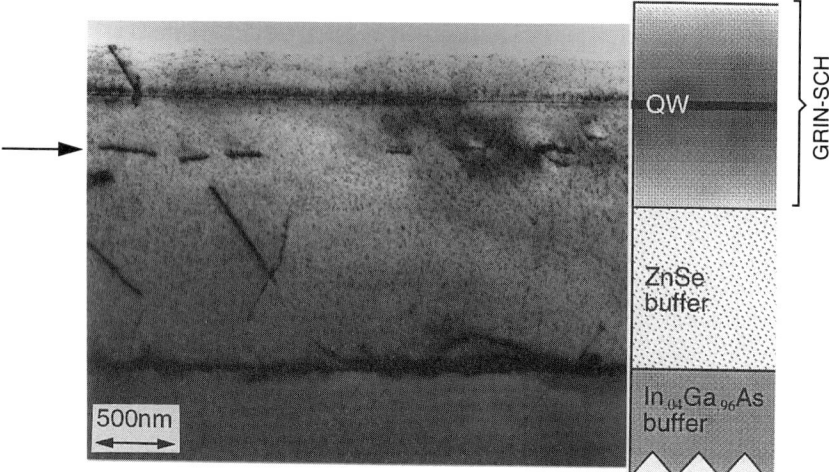

Fig. 1   Cross-sectional view in bright field of sample 294 near the [110] zone axis, with a scaled scheme of the structure.

First of all, there is a very high density of misfit defects (very probably stacking faults bounded by partial dislocations) originating at the substrate-InGaAs buffer interface, which means that the buffer layer is relaxed. However, only a small number (<5%) of them propagates beyond the InGaAs-ZnSe interface. The blockage of the misfit defects gives raise to a high density of dislocations in the interface plane, as suggested by the contrasts around the interface. Some more work is needed to understand this unexpected behaviour, but it is clear from Fig. 1 that it allows the growth of subsequent high quality II-VI layers. The quantum well is readily visible at the top of the micrograph, as are dislocations in the first half of the GRIN-SCH (indicated by an arrow).

An analysis in plan-view, shown in Fig. 2, reveals that there is an array of dislocations running approximately along the [100] and [010] directions.

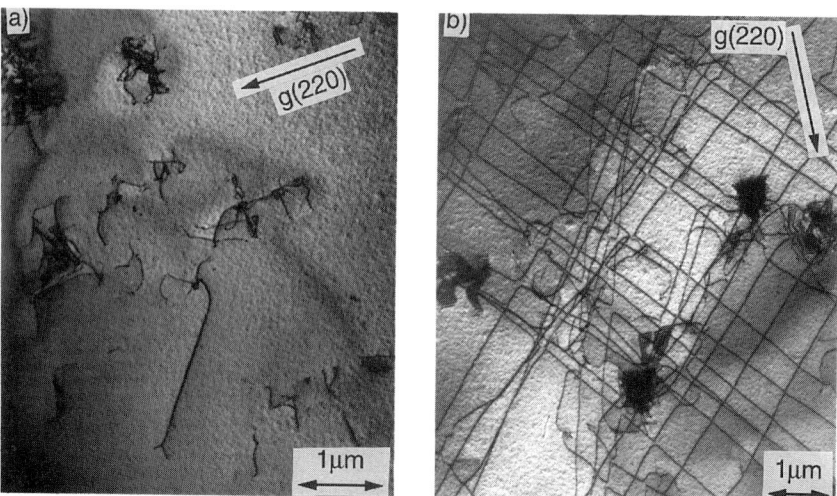

Fig. 2   Plan-views of sample 292, bright field in two-wave excitation near the [001] zone axis; foil thicknesses are (a) 500nm and (b) 1μm.

The foil thicknesses vary between Fig. 2a and Fig. 2b from 500nm to 1μm, as determined by the extent of the stacking fault contrast (which show vanishing contrast in the diffraction conditions of Fig. 2a). We can conclude that the array is located in the first half of the GRINSCH, and that the dislocations observed in cross-section ~350nm below the QW (indicated by an arrow in Fig. 1) are part of this array. They are typical of misfit relaxation induced by compressive strain (Jouneau 1993 and references therein), which is present in our structures since the lattice parameter of ZnCdSe increases with increasing Cd content.

The plan-views show furthermore the presence of stacking faults, threading dislocations, and pyramidal growth (the last on two samples only). It is also interesting to note that dissociated 60° dislocations (Chen et al. 1994) are present everywhere in the II-VI layers, whereas none has been detected in the InGaAs layers.

4. CATHODOLUMINESCENCE PROPERTIES

4.1. Spectrally resolved CL

The CL of the four samples was studied at various temperatures between 10K and 300K. All examined structures show strong cathodoluminescence (with typical beam powers of 2×10^{-5} W) up to 300K. In order to allow direct comparison between the samples and to reduce degradation effects (Bonard et al 1995), the spectra are acquired in scan mode at 5000x magnification (scanned surface of $384\mu m^2$) at TV- rate. Under these conditions and at 20keV, the required current to match the typical threshold power for this kind of structures ($\sim$10kW cm^{-2} at 83K) would therefore be 1.9μA.

The following figures show typical spectra taken on the surface of samples 295 and 296 (samples 292 and 294 show similar features):

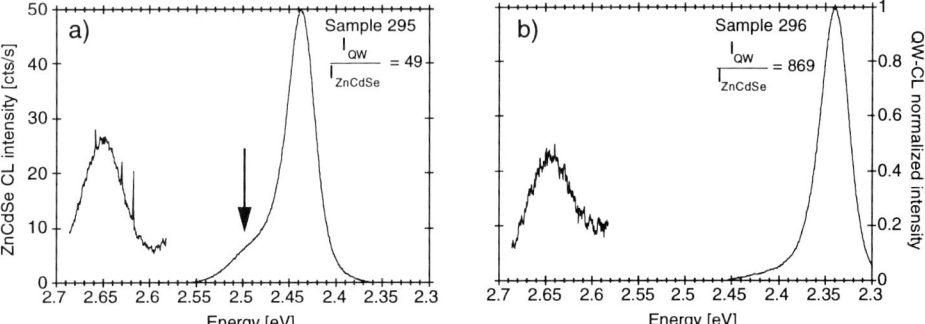

Fig. 3   Spectra taken at 300K, 20keV, 1nA on samples (a) 295 and (b) 296, scan mode. The QW-CL is normalized to the peak intensity, the ZnCdSe signal (at 2.65eV) is as acquired.

While the CL intensity from the ZnCdSe (arising from recombinations in the GRINSCH) is comparable for all samples once the background is removed, the wells show considerable differences. The actual lasers (292, 296) emit at least twice as much luminescence as the two other samples under similar excitation conditions (this disparity is present for temperatures greater than 80K). As we can see in the table of section 4.2, the ratio between QW and ZnCdSe CL intensities (for the conditions used in the acquisition of Fig. 3) provides a good indication for the possibility of light amplification in a structure.

For both non-lasing structures, a second transition is easily visible (indicated by arrows on Fig. 3a), located on the high energy side of the QW peak: it is present above 250K under a large range of injection conditions (10pA to 100nA at 20keV). We observe this feature on all samples between 10K and 300K, but it becomes apparent only under extreme injection conditions (more than 100nA at 20kV).

4.2. Defect statistics

Polychromatic maps show non-uniform CL emission; the defects affecting the CL are visible mainly as dark spots, with a few dark-line defects (DLD) oriented along [100] and [010].

The recorded densities of defects in TEM and CL, along with the principal characteristics of the four analysed samples are listed in the following table:

Sample Nr.	achieves lasing ?	threshold [kWcm^{-2}]	QW parameters [Å]	x	TEM [cm^{-2}] [100] disl.	threading	CL [cm^{-2}] DLD	point+pyr	$\frac{I_{QW}}{I_{ZnCdSe}}$
292	yes	11@83K	80	0.2	9×10^7	2×10^8	1×10^5	1×10^7	202
294	no	—	120	0.17	7×10^7	6×10^7	$<2 \times 10^4$	2×10^7	66
295	no	—	80	0.2	3×10^7	1×10^7	1×10^6	8×10^6	49
296	yes	8.3@83K	50	0.25	3×10^7	4×10^7	$<2 \times 10^4$	2×10^7	869

We see that the defects observed in CL do not readily correspond to defects observed in TEM, since the densities match neither for DLDs nor for point defects. The CL contrast is probably produced by dislocations crossing the well, but other non-radiative recombination centres, such as clusters of point defects, may also play a part. Most important, the density of defects does not have a significant influence on the lasing/non-lasing behaviour.

5. CONCLUSION

The raw luminescence intensity emitted by the well is a crucial parameter for a structure in order to attain laser amplification. The ratio of QW and ZnCdSe intensities turns out to be a good estimator for the threshold power; CL is a valuable tool for providing a first quantitative estimate of a structure's quality, prior to any processing.

Furthermore, crystallographic defects have little or no part at all in the lasing/non-lasing behaviour, but preliminary results shows that they may influence significantly the resistance to degradation of the structures.

TEM observations show a different dislocation behaviour between III-V and II-VI materials. Misfit defects initiating at the substrate- buffer interface are blocked at the III-V - II-VI interface, allowing so the subsequent growth of II-VI layers of good crystallographic quality. An array of dislocations is present at the beginning of the GRINSCH: their nature indicates a possible relaxation of the misfit. However, this needs confirmation, through double X-Ray diffraction studies for example, so raising the need of a multi-technique study for further understanding of the structures.

We wish to thank heartily our colleagues from the Centre Interdépartemental de Microscopie Électronique (CIME-EPFL) for their collaboration, in particular B. Garoni, G. Peter and B. Senior for the expert technical support with the microscopes, as well as P.-A. Buffat and P.-H. Jouneau for valuable discussions.

Jean-Marc Bonard acknowledges the financial support of the "OPTIQUE" priority program of the board of the Swiss Federal Institutes for Technology.

REFERENCES

Bonard J-M, Ganière J-D, Vanzetti L, Sorba L, Franciosi A, Hervé D, Molva E 1995 (unpublished results)
Hervé D, Molva E, Vanzetti L, Sorba L and Franciosi A, 1995 Electron. Lett. (to be published)
Lozykowski H J, Shastri V K, 1991 J. Appl. Phys. 69(5) 3235
Chen Y, Liu X, Weber E, Bourret E D, Liliental-Weber Z, Haller E E, Washburn J, Olego D J, Dorman D R, Gaines J M and Tasker N R 1994 Appl. Phys. Lett. 65(5) 549
Jouneau P-H 1993 PhD thesis (Institut National Polytechnique de Grenoble, France) pp 53-127

Inst. Phys. Conf. Ser. No 146
Paper presented at Microsc. Semicond. Mater. Conf., Oxford, 20–23 March 1995

Characterization of semiconductor materials using a cryogenic Fourier transform SEM-CL system

V Higgs, E C Lightowlers, A T Collins and A Mainwood

Department of Physics, Kings College London, Strand WC2R 2LS, UK

ABSTRACT: A cryogenic Fourier transform facility has been developed to obtain high resolution CL spectra and CL images from semiconductor materials, throughout the spectral region of 5600-45000 cm^{-1}. High resolution (0.5-2 cm^{-1}) CL spectra were recorded from individual SiGe quantum wires, and GaAs and CdSe quantum well structures using a focused electron beam. However, in contrast, it was very difficult to obtain high resolution spectra from synthetic diamond, CVD diamond films, bulk ZnSe, and epitaxial films of GaN because of the low total light output. By defocusing the electron beam (from 300 Å -300 µm) it was found that the luminescence emission increased between a factor of 10^3-10^5, enabling high resolution spectra to be obtained, but only from a larger excited area.

1. INTRODUCTION

Fourier transform (FT) spectroscopy has long been accepted as a standard technique for mid- to far-infrared absorption spectroscopy (Giffiths and deHaseth 1986). The two main advantages of FT spectroscopy, the Fellgett and Jacquinot advantages, result in an improvement of the signal-to-noise ratio and faster data acquisition from a wide spectral range (100-10,000 cm^{-1}). FT absorption spectroscopy has been applied to the determination of impurities in Si and GaAs (e.g. carbon, hydrogen) and to monitoring the complex point defect chemistry in these materials. Complementary to this type of analysis, Fourier transform photoluminescence spectroscopy (FT-PLS) has been developed to characterize defects in Si (McL. Colley and Lightowlers 1987), and has also been used to investigate the optical properties of bulk Ge and SiGe heterostructures. Throughout this spectral region (3000-9500 cm^{-1}), very sensitive photomultipliers are as yet not available and it has been recognised that FT-PLS offers certain advantages over conventional dispersive spectroscopy. However, only a limited number of investigations using FT-PLS spectroscopy have been carried out in the spectral region above 9500 cm^{-1} in spite of the availability of commercial dedicated FT-PLS systems. Similarly, nearly all the applications of cathodoluminescence (CL) spectroscopy in the shorter wavelength region utilise very sensitive photon counting detectors coupled with a standard monochromator. More recently, CCD detectors have become available and are starting to be incorporated with existing CL systems.

In the present investigation, a FT-CL system has been developed to record high resolution CL spectra and CL images throughout the near-infrared, visible and ultraviolet regions (5600-45000 cm^{-1}). The CL is generated in a SEM fitted with a variable temperature cold stage (5-300 K). Using this equipment, in this paper we present results obtained from a range of semiconducting materials and demonstrate that in some cases high resolution spectra can be obtained (0.5 -2 cm^{-1}) using a tightly focused electron beam. However for other materials the luminescence emission is very low when using a focused beam, and the electron beam has to be defocused to obtain the maximum light output. This effect makes it difficult to obtain high resolution spectra from a highly localised area . A preliminary investigation into this phenomenon and the materials issues are discussed below.

2. EXPERIMENTAL

CL measurements were made at temperatures between T=5-300 K using the CL mode of a Jeol 35 C SEM. A range of excitation conditions was used including beam energies between 5-35 keV and beam currents of 0.1 nA- 0.1 μA. CL spectra were recorded using both focused and defocused electron beams during acquisition. Specially designed mirror optics were used to collect the luminescence and either direct a parallel beam into a Bruker IFS 66 Fourier transform spectrometer or direct a focused beam on the detector through narrow band pass filters for monochromatic imaging. A cooled Ge photodiode detector was used for the CL experiments carried out on bulk Si, SiGe and GaAs. In the visible region, a blue-enhanced Si diode was employed, and for comparison some spectra were recorded using a photomultiplier with an S20 photocathode. To further investigate the defocusing phenomena, additional luminescence measurements were made using a dedicated electron microprobe (Jeol 8600) which has a calibrated variable defocusing control (up to 300 μm beam diameter).

3. RESULTS AND DISCUSSION

CL spectra were recorded (resolution = 2 cm^{-1}) with a focused electron beam from SiGe quantum wires grown by gas source molecular beam epitaxy (GS-MBE) on V-groove patterned Si substrates. The low temperature CL spectra (T= 5K) contained three distinct SiGe excitonic features and their transverse optic phonon replicas (X_{TO}). A typical CL spectrum from one SiGe quantum wire is shown in Fig.1a, in which the three X_{TO} phonon replicas are identified in the CL spectrum. Monochromatic images (Higgs et al 1995) revealed that these features are related to recombination in the SiGe quantum wires and the SiGe quantum wells in the (100) and (111) directions. CL spectra were also readily obtained from SiGe quantum wells and bulk Si. The spectral resolution obtained is better than that obtained with conventional dispersive photoluminescence spectra.

Higher resolution (0.5 cm^{-1}) CL spectra and high quality CL images were obtained from GaAs/AlGaAs and CdSe quantum well structures. However, in contrast, it was very difficult to obtain high resolution spectra from CVD diamond films, synthetic diamond, bulk and epitaxial films of ZnSe and epitaxial films of GaN. This is because the total luminescence emission using a focused electron beam (beam currents= 0.1 nA-1 μA) is extremely small. By simply defocusing the electron beam up to a spot size of 300 μm the luminescence increased by a factor of 10^4-10^5. From a selected number of CVD diamond films it was possible to obtain CL spectra in the focused beam mode with a much higher spectral resolution (16 cm^{-1}) than that obtained using a conventional dispersive CL system (see Fig. 1b). To investigate these effects further, a selection of samples was characterized under the same defocusing conditions.

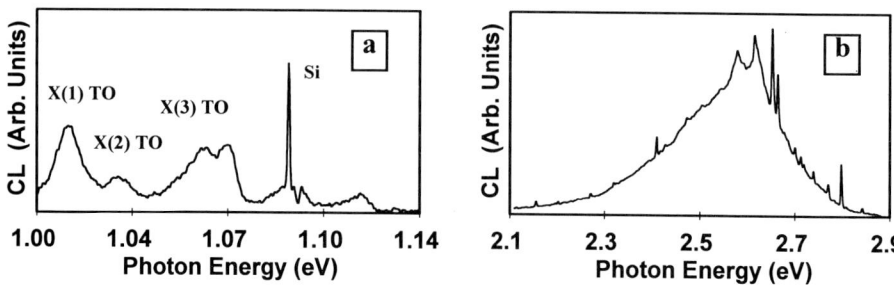

Fig. 1. CL spectra recorded from a) SiGe quantum wire, b) CVD diamond film.

Fig. 2a shows the variation of the integrated peak CL intensity as a function of beam defocus at T= 5 K. For bulk FZ Si (curve A) and bulk GaAs (curve B) the bound exciton features did not vary with beam defocus; the same results were obtained for other Si and GaAs samples with different doping levels and different dopants. In addition, the free exciton features observed in the Si and GaAs samples were also unaffected by defocusing the electron beam, whereas in the case of bulk ZnSe (curve C) the bound exciton features increased by a factor of ≈ 800. Similarly, for CVD diamond (curve D), the dislocation-related blue band A (Collins 1990), commonly observed in diamond films, increased by ≈ 300 times. CL spectra containing the intrinsic free exciton features could be obtained for this sample only using the maximum defocused beam diameter. Because of the experimental configuration of the CL system we could only defocus the spot size to a maximum of ≈ 20 μm. To further examine the defocusing effect, the samples were also analysed using a Jeol 8600 dedicated X-ray microprobe. Fig. 2b shows the effect of a larger defocus on the CL emission from a natural type IIa diamond; the different traces show the defocusing effects at three different points chosen at random. The intensity of the light output increased by a factor of up to 10^5 at the maximum defocused beam. Also the rate of change of CL intensity (gradient) before a constant light emission level is obtained (the flat horizontal section of the curves in Fig. 2b) varies depending on the position on the sample where the data were recorded. Similar behaviour was observed on other diamonds and other materials including ZnSe and GaN.

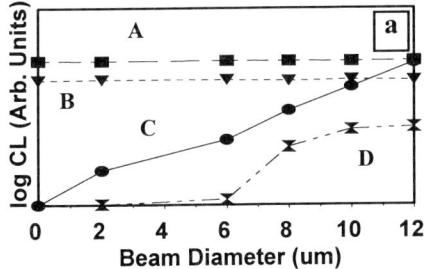 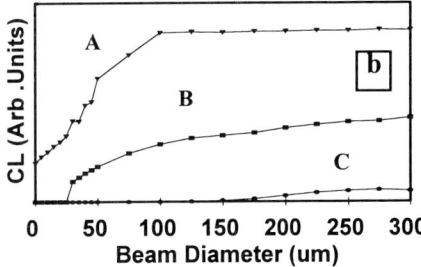

Fig. 2. The Influence of beam defocus a) Si, GaAs, ZnSe and CVD diamond, b) type IIa diamond.

The reduction in the luminescence intensity in the focused electron beam could be due to many different effects: thermal dissociation, ionisation by the Stark effect, impact ionisation by hot carriers, or saturation of the radiative centres. An electron beam blanking system (2 ns switching time) was employed to modulate the beam to eliminate heating or electric field effects. However, no enhancement of the luminescence was observed during modulation for any of these materials. Under these high levels of excitation produced in the focused beam we suggest that the trapping of electrons, holes and excitons at non-radiative centres is dominant. When a high concentration of electrons, holes and excitons is produced which greatly exceeds the density of radiative centres, then most are trapped by non-radiative centres and cannot migrate. This implies a relatively high concentration of non-radiative trapping centres, though trapping cross-section and lifetime are also important parameters. When the beam is defocused and the electron, hole and exciton density is less than the concentration of radiative centres, a much higher proportion can recombine radiatively. In Si and GaAs the trap density is low, so the diffusion length and free exciton lifetime are long. In the focused beam mode of operation the excitons, electrons and holes can readily diffuse away from the point of excitation, and the radiative centres are not saturated. Whereas, in synthetic diamond, ZnSe and GaN the dislocation levels and trap levels are very high, and therefore the diffusion length and lifetime are shorter, and in the focused beam mode the excitons, holes electrons become trapped at non-radiative centres and cannot migrate to other radiative centres.

We have constructed a simple finite element computer program to examine these mechanisms. The variables which are crucial are the radiative and non-radiative lifetimes of the exciton, and the

762

density of radiative traps. In this simple model, we have assumed a distribution of radiative traps with a lifetime of 10 ns with a large capture cross section together and a non-radiative recombination mechanism with a lifetime of 1 ns. The beam energy is 30 keV and the beam current is 5×10^{-7} A. The results for diamond are shown in Fig. 3, with the numbers labelling the curves giving the density of defects per cubic micron, so that 100 corresponds to 10^{14} defects cm^{-3}. There is qualitative agreement with the experimental dependence on beam radius (compare Fig. 3 with Fig. 2b) and radiative trap densities of 10^{14} -10^{15} defects cm^{-3} are plausible estimates. This model can be applied to other semiconductor materials and by direct comparison with CL defocusing measurements maybe useful as a qualitative indication of material quality.

Fig. 3. The Influence of beam defocus for diamond with different concentrations of radiative centres

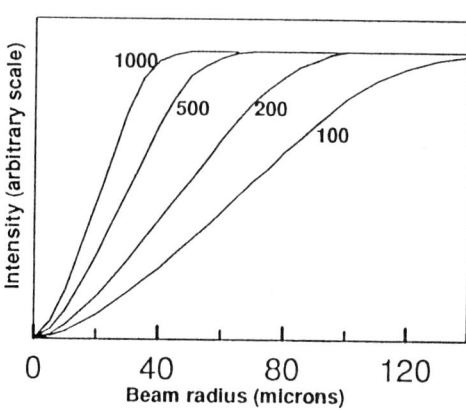

It seems likely that the saturation of the radiative centres is a possible explanation for the weak bound exciton emission in the more defective materials using a highly focused beam . However it does not rule out the possibility that the other effects (e. g. electric field) may be more dominant in different materials. Also, the reason for the quenching of the free exciton luminescence in diamond as the diameter of the electron beam is reduced, is still under investigation. Finally we note that there are more drawbacks in using a FT interferometer with a photomultiplier (PM) detector. A PM detector is very sensitive, but the noise increases with the square root of the incident signal, and the Fellgett advantage favouring FT spectroscopy is lost. Therefore, under the same excitation conditions, there will be certain cases where it would be possible to measure a CL spectrum in the spot mode using a conventional CL dispersive system and not in a FT-CL system. However, it is clear that in the cases where the material quality is good, high resolution CL spectra can be obtained in the focused beam mode using FT-CL, whereas in the more defective material, CL spectra have to be obtained from a larger local excited area .

ACKNOWLEDGEMENTS

This work was supported by the Science and Engineering Research Council.

REFERENCES

Collins AT 1990 Dia.. and Rel. Mater. 1 457
Griffiths PR and de Haseth JA Fourier transform spectroscopy (Wiley, New York 1986)
Higgs V Lightowlers EC Usami N Shiraki Y Mine T Fukatsu S to be published in
    J. Crys. Growth 1995
McL. Colley P and Lightowlers EC 1987 Semi. Sci. Tech. 2 187

Inst. Phys. Conf. Ser. No 146
Paper presented at Microsc. Semicond. Mater. Conf., Oxford, 20–23 March 1995

Multiple scattering simulation of electron channelling contrast images of dislocations at interfaces

S L Dudarev[†], J T Czernuszka[†], L-M Peng[‡], A J Wilkinson[†] and M J Whelan[†]

[†]Department of Materials, University of Oxford, Parks Road, Oxford OX1 3PH, England

[‡]Beijing Laboratory of Electron Microscopy, Chinese Academy of Sciences, P.O. Box 2724, Beijing 100080, People's Republic of China

ABSTRACT: Quantum-mechanical multiple scattering theory of electron backscattering from crystalline materials is applied to simulate channelling images of dislocations situated near the surface of a crystal. A computational algorithm is developed which uses perturbation expansion of the solution of an inhomogeneous transport equation and which makes it possible to generate two-dimensional images of dislocations for various orientations of the Burgers vector. We compare the images simulated numerically with the results of experimental observations performed at inclined incidence using a conventional SEM and recently developed detector of backscattered electrons. The origin of the oscillations of the image intensity is discussed.

1. INTRODUCTION

The possibility of imaging of dislocations using the so-called channelling mode in a conventional scanning electron microscope (SEM) was proposed by Booker el al (1967) soon after the experimental discovery of electron channelling patterns by Coates (1967). Channelling patterns arise in SEM as a result of variation of the coefficient of backscattering of electrons with change in the orientation of the incident beam with respect to the atomic planes of the crystal. A dislocation becomes visible due to local bending of the atomic planes and the associated change of the diffraction conditions for incident electrons, which results in a variation of the coefficient of backscattering when the incident beam scans over the region corresponding to the distortion of the lattice. The theory explaining the mechanism of formation of electron channeling patterns was first formulated by Hirsch and Humphreys (1970) for perfect crystals and later generalized by Spencer et al (1972) to the case of crystals containing defects. Another theoretical formulation suitable for calculation of electron channelling images of defects in crystals was developed by Clark and Howie (1971a). Later Clark (1971b) published the first images of dislocations observed in SEM using thin specimens and backscattered electrons. The problem of imaging dislocations in bulk specimens has been addressed by Morin et al (1979) and Fontaine et al (1983) who obtained images of dislocations using an extensively modified SEM. A field-emission gun (FEG) was installed in order to increase the brightness of the image, and a retarding field detector of backscattered electrons was used to collect only low-loss electrons. However, this technique was not used by other groups because of the complexity of the detector system which was employed. More recently .Czernuszka et al (1990) has shown how dislocations can be observed using a conventional FEG SEM with very little modification. The contrast of the channelling image can be improved considerably using the inclined geometry of observations shown schematically in Fig. 1 and collecting all the backscattered electrons without energy separation. This technique was later employed by Wilkinson et al (1993) for imaging of interfacial defects in strained silicon-germanium layers on a silicon substrate. The new experimental development has been in advance of existing computational techniques for simulation of the energy and angle distribution of high-energy electrons backscattered from crystalline materials. To remedy this we describe here a new theoretical approach to the problem of electron backscattering from crystals which has been recently developed by Dudarev, Rez and Whelan (1995), and show how it can be applied to the problem of calculation of channelling contrast (CC) images of dislocations.

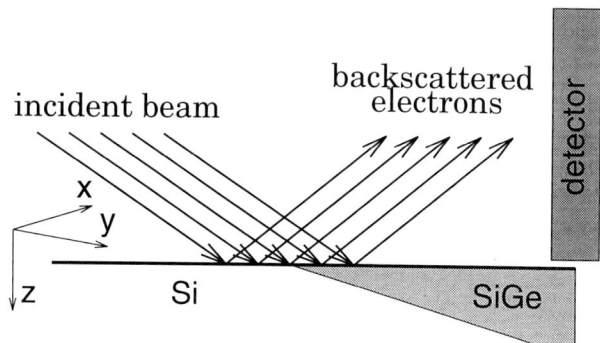

Figure 1. The geometry of scattering

2. EFFECT OF DIFFRACTION ON ELECTRON BACKSCATTERING

In our theoretical approach we follow the idea proposed by Spencer and Humphreys (1980) that the quantitative treatment of electron backscattering from crystals requires solving a particular type of inhomogeneous transport equation. In this equation the source function depends on the diffraction conditions for the incident beam. Developing this approach, Dudarev, Rez and Whelan (1995) have shown that this inhomogeneous transport equation can be derived from the kinetic equation for

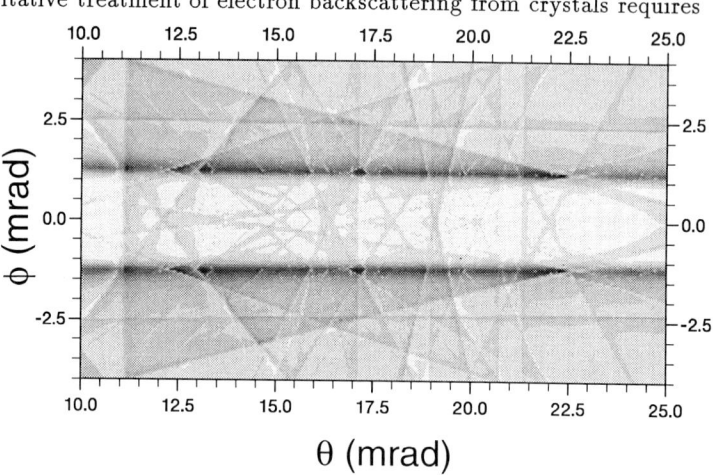

Figure 2. A simulated channelling pattern

the so-called density matrix [see Dudarev, Peng and Whelan (1993) and Dudarev, Vvedensky and Whelan (1994)], and have formulated an algorithm suitable for solving this equation taking multiple quasi-elastic phonon scattering and energy losses into account. In Fig. 2 there is shown a two-dimensional channelling pattern simulated for 25 keV electrons and the Si (100) surface using the method developed by Dudarev, Rez and Whelan (1995). The Si (220) channelling band employed to obtain experimental images of dislocations is seen in the centre of the pattern shown in Fig. 2.

3. PERTURBATION TREATMENT OF THE IMAGE CONTRAST

In 1974 Howie proposed the idea that the contrast of channelling patterns can be treated by perturbation, and recently it has been shown [see Dudarev, Rez and Whelan (1995)] how perturbation calculations can be performed practically by transforming the transport equation into integral form and solving it by iteration. The contrast of the channelling image can then be calculated using the following equation:

$$\delta R(\cos\theta, \phi, E) = R(\cos\theta, \phi, E) - R_{\text{rand}}(\cos\theta, \phi, E) = K(\cos\theta, \phi, E \mid \cos\theta_0, \phi_0, E_0)$$

$$\times \frac{1}{S} \int\int dx\, dy \int_0^\infty dz\, z \left[\frac{1}{\hbar} \sum_{h,l} \Phi_h(\mathbf{r})\Phi_l^*(\mathbf{r})\gamma_{lh} - v_0 w_{\text{el}}^{(tot)}(E_0) \exp\left(-w_{\text{el}}^{(tot)}(E_0)z/\cos\theta_0\right) \right], (1)$$

where $w_{\text{el}}(\cos\psi, E)$ is proportional to the differential cross-section of phonon scattering, and θ_0 and ϕ_0 are polar and azimuthal angles of incidence respectively. The amplitudes $\Phi_h(\mathbf{r})$ entering (1) satisfy the equation of the form (the so-called Takagi's equation)

$$i\frac{\hbar^2}{m}(\mathbf{k}_0 + \mathbf{G}_h)\frac{\partial}{\partial \mathbf{r}}\Phi_h(\mathbf{r}) = -\Phi_h(\mathbf{r})\frac{\hbar^2}{m}\left\{(\mathbf{k}_0 + \mathbf{G}_h)\frac{\partial}{\partial \mathbf{r}}\right\}(\mathbf{G}_h \cdot \mathbf{R}(\mathbf{r}))$$

$$+\frac{\hbar^2}{2m}[(\mathbf{k}_0 + \mathbf{G}_h)^2 - \mathbf{k}_0^2]\Phi_h(\mathbf{r}) + \sum_t (U_{ht} - \frac{i}{2}\gamma_{ht})\Phi_t(\mathbf{r}). \quad (2)$$

where the Fourier components of the potential U_{ht} and $-[i/2]\gamma_{ht}$ are independent of the local displacement of the lattice, and the boundary condition on (2) at $z = 0$ is

$$\Phi_h(\mathbf{r})|_{z=0} = \delta(x - x_0)\delta(y - y_0)\delta_{h0},$$

where (x_0, y_0) are the coordinates of the point where the incident beam enters the crystal. The observed channelling contrast is associated with variation of $\delta R(\cos\theta, \phi, E)$ as a function of (x_0, y_0). Note that the mechanism giving rise to channelling contrast images of defects situated near the surface differs from that responsible for reflection electron microscope (REM) or X-ray images [see Peng et al (1989) and Kaganer et al (1991)]. In the latter case the origin of the contrast is associated with bending of the surface and the resulting variation of the *direction* of the specular reflection while in the former case the observed variation of the coefficient of backscattering results from the local variations of the orientation of the lattice and corresponding changes in the effective depth of penetration of

Figure 3. Channelling image of an interface dislocation

electrons in the crystal bulk. In Fig. 3 there is shown a channelling image of a dislocation observed using the (220) reflection in a commercial SEM described by Czernuszka et al (1990) at inclined incidence ($\theta_0 = 37°$). The left (dark) part of the image corresponds to electron backscattering from pure Si, while the brighter area corresponds to backscatte-

766

ring from SiGe alloy. The bright horizontal line near the centre of the image originates from scattering of electrons by a dislocation line lying in the plane of the interface between Si and SiGe. The theoretical simulations of channelling images are based on direct numerical integration of equation (2) using displacement fields of a general dislocation evaluated by Shaibani and Hazzledine (1981). From the analysis of previous experimental observations it was expected that the dislocations situated at the inteface between the Si and SiGe crystals are 60° misfit dislocations, the Burgers vector of which makes angles of 60° or 120° with the $[01\bar{1}]$ direction. There are eight possible Burgers vectors satisfying this condition, namely, $\mathbf{b}_{1,2,3,4} = \{a/2\sqrt{2}\}(\pm1, 0, \pm1)$ and $\mathbf{b}_{5,6,7,8} = \{a/2\sqrt{2}\}(\pm1, \pm1, 0)$. Numerical simulations show that there are only *two* distinct images corresponding to these eight possible orientations of $\mathbf{b}$. These two images are shown in Figs. 4 and 5. As follows from our analysis, a series of observations using different $\mathbf{G}$ vectors is required to determine the orientation of the Burgers vector unambiguously.

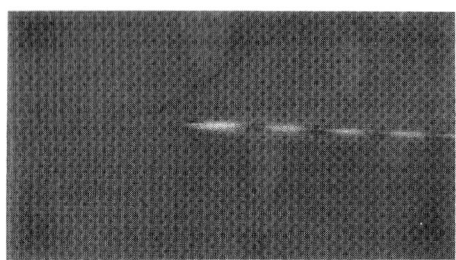   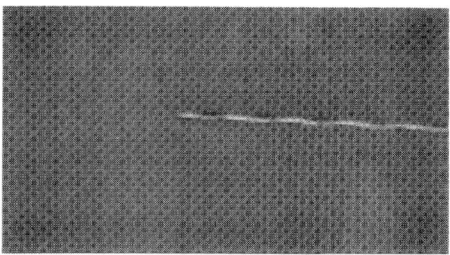

Figure 4. CC image simulated for $\mathbf{b}\|$ [110]   Figure 5. CC image simulated for $\mathbf{b}\|$ [$\bar{1}$10]

REFERENCES

Booker G R, Shaw A M B, Whelan M J and Hirsch P B , 1967 Philos. Mag. 16 1185
Coates D G, 1967 Philos. Mag. 16 1179
Clarke D R and Howie A, 1971a Philos. Mag. 24 959
Clarke D R, 1971b Philos. Mag. 24 973
Czernuszka J T, Long N J, Boyes E D and Hirsch P B, 1990 Philos. Mag. Lett. 62 227
Dudarev S L, Peng L-M and Whelan M J, 1993 Phys. Rev. B 48 13408
Dudarev S L, Vvedensky D D and Whelan M J, 1994 Phys. Rev. B 50 14525
Dudarev S L, Rez P and Whelan M J, 1995 Phys. Rev. B 51 3397
Fontaine G, Morin P and Pitaval M, in: *Microscopy of Semiconducting Materials*, 1983, ed. by A G Cullis, S M Davidson and G R Booker, Inst. of Phys. Conf. Ser. No 67 (IOP, Bristol, 1983), p. 213
Hirsch P B and Humphreys C J, 1970 in: *Scanning Electron Microscopy 1970*, ed. by O. Johari, Proc. 3rd Annual Sympos. (IIT Res. Inst., Chicago, 1970) p. 451
Howie A, 1974 in: *Quantitative Scanning Electron Microscopy*, ed. D B Holt, M D Muir, P R Grant and I M Boswarva (Academic Press, London, 1974), p. 183
Kaganer V M and Möhling W, 1991 Phys. Stat. Sol. (a) 123 379
Morin P, Pitaval M, Besnard D and Fontaine G, 1979 Philos. Mag. 40 511
Peng L-M, Cowley J M and Hsu T, 1989 Ultramicroscopy 29 135
Shaibani S J and Hazzledine P M, 1981 Philos. Mag. 44 657
Spencer J P, Humphreys C J and Hirsch P B, 1972 Philos. Mag. 26 193
Spencer J P and Humphreys C J, 1980 Philos. Mag. 42 433
Wilkinson A J, Anstis G R, Czernuszka J T, Long N J and Hirsch P B, 1993 Philos. Mag. A 68 59

Inst. Phys. Conf. Ser. No 146
Paper presented at Microsc. Semicond. Mater. Conf., Oxford, 20–23 March 1995

The effect of spherical aberration and surface reflection on the scanning infra-red microscope imaging of oxide particles in Si

Z Laczik, P Török* and G R Booker

University of Oxford, Department of Materials, Parks Road, Oxford OX1 3PH, UK

ABSTRACT: The two major factors limiting the reflection confocal mode scanning infra-red micro-scope imaging of small oxide particles in bulk silicon are a) the decrease in signal intensity due to the spherical aberration introduced by the specimen, and b) the presence of a background signal caused by specular reflection from the front specimen surface. An analysis of these two effects is presented. The results show that when using the standard reflection confocal mode, small oxide particles can only be imaged for depths of ~50 to ~150 μm below the surface. However, it is also shown that this range can be extended to 0 to ~600 μm by correction of the spherical aberration and by the use of the dark-field imaging mode of the scanning infra-red microscope.

1. INTRODUCTION

One of the important applications of scanning infra-red microscopy (SIRM) is the imaging of oxide particles present in commercial Czochralski silicon wafers heat-treated for internal oxide gettering. In general, as-grown Czochralski silicon wafers are annealed to precipitate some of the solute oxygen to produce oxide particles with number densities 10^8 to 10^{10} cm^{-3} and with sizes in the range of a few nanometers to a few hundred nanometers (Falster et al. 1992). For the effective operation of internal oxide gettering the three dimensional distribution of the oxide particles needs to be carefully controlled, the main parameters being the bulk number density of the oxide particles and the depth of the surface denuded zone, where no particles are present. The two major factors limiting the reflection confocal mode scanning infra-red microscope imaging of such small oxide particles are a) the decrease in signal intensity and reduced resolution due to the spherical aberration (SA) introduced by the specimen as the focusing depth increases, and b) the presence of a background signal caused by specular reflection from the front specimen surface as the focusing depth decreases. It is demonstrated in the following that correction for the SA completely eliminates a) and that the use of dark-field imaging completely elimi-nates b). Simultaneous use of the two extends the range of operation to 0 to ~600 μm focusing depths.

In a reflection mode scanning infra-red microscope (RC-SIRM) the light of a laser is brought to a focus (the 'probe') by a probe-forming lens within the specimen. Either the light beam or the specimen is then raster scanned and the reflected/scattered light is collected by the same lens and after passing through a pin-hole is detected by a photodiode. An image is then built up pixel-by-pixel from the detector signal. As shown in Fig. 1, the confocal arrangement ensures that mainly light from the in-focus regions only reaches the detector. For further general details on SIRM and scanning confocal microscopy the reader is referred to Booker et al. (1992) and Wilson and Sheppard (1984).

For the present experiments a high-brightness, low-noise semiconductor laser with 1300 nm wave-length was used as the light source and a high numerical aperture (NA 0.85) lens with variable SA correction for 0 to ~600 μm focusing depths was used as the probe-forming lens. Confocal detection was realised using a fiber optic arrangement which was equivalent to using a pin-hole of optimum size.

2. EFFECT OF SPHERICAL ABERRATION

The detected RC-SIRM signal is directly related to the electric energy density in the probe at the

---

*Now at the Multi Imaging Centre, Univ. of Cambridge, Downing Street, Cambridge CB2 3DY, UK

768

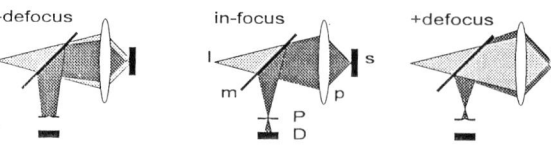

Fig. 1. Schematic diagram illustrating confocal detection in reflection mode SIRM imaging. [l light source, m beam-splitter, p probe-forming lens, s specimen, P pin-hole, D detector]

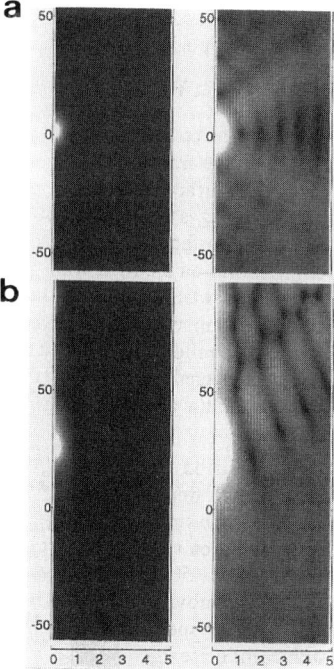

position where the object being examined is present. The peak electric energy density corresponds to the peak RC-SIRM signal, i.e. to the sensitivity of the SIRM, which is less than 50 nm (under optimum conditions) in terms of minimum particle size that can be imaged. The size/shape (e.g. half-maximum iso-surface) of the electric energy distribution corresponds to the image size/shape for point objects, i.e. to the resolution of the SIRM, which is 0.7 μm and 7 μm for the spatial and depths resolutions respectively under optimum conditions. Fig. 2a shows the calculated electric energy density distribution that a point object situated 10 μm below the front specimen surface (10 μm focusing depth) would experience as the probe was scanned across it in the X-Z plane (Török et al. 1995) where X and Z are the radial and axial directions respectively. The left and right panels show the distribution on linear and logarithmic scales respectively. For this depth below surface position of the point object there is virtually no SA caused by the specimen present. Fig. 2b shows similar plots for 300 μm focusing depth. The electric energy density distribution is significantly broadened and additional secondary peaks appear both at axial positions and off axis. Fig. 3 shows

Fig. 2. Calculated time averaged electric energy density distributions for 10 and 300 μm focusing depths. The scale marks are in microns.

how the peak intensity of the electric energy distributions decreases with increasing focusing depth. Both the increased probe size and the decreased intensity in the probe are due to the SA caused by the specimen.

Fig. 4 shows measured SIRM peak signal values for point and plane objects (small oxide particles and the back specimen surface respectively) as a function of the SA correction settings on the lens. The curve for the point object was recorded by keeping an individual oxide particle located at a depth of 250 μm below the front surface exactly in focus and only changing the SA correction setting on the lens. For the plane object curve the same procedure was used but for the back surface of the specimen. The behaviour predicted from the calculated distributions can indeed be observed, as illustrated by, for

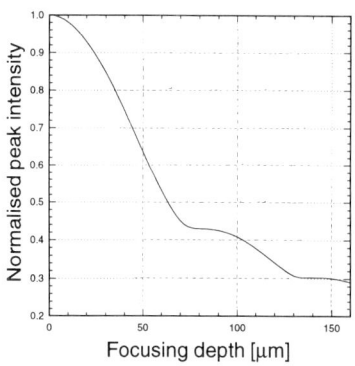

Fig. 3. Calculated normalised peak intensity in the focused probe for 0 to 300 μm depths.

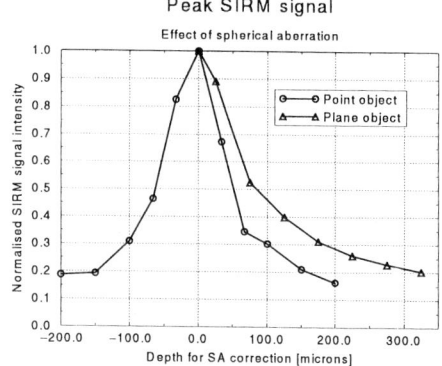

Fig. 4. Measured SIRM peak signal for point and plane objects as a function of SA.

example, the ~85% reduction in the RC-SIRM signal as the SA correction setting is changed by 200 μm from its optimum position (marked 0 μm in Fig. 4) .

3. EFFECT OF SURFACE REFLECTION

To assess the combined effects of reduced signal due to SA and increased noise due to specular reflection from the front specimen, XY RC-SIRM images were recorded at different depths below the front specimen surface with the SA correction adjusted for zero depth (Fig. 5) and 200 μm depth (Fig. 6). In the X-Z images the specimen surface appears as a white band at the bottom, and increasing depths below the front surface correspond to positions further towards the top. For both Figs. 5 and 6 the XY images are shown on the right with their depth positions marked on the XZ image on the left. The XZ images were recorded with a 100x lower signal amplification setting (compared to the XY images) to prevent the saturation of the detector when the front surface was in focus. Comparison of the signal from the surface reflection in the XZ image and the signal from the particles in the XY images showed that the surface signal is ~10^2 to 10^4 times larger than the particle signal. Particles would be observed in the XZ images if the amplification were increased.

For Fig. 5, i.e. SA correction adjusted for zero depth, the depth range in which the small oxide particles can be clearly imaged is ~50 to ~150 μm. On moving closer to the front specimen surface the noise signal due to the surface reflection increases. For depths <50 μm, e.g. the bottom XY image, the noise is comparable with the signal from the small particles, and the particles are not clearly observed. For depths >150 μm, as the depth increases (top XY images) the particle spot sizes increase (poorer spatial resolution), the particle spot number density increases (poorer depth resolution), and the particle signal decreases (as in Figs. 3 and 4), and so the particles are not clearly observed. These three effects are due to the increased SA at the larger focusing depths.

For Fig. 6, i.e. SA correction adjusted for 200 μm depth, the band corresponding to the specimen surface in the

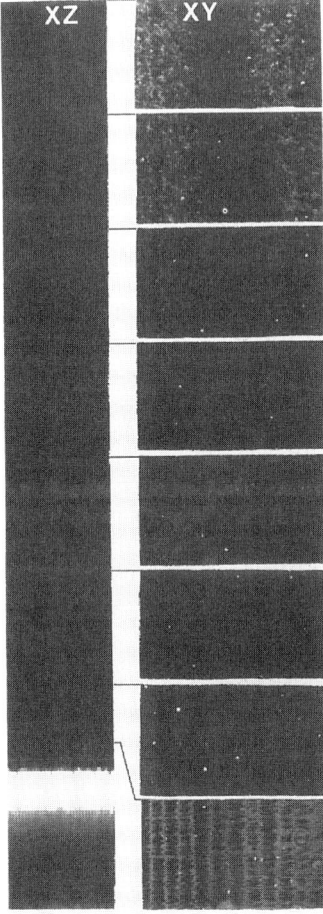

Figure 5. XZ and XY RC-SIRM images of oxide particles in Cz Si with SA correction set for 0 μm depth below the front surface.

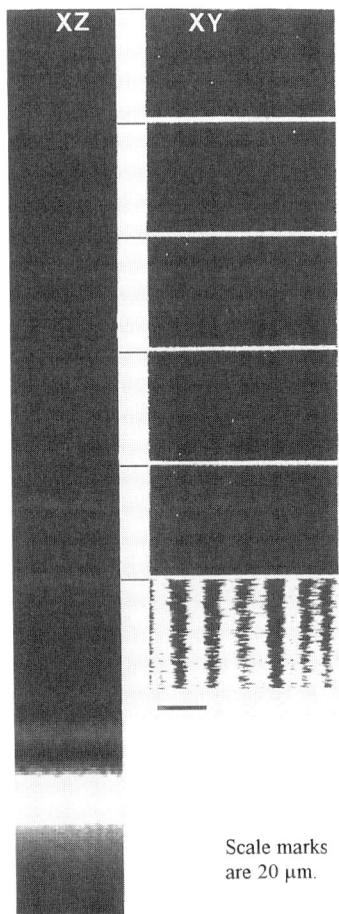

Scale marks are 20 μm.

Figure 6. XZ and XY RC-SIRM images of oxide particles in Cz Si with SA correction set for 200 μm depth below the front surface.

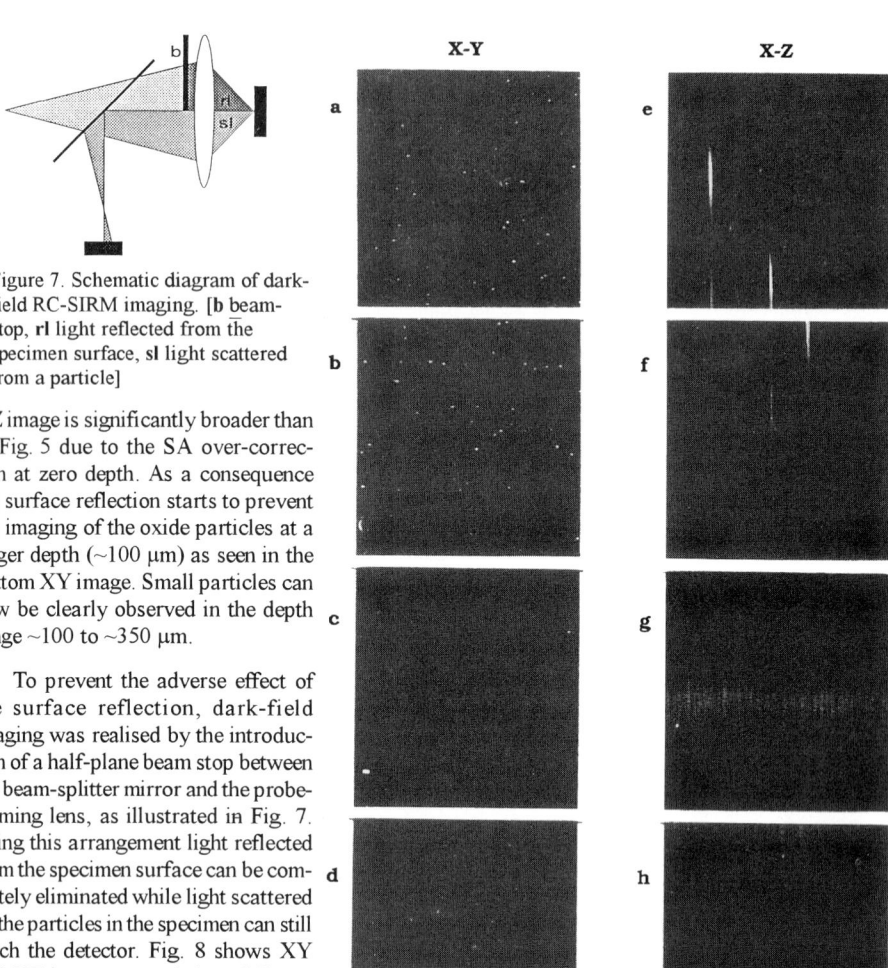

Figure 7. Schematic diagram of dark-field RC-SIRM imaging. [b beam-stop, rl light reflected from the specimen surface, sl light scattered from a particle]

XZ image is significantly broader than in Fig. 5 due to the SA over-correction at zero depth. As a consequence the surface reflection starts to prevent the imaging of the oxide particles at a larger depth (~100 μm) as seen in the bottom XY image. Small particles can now be clearly observed in the depth range ~100 to ~350 μm.

To prevent the adverse effect of the surface reflection, dark-field imaging was realised by the introduction of a half-plane beam stop between the beam-splitter mirror and the probe-forming lens, as illustrated in Fig. 7. Using this arrangement light reflected from the specimen surface can be completely eliminated while light scattered by the particles in the specimen can still reach the detector. Fig. 8 shows XY and XZ images recorded at different depths using the arrangement shown in Fig. 7 with the beam stop aligned so that the surface signal is smaller than the signal from the particles but is not reduced to zero (Fig. 8g). It is clearly demonstrated that the adverse effect of surface reflection can be eliminated using this technique while the imaging of the oxide particles is only slightly affected.

Figure 8. Dark-field RC-SIRM images of oxide particles in Cz Si. The four rows correspond to 350, 175, 10 and -40 μm depths.

ACKNOWLEDGEMENTS

We are grateful to Professor DG Pettifor for provision of laboratory facilities and access to the computer imaging facilities in the Materials Modelling Laboratory, which is partially funded by SERC Grant No. GR/H58278. We are also thankful to Dr R Falster for his interest in the work.

REFERENCES

Booker G R, Laczik Z, Kidd P, 1992, Semiconductor Sci. and Technology 7 A110.
Falster R, Laczik Z, Booker G R, Bhatti A R and Török P, 1992, MRS Symp. Proc. 262 945.
Török P, Varga P and Booker G R, 1995, J. Opt. Soc. Am. A, in press.
Wilson T and Sheppard C, 1984, 'Theory and Practice of Scanning Optical Microscopy', Acad. Press

Inst. Phys. Conf. Ser. No 146
Paper presented at Microsc. Semicond. Mater. Conf., Oxford, 20–23 March 1995

Imaging of oxide particles in Czochralski silicon wafers using high-performance reflection confocal scanning infra-red microscopy and transmission electron microscopy

P Török[*], B Pécz[1], Z Laczik, GR Booker, G Radnóczi[1] and R Falster[2]

University of Oxford, Department of Materials, Parks Road, Oxford OX1 3PH, UK
[1] Research Institute for Technical Physics, H-1325 Budapest, PO Box 76, Hungary
[2] MEMC Electronic Materials, Novara, Italy

ABSTRACT: (001) Cz Si wafers containing 5×10^{17} cm^{-3} oxygen atoms were heat treated to produce oxide precipitate particles of number density either $\sim 10^8$ or $\sim 10^9$ cm^{-3}, and subsequently annealed at 1000°C for 2 to 160 hr. Reflection confocal scanning infra-red microscope (RC-SIRM) and transmission electron microscope (TEM) examinations were made of the oxide particles and associated dislocations. The oxide particles occurred as plates ~ 100 to 200 nm across and ~ 10 to 25 nm thick. Initially single isolated plates occurred but further annealing caused the plates to generate dislocations and further plates, thereby producing defect clusters. Calculation of the amount of oxygen in the precipitated oxide plates from their size and number density showed that almost all of the oxygen in the wafers was precipitated after the 160 hr anneal. The RC-SIRM and TEM methods provided complementary data for such precipitation studies.

1. INTRODUCTION

Oxide particles in Czochralski (Cz) silicon wafers (Shimura 1989) are important because they can getter fast diffusing metal impurities during the subsequent device processing (Bhatti et al. 1991) and hence improve device performance and yields. Infra-red microscope methods are increasingly being used to image the particles within the bulk wafers because these methods are rapid and non-destructive and enable number densities and distributions to be directly determined. However, the image contrast arises mainly by light scattering and so these methods do not in general image dislocations. Conversely, transmission electron microscope (TEM) methods are slow and destructive and in general do not give good statistically significant data concerning number densities. However, they provide detailed information concerning particle sizes, shapes and local distributions, and dislocations can be imaged. In the present work two series of heat treated Cz silicon wafers were examined by scanning infra-red microscopy (SIRM) (Booker et al. 1992, Booker et al. 1995) and one of the two series was examined by TEM.

2. EXPERIMENTAL

The Cz silicon wafers were (001) n-type, 650 μm thick and contained $\sim 5 \times 10^{17}$ cm^{-3} oxygen atoms. A wafer series I was heat treated at 800°C/4 hr and 1000°C/4 hr and a wafer series II at 900°C/15 min, 800°C/4 hr and 1000°C/4 hr. Wafers from each series were then multi-stage annealed at 1000°C for either 2, 4, 16, 32, 85 or 160 hr. The wafers were examined using our high-performance reflection confocal SIRM (Török et al. 1993) with a lateral resolution of ~ 0.7 μm, a depth resolution of ~ 5 μm and a sensitivity such that individual particles down to < 50 nm across could be imaged (diameter of spherical particle of equivalent volume). Thin foil TEM plan-view specimens were examined using either a Philips CM20 at 200 kV or a JEOL 4000EX at 400 kV. For both the SIRM and TEM studies, the regions of the wafers imaged corresponded to the 'middle' depth, i.e. to the bulk rather the surface regions of the wafers.

---

[*] Now at Multi-Imaging Centre, Univ. of Cambridge, Downing Street, Cambridge CB2 3DY, UK

3. RESULTS

The SIRM examinations of series I showed single bright spots corresponding to oxide particles for all of the wafers (Fig. 1). The in-focus bright spots were ~1 μm across. The smaller and less intense bright spots arise from particles slightly out-of-focus. The depth range for which particles appear in such images is ~5 μm. The spot number density ranged from $4x10^8$ to $1.1x10^9$ cm^{-3}, with local variations occurring both within individual wafers and on going from wafer to wafer. However, there was no significant change in mean number density as the annealing time increased. Secco etching on cleaved cross-sections followed by conventional optical microscopy was performed on some of these wafers and pits were observed corresponding to a bulk particle number density of ~10^9 cm^{-3}. SIRM bright spot signal intensities above the background level were measured from line traces taken across many in-focus individual spots and these were averaged for the individual wafers. As the annealing time increased from 2 to 16 hr, the intensity increased by 3.4x; from 16 to 32 hr, it increased by 11x; and from 32 to 160 hr, it remained closely the same (Table, row 1). Increases in such spot signal intensities are generally taken to indicate increases in particle size. (These increases in spot intensity are not apparent on going from Figs. 1a to f because a software contrast normalisation process was used in which the highest intensity present in each image is 'stretched' to the maximum brightness before saturation occurs. Consequently, the only apparent change in the images is a more pronounced background when the spot signal intensity is low, e.g. Fig. 1a).

The SIRM results for series II were somewhat similar to those of series I. The bright spot number density ranged from $6 x 10^7$ to $1.6 x 10^8$ cm^{-3} and the mean number density did not depend significantly on the annealing time. However, for the 160 hr wafer, many of the bright 'spots' were in the form of irregular lines up to ~15 μm long consisting of several bright spots, often overlapping (Fig. 2). As the annealing time increased from 2 to 16 hr, the bright spot signal intensity increased by 4.5x; from 16 to 32 hr, it increased by 8.9x; and from 32 to 160 hr, it increased by 4.3x, and so there was again a large signal increase on going from 16 to 32 hr.

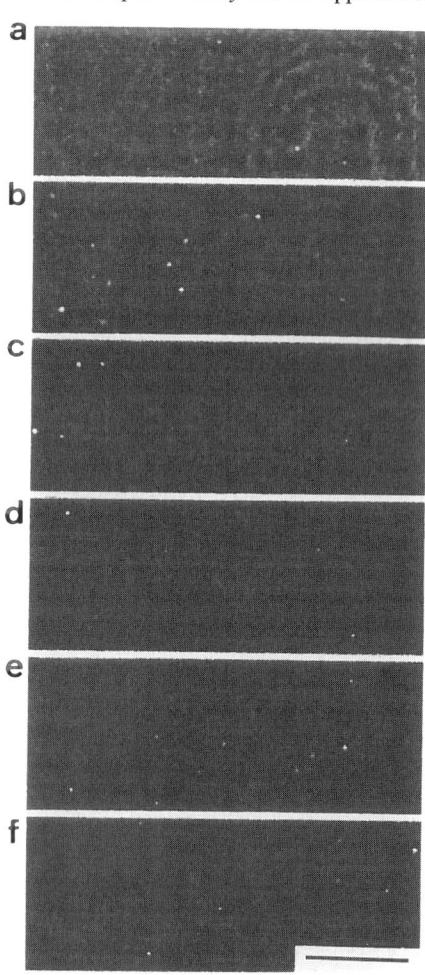

Fig. 1. RC-SIRM images showing oxide particles in series I wafers. a) to f) correspond to anneal times of 2, 4, 16, 32, 85 and 160 hr (Scale bar 20 μm).

TEM examinations were made of the wafers in series I. For all of the wafers, the oxide particles mostly occurred as approximately square plates ~100 to 200 nm across and ~10 to 25 nm thick, oriented on (001) planes and with the plate edges along <110> directions. HRTEM examinations showed that the plates were amorphous (Fig. 3) and energy dispersive x-ray analysis (EDX) showed that the plate composition corresponded approximately to SiO_2. For the 2, 4 and 16 hr wafers, mostly single isolated plates occurred. Some of these plates were associated with local strain fields (Fig. 4a) and some with small tangles of dislocations (Fig. 4b). For the 32, 85 and 160 hr wafers, single isolated plates were still present.

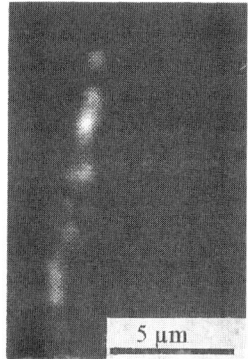

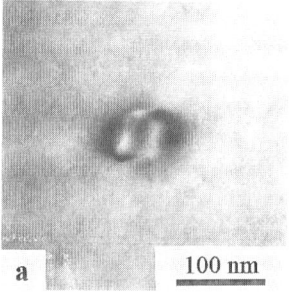

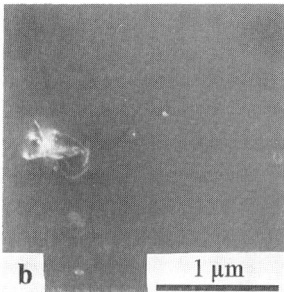

Fig. 2. RC-SIRM image showing cluster of oxide particles in the series II wafer annealed for 160 hr.

Fig. 3. HRTEM image showing amorphous oxide particle in the series I wafer annealed for 160 hr.

However, most of the plates occurred in clusters together with dislocation loops typically 200 nm across which had been punched-out along <110> directions and sometimes with dislocation tangles (Figs. 5a and b). There were up to ~40 plates per cluster. Also occasionally observed in some of the wafers were oxide plates up to ~0.5 to 1.0 μm across (Figs. 6a and b) and stacking fault loops (Fig. 7) up to ~10 μm across.

Fig. 4. TEM images of single oxide particles in the series I wafer annealed for 2 hr. a) with associated strain field, b) with dislocation tangles

Wafer	Additional Annealing Time (hr)					
Series	2	4	16	32	85	160
I	1.0	1.6	3.4	39	39	40
II	1.0	2.5	4.5	40	118	170

Table: For each wafer series, the signal was normalised to that of the wafer annealed for 2 hr.

4. DISCUSSION

The SIRM and TEM methods provided complementary information concerning the oxide precipitation process occurring and our interpretation of the results is as follows. For the series I wafers, the initial precipitation was mainly in the form of single isolated oxide plates of number density ~10^9 cm^{-3}. As the annealing time increased (2 to 16 hr), the plates slowly increased in size (Table, row 1). The local strain fields associated with the individual plates increased and this eventually caused local generation of dislocations, both loops by the punching-out mechanism and tangles by combined glide and climb. Oxygen diffused to these stress centres and further oxide plates were formed, either on or adjacent to the existing dislocations. Consequently, clusters consisting of plates and dislocations formed and the process increased in rate once initiated (16 to 32 hr). At this stage, if the mean plate size is taken as 200 x 200 x 20 nm, the cluster number density as 10^9 cm^{-3} and the number of plates per cluster as 25, calculation shows that the amount of oxygen in the precipitated plates is ~$4x10^{17}$ cm^{-3}, which corresponds to most of the oxygen present in the wafers. Consequently, on further annealing (32 to 160 hr), little additional precipitation occurred (Table, row 1). For the individual clusters, most of the plates occurred in the central region of the clusters, typically ~2 μm across. Hence, in the SIRM images, the individual plates in the individual clusters were not resolved and so each cluster appeared as a single bright spot.

774

For the series II wafers, although only SIRM results were obtained, these can now be interpreted as follows. The initial precipitation was mainly as single isolated oxide plates of number density $\sim 10^8$ cm^{-3}. The lower value compared with series I occurred because of the three-stage, rather than two-stage, initial heat treatment that was given. As the annealing time increased (2 to 16 hr), the plates slowly increased in size (Table, row 2) and these again generated clusters consisting of plates and dislocations (16 to 32 hr). At this stage, because the number density was

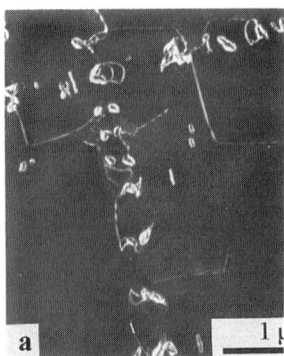 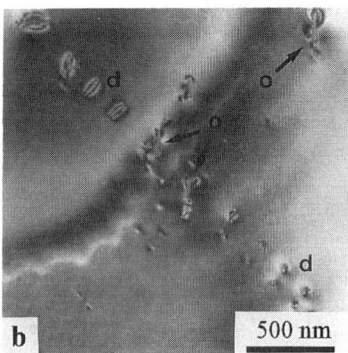

Fig. 5 TEM images of oxide particles (**o**) with punched-out dislocation loops (**d**) and dislocation tangles in the series I wafers annealed for a) 32 hr and b) 160 hr.

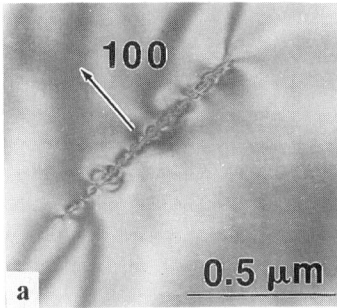 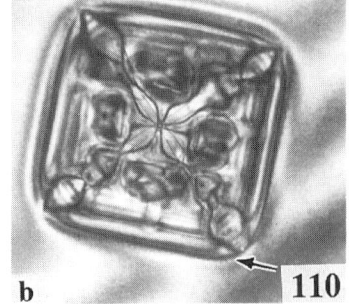

Fig. 6. TEM images of large oxide particles in the series I wafers annealed for a) 16 hr and b) 85 hr.

~ 10x smaller than for series I, much of the oxygen present in the wafers had not yet precipitated as oxide plates. Consequently, on further annealing (32 to 160 hr), more oxygen diffused to the clusters and more plates formed (Table, row 2). Hence, compared with series I, for the 160 hr anneal, there were fewer clusters but they were larger and contained more plates and dislocations. In particular, the lines of punched-out loops and associated plates were now significantly longer. In the SIRM images for the 160 hr anneal, small groups of plates within individual clusters appeared as irregular lines of single or overlapping bright spots (Fig. 2).

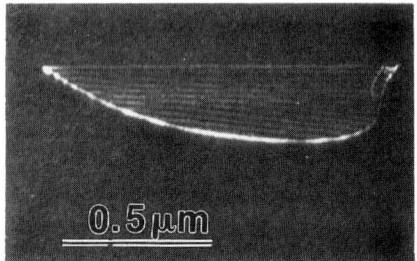

Fig. 7. TEM image for stacking fault in the series I wafers annealed for 2 hr.

ACKNOWLEDGEMENTS

The authors wish to thank the Engineering and Physical Sciences Research Council (EPSRC) UK for support.

REFERENCES

Bhatti A R, Falster R and Booker G R, 1991 Solid State Phenomena 19-20 51.
Booker G R, Laczik Z and Kidd P, 1992 Semiconductor Sci. & Tech 1 A110.
Booker G R, Laczik Z and Török P, 1995 Inst. Phys. Conf. Ser. (in press).
Shimura F, 1989 Semiconductor Silicon Crystal Technology (Academic Press) 279.
Török P, Booker G R, Laczik Z and Falster R, 1993 Inst. Phys. Conf. Ser. No 134 771.

Inst. Phys. Conf. Ser. No 146
Paper presented at Microsc. Semicond. Mater. Conf., Oxford, 20–23 March 1995
© 1995 IOP Publishing Ltd

Mid-IR-laser microscopy as a tool for defect investigation in bulk semiconductors

O V Astafiev, V P Kalinushkin and V A Yuryev

General Physics Institute of RAS, 38, Vavilov Street, Moscow, 117942, Russia

ABSTRACT: A non-destructive optitcal technique described in this paper is an effective new tool for the investigation of defects in semiconductors. The basic instrument for this technique — a mid-IR-laser microscope — being sensitive to accumulations of free carriers enables the study of both accumulations of electrically-active defects or impurities in bulk semiconductors and doped domains in semiconductor structures. The optical beam induced scattering mode of this microscope is designed for the investigation of recombination-active defects but unlike EBIC it requires neither Schottky barrier or p–n junction nor special preparation of samples.

1. INTRODUCTION

At present, many techniques have been developed to study defects on the micron and sub-micron scale in semiconductors. Unfortunately most of them either break down the studied sample or require rather complicated procedures of sample preparation that in some cases may modify the properties of the defects studied. Moreover, practically all of the modern methods for investigating such defects enable at best to observe defects but cannot give information about their composition. The specificity of the techniques which may be applicable for analysis of defect composition, such as e.g. X-ray microanalysis, microcathodoluminescence or photoluminescence, does not allow one to investigate defects situated in crystal bulk. Even the defects located in near-surface regions are hard to be studied by these methods because of insufficient sensitivity or, like in the case of X-ray microanalysis, because the defects are often composed of intrinsic defects. So despite the great success that now has been achieved in the understanding of defect formation mechanisms, their composition and properties, new methods for defect investigations, which might appreciably accelerate the progress in these issues, are presently required like they never were in the past. Non-destructive methods for investigation of defects in crystal bulk might e.g. give opportunities for direct studies of the processes of inner gettering especially if they are combined with techniques for studies of defects located in near-surface layers. Considerable progress may be achieved in the investigation of defects located in the vicinity of semiconductor interfaces and surfaces coated with dielectric layers and so on. We could give many examples of other applications of such techniques (see e.g. Kalinushkin *et al* 1995b). From our viewpoint, the most attractive methods for such investigations are those which enables the study of the crystal domains enriched with electrically-active defects and/or defects interacting with non-equilibrium carriers (we call them large-scale electrically-active defect accumulations, LSDAs, and large-scale recombination-active defects, LSRDs). The most appropriate methods for investigating these defects are those sensitive directly to micron-scale variations of the free carrier concentration. This is because many of the procedures for determining the point-defect structure of materials, e.g. such as measurements of temperature dependencies of conductivity or photocurrent measurements, may be suited for these techniques.

It seems, however, that until recently, the low-angle mid-IR-laser light scattering technique, LALS (see Kalinushkin 1988, Voronkov *et al* 1990, Kalinushkin *et al* 1995b and many references therein), was the only method that satisfied all these requirements, and many studies of LSDAs and LSRDs in different materials were carried out by this method — see e.g. references cited by Kalinushkin *et al* (1995b) and Kalinushkin (1988). Nonetheless LALS in its standard form (or LALS with angular resolution) has two main disadvantages: firstly, it does not permit the study of each individual defect and gives information averaged over a group of defects with close parameters which are located within the probe beam, and secondly, using only LALS one cannot estimate the concentrations of LSDAs which are necessary to evaluate such parameters as free carrier concentrations in LSDAs (and, hence, the energy locations of point centers constituting them). Therefore one must use other methods to evaluate the concentrations of LSDAs which decreases the reliability of the information obtained (however, the data obtained e.g. by Kalinushkin *et al* (1991) for InP and GaAs were confirmed by Yuryev and Kalinushkin (1995) and Yuryev *et al* (1995) by direct observations).

To overcome the mentioned shortcomings of LALS, a new method for visualization of free carrier accumulations was recently proposed (Astafiev *et al* 1994b, Kalinushkin *et al* 1995b, Astafiev *et al* 1995a, Astafiev *et al* 1995b and Yuryev *et al* 1995), which is a kind of scanning laser microscopy working in the mid-infrared wavelength range. Being an evolution of LALS (it is often referred to as scanning LALS or SLALS), this method possesses all its merits and admits all modifications which make LALS so convenient for investigation of LSDAs. Before long, the optical beam induced LALS (OLALS) mode of the mid-IR-laser microscopy was developed (Astafiev *et al* 1995c, Astafiev *et al* 1995d and Kalinushkin *et al* 1995b) on the basis of LALS with surface photoexcitation (Kalinushkin *et al* 1994a and Kalinushkin *et al* 1995a) for investigation of LSRDs in near-surface and near-interface layers of semiconductors. The latter technique is a non-destructive contactless optical analog of EBIC or, more exactly, OBIC but in contrast to these methods OLALS does not require Schottky barrier or p–n junction and any special surface preparation.

The present paper is devoted to the description of the developed techniques of mid-IR-laser microscopy and illustration of their serviceability by the experimental results obtained for bulk Si single crystals and their near-surface regions.

2. SLALS MODE OF MID-IR-LASER MICROSCOPE

2.1 Basic Instrument

The SLALS technique is shown schematically in Fig.1. The plane wave of the laser source illuminates a thin semiconductor parallel-sided crystal with polished surfaces (usually standard technological wafer before structure production). The wafer is located in the focal plane of a lens L1. Let a defect be in the crystal bulk in the front focal plane of L1. It scatters the probe wave producing an additional scattered wave, which diverges in an angle of an order of λ/a where λ is the wave-length, a is the defect's characteristic size. A resultant wave after the defect is the sum of the undisturbed plane wave and that scattered by the defect. The lens L1 condenses the plane wave in the back focus to a spot with the size of about $\lambda f_1/D1$ where $D1$ is the diameter of the probe plane-wave beam, f_1 is the focal length of the lens L1. A small mirror turned to the angle of 45° to the focal plane or an absorbing screen is positioned in the back focal plane to remove the probe wave radiation. The scattered wave, being almost a plane

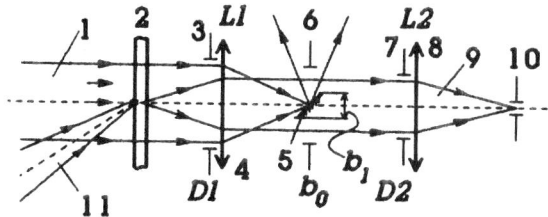

Figure 1: Optical diagram of the central dark ground mid-IR-laser microscope: (1) mid-IR probe wave; (2) sample; (3,6,7) diaphragms; (4,8) lenses; (5) mirror or opaque screen; (9) scattered wave; (10) IR photodetector; (11) exciting light beam (used in OLALS).

wave with characteristic beam diameter of $\lambda f_1/a$ after L1, passes to the second lens L2 almost without losses if the screen size is smaller than $\lambda f_1/a$. So the scattered wave without probe radiation reaches the lens L2 and the image of the defect is formed in the back focal plane of L2 in scattered rays.

The detailed analysis of the SLALS technique with application to the the central dark field method of spatial-frequency signal filtering, which is schematically presented in Fig.1, is made in the paper by Astafiev *et al* (1995b).

2.2 LSDAs in CZ Si:B

The single crystals of standard industrial Si:B studied in this work were grown by the Czochralski (CZ) process and had the specific resistivity of 12 Ω·cm. The thickness of the samples was 300 μm. The large-scale electrically active defects with the dimensions ranging from 3–5 μm to 40–50 μm were observed previously in analogous crystals in the works by Buzynin *et al* (1990) and Astafiev et al (1994a). Their concentrations were estimated as 10^5–10^7 cm^{-3}, the values of the relative variations of the dielectric constants $\delta\epsilon_m$ in them were evaluated as 10^{-3}–10^{-4}.

The specimens sharply differing in the concentration of defects as determined by selective etching were studied in the present work: their concentrations were about 2×10^5 cm^{-3} in the sample 1 and about 2×10^4 cm^{-3} in the sample 2.

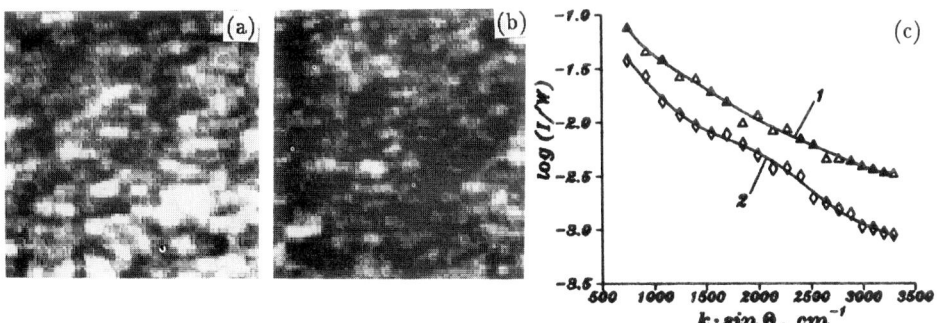

Figure 2: LSDAs in CZ Si:B (left to right): SLALS images of the samples 1 (*a*) and 2 (*b*), 1×1 mm^2; LALS diagrams (*c*) for the samples 1 (1) and 2 (2); λ=10.6 μm.

Fig.2(*a, b*) presents the microphotographs of these samples obtained with the use of SLALS (the areas of 1×1mm^2 are depicted in the photographs). Like in the photographs of indium phosphide and gallium arsenide (Yuryev and Kalinushkin 1995), the white spots in the pictures are the images of LSDAs. The mean value of the light scattering intensity by the defects in Fig.2(a) is by around 3 times greater than that in Fig.2(b), that is completely in agreement with the results obtained by LALS — the ratio of the light scattering intensities by the samples 1 and 2 in the LALS diagrams given in Fig.2(c) is also nearly equal to 3. The concentration of defects determined in the sample 1 was about 10^6 cm^{-3}, whereas that in the sample 2 was about $(4-5)\times10^5$ cm^{-3}.

Note that the values of the defect concentrations revealed in the analogous samples by EBIC with the special surface preparation (Buzynin et al 1989) appeared to be nearly equal to the above values obtained by SLALS.

The results obtained allow us to conclude the following: 1) using the SLALS-microscope we visualized the defects which were previously investigated by LALS — the so called "weak impurity accumulations" (Buzynin et al 1990 and Astafiev et al 1994a); 2) using EBIC with the special surface preparation — the data obtained by this method were used by Buzynin et al (1990) and Astafiev et al (1994a) — we also revealed namely "weak impurity accumulations", i.e. the estimations of LSDAs parameters and thermal activation energies of the centers constituting LSDAs made by Buzynin et al (1990) and Astafiev et al (1994a) are valid.

The radii of the accumulations calculated from the LALS diagrams given in Fig.2(c) are $a \sim 10 - 14\,\mu$m. Measuring the zero-angle light scattering intensity I_0/W from LALS and determining the concentration of LSDAs from SLALS one can easily estimate the values of $\delta\epsilon_m$ in LSDAs. They are $(6.5-10)\times10^{-4}$ and $(11-14)\times10^{-4}$ for the samples 1 and 2, respectively.

Note also that the correlation between the concentrations of defects revealed by selective etching and SLALS is purely qualitative. This is also characteristic for most of the comparative experiments on etching and LALS. However, enquiring into the reasons for the discrepancies observed is beyond the scope of this paper.

3. OPTICAL BEAM INDUCED LALS

3.1 Description of the Method

This mode of SLALS was recently proposed by us (Kalinushkin et al 1995b and Astafiev et al 1995c, 1995d) as a scanning modification of LALS with sample photoexcitation (Kalinushkin 1988, Kalinushkin et al 1994 and Kalinushkin et al 1995a). Its principle diagram is given in Fig.1. The method, as well as LALS, can work in two modes: in the mode of bulk photoexcitation and in the mode of surface photoexcitation. The regimes are different only by the choice of pumping laser: the first regime uses a laser with quantum energy less than the studied sample bandgap, whereas the second one uses a laser with quantum energy greater than the bandgap, and in general both regimes are quite analogous.

The essence of the method, say for surface excitation, is as follows. A highly focused beam (in contrast to LALS with photoexcitation where a wide beam is used) generates electron-hole pairs in the sample (in the chosen case, in its near-surface region). If the characteristic dimensions of the non-uniformity of the generated electron-hole pair distribution are as small as it is required in the paper by Astafiev et al (1995b), the scattered mid-IR-laser light of the SLALS microscope starts reaching the photodetector. Its intensity is proportional to the square of generated carrier concentration in the spot. The characteristic sizes of the non-uniformity are controlled by the sizes of the exciting laser spot, the carrier diffusion length and the surface recombination velocity. Even if the diffusion length is large (e.g. in Si), the inhomogeneity with small enough characteristic dimensions remains in the carrier distribution because of the surface recombination. This inhomogeneity is detected by the SLALS microscope. The carrier concentration in such a "droplet" is controlled with the electron-hole pair life-time in a given area of the sample.

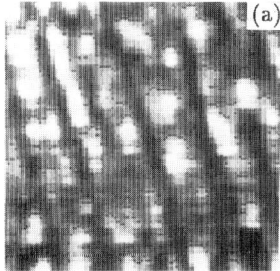 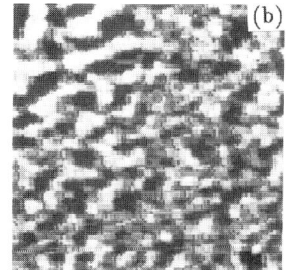 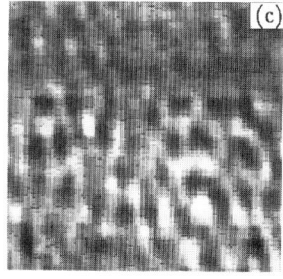

Figure 3: OLALS images, left to right: FZ Si:P, chemico-dynamic (a) and mechanical (b) polishing; CZ Si:B coated with 1200 Å thick SiO_2 film (c); 1×1 mm^2, $\lambda = 10.6\,\mu$m, $\lambda_{ex} = 633$ nm.

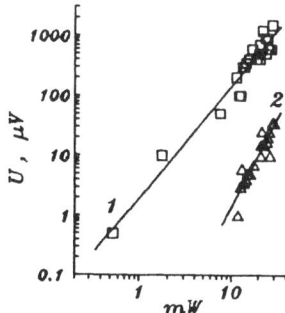

Figure 4: Dependences of MCT photodetector signal on the absorbed power of He-Ne laser radiation for chemico-dynamically (1) and mechanically (2) polished sides of FZ Si:P wafer depicted in Fig.3; $\lambda = 10.6\,\mu$m, $\lambda_{ex} = 633$ nm.

It is clear that the method is a very close analog of the electron or optical beam induced current (EBIC or OBIC) methods, but is different in that OLALS requires neither a Schottky barrier nor a p–n junction. It also does not require any special preparation of surfaces.

But the most important property of the developed method is its ability to obtain information from interfaces and surfaces covered with coatings and epilayers until the wafer is metallized.

It is also possible to create a kind of tomographic microscopy on the basis of OLALS, and this problem does not seem to be very difficult.

Note that modulated 50 mW He-Ne laser radiation at the wavelength of 0.63 μm was used in this work in the OLALS experiments. The signal was detected with lock-in nanovoltmeter at the modulation frequency.

3.2 OLALS Images of Silicon Wafers

Fig.3 demonstrates OLALS images of FZ Si:P wafer surfaces (1×1mm^2). The pictures (a) and (b) present two sides of the same wafer, one of which was polished in chemico-dynamic way ("finished side") and the other was polished mechanically up to optical precision grade. The darker the areas in the photographs, the shorter the carrier life is. So the dark stripes in the pictures of the finished side correspond, from our viewpoint, to tracks of underpolished and/or underetched scratches. A very badly damaged layer is registered on the mechanically polished side (the signal from this side was by around 1000 times lower than that from the finished side). The picture (c) gives the micrograph of CZ Si:B wafer surface coated with 1200 Å thick layer of SiO_2 (this wafer was taken directly from the technological line of CCD production). The dark spots are likely the images of defective regions.

Fig.4 demonstrates the dependence of IR-photodetector signal on the power of the exciting He-Ne laser for the FZ Si:P wafer depicted in Fig.3. Two cases are shown: for the finished side (marked as 1) and for the mechanically polished side (marked as 2). It is seen that in the case (1) the signal is proportional to the square of the photoexciting laser intensity whereas in the case (2) the signal is proportional to the third power of the He-Ne laser intensity. These lines confirm that the SLALS microscope works with scattered rays and the scattering by the domain with generated non-equilibrium carriers allows us to make imaging in the OLALS mode. The cubic dependence (2) has not been explained yet. Note that the same depence was obtained for the mechanically polished Ge sample in the work by Kalinushkin *et al* (1995a).

4. CONCLUSION

It was shown in this paper that the mid-IR-laser microscopy may become a useful tool for defect investigations in semiconductors. This method can be easily complemented with such well developed techniques for defect composition analysis as measurements of temperature dependencies of LALS (see e.g. Kalinushkin *et al* 1995 and Kalinushkin *et al* 1991). The measurements of LALS photoexcitation spectra might be also very useful for this purpose. Such well known technique as photoluminescence mapping might be used as a complementary method to the OLALS mode, with the same exciting laser being used for both techniques and the measurements being made simultaneously. The tomographic SLALS microscope is also under development now using the principles of laser heterodyning (Sawatari 1973 and Protopopov and Ustinov 1985). The methods of phase contrast and interference microscopies (Françon 1954) also might appear to be useful.

REFERENCES

Astafiev O V, Buzynin A N, Buvaltsev A I *et al* 1994a Semicond. <u>28</u> (3) 407
Astafiev O V, Kalinushkin V P and Yuryev V A 1994b Proc. SPIE <u>2332</u> 138
Astafiev O V, Kalinushkin V P and Yuryev V A 1995a Mater. Sci. Eng. (B) in press
Astafiev O V, Kalinushkin V P and Yuryev V A 1995b Microelectronics in press
Astafiev O V, Kalinushkin V P and Yuryev V A 1995c Microelectronics in press
Astafiev O V, Kalinushkin V P and Yuryev V A 1995d J. Tech. Phys. Lett. in press
Buzynin A N, Butylkina N A, Kalinushkin V P *et al* 1989 USSR Patent No.1531766
Buzynin A N, Zabolotskiy S E, Kalinushkin V P *et al* 1990 Sov. Phys.–Semicond. <u>24</u> (2) 264
Françon M 1954 Le Microscope á Contraste de Phase et Microscope Interferentiel (Paris: Centre National de la Recherche Scientifique)
Kalinushkin V P 1988 Proc. Inst. Gen. Phys. Acad. Sci. USSR Vol.4 Laser Methods of Defect Investigations in Semiconductors and Dielectrics (New York: Nova) pp.1–75
Kalinushkin V P, Murin D I, Astafiev O V *et al* 1995a Phys. Chem. Mech. Surf. (4) in press
Kalinushkin V P, Murin D I, Yuryev V A *et al* 1994 Proc. SPIE <u>2332</u> 146
Kalinushkin V P, Yuryev V A and Astafiev O V 1995b Mater. Sci. Technol. in press
Kalinushkin V P, Yuryev V A, Murin D I *et al* 1991 Semicond. Sci. Technol. <u>7</u> A255
Protopopov V V and Ustinov N D 1985 Laser Heterodyning (Moscow: Nauka)
Sawatari T 1973 Appl.Opt. <u>12</u> 2768
Voronkov V V, Zabolotskiy S E, Kalinushkin V P *et al* 1990 J. Cryst. Growth <u>103</u> 126
Yuryev V A, Astafiev O V and Kalinushkin V P 1995 Semicond. in press
Yuryev V A and Kalinushkin V P 1995 Mater.Sci.Eng. (B) in press

Subject Index

Author Index